BIOGEOCHEMISTRY

THIRD EDITION

BIOGEOCHEMISTRY
An Analysis of Global Change

THIRD EDITION

WILLIAM H. SCHLESINGER
Cary Institute of Ecosystem Studies
Millbrook, NY

EMILY S. BERNHARDT
Department of Biology
Duke University
Durham, NC

GEORGE GREEN LIBRARY OF
SCIENCE AND ENGINEERING

AMSTERDAM • BOSTON • HEIDELBERG • LONDON
NEW YORK • OXFORD • PARIS • SAN DIEGO
SAN FRANCISCO • SINGAPORE • SYDNEY • TOKYO
Academic Press is an imprint of Elsevier

Library of Congress Cataloging-in-Publication Data
Schlesinger, William H.
  Biogeochemistry : an analysis of global change / William H. Schlesinger,
Emily S. Bernhardt. — 3rd edition.
      pages   cm
  Includes bibliographical references and index.
  ISBN 978-0-12-385874-0
1. Biogeochemistry. I. Bernhardt, Emily S.   II. Title.
  QH343.7.S35 2012
  572–dc23
                                                          2012030107

British Library Cataloguing-in-Publication Data
A catalogue record for this book is available from the British Library.

For information on all Academic Press publications
visit our website at *http://store.elsevier.com*

To
Lisa and Justin, for supporting our obsession with biogeochemistry
and to Hannah and Gwyneth, with hopes for a more sustainable future.

Contents

Part II

GLOBAL CYCLES

Preface

This is a textbook about the chemistry of the surface of the Earth—the arena of life that is now increasingly affected by human activities. It is remarkable how many of our current environmental problems, ranging from climate change to ocean acidification, stem from changes in the Earth's chemistry wrought by humans. We hope that students using this book will come to appreciate how the Earth functions naturally as a chemical system, what events have caused changes in Earth's surface chemistry in the past, and what is causing our planet to change rapidly today. The book melds a wide range of disciplines from astrophysics to molecular biology and scales of time, from the origin of the Earth to the coming decades.

Like its previous editions, this book's organization follows the structure of a class in biogeochemistry that we have taught for many years at Duke University—first by Schlesinger, then by both of us, and more recently by Bernhardt. Following our class syllabus, we have organized the book into two sections: Part I covers the microbial and chemical reactions that occur in the atmosphere, on land, in freshwaters, and in the sea. Part II is a set of shorter chapters that link the mechanistic understanding of the earlier chapters to a large-scale, synthetic view of global biogeochemical cycles.

Every section of this book has been revised from the most recent edition (1997), with special emphasis given to expand our treatment of freshwaters and aquatic ecosystems, to include satellite and model-derived global maps of Earth's chemical characteristics, and to provide new global budgets for the major biogeochemical elements and mercury.

Throughout this book, we give special emphasis to the chemical reactions that link the elements that are important to life. The coupling of element cycling begins in biochemistry and constrains element cycling at the global level. In several locations, we show how computer models can be used to help understand and predict elemental cycling and ecosystem function. Many of these models are based on biochemistry and interactions among the biochemical elements. The models are useful in extrapolating small-scale observations to the global level. The models are often validated by observations taken from satellites, particularly since the deployment of NASA's Earth Observing System (EOS). We hope that this book will link disparate fields ranging from microbiology to global change ecology—all of which are now part of the science of biogeochemistry.

This text provides the framework for a class in biogeochemistry. It is meant to be supplemented by readings from the current literature, so that areas of specific interest or recent progress can be explored in more detail. Although not encyclopedic, the text includes more than 4500 references to aid students and others who wish to explore any topic more thoroughly. Reflecting the book's interdisciplinary nature, we have made a special effort to provide abundant cross-referencing of chapters, figures, and tables throughout.

As with this book's first and second editions, we hope that this one will stimulate a new generation of students to address the science and policy of global change. Denial is not an option; our planet needs attentive stewardship.

WHS and ESB

Millbrook, NY, and Durham, NC

Acknowledgments

A huge number of people have helped to produce this volume, both directly through their efforts to supply us with data, references, reviews, and figures, and indirectly through their early influence on our careers. The latter include (for WHS) Jim Eicher, Joe Chadbourne, John Baker, Russ Hansen, Bill Reiners, Noye Johnson, Bob Reynolds, and Peter Marks; and (for ESB) Gene Likens, Barbara Peckarsky, Lars Hedin, Alex Flecker, Margaret Palmer, Pat Mulholland, and Bob Hall.

A number of friends helped immensely in the preparation of this edition by providing reviews of early drafts of our efforts, including Ron Kiene, Susan Lozier, Elise Pendall, Emma Rosi-Marshall, Dan Richter, Lisa Dellwo Schlesinger, Jim Siedow, Dave Stevenson, Mike Tice and Paul Wennberg. Several of them tested the drafts of certain chapters in their classes. Greg Okin and Daniel Giammar helped test the problem sets.

For this edition, we thank the following for references, figures, and comments—large and small—who have helped to produce a better book: Geoff Abers, Andy Andreae, Alison Appling, Dennis Baldocchi, Mike Behrenfeld, Neil Bettez, Jim Brown, Amy Burgin, Oliver Chadwick, Terry Chapin, Ben Colman, Jim Clark, Jon Cole, Bruce Corliss, Randy Dahlgren, Paul Falkowski, Ian Faloona, Jack Fishman, Jacqueline Flückiger, Wendy Freeman, Jim Galloway, Nicholas Gruber, Jim Hansen, Kris Havstad, James Heffernan, Ashley Helton, Kirsten Hofmockel, Ben Houlton, Dan Jacob, Steve Jasinski, Jason Kaye, Gabriel Katul, Ralph Keeling, Emily Klein, George Kling, Jean Knops, Arancha Lana, Steve Leavitt, Lance Lesack, Gene Likens, Gary Lovett, George Luther, Brian Lutz, John Magnuson, Pat Megonigal, Patrick Mitchell, Scott Morford, Karl Niklas, Ram Oren, Steve Piper, Jim Randerson, Sasha Reed, Bill Reiners, Joan Riera, Phil Robertson, Jorge Sarmiento, Noelle Selin, Gus Shaver, Hank Shugart, John Simon, Emily Stanley, Phil Taylor, Eileen Thorsos, Kevin Trenberth, Remco van den Bos, Peter Vitousek, Mark Walbridge, Matt Wallenstein, Kathie Weathers, and Charlie Yocum. In addition, Deb Fargione helped to keep more than 4500 references easily accessible.

We would also like to thank the many students of Duke's Biogeochemistry course over the years for helping us refine our own understanding of biogeochemistry.

As always, any errors are our own, and we welcome hearing from you with comments, both large and small. You can contact us at schlesingerw@caryinstitute.org or emily.bernhardt@duke.edu.

BIOGEOCHEMISTRY

PROCESSES AND REACTIONS

Introduction

WHAT IS BIOGEOCHEMISTRY?

Today life is found from the deepest ocean trenches to the heights of the atmosphere above Mt. Everest; from the hottest and driest deserts in Chile to the coldest snows of Antarctica; from acid mine drainage in California, with pH < 1.0, to alkaline groundwaters in South Africa. More than 3.5 billion years of life on Earth has allowed the evolutionary process to fill nearly all habitats with species, large and small. And collectively these species have left their mark on the environment in the form of waste products, byproducts, and their own dead remains. Look into any shovel of soil and you will see organic materials that are evidence of life—a sharp contrast to what we see on the barren surface of Mars. Any laboratory sample of the atmosphere will contain nearly 21% oxygen, an unusually high concentration given that the Earth harbors lots of organic materials, such as wood, that are readily consumed by fire. All evidence suggests that the oxygen in Earth's atmosphere is derived and maintained from the photosynthesis of green plants. In a very real sense, O_2 is the signature of life on Earth (Sagan et al. 1993).

The century-old science of biogeochemistry recognizes that the influence of life is so pervasive that there is no pure science of geochemistry at the surface of Earth (Vernadsky 1998). Indeed, many of the Earth's characteristics are only hospitable to life today because of the current and historic abundance of life on this planet (Reiners 1986). Granted some Earthly characteristics, such as its gravity, the seasons, and the radiation received from the Sun,

3

are determined by the size and position of our planet in the solar system. But most other features, including liquid water, climate, and a nitrogen-rich atmosphere, are at least partially due to the presence of life. Life is the *bio* in biogeochemistry.

At present, there is ample evidence that our species, *Homo sapiens,* is leaving unusual imprints on Earth's chemistry. The human combustion of fossil fuels is raising the concentration of carbon dioxide in our atmosphere to levels not seen in the past 20 million years (Pearson and Palmer 2000). Our release of an unusual class of industrial compounds known as chlorofluorocarbons has depleted the concentration of ozone in the upper atmosphere, where it protects the Earth's surface from harmful levels of ultraviolet light (Rowland 1989). In our effort to feed 7 billion people, we produce vast quantities of nitrogen and phosphorus fertilizers, resulting in the runoff of nutrients that pollute surface and coastal waters (Chapter 12). As a result of coal combustion and other human activities, the concentrations of mercury in freshly caught fish are much higher than a century ago (Monteiro and Furness 1997), rendering many species unfit for regular human consumption. Certainly we are not the first species that has altered the chemical environment of planet Earth, but if our current behavior remains unchecked, it is well worth asking if we may jeopardize our own persistence.

UNDERSTANDING THE EARTH AS A CHEMICAL SYSTEM

Just as a laboratory chemist attempts to observe and understand the reactions in a closed test tube, biogeochemists try to understand the chemistry of nature, where the reactants are found in a complex mix of materials in solid, liquid, and gaseous phases. In most cases, biogeochemistry is a nightmare to a traditional laboratory chemist: the reactants are impure, their concentrations are low, and the temperature is variable. About all you can say about the Earth as a chemical system is that it is closed with respect to mass, save for a few meteors arriving and a few satellites leaving our planet. This closed chemical system is powered by the receipt of energy from the Sun, which has allowed the elaboration of life in many habitats (Falkowski et al. 2008).

Biogeochemists often build models for what controls Earth's surface chemistry and how Earth's chemistry may have changed through the ages. Unlike laboratory chemists, we have no replicate planets for experimentation, so our models must be tested and validated by inference. If our models suggest that the accumulation of organic materials in ocean sediments is associated with the deposition of gypsum ($CaSO_4 \cdot 2H_2O$), we must dig down through the sedimentary layers to see if this correlation occurs in the geologic record (Garrels and Lerman 1981). Finding the correlation does not prove the model, but it adds a degree of validity to our understanding of how Earth works—its biogeochemistry. Models must be revised when observations are inconsistent with their predictions.

Earth's conditions, such as the composition of the atmosphere, change only slowly from year to year, so biogeochemists often build steady-state models. As an example, in a steady-state model of the atmosphere, the inputs and losses of gases are balanced each year; the individual molecules in the atmosphere change, but the total content of each stays relatively constant. The assumption of a steady-state brings a degree of tidiness to our models of Earth's chemistry, but we should always be cognizant of the potential for nonlinear and cyclic behavior in Earth's characteristics. Indeed, some cycles, such as the daily rotation of

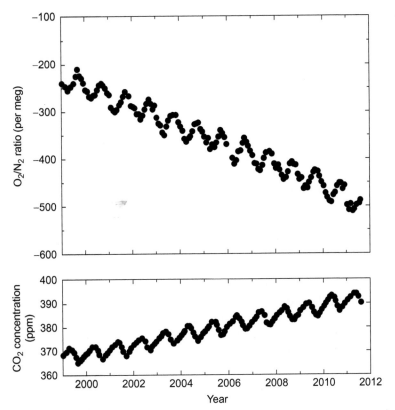

FIGURE 1.1   Annual cycles of CO_2 and O_2 in the atmosphere. Changes in the concentration of O_2 are expressed relative to concentrations of nitrogen (N_2) in the same samples. Note that the peak of O_2 in the atmosphere corresponds to the minimum CO_2 in late summer, presumably due to the seasonal course of photosynthesis in the Northern Hemisphere. *Source: From Ralph Keeling, unpublished data used by permission.*

the Earth around its axis and its annual rotation about the Sun, are now so obvious that it seems surprising that they were mysterious to philosophers and scientists throughout much of human history.

Steady-state models often are unable to incorporate the cyclic activities of the biosphere, which we define as the sum of all the live and dead materials on Earth.[1] During the summer, total plant photosynthesis in the Northern Hemisphere exceeds respiration by decomposers. This results in a temporary storage of carbon in plant tissues and a seasonal decrease in atmospheric CO_2, which is lowest during August of each year in the Northern Hemisphere (Figure 1.1). The annual cycle is completed during the winter months, when atmospheric CO_2 returns to higher levels, as decomposition continues when many plants are dormant or leafless.

[1] Some workers use the term biosphere to refer to the regions or volume of Earth that harbor life. We prefer the definition used here, so that the oceans, atmosphere, and surface crust can be recognized separately. Our definition of the biosphere recognizes that it has mass, but also functional properties derived from the species that are present.

Certainly, it would be a mistake to model the activity of the biosphere by considering only the summertime conditions, but a steady-state model can ignore the annual cycle if it uses a particular time each year as a baseline condition to examine changes over decades.

Over a longer time frame, the size of the biosphere has decreased during glacial periods and increased during post-glacial recovery. Similarly, the storage of organic carbon increased strongly during the Carboniferous Period—about 300 million years ago, when most of the major deposits of coal were laid down. The unique conditions of the Carboniferous Period are poorly understood, but it is certainly possible that such conditions are part of a long-term cycle that might return again. Significantly, unless we recognize the existence and periodicity of cycles and nonlinear behavior and adjust our models accordingly we may err in our assumption of a steady state in Earth's biogeochemistry.

All current observations of global change must be evaluated in the context of underlying cycles and potentially non-steady-state conditions in the Earth's system. The current changes in atmospheric CO_2 are best viewed in the context of cyclic changes seen during the last 800,000 years in a record obtained from the bubbles of air trapped in the Antarctic ice pack. These bubbles have been analyzed in a core taken near Vostok, Antarctica (Figure 1.2). During the entire 800,000-year period, the concentration of atmospheric CO_2 appears to have oscillated between high values during warm periods and lower values during glacial intervals. Glacial cycles are linked to small variations in Earth's orbit that alter the receipt of radiation from the Sun (Berger 1978; Harrington 1987). During the peak of the last glacial epoch (20,000 years ago), CO_2 ranged from 180 to 200 ppm in the atmosphere. CO_2 rose dramatically at the end of the last glacial (10,000 years ago) and was relatively stable at

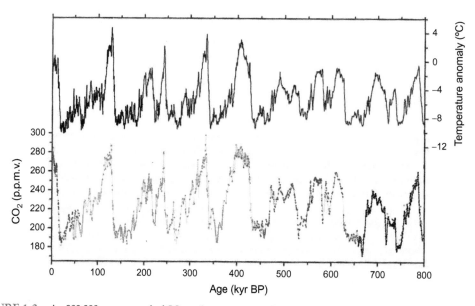

FIGURE 1.2   An 800,000-year record of CO_2 and temperature, showing the minimum temperatures correspond to minimum CO_2 concentrations seen in cycles of ~120,000 periodicity, associated with Pleistocene glacial epochs. *Source: From Luthi et al. (2008)*

280 ppm until the Industrial Revolution. The rapid increase in CO_2 at the end of the last glacial epoch may have amplified the global warming that melted the continental ice sheets (Sowers and Bender 1995, Shakun et al. 2012).

When viewed in the context of this cycle, we can see that the recent increase in atmospheric CO_2 to today's value of about 400 ppm has occurred at an exceedingly rapid rate, which carries the planet into a range of concentrations never before experienced during the evolution of modern human social and economic systems, starting about 8000 years ago (Flückiger et al. 2002). If the past is an accurate predictor of the future, higher atmospheric CO_2 will lead to global warming, but any observed changes in global climate must also be evaluated in the context of long-term cycles in climate with many possible causes (Crowley 2000; Stott et al. 2000).

The Earth has many feedbacks that buffer perturbations of its chemistry, so that steady-state models work well under many circumstances. For instance, Robert Berner and his coworkers at Yale University have elucidated the components of a carbonate–silicate cycle that stabilizes Earth's climate and its atmospheric chemistry over geologic time (Berner and Lasaga 1989). The model is based on the interaction of carbon dioxide with Earth's crust. Since CO_2 in the atmosphere dissolves in rainwater to form carbonic acid (H_2CO_3), it reacts with the minerals exposed on land in the process known as rock weathering (Chapter 4). The products of rock weathering are carried by rivers to the sea (Figure 1.3).

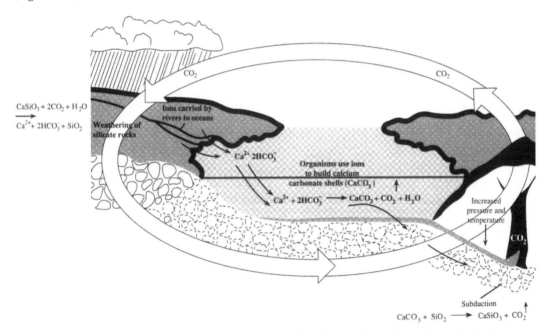

FIGURE 1.3   The interaction between the carbonate and the silicate cycles at the surface of Earth. Long-term control of atmospheric CO_2 is achieved by dissolution of CO_2 in surface waters and its participation in the weathering of rocks. This carbon is carried to the sea as bicarbonate (HCO_3^-), and it is eventually buried as part of carbonate sediments in the oceanic crust. CO_2 is released back to the atmosphere when these rocks undergo metamorphism at high temperature and pressures deep in Earth. *Source: Modified from Kasting et al. (1988).*

In the oceans, limestone (calcium carbonate) and organic matter are deposited in marine sediments, which in time are carried by subduction into Earth's upper mantle. Here the sediments are metamorphosed; calcium and silicon are converted back into the minerals of silicate rock, and the carbon is returned to the atmosphere as CO_2 in volcanic emissions. On Earth, the entire oceanic crust appears to circulate through this pathway in <200 million years (Muller et al. 2008). The presence of life on Earth does not speed the turning of this cycle, but it may increase the amount of material moving in the various pathways by increasing the rate of rock weathering on land and the rate of carbonate precipitation in the sea.

The carbonate–silicate model is a steady-state model, in the sense that it shows equal transfers of material along the flow-paths and no change in the mass of various compartments over time. In fact, such a model suggests a degree of self-regulation of the system, because any period of high CO_2 emissions from volcanoes should lead to greater rates of rock weathering, removing CO_2 from the atmosphere and restoring balance to the system. However, the assumption of a steady state may not be valid during transient periods of rapid change. For example, high rates of volcanic activity may have resulted in a temporary increase in atmospheric CO_2 and a period of global warming during the Eocene, 40 million years ago (Owen and Rea 1985). And clearly, since the Industrial Revolution, humans have added more carbon dioxide to the atmosphere than the carbonate–silicate cycle or the ocean can absorb each year (Chapter 11).

Because the atmosphere is well mixed, changes in its composition are perhaps our best evidence of human alteration of Earth's surface chemistry. Concern about global change is greatest when we see increases in atmospheric content of constituents such as carbon dioxide, methane (CH_4), and nitrous oxide (N_2O), for which we see little or no precedent in the geologic record. These gases are produced by organisms, so changes in their global abundance must reflect massive changes in the composition or activity of the biosphere.

Humans have also changed other aspects of Earth's natural biogeochemistry. For example, when human activities increase the erosion of soil, we alter the natural rate of sediment delivery to the oceans and the deposition of sediments on the seafloor (Wilkinson and McElroy 2007, Syvitski et al. 2005). As in the case of atmospheric CO_2, evidence for global changes in erosion induced by humans must be considered in the context of long-term oscillations in the rate of crustal exposure, weathering, and sedimentation due to changes in climate and sea level (Worsley and Davies 1979, Zhang et al. 2001).

Human extraction of fossil fuels and the mining of metal ores substantially enhance the rate at which these materials are available to the biosphere, relative to background rates dependent on geologic uplift and surface weathering (Bertine and Goldberg 1971). For example, the mining and industrial use of lead (Pb) has increased the transport of Pb in world rivers by about a factor of 10 (Martin and Meybeck 1979). Recent changes in the content of lead in coastal sediments appear directly related to fluctuations in the use of Pb by humans, especially in leaded gasoline (Trefry et al. 1985)—trends superimposed on underlying natural variations in the movements of Pb at Earth's surface (Marteel et al. 2008, Pearson et al. 2010).

Recent estimates suggest that the global cycles of many metals have been significantly increased by human activities (Table 1.1). Some of these metals are released to the atmosphere and deposited in remote locations (Boutron et al. 1994). For example, the combustion of

TABLE 1.1   Movement of Selected Elements through the Atmosphere

| Element | Natural | | | Anthropogenic | | Ratio anthropogenic: natural |
| | Continental dust[a] | Volcanic | | Industrial particles | Fossil fuel | |
		Dust	Gas			
Al	356,500	132,750	8.4	40,000	32,000	0.15
Fe	190,000	87,750	3.7	75,000	32,000	0.38
Cu	100	93	0.012	2,200	430	13.63
Zn	250	108	0.14	7,000	1,400	23.46
Pb	50	8.7	0.012	16,000	4,300	345.83

[a] All data are expressed in 10^8 g/yr.
Source: From Lantzy and MacKenzie (1979). Used with permission.

coal has raised the concentration of mercury (Hg) deposited in Greenland ice layers in the past 100 years (Weiss et al. 1971). Recognizing that the deposition of Hg in the Antarctic ice cap shows large variations over the past 34,000 years (Vandal et al. 1993), we must evaluate any recent increase in Hg deposition in the context of past cyclic changes in Hg transport through the atmosphere. Again, human-induced changes in the movement of materials through the atmosphere must be placed in the context of natural cycles in Earth system function (Nriagu 1989).

SCALES OF ENDEAVOR

The science of biogeochemistry spans a huge range of space and time, spanning most of the geologic epochs of Earth's history (see inside back cover). Molecular biologists contribute their understanding of the chemical structure and spatial configuration of biochemical molecules, explaining why some biochemical reactions occur more readily than others (Newman and Banfield 2002). Increasingly, genomic sequencing allows biogeochemists to identify the microbes that are active in soils and sediments and what regulates their gene expression (Fierer et al. 2007). Physiologists measure variations in the activities of organisms, while ecosystem scientists measure the movement of materials and energy through well-defined units of the landscape.

Geologists study the chemical weathering of minerals in rocks and soils and document Earth's past from sedimentary cores taken from lakes, oceans, and continental ice packs. Atmospheric scientists provide details of reactions between gases and the radiative properties of the planet. Meanwhile, remote sensing from aircraft and satellites allows biogeochemists to see the Earth at the largest scale, measuring global photosynthesis (Running et al. 2004) and following the movement of desert dusts around the planet (Uno et al. 2009). Indeed the skills needed by the modern biogeochemist are so broad that many students find their entrance to this new field bewildering. But the fun of being a biogeochemist stems from the challenge of integrating new science from diverse disciplines. And luckily, there are a few basic rules that guide the journey, as described in the next few subsections.

Thermodynamics

Two basic laws of physical chemistry, the laws of thermodynamics, tell us that energy can be converted from one form to another and that chemical reactions should proceed spontaneously to yield the lowest state of free energy, G, in the environment. The lowest free energy of a chemical reaction represents its equilibrium, and it is found in a mix of chemical species that show maximum bond strength and maximum disorder among the components. In the face of these basic laws, living systems create non-equilibrium conditions; life captures energy to counteract reactions that might happen spontaneously to maximize disorder.

Even the simplest cell is an ordered system; a membrane separates an inside from an outside, and the inside contains a mix of very specialized molecules. Biological molecules are collections of compounds with relatively weak bonds. For instance, to break the covalent bonds between two carbon atoms requires 83 kcal/mole, versus 192 kcal/mole for each of the double bonds between carbon and oxygen in CO_2 (Davies 1972, Morowitz 1968). In living tissue most of the bonds between carbon (C), hydrogen (H), nitrogen (N), oxygen (O), phosphorus (P), and sulfur (S), the major biochemical elements, are reduced or "electron-rich" bonds that are relatively weak (Chapter 7). It is an apparent violation of the laws of thermodynamics that the weak bonds in the molecules of living organisms exist in the presence of a strong oxidizing agent in the form of O_2 in the atmosphere. Thermodynamics would predict a spontaneous reaction between these components to produce CO_2, H_2O, and NO_3^-—molecules with much stronger bonds. In fact, after the death of an organism, this is exactly what happens! Living organisms must continuously process energy to counteract the basic laws of thermodynamics that would otherwise produce disordered systems with oxidized molecules and stronger bonds.

During photosynthesis, plants capture the energy in sunlight and convert the strong bonds between carbon and oxygen in CO_2 to the weak, reduced biochemical bonds in organic materials. As heterotrophic organisms, herbivores eat plants to extract this energy by capitalizing on the natural tendency for electrons to flow from reduced bonds back to oxidizing substances, such as O_2. Heterotrophs oxidize the carbon bonds in organic matter and convert the carbon back to CO_2. A variety of other metabolic pathways have evolved using transformations among other compounds (Chapters 2 and 7), but in every case metabolic energy is obtained from the flow of electrons between compounds in oxidized or reduced states. Metabolism is possible because living systems can sequester high concentrations of oxidized and reduced substances from their environment. Without membranes to compartmentalize living cells, thermodynamics would predict a uniform mix, and energy transformations, such as respiration, would be impossible.

Free oxygen appeared in Earth's surface environments sometime after the appearance of autotrophic, photosynthetic organisms (Chapter 2). Free O_2 is one of the most oxidizing substances known, and the movement of electrons from reduced substances to O_2 releases large amounts of free energy. Thus, large releases of free energy are found in aerobic metabolism, including the efficient metabolism of eukaryotic cells. The appearance of eukaryotic cells on Earth was not immediate; the fossil record suggests that they evolved nearly 1.5 billion years after the appearance of the simplest living cells (Knoll 2003). Presumably the evolution of eukaryotic cells was possible only after the accumulation of sufficient O_2 in the environment to sustain aerobic metabolic systems. In turn, aerobic metabolism offered large amounts of energy that could allow the elaborate structure and activity of higher organisms. Here some humility is important: eukaryotic cells may perform biochemistry faster and more efficiently, but the full range of known biochemical transformations is found amongst the members of the prokaryotic kingdom.

Stoichiometry

A second organizing principle of biogeochemistry stems from the coupling of elements in the chemical structure of the molecules of which life is built—cellulose, protein, and the like. Redfield's (1958) observation of consistent amounts of C, N, and P in phytoplankton biomass is now honored by a ratio that carries his name (Chapter 9). Reiners (1986) carried the concept of predictable stoichiometric ratios in living matter to much of the biosphere, allowing us to predict the movement of one element in an ecosystem by measurements of another. Sterner and Elser (2002) have presented stoichiometry as a major control on the structure and function of ecosystems. The growth of land plants is often determined from the nitrogen content of their leaves and the nitrogen availability in the soil (Chapter 6), whereas phosphorus availability explains much of the variation in algal productivity in lakes (Chapter 8). The population growth of some animals may be determined by sodium—an essential element that is found at a low concentration in potential food materials, relative to its concentration in body tissues.

Although the stoichiometry of biomass allows us to predict the concentration of elements in living matter, the expected ratio of elements in biomass is not absolute, such as the ratio of C to N in a reagent bottle of alanine. For instance, a sample of phytoplankton will contain a mix of species that vary in individual N/P ratios, with the weighted average close to that postulated by Redfield (Klausmeier et al. 2004). And, of course, a large organism will contain a mix of metabolic compounds (largely protein) and structural components (e.g., wood or bone) that differ in elemental composition (Reiners 1986, Arrigo et al. 2005; Elser et al. 2010). In some sense, organisms are what they eat, but decomposers can adjust their metabolism (Manzoni et al. 2008) and enzymatic production (Sinsabaugh et al. 2009) to feed on a wide range of substrates, even as they maintain a constant stoichiometry in their own biomass.

In some cases, trace elements control the cycle of major elements, such as nitrogen, by their role as activators and cofactors of enzymatic synthesis and activity. When nitrogen supplies are low, signal transduction by P activates the genes for N fixation in bacteria (Stock et al. 1990). The enzyme for nitrogen fixation, nitrogenase, contains iron (Fe) and molybdenum (Mo). Over large areas of the oceans, Falkowski et al. (1998) show that iron, delivered to the surface waters by the wind erosion of desert soils, controls marine production, which is often limited by N fixation. Similarly, when phosphorus supply is low, plants and microbes may produce alkaline phosphatase, containing zinc, to release P from dead materials (Shaked

	Oxidized ⟶ Reduced			
	H_2O/O_2	C	N	S
Oxidized H_2O/O_2	X	Photosynthesis $CO_2 \longrightarrow C$ $H_2O \longrightarrow O_2$		
C	Respiration $C \longrightarrow CO_2$ $O_2 \longrightarrow H_2O$	X	Denitrification $C \longrightarrow CO_2$ $NO_3 \longrightarrow N_2$	Sulfate-Reduction $C \longrightarrow CO_2$ $SO_4 \longrightarrow H_2S$
N	Heterotrophic Nitrification $NH_4 \longrightarrow NO_3$ $O_2 \longrightarrow H_2O$	Chemoautotrophy (Nitrification) $NH_4 \longrightarrow NO_3$ $CO_2 \longrightarrow C$	Anammox $NH_4 + NO_2 \longrightarrow N_2 + 2H_2O$	?
Reduced S	Sulfur Oxidation $S \longrightarrow SO_4$ $O_2 \longrightarrow H_2O$	Chemoautotrophy (Sulfur-based Photosynthesis) $S \longrightarrow SO_4$ $CO_2 \longrightarrow C$	Autotrophic Denitrification $S \longrightarrow SO_4$ $NO_3 \longrightarrow N_2/NH_4$	X

FIGURE 1.4   A matrix showing how cellular metabolisms couple oxidation and reduction reactions. The cells in the matrix are occupied by organisms or a consortium of organisms that reduce the element at the top of the column, while oxidizing an element at the beginning of the row. *Source: From Schlesinger et al. (2011).*

et al. 2006). Thus, the productivity of some ecosystems can be stimulated either by adding the limiting element itself or by adding a trace element that facilitates nutrient acquisition (Arrigo et al. 2005).

The elements of life are also coupled in metabolism, since organisms employ some elements in energy-yielding reactions, without incorporating them into biomass. Coupled biogeochemistry of the elements in metabolism stems from the flow of electrons in the oxidation/reduction reactions that power all of life (Morowitz 1968, Falkowski et al. 2008). Coupled metabolism is illustrated by a matrix, where each element in a column is reduced while the element in an intersecting row is oxidized (Figure 1.4). All of Earth's metabolisms can be placed in the various cells of this matrix and in a few adjacent cells that would incorporate columns and rows for Fe and other trace metals. The matrix incorporates the range of metabolisms possible on Earth, should the right conditions exist (Bartlett 1986).

Large-Scale Experiments

Biogeochemists frequently conduct large-scale experiments to assess the response of natural systems to human perturbation. Schindler (1974) added phosphorus to experimental lakes in Canada to show that it was the primary nutrient limiting algal growth in those ecosystems (Figure 1.5). Bormann et al. (1974) deforested an entire watershed to demonstrate the importance of vegetation in sequestering nutrients in ecosystems. Several experiments have exposed replicated plots of forests, grasslands, and desert ecosystems to

FIGURE 1.5   An ecosystem-level experiment in which a lake was divided and one half (distant) fertilized with phosphorus, while the basin in the foreground acted as a control. The phosphorus-fertilized basin shows a bloom of nitrogen-fixing cyanobacteria. *Source: From Schindler (1974); www.sciencemag.org/content/184/4139/897.short. Used with permission.*

high CO_2 to simulate plant growth in the future environments on Earth (Chapter 5). And oceanographers have added Fe to large patches of the sea to ascertain whether it normally limits the growth of marine phytoplankton (Chapter 9). In many cases these large experiments and field campaigns are designed to test the predictions of models and to validate them.

Models

With sufficient empirical observations, biogeochemists can often build mathematical models for how ecosystems function. Equations can express what controls the movement of energy and materials through organisms or individual compartments of an ecosystem, such as the soil. The equations often incorporate the constraints of thermodynamics and stoichiometry. These models allow us to determine which processes control the productivity and biogeochemical cycling in ecosystems, and where our understanding is incomplete. Models that are able to reproduce past dynamics reliably allow us to explore the behavior of ecosystems in response to future perturbations that may lie outside the natural range of environmental variation.

LOVELOCK'S GAIA

In a provocative book, *Gaia*, published in 1979, James Lovelock focused scientific attention on the chemical conditions of the present-day Earth, especially in the atmosphere, that are extremely unusual and in disequilibrium with respect to thermodynamics. The 21% atmospheric content of O_2 is the most obvious result of living organisms, but other gases, including NH_3 and CH_4, are found at higher concentrations than one would expect in an O_2-rich atmosphere (Chapter 3). This level of O_2 in our atmosphere is maintained despite known reactions that should consume O_2 in reaction with crustal minerals and organic carbon. Further, Lovelock suggested that the *albedo* (reflectivity) of Earth must be regulated by the biosphere, because the planet has shown relatively small changes in surface temperature despite large fluctuations in the Sun's radiation during the history of life on Earth (Watson and Lovelock 1983).

Lovelock suggested that the conditions of our planet are so unusual that they could only be expected to result from activities of the biosphere. Indeed, *Gaia* suggests that the biosphere evolved to regulate conditions within a range favorable for the continued persistence of life on Earth. In Lovelock's view, the planet functions as a kind of "superorganism," providing planetary homeostasis. Reflecting the vigor and excitement of a new scientific field, other workers have strongly disagreed—not denying that biotic factors have strongly influenced the conditions on Earth, but not accepting the hypothesis of purposeful self-regulation of the planet (Lenton 1998).

Like all models, *Gaia* remains as a provocative hypothesis, but the rapid pace at which humans are changing the biosphere should alarm us all. Some ecologists see the potential for critical transitions in ecosystem function; points beyond which human impacts would not allow the system to rebound to its prior state, even if the impacts ceased (Scheffer et al. 2009). Others have attempted to quantify these thresholds, so that we may recognize them in time (Rockstrom et al. 2009). In all these endeavors, policy makers are desperate for biogeochemists to deliver a clear articulation of how the world works, the extent and impact of the human perturbation, and what to do about it.

Recommended Readings

Gorham, E. 1991. Biogeochemistry: Its origins and development. *Biogeochemistry 13*:199–239.
Kump, L.R., J.F. Kasting, and R.G. Crane. 2010. *The Earth System*, second ed. Prentice Hall.
Lovelock, J.E. 2000. *The Ages of Gaia*. Oxford University Press.
Sterner, R.W., and J.J. Elser. 2002. *Ecological Stoichiometry*. Princeton University Press.
Smil, V. 1997. *The Cycles of Life*. Scientific American Press.
Volk, T. 1998. *Gaia's Body*. Springer/Copernicus.
Williams, G.R. 1996. *The Molecular Biology of Gaia*. Columbia University Press.

Origins

INTRODUCTION

Six elements, H, C, N, O, P, and S, are the major constituents of living tissue and account for 95% of the mass of the biosphere. At least 25 other elements are known to be essential to at least one form of life, and it is possible that this list may grow slightly as we improve our understanding of the role of trace elements in biochemistry (Williams and Fraústo da Silva 1996).[1] In the periodic table (see inside front cover), nearly all the elements essential to life are found at atomic numbers lower than that of iodine at 53. Even though living organisms affect the distribution and abundance of some of the heavier elements, the biosphere is built from the "light" elements (Deevey 1970, Wackett et al. 2004). Ultimately, the environment in which life arose and the arena for biogeochemistry today was determined by the relative abundance of chemical elements in

---

[1] Arsenic is known to be an essential trace element for some species, but recent reports of bacteria that are able to grow using arsenic as a substitute for phosphorus (Wolfe-Simon et al. 2011) are largely discounted (Erb et al. 2012).

15

our galaxy and by the subsequent concentration and redistribution of those elements on Earth's surface.

In this chapter we will examine models that astrophysicists suggest for the origin of the elements. Then we will examine models for the formation of the solar system and the planets. There is good evidence that the conditions on the surface of the Earth changed greatly during the first billion years or so after its formation—before life arose. Early differentiation of the Earth, the cooling of its surface, and the composition of the earliest oceans determined the arena for the origins of life. Later changes caused by the evolution and proliferation of life strongly determined the conditions on our planet today. In this chapter, we will consider the origin of the major metabolic pathways that characterize life and affect Earth's biogeochemistry. The chapter ends with a discussion of the planetary evolution that has occurred on Earth compared to its near neighbors—Mars and Venus.

ORIGINS OF THE ELEMENTS

Any model for the origin of the chemical elements must account for their relative abundance in the Universe. Estimates of the cosmic abundance of elements are made by examining the spectral emission from the stars in distant galaxies as well as the emission from our Sun (Ross and Aller 1976). Analyses of meteorites also provide important information on the composition of the solar system (Figure 2.1). Two points are obvious: (1) with three exceptions—lithium (Li), beryllium (Be), and boron (B)—the light elements, that is, those with an atomic number <30, are far more abundant than the heavy elements; (2) especially among the light elements, the even-numbered elements are more abundant than the odd-numbered elements of similar atomic weight.

A central theory of astrophysics is that the Universe began with a gigantic explosion, "the Big Bang," about 13.7 billion years ago (Freedman and Madore 2010). The Big Bang initiated the fusion of hypothetical fundamental particles, known as quarks, to form protons (^{1}H) and neutrons, and it allowed the fusion of protons and neutrons to form some simple atomic nuclei (^{2}H, ^{3}He, ^{4}He, and a small amount of ^{7}Li). See Malaney and Fowler (1988), Pagel (1993), and Copi et al. (1995). After the Big Bang, the Universe began to expand outward, so there was a rapid decline in the temperatures and pressures that would be needed to produce heavier elements by fusion in interstellar space. Moreover, the elements with atomic masses of 5 and 8 are unstable, so no fusion of the abundant initial products of the Big Bang (i.e., ^{1}H and ^{4}He) could yield an appreciable, persistent amount of a heavier element. Thus, the Big Bang can explain the origin of elements up to ^{7}Li, but the origin of heavier elements had to await the formation of stars in the Universe—about 1 billion years later.

A model for the synthesis of heavier elements in stars was first proposed by Burbidge et al. (1957), who outlined a series of pathways that could occur in the interior of massive stars during their evolution (Fowler 1984, Wallerstein 1988, Trimble 1997). As a star ages, the abundance of hydrogen (H) in the core declines as it is converted to helium (He) by fusion. As the heat from nuclear fusion decreases, the star begins to collapse inward under its own gravity. This collapse increases the internal temperature and pressure until He begins

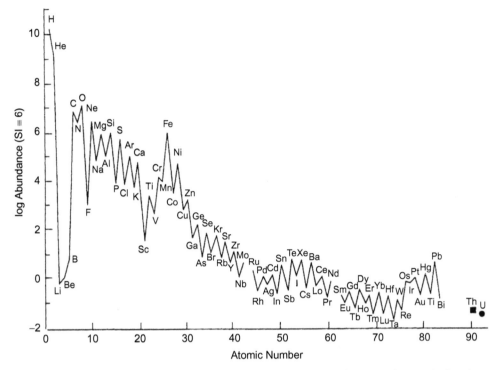

FIGURE 2.1   The relative abundance of elements in the solar system, also known as the cosmic abundance, as a function of atomic number. Abundances are plotted logarithmically and scaled so that silicon (Si) = 1,000,000. *Source: From a drawing in Brownlee (1992) based on the data of Anders and Grevesse (1989).*

to be converted via fusion reactions to form carbon (C) in a two-step reaction known as the triple-alpha process. First,

$$^4He + \, ^4He \rightleftarrows \, ^8Be. \tag{2.1}$$

Then, while most ^{8}Be decays spontaneously back to helium, the momentary existence of small amounts of ^{8}Be under these conditions allows reaction with helium to produce carbon:

$$^8Be + \, ^4He \rightarrow \, ^{12}C. \tag{2.2}$$

The main product of this so-called helium "burning" reaction is ^{12}C, and the rate of this reaction determines the abundance of C in the Universe (Oberhummer et al. 2000). ^{16}O is built by the addition of ^{4}He to ^{12}C, and nitrogen by the successive addition of protons to ^{12}C. As the supply of helium begins to decline, a second phase of stellar collapse is followed by the initiation of a sequence of further fusion reactions in massive stars (Fowler 1984). First, fusion of two ^{12}C forms ^{24}Mg (magnesium), some of which decays to ^{20}Ne (neon) by loss of an alpha (^{4}He) particle. Subsequently, oxygen burning produces ^{32}S, which forms an appreciable amount of ^{28}Si (silicon) by loss of an alpha particle (Woosley 1986).

I. PROCESSES AND REACTIONS

A variety of fusion reactions in massive stars are thought to be responsible for the synthesis—known as stellar nucleosynthesis—of the even-numbered elements up to iron (Fe) (Fowler 1984, Trimble 1997). (Smaller stars, like our Sun, do not go through all these reactions and burn out along the way, becoming white dwarfs.) These fusion reactions release energy and produce increasingly stable nuclei (Friedlander et al. 1964). However, to make a nucleus heavier than Fe requires energy, so when a star's core is dominated by Fe, it can no longer burn. This leads to the catastrophic collapse and explosion of the star, which we recognize as a supernova. Heavier elements are apparently formed by the successive capture of neutrons by Fe, either deep in the interior of stable stars (s-process) or during the explosion of a supernova (r-process; Woosley and Phillips 1988, Burrows 2000, Cowan and Sneden 2006). A supernova casts all portions of the star into space as hot gases (Chevalier and Sarazin 1987).

This model explains a number of observations about the abundance of the chemical elements in the Universe. First, the abundance of elements declines logarithmically with increasing mass beyond hydrogen and helium, the original building blocks of the Universe. However, as the Universe ages, more and more of the hydrogen will be converted to heavier elements during the evolution of stars. Astrophysicists can recognize younger, second-generation stars, such as our Sun, that have formed from the remnants of previous supernovas because they contain a higher abundance of iron and heavier elements than older, first-generation stars, in which the initial hydrogen-burning reactions are still predominant (Penzias 1979). We should all be thankful for the fusion reactions in massive stars which have formed most of the chemical elements of life.

Second, because the first step in the formation of all the elements beyond lithium is the fusion of nuclei with an even number of atomic mass (e.g., ^{4}He, ^{12}C), the even-numbered light elements are relatively abundant in the cosmos. The odd-numbered light elements are formed by the addition of neutrons to nuclei in the interior of massive stars (s-process) and by the fission of heavier even-numbered nuclei. In most cases an odd-numbered nucleus is slightly less stable than its even-numbered "neighbors," so we should expect odd-numbered nuclei to be less abundant. For example, phosphorus is formed in the reaction

$$^{16}O + {}^{16}O \rightarrow {}^{32}S \rightarrow {}^{31}P + {}^{1}H. \tag{2.3}$$

Thus, phosphorus is much less abundant than the adjacent elements in the periodic table, Si (silicon) and S (sulfur) (Figure 2.1). It is interesting to speculate that the low cosmic abundance of P (phosphorus) formed by this and other reactions of nucleosynthesis may account for the fact that P is often in short supply for the biosphere on Earth today (Macia et al. 1997).

Finally, the low cosmic abundance of Li, Be, and B is due to the fact that the initial fusion reactions pass over nuclei with atomic masses of 5 to 8, forming ^{12}C, as shown in Eqs. 2.1 and 2.2. Apparently, most Li, Be, and B are formed by spallation—the fission of heavier elements that are hit by cosmic rays in interstellar space (Olive and Schramm 1992, Reeves 1994, Chaussidon and Robert 1995).

This model for the origin and cosmic abundance of the elements offers some initial constraints for biogeochemistry. All things being equal, we might expect that the chemical environment in which life arose would approximate the cosmic abundance of elements. Thus, the evolution of biochemical molecules might be expected to capitalize on the light elements that were abundant in the primordial environment. It is then of no great surprise that no element heavier than Fe is more than a trace constituent in living tissue and that among the light elements, no Li or Be, and only traces of B, are essential components of biochemistry (Wackett et al. 2004). The composition of life is remarkably similar to the composition of the Universe; as put by Fowler (1984), we are all "a little bit of stardust."

ORIGIN OF THE SOLAR SYSTEM AND THE SOLID EARTH

The Milky Way galaxy is about 12.5 billion years old (Dauphas 2005), indicating that the first stars and galaxies had formed within a billion years after the Big Bang (Cayrel et al. 2001). By comparison, as a second-generation star, our Sun appears to be only about 4.57 billion years old (Baker et al. 2005, Bonanno et al. 2002, Bouvier and Wadhawa 2010). Current models for the origin of the solar system suggest that the Sun and its planets formed from a cloud of interstellar gas and dust, possibly including the remnants of a supernova (Chevalier and Sarazin 1987). This cloud of material would have the composition of the cosmic mix of elements (Figure 2.1). As the Sun and the planets began to condense, each developed a gravitational field that helped capture materials that added to its initial mass. The mass concentrated in the Sun apparently allowed condensation to pressures that reinitiated the fusion of hydrogen to helium.

The planets of our solar system appear to have formed from the coalescing of dust to form small bodies, known as planetesimals, within the primitive solar cloud (Beckwith and Sargent 1996, Baker et al. 2005). Collisions among the planetesimals would have formed the planets. The process is likely to have been fairly rapid. Several lines of evidence show planetesimals forming during the first million years of the solar system (Srinivasan et al. 1999, Yin et al. 2002, Alexander et al. 2001), and most stars appear to lose their disk of gases and dust within 400 million years after their formation (Habing et al. 1999). Recent observations suggest that a similar process is now occurring around another star in our galaxy, ß Pictoris (Lagage and Pantin 1994, Lagrange et al. 2010), and Earth-size and larger planets have been detected around numerous other stars in our galaxy (Gaidos et al. 2007, Borucki et al. 2010, Lissauer et al. 2011).

Overall, the original solar nebula is likely to have been composed of about 98% gaseous elements (H, He, and noble gases), 1.5% icy solids (H_2O, NH_3, and CH_4), and 0.5% rocky solid materials, but the composition of each planet was determined by its position relative to the Sun and the rate at which the planet grew (McSween 1989). The "inner" planets (Mercury, Venus, Earth, and Mars) seemed to have formed in an area where the solar nebula was very hot, perhaps at a temperature close to 1200 K (Boss 1988). Venus, Earth, and Mars are all depleted in light elements compared to the cosmic abundances, and they are dominated by silicate minerals that condense at high temperatures and contain large amounts of FeO (McSween 1989). The mean density of Earth is about 5.5 g/cm^3. The high density of the inner planets contrasts with the lower average density of the larger, outer planets, known as gas giants, which captured a greater fraction of lighter constituents from the initial solar cloud (Table 2.1). Jupiter contains much hydrogen and helium. The average density of Jupiter is 1.3 g/cm^3, and its overall composition does not appear too different from the solar abundance of elements (Lunine 1989, Niemann et al. 1996). Some astronomers have pointed out that the hydrogen-rich atmosphere on Jupiter is similar to the composition of "brown dwarfs"—stars that never "ignited" (Kulkarni 1997).

From the initial solar cloud of elements, the chemical composition of the Earth is a selective mix, peculiar to the orbit of the incipient planet. The majority of the mass of the Earth seems

TABLE 2.1   Characteristics of the Planets

Planet Name[a]	Radius 10^8 cm	Volume 10^{26} cm^3	Mass 10^{27} gm	Density gm/cm^3	Corrected density[b] gm/cm^3
Mercury	2.44	0.61	0.33	5.42	5.4
Venus	6.05	9.3	4.9	5.25	4.3
Earth	6.38	10.9	6.0	5.52	4.3
Mars	3.40	1.6	0.64	3.94	3.7
Jupiter	71.90	15,560	1900	1.31	<1.3
Saturn	60.20	9130	570	0.69	<0.7
Uranus	25.40	690	88	1.31	<1.3
Neptune	24.75	635	103	1.67	<1.7

[a] *The mass of the Sun is 1.99×10^{33} gm, 1000× the mass of Jupiter.*
[b] *Density a planet would have in the absence of gravitational squeezing.*
Source: From Broecker (1985, p. 73). Published by Lamont Dougherty Laboratory, Columbia University. Used with permission.

likely to have accreted by about 4.5 bya—within about 100 million years of the origin of the solar system (Allègre et al. 1995, Kunz et al. 1998, Yin et al. 2002, Touboul et al. 2007, Jackson et al. 2010). Several theories account for the origin and differentiation of Earth. One suggests that Earth may have grown by homogeneous accretion; that is, throughout its early history, Earth may have captured planetesimals that were relatively similar in composition (Stevenson 1983, 2008).

Kinetic energy generated during the collision of these planetesimals (Wetherill 1985), as well as the heat generated from radioactive decay in its interior (Hanks and Anderson 1969), would heat the primitive Earth to the melting point of iron, nickel, and other metals, forming a magma ocean. These heavy elements were "smelted" from the materials arriving from space and sank to the interior of the Earth to form the core (Agee 1990, Newsom and Sims 1991, Wood et al. 2006).

As Earth cooled, lighter minerals progressively solidified to form a mantle dominated by perovskite ($MgSiO_3$), with some complement of olivine ($FeMgSiO_4$), and a crust dominated by aluminosilicate minerals of lower density and the approximate composition of feldspar (Chapter 4). Thus, despite the abundance of iron in the cosmos and in the Earth as a whole, the crust of the Earth is largely composed of Si, Al, and O (Figure 2.2). The aluminosilicate rocks of the crust "float" on the heavier semifluid rocks of the mantle (Figure 2.3; Bowring and Housh 1995).

An alternative theory for the origin of Earth suggests that the characteristics of planetesimals and other materials contributing to the growth of the planet were not uniform through time. Theories of heterogeneous accretion suggest that materials in the Earth's mantle arrived later than those of the core (Harper and Jacobsen 1996, Schönbächler et al. 2010), and that a late veneer delivered by a class of meteors known as carbonaceous chondrites was responsible for most of the light elements and volatiles on Earth (Anders and Owen 1977, Wetherill 1994, Javoy 1997, Kramers 2003). The two accretion theories are not mutually exclusive; it is

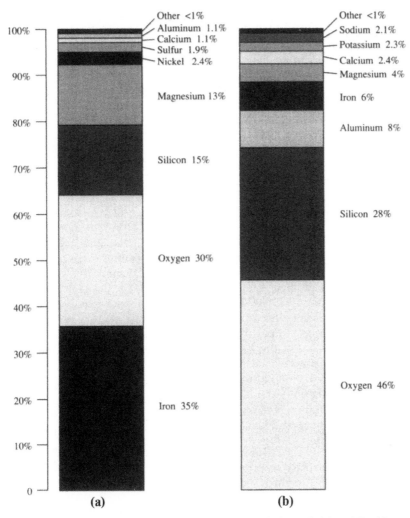

FIGURE 2.2 Relative abundance of elements by weight in the whole Earth (a) and Earth's crust (b). *Source: From Earth (fourth ed.) by Frank Press and Raymond Siever. Copyright 1986 by W.H. Freeman and Company. Used with permission.*

possible that a large fraction of the Earth mass was delivered by homogeneous accretion, followed by a late veneer of chondritic materials (Willbold et al. 2011).

It is likely that during its late accretion, Earth was impacted by a large body—known as Theia—which knocked a portion of the incipient planet into an orbit about it, forming the Moon (Lee et al. 1997). The Moon's age is estimated at 4.527 billion years (Kleine et al. 2005). Earth's early history was probably dominated by frequent large impacts, but based on the age distribution of craters on the Moon, it is postulated that most of the large impacts occurred before 2.0 bya (Neukum 1977, Cohen et al. 2000, Bottke et al. 2012). The present-day

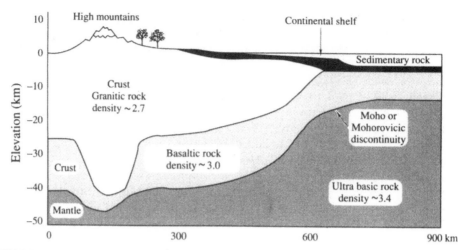

FIGURE 2.3   A geologic profile of the Earth's surface. On land the crust is dominated by granitic rocks, largely composed of Si and Al (Chapter 4). The oceanic crust is dominated by basaltic rocks with a large proportion of Si and Mg. Both granite and basalt have a lower density than the upper mantle, which contains ultrabasic rocks with the approximate composition of olivine ($FeMgSiO_4$). *Source: From Howard and Mitchell (1985).*

receipt of extraterrestrial materials (8 to 38×10^9 g/yr; Taylor et al. 1998, Love and Brownlee 1993, Cziczo et al. 2001) is much too low to account for Earth's mass (6×10^{27} g), even if it has continued for all of Earth's history.

    Consistent with either theory are several lines of evidence that the primitive Earth was devoid of an atmosphere derived from the solar nebula—that is, a primary atmosphere. During its early history, the gravitational field on the small, accreting Earth would have been too weak to retain gaseous elements, and the incoming planetesimals were likely to have been too small and too hot to carry an envelope of volatiles. The impact of Theia is also likely to have blown away any volatiles that had accumulated in Earth's atmosphere by that time. Today, volcanic emissions of some inert (noble) gases, such as ^3He, ^{20}Ne (neon), and ^{36}Ar (argon), which are derived from the solar nebula, result from continuing degassing of primary volatiles that must have been delivered to the primitive Earth trapped in pockets (fluid inclusions) in accreting chondrites (Lupton and Craig 1981, Burnard et al. 1997, Jackson et al. 2010). Otherwise, the Earth's atmosphere appears to be of secondary origin.

    If a significant fraction of today's atmosphere were derived from the original solar cloud, we might expect that its gases would exist in proportion to their solar abundances (refer to Figure 2.1). Here, ^{20}Ne is of particular interest because it is not produced by any known radioactive decay, it is too heavy to escape from Earth's gravity, and as an inert gas it is not likely to have been consumed in any reaction with crustal minerals (Walker 1977).[2] Thus, the

---

[2] ^{20}Ne is one of the isotopes of neon. Isotopes of an element have the same number of protons in the nucleus, but differ in the number of neutrons, so they differ in atomic weight. Naturally occurring chemical elements are usually mixtures of isotopes, and their listed atomic weights are average values for the mixture. Most of the elements in the periodic table have two or more isotopes, with 254 stable (i.e. nonradioactive) isotopes for the first 80 elements.

present-day abundance of ^{20}Ne in the atmosphere is likely to represent its primary abundance—that derived from the solar nebula. Assuming that other solar gases were delivered to the Earth in a similar manner, we can calculate the total mass of the primary atmosphere by multiplying the mass of ^{20}Ne in today's atmosphere by the ratio of each of the other gases to ^{20}Ne in the solar abundance. For example, the solar ratio of nitrogen to neon is 0.91 (Figure 2.1). If the present-day atmospheric mass of neon, 6.5×10^{16} g, is all from primary sources, then $(0.91) \times (6.5 \times 10^{16}$ g$)$ should be the mass of nitrogen that is also of primary origin. The product, 5.9×10^{16} g, is much less than the observed atmospheric mass of nitrogen, 39×10^{20} g. Thus, most of the nitrogen in today's atmosphere must be derived from other sources.

ORIGIN OF THE ATMOSPHERE AND THE OCEANS

Much of the Earth's inventory of "light" elements is likely to have been delivered to the planet as constituents of the silicate minerals in carbonaceous chondrites, perhaps in a late veneer of accretion (Javoy 1997). Even today, many silicate minerals in the Earth's mantle carry elements such as oxygen and hydrogen as part of their crystalline structure (Bell and Rossman 1992, Meade et al. 1994). Of particular interest to biogeochemistry, carbonaceous chondrites typically contain from 0.5 to 3.6% C (carbon) and 0.01 to 0.28% N (nitrogen) (Anders and Owen 1977), which may represent the original source of these elements for the biosphere.

The origin of the Earth's atmosphere is closely tied to the appearance and evolution of its crust, which differentiated from the mantle by melting and by density separation under the heat generated by large impacts and internal radioactive decay (Fanale 1971, Stevenson 1983, Kunz et al. 1998). During melting, elements such as H, O, C, and N would have been released from the mantle as volcanic gases. Several lines of evidence point to the existence of a continental crust by 4.4 bya (Wilde et al. 2001, Watson and Harrison 2005, O'Neil et al. 2008), and its volume appears to have grown through Earth's history (Collerson and Kamber 1999, Abbott et al. 2006, Hawkesworth and Kemp 2006). Thus, the accumulation of a secondary atmosphere began early in Earth's history (Kunz et al. 1998).

Today, a variety of gases are released during volcanic eruptions at the Earth's surface. The emissions associated with the eruption of island basalts, such as on Hawaii, offer a good indication of the composition of mantle degassing, since they are less likely to be contaminated by younger crustal materials that have been subducted (Marty 2012). Table 2.2 gives the composition of gases emitted from various volcanoes. Characteristically, water vapor dominates the emissions, but small quantities of C, N, and S gases are also present (Tajika 1998). Volcanic emissions, representing degassing of Earth's interior, are consistent with the observation that Earth's atmosphere is of secondary origin—largely derived from solid materials. The earliest atmosphere is likely to have been dominated by H_2, H_2O, CO, and CO_2 depending on the mix of chondritic materials (Schaefer and Fegley 2010). Some elements, including H, can dissolve in magma so that a significant proportion of the Earth's inventory of water, carbon, and nitrogen may still reside in the mantle (Bell and Rossman 1992, Murakami et al. 2002, Marty 2012). The total extent of mantle degassing through geologic time is unknown, but perhaps is 50% based on the content of ^{40}Ar in Earth's mantle (Marty 2012).

TABLE 2.2 Composition of Gases Emitted by Volcanoes

Volcano	Units	H_2O	H_2	CO_2	SO_2	H_2S	HCl	HF	N_2	NH_3	O_2	Ar	CH_4	References
Kudryavy, Russia	mole %	95.00	0.56	2.00	1.32	0.41	0.3700	0.030	0.21	–	0.03	0.002	0.002	Taran et al. (1995)
Nevado del Ruiz, Colombia	wgt. %	94.90		2.91	2.74	0.80	0.0052							Williams et al. (1986)
Kamchatka, Russia	vol. %	78.60	3.01	4.87	0.03	0.16	0.5700	0.056	11.87	0.11	0.01	0.060	0.440	Dobrovolsky (1994)

The isotopic ratio of argon gas on Earth (i.e., $^{40}Ar/^{36}Ar$) is suggestive of the proportion of our present atmosphere that is derived from mantle degassing. On Earth the isotope ^{40}Ar appears to be wholly the result of the radioactive decay of ^{40}K in the mantle (Farley and Neroda 1998), while the isotope ^{36}Ar was delivered intact from the original solar nebula. Like ^{20}Ne, this noble element is too heavy to escape the gravity of the Earth.[3] Thus, the atmospheric content of ^{36}Ar should represent the proportion that is due to the residual primary atmosphere (i.e. the solar nebula), whereas the content of ^{40}Ar is indicative of the proportion due to crustal degassing. The ratio of $^{40}Ar/^{36}Ar$ on Earth is nearly 300, suggesting that 99.7% of the Ar in our present atmosphere is derived from the interior of the Earth. In contrast, the Viking spacecraft measured a much higher ratio of 2750 for ^{40}Ar versus ^{36}Ar in the atmosphere on Mars (Owen and Biemann 1976). This observation supports the emerging belief that Mars lost a large portion of its primary atmosphere, and that most of the atmosphere on Mars today is derived from degassing (Carr 1987).

The ratio of N_2 to ^{40}Ar in the Earth's mantle is close to that of today's atmosphere ($\sim$80)—implying a common source for both elements at the Earth's surface (Marty 1995). Volcanic emissions and mantle degassing were undoubtedly greatest in early Earth history; the present-day flux of nitrogen from volcanoes (0.78 to 1.23×10^{11} g/yr; Sano et al. 2001, Tajika 1998) is too low to account for the current inventory of N at the Earth's surface ($\sim$50 $\times 10^{20}$ g; Table 2.3) even if it has continued for all 4.5 billion years of Earth's history. Moreover, some nitrogen has also returned to the mantle by subduction (Zhang and Zindler 1993).

It is possible that a late impact of comets contributed to the gaseous inventory on Earth (Chyba 1990a). If so, the proportion must be small, because the isotopic ratio of H measured in the ices of the Hale Bopp and other comets does not match the ratio in the present inventory of water on Earth (Meier et al. 1998). Nevertheless, it is difficult to explain the presence of ice on the surface of the Moon, which has undergone only slight degassing other than by delivery in a late arrival of comets impacting its surface (Clark 2009, Sunshine et al. 2009, Colaprete et al. 2010, Zuber et al. 2012).

As long as the Earth was very hot, volatiles remained in the atmosphere, but when the surface temperature cooled to the condensation point of water, water could condense out of the primitive atmosphere to form the oceans. This must have been a rainstorm of true global proportion! Several lines of evidence point to the existence of liquid water on the Earth's surface 4.3 bya (Mojzsis et al. 2001, Wilde et al. 2001). Although heat from large, late-arriving meteors may have caused temporary revaporization of some of the earliest oceans (Sleep et al. 1989, Abramov and Mojzsis 2009), the geologic record suggests that liquid water has been present on the Earth's surface continuously for the past 3.8 billion years. Indeed, despite a few lingering meteor impacts, Earth may have harbored a temperate climate nearly 3.4 bya (Hren et al. 2009, Blake et al. 2010).

Various other gases would have quickly entered the primitive ocean as a result of their high solubility in water; for example:

$$CO_2 + H_2O \rightarrow H_2CO_3 \rightarrow H^+ + HCO_3^- \tag{2.4}$$
$$HCl + H_2O \rightarrow H_3O^+ + Cl^- \tag{2.5}$$
$$SO_2 + H_2O \rightarrow H_2SO_3. \tag{2.6}$$

---

[3] A small amount of ^{40}Ar has been destroyed by cosmic rays, producing ^{35}S in the atmosphere (Tanaka and Turekian 1991).

TABLE 2.3   Total Inventory of Volatiles at the Earth's Surface[a]

Reservoir	H$_2$O	CO$_2$	C	O$_2$	N	S	Cl	Ar	Total (rounded)
Atmosphere (see Table 3.1)	1.3	0.31	—	119	387	—	—	6.6	514
Oceans	135,000	19.3[b]	0.07	256[c]	2[d]	128[e]	2610	—	138,000
Land plants	0.1	—	0.06	—	0.0004	—	—	—	0.16
Soils	12	0.40[f,g]	0.15	—	0.0095	—	—	—	12.6
Freshwater (including ice and groundwater)	4850	—	—	—	—	—	—	—	4,850
Sedimentary rocks	15,000[h]	30000[g]	1560	4745[i]	200[j]	744[k]	500[h]	—	52,750
Total (rounded)	155,000	30,000	1560	5120	590	872	3100	7	196,000
See also	Fig. 10.1	Fig. 11.1	Fig. 11.1	Fig. 2.8		Table 13.1			

[a] All data are expressed as 10^{19} g, with values derived from this text unless noted otherwise.
[b] Assumes the pool of inorganic C is in the form of HCO$_3^-$.
[c] Oxygen content of dissolved SO$_4^{2-}$.
[d] Dissolved N$_2$.
[e] S content of SO$_4^{2-}$.
[f] Desert soil carbonates.
[g] Assumes 60% of CaCO$_3$ is carbon and oxygen.
[h] Walker (1977).
[i] O$_2$ held in sedimentary Fe$_2$O$_3$ and evaporites CaSO$_4$.
[j] Goldblatt et al. (2009).
[k] S content of CaSO$_4$ and FeS$_2$.

These reactions removed a large proportion of reactive water-soluble gases from the atmosphere, as predicted by Henry's Law for the partitioning of gases between gaseous and dissolved phases:

$$S = kP, \tag{2.7}$$

where S is the solubility of a gas in a liquid, k is the solubility constant, and P is the overlying pressure in the atmosphere. Under one atmosphere of partial pressure, the solubilities of CO_2, HCl, and SO_2 in water are 1.4, 700, and 94.1 g/liter, respectively, at 25°C. When dissolved in water, all of these gases form acids, which would be neutralized by immediate reaction with the surface minerals on Earth. Thus, after cooling, the Earth's earliest atmosphere is likely to have been dominated by N_2, which has relatively low solubility in water (0.018 g/liter at 25°C).

Because many gases dissolve so readily in water, an estimate of the total extent of crustal degassing through geologic time must consider the mass of the atmosphere, the mass of oceans, and the mass of volatile elements that are now contained in sedimentary minerals, such as $CaCO_3$, which have been deposited from seawater (Li 1972). By this accounting, the mass of the present-day atmosphere (5.14×10^{21} g; Trenberth and Guillemot 1994) represents less than 1% of the total degassing of the Earth's mantle over geologic time (refer to Table 2.3). The oceans and various marine sediments contain nearly all of the remainder, and some volatiles have returned to the upper mantle by subduction of Earth's oceanic crust (Zhang and Zindler 1993, Plank and Langmuir 1998, Kerrick and Connolly 2001).

Despite uncertainty about the exact composition of the Earth's earliest atmosphere, several lines of evidence suggest that by the time life arose, the atmosphere was dominated by N_2, CO_2, and H_2O (Holland 1984), which were in equilibrium with the oceans, and by trace quantities of other gases from volcanic emissions that were continuing at that time (Hunten 1993, Yamagata et al. 1991). There was certainly no O_2; the small concentrations produced by the photolysis of water in the upper atmosphere would rapidly be consumed in the oxidation of reduced gases and crustal minerals (Walker 1977, Kasting and Walker 1981).

During its early evolution as a star, the Sun's luminosity was as much as 30% lower than at present. We might expect that primitive Earth was colder than today, but the fossil record indicates a continuous presence of liquid water on the Earth's surface since 3.8 bya. One explanation is that the primitive atmosphere contained much higher concentrations of water vapor, CO_2, CH_4, and other greenhouse gases than today (Walker 1985). These gases would trap outgoing infrared radiation and produce global warming through the "greenhouse" effect (refer to Figure 3.2). In fact, even today the presence of water vapor and CO_2 in the atmosphere creates a significant greenhouse effect on Earth—about 75% due to water vapor and 25% from CO_2 (Lacis et al. 2010, Schmidt et al. 2010a,b). Without these gases, Earth's temperature would be about 33°C cooler, and the planet would be covered with ice (Ramanathan 1988).

There are few direct indications of the composition of the earliest seawater. Like today's seawater, the Precambrian ocean is likely to have contained a substantial amount of chloride. HCl and Cl_2 emitted by volcanoes would dissolve in water, forming Cl^- (Eq. 2.5). The acids produced by the dissolution of these and other gases in water (Eqs. 2.4–2.6) would have reacted with minerals of the Earth's crust, releasing Na^+, Mg^{2+}, and other cations by chemical weathering (Chapter 4). Carried by rivers, these cations would accumulate in seawater until their concentrations increased to levels that would precipitate secondary minerals.

For instance, sedimentary accumulations of $CaCO_3$ of Precambrian age indicate that the primitive oceans had substantial concentrations of Ca^{2+} (Walker 1983). Thus, it is likely that the dominant cations (Na, Mg, and Ca) and the dominant anion (Cl) in Precambrian seawater were similar to those in seawater today (Holland 1984, Morse and MacKenzie 1998). Only SO_4^{2-} seems to have been less concentrated in the Precambrian ocean (Grotzinger and Kasting 1993, Habicht et al. 2002).

ORIGIN OF LIFE

Fundamental to all considerations of the origins of life are the characteristics of living systems as we know them today. To develop theories and a timeline for the evolution of life on primitive Earth, we need to be able to recognize some or all of these traits in fossil sediments and in the products of laboratory synthesis of organic materials. These characteristics include the presence of a physical membrane, metabolic machinery for obtaining energy from the environment, and genetic material allowing heritability. These fundamental characteristics separate life from abiotic organic materials. A surrounding or *plasma membrane* allows segregation of the building blocks of biochemistry, up to 30 elements, at concentrations and proportions that typically diverge substantially from the surrounding environment.

Internal membranes, such as the mitochondrial membrane, allow separation of materials within cells, facilitating the capture of energy as electrons flow from electron-rich (reduced) to electron-poor (oxidized) substances that are obtained from the environment (Figure 1.4). Autotrophic organisms produce their own organic materials by capturing energy from the Sun (photoautotrophy) or other external sources (chemoautotrophy); heterotrophic organisms consume the organic materials produced by others. Generic material allows these structural innovations to be repeatable and heritable so that organisms can grow and reproduce.

An initial constraint on the evolution of life was a lack of organic molecules on primitive Earth. In 1871, Darwin postulated that the interaction of sunlight with marine salts under a primitive atmosphere might have created these primordial organic building blocks.[4] Working with Harold Urey in the early 1950s, Stanley Miller carried out this experiment by adding the probable constituents of the primitive atmosphere and oceans to a laboratory flask and subjecting the mix to an electric discharge to represent the effects of lightning. After several days, Miller found that simple, reduced organic molecules had been produced (Miller 1953, 1957). This experiment, possibly simulating the conditions on early Earth, suggested that the organic constituents of living organisms could be produced abiotically.

This experiment has been repeated in many laboratories, and under a wide variety of conditions (Chang et al. 1983). Ultraviolet light can substitute for electrical discharges as an energy source; a high flux of ultraviolet light would be expected on primitive Earth in the absence of an ozone (O_3) shield in the stratosphere (Chapter 3). Additional energy for abiotic

---

[4] From a letter from Charles Darwin to Joseph Hooker, 1871: "but if (and oh! what a big if!) we could conceive in some warm little pond, with all sorts of ammonia and phosphoric salts, light, heat, electricity, etc., present, that a protein compound was chemically formed ready to undergo still more complex changes, at the present day such matter would be instantly devoured or absorbed, which would not have been the case before living creatures were formed."

synthesis may have been derived from the impact of late-arriving meteors and comets passing through the atmosphere (Chyba and Sagan 1992, McKay and Borucki 1997) or at hydrothermal vents in the deep sea (Russell 2006).

The mix of atmospheric constituents taken to best represent the primitive atmosphere is controversial. H_2 may have been an important component of Earth's earliest atmosphere (Tian et al. 2005), and the yield of organic molecules is greatest in such highly reducing conditions. Nevertheless, an acceptable yield of simple organic molecules is found in experiments using mildly reducing atmospheres, composed of CO_2, H_2O, and N_2 (Pinto et al. 1980), which are the more probable conditions on the primitive Earth (Trail et al. 2011). The experiments are never successful when free O_2 is included; O_2 rapidly oxidizes the simple organic products before they can accumulate.

Interstellar dust particles and cometary ices also contain a wide variety of simple organic molecules (Busemann et al. 2006, Carr and Najita 2008, Sloan et al. 2009), and various amino acids are found in carbonaceous chondrites (Kvenvolden et al. 1970, Cooper et al. 2001, Pizzarello et al. 2001, Herd et al. 2011), suggesting that abiotic synthesis of organic molecules may be widespread in the galaxy (Orgel 1994, Irvine 1998, Ciesla and Sandford 2012). Significantly, it is possible that a small fraction of the organic molecules in chondrites and comets survives passage through the Earth's atmosphere, contributing to the inventory of organic molecules on its surface (Anders 1989, Chyba and Sagan 1992). Even if the total mass received is small, exogenous sources of organic molecules are important, for they may have served as chemical templates, speeding the rate of abiotic synthesis and the assembly of organic molecules on Earth.

A wide variety of simple organic molecules have now been produced under abiotic conditions in the laboratory (Dickerson 1978). In many cases hydrogen cyanide and formaldehyde are important initial products that polymerize to produce simple sugars such as ribose and more complex molecules such as amino acids and nucleotides. Even methionine, a sulfur-containing amino acid, has been synthesized abiotically (Van Trump and Miller 1972). The volcanic gas carbonyl sulfide (COS) can catalyze the binding of amino acids to form polypeptides (Leman et al. 2004), and short chains of amino acids have been linked by condensation reactions involving phosphates (Rabinowitz et al. 1969, Lohrmann and Orgel 1973). An early abiotic role for organic polyphosphates in synthesis speaks strongly for the origin of adenosine triphosphate (ATP) as the energizing reactant in virtually all biochemical reactions that we know today (Dickerson 1978).

Clay minerals, with their surface charge and repeating crystalline structure, may have acted to concentrate simple, polar organic molecules from the primitive ocean, making assembly into more complicated forms, such as RNA and protein, more likely (Cairns-Smith 1985, Ferris et al. 1996, Hanczyc et al. 2003). Metal ions such as zinc and copper can enhance the binding of nucleotides and amino acids to clays (Lawless and Levi 1979, Huber and Wachtershauser 1998, 2006). It is interesting to speculate why nature incorporates only the "left-handed" forms of amino acids in proteins, when equal forms of L- and D-enantiomers are produced by abiotic synthesis (Figure 2.4). Apparently, the light of stars is polarizing, creating an abundance of L-enantiomers during organic synthesis in the interstellar environment (Engel and Macko 2001). If the organic molecules in meteorites served as a chemical template for abiotic synthesis on Earth, meteors may have carried the preference for L-enantiomers in organic synthesis at Earth's surface (Engel and Macko 1997, Bailey et al. 1998, Pizzarello and Weber 2004).

I. PROCESSES AND REACTIONS

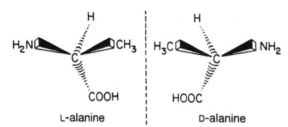

FIGURE 2.4 The left-handed (L) and right-handed (D) forms, known as enantiomers, of the amino acid alanine. No rotation of these molecules allows them to be superimposed. Although both forms are found in the extraterrestrial organic matter of carbonaceous chondrites, all life on Earth incorporates only the L form in proteins. *Source: From Chyba (1990b).*

Recently, scientists studying the origins of life have focused on submarine hydrothermal vent systems, which today harbor a diversity of life forms, as the arena for Earth's earliest life (Kelley et al. 2002, Russell 2006). Hydrothermal vents appear to support the abiotic synthesis of simple organic molecules, including formate, acetate (Lang et al. 2010), pyruvate (Cody et al. 2000), and amino acids (Huber and Wächtershäuser 2006). Indeed, the energetics of amino acid synthesis is favorable in these environments (Amend and Shock 1998). An origin of life in the high temperature, extreme pH, and high salinity of these habitats may explain how life persists in such a wide range of extreme habitats today (Rothschild and Mancinelli 2001, Marion et al. 2003).

Just as droplets of cooking oil form "beads" on the surface of water, it has long been known that some organic polymers will spontaneously form coacervates, which are colloidal droplets small enough to remain suspended in water. Coacervates are perhaps the simplest systems that might be said to be "bound," as if by a membrane, providing an inside and an outside. Yanagawa et al. (1988) describe several experiments in which protocellular structures with lipoprotein envelopes were constructed in the laboratory. In such structures, the concentration of substances will differ between the inside (hydrophobic) and the outside (hydrophilic) as a result of the differing solubility of substances in an organic medium and water, respectively. Mansy et al. (2008) show how primitive membranes may have allowed the transport of charged substances to the interior of protocells, allowing the evolution of heterotrophic metabolism.

Some organic molecules produced in the laboratory will self-replicate, suggesting potential mechanisms that may have increased the initial yield of organic molecules from abiotic synthesis (Hong et al. 1992, Orgel 1992, Lee et al. 1996). Other laboratories have produced simple organic structures, known as micelles, that will self-replicate their external framework (Bachmann et al. 1992). There is good reason to believe that the earliest genetic material controlling replication may not have been DNA but a related molecule, RNA, which can also perform catalytic activities (de Duve 1995, Robertson and Miller 1995).

Recent reports indicate some success in the abiotic synthesis of RNA precursors, which could subsequently support the abiotic synthesis of lengthy RNA molecules (Unrau and Bartel 1998, Powner et al. 2009). Vesicles that form around clay particles are found to enhance the polymerization of RNA (Hanczyc et al. 2003). Recently Gibson et al. (2010) inserted synthetic DNA into bacteria, where it replaced the native DNA and began reproducing. This work brings us one step closer to replicating the assembly of simple organic molecules into a complete self-replicating, metabolizing, and membrane-bound form that we might call life, with its origins in the laboratory.

A traditional view holds that life arose in the sea, and that biochemistry preferentially incorporated constituents that were abundant in seawater. For example, Banin and Navrot (1975) point out the striking correlation between the abundance of elements in today's biota and the solubility of elements in seawater. Elements with low ionic potential (i.e., ionic charge/ionic radius) are found as soluble cations (Na^+, K^+, Mg^{2+}, and Ca^{2+}) in seawater and as important components of biochemistry. Other elements, including C, N, and S, that form soluble oxyanions in seawater (HCO_3^-, NO_3^-, and SO_4^{2-}), are also abundant biochemical constituents. Molybdenum is much more abundant in biota than one might expect based on its crustal abundance; molybdenum forms the soluble molybdate ion (MoO_4^{2-}) in ocean water. In contrast, aluminum (Al) and silicon (Si) form insoluble hydroxides in seawater. They are found at low concentrations in living tissue, despite relatively high concentrations in the Earth's crust (Hutchinson 1943). Indeed, many elements that are rare in seawater are familiar poisons to living systems (e.g., Be, As, Hg, Pb, and Cd).

Although phosphorus forms a soluble oxyanion, PO_4^{3-}, it may never have been particularly abundant in seawater, owing to its tendency to bind to other minerals (Griffith et al. 1977). Unique properties of phosphorus may account for its major role in biochemistry, despite its relatively low geochemical abundance on Earth. With three ionized groups, phosphoric acid can link two nucleotides in DNA, with the third negative site acting to prevent hydrolysis and maintain the molecule within a cell membrane (Westheimer 1987). These ionic properties also allow phosphorus to serve in intermediary metabolism and energy transfer in ATP.

In sum, if one begins with the cosmic abundance of elements as an initial constraint, and the partitioning of elements during the formation of the Earth as subsequent constraints, then solubility in water appears to be a final constraint in determining the relative abundance of elements in the geochemical arena in which life arose. Those elements that were abundant in seawater are important biochemical constituents. Phosphorus appears as an important exception—an important biochemical constituent that has been in short supply for much of the Earth's biosphere through geologic time.

EVOLUTION OF METABOLIC PATHWAYS

In 1983, Awramik et al. reported that 3.5-billion-year-old rocks collected in Western Australia contained microfossils. While these observations were not without controversy (Brasier et al. 2004, Garcia-Ruiz et al. 2003), these and other specimens of about the same age may contain evidence of the first life on Earth (compare with Schopf et al. 2002). The earliest organisms on Earth may have resembled the methanogenic archaea that survive today in anaerobic hydrothermal (volcanic) environments at pH ranging from 9 to 11 and temperatures above 90°C (Rasmussen 2000, Huber et al. 1989, Kelley et al. 2005). Archaea are distinct from bacteria due to a lack of a muramic acid component in the cell wall and a distinct r-RNA sequence (Fox et al. 1980). Halophilic (salt-tolerant), acidophilic (acid-tolerant), and thermophilic (heat-tolerant) forms of archaea are also known (Figure 2.5). Kashefi and Lovley (2003) describe iron-reducing archaea growing at 121°C near a deep sea hydrothermal vent of the North Pacific—a potential analog of one of the earliest habitats for life on Earth.

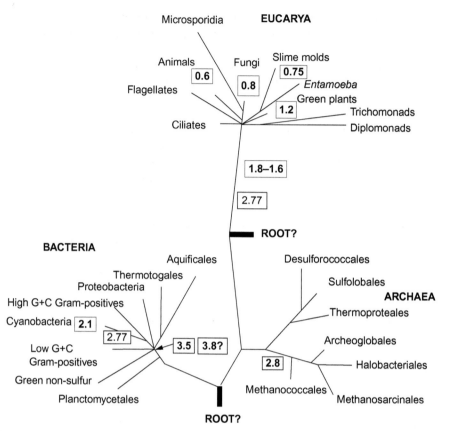

FIGURE 2.5   Relationship of the three domains of the tree of life, with boxes showing the estimated time (billions of years ago) for the first appearance of various forms. *Source: From Javaux (2006).*

The most primitive metabolic pathway probably involved the production of methane by splitting simple organic molecules, such as acetate, that would have been present in the oceans from abiotic synthesis:

$$CH_3COOH \rightarrow CO_2 + CH_4. \tag{2.8}$$

Organisms using this metabolism were scavengers of the products of abiotic synthesis and obligate heterotrophs, sometimes classified as chemoheterotrophs. The modern fermenting bacteria in the order Methanobacteriales may be our best present-day analogs.

Longer pathways of anaerobic metabolism, such as glycolysis, probably followed with increasing elaboration and specificity of enzyme systems. Oxidation of simple organic molecules in anaerobic respiration was coupled to the reduction of inorganic substrates from the environment. For example, sometime after the appearance of methanogenesis from acetate splitting, methanogenesis by CO_2 reduction,

$$CO_2 + 4H_2 \rightarrow CH_4 + 2H_2O, \tag{2.9}$$

probably arose among early heterotrophic microorganisms. Generally this reaction occurs in two steps: fermenting bacteria convert organic matter to acetate, H_2, and CO_2, and then archaea transform these to methane, following Eq. 2.9 (Wolin and Miller 1987, Kral et al. 1998). Note that this methanogenic reaction is more complicated than that from acetate splitting and would require a more complex enzymatic catalysis.

Evidence for the first methanogens is found in rocks more than 3.5 billion years old (Ueno et al. 2006). Both pathways of methanogenesis are found among the fermenting bacteria that inhabit wetlands and coastal ocean sediments today (see Chapters 7 and 9). Without O_2 in the atmosphere, these early microbial metabolisms may have led to large accumulations of methane and an enhanced greenhouse effect on Earth (Catling et al. 2001).

Today, microbial communities performing methanogenesis by CO_2 reduction are also found deep in the Earth, where H_2 is available from geologic sources (Stevens and McKinley 1995, Chapelle et al. 2002). These microbial populations are functionally isolated from the rest of the biosphere, and indicate another potential habitat for the first life on Earth. Indeed, a vast elaboration of prokaryotes is found at great depths on land and in ocean sediments worldwide, where they persist with extremely low rates of metabolism (Whitman et al. 1998, Parkes et al. 2005, Krumholz et al. 1997, Fisk et al. 1998, Schippers et al. 2005, Lomstein et al. 2012, Røy et al. 2012).

Before the advent of atmospheric O_2, the primitive oceans are likely to have contained low concentrations of available nitrogen—largely in the form of nitrate (NO_3^-; Kasting and Walker 1981; but see also Yung and McElroy 1979). Thus, the earliest organisms had limited supplies of nitrogen available for protein synthesis. There is little firm evidence that dates the origin of nitrogen fixation, in which certain bacteria break the inert, triple bond in N_2 and reduce the nitrogen to NH_3, but today this reaction is performed by bacteria that require strict local anaerobic conditions. The reaction

$$N_2 + 8H^+ + 8e^- + 16ATP \rightarrow 2NH_3 + H_2 + 16ADP + 16Pi \qquad (2.10)$$

is catalyzed by the enzyme complex known as *nitrogenase*, which consists of two proteins incorporating iron and molybdenum in their molecular structure (Georgiadis et al. 1992, Kim and Rees 1992, Chan et al. 1993). The modern form of nitrogenase, containing molybdenum, may have appeared only 1.5 to 2.2 bya, having evolved from earlier forms in methanogenic bacteria (Boyd et al. 2011). A cofactor, vitamin B_{12} that contains cobalt, is also essential (Palit et al. 1994, O'Hara et al. 1988). Nitrogen fixation requires the expenditure of large amounts of energy; breaking the N_2 bond requires 226 kcal/mole (Davies 1972). Modern nitrogen-fixing cyanobacteria couple nitrogen fixation to their photosynthetic reaction; other nitrogen-fixing organisms are frequently symbiotic with higher plants (Chapter 6).

Photosynthesis: The Origin of Oxygen on Earth

Despite various pathways of anaerobic metabolism, the opportunities for heterotrophic organisms must have been quite limited in a world where organic molecules were only available as a result of abiotic synthesis. Natural selection would strongly favor autotrophic systems that could supply their own reduced organic molecules for metabolism. Some of the earliest autotrophic metabolisms may have depended on H_2 (Schidlowski 1983, Tice and Lowe 2006, Canfield et al. 2006), namely:

$$2H_2 + CO_2 \rightarrow CH_2O + H_2O. \qquad (2.11)$$

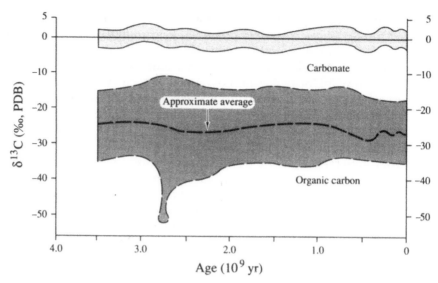

FIGURE 2.6   The isotopic composition of carbon in fossil organic matter and marine carbonates through geologic time, showing the range (*shaded*) among specimens of each age. The isotopic composition is shown as the ratio of ^{13}C to ^{12}C, relative to the ratio in an arbitrary standard (PDB belemite), which is assigned a ratio of 0.0. Carbon in organic matter is 2.8% less rich in ^{13}C than the standard, and this depletion is expressed as $-28‰$ $\delta^{13}C$ (see Chapter 5). *Source: From Schidlowski (1983).*

And we might also expect that one of the early photosynthetic reactions might have been based on the oxidation of the highly reduced gas, hydrogen sulfide (H_2S) (Schidlowski 1983, Xiong et al. 2000). For H_2S, the reaction

$$CO_2 + 2H_2S \rightarrow CH_2O + 2S + H_2O \qquad (2.12)$$

was probably performed by sulfur bacteria, not unlike the anaerobic forms of green and purple sulfur bacteria of today. These bacteria could have been particularly abundant around shallow submarine volcanic emissions of reduced gases, including H_2S.

Several indirect lines of evidence suggest that photosynthesis occurred in ancient seas of 3.8 bya. Photosynthesis produces organic carbon in which ^{13}C is depleted relative to its abundance in dissolved bicarbonate (HCO_3^-), and there are no other processes known to produce such strong fractionations between the stable isotopes of carbon.[5] Indeed, the carbon in carbonaceous chondrites is enriched in ^{13}C (Engel et al. 1990, Herd et al. 2011). Fossil organic matter with ^{13}C depletion is found in rocks from Greenland dating back to at least 3.8 bya (Mojzsis et al. 1996, Rosing 1999, Schidlowski 2001; Figure 2.6). This discrimination, which is about -2.8% ($-28‰$) in the dominant form of present-day photosynthesis, is based on

[5] When two isotopes are available, here ^{12}C and ^{13}C, most biochemical pathways proceed more rapidly with the lighter isotope, which is often more abundant. This preferential use is known as mass-dependent fractionation, and it leaves metabolic products with a different ratio of isotopes than found in the surrounding environment.

the slower diffusion of $^{13}CO_2$ relative to $^{12}CO_2$ and the greater affinity of the carbon-fixation enzyme, ribulose bisphosphate carboxylase, for the more abundant $^{12}CO_2$ (see Chapter 5).

Some workers have suggested that the C-isotope depletion seen in these rocks is an artifact of metamorphism (Fedo and Whitehouse 2002, van Zuilen et al. 2002; but see Dauphas et al. 2004). But other evidence for the existence of oxygen-producing photosynthesis is also found in these and other deposits, which are known as *banded iron formations* (BIF; Figure 2.7). Under the anoxic conditions of primitive Earth, Fe^{2+} released during rock weathering and from submarine hydrothermal emissions would be soluble and accumulate in seawater. With the advent of oxygenic photosynthesis, O_2 would be available to oxidize Fe^{2+} and deposit Fe_2O_3 in the sediments of the primitive ocean, namely:

$$4Fe^{2+} + O_2 + 4H_2O \rightarrow 2Fe_2O_3 \downarrow + 8H^+. \tag{2.13}$$

Massive worldwide deposits of the Fe_2O_3 in the banded iron formation are often taken as evidence for the presence of oxygen-producing photosynthesis based on the photochemical splitting of water in sunlight:

$$H_2O + CO_2 \rightarrow CH_2O + O_2. \tag{2.14}$$

Despite the relatively large energy barrier inherent in this photosynthetic reaction, there must have been strong selection for photosynthesis based on the splitting of water, particularly as the limited supplies of H_2S in the primitive ocean were removed by sulfur bacteria (Schidlowski 1983). Water offered an inexhaustible supply of substrate for photosynthesis.

Banded iron formations reach a peak occurrence in sediments deposited 2.5 to 3.0 bya (Walker et al. 1983). Most of the major deposits of iron ore in the United States (Minnesota), Australia, and South Africa are found in formations of this age (Meyer 1985). Presumably the deposition of BIF ended when the Earth's primitive oceans were swept clear of Fe^{2+} and excess oxygen could diffuse to the atmosphere.

Rather than taking banded iron formation as evidence of oxygenic photosynthesis, some workers have pointed out that it could also have been deposited by anaerobic, Fe^{2+}-oxidizing bacteria (Kappler et al. 2005, Widdel et al. 1993), namely:

$$4Fe^{2+} + HCO_3^- + 10H_2O \rightarrow CH_2O + 4Fe(OH)_3 + 7H^+. \tag{2.15}$$

FIGURE 2.7 Banded iron formation from the 3.25-billion-year-old Barberton Greenstone Belt, South Africa. *Sources: Collected by M.M. Tice (Texas A&M University); photo © 2010, Lisa M. Dellwo.*

Invoking this reaction as an alternative, one might conclude that oxygen-evolving photosynthesis dates only to the first appearance of cyanobacteria, which are known oxygenic forms today. Evidence for the presence of cyanobacteria is found in sediments deposited 2.5 to 2.7 bya (Summons et al. 1999, Brocks et al. 1999) or younger (Rasmussen et al. 2008). Indeed, some workers believe that anoxygenic Fe-photosynthesis may have dominated primary production in the primitive ocean (Table 2.4). Note that these two pathways for the deposition of BIF (Eqs. 2.13 and 2.15) are not mutually exclusive; both Fe-photosynthesizers and cyanobacteria can precipitate iron oxides in marine environments today (Trouwborst et al. 2007).

Despite some doubts regarding an early evolution of oxygenic photosynthesis, there is strong evidence that some forms of photosynthetic microbes existed at least 3.4 bya (Tice and Lowe 2004), and robust reports of microfossils, 3.2 to 3.4 bya, are derived from rocks of South Africa (Javaux et al. 2010, Fliegel et al. 2010) and Australia (Wacey et al. 2011). Thus, life seems to have appeared about 500 million years after the last great impacts of Earth's accretion. Whenever it first appeared, oxygenic photosynthesis offered a higher energy yield than other forms of photosynthesis, so these autotrophs might be expected to proliferate rapidly in the competitive arena of nature (Table 2.4).

Even in the face of an early evolution of oxygen-producing photosynthesis, many lines of evidence indicate that the Earth's atmosphere seems to have remained anoxic until about 2.45 to 2.32 billion years ago (Farquhar et al. 2000, 2011, Bekker et al. 2004, Sessions et al. 2009). Until recently, most researchers attributed the lack of oxygen solely to its reaction with reduced iron (Fe^{2+}) in seawater and the deposition of Fe_2O_3 in banded iron formations (Cloud 1973). Oxidation of other reduced species, perhaps sulfide (S^{2-}), may have also played a role, accounting for the slow buildup of SO_4^{2-} in Precambrian seawater (Walker and Brimblecombe 1985, Habicht et al. 2002). It is also possible that the early deposition of iron oxides in the banded iron formation held phosphorus concentrations at low levels, slowing the proliferation of photosynthetic organisms (Bjerrum and Canfield 2002, but see also Konhauser et al. 2007).

Several recent papers postulate an early evolution of aerobic respiration, closely coupled to local sites of O_2 production, which may have held the concentration of O_2 at low levels (Towe 1990, Castresana and Saraste 1995). Aerobic oxidation of methane may have kept oxygen at low levels until 2.7 bya (Konhauser et al. 2009). Only when the oceans were swept clear of reduced substances such as Fe^{2+}, S^{2-}, and CH_4 could excess O_2 accumulate in seawater and diffuse to the atmosphere. Thus, what is known as the Great Oxidation Event began about 2.4 bya, achieving 1% of the present level of O_2 in the atmosphere about 2.0 bya (Kump et al. 2011).

TABLE 2.4   Estimates of Marine Primary Production about 3.5 Billion Years Ago

Process	Annual rate	See Equation
H_2-based anoxygenic photosynthesis	0.35×10^{15} gC/yr	2.11
Sulfur-based anoxygenic photosynthesis	0.03	2.12
Fe-based anoxygenic photosynthesis	4.0	2.15
Present day	~50.0	2.14; Chapter 9

Source: Modified from Canfield et al. (2006).

Chemoautotrophy

Oxygen also enabled the evolution of several new biochemical pathways of critical significance to the global cycles of biogeochemistry (Raymond and Segre 2006). Two forms of aerobic biochemistry constitute chemoautotrophy. One based on sulfur or H_2S,

$$2S + 2H_2O + O_2 \rightarrow 2SO_4^{2-} + 4H^+, \tag{2.16}$$

is performed by various species of *Thiobacilli* (Ralph 1979). The protons generated are coupled to energy-producing reactions, including the fixation of CO_2 into organic matter (refer to Figure 1.4). On primitive Earth, these organisms could capitalize on elemental sulfur deposited from anaerobic photosynthesis (Eq. 2.12), and today they are found in local environments where elemental sulfur or H_2S is present, including some deep-sea hydrothermal vents (Chapter 9), caves (Sarbu et al. 1996), wetlands (Chapter 7), and lake sediments (Chapter 8).

Also important are the chemoautotrophic reactions involving nitrogen transformations by *Nitrosomonas* and *Nitrobacter* bacteria:

$$2NH_4^+ + 3O_2 \rightarrow 2NO_2^- + 2H_2O + 4H^+ \tag{2.17}$$

and

$$2NO_2^- + O_2 \rightarrow 2NO_3. \tag{2.18}$$

These reactions constitute *nitrification*, and the energy released is coupled to low rates of carbon fixation; that is, nitrifying bacteria are chemoautotrophs. Evidence for the first occurrence of sulfide oxidation and nitrification is indirect evidence for the presence of O_2 on Earth.

Anaerobic Respiration

With the appearance of SO_4^{2-} and NO_3^- as products of chemoautotrophic reactions, other metabolic pathways could evolve. The sulfate-reducing pathway, which depends on SO_4^{2-},

$$2CH_2O + 2H^+ + SO_4^{2-} \rightarrow H_2S + 2CO_2 + 2H_2O, \tag{2.19}$$

is found in archaea dating to 2.4 to 2.7 bya, on the basis of the S-isotope ratios in preserved sediments (Cameron 1982, Parnell et al. 2010). The late appearance of sulfate-reduction relative to photosynthesis may be related to the time needed to accumulate sufficient SO_4^{2-} in ocean waters, from the oxidation of sulfides, to make this an efficient means of metabolism (Habicht et al. 2002, Kah et al. 2004).

This biochemical pathway has been found in a group of thermophilic archaea isolated from the sediments of hydrothermal vent systems in the Mediterranean Sea, where a hot, anaerobic, and acidic microenvironment may resemble the conditions of primitive Earth (Stetter et al. 1987, Jorgensen et al. 1992, Elsgaard et al. 1994). In South Africa, simple microbial communities isolated at 2.8-km depth consist of sulfate-reducing archaea that fix nitrogen and are completely isolated from energy inputs from the Sun at the Earth's surface (Lin et al. 2006, Chivian et al. 2008).

Similarly, today an anaerobic, heterotrophic reaction called *denitrification* is performed by bacteria, commonly of the genus *Pseudomonas*, found in soils and wet sediments (Knowles 1982), namely:

$$5CH_2O + 4H^+ + 4NO_3^- \rightarrow 2N_2 + 5CO_2 + 7H_2O. \qquad (2.20)$$

The denitrifying reaction requires NO_3^-, and its preferential use of $^{14}NO_3^-$ over $^{15}NO_3^-$ leaves the ocean enriched in $^{15}NO_3^-$. Rocks showing this enrichment are dated to at least 2.0 bya (Beaumont and Robert 1999, Papineau et al. 2005) and perhaps earlier (Garvin et al. 2009, Godfrey and Falkowski 2009). At that time, nitrate must have been present as products of nitrification reactions (Eqs. 2.17 and 2.18), providing another indirect line of evidence for the presence of O_2 on Earth.

Although the denitrification reaction requires anoxic environments, denitrifiers are facultatively aerobic—that is, switching to aerobic respiration when O_2 is present. This is consistent with several lines of evidence that suggest that denitrification may have appeared later than the strictly anaerobic pathways of methanogenesis and sulfate reduction (Betlach 1982). Denitrification would have been efficient only after relatively high concentrations of NO_3^- had accumulated in the primitive ocean, which is likely to have contained low NO_3^- at the start (Kasting and Walker 1981). Thus, the evolution of denitrification may have been delayed until sufficient O_2 was present in the environment to drive the nitrification reactions (Eqs. 2.17 and 2.18). It is interesting to note that having evolved in a world dominated by O_2, the enzymes of today's denitrifying organisms are not destroyed, but merely inactivated, by O_2 (Bonin et al. 1989, McKenney et al. 1994).

The first O_2 that reached the atmosphere was probably immediately involved in oxidation reactions with reduced atmospheric gases and with exposed crustal minerals of the barren land (Holland et al. 1989, Kump et al. 2011). Oxidation of reduced minerals, such as pyrite (FeS_2), would transfer SO_4^{2-} and Fe_2O_3 to the oceans in riverflow (Konhauser et al. 2011). Deposits of Fe_2O_3 that are found in alternating layers with other sediments of terrestrial origin constitute *red beds*, which are found beginning at 2.0 bya and indicative of aerobic terrestrial weathering (Van Houten 1973). It is noteworthy that the earliest occurrence of red beds roughly coincides—with little overlap—with the latest deposition of banded iron formation, further evidence that the oceans were swept clear of reduced Fe before O_2 began to diffuse to the atmosphere.

Canfield (1998) suggested that with O_2 in Earth's atmosphere, the oceans about 2 bya may have consisted of oxic surface waters and anoxic bottom waters, which only later became oxic throughout the water column (Reinhard et al. 2009). In this model, the bottom waters may have contained substantial concentrations of sulfide, produced by sulfate-reducing bacteria using SO_4^{2-} mixing down from above. Since most metals are scarcely soluble in the presence of sulfide (Anbar and Knoll 2002), Fe^{2+} would have been removed from these bottom waters. Deposition of the banded iron formation would cease because Fe^{2+} was precipitated as FeS_2 rather than Fe_2O_3 (but see Planavsky et al. 2011).

Oxygen began to accumulate to its present-day atmospheric level of 21% when the rate of O_2 production by photosynthesis exceeded its rate of consumption by the oxidation of reduced substances. Atmospheric oxygen may have reached 21% as early as the Silurian—about 430 mya (see inside back cover), and it is not likely to have fluctuated outside the range of 15 to 35% ever since (Berner and Canfield 1989, Scott and Glasspool 2006). What maintains the concentration at such stable levels? Walker (1980) examined all the oxidation/reduction reactions affecting atmospheric O_2, and suggested that the balance is due to the negative feedback between O_2 and the long-term net burial of organic matter in sedimentary rocks. When O_2 rises, less organic matter escapes decomposition, stemming a further rise in O_2. We will examine these processes in more detail in Chapters 3 and 11, but here it is interesting to note the significance of an atmosphere with 21% O_2. Lovelock (1979) points out that with <15% O_2, fires would not burn, and at >25% O_2, even wet organic matter would burn freely (Watson et al. 1978, Belcher and McElwain 2008). Either scenario would result in a profoundly different world than that of today.

The release of O_2 by photosynthesis is perhaps the single most significant effect of life on the geochemistry of the Earth's surface (Raymond and Segre 2006). The accumulation of free O_2 in the atmosphere has established the oxidation state for most of the Earth's surface for the last 2 billion years. However, of all the oxygen ever evolved from photosynthesis, only about 2% resides in the atmosphere today; the remainder is buried in various oxidized sediments, including banded iron formations and red beds (see Figure 2.8 and refer to Table 2.3). The total inventory of free oxygen that has ever been released on the Earth's surface is, of course, balanced stoichiometrically by the storage of reduced carbon in the Earth's crust, including coal, oil, and other reduced compounds of biogenic origin (e.g., sedimentary pyrite). The sedimentary storage of organic

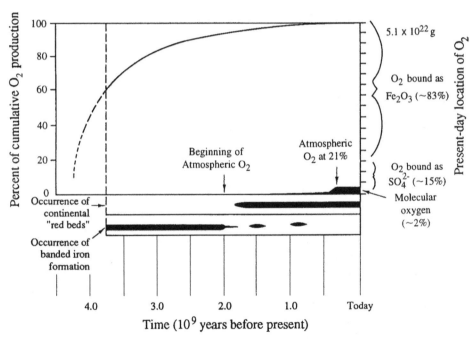

FIGURE 2.8   Cumulative history of O_2 released by photosynthesis through geologic time. Of more than 5.1×10^{22} g of O_2 released, about 98% is contained in seawater and sedimentary rocks, beginning with the occurrence of banded iron formation beginning at least 3.5 bya. Although O_2 was released to the atmosphere beginning about 2.0 bya, it was consumed in terrestrial weathering processes to form red beds, so that the accumulation of O_2 to present levels in the atmosphere was delayed to 400 mya. *Source: Modified from Schidlowski (1980).*

carbon is now estimated at 1.56×10^{22} g (Des Marais et al. 1992; see also Table 2.3), representing the cumulative net production of *bio*geochemistry since the origin of life.

The release of free oxygen as a byproduct of photosynthesis also dramatically altered the evolution of life on Earth. The pathways of anaerobic respiration by methanogenic bacteria and photosynthesis by sulfur bacteria are poisoned by O_2. These organisms generally lack catalase and have only low levels of superoxide dismutase—two enzymes that protect cellular structures from damage by highly oxidizing compounds such as O_2 (Fridovich 1975). Today, these metabolisms are confined to local anoxic environments. Alternatively, eukaryotic metabolism is possible at O_2 levels that are about 1% of present levels (Berkner and Marshall 1965, Chapman and Schopf 1983). Fossil evidence of eukaryotic organisms is found in rocks formed 1.7 to 1.9 bya (Knoll 1992; Figure 2.5), and perhaps even as much as 2.1 billion years ago (Han and Runnegar 1992). Large colonial organisms are reported from rocks 2.1 bya (El Albani et al. 2010). The rate of evolution of amino acid sequences among major groups of organisms suggests that prokaryotes and eukaryotes diverged 2 bya (Doolittle et al. 1996). All these dates are generally consistent with the end of deposition of the banded iron formation and the presence of O_2 in the atmosphere as indicated by red beds (Table 2.5).

TABLE 2.5   Milestones in the Deep History of the Earth

Milestone	When occurred (bya)
Origin of the Universe	13.7
Origin of the Milky Way Galaxy	12.5
Origin of the Sun	4.57
Accretion of the Earth largely complete[a]	4.5
Liquid water on Earth	4.3
Last of the great impacts	3.8
Earliest evidence of photosynthesis	
Depleted ^{13}C and banded iron formations	3.8
Earliest evidence of cellular structures	3.5
First evidence of cyanobacteria	2.7
First evidence of O_2 in the atmosphere	2.45
Evidence of seawater SO_4^{2-}, thus O_2	2.4
Evidence of denitrification, hence NO_3^- and O_2	2.5
Evidence of aerobic rock weathering (red beds)	2.0
First evidence of eukaryotes	2.0
End of banded iron formation	1.8
Land plants	0.43
Genus *Homo*	0.002

[a] *Impact of Theia at 4.527 bya forms the Moon.*

FIGURE 2.9   Cyanobacteria inhabit the space beneath quartz stones in the Mojave Desert, California, where they photosynthesize on the light passing through these translucent rocks. *Source: Schlesinger et al. (2003). Photo © 2010, Lisa M. Dellwo.*

O_2 in the environment allowed eukaryotes to localize their heterotrophic respiration in mitochondria, providing an efficient means of metabolism and allowing a rapid proliferation of higher forms of life. Similarly, more efficient photosynthesis in the chloroplasts of eukaryotic plant cells presumably enhanced the production and further accumulation of atmospheric oxygen.

O_2 in the stratosphere is subject to photochemical reactions leading to the formation of ozone (Chapter 3). Today, stratospheric ozone provides an effective shield for much of the ultraviolet radiation from the Sun that would otherwise reach the Earth's surface and destroy most life. Before the O_3 layer developed, the earliest colonists on land may have resembled the microbes and algae that inhabit desert rocks of today (e.g., Friedmann 1982, Bell 1993, Schlesinger et al. 2003, Phoenix et al. 2006; Figure 2.9). Although there is some fossil evidence for the occurrence of extensive microbial communities on land during the Precambrian (Horodyski and Knauth 1994, Knauth and Kennedy 2009, Strother et al. 2011), it is unlikely that higher organisms were able to colonize land abundantly until the ozone shield developed. Multicellular organisms are found in ocean sediments dating to about 680 mya, but the colonization of land by higher plants was apparently delayed until the Silurian (Gensel and Andrews 1987, Kenrick and Crane 1997). A proliferation of plants on land followed the development of lignified, woody tissues (Lowry et al. 1980) and the origin of effective symbioses with mycorrhizal fungi that allow plants to obtain phosphorus from unavailable forms in the soil (Pirozynski and Malloch 1975, Simon et al. 1993, Yuan et al. 2005; Chapter 6). Even primitive land plants are likely to have speeded the formation of clay minerals, which help preserve organic matter from degradation and thus further the accumulation of oxygen in the atmosphere (Kennedy et al. 2006, Chapter 4).

COMPARATIVE PLANETARY HISTORY: EARTH, MARS, AND VENUS

In the release of free O_2 to the atmosphere, life has profoundly affected the conditions on the surface of the Earth. But what might have been the conditions on Earth in the absence of life? Some indications are given by our neighboring planets Mars and Venus, which are the best replicates we have for the underlying geochemical arena on Earth. We are fairly

FIGURE 2.10 The surface of Mars as seen from the *Viking 2 Lander* in 1976.

confident that there has never been life on these planets, so their surface composition represents the cumulative effect of 4.5 billion years of abiotic processes. Our understanding of the atmosphere of Mars has improved markedly since the Viking landing in 1976, followed by landings in 1997, 2004, and 2007 that further explored its atmosphere and surface properties with robotic instruments (Figure 2.10).[6]

Table 2.6 compares a number of properties and conditions on Earth, Mars, and Venus. Two properties characterize the atmosphere of these planets: the total mass (or pressure) and the proportional abundance of constituents. The present atmosphere on Mars is only about 0.76% as massive as that on Earth (Hess et al. 1976). We should expect a less massive atmosphere on Mars than on Earth because the gravitational field is weaker on a smaller planet. Mars probably began with a smaller allocation of the solar nebula during planetary formation, and we should expect that a small planet would have retained less internal heat to drive tectonic activity and outgassing of its mantle after its formation (Anders and Owen 1977, Owen and Biemann 1976). Estimates of the cumulative generation of magma are substantially lower for Mars (0.17 km^3/yr) than for Earth (26–34 km^3/yr) or Venus (<19 km^3/yr) (Greeley and Schneid 1991).

We should also expect that the surface temperature on Mars would be colder than that on Earth because the planet is much farther from the Sun. The average temperature on Mars, $-53°C$ at the site of the Viking landing (Kieffer 1976), ensures that water is frozen on most of the Martian surface at all seasons. Ice is found at both poles of Mars and in Martian soils in other areas (Titus et al. 2003, Mustard et al. 2001, Smith et al. 2009). In the absence of liquid water, we would expect that the atmosphere on Mars would be mostly dominated by CO_2, which readily dissolves in seawater on Earth (Eq. 2.4). Indeed, CO_2 constitutes a major

---

[6] The NASA lander, *Curiosity*, arrived safely on Mars on August 6, 2012.

TABLE 2.6 Some Characteristics of the Inner Planets

	Mars[a]	Earth	Venus[b]
Distance to the Sun (10^6 km)	228	150	108
Surface temperature (°C)	−53	16	474
Radius (km)	3390	6371	6049
Atmospheric pressure (bars)	0.007	1	92
Atmospheric mass (g)	2.4×10^{19}	5.3×10^{21}	5.3×10^{23}
Atmospheric composition (% wt.)			
CO_2	95	0.04	98
N_2	2.5	78	2
O_2	0.25	21	0
H_2O	0.10	1	0.05
$^{40}Ar/^{36}Ar$[c]	2840	296	1

[a] From Owen and Biemann (1976).
[b] From Nozette and Lewis (1982).
[c] Wayne (1991).

proportion of the thin atmosphere of Mars, and the observed fluctuations of the ice cap at the south pole of Mars appear due to seasonal variations in the amount of CO_2 that is frozen out of its atmosphere (Leighton and Murray 1966, James et al. 1992, Phillips et al. 2011a).

Several attributes of Mars are anomalous. First, with most of the water and CO_2 now trapped on the surface, why is N_2 such a minor component of the atmosphere on Mars? Second, why do the surface conditions on Mars indicate a period when small amounts of liquid water may have been present on its surface (Malin and Edgett 2000, 2003; Squyres et al. 2004; Solomon et al. 2005)—perhaps even recently (Malin et al. 2006, Christensen 2003, McEwen et al. 2011). Could it be that a more massive early atmosphere may once have allowed a significant "greenhouse effect" on Mars and warmer surface temperatures than today (Pollack et al. 1987)? Such a scenario could explain an early sporadic occurrence of liquid water on Mars, but if it is correct, why did Mars lose its atmosphere and cool to its present surface temperature of −53°C?

Losses of atmospheric gases from Mars may have resulted from several processes. A thick atmosphere on Mars may have been lost to space as a result of catastrophic impacts during its early history (Carr 1987, Melosh and Vickery 1989) or by a process known as "sputtering" driven by solar wind (Kass and Yung 1995, Hutchins and Jakosky 1996). Impacts are consistent with the low abundance of noble gases on Mars relative to the concentrations in carbonaceous chondrites and the Sun (Hunten 1993, Pepin 2006). Catastrophic loss of an early atmosphere is also consistent with the observation that nearly the entire atmosphere on Mars is secondary, as evidenced by a $^{40}Ar/^{36}Ar$ ratio of 2750 or more (Owen and Biemann 1976, Biemann et al. 1976, Wayne 1991, Yung and DeMore 1999). Mars shows some evidence of recent volcanic activity (Neukum et al. 2004), and a significant proportion of its thin atmosphere may have been derived <1 bya (Gillmann et al. 2011, Niles et al. 2010).

Loss of water from Mars may have also occurred as the water vapor in its atmosphere underwent photolysis by ultraviolet light. Observations of analogous processes on Earth are instructive. In the upper atmosphere on Earth, small amounts of water vapor are subject to photodisassociation, with the loss of H_2 to space. However, because the upper atmosphere is cold, little water vapor is present, and the process has been minor throughout Earth's history (Chapter 10). If this process were significant on Mars, we would expect that the loss of ^1H would be more rapid than that of ^2H, leaving a greater proportion of ^2H_2O in the planetary inventory. Owen et al. (1988) found that the ratio of ^2H (deuterium) to ^1H on Mars is about $6\times$ that on Earth, suggesting that Mars may have once possessed a large inventory of water that has been lost to space (de Bergh 1993, Krasnopolsky and Feldman 2001). The relative abundance of ^2H has potentially increased through time, as seen in measurements of putative Martian meteorites of various ages that have been collected on Earth (Greenwood et al. 2008). Although a small amount of O_2 is found in the Martian atmosphere (Table 2.6), most of the oxygen produced from the photolysis of water has probably oxidized minerals of the crust, namely:

$$4FeO + O_2 \rightarrow 2Fe_2O_3, \tag{2.21}$$

giving Mars its reddish color (Figure 2.10).

Nitrogen may have also been lost from Mars as N_2 underwent photodisassociation in the upper atmosphere, forming monomeric N. This process occurs on Earth as well, but even N is too heavy to escape the Earth's gravitational field and quickly recombines to form N_2. With its smaller size, Mars allows the loss of N. Relative to the Earth, a higher proportion of $^{15}N_2$ in the Martian atmosphere is suggestive of this process, since the escape of ^{15}N would be slower than that of ^{14}N, which has a lower atomic weight (McElroy et al. 1976, Murty and Mohapatra 1997). Both the Earth and Mars have a higher relative proportion of ^{15}N than the solar composition—Mars more so (Marty et al. 2011).

With losses of H_2O and N_2 to space, it is not surprising that the Martian atmosphere is dominated by CO_2. What is surprising is that the atmospheric mass is so low. As much as 3 bars of CO_2 may have been degassed from the interior of Mars, but only about 10% of that amount appears to be frozen in the polar ice caps and the soil (Kahn 1985). Some CO_2 may have been lost to space (Kass and Yung 1995), but during an earlier period of moist conditions, CO_2 may have also reacted with the crust of Mars, weathering rocks and forming carbonate minerals on its surface (Bandfield et al. 2003, Ehlmann et al. 2008, Boynton et al. 2009, Morris et al. 2010). With the loss of tectonic activity on Mars, there was no mechanism to release this CO_2 back to the atmosphere, as there is on Earth (refer to Figure 1.3).

In sum, various lines of evidence suggest that Mars had a higher inventory of volatiles early in its history, but most of the atmosphere has been lost to space or in reactions with its crust. The presence of water on Mars may have once offered an environment conducive to the evolution of life, especially in light of the relatively rapid appearance of life on Earth. Subsurface and hydrothermal environments are also possible sites for early life on Mars (Squyres et al. 2008), and today, intriguing emissions of CH_4 are observed from the Martian surface (Formisano et al. 2004, Mumma et al. 2009). Nevertheless, evidence for past life on Mars is rather scant (McKay et al. 1996), and there is no evidence of liquid water or life on Mars today. Even organic molecules received from meteoric impacts on the surface of Mars have been oxidized by the Sun's ultraviolet light (Kminek and Bada 2006).

Through geologic time, the loss of water from Mars would remove a large component of greenhouse warming from the planet. The thin atmosphere that remains is dominated by CO_2, but it offers little greenhouse warming—raising the temperature of Mars only about 10°C over what might be seen if Mars had no atmosphere at all (Houghton 1986). Our best estimate suggests that the volume of CO_2 now frozen in the polar ice caps (100 mbar) and soils (300 mbar) on Mars is insufficient to supply the 2 bars of atmospheric CO_2 that would be necessary for the greenhouse effect to raise the temperature of the planet above the freezing point of water (Pollack et al. 1987). Thus, it would be difficult to use planetary-level engineering to establish a large, self-sustained greenhouse effect on Mars, allowing humans to colonize the planet (McKay et al. 1991).

On Venus, the ratio of the mass of the atmosphere to the mass of the planet (1.09×10^{-4}) is only slightly less than the ratio of the total mass of volatiles on Earth (see Table 2.3) to the mass of the Earth (3.3×10^{-4}). These values suggest a similar degree of crustal degassing on these planets. Indeed, volcanism is observed on Venus today (Smrekar et al. 2010, Bondarenko et al. 2010), and the atmosphere on Venus is remarkably similar to the average composition of carbonaceous chondrites (Pepin 2006). Despite the evidence of outgassing, the $^{40}Ar/^{36}Ar$ ratio on Venus is close to 1.0, implying that, relative to the Earth, Venus may have also retained a greater fraction of gases from the solar nebula during its accretion (Pollack and Black 1982). Unlike the Earth, the high surface temperature of 474°C on Venus ensures that its present inventory of volatiles resides entirely in its atmosphere. The atmospheric pressure on Venus is nearly 100× that of Earth (Table 2.6). The hot, high-pressure conditions have made it difficult to land spacecraft for the exploration of the surface of Venus.

The massive atmosphere on Venus is dominated by CO_2, conferring a large greenhouse warming and surface temperatures well in excess of that predicted for a nonreflective body at the same distance from the Sun (54°C; Houghton 1986).[7] The relative abundance of CO_2 and N_2 in the atmosphere of Venus is roughly similar to that in the total inventory of volatiles on Earth (Oyama et al. 1979, Pollack and Black 1982, Lecuyer et al. 2000). What is unusual about Venus is the low abundance of water in its atmosphere. Was Venus wet in the past?

The ratio of ^2H (deuterium) to ^1H on Venus is >100× higher than that on Earth (Donahue et al. 1982, McElroy et al. 1982, de Bergh et al. 1991), suggesting that Venus, like Mars, may have possessed a large inventory of water in the past, but lost water through a process that differentiates between the isotopes of hydrogen. With the warm initial conditions on Venus, a large amount of the water vapor in the atmosphere may have been subject to photodisassociation, causing the planet to dry out through its history (Kasting et al. 1988, Lecuyer et al. 2000). The oxygen released during the photodisassociation of water has probably reacted with crustal minerals (Donahue et al. 1982, McGill et al. 1983, p. 87).

At the surface temperatures found on Venus, little CO_2 can react with its crust (compare with Figure 1.3), so high concentrations of CO_2 remain in the atmosphere (Nozette and Lewis 1982). Various other gases, such as SO_2, that are found dissolved in seawater on Earth also

---

[7] One of the widely cited reports of the Intergovernmental Panel on Climate Change (IPCC) indicates that the surface temperature on Venus in the absence of a greenhouse effect would be −47°C (Houghton et al. 1990). This is lower than the value given here because the IPCC report accounts for the reflectivity of the thick cloud layer on Venus, whereas our value considers the equilibrium temperature for a black body absorber in the orbit of Venus. See Lewis and Prinn (1984, p. 97).

reside as gases in the atmosphere on Venus (Oyama et al. 1979, Svedhem et al. 2007, Marcq et al. 2011). Continuing volcanic releases of CO_2 have accumulated in the atmosphere to produce a runaway greenhouse effect in which increasing temperatures allow an increasing potential for the atmosphere to hold CO_2 and other gases (Walker 1977). Thus, the current temperature on Venus, 474°C, is much greater than we would predict if Venus had no atmosphere and is not conducive to life as we know it.

The proliferation of known planets around other stars and observations of organic molecules in interstellar dusts and meteorites beg for further exploration for the presence of extraterrestrial life in our galaxy and beyond (Lissauer et al. 2011, Cassan et al. 2012). At least one planet, Kepler-22b, which is 600 light years from Earth, shows a size and an orbit around its sun that might be conducive to life. The massive atmospheres on the outer planets of our solar system are not conducive to life, but Europa and the other moons of Jupiter show evidence of subsurface oceans beneath surface ice (Carr et al. 1998, Kivelson et al. 2000). Some of these habitats may be subjected to hydrothermal activity, providing a submarine habitat for the abiotic synthesis of organic materials and perhaps modest forms of metabolism (Gaidos et al. 1999, Marion et al. 2003). A small amount of O_2 is detected in the atmosphere of Europa (10^{-11} of that on Earth), where it may originate from a photolytic process similar to that leading to the loss of water from Mars and Venus (Hall et al. 1995).

The atmosphere of Saturn's moon Titan is dominated by nitrogen (96%) and methane (3%), with an $^{40}Ar/^{36}Ar$ ratio of 154, suggesting a secondary origin from outgassing (Niemann et al. 2005, Yung and DeMore 1999, p. 202). Huge lakes of liquid methane are potentially stable on its surface (Stofan et al. 2007, Lorenz et al. 2008), and liquid methane falls from its atmosphere as rain (Hueso and Sanchez-Lavega 2006, Turtle et al. 2011). On Saturn's moon, Enceladus, frozen water-ice covers much of the surface (Brown et al. 2006a), and geysers appear to spew water vapor into the atmosphere (Waite et al. 2006, Postberg et al. 2011). Habitats with liquid water, hydrothermal activity, and reduced gases (especially methane) on planets and planetary moons are likely places to look for extraterrestrial life within and outside our solar system (Swain et al. 2008). Subglacial lakes in Antarctica are analogous terrestrial habitats on Earth (Priscu et al. 1998).

Certainly the most unusual characteristic of the Earth's atmosphere is the presence of large amounts of O_2, which is an unequivocal indication of life on this planet (Sagan et al. 1993). Having examined the conditions on Mars, Venus, and other bodies of our solar system, we can now offer some speculation on the conditions that might exist on a lifeless Earth. At a distance of 150×10^6 km from the Sun, the surface temperature on the Earth, assuming no reflectivity to incoming solar radiation, would be close to the freezing point of water (Houghton 1986). Such cold conditions would seem to ensure that the atmosphere on the Earth has never contained much water vapor, so relatively little water has been lost to space as a result of photolysis in the upper atmosphere.

Despite the small amount of H_2O in Earth's atmosphere, the atmosphere confers enough greenhouse warming to the planet to have maintained liquid oceans for most of its history. Thus, even on a lifeless Earth, most of the inventory of volatiles would reside in the oceans. The atmosphere on a lifeless Earth would be dominated by N_2, which is only slightly soluble in water. The size of Earth and its gravitational field ensure that photolysis of N_2 does not result in the loss of N from the planet. Moreover, the rate of fixation of nitrogen by lightning in an atmosphere without O_2 appears too low to transfer a significant portion of N_2 from the atmosphere to the oceans (Kasting and Walker 1981; Chapter 12). Thus, the main effect of life has been to dilute the initial nitrogen-rich atmosphere on Earth with a large quantity of O_2 (Walker 1984).

How long will the Earth be hospitable to life? Barring some unforeseen catastrophe, we can speculate that the biosphere will persist as long as our planet harbors liquid water on its surface. Eventually, however, a gradual increase in the Sun's luminosity will warm the Earth, causing a photolytic loss of water from the upper atmosphere, irreversible oxidation of the Earth's surface, and the demise of life—perhaps after another 2.5 billion years (Lovelock and Whitfield 1982, Caldeira and Kasting 1992). The Sun itself will burn out in 10 billion years. If we manage the planet well, studies of biogeochemistry have a long future.

SUMMARY

In this chapter we have reviewed theories for the formation and differentiation of early Earth. In the process of planetary formation, certain elements were concentrated near its surface and only some elements were readily soluble in seawater. Thus, the environment in which life arose is a special mix taken from the geochemical abundance of elements that were available on Earth. Simple organic molecules can be produced by physical processes in the laboratory; presumably similar reactions occurred in high-energy habitats on primitive Earth. Life may have arisen by the abiotic assembly of these constituents into simple forms, resembling the most primitive bacteria that we know of today. Essential to living systems is the processing of energy, which is likely to have begun with the heterotrophic consumption of molecules found in the environment. A persistent scarcity of such molecules is likely to have led to selection for the autotrophic production of energy by various pathways, including photosynthesis. Autotrophic photosynthesis appears to be responsible for nearly all the production of O_2, which has accumulated in the Earth's atmosphere over the last 2 billion years. The major biogeochemical cycles on Earth are mediated by organisms whose metabolic activities couple the oxidation and reduction of substances isolated from the environment.

Recommended Readings

Broecker, W.S. 1985. *How to Build a Habitable Planet*. LaMont–Doherty Earth Observatory.

Brown, G.C. and A.E. Mussett. 1981. *The Inaccessible Earth*. Unwin Hyman.

Cox, P.A. 1989. *The Elements*. Oxford University Press.

Fraústo da Silva, J.J.R. and R.J.P. Williams. 1991. *The Biological Chemistry of Life*. Oxford University Press.

Hazen, R.M. 2012. *The Story of Earth: The First 4.5 Billion Years, from Stardust to Living Planet*. Viking Books.

Jakosky, B. 1998. *The Search for Life on Other Planets*. Cambridge University Press.

Kasting J.F. 2010. *How to Find a Habitable Planet*. Princeton University Press.

Knoll, A.H. 2003. *Life on a Young Planet*. Princeton University Press.

Williams, G.R. 1996. *The Molecular Biology of Gaia*. Columbia University Press.

Yung, Y.L. and W.B. DeMore. 1999. *Photochemistry of Planetary Atmospheres*. Oxford University Press.

PROBLEMS

2.1. If the average carbon content of a carbonaceous chondrite is 3.5% by mass (compare with Anders and Grevesse 1989), what mass of such meteorites would have to be received to account for the total mass of carbon on Earth (Table 2.3)? How does this compare to the total mass of the Earth? What would be the % nitrogen content in these chondrites to account for the nitrogen inventory on Earth?

2.2. Assuming that the net primary production of today's biosphere (see Tables 5.3 and 9.2) has pertained to the Earth for the past 400,000,000 years, what is the cumulative amount of carbon that has at one time or another been held in living tissue over this interval? How does this compare to the carbon inventory on Earth and to the mass of the Earth's crust (estimate or cite a source for this number)? What is the significance of your conclusion?

2.3. On today's Earth, water at sea level (that is, under 1 atmosphere of pressure) boils at 100°C. If the total inventory of volatiles (Table 2.3) was present in the atmosphere of the Earth about 4.0 bya, at what temperature would liquid water have first condensed to form the oceans? Based on your calculation, what is the maximum fraction of today's inventory that could have been in the atmosphere when the condensation began? *Hint*: You will need to consult a table for the vapor pressure of water as a function of temperature.

2.4. The best estimates of the total amount of carbon (i.e., CO_2) released from volcanoes worldwide is about 0.05×10^{15} gC/yr. Using the data of Table 2.3, what does this suggest for the mean residence time for carbon in marine sediments? Why is this older than the oldest oceanic crust?

2.5. What is the N_2/Ar molar ratio in the volcanic emissions compiled in Table 2.2 and in the total inventory of volatiles on Earth (Table 2.3)? Why might you expect the N_2/Ar ratio to be higher in volcanic emissions than in the inventory?

INTRODUCTION

There are several reasons to begin our treatment of biogeochemistry with a consideration of the atmosphere. The atmosphere has evolved as a result of the history of life on Earth (Chapter 2), and there is good evidence that it is now changing rapidly as a result of human activities. The atmosphere controls Earth's climate and ultimately determines the conditions in which we live—our supplies of food and water, our health, and our economy. Further, the atmosphere is relatively well mixed, so changes in its composition can be taken as a first index of changes in biogeochemical processes at the global level. The circulation of the atmosphere transports biogeochemical constituents between the oceans and land, resulting in a global circulation of elements.

We begin our discussion with a brief consideration of the structure, circulation, and composition of the atmosphere. Then we examine reactions that occur among various gases, especially in the lower atmosphere. Many of these reactions remove constituents from the atmosphere, depositing them on the surface of the land and sea. In the face of constant losses, the composition of the atmosphere is maintained by biotic processes that supply gases to the atmosphere. We mention the sources of atmospheric gases here briefly, but they will be treated in more detail in later chapters of this book, especially as we examine the microbial reactions that occur in soils, wetlands, and ocean sediments. Finally, we discuss human impacts on the global atmosphere, as seen in ozone depletion and climate change.

STRUCTURE AND CIRCULATION

The atmosphere is held on Earth's surface by the gravitational attraction of the Earth. At any altitude, the downward force (F) is related to the mass (M) of the atmosphere above that point:

$$F = M(g),   (3.1)$$

where g is the acceleration due to gravity (980 cm/sec^2 at sea level). Pressure (force per unit area) decreases with increasing altitude because the mass of the overlying atmosphere is smaller (Walker 1977). Decline in atmospheric pressure (P in bars) with altitude (A in km) is approximated by the logarithmic relation:

$$\log P = -0.06(A),   (3.2)$$

over the whole atmosphere (Figure 3.1).

Although the chemical composition of the atmosphere is relatively uniform, when we visit high mountains, we often say that the atmosphere seems "thinner" than at sea level. The abundance of molecules in each volume of the atmosphere is greater at sea level, because it is compressed by the pressure of the overlying atmosphere. Thus, the lower atmosphere, the *troposphere*, contains about 80% of the atmospheric mass (Warneck 2000), and jet aircraft flying at high altitudes require cabin pressurization for their passengers.

Certain atmospheric constituents, such as ozone, aerosols, and clouds absorb and reflect portions of the radiation that the Earth receives from the Sun, so only about half of the Sun's radiation penetrates the atmosphere to be absorbed at the Earth's surface (Figure 3.2). The overall reflectivity or *albedo* of the Earth, as measured by changes in "earthshine" received by the Moon, is about 30% (Goode et al. 2001). The Earth's albedo has apparently increased slightly in recent years, presumably due to particulate air pollutants (Pallé et al. 2009, Wang et al. 2009). Greater albedo reduces the radiation reaching Earth's surface (i.e., global dimming).

The land and ocean surfaces reradiate long wave (heat) radiation to the atmosphere, so the atmosphere is heated from the bottom and is warmest at the Earth's surface (Figure. 3.1). Because warm air is less dense and rises, the troposphere is well mixed. The top of the troposphere extends to 8 to 17 km, varying seasonally and with latitude. The temperature of the upper troposphere is about −60°C, which ensures that the atmosphere above 10 km contains only small amounts of water vapor.

Above the troposphere, the stratosphere is defined by the zone in which temperatures increase with altitude, extending to about 50 km (Figure 3.1). The increase is largely due to the absorption of ultraviolet light by ozone. Vertical mixing in the stratosphere is limited, as is exchange across the boundary between the troposphere and the stratosphere, the *tropopause*.

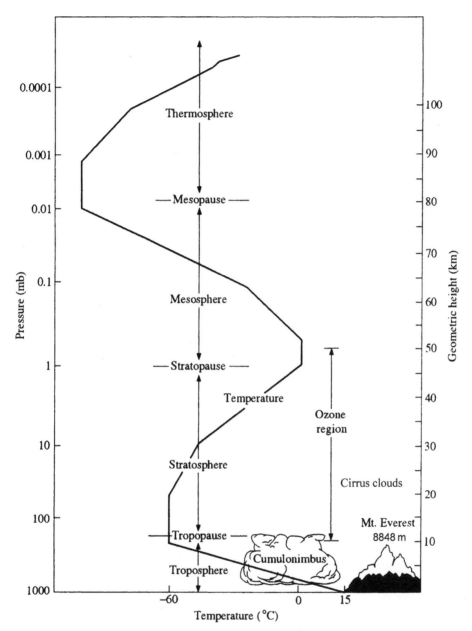

FIGURE 3.1 Vertical structure and zonation of the atmosphere, showing the temperature profile to 100-km altitude. Note the logarithmic decline in pressure (*left axis*) as a function of altitude.

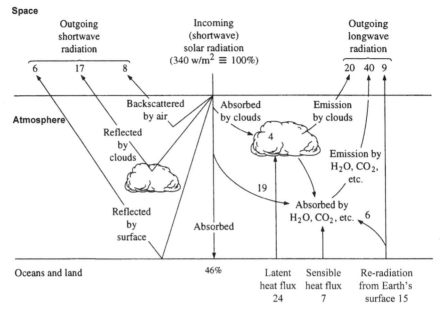

FIGURE 3.2   The radiation budget for Earth, showing the proportional fate of the energy that Earth receives from the Sun, about $340\,W/m^2$ largely in short wavelengths. About one-third of this radiation is reflected back to space and the remaining is absorbed by the atmosphere (23%) or the surface (46%). Long-wave radiation (infrared) is emitted from the Earth's surface, some of which is absorbed by atmospheric gases, warming the atmosphere (the greenhouse effect). The atmosphere emits long-wave radiation, so that the total energy received is balanced by the total energy emitted from the planet. *Source: Modified from MacCracken (1985).*

Thus, materials that enter the stratosphere remain there for long periods, allowing for high-altitude transport around the globe.

The thermal mixing of the troposphere is largely responsible for the global circulation of the atmosphere, as well as local weather patterns (Figure. 3.3). The large annual receipt of solar energy at the equator causes warming of the atmosphere (*sensible heat*) and the evaporation of large amounts of water, carrying *latent heat*, from tropical oceans and rainforests. As this warm, moist air rises, it cools, producing a large amount of precipitation in equatorial regions. Having lost its moisture, the rising air mass moves both north and south, away from the equator. In a belt centered on approximately 30° N or S latitude, these dry air masses sink to the Earth's surface, undergoing compressional heating. Most of the world's major deserts are associated with the downward movement of hot, dry air at this latitude. A similar, but much weaker, circulation pattern is found at the poles, where cold air sinks and moves north or south along the Earth's surface to lower latitudes. Known as *direct Hadley cells*, the tropical and polar circulation patterns drive an indirect circulation in each hemisphere between 40° and 60° latitude, producing regional storm systems and the prevailing west winds that we experience in the temperate zone (Figure 3.3).

The tropospheric air in each hemisphere mixes on a time scale of a few months (Warneck 2000), allowing for regional transport of air pollutants that persist for more than a few days. For instance, in 1995 carbon monoxide (CO) from Canadian forest fires contributed to air

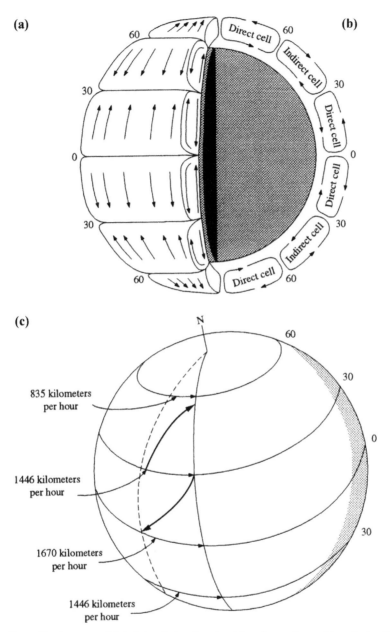

FIGURE 3.3   Generalized pattern of global circulation showing (a) surface patterns, (b) vertical patterns, and (c) the origin of the Coriolis force. As air masses move across different latitudes, they are deflected by the Coriolis force, which arises because of the different speeds of the Earth's rotation at different latitudes. For instance, if you were riding on an air mass moving at a constant speed south from 30° N latitude, you would begin your journey seeing 1446 km of the Earth's surface pass to the east every hour. By the time your air mass reached the equator, 1670 km would be passing to the east each hour. While moving south at a constant velocity, you would find that you had traveled 214 km west of your expected trajectory. The Coriolis force means that all movements of air in the Northern Hemisphere are deflected to the right; those in the Southern Hemisphere are deflected to the left. *Source: Modified from Oort (1970) and Gross (1977).*

pollutant loads in the eastern United States (Wotawa and Trainer 2000). The eruption of the Eyjafjallajökull volcano in Iceland on April 13 and 14, 2010 produced a cloud of volcanic ash over Poland several days later (Pietruczuk et al. 2010, Langmann et al. 2012) and disrupted airplane travel over much of Europe for several weeks. Vertical mixing in the troposphere is driven by convection, especially in thunderstorms, so that much of the air in the upper troposphere is less than a week old (Brunner et al. 1998, Bertram et al. 2007). Each year, there is also complete mixing of tropospheric air between the Northern and the Southern Hemispheres across the intertropical convergence zone (ITCZ). If a gas shows a higher concentration in one hemisphere, we can infer that a large natural or human source must exist in that hemisphere, overwhelming the tendency for atmospheric mixing to equalize the concentrations (Figure 3.4).

Exchange between the troposphere and the stratosphere is driven by several processes (Warneck 2000). In the tropical Hadley cells, rising air masses carry some tropospheric air to the stratosphere (Holton et al. 1995, Fueglistaler et al. 2004). The strength of the updraft varies seasonally, as a result of variations in the radiation received from the Sun. When the height of the tropopause drops, tropospheric air is trapped in the stratosphere, or vice versa. There is also exchange across the tropopause due to large-scale wind movements (Appenzeller and Davies 1992, Hocking et al. 2007), thunderstorms (Dickerson et al. 1987, Randel et al. 2010), and eddy diffusion (Warneck 2000).

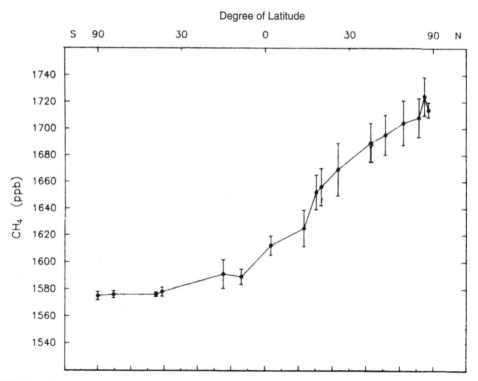

FIGURE 3.4   The latitudinal variation in the mean concentration of methane (CH_4) in Earth's atmosphere. *Source: From Steele et al. (1987). Used with permission of Reidel Publishing.*

Atmospheric scientists have examined the exchange of air mass between the troposphere and the stratosphere by following the fate of industrial pollutants released to the troposphere and radioactive contaminants released to the stratosphere in tests of atomic weapons during the 1950s and early 1960s (Warneck 2000). In these considerations, the concept of mean residence time is useful. For any reservoir that is in steady state, *mean residence time* (MRT) is defined as

$$MRT = Mass/flux, \tag{3.3}$$

where flux may be either the input or the loss from the reservoir.[1] Since the stratosphere is not well mixed vertically, the mean residence time of stratospheric air increases with altitude (Waugh and Hall 2002). However, the return of stratospheric air to the troposphere, about 4×10^{17} kg/yr (Seo and Bowman 2002), amounts to about 40% of the stratospheric mass each year, leading to an overall mean residence time of 2.6 years for stratospheric air. Thus, when a large volcano injects sulfur dioxide into the stratosphere, about half of it will remain after 2 years and about 5% will remain after 7.5 years.

ATMOSPHERIC COMPOSITION

Gases

Table 3.1 gives the globally averaged concentration of some important gases in the atmosphere. Three gases—nitrogen, oxygen, and argon—make up 99% of the atmospheric mass of 5.14×10^{21} g (Trenberth and Guillemot 1994). The mean residence times of these gases are much longer than the rate of atmospheric mixing. Thus, the concentrations of N_2, O_2, and all noble gases (He, Ne, Ar, Kr, and Xe) are globally uniform and time-invariant.

Several hundred trace gases, including a wide variety of volatile organic compounds (VOCs), are also found in the atmosphere. The most abundant volatile organic compound from vegetation is isoprene, which is commonly emitted from many coniferous forest species (Guenther et al. 2000, Fuentes et al. 2000). For comparison, Table 3.2 shows the volatile emission from several species of desert shrubs in the southwestern United States, where monoterpene compounds dominate the flux (Geron et al. 2006). The various volatile organic compounds derived from vegetation are known as nonmethane hydrocarbons (NMHC). Humans also add a wide variety of trace gases to the atmosphere, including oxygenated organic gases, such as acetone and alcohols (Piccot et al. 1992, Chameides et al. 1992, Singh et al. 2001).

Most trace gases are highly reactive and thus have short mean residence times, so it is not surprising that they are minor constituents in the atmosphere (Atkinson and Arey 2003). The concentration of such gases varies in space and time. For instance, we expect high concentrations of certain pollutants (ozone, carbon monoxide, etc.) over cities (e.g., Idso et al. 2001) and high concentrations of some reduced gases (methane and hydrogen sulfide) over swamps and other areas of anaerobic decomposition (e.g., Harriss et al. 1982, Steudler and Peterson

---

[1] Assuming exponential decay of a tracer from a reservoir that is in steady state, the fractional loss per year $(-k)$ is equal to the reciprocal of the mean residence time in years (i.e., $1/MRT$). The amount remaining in the reservoir at any time t (in years) as a fraction of the original content is equal to e^{-kt}, the half-life of the reservoir in years is $0.693/k$, and 95% will have disappeared from the reservoir after $3/k$ years.

TABLE 3.1   Global Average Concentration of Well-Mixed Atmospheric Constituents[a]

Compounds	Formula	Concentration	Total mass (g)
Major constituents (%)			
Nitrogen	N_2	78.084	3.87×10^{21}
Oxygen	O_2	20.946	1.19×10^{21}
Argon	Ar	0.934	6.59×10^{19}
Parts-per-million constituents (ppm = 10^{-6} or $\mu l/l$)			
Carbon dioxide	CO_2	400	3.11×10^{18}
Neon	Ne	18.2	6.49×10^{16}
Helium	He	5.24	3.70×10^{15}
Methane	CH_4	1.83	5.19×10^{15}
Krypton	Kr	1.14	1.69×10^{16}
Parts-per-billion constituents (ppb = 10^{-9} or nl/l)			
Hydrogen	H_2	510	1.82×10^{14}
Nitrous oxide	N_2O	320	2.49×10^{13}
Xenon	Xe	87	2.02×10^{15}
Parts-per-trillion constituents (ppt = 10^{-12})			
Carbonyl sulfide	COS	500	5.30×10^{12}
Chlorofluorocarbons			
CFC 11	CCl_3F	280	6.79×10^{12}
CFC 12	CCl_2F_2	550	3.12×10^{13}
Methylchloride	CH_3Cl	620	5.53×10^{12}
Methylbromide	CH_3Br	11	1.84×10^{11}

[a] Those with a mean residence time >1 year. Assuming a dry atmosphere with a molecular weight of 28.97, the overall mass of the atmosphere sums to 514×10^{19} g.

Source: Updated from Trenberth and Guillemot (1994).

1985). Winds mix the concentrations of these gases to their average tropospheric background concentration within a short distance downwind of local sources. While spatial heterogeneity in atmospheric concentrations can help us identify important source areas, we can best perceive global changes in atmospheric composition, such as the current increase in CH_4, by making long-term measurements in remote locations.

Junge (1974) related geographic variations in the atmospheric concentration of various gases to their estimated mean residence time in the atmosphere (Figure 3.5). Gases that have short mean residence times are highly variable from place to place, whereas those that have long mean residence times relative to atmospheric mixing show relatively little spatial variation. For example, the average volume of water in the atmosphere is equivalent to about 13,000 km^3 at any time, or 24.6 mm above any point on the Earth's surface (Trenberth 1998). The average daily precipitation would be about 2.73 mm if it were deposited evenly around the globe. Thus, the mean residence time for water vapor in the atmosphere is

$$24.6 \text{ mm}/2.73 \text{ mm day}^{-1} = 9.1 \text{ days}. \qquad (3.4)$$

This is a short time compared to the circulation of the troposphere, so we should expect water vapor to show highly variable concentrations in space and time (Figure 3.5). This relationship between variation in concentration and residence time in the atmosphere

TABLE 3.2 Emission of Volatile Organic Compounds ($\mu gC\ g^{-1}\ h^{-1}$) from Desert Shrubs of the U.S. Southwest

Species	Isoprene	α-Pinene	β-Pinene	Camphene	Myrcene	d-Limonene	Γ-Monoterpenes
Ambrosia deltoidea	<0.1	0.06 (0.02)	0.31 (0.18)	0.51 (0.18)	2.3 (1.1)	1.0 (0.32)	4.1 (1.6)
Ambrosia dumosa	<0.1	1.6 (0.93)	3.0 (1.5)	0.06 (0.03)	1.1 (0.82)	2.0 (0.87)	7.9 (3.4)
Atriplex canescens	<0.1	0	0	0.17 (0.17)	0.13 (0.12)	0	0.31 (0.29)
Chrysothamnus nauseosus	<0.1	0.28 (0.28)	0	0	0.16 (0.16)	0.21 (0.02)	0.65 (0.46)
Ephedra nevadensis	10 (4.0)	0.05 (0.01)	0.03 (0.017)	0.01 (0.006)	0.09 (0.06)	0.11 (0.06)	0.30 (0.04)
Hymenoclea salsola	<0.1	1.4 (0.31)	0.06 (0.06)	0.02 (0.02)	0.35 (0.26)	0.30 (0.30)	2.6 (0.57)
Krameria erecta	<0.1	0.02 (0.02)	0.06 (0.001)	0.03 (0.003)	0.14 (0.05)	0.05 (0.003)	0.30 (0.03)
Larrea tridentata	<0.1	0.37 (0.18)	0.12 (0.04)	0.44 (0.18)	0.30 (0.13)	0.74 (0.31)	2.0 (0.48)
Lycium andersonii	<0.1	0.10 (0.03)	0.27 (0.10)	0.11 (0.01)	0.39 (0.02)	0.27 (0.06)	1.1 (0.18)
Psorothamnus fremontii	35 (10)	0.50 (0.19)	0	0	1.0 (0.24)	0.50 (0.18)	2.0 (0.2)

Source: From Geron et al. (2006).

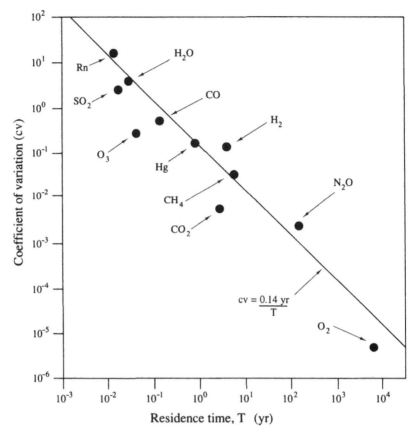

FIGURE 3.5   Variability in the concentration of atmospheric gases, expressed as the coefficient of variation among measurements, as a function of their estimated mean residence times in the atmosphere. *Source: Modified from Junge (1974), as updated by Slinn (1988).*

extends to trace organic species (e.g., propane), which have residence times of a few days (Jobson et al. 1999).

The mean residence time for carbon dioxide is about 5 years—only slightly longer than the mixing time for the atmosphere. Owing to the seasonal uptake of CO_2 by plants, CO_2 shows a minor seasonal and latitudinal variation ($\pm$ about 1%) in its global concentration of ~400 ppm (Figures 1.1 and 3.6). In contrast, painstaking analyses are required to show *any* variation in the concentration of O_2 because the amount in the atmosphere is so large and its mean residence time, 4000 years, is so much longer than the mixing time of the atmosphere (Keeling and Shertz 1992).

Gases with mean residence times of <1 year in the troposphere do not persist long enough for appreciable mixing into the stratosphere. Indeed, one of the most valuable, but dangerous, industrial properties of the chlorofluorocarbons is that they are chemically inert and thus long-lived in the troposphere (Rowland 1989). This allows chlorofluorocarbons to mix into the stratosphere, where they lead to the destruction of ozone by ultraviolet light (see pp. 82–83).

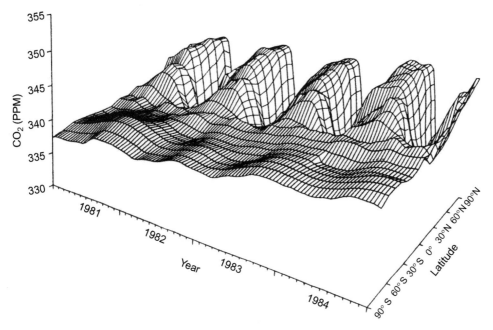

FIGURE 3.6 Seasonal fluctuations in the concentration of atmospheric CO_2 (1981–1984), shown as a function of 10° latitudinal belts (Conway et al. 1988). Note the smaller amplitude of the fluctuations in the Southern Hemisphere, reaching peak concentrations during the Northern Hemisphere's winter.

Aerosols

In addition to gaseous components, the atmosphere contains particles, known as aerosols, that arise from a variety of sources (Table 3.3). Soil particles are dispersed by wind erosion (also known as deflation weathering or eolian transport) from arid and semiarid regions (Pye 1987, Engelstaedter et al. 2003, Ravi et al. 2011). Particles with diameter <1.0 μm are held aloft by turbulent motion and are subject to long-range transport. Current estimates suggest that up to 2×10^{15} g/yr of soil particles enter the atmosphere from arid and barren agricultural soils (Zender et al. 2004), and about 20% of these particles are involved in long-range transport. The flux of soil dust has increased due to human cultivation, especially in semiarid lands (Tegen et al. 2004, Mulitza et al. 2010) and increases during droughts. Dust from the deserts of central Asia falls in the Pacific Ocean (Duce et al. 1980), where it contributes much of the iron needed by oceanic phytoplankton (Mahowald et al. 2005b, Chapter 9). Similarly, dust from the Sahara supplies nutrients to phytoplankton in the Atlantic Ocean (Talbot et al. 1986, Wu et al. 2000, Jickells et al. 2005) and phosphorus to Amazon rainforests (Swap et al. 1992, 1996; Perry et al. 1997; Okin et al. 2004). Dust from desert soils is monitored by several satellites, including NASA's MODIS satellite (Tanre et al. 2001, Kaufman et al. 2002; see Figure 3.7). Typically, while it is in transit, soil dust warms the atmosphere over land and cools the atmosphere over the oceans, which have lower surface albedo (reflectivity) (Ackerman and Chung 1992, Kellogg 1992, Yang et al. 2009).

TABLE 3.3　Global Production and Atmospheric Burden of Aerosols from Natural and Human-Derived Sources

	Mass emission 10^{12}g/yr	Mass Burden Tg	Number Produced per year	Number Burden
Carbonaceous aerosols				
Primary organic (0-2 μm)	95	1.2	–	$310 \cdot 10^{24}$
Biomass burning	54	–	$7 \cdot 10^{27}$	–
Fossil fuel	4	–	–	–
Biogenic	35	0.2	–	–
Black carbon (0–2 μm)	10	0.1	–	$270 \cdot 10^{24}$
Open burning and biofuel	6	–	–	–
Fossil fuel	4.5	–	–	–
Secondary organic	28	0.8	–	–
Biogenic	25	0.7	–	–
Anthropogenic	3.5	0.08	–	–
Sulfates	200	2.8	$2 \cdot 10^{28}$	–
Biogenic	57	1.2	–	–
Volcanic	21	0.2	–	–
Anthropogenic	122	1.4	–	–
Nitrates	18	0.49	–	–
Industrial dust, etc.	100	1.1	–	–
Sea salt				
$d < 1\ \mu$m	180	3.5	$7.4 \cdot 10^{26}$	–
$d = 1\text{–}16\ \mu$m	9940	12	$4.6 \cdot 10^{26}$	–
Total	10,130	15	$1.2 \cdot 10^{27}$	$27 \cdot 10^{24}$
Mineral (soil) dust				
$<1\ \mu$m	165	4.7	$4.1 \cdot 10^{25}$	–
$1\text{–}2.5\ \mu$m	496	12.5	$9.6 \cdot 10^{25}$	–
$2.5\text{–}10\ \mu$m	992	6	–	–
Total	1600	18 ± 5	$1.4 \cdot 10^{26}$	$11 \cdot 10^{24}$

Source: From Andreae and Rosenfeld (2008).

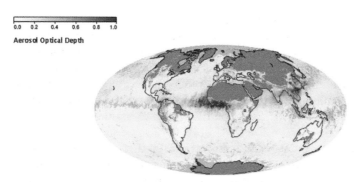

FIGURE 3.7　Aerosols in Earth's atmosphere, measured as AOD by the NASA MODIS satellite during March 2010. Optical depth is the fraction of light absorbed by aerosols in a column of air. Note high amounts of aerosols exiting the southern Sahel region of Africa, blowing westward to the Amazon, and high concentrations of aerosols emitted from the deserts of China, blowing eastward across the Pacific Ocean. *Source: From http://earthobservatory.nasa.gov/GlobalMaps/ view.php?d1=MODAL2_M_AER_OD.*

An enormous quantity of particles enter the atmosphere from the ocean as a result of tiny droplets that become airborne with the bursting of bubbles at the surface (MacIntyre 1974, Wu 1981). As the water evaporates from these bubbles, the salts crystallize to form seasalt aerosols, which carry the approximate chemical composition of seawater (Glass and Matteson 1973, Möller 1990). As in the case of soil dust, most seasalt aerosols are relatively large and settle from the atmosphere quickly, but a significant proportion remains in the atmosphere for global transport. Möller (1990) estimates a total seasalt production of 10×10^{15} g/yr, which carries about 200×10^{12} g of chloride from sea to land (see Figure 3.16 later). Other global estimates of seasalt production are similar (Erickson and Duce 1988, Gong et al. 1997, Sofiev et al. 2011).

Organic particles are produced from a wide variety of sources, including pollen, plant fragments, and bacteria (Després et al. 2012). Forest fires produce particles of charcoal that are carried throughout the troposphere, and small organic particles (soot) are produced by the condensation of volatile hydrocarbons from the smoke of forest fires (Hahn 1980, Cachier et al. 1989). Forest fires in the Amazon are thought to release as much as 1×10^{13} g of particulate matter to the atmosphere each year (Kaufman et al. 1990). It is likely that the global production of aerosols from forest fires has increased markedly as a result of higher rates of biomass burning in the tropics (Andreae 1991, Cahoon et al. 1992). Aerosols from these fires may affect regional patterns of rainfall (Cachier and Ducret 1991) and global climate (Penner et al. 1992). At the same time, in the temperate zone, control of forest fires has reduced the aerosol loading to the atmosphere over the last century (Clark and Royall 1994).

Volcanoes inject finely divided rock material—volcanic ash—into the atmosphere where it is deposited over large areas (Table 3.4), contributing to soil development in regions that are downwind from major eruptions (Watkins et al. 1978, Dahlgren et al. 1999, Zobel and Antos 1991). Volcanic gases and ash that are transported to the stratosphere by violent eruptions undergo global transport, affecting climate for several years (Langway et al. 1995, McCormick et al. 1995, Briffa et al. 1998).

Small particles, known as secondary aerosols, are also produced by reactions between gases in the atmosphere. For instance, when SO_2 is oxidized to sulfuric acid (H_2SO_4) in the atmosphere, particles rich in $(NH_4)_2SO_4$ may be produced by a subsequent reaction with atmospheric ammonia (NH_3; Behera and Sharma 2011):

$$2NH_3 + H_2SO_4 \rightarrow (NH_4)_2SO_4 \tag{3.5}$$

(Ammonia is derived from a variety of sources, primarily associated with agricultural activities; Chapter 12.) Sulfate aerosols are also produced during the oxidation of dimethylsulfide released from the ocean (Chapter 9). Sulfate aerosols increase the albedo of the Earth's atmosphere, so estimates of the abundance of sulfate aerosols are an important component of global climate models (Kiehl and Briegleb 1993, Mitchell et al. 1995). Secondary aerosols are also produced from volatile organic compounds, such as isoprene, that are released from plants (Kavouras et al. 1998, O'Dowd et al. 2002, Henze and Seinfeld 2006, Jimenez et al. 2009, Pöschl et al. 2010).

Finally, a wide variety of particles are produced from human industrial processes, especially the burning of coal (Hulett et al. 1980, Shaw 1987). Globally, the release of particles during the combustion of fossil fuels rivals the mobilization of elements by rock weathering

TABLE 3.4   Composition of Airborne Particulate Volcanic Ash Sample Collected during Mt. St. Helens Eruption in Washington State on May 19, 1980

Constituent	Particulate sample	Average ash
Major elements (%)		
SiO_2	$\equiv$65.0	65.0
Fe_2O_3	6.7	4.81
CaO	3.0	4.94
K_2O	2.0	1.47
TiO_2	0.42	0.69
MnO	0.054	0.077
P_2O_5	–	0.17
Trace elements (ppm)		
S	3220	940
Cl	1190	660
Cu	61	36
Zn	34	53
Br	<8	~1
Rb	<17	32
Sr	285	460
Zr	142	170
Pb	36	8.7

Source: From Fruchter et al. (1980); and Hooper et al. (1980), used with permission of American Association for the Advancement of Science.

at the Earth's surface (Bertine and Goldberg 1971). Fine particulate air pollution has significant human health effects (Samet et al. 2000, Pope et al. 2009, Anenberg et al. 2010). Fortunately, the mass of industrial aerosols has declined in many developed countries where pollution controls have been instituted (Renberg and Wik 1984). One of the most widespread anthropogenic aerosols, particles of lead from automobile exhaust, has declined in global abundance over the past 30 years due to a reduction in the use of leaded gasoline (Boutron et al. 1991). In other regions, where air pollution is unregulated, concentrations of aerosols have increased in recent years (Streets et al. 2008, Dey and Girolamo 2011), contributing to observations of global dimming. Overall, human activities probably account for about 10% of the burden of aerosols in today's atmosphere (Table 3.3).

Small particles (<1.0 μm)[2] are much more numerous in the atmosphere than large particles, but it is the large particles that contribute the most to the total airborne mass (Warneck 2000, Raes et al. 2000). Nanoparticles (<0.3 μm), derived from a variety of natural and

[2] The U.S. EPA designates small aerosols, those <2.5 μm, as $PM_{2.5}$.

manufactured sources (Kumar et al. 2010, Hendren et al. 2011), are of particular concern to human health. The mass of aerosols declines with increasing altitude from values ranging between 1 and 50 $\mu g/m^3$ near unpolluted regions of the Earth's surface. Although there is an inverse relation between the size of particles and their persistence in the atmosphere, the overall mean residence time for tropospheric aerosols is about 5 days (Warneck 2000). Thus, aerosols are not uniform in their distribution in the atmosphere. As a result of their longer mean residence time, small particles have the greatest influence on Earth's climate, and they carry the largest mass of material through the atmosphere.

The composition of tropospheric aerosols varies greatly depending on the proximity of continental, maritime, or anthropogenic sources (Heintzenberg 1989, Murphy et al. 1998). Over land, aerosols are often dominated by soil minerals and human pollutants (Shaw 1987, Gillette et al. 1992). Over the ocean, the composition of aerosols is a mixture of contributions from silicate minerals of continental origin and seasalt from the ocean (Andreae et al. 1986). Various workers have used ratios among the elemental constituents of aerosols to deduce the relative contribution of different sources (e.g., Moyers et al. 1977, Rahn and Lowenthal 1984).

Aerosols are important in reactions with atmospheric gases and as nuclei for the condensation of raindrops. The latter are known as cloud condensation nuclei, often abbreviated CCN. Raindrops are formed when water vapor begins to condense on aerosols >0.1 μm in diameter. As raindrops enlarge and fall to the ground, they collide with other particles and absorb atmospheric gases. Soil dusts often contain a large portion of insoluble material (Reheis and Kihl 1995), but seasalt aerosols and those derived from pollution sources are readily soluble and contribute to the dissolved chemical content of rainwater. Reactions of atmospheric gases with aerosols or raindrops are known as *heterogeneous* or *multiphase reactions* (Ravishankara 1997). Such reactions are responsible for the ultimate removal of many reactive gases from the atmosphere.

BIOGEOCHEMICAL REACTIONS IN THE TROPOSPHERE

Major Constituents—Nitrogen and Oxygen

It is perhaps not surprising that the major constituents of the atmosphere, N_2, O_2, and Ar, have nearly uniform concentrations and long mean residence times in the atmosphere. Argon is inert and has accumulated in the Earth's atmosphere since the earliest degassing of its crust (Chapter 2). From a biogeochemical perspective, N_2 is practically inert; reactive N is found only in molecules such as NH_3 and NO. Collectively the reactive nitrogen gases are sometimes called "odd" nitrogen, because the molecules have an odd number of N atoms (versus N_2 or N_2O).[3] Despite its abundance in the atmosphere, N_2 is so inert that the rate of formation of odd, or reactive, nitrogen is the primary factor that limits the growth of plants on land and in the oceans (Delwiche 1970, LeBauer and Treseder 2008). Among atmospheric gases, only argon and the other noble gases are less reactive.

---

[3] The term "odd nitrogen" is not ideal. In practice it refers only to the various oxidized forms of nitrogen in the atmosphere, including N_2O_5, but not to NH_3, which also has an odd number of N atoms.

Conversion of N_2 to reactive compounds, *N fixation*, occurs in lightning bolts, but the estimated global production of NO by lightning ($<5 \times 10^{12}$ g N/yr; Chapter 12) is too low to account for a significant turnover of N_2 in the atmosphere. By far the most important source of fixed nitrogen for the biosphere derives from the bacteria that convert N_2 to NH_3 in the process of biological nitrogen fixation (Eq. 2.10). The global rate of biological N fixation is poorly known because it must be extrapolated from small-scale measurements to the entire surface of the Earth (Chapters 6 and 9). Including human activities, global nitrogen fixation is not likely to exceed 450×10^{12} g N/yr, with the production of synthetic nitrogen fertilizer now accounting for about one-third of the total (Chapter 12).

The natural rate of nitrogen fixation would remove the pool of N_2 from the atmosphere in about 40 million years.[4] Fortunately, denitrification (Eq. 2.20) returns N_2 to the atmosphere. At present, we have little evidence that the rate of either N fixation or denitrification changes significantly in response to changes in the concentration of N_2 in the atmosphere. While the biosphere is responsible for the maintenance of N_2 in Earth's atmosphere over geologic time, it plays a minor role in stabilizing the concentration of atmospheric N_2 over shorter periods, since the pool of N_2 in the atmosphere is so large (Walker 1984).

In Chapter 2 we discussed the accumulation of O_2 in the atmosphere during the evolution of life on Earth. The atmosphere now contains only a small portion of the total O_2 released by photosynthesis through geologic time (Figure 2.8). However, the atmosphere contains much more O_2 than can be explained by the storage of carbon in land plants today. The instantaneous combustion of all the organic matter now stored on land would reduce the pool of atmospheric oxygen content by only 0.45% (Chapter 5). The accumulation of O_2 in the atmosphere is the result of the long-term burial of reduced carbon in ocean sediments (Berner 1982), which contain nearly all of the reduced, organic carbon on Earth (Table 2.3). The rate of burial is determined by the area and depth of the ocean floor that is subject to anoxic conditions (Walker 1977, Hartnett et al. 1998). Because the area and depth vary inversely with the concentration of atmospheric O_2, the balance between the burial of organic matter and its oxidation maintains O_2 at a steady-state concentration of about 21% (see also Chapters 9 and 11).

A large amount of O_2 has been consumed in weathering of reduced crustal minerals, especially Fe and S, through geologic time (Figure 2.8); the current rate of exposure of these minerals would consume all atmospheric oxygen in about 70 million years (Lenton 2001; see Figure 11.8). However, the rate of exposure is not likely to vary greatly in response to changes in atmospheric O_2, so weathering is not the major factor controlling O_2 in the atmosphere. In sum, despite the potential reactivity of O_2, its rate of reaction with reduced compounds is rather slow, and O_2 is a stable component of the atmosphere. The mean residence time of O_2 in the atmosphere is on the order of 4000 years, largely determined by exchange with the biosphere (compare Figures 3.5 and 11.8). As such O_2 is well mixed and uniform in the atmosphere. Annual photosynthesis and respiration cause seasonal variations in O_2 concentration of about $\pm0.002\%$ (Figure 1.1).

---

[4] Mass of N in the atmosphere (Table 3.1) divided by the global rate of biological nitrogen fixation (270×10^{12} gN/yr) gives a mean residence time of 14,300,000 years for N_2 in the atmosphere. Thus, $k = 7.0 \times 10^{-8}$ and $3/k =$ 43 million years.

Carbon Dioxide

Carbon dioxide is not reactive with other gases in the atmosphere. The concentration of CO_2 is affected by interactions with the Earth's surface, including the reactions of the carbonate–silicate cycle (Figure 1.3), gas exchange with seawater following Henry's Law (Eq. 2.4), and annual cycles of photosynthesis and respiration by land plants (Figures 1.1 and 3.6). For the Earth's land surface, our best estimates of plant uptake (60×10^{15} g C/yr; Chapter 5) suggest a mean residence time of about 12.5 years before a hypothetical molecule of CO_2 in the atmosphere is captured by photosynthesis. The annual exchange of CO_2 with seawater, particularly in areas of cold, downwelling water and high productivity (Chapter 9), is about $1.5\times$ as large as the annual uptake of CO_2 by land plants. Both plant and ocean uptake are likely to increase with increasing concentrations of atmospheric CO_2, potentially buffering fluctuations in its concentration (Chapters 5, 9, and 11). Following Eq. 3.3, the mean residence time for CO_2, determined by the total flux from the atmosphere (the sum of land and ocean uptake), is about 5 years, so CO_2 shows some seasonal and latitudinal variation in the atmosphere (Figures 1.1 and 3.6).

The carbonate–silicate cycle (Figure 1.3) also buffers the concentration of CO_2 in the atmosphere, but does not affect the concentration of atmospheric CO_2 significantly in periods of less than ~10,000 years (Hilley and Porder 2008). We will compare the relative importance of these processes in more detail in Chapter 11, which examines the global carbon cycle. The current increase in atmospheric CO_2 is a non-steady-state condition, caused by the combustion of fossil fuels and destruction of land vegetation. CO_2 is released by these processes faster than it can be taken up by land vegetation and the sea. If these activities were to cease, atmospheric CO_2 would return to a steady state, and after several hundred years nearly all of the CO_2 released by humans would reside in the oceans. In the meantime, higher concentrations of CO_2 are likely to cause significant atmospheric warming through the "greenhouse effect" (refer to Figure 3.2).

Trace Biogenic Gases

Volcanoes are the original source of volatiles in the Earth's atmosphere (Chapter 2) and a small continuing source of some of the reduced gases (H_2S, H_2, NH_3, CH_4) that are found in the atmosphere today (Table 2.2). However, in most cases, the concentrations of these gases in today's atmosphere are dominated by supply from the biosphere, particularly by microbial activity (Monson and Holland 2001). Methane is largely produced by anaerobic decomposition in wetlands (Chapters 7 and 11), nitrogen oxides by soil microbial transformations (Chapters 6 and 12), carbon monoxide by combustion of biomass and fossil fuels (Chapters 5 and 11), and volatile hydrocarbons, especially isoprene, by vegetation and human industrial activities (Chapter 5). The production of trace gases containing N and S contributes to the global cycling of these elements, which is controlled by the biosphere (Crutzen 1983). These and other trace gases are found at concentrations well in excess of what is predicted from equilibrium geochemistry in an atmosphere with 21% O_2 (Table 3.5).

Unlike major atmospheric constituents, many of the trace biogenic gases in the atmosphere are highly reactive, showing short mean residence times and variable concentrations in space and time (refer to Figure 3.5). Concentrations of these gases in the atmosphere are determined by the balance between local sources and chemical reactions—known as *sinks*—that remove

TABLE 3.5   Some Trace Biogenic Gases in the Atmosphere

Compound	Formula	Concentration (ppb) Expected[a]	Actual[b]	Mean residence time	Percentage of sink due to OH
Carbon compounds					
Methane	CH_4	10^{-148}	1830	9 years	90
Carbon monoxide	CO	10^{-51}	45–250	60 days	80
Isoprene	$CH_2=C(CH_3)-CH=CH_2$		0.2–10.0	<1 day	100
Nitrogen compounds					
Nitrous oxide	N_2O	10^{-22}	320	120 years	0
Nitric oxides	NO_x	10^{-13}	0.02–10.0	1 day	100
Ammonia	NH_3	10^{-63}	0.08–5.0	5 days	<2
Sulfur compounds					
Dimethylsulfide	$(CH_3)_2S$		0.004–0.06	1 day	50
Hydrogen sulfide	H_2S		<0.04	4 days	100
Carbonyl sulfide	COS	0	0.50	5 years	20
Sulfur dioxide	SO_2	0	0.02–0.10	3 days	50

[a] *Approximate values in equilibrium with an atmosphere containing 21% O_2 (Chameides and Davis 1982).*
[b] *For short-lived gases, the value is the range expected in remote, unpolluted atmospheres.*

these gases from the atmosphere. Sinks are largely driven by oxidation reactions and the capture of the reaction products by rainfall. Currently the concentration of nearly all these constituents is increasing as a result of human activities, suggesting that humans are affecting biogeochemistry at the global level (Prinn 2003).

Despite its abundance, O_2 does not directly oxidize reduced gases in the atmosphere. Instead, a small proportion of the oxygen is converted to the powerful atmospheric oxidants ozone (O_3) and hydroxyl radical (OH) through a series of reactions driven by sunlight (Logan 1985, Thompson 1992). Ozone and OH are the primary gases that oxidize many of the trace gases to CO_2, HNO_3, and H_2SO_4.

It is important to understand the natural production, occurrence, and reactions of ozone in the atmosphere. Nearly daily we read seemingly contradictory reports of the harmful effects of ozone depletion in the stratosphere and harmful effects of ozone pollution in the troposphere. In each case, human activities are upsetting the natural concentrations of ozone that are critical to atmospheric biogeochemistry.

Most ozone is produced by the reaction of sunlight with O_2 in the stratosphere, as described in the next section. Some of this ozone is transported to the Earth's surface by the mixing of stratospheric and tropospheric air (e.g., Hocking et al. 2007), where it contributes to the budget of ozone in the troposphere (Table 3.6). However, observations of high ozone

TABLE 3.6   Tropospheric Ozone Budget

Source or sink	Pre-industrial	Present	Human impact
Stratospheric injection	+696	+696	0
Tropospheric production (i.e., dirty atmosphere reactions)	+199	+686	+487
Tropospheric sink (i.e., clean atmosphere reactions)	−435	−558	−123
Dry deposition	−459	−825	−366

Note: All values in Tg (10^{12} g) of O_3/year.
Source: From Levy et al. (1997).

concentrations in the smog of polluted cities (e.g., Los Angeles) alerted atmospheric chemists to reactions by which ozone is produced in the troposphere (Warneck 2000).

When NO_2 is present in the atmosphere, it is dissociated by sunlight ($h\nu$),

$$NO_2 + h\nu \rightarrow NO + O, \tag{3.6}$$

followed by a reaction producing ozone:

$$O + O_2 \rightarrow O_3. \tag{3.7}$$

This reaction sequence is an example of a *homogeneous gas reaction*, that is, a reaction between atmospheric constituents that are all in the gaseous phase. The net reaction is

$$NO_2 + O_2 \leftrightarrow NO + O_3, \tag{3.8}$$

which is an equilibrium reaction, so high concentrations of NO tend to drive the reaction backward. Sunlight is essential to form ozone by these pathways, so they are known as photochemical reactions. At night, ozone is consumed by reactions with NO_2 to form nitric acid (Brown et al. 2006b).

Both NO_2 and NO, collectively known as NO_x, are found in polluted air, in which they are derived from industrial and automobile emissions.[5] Small concentrations of both of these constituents are also found in the natural atmosphere, where they are derived from forest fires, lightning discharges, and microbial processes in the soil (Chapter 6). Thus, the production of ozone from NO_2 has probably always occurred in the troposphere, and the present-day concentrations of tropospheric ozone have simply increased as industrial emissions have raised the concentration of NO_2 and other precursors to O_3 formation (Volz and Kley 1988, Lelieveld et al. 2004, Cooper et al. 2010).

Ozone is subject to further photochemical reaction in the troposphere,

$$O_3 + h\nu \rightarrow O_2 + O(^1D), \tag{3.9}$$

where $h\nu$ is ultraviolet light with wavelengths <318 nm and $O(^1D)$ is an excited atom of oxygen. Reaction of $O(^1D)$ with water yields hydroxyl radicals:

$$O(^1D) + H_2O \rightarrow 2OH. \tag{3.10}$$

---

[5] NO_x (pronounced "knocks") refers to the sum of NO + NO_2. NO_y is used to refer to the sum of NO_x plus all other oxidized forms of nitrogen—for example, HNO_3 and $CH_3C(O)O_2NO_2$ (peroxyacetyl nitrate or PAN).

The formation of hydroxyl radicals is strongly correlated with the amount of ultraviolet radiation (Rohrer and Berresheim 2006). Hydroxyl radicals may further react to produce HO_2 and H_2O_2,

$$2OH + 2O_3 \rightarrow 2HO_2 + 2O_2 \tag{3.11}$$

$$2HO_2 \rightarrow H_2O_2 + O_2, \tag{3.12}$$

which are other short-lived oxidizing compounds in the atmosphere (Thompson 1992, Crutzen et al. 1999).

Hydroxyl radicals exist with a mean concentration of about 1×10^6 molecules/cm^3 (Prinn et al. 1995). The highest concentrations occur in daylight (Platt et al. 1988, Mount 1992) and at tropical latitudes, where the concentration of water vapor is greatest (Hewitt and Harrison 1985). The average OH radical persists for only a few seconds in the atmosphere, so concentrations of OH are highly variable. Local concentrations can be measured using beams of laser-derived light, which is absorbed as a function of the number of OH radicals in its path (Dorn et al. 1988, Mount et al. 1997).

Because of its short mean residence time, the global mean concentration of OH radicals must be estimated indirectly. For this purpose, atmospheric chemists have relied on methyl-chloroform (trichloroethane), a gas that is known to result only from human activity. Methyl-chloroform has a mean residence time of about 4.8 years (Prinn et al. 1995), so it is reasonably well mixed in the atmosphere. In the laboratory, it reacts with OH,

$$OH + CH_3CCl_3 \rightarrow H_2O + CH_2CCl_3, \tag{3.13}$$

and the rate constant, K, for the reaction is 0.85×10^{-14} cm^3 molecule^{-1} sec^{-1} at 25°C (Talukdar et al. 1992). Then, knowing the industrial production of CH_3CCl_3, its accumulation in the atmosphere and K, one can calculate the concentration of OH that must be present, namely,

$$OH = (\text{Production} - \text{Accumulation})/K. \tag{3.14}$$

Hydroxyl radicals are the major source of oxidizing power in the troposphere. For example, in an unpolluted atmosphere, hydroxyl radicals destroy methane in a series of reactions,

$$CH_4 + OH \rightarrow CH_3 + H_2O \tag{3.15}$$

$$CH_3 + O_2 \rightarrow CH_3O_2 \tag{3.16}$$

$$CH_3O_2 + HO_2 \rightarrow CH_3O_2H + O_2 \tag{3.17}$$

$$CH_3O_2H \rightarrow CH_3O + OH \tag{3.18}$$

$$CH_3O + O_2 \rightarrow CH_2O + HO_2, \tag{3.19}$$

for which the net reaction is

$$CH_4 + O_2 \rightarrow CH_2O + H_2O. \tag{3.20}$$

Note that the hydroxyl radical has acted as a catalyst to initiate the oxidation of CH_4 and its byproducts by O_2. Other volatile organic compounds are also oxidized through this pathway, which yields formaldehyde (CH_2O; Atkinson 2000, Atkinson and Arey 2003).

The formaldehyde that is produced in these reactions is further oxidized to carbon monoxide,

$$CH_2O + OH + O_2 \rightarrow CO + H_2O + HO_2, \tag{3.21}$$

and CO is oxidized by OH to produce CO_2,

$$CO + OH \rightarrow H + CO_2 \tag{3.22}$$

$$H + O_2 \rightarrow HO_2 \tag{3.23}$$

$$HO_2 + O_3 \rightarrow OH + 2O_2, \tag{3.24}$$

for which the net reaction is

$$CO + O_3 \rightarrow CO_2 + O_2. \tag{3.25}$$

Thus, OH acts to scrub the atmosphere of a wide variety of reduced carbon gases, ultimately oxidizing their carbon atoms to carbon dioxide.

Hydroxyl radicals can also react with NO_2 and SO_2 in homogeneous gas reactions:

$$NO_2 + OH \rightarrow HNO_3 \tag{3.26}$$

$$SO_2 + OH \rightarrow SO_3 + HO_2, \tag{3.27}$$

and the latter reaction is followed by a heterogeneous reaction with raindrops:

$$SO_3 + H_2O \rightarrow H_2SO_4, \tag{3.28}$$

which removes sulfur dioxide from the atmosphere, causing acid rain. Sulfur dioxide is also oxidized by hydrogen peroxide, according to Chandler et al. (1988):

$$SO_2 + H_2O_2 \rightarrow H_2SO_4. \tag{3.29}$$

The reaction of OH with NO_2 is very fast, and it produces nitric acid that is removed from the atmosphere by a heterogeneous interaction with raindrops (Munger et al. 1998). The reactions with SO_2 are much slower, accounting for the long-distance transport of SO_2 as a pollutant in the atmosphere (Rodhe 1981). Hydrogen sulfide (H_2S) and dimethylsulfide ($(CH_3)_2S$, released from anaerobic soils (Chapter 7) and the ocean surface (Chapter 9), are also removed by reactions with OH and other oxidizing compounds, leading to the deposition of H_2SO_4 (Toon et al. 1987). Thus, OH radicals cleanse the atmosphere of trace N and S gases by converting them to *acid anions* (NO_3^-, SO_4^{2-}) in the atmosphere.

The vast majority of OH radicals in the atmosphere is consumed in reactions with CO and CH_4. Although the concentration of methane is much higher than that of carbon monoxide in unpolluted atmospheres, the reaction of OH with CO is much faster. The speed of reaction of CO with OH accounts for the short mean residence time of CO in the atmosphere (Table 3.5). The mean residence time for methane is much longer, accounting for its more uniform distribution (Figure 3.5). One explanation for the current increase in methane in the atmosphere is that the anthropogenic release of CO consumes OH radicals previously available for the oxidation of methane (Khalil and Rasmussen 1985), but other measurements suggest that OH concentrations have declined only slightly (or perhaps not at all) in recent years (Prinn et al. 1995, 2005; Montzka et al. 2011b). Increasing deposition of formaldehyde in the Greenland snowpack indicates increasing oxidation of methane in the atmosphere (Eq. 3.20; Staffelbach et al. 1991).

In unpolluted atmospheres, all these reactions consume OH. In "dirty" atmospheres, a different set of reactions pertains, in which there can be a net *production* of O_3, and thus OH, during the oxidation of reduced gases (Jenkin and Clemitshaw 2000, Sillman 1999). When the concentration of NO is >10 ppt, which we will define as a "dirty" atmosphere (Jacob and Wofsy 1990), the oxidation of carbon monoxide begins by reaction with hydroxyl radical and proceeds as follows (Crutzen and Zimmermann 1991):

$$CO + OH \rightarrow CO_2 + H \tag{3.30}$$

$$H + O_2 \rightarrow HO_2 \tag{3.31}$$

$$HO_2 + NO \rightarrow OH + NO_2 \tag{3.32}$$

$$NO_2 + hv \rightarrow NO + O \tag{3.33}$$

$$O + O_2 \rightarrow O_3. \tag{3.34}$$

The net reaction is

$$CO + 2O_2 \rightarrow CO_2 + O_3. \tag{3.35}$$

Similarly, the oxidation of methane in the presence of high concentrations of NO proceeds through a large number of steps, yielding a net reaction of

$$CH_4 + 4O_2 \rightarrow CH_2O + H_2O + 2O_3. \tag{3.36}$$

In both cases, NO acts as a catalyst leading to the oxidation of reduced gases by oxygen.

Figure 3.8 shows the contrasting pathways of carbon monoxide oxidation in clean and dirty atmospheres. Crutzen (1988) points out that the oxidation of one molecule of CH_4 could consume up to 3.5 molecules of OH and 1.7 molecules of O_3 when the NO concentration is low, whereas it would yield a net gain of 0.5 OH and 3.7 O_3 in polluted environments (see also Wuebbles and Tamaresis 1993). Although they were first discovered in urban areas, the reactions of dirty atmospheres are likely to be relatively widespread in nature. NO is produced naturally by soil microbes (Chapter 6) and forest fires. Concentrations of NO >10 ppt are present over most of the Earth's land surface (Chameides et al. 1992, Levy et al. 1999). In the presence of NO, oxidation of volatile hydrocarbons emitted from vegetation, and CO emitted from both vegetation and forest fires, can account for unexpectedly high concentrations of O_3 over rural areas of the southeastern United States (Figure 3.9) (Jacob et al. 1993, Kleinman et al. 1994, Kang et al. 2003) and in remote tropical regions (Crutzen et al. 1985, Zimmerman et al. 1988, Jacob and Wofsy 1990, Andreae et al. 1994a). In urban areas, where the concentration of NO_x is especially high due to industrial pollution, effective control of atmospheric O_3 levels may also depend on the regulation of volatile hydrocarbons (Chameides et al. 1988, Seinfeld 1989). In rural areas, ozone formation is usually limited

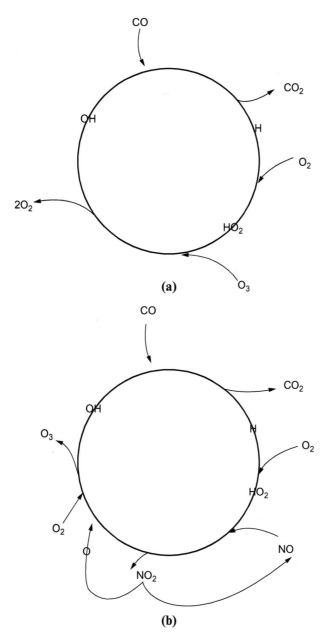

FIGURE 3.8 Reaction chain for the oxidation of CO in (a) clean and (b) dirty atmospheric conditions.

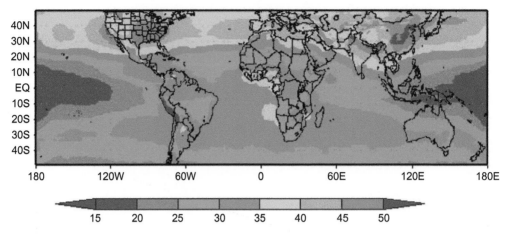

FIGURE 3.9   Distribution of ozone in Earth's atmosphere, for summer months, averaged over 1979–1991. Note high ozone concentrations over the eastern United States and China. Data are in Dobson units (see page 83). *Source: From Fishman et al. (2003). Used with permission of European Geosciences Union.*

by the concentration of NO_x, especially during the growing season, when vegetation actively emits volatile organic compounds (Figure 3.10; Aneja et al. 1996).

Understanding changes in the concentration of OH and other oxidizing species in the atmosphere is critical to predicting future trends in the concentration of trace gases, such as CH_4, that can contribute to greenhouse warming. Some models predict an increase in O_3 (Isaksen and Hov 1987, Hough and Derwent 1990, Thompson 1992, Prinn 2003) in the atmosphere as a result of increasing human emissions of NO, creating dirty atmosphere conditions over much of the planet. Indeed, measurements in Europe in the late 1800s indicate lower concentrations of tropospheric ozone than today (Volz and Kley 1988, Marenco et al. 1994). The models are also consistent with indirect observations that the global concentration of OH has remained fairly stable in recent years, despite increasing emissions of reduced gases that should scrub OH from the atmosphere (Prinn et al. 1995, 2005; Montzka et al. 2011b). Concentrations of H_2O_2, derived from OH (Eqs. 3.11 and 3.12), have increased in layers of Greenland ice deposited during the last 200 years, suggesting a greater oxidizing capacity in the northern hemisphere as a result of human activities (Figure 3.11). Several recent papers suggest additional pathways leading to the formation of OH (Li et al. 2008, Hofzumahaus et al. 2009), so that its production may not be restricted to the photochemical reactions outlined in Eqs. 3.8 through 3.10. Soil microbes that produce nitrite (NO_2^-) are a potential source of nitrous acid (HONO) in the atmosphere and OH radicals (Su et al. 2011).

Some of the O_3 produced over the continents undergoes long-distance transport (Jacob et al. 1993, Parrish et al. 1993, Cooper et al. 2010, Brown-Steiner and Hess 2011), resulting in the appearance of O_3 and its byproducts at considerable distances from their source (Figure 3.9). In some rural areas, concentrations of O_3 from local production and transport from nearby cities inhibit the growth of agricultural crops and trees (Chameides et al. 1994). Other workers disagree, finding that local atmospheric conditions, rather than global changes in transport from polluted areas, determine the oxidizing capacity of the atmosphere over much of the planet (Oltmans and Levy 1992, Ayers et al. 1992, Kang et al. 2003).

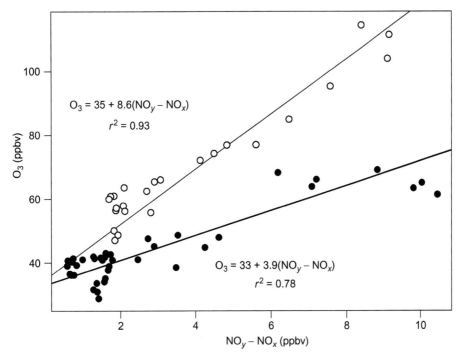

FIGURE 3.10 Ambient O_3 versus NO_y–NO_x concentrations in the atmosphere at Harward forest in northern Massachusetts in the United States, May 6–12, 1990 (●) and August 24–30, 1992 (○). *Source: From Hirsch et al. (1996). Used with permission of American Geophysical Union.*

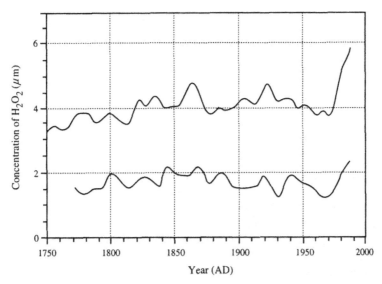

FIGURE 3.11 Variation in the mean annual H_2O_2 concentration over the past 200 years as seen in two cores from the Greenland ice pack. *Source: Modified with permission of Macmillan from Sigg and Neftel (1991).*

I. PROCESSES AND REACTIONS

ATMOSPHERIC DEPOSITION

Elements of biogeochemical interest are deposited on the Earth's surface as a result of rainfall, dry deposition, and the direct absorption of gases from the atmosphere. The importance of each of these processes differs for different regions and for different elements (Gorham 1961).

Processes

In many forests, a large fraction of the annual uptake and circulation of nutrient elements in vegetation may be derived from the atmosphere (Miller et al. 1993b, Kennedy et al. 2002a, Avila et al. 1998). The atmosphere accounts for nearly all of the nitrogen and sulfur that circulates in terrestrial ecosystems, with rock weathering providing only smaller amounts (see Table 4.5 and Chapter 6).

The chemical composition of rainfall has received great attention, as a result of widespread concern about dissolved constituents that lead to "acid rain." The dissolved constituents in rainfall are often separated into two fractions. The *rainout* component consists of constituents derived from cloud processes, such as the nucleation of raindrops. The *washout* component is derived from below cloud level, by scavenging of aerosol particles and the dissolution of gases in raindrops as they fall (Brimblecombe and Dawson 1984, Shimshock and de Pena 1989). The dissolved content in both fractions represents the results of heterogeneous reactions between gases and raindrops in the atmosphere.

The relative contribution of rainout and washout varies depending on the length of the rainstorm. As washout cleanses the lower atmosphere, the content of dissolved materials in rainfall declines. Thus, the concentration of dissolved constituents in precipitation is inversely related to the rate of precipitation (Gatz and Dingle 1971) and to the total volume that has fallen (Likens et al. 1984, Lesack and Melack 1991, Minoura and Iwasaka 1996). The concentration of dissolved constituents also varies inversely as a function of mean raindrop size (Georgii and Wötzel 1970, Bator and Collett 1997). This inverse relation explains why extremely high concentrations of dissolved constituents are found in fog waters (Weathers et al. 1986, Waldman et al. 1982, Clark et al. 1998, Elbert et al. 2000). Capture of fog and cloud water by vegetation is an important component of the deposition of nutrient elements from the atmosphere in some high-elevation and coastal ecosystems (Lovett et al. 1982, Waldman et al. 1985, Ewing et al. 2009, Weathers et al. 2000).

The relative efficiency of scavenging by rainwater is often expressed as the washout ratio:

$$\text{Washout} = \frac{\text{Ionic concentration in rain}(\text{mg/liter})}{\text{Ionic concentration in air}(\text{mg/m}^3)}. \tag{3.37}$$

With units of m^3/liter, this ratio gives an indication of the volume of atmosphere cleansed by each liter of rainfall as it falls. Large ratios are generally found for ions that are derived from relatively large aerosols or from highly water-soluble gases in the atmosphere. Snowfall is generally less efficient at scavenging than rainfall.

The deposition of nutrients by precipitation is often called wetfall; dryfall is the result of gravitational sedimentation of particles during periods without rain (Hidy 1970, Wesely and

Hicks 1999). Dryfall of dusts downwind of arid lands is often spectacular; Liu et al. (1981) reported 100 g m^{-2} hr^{-1} of dustfall in Beijing, China, as a result of a single dust storm on April 18, 1980. Enormous deposits of wind-deposited soil, known as loess, were laid down during glacial periods, when large areas of semiarid land were subject to wind erosion (Pye 1987, Simonson 1995, Muhs et al. 2001). Today, various elements necessary for plant growth are released by chemical weathering of soil minerals in these deposits (Chapter 4).

The dryfall received in many areas contains a significant fraction that is easily dissolved by soil waters and immediately available for plant uptake. Despite the high rainfall found in the southeastern United States, Swank and Henderson (1976) reported that 19 to 64% of the total annual atmospheric deposition of ions such as Ca, Na, K, and Mg, and up to 89% of the deposition of P, was derived from dryfall. Dryfall inputs of P may assume special significance to plant growth in areas where the release of P from rock weathering is very small (Newman 1995, Chadwick et al. 1999, Okin et al. 2004). Dry deposition contributes about 30 to 60% of the deposition of sulfur in New Hampshire (Likens et al. 1990; compare Tanaka and Turekian 1995). Similarly, 34% of the atmospheric inputs of nitrogen to Harvard Forest (Massachusetts) are derived from dry deposition (Munger et al. 1998). Organic nitrogen compounds deposited from the atmosphere are decomposed by soil microbes, providing additional plant nutrients (Neff et al. 2002a, Mace et al. 2003, Zhang et al. 2012b).

Dryfall is often measured in collectors that are designed to close during rainstorms. When open to the atmosphere, these instruments capture particles that are deposited vertically, known as *sedimentation*. In natural ecosystems, dryfall is also derived by the capture of particles on vegetation surfaces. When vegetation captures particles that are moving horizontally in the airstream, the process is known as *impaction* (Hidy 1970). Impaction is a particularly important process in the capture of seasalt aerosols near the ocean (Art et al. 1974, Potts 1978).

In addition to the uptake of CO_2 in photosynthesis, vegetation also absorbs N– and S– containing gases directly from the atmosphere (Hosker and Lindberg 1982, Lindberg et al. 1986, Sparks et al. 2003, Turnipseed et al. 2006). Uptake of pollutant O_3, SO_2, and NO_2 by vegetation is particularly important in humid regions (McLaughlin and Taylor 1981, Rondón and Granat 1994), where plant stomata remain open for long periods. Lovett and Lindberg (1986, 1993) found that uptake of HNO_3 vapor accounted for 75% of the annual dry deposition of nitrogen (4.8 kg/ha) in a deciduous forest in Tennessee, where dry deposition was nearly half of the total annual deposition of nitrogen from the atmosphere. Vegetation can also be a source or a sink for atmospheric NH_3, depending on the ambient concentration in the atmosphere (Langford and Fehsenfeld 1992, Sutton et al. 1993, Pryor et al. 2001). Plants can also remove volatile organic compounds from the air (Simonich and Hites 1994).

The total capture of dry particles and gases by land plants is difficult to measure. When rainfall is collected inside a forest, it contains materials that have been deposited on the plant surfaces, but also large quantities of elements that are derived from the plants themselves (Parker 1983, Chapter 6). Artificial collectors (surrogate surfaces) are often used to approximate the capture by vegetation (White and Turner 1970, Vandenberg and Knoerr 1985, Lindberg and Lovett 1985). The capture on known surfaces can be compared to the airborne concentrations to calculate a deposition velocity (Sehmel 1980):

$$\text{Deposition velocity} = \frac{\text{Rate of dryfall}(\text{mg/cm}^2/\text{sec})}{\text{Concentration in air}(\text{mg/cm}^3)}. \tag{3.38}$$

In units of cm/sec, these velocities can be multiplied by the estimated surface area of vegetation (cm²) and the concentration in the air to calculate total deposition for an ecosystem. For example, Lovett and Lindberg (1986) used a deposition velocity of 2.0 cm/sec to calculate a nitrogen deposition of 3.0 kg/ha/yr in a forest with a leaf area index of 5.8 m²/m² and an ambient concentration of 0.82 μg N m^{-3} in the form of nitric acid vapor. It is often unclear if deposition velocities measured using artificial surfaces apply to natural surfaces (e.g., bark), and accurate estimates of the surface area of vegetation are difficult (Whittaker and Woodwell 1968). Clearly, further work on dry deposition is needed (Lovett 1994, Petroff et al. 2008).

Atmospheric deposition on the surface of the sea is often estimated from collections of wetfall and dryfall on remote islands (Duce et al. 1991). The surface of the sea can also exchange gases with the atmosphere (Liss and Slater 1974), often acting as a sink for atmospheric CO_2 (Sabine et al. 2004) and SO_2 (Beilke and Lamb 1974) and as a source of NH_3 (Quinn et al. 1987, 1988).

Regional Patterns and Trends

Regional patterns of rainfall chemistry in the United States reflect the relative importance of different constituent sources and deposition processes in different areas (Munger and Eisenreich 1983). Coastal areas are dominated by atmospheric inputs from the sea, with large inputs of Na, Mg, Cl, and SO_4 that are the major constituents in the seasalt aerosol (Junge and Werby 1958, Hedin et al. 1995). Areas of arid and semiarid land show high concentrations of soil-derived constituents, such as Ca, in rainfall (Figure 3.12; Young et al. 1988, Sequeira 1993,

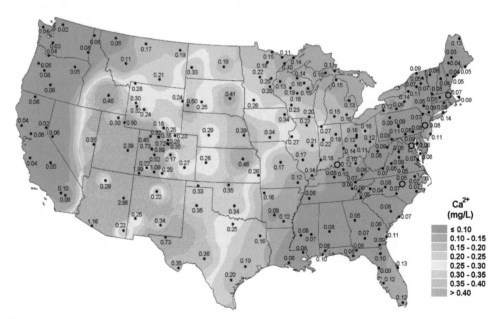

FIGURE 3.12 Mean calcium concentration (mg/l) in wetfall precipitation in the United States for 2009. *Source: From the National Atmospheric Deposition Program/National Trends Network (2009); http://nadp.sws.uiuc.edu.*

Gillette et al. 1992). Areas downwind of regional pollution show exceedingly low pH and high concentrations of SO_4^{2-} and NO_3^- (Schwartz 1989, Ollinger et al. 1993).

The ratio among ionic constituents in rainfall can be used to trace their origin. Except in unusual circumstances, nearly all the sodium (Na) in rainfall is derived from the ocean. When magnesium is found in a ratio of 0.12 with respect to Na—the ratio in seawater (refer to Table 9.1)—we may presume that the Mg is also of marine origin. In the southeastern United States, however, Mg/Na ratios in wetfall range from 0.29 to 0.76 (Swank and Henderson 1976). Here the Mg content has increased relative to Na, presumably because the airflow that brings precipitation to this region has crossed the United States, picking up Mg from soil dust and other sources. Schlesinger et al. (1982b) used this approach to deduce nonmarine sources of Ca and SO_4 in the rainfall in coastal California (Figure 3.13).

Iron (Fe) and aluminum (Al) are largely derived from the soil, and ratios of various ions to these elements in soil can be used to predict their expected concentrations in rainfall when soil dust is a major source (Lawson and Winchester 1979, Warneck 2000). High concentrations of Al in dryfall on Hawaii were traced to springtime dust storms on the central plains of China (Parrington et al. 1983). Soil mineralogy in dusts from the 1930s' Great Plains dust bowl can be identified in layers of the Greenland ice pack (Donarummo et al. 2003). Windborne particles of soil and vegetation contribute significantly to the global transport of trace metals in the atmosphere (Nriagu 1989).

In many areas downwind of pollution, a strong correlation between H^+ and SO_4^{2-} is the result of the production of H_2SO_4 during the oxidation of SO_2 and its dissolution in rainfall (Eqs. 3.27 and 3.28; Cogbill and Likens 1974, Irwin and Williams 1988). Nitrate (NO_3^-) also contributes to the strong acid content in rainfall (HNO_3). These constituents depress the pH of rainfall below 5.6, which would be expected for water in equilibrium with atmospheric CO_2 (Galloway et al. 1976). In contrast, ammonia (NH_3) is a net source of alkalinity in rainwater, since its dissolution produces OH^-:

$$NH_3 + H_2O \rightarrow NH_4^+ + OH^-. \tag{3.39}$$

The pH of rainfall is determined by the concentration of strong acid anions that are not balanced by NH_4^+ and Ca^{2+} (from $CaCO_3$), namely (from Gorham et al. 1984),

$$H^+ = \left[NO_3^- + 2SO_4^{2-}\right] - \left[NH_4^+ + 2Ca^{2+}\right]. \tag{3.40}$$

In Kanpur, India, ammonia dominated the neutralization of acidity in rainfall during the wet season, while Ca played a similar role in the dry season, when more soil dust was present in the atmosphere (Shukla and Sharma 2010).

Globally, about 40% of the atmosphere's acidity is neutralized by NH_3 (Chapter 13), with a higher proportion in the Southern Hemisphere where there is less industrial pollution (Savoie et al. 1993). In the eastern United States, the acidity of rainfall is often directly correlated to the concentration of SO_4^{2-}, which is related to pollutant emissions of SO_2 in areas upwind (Likens et al. 2005). Similar relationships are seen between emissions of NO_x and the NO_3^- content of rain (Butler et al. 2003, 2005). In the western United States, the relationship between acidity and the acid-forming anions is less clear because they have often reacted with soil aerosols containing $CaCO_3$ (Epstein and Oppenheimer 1986, Oppenheimer et al. 1985, Young et al. 1988, Reheis and Kihl 1995).

I. PROCESSES AND REACTIONS

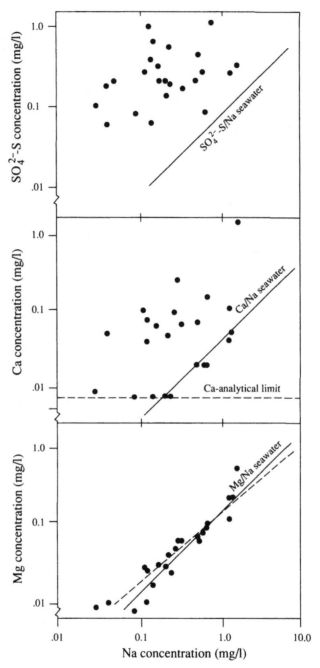

FIGURE 3.13   Concentrations of SO$_4$, Ca, and Mg in wetfall precipitation near Santa Barbara, California, plotted as a logarithmic function of Na concentration in the same samples (Schlesinger et al. 1982b). The *solid line* represents the ratio of these ions to Na in seawater. Ca and SO$_4$ are enriched in wetfall relative to seawater, whereas Mg shows a correlation (*dashed line*) that is not significantly different from the ratio expected in seawater.

In the past 100 years, increases in the concentration of NO_3 and SO_4 in the Greenland and Tibetan snowpacks have reflected the changes in the abundance of anthropogenic pollutants due to industrialization in the Northern Hemisphere (Mayewski et al. 1986, 1990; Thompson et al. 2000). There are no apparent changes in the deposition of these ions in the Southern Hemisphere as recorded by Antarctic ice (Langway et al. 1994). Similarly, the uppermost sediments in lakes of the Northern Hemisphere contain higher concentrations of many trace metals, presumably from industrial sources (Galloway and Likens 1979, Swain et al. 1992).

Long-term records of precipitation chemistry are rare, but the collections at the Hubbard Brook Ecosystem in central New Hampshire and eastern Tennessee suggest a recent decline in the concentrations of SO_4 that may reflect improved control of emissions (Likens et al. 1984, 2002; compare Kelly et al. 2002, Zbieranowski and Aherne 2011, Lutz et al. 2012b). Improvements in air quality as a result of the implementation of the Clean Air Act in 1990 have resulted in significant decreases in the acidity of rainfall over the eastern United States and Canada (Figure 3.14; Hedin et al. 1987). Similarly, the uppermost layers of ice in glaciers of the French Alps and Greenland show lower concentrations of SO_4^{2-} and NO_3^-, presumably due to control of pollutant emissions in upwind sources in recent years (Preunkert et al. 2001, 2004, Fischer et al. 1998).

Even with pollutant abatement, long-term records suggest that many natural ecosystems currently receive a greater input of N, S, and other elements of biogeochemical importance than before widespread emissions from human activities. Pollutant emissions have more than doubled the annual input of S-containing gases to the atmosphere globally (Chapter 13). Excess deposition of nitrogen might be expected to enhance the growth of forests, but in combination with acidity, this fertilization effect may lead to deficiencies of P, Mg, and other plant nutrients (Chapters 4 and 6). Atmospheric deposition of nitrogen makes a significant contribution to the nutrient load and eutrophication of lakes (Bergström and Jansson 2006), estuaries (Nixon et al. 1996, Latimer and Charpentier 2010), and coastal waters (Paerl et al. 1999). The western North Atlantic Ocean receives about 20 to 40% of the sulfur and nitrogen oxides emitted in eastern North America (Galloway and Whelpdale 1987, Liang et al. 1998). Although pollutant emissions have declined in North America and Europe, increasing emissions are seen from India and China (Lelieveld et al. 2001, Richter et al. 2005, Stern 2006). The airborne concentrations of many air pollutants can now be measured using satellite technology (Richter et al. 2005, Clarisse et al. 2009, Martin 2008).

BIOGEOCHEMICAL REACTIONS IN THE STRATOSPHERE

Ozone

Ozone is produced in the stratosphere by the disassociation of oxygen atoms that are exposed to shortwave solar radiation. The reaction accounts for most of the absorption of ultraviolet sunlight ($h\nu$) at wavelengths of 180 to 240 nm and proceeds as follows:

$$O_2 + h\nu \rightarrow O + O \tag{3.41}$$

$$O + O_2 \rightarrow O_3. \tag{3.42}$$

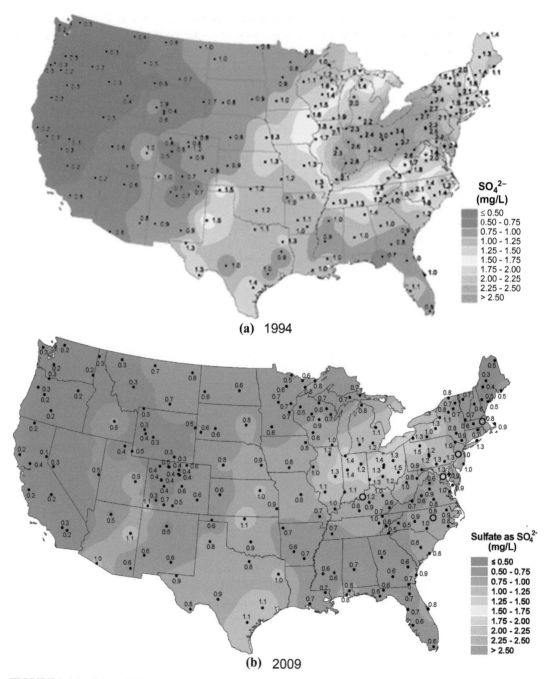

FIGURE 3.14   Sulfate (SO$_4$) concentration (mg/l) measured in samples of wetfall precipitation across the United States, showing the effect of the Clean Air Act in reducing SO$_2$ emissions and thus SO$_4$ deposition between (a) 1994 and (b) 2009. *Source: From the National Atmospheric Deposition Program/National Trends Network (2009); http://nadp.sws. uiuc.edu.*

Some ozone from the stratosphere mixes down into the troposphere, where the production of O_3 by these reactions is limited because there is less ultraviolet light (refer to Table 3.6). Most of the remaining ozone is destroyed by a variety of reactions in the stratosphere.

Absorption of ultraviolet light at wavelengths between 200 and 320 nm destroys ozone:

$$O_3 + h\nu \rightarrow O_2 + O \tag{3.43}$$

$$O + O_3 \rightarrow O_2 + O_2. \tag{3.44}$$

This absorption warms the stratosphere (refer to Figure 3.1) and protects the Earth's surface from the ultraviolet portion of the solar spectrum that is most damaging to living tissue (uvB). Stratospheric ozone is also destroyed by reaction with OH (Wennberg et al. 1994),

$$O_3 + OH \rightarrow HO_2 + O_2 \tag{3.45}$$

$$HO_2 + O_3 \rightarrow OH + 2O_2, \tag{3.46}$$

and by reactions stemming from the presence of nitrous oxide (N_2O), which mixes up from the troposphere. Tropospheric N_2O is produced in a variety of ways (Chapters 6 and 12), but it is inert in the lower atmosphere. The only significant sink for N_2O is photolysis in the stratosphere. About 80% of the N_2O reaching the stratosphere is destroyed in a reaction producing N_2 (Warneck 2000),

$$N_2O \rightarrow N_2 + O(^1D), \tag{3.47}$$

and about 20% in reactions with the $O(^1D)$ produced in Eq. 3.47, mainly

$$N_2O + O(^1D) \rightarrow N_2 + O_2, \tag{3.48}$$

but with a small amount forming NO:

$$N_2O + O(^1D) \rightarrow 2NO. \tag{3.49}$$

The nitric oxide (NO) produced in reaction 3.49 destroys ozone in a series of reactions,

$$NO + O_3 \rightarrow NO_2 + O_2 \tag{3.50}$$

$$O_3 \rightarrow O + O_2 \tag{3.51}$$

$$NO_2 + O \rightarrow NO + O_2, \tag{3.52}$$

for which the net reaction is

$$2O_3 \rightarrow 3O_2. \tag{3.53}$$

Note that the mean residence time of NO in the troposphere is too short for an appreciable amount to reach the stratosphere, where it might contribute to the destruction of ozone. Nearly all the NO in the stratosphere is produced in the stratosphere from N_2O; only a small amount is contributed by high-altitude aircraft. Eventually NO_2 is removed from the strato-sphere by reacting with OH to produce nitric acid (Eq. 3.26), which mixes down to the tro-posphere and is removed by the heterogeneous reaction with raindrops.[6]

Finally, stratospheric ozone is destroyed by chlorine, which acts as a catalyst in the reaction

$$Cl + O_3 \rightarrow ClO + O_2 \tag{3.54}$$

$$O_3 + h\nu \rightarrow O + O_2 \tag{3.55}$$

$$ClO + O \rightarrow Cl + O_2, \tag{3.56}$$

for a net reaction of

$$2O_3 + h\nu \rightarrow 3O_2. \tag{3.57}$$

Although each Cl produced may cycle through these reactions and destroy many mole-cules of O_3, Cl is eventually converted to HCl and removed from the stratosphere by down-ward mixing and heterogeneous interaction with cloud drops in the troposphere (Rowland 1989, Solomon 1990).

The balance between ozone production (refer to Eqs. 3.41 and 3.42) and the various reac-tions that destroy ozone maintains a steady-state concentration of stratospheric O_3 with a peak of approximately 7×10^{18} molecules/m^3 at 30 km altitude (Warneck 2000). Although the photochemical production of O_3 is greatest at the equator, the density of the ozone layer is normally thickest at the poles.

Since the mid-1980s, field measurements have indicated that the total density of ozone mol-ecules in the atmospheric column has declined significantly over Antarctica (Farman et al. 1985; Figure 3.15) and globally (Herman 2010). The decline, as much as 0.3%/yr, was unprecedented and represents a perturbation of global biogeochemistry. Destruction of ozone is likely to lead to an increased flux of ultraviolet radiation to the Earth's surface (Correll et al. 1992, Kerr and McEl-roy 1993, McKenzie et al. 1999) and increased incidence of skin cancer and cataracts in humans (Norval et al. 2007). Greater uvB radiation at the Earth's surface is likely to reduce marine pro-duction in the upper water column of the Southern Ocean around Antarctica (Smith et al. 1992b, Arrigo et al. 2003). Ultraviolet radiation also causes deleterious effects on land plants (Caldwell and Flint 1994, Day and Neale 2002). Because previous, steady-state ozone concentrations were maintained in the face of natural photochemical reactions that produce and consume ozone, at-tention focused on how this balance might have been disrupted by human activities (Cicerone 1987, McElroy and Salawitch 1989, Rowland 1989).

Chlorofluorocarbons (freons), which are produced as aerosol propellants, refrigerants, and solvents, have no known natural source in the atmosphere (Prather 1985). These compounds are chemically inert in the troposphere, so they eventually mix into the stratosphere where they are decomposed by photochemical reactions producing active chlorine (Molina and Rowland 1974, Rowland 1989, 1991):

$$CCl_2F_2 \rightarrow Cl + CClF_2, \tag{3.58}$$

---

[6] Atmospheric chemists refer to this reaction as denitrification. It is not to be confused with the denitrification performed by certain bacteria, which remove NO_3^- and produce N_2 in anaerobic soils and sediments (Chapter 7).

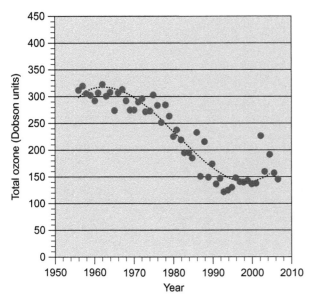

FIGURE 3.15 The decline in ozone (O_3) over Antarctica since the 1950s, and its recovery in recent years as a result of the Montreal Protocol, instituted in 1989. The customary unit for the total number of ozone molecules in an atmospheric column, the Dobson unit, is equivalent to 2.69×10^{16} molecules/cm^2 of the Earth's surface. *Source: From Kump et al. (2010). Used with permission of Pearson/Prentice Hall.*

which can destroy ozone by the reactions of Eqs. 3.54 through 3.56. These reactions are greatly enhanced in the presence of ice particles, which accounts for the first observations of the O_3 "hole" in the springtime over Antarctica (Farman et al. 1985, Solomon et al. 1986). In a dry atmosphere, ClO reacts with NO_2 to form $ClONO_2$, an inactive compound that removes both gases from O_3 destruction. In the presence of ice clouds, $ClONO_2$ breaks down:

$$ClONO_2 + HCl \rightarrow Cl_2 + HNO_3 \tag{3.59}$$

$$Cl_2 + h\nu \rightarrow 2Cl, \tag{3.60}$$

producing active chlorine for ozone destruction (Molina et al. 1987, Solomon 1990). Significantly, during the last 40 years, levels of active chlorine have increased in a mirror image to the loss of ozone from the stratosphere (Solomon 1990).

The relative importance of chlorofluorocarbons versus natural sources of chlorine in the stratosphere is apparent in a global budget for atmospheric chlorine (Figure 3.16). Seasalt aerosols are the largest natural source of chlorine in the troposphere, but they have such a short mean residence time that they do not contribute Cl to the stratosphere. There is also no good reason to suspect that seasalt aerosols have increased in abundance in the last few decades. Similarly, industrial emissions of HCl are rapidly removed from the troposphere by rainfall. Especially violent volcanic eruptions can inject gases directly into the stratosphere, sometimes adding to stratospheric Cl (Johnston 1980, Mankin and Coffey 1984). However, in most cases only a small amount of Cl reaches the stratosphere, because various processes remove HCl from the rising volcanic plume (Tabazadeh and Turco 1993, Textor et al. 2003). After the Mount Pinatubo eruption, which released 4.5×10^{12} g of HCl, stratospheric Cl increased by <1% (Mankin et al. 1992).

The only significant natural source of Cl in the stratosphere stems from the production of methylchloride by the ocean surface, by plants in coastal salt marshes and upland forests, by

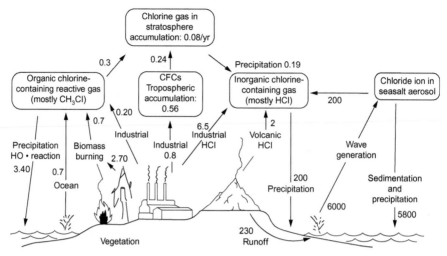

FIGURE 3.16   A global budget for Cl in the troposphere and the stratosphere. All data are given in 10^{12} g Cl/yr. *Sources: Modified and updated from Möller (1990), Graedel and Crutzen (1993), and Graedel and Keene (1995), with new data from McCulloch et al. (1999) and other sources listed in Table 3.7.*

litter decomposition (Keppler et al. 2000, Hamilton et al. 2003), and by forest fires (Table 3.7). There is no strong industrial source that might be indicated by an increase in the concentration of CH_3Cl in the Antarctic ice pack during the past 100 years (Butler et al. 1999, Saltzman et al. 2009), or by a significant difference in the concentration of methylchloride between the Northern and Southern Hemispheres (Beyersdorf et al. 2010). The current budget for CH_3Cl is imbalanced, with sources slightly exceeding sinks, but methylchloride has a mean residence time of about 1.3 years in the atmosphere, so a small portion mixes into the stratosphere.

In the global Cl budget, the relatively small industrial production of chlorofluorocarbons, which are inert in the troposphere, is the dominant source of Cl delivered to the stratosphere (Figure 3.16; Russell et al. 1996). Increasing concentrations of these compounds have been strongly implicated in ozone destruction (Rowland 1989, Butler et al. 1999). Happily, with the advent of the Montreal Protocol in 1989, which limits the use of these compounds worldwide, there is already some evidence that the growth rate of these compounds in the atmosphere is slowing (Elkins et al. 1993, Montzka et al. 1996, Solomon et al. 2006) and that the ozone hole may be starting a slow recovery (Figure 3.15; Yang et al. 2008, Newman et al. 2006). Indeed, the main cause of continued human impacts on stratospheric ozone may stem from our continuing contributions to the rise in N_2O in Earth's atmosphere (Ravishankara et al. 2009; Chapter 12).

Similar reactions are possible with compounds containing bromine; in fact, Br compounds may be even more potent in the destruction of stratospheric O_3 than Cl (Wennberg et al. 1994). Industry is a source of methylbromide (CH_3Br), which is used as an agricultural fumigant (Yagi et al. 1995). Methylbromide is also released from the ocean's surface, coastal vegetation, and biomass burning (Table 3.7). Sinks of CH_3Br include uptake by the oceans and soils and oxidation by OH radical. After rising throughout the Industrial Revolution (Saltzman et al. 2008, Khalil et al. 1993a), the atmospheric concentration of methylbromide appears to have

TABLE 3.7   Budgets of CH_3Cl and CH_3Br in the Atmosphere (Tg/yr)

Sources	CH_3Cl	CH_3Br	References
Ocean surface	0.65–0.76	0.10	Thompson et al. 2002, Yoshida et al. 2006, Anbar et al. 1996
Coastal vegetation	0.05–0.17	0.0014–0.014	Rhew et al. 2000, Hu et al. 2010, Manley et al. 2006
Tropical forests	0.91–1.5		Yokouchi et al. 2002a, Saito et al. 2008, Blei et al. 2010
Other vegetation (by difference)	1.4		Yoshida et al. 2006
Biomass burning	0.55–0.90	0.029	Andreae and Merlet 2001, Yoshida et al. 2006, Thompson et al. 2002
Industrial uses	0.11–0.16	0.05	McCulloch et al. 1999, Thompson et al. 2002
Total of sources (best estimates)	**4.28**	**0.19**	
Sinks			
Ocean uptake		0.12	Anbar et al. 1996
Soil uptake	0.25	0.022–0.042	Shorter et al. 1995, Serca et al. 1998
Reaction with OH	3.37	0.09	Thompson et al. 2002
Loss to stratosphere	0.28	0.006	Thompson et al. 2002
Total of sinks (best estimates)	**3.90**	**0.24**	

declined in recent years (Yokouchi et al. 2002b, Yvon-Lewis et al. 2009). The global budget of CH_3Br and its mean residence time (about 0.8 years; Colman et al. 1998) in the atmosphere are poorly constrained—sinks exceed sources (Table 3.7). However, some CH_3Br persists long enough to reach the stratosphere, where it can lead to ozone destruction.

Among other halogen-containing gases, the lifetimes of bromoform ($CHBr_3$; Quack and Wallace 2003) and methyliodide (CH_3I; Campos et al. 1996, Bell et al. 2002, Butler et al. 2007, Yokouchi et al. 2008), both produced by marine phytoplankton, and various inorganic fluoride compounds (e.g., CH_3F) are too short for appreciable mixing into the stratosphere. The observed increase of fluoride in the stratosphere appears solely due to the upward transport of chlorofluorocarbons, and it is an independent verification of their destruction in the stratosphere by ultraviolet light (Russell et al. 1996). However, F is ineffective as a catalyst for ozone destruction.

Satellite observations have greatly aided our understanding of changes in stratospheric ozone. The loss of ozone from the atmosphere has been monitored since 1979 when the first Total Ozone Mapping Spectrometer (TOMS) began records of the abundance of O_3 in a column extending from the bottom to the top of the atmosphere (Figure 3.17). Both the

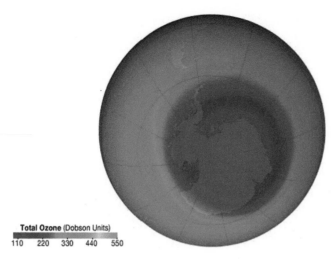

FIGURE 3.17   The average abundance of ozone in the atmosphere of the Southern Hemisphere during October 2006. The ozone "hole," seen in blue and purple, is actually an area where the abundance of ozone in the stratosphere is reduced—perhaps better described as a thinning rather than a hole. *Source: From http://ozonewatch.gsfc.nasa.gov/ monthly/monthly_2006-10.html.*

Total Ozone (Dobson Units)

110   220   330   440   550

area and the minimum column abundance of ozone appear to have stabilized in recent years, after strong declines at the beginning of the record (refer to Figure 3.15).[7] Similar, though less extensive, ozone losses are reported for the Arctic (Solomon et al. 2007, Manney et al. 2011).

Stratospheric Sulfur Compounds

Sulfate aerosols in the stratosphere are important to the albedo of the Earth (Warneck 2000). A layer of sulfate aerosols, known as the Junge layer, is found in the stratosphere at about 20 to 25 km altitude. Its origin is twofold. Large volcanic eruptions can inject SO_2 into the stratosphere, where it is oxidized to sulfate (Eqs. 3.27 and 3.28). Large eruptions have the potential to increase the abundance of stratospheric sulfate 100-fold (Arnold and Bührke 1983, Hofmann and Rosen 1983), and the sulfate aerosols persist in the stratosphere for several years, cooling the planet (McCormick et al. 1995, Briffa et al. 1998). During periods without volcanic activity, the dominant source of stratospheric sulfate derives from carbonyl sulfide (COS)[8] that mixes up from the troposphere, where it originates from a variety of sources (Chapter 13). Most sulfur gases are so reactive that they do not reach the stratosphere, but COS has a mean residence time of about 5 years in the atmosphere (refer to Table 3.5), so about one-third of the annual production mixes to the stratosphere. Additional COS may be lofted to the stratosphere in the smoke plumes of large wildfires (Notholt et al. 2003).

Carbonyl sulfide that reaches the stratosphere is oxidized by photolysis, forming sulfate aerosols which contribute to the Junge layer (Chin and Davis 1993). Eventually, these aerosols are removed from the stratosphere by downward mixing of stratospheric air.

[7] See *http://ozoneaq.gsfc.nasa.gov/* and *http://ozonewatch.gsfc.nasa.gov/*.

[8] Also abbreviated OCS.

The concentration of COS in the atmosphere today is higher than in preindustrial times (Montzka et al. 2004), although it seems to have declined slightly in recent years (Rinsland et al. 2002, Sturges et al. 2001).

MODELS OF THE ATMOSPHERE AND GLOBAL CLIMATE

A large number of models have been developed to explain the physical properties and chemical reactions in the atmosphere. When these models attempt to predict the characteristics in a single column of the atmosphere, they are known as one-dimensional (1D) and radiative-convective models. For example, Figure 3.2 is a simple 1D model for the greenhouse effect, which assumes that the behavior of the Earth's atmosphere can be approximated by average values applied to the entire surface. Two-dimensional (2D) models can be developed using the vertical dimension and a single horizontal dimension (e.g., latitude) to examine the change in atmospheric characteristics across a known distance of the Earth's surface (e.g., Brasseur and Hitchman 1988, Hough and Derwent 1990). On a regional scale, these are particularly useful in following the fate of pollution emissions (e.g., Rodhe 1981, Asman and van Jaarsveld 1992, Berge and Jakobsen 2002). Three-dimensional (3D) models attempt to follow the fate of particular parcels of air as they move both horizontally and vertically in the atmosphere. These dynamic 3D models are known as *general circulation models* (GCMs) for the globe (Figure 3.18).

Many models are constructed to include both chemical reactions and physical phenomena, such as the circulation of the atmosphere due to temperature differences. Chemical transformations are parameterized using the rate and equilibrium coefficients for the reactions that we have examined in this chapter. Because there are a large number of reactions, most of these models are quite complex (e.g., Logan et al. 1981, Isaksen and Hov 1987, Lelieveld and Crutzen 1990), but they give useful predictions of future atmospheric composition when the input of several constituents is changing simultaneously.

Nearly all climate models predict that a substantial warming of the atmosphere (2–4.5°C) will accompany increasing concentrations of CO_2, N_2O, CH_4, and chlorofluorocarbons in the atmosphere (IPCC 2007).[9] Largely stemming from fossil fuel combustion, the concentration of CO_2 is now higher than at any time in the past 20 million years (Pearson and Palmer 2000). During the past 150 years, the concentrations of CO_2, CH_4, and N_2O have risen above levels seen at any time during the past 10,000 years—spanning the entire history of human civilization (Flückiger et al. 1999). Atmospheric warming, resulting from the absorption of infrared (heat) radiation by these gases emitted from the Earth's surface, is known as the greenhouse effect or radiative *forcing* (Figure 3.2). The predicted warming of future climate is greatest near the poles, where there is normally the greatest net loss of infrared radiation relative to incident sunlight (Manabe and Wetherald 1980). Dramatic, recent declines in the extent of Arctic sea ice suggest that these predictions are already proving correct (Serreze et al. 2007; Chapter 10). For the same reason, future nighttime and wintertime temperatures worldwide are likely to show large changes relative to today's conditions. Presumably the oceans will warm more

---

[9] Through its industrial use as a solvent, nitrogen trifluoride (NF_3) is potentially an important contributor to Earth's greenhouse effect, but it is not included in most assessments of climate change (Prather and Hau 2008, Weiss et al. 2008).

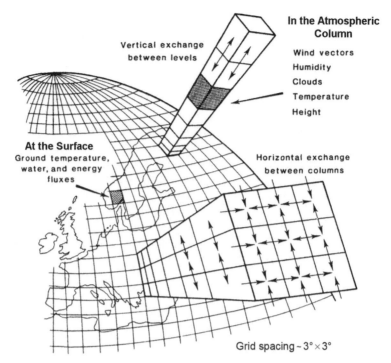

FIGURE 3.18   Conceptual structure of a dynamic three-dimensional general circulation model for the Earth's atmosphere, indicating the variables that must be included for a global model to function properly. *Source: From Henderson-Sellers and McGuffie (1987). Copyright © 1987. Reprinted by permission of John Wiley & Sons, Ltd.*

slowly than the atmosphere, but eventually warmer ocean waters will allow greater rates of evaporation, increasing the circulation of water in the global hydrologic cycle (Chapter 10). Water vapor also absorbs infrared radiation, so it is likely to further accelerate the potential greenhouse effect (Raval and Ramanathan 1989, Rind et al. 1991, Soden et al. 2005, Willett et al. 2007). Thus, most models predict that higher concentrations of CO_2 and other trace gases in the atmosphere will make the Earth a warmer and more humid planet.

Incoming solar radiation delivers about 340 W/m^2 to the Earth (Figure 3.2).[10] The natural greenhouse effect warms the planet about 33°C by trapping 153 W/m^2 of outgoing radiation (Ramanathan 1988).[11] For the past 30 years or so, there has been a small increase in the Sun's luminosity (+0.12 to +0.16 W/m^2) (IPCC 2007, Foukal et al. 2006, Pinker et al. 2005), but the human impact on radiative forcing due to increasing concentrations of atmospheric trace gases currently adds about 2.3 W/m^2 to the natural greenhouse effect (IPCC 2007), causing measurable

[10] The Sun's radiation, measured outside Earth's atmosphere, delivers 1379 W/m^2, known as the solar constant (McElroy 2002). A one-dimensional model for the Earth's radiation budget shows an annual input of ~340 W/m^{-2}, because only ¼ of the Earth's surface is exposed to sunlight at any moment.

[11] The Earth's greenhouse effect is dominated by H_2O and CO_2. O_2 and N_2 provide only a trivial contribution to Earth's radiation balance, together adding 0.28 W/m^2 to the greenhouse effect (Hopfner et al. 2012).

changes in the spectral distribution of radiation leaving the Earth (Harries et al. 2001). Aerosols tend to cool the atmosphere, and increases in aerosols due to human activities are thought to reduce the global radiative forcing by about 1.2 W/m^2 (i.e., global dimming; IPCC 2007, Bellouin et al. 2005, Mahowald 2011). It is interesting to note that aerosol concentrations were higher (Patterson et al. 1999, Lambert et al. 2008) and CO_2 concentrations and temperatures were lower (Figure 1.2) during the last glacial period.

Long-term records from tree rings and ice cores suggest that substantial warming of the global climate is now occurring, coincident with rising CO_2 (Mann et al. 1999; Thompson et al. 2000). Recent temperatures in Europe are greater than at any point in the past 500 years (Luterbacher et al. 2004). These changes in climate cannot be explained by natural phenomena alone (Crowley 2000, Stott et al. 2000). How rapidly these changes in climate occur will be moderated by the thermal buffer capacity of the world's oceans, which can absorb enormous quantities of heat. Already, several long-term records suggest increases in the ocean's temperature worldwide (Barnett et al. 2005, Levitus et al. 2001).

Differential warming of the atmosphere and oceans will also change global patterns of precipitation and evapotranspiration (Manabe and Wetherald 1986, Rind et al. 1990, Zhang et al. 2007), causing substantial changes in soil moisture of most areas outside the tropics. Arid regions, such as the southwestern United States, are especially likely to experience increased drought (Cook et al. 2004, Seager et al. 2007), consistent with recent trends in rainfall in this region (Milly et al. 2005).

Climate change affects biogeochemistry (and vice versa) in a variety of ways. Land plants and ocean waters take up substantial quantities of CO_2 from the atmosphere, potentially slowing the rate of climate change (Chapters 5, 9, and 11). Clearing vegetation alters the albedo of the Earth's land surface, potentially altering radiative forcing. Warmer temperatures may increase the rate of decomposition of soil carbon now frozen in Arctic permafrost (Dorrepaal et al. 2009, Schuur et al. 2009), and trigger the release of methane now frozen in ocean sediments, resulting in positive feedbacks that further exacerbate global warming (Chapter 11). Changes in climate are likely to affect the distribution of many plants and animals, potentially causing many extinctions (Thomas et al. 2004a); they also impact a wide range of conditions affecting human health and economic activity.

SUMMARY

In this chapter we have examined the physical structure, circulation, and composition of the atmosphere. Major constituents, such as N_2, are rather unreactive and have long mean residence times in the atmosphere. CO_2 is largely controlled by plant photosynthetic uptake and by its dissolution in waters on the surface of the Earth. The atmosphere contains a variety of minor constituents, many of which are reduced gases. These gases are highly reactive in homogeneous reactions with hydroxyl (OH) radicals and heterogeneous reactions with aerosols and cloud droplets, which scrub them from the atmosphere. Changes in the concentration of many trace gases are indicative of global change, perhaps leading to future climatic warming and higher surface flux of ultraviolet light. The oxidized products of trace gases are deposited in land and ocean ecosystems, resulting in inputs of N, S, and other elements of biogeochemical significance. Pollution of the atmosphere by the release of oxidized gases containing N and S as a result of human activities results in acid deposition in downwind ecosystems. The enhanced deposition of N and S represents altered biogeochemical cycling on a regional and global basis. Changes in stratospheric ozone and global climate are early warnings of the human impact on the atmosphere of our planet.

Recommended Readings

Brasseur, G.P., J.J. Orlando, and G.S. Tyndall. 1999. *Atmospheric Chemistry and Global Change*. Oxford University Press.
Graedel, T.E. and P.J. Crutzen. 1993. *Atmospheric Change*. Freeman.
Henderson-Sellers, A. and K. McGuffie. 1987. *A Climate Modelling Primer*. Wiley.
Jacob, D.J. 1999. *Introduction to Atmospheric Chemistry*. Princeton University Press.
McElroy, M.B. 2002. *The Atmospheric Environment*. Princeton University Press.
Seinfeld, J.H. and S.N. Pandis. 2006. *Atmospheric Chemistry and Physics*. Wiley.
Walker, J.C.G. 1977. *Evolution of the Atmosphere*. Macmillan.
Warneck, P. 2000. *Chemistry of the Natural Atmosphere* (second ed). Academic Press/Elsevier.
Wayne, R.P. 1991. *Chemistry of Atmospheres*. Clarendon Press.

PROBLEMS

3.1 If you are vacationing in Aspen, Colorado (8000 ft elevation), what is the mass of the atmosphere above you, relative to that at sea level?

3.2 For any planet, the equilibrium surface temperature is determined by the balance between the radiation received and the radiation lost to space. The input is determined by

$$Input = [F(1 - A)\pi(R^2)]/4,$$

where
F is the solar constant; for Earth $F = 1379$ W/m^2
A is the albedo of the surface
R = radius of the planet
The loss of radiation is determined by the emission of infraradiation:

$$Loss = (\sigma)T^4\pi R^2,$$

where
σ is the Stefan-Boltzmann constant (5.673×10^{-8} W/m^2/K^{-4})
T is the temperature in degrees Kelvin
R = radius of the planet
Thus, if the input of radiation increases, the temperature will rise. But with a rise in temperature there is a dramatic increase in loss (i.e., temperature to the 4th power), so eventually the input and loss will again be equal. At that time, the planet will be in thermal equilibrium but at a higher surface temperature. This is essentially the origin of the "greenhouse effect" on Earth, which is driven not by changes in the Sun's luminosity but by the tendency for radiatively active gases (such as CO_2 and H_2O) to reduce the loss of infrared radiation from Earth's surface.
 (a) If the albedo of Earth was 0 (i.e., if the Earth were a perfect black-body absorber), what would be its equilibrium temperature?
 (b) Actually, the albedo of Earth is about 0.30. What should be its equilibrium temperature? Compared to the mean temperature on Earth, what is the "greenhouse effect"?

3.3 The mass of tropospheric air that is injected into the stratosphere results in a mean residence time for stratospheric air of 2.6 years. Presumably, each year, the same volume of air mixes down from the stratosphere into the troposphere. What is the mean residence time of tropospheric air with respect to the stratosphere? Given their 100-year mean residence time in Earth's atmosphere, what does this say about the circulation of CFCs in our atmosphere?

3.4 What percent of the emissions of a gas with constant emissions and a mean residence time of 120 days in the troposphere will mix into the stratosphere?

3.5 The deposition velocity of NH_3 in forests and other natural ecosystems is often about 10 to 50 mm/sec (Sutton et al. 1993). Given the range of concentrations of NH_3 in the atmosphere (Table 3.5), what is the expected range of dry deposition of NH_3 in terrestrial ecosystems?

4

The Lithosphere

INTRODUCTION

Since early geologic time, the atmosphere has interacted with the exposed crust of the Earth, causing rock weathering. Many of the volcanic gases in the Earth's earliest atmosphere dissolved in water to form acids that could react with surface minerals (Chapter 2). Later, as oxygen accumulated in the atmosphere, rock weathering occurred as a result of the oxidation of reduced minerals, such as pyrite, that were exposed at the Earth's surface. At least since the advent of land plants, rock minerals have also been exposed to high concentrations of carbon dioxide in the soil as a result of the metabolic activities of soil microbes and plant roots. Today, carbonic acid (H_2CO_3), derived by reaction of CO_2 with soil water, is a major cause of rock weathering in most ecosystems. In recent years, humans have added large quantities of NO_x and SO_2 to the atmosphere, causing acid rain (Chapter 3) and increasing the rate of rock weathering in many areas (Cronan 1980).

Siever (1974) proposed a basic equation to summarize the close linkage between the Earth's atmosphere and its crust:

$$\text{Igneous Rocks} + \text{Acid Volatiles} = \text{Sedimentary Rocks} + \text{Salty Oceans.} \qquad (4.1)$$

This formula recognizes that through geologic time the primary minerals of the Earth's crust have been exposed to reactive, acid-forming C, N, and S gases of the atmosphere. The products of the reaction are carried to the oceans, where they accumulate as dissolved salts or in ocean sediments (Li 1972). Large amounts of sedimentary rock have formed through geologic time; indeed, about two-thirds of the rocks now exposed on land are sedimentary rocks that have been uplifted by tectonic activity (Durr et al. 2005, Suchet et al. 2003, Wilkinson et al. 2010). With their uplift, these sedimentary rocks are subject to further weathering reactions with acid volatiles, in accord with Siever's basic equation. Eventually, geologic processes, known as subduction, carry sedimentary rocks to the deep Earth, where CO_2 is released and the solid constituents are converted back to primary minerals under great heat and pressure (Figure 1.3; see also Siever 1974).

This chapter reviews the basic types of rock weathering on land and the processes that drive these weathering reactions. Rock weathering is especially important to the bioavailability of elements that have no gaseous forms (e.g., Ca, K, Fe, and P; Table 4.1). Weathering is basic to soil fertility, biological diversity, and agricultural productivity (Huston 1993). Reactions between soil waters and the solid materials in soil determine the availability of essential elements to biota and the losses of these elements in runoff. Conversely, land plants and soil microbes affect rock weathering and soil development, potentially regulating global biogeochemistry and the Earth's climate. In this chapter, we examine soil development in the major ecosystems on Earth. Finally, we will estimate the global rate of rock weathering in an attempt to determine the annual new supply of biochemical elements on land and the total delivery of weathering products to rivers and the sea.

TABLE 4.1   Approximate Mean Composition of Earth's Continental Crust

Constituent	Percentage composition
Si	28.8
Al	7.96
Fe	4.32
Ca	3.85
Na	2.36
Mg	2.20
K	2.14
Ti	0.40
P	0.076
Mn	0.072
S	0.070

Source: Data from Wedepohl (1995).

ROCK WEATHERING

Following geologic uplift and exposure at the Earth's surface, all rocks undergo weathering, a general term that encompasses a wide variety of processes that decompose rocks. Mechanical weathering is the fragmentation or loss of materials without a chemical reaction; in laboratory terminology it is equivalent to a physical change. Mechanical weathering is important in extreme and highly seasonal climates and in areas with much exposed rock. Wind abrasion is a form of mechanical weathering in arid environments, whereas the expansion of frozen water in rock crevices—often resulting in fractured rocks—is an important form of mechanical weathering in cold climates. Plant roots also fragment rock when they grow in crevices.

Where the products of mechanical weathering are not rapidly removed by erosion, thick soils develop, and the landscape is said to be "transport-limited." In other areas, finely divided rock and soil are removed by erosion—the loss of particulate solids from the ecosystem. The products of mechanical weathering may also be lost in catastrophic events such as landslides (Swanson et al. 1982). Indicative of the high rates of mechanical weathering and erosion at high elevations, rivers draining mountainous regions often carry exceptionally large sediment loads to the sea (Milliman and Syvitski 1992).

Chemical Weathering

Chemical weathering occurs when the minerals in rocks and soils react with acidic and oxidizing substances. Usually chemical weathering involves water, and various constituents are released as dissolved ions that are available for uptake by biota or that are lost in stream waters. In many cases, mechanical weathering is important in exposing the minerals in rocks to chemical attack (e.g., Miller and Drever 1977, Anbeek 1993, Nugent et al. 1998, Hilley et al. 2010). If the rate of mechanical weathering is slow, it may limit the rate of chemical weathering and lower the fertility of soils (Gabet and Mudd 2009).

The rate of chemical weathering depends on the mineral composition of rocks. Igneous and metamorphic rocks contain primary minerals (e.g., olivine and plagioclase) that were formed under conditions of high temperature and pressure deep in the Earth. These primary silicate minerals are crystalline in structure, and they are found in two classes—the ferromagnesian or *mafic* series and the plagioclase or *felsic* series—depending on the crystal structure and the presence of magnesium versus aluminum in the crystal lattice (Figure 4.1). Among these minerals, the rate of weathering tends to follow the reverse order of the sequence of mineral formation during the original cooling and crystallization of rock; that is, minerals that condensed first are the most susceptible to weathering reactions (Goldich 1938).

Minerals formed during rapid, early crystallization of magma at high temperatures contain few bonds that link the units of their crystalline structure. They also have frequent substitutions of various cations—for example, Ca, Na, K, Mg, and trace metals (Fe and Mn)—in their crystal lattice, distorting its shape and increasing its susceptibility to weathering. Thus, for example, olivine, which is formed under conditions of great heat and pressure deep within the Earth, is most likely to weather rapidly when exposed at the Earth's surface.

I. PROCESSES AND REACTIONS

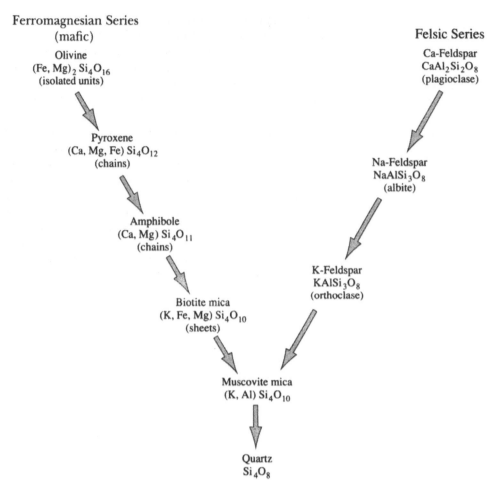

Ferromagnesian Series
(mafic)

Olivine
$(Fe, Mg)_2 Si_4O_{16}$
(isolated units)

Pyroxene
$(Ca, Mg, Fe) Si_4O_{12}$
(chains)

Amphibole
$(Ca, Mg) Si_4O_{11}$
(chains)

Biotite mica
$(K, Fe, Mg) Si_4O_{10}$
(sheets)

Felsic Series

Ca-Feldspar
$CaAl_2Si_2O_8$
(plagioclase)

Na-Feldspar
$NaAlSi_3O_8$
(albite)

K-Feldspar
$KAlSi_3O_8$
(orthoclase)

Muscovite mica
$(K, Al) Si_4O_{10}$

Quartz
Si_4O_8

FIGURE 4.1   Silicate minerals are divided into two classes, the ferromagnesian series and the felsic series, based on the presence of Mg or Al in the crystal structure. Among the ferromagnesian series, minerals that exist as isolated crystal units (e.g., olivine) are most susceptible to weathering, while those showing linkage of crystal units and a lower ratio of oxygen to silicon are more resistant. Among the felsic series, Ca-feldspar (plagioclase) is more susceptible to weathering than Na-feldspar (albite) and K-feldspar (orthoclase). Quartz is the most resistant of all. This weathering series is the reverse of the order in which these minerals are precipitated during the cooling of magma.

In rocks or soils of mixed composition, chemical weathering is concentrated on the relatively labile minerals, while other minerals may be unaffected (April et al. 1986, White et al. 1996, 2001). The dissolution and loss of some constituents may result in a reduction of the density of bedrock, with little collapse or loss of the initial rock volume—a process known as isovolumetric weathering. The product, "rotten" rock known as *saprolite*, comprises the lower soil profile of many regions, especially in the southeastern United States (Gardner et al. 1978, Velbel 1990, Stolt et al. 1992, Oh and Richter 2005). In other cases, the removal of some constituents is accompanied by the collapse of the soil profile and an apparent increase in the volumetric concentration of the elements that remain (e.g., Zr, Ti, and Fe; Brimhall et al. 1991).

Quartz is very resistant to chemical weathering and often remains when other minerals are lost (Figure 4.1). Quartz is a relatively simple silicate mineral consisting only of silicon and oxygen in tetrahedral crystals that are linked in three dimensions. In many cases, the sand fraction of soils is largely composed of quartz crystals that remain following the chemical weathering and loss of other constituents during soil development (Brimhall et al. 1991).

In addition to mineralogy, rock weathering also depends on climate (White and Blum 1995, Gislason et al. 2009). Chemical weathering involves chemical reactions, so it is not surprising that it occurs most rapidly under conditions of higher temperature and precipitation (White et al. 1998, West et al. 2005). Chemical weathering is more rapid in tropical forests than in temperate forests, and more rapid in most forests than in grasslands or deserts. Chemical weathering proceeds, albeit at low rates, in cold Antarctic environments (Hodson et al. 2010). White and Blum (1995) show that the loss of Si in stream water, which is often a good index of chemical weathering, is directly related to precipitation and temperature over much of the Earth's surface (Figure 4.2). With the ongoing global changes in climate, the likelihood of higher future temperatures and precipitation should increase the rate of chemical weathering worldwide.

The dominant form of chemical weathering is the carbonation reaction, driven by the formation of carbonic acid, H_2CO_3, in the soil solution:

$$H_2O + CO_2 \leftrightarrow H^+ + HCO_3^- \leftrightarrow H_2CO_3. \tag{4.2}$$

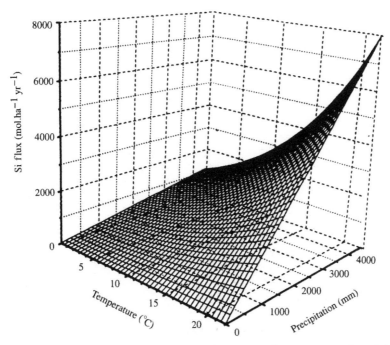

FIGURE 4.2   Loss of silicon (SiO_2) in runoff as a function of mean annual temperature and precipitation in various areas of the world. *Source: Modified from White and Blum (1995).*

Because plant roots and soil microbes release CO_2 to the soil, the concentration of H_2CO_3 in soil waters is often much greater than that in equilibrium with atmospheric CO_2 at 400 ppm (i.e., 0.04%) (Castelle and Galloway 1990, Amundson and Davidson 1990, Pinol et al. 1995). Buyanovsky and Wagner (1983) report seasonal CO_2 concentrations of greater than 7% in the soil beneath wheat fields in Missouri. Such high concentrations of CO_2 can extend to considerable depths in the soil profile, affecting the weathering of underlying rock (Sears and Langmuir 1982, Richter and Markewitz 1995). Wood and Petraitis (1984) found CO_2 concentrations of 1.0% at 36 m, which they link to the downward transport of organic materials that subsequently decompose at depth. Solomon and Cerling (1987) found that high concentrations of CO_2 accumulated in the soil under a mountain snowpack, potentially leading to significant weathering during the winter (see also Berner and Rao 1997).

Plant growth is greatest in warm and wet climates. As a result of root growth and activity, these areas maintain the highest levels of soil CO_2 and the greatest rates of carbonation weathering (Johnson et al. 1977). Examining data from a variety of ecosystems, Brook et al. (1983) found that the average concentration of soil CO_2 varies as a function of actual evapotranspiration, a composite measure of temperature and available soil moisture, at the site (Figure 4.3). However, rock weathering appears to be controlled by carbonation weathering even in arid regions (e.g., Routson et al. 1977). By maintaining high concentrations of CO_2 in the soil, plants and soil microbes control the process of rock weathering—a good example of *biogeochemistry* at the Earth's surface.

Carbonic acid attacks silicate rocks. For example, weathering of the Na-feldspar, albite, proceeds as

$$2NaAlSi_3O_8 + 2H_2CO_3 + 9H_2O \rightarrow$$
$$2Na^+ + 2HCO_3^- + 4H_4SiO_4 + Al_2Si_2O_5(OH)_4. \tag{4.3}$$

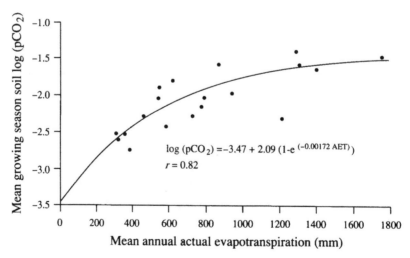

FIGURE 4.3   The relationship between the mean concentration of CO_2 in the soil pore space and the actual evapotranspiration of the site for various ecosystems of the world. *Source: From Brook et al. (1983).*

During this process, a primary mineral is converted to a secondary mineral, kaolinite, by the removal of Na^+ and soluble silica. A sign that carbonation weathering has occurred is the observation that HCO_3^- is the dominant anion in runoff waters (Ohte and Tokuchi 1999). Formation of kaolinite involves hydration with H^+ and water. The secondary mineral has a lower ratio of Si to Al as a result of the loss of some Si to stream waters. Because only some of the constituents of the primary mineral are released, this type of weathering reaction is known as an *incongruent dissolution*. Under conditions of high rainfall, as in the humid tropics, kaolinite may undergo a second incongruent dissolution to form another secondary mineral, gibbsite:

$$Al_2Si_2O_5(OH)_4 + 5H_2O \rightarrow 2H_4SiO_4 + Al_2O_3 \cdot 3H_2O. \tag{4.4}$$

Some weathering reactions involve congruent dissolutions. In moist climates, limestone undergoes a relatively rapid congruent dissolution during carbonation weathering:

$$CaCO_3 + H_2CO_3 \rightarrow Ca^{2+} + 2HCO_3^-. \tag{4.5}$$

Olivine ($FeMgSiO_4$) also undergoes congruent dissolution in water, releasing Fe, Mg, and Si (Grandstaff 1986). The magnesium and silicon are lost in runoff waters, but usually the Fe reacts with oxygen, resulting in the precipitation of Fe_2O_3 in the soil profile. Similarly, pyrite (FeS_2) undergoes a congruent reaction during its oxidation:

$$4FeS_2 + 8H_2O + 15O_2 \rightarrow 2Fe_2O_3 + 16H^+ + 8SO_4^{2-}. \tag{4.6}$$

The H^+ produced in this reaction accounts for the acidity of runoff from many mining operations. As with the weathering of olivine, the Fe from weathered pyrite is subsequently precipitated as Fe_2O_3 in the soil profile or streambed (Bloomfield 1972, Johnson et al. 1992). Often this reaction is mediated by the chemoautotrophic bacterium *Thiobacillus ferrooxidans* (Eq. 2.16; Temple and Colmer 1951, Ralph 1979, Schrenk et al. 1998).

In addition to carbonic acid, organisms release a variety of organic acids to the soil solution that can be involved in the weathering of silicate minerals (Ugolini and Sletten 1991). Many simple organic compounds, including acetic and citric acids, are released from plant roots (Smith 1976, Tyler and Ström 1995, Jones 1998). Phenolic acids (i.e., tannins) are released during the decomposition of plant litter, and soil microbes produce fulvic and humic acids during decomposition of plant remains (Chapter 5). Many fungi release oxalic acid that results in chemical weathering (Cromack et al. 1979, Lapeyrie et al. 1987, Welch and Ullman 1993, Cama and Ganor 2006). Organic acids from plant roots and microbes can weather biotite mica, releasing K (Boyle and Voigt 1973, April and Keller 1990). Soil microbes appear to preferentially colonize and weather the surface of minerals that contain elements, such as phosphorus, that are otherwise in short supply for their growth and reproduction (Rogers et al. 1998, Banfield et al. 1999).

In addition to their contributions to total acidity, organic acids speed weathering reactions in the soil by combining with some weathering products, in a process called *chelation*.[1] When Fe and Al combine with fulvic acid, they are mobile and move to the lower soil profile in percolating water (Dahlgren and Walker 1993, Lundström 1993). When these elements are involved in chelation, their inorganic concentration in the soil solution remains low, the

---

[1] Derived from the Greek word for claw, chelation applies when an organic molecule forms two or more separate covalent bonds with an ion, usually a metal, that keeps it in solution.

equilibrium between dissolved products and mineral forms is not achieved, and chemical weathering may continue unabated (Berggren and Mulder 1995, Zhang and Bloom 1999). Grandstaff (1986) found that additions of small concentrations of EDTA (an organic chelation agent) to weathering solutions increased the dissolution of olivine by 110 times over inorganic conditions. Fulvic and humic acids increase the weathering of a variety of silicate minerals, including quartz, particularly when the soil solution is neutral or slightly acid (Tan 1980, Bennett et al. 1988, Wogelius and Walther 1991, Welch and Ullman 1993).

Organic acids often dominate the acidity of the upper soil profile, while carbonic acid is important below (Ugolini et al. 1977). In general, organic acids dominate the weathering processes in cool temperate forests where decomposition processes are slow and incomplete, whereas carbonic acid drives the chemical weathering in tropical forests where lower concentrations of fulvic acids remain after the decomposition of plant debris (Johnson et al. 1977, Ohte and Tokuchi 1999).

There is much debate about changes in the rate of chemical weathering through geologic time, especially as a result of the evolution of vascular land plants (Berner and Kothavala 2001, Drever 1994). Greater rock weathering, beginning with the initial colonization of land by plants, is likely to have been responsible for a decline in atmospheric CO_2 300 to 400 million years ago, owing to the consumption of CO_2 in carbonation weathering (Knoll and James 1987, Berner 1997, Mora et al. 1996). Various workers have found evidence for higher rates of rock weathering under areas of vegetation (Moulton et al. 2000), suggesting that higher plants add large amounts of CO_2 to the soil from root metabolism and from the decomposition of plant debris by microbes (Kelly et al. 1998). This view holds that the process of photosynthesis acts to speed the transfer of an acid volatile, CO_2, from the atmosphere to the soil profile. Even before the advent of vascular land plants, the Earth's surface may have been covered with algae and lichens, producing relatively high levels of CO_2 in the soil (Keller and Wood 1993, Retallack 1997).

Other workers disagree, suggesting that tectonic uplift and erosion—processes that stimulate mechanical weathering—are the most important determinants of the rate of chemical weathering (Hilley and Porder 2008, Gabet and Mudd 2009, Riebe et al. 2003, Dixon et al. 2012). Still others argue that temperature and precipitation are the dominant factors controlling chemical weathering, with plants playing a lesser role (Gislason et al. 2009, Ohte and Tokuchi 1999, Kump et al. 2000). In sum, a variety of factors—tectonics, mineralogy, climate plants, and soil microbes—play important roles in determining the rate of rock weathering, and it may be impossible to attribute a predominant role to any one of them (Gaillardet et al. 1999, Gabet and Mudd 2009, Anderson et al. 2002).

Because carbonation weathering consumes atmospheric CO_2, the rate of rock weathering on Earth has a major long-term effect on global climate. Schwartzman and Volk (1989) suggest that without CO_2-enhanced chemical weathering, the temperature of the Earth would be too hot for all but the most primitive microbes. As atmospheric CO_2 increases, we can expect rates of rock weathering to increase. This weathering will be mediated directly by increasing soil CO_2 concentrations and indirectly by rising global temperatures (compare Figure 4.2). Several experiments show that the amount of CO_2 in the soil profile increases when plants are grown at high CO_2 (Andrews and Schlesinger 2001, Bernhardt et al. 2006, Williams et al. 2003, Karberg et al. 2005; Figure 4.4). And in soil-warming experiments, greater microbial activity leads to higher concentrations of CO_2 in the soil pore space, and presumably greater carbonation weathering (Figure 4.5).

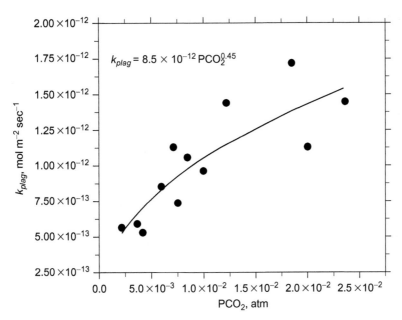

FIGURE 4.4   Dissolution of Ca-feldspar (plagioclase) as a function of soil CO_2 concentrations in watersheds of the Sierra Nevada (California), subject to differential hydrothermal activity. *Source: From Navarre-Sitchler and Thyne (2007).*

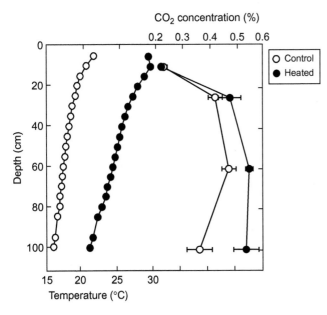

FIGURE 4.5   Soil temperature and pore-space CO_2 concentration (mean ±S.E.) as a function of depth in control and experimentally heated (+5°C) plots in a hardwood forest in Massachusetts. *Source: From unpublished work of Megonigal.*

Secondary Minerals

Secondary minerals are formed as byproducts of weathering at the Earth's surface. Usually the formation of secondary minerals begins near the site where primary minerals are being attacked, perhaps even originating as coatings on the crystal surfaces (Casey et al. 1993, Nugent et al. 1998). Although weathering leads to the loss of Si as a dissolved constituent in stream water (Figure 4.2), some Si is often retained in the formation of secondary minerals (see Eq. 4.3).

Many types of secondary minerals can form in soils during chemical weathering. The secondary minerals in temperate forest soils are often dominated by layered silicate or "clay" minerals. These exist as small (<0.002-mm) particles that control the structural and chemical properties of soils. In general, two types of layers characterize the crystalline structure of secondary, aluminosilicate clay minerals—Si layers and layers dominated by Al, Fe, and Mg. These layers are held together by shared oxygen atoms. Clay minerals and the size of their crystal units are recognized by the number, order, and ratio of these layers (Birkeland 1984). Moderately weathered soils are often dominated by secondary minerals such as montmorillonite and illite, which have a 2:1 ratio of Si- to Al-dominated layers. More strongly weathered soils, such as in the southeastern United States, are dominated by kaolinite clays with a 1:1 ratio of layers, reflecting a greater loss of Si.

Because secondary minerals may incorporate elements important to biochemistry, one cannot assume that the release of those elements from primary minerals leads to an immediate increase in the pool of ions available for uptake by plants. Magnesium is often fixed in the crystal lattice of montmorillonite, whereas illite contains K (Martin and Sparks 1985, Harris et al. 1988). These are common secondary minerals in temperate soils. Similarly, although little nitrogen is contained in primary minerals, some 2:1 clay minerals incorporate N as ammonium (NH_4) in their crystal lattice (Holloway and Dahlgren 2002). Ammonium contained in clay minerals can represent more than 10% of the total N in some soils (Stevenson 1982, Smith et al. 1994, Johnson et al. 2012). The weathering of sedimentary rocks containing ancient clay minerals with "fixed" ammonium can release large quantities of nitrogen to stream waters (Holloway et al. 1998). Recognizing the widespread nitrogen limitation on land (Chapter 6), the release of nitrogen from rock weathering may play an important role in determining the availability of N for plant growth (Mengel and Scherer 1981, Baethgen and Alley 1987, Green et al. 1994, Morford et al. 2011).

In contrast to the loss of Si and other cations (e.g., Ca and Na) to runoff waters, Al and Fe are relatively insoluble in soils unless they are involved in chelation relations with organic matter (Huang 1988, Alvarez et al. 1992, Allan and Roulet 1994, Ross and Bartlett 1996). In the absence of chelation reactions, these elements tend to accumulate in the soil as oxides. Initially, free Fe accumulates in amorphous and poorly crystallized forms, known as ferrihydrite, which are often quantified by extraction in a weak oxalate solution (Shoji et al. 1993, Birkeland 1984). With increasing time, most Fe is found in crystalline oxides and hydroxides, which are traditionally extracted using a reducing solution of citrate-dithionate (Chorover et al. 2004). Some of these mineral transformations involve bacteria and thus are biogeochemical in nature (Fassbinder et al. 1990).

Crystalline oxides and hydrous oxides of Fe (e.g., goethite and hematite) and Al (e.g., gibbsite and boehmite) are common in many tropical soils, where high temperatures and rainfall

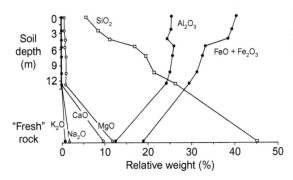

FIGURE 4.6 Content of different rock-forming minerals (% mass) above basalt bedrock in Kauai, Hawaii. *Source: From Bluth and Kump (1994).*

cause relatively rapid decomposition of plant debris and few organic acids remain to chelate and mobilize Fe and Al. Under these climatic conditions, the secondary clay minerals typical of temperate zone soils are subject to weathering, with the near-complete removal of Si, Ca, K, and other basic cations in stream water (Figure 4.6). However, in an interesting example of the importance of biota in soil development, Lucas et al. (1993) show that kaolinite may persist in the upper horizons of some rainforest soils due to the plant uptake of Si from the lower soil profile and the return of Si to the soil surface in plant debris (see also, Alexandre et al. 1997, Markewitz and Richter 1998, Gerard et al. 2008, Conley 2002). Similar plant "pumping" also maintains higher levels of K at the surface of many soils (Jobbagy and Jackson 2004, Barre et al. 2009).

SOIL CHEMICAL REACTIONS

Following their release by weathering, the availability of essential biochemical elements to biota is controlled by a number of reactions that determine the equilibrium between concentrations in the soil solution and contents that are associated with soil minerals or organic materials. In contrast to the kinetics of weathering reactions, soil exchange reactions occur relatively rapidly (Furrer et al. 1989, Maher 2010). The specific soil reactions differ depending on how soil development has progressed under the influence of climate, time, biota, topographic position, and the parent material of the site (Jenny 1980).

Cation Exchange Capacity

The layered silicate clay minerals that dominate temperate zone soils possess a net negative charge that attracts and holds cations dissolved in the soil solution. This negative charge has several origins; most of it arises from ionic substitutions within silicate clays, especially in 2:1 clays. For example, when Mg^{2+} substitutes for Al^{3+} in montmorillonite, there is an unsatisfied negative charge in the internal crystal lattice. This negative charge is permanent in the sense that it arises inside the crystal structure and cannot be neutralized by covalent bonding of cations from the soil solution. Permanent charge is expressed as a zone or "halo" of negative charge surrounding the surface of clay particles in the soil.

A second source of negative charge is found at the edges of clay particles, where hydroxide ($-OH$) groups are often exposed to the soil solution. Depending on the pH of the solution, the H^+ ion may be more or less strongly bound to this group. In most cases, a considerable number of the H^+ are dissociated, leaving negative charges ($-O^-$) that can attract and bind cations (e.g., Ca^{2+}, K^+, and NH_4^+). This cation exchange capacity is known as *pH-dependent charge*. The binding is reversible and exists in equilibrium with ionic concentrations in the soil solution.

In many temperate soils, a large amount of cation exchange capacity is also contributed by soil organic matter (Yuan et al. 1967). These are pH-dependent charges originating from the phenolic ($-OH$) and organic acid ($-COOH$) groups of soil humic materials. In some sandy soils, as in central Florida, and in many highly weathered soils, nearly all cation exchange is the result of soil organic matter (e.g., Daniels et al. 1987, Richter et al. 1994). Organic matter is also the major source of cation exchange in desert soils that contain relatively small amounts of secondary clay minerals as a result of limited chemical weathering.

The total negative charge in a soil is expressed as mEq/100 g or cmol(+)/kg of soil, which constitutes the cation exchange capacity (CEC). Exchange of cations occurs as a function of chemical mass balance with the soil solution. Elaborate models of ion exchange have been developed by soil chemists (Sposito 1989). In general, cations are held on the exchange sites and displace one another in the sequence

$$Al^{3+} > H^+ > Sr^{2+} > Ca^{2+} > Mg^{2+} > Rb^+ > K^+ > NH_4^+ > Na^+ > Li^+ \qquad (4.7)$$

(Sparks 2004, Sahai 2000). This sequence assumes equal molar concentrations in the initial soil solution and can be altered by the presence of large quantities of the more weakly held ions. Agricultural liming, for example, is an attempt to displace Al^{3+} ions from the exchange sites by "swamping" the soil solution with excess Ca^{2+}. In most cases, few cation exchange sites are actually occupied by H^+, which quickly weathers soil minerals, releasing Al and other cations.

Cations other than Al and H are informally known as base cations, since they tend to form bases—for example, $Ca(OH)_2$—when they are released to the soil solution (Birkeland 1984, p. 23). The percentage of the total cation exchange capacity occupied by base cations is termed *base saturation*. Both cation exchange capacity and base saturation increase during initial soil development on newly exposed parent materials. However, as the weathering of soil minerals continues, cation exchange capacity and base saturation decline (Bockheim 1980). Temperate forest soils dominated by 2:1 clay minerals have greater cation exchange capacity than those dominated by 1:1 clay minerals such as kaolinite. Highly weathered soils in the humid tropics are dominated by aluminum hydroxide minerals, which offer essentially no cation exchange capacity in the mineral fraction at their natural soil pH.

Soil Buffering

Cation exchange capacity acts to buffer the acidity of many temperate soils. When H^+ is added to the soil solution, it exchanges for cations, especially Ca, on clay minerals and organic matter (Bache 1984, James and Riha 1986). Over a wide range of pH, temperate soils maintain a constant value (k) for the expression

$$pH - 1/2(pCa) = k, \qquad (4.8)$$

which is known as the lime potential. This expression suggests that when H^+ is added to the soil solution (lower pH), the concentration of Ca^{2+} increases in the soil solution (lower pCa), so that k remains constant. The 1/2 reflects the valence of Ca^{2+} versus H^+. As long as there is sufficient base saturation (e.g., >15%), buffering by CEC explains why the pH of many temperate soils may show relatively little change when they are exposed to acid rain (Federer and Hornbeck 1985, David et al. 1991, Johnson et al. 1994, Likens et al. 1996).

In strongly acid soils, as in the humid tropics, there is little CEC to buffer the soil solution. These soils are buffered by various geochemical reactions involving aluminum (Figure 4.7). Aluminum is not a base cation inasmuch as its release to the soil solution leads to the formation of H^+ when Al^{3+} is precipitated as aluminum hydroxide:

$$Al^{3+} + H_2O \leftrightarrow Al(OH)^{2+} + H^+ \tag{4.9}$$

$$Al(OH)^{2+} + H_2O \leftrightarrow Al(OH)_2^+ + H^+ \tag{4.10}$$

$$Al(OH)_2^+ + H_2O \leftrightarrow Al(OH)_3 + H^+. \tag{4.11}$$

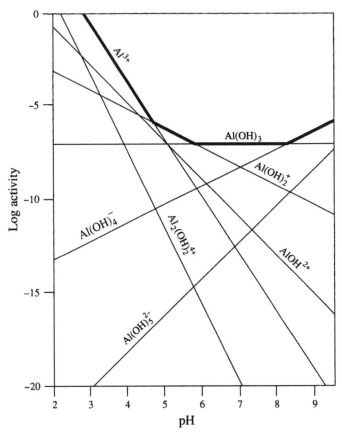

FIGURE 4.7   The solubility of aluminum as a function of pH. For pH in the neutral range, gibbsite [$Al(OH)_3$] controls aluminum solubility, and there is little Al^{+3} in solution. Al^{+3} becomes more soluble at pH < 4.7. *Source: From Lindsay (1979).*

These reactions account for the acidity of many soils in the humid tropics (Sanchez et al. 1982a). Note that the reactions are reversible, so that the soil solution is also buffered against additions of H^+ by the dissolution of aluminum hydroxide. The acid rain received by the northeastern United States appears to dissolve gibbsite [$Al(OH)_3$] from many forest soils, leading to high concentrations of Al^{3+} that are toxic to fish in lakes and streams at high elevations. As stream waters flow to lower elevations, H^+ is consumed in weathering reactions with various silicate minerals, stream water pH increases, and aluminum hydroxides are precipitated (Johnson et al. 1981a). In other regions, dissolution of Al-organic complexes (rather than gibbsite) appears to control the concentration of dissolved Al^{3+} in the soil solution (Mulder and Stein 1994, Allan and Roulet 1994).

Anion Adsorption Capacity

In contrast to the permanent, negative charge in soils of the temperate zone, tropical soils dominated by oxides and hydrous oxides of iron and aluminum show variable charge, depending on soil pH (Uehara and Gillman 1981, Sollins et al. 1988, Arai and Sparks 2007). Under acid conditions these soils possess positive charge, as a result of the association of H^+ with the surface hydroxide groups (Figure 4.8). With an experimental increase in pH, a soil sample is observed to pass through a zero point of charge (ZPC), where the number of cation and anion exchange sites is equal. These soils develop cation exchange capacity at high pH. For gibbsite, the ZPC occurs around pH 9.0, so significant anion adsorption capacity (AAC) is present in acid tropical soils in most field situations.

It is important to recognize that these reactions occur on soil constituents everywhere, but the ZPC of layered silicate minerals or soil organic matter occurs at pH < 2.0, so they offer little anion adsorption capacity in most conditions (Sposito 1989, Polubesova et al. 1995). The ZPC of a bulk soil sample will depend on the relative mix of various minerals and organic matter (Chorover and Sposito 1995). Tropical soils in Costa Rica show a ZPC at a pH of about 4.0, as a result of their mixture of soil organic matter and gibbsite (Sollins et al. 1988). Some anion adsorption capacity is found in temperate soils when iron and aluminum oxides and hydroxides occur in the lower soil profile (Johnson et al. 1981a, 1986a).

Anion adsorption capacity is typically greater on poorly crystalline forms of Fe and Al (oxalate-extractable), which have greater surface area than crystalline forms (dithionate-extractable) (Parfitt and Smart 1978, Johnson et al. 1986a, Chorover et al. 2004). Potential adsorption of various anions, including sulfate from acid rain, is positively correlated to the oxalate-extractable Al in a variety of soils (Harrison et al. 1989, Courchesne and Hendershot 1989, MacDonald and Hart 1990, Walbridge et al. 1991).

Anion adsorption follows the sequence

$$PO_4^{3-} > SO_4^{2-} > Cl^- > NO_3^-, \tag{4.12}$$

which accounts for the low availability of phosphorus in many tropical soils (Strahm and Harrison 2007). Frequently anion exchange is described using a Langmuir model, in which the content of anions held on exchange sites is expressed as a function of the concentration in the solution (Travis and Etnier 1981, Reuss and Johnson 1986, Autry and Fitzgerald 1993). Phosphorus, sulfate, and selenate (SeO_4^{2-}) are so strongly held that the binding is known

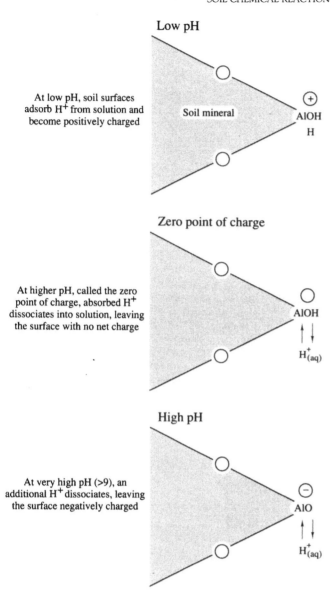

Low pH

At low pH, soil surfaces adsorb H^+ from solution and become positively charged

Soil mineral

AlOH
H

$\oplus$

Zero point of charge

At higher pH, called the zero point of charge, absorbed H^+ dissociates into solution, leaving the surface with no net charge

AlOH

$H^+_{(aq)}$

High pH

At very high pH (>9), an additional H^+ dissociates, leaving the surface negatively charged

AlO

$H^+_{(aq)}$

$\ominus$

FIGURE 4.8   Variation in surface charge on iron and aluminum hydroxides as a function of pH of the soil solution. *Source: From Johnson and Cole (1980).*

as *specific adsorption* or ligand exchange, which is thought to replace $-OH$ groups on the surface of the minerals (Figure 4.9; Hingston et al. 1967, Guadalix and Pardo 1991, Bhatti et al. 1998). Here, the adsorption of SO_4^{2-} from acid rain is associated with an increase in soil pH, a decline in apparent ZPC, and higher cation exchange capacity (e.g., Marcano-Martinez and McBride 1989, David et al. 1991). All these anions are also involved in nonspecific adsorption, which is more readily reversible with changes in their concentration in the soil solution.

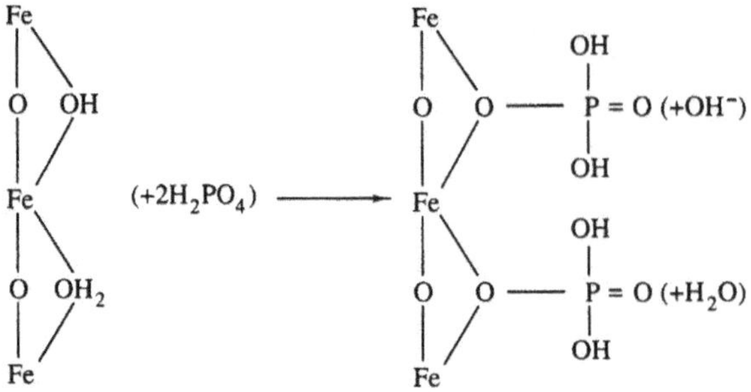

FIGURE 4.9 The specific adsorption of phosphate by iron sesquioxides may release OH or H_2O to the soil solution. *Source: From Binkley (1986).*

Anion adsorption capacity is inhibited by soil organic matter, especially organic anions, which tend to bind to the reactive surfaces of Fe and Al minerals (Johnson and Todd 1983, Hue 1991, Karltun and Gustafsson 1993, Gu et al. 1995). Soils or soil layers that are rich in organic matter are less efficient in anion adsorption than those dominated solely by Fe and Al oxide and hydroxide minerals. Percolating waters often carry SO_4^{2-} from the upper, organic layers of the soil to lower depths, where it is captured by Fe and Al minerals (Dethier et al. 1988, Vance and David 1991). Thus, anion adsorption capacity is determined by the effects of soil organic matter on a variety of soil properties: increasing AAC by inhibiting the crystallization of Fe and Al minerals, but reducing AAC by binding to the anion exchange sites (Johnson et al. 1986a, Kaiser and Zech 1998).

Phosphorus Minerals

Phosphorus (P) deserves special attention, since it is often in limited supply for plants. The only primary mineral with significant phosphorus content is apatite, which can undergo carbonation weathering in a congruent reaction, releasing P:

$$Ca_5(PO_4)_3OH + 4H_2CO_3 \rightarrow 5Ca^{2+} + 3HPO_4^{2-} + 4HCO_3^- + H_2O. \tag{4.13}$$

Although this phosphorus may be accumulated by biota (organic-P), a large proportion of the available P is involved in reactions with other soil minerals, leading to its precipitation in unavailable forms.

As seen in Figure 4.10, the maximum level of available phosphorus in the soil solution is found at a pH of about 7.0. In acid soils, P availability is controlled by direct precipitation with iron and aluminum (Lindsay and Moreno 1960, Arai and Sparks 2007), whereas in alkaline soils phosphorus is often precipitated with calcium minerals (Cole and Olsen 1959, Lajtha and Bloomer 1988), and therefore may be deficient for optimal plant growth (Tyler 1994). Binding to iron and aluminum oxides accounts for the low availability of phosphorus in many tropical soils (Sanchez et al. 1982b, Smeck 1985, Agbenin 2003). When phosphorus is captured in the interior of crystalline Fe and Al oxides, it is known as occluded P, which is essentially unavailable to biota.

Walker and Syers (1976) diagram the general evolution of phosphorus availability during the weathering of rocks containing apatite (Figure 4.11). Apatite weathers rapidly, giving rise to phosphorus contained in various other forms and to a decline of total phosphorus in the system due to losses in runoff (Singleton and Lavkulich 1987a, Schlesinger et al. 1998,

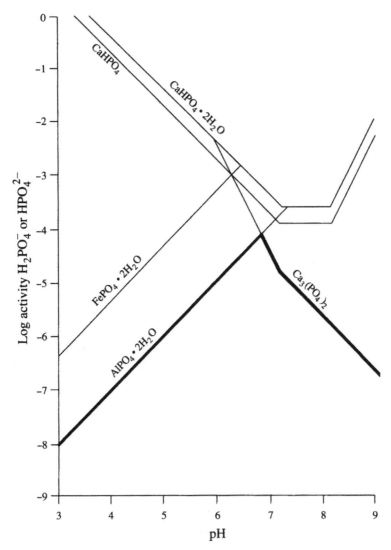

FIGURE 4.10 The solubility of phosphorus in the soil solution as a function of pH. Precipitation with Al sets the upper limit on dissolved phosphate at low pH (bold line); precipitation with Ca sets a similar limit at high pH. Phosphorus is most available at a pH of about 7.0. *Modified from Lindsay and Vlek (1977).*

Filippelli and Souch 1999, Selmants and Hart 2010). Nonoccluded phosphorus includes forms that are held on the surface of soil minerals by a variety of reactions, including anion adsorption. With time, crystalline oxide minerals accumulate, and phosphorus accumulates in occluded forms. At the later stages of weathering and soil development, occluded and organic P dominate the forms of P remaining in the system (Cross and Schlesinger 1995, Crews et al. 1995, Richardson et al. 2004, Yang and Post 2011). At this stage, almost all available phosphorus may be found in organic forms in the upper soil profile, while phosphorus

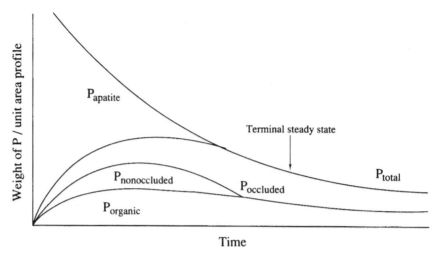

FIGURE 4.11   Changes in the forms of phosphorus found during soil development on sand dunes in New Zealand. *Source: Modified from Walker and Syers (1976).*

found at lower depths is largely bound with secondary minerals (Yanai 1992). Plant growth may depend almost entirely on the release of phosphorus from dead organic matter, defining a biogeochemical cycle of phosphorus in the upper soil horizons (Wood et al. 1984).

As seen for the weathering of silicate minerals, organic acids can influence the release of phosphorus during rock weathering. Jurinak et al. (1986) show how the production of oxalic acid (CH_2O_4) by plant roots can lead to the weathering of P from apatite. Oxalate production is directly related to soil phosphorus availability in sandy soils of Florida (Figure 4.12).

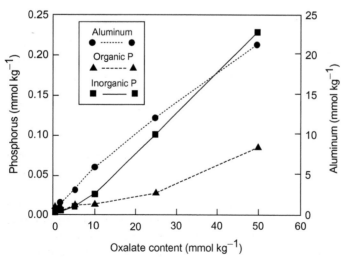

FIGURE 4.12   Release of inorganic P, organic P, and Al from a spodic horizon of a pine forest in Florida, following a single oxalate addition at different levels. *Source: From Fox and Commerford (1992a). Used with permission of American Society of Agronomy.*

Organic acids can inhibit the crystallization of Al and Fe oxides, reducing the rate of phosphorus occlusion (Schwertmann 1966, Kodama and Schnitzer 1977, 1980), and allowing noncrystalline (amorphous) forms to dominate the P-adsorption capacity of the soil (Walbridge et al. 1991, Yuan and Lavkulich 1994).

In addition, phosphorus may be more available in the presence of organic acids, such as oxalate, which remove Fe and Ca from the soil solution by chelation and precipitation (Graustein et al. 1977, Welch et al. 2002, Wang et al. 2008). The production and release of oxalic acid by mycorrhizal fungi (Chapter 6) explains their importance to the phosphorus nutrition of higher plants (Bolan et al. 1984, Cromack et al. 1979) and the greater availability of phosphorus under fungal mats (Fisher 1972, 1977). Some workers believe that the mobilization of phosphorus by symbiotic fungi was a precursor to the successful establishment of plants on land (Chapter 2).

SOIL DEVELOPMENT

Soils usually consist of a number of layers, or horizons, that collectively constitute the complete soil profile, or *pedon*. Rock weathering, water movement, and organic decomposition all influence the development of a soil profile under varying climatic conditions. Indeed, Jenny (1941, 1980) suggested that soil profile development could only be understood in the context of the interactions of climate, biota, topography, parent material, and time. Today, humans are one of the organisms that have dramatic effects on soil development in many regions (Amundson and Jenny 1991, Richter 2007). Recognition of the processes that occur in the different layers of soil is an essential part of understanding biogeochemical cycles on land. Chemical weathering occurs in the upper soil profile as well as in the underlying layers of fractured rock. In this section, we consider soil development in forests, grasslands, and deserts.

Forests

In forests it is often easy to separate an organic layer, the forest floor or "O-horizon," from the underlying layers of mineral soil. The thickness and presence of the forest floor varies throughout the year, especially in regions where plant litterfall is strongly seasonal. In some tropical forests decomposition of fresh litter is so rapid that there is little surface litter (Olson 1963, Vogt et al. 1986). On the other hand, slow decomposition in coniferous forests, especially in the boreal zone, results in the accumulation of a thick forest floor, known as a *mor*, that is sharply differentiated from the underlying soil (Romell 1935). Much of the arctic zone is characterized by waterlogged soils, in which the entire rooting zone is composed of organic materials. Such peatland soils are known as *Histosols*. We will treat the special properties of waterlogged organic soils in Chapter 7.

The upper mineral soil is designated as the A-horizon. Soil water percolating through the forest floor often contains a variety of organic acids derived from the microbial decomposition of litter (Vance and David 1991, Strobel 2001). These organic acids dominate the weathering of soil minerals in the A-horizon. Solutions collected beneath the A-horizon carry cations and

TABLE 4.2   Chemical Composition of Precipitation, Soil Solutions, and Groundwater
in a 175-year-old *Abies amabilis* Stand in Northern Washington

Solution	pH	Total cations (mEq/liter)	Soluble ions (mg/liter)			Total (mg/liter)	
			Fe	Si	Al	N	P
Precipitation							
Above canopy	5.8	0.03	<0.01	0.09	0.03	0.60	0.01
Below canopy	5.0	0.10	0.02	0.09	0.06	0.40	0.05
Forest floor	4.7	0.14	0.04	3.50	0.79	0.54	0.04
Soil							
15 cm E	4.6	0.12	0.04	3.55	0.50	0.41	0.02
30 cm B_s	5.0	0.08	0.01	3.87	0.27	0.20	0.02
60 cm B3	5.6	0.25	0.02	2.90	0.58	0.37	0.03
Groundwater	6.2	0.26	0.01	4.29	0.02	0.14	0.01

*Source: Data from Ugolini et al. (1977), Soil Science **124**: 291–302. Copyright (1977) Williams and Wilkins.*

silicate, derived from weathering reactions (Table 4.2). Iron and Al are removed from the A-horizon by chelation with fulvic acid and low-molecular-weight (LMW) organic acids that percolate downward from the forest floor (Antweiler and Drever 1983, Driscoll et al. 1985, Zysset et al. 1999, Fuss et al. 2011). The removal of mineral components from the A-horizon is known as *eluviation*, while the downward transport of Fe and Al in conjunction with organic acids is known as *podzolization* (Chesworth and Macias-Vasquez 1985, Lundström et al. 2000a).

Although it occurs throughout the world, podzolization is particularly intense in subarctic (boreal) and cool temperate forests (e.g., Ugolini et al. 1987, De Kimpe and Martel 1976, Langley-Turnbaugh and Bockheim 1998). Many of these areas are characterized by coniferous forests that produce litterfall that is rich in phenolic compounds and organic acids (Cronan and Aiken 1985, Strobel 2001). In these ecosystems, decomposition is slow and incomplete, and large quantities of organic acids are available to percolate from the forest floor into the underlying A-horizon. The pH of the soil solution is often as low as 4.0 (Dethier et al. 1988, Vance and David 1991).

When the removal of Fe, Al, and organic matter is very strong, a whitish layer is easily recognized beneath the forest floor (Figure 4.13). This horizon is sometimes designated as an A_e- or E- (eluvial) horizon, which may consist entirely of quartz grains that are resistant to weathering and relatively insoluble under acid conditions (Pedro et al. 1978). These eluvial horizons reflect the importance of biota in soil development; in the absence of organic chelation, one would expect Fe and Al to accumulate as weathering products and for Si to dominate the losses from the upper soil profile.

During soil development, substances leached from the A- and E-horizons are deposited in the underlying B-horizons (Jersak et al. 1995, Langley-Turnbaugh and Bockheim 1998). These are defined as the zone of deposition or the *illuvial* horizons, where secondary clay minerals

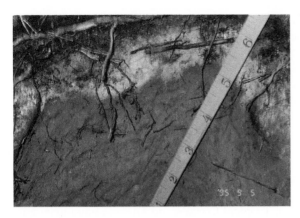

FIGURE 4.13 Soil profile, Nyanget, Svartberget Forest Research Station, north Sweden, showing the E- (whitish) and B-horizons. The scale is in decimeters. The role of biology in the process of podzolization is exemplified by the removal of Al and Fe minerals from the E-horizon in complexes with organic acids and their precipitation as oxides in the B-horizon. *Source: From Lundström et al. (2000b).*

accumulate. Clay minerals arrest the downward movement of dissolved organic compounds that are carrying Fe and Al (Greenland 1971, Chesworth and Macias-Vasquez 1985, Cronan and Aiken 1985, Schulthess and Huang 1991, Jansen et al. 2005). Typically, Fe oxides and hydrous-oxides precipitate first, and Al moves lower in the profile (Adams et al. 1980, Olsson and Melkerud 1989, Law et al. 1991a), but the mechanisms underlying this difference are not entirely clear (Lundström et al. 2000a).

Strongly podzolized soils, *Spodosols*, are characterized by a dark *spodic* horizon-designated B_{hs} that is rich in Fe and organic matter. On fresh parent materials, a spodic horizon can develop in 350 to 1000 years (Singleton and Lavkulich 1987b; Protz et al. 1984, 1988; Barrett and Schaetzl 1992). In New England forests, the accumulation of organic matter in the spodic horizon appears to limit the loss of dissolved organic carbon in streams (McDowell and Wood 1984). Most of the Fe and Al in the B-horizon is found in crystalline oxides (e.g., Fe_2O_3), but near root channels and pockets of buried organic matter, Fe^{3+} may be reduced to Fe^{2+} and leached away (Fimmen et al. 2008, Dubinsky et al. 2010, Fuss et al. 2011). Phosphorus can be mobilized from these areas of the B-horizon, which often appear grayish in the field (Chacon et al. 2006).

In warmer climates, decomposition is more rapid, smaller quantities of fulvic acids remain to percolate through the A-horizon, podzolization is less intense, and there is no sharply defined E-horizon (Pedro et al. 1978). In most areas of the tropics, decomposition is so complete that there is almost no soluble organic acid percolating through the soil profile (Johnson et al. 1977). In the absence of podzolization, Fe and Al are not removed from the upper soil profile by chelation; rather, they are precipitated as oxides and hydroxides that accumulate in the zone of active weathering. Unless they cycle through vegetation, cations and Si are lost to runoff. Podzolization is found in only a few tropical soils, usually those that develop on sandy parent materials (Bravard and Righi 1989).

Soils in tropical forests may be many meters in depth, because in many areas they have developed over millions of years without disturbances such as glaciation (Birkeland 1984). In the absence of clear zones of eluviation and illuviation, the distinction of A- and B-horizons is difficult. Long periods of intense weathering have removed cations and silicon from the entire soil profile. The climate of most tropical regions includes high precipitation, and the solubility of Si increases with increasing temperature (Figure 4.2). Many tropical soils are classified as *Oxisols*, on the basis of high contents of Fe and Al oxides throughout the soil profile (Richter and Babbar 1991). Over large portions of lowland tropical rainforests, soils are acid (with Al buffering), low in base cations, and infertile with P deficiency (Sanchez et al. 1982b). Under extreme conditions, these soils are known informally as *laterite*.

A comparative index of soil formation and the degree of podzolization is seen in the ratio of Si to sesquioxides (Fe and Al) in the soil profile (Table 4.3). In boreal forest soils, Si is relatively immobile and Fe and Al are removed, which results in high values for this ratio in the A-horizon. The accumulation of secondary minerals such as montmorillonite in the moderately weathered soils of the glaciated portion of the United States yields Si/sesquioxide ratios of two to four as a result of the ratio of Si to Al in the crystal lattice. Silicon/sesquioxide ratios are lower in more highly weathered soils. During soil development in the southeastern United States, low ratios characterize older soils in which kaolinite (a 1:1 mineral) has accumulated as a secondary mineral (Figure 4.14). Tropical Oxisols and Ultisols have very low values for this ratio in all horizons because they are dominated by kaolinite and iron and aluminum minerals.

Below the B-horizons, the C-horizon consists of coarsely fragmented soil material with little organic content. Carbonation weathering tends to predominate in the C-horizon, which may consist of a thick saprolite (Ugolini et al. 1977). Weathering can extend to a depth of hundreds of meters in the soil profile (Richter and Markewitz 1995), and it controls the chemistry of groundwater. For a rainforest in Puerto Rico, nearly all of the Al was released by weathering in the upper soil profile, but 50% of the Si in stream water was derived from weathering in the C-horizon (Riebe et al. 2003). When the soil has developed from local materials, the C-horizon shows mineralogical similarity to the underlying parent rock. In contrast, when the parent materials have been deposited by transport, there may be little correspondence between the C-horizon and the underlying bedrock. Some of the world's richest agricultural soils are found on transported materials, such as glacial till, floodplain deposits, and volcanic ash.

The distribution of soils forms a continuous gradient over broad geographic regions. For example, the degree of podzolization varies beneath deciduous and coniferous forests in the same geographic region (Stanley and Ciolkosz 1981, De Kimpe and Martel 1976). Soil profile

TABLE 4.3   Silicon/Sesquioxide ($Al_2O_3 + Fe_2O_3$) Ratios for the A- and B-Horizons of Some Soils in Different Climatic Regions

Region	Number of sites	Mean Si/sesquioxide ratio		References
		A-horizon	B-horizon	
Boreal	1	12.0	8.1	Wright et al. (1959)
Boreal	1	9.3	6.7	Leahey (1947)
Cool-temperate	4	4.07	2.28	Mackney (1961)
Warm-temperate	6	3.77	3.15	Tan and Troth (1982)
Tropical	5	1.47	1.61	Tan and Troth (1982)

Note that the removal of Al and Fe results in high values in boreal and cool temperate soils, especially in the A-horizon. Lower values characterize tropical soils, where there is little differentiation between horizons as a result of the removal of Si from the entire profile in long periods of weathering.

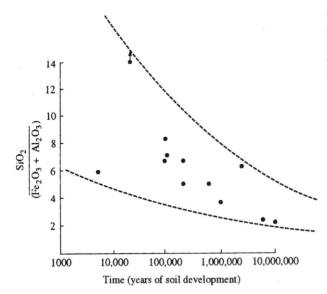

FIGURE 4.14 Changes in Si-to-sesquioxide ratio during the long-term development of soils in the southeastern United States (Markewitz and Pavich 1991). The data are recalculated from Markewitz et al. (1989)

development on steep slopes is often incomplete as a result of landslides and other mechanical weathering events. Soils in floodplain areas and those that have received deposits of volcanic ash may have "buried" horizons (Dahlgren et al. 1999). In areas of recent disturbance, soils with little or no profile development are known as *Inceptisols* and *Entisols*, respectively. In all cases one must remember that soil profile development is slow compared to changes in vegetation.

Soil profile development is also affected by human activities. In the Piedmont region of the southeastern United States, the forest floor often resides directly on top of the B-horizon. Here, the A-horizon was lost by erosion during past agricultural use. In the northeastern United States, forest Spodosols are now exposed to acid rain. At high elevations, the acidity of the solution percolating through the forest floor and A-horizon is not dominated by organic or carbonic acids, but by "strong" acids such as H_2SO_4 (see Chapter 13). At normal levels of acidity, aluminum is mobile only as an organic chelate, and it is precipitated in the B-horizon. Under the present conditions of higher acidity, Al is mobile as Al^{3+}, a potentially toxic form that is carried through the lower profile to stream waters, with SO_4^{2-} as a balancing anion (Figure 4.7; Johnson et al. 1972, Cronan 1980, Reuss et al. 1987). The overall rate of chemical weathering has increased (Cronan 1980, April et al. 1986). Fortunately, with the implementation of the Clean Air Act in the early 1990s, the deposition of acidity and SO_4^{2-} and the stream water losses of Al^{3+} are beginning to decline (Palmer and Driscoll 2002, Clow and Mast 2002, Likens et al. 2002).

Grasslands

In contrast to soil development in forests, where precipitation greatly exceeds evapotranspiration and excess water is available for soil leaching and runoff, soil development in grasslands proceeds under conditions of relative drought. The products of chemical weathering

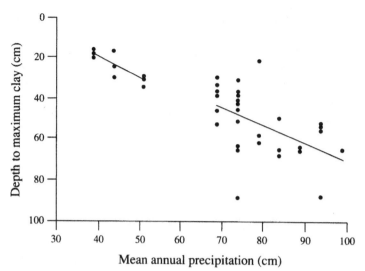

FIGURE 4.15   Depth-to-peak content of clay in the soil profile, an index of weathering and soil development, decreases from east to west across the Great Plains of the United States as a function of the decrease in mean annual precipitation. *Source: From Honeycutt et al. (1990). Used with permission of American Society of Agronomy.*

are not rapidly leached from the soil profile, so soils remain near neutral pH with high base saturation. High contents of Ca and other cations tend to flocculate[2] clay minerals and fulvic acids in the upper profile (Oades 1988), limiting podzolization and the downward movement of weathering products. Often there is little development of a clay-rich or B_t-horizon in grassland soils until the upper profile has been leached free of Ca. Overall, the intensity of chemical weathering and podzolization in grassland soils is much less than in forests (Madsen and Nornberg 1995).

Trends in the development of grassland soils are best seen by examining a transect across the midportion of the United States. Mean annual precipitation decreases westward from the tallgrass prairie near the Mississippi River valley to the shortgrass prairie at the base of the Rocky Mountains. Honeycutt et al. (1990) show that the thickness of the soil profile—measured by the depth to the peak content of clay—decreases from east to west along this gradient (Figure 4.15). Along the same gradient, base saturation, pH, and the content of calcium increase (Ruhe 1984, Gunal and Ransom 2006). Similar trends are seen along a gradient of decreasing precipitation in tropical grasslands of eastern Africa (Scott 1962). In the western Great Plains, the leaching of the soil profile is so limited that Ca precipitates as $CaCO_3$, which accumulates in the lower profile in calcic horizons, designated B_k and informally known as *caliche.*

The climatic regime of grasslands results in lower levels of plant growth than in forests, and smaller quantities of plant residues are added to the soil each year. Nevertheless, grassland soils contain large stores of organic matter because the limited availability of water also results in slower rates of decomposition (Chapter 5). Most grassland soils in the temperate zone are classified as *Mollisols,* on the basis of a high content of organic carbon and high base saturation in the surface layers.

---

[2] Flocculation occurs when materials that are moving in water as colloids form insoluble aggregates, usually by the neutralization of hydrophyllic surface charges.

As seen for forest ecosystems, soil properties in grasslands vary locally as a result of differences in underlying parent materials and hillslope positions. Schimel et al. (1985) showed how the downslope movement of materials results in thin soils on hillslopes and accumulations of organic matter and nitrogen in local depressions. Similarly, Aguilar and Heil (1988) found that grassland soils derived from sandstone have lower contents of organic carbon, nitrogen, and total P than soils derived from fine-textured materials, such as shale, which have a higher content of clay minerals. The proportion of P contained in organic forms is greater for soils derived from sandstone. These differences in soil properties can strongly affect the local productivity of grasslands.

Deserts

The trends in soil development that are seen with increasing aridity in grasslands reach their extreme expression in deserts. Desert soils often contain clay minerals and horizons of clay accumulations derived from eolian dust, giving the appearance that substantial weathering has occurred (Singer 1989, Reheis and Kihl 1995). In fact, soil development in deserts occurs slowly because of limited weathering and leaching of the soil profile. In the face of limited weathering and runoff, many desert soils show a net gain of materials from atmospheric deposition (Ewing et al. 2006, Michalski et al. 2004a, Schlesinger et al. 2000).

Despite the limited chemical weathering in deserts, the small amount of water percolating through these soils transports substances both vertically in the profile and horizontally across the landscape. As water is removed by plant uptake, soluble substances precipitate. Typically, well-developed desert soils contain $CaCO_3$ horizons that show progressive development and cementation through time (Gile et al. 1966). When desert soils contain distinct horizons, they are classified as *Aridisols* (Dregne 1976).

Because $CaCO_3$ precipitates in desert soils in equilibrium with CO_2 derived from plant root respiration,

$$2CO_2 + 2H_2O \rightarrow 2H^+ + 2HCO_3^- \tag{4.14}$$

$$Ca^{2+} + 2HCO_3^- \rightarrow CaCO_3\downarrow + H_2O + CO_2, \tag{4.15}$$

the carbonate carries a carbon isotopic signature that can be traced to photosynthesis (Chapter 5) and used as an index of past climate and vegetation (Quade et al. 1989a). Seasonal variations in the activity of plant roots (Schlesinger 1985) and soil microbes (Monger et al. 1991) affect $CaCO_3$ deposition and other aspects of soil development in deserts, despite the outward appearance that abiotic processes should be dominant. For instance, beneath desert shrubs, the deposition of $CaCO_3$ is inhibited by the presence of dissolved organic materials in the soil solution (Inskeep and Bloom 1986, Reddy et al. 1990, Suarez et al. 1992), and calcic horizons are found deeper in the soil profile.

Depth to the $CaCO_3$ horizon in deserts shows a direct relation to mean annual rainfall and wetting of the soil profile (Arkley 1963). Beneath the $CaCO_3$, one may find horizons in which $CaSO_4 \cdot 2H_2O$ (gypsum) or NaCl are dominant, reflecting the greater solubility and downward movement of these salts (Yaalon 1965, Marion et al. 2008). Similar patterns are seen across the landscape, where Na, Cl, and SO_4 are carried to intermittent lakes in basin lows, while Ca remains in the upland soils of the adjacent piedmont (Drever and Smith 1978, Eghbal et al. 1989, Amundson et al. 1989a).

Despite sparse plant cover, much of the nutrient cycling in desert ecosystems is controlled by biota. With widespread root systems, desert shrubs accumulate nutrients from a large area and concentrate dead organic matter in the local area beneath their canopy. Most of the annual turnover of N, P, and other elements is controlled by biogeochemical processes in these "islands of fertility" (Schlesinger et al. 1996, Titus et al. 2002).

Models of Soil Development

The processes underlying soil profile development are conducive to simulation modeling. Models of soil chemistry include the weathering reactions described earlier in this chapter and equilibrium constants for the exchange of cations and anions between the soil solution and the mineral phases (Furrer et al. 1989). Depending on the time scale of the simulation, a model of soil chemistry usually routes daily or annual precipitation sequentially through the soil profile, where the solution achieves an equilibrium with the soil minerals in each horizon. Removal of water from the soil profile is calculated from estimates of evaporation from the soil surface, plant uptake, and runoff to streams. Simulation models of soil processes have been constructed to predict soil profile development and to calculate losses of dissolved constituents in forested regions subject to acid rain (Reuss 1980; Cosby et al. 1985, 1986; David et al. 1988).

The long-term development of soils in arid regions is simulated in a model, CALDEP, developed by Marion et al. (1985), in which daily precipitation achieves equilibrium with carbonate biogeochemistry as it percolates through the soil profile. Plant root respiration is explicitly included in the calculation, and it varies seasonally and with depth in the profile. Plants also control the loss of water from the soil surface by evapotranspiration. The model suggests that the $CaCO_3$ horizon will be deeper in a soil profile developed from coarse-textured parent materials, which allow greater percolation of water, and when plant root respiration varies seasonally, showing high values of soil CO_2 during the growing season.

Using current climatic conditions to parameterize precipitation and evaporation, the model was run to simulate 500 years of soil profile development (Figure 4.16). The model predicted mean depths to the $CaCO_3$ that were much shallower than observed in a sample of 16 desert soils from Arizona. When the model was reparameterized using the cool, wet conditions that are thought to have been widespread in this region during the latest Pleistocene glaciation, the predicted depth of $CaCO_3$ closely matched that found in the field.

These conditions produced greater percolation of soil moisture and lower rates of evaporation from the soil surface. Such models are only as good as the data used in the simulations, and rarely can models establish the importance of processes unequivocally. Nevertheless, models are useful for hypothesis development and for organizing research priorities. CALDEP suggests that most $CaCO_3$ horizons were formed during the Pleistocene, when deserts received more precipitation than today—consistent with the age of many desert soil carbonates (Schlesinger 1985). In the southwestern United States, most calcic horizons are >10,000 years old, and the $CaCO_3$ has accumulated at rates of 1.0 to 5.0 g m^{-2} yr^{-1} from the downward transport of Ca-rich minerals deposited from the atmosphere (Schlesinger 1985, Capo and Chadwick 1999, Van der Hoven and Quade 2002).

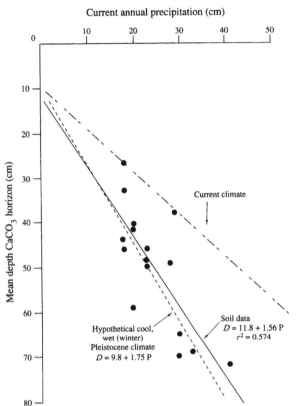

Current annual precipitation (cm)

FIGURE 4.16 Depth to $CaCO_3$ in desert soils of Arizona, as a function of mean annual precipitation. The dashed line shows the prediction from the CALDEP model using current precipitation regimes. The solid line shows the best fit to actual data reported from the field. The dotted line shows the predictions when the model is run with postulated climatic data from the latest Pleistocene pluvial period. *Source: Modified from Marion et al. (1985).*

Current climate

Soil data
$D = 11.8 + 1.56\ P$
$r^2 = 0.574$

Hypothetical cool,
wet (winter)
Pleistocene climate
$D = 9.8 + 1.75\ P$

WEATHERING RATES

Rock weathering and soil development are difficult to study because the processes occur slowly and the soil profile is nearly impossible to sample without disturbing many of the chemical reactions of interest. Nevertheless, estimates of weathering are needed to understand the biogeochemistry of local watersheds, where essential elements for biota are derived from the underlying rock (Chapter 6). Often we must infer weathering rates from what remains in the soil profile and what is lost to stream water. Estimates of the dissolved and suspended load of rivers allow us to calculate a global rate of chemical weathering, which supplies nutrient elements to land and marine biota.

Chemical Weathering Rates

Variations in the concentration of dissolved ions in streams can be related to the rate of discharge, the origin of their waters, and chemical weathering (Johnson et al. 1969, Maher 2011). In a simple geochemical system, we might expect that stream water concentrations would be highest at periods of low flow, because most of the water would be derived from drainage of the soil profile where it would be in equilibrium with various rock weathering and ion-exchange reactions. As stream flow increases, we might expect concentrations to decline as an increasing proportion of the flow is derived from precipitation and surface runoff, with little or no equilibration with the soil mineral phases. This simple geochemical model often explains the behavior of major ions in stream water (Ca, Mg, Na, Si, Cl, SO_4, and HCO_3), which are readily soluble in water and available in excess of biotic demand (Meyer et al. 1988).

During rainfall or seasonal flooding, the concentrations of dissolved ions in stream waters are often higher as the waters are rising than during the equivalent flows during the receding period (Whitfield and Schreier 1981, McDiffett et al. 1989). The effect, known as *hysteresis*, is thought to result from an initial flushing of the highly concentrated waters that accumulate in the soil pores during periods of low flow. Not all ions show consistent hysteresis patterns, so to calculate the total annual loss of dissolved ions from a watershed, the stream flow discharge for each day must be multiplied by the concentration measured at that discharge and the products summed for all 365 days.

Even for elements that show lower concentrations at greater discharge, the total removal from the landscape is greatest during years of high stream flow (Figure 4.17)—that is, the increase in flow predominates over the expected dilution of dissolved materials. In comparisons made over large geographic regions, concentrations are greatest in rivers that drain areas with limited runoff, but total transport is greater in rivers with greater discharge (e.g., Figure 4.18; Bluth and Kump 1994, Gaillardet et al. 1999).

Jennings (1983) shows a dilution of Ca and Mg with increasing discharge in areas of limestone in New Zealand, but the slope of the relationship changes slightly from summer to winter, reflecting a greater weathering of limestone by the more active respiration of roots during the summer (see also, Laudelout and Robert 1994). Plants also exert control on stream water chemistry by the uptake of essential nutrient elements. In most areas, the concentrations of plant-essential nutrients, such as potassium and nitrate, show little relation to stream water

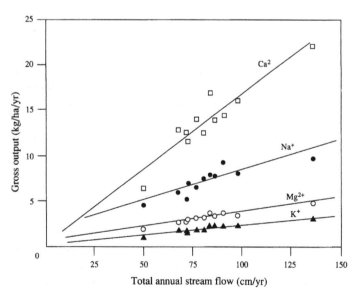

FIGURE 4.17   Annual stream water loss of major cations as a function of the stream discharge in different years in the Hubbard Brook Experimental Forest in New Hampshire. *Source: From Likens and Bormann 1995a).*

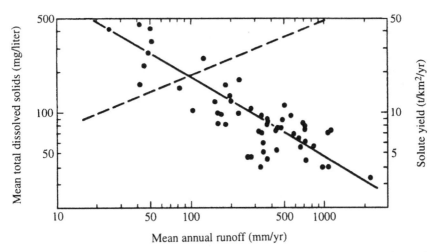

FIGURE 4.18   Variation in the concentration of total dissolved solids (*solid line*) and the total annual transport of dissolved substances (*dashed line*, shown without data) for various streams in Kenya as a function of mean annual runoff. *Source: From Dunne and Leopold (1978). Used with permission of Springer-Verlag.*

discharge because they are regulated by microbial activity and root uptake (Lewis and Grant 1979, Feller and Kimmins 1979, McDowell and Asbury 1994). In a deciduous forest in New Hampshire, the lowest potassium and nitrate concentrations in stream water are found during the low-flow periods of summer, when biotic demands are greatest (Johnson et al. 1969, Likens et al. 1994).

Plant uptake of essential elements and the retention of cations on soil minerals complicate the use of stream water concentrations to calculate the rate of chemical weathering over short periods of time (Taylor and Velbel 1991, Gardner 1990). However the eventual loss of cations to riverflow explains the decline in base saturation and pH during soil development (Bockheim 1980). Losses of dissolved constituents from terrestrial ecosystems represent the products of chemical weathering and constitute *chemical denudation* of the landscape.

One of the best-known attempts to calculate the rate of chemical weathering began in 1963, when Gene Likens, Herbert Bormann, and Noye Johnson quantified the chemical budgets for the Hubbard Brook forest in New Hampshire (Likens and Bormann 1995a). Here, a number of comparable watersheds are underlain by impermeable bedrock with no apparent subsurface drainage to groundwater. These workers reasoned that if the atmospheric inputs of chemical elements are subtracted from the stream water losses, the difference should reflect the annual release from rock weathering. They estimated the rate of rock weathering using the equation

$$\text{Weathering} = \frac{(\text{Ca lost in stream water}) - (\text{Ca received in precipitation})}{(\text{Ca in parent material}) - (\text{Ca in residual material in soil})} \qquad (4.16)$$

The solution of this equation shows rather different amounts of weathering when the calculations are performed using other rock-forming elements (Table 4.4). The observed losses of calcium and sodium in stream water imply higher rates of weathering than what is calculated using potassium and magnesium in the same equation. Johnson et al. (1968) suggest that the latter elements are accumulating in secondary minerals (illite and vermiculite) in the soil.

TABLE 4.4   Calculation of the Rate of Primary Mineral Weathering, Using the Stream Water Losses and Mineral Concentrations of Cationic Elements

Element	Annual net loss (kg ha^{-1} yr^{-1})	Concentration in rock (kg/kg of rock)	Concentration in soil (kg/kg of soil)	Calculated rock weathering (kg ha^{-1} yr^{-1})
Ca	8.0	0.014	0.004	800
Na	4.6	0.016	0.010	770
K	0.1	0.029	0.024	20
Mg	1.8	0.011	0.001	180

Source: Data from Johnson et al. (1968).

In addition, trees may take up and store essential elements in long-lived tissues (e.g., wood growth), temporarily reducing the loss of some elements in stream water (Taylor and Velbel 1991; see also Chapter 6).

In calculating weathering rates, it is often useful to examine the watershed budgets for Cl and Si. In nearly all cases, the atmospheric inputs of Si are trivial, so Si in stream water is an index of net chemical weathering, which is its only source (e.g., Figure 4.2). In contrast, the Cl content of rocks is normally very low. Nearly all Cl$^-$ in stream water is derived from the atmosphere, passing through the system relatively unimpeded by biotic activity or geochemical reactions (but see Lovett et al. 2005, Oberg et al. 2005). A balanced budget for Cl is often indicative of an accurate hydrologic budget for the watershed (Juang and Johnson 1967, Svensson et al. 2012).

Release from rock weathering is the dominant source of Ca, Mg, K, Fe, and P in stream waters draining the Hubbard Brook Experimental Forest, whereas deposition from the atmosphere is the dominant input for Cl, S, and N, which have a small content in rocks (Table 4.5).

TABLE 4.5   Inputs and Outputs of Elements from Hubbard Brook Experimental Forest

	Inputs (%)		Output as percent of input
	Atmosphere	Weathering	
Ca	9	91	59
Mg	15	85	78
K	11	89	24
Fe	0	100	25
P	1	99	1
S	96	4	90
N	100	0	19
Na	22	78	98
Cl	100	0	74

Source: Data from Likens et al. (1981).

In forests not subject to regional inputs of acid rain, the proportion of sulfur that is derived from the atmosphere is often somewhat lower than that at Hubbard Brook (e.g., Mitchell et al. 1986). A large number of watershed studies have been conducted, allowing similar calculations of weathering rates for a variety of ecosystems (Table 4.6; Likens and Bormann 1995b, Henderson et al. 1978, Feller and Kimmins 1979, Velbel 1992).

A relative index of chemical weathering is calculated by summing the annual losses of various elements in stream water. In comparisons of ecosystems of the world, chemical denudation is found to increase with increasing runoff (Figure 4.2). Total dissolved transport ranges from 47.6 to 372.7 kg ha^{-1} yr^{-1} among the watersheds in Table 4.6, and Alexander (1988) found that chemical denudation ranged from 20 to 200 kg ha^{-1} yr^{-1} among 18 undisturbed ecosystems in a variety of climatic regimes. Often there is much regional variability in calculated weathering rates (Nezat et al. 2004, Schaller et al. 2010), but globally, rivers transport about 4×10^{15} g of dissolved substances to the oceans each year—an average of about 267 kg ha^{-1} yr^{-1} for the Earth's land surface (refer to Table 4.8 later in chapter).

In most areas underlain by silicate rocks, the loss of elements in stream water relative to their concentration in bedrock follows the order

$$Ca > Na > Mg > K > Si > Fe > Al, \tag{4.17}$$

but the order is affected by the specific composition of bedrock and the secondary minerals that are formed in the soil profile (Holland 1978, Harden 1988, Hudson 1995). This general order reflects the tendency for Ca- and Na-silicates to weather easily and for the

TABLE 4.6   Net Transport (export minus atmospheric deposition) of Major Ions, Soluble Silica, and Suspended Solids from Various Watersheds of Forested Ecosystems

Watershed characteristics	Caura River, Venezuela	Gambia River, West Africa	Catoctin Mtns., Maryland	Hubbard Brook, New Hampshire
Size (km^2)	47,500	42,000	5.5	2
Precipitation (cm)	450	94	112	130
Vegetation	Tropical	Savanna	Temperate	Temperate
Net dissolved transport (kg ha^{-1} yr^{-1})				
Na	19.4	3.9	7.3	5.9
K	13.6	1.4	14.1	1.5
Ca	14.2	4.0	11.9	11.7
Mg	5.7	2.0	15.6	2.7
HCO$_3^-$	124.0	20.3	78.1	7.7
Cl$^-$	−1.4	0.6	16.6	−1.6
SO$_4^{2-}$	1.5	0.4	21.2	14.8
SiO$_2$	195.7	15.0	56.1	37.7
Total transport (kg ha^{-1} yr^{-1})	372.7	47.6	220.9	80.4

Source: Modified from Lewis et al. (1987). Used with permission of Springer.

limited incorporation of Ca and Na in secondary minerals. The composition of individual streams may differ strongly from these patterns depending on local conditions. For instance, streams draining areas of carbonate terrain are dominated by Ca and HCO_3 (e.g., Laudelout and Robert 1994), and stream waters may contain high concentrations of Na, Cl, and SO_4 where evaporite minerals are exposed (e.g., Stallard and Edmond 1983). In most cases, Fe and Al are retained in the lower soil profile as oxides and hydroxides, which are essentially immobile (Chesworth et al. 1981, Olsson and Melkerud 2000, Lichter 1998).

Because temperate forest soils are dominated by clay minerals with permanent negative charge, the loss of cations is determined by the availability of anions that "carry" cations through the soil profile to stream waters (Gorham et al. 1979, Johnson and Cole 1980, Terman 1977, Christ et al. 1999). In most cases, the dominant anion in soil water is bicarbonate (HCO_3^-); thus, the activities of plant roots and soil biota control chemical weathering and the composition of stream water. The high stream water concentration of HCO_3^- in the rainforest of Venezuela (Table 4.6) reflects the importance of carbonation weathering in tropical ecosystems (refer to Figures 4.3 and 4.4). The total mobilization of cations and silicon in the Venezuelan study is also high, consistent with our expectations of rapid chemical weathering in tropical climates (Figure 4.2). In Chapter 6, we will see how an increase in the availability of NO_3^-—a mobile anion—increases the loss of cations following disturbances, such as forest cutting (Figure 6.12).

In both the northeastern United States and much of Europe, losses of base cations appear to have increased in recent years because of the large amounts of H^+ and SO_4^{2-} delivered to the soil by acid rain (Tamm and Hällbäcken 1986, Sjöström and Qvarfort 1992, Wright et al. 1994, Fernandez et al. 2003, Lapenis et al. 2004, Blake et al. 1999, Warby et al. 2009). Some of the cations are lost from the cation exchange capacity, while others are derived from increased rates of chemical weathering under the influence of strong acids (Miller et al. 1993, Likens et al. 1996). The cations move through the soil to stream water, with SO_4^{2-} acting as a balancing anion. In contrast, in the more highly weathered soils of the southeastern United States, increased losses of cations from acid rain have been less dramatic because the soils in this region have greater anion adsorption capacity, so SO_4^{2-} is retained on soil minerals and cannot act as a mobile anion (Reuss and Johnson 1986, Harrison et al. 1989, Cronan et al. 1990). As a historical index of anion adsorption in Tennessee, Johnson et al. (1981a) found a lower content of SO_4^{2-} in the soils beneath a house built in 1890 (67 mg/kg) compared to that in adjacent field soils (195 mg/kg) that have been exposed to acid rain throughout the twentieth century.

In soils of the humid tropics, dominated by variable-charge minerals, one might expect that the abundance of mobile *cations* might determine the loss of anions from adsorption sites. Indeed, the loss of nitrate is retarded as water passes through experimental columns of tropical and volcanic soils (Wong et al. 1990, Bellini et al. 1996, Maeda et al. 2008, Strahm and Harrison 2006), and the loss of NO_3^- from forests in Costa Rica is reduced by a high soil anion adsorption capacity (Matson et al. 1987, Reynolds-Vargas et al. 1994, Ryan et al. 2001). In many cases, the adsorption of SO_4^{2-} in tropical soils reduces its importance as a mobile anion in the stream waters and groundwaters draining these regions (Szikszay et al. 1990; Table 4.6).

The organic compounds in river water, especially the fulvic acids, are important in the dissolved transport of Fe and Al. Because, these metals form complexes with organic acids, they are carried at concentrations well in excess of the solubility of Fe and Al hydroxides in river water (Perdue et al. 1976). The importance of dissolved organic acids in the transport of metals to the sea is a good example of the influence of terrestrial biota over simple geochemical processes that might otherwise determine the movement of materials on the surface of the Earth.

The composition of "average" river water was calculated by Livingstone (1963a) from measurements on a large number of rivers (Table 4.7; see also Table 9.1). Livingstone's estimate of total dissolved transport, 3.76×10^{15} g/yr, is largely confirmed by more recent work (e.g., Meybeck 1979; see also Table 4.8). Nearly all the Ca, Mg, and K in river water is derived from rock weathering (Table 4.9). Weathering of carbonates is the dominant source for Ca, while silicates are the dominant source for Mg and K (Holland 1978). Primary minerals in igneous rocks account for 30% of the dissolved constituents from chemical weathering delivered to the ocean—slightly less than the proportional exposure of igneous rocks at the Earth's surface (Durr et al. 2005). The chemical weathering of sedimentary rocks, especially carbonates, accounts for the remainder (Li 1972, Blatt and Jones 1975, Suchet et al. 2003).

Not all of the constituents in rivers are derived from rock weathering. Reflecting the importance of carbonation weathering, about two-thirds of the HCO_3^- in rivers is derived from the atmosphere, either directly from CO_2 or indirectly via organic decomposition and root respiration that contribute CO_2 to the soil profile (Holland 1978, Meybeck 1987, Moosdorf et al. 2011). Because chemical weathering involves the reaction between atmospheric constituents and rock minerals, weathering of 100 kg of igneous rock results in 113 kg of sediments that are deposited in the ocean and about 2.5 kg of salts that are added to seawater (Li 1972). Thus, a significant fraction of the transport of total dissolved substances in rivers is derived from the atmosphere and does not represent true chemical denudation of the continents (Berner and Berner 1987).

TABLE 4.7   Mean Composition of Dissolved Ions in River Waters of the World

Continent	HCO_3^-	SO_4^{2-}	Cl^-	NO_3^-	Ca^{2+}	Mg^{2+}	Na^+	K^+	Fe	SiO_2	Sum
North America	68	20	8	1	21	5	9	1.4	0.16	9	142
South America	31	4.8	4.9	0.7	7.2	1.5	4	2	1.4	11.9	69
Europe	95	24	6.9	3.7	31.1	5.6	5.4	1.7	0.8	7.5	182
Asia	79	8.4	8.7	0.7	18.4	5.6	9.3		0.01	11.7	142
Africa	43	13.5	12.1	0.8	12.5	3.8	11		1.3	23.2	121
Australia	31.6	2.6	10	0.05	3.9	2.7	2.9	1.4	0.3	3.9	59
World	58.4	11.2	7.8	1	15	4.1	6.3	2.3	0.67	13.1	120
Anions[a]	0.958	0.233	0.220	0.017							1.428
Cations[a]					0.750	0.342	0.274	0.059			1.425

[a] *Millequivalents of strongly ionized components.*
Source: Livingstone (1963b); concentrations in mg/liter.

TABLE 4.8   Chemical and Mechanical Denudation of the Continents

| Continent | Chemical denudation[a] | | Mechanical denudation[b] | | Ratio mechanical/ chemical |
	Total (10^{14} g/yr)	Per unit area (kg ha^{-1} yr^{-1})	Total (10^{14} g/yr)	Per unit area (kg ha^{-1} yr^{-1})	
N. America	7.0	330	14.6	840	2.1
S. America	5.5	280	17.9	1000	3.3
Asia	14.9	320	94.3	3040	6.3
Africa	7.1	240	5.3	350	0.7
Europe	4.6	420	2.3	500	0.5
Australia	0.2	20	0.6	280	3.0
Total	39.3	267	135.0	918	3.4

[a] Source: From Garrels and MacKenzie (1971).
[b] Source: From Milliman and Meade (1982).

TABLE 4.9   Sources of Major Elements in World River Waters (in percent of actual concentrations)

| Element | Atmospheric cyclic salt | Weathering | | | Pollution |
		Carbonates	Silicates	Evaporites	
Ca^{2+}	0.1	65	18	8	9
HCO_3^-	≪1	61	37	0	2
Na^+	8	0	22	42	28
Cl^-	13	0	0	57	30
SO_4^{2-}[a]	2	0	0	22	43
Mg^{2+}	2	36	54	≪1	8
K^+	1	0	87	5	7
H_4SiO_4	≪1	0	99+	0	0

[a] SO_4^{2-} is also derived from the weathering of pyrite.
Source: From Berner and Berner (1987). Used with permission of Prentice Hall.

A significant fraction of the Na, Cl, and SO_4 in riverflow is also derived from the atmosphere—from marine aerosols ("cyclic salts"; Chapter 3) that are deposited on land,[3] but at least some Na is also derived from weathering, since its content in river water is in excess of the molar equivalent of Cl, which would be expected if seasalt were the sole source.

[3] The amount of Cl in rivers that is derived from the atmosphere is the subject of some controversy (Berner and Berner 1987). Some budgets (e.g., Table 4.9) suggest that only a small amount of Cl is "cyclic," based on the observations by Stallard and Edmond (1981, 1983) in the Amazon Basin. Other workers have assumed that a larger fraction of the Cl in global riverflow is derived from the sea, with values ranging from 85% (Dobrovolsky 1994, p. 83) to nearly 100% (Möller 1990; compare Figure 3.16).

The river transport of some dissolved ions has been increased by human activities, such as mining, which accelerate the natural rate of crustal exposure and rock weathering on Earth (Bertine and Goldberg 1971, Martin and Meybeck 1979). And the widespread use of roadsalt as a de-icer has dramatically increased the concentration of Cl in runoff waters (Kaushal et al. 2005). Humans have enhanced the atmospheric deposition of NO_3 and SO_4, accounting for the relatively high concentrations of these ions in the runoff from industrialized continents (refer to Tables 4.7 and 4.9). The flux of HCO_3^- in the Mississippi River appears to have increased significantly during the past century as a result of a variety of human activities such as agricultural liming and rising CO_2 in Earth's atmosphere (Raymond et al. 2008).

Gibbs (1970) used the concentrations of ions in major world rivers to examine the origins of the dissolved constituents in their waters. Rivers dominated by precipitation show low concentrations of dissolved substances, and a high ratio of Cl to the total of $Cl + HCO_3^-$, reflecting the importance of Cl from rainfall (Zone A in Figure 4.19). Rivers in which the dissolved load is largely derived from chemical weathering show higher concentrations of dissolved substances, and HCO_3^- is the dominant anion (Zone B), reflecting the importance of carbonation weathering in most soils. Rivers that pass through arid regions lose a significant amount of water to evaporation before reaching the ocean. These rivers (Zone C) show the greatest concentrations of dissolved ions and high ratios of $Cl/(Cl + HCO_3^-)$, because HCO_3^- has been removed by the chemical precipitation of minerals such as $CaCO_3$ on the streambed (Holland 1978). In this scheme, seawater represents the endpoint of the evaporative concentration of river waters. These relationships are also seen when Na and Ca are used to scale the x axis in Figure 4.19, with the relative concentration of Na as an index of rainfall and Ca as an index of chemical weathering (Gaillardet et al. 1999).

Mechanical Weathering

In addition to chemical denudation, a large amount of material derived from mechanical weathering is eroded from land and carried in rivers as the particulate or suspended load. These materials have received less attention from biogeochemists because their elemental contents are not immediately available to biota.

The concentration of suspended sediment often shows a curvilinear relationship with stream flow, increasing exponentially at high flows (Parker and Troutman 1989; Figure 4.20). At low flows, suspended sediments are dominated by organic materials, but the proportion of organic matter declines as the amount of suspended sediment increases during high flows, when soil erosion is greatest (Meybeck 1982, Ittekkot and Arain 1986, Paolini 1995). Long-term records show that the sediment transport during occasional extreme events often exceeds the total transport during long periods of more normal conditions (Van Sickle 1981, Swanson et al. 1982). Large concentrations of suspended sediments are found during flash floods in deserts (Baker 1977, Fisher and Minckley 1978, Laronne and Reid 1993).

Transport of suspended sediments in world rivers is affected by many factors, including elevation, topographic relief, and runoff from the watershed (Gaillardet et al. 1999). Although the rivers draining arid regions show high concentrations of suspended sediments, their total flow is limited, so the loss of soil materials per unit of landscape is rather low (Milliman and

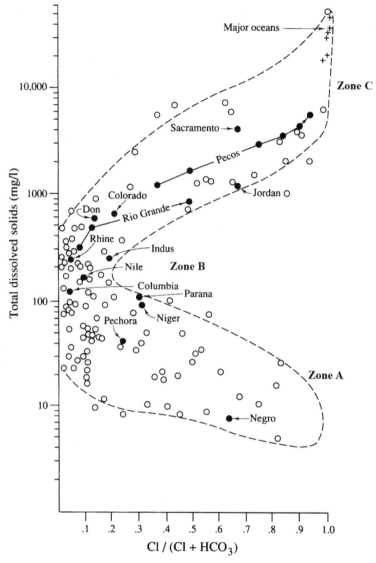

FIGURE 4.19   Variations in the total dissolved solids in rivers and lakes as a function of the ratio of Cl/(Cl + HCO₃) in their waters. *Source: From Gibbs (1970). Used with permission of the American Association for the Advancement of Science.*

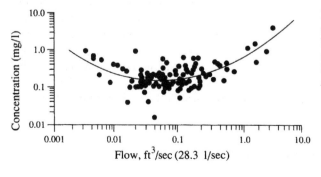

FIGURE 4.20   Concentration of particulate matter as a function of stream flow in the Hubbard Brook Experimental Forest of New Hampshire (U.S.A.). *Source: From Bormann et al. (1974). Used with permission of the Ecological Society of America.*

Meade 1983). Rivers draining southern Asia carry 70% of the global transport of suspended sediments, 13.5×10^{15} g/yr (refer to Table 4.8). Large sediment loads in the rivers of China are derived from the erosion of massive deposits of wind-derived soils, loess, in their drainage basin. In contrast, despite carrying 20% of the world's riverflow, the Amazon River carries only about 600 to 900×10^{12} g of suspended sediment each year, about 5% of the world's total (Mikhailov 2010, Wittmann et al. 2011). Most of the Amazon Basin is situated at low elevations with limited topographic relief, which accounts for its relatively low yield of suspended sediments (Meybeck 1977).

Much of the sediment removed from uplands is deposited in stream channels and flood-plains in the lower reaches of rivers (Trimble 1977, Longmore et al. 1983, Gurtz et al. 1988). Thus, the sediment yield per unit area of watershed declines with increasing watershed area (Milliman and Meade 1983). Despite large seasonal variations in volume, the daily sediment transport of the Amazon is rather constant as a result of storage of sediment in the floodplain during periods of rising waters and remobilization during falling waters (Meade et al. 1985).

Erosion increases when vegetation is removed (Bormann et al. 1974), and global land cleared by humans for agriculture, mining, and construction has dramatically increased the transport of suspended sediments in many parts of the world (Wilkinson and McElroy 2007, Hooke et al. 2000). For instance, in the eastern United States, coal mining now vastly exceeds all natural processes that mobilize materials from the Earth's crust (Figure 4.21). Worldwide, human activities have simultaneously nearly doubled the transport of suspended materials in rivers, but reduced the amount of the mobilized material reaching the oceans owing to the construction of dams and reservoirs (Syvitski et al. 2005). Large amounts of suspended sediment are captured in lakes and floodplains and behind dams and other human structures (Dynesius and Nilsson 1994, Vorosmarty et al. 2003, Walling and Fang 2003).

Total Denudation Rates

The total denudation of land is dominated by the products of mechanical weathering, which exceeds chemical weathering by three to four times, worldwide (Table 4.8). The mean rate of total continental denudation is about 1000 kg ha^{-1} yr^{-1}, with approximately 75% carried in the suspended sediments in rivers (Alexander 1988, Tamrazyan 1989, Wakatsuki and Rasyidin 1992, Gaillardet et al. 1999). The importance of mechanical weathering increases with elevation (e.g., Reiners et al. 2003, Gaillardet et al. 1999); differences in mean elevation among the continents explain much of the variation in mechanical weathering in Table 4.8. Recent estimates of the total transport of suspended materials in all rivers of the world range from 12.6 to 13.5×10^{15} g/yr (Milliman and Meade 1983, Syvitski et al. 2005).

Assuming that the specific gravity of suspended sediment is 2.5 g/cm^3, these estimates are 4 to 5 times higher than estimates (1.27 km^3/yr) of the volume of deep ocean sediments derived from land (Howell and Murray 1986) and the current volume of sedimentary materials being subducted into the mantle (0.73 km^3/yr; Plank and Langmuir 1998). Presumably, most sediment is deposited near the shore, in continental shelf deposits (Chapter 9). Gregor (1970) suggests that the global rate of sediment transport, an overall measure of mechanical weathering, may have been about four times greater before the land surface was colonized by plants (compare Wilkinson and McElroy 2007). Indeed, today, especially high concentrations of

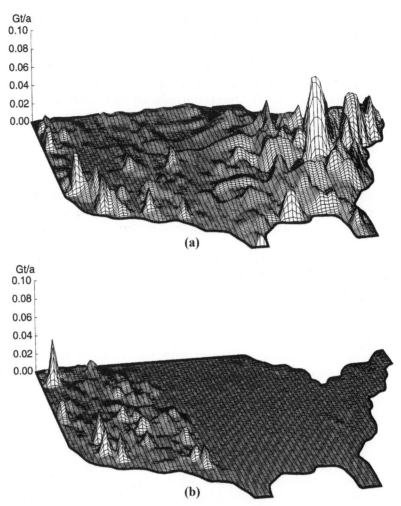

FIGURE 4.21   Rate at which surficial materials are mobilized from the Earth's crust (Gt/yr = 10^{15} g/yr) in grid cells measuring $1 \times 1°$ latitude and longitude by humans (A) and rivers (B). The massive mobilization of crustal materials in the eastern United States is largely the result of surface mining for coal (Hooke 1999).

suspended sediment are seen in rivers draining arid and semi-arid regions where vegetation is sparse (Milliman and Meade 1983).

Because Fe, Al, and Mn are only slightly soluble in water, particulate and suspended sediments account for most of the removal of these elements from terrestrial ecosystems (Table 4.10; Benoit and Rozan 1999). The numbers shown in the table are estimated by comparisons of the concentrations in bedrock to those in weathered residues (river particulates); both are normalized to Al. Suspended sediments are also enriched in phosphorus, owing to chemical reactions between dissolved P and various soil minerals (Avnimelech and McHenry 1984, Sharpley 1985). As a result of a variety of human activities, the river transport of many metals (e.g., Cu, Pb, and Zn) is now greater than the transport under preindustrial

Meade 1983). Rivers draining southern Asia carry 70% of the global transport of suspended sediments, 13.5×10^{15} g/yr (refer to Table 4.8). Large sediment loads in the rivers of China are derived from the erosion of massive deposits of wind-derived soils, loess, in their drainage basin. In contrast, despite carrying 20% of the world's riverflow, the Amazon River carries only about 600 to 900×10^{12} g of suspended sediment each year, about 5% of the world's total (Mikhailov 2010, Wittmann et al. 2011). Most of the Amazon Basin is situated at low elevations with limited topographic relief, which accounts for its relatively low yield of suspended sediments (Meybeck 1977).

Much of the sediment removed from uplands is deposited in stream channels and flood-plains in the lower reaches of rivers (Trimble 1977, Longmore et al. 1983, Gurtz et al. 1988). Thus, the sediment yield per unit area of watershed declines with increasing watershed area (Milliman and Meade 1983). Despite large seasonal variations in volume, the daily sediment transport of the Amazon is rather constant as a result of storage of sediment in the floodplain during periods of rising waters and remobilization during falling waters (Meade et al. 1985).

Erosion increases when vegetation is removed (Bormann et al. 1974), and global land cleared by humans for agriculture, mining, and construction has dramatically increased the transport of suspended sediments in many parts of the world (Wilkinson and McElroy 2007, Hooke et al. 2000). For instance, in the eastern United States, coal mining now vastly exceeds all natural processes that mobilize materials from the Earth's crust (Figure 4.21). Worldwide, human activities have simultaneously nearly doubled the transport of suspended materials in rivers, but reduced the amount of the mobilized material reaching the oceans owing to the construction of dams and reservoirs (Syvitski et al. 2005). Large amounts of suspended sediment are captured in lakes and floodplains and behind dams and other human structures (Dynesius and Nilsson 1994, Vorosmarty et al. 2003, Walling and Fang 2003).

Total Denudation Rates

The total denudation of land is dominated by the products of mechanical weathering, which exceeds chemical weathering by three to four times, worldwide (Table 4.8). The mean rate of total continental denudation is about 1000 kg ha^{-1} yr^{-1}, with approximately 75% carried in the suspended sediments in rivers (Alexander 1988, Tamrazyan 1989, Wakatsuki and Rasyidin 1992, Gaillardet et al. 1999). The importance of mechanical weathering increases with elevation (e.g., Reiners et al. 2003, Gaillardet et al. 1999); differences in mean elevation among the continents explain much of the variation in mechanical weathering in Table 4.8.

Recent estimates of the total transport of suspended materials in all rivers of the world range from 12.6 to 13.5×10^{15} g/yr (Milliman and Meade 1983, Syvitski et al. 2005). Assuming that the specific gravity of suspended sediment is 2.5 g/cm^3, these estimates are 4 to 5 times higher than estimates (1.27 km^3/yr) of the volume of deep ocean sediments derived from land (Howell and Murray 1986) and the current volume of sedimentary materials being subducted into the mantle (0.73 km^3/yr; Plank and Langmuir 1998). Presumably, most sediment is deposited near the shore, in continental shelf deposits (Chapter 9). Gregor (1970) suggests that the global rate of sediment transport, an overall measure of mechanical weathering, may have been about four times greater before the land surface was colonized by plants (compare Wilkinson and McElroy 2007). Indeed, today, especially high concentrations of downward diffusion of atmospheric CO_2 through the soil profile. However, at periods in Earth's early history, the concentration of atmospheric CO_2 was most certainly higher than today, presumably yielding high rates of carbonation weathering. Weathering is also driven by the availability of water. The high concentration of CO_2 on Venus (Table 2.6) is ineffective in weathering because the surface of the planet is dry (Nozette and Lewis 1982).

By mining minerals from its crust and extracting buried fossil fuels from the Earth, humans have increased the global rates of chemical and mechanical weathering and added significant quantities of dissolved materials to global riverflow. Currently, the removal of fossil fuels from the Earth's crust mobilizes and oxidizes carbon at a rate (9×10^{15} g/yr), which exceeds the total of all natural chemical weathering (Table 4.8). Human exposure and erosion of soils in agricultural use have increased the global denudation due to mechanical weathering by a factor of about 2 (Syvitski et al. 2005), leading to increases in the rate of sediment accumulation in estuaries and river deltas (Chapter 8).

Chemical weathering is a source of essential elements for the biochemistry of life, but stream water runoff removes these elements from the land surface. Chemical reactions among soil constituents and uptake by biota determine the rate of loss, but the inevitable removal of cations results in lower soil pH and base saturation through time (Bockheim 1980). Phosphorus is particularly critical as a soil nutrient because it is not abundant in crustal rocks and is easily precipitated in unavailable forms in the soil. Old soils in highly weathered landscapes are composed of resistant, residual Fe and Al oxide minerals. In these soils, P is often deficient and even the small atmospheric inputs of P are important for plant growth (Chadwick et al. 1999; Chapter 6).

Recommended Readings

Birkeland, P.W. 1989. *Soils and Geomorphology* (third ed.). Oxford University Press.
Bleam, W.F. 2012. *Soil and Environmental Chemistry*. Elsevier.
Drever, J.I. (Ed.). 2004. *Surface and Groundwater, Weathering and Soils*. Volume 5 of *Treatise on Geochemistry*. Elsevier.
Garrels, R.M. and F.T. MacKenzie. 1971. *Evolution of Sedimentary Rocks*. W.W. Norton.
Huang, P.M., Y. Li, and M.E. Summer. 2011. *Handbook of Soil Science* (second ed.). CRC Press.
Jenny, H. 1980. *The Soil Resource*. Springer-Verlag.
Likens, G.E. and F.H. Bormann. 1995. *Biogeochemistry of a Forested Ecosystem* (second ed.). Springer-Verlag.
Lindsay, W.L. 1979. *Chemical Equilibria in Soils*. Wiley.
Reiners, W.A. and K.L. Driese. 2004. *Transport Processes in Nature*. Cambridge University Press.
Reuss, J.O. and D.W. Johnson. 1986. *Acid Deposition and the Acidification of Soils and Waters*. Springer-Verlag.
Sparks, D.L. 1995. *Environmental Soil Chemistry*. Elsevier/Academic Press.
Sposito, G. 1989. *The Surface Chemistry of Natural Particles*. Oxford University Press.

PROBLEMS

4.1. Consider a forested watershed in which evapotranspiration removes about 50% of the annual rainfall and there are no losses of water to groundwater. A scientist makes the following observations:

Precipitation:
- Amount: 100 cm/year
- Mg concentration: 0.05 mg/l

Stream water:
- Volume 1.5×10^8 l/year
- Mg concentration: 0.40 mg/l

Bedrock:
- Al concentration: 10 g/kg
- Weathering rate: 800 kg/ha/year
- Al/Mg ratio: 5

Diagram the Mg cycle in this system, expressing the inputs and outputs in terms of kilograms per hectare (10,000 m²) per year. Assuming all these measurements are accurate, how might you explain the small discrepancy this scientist encounters in the budget for Mg?

4.2. Using the rates of chemical and mechanical denudation in Table 4.8, and assuming that the Earth's continental crust is in steady state, estimate the mean rate of uplift of the continental crust in cm per year.

4.3. Calculate the transport of major ions to the sea, using the data of Table 9.1 and a global riverflux of 4×10^{16} l/yr. Using the data given in Table 4.9 what is the fraction of the annual flux of HCO_3 in rivers that is derived from the atmosphere? How much should the global rate of chemical denudation (Table 4.8) be reduced to account for the HCO_3 that is derived from the atmosphere?

4.4. What is the relative transport of crustal materials from land to sea in soil dust versus in rivers?

The Biosphere
The Carbon Cycle of Terrestrial Ecosystems

INTRODUCTION

Photosynthesis is the biogeochemical process that transfers carbon from its oxidized form, CO_2, to the reduced (organic) forms that result in plant growth. Directly or indirectly, photosynthesis provides the energy for all other forms of life in the biosphere, and the use of plant products for food, fuel, and shelter brings photosynthesis into our daily lives. The fossil fuels that power modern society are derived from plant photosynthesis in the geologic past (Dukes 2003). Plant growth affects the composition of the atmosphere (Chapter 3) and the development

135

of soils (Chapter 4), linking photosynthesis to other aspects of global biogeochemistry. Indeed, the presence of organic carbon in soils and sediments and O_2 in our atmosphere provides a striking contrast between the *bio*geochemistry on Earth and the simple geochemistry that characterizes our neighboring planets.

In this chapter we consider the measurement of net primary production—the rate of accumulation of organic carbon in land plants. A similar treatment of photosynthesis in the world's oceans is given in Chapter 9. The rate of plant growth varies widely over the land surface. Deserts and continental ice masses may have little or no net primary production (NPP), while tropical rainforests can show annual production of $>1000 \, g \, C/m^2$.

Various environmental factors affect the rate of net primary productivity on land and the total storage of organic carbon in plant tissues (biomass), dead plant parts (detritus), and soil organic matter. As any home gardener knows, light and water are important, but plant growth is also determined by the stock of available nutrients in the soil. These nutrients are ultimately derived from the atmosphere or from the underlying bedrock (Table 4.5). The overall storage of carbon on land is determined by the balance between primary production and decomposition, which returns carbon to the atmosphere as CO_2.

PHOTOSYNTHESIS

Containing a central atom of magnesium, the chlorophyll molecule is a prime example of how plants have incorporated an abundant product of rock weathering as an essential element in their biochemistry (Figure 5.1). When photosynthetic pigments absorb sunlight, a few of the chlorophyll molecules are oxidized—passing an electron to a sequence of proteins

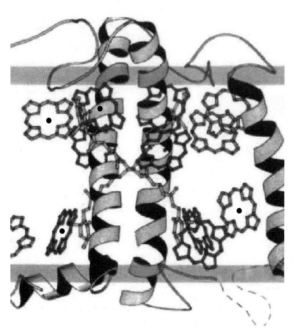

FIGURE 5.1 Molecular structure of the light-harvesting complex of photosystem II, showing the position of the Mg-porphyrin groups, shown with black dots. *Source: From Kuhlbrandt et al. (1994).*

that ultimately add the electron to a high-energy molecule, known as nicotinamide adenine dinucleotide phosphate ($NADP^+$), which is thus reduced to form NADPH. The chlorophyll molecule regains an electron from a water molecule, which is split by an enzyme containing manganese, calcium, and chlorine, in a complex three-dimensional structure (Yano et al. 2006, Guskov et al. 2009, Umena et al. 2011). This reaction is the origin of O_2 in the Earth's atmosphere:

$$2H_2O \rightarrow 4H^+ + 4e^- + O_2 \uparrow \tag{5.1}$$

In all cases, the photosynthetic pigments and proteins are embedded in a cell membrane, which allows protons (e.g., H^+ of Eq. 5.1) to build up to high concentrations on one side of the membrane and for this potential energy to be captured in a high-energy compound, adenosine triphosphate or ATP. In higher plants, the accumulation of protons occurs within the chloroplasts of leaf cells, whereas in photosynthetic bacteria, the reaction is conducted across the external cell membrane.

The high-energy compounds NADPH and ATP are then used by a suite of enzymes to reduce CO_2 and build carbohydrate molecules. The reaction begins with the enzyme ribulose bisphosphate carboxylase, also known as *Rubisco*, which adds CO_2 to the basic carbohydrate unit.[1] The overall reaction for photosynthesis is shown in the following equation,

$$CO_2 + H_2O \rightarrow CH_2O + O_2, \tag{5.2}$$

but we should remember that the process occurs in two stages.

First, the capture of light energy allows water molecules to be split and high-energy molecules to form. This drives a second reaction in which CO_2 is converted to carbohydrate. The efficiency of photosynthesis relative to the sunlight absorbed by chlorophyll is known as the *quantum yield efficiency*, which is normally close to 0.081 moles of CO_2 captured in carbohydrate per mole of photons absorbed, when soil water and other environmental factors are optimal (Singsaas et al. 2001). Photosynthesis can occur in markedly low-light environments, such as beneath snow and ice and under translucent rocks (Starr and Oberbauer 2003, Gradinger 1995, Schlesinger et al. 2003); moonlight, however, is not enough (Raven and Cockell 2006).

The carbon dioxide used in photosynthesis diffuses into plant leaves through pores, known as *stomates*, which are generally found on the lower surface of broad-leaf plants. One factor that determines the rate of photosynthesis is the stomatal aperture, which plant

---

[1] For understanding global biogeochemistry, we focus on the photosynthesis of C3 plants, which account for the overwhelming proportion of plant biomass and net primary productivity on Earth. C3 plants are so named because the first product of the photosynthetic reaction is a carbohydrate containing three carbon atoms. However, some plant species, largely warm-climate grasses, conduct photosynthesis by another biochemical pathway, known as C4 photosynthesis (Ehleringer and Monson 1993). C4 plants may account for up to 23% of global net primary production (Lloyd and Farquhar 1994, Still et al. 2003), but their contribution to global biomass is small because most species are not woody. The overall photosynthetic reaction is identical to Eq. 5.2, but C4 plants have different water-use efficiency and a different C-isotopic fractionation in their tissues (average −12 ‰). The isotopic ratio of plant debris preserved in soils can be used to trace changes in the past distribution of C3 and C4 plants (e.g., Quade et al. 1989a, Ambrose and Sikes 1991).

physiologists express as stomatal *conductance* in units of cm/sec. Stomatal conductance is controlled primarily by the availability of light and water to the plant and the concentration of CO_2 inside the leaf, where it is consumed by photosynthesis. When well-watered plants are actively photosynthesizing, internal CO_2 is relatively low and stomates show maximum conductance. Under such conditions, the amount and activity of Rubisco may determine the rate of photosynthesis (Sharkey 1985).

Water-Use Efficiency

There is a trade-off in photosynthesis; when plant stomates are open, allowing CO_2 to diffuse inward, O_2 and H_2O diffuse outward to the atmosphere. The loss of water through stomates, *transpiration*, is the major mechanism by which soil moisture is returned to the atmosphere (Chapter 10). In the Hubbard Brook Experimental Forest in New Hampshire (see Chapter 4), about 25% of the annual precipitation is lost by plant uptake and transpiration; stream flow increased by 26 to 40% when the forest was cut (Pierce et al. 1970). Because water for plant growth is often in short supply (Kramer 1982), these large losses of water by plants are somewhat surprising. One might expect natural selection for more efficient use of water by plants.

Plant physiologists express the loss of water relative to photosynthesis as water-use efficiency (WUE), namely,

$$\text{WUE} = \text{mmoles of } CO_2 \text{ fixed/moles of } H_2O \text{ lost.} \tag{5.3}$$

For most plants, water-use efficiency typically ranges from 0.86 to 1.50 mmol/mol, depending on environmental conditions (Osmond et al. 1982). Water-use efficiency is higher at lower stomatal conductance.

Rising concentrations of CO_2 in the atmosphere allow the same rate of photosynthesis to occur at lower stomatal conductance, thus increasing WUE (Bazzaz 1990, Ceulemans and Mousseau 1994). There is also some evidence that the number of stomates per unit of leaf surface has declined as atmospheric CO_2 rose during the Industrial Revolution (Woodward 1987, 1993; Peñuelas and Matamala 1990). The olive leaves preserved in King Tut's tomb (1327 B.C.) have a higher density of stomates than the leaves of the same species now growing in Egypt (Beerling and Chaloner 1993a).

Equation 5.3 largely applies to short-term experiments in the laboratory. For the biogeochemist, long-term average WUE can be estimated from the carbon isotope composition of plant tissues, especially tree rings. This method is based on the observation that the diffusion of $^{12}CO_2$ is more rapid than that of $^{13}CO_2$, which has a slightly higher molecular weight. Thus, over any time period, a greater proportion of $^{12}CO_2$ enters the leaf than $^{13}CO_2$. Inside the leaf, ribulose bisphosphate carboxylase also has a higher affinity for $^{12}CO_2$. As a result of these factors, plant tissue contains a lower proportion of $^{13}CO_2$ than the atmosphere by about 2% (= 20‰) (O'Leary 1988). The discrimination (*fractionation*) between carbon isotopes is expressed relative to an accepted standard as

$$\delta^{13}C = \left[\frac{^{13}C/^{12}C_{\text{sample}} - {}^{13}C/^{12}C_{\text{standard}}}{^{13}C/^{12}C_{\text{standard}}} \right] \times 1000 \tag{5.4}$$

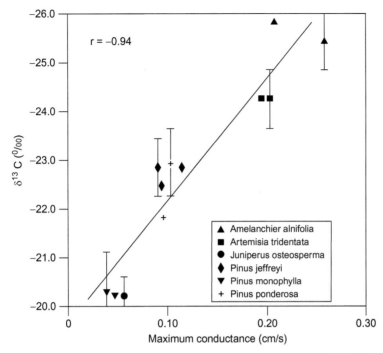

FIGURE 5.2 Relationship between the content of 13C in plant tissues (expressed as $\delta^{13}C$) and stomatal conductance for a variety of plant species in western Nevada. *Source: Modified from DeLucia et al. (1988).*

using the units of parts per thousand parts (‰). Because atmospheric CO_2 shows an isotopic ratio of −8.0‰ versus the standard, most plant tissues show $\delta^{13}C$ of about −28‰—that is, (−8‰) + (−20‰). Sedimentary organic carbon with this isotopic signature is useful in determining the antiquity of photosynthesis as a biochemical process (Figure 2.6).

The discrimination between $^{12}CO_2$ and $^{13}CO_2$ during photosynthesis is greatest (most negative $\delta^{13}C$) when stomatal conductance is high (Figure 5.2). When stomates are partially or completely closed, nearly all of the CO_2 inside the leaf reacts with ribulose bisphosphate carboxylase and there is less fractionation of the isotopes. Therefore, the isotopic ratio of plant tissue is directly related to the average stomatal conductance during its growth, providing a long-term index of water-use efficiency (Farquhar et al. 1989). Significantly, $\delta^{13}C$ values of preserved plant materials and tree rings indicate that WUE of plants increased as the concentration of atmospheric CO_2 rose at the end of the last glacial period (Van de Water et al. 1994) and during the last several hundred years (Peñuelas and Azcón-Bieto 1992, Feng 1999, Saurer et al. 2004, Watmough et al. 2001, Kohler et al. 2010).

Nutrient-Use Efficiency

Over a broad range of plant species, the rate of photosynthesis is directly correlated to leaf nitrogen content when both are expressed on a mass basis (Reich et al. 1992, 1999, Atkinson et al. 2010; Figure 5.3). Most leaf nitrogen is contained in enzymes; by itself, ribulose bisphosphate carboxylase usually accounts for 20 to 30% of leaf nitrogen (Evans 1989). Seemann et al.

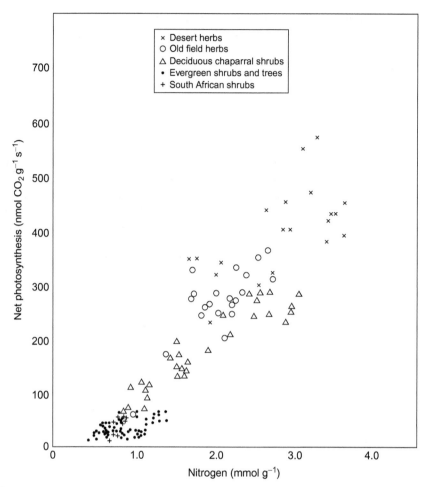

FIGURE 5.3   Relationship between net photosynthesis and leaf nitrogen content among 21 species from different environments. *Source: From Field and Mooney (1986). Used with permission of Cambridge University Press.*

(1987) found that photosynthetic potential is directly related to the content of ribulose bisphosphate carboxylase and leaf nitrogen in several species, suggesting that the availability of nitrogen determines leaf enzyme content and, thus, the rate of photosynthesis in land plants. In addition to nitrogen, leaf phosphorus content may be an important determinant of photosynthetic capacity in some species (Reich and Schoettle 1988, DeLucia and Schlesinger 1995, Raaimakers et al. 1995), and adequate P often determines the relationship of photosynthesis to N (Reich et al. 2009). Despite their central role in the biochemistry of photosynthesis, magnesium and manganese are seldom in short supply for plant growth.

Because most land plants grow under conditions of nitrogen deficiency, we might expect adjustments in nutrient use to maximize photosynthesis under varying conditions of soil fertility. The rate of photosynthesis per unit of leaf nitrogen—the slope of the line in Figure 5.3—would be one measure of nutrient-use efficiency (NUE) (Evans 1989). Overall, the data of this figure would seem to indicate that most species have similar photosynthetic

NUE, but subtle variations in NUE are seen among different types of plants (Reich et al. 1995) and among plants grown at different levels of fertility (Reich et al. 1994). For many plant species, when leaf nutrient content increases (by fertilization), NUE declines (Ingestad 1979a, Lajtha and Whitford 1989). Nutrient-use efficiency also appears inversely correlated to WUE across many species (Field et al. 1983, DeLucia and Schlesinger 1991).

RESPIRATION

Photosynthesis is usually measured by placing leaves or whole plants in closed chambers and measuring the uptake of CO_2 or release of O_2. The rates are a measure of *net* photosynthesis by the plant—that is, the fixation of carbon in excess of the simultaneous release of CO_2 by plant metabolism. Plant metabolism, known as respiration, is largely the result of mitochondrial activity in plant cells, and it is correlated to the nitrogen content, which is a good index of metabolic activity in most plant tissues (Figure 5.4; see also Ryan 1995, Vose and Ryan 2002, Reich et al. 2006). In woody plants, a large fraction of the respiration is contributed by stems and roots owing to their large contribution to total plant biomass (Amthor 1984, Ryan et al. 1994). For leaf tissues, rates of respiration are higher in the daytime than during the night as a result of the additional process of photorespiration.[2]

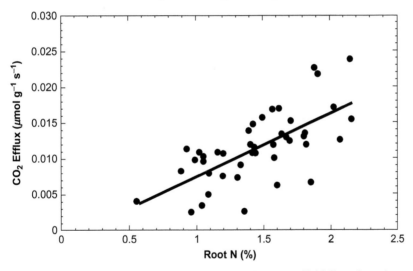

FIGURE 5.4   Root respiration as a function of nitrogen content (%) in roots of loblolly and ponderosa pine. *Source: From Griffin et al. (1997). Used with permission of Springer.*

[2] Photorespiration is not simply respiration during the day, but a higher level of plant respiration that is observed in sunlight as a result of a competitive reaction of Rubisco with O_2, which the plant must expend energy to reverse (Sharkey 1988). The reaction with O_2 is a function of the O_2/CO_2 ratio in the chloroplast, which is greater at high temperatures and during drought. Although photorespiration has generally been regarded as detrimental to plant growth, there is some evidence that the process is important to nitrate assimilation in land plants (Rachmilevitch et al. 2004) and to the protection of the photosynthetic mechanism at high light during periods of drought (Mahall and Schlesinger 1982).

About one-half of the carbon fixation by photosynthesis is respired by plants, so the gross rate of photosynthesis is often twice the measured rate of carbon uptake (Farrar 1985, Amthor 1989). For long-lived woody plants, maintenance respiration increases with stand age, consuming an increasing fraction of the gross photosynthesis and contributing to the reduction in the rate of plant growth with age (Kira and Shidei 1967, DeLucia et al. 2007, Piao et al. 2010). Plant respiration generally increases with increasing temperature, accounting for high rates of respiration in tropical forests and potentially higher rates of plant respiration with global warming (Ryan 1991; Ryan et al. 1994, 1995). In desert environments, total respiration in trees may increase as plants allocate more tissue to sapwood (Callaway et al. 1994). The increase in plant respiration in warmer conditions, versus the relatively insensitive response of photosynthesis to temperature, has important implications for global net primary production in the face of ongoing global climate change (Dillaway and Kruger 2010, Cai et al. 2009, Piao et al. 2010).

NET PRIMARY PRODUCTION

The rate of plant growth, *net primary production* (NPP), measured by ecologists in the field is analogous to the rate of net photosynthesis measured by plant physiologists in the laboratory. For plants in nature, we say that

$$\text{Gross primary production} - \text{plant respiration} = \text{net primary production} \qquad (5.5)$$

NPP is, however, not directly equivalent to plant growth as measured by foresters, ranchers, and farmers. Some fraction of NPP is lost to herbivores, fires, and in the death and loss of tissues, known collectively as *litterfall*. Foresters frequently call the NPP that remains the *true increment*, which may add to the accumulation of biomass over many years. When mortality occurs during forest development, the true increment is the net increase in the mass of woody tissue in living plants, after subtracting the mass of individuals that die over the same interval (Clark et al. 2001).

The annual accumulation of organic matter per unit of land is a measure of NPP, often expressed in units of $g\ m^{-2}\ yr^{-1}$. Plant tissue typically contains about 45 to 50% carbon, so division by two is a convenient way to convert the accumulation of organic matter to carbon fixation (Reichle et al. 1973a). Net primary production can also be expressed in units of energy, by measurements of the caloric content of various plant tissues (Paine 1971, Darling 1976). Calories are particularly useful for expressing the efficiency of photosynthesis relative to the receipt of sunlight energy. Net primary production typically increases as a function of intercepted radiation (e.g., Runyon et al. 1994), but even in forests, photosynthesis usually captures only about 1% of the total energy received from sunlight (Botkin and Malone 1968, Reiners 1972, Schulze et al. 2010). Even crop plants show maximum efficiencies of <5% of incident sunlight (Amthor 2010).

Measurement and Allocation of NPP

Traditional methods for measuring NPP in forests and shrublands involve the harvest of vegetation and calculation of the annual growth of wood and the mass of foliage at the peak of annual leaf display (Clark et al. 2001). The harvest data of a few individuals are used to calculate the mass and increment in plants of varying size by virtue of tightly constrained relationships between diameter, height, mass, and growth of individual plants—known as *plant allometry* (Whittaker and Marks 1975, Enquist et al. 2007, Niklas and Enquist 2002).

Independent estimates of the seasonal loss of plant parts can be obtained from collections of plant litterfall throughout the year. In grasslands, where there is little or no true increment, estimates of NPP generally involve the difference between the mass of tissue harvested from small plots at the beginning and the end of the growing season (e.g.,Wiegert and Evans 1964, Lauenroth and Whitman 1977, Singh et al. 1975). These estimates must be corrected for the consumption and loss of tissues during the same period. Similar approaches are used to measure the growth of roots by sequential coring throughout the growing season (Neill 1992, Vogt et al. 1998, Makkonen and Helmisaari 1999).

Allocation of net primary production varies with vegetation type and age. In forests, 25 to 35% of aboveground production is found in leaves (Whittaker 1974), with this percentage tending to decrease with stand age. Allocation to foliage in shrublands is generally greater, ranging from 35 to 60% in desert and chaparral shrubs (Whittaker and Niering 1975, Gray 1982). In grassland communities, essentially all aboveground net primary production is found in photosynthetic tissues. Across a broad range of species, allocation of photosynthate to leaf growth relative to stem growth is 0.53 (Niklas and Enquist 2002).

As seen in laboratory studies of photosynthesis, plant respiration accounts for about half of GPP in ecosystem studies (Waring et al. 1998, Litton et al. 2007; Figure 5.5). As a result of their massive structure and high environmental temperatures, tropical forests may expend a greater percentage of their GPP in respiration (Whittaker and Marks 1975, Ryan et al. 1994, Luyssaert et al. 2007, Piao et al. 2010), leaving less for wood growth. Comparing plant communities in different regions, Jordan (1971) found that the proportional allocation of NPP to wood growth was greater in boreal forests than in tropical forests—that is, there is greater wood production per unit of foliage in boreal forests. Webb et al. (1983) found a logarithmic relationship between total aboveground NPP and foliage biomass for a variety of plant communities in North America, with some deserts showing exceptionally high values of this ratio

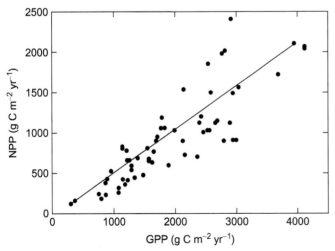

FIGURE 5.5 Relationship between net primary production (NPP) and gross primary production (GPP) in different forest types. *Source: From DeLucia et al. (2007).*

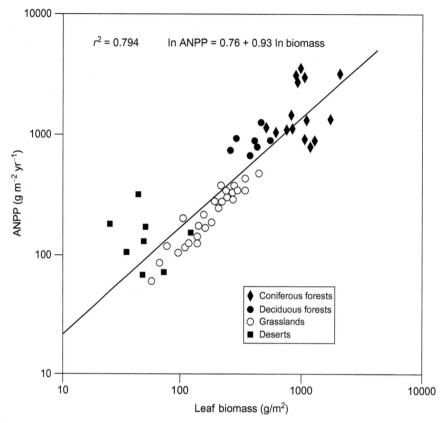

FIGURE 5.6   Using data from a variety of ecosystems in North America, Webb et al. (1983) found a strong relation between the annual aboveground NPP and leaf biomass.

(Figure 5.6). Compared to communities with abundant precipitation, however, desert shrublands show relatively low allocation of NPP to wood production (Jordan 1971), perhaps as a result of a large allocation to roots (Wallace et al. 1974, Mokany et al. 2006).

Because measurements of roots are difficult, many studies of NPP include data only for the aboveground tissues. Root biomass ranges between 20 and 40% of total biomass in forest ecosystems (Vogt et al. 1996, Poorter et al. 2012), and the annual growth of root tissues accounts for a significant fraction of the total NPP in most communities—averaging about 15 to 25% across a broad range of plant size (Pan et al. 2006; Niklas and Enquist 2002; K. Niklas, personal communication, 2010). Some NPP is lost in soluble organic compounds released from roots, perhaps as much as 20% in some circumstances (van Hees et al. 2005, Fahey et al. 2005).[3] Trees allocate proportionally more photosynthate to root growth in low-fertility soils (Axelsson

[3] Various workers use "total belowground allocation" to describe the total allocation of GPP to root and mycorrhizal respiration, to leakage of root exudates, and to root production—only the latter has traditionally been considered a component of NPP (Raich and Nadelhoffer 1989).

1981, Gower et al. 1992, Powers et al. 2005), although the absolute amount of root growth is greatest on sites with high NPP (Raich and Nadelhoffer 1989, Aerts and Chapin 2000).

Edwards and Harris (1977) reported that the growth and death of roots in a forest in Tennessee delivered 733 g C m^{-2} yr^{-1} to the soil, whereas the aboveground production was 685 g C m^{-2} yr^{-1} (Reichle et al. 1973a). Similarly, roots composed more than half of the NPP in coniferous forests in Washington (Table 5.1) and in the deciduous forest at Hubbard Brook (Fahey and Hughes 1994). An even larger proportion of total NPP is allocated to root growth in many grassland ecosystems (Lauenroth and Whitman 1977, Warembourg and Paul 1977).

Although there are strong relations between above- and belowground *biomass* (Cairns et al. 1997, Enquist and Niklas 2002, Mokany et al. 2006, Cheng and Niklas 2007), there are no obvious generalizations that allow us to predict the allocation of NPP to the *growth* of shoots and roots worldwide (Nadelhoffer and Raich 1992, Gower et al. 1996, Litton et al. 2007).

Much of the plant photosynthate that is allocated belowground supports the growth of roots <2 mm diameter, known as fine roots. In grasslands, the total length of fine roots may exceed 100 km beneath each square meter of the soil (Jackson et al. 1997). In most ecosystems,

TABLE 5.1   Net Primary Production in 23- and 180-yr-old *Abies amabilis* Forests in the Cascade Mountains in Washington

	23-yr-old		180-yr-old	
Aboveground	**g m^{-2}yr^{-1}**	**% of total**	**g m^{-2}yr^{-1}**	**% of total**
Biomass increment				
Tree total	426		232	
Shrub stems	6		<1	
Total	432	18.37	232	9.33
Detritus production				
Litterfall	151		218	
Mortality	30			
Herb layer turnover	32		5	
Total	213	9.06	223	8.97
Total aboveground	**645**	**27.42**	**455**	**18.30**
Belowground				
Roots				
Fine (≤2 mm)	650	27.64	1290	51.87
Fibrous-textured	571		1196	
Mycorrhizal	79		94	
Coarse (>2 mm)	358		324	
Angiosperm fine root turnover	373		44	
Total root turnover	1381	58.72	1658	66.67
Mycorrhizal fungal component	326	13.86	374	15.04
Total below ground	1707	72.58	2032	81.70
Ecosystem total	**2352**	**2352**	**2487**	

Source: Form Vogt et al. (1982). Used with permission of Ecological Society of America.

about half of these roots die each year (Gill and Jackson 2000), consistent with observations of roots using transparent soil tubes, known as *minirhizotrons* (Strand et al. 2008, Eissenstat and Yanai 1997). However, studies following the disappearance of carbon from isotopic-labeled root systems indicate considerable longevity of some fine roots—sometimes exceeding 5 years (Matamala et al. 2003, Gaudinski et al. 2001, Riley et al. 2009). It is likely that many root systems consist of a large fraction of roots with relatively short longevity and a smaller population that lasts several years (Joslin et al. 2006, Gaudinski et al. 2010).

NET ECOSYSTEM PRODUCTION AND EDDY-COVARIANCE STUDIES

As long as they are growing, plants allocate some NPP to the accumulation of biomass, with the remaining NPP passing to herbivores or decomposers, which convert organic carbon to CO_2 and return it to the atmosphere. We define *net ecosystem production* (NEP) as

$$NEP = NPP - (R_h + R_d); \qquad (5.6)$$

thus,

$$NEP = GPP - (R_p + R_h + R_d), \qquad (5.7)$$

where R_p, R_h, and R_d represent the respiration of plants, herbivores, and decomposers, respectively.[4] Except in unusual circumstances, NEP will be a partial fraction of NPP. In young forests, NEP may be 50% of NPP, but in older stands, when vegetation is not accumulating biomass, nearly all the NEP will be found in small increments to soil organic matter (Schlesinger 1990, Law et al. 2003, Pregitzer and Euskirchen 2004).

In recent years, atmospheric scientists have made indirect measurement of net ecosystem production of whole ecosystems by calculating the net uptake of CO_2 within a hypothetic column of the atmosphere with a small ground "footprint," typically 1 m^2. Substantial theory underlies this approach. Namely, if there is no carbon exchange between the atmosphere and the biosphere, such as over the surface of a parking lot, we would expect the atmosphere to show a uniform concentration of CO_2 at all heights above the surface—with values close to the global average in the troposphere ($\sim$400 ppm; Table 3.1). In contrast, inside the canopy of a forest, photosynthesis will deplete CO_2 during the day, while CO_2 will be enriched in samples taken near the soil surface, where it is emitted by the activity of decomposers. These differences in CO_2 concentration with height persist in the face of winds that might transport fresh air from outside the ecosystem or mix air within the forest, resulting in a uniform concentration of CO_2. Thus, if we measure the concentration of CO_2 at various heights within the forest and the delivery of fresh air at each height, then we can estimate the carbon uptake, or CO_2 release, necessary to maintain the differences in CO_2 at different heights. Integrating over height and time, these measurements would indicate the net exchange of CO_2—net ecosystem production—in a forest or other types of vegetation.

---

[4] When calculated for large areas, NEP is sometimes called net biome production (NBP) (Randerson et al. 2002).

FIGURE 5.7   An eddy-covariance (flux) tower in a deciduous forest in North Carolina. *Photo courtesy of G. Katul, Duke University.*

Now consider the height-profile of carbon dioxide in the middle of a uniform expanse of vegetation. The air moving through the column above a hypothetical 1-m^2 ground foot-print will arrive and leave with the same distribution of CO_2 with height. The only net exchange with the atmosphere will occur from above the canopy as a result of wind- and eddy-driven transport into or out of the vegetation. *Eddy-covariance measurements* of net carbon exchange, so named because they trace the simultaneous variation in CO_2 con-centrations and vertical wind velocity, have been made in a large number of sites world-wide. The method requires a tower with wind speed and CO_2 analyzers at varying heights (Figure 5.7), and works best in large areas of relatively uniform vegetation and flat topog-raphy, where the effects of turbulence are minimized (Baldocchi 2003). Eddy-covariance studies can provide simultaneous measurements of the net flux of several gases, such as CO_2 and H_2O, providing an estimate of the WUE of photosynthesis at the ecosystem level (Figure 5.8)[5]. At night, the outward flux of CO_2 would represent the total respiration

[5] Developed by atmospheric scientists who were interested in the disappearance of CO_2 from the atmosphere, eddy-covariance studies assign plant uptake of CO_2 a negative value. Ecologists, using harvests to estimate increases in carbon storage in an ecosystem, express NEP as a positive value. Thus, NEP of -100 g C m^{-2} yr^{-1} reported by an eddy-covariance study is equivalent to $+100$ g C m^{-2} yr^{-1} reported by foresters, and both indicate net carbon storage in the ecosystem. In this book, we follow the latter convention and assign a positive value to all NPP and NEP estimates that indicate net carbon uptake by vegetation.

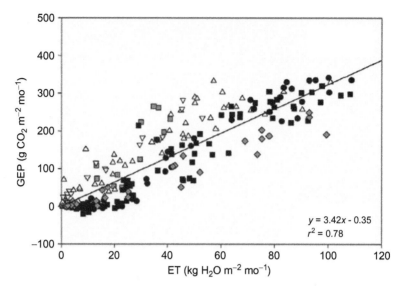

FIGURE 5.8   Monthly gross primary production and evaporation in various temperate deciduous forests, measured by eddy-covariance techniques. The slope of the line is an estimate of water-use efficiency, here equivalent to 1.4 mmol/mol (see Eq. 5.3). *Source: From Law et al. (2002).*

in the ecosystem. GPP can be calculated from NEP using Eq. 5.7, with the assumption that the respiration flux measured during the night applies during the 24-hour period.[6]

It is interesting to compare traditional harvest estimates to NEP obtained from eddy-covariance studies (Table 5.2). Barford et al. (2001) used eddy-covariance to estimate GPP of 1300 g C m^{-2} yr^{-1} during an 8-year study in a deciduous woodland at Harvard Forest in Massachusetts. Total respiration was 1100 g C m^{-2} yr^{-1}, resulting in NEP of 200 g C m^{-2} yr^{-1}—a preliminary estimate of net carbon sequestration in wood and soil organic matter. Independent measurements of NEP from traditional harvests indicate net carbon storage of 160 g C m^{-2} yr^{-1} showing agreement within 20 to 25%.

In an experimental forest in Michigan, Gough et al. (2008) concluded that the differences between biometric (harvest) and meteorological (eddy-covariance) estimates of NPP were related to late-season photosynthesis which was allocated to storage rather than growth. NPP measured by these techniques agreed within 1% when data were averaged over 5 years. Unfortunately, many investigators using eddy-covariance techniques have not simultaneously used traditional methods to validate the carbon accumulation at their field sites (Luyssaert et al. 2009).

Eddy-covariance studies of carbon uptake can be applied in a wide variety of situations, including ecosystem-level studies of the net carbon balance of cities. Despite harboring

---

[6] Net carbon uptake in eddy-covariance studies is sometimes known as net ecosystem exchange (NEE).

TABLE 5.2 GPP, NPP, and NEP for Some Young Temperate and Boreal Forest Ecosystems Measured by Harvest (H) and Eddy Covariance (CV) Methods

Ecosystem type	Age	Method	GPP	NPP	NEP	References
Pinus sylvestris (Finland)	40	CV H	1005		185 228	Kolari et al. 2004
Picea rubens (Maine)	90	CV	1339		174	Hollinger et al. 2004
Pinus taeda (North Carolina)	16	CV H	2238	986	433 428	Juang et al. 2006 Hamilton et al. 2002 McCarthy et al. 2010
Pinus elliottii (Florida)	24	CV H	2606		675 745	Clark et al. 2004
Pinus ponderosa (Oregon)	56–89	CV H	1208	400	324 118	Law et al. 2000 Law et al. 2003
Mixed deciduous (Massachusetts)	60	CV H	1300		200 160	Barford et al. 2001
Mixed deciduous (Michigan)	85	CV H	654		151 153	Gough et al. 2008

Note: All data in g C m^{-2} yr^{-1}

considerable forest cover, suburban areas in Baltimore show a net release of 361 g C m^{-1} yr^{-1} or 241 kg C yr^{-1} for each inhabitant (Crawford et al. 2011). Other world cities have even higher rates of net efflux, reflecting the balance between CO_2 uptake by vegetation vs. CO_2 release by fossil fuel combustion for heating and transport (Crawford et al. 2011, Bergeron and Strachan 2011).

THE FATE OF NET PRIMARY PRODUCTION

As defined in Eq. 5.6, NEP would seem equivalent to the incremental accumulation of organic matter in an ecosystem—largely in wood growth and increments of soil organic matter. Even old forests that have stopped growing continue to store some organic matter in soils (Luyssaert et al. 2008, Law et al. 2003, Schlesinger 1990, Zhou et al. 2006). Only a small amount of carbon from photosynthesis accumulates in ecosystems in other forms, including calcium oxalate and calcium carbonate in plant tissues (Stone and Boonkird 1963, Braissant et al. 2004, Cailleau et al. 2004) and calcium carbonate in soils (Chapter 4). If it survives the appetite of herbivores, most of the remaining NPP passes to the decomposers, where it is respired to CO_2. Terrestrial vegetation is also subject to fire, with a yearly frequency in some grasslands and century-long return intervals for fire in some forests. Fires return carbon to the atmosphere, largely as CO_2, analogous to a large, generalist herbivore.

Although herbivory may play a role in controlling forest productivity and nutrient cycling (Chapter 6), the consumption of plant tissues by herbivores is nearly always <20% of terrestrial NPP (e.g., Mispagel 1978, McNaughton et al. 1989, Cyr and Pace 1993, Cebrian and Lartigue 2004). Higher values are nearly always associated with insect outbreaks (e.g., Kurz et al. 2008). By consuming leaf area and root biomass, herbivores may have an indirect effect on NPP that is larger than their direct consumption (Reichle et al. 1973b, Llewellyn 1975, Ingham and Detling 1990). Globally, herbivores consume about 5% of terrestrial NPP (Whittaker and Likens 1973). Respiration by decomposers consumes most of the rest (Chapter 11).

A small amount of carbon is lost from ecosystems in organic materials that are carried by streamwaters and groundwaters (Chapter 8) and respired to CO_2 outside the boundaries of the ecosystem, typically amounting to 1 to 10 g C m^{-2} yr^{-1}, or less than 1% of global terrestrial NPP (Schlesinger and Melack 1981). Similarly, when volatile organic compounds (VOCs) are produced by plants and lost to the atmosphere, they represent a small portion of NPP that is oxidized by hydroxyl radicals outside of the ecosystem (Chapter 3; Kesselmeier et al. 2002). Emissions of isoprene have been measured using eddy-covariance methods (Rinne et al. 2000).

Globally, the emission of reduced carbon compounds from natural vegetation may exceed 1×10^{15} g C/yr, or about 2% of NPP (Guenther 2002, Laothawornkitkul et al. 2009). This small fraction accounts for much of the CO and CH_4 in the atmosphere (Chapter 11). Losses of organic carbon compounds in streamwaters and as volatile organics explain why NEP is not always directly equivalent to new incremental storage of organic matter in the ecosystem (Lovett et al. 2006, Chapin et al. 2006, Kindler et al. 2011).

REMOTE SENSING OF PRIMARY PRODUCTION AND BIOMASS

Harvest measurements and eddy-covariance studies of NPP are labor intensive and necessarily applied only to small areas. Since the productivity of vegetation may vary greatly over the landscape, regional estimates of productivity by harvest are often prohibitively expensive. For the past decade, a NASA satellite, known as Moderate Resolution Imaging Spectroradiometer (MODIS), has provided integrated estimates of GPP over large areas for studies of global change (Running et al. 2004). MODIS replaces a number of older satellites, including LANDSAT and NOAA-AVHRR, which provided early estimates of global NPP using similar approaches (Box et al. 1989, Field et al.1998).

The basis of satellite measurements of GPP is the differential absorption of light by chlorophyll and other leaf pigments. Green plants look green because chlorophyll preferentially absorbs light in the blue and red portions of the solar spectrum, reflecting a large portion of the green light to our eyes. Despite its strong absorption of red light (760 nm), chlorophyll shows little absorption of infrared light at wavelengths of 800 to 1200 nm. Thus, to provide an index of the underlying "greenness" of the Earth's surface, satellites measure the surface reflectance in discrete portions of the visible and infrared spectrum (Figure 5.9). Bare soil

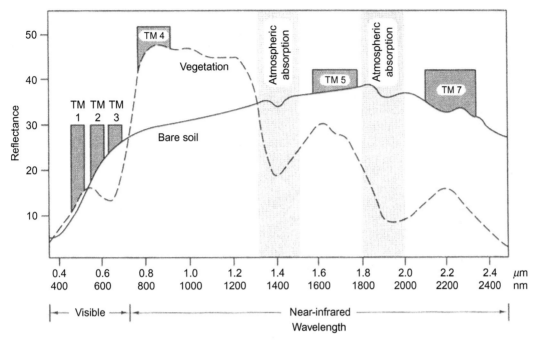

FIGURE 5.9 A portion of the solar spectrum showing the typical reflectance from soil (−) and leaf (−−−) surfaces and the portions of the spectrum that are measured by the LANDSAT satellite.

shows similar reflectance in the infrared and red wavebands, whereas vegetation shows an infrared/red ratio ≫1.0 as a result of the absorption of red light by chlorophyll.

The normalized difference vegetation index (NDVI) is calculated as:

$$NDVI = (NIR - VIS)/(NIR + VIS), \qquad (5.8)$$

where NIR is reflectance in the near-infrared and VIS is reflectance in the visible red wavebands, respectively. This index minimizes the effects of variations in background reflectance and emphasizes variations in the data that occur because of the density of green vegetation. NDVI allows global mapping of a greenness index for the Earth's land surface, and satellite measurements of greenness can provide estimates of NPP, assuming that greenness is directly related to leaf area[7] and that LAI is a good predictor of NPP (Gholz 1982; Figures 5.6 and 5.10).

For the past decade, the MODIS satellite has provided an estimate of GPP by assuming that

$$GPP = \varepsilon \times NDVI \times PAR, \qquad (5.9)$$

where ε is a measured coefficient expressing the efficiency of conversion of sunlight energy into plant growth in various ecosystems (Field et al. 1995), and PAR is a measure of

---

[7] Ecosystem ecologists often express leaf area as the area of leaves exposed above 1 m^2 of ground surface. The term, *leaf area index* or LAI, has the units of m^2/m^2.

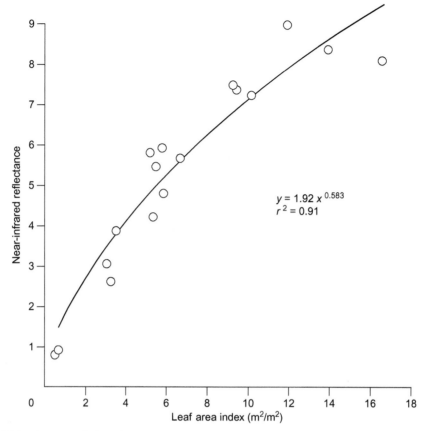

FIGURE 5.10   The ratio of light reflected in the near-infrared and red spectral bands (wavebands TM4 and TM3 of the LANDSAT satellite; see Figure 5.9) is related to LAI for forest stands in the northwestern United States. *Source: From Peterson et al. (1987).*

photosynthetically active radiation. The estimate is computed every 8 days for the Earth's land surface at 1-km spatial resolution (Running et al. 2004). Currently, the satellite measurements of NDVI are coupled to independent measurements of surface climate conditions that affect ε. It is possible that direct satellite estimates of ε may be possible in the near future (Grace et al. 2007, Hilker et al. 2011, Frankenberg et al. 2011).

Remote sensing of biomass is more difficult than for LAI and NPP. Synthetic aperture radar (SAR) is used to measure vegetation biomass based on the absorption of microwave radiation by the water held in woody biomass (Le Toan et al. 2011; Figure 5.11). Biomass estimates have used radar or LiDAR to estimate forest height, which is often directly related to biomass (Treuhaft et al. 2004, 2010; Shugart et al. 2010). Boudreau et al. (2008) used a combination of field measurements and aircraft and satellite LiDAR systems to estimate forest biomass from measurements of its height in Quebec. Similar techniques have been applied to tropical rainforests in Costa Rica and Brazil (Dubayah et al. 2010, Drake et al. 2003, Asner et al. 2010, Clark et al. 2011).

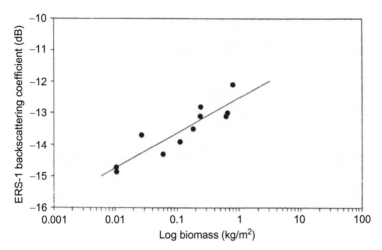

FIGURE 5.11   The reflected microwave radiation (backscattering coefficient) measured by an airborne SAR for stands of young loblolly pine (*Pinus taeda*) in central North Carolina. *Source: Modified from Kasischke et al. (1994).*

GLOBAL ESTIMATES OF NET PRIMARY PRODUCTION AND BIOMASS

Beer et al. (2010) estimate 123×10^{15} g C/yr of global GPP based on 8 years of observations of the land surface by the MODIS satellite (compare Jung et al. 2011). Since one-half of the GPP is normally consumed by respiration, the indicated global terrestrial net primary production is about 60×10^{15} g C/yr—near the high end of the range of earlier estimates based on the modeling and aggregation of harvest data ($45-65 \times 10^{15}$ g C/yr) of the following authors in order: Whittaker and Likens (1973), Lieth (1975), Melillo et al. (1993), Field et al. (1998), Saugier et al. (2001), Del Grosso et al. (2008), and Ito (2011). As expected, a global map of terrestrial NPP shows the highest values in tropical rainforests and the lowest values in areas of extreme desert and ice (Figure 5.12).

Aggregations of data from harvest and eddy-covariance studies also suggest that the primary productivity of forests is greatest in the tropics and declines with increasing latitude to low values in boreal forests and shrub tundra (Table 5.3). In seasonal environments, photosynthetic rates often acclimate to changes in temperature (Lange et al. 1974, Gunderson et al. 2010). Thus, daily values for NPP are relatively similar in many ecosystems; it is the length of the growing season, as determined by temperature and moisture, that determines annual NPP (Kerkhoff et al. 2005). Among European forests, net ecosystem production is lower in northern forests as a result of a greater effect of low temperatures on the length of the growing season, reducing GPP relative to ecosystem respiration (Valentini et al. 2000, Janssens et al. 2001, Van Dijk and Dolman 2004). Along a gradient of decreasing precipitation, NPP declines from forests to grasslands, showing very low values in most deserts (Knapp and Smith 2001). In all biomes, rain-use efficiency by vegetation is greatest during dry years, when it approaches a value of 0.21 g C in aboveground NPP per mm of precipitation (WUE = 0.315 mmol/mol) across a broad range of ecosystems (see Eq. 5.3; Huxman et al. 2004).

Evidence for the importance of temperature and moisture as controls of NPP is seen in regional comparisons of productivity, especially patterns along gradients of elevation. Whittaker (1975) found that net primary production declined with increasing elevation in the forested

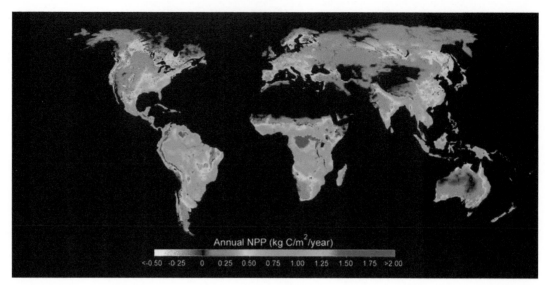

Annual NPP (kg C/m²/year)

<-0.50   -0.25   0   0.25   0.50   0.75   1.00   1.25   1.50   1.75   >2.00

FIGURE 5.12   Distribution of global NPP on land for 2002, computed from MODIS data. *Source: From Running et al. 2004, Figure 5 in BioScience, June 2004; used with permission.*

TABLE 5.3   Biomass and Net Primary Production in Terrestrial Ecosystems

Biome	Area (10^6 km²)	NPP (g C m^{-2} yr^{-1})	Total NPP (10^{15}g C yr^{-1})	Biomass (g C m^{-2})	Total plant C pool (10^{15}g C)
Tropical forests	17.5	1250	20.6	19,400	320
Temperate forests	10.4	775	7.6	13,350	130
Boreal forests	13.7	190	2.4	4150	54
Mediterranean shrublands	2.8	500	1.3	6000	16
Tropical savannas/grasslands	27.6	540	14.0	2850	74
Temperate grasslands	15.0	375	5.3	375	6
Deserts	27.7	125	3.3	350	9
Arctic tundra	5.6	90	0.5	325	2
Crops	13.5	305	3.9	305	4
Ice	15.5				
Total	**149.3**		**58.9**		**615**

From data compiled by Saugier et al. 2001, assuming a 50% carbon content in plant tissues.

mountains of the eastern United States, presumably reflecting the influence of declining temperatures (i.e., a shorter growing season). In the southwestern United States, where precipitation is more limited, NPP tends to increase with elevation in communities ranging from desert shrublands to montane forests (Whittaker and Niering 1975). Sala et al. (1988) show a direct relationship between NPP and precipitation within the grasslands of the central United

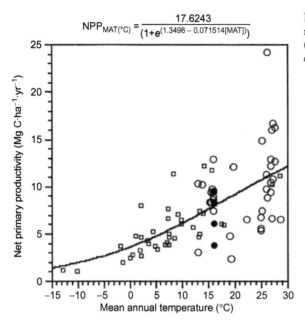

$$NPP_{MAT(°C)} = \frac{17.6243}{(1+e^{(1.3496 - 0.071514[MAT])})}$$

FIGURE 5.13 NPP in world forests versus mean annual temperature. *Soruce: From Schuur (2003). Used with permission of the Ecological Society of America.*

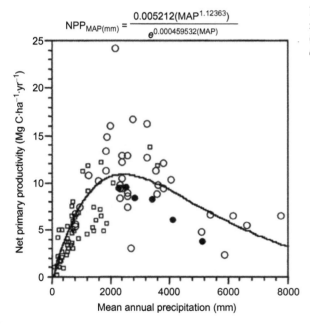

$$NPP_{MAP(mm)} = \frac{0.005212(MAP^{1.12363})}{e^{0.000459532(MAP)}}$$

FIGURE 5.14 NPP in world forests versus mean annual precipitation. *Source: From Schuur (2003). Used with the permission of the Ecological Society of America.*

States. Compilations of data from various world biomes show strong relations of NPP to temperature and mean annual precipitation (Scurlock and Olson 2002; Figures 5.13 and 5.14).

In forests of the northwestern United States, NPP and LAI are directly related to site water balance, which is the difference between precipitation inputs and losses of soil moisture by runoff and evapotranspiration during the growing season (Grier and Running 1977, Gholz 1982). Rosenzweig (1968) combined temperature and precipitation to calculate actual evapotranspiration, which shows a positive correlation to NPP in temperate zone ecosystems (compare Webb et al. 1978). The overall strength of the relationship may partially derive from the influence of these variables on microbial processes that speed nutrient turnover in the soil (Chapter 6). Nutrient availability often determines local differences in net primary productivity among sites within the temperate zone (e.g., Pastor et al. 1984). In tropical rainforests, where both light and moisture are abundant, the relationship of NPP to these variables is weak, and local soil conditions determining fertility are potentially more important (Brown and Lugo 1982, Cleveland et al. 2011).

Estimates of the total biomass of land plants range from 560 to 615×10^{15} g C, derived from the aggregation of harvest data worldwide (Olson et al. 1983, Saugier et al. 2001). Based on a random sampling of land areas, total biomass in U.S. forests is about 18×10^{15} g C (Blackard et al. 2008), whereas forests in China and India contain about 4.75×10^{15} g C and 2.9×10^{15} g C, respectively (Fang et al. 2011, Kaul et al. 2011). By comparison, total biomass in the tropical forests of Brazil may be as large as 50×10^{15} g C (Nogueira et al. 2008). The ratio of biomass/NPP is an estimate of the mean residence time for an atom of carbon in plant tissues (compare to Eq. 3.3). The global values yield an overall mean residence time of about 10 years, but this value varies from about 4 in deserts to >20 in some forests (see Table 5.3; compare to Fahey et al. 2005). For the United States, the mean residence time of C in vegetation is about 5 years—that is, forest biomass of 18×10^{15} g C divided by NPP of 3.5×10^{15} g C/yr, which is calculated from Xiao et al. (2010). Of course, we must remember that these are weighted averages. In forests some tissues, such as leaves, may last only a few months, while wood may last for centuries.

Estimates, such as those in Table 5.3, are calculated by classifying the land vegetation into a small number of categories and by assigning a mean value to the NPP and biomass of each category based on data from the widest possible number of field studies. The classification of vegetation is arbitrary, and estimates of the land area in each category often vary considerably (Golley 1972). Moreover, the NPP data often do not reflect average values because ecologists tend to select mature, well-developed stands for study. Random site selection often produces lower regional values (Botkin and Simpson 1990, Botkin et al. 1993, Jenkins et al. 2001). Remote sensing estimates of NPP and biomass have the advantage of including the full range of variation seen in the field (Zhang and Kondragunta 2006). MODIS offers continuous, realistic estimates of NPP over large scales, with some loss of accuracy at local sites (Pan et al. 2006).

NET PRIMARY PRODUCTION AND GLOBAL CHANGE

Since the beginning of civilization, humans have harvested the Earth's net primary production for food, fuel, and fiber. Indeed, the pages of this book were once part of a living tree. Cultivated lands and pasture now occupy about 40% of the world's land surface (Ellis et al. 2010, Goldewijk et al. 2010, Ramankutty and Foley 1998, Sterling and Ducharne

2008), and rates of tropical deforestation, largely for new cultivation, are estimated at 5.6 to 5.8×10^6 ha/yr (Archard et al. 2002, DeFries et al. 2002).[8] Directly and indirectly humans now use 11 to 24% of potential NPP on the land surface (Haberl et al. 2007, Imhoff et al. 2004b), with most of the carbon released to the atmosphere as CO_2. Ancillary human impacts on the land surface may raise our total appropriation of photosynthesis to 40% annually (Vitousek et al. 1986). These high values for the consumption of NPP by a single species do not bode well for the future of other species on the planet.

The human harvest of natural vegetation is not uniform across the planet. High rates of harvest in the tropics are balanced by the abandonment and regrowth of cultivated land elsewhere (Imhoff et al. 2004b). In the southeastern U.S. coastal plain, young forests are storing 90,000 tC/yr, which is equivalent to NEP of 100 g C/m^2/yr (90×10^9g/yr) (Binford et al. 2006, Delcourt and Harris 1980). Similarly, in Europe regrowing forests provide a net storage of carbon (Peters et al. 2010, Luyssaert et al. 2010). Nevertheless, as a result of expanding urban areas, overall NPP has declined by 0.4% in the Southeast (Milesi et al. 2003). Imhoff et al. (2004a) estimate a 1.6% loss of NPP due to urbanization across the entire United States.

In the Great Plains of the United States, irrigated and fertilized lands have potentially increased NPP by 10% above the level of native ecosystems in that region (Bradford et al. 2005); however, in most areas, agricultural lands are a net source of CO_2 to the atmosphere (i.e., negative NEP), especially when ancillary CO_2 emissions, such as those from diesel fuel, are included (West et al. 2010).

Fires are a normal part of the Earth's terrestrial ecosystems, especially in areas of tropical savanna (Cahoon et al. 1992). However, much of the land-clearing by humans has increased the global extent of fire (Bowman et al. 2009, Mouillot and Field 2005). Global estimates of CO_2 emissions from fires range from 1.4 to 3.6×10^{15} g C/yr (van der Werf et al. 2003, Mouillot et al. 2006, Schultz et al. 2008, Mieville et al. 2010); fires also contribute to global sources of a variety of trace gases (Andreae and Merlet 2001, Jain et al. 2006; Chapter 3 and 6). Large fires in Kalimantan, Indonesia, in 1997 are estimated to have released between 0.81 and 2.57×10^{15} g C into the atmosphere in a single year (Page et al. 2002). Large fires have also resulted in major losses of carbon from boreal forests in recent years (Kasischke et al. 1995, Bond-Lamberty et al. 2007). Many ecologists anticipate a greater frequency of fires as a result of global climate change.

The rising concentration of carbon dioxide in the atmosphere from fossil fuel combustion and biomass burning increases the availability of a basic reactant for photosynthesis (Eq. 5.2). Early studies of plant response to high CO_2 showed an average 31% increase in growth with a doubling of CO_2 concentrations for woody plants in controlled experiments (Curtis and Wang 1998, Wang et al. 2012). When it was noted that the growth responses were much lower in the absence of fertilization (Thomas et al. 1994, Hattenschwiler et al. 1997, Poorter and Perez-Soba 2001),

---

[8] Country-specific statistics for the extent of forests are given in The Global Forest Resources Assessment of the Food and Agriculture Organization. www.fao.org/forestry/fra/fra2010/en/. The gross rate of forest cover loss is approximately 4 times larger than the net rate as a result of the regrowth of forests on some disturbed lands (Hansen et al. 2010). In the Amazon basin, the rate of deforestation declined significantly between 2004 and 2011 (Davidson et al. 2012). For the United States, the gross rate of deforestation is about 1×10^6 ha/yr (Masek et al. 2011), but as a result of reforestation and afforestation, there has been a net increase in forest area during the last decade.

FIGURE 5.15 The Free-Air CO_2 Enrichment experiment in Duke Forest in central North Carolina. Each plot is 30 m in diameter and surrounded by 16 towers, which emit CO_2 so as to maintain a specified concentration in the cylindrical volume of the plot to the height of the forest canopy. All other factors, including soil fertility, are allowed to vary naturally in the control and experimental plots.

investigators established large-scale, long-term experiments in a variety of ecosystems using Free-Air CO_2 Enrichment (FACE) technology (see Figure 5.15 and Hendrey et al. 1999).

The results from various forest FACE experiments show an 18% increase in NPP with growth at $+200$ ppm CO_2—the atmospheric levels expected globally in 2050 (Norby et al. 2005). Crop plants show growth stimulations ranging from 12 to 14% in rice, wheat, and soybeans (Long et al. 2006). In all species, growth at high CO_2 stimulates the proportional allocation of NPP to roots, which may increase the delivery of carbon to soils (Rogers et al. 1994, Jackson et al. 2009). There seems little doubt that the rise of CO_2 in Earth's atmosphere and thus the rate of anticipated climate change would be greater if it were not for CO_2 uptake in areas of undisturbed vegetation and the regrowth of vegetation on previously cleared lands.

Along with exposure to high CO_2, humans have increased the exposure of forests to ozone, acid rain, and various forms of reactive nitrogen, with variable effects on growth. Nitrogen deposition from the atmosphere, producing a number of changes in soil biogeochemistry (Chapter 6), is likely to increase carbon uptake and storage in forests through its effect as a plant fertilizer (Magnani et al. 2007, Thomas et al. 2010). On the other hand, ozone reduces the growth of most plants when concentrations exceed 100 ppb (Richardson et al. 1992, Gregg et al. 2003), although growth at high CO_2 partially compensates for the ozone effect (Reid and Fiscus 1998, King et al. 2005, Poorter and Perez-Soba 2001, Penuelas et al. 2011). Losses of photosynthetic efficiency from air pollutants are noted even among plants in high northern latitudes, where one might expect relatively low pollution loads (Odasz-Albrigtsen et al. 2000, Savva and Berninger 2010).

With the onslaught of human effects on terrestrial NPP, it is worth asking whether we have produced measureable changes in global NPP in recent years. Tree-ring records are equivocal—some showing recent increases in growth (Soule and Knapp 2006), others showing little or no effect, perhaps due to concurrent drought (Barber et al. 2000, Gedalof and Berg 2010, Andreu-Hayles et al. 2011, Penuelas et al. 2011). Using satellite data to monitor NDVI (Eq. 5.8), several studies noted increases in global NPP during the 1980s and 1990s, largely through changes in high northern latitudes (Myneni et al. 1997), and perhaps globally (Nemani et al. 2003). Most of the change was attributed to changes in temperature, which determines the

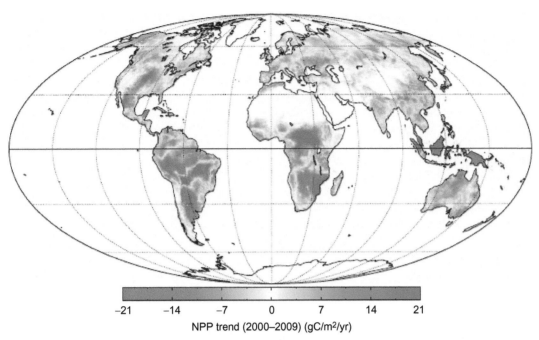

-21   -14   -7   0   7   14   21

NPP trend (2000–2009) (gC/m2/yr)

FIGURE 5.16   Change in terrestrial NPP from 2000 to 2009 from MODIS. *Source: From Zhao and Running 2010. Used with permission of the American Association for the Advancement of Science.*

length of the growing season. Strangely, this effect seems to be reversed in the MODIS-derived NPP record from 2000 to 2009 which shows a 1% decline in global NPP attributed to increasing drought in the Southern Hemisphere (Figure 5.16; Piao et al. 2011). Eddy-covariance studies in Europe show a severe reduction in NPP during 2003 due to heat and drought, so that the region became a source of CO_2 to the atmosphere (i.e., negative NEP; Ciais et al. 2005).

Certainly, global changes in NPP and biomass accompanied past climate changes associated with glacial intervals. At the last glacial maximum 19,000 years ago, carbon storage in land plants and soils was 30 to 50% lower than today (Bird et al. 1994, Beerling 1999, Kohler and Fischer 2004). NPP on the world's land surface was presumably depressed as well, because of lower plant cover; cold, dry climates; and low atmospheric CO_2 (Gerhart and Ward 2010). At the last glacial maximum, Landais et al. (2007) estimate that terrestrial NPP was only 65 to 70% of today's value. With the future climate changes due to greenhouse warming, plant biomass may increase up to 10% over present-day conditions (Smith et al. 1992c), although a transient period of drought may reduce terrestrial productivity during the next few decades (Rind et al. 1990, Smith and Shugart 1993). Overall, we might expect higher terrestrial NPP on a warmer, wetter world in the future (Wu et al. 2011).

DETRITUS

The largest fraction of NPP is delivered to the soil as dead organic matter. Global patterns in the deposition of plant litterfall are similar to global patterns in NPP (Matthews 1997). The deposition of litterfall declines with increasing latitude from tropical to boreal forests (Vogt et al.

1986, Lonsdale 1988, Berg et al. 1999). Leaf tissues account for about 70% of litterfall in forests (O'Neill and De Angelis 1981, Meentemeyer et al. 1982), but the deposition of woody litter tends to increase with forest age, and fallen logs may be a conspicuous component of the forest floor in old-growth forests (Lang and Forman 1978, Harmon et al. 1986). In grassland ecosystems, where little of the aboveground production is contained in perennial tissues, the annual litterfall is nearly equal to annual net primary production.

In most areas, the annual growth and death of fine roots contributes a large amount of detritus to the soil, which has been overlooked by studies that only consider aboveground litterfall (Vogt et al. 1986, Nadelhoffer and Raich 1992). Using actual evapotranspiration to predict global patterns of litterfall; Meentemeyer et al. (1982) estimated 27×10^{15} g C for the annual production of aboveground litterfall worldwide. Matthews (1997) indicates total detritus production of $\sim 50 \times 10^{15}$ g C/yr, suggesting that about half of global NPP occurs belowground. Her value is slightly smaller than current estimates of terrestrial NPP, which allows us to accommodate ancillary losses of organic carbon to the atmosphere and to streams and groundwater.

The Decomposition Process

Most detritus, whether from litterfall or root turnover, is delivered to the upper layers of the soil where it is subject to the decomposition by microfauna, bacteria, and fungi (Swift et al. 1979, Schaefer 1990). Decomposition leads to the release of CO_2, H_2O, and nutrient elements, and to the microbial production of highly resistant organic compounds known as *humus*. Humus compounds accumulate in the lower soil profile (Chapter 4) and compose the bulk of soil organic matter (Schlesinger 1977, Rumpel and Kogel-Knabner 2011). The dynamics of the pool of carbon in soils is best viewed in two stages—processes leading to rapid turnover of the majority of litter at the surface and processes leading to the slower production, accumulation, and turnover of humus at depth.

The litterbag approach is widely used to study decomposition at the surface of the soil. Fresh litter is confined in mesh bags that are placed on the ground and collected for measurements at periodic intervals (Singh and Gupta 1977). Simple models of decay are based on an exponential pattern of loss, where the fraction remaining after 1 year is given by

$$X/Xo = e^{-k} \tag{5.10}$$

An alternative, the mass-balance approach, suggests that the annual decomposition should equal the annual input of fresh debris so that the mass of detritus stays constant. Under these assumptions, a constant fraction, k, of the detrital mass decomposes, so that

$$\text{litterfall} = k(\text{detrital mass}), \tag{5.11}$$

or

$$\text{litterfall/detrital mass} = k. \tag{5.12}$$

When the detritus is in steady state, the values for k calculated from the litterbag and mass-balance approaches should be equivalent, and mean residence time for plant debris is $1/k$ (Olson 1963; see also footnote on p. 55). For a forest in the Pacific Northwest, Vogt et al. (1983) shows the importance of fine roots in the calculation of mean residence times by the mass-balance approach. When root turnover was included, the mean residence time for organic matter in the forest floor was 8.2 to 15.6 years, compared to 31.7 to 68.6 years calculated from aboveground litter alone.

With either approach, when decomposition rates are rapid, values for k are greater than 1.0, and there is little surface accumulation (e.g., in tropical rainforests; Cuevas and Medina 1988, Gholz et al. 2000, Powers et al. 2009). In such systems, decomposition has the potential to respire more than the annual input of organic carbon in litterfall. In contrast, in some peatlands values for k are very small (e.g., 0.001; Olson 1963). Decomposition in grasslands shows a range of 0.20 to 0.60 in values for k (Vossbrinck et al. 1979, Seastedt 1988), but values for deserts may be as high as 1.00 due to the action of termites (Schaefer and Whitford 1981) and photooxidation of litter by ultraviolet light (Austin and Vivanco 2006, Gallo et al. 2009). In many ecosystems, decomposition shows a rapid initial phase of decomposition, followed by a slower phase in which some material may persist for decades (Harmon et al. 2009). Two- or three-phase exponential models are often best to describe this pattern of decomposition and for the most accurate estimates of k (Minderman 1968, Adair et al. 2008).

Decomposition rates vary as a function of temperature, moisture, and the chemical composition of the litter material. Microbial activity increases exponentially with increasing temperature (e.g., Edwards 1975). For plant litter, this relation often shows a Q_{10} of ≥ 2.0, that is a doubling in activity per 10°C increase in temperature (Raich and Schlesinger 1992, Kirschbaum 1995, Katterer et al. 1998). Van Cleve et al. (1981) found that the thickness of the forest floor in black spruce forests in Alaska was inversely related to the cumulative degree days favorable to decomposition each year. In contrast, soil moisture often limits the rate of decomposition in arid and semiarid regions (Strojan et al. 1987, Amundson et al. 1989, Epstein et al. 2002), and moisture assumes increasing importance when temperate forest soils are subject to experimental warming, which dries them (Peterjohn et al. 1994).

Meentemeyer (1978a) compiled data from various decomposition studies to relate surface decomposition to actual evapotranspiration, and used the resulting equation to predict regional patterns of decomposition (Figure 5.17). His predictions are consistent with observations of surface litter in much of the United States (e.g., Lang and Forman 1978). Actual evapotranspiration is also a good predictor of decomposition in Europe (Berg et al. 1993), but less successful in predicting the decomposition of fine roots (Silver and Miya 2001). Improvements in these predictions are found when chemical parameters, such as lignin and nitrogen, are also considered (Meentemeyer 1978b, Melillo et al. 1982), but we defer a discussion of the dynamics of nutrient elements during decomposition until Chapter 6.

Humus Formation and Soil Organic Matter

Plant litter and soil microbes constitute the cellular fraction of soil organic matter. As decomposition proceeds, there is an increasing content of noncellular organic matter (i.e., humus) which appears to result from microbial activity. The structure of humus is poorly known, but it contains numerous aromatic rings with phenolic (–OH) and organic acid (–COOH) groups (Flaig et al. 1975, Stevenson 1986). As we saw in Chapter 4, these groups offer a major source of cation exchange capacity in many soils. Some tentative models have been proposed for the complete molecular structure of humus (Schulten and Schnitzer 1993, 1997), but many scientists believe that a large portion of soil humus is amorphous, with no consistent molecular weight or repeating units.

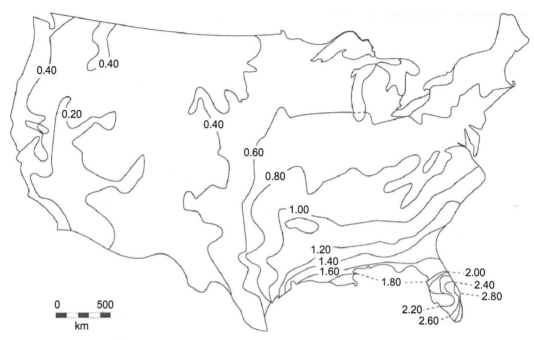

FIGURE 5.17   Rates of decomposition of fresh litter in the United States predicted by a stimulation model using actual evapotranspiration as a predictive variable. Isopleth values are the fractional loss rate (*k*) of mass from fresh litter during the first year of decomposition. *Source: From Meentemeyer (1978a.)*

The structure of humic substances also changes dramatically in response to changes in the pH, ion strength and complexing ions in the soil solution (Myneni et al. 1999). Recent progress in elucidating the chemical structure of humus has been made using ^{13}C nuclear magnetic resonance (NMR) spectroscopy (Mahieu et al. 1999, Baldock et al. 2004) and pyrolysis-field ionization mass spectrometry (Py-FIMS) (Schnitzer and Schulten 1992). The most recalcitrant fractions of soil humus appear to have a large component of polymethylene ($C=C=C$) groups that are synthesized by microorganisms (Baldock et al. 1992).

Traditional chemical characterizations of humus have been based on the solubility of humic and fulvic acid components in alkaline and acid solutions, respectively (Figure 5.18). The acid-soluble component of humus, primarily fulvic acid, controls the downward movement of Fe and Al in soils (Chapter 4). Percolating downward from the forest floor and A-horizon, fulvic acids often account for a large fraction of the soil organic matter in the lower soil profile, where they are complexed with clay minerals and calcium (Beyer et al. 1993, Oades 1988, Kalbitz et al. 2000). Noncrystalline forms of Fe- and Al-oxides are particularly effective in preserving organic matter with surface adsorption (Torn et al. 1997, Mikutta et al. 2006, Powers and Veldkamp 2005). This humus is very resistant to microbial attack; extracted humic materials from forest soil in Saskatchewan had a measured mean ^{14}C age of 250 to 940 years (Campbell et al. 1967).

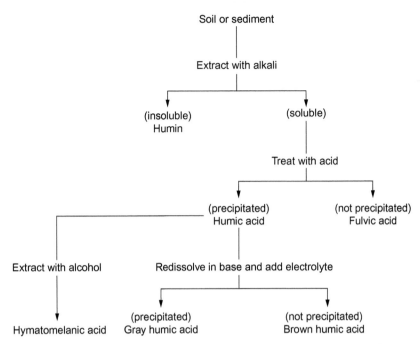

Soil or sediment

Extract with alkali

(insoluble)
Humin

(soluble)

Treat with acid

(precipitated)
Humic acid

(not precipitated)
Fulvic acid

Extract with alcohol

Redissolve in base and add electrolyte

(precipitated)
Gray humic acid

(not precipitated)
Brown humic acid

Hymatomelanic acid

FIGURE 5.18   Fractionation of fulvic and humic acid components from soil organic matter. *Source: From Stevenson (1986).*

Under most vegetation, the mass of humus in the soil profile exceeds the combined content of organic matter in the forest floor and aboveground biomass. Globally the pool of organic carbon in world soils amounts to about 1500×10^{15} g to 1-m depth (Schlesinger 1977, Batjes 1996, Amundson 2001). Many tropical soils contain small amounts of soil organic matter dispersed in the lower profile. Table 5.4 provides a global inventory of plant detritus and soil organic matter, totaling 2344×10^{15} g C to 3-m depth (Jobbágy and Jackson 2000). Even that value may underestimate the total mass of organic material stored in regions of permafrost (Tarnocai et al. 2009).

The global estimate of soil organic matter, divided by the estimate of global litterfall, suggests a mean residence time of about 50 years for the total pool of organic carbon in soils, but the mean residence time varies over several orders of magnitude between the surface litter and the various humus fractions (Figure 5.19). In the temperate zone, the mass of soil organic matter and its mean residence time increase from warm-temperate to boreal forests (Schlesinger 1977, Garten 2011, Frank et al. 2012). Regional inventories of the distribution and abundance of soil organic carbon (0–100 cm) are available for the United States (74×10^{15} g C; Guo et al. 2006), China (84 to 89×10^{15} g C; Yu et al. 2007, Li et al. 2007), India (63×10^{15} g C; Lal 2004), and other nations as part of recent national accounts of carbon storage in vegetation and soils.

Turnover

The incorporation of nuclear-bomb-derived radiocarbon (^{14}C) into different fractions of soil organic matter shows promise as a means of estimating their turnover (Trumbore 1993). O'Brien and Stout (1978) used radiocarbon dating to find that 16% of the organic matter

TABLE 5.4   Distribution of Soil Organic Matter by Ecosystem Types

Biome	World area (10^6 km^2)	Mean soil profile carbon (kgC/m^2) 0–100 cm	0–300 cm	Total soil carbon pool (10^{15} gC) 0–300 cm
Tropical forests				
Deciduous	7.5	15.8	29.1	218
Evergreen	17.0	18.6	27.9	474
Temperate forests				
Deciduous	7	17.4	22.8	160
Evergreen	5	14.5	20.4	102
Boreal forests	12	9.3	12.5	150
Mediterranean shrublands	8.5	8.9	14.6	124
Tropical savannas/grasslands	15	13.2	23.0	345
Temperate grasslands	9	11.7	19.1	172
Deserts	18	6.2	11.5	208*
Arctic tundra	8	14.2	18.0	144
Crops	14	11.2	17.7	248
Extreme desert, rock and ice	15.5			
Total	**136.5**			**2344**

Note: Excludes soil carbonates, which may contain an additional 930 x 10^{15} gC (Schlesinger 1985).
Source: From Jobbágy and Jackson (2000). Used with permission of the Ecological Society of America.

in a pasture soil had a minimum age of 5700 years, while the rest was of recent origin and concentrated near the surface. In British deciduous woodlands, the distribution of radio-carbon is also compatible with two pools of carbon, each with about 3.5 kg/m^2 to 15-cm depth (Tipping et al. 2009). Because of different turnover times, there is no universal decomposition constant, k, that can be applied to the entire mass of organic matter in the soil profile (Trumbore 1997, Gaudinski et al. 2000).

    Field measurements of the flux of CO_2 from the soil surface provide an estimate of the total respiration in the soil. Most of the production of CO_2 occurs in the surface litter where decomposition is rapid and a large proportion of the fine root biomass is found (Bowden et al. 1993). Edwards and Sollins (1973) found that only 17% of the annual production of CO_2 in a temperate forest soil was contributed by soil layers below 15 cm. Flux of CO_2 from the deeper soil layers is presumably due to the decomposition of humus substances. Production of CO_2 in the soil leads to the accumulation of CO_2 in the soil pore space, which drives carbonation weathering in the lower profile (Chapter 4). Geologic sources of CO_2 diffusing upward to the soil surface are normally very small (Keller and Bacon 1998).

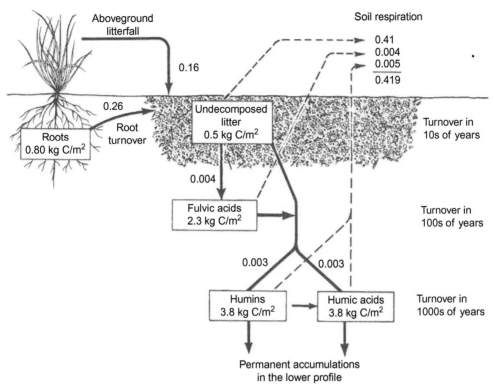

FIGURE 5.19   Turnover of detritus and soil organic fractions in a grassland soil, in units of kgC m^{-2} yr^{-1}. Note that mean residence time can be calculated for each fraction from measurements of the quantity in the soil and the annual production or loss (respiration) from that fraction. *Source: From Schlesinger (1977).*

Unfortunately, the respiration of living roots makes it impossible to use estimates of CO_2 flux from the soil surface to calculate turnover of the soil organic pool (Fahey et al. 2005). In a compilation of values, Schlesinger (1977) found that CO_2 evolution exceeded the deposition of aboveground litter by a factor of about 2.5 (Figure 5.20). The additional CO_2 is presumably derived from root and mycorrhizal metabolism and the decomposition of root detritus (Raich and Nadelhoffer 1989, Subke et al. 2011).

In a field experiment using girdling of trees to eliminate the transport of new photosynthate to roots, Hogberg et al. (2001) reported a 54% decline in soil respiration; presumably the remaining respiration was due to decomposers in the soil (compare to Andrews et al. 1999, Hanson et al. 2000). Globally, soil respiration is 80 to 100×10^{15} g C/yr, with about half derived from the respiration of live roots and the remainder from decomposition (Raich et al. 2002, Subke et al. 2006, Bond-Lamberty and Thomson 2010). Soil respiration shows a strong correlation with NPP and detritus inputs in world ecosystems (Raich and Tufekcioglu 2000, Bond-Lamberty et al. 2004, Hibbard et al. 2005).

The global distribution of soil organic matter shows how moisture and temperature control the balance between primary production and decomposition in surface and lower soil layers (Amundson 2001). Among forests, accumulations in the forest floor increase from tropical to

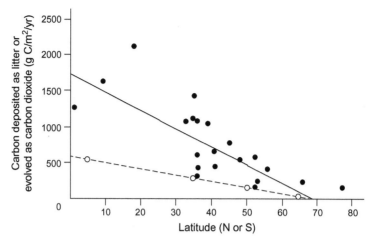

FIGURE 5.20   Latitudinal trends for carbon dynamics in forest and woodland soils of the world. The *dashed line* shows the mean annual input of organic carbon to the soil by litterfall. The *solid line* shows the loss of carbon, measured as the flux CO_2 from the surface. The difference between these lines represents the loss of CO_2 from root and mycorrhizae respiration and from the decomposition of root detritus and exudates. *Source: From Schelesinger (1977).*

boreal climates. Net primary productivity shows the opposite trend, so the accumulation of soil organic matter is largely due to differences in decomposition. Thus, compared to the process of primary production, soil microbes are more sensitive to regional differences in temperature and moisture (Figure 5.20). Worldwide, the accumulation of organic matter in surface litter seems more related to factors controlling decomposition than to the NPP of terrestrial ecosystems (Cebrián and Duarte 1995, Valentini et al. 2000; but see Frank et al. 2012).

Parton et al. (1987) developed a model based on the differential turnover of soil organic fractions to predict the accumulation of soil organic matter in grassland ecosystems. Accurate predictions were achieved when temperature, moisture, soil texture, and plant lignin content were included as variables. Despite relatively low NPP, soils of temperate grasslands contain large amounts of soil organic matter (Sanchez et al. 1982b) due to relatively low rates of decomposition and a larger fraction of plant debris that is derived from root turnover (Oades 1988). In contrast, tropical grasslands and savannas have relatively small accumulations of surface litter, perhaps due to frequent fire (Kadeba 1978, Jones 1973).

Storage of soil organic matter represents a component of net ecosystem production (NEP) in terrestrial ecosystems. Studies of soil chronosequences show that soil organic matter accumulates rapidly on disturbed sites, but rates decline to 1 to 12 $g\,C\,m^{-2}\,yr^{-1}$ during long-term soil development (Figure 5.21; Schlesinger 1990, Chadwick et al. 1994), with the highest rates under cool, wet conditions. Many wetland soils also show large rates of organic accumulation due to anoxic conditions in their sediments (Chapter 7). The low rate of accumulation of soil organic matter in upland soils speaks strongly for the efficiency of decomposers using aerobic metabolic pathways of degradation (Gale and Gilmour 1988). With relatively high nutrient content, humic substances are not inherently resistant to decomposition, but they are stabilized by interactions with soil minerals (Schmidt et al. 2011, Allison 2006). Globally, the annual

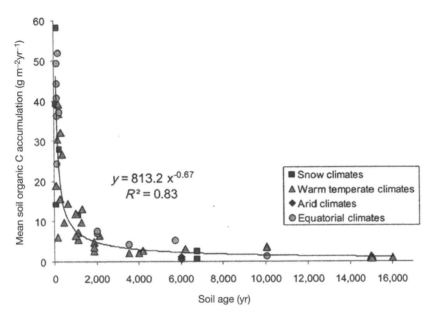

FIGURE 5.21   The rate of accumulation of organic matter in soil chronosequences of different age and climate zones, all derived from volcanic materials. *Source: From Zehetner (2010).*

net production of humus substances is $<0.4 \times 10^{15}$ g C/yr or only about 0.7% of NPP (Schlesinger 1990).

The mass of soil organic matter in most upland ecosystems is likely to have been fairly constant before widespread human disturbance of soils. When soils show a steady state in organic content, the production of humic compounds must be equal to their removal from soils by erosion. Coincidentally, estimates of the global transport of organic carbon in rivers are also about 0.4×10^{15} g C/yr (Schlesinger and Melack 1981, Meybeck 1982), suggesting that before human impacts, NEP for the Earth's land surface was essentially zero.

For areas covered by the last continental glaciation, the total accumulation of soil organic matter represents NEP for the last 10,000 years. The maximum extent of the last glacial, covering 29.5×10^6 km^2 of the present land area (Flint 1971), now contains roughly 300×10^{15} g C—that is, more than 10% of the organic carbon contained in all the world's soils (Table 5.4). In these areas, soil organic matter has accumulated at rates of about 1.35 g C m^{-2} yr^{-1} during the Holocene period. The current rate of storage in northern ecosystems (0.015 to 0.035×10^{15} g C/yr) is too small to be a significant sink for human releases of CO_2 to the atmosphere from fossil fuels, nor is it likely to have increased significantly during the last century (Gorham 1991, Harden et al. 1992).

Total storage of carbon in soils, 1456×10^{15} g or 121×10^{15} moles, can account for only 0.3% of the O_2 content of the atmosphere, given that the storage of organic carbon and the release of O_2 occur on a mole-for-mole basis during photosynthesis (Eq. 5.2). Thus, accumulations of atmospheric O_2 cannot be the result of the storage of organic carbon on land. Long-term storage of organic carbon appears to be dominated by accumulations in anoxic marine sediments (Chapter 9).

SOIL ORGANIC MATTER AND GLOBAL CHANGE

When soils are brought under cultivation, the content of organic matter in soil declines. Most cultivation reduces the inputs of fresh plant debris and increases the rate of decomposition of soil organic matter as a result of better soil moisture and aeration conditions. Losses from many soils are typically 20 to 30% within the first few decades of cultivation (e.g., Figure 5.22, Don et al. 2011). The loss is greatest during the first years of cultivation. Eventually a new, lower level of soil organic matter is achieved that is in equilibrium with the plant production and decomposition in cropland (Jenkinson and Rayner 1977). Some of the soil organic matter is lost as a result of erosion and buried elsewhere, but most is probably oxidized to CO_2 and released to the atmosphere (Van Oost et al. 2007).

With about 10% of the world's soils under cultivation (refer to Table 5.4), losses of organic matter from agricultural soils have been a major component of the increase in atmospheric CO_2 during the past several centuries (Schlesinger 1984). The current rate of release from soils, as much as 0.8×10^{15} g C/yr, is largely dependent on the rate at which natural ecosystems, especially in the tropics, are being converted to agriculture (Maia et al. 2010, Don et al. 2011). Especially large losses of soil carbon are seen when organic soils in wetlands and peatlands are drained (Armentano and Menges 1986, Hutchinson 1980; Chapter 7).

The dynamics of soil organic matter are illustrated by the pattern of loss after land is converted to agriculture. Recall that soil organic matter consists of a labile and a resistant

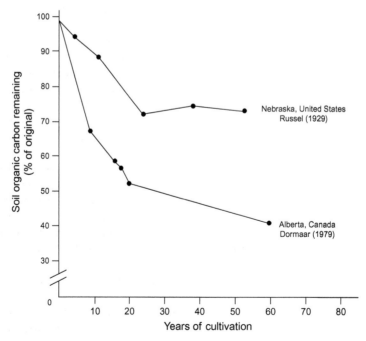

FIGURE 5.22   Decline in soil organic matter following conversation of native soil to agriculture in two grassland soils. *Source: From Schlesinger (1986).*

fraction. The labile fraction is composed of fresh plant materials that are subject to rapid decomposition, whereas the resistant fraction is composed of humic materials that are often complexed with clay minerals. Rather than using biochemical fractionations (refer to Figure 5.18), some workers have used size or density fractionation to quantify the labile and resistant organic matter. Density fractionations are performed by adding soil samples to solutions of increasing specific gravity and collecting the material that floats to the surface (Spycher et al. 1983).

In size fractionation, soils are passed through screens of varying mesh (Tisdall and Oades 1982, Elliott 1986). Most of the turnover of soil organic matter is in the "light" or large fractions that represent fresh plant materials (Tiessen and Stewart 1983). The "heavy" fraction is composed of polysaccharides (sugars) and humic materials that are complexed with clay minerals to form microaggregates of relatively high specific gravity (Tisdall and Oades 1982, Tiessen and Stewart 1988). The radiocarbon age of the different size or weight fractions indicates their rate of turnover. Anderson and Paul (1984) reported a ^{14}C age of 1255 years for organic matter in the clay fraction of a soil for which the overall age was 795 years. The decline in soil organic matter in agricultural soils is largely the result of losses from the light fraction (Buyanovsky et al. 1994, Cambardella and Elliott 1994).

Losses of soil organic matter with cultivation contribute to the rise of atmospheric CO_2, but also to the hope that better management of agricultural soils might restore their native soil carbon and act to store CO_2 that would otherwise accumulate in the atmosphere. A reduced frequency of tillage results in the accumulation of soil carbon in many circumstances (West and Marland 2002). Successful management of agricultural soils may benefit from the preservation of their organic microaggregate structure by reduced tillage.

Greater returns of crop residues, stimulated by fertilizer and irrigation, can also result in accumulations of soil organic matter, although in most circumstances the enhanced soil carbon storage is less than the CO_2 emitted during the manufacture or delivery of these amendments (Schlesinger 2000, Russell et al. 2005, Khan et al. 2007, Townsend-Small and Czimczik 2010). Additions of biochar also increase soil organic matter in some circumstances (Woolf et al. 2010). When ancillary and offsite emissions are considered, most agricultural lands in the United States show negative NEP (West et al. 2010). Indeed, allowing vegetation to regrow on abandoned agricultural lands seems the only certain way to increase soil carbon (Vuichard et al. 2008, McLauchlan et al. 2006).

Significantly, soil organic matter can accumulate fairly rapidly when agricultural soils are abandoned, with rates averaging 33 g C m^{-2} yr^{-1} in a wide review of published values (Post and Kwon 2000, Guo and Gifford 2002, Clark and Johnson 2011)—much higher than the accumulation under undisturbed vegetation (see Figure 5.21). Soils in urban environments—lawns, parks, and golf courses—can also show increases in soil carbon storage with intensive management (Golubiewski 2006, Pouyat et al. 2009, Raciti et al. 2011). In all cases of natural succession, the accumulation of carbon in soil organic matter is dwarfed by the accumulation in regrowing woody vegetation (Richter et al. 1996, Johnson et al. 2003, Hooker and Compton 2003).

Growth at high CO_2 increases the productivity of vegetation, especially belowground, so it is natural to expect increasing storage of soil carbon as a result. Nevertheless, various high CO_2 experiments using FACE technology in forests report only small differences in soil carbon storage (Jastrow et al. 2005, Hagedorn et al. 2003, Lichter et al. 2008, Hungate et al. 2009).

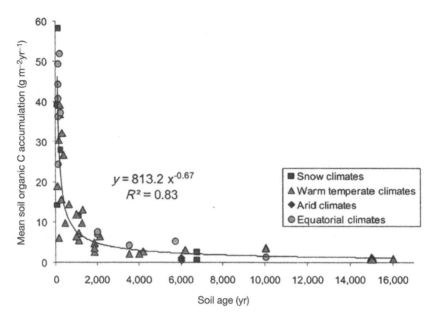

FIGURE 5.21   The rate of accumulation of organic matter in soil chronosequences of different age and climate zones, all derived from volcanic materials. *Source: From Zehetner (2010).*

net production of humus substances is $<0.4 \times 10^{15}$ g C/yr or only about 0.7% of NPP (Schlesinger 1990).

The mass of soil organic matter in most upland ecosystems is likely to have been fairly constant before widespread human disturbance of soils. When soils show a steady state in organic content, the production of humic compounds must be equal to their removal from soils by erosion. Coincidentally, estimates of the global transport of organic carbon in rivers are also about 0.4×10^{15} g C/yr (Schlesinger and Melack 1981, Meybeck 1982), suggesting that before human impacts, NEP for the Earth's land surface was essentially zero.

For areas covered by the last continental glaciation, the total accumulation of soil organic matter represents NEP for the last 10,000 years. The maximum extent of the last glaciations, covering 29.5×10^{6} km^2 of the present land area (Flint 1971), now contains roughly 300×10^{15} g C—that is, more than 10% of the organic carbon contained in all the world's soils (Table 5.4). In these areas, soil organic matter has accumulated at rates of about 1.35 g C m^{-2} yr^{-1} during the Holocene period. The current rate of storage in northern ecosystems (0.015 to 0.035×10^{15} g C/yr) is too small to be a significant sink for human releases of CO_2 to the atmosphere from fossil fuels, nor is it likely to have increased significantly during the last century (Gorham 1991, Harden et al. 1992).

Total storage of carbon in soils, 1456×10^{15} g or 121×10^{15} moles, can account for only 0.3% of the O_2 content of the atmosphere, given that the storage of organic carbon and the release of O_2 occur on a mole-for-mole basis during photosynthesis (Eq. 5.2). Thus, accumulations of atmospheric O_2 cannot be the result of the storage of organic carbon on land. Long-term storage of organic carbon appears to be dominated by accumulations in anoxic marine sediments (Chapter 9).

energy. Most of the remaining energy evaporates water and heats the air, resulting in the global circulation of the atmosphere (Chapters 3 and 10). Thus, the terrestrial biosphere is fueled by a relatively inefficient initial process.

During photosynthesis, plants take up moisture from the soil and lose it to the atmosphere in the process of transpiration. Available moisture appears to be a primary factor determining global variation in leaf area and NPP. Among communities with adequate soil moisture, net primary production is determined by the length of the growing season and mean annual temperature; both are an index of the receipt of solar energy. Soil nutrients appear to be of secondary importance to NPP on land, perhaps because plants have various adaptations for obtaining and recycling nutrients efficiently when they are in short supply (Chapter 6).

Most net primary production is delivered to the soil, where it is decomposed by a variety of organisms. The decomposition process is remarkably efficient, so only small amounts of NPP are added to the long-term storage of soil organic matter or humus each year. Soil organic matter consists of a dynamic pool near the surface in which there is rapid turnover of fresh plant detritus and little long-term accumulation, and a large refractory pool of humic substances that are dispersed throughout the soil profile. Thus, the turnover time of organic carbon in the soil ranges from about 3 years for the litter to thousands of years for humus. For the United States, the mean residence time of carbon in terrestrial ecosystems is about 46 years (Zhou and Luo 2008).

Humans have altered the processes of net primary production and decomposition on land, resulting in the transfer of organic carbon to the atmosphere, and perhaps a permanent reduction in the global rate of NPP. This disruption has produced global changes in the biogeochemical cycle of carbon, but little change in the atmospheric concentration of O_2.

Recommended Readings

Chapin, F.S., P.A. Matson, and P.M. Vitousek. 2012. *Principles of Terrestrial Ecosystem Ecology*. Springer.

Fahey, T.J., A.K. Knapp. 2007. *Principles and Standards for Measuring Primary Production*. Oxford University Press.

Reichle, D.E. (ed.). 1981. *Dynamic Properties of Forest Ecosystems*. Cambridge University Press.

Roy, J., B. Saugier, and H.A. Mooney. 2001. *Terrestrial Global Productivity*. Academic Press.

Swift, M.J., O.W. Heal, and J.M. Anderson. 1979. *Decomposition in Terrestrial Ecosystems*. University of California Press.

Waring, R.H., and S.W. Running. 2007. *Forest Ecosystems: Analysis at Multiple Scales* (third ed.). Elsevier.

PROBLEMS

1. Fick's law of diffusion states that the Flux=Gradient/Resistance. For a plant leaf, the concentration of CO_2 typically ranges from about 100 $\mu l/l$ inside the leaf to 400 $\mu l/l$ in the atmosphere. For water vapor, the gradient of concentration ranges from saturation inside the leaf to the vapor pressure of the free atmosphere, as determined by relative humidity and temperature outside the leaf. Assume stomatal resistance is 5 sec/cm. What is the flux of CO_2 into the leaf and water vapor out of the leaf at 25 °C and 30% relative humidity? (Remember to consider the different value of diffusivity for each of these gases in air). If the stomatal aperture decreases, so that the resistance increases to 10 sec/cm, what is the flux of these two gases? What do these calculations suggest as to why the water-use efficiency of vegetation is so low?

2. Assume that the shape of a tree trunk approximates that of a cone and that only that outermost ring of wood, with 2-mm thickness, is metabolically active (i.e., it respires at a rate of 0.005 μmol $CO_2/$ cm^3/sec) and functional in water transport. In a cross-section of the trunk at the base, if each cm^2 of these outer layers of wood support 0.50 m^2 of leaf area, and each unit of leaf area fixes carbon at a rate of 10 μmol $CO_2/m^2/sec$, what fraction of the photosynthate generated by a tree of 50-cm diameter will go to the support of wood tissue when it is 15-m tall? (Assume a 12-hour daylight priod, each day for a 365-day growing season.)

3. Physiologists often use the expression Q_{10} to express the change in a metabolic process that occurs with a 10°C rise in temperature. In many areas, the Q_{10} for soil respiration is about 2.4. What will be the annual global increase in soil respiration with a 3°C rise in global temperature during the next century? How does this compare to fossil fuel emissions?

4. There is a total of 10.4 kg C/m^2 in the grassland soil pictured in Figure 5.19. Calculate the turnover of the soil organic carbon with respect to each of the three components of soil respiration and with respect to total soil respiration. Then, calculate the overall turnover time with respect to total inputs to the soil. How do these compare? Is the overall turnover time controlled by the pools with rapid or slow turnover?

The Biosphere
Biogeochemical Cycling on Land

INTRODUCTION

Living tissue is primarily composed of carbon, hydrogen, and oxygen in the approximate proportion of CH$_2$O, but more than 25 other elements are necessary for biochemical reactions and for the growth of structural biomass. For example, phosphorus (P) is required for adenosine triphosphate (ATP), the universal molecule for energy transformations in organisms, and calcium (Ca) is a major structural component in both plants and animals. The enzymes and structural proteins found in plants and animals contain about 16% nitrogen (N) by weight. Earlier we saw that the enzyme ribulose bisphosphate carboxylase determines the rate of photosynthesis—carbon uptake—by many plant species (Chapter 5). The link between C and N that begins in cellular biochemistry extends to the global biogeochemical cycles of these elements.

The various elements essential to biochemical structure and function are often found in predictable proportions in living tissues (e.g., wood, leaf, bone, and muscle) (Reiners 1986, Sterner and Elser 2002). For instance, the ratio of C to N in leaf tissue ranges from 25 to 50 (i.e., 1–2% N). At the global level, our estimate of net primary production (NPP), 60×10^{15} g C/yr, implies that at least 1200×10^{12} g of nitrogen must be supplied to plants each year through biogeochemical cycling to achieve the level of NPP that we observe. As we shall see, the availability of some elements, such as N and P, is often limited, and the supply of these elements controls the rate of net primary production in many terrestrial ecosystems (Reich et al. 1997, LeBauer and Treseder 2008, Elser et al. 2007, Xia and Wan 2008).

Conversely, for elements that are typically available in greater quantities (e.g., Ca and S) the rate of net primary production often determines the rate of cycling in the ecosystem and losses to stream waters. In every case, the biosphere exerts a strong control on the geochemical behavior of the major elements of life. Much less biological control is seen in the cycling of elements such as sodium (Na) and chloride (Cl), which are less important constituents of biomass (Gorham et al. 1979).

In earlier chapters, we found that the atmosphere is the dominant source of C, N, and S in terrestrial ecosystems. Except on old, highly weathered soils, rock weathering is the major source for most of the remaining biochemical elements (e.g., Ca, Mg, K, Fe, and P). In any terrestrial ecosystem, the receipt of elements from the atmosphere and the lithosphere represents an input of new quantities of nutrients for plant growth.[1] However, as a result of internal cycling and retention of past inputs, plant growth is not solely dependent on new inputs to the system. In fact, the annual circulation of important elements such as N within an ecosystem is often 10 to $20\times$ greater than the amount received from outside the system (Table 6.1).[2] This large internal, or *intrasystem, cycle* is achieved by the long-term retention and recycling of elements derived from the atmosphere and the lithosphere. Important biochemical elements are accumulated in terrestrial ecosystems by biotic uptake, whereas nonessential elements pass through these systems under simple geochemical control (Vitousek and Reiners 1975).

In this chapter, we analyze the cycles of biochemical elements in terrestrial ecosystems. We begin by examining aspects of plant uptake, allocations during growth, and losses due to the death of plants and plant tissues. Then we see how elements such as N, P, and S in dead organic matter are transformed in the soil, leading to their release for plant uptake or for loss from the ecosystem. The yearly uptake, allocation, return, and release of nutrient elements in an ecosystem constitute the nutrient cycle. Throughout, we stress interactions between

---

[1] Unlike the well-known models developed for the Hubbard Brook Ecosystem (e.g., Likens and Bormann 1995), the nutrient budgets in this book consider rock weathering as an *external* source of nutrients that enter a terrestrial ecosystem each year (Gorham et al. 1979).

[2] Volk (1998) defines the recycling ratio as the ratio of the amount cycling in a system to the amount exiting it, which is 6 for N and 4 for P in terrestrial ecosystems (see Chapter 12).

TABLE 6.1 Percentage of the Annual Requirement of Nutrients for Plant Growth in the Northern Hardwoods Forest at Hubbard Brook, New Hampshire, Which Could Be Supplied by Various Sources of Available Nutrients

Process	N	P	K	Ca	Mg
Growth requirement (Kg ha^{-1} yr^{-1})	115.4	12.3	66.9	62.2	9.5
Percentage of the requirement that could be supplied by:					
Intersystem inputs					
Atmospheric	18	0	1	4	6
Rock weathering	0	1	11	34	37
Intrasystem transfers					
Reabsorptions	31	28	4	0	2
Detritus turnover (includes return in throughfall and stemflow)	69	67	87	85	87

Note: Calculated using Eqs. 6.2 and 6.3.

Source: Reabsorption data are from Ryan and Bormann (1982). Data for N, K, Ca, and Mg are from Likens and Bormann (1995) and for P from Yanai (1992).

carbon and other biochemical elements and examine how land plants have adapted to the widespread limitations of N and P in terrestrial ecosystems. We deduce changes in the sources of nutrients that determine plant growth during ecosystem development.

BIOGEOCHEMICAL CYCLING IN LAND PLANTS

Nutrient Uptake

It is easy to forget the essential, initial role played by plants in all of biochemistry. Plants obtain essential elements from the soil (e.g., N from NO_3^-) and incorporate them into biochemical molecules (e.g., amino acids) (Oaks 1994). Animals may eat plants, and each other, and synthesize new amino acids, but the building blocks of the amino acids in animal protein are those originally synthesized in plants. Only in isolated instances, for example, in animals at natural salt licks, do we find a direct transfer of elements from inorganic form to animal biochemistry—geophagy (Jones and Hanson 1985). There are few vitamin pills in the natural biosphere!

Soil chemical characteristics, including mineralogy and ion exchange, set the initial constraints on the availability of essential elements for plant uptake. However, when plant uptake of an element is rapid, plants can release organic compounds that enhance the solubility of elements, such as P, from soil minerals (Chapter 4). Thus, plants can affect the availability of nutrients needed for their own growth and adapt to a wide range of soil fertility (Forde and Lorenzo 2001). Although foliar uptake is known, the vast majority of plant nutrient uptake passes through roots. A few unusual, insectivorous plants obtain N and P by digesting captured organisms (Adamec 1997, Wakefield et al. 2005). For instance, Dixon et al. (1980) found that 11 to 17% of the annual uptake of N in *Drosera erythrorhiza* (sundew) can be obtained from captured insects.

Delivery of ions to plant roots can occur by several pathways (Barber 1962). The concentration of some elements in the soil solution is such that their passive uptake with water is adequate for plant nutrition (Turner 1982). In some cases, the delivery is excessive, and the ions must be actively excluded at the root surface. For example, it is not unusual to see accumulations of Ca, as $CaCO_3$, surrounding the roots of desert shrubs growing in calcareous soils (Klappa 1980, Wullstein and Pratt 1981). In contrast, for N, P, and K, the concentration in the soil solution is often much too low for adequate delivery in the transpiration stream, and plant uptake is enhanced by enzymes—transporters—that carry ions through channels in the root membrane using active transport (Hirsch et al. 1998, Khademi et al. 2004, Williams and Miller 2001). Indeed, in a process known as signal transduction, NO_3^- appears to activate the enzymes that promote its own uptake (Zhang and Forde 1998, Tischner 2000). These ion transporters account for the large portion of root respiration that is associated with nutrient uptake (Chapter 5).

The transporter systems embedded in root membranes achieve increasing rates of nutrient uptake as a function of increasing concentrations in the soil solution until the activity of the enzyme system is saturated. Chapin and Oechel (1983) found that populations of the arctic sedge, *Carex aquatilis*, from colder habitats have higher rates of uptake than those from warmer habitats, presumably reflecting adaptation to the lower availability of phosphorus in cold soils (Figure 6.1). In comparative studies, root physiologists define the specific absorption rate (SAR) as the rate of uptake of a nutrient from the soil per unit of root mass over a specified period of time.

Normally, the uptake of N and P is so rapid and the concentrations in the soil solution are so low that these elements are effectively absent in the soil solution surrounding roots,

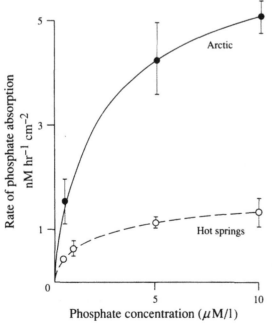

FIGURE 6.1 Rate of phosphate absorption per unit of root surface area in populations of *Carex aquatilis* from cold (arctic) and warm (hot springs) habitats, measured at 5°C. *Source: From Chapin (1974). Used with permission of the Ecological Society of America.*

and the rate of uptake is determined by diffusion to the root from other areas (Nye 1977). Phosphate is particularly immobile in most soils, and the rate of diffusion strongly limits P supply to plant roots (Robinson 1986). Although adaptations for more efficient root enzymes are seen in some species (Pennell et al. 1990), the most apparent response of plants to low nutrient concentrations is an increase in the root/shoot ratio, which increases the volume of soil exploited and decreases diffusion distances (Aerts and Chapin 2000, Clarkson and Hanson 1980, Robinson 1994). In many species, the relative growth rate of roots determines the uptake of nitrogen and phosphorus (Newman and Andrews 1973; Figure 6.2). Enhanced root growth is found in low phosphorus soils (Bates and Lynch 1996, Ma et al. 2001), and roots rapidly proliferate in nutrient-rich patches (Jackson et al. 1990, Black et al. 1994).

Plants and soil microbes exude enzymes into the soil that can release inorganic phosphorus from organic matter. These extracellular enzymes are known as *phosphatases*, which have different forms in acid and alkaline soils (Malcolm 1983, Tarafdar and Claassen 1988, Dinkelaker and Marschner 1992, Duff et al. 1994). In many cases, root phosphatase activity is inversely proportional to available soil P (Fox and Comerford 1992a, Treseder and Vitousek 2001). For example, phosphatase activity rises with the accumulation of organic matter in soils during the development of *Eucalyptus* plantations after fire (Polglase et al. 1992). Phosphatase activity associated with root surfaces is particularly significant to plants in phosphorus-poor habitats, and it may supply up to 75% of the annual phosphorus

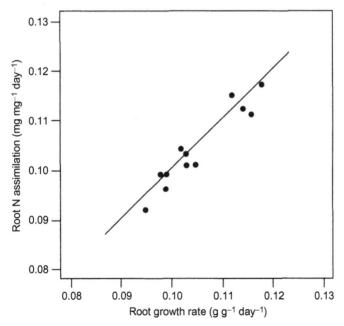

FIGURE 6.2 The rate of N uptake in tobacco as a function of the relative growth rate of roots. *Source: From Raper et al. (1978). Used with permission of the University of Chicago Press.*

I. PROCESSES AND REACTIONS

demand of some tundra and boreal forest species (Kroehler and Linkins 1991, Firsching and Claassen, 1996).

Nutrient Balance

In addition to an adequate supply of nutrient elements, plant growth is affected by the balance of nutrients in the soil (Shear et al. 1946). For seedlings of several tree species, Ingestad (1979b) found that a solution containing 100 parts N, 15 parts P, 50 parts K, 5 parts Ca and Mg, and 10 parts S was ideal for maximum growth. In a compilation of data from nearly 10,000 species, leaf N and P contents were related by a 2/3-power-law function, with a mean N/P ratio of 10.9 (by mass; Reich et al. 2010; compare Kerkhoff et al. 2005). Despite wide variations in nutrient availability in the environment, most plants show an N:P ratio of about 14 to 15 (by mass) in leaf tissues (Gusewell 2004, McGroddy et al. 2004, Han 2005, Koerselman and Meuleman 1996), with N deficiency at lower and P deficiency at higher values. However, unless the supply of a nutrient reaches very low levels, plants usually do not show deficiency symptoms; they simply grow more slowly (Clarkson and Hanson 1980). Inherent slow growth is a characteristic of plants adapted to infertile habitats, and it often persists even when nutrients are added experimentally (Chapin et al. 1986a).

Because more nutrients occur as positively charged ions than as negatively charged ions in the soil solution, one might expect that plant roots would develop a charge imbalance as a result of nutrient uptake. When ions such as K^+ are removed from the soil solution in excess of the uptake of negatively charged ions, the plant releases H^+ to maintain an internal balance of charge (Maathuis and Sanders 1994). This H^+ may, in turn, replace K^+ on a cation exchange site, driving another K^+ into the soil solution. The high concentration of N in plant tissues causes the form in which N is taken up to dominate this process (Table 6.2). Oaks (1992) has shown how plants that use NH_4^+ as an N source tend to acidify the immediate zone around their roots (Figure 6.3). The uptake of NO_3^- has the opposite effect as a result of plant releases of HCO_3^- and organic anions to balance the negative charge (Nye 1981, Hedley et al. 1982a, Schöttelndreier and Falkengren-Grerup 1999).

TABLE 6.2   Chemical Composition and Ionic Balance for Perennial Ryegrass

	N	P	S	Cl	K	Na	Mg	Ca
Percent in leaf tissue	4.00	0.40	0.30	0.20	2.50	0.20	0.25	1.00
Equivalent weight (g)	14.00	30.98	16.03	35.46	39.10	22.99	12.16	20.04
mEq present	285.7	12.9	18.7	5.6	63.9	8.8	20.6	49.9
Sum of mEq	±285.7		−37.2			+143.1		

Imbalance in mEq %
  (a) where ammonium nitrogen is taken up: 285.7 + 143.1 − 37.2 = +391.6
  (b) where nitrate nitrogen is taken up: 143.1 − 285.7 − 37.2 = −179.8

Source: From Middleton and Smith (1979). Used with permission of Springer.

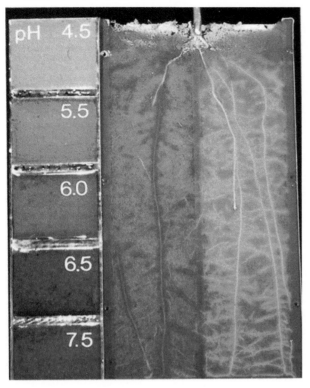

FIGURE 6.3   The pH of the soil in plants fertilized with nitrate (left) and ammonium (right), shown with a dye that changes color as a function of acidity. *Source: From Oaks (1994). Used with permission of NRC Research Press.*

Nitrogen Assimilation

Among various habitats, the availability of soil nitrogen as NH_4^+ or NO_3^- differs largely depending on the environmental conditions that affect the conversion of NH_4^+ to NO_3^- in the microbial process known as nitrification (Eqs. 2.17 and 2.18). For example, in waterlogged soils, almost all nitrogen is found as NH_4^+ (Barsdate and Alexander 1975), whereas in some deserts and forests, nearly all mineralized NH_4 is converted to NO_3^- (Virginia and Jarrell 1983, Nadelhoffer et al. 1984). Many species show a preference for NO_3^-, although species occurring in sites where nitrification is slow or inhibited often tend to show superior growth with ammonium (Haynes and Goh 1978, Adams and Attiwill 1982, Falkengren-Grerup 1995, Kronzucker et al. 1997, Wang and Macko 2011).

Derived from the breakdown of proteins, amino acids are found in many soils (Yu et al. 2002, Hofmockel et al. 2010) and used as a source of N by plants in a wide range of habitats, including tundra (Kielland 1994, Schimel and Chapin 1996, Nordin et al. 2004), boreal and temperate forest (Nasholm et al. 1998, Finzi and Berthrong 2005), and desert ecosystems (Jin and Evans 2010). Direct uptake of amino acids has been demonstrated using isotopically labeled amino acids (Nasholm et al. 1998) and nanoscale labels known as "quantum dots," which are attached to amino acids (Whiteside et al. 2009). Generally, the uptake of amino acids is greatest when the availability of inorganic N is low (Finzi and Berthrong 2005). In a British grassland, most species showed preferential uptake of inorganic N over amino acids (Harrison et al. 2007).

Once inside the plant, NO_3^- and NH_4^+ are converted to amino groups $(-NH_2)$ that are attached to soluble organic compounds. In many woody species these conversions occur in the roots, and N is transported to the shoot as amides, amino acids, and ureide compounds in the xylem stream (Andrews 1986, Tischner 2000). However, in some species, N in the xylem is found as NO_3^-, and the reduction of NO_3^- to $-NH_2$ occurs in leaf tissues (Smirnoff et al. 1984). Eventually, most plant N is incorporated into protein, and the amino acid arginine is used for storage of excess N (Llácer et al. 2008).

The conversion of NO_3^- to $-NH_2$ is a biochemical reduction reaction that requires metabolic energy and is catalyzed by an enzyme, *nitrate reductase*. One might puzzle why most plants do not show a clear preference for NH_4^+, which is assimilated more easily. Several explanations have been offered. Recall that NH_4^+ interacts with soil cation exchange sites, whereas NO_3^- is extremely mobile in most soils. The rate of delivery of NO_3^- to the root by diffusion or mass flow is much higher than that of NH_4^+ under otherwise equivalent conditions (Raven et al. 1992). Plants that utilize NH_4^+ may have to compensate for the differences in diffusion by investing more energy in root growth (Gijsman 1990, Oaks 1992, Bloom et al. 1993).

Uptake of NO_3^- also avoids the competition that occurs in root enzyme carriers between NH_4^+ and other positively charged nutrient ions. For example, the presence of large amounts of K^+ in the soil solution can reduce the uptake of NH_4^+ (Haynes and Goh 1978). Finally, relatively low concentrations of NH_4^+ are potentially toxic to plant tissues. These potential disadvantages in the uptake of NH_4^+ may explain why many plants take up NO_3^- when thermodynamic calculations suggest that the metabolic costs of reducing NO_3^- are greater than for plants that assimilate NH_4^+ or amino acids directly (Middleton and Smith 1979, Gutschick 1981, Bloom et al. 1992, Zerihun et al. 1998).

It is unclear why so many woody species concentrate nitrate reductase in their roots, when the same reaction performed in leaf tissues, where it can be coupled to the photosynthetic reaction, is energetically much less costly (Gutschick 1981, Andrews 1986). Addition of NO_3^- to the soil often induces the production of root enzymes for NO_3^- uptake and the synthesis of more nitrate reductase in plant tissues (Lee and Stewart 1978, Hoff et al. 1992, Oaks 1994, Tischner 2000). There is some evidence that the proportion of nitrate reductase in the shoot increases at high levels of available NO_3^- (Andrews 1986). Both photosynthetic rates and nitrate uptake increase when plants are grown at high CO_2, but the response is not universal (Bassirirad 2000).

Nitrogen Fixation

Several types of bacteria possess the enzyme *nitrogenase*, which converts atmospheric N_2 to NH_3 in local conditions of cellular anoxia (see Eq. 2.10). Some of these exist as free-living (asymbiotic) forms in soils, but others, such as *Rhizobium* and *Frankia*, form symbiotic associations with the roots of higher plants. The symbiotic bacteria reside in root nodules that are easily recognized in the field. Nitrogen fixation is especially well known among species of legumes (Leguminosae) (Bryan et al. 1996).

Nitrogen that enters terrestrial ecosystems by fixation is a "new" input in the sense that it is derived from outside the boundaries of the ecosystem—that is, from the atmosphere. The reduction of N_2 to NH_3 has large metabolic costs that require the respiration of organic carbon.

Nevertheless, Gutschick (1981) suggests that symbiotic fixation in higher plants is not greatly less efficient than the uptake of NO_3^- for those species in which the nitrate reductase is located in plant roots. Only a few land plants support symbiotic nitrogen fixation, and it is interesting to speculate why nitrogen fixation is not more widespread, when nitrogen limitations of net primary production are so frequent (Vitousek and Howarth 1991, Crews 1999). Globally, plants "spend" only about 2.5% of NPP on nitrogen fixation (Gutschick 1981).

The energy cost of nitrogen fixation links this biogeochemical process to the availability of organic carbon, provided by net primary production. In plants with symbiotic nitrogen fixation, the rate of N fixation is often directly related to the rate of photosynthesis and the efficiency of plant growth (Bormann and Gordon 1984). N fixation is stimulated in seedlings of various species grown at high CO_2 (Tissue et al. 1997, Millett et al. 2012), but it is unclear if this initial effect is persistent in long-term field experiments (Hungate et al. 2004). Free-living heterotrophic bacteria that conduct asymbiotic nitrogen fixation are usually found in organic soils or local areas with high levels of organic matter that provide a ready source of energy (Granhall 1981, Billings et al. 2003). Organic-rich environments also foster the development of cellular anaerobiosis required by the nitrogenase enzyme (Marchal and Vanderleyden 2000). For instance, nitrogen fixation is frequently observed in rotten logs (Roskoski 1980, Silvester et al. 1982, Griffiths et al. 1993), where it is probably associated with anaerobic cellulolytic bacteria (Leschine et al. 1988). N-fixing symbioses are found in a wide variety of local microenvironments where other organisms provide abundant organic matter, such as the root zone of desert grasses (Herman et al. 1993), the hind-gut of termites (Breznak et al. 1973, Yamada et al. 2006, Hongoh et al. 2008), the interior of pineapples (Tapia-Hernandez 2000), feathermoss carpets in the boreal forest (DeLuca et al. 2002), and the fungus gardens of leaf-cutter ants in tropical rainforests (Pinto-Tomas 2009). Studies of nitrogen fixation in these habitats are often aided by the identification of the genes coding for nitrogenase (nifH) using molecular techniques (Widmer et al. 1999, Reed et al. 2010).

In both symbiotic and asymbiotic forms, nitrogen fixation is generally inhibited at high levels of available nitrogen (Cejudo et al. 1984). In many cases, the rate of fixation appears to be controlled by the N:P ratio in the soil (Chapin et al. 1991, Smith 1992a), and added phosphorus stimulates asymbiotic N fixation (Figure 6.4). In bacteria, phosphorus appears to activate the gene for the synthesis of nitrogenase (Stock et al. 1990), illustrating how the linkage between the global cycles of nitrogen and phosphorus has a basis in molecular biology. Requirements for Mo and Fe as structural components of nitrogenase also link nitrogen fixation to the availability of these elements in natural ecosystems (Kim and Rees 1994, O'Hara et al. 1988). Low availability of Mo may limit asymbiotic N fixation in many forests (Silvester 1989, Barron et al. 2009). Some plants with symbiotic N-fixing bacteria appear to acidify their rooting zone to make Fe and P more available (Ae et al. 1990, Raven et al. 1990, Gillespie and Pope 1990), and legumes appear to have several mechanisms to aid the uptake and retention of P in phosphorus-poor soils (He et al. 2011, Venterink 2011).

The isotopic ratio of N in plant tissues is expressed as $\delta^{15}N$, using a calculation analogous to what we saw for the isotopes of carbon in Chapter 5 (Robinson 2001). In the case of nitrogen, the standard is the atmosphere, which contains 99.63% ^{14}N and 0.37% ^{15}N. Nitrogenase shows only a slight discrimination between the isotopes of N, that is, between $^{15}N_2$ and $^{14}N_2$ (Handley and Raven 1992, Hogberg 1997), so differences in the isotopic ratio of nitrogen among plant species growing in the same soil can be used to suggest which species may be

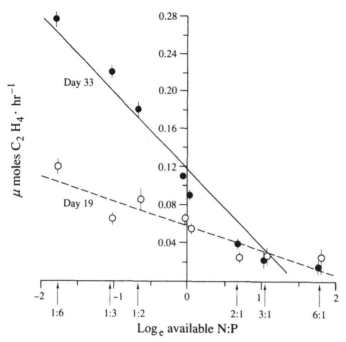

FIGURE 6.4 Acetylene reduction as an index of nitrogen fixation in asymbiotic N-fixing bacteria as a function of the ratio of N to P in the soil. *Source: From Eisele et al. (1989). Used with permission of Springer.*

involved in nitrogen fixation (Virginia and Delwiche 1982, Yoneyama et al. 1993). Nitrogen-fixing species typically show values of $\delta^{15}N$ that are slightly negative or close to the atmospheric ratio ($\delta^{15}N = 0$), whereas nonfixing species show a wide range of values (usually positive) depending on various N transformations in the soil (Garten and Van Miegroet 1994, Hogberg 1997; Figure 6.5) (see also pp. 202–203).

Shearer et al. (1983) used the difference in isotopic ratio between *Prosopis* grown in the laboratory without added N (i.e., all nitrogen was derived from fixation) and the same species in the field to estimate that the field plants derived 43 to 61% of their nitrogen from fixation. Of course, when nitrogen-fixing plants die, their nitrogen content is available for other species in the ecosystem (Huss-Danell 1986, van Kessel et al. 1994). Lajtha and Schlesinger (1986) found that the desert shrub *Larrea tridentata*, growing adjacent to nitrogen-fixing *Prosopis*, had lower $\delta^{15}N$ than when *Larrea* was growing alone.

Nitrogenase activity can be measured using the acetylene-reduction technique, which is based on the observation that this enzyme also converts acetylene to ethylene under experimental conditions. Plants or nodules are placed in small chambers or small chambers are placed over field plots, and the conversion of injected acetylene to ethylene over a known time period is measured using gas chromatography. The conversion of acetylene (in moles) is not exactly equivalent to the potential rate of fixation of N_2 because the enzyme has different affinities for these substrates. However, appropriate conversion ratios can be determined using other techniques (Schwintzer and Tjepkema 1994, Liengen 1999). For instance, investigators have added $^{15}N_2$, the heavy stable isotope of N, to closed chambers and used the increase in organic compounds containing ^{15}N in test plants or soil as a measure of nitrogen fixation (e.g., Silvester et al. 1982, Zechmeister-Boltenstern and Kinzel 1990).

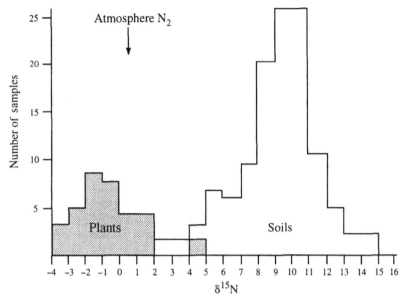

FIGURE 6.5   Frequency distribution of $\delta^{15}N$ in the tissues of 34 nitrogen-fixing plants and in the organic matter of 124 soils from throughout the United States. *Source: Plotted using data from Shearer and Kohl (1988, 1989).*

Heterotrophic N-fixing bacteria and cyanobacteria (blue-green algae) are widespread, and their nitrogen fixation can be an important source of N for some terrestrial ecosystems (Reed et al. 2011). Exceptionally high rates of fixation have been recorded in cyanobacterial crusts that cover the soil surface in some desert ecosystems (Rychert et al. 1978), but in most cases, the total input from these sources of asymbiotic N fixation is in the range of 1 to 5 kg N ha^{-1} yr^{-1} (Boring et al. 1988, Cushon and Feller 1989, Son 2001, Cleveland et al. 2010). This input is usually less than the annual deposition of nitrogen in wetfall and dryfall from the atmosphere (Schwintzer and Tjepkema 1994).

The importance of symbiotic nitrogen fixation in terrestrial ecosystems varies widely depending on the presence of species that harbor symbiotic bacteria (Reed et al. 2011). Grazed pastures with clover routinely show N fixation at rates of 100 to 200 kg N ha^{-1} yr^{-1} (Bolan et al. 2004). In natural ecosystems, some of the greatest rates of N fixation are seen in species that invade after disturbance, where high light levels allow maximum photosynthesis (Vitousek and Howarth 1991). For example, in the recovery of Douglas fir forests after fire, Youngberg and Wollum (1976) found that the nodulated shrub *Ceanothus velutinus* contributed up to 100 kg N ha^{-1} yr^{-1} on some sites. Invasion of the exotic, nitrogen-fixing tree *Myrica fay*, provides important inputs of nitrogen (18 kg ha^{-1} yr^{-1}) on fresh volcanic ashflows in Hawaii (Vitousek et al. 1987). In most cases the importance of plants with symbiotic nitrogen fixation declines with the recovery of mature vegetation, and their occurrence in undisturbed communities is limited. In this regard, the widespread occurrence of leguminous species in tropical forests is deserving of further study (Kreibich et al. 2006, Vitousek et al. 2002). The sporadic occurrence of symbiotic nitrogen fixation in terrestrial ecosystems makes it difficult to extrapolate from local studies to provide a global estimate of its importance.

The global rate of N fixation (asymbiotic + symbiotic) in natural ecosystems may supply 60 to 100×10^{12} g N/yr, or about 10% of the annual plant demand for nitrogen on land (Chapter 12).

Mycorrhizal Fungi

Symbiotic associations between fungi and higher plants are found in most ecosystems (Allen 1992). This symbiosis is important for the nutrition of plants and may have even determined the origin of land plants (Simon et al. 1993, Courty et al. 2010). There are several forms of symbiosis. In temperate regions, many trees harbor ectotrophic mycorrhizal fungi. These fungi form a sheath around the active fine roots and extend additional hyphae into the surrounding soil. In many areas, especially the tropics, plants possess endotrophic mycorrhizal fungi in which the hyphae actually penetrate cells of the root.

By virtue of their large surface area and efficient absorption capacity, mycorrhizal fungi are able to obtain soil nutrients and transfer these to the higher plant root. Recent work has elucidated the genetic development of the symbiosis and the molecular structure of a transporter protein in mycorrhizae that moves phosphorus into the roots (Harrison and van Buuren 1995, Bucher 2007). Mycorrhizal fungi are directly involved in the decomposition of soil organic materials through the release of extracellular enzymes such as cellulases and phosphatases (Antibus et al. 1981, Dodd et al. 1987, Hodge et al. 2001) and in the weathering of soil minerals through the release of organic acids (Bolan et al. 1984, Illmer et al. 1995, Van Breemen et al. 2000, van Scholl et al. 2008, Blum et al. 2002; see also Chapter 4). It is important to remember that most of these reactions are also associated with plant roots; mycorrhizae simply enhance their occurrence in the rhizosphere, increasing the overall rate of plant nutrient uptake (Bolan 1991). In return, mycorrhizal fungi depend on the host plant for supplies of carbohydrate.

The importance of mycorrhizae in infertile sites is well known. Treseder and Vitousek (2001) show greater mycorrhizal colonization, greater extracellular phosphatase, and greater phosphorus uptake by plants on P-deficient soils in Hawaii. Many species of pine require ectotrophic mycorrhizae, which perhaps accounts for the success of pines in nutrient-poor soils. Most tropical trees appear to require endotrophic mycorrhizal associations for proper growth (Janos 1980), and mycorrhizal fungi are widespread among the *Eucalyptus* species growing in the low-phosphorus soils of Australia. Berliner et al. (1986) report the complete exclusion of *Cistus incanus* from basaltic soils in Israel due to a failure of mycorrhizal development. The same species grows well on adjacent calcareous soils or on basaltic soils supplied with fertilizer.

Mycorrhizal fungi are especially important in the transfer of those soil nutrients with low diffusion rates in the soil. A large number of studies document the importance of mycorrhizae in P nutrition (Koide 1991), but absorption of N and other nutrients is also known (Bowen and Smith 1981, Ames et al. 1983, Govindarajulu et al. 2005). In forests, mycorrhizae mediate the uptake of amino acids by trees. Mycorrhizae appear responsible for 61 to 86% of N uptake in Arctic tundra (Hobbie and Hobbie 2006).

Some plants with mycorrhizal fungi show higher levels of various nutrients in foliage, but frequently the enhanced uptake of nutrients results in higher rates of growth (Schultz et al. 1979). Rose and Youngberg (1981) provide an insightful experiment with *Ceanothus velutinus* growing in nitrogen-deficient soils with and without mycorrhizae and symbiotic nitrogen-fixing bacteria (Table 6.3). The highest rates of growth were seen when both of these symbiotic associations were present, which also allowed a decrease in the root/shoot ratio. Nitrogen fixation enhanced the uptake of phosphorus by mycorrhizal fungi. These results illustrate the strong interactions between N, P, and C in the nutrition of higher plants.

Under conditions of nutrient deficiency, plant growth usually slows, whereas photosynthesis continues at relatively high rates (Chapin 1980) and the content of soluble carbohydrate in the plant increases. Marx et al. (1977) found that high concentrations of carbohydrate in root tissues of loblolly pine (*Pinus taeda*) stimulated mycorrhizal infection (Figure 6.6). Plants grown at high CO_2 also appear to allocate additional carbohydrate to fine roots to support mycorrhizal development (DeLucia et al. 1997, Pritchard et al. 2008a, Phillips et al. 2011a). Thus, internal plant allocation of carbohydrates to roots may result in increased nutrient uptake by mycorrhizae and an alleviation of nutrient deficiencies (Bucking and Shachar-Hill 2005, Ryan et al. 2012).

Mycorrhizal fungi use a fraction of the carbon fixed by the host plant, representing a drain on net primary production that might otherwise be allocated to growth (Rygiewicz and Andersen 1994). That the cost of symbiotic fungi is significant is underscored by experiments in which the degree of colonization declined and plant growth increased when plants were fertilized (e.g., Blaise and Garbaye 1983). Across a wide variety of species, the carbon allocation to mycorrhizae appears to lower net primary productivity by about 15% (Hobbie 2006; see also Table 5.1).

TABLE 6.3   Effects of Mycorrhizae and N-Fixing Nodules on Growth and Nitrogen Fixation in *Ceanothus velutinus* Seedlings

	Control	+ Mycorrhizae	+ Nodules	+ Mycorrhizae and nodules
Mean shoot dry weight (mg)	72.8	84.4	392.9	1028.8
Mean root dry weight (mg)	166.4	183.4	285.0	904.4
Root/shoot	2.29	2.17	0.73	0.88
Nodules per plant	0	0	3	5
Mean nodule weight (mg)	0	0	10.5	44.6
Acetylene reduction (mg/nodule/hr)	0	0	27.85	40.46
Percent mycorrhizal colonization	0	45	0	80
Nutrient concentration (in shoot, %)				
N	0.32	0.30	1.24	1.31
P	0.08	0.07	0.25	0.25
Ca			1.07	1.15

Source: From Rose and Youngberg (1981). Used with permission of NRC Research Press.

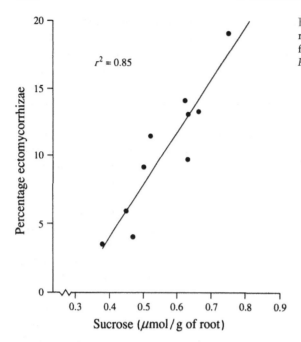

FIGURE 6.6   Relationship between infection of the roots of loblolly pine (*Pinus taeda*) by ectomycorrhizal fungi and the sucrose concentration in the root. *Source: From Marx et al. (1977).*

NUTRIENT ALLOCATIONS AND CYCLING IN LAND VEGETATION

The Annual Intrasystem Cycle

The plant uptake of nutrients from the soil is allocated to the growth of new tissues. Although short-lived tissues (leaves and fine roots) compose a small fraction of total plant biomass, they receive the largest proportion of the annual nutrient uptake (Pregitzer et al. 2010). Growth of leaves and roots received 87% of the N and 79% of the P allocated to new tissues in a deciduous forest in England (Cole and Rapp 1981, p. 404). In a perennial grassland dominated by *Bouteloua gracilis* in Colorado, new growth of aboveground tissues accounted for 67% of the annual uptake of N (Woodmansee et al. 1978).

When leaf buds break and new foliage begins to grow, the leaf tissues often have high concentrations of N, P, and K. As the foliage matures, these concentrations often decrease (van den Driessche 1974). Some of these changes are due to the increasing accumulation of photosynthetic products with time and to leaf thickening during development. Leaf mass per unit area (mg/cm^2) may increase as much as 50% during the growing season and then decline as the leaf senesces (Smith et al. 1981). The initial concentrations of N and P are diluted as the leaf tissues accumulate carbohydrates and cellulose. In contrast, the concentrations of some nutrients (e.g., Ca, Mg, and Fe) often increase with leaf age (van den Driessche 1974). Increases in calcium concentration with leaf age result from secondary thickening, including calcium pectate deposition in cell walls, and from increasing storage of calcium oxalate in cell vacuoles.

Although there are variations among species, nutrient concentrations in mature foliage are related to the rate of photosynthesis (Chapter 5) and plant growth (e.g., Tilton 1978), and analysis of foliage is often a good index of site fertility (van den Driessche 1974, Ordonez et al. 2009). Leaf concentrations of trace metals often reflect the content of the underlying soil, such that leaf tissues are useful for mineral prospecting in some areas (Cannon 1960, Brooks 1973). Among tropical forests, concentrations of major nutrients in leaves are significantly higher on more fertile soils (Vitousek and Sanford 1986). Yin (1993) found that concentration of N and P in the foliage of deciduous trees varied systematically with higher values among species in colder habitats than in the tropics. The higher leaf nutrient contents in colder climates may allow for higher photosynthetic rates and rapid growth of these species in response to a short growing season (Mooney and Billings 1961, Körner 1989, Reich and Oleksyn 2004).

After fertilization with a specific nutrient, the concentrations of other leaf nutrients can show surprising changes. For example, leaf N increased when Miller et al. (1976) fertilized Corsican pine (*Pinus nigra*) with N, but in the same samples, concentrations of P, Ca, and Mg declined. When nitrogen fertilization of N-deficient stands stimulates photosynthesis, the concentrations of other nutrients in foliage may be diluted by new accumulations of carbohydrate (Fowells and Krauss 1959, Timmer and Stone 1978, Jarrell and Beverly 1981). In some cases, uptake of P from the soil may fall behind the rates needed for maximum growth at the newly established levels of N availability. In other cases, improvements in plant nitrogen status enhance the uptake of other elements as well (e.g., Table 6.3). Plant response to multiple element fertilizations are illustrative of the importance of considering balanced nutrition for maximum plant growth.

Once leaves are fully expanded, changes in the nutrient content per unit of leaf area indicate movements of nutrients between the foliage and the stem. Woodwell (1974) found that oak leaves rapidly accumulated N during the early summer, presumably as a component of photosynthetic enzymes. The leaf content of N, P, and K remained relatively constant at high levels throughout the growing season, but concentrations of all three elements declined rapidly in autumn prior to leaf abscission. Such losses often represent active withdrawal of nutrients from foliage for reuse during the next year. Some trace micronutrients are also withdrawn before leaf-fall (Killingbeck 1985), but usually reabsorption of foliar Ca and Mg is limited. Fife and Nambiar (1984) observed that reabsorption of N, P, and K was not just related to leaf senescence in *Pinus radiata*; these nutrients could also move from tissues produced early and later during the same growing season.

Leaf nutrient contents are also affected by rainfall that leaches nutrients from the leaf surface (Tukey 1970, Parker 1983). In particular, seasonal changes in the content of K, which is highly soluble and especially concentrated in cells near the leaf surface, may represent leaching. The losses of nutrients in leaching often follow the order

$$K \gg P > N > Ca. \qquad (6.1)$$

Leaching rates generally increase as foliage senesces before abscission; thus, care must be taken to recognize changes due to leaching versus changes due to active nutrient withdrawals (Ostman and Weaver 1982).

Nutrient losses by leaching differ among leaf types. Luxmoore et al. (1981) calculated lower rates of leaching loss from pines than from broad-leaf deciduous species in a forest in

Tennessee. Such differences may be the result of variation in leaf nutrient concentration, surface-area-to-volume ratio, surface texture, and leaf age. Among the trees of the humid tropics, the smooth surface of broad sclerophylls may be an adaptive response to reducing leaching by minimizing the length of time that rainwater is in contact with the leaf surface (Dean and Smith 1978). Species-specific differences in rates of leaching from potential host trees may explain differences in their epiphyte loads (Benzing and Renfrow 1974, Awasthi et al. 1995), with many epiphytes showing P deficiency (Wanek and Zotz 2011).

Rainwater that passes through a vegetation canopy is called *throughfall*, which is usually collected in funnels or troughs placed on the ground. Throughfall contains nutrients leached from leaf surfaces and is most important in the cycling of nutrients such as K (Parker 1983, Schaefer and Reiners 1989). In forests, rainwater that travels down the surface of stems is called *stemflow* (Levia and Frost 2003). The concentrations of nutrients in stemflow waters are high, but usually much more water reaches the ground as throughfall. Stemflow is significant to the extent that it returns highly concentrated nutrient solutions to the soil at the base of plants (Gersper and Holowaychuk 1971).

Leaching varies seasonally depending on forest type and climate. Not surprisingly, in temperate deciduous forests, the greatest losses are during the summer months (Lindberg et al. 1986). Some of the nutrient content in throughfall is derived from aerosols that are deposited on leaf surfaces (Chapter 3). Indeed, Lindberg and Garten (1988) found that about 85% of the apparent loss of sulfate from a forest canopy was due to dry deposition on leaf surfaces, and some studies have used the SO_4^{2-} content of throughfall to estimate dry deposition in the canopy (Garten et al. 1988, Ivens et al. 1990, Neary and Gizyn 1994).

For most elements, however, leaching of nutrients from vegetation makes it difficult to use nutrient concentrations in the rainfall collected under a canopy to calculate dry deposition on leaf surfaces (Chapter 3). In some cases, leaves appear to take up nutrients from rainfall, particularly soluble forms of N (Carlisle et al. 1966, Miller et al. 1976, Garten and Hanson 1990, Lovett and Lindberg 1993). Various reactive nitrogen gases (e.g., NH_3, NO_x, and peroxyacetyl nitrate) are also absorbed at the leaf surface (Gessler et al. 2000, Sparks et al. 2003).

Litterfall

When the biomass of vegetation is not changing, the annual production of new tissues is balanced by the senescence and loss of plant parts (Chapter 5). In the intrasystem cycle, plant litterfall is the dominant pathway for nutrient return to the soil, especially for N and P (Figure 6.7). Below ground, root death also makes a major contribution of nutrients to the soil each year (Cox et al. 1978, Vogt et al. 1983, Burke and Raynal 1994).

The nutrient concentrations in litterfall differ from the nutrient concentrations in mature foliage by the reabsorption of constituents during leaf senescence (Killingbeck 1996). In the tundra shrub *Eriophorum vaginatum*, Chapin et al. (1986b) found that all organic N and P compounds decreased to a similar extent during leaf senescence, suggesting that reabsorption is not simply limited to certain biochemical compounds that are particularly susceptible to hydrolysis. Nutrient reabsorption is also known during the senescence of fine roots (Freschet et al. 2010) and during the aging of sapwood to heartwood in trees (Laclau et al. 2001).

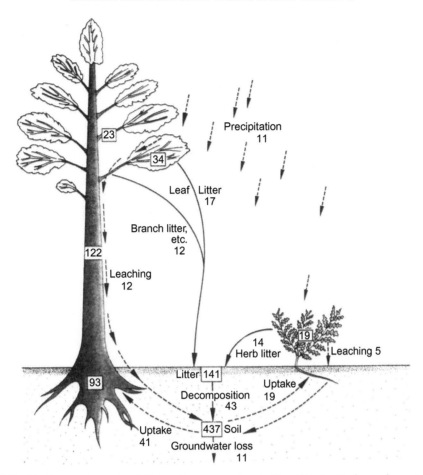

FIGURE 6.7   The intrasystem cycle of Ca in a forest ecosystem in Great Britain. Pools are shown in kg/ha and annual flux in kg ha^{-1} yr^{-1}. *Source: From Whittaker, R.H., Communities and Ecosystems (1970, p. 110). Reprinted by permission of Prentice Hall, Upper Saddle River, New Jersey.*

Nutrient reabsorption potentially confers a second type of nutrient-use efficiency on vegetation (see Chapter 5 for a discussion of nutrient-use efficiency in photosynthesis). Nutrients that are reabsorbed can be used in net primary production in future years, increasing the carbon fixed per unit of nutrient uptake (Salifu and Timmer 2003).

Compiling data from a wide range of species, Aerts (1996) found a mean fractional reabsorption of 50% N and 52% P during leaf senescence. A recent report by Vergutz et al. (2012) suggests that N and P resorption from leaves frequently may exceed 60%. Somewhat lower values are seen in a California shrubland (Table 6.4), in the Hubbard Brook forest (Table 6.1), and in grassland ecosystems (Woodmansee et al. 1978). Lajtha (1987) found exceptionally high values for P reabsorption (72–86%) in the desert shrub *Larrea tridentata* growing in calcareous soils in which P availability is limited due to the precipitation of calcium phosphate

TABLE 6.4 Nutrient Cycling in a 22-Year-Old Stand of the Chaparral Shrub (*Ceanothus megacarpus*) near Santa Barbara, California

	Biomass	N	P	K	Ca	Mg
Atmospheric input (g m^{-2} yr^{-1})						
Deposition		0.15		0.06	0.19	0.10
N-fixation		0.11				
Total input		0.26		0.06	0.19	0.10
Compartment pools (g/m^2)						
Foliage	553	8.20	0.38	2.07	4.50	0.98
Live wood	5929	32.60	2.43	13.93	28.99	3.20
Reproductive tissues	81	0.92	0.08	0.47	0.32	0.06
Total live	6563	41.72	2.89	16.47	33.81	4.24
Dead wood	1142	6.28	0.46	2.68	5.58	0.61
Surface litter	2027	20.5	0.6	4.7	26.1	6.7
Annual flux (g m^{-2} yr^{-1})						
Requirement for production						
Foliage	553	9.35	0.48	2.81	4.89	1.04
New twigs	120	1.18	0.06	0.62	0.71	0.11
Wood increment	302	1.66	0.12	0.71	1.47	0.16
Reproductive tissues	81	0.92	0.08	0.47	0.32	0.07
Total in production	1056	13.11	0.74	4.61	7.39	1.38
Reabsorption before abscission		4.15	0.29	0	0	0
Return to soil						
Litterfall	727	6.65	0.32	2.10	8.01	1.41
Branch mortality	74	0.22	0.01	0.15	0.44	0.02
Throughfall		0.19	0	0.94	0.31	0.09
Stemflow		0.24	0	0.87	0.78	0.25
Total return	801	7.30	0.33	4.06	9.54	1.77
Uptake (= increment + return)		8.96	0.45	4.77	11.01	1.93
Stream-water loss (g m^{-2} yr^{-1})		0.03	0.01	0.06	0.09	0.06
Comparisons of turnover and flux						
Foliage requirement/total requirement (%)		71.3	64.9	61.0	66.2	75.4
Litter fall/total return (%)		91.1	97.0	51.7	84.0	79.7
Uptake/total live pool (%)		21.4	15.6	29.0	32.6	45.5
Return/uptake (%)		81.4	73.3	85.1	86.6	91.7
Reabsorption/requirement (%)		31.7	39.0	0	0	0
Surface litter/litter fall (yr)	2.8	3.1	1.9	1.2	3.3	4.8

Source: Modified from Gray (1983) and Schlesinger et al. (1982b).

minerals (see Figure 4.10). DeLucia and Schlesinger (1995) report 94% reabsorption of leaf P in *Cyrilla racemiflora* in a P-limited bog in the southeastern United States.

Plants grown with low nutrient availability or occurring on infertile sites tend to have low nutrient concentrations in mature leaves and litter; they generally reabsorb a smaller *amount* but a larger *proportion* of the nutrient pool in senescent leaves, compared to individuals of the same species under conditions of greater nutrient availability (Chapin 1988, Boerner 1984,

Pugnaire and Chapin 1993, Killingbeck 1996, Kobe et al. 2005). In most cases, however, species appear to have only a limited ability to adjust the efficiency of reabsorption of leaf nutrients as a function of site fertility (Chapin and Kedrowski 1983, Birk and Vitousek 1986, Chapin and Moilanen 1991, Enoki and Kawaguchi 1999).

Differences in nutrient-use efficiency in reabsorption between nutrient-rich and nutrient-poor sites are not as likely to be due to a direct response of plants as to the tendency for species with higher inherent capabilities for nutrient reabsorption to dominate nutrient-poor sites (Pastor et al. 1984, Chapin et al. 1986b, Schlesinger et al. 1989). Among tropical forests, reabsorption of P varies as an inverse function of site fertility (Kitayama et al. 2000; Figure 6.8). As a result of mycorrhizal associations and internal conservation of P, it appears that tropical trees are well adapted to P-deficient soils, which are widespread in these regions (Cuevas and Medina 1986, Paoli et al. 2005, Cleveland et al. 2011).

Mass Balance of the Intrasystem Cycle

The annual circulation of nutrients in land vegetation, the intrasystem cycle (Figure 6.7), can be modeled using the mass-balance approach. Nutrient requirement is equal to the peak nutrient content in newly produced tissues during the growing season (refer to Table 6.1). Nutrient uptake cannot be measured directly, but uptake must equal the annual storage in

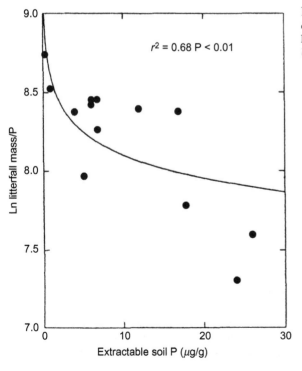

$r^2 = 0.68\ P < 0.01$

FIGURE 6.8   Litterfall mass/P ratio as a function of soil phosphorus availability measured in 13 humid tropical forests. *Source: From Silver (1994). Used with permission of Springer.*

perennial tissues, such as wood, plus the replacement of losses in litterfall and leaching; thus, the following equation:

$$\text{Uptake} = \text{Retained} + \text{Returned}. \tag{6.2}$$

Uptake is less than the annual requirement by the amount reabsorbed from leaf tissues before abscission; namely:

$$\text{Requirement} = \text{Uptake} + \text{Retranslocation}. \tag{6.3}$$

The requirement is the nutrient flux needed to complete a mass balance; it should not be taken as indicative of biological requirements. In fact, this equation can be solved for nonessential elements such as Na. For a forest in Tennessee, the mass-balance approach was used to show that net accumulations of Ca and Mg in vegetation were directly related to decreases in the content of exchangeable Ca and Mg in the soil during 11 years of growth (Johnson et al. 1988). Similarly, during 100 years of forest growth on abandoned agricultural land, Hooker and Compton (2003) showed that the nitrogen accumulation by vegetation was largely supplied from the soil N pool—not new inputs. Mass-balance studies also show that some dissolved silicon (Si), which is often used as an index of rock weathering, is retained and recycled by terrestrial vegetation (Markewitz and Richter 1998, Conley 2002, Derry et al. 2005, Cornelis et al. 2010).

The mass-balance approach was used to analyze the internal storage and the annual transfers of nutrients in the aboveground portion of a California shrubland (Table 6.4). These data serve to summarize many aspects of the intrasystem cycle. Note that 71% of the annual requirement of N is allocated to foliage, whereas much less is allocated to stem wood. Nevertheless, total nutrient storage in short-lived tissues is small compared to storage in wood, which has lower nutrient concentrations than leaf tissue but has accumulated during 22 years of growth. For most nutrients in this ecosystem, the storage in wood increases by about 5% each year.

In this community the nutrient flux in stemflow is unusually large but the total annual return in leaching is relatively small, except for K. Despite substantial reabsorption of N and P before leaf abscission, litterfall is the dominant pathway of return of these elements to the soil from the aboveground vegetation. Ca is actively exported to the leaves before abscission (i.e., requirement < uptake). In this shrubland, annual uptake is 16 to 46% of the total storage in vegetation, but 73 to 92% of the uptake is returned each year. As in most studies, some of these calculations would be revised if belowground transfers were better understood.

There are changes in the nutrient cycles in vegetation, which accompany changes in the allocation of net primary production with stand age. During forest regrowth after disturbance, leaf area develops rapidly and the nutrient movements dependent on leaf area (i.e., litterfall and leaching) are quickly reestablished (Marks and Bormann 1972, Boring et al. 1981, Davidson et al. 2007). Gholz et al. (1985) found that the proportion of the annual requirement met by internal cycling (i.e., nutrient reabsorption from leaves) increased with time during the development of pine forests in Florida. Nutrients are accumulated most rapidly during the early development of forests and more slowly as the aboveground biomass reaches a steady state (Gholz et al. 1985, Pearson et al. 1987, Reiners 1992). Percentage turnover in vegetation declines as the mass and nutrient storage in vegetation increase. In mature forests, leaf biomass is <5% of the total and leaves contain only 5 to 20% of the total nutrient pool in vegetation (Waring and Schlesinger 1985).

TABLE 6.5 Biomass and Element Accumulations in Biomass of Mature Forests

Forest biome	Number of stands	Total biomass (t/ha)	Percent of total biomass				Mass ratio		
			Leaf	Branch	Bole	Roots	C/N	C/P	N/P
Northern/subalpine conifer	12	233	4.5	10.2	62.8	22.6	143	1246	8.71
Temperate broadleaf deciduous	13	286	1.1	16.2	63.1	19.5	165	1384	8.40
Giant temperate conifer	5	624	2.5	10.2	66.4	20.8	158	1345	8.53
Temperate broadleaf evergreen	15	315	2.7	14.7	66.2	16.5	159	1383	8.73
Tropical/subtropical closed forest	13	494	1.9	21.8	59.8	16.4	161	1394	8.65
Tropical/subtropical woodland and savanna	13	107	3.6	19.1	60.4	16.9	147	1290	8.80

Source: From Vitousek et al. (1988). Used with permission of Springer.

Vitousek et al. (1988) have compiled data showing the proportions of biomass (i.e., carbon) and major nutrient elements in various types of mature forest (Table 6.5). The nutrient ratios vary over a surprisingly small range, so the global pattern of element stocks in vegetation is similar to that for biomass; that is, tropical > temperate > boreal forests (Table 5.3). It is important to remember that these ratios are calculated for the total plant biomass; the concentrations of nutrients in leaf tissues are higher, and C/N and C/P ratios in leaves are correspondingly smaller. Thus, nutrient ratios for whole-plant biomass increase with time as the vegetation becomes increasingly dominated by structural tissues with lower nutrient content (Vitousek et al. 1988, Reiners 1992).

Nutrient Use Efficiency

A mass balance for the intrasystem cycle of vegetation allows us to calculate an integrated measure of nutrient-use efficiency by vegetation—net primary production per unit nutrient uptake (Pastor and Bridgham 1999). This measure is affected by various factors that we have examined individually, including the rate of photosynthesis per unit leaf nutrient (Chapter 5), uptake per unit of root growth (refer to Figure 6.2), and leaching and nutrient retranslocations from leaves. As a result of changes in these factors, net primary production per unit of nitrogen or phosphorus taken from the soil increased by factors of 5 and 10, respectively, during the growth of pine forests in central Florida (Gholz et al. 1985).

Among temperate forests, the annual circulation of nutrients in coniferous forests is much lower than the circulation in deciduous forests, largely as a result of lower leaf nutrient concentrations and lower leaf turnover in coniferous forest species (Cole and Rapp 1981, Aerts 1996). The foliage of some coniferous species persists for 8 to 10 years. Also, leaching losses are lower in coniferous forests (Parker 1983), and photosynthesis per unit of leaf nitrogen tends to be greater in coniferous species (Reich et al. 1995). Together these mechanisms result in greater nutrient-use efficiency in coniferous forests compared to deciduous forests of the

TABLE 6.6   Net Primary Production ($kg\ ha^{-1}\ yr^{-1}$) per Unit of Nutrient Uptake, as an Index of Nutrient-Use Efficiency in Deciduous and Coniferous Forests

Forest type	Production per unit nutrient uptake				
	N	P	K	Ca	Mg
Deciduous	143	1859	216	130	915
Coniferous	194	1519	354	217	1559

Source: From Cole and Rapp (1981).

world (Table 6.6). Higher nutrient-use efficiency in coniferous species may explain their frequent occurrence on nutrient-poor sites and in boreal climates where soil nutrient turnover is slow. Significantly, larch (*Larix* sp.), one of the few deciduous species in the boreal forest, has exceptionally high fractional reabsorption of foliar nutrients (Carlyle and Malcolm 1986).

The high nutrient-use efficiency of most conifers may also extend to the occurrence of broad-leaf evergreen vegetation on nutrient-poor soils in other climates (Monk 1966; Beadle 1966; Goldberg 1982, 1985; DeLucia and Schlesinger 1995). Escudero et al. (1992) suggest that leaf longevity was the most important factor increasing nutrient-use efficiency among various trees and shrubs in central Spain (compare Reich et al. 1992), since deciduous and evergreen species have roughly similar amounts of nutrient reabsorption during leaf senescence (del Arco et al. 1991, Aerts 1996, Eckstein et al. 1999).

For biogeochemical cycling in vegetation, we have seen that the leaves and fine roots contain only a small portion of the nutrient content in biomass, but the growth, death, and replacement of these tissues largely determine the annual intrasystem cycle of nutrients. Net primary production is positively correlated to soil N availability in both coniferous and deciduous forests (Zak et al. 1989, Reich et al. 1997), but differences in nutrient-use efficiency tend to weaken the correlation, so that light and moisture are the primary determinants of net primary production on a global basis (Figures 5.13 and 5.14). When nutrient concentrations in litter are low, as might be expected after reabsorption of nutrients, decomposition is slower (Scott and Binkley 1997, Lovett et al. 2004). Thus, intrasystem cycling contains a positive feedback to the extent that an increase in nutrient-use efficiency by vegetation may reduce the future availability of soil nutrients for plant uptake (Shaver and Melillo 1984).

Because of the uptake of nutrients from the soil and the intrasystem cycling of nutrients, terrestrial vegetation leaves a marked imprint on the nutrient distribution in soils. Some nutrients are actively accumulated from deep in the soil profile and deposited at the surface (Marsh et al. 2000, Lawrence and Schlesinger 2001, Jobbágy and Jackson 2004). This effect is most pronounced for nutrients that are strongly recycled within the plant community, while others are more evenly distributed through the soil profile. As a result, a global compilation of 10,000 soil profiles shows the following rank-order of nutrient concentrations with depth (shallow to deep) (Jobbágy and Jackson 2001):

$$P > K > Ca > Mg > Na = Cl = SO_4. \tag{6.4}$$

Nutrient uplift by plants is mediated by precipitation—if the site is very wet, leaching tends to dominate over plant uplift (Porder and Chadwick 2009). In deserts and other areas where there is patchy vegetation, plant nutrients are strongly concentrated under shrubs, whereas nonlimiting or nonessential nutrients accumulate in the barren spaces between shrubs (Schlesinger et al. 1996, Gallardo and Parama 2007). Even in forests individual trees can leave an imprint on soil chemistry (Boettcher and Kalisz 1990, Rodriguez et al. 2011).

BIOGEOCHEMICAL CYCLING IN THE SOIL

Despite new inputs from the atmosphere and from rock weathering, and adaptations in plants to minimize their loss of nutrients, most of the annual nutrient requirement of land plants is supplied from the decomposition of dead materials in the soil (refer to Table 6.1). Decomposition of dead organic matter completes the intrasystem cycle by returning nutrient elements to the soil where they are available for plant uptake.

Soil Microbial Biomass and the Decomposition Process

Decomposition is a general term that refers to the breakdown of organic matter. *Mineralization* is a more specific term[3] that refers to processes that release carbon as CO_2 and nutrients in inorganic form (e.g., N as NH_4^+ and P as PO_4^{3-}).

A variety of soil animals, including earthworms, fragment and mix fresh litterfall (Swift et al. 1979, Wolfe 2001); however, the main biogeochemical transformations are performed by fungi and bacteria in the soil. Most of the mineralization reactions are the result of the activity of extracellular degradative enzymes, released by soil microbes and mycorrhizae (Burns 1982; Linkins et al. 1990; Sinsabaugh et al. 1993, 2002, 2008). The release of a wide variety of extracellular enzymes, including cellulases and proteases, increases in response to the amount of freshly deposited organic matter available for decomposition.

Microbial biomass (bacteria + fungi) typically comprises <3% of the organic carbon found in soils (Wardle 1992, Zak et al. 1994). High levels of microbial biomass are found in most forest soils and lower levels in deserts (Insam 1990, Gallardo and Schlesinger 1992). Fungi dominate over bacteria in most well-drained, upland soils (Anderson and Domsch 1980). Ruess and Seagle (1994) found a direct correlation between soil microbial biomass and soil respiration in the grasslands of the African Serengeti, and microbial biomass is often measured as an index of its activity (Andersson et al. 2004a, Booth et al. 2005).

Determination of microbial biomass is often performed by one of several techniques involving fumigation with chloroform (Jenkinson and Powlson 1976, Martens 1995, Joergensen et al. 2011). For instance, in a subdivided soil sample, total soluble nitrogen (NH_4^+, NO_3^-, and dissolved organic N) is measured before and after fumigation with chloroform. A higher content in the fumigated sample is assumed to result from the lysis of microbes that were killed by chloroform (Brookes et al. 1985, Joergensen 1996). Microbial biomass is then calculated by assuming a standard nitrogen content in microbial tissue and a correction factor, Kn, to account for microbial N that is not immediately released by fumigation (Voroney and Paul 1984, Shen et al. 1984, Joergensen and Mueller 1996). The technique is justified by the observation of relatively constant C/N and C/P ratios in soil microbial biomass from many

---

[3] This use of the term *mineralization* differs from its common usage in the literature of geology, in which mineralization refers to various processes (e.g., precipitation from hydrothermal fluids) that result in the deposition of metals in an ore deposit of economic significance.

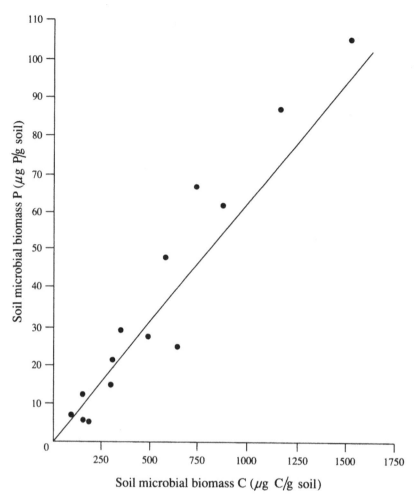

FIGURE 6.9   Relationship between the phosphorus and carbon contained in the microbial biomass of 14 soils. *Source: From Brookes et al. (1984).*

different environments (e.g., Figure 6.9). Microbial biomass is also measured by changes in respiration on the addition of glucose (i.e., substrate-induced respiration; Anderson and Domsch 1978, Lin and Brookes 1999) and extractions of the phospholipid fatty acid (PLFA) content from soil samples (Leckie et al. 2004, Bailey et al. 2002).

Soil microbes have high nutrient concentrations relative to the organic matter they decompose (Diaz-Ravina et al. 1993, Cleveland and Liptzin 2007). Microbial biomass contained 2.5 to 5.6% of the organic carbon but up to 19.2% of the organic phosphorus in tropical soils of central India (Srivastava and Singh 1988). During the decomposition of plant material, respiration of soil microbes converts organic carbon to CO_2, while the N and P content are initially retained in microbial biomass. When the decomposition of fresh litter is observed in litterbags (Chapter 5), the C/N and C/P ratios decline as decomposition proceeds and as the remaining

TABLE 6.7   Ratios of Nutrient Elements to Carbon in the Litter of Scots Pine (*Pinus sylvestris*) at Sequential Stages of Decomposition

	C/N	C/P	C/K	C/S	C/Ca	C/Mg	C/Mn
Needle litter							
Initial	134	2630	705	1210	79	1350	330
After incubation of:							
1 year	85	1330	735	864	101	1870	576
2 year	66	912	867	ND	107	2360	800
3 year	53	948	1970	ND	132	1710	1110
4 year	46	869	1360	496	104	704	988
5 year	41	656	591	497	231	1600	1120
Fungal biomass							
Scots pine forest	12	64	41	ND	ND	ND	ND

Note: C/N and C/P ratios decline with time, which indicates retention of these nutrients as C is lost, whereas C/Ca and C/K ratios increase, which indicates that these nutrients are lost more rapidly than carbon.
Source: From Staaf and Berg (1982). Used with permission of NRC Research Press.

materials are progressively dominated by microbial biomass that has colonized and grown on the substrate (Table 6.7; Sinsabaugh et al. 1993, Manzoni et al. 2010, Fahey et al. 2011).

The accumulation of N, P, and other nutrients in soil microbes is known as *immobilization*. Immobilization is most significant for N and P, which are limiting to microbial growth and usually less obvious for Mg and K, which are available in greater quantities (Jorgensen et al. 1980, Staaf and Berg 1982). In the process of immobilization, soil microbes not only retain the nutrients released from their substrate but also accumulate nutrients from the soil solution— net immobilization (Figure 6.10; Drury et al. 1991). Microbial uptake of NH_4^+ is rapid, sequestering available NH_4^+ that might otherwise be available for plant uptake or nitrifying bacteria

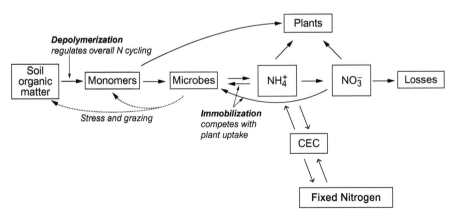

FIGURE 6.10   A conceptual model for the soil nitrogen cycle. *Source: Modified from Schimel and Bennett (2004); see also Drury et al. (1991).*

(Jackson et al. 1989, Schimel and Firestone 1989). In a beech forest near Vienna, Austria, microbial uptake exceeded mineralization during the winter (Kaiser et al. 2011). In cases of net accumulation by microbes, the total nutrient content of the substrate appears to increase during the initial phases of decomposition (e.g., Aber and Melillo 1980, Berg 1988).

Mineralization slows when fresh residues with high C/N ratio are added to the soil (Turner and Olson 1976, Gallardo and Merino 1998). Fallen logs have low N contents, and long-term immobilization of N is especially evident during log decay (Lambert et al. 1980, Fahey 1983, Schimel and Firestone 1989). In contrast, leaf tissues decompose more readily. Ecologists have long used the C/N ratio of litterfall as an index of its potential rate of decomposition (Taylor et al. 1989, Enriquez et al. 1993). Lignin/nitrogen (Melillo et al. 1982) and lignin/phosphorus (Wieder et al. 2009) ratios in litter are also good predictors of the rate of decomposition in various ecosystems (Figure 6.11). These relationships allow us to predict the rate of decomposition and mineralization over large regions using remote sensing (Fan et al. 1998a, Ollinger et al. 2002).

When microbial growth slows, there is little further nutrient immobilization. Thus, immobilization of nutrients predominates in the layer of fresh litter on the soil surface, while net mineralization of N, P, and S is usually greatest in the lower forest floor (Federer 1983). Net mineralization of N often begins with C/N ratios near 30:1, but this can vary depending on the substrate and the assimilation efficiency of the decomposer (Manzoni et al. 2010). Immobilization is least rapid and the release of nitrogen is most rapid in substrates of greatest initial nitrogen content (Parton et al. 2007, Manzoni et al. 2008).

During the decomposition process, fulvic acids and other dissolved organic compounds carry nutrients to the lower soil horizons (Schoenau and Bettany 1987, Qualls and Haines 1991), where they add to the nutrient pool in humus. Sollins et al. (1984) found that the "light" fraction of soil organic matter, representing fresh plant residues, had a higher C/N ratio and lower mineralization than the "heavy" fraction, composed of humic substances. A small fraction of the substrate is converted to humic compounds (Chapter 5) that paradoxically have high N content (Schulten and Schnitzer 1993) but long-term stability in the soil profile (Chapter 5). Decaying plant litter appears to adsorb Al and Fe (Rustad 1994,

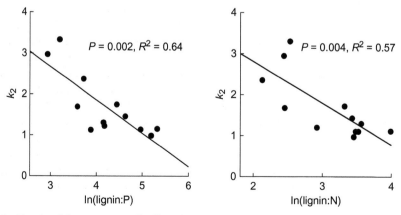

FIGURE 6.11   Fractional decomposition of leaf litter from wet tropical forests as a function of the lignin-to-P and lignin-to-N ratio in newly fallen litter. *Source: From Wieder et al. (2009). Used with permission of the Ecological Society of America.*

TABLE 6.8   Mean Residence Time, in Years, of Organic Matter and Nutrients in the Surface Litter of Forest and Woodland Ecosystems

Region	Mean residence time (year)					
	Organic matter	N	P	K	Ca	Mg
Boreal forest	353	230	324	94	149	455
Temperate forest						
Coniferous	17	17.9	15.3	2.2	5.9	12.9
Deciduous	4	5.5	5.8	1.3	3.0	3.4
Mediterranean	3.8	4.2	3.6	1.4	5.0	2.8
Tropical rainforest	0.4	2.0	1.6	0.7	1.5	1.1

Note: Values are calculated by dividing the forest floor mass by the mean annual litterfall.
Sources: Boreal and temperate values are from Cole and Rapp (1981); tropical values are from Edwards and Grubb (1982) and Edwards (1977, 1982); Mediterranean values are from Gray and Schlesinger (1981).

Laskowski et al. 1995), perhaps in compounds that are precursors to fulvic acids that carry Al and Fe to the lower soil profile in the process of podzolization (Chapter 4).

Differential losses of nutrients and nutrient immobilizations mean that the loss of mass from litterbags cannot be directly equated with the proportional release of its original nutrient content (Jorgensen et al. 1980, Rustad 1994). Table 6.8 shows the mean residence time for organic matter and its nutrient content in the surface litter of various ecosystems. Some nutrients, such as K, that are easily leached from litter may show mineralization rates in excess of the loss of litter mass. Others, such as N, turn over more slowly due to immobilization in microbial tissues. Pregitzer et al. (2010) used experimental additions of $^{15}NO_3$ to a sugar maple forest in Michigan to measure a mean residence time of 6.5 years for N in the forest floor.

Vogt et al. (1986) suggest that immobilization of N is greatest in temperate and boreal forests, whereas the immobilization of P is more important in tropical forests. Given the nutrient limitations facing soil microbes, it is perhaps not surprising that their activity increases with nutrient additions (e.g., Cleveland and Townsend 2006, Allen and Schlesinger 2004), leading to losses of organic matter from the soil profile (Neff et al. 2002b, Mack et al. 2004). Paradoxically, most nutrient-addition experiments show decreased soil microbial biomass (Wallenstein et al. 2006, Treseder 2008, Lu et al. 2010, Liu and Greaver 2010).

In Chapter 5 we saw that the pool of soil organic matter greatly exceeds the mass of vegetation in most ecosystems. As a result of its high nutrient content, humus also dominates the storage of biogeochemical elements in most ecosystems. Aboveground biomass contains only 4 to 8% of the total quantity of N in temperate forests (Cole and Rapp 1981) and 3 to 32% in tropical forests (Edwards and Grubb 1982). Generally, the ratio of C, N, P, and S in humus is close to 140:10:1.3:1.3 (Stevenson 1986, Schulten and Schnitzer 1993; compare Cleveland and Liptzin 2007), so the global pool of nitrogen in soil, estimated at 95 to 140 × 10^{15} g (Post et al. 1985, Batjes 1996), dwarfs the pool of nitrogen in vegetation, 3.8 × 10^{15} g.[4] Owing to the

---

[4] Calculated using the global biomass of 615 × 10^{15} g C (Table 5.3) and a C/N ratio in vegetation of 160 (Table 6.5).

stability of humus substances in the soil, the large nutrient pool in humus turns over very slowly. Typically soil microbes mineralize about 1 to 3% of the pool of nitrogen in the soil each year (Connell et al. 1995).

Simple measurements of extractable nutrients, such as NH_4^+ or PO_4^{3-}, are unlikely to give a good index of nutrient availability in terrestrial ecosystems. These nutrients are subject to active uptake by plant roots, immobilization by soil microbes, and a variety of other processes that rapidly remove available forms from the soil solution. At any moment, the quantity extractable from a soil sample may be only a small fraction of what is made available by mineralization during the course of a growing season (Davidson et al. 1990). Thus, studies of biogeochemical cycling in the soil need to be based on measurements that record the dynamic nature of nutrient turnover.

Nitrogen Cycling

The mineralization of N from decomposing materials begins with the release of amino acids and other simple organic-N molecules by the microbes involved in decomposition (Schimel and Bennett 2004, Geisseler et al. 2010; Figure 6.10). Some of the amino acids are taken up directly by plants and soil microbes. Using ^{15}N as a tracer, Marumoto et al. (1982) have shown that much of the N mineralized in the soil is released from dead microbes. The presence of soil animals that feed on bacteria and fungi can increase the rates of release of N and P from microbial tissues (Cole et al. 1978, Anderson et al. 1983). The release, or mineralization, of NH_4^+ from organic forms is known as *ammonification*.

Subsequently, a variety of abiotic and biotic processes may remove NH_4^+ from the soil solution, including uptake by plants, immobilization by microbes, and fixation in clay minerals (Johnson et al. 2000a). Some of the remaining NH_4^+ may undergo nitrification, in which the oxidation of NH_4^+ to NO_3^- is coupled to the fixation of carbon by chemoautotrophic bacteria, traditionally classified within the genera *Nitrosomonas* and *Nitrobacter* (Meyer 1994; see Eqs. 2.17 and 2.18). Recent work suggests that nitrification is also performed by prokaryotes in the more primitive group archaea, although it is unclear whether they normally achieve the same level of activity in soils as bacteria (Leininger et al. 2006, Di et al. 2009). In some cases, NH_4^+ is also oxidized by heterotrophic nitrification, producing NO_3^- (Schimel et al. 1984, Duggin et al. 1991, Brierley and Wood 2001, Pedersen et al. 1999).

Nitrate may be taken up by plants and microbes or lost from the ecosystem in runoff waters or in emissions of N-containing gases during denitrification. An intermediate product in the nitrification reaction (see Eq. 2.17), NO_2^- appears to bind to soil organic matter by abiotic processes (Dail et al. 2001, Fitzhugh et al. 2003, Davidson et al. 2003). Nitrate taken up by soil microbes (immobilization) is reduced to NH_4^+ by nitrate reductase and used in microbial growth (Davidson et al. 1990, DeLuca and Keeney 1993, Downs et al. 1996). This process is known as *assimilatory reduction*. Nitrate can also be utilized by dissimilatory nitrate-reducing bacteria to produce NH_4^+, which can cycle back through these pathways. Dissimilatory nitrate reduction to ammonium (DNRA) is best known from its occurrence in wet tropical soils, where it can exceed denitrification—the conversion of NO_3^- to N_2 (Silver et al. 2001, Rutting et al. 2008, Templer et al. 2008). DNRA potentially reduces the loss of NO_3^- to stream waters.

At any time, the extractable quantities of NH_4^+ and NO_3^- in the soil represent the net result of all of these processes. A low concentration of NH_4^+ is not necessarily an indication of low mineralization rates; it can also indicate rapid nitrification or plant uptake (Rosswall 1982, Davidson et al. 1990). A variety of techniques are available to study the transformations of nitrogen in the soil (Binkley and Hart 1989). Many workers have used the "buried-bag" approach to examine net mineralization. A soil sample is subdivided and part is extracted immediately, usually with KCl, to measure the available NH_4^+ and NO_3^-. The remaining soil is replaced in the field in a polyethylene bag, which is permeable to O_2 but not to H_2O. After a short period, usually 30 days, the bag is retrieved and its contents analyzed for the forms of available N. An increase in the quantity of available N in the buried bag is taken to represent *net* mineralization (i.e., the mineralization in excess of microbial immobilization) in the absence of plant uptake. Repeated samples taken through an annual cycle allow an estimate of annual net mineralization, which can be correlated with plant uptake and cycling (Pastor et al. 1984).

Field measurements can also be performed in tubes (Raison et al. 1987) or trenched plots (Vitousek et al. 1982). In the latter, a block of soil, often 1 m^2, is isolated on all sides by trenching and the trenches are lined with plastic to prevent the invasion of roots. Plants rooted in this plot are removed, but the area is not otherwise disturbed. Periodic measurements of NH_4^+ and NO_3^- indicate rates of mineralization and nitrification in the absence of plant uptake. Trenching also eliminates the plant uptake of water, so this approach measures microbial activity at artificially high soil moisture content, with potentially greater losses from the ecosystem due to leaching and denitrification.

A more expensive approach involves the use of $^{15}NH_4^+$ to label the initial pool of available NH_4^+ in the soil (Van Cleve and White 1980, Davidson et al. 1991a, Di et al. 2000). After a period of time, the pool is remeasured for the ratio of $^{15}NH_4^+$ to $^{14}NH_4^+$. A decline in proportion of $^{15}NH_4^+$ is assumed to result from the microbial mineralization of NH_4^+ from the pool of N in soil organic matter. This technique gives a measure of total (gross) mineralization under natural field conditions. Using this approach, Davidson et al. (1992) found that net mineralization was only 14% of the total in a coniferous forest; the remainder was immobilized by the microbial community. Known as the isotope-pool dilution technique, this approach has also been applied to determine turnover of amino acids (Wanek et al. 2010) and NO_3^- (Stark and Hart 1997). For instance, $^{15}NO_3$ is used to label the pool of NO_3^- in the soil, and gross nitrification is measured as the label is diluted by $^{14}NO_3^-$. Net nitrification can also be studied by measuring changes in the concentration of NH_4^+ and NO_3^- after application of compounds that specifically inhibit nitrification, including nitrapyrin (Bundy and Bremner 1973), acetylene (Berg et al. 1982), or chlorate (Belser and Mays 1980).

Mineralization and nitrification have been studied in a wide variety of ecosystems (Vitousek and Melillo 1979, Robertson 1982a, Vitousek and Matson 1988, Davidson et al. 1992). Net mineralization typically ranges from 20 to 120 kg N ha^{-1} yr^{-1} in forests (Pastor et al. 1984, Fan et al. 1998b, Perakis and Sinkhorn 2011), 40 to 90 kg N ha^{-1} yr^{-1} in grasslands (Hatch et al. 1990), and 10 to 30 kg N ha^{-1} yr^{-1} in deserts (Schlesinger et al. 2006). Generally, net mineralization is directly related to the total content of organic nitrogen in the soil (e.g., Marion and Black 1988, McCarty et al. 1995, Accoe et al. 2004, Perakis and Sinkhorn 2011), but mineralization is also closely linked to the availability of carbon (Booth et al. 2005). Vegetation with a high C/N ratio in litterfall often shows low rates of mineralization in the soil (Gosz 1981, Vitousek et al. 1982). When field plots are fertilized with sugar, net mineralization and nitrification slow because of increased immobilization of NH_4^+ by soil microbes

(DeLuca and Keeney 1993, Zagal and Persson 1994). Fertilization of a Douglas fir forest with sugar resulted in lower N content in leaves and greater nutrient reabsorption before leaf-fall (Turner and Olson 1976), showing a direct link between microbial processes in the soil and the nutrient-use efficiency of vegetation.

Although soil microbial populations may adapt to a wide variety of field conditions, nitrification is generally lower at low pH, low O_2, low soil moisture content, and high litter C/N ratios (Wetselaar 1968, Rosswall 1982, Robertson 1982b, Bramley and White 1990, Booth et al. 2005). Compared to mineralization, nitrification is more sensitive to low soil water content, so NH_4^+ accumulates in seasonally dry desert soils (Hartley and Schlesinger 2000). Generally, however, nitrification rates are high whenever NH_4^+ is readily available (Robertson and Vitousek 1981, Vitousek and Matson 1988).

A large amount of effort has been directed toward understanding the control of nitrification following disturbances, such as forest harvest or fire (Vitousek and Melillo 1979, Vitousek et al. 1982). When vegetation is removed, soil temperature and moisture content are generally higher, and rapid ammonification increases the availability of NH_4^+. Subsequently, nitrification may be so rapid that uptake by regrowing vegetation and immobilization by soil microbes are insufficient to prevent large losses of NO_3^- in stream water. However, not all disturbed sites show large losses of NO_3^-. In pine forests in the southeastern United States, microbial immobilization in harvest debris accounted for 83% of the uptake of ^{15}N that was applied as an experimental tracer following forest harvest (Vitousek and Matson 1984). Immobilization of nitrate accounts for a large fraction of the nitrogen turnover in coniferous forests (Stark and Hart 1997), and microbial immobilization also retards the loss of nitrate following burning of tallgrass prairie (Seastedt and Hayes 1988).

In general, nitrification and losses of NO_3^- in stream water after disturbance are greatest in forests with high nitrogen availability (Krause 1982, Vitousek et al. 1982). Rates of nitrification decline during the early recovery of vegetation, and only minor differences are seen between middle- and old-age forests (Robertson and Vitousek 1981, Christensen and MacAller 1985, Davidson et al. 1992). There is some evidence that nitrification is inhibited by terpenoid and tannin compounds released by some types of vegetation (Olson and Reiners 1983, White 1988, Subbarao et al. 2009).

Increases in nitrification following disturbance affect other aspects of ecosystem function. Nitrification generates acidity (Eq. 2.17), so losses of NO_3^- in stream water are often accompanied by increased losses of cations, which are removed from cation exchange sites in favor of H^+ (Likens et al. 1970b). Stream-water losses of nearly all biogeochemical elements increased following harvest of the Hubbard Brook forest in New Hampshire. Sulfate was a curious exception (Figure 6.12). Nodvin et al. (1988) showed that the decline in stream water SO_4^{2-} concentrations after forest harvest was a result of the acidity generated from nitrification, which increased soil anion adsorption capacity (Chapter 4; Mitchell et al. 1989). These observations are a good example of a link between the biogeochemical cycles of N and S in terrestrial ecosystems.

Since the various transformations of N in the soil favor ^{14}N over ^{15}N (Hogberg 1997), ^{15}N increases in the undecomposed residues (Nadelhoffer and Fry 1988), with depth in the soil profile (Koba et al. 1998, Hobbie and Ouimette 2009, Piccolo et al. 1996, Kramer et al. 2003), and with time of ecosystem development (Brenner et al. 2001, Billings and Richter 2006). In contrast, nitrate in runoff waters is depleted in ^{15}N (Spoelstra et al. 2007). Nearly all soils

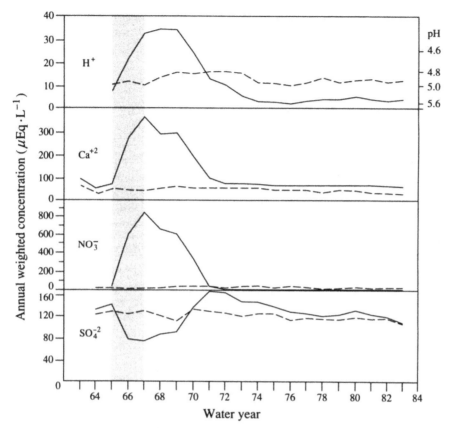

FIGURE 6.12   Concentration of H^+, Ca^{2+}, NO_3^-, and SO_4^{2-} in streams draining the Hubbard Brook Experimental Forest for the years 1964–1984. Streams draining undisturbed forest are shown with the dashed line. The solid line depicts the concentrations in a stream draining a single watershed that was disturbed between 1965 and 1967 (*shaded*). Losses of Ca and NO_3 increased strongly during the period of disturbance, and then recovered to normal values as the vegetation regenerated. The budget for SO_4 shows greater retention during and after the period of disturbance, presumably as a result of increased acidity and anion adsorption in the soil. *Source: Modified from Nodvin et al. (1988).*

show $\delta^{15}N > 0$, with the greatest values seen in ecosystems with rapid nitrogen cycling (Templer et al. 2007) and with substantial losses of NO_3^- in runoff and emissions of nitrogen gases from the ecosystem (Amundson et al. 2003, Pardo et al. 2002). Since plants depend on nitrogen mineralization for uptake, plant $\delta^{15}N$ is usually lower than that of soil organic matter, but $\delta^{15}N$ in plants increases as a function of soil nitrogen cycling, which progressively enriches ^{15}N remaining in the soil (Garten and Van Miegroet 1994, Templer et al. 2007).

Emission of Nitrogen Gases from Soils

During transformations of nitrogen in the soil, a variety of nitrogen gases, including NH_3, NO, N_2O, and N_2, are produced as products and byproducts of microbial activity (Figure 6.13). Some of these may escape from the ecosystem, contributing to a loss of local

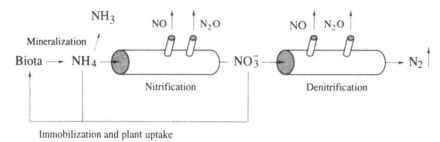

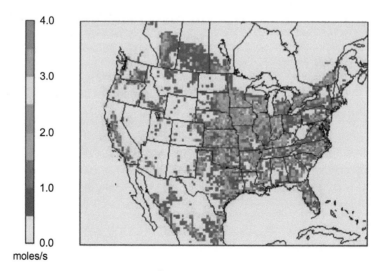

FIGURE 6.13   Microbial processes that yield nitrogen gases during nitrification and denitrification in the soil. *Source: Modified from Firestone and Davidson (1989).*

soil fertility. More significantly, terrestrial ecosystems are a significant source of these gases in the atmosphere (Chapters 3 and 12).

In soils, ammonium may be converted to ammonia gas (NH_3), which is lost to the atmosphere. The reaction

$$NH_4^+ + OH^- \rightarrow NH_3\uparrow + H_2O \tag{6.5}$$

is favored in dry soils and deserts, where accumulations of $CaCO_3$ maintain alkaline pH. Low cation exchange capacity and low rates of nitrification also maximize the production and loss of NH_3 (Nelson 1982, Freney et al. 1983). Small losses of NH_3 have been measured in a variety of natural forest and grassland soils worldwide (Schlesinger and Hartley 1992; Figure 6.14). Losses of NH_3 are greatest in fertilized soils and during the decomposition of urea excreted by wild and domestic animals (Terman 1979). During the loss of NH_3

FIGURE 6.14   Soil emissions of NH_3 moles/m^2/sec for the continental United States, showing relatively high emissions from various agricultural regions and lower emissions from undisturbed ecosystems. *Source: From Gilliland et al. (2006).*

from soils, isotopic fraction occurs, leaving soils enriched in ^{15}N (Mizutani et al. 1986, Mizutani and Wada 1988).

Vegetation can also be a source of NH_3 during leaf senescence (Whitehead et al. 1988, Heckathorn and DeLucia 1995), but some of the NH_3 emitted by soils may be taken up by plants, so in many cases terrestrial ecosystems are only a small net source of NH_3 to the atmosphere (Langford and Fehsenfeld 1992, Sutton et al. 1993, Pryor et al. 2001). The loss of ammonia is typically <1 kg ha^{-1} yr^{-1} in soils that are not impacted by fertilizers or domestic animals. The global flux from natural soils is about 2.4×10^{12} g N/yr (Chapter 12). This flux to the atmosphere is significant inasmuch as NH_3 is the only substance that is a net source of alkalinity in the atmosphere, where it can reduce the acidity of rain (Eqs. 3.5 and 3.40). Extremely high NH_3 volatilization from barns and animal feedlots may result in enhanced atmospheric deposition of NH_4^+ in areas immediately downwind (Draaijers et al. 1989, Aneja et al. 2003, Theobald et al. 2006). Somewhat counterintuitively, these large inputs of NH_4^+ may acidify soils, as the NH_4^+ is nitrified and the nitrate is taken up by vegetation (van Breemen et al. 1982, Verstraten et al. 1990).

Both nitric oxide (NO) and nitrous oxide (N_2O) are generated as microbial byproducts of nitrification, with NO generally being the more abundant (Williams et al. 1992).[5] Specifically, NO is released during the oxidation of NO_2^- to NO_3^- by chemoautotrophic bacteria (Eq. 2.18; Venterea and Rolston 2000). Typically, about 1 to 3% of the nitrogen passing through the nitrification pathway is volatilized as NO each year (Baumgärtner and Conrad 1992, Hutchinson et al. 1993), and the net flux from soils to the atmosphere is estimated to be about 12×10^{12} g N/yr globally (Ganzeveld et al. 2002). Davidson et al. (1998) indicate that soils supply 10% of the NO_x emitted in the southeastern United States, with industrial and transportation sources accounting for the rest. (Recall that NO plays a major role in the chemistry of ozone in the troposphere—see Chapter 3.)

The flux of NO from soils is highest under conditions that stimulate nitrification, including fertilization with NH_4^+ (Skiba et al. 1993, Roelle et al. 1999). Soil NO efflux is directly related to nitrification in Chihuahuan desert soils (Figure 6.15). In various ecosystems, the flux

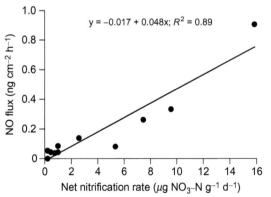

FIGURE 6.15 Emission of NO as a function of nitrification rates in soils of the Chihuahuan Desert, New Mexico. *Source: From Hartley and Schlesinger (2000).*

---

[5] The production of N_2O by several pathways during nitrification is sometimes called nitrifier denitrification (Wrage et al. 2001, Kool et al. 2011).

appears to increase as a function of soil temperature (Williams et al. 1992, Roelle et al. 1999, Van Dijk and Duyzer 1999) and immediately following the wetting of dry soils (Davidson et al. 1991b, 1993; Ghude et al. 2010; Hartley and Schlesinger 2000). When the atmospheric concentration of nitric oxide is high, some NO is taken up by plants and soils, reducing the net flux to the atmosphere (Rondón and Granat 1994, Slemr and Seiler 1991, Ganzeveld et al. 2002). The atmospheric concentration that produces no net uptake or loss is known as the *compensation point*. In most cases, the background concentration in the atmosphere, about 10 ppbv (refer to Table 3.5), is below the compensation point, so terrestrial ecosystems are a net source of NO to the atmosphere (Kaplan et al. 1988, Duyzer and Fowler 1994, Ludwig et al. 2001). In remote areas, the concentration of NO is greatest in the lower atmosphere, and it declines with altitude (Luke et al. 1992).

Losses of NO and N_2O increase with factors that increase the rate of nitrification in soils, including the clearing, cultivation, and fertilization of agricultural soils (Conrad et al. 1983, Mosier et al. 1991, Clayton et al. 1994, Bouwman et al. 2002a). Shepherd et al. (1991) report that 11% of fertilizer N was lost as NO and 5% as N_2O in some cultivated fields in Ontario. When tropical forests are cleared, the losses of NO and N_2O from soils increase dramatically (Sanhueza et al. 1994, Keller et al. 1993, Weitz et al. 1998), but older pastures often have lower N_2O emissions than uncut forest (Melillo et al. 2001, Verchot et al. 1999). Thus, fertilized and newly cleared fields may be responsible for the rising concentration of N_2O in Earth's atmosphere (Chapter 12).

Nitrate is also converted to NO, N_2O, and N_2 in the process of denitrification (Knowles 1982, Firestone 1982, Ye 1994, Goregues et al. 2005). This reaction (Eq. 2.20) is performed by soil bacteria that are aerobic heterotrophs in the presence of O_2 but facultative anaerobes at low concentrations of O_2. During anoxia, heterotrophic activity continues, with nitrate serving as the terminal electron acceptor in metabolism. The structure of the various denitrification enzymes, *nitrite reductases*, contains Fe and Cu (Godden et al. 1991, Tavares et al. 2006, Hino et al. 2010, Pomowski et al. 2011). Because NO_3^- is reduced, but not incorporated into microbial tissue, denitrification is also known as *dissimilatory nitrate reduction*. Bacteria in the genus *Pseudomonas* are the best known denitrifiers, but many others are reported (Knowles 1982, Tiedje et al. 1989).

For a long time, denitrification was thought to occur only in flooded, anoxic soils (Chapter 7), and its importance in upland ecosystems was overlooked. Indeed, the activity of denitrification enzymes is often greatest at low soil O_2 concentrations (Burgin et al. 2010). Now, soil scientists have shown that oxygen diffusion to the center of soil aggregates is so slow that anoxic microsites are common, even in well-drained soils (Figure 6.16; Tiedje et al. 1984, Sexstone et al. 1985a, van der Lee et al. 1999, Ju et al. 2011). Thus, denitrification is widespread in terrestrial ecosystems, especially those in which organic carbon and nitrate are readily available (Burford and Bremner 1975, Carter et al. 1995, Wagner et al. 1996, Wolf and Russow 2000).

Davidson and Swank (1987) found that additions of NO_3^- stimulated denitrification in the surface litter of forests in western North Carolina, and the addition of organic carbon stimulated denitrification in the mineral soil. Additions of organic carbon stimulated the expression of genes related to denitrification in some cultivated soils (Miller et al. 2012). Rainfall generally increases the rate of denitrification because the diffusion of oxygen is slower in wet soils (Sexstone et al. 1985b, Smith and Tiedje 1979, Rudaz et al. 1991, Peterjohn and Schlesinger 1991).

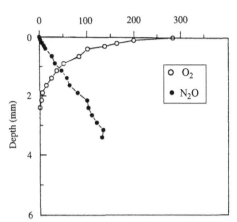

FIGURE 6.16   Concentration of O_2 and N_2O (μM) determined in a soil aggregate as a function of the depth of penetration of a microelectrode. *Source: From Hojberg et al. (1994).*

Typically the production of both NO and N_2O produced by nitrification declines with increasing soil moisture content (or declining O_2), and in anoxic conditions N_2O is entirely due to denitrification (Khalil et al. 2004, Wolf and Russow 2000, Wrage et al. 2001). In Germany, well-drained soils with near-neutral pH produced NO only from nitrification, whereas in acid, anoxic soils, NO was produced from denitrification (Remde and Conrad 1991). In studies of a semidesert ecosystem, Mummey et al. (1994) found that nitrification accounted for 61 to 98% of the N_2O produced in moist soils, but denitrification was the predominant reaction in saturated conditions. In the wet soils of Amazon rainforests, N_2O appeared to be mostly from denitrification (Livingston et al. 1988, Keller et al. 1988).

The relative importance of NO, N_2O, and N_2 as products of denitrification varies depending on environmental conditions (Firestone and Davidson 1989, Bonin et al. 1989). Typically, in denitrification, the production of N_2O dwarfs the production of NO, so the proportional and total loss (nitrification + denitrification) of NO from soils declines with increasing moisture content, while the flux of N_2O increases (Figure 6.17; Drury et al. 1992, Bollmann and

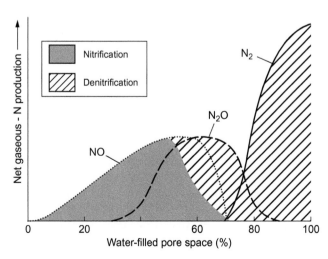

FIGURE 6.17   Relative emission of NO, N_2O, and N_2 from nitrification and denitrification as a function of the water content of soils. *Source: From Davidson et al. (2000). Used with permission of the American Institute of Biological Sciences. All rights reserved.*

Conrad 1998, Wolf and Russow 2000). Factors affecting the relative loss of N_2O and N_2 by denitrification are poorly understood, but they include soil pH and the relative abundance of NO_3^- and O_2 as oxidants and organic carbon as a reductant (Firestone et al. 1980, McKenney et al. 1994, Chen et al. 1995, Morley and Baggs 2010, Burgin and Groffman 2012, Zhang et al. 2009). When NO_3^- is abundant relative to the supply of organic carbon, N_2O can be an important product (Firestone and Davidson 1989, Mathieu et al. 2006, Huang et al. 2004). In studies of forest soils in New Hampshire (Melillo et al. 1983) and Michigan (Merrill and Zak 1992), N_2O was the only significant product of denitrification. In some circumstances, N_2O can diffuse into soils, where it is consumed by dentrifiers, producing N_2 (Frasier et al. 2010, Goldberg and Gebauer 2009).

The ratio of N_2O to N_2 produced in denitrification varies widely (Weier et al. 1993). The median ratio for the overall emission of these gases from upland soils appears to be about 1:1 (Schlesinger 2009), but the proportional emission of N_2O from wetter soils is much lower. The total loss of N from soils by denitrification is typically $<2\,\mathrm{kg\,ha^{-1}\,yr^{-1}}$ in forests and grasslands. N_2O losses from croplands are typically 2 to $4\,\mathrm{kg\,N\,ha^{-1}\,yr^{-1}}$ (Roelandt et al. 2005), but occasional values as high as $13\,\mathrm{kg\,N\,ha^{-1}\,yr^{-1}}$ are reported from agricultural soils (Barton et al. 1999). Globally, the total loss of N_2 from soils ($\sim 44 \times 10^{12}\,\mathrm{g\,N/yr}$; Chapter 12) dwarfs the loss of N_2O ($<10 \times 10^{12}\,\mathrm{g\,N/yr}$) or NO ($12 \times 10^{12}\,\mathrm{g\,N/yr}$). Denitrification is a major process that returns N_2 to the atmosphere, completing the global biogeochemical cycle of nitrogen (Chapter 12). The flux of N_2O is significant inasmuch as its concentration in the atmosphere is increasing, and N_2O plays an important role as a greenhouse gas and as a catalyst of ozone chemistry in the stratosphere (Chapters 3 and 12).

Field measurements of denitrification are problematic, since the production of N_2 is difficult to measure against the background concentration of 78% in Earth's atmosphere. Recently, various workers have measured denitrification using an inert gas, such as helium or argon, to fill collection chambers in the field, allowing the efflux of N_2 from the soil to be observed more easily (Scholefield et al. 1997, Butterbach-Bahl et al. 2002, Dannenmann et al. 2008, Burgin et al. 2010).

Many laboratory studies have also measured denitrification using acetylene to block the conversion of the intermediate denitrification product, N_2O, to the final product, N_2 (refer to Figure 6.13; Yoshinari and Knowles 1976, Burton and Beauchamp 1984, Davidson et al. 1986, Tiedje et al. 1989). With the application of acetylene to laboratory soils or field plots, the sole product of denitrification is N_2O, which is easy to measure with gas chromatography against its background concentration of about 320 ppb in the atmosphere. Alternatively, many field workers have measured denitrification by the application of $^{15}NO_3^-$ to field plots and by measurements of the release of ^{15}N gases or the decline in $^{15}NO_3^-$ remaining in the soil (Parkin et al. 1985, Mosier et al. 1986, Mathieu et al. 2006, Zhang et al. 2009).

Field estimates of denitrification are complicated by high spatial variability. At the local scale, a large portion of the total variability is found at distances of $<10\,\mathrm{cm}$, which Parkin (1987) link to the local distribution of soil aggregates that provide anaerobic microsites. In one case, Parkin (1987) found that 85% of the total denitrification in a 15-cm-diameter soil core was located under a $1\text{-}\mathrm{cm}^2$ section of decaying pigweed (*Amaranthus*) leaf! In desert ecosystems, soil nitrogen content and nitrification rates are localized under shrubs, and denitrification is largely confined to those areas (Virginia et al. 1982, Peterjohn and Schlesinger 1991). Differences in the microbial communities between natural and disturbed ecosystems are also likely to contribute to variation in denitrification (Cavigelli and Robertson 2000).

Robertson et al. (1988) documented the pattern of mineralization, nitrification, and denitrification in a field in Michigan. All these processes showed large spatial variation, but the coefficient-of-variation for denitrification, 275%, was the largest measured. Significant correlations were found among these processes. Soil respiration and potential nitrification explained 37% of the variation in denitrification, presumably due to the dependence of denitrification on organic carbon and NO_3^- as substrates.

The high variability in these processes makes it difficult to use measurements from a few sample chambers to calculate a mean or total flux from an ecosystem (Ambus and Christensen 1994, Mathieu et al. 2006). High rates of denitrification are often confined to particular landscape positions, where conditions are favorable. For example, Peterjohn and Correll (1984) suggested that the runoff of nitrate from agricultural fields was largely denitrified in streamside forests, minimizing the losses in rivers (see also, Pinay et al. 1993, Schipper et al. 1993, Jordan et al. 1993, Ettema et al. 1999). In calculating regional averages for denitrification, investigators must evaluate the relative contributions from local areas of high and low activity (e.g., Groffman and Tiedje 1989, Matson et al. 1991, Yavitt and Fahey 1993, Morse et al. 2012).

As in the case of NH_3 volatilization, the losses of N gases as products and byproducts of nitrification and denitrification leave soils enriched in ^{15}N. Denitrifying bacteria fractionate among the isotopes of available nitrogen—that is, between $^{14}NO_3^-$ and $^{15}NO_3^-$ (Handley and Raven 1992, Robinson 2001, Snider et al. 2009). Preference for $^{14}NO_3^-$ leads to positive $\delta^{15}N$ in most soils (refer to Figure 6.5), as $^{14}N_2$ is lost from the soil by denitrification (Shearer and Kohl 1988, Knöller et al. 2011). Evans and Ehleringer (1993) show a strong inverse relation between $\delta^{15}N$ and the nitrogen content in soils (Figure 6.18), suggesting that soils with low nitrogen are enriched in ^{15}N as a result of the loss of N gases (Garten 1993). Strong enrichments in ^{15}N are also seen in saturated soils with low redox potential (Chapter 7) compared to

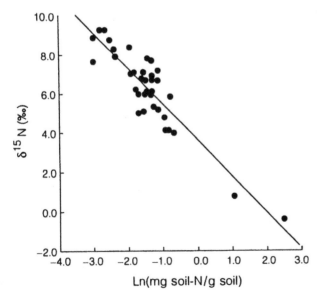

FIGURE 6.18 $\delta^{15}N$ of soil organic matter as a function of the total N content of the soil in juniper woodlands of Utah. *Source: From Evans and Ehleringer (1993).*

adjacent well-drained soils in the same field (Sutherland et al. 1993). Strong enrichments in tropical soils suggest that gaseous losses may be a major pathway of nitrogen loss (Martinelli et al. 1999, Houlton et al. 2006, Koba et al. 2012).

Soil Phosphorus Cycling

Transformations of organic phosphorus in the soil are difficult to study because of the reaction of available phosphorus with various soil minerals (Figures 4.10 and 6.19). A few workers have examined phosphorus mineralization in soils using the buried-bag approach (e.g., Pastor et al. 1984), but in many cases there is no apparent mineralization because of the immediate complexation of P with soil minerals. Thus, most studies of phosphorus cycling have followed the decay of radioactively labeled plant materials (Harrison 1982) or measured the dilution of radioactive ^{32}P that is applied to the soil pool as a tracer (Walbridge and Vitousek 1987, Lopez-Hernandez et al. 1998). With the isotope-dilution technique, one must assume that ^{32}P equilibrates with all the chemical pools in the soil and that the only dilution of its concentration is by the mineralization of organic phosphorus (Kellogg et al. 2006). Unfortunately, these assumptions are not always valid, making the technique difficult to apply in many instances (Walbridge and Vitousek 1987, Di et al. 1997, Bunemann et al. 2007).

In the face of difficulty measuring P mineralization directly, many workers have used sequential extractions to quantify phosphorus availability in the soil (Hedley et al. 1982b, Stevenson 1986, Tiessen et al. 1984). Extraction with 0.5 M NaHCO$_3$ is a convenient index of labile inorganic and soluble organic phosphorus in many soils (Olsen et al. 1954, Sharpley

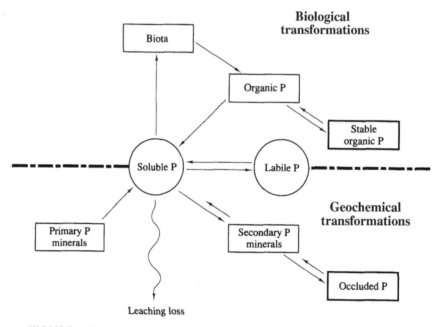

FIGURE 6.19   Phosphorus transformations in the soil. *Source: From Smeck (1985).*

et al. 1987). Organic P is often determined as the difference between PO_4 in a sample that has been combusted at high temperatures and an untreated sample (Stevenson 1986); microbial P, by the change in extractable phosphorus after fumigation with chloroform (Brookes et al. 1982, 1984).

Extraction with NaOH (to raise pH and lower anion adsorption capacity) indicates the amount of P that is held on Fe and Al minerals, while extraction with HCl releases P from many Ca-bound forms, including $CaCO_3$ (Tiessen et al. 1984, Cross and Schlesinger 1995). Acid-extractable phosphorus also includes P derived from apatite (Chapter 4), including secondary hydroxyapatite—$Ca_5OH(PO_4)_3$—from bones and fluoroapatite—$Ca_5F(PO_4)_3$—in teeth. These biominerals in soils are sometimes used by archeologists to determine the location of past human activity and settlements (Sjöberg 1976, Vitousek et al. 2004).

In most ecosystems, much of the phosphorus available for biogeochemical cycling is held in organic forms (Chapin et al. 1978, Wood et al. 1984, Yanai 1992, Gressel et al. 1996), especially inositol phosphates (Turner and Millward 2002). These organic forms can be isolated and identified using phosphorus-31 nuclear magnetic resonance spectroscopy (Turner et al. 2007, Turner and Engelbrecht 2011). Earlier we discussed the ability of soil microbes, mycorrhizae, and plant roots to release phosphatase enzymes and organic acids that mineralize P from organic and inorganic forms (see also Chapter 4).

Much of the P in decomposing materials is found in ester linkages (i.e., -C-O-P). These groups may be mineralized by the release of extracellular enzymes (e.g., phosphatases) in response to specific microbial demand for P (McGill and Cole 1981). Release of acid phosphatases by soil microbes is directly related to levels of soil organic matter (Tabatabai and Dick 1979, Polglase et al. 1992). During forest development, the phosphorus taken up from labile pools in the soil is replenished by P released from the anion adsorption and nonoccluded pools, which presumably equilibrate with the soil solution over longer periods (Richter et al. 2006).

Walbridge et al. (1991) found that up to 35% of the organic P in the undecomposed litter in a warm-temperate forest was held in microbial biomass, and Gallardo and Schlesinger (1994) found that additions of inorganic P increased the microbial biomass in the lower horizons of a forest soil in North Carolina, where the mineralogy is dominated by Fe and Al-oxide minerals with strong phosphorus adsorption capacity. Similar results are reported for tropical forests (Cleveland et al. 2002, Liu et al. 2012). P immobilization in microbial biomass also dominated P cycling in some European grasslands (Banemann et al. 2012). In the course of decomposition, organic phosphorus compounds move from the forest floor to the lower soil profile, where they accumulate in humus (Schoenau and Bettany 1987, Qualls and Haines 1991, Kaiser et al. 2003, Turner and Haygarth 2000).

Sulfur Cycling

Similar to phosphorus, the cycle of sulfur in the soil is affected by both chemical and biological reactions. Sulfur is derived from atmospheric deposition (Chapter 3) and from the weathering of sulfur-bearing minerals in rocks (Chapter 4), and the proportion from each source varies with location and soil development (Novak et al. 2005, Bern and Townsend 2008, Mitchell et al. 2011b). The concentration of SO_4^{2-} in the soil solution exists in equilibrium with sulfate adsorbed on soil minerals (Chapter 4). Plant uptake of SO_4^{2-} is followed by assimilatory reduction and incorporation of sulfur into glutathione (Kostner et al. 1998) and the amino acids cysteine and methionine, which are incorporated into protein (Johnson 1984). The molecular structure of the S-reducing enzyme contains Fe as a cofactor (Crane et al. 1995). A small quantity of sulfur in plants is found in ester-bonded sulfates (-C-O-SO$_4$), and when soil sulfate concentrations are high, plants may accumulate SO$_4$ in leaf tissues (Turner et al. 1980).

In most soils, the majority of the S is held in organic forms (Bartel-Ortiz and David 1988, Mitchell et al. 1992, Houle and Carignan 1992) in a variety of compounds (Zhao et al. 2006, Schroth et al. 2007). Decomposition of plant tissues is accompanied by microbial immobilization of S (Saggar et al. 1981, Staaf and Berg 1982, Fitzgerald et al. 1984). Using ^{35}S as a radio-isotopic tracer, Wu et al. (1995) found high rates of microbial immobilization of ^{35}SO$_4^{2-}$ when glucose was added to soils (compare Houle et al. 2001). Downward movement of fulvic acids appears to transport organic sulfur compounds to the lower soil profile (Schoenau and Bettany 1987, Kaiser and Guggenberger 2005), where they are mineralized (Houle et al. 2001, Dail and Fitzgerald 1999). Typically, mineralization of SO$_4^{2-}$ begins at C/S ratios <200 (Stevenson 1986). Sulfur in soil organic matter shows higher δ^{34}S than soil sulfate, suggesting that soil microbes discriminate against the heavy isotope of S in favor of ^{32}S during mineralization (Mayer et al. 1995). Most of the SO$_4^{2-}$ in runoff waters appears to have passed through the organic pool (Likens et al. 2002, Novak et al. 2005).

In forest soils, the microbial immobilization of added SO$_4^{2-}$ is greatest in the upper soil profile, and anion adsorption of inorganic SO$_4^{2-}$ dominates the B horizons, where sesquioxide minerals are present (Schindler et al. 1986, Randlett et al. 1992, Houle et al. 2001). In most cases, the majority of microbial S is found in carbon-bonded forms (David et al. 1982, Watwood et al. 1988, Schindler et al. 1986, Mitchell et al. 1986, Dhamala and Mitchell 1995). Organic S appears to accumulate in areas of high inputs of SO$_4^{2-}$ from acid rain (Likens et al. 2002, Armbruster et al. 2003). However, at the Coweeta Experimental Forest in North Carolina, a large portion of the immobilization of sulfur by soil microbes was accumulated as ester sulfates (Fitzgerald et al. 1985, Watwood and Fitzgerald 1988), yielding a significant sink for SO$_4^{2-}$ deposited from the atmosphere (Swank et al. 1984). Despite the predominance of organic forms, the pool of SO$_4^{2-}$ in most soils is not insignificant. In the study of a forest in Tennessee, Johnson et al. (1982) found that the pool of adsorbed SO$_4^{2-}$ was larger than the total pool of S in vegetation by a factor of 15.

To maintain a charge balance, plant uptake and reduction of SO$_4^{2-}$ consumes H$^+$ from the soil, whereas the mineralization of organic sulfur returns H$^+$ to the soil solution, producing no net increase in acidity (Binkley and Richter 1987). In contrast, reduced inorganic sulfur is found in association with some rock minerals (e.g., pyrite), and the oxidative weathering of reduced sulfide minerals accounts for highly acidic solutions draining mine tailings (Eqs. 2.16 and 4.6). This oxidation is performed by chemoautotrophic bacteria, generally in the genus *Thiobacillus*.

Production of reduced sulfur gases such as H$_2$S, COS (carbonyl sulfide), and (CH$_3$)$_2$S (dimethylsulfide) is largely confined to wetland soils, since highly reducing, anaerobic conditions are required (Chapter 7). Globally, upland soils are only a small source of sulfur gases in the atmosphere (Lamb et al. 1987, Goldan et al. 1987, Staubes et al. 1989, Yi et al. 2010). However, many plants (e.g., garlic) produce a variety of volatile organic sulfur compounds that activate sensory receptors in humans and presumably other herbivores

(Bautista et al. 2005).[6] The smell of CS_2 (carbon disulfide) is often found when excavating the roots of the tropical tree *Stryphnodendron excelsum* (Haines et al. 1989), and many plant leaves are known to release sulfur gases during photosynthesis (Winner et al. 1981, Garten 1990, Kesselmeier et al. 1993).

Often the total net flux of sulfur gases from an ecosystem (soil + plant) is estimated by examining the vertical profile of gas concentrations in the atmosphere (e.g., Andreae and Andreae 1988). Hydrogen sulfide appears to dominate the release of sulfur gases from plants (Delmas and Servant 1983, Andreae et al. 1990, Rennenberg 1991). Terrestrial ecosystems also appear to be a source of $(CH_3)_2S$ during the day (Andreae et al. 1990, Berresheim and Vulcan 1992, Kesselmeier et al. 1993), but vegetation is a major sink for COS globally (Chapter 13).

Transformations in Fire

During fires, nutrients are lost in gases and in the particles of smoke (Andreae and Merlet 2001), and soil nutrient availability increases with the addition of ash to soil (Raison 1979, Giardina et al. 2000). Following fire, there is often increased runoff and erosion from bare, ash-covered soils. High rates of nitrification in these nutrient-rich soils can stimulate the loss of NO and N_2O after fire (Anderson et al. 1988, Levine et al. 1988).

Increases in the rate of forest burning worldwide have the potential to deplete the nutrient content of soils and add trace gases to the atmosphere (Mahowald et al. 2005a). However, before human intervention, fires were a natural part of the environment in many regions; thus, nutrient losses as a result of fire occurred at infrequent but somewhat regular intervals (Clark 1990). Using a mass-balance approach we can estimate the length of time it takes to replace the nutrients that are lost in a single fire. For instance, 11 to 40 kg/ha of N are lost in small ground fires in southeastern pine forests (Richter et al. 1982), equivalent to 3 to 12 times the annual deposition of N from the atmosphere in this region (Swank and Henderson 1976). In contrast, periodic losses of N in fires may dominate the long-term nitrogen budget in semi-arid forests—requiring hundreds of years of new inputs to replace the losses from a single fire (Johnson et al. 1998).

When leaves and twigs are burned under laboratory conditions, up to 90% of their N content can be lost, presumably as N_2 or as one or more forms of nitrogen oxide gases (DeBell and Ralston 1970, Lobert et al. 1990). Forest fires volatilize nitrogen in proportion to the heat generated and the organic matter consumed (DeBano and Conrad 1978, Raison et al. 1985, McNaughton et al. 1998); the rate of loss declines dramatically as fires pass from flaming to smoldering phases (Crutzen and Andreae 1990). Typically N losses in forest fires range from 100 to 600 kg/ha, or 10 to 40% of the amount in aboveground vegetation and surface litter (Johnson et al. 1998). Especially large losses are reported from slash fires in the Amazon rainforest (Kauffman et al. 1993).

Studies of the gaseous products of fires are often conducted by flying aircraft through the smoke plume to gather gas samples (e.g., Cofer et al. 1990, Nance et al. 1993, Hurst et al. 1994). The enrichment of CO_2 and CO over the atmospheric background is measured, as well as the

---

[6] The tasty compound in garlic is diallyl thiosulfinate or diallyl disulfide.

ratio of other gases to CO_2 in the smoke (e.g., NH_3/CO_2). Assuming that the carbon in the fuel is all converted to CO_2 and CO, the loss of other fuel constituents as gases and particles can be calculated from estimates of the carbon in the biomass consumed by fire and the ratio of the constituent in question to the total carbon ($CO_2 + CO$) in smoke (Laursen et al. 1992, Delmas et al. 1995). Thus, global estimates of the volatilization of nitrogen from forest fires can be calculated from global estimates of the amount of carbon lost in forest fires each year (Andreae and Merlet 2001, Schultz et al. 2008). N_2 dominates the gaseous loss of nitrogen (Kuhlbusch et al. 1991), constituting a form of "pyrodenitrification" that removes fixed nitrogen from the biosphere (Chapters 3 and 12).

The losses of other nitrogen gases in forest fires account for 6% of the N_2O, 15% of the NH_3, and 18% of the NO_x emitted annually to the atmosphere (Chapter 12). Tropospheric circulation can carry the plume of NO_x from fires in the boreal forest of Canada to Europe (Spichtinger et al. 2001). Forest fires are also a major global source of CO (Seiler and Conrad 1987; refer to Table 11.3) and smaller sources of CH_4 (Delmas et al. 1991, Quay et al. 1991), CH_3Br, CH_3Cl (refer to Table 3.7), and SO_2 in the atmosphere (Sanborn and Ballard 1991, Crutzen and Andreae 1990).

Air currents and updrafts during fire carry particles of ash that remove other nutrients from the site. These losses are usually much smaller than gaseous losses (Arianoutsou and Margaris 1981, Gaudichet et al. 1995). Expressed as a percentage of the amount present in aboveground vegetation and litter before fire, the total loss of plant nutrients in gases and particulates often follows the order $N \gg K > Mg > Ca > P > 0\%$. Differential rates of loss change the balance of nutrients available in the soil after fire (Raison et al. 1985), and nutrient losses to the atmosphere in fire may enhance the atmospheric deposition of nutrients in adjacent locations (Clayton 1976, Lewis 1981).

Depending on intensity, fire kills aboveground vegetation and transfers varying proportions of its mass and nutrient content to the soil as ash. There are a large number of changes in chemical and biological properties of soil as a result of the addition of ash (Raison 1979). Cations and P may be readily available in ash, which usually increases soil pH (Woodmansee and Wallach 1981). Burning increases extractable P, but reduces the levels of organic P and phosphatase activity in soil (DeBano and Klopatek 1988, Saa et al. 1993, Serrasolsas and Khanna 1995). Nitrogen in the ash is subject to rapid mineralization and nitrification (Christensen 1977, Dunn et al. 1979, Matson et al. 1987), so available NH_4^+ and NO_3^- usually increase after fire, even though total soil N may be unchanged (Wan et al. 2001). The increase in available nutrients as a result of ashfall is usually short-lived, as nutrients are taken up by vegetation or lost to leaching and erosion (Lewis 1974, Christensen 1977, Uhl and Jordan 1984). Enhanced emissions of NO and accumulations of nitrate after fire stimulate the germination of post-fire species (Keeley and Fotheringham 1997).

Stream-water runoff typically increases following fire because of reduced water losses in transpiration. High nutrient availability in the soil coupled with greater runoff can lead to large hydrologic losses of nutrients from ecosystems following burning. The loss of nutrients in runoff depends on many factors, including the season, rainfall pattern, and the growth of postfire vegetation (Dyrness et al. 1989). Wright (1976) noted significant increases in the loss of K and P from burned forest watersheds in Minnesota. These losses were greatest in the first 2 years after fire; by the third year there was actually less P lost from burned watersheds than from adjacent mature forests, presumably due to uptake by regrowing vegetation (McColl and Grigal 1975; compare Saá et al. 1994). Although there are exceptions, the relative increase in the loss of Ca, Mg, Na, and K in runoff waters after fire often exceeds that of N and P (Chorover et al. 1994).

Ice cores and sediments contain a historical record of biomass burning. Layers of ice with buried ash often contain especially high concentrations of NH_4^+, indicating substantial NH_3 volatilization during fires (Legrand et al. 1992, Whitlow et al. 1994). Cores taken from the Greenland ice cap show episodes of increased biomass burning that appear to be related to the European colonization of North America (Whitlow et al. 1994, Savarino and Legrand 1998). Lake and ocean sediments contain layers of buried ash that indicate the frequency of fires (Mensing et al. 1999, Clark et al. 1996). It is likely that humans have significantly increased the rate of biomass burning worldwide, especially as a result of tropical deforestation (Crutzen and Andreae 1990, Cahoon et al. 1992).

The Role of Animals

Discussions of terrestrial biogeochemistry usually center on the role of plants and soil microbes. Having seen that animals harvest about 5% of terrestrial net primary production (Chapter 5), it is legitimate to ask if they might play a significant role in nutrient cycling. Certainly an impressive nutrient influx is observed in the soils below roosting birds (Gilmore et al. 1984, Mizutani and Wada 1988, Lindeboom 1984, Simas et al. 2007, Maron et al. 2006). In Yellowstone National Park, elk appear to redistribute plant materials among habitats in the landscape, increasing the nitrogen content and nitrogen mineralization in soils where they congregate (Frank et al. 1994, Frank and Groffman 1998). Soil $\delta^{15}N$ is higher in grazed soils, indicating greater N losses from the ecosystem (Frank and Evans 1997). Grazing in the Serengeti of Africa appears to stimulate nutrient cycling and plant productivity, providing better habitat for the animals (McNaughton et al. 1997).

Various workers have suggested that the grazing of vegetation, especially by insects, stimulates the intrasystem cycle of nutrients and might even be advantageous for terrestrial vegetation (Owen and Wiegert 1976). Trees that are susceptible to herbivory are often those that are deficient in minerals or otherwise stressed (Waring and Schlesinger 1985). Periodic herbivory may stimulate nutrient return to the soil via insect frass and alleviate nutrient deficiencies (Mattson and Addy 1975, Yang 2004). Risley and Crossley (1988) also noted significant premature leaf-fall in a forest that was subject to insect grazing. These leaves delivered large quantities of nutrients to the soil, since nutrient reabsorption had not yet occurred. In the same forest, Swank et al. (1981) noted an increase in stream-water nitrate when the trees were defoliated by grazing insects.

An enormous literature exists on the characteristics of plant tissues that are selected for food. Seasonal variations in the plants selected as food by large mammals may help these grazing animals avoid mineral deficiency (McNaughton 1990, Ben-Shahar and Coe 1992, Grasman and Hellgren 1993). Many studies report that herbivory is centered on plants with high nitrogen contents (Mattson 1980, Lightfoot and Whitford 1987, Griffin et al. 1998), suggesting that animal populations might be limited by N. However, the preference for such tissues may be related more to their high water (Scriber 1977) and low phenolic contents (Jonasson et al. 1986) than to a specific search for leaves with high amino acid content.

Grazing often reduces plant photosynthesis while nutrient uptake continues, resulting in high nutrient contents in the aboveground tissues that remain (McNaughton and Chapin 1985). Grazing may even enhance nitrogen uptake in some species (Jaramillo and Detling 1988). Thus, consumers sometimes increase the nutritional quality of the forage available for future consumption, although the quantity of defensive compounds may also increase (White 1984, Seastedt 1985).

In extreme cases, defoliations may be the dominant form of nutrient turnover in the ecosystem (Hollinger 1986). Usually, however, the role of grazing animals in terrestrial ecosystems is rather minor (Gosz et al. 1978, Woodmansee 1978, Pletscher et al. 1989), and certainly of limited benefit to plants (Lamb 1985). In fact, plants often show large allocations of net primary production to defensive compounds (Coley et al. 1985) and higher net primary production when they are relieved of pests (Cates 1975, Morrow and LaMarche 1978, Marquis and Whelan 1994).

At higher trophic levels, predators may also affect nutrient cycling in terrestrial ecosystems, inasmuch as they create local patches of nutrient-rich soil with the delivery of dead animals (Schmitz et al. 2010, Carter et al. 2007). Spawning salmon deliver nitrogen from the marine environment to streamside forests (Helfield and Naiman 2001, Ben-David et al. 1998), especially when bears prey on them (Hilderbrand et al. 1999). Similarly, sea turtles return nutrients to dune habitats, where they lay eggs (Bouchard and Bjorndal 2000, Hannan et al. 2007). When predators—foxes—were introduced to an Aleutian island, they disrupted nesting seabirds, which normally delivered nutrients from the marine environment to the soil (Maron et al. 2006).

The role of animals in litter decomposition is much more significant (Swift et al. 1979, Hole 1981, Seastedt and Crossley 1980). Nematodes, earthworms, and termites are particularly widespread and important in the initial breakdown of litter and the turnover of nutrients in the soil. Schaefer and Whitford (1981) found that termites were responsible for the turnover of 8% of litter N in a desert soil (Figure 6.20). An additional 2% of the pool of nitrogen in surface litter was transported belowground by their burrowing activities. When termites were excluded by the application of pesticides, decomposition slowed and surface litter accumulated. Because soil animals have short lifetimes, their nutrient contents are rapidly decomposed and returned to the intrasystem cycle (Seastedt and Tate 1981).

It is interesting to view the biogeochemistry of animals from another perspective: What is the role of biogeochemistry in determining the distribution and abundance of animals? The death of ducks and cattle feeding in areas of high soil selenium (Se) suggests that such interactions might be of widespread significance.

Plants have no essential role for sodium in their biochemistry, and naturally have low Na contents due to limited uptake and exclusion at the root surface (Smith 1976). On the other hand, sodium is an important, essential element for all animals. The wide ratio between the Na content of herbivores to that in their foodstuffs suggests that Na might limit mammal populations generally. Observations of Na deficiency are supported by the interest that many animals show in natural salt licks (Jones and Hanson 1985, Freeland et al. 1985, Smedley and Eisner 1995) and Na-rich plants (Botkin et al. 1973, Rothman et al. 2006). Weir (1972) suggested that the distribution of elephants in central Africa was at least partially dependent on sodium in seasonal waterholes, and McNaughton (1988) found that the abundance of ungulates in the Serengeti area was linked to Na, P, and Mg in plant tissues available

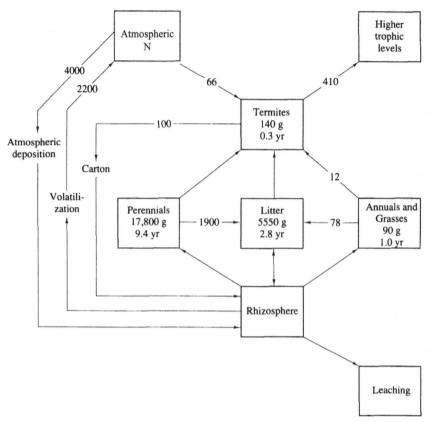

FIGURE 6.20 Nitrogen cycle in the Chihuahuan Desert of New Mexico, showing the role of termites in nitrogen transformations. Flux of nitrogen is shown along arrows in g N ha^{-1} yr^{-1}; nitrogen pools are shown in boxes with turnover time in years. *Source: From Schaefer and Whitford (1981). Used with permission of Springer.*

for grazing. Thus, animal populations may be affected by the availability of Na in natural ecosystems. Aumann (1965) found high rodent populations in areas of Na-rich soils, and speculated that the increased abundance of rodents in the eastern United States during the 1930s might have been due to a large deposition of Na-rich soil dust that was derived from the prairies during the "Dust Bowl." Such a case would link the abundance of animals to the biogeochemistry of soils and to soil erosion by wind in a distant region. Kaspari et al. (2009) suggest that the decomposition of litter by termites in tropical forests is mediated by the availability of sodium, derived from atmospheric deposition, which decreases inland from coastal areas.

CALCULATING LANDSCAPE MASS BALANCE

Elements are retained in terrestrial ecosystems when they play a functional role in biochemistry or are incorporated into organic matter. The pool of nutrients held in the soil and vegetation is many times larger than the annual receipt of nutrients from the atmosphere and rock weathering (e.g., Table 6.4). In the Hubbard Brook Experimental Forest in New Hampshire, turnover times (mass/input) range from 21 years for Mg to >100 years for P in the vegetation and forest floor (Likens and Bormann 1995b, Yanai 1992). In contrast, for a nonessential element, sodium (Na), the turnover time is rapid (1.2 years), because Na is not retained by biota or incorporated into humus.

Because chlorine (Cl) is highly soluble, not strongly involved in soil chemical reactions (Chapter 4), and only a trace element in plant nutrition (White and Broadley 2001), chloride (Cl^-) has traditionally been used as a tracer of hydrologic flux through ecosystems (Juang and Johnson 1967). However, several studies (Oberg et al. 2005, Bastviken et al. 2007, Leri and Myneni 2010) show that some Cl is incorporated and retained in soil organic matter, partially compromising its use as a conservative tracer of geochemical processes when the overall Cl flux is small (Svensson et al. 2012). Some nonessential—even toxic—elements such as lead (Pb) that bind to organic matter may also accumulate in soils (Smith and Siccama 1981, Friedland and Johnson 1985, Dörr and Münnich 1989, Kaste et al. 2005). Even though Pb is not involved in biochemistry, its retention in the ecosystem is the result of the presence of biotic processes. Studies of the movement of Si, Cl, Pb, and mercury (Hg) in the Earth's terrestrial ecosystems all fall into the realm of biogeochemistry.

Annual mineralization, plant uptake, and litterfall result in a large internal cycle of elements in most ecosystems. Annual nitrogen inputs are typically 1 to 5 kg ha^{-1} yr^{-1}, while mineralization of soil nitrogen is 50 to 100 kg ha^{-1} yr^{-1} (Bowden 1986). Despite such large movements of available nutrients within the ecosystem, there are usually only small losses of N in streams draining forested landscapes (~3 kg N ha^{-1} yr^{-1}; Lewis 2002). The minor loss of nitrogen in stream water speaks strongly for the efficiency of biological processes that retain elements essential to biochemistry. Where plants are present, most nitrogen deposited from the atmosphere is taken up and recycled in the ecosystem (Durka et al. 1994), whereas in deserts, a substantial portion is lost (Michalski et al. 2004b).

Relatively few studies have included measurements of gaseous flux in ecosystem nutrient budgets (Schlesinger 2009). Losses of nitrogen in denitrification may explain why the retention of N applied in fertilizer is often somewhat lower than that of other elements (e.g., P and K), which have no gaseous phase (Stone and Kszystyniak 1977). Globally, denitrification may explain the tendency for the growth of most vegetation to be N limited, despite efficient plant uptake of N from the soil and only minor losses in stream water (Houlton et al. 2006; Chapter 12).

Allan et al. (1993) compared the mass balance of elements in small patches of forest occupying rock outcrops in Ontario. Areas of bare rock showed net losses of various elements, whereas adjacent patches of forest showed accumulations of N, P, and Ca in vegetation. We should not, however, expect that the essential biological nutrients will accumulate indefinitely in all ecosystems. The incorporation of N and P in biomass should be greatest when structural biomass and soil organic matter are accumulating rapidly—that is, in young ecosystems where there is positive net ecosystem production (Chapter 5). Losses of N should be higher in mature, steady-state ecosystems where the total biomass is stable (Vitousek and Reiners 1975, Davidson et al. 2007). The extent to which N is incorporated into biota may depend on its availability relative to other elements. For instance, lowland tropical rainforests forests, where vegetation growth is generally limited by P, appear to be "leaky" with respect to N relative to temperate forests (Martinelli et al. 1999, Brookshire et al. 2012).

Using the mass-balance approach, where

$$\text{Input} - \text{Output} = \Delta\text{Storage}, \tag{6.6}$$

Vitousek (1977) found greater losses of available N from old-growth forests than from younger sites in New Hampshire. Hedin et al. (1995b) confirmed high nutrient losses in old-growth forests of Chile, with forms of dissolved organic nitrogen being an important fraction of the total loss. The relatively high losses of N and P from the Caura River in Venezuela (Table 6.9) are consistent with the mature vegetation covering most of its watershed (Lewis 1986, compare Davidson et al. 2007).

In seasonal climates, losses of N and K in stream waters are usually minor during the growing season and greater during the winter period of plant dormancy (Likens and Bormann 1995, Likens et al. 1994). Often there is little seasonal variation in the loss of Na and Cl, which pass through the system under simple geochemical control (Johnson et al. 1969, Belillas and Rodà 1991b). Stream-water losses of nutrients give old-growth ecosystems the appearance of being "leaky," but it is important to recognize that outputs represent the excess of inputs over the seasonal demand for nutrients by vegetation and soil microbes (Gorham et al. 1979).

During long periods of soil development, chemical weathering depletes soils of the essential nutrients traditionally thought to be derived from bedrock, especially Ca and P (Figure 4.11). Thus, studying a 4-million-year-old sequence of soils in Hawaii, Chadwick et al. (1999) showed that small atmospheric inputs of phosphorus assume great importance to vegetation growing on ancient soils (Figure 6.21). Vegetation progressively shifts from N limitation to P limitation with soil age (Vitousek 2004, Richardson et al. 2004). Atmospheric inputs of phosphorus from the long-distance transport of desert dust seem essential to the continued productivity of tropical rainforests in Hawaii and the Amazon Basin (Gardner 1990, Okin et al. 2004, Bristow et al. 2010).

Indeed, using strontium (Sr) as a tracer, some workers have suggested that the vegetation in many regions is dependent on atmospheric inputs of elements traditionally associated with rock weathering (Graustein and Armstrong 1983, Miller et al. 1993, Kennedy et al. 1998). These observations are not universal; in areas of rapid geologic uplift and erosion, weathering is a persistent source of phosphorus and other plant nutrients that are derived from bedrock

TABLE 6.9   Annual Chemical Budgets for Undisturbed Forests in Various World Regions

Location and reference	Precipitation (cm)	Chemical (kg ha^{-1} yr^{-1})			
		Ca	Cl	N	P
British Columbia (Feller and Kimmins 1979)	240	15.8	2.9	−2.6	0
Oregon (Martin and Harr 1988)	219	41.2	—	−1.2	0.3
New Hampshire (Likens and Bormann 1995)	130	11.7	−1.6	−16.7	0
North Carolina (Swank and Douglass 1977)	185	3.9	1.7	−5.5	−0.1
Venezuela (Lewis et al. 1987, Lewis 1988)	450	14.2	−1.4	8.5	0.32
Brazil (Lesack and Melack 1996)	240	−0.52	3.58	−2.4	−0.04

Note: Total stream-water losses minus atmospheric deposition.

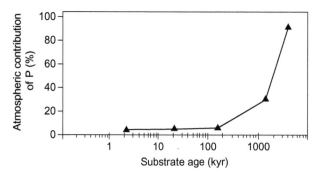

FIGURE 6.21 The relative importance of atmospheric inputs of phosphorus to ecosystems on the Hawaiian islands as a function of the age of the landscape. *Source: From Chadwick et al. (1999).*

(Bern et al. 2005, Porder et al. 2006). Generally, the long-term productivity of terrestrial vegetation may depend on periodic renewal of weatherable minerals (Wardle et al. 2004). Thus, biogeochemists must include both atmospheric and bedrock sources in ecosystem nutrient budgets, especially when these budgets are used to evaluate the impacts of changing levels of air pollution and atmospheric deposition on forest growth (e.g., Drouet et al. 2005, Mitchell et al. 2011a).

Among elements in short supply to biota, nitrogen is unique in that it is largely derived from the atmosphere (Table 4.5).[7] Net primary production in some temperate forests appears to show a correlation to N inputs in precipitation (Cole and Rapp 1981). Comparing forests from Oregon, Tennessee, and North Carolina, Henderson et al. (1978) noted strong N retention in each, despite a tenfold difference in N input from the atmosphere. The data suggest that plant growth is limited by N in each region. In contrast, losses of Ca were always a large percentage of the amount cycling in these forests. Especially on limestone soils, ample supplies of Ca were derived from rock weathering and Ca was not in short supply. Thus, abundant (e.g., Ca) and nonessential (e.g., Na) elements are most useful in estimating the rate of rock weathering (Chapter 4), whereas biogeochemistry controls the loss of scarce elements that are essential to life.

While most studies of ecosystem mass balance have considered watersheds, Baker et al. (2001) developed a nitrogen budget for the metropolitan ecosystem of Phoenix, Arizona, in which anthropogenic inputs (from food and pet food, combustion, and fertilizer) and gaseous outputs (NO_x and N_2) composed the largest movements of nitrogen (Table 6.10). Metson et al. (2011) formulated a similar assessment of the phosphorus budget for Phoenix. The ecosystem concept can even be applied to individual households to produce biogeochemical budgets for their inputs and outputs (Fissore et al. 2011). At the other extreme, Ti et al. (2012) compiled an input-output budget for nitrogen in all of mainland China. They found that chemical fertilizers, N fixation, and precipitation dominated the sources of nitrogen in this region, while denitrification, ammonia volatilization, and hydrologic export dominated the losses.

---

[7] Sedimentary and metasedimentary rocks usually contain a small amount of nitrogen, which can be released on weathering (Holloway and Dahlgren 2002). Nitrogen derived from bedrock can make a significant contribution to the nitrogen budget of some forests (Morford et al. 2011) and to the global sources of nitrogen for land plants (Chapter 12).

TABLE 6.10 Nitrogen Budget for the Phoenix Metropolitan Area

Inputs	N flux (G g y^{-1})
Surface water	1.2
Wet deposition	3.0
Human food	9.9
N-containing chemicals	5.8
Food for dairy cows	0.8
Pet food	2.7
Commercial fertilizer	24.3
Biological fixation	
Alfalfa	7.5
Desert plants	7.1
Fixation by combustion	36.3
Total inputs of fixed N	**98.4**
Outputs	
Surface water	2.6
Cows for slaughter	0.1
Milk	2.4
Atmospheric NO_x	17.1
Atmospheric NH_3	3.8
N_2O from denitrification	4.6
N_2 from denitrification	46.9
Total outputs	**77.5**
Accumulation (inputs-outputs)	20.9
Net subsurface storage	8.3
Landfills	8.6
Increase in human biomass	0.2

Source: Modified from Baker et al. 2001.

Many of the transformations in biochemistry involve oxidation and reduction reactions that generate or consume acidity (H^+). For instance, H^+ is produced during nitrification and consumed in the plant uptake and reduction of NO_3^-. Binkley and Richter (1987) review these processes and show how ecosystem budgets for H^+ may be useful as an index of net change in ecosystem function, particularly as soils acidify during ecosystem development (Chapter 4). H^+-ion budgets are also useful as an index of human impact, especially from acid rain and excess nitrogen deposition (Driscoll and Likens 1982). For example, a net increase in acidity is expected when excess NH_4^+ deposition is subject to nitrification, with the subsequent loss of NO_3^- in stream water (van Breemen et al. 1982). H^+ budgets are analogous to measurements of human body temperature. When we see a change, we suspect that the ecosystem is stressed, but we must look carefully within the system for the actual diagnosis.

HUMAN IMPACTS ON TERRESTRIAL BIOGEOCHEMISTRY

Acid Rain

Forest growth has declined in many areas that are downwind of air pollution (Savva and Berninger 2010). In addition to the direct effects of ozone and other gaseous pollutants on their growth, plants in these areas are subject to "acid rain"—perhaps better named *acid deposition*, since some of the acidity is delivered as dryfall (Chapter 3). While all rain is naturally somewhat acidic (Chapter 13), human activities can produce rain of exceptional low pH as a result of NO_3^- and SO_4^{2-} that are derived from the dissolution of gaseous pollutants in raindrops (Eqs. 3.26–3.30). The chemical inputs in acid rain affect several aspects of soil chemistry and plant nutrition, leading to changes in plant growth rates.

Inputs of H^+ in acid rain increase the rate of weathering of soil minerals, the release of cations from cation exchange sites, and the movement of Al^{3+} into the soil solution (Chapter 4). At the Rothamsted Experimental Station in the United Kingdom, soil pH declined from 6.2 to 3.8 between 1883 and 1991, in association with acid rainfall (Blake et al. 1999). Depending on the underlying parent rocks, the forest floor and soil exchange capacity may be substantially depleted of Ca^{2+} in areas of acid deposition (Miller et al. 1993, Wright et al. 1994, Likens et al. 1996, Johnson et al. 2008a). Forests in the Adirondack Mountains of New York lost 64% of their soil Ca between 1930 and 2006 (Bedison and Johnson 2010).

Field experiments simulating acid rain show the depletion of cations on soil cation exchange sites and mobilization of Al^{+3} (Fernandez et al. 2003). Between 1984 and 2001 the loss of Ca from soils in the northeastern United States was closely balanced with an increase in Al on the cation exchange sites (Warby et al. 2009). High concentrations of Al^{3+} may reduce the plant uptake of Ca^{2+} and other cations (Godbold et al. 1988, Bondietti et al. 1989), and in the northeastern United States forest growth appears to decline as a result of a decreased Ca/Al ratio in the soil solution (Shortle and Smith 1988, Cronan and Grigal 1995). In a spruce forest of New England, Bullen and Bailey (2005) document decreasing Ca, Sr, and other cations and increasing Al in tree rings during the past century of acid inputs. Losses of Al from the mineral soil horizons are also associated with the mobilization of Al-bound phosphorus (SanClements et al. 2010).

Bernier and Brazeau (1988a, 1988b) link dieback of sugar maple to deficiencies of K on areas of low-K rocks and to deficiencies of Mg on low-Mg granites in southeastern Quebec. Magnesium deficiencies are also seen in the forests of central Europe, where forest decline is linked to an imbalance in the supply of Mg and N to plants (Oren et al. 1988, Berger and Glatzel 1994). Graveland et al. (1994) suggest that the effects of acid rain are also seen at higher trophic levels; in the Netherlands, birds showed poor reproduction in forests subject to acid rain as a result of a decline in the abundance of snails, which are the main source of Ca for eggshell development. Similar effects are potentially related to the decline of the Wood Thrush in eastern North America (Hames et al. 2002).

With an abatement of air pollution, the acidity of rainfall has declined in many areas of the eastern United States and Europe, commencing a slow recovery of soils and vegetation, especially in sites where Ca has been strongly depleted from the soil (Palmer et al. 2004). Soil pH increased in England and Wales between 1978 and 2003, presumably as a result of lower emissions of sulfur dioxide and lower levels of acid rain (Kirk et al. 2010). Atmospheric inputs of Ca in dryfall can assume special significance in the rejuvenation of soil cation exchange capacity (Drouet et al. 2005). When researchers added 1.2 tons Ca per hectare as the mineral wollastonite ($CaSiO_3$) to the forest at Hubbard Brook, New Hampshire, the addition restored soil Ca (Cho et al. 2010), ameliorated the effects of acidity on the growth of sugar maple (Juice et al. 2006), and reduced losses of nitrogen (Groffman and Fisk 2011).

It is important to note that soil acidity can derive from a number of causes. During forest growth, the accumulation of cations in biomass leads to greater soil acidity (Berthrong et al. 2009). For a U.S. forest in South Carolina, Markewitz et al. (1998) attribute 62% of soil acidity to plant uptake and 38% to atmospheric deposition. Permanent increases in soil acidity will result if the new inputs of cations during a forest growth cycle are less than the removals in harvest. Similarly, the use of NH_4^+ fertilizers, which generate H^+ during nitrification, appears responsible for the acidification of agricultural soils in China (Guo et al. 2010).

Nitrogen Saturation

Currently, the deposition of available nitrogen from the atmosphere in the northeastern United States and western Europe (~ 10 kg N ha^{-1} yr^{-1}) is about five times greater than that recorded under pristine conditions.[8] The excess nitrogen derives from combustion of fossil fuels (releasing NO_x) and agricultural activities (releasing NH_3) upwind (Fang et al. 2011). Many workers have speculated that excess nitrogen deposition may act as fertilizer—stimulating the growth of trees. Indeed, the added nitrogen could lead to significant enhanced growth and carbon storage in forests, enhancing the removal of CO_2 from the atmosphere (Thomas et al. 2010, Magnani et al. 2007; but see Sutton et al. 2008, Hogberg 2012). Fertilizer experiments show that some of the added nitrogen also accumulates in soil organic matter (Nave et al. 2009, Gardner and Drinkwater 2009), enhancing carbon storage in soils (Nadelhoffer et al. 2004, Pregitzer et al. 2008, Hyvonen et al. 2008, Liu and Greaver 2010). Surprisingly, even though most forests are nitrogen-limited, plant uptake of exogenous nitrogen is usually only 10 to 30% of that applied (Schlesinger 2009, Pregitzer et al. 2010).

In some areas of high nitrogen deposition, particularly at high elevations, forest decline is observed as the ecosystem becomes saturated with nitrogen (Aber et al. 1998, 2003; McNulty et al. 2005; Lovett and Goodale 2011). In these sites, nitrification rates increase dramatically, yielding higher losses of NO_3^- in stream waters (Peterjohn et al. 1996, Corre et al. 2003, Lu et al. 2010) and greater emissions of N_2O to the atmosphere (Brumme and Beese 1992, Peterjohn et al. 1998, Venterea et al. 2003). While the losses of nitrogen to stream waters

---

[8] See, for example, *http://nadp.sws.uiuc.edu/maplib/pdf/2010/TotalN_10.pdf.*

are normally dominated by dissolved organic nitrogen (DON) in areas of excessive nitrogen deposition, NO_3^- becomes increasingly dominant (Perakis and Hedin 2002, Lovett et al. 2000, Lutz et al. 2011). Forests receiving high nitrogen deposition show higher $\delta^{15}N$ in canopy foliage, indicative of a high rate of nitrification, and NO_3^- losses in streams increase (Pardo et al. 2007).

Along three gradients of increasing air pollution in southern California, Zinke (1980) showed that N content in the foliage of Douglas fir increased from 1% to more than 2%, while P content decreased abruptly, changing the ratios of N to P from about 7 in relatively pristine areas to 20 to 30 in polluted areas. Such an imbalance in leaf N/P ratios is also seen in the Netherlands, in areas of excessive inputs of NH_4^+ from the atmosphere (Mohren et al. 1986). Historical collections show increasing nitrogen concentrations in some plants during the past century (Peñuelas and Filella 2001), perhaps indicating a shift of the terrestrial biosphere away from N deficiency (Elser et al. 2007). Nevertheless, in areas of high N deposition, forests show only scattered evidence of P deficiency (Gress et al. 2007, Finzi et al. 2009, Weand et al. 2010).

The symptoms of nitrogen saturation vary as a function of underlying site fertility, species composition, and other factors. Lovett and Goodale (2011) stress the importance of the rate of N input to the rate of N uptake by plants and soil microbes in controlling the appearance of nitrogen saturation and enhanced N loss. Low fertility sites may show only small changes in nitrification because plants take up the excess N deposition from the atmosphere (Fenn et al. 1998, Lovett et al. 2000). Without specific field experiments, it is often difficult to separate the effects of acid rain from those of excess nitrogen, since a large fraction of the nitrogen deposited from the atmosphere arrives as nitric acid, and inputs of NH_4^+ generate acidity if they are nitrified (Stevens et al. 2011). Temperate forests have been studied most extensively, but it is likely that these effects will be increasingly found in tropical regions, where P limitation predominates (Hall and Matson 1999, Koehler et al. 2009, Corre et al. 2010, Cusack et al. 2011, Hietz et al. 2011).

Nitrogen saturation is reversible. With experimental reductions of nitrogen inputs in areas of high deposition, the rates of nitrification and the loss of NO_3^- to stream waters decline (Quist et al. 1999, Corre and Lamerdorf 2004, Lutz et al. 2012b).

Rising CO_2 and Global Warming

Rising concentrations of CO_2 in Earth's atmosphere appear to stimulate the growth and carbon storage of land plants by enhancing plant photosynthesis (Chapter 5). Early greenhouse studies suggested that this response might be short-lived because of soil nutrient limitations (Thomas et al. 1994a). Several workers postulated a progressive nutrient limitation of field plants grown at high CO_2 (Luo and Reynolds 1999, Luo et al. 2004); however, some experiments that exposed intact forests to high CO_2 show a positive response to CO_2 that lasts up to a decade (Finzi et al. 2006). Some of the greater nutrient demand by faster growing plants is met by greater nutrient-use efficiency in photosynthesis (Springer et al. 2005), greater nutrient reabsorption before leaf abscission (Finzi et al. 2002, Norby et al. 2001), and greater allocation of carbon to root exudates that stimulate the decomposition of soil organic matter and nutrient mineralization (Drake et al. 2011, Phillips et al. 2011b).

Plants also respond to elevated CO_2 with greater root growth, which appears to explore the soil nutrient pool more fully (Norby and Iversen 2006, Hungate et al. 2006, Pritchard et al. 2008b, Finzi et al. 2007, Jackson et al. 2009). The duration of the positive growth response of plants to high CO_2 in field experiments is surprising; eventually, stoichiometric constraints, such as the C/N ratio in plant biomass, will limit the amount of carbon that can be sequestered in woody biomass and soils in the absence of exogenous inputs of N (Johnson 2006, van Groenigen et al. 2006). Indeed, nitrogen appears to constrain long-term growth response at the FACE experiment at Oak Ridge, Tennessee (Norby et al. 2010, Garten et al. 2011).

Soil-warming experiments, designed to simulate ongoing climate change, typically show an increase in soil nitrogen mineralization (Van Cleve et al. 1990, Rustad et al. 2001, Shaw et al. 2001, Melillo et al. 2002). The change in soil microbial activity mobilizes nitrogen for plant uptake, potentially enhancing plant growth and carbon uptake. In wet tundra, soil warming stimulated the rate of decomposition, but caused only a small increase in plant growth in field experiments (Johnson et al. 2000b, Mack et al. 2004, Shaver et al. 2006). In contrast, in temperate forest ecosystems, soil warming stimulates nitrogen mineralization, plant carbon uptake, and net carbon sequestration in aboveground tissues (Melillo et al. 2011). In some areas, where the loss of the insulating effect of a winter snow pack results in frozen soils, nitrogen mineralization rates are likely to decline under warmer, future climatic conditions (Groffman et al. 2009).

SUMMARY

Interactions between plants, animals, and soil microbes link the internal biogeochemistry of terrestrial ecosystems. Plants adapted to low nutrient availability have low nutrient contents and higher nutrient reabsorption before leaf-fall, yielding higher nutrient-use efficiency (Figure 6.22). In some cases these characteristics can be induced by experimental treatments that reduce nutrient availability. For instance, when Douglas fir were fertilized with sugar, which increases the C/N ratio of the soil and the immobilization of N by microbes, reabsorption of foliar N increased, implying greater nutrient-use efficiency by the trees (Turner and

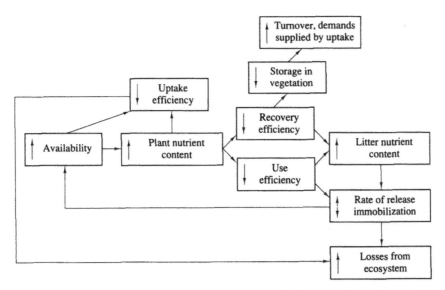

FIGURE 6.22 Changes in internal nutrient cycling that are expected with changes in nutrient availability. *Source: From Shaver and Melillo (1984). Used with permission of the Ecological Society of America.*

Olson 1976). Internal cycling by the vegetation may partially alleviate nutrient deficiencies, but decomposition of nutrient-poor litterfall is slow, further exacerbating the low availability of nutrients in the soil (Hobbie 1992, Lovett et al. 2004). Thus, nutrient-poor sites are likely to be occupied by vegetation that is specially adapted for long-term persistence under such conditions (Chapin et al. 1986b). In turn, the vegetation leaves its imprint on microbial activity and soil properties (Lovett et al. 2004, Reich et al. 2005).

Biogeochemistry controls the distribution and characteristics of vegetation at varying scales. Continental distributions of vegetation, such as the widespread dominance of conifers in the boreal regions, are likely to be related to the higher nutrient-use efficiency of evergreen vegetation under conditions of limited nutrient turnover in the soil. The effect of soil properties on the regional distribution of vegetation is seen in the occurrence of evergreen vegetation on nutrient-poor, hydrothermally altered soils in arid and semiarid climates (Figure 6.23). Fine-scale spatial heterogeneity of soil properties, as recorded by Robertson et al. (1988) for a field in Michigan, has been linked to the maintenance of diversity in land plant communities (Tilman 1985), and several studies show the importance of local soil conditions to the distribution and abundance of forest and grassland herbs (Snaydon 1962, Pigott and Taylor 1964, Lechowicz and Bell 1991, John et al. 2007). Additions of fertilizer tend to reduce the species diversity of plant communities (Huenneke et al. 1990, Wedin and Tilman 1996, Stevens et al. 2006, Cleland and Harpole 2010).

Linkages among components of the intrasystem cycle suggest that an integrative index of terrestrial biogeochemistry might be derived from the measure of a single component, such as the chemical characteristics of the leaf canopy (Matson et al. 1994). Wessman et al. (1988b) analyzed the spectral reflectance of leaf tissues in the laboratory as a first step toward developing an index of forest canopies by remote sensing. Their data show a strong correlation between nitrogen and lignin measured by infrared reflectance and by traditional laboratory analyses. Several workers now use satellite measurements of reflectance to characterize canopy properties (Martin and Aber 1997, Asner and Vitousek 2005, Kokaly et al. 2009, Ollinger 2011).

In the White Mountains of New Hampshire, forest productivity appears related to canopy nitrogen content, as measured by remote sensing (Figure 6.24; Ollinger and Smith 2005,

FIGURE 6.23 Occurrence of *Pinus ponderosa* and *Pinus jeffreyi* on acid, nutrient-poor hydrothermally altered andesites in the Great Basin Desert of Nevada, with *Artemisia tridentata* occurring on adjacent desert soils, of higher pH and phosphorus availability. *Sources: Schlesinger et al. (1989) and Gallardo and Schlesinger (1996).*

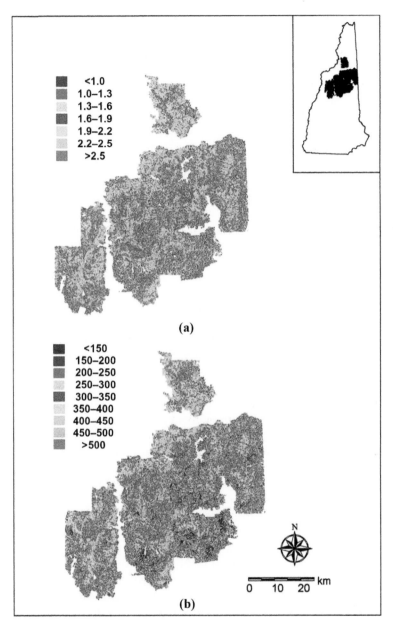

FIGURE 6.24 Spatial variation of canopy nitrogen and forest wood production in central New Hampshire, as estimated from the AVIRIS satellite. (a) Whole-canopy nitrogen concentration (%). (b) Aboveground woody biomass production (g·M^{-2}·yr^{-1}). *Source: From Smith et al. (2002). Used with permission of the Ecological Society of America.*

Ollinger et al. 2008), and variations of leaf C/N ratio across sites provide a convenient index of the intrasystem cycle of nutrients (Ollinger et al. 2002). Recognizing that decomposition is frequently controlled by the lignin and nitrogen content of litter (refer to Figure 6.11), remote sensing of canopy characteristics has potential for comparative regional studies of nutrient cycling in different plant communities (Myrold et al. 1989). Canopy lignin, measured by aircraft remote sensing, was highly correlated to soil nitrogen mineralization in Wisconsin forests (Wessman et al. 1988b, Pastor et al. 1984). These studies reinforce our appreciation of the linkage between vegetation and soil characteristics, as outlined in Figure 6.22.

Various models demonstrate other linkages between plant and soil processes in terrestrial biogeochemistry. Walker and Adams (1958) suggested that the level of available phosphorus during soil development was the primary determinant of terrestrial net primary production, since nitrogen-fixing bacteria depend on a supply of organic carbon and available phosphorus. They use the level of organic carbon in the soil as an index of terrestrial productivity and suggest that organic carbon peaks midway during soil development and then declines as an increasing fraction of the phosphorus is rendered unavailable by precipitation with secondary minerals (Figure 4.11). The model is consistent with observations of the increasing limitation of NPP by phosphorus during soil development (Chadwick et al. 1999, Richardson et al. 2004).

Numerous workers have examined the Walker and Adams (1958) hypothesis in various ecosystems. Tiessen et al. (1984) found that available phosphorus explained 24% of the variability of organic carbon in a collection of 168 soils from eight different soil orders. Roberts et al. (1985) found a similar relationship between bicarbonate-extractable P and organic carbon in several grassland soils of Saskatchewan. Raghubanshi (1992) found that phosphorus was well correlated to soil organic matter, soil nitrogen, and nitrogen mineralization rates in dry tropical forests of India. Thus, available phosphorus explains some, but not all, of the variation in soil organic carbon, which is ultimately derived from the production of vegetation. The linkage of phosphorus and carbon is likely to be strongest during early soil development, when both organic phosphorus and carbon are accumulating. The importance of organic phosphorus increases during soil development, and through the release of phosphatase enzymes, vegetation interacts with the soil pool to control the mineralization of P.

Parton et al. (1988) present a model linking the cycling of C, N, P, and S in grassland ecosystems. The flow of carbon is shown in Figure 6.25. The nitrogen cycling submodel has a similar structure, since the model assumes that most nitrogen is bonded directly to carbon in amino groups (McGill and Cole 1981). Lignin controls decomposition rates and nitrogen is mineralized from soil pools when critical C/N ratios are achieved during the respiration of carbon. Phosphorus availability is controlled by a modification of a model first presented by Cole et al. (1977), which includes C/P control over mineralization of organic pools and geochemical control over the availability of inorganic forms as in Figure 6.19. However, unlike N, C/P ratios in plant tissues and soil organic matter are allowed to vary widely as a function of P availability.

The complete model was used to predict patterns of primary production and nutrient mineralizations during 10,000 years of soil development (Figure 6.26). Net primary production and accumulations of soil organic matter are strongly linked to P availability during the

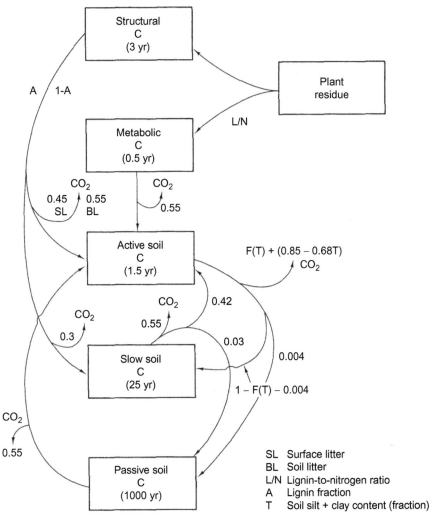

FIGURE 6.25 Flow diagram for carbon in the CENTURY model. The proportion of carbon moving along each flowpath is shown as a fraction, and turnover times for reservoirs are shown in parentheses. *Source: From Parton et al. (1988). Used with permission of Springer.*

first 800 years, after which increases in plant production are related to increases in soil N mineralization. Organic P increases throughout the 10,000-year sequence. In simulations of the response of native soils to cultivation, the model predicted a correlated decline in the native levels of organic carbon and nitrogen in the soil, but a relatively small decline in P. Validation of the model is seen in the data of Tiessen et al. (1982), who found declines of 51% for C and 44% for N, but only 30% for P in a silt loam soil cultivated for 90 years in Saskatchewan.

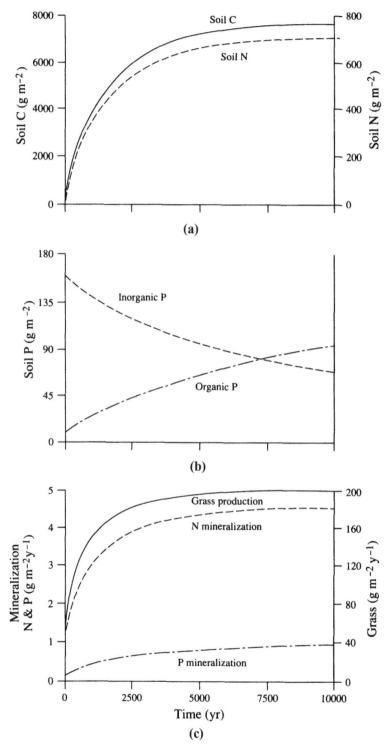

FIGURE 6.26 Simulated changes in soil C, N, and P during 10,000 years of soil development in a grassland, using the CENTURY model. *Source: Parton et al. (1988). Used with permission of Springer.*

Recommended Readings

Aber, J.D., and J.M. Melillo. 2001. *Terrestrial Ecosystems*. Academic Press/Elsevier.

Agren, G.L. and F.G. Andersson. 2011. *Terrestrial Ecosystem Ecology*. Cambridge University Press.

Dobrovolsky, V.V. 1994. *Biogeochemistry of the World's Land*. CRC Press.

Johnson, D.W., and S.E. Lindberg (Eds.). 1992. *Atmospheric Deposition and Nutrient Cycling*. Springer-Verlag.

Likens, G.E., and F.H. Bormann. 1995. *Biogeochemistry of a Forested Ecosystem*, second ed. Springer-Verlag.

Marschner, P., and Z. Rengel (Eds.). 2007. *Soil Biology: Nutrient Cycling in Terrestrial Ecosystems*. Springer-Verlag.

Paul, E.A. 2007. *Soil Microbiology, Ecology and Biochemistry*, third ed. Academic Press/Elsevier.

Sterner, R.W., and J.J. Elser. 2002. *Ecological Stoichiometry*. Princeton University Press.

Vitousek, P.M. 2004. *Nutrient Cycling and Limitation*. Princeton University Press.

Wolfe, D.W. 2001. *Tales from the Underground*. Perseus.

PROBLEMS

1. For ryegrass (Table 6.2), what ratio of plant uptake of NH_4 and NO_3 will produce no change in the acidity of soil in the rooting zone?

2. Using data for the chaparral shrubland (Table 6.4), calculate nutrient-use efficiency (NPP/unit nutrient uptake) for each element. Speculate on why nutrient-use efficiency in this shrubland differs from that in the forests listed in Table 6.6.

3. What is the C/N ratio of carbonaceous chondrites, the Earth's crust, land plants, and terrestrial animals? What does this suggest about nitrogen availability to the biosphere?

4. If global net primary productivity on land is 60×10^{15} g C/yr, and the average C/N ratio of plant biomass is 100, what is the upper limit (i.e., maximum estimate) of oxygen used every year in microbial nitrification?

5. Assume that a forest soil contains 8000 kg N/ha and that each year 80 kg N/ha/yr are mineralized and half of those are nitrified and lost. There is very little isotopic fractionation during mineralization, but during nitrification ^{14}N molecules react up to $1.035\times$ faster than ^{15}N molecules. Assuming no new inputs to the soil, what is the change in $\delta^{15}N$ in the soil pool after one year? (Do not round the numbers as you perform this calculation.)

Wetland Ecosystems

INTRODUCTION

Aquatic and terrestrial ecosystems differ in the relative importance of hydrology as a critical factor in their biogeochemistry. In terrestrial ecosystems water may limit autotrophic or heterotrophic activity directly. In aquatic ecosystems where water is abundant, hydrology plays other fundamental roles. First, because oxygen diffuses 10^4 more slowly in water than it does in air, water indirectly limits biogeochemical activity by constraining oxygen supply. As a result there are many places in aquatic ecosystems where oxygen consumption exceeds rates of delivery, so anoxia is the typical condition of most sediments in wetlands, lakes, streams, and oceans. The plants and microbes controlling the biogeochemistry of aquatic

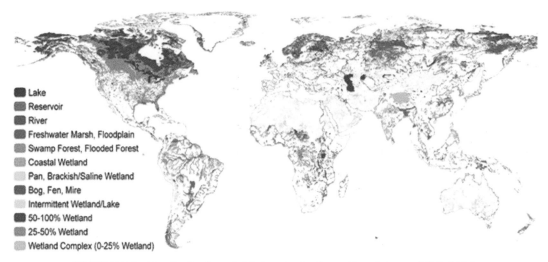

Lake
Reservoir
River
Freshwater Marsh, Floodplain
Swamp Forest, Flooded Forest
Coastal Wetland
Pan, Brackish/Saline Wetland
Bog, Fen, Mire
Intermittent Wetland/Lake
50–100% Wetland
25–50% Wetland
Wetland Complex (0–25% Wetland)

FIGURE 7.1   The distribution of global wetlands. *Source: From Lehner and Doll (2004).*

ecosystems must cope with limited oxygen supplies. Second, because of their low topo-graphic positions where surface waters collect or groundwaters emerge, aquatic ecosystems receive substantial inputs from the surrounding terrestrial catchments.

The importance of these subsidies from terrestrial ecosystems depends to a great extent on the ratio of shoreline to the volume of the ecosystem. In many aquatic ecosystems these *allochthonous* (i.e., externally derived) inputs of energy and elements can exceed *autochthonous* (in situ) inputs by deposition, photosynthesis, or fixation, such that the biogeochemical cycles of many (perhaps most) aquatic ecosystems are net heterotrophic, or reliant on surrounding terrestrial ecosystems to provide the majority of their annual supply of organic matter and essential elements.

For many aquatic ecosystems, ecosystem boundaries can be difficult to assess on the ground, and even more difficult to estimate using remote sensing. All estimates of wetland area are thus fraught with uncertainty. Improvements in remote sensing will offer refined estimates of the areal extent of aquatic systems and the contribution of aquatic ecosystems to global biogeochemical cycles. The current global estimates of lakes, rivers, and wetlands are shown in Figure 7.1 and Table 7.1.

Although there are a wide variety of wetland types (e.g., marshes, bogs, fens, swamps), all wetlands are characterized by the unique features of their hydrology, vegetation, and soils. All wetlands have water at or near the surface for at least some portion of the year; as a result they have hydric soils[1] that exhibit intermittent to permanent anoxia. Hydrophytic (water-loving) plants capable of living in saturated soils dominate wetland vegetation. Estimates of the global wetland area are extremely variable, ranging from 5.3 to 12.8×10^6 km^2 depend-ing on the stringency of the wetland definition applied (Matthews and Fung 1987, Aselmann and Crutzen 1989, Lehner and Doll 2004, Mitsch and Gosselink 2007). Despite their relatively

[1] Soils that are formed under conditions of periodic or continuous saturation sufficient to develop anoxic conditions in the upper horizons.

TABLE 7.1   Estimated Global Spatial Extent of Inland Waters

Class	Global Area	
	10^3 km^2	%
1. Lake	2428	1.8
2. Reservoir	251	0.2
3. River	360	0.3
4. Freshwater Marsh, Floodplain	2529	1.9
5. Swamp Forest, Flooded Forest	1165	0.9
6. Coastal Wetland	660	0.5
7. Pan, Brackish/Saline Wetland	435	0.3
8. Bog, Fen, Mire	708	0.5
9. Intermittent Wetland/Lake	690	0.5
10. Wetland Complexes		
50–100% Wetland	882–1764	0.7–1.3
35–50% Wetland	790–1580	0.6–1.2
0–25% Wetland	0–228	0–0.2
Total lakes and reservoirs (1–3)	2679	2.0
Total Wetlands (4–10)	8219–10,119	6.2–7.6

Source: Data from Lehner and Doll (2004). In these analyses, they assumed a total global land surface area (excluding Antarctica and glaciated Greenland) of 133 million km².

small proportional area, wetlands play an important role in global carbon cycling. In some instances, wetlands have the highest average productivity of any ecosystem type ($1300 \text{ g C m}^{-2} \text{ yr}^{-1}$) (Houghton and Skole 1990). Wetlands contribute 7 to 15% of global terrestrial productivity, and collectively store more than half of all the soil carbon on Earth (peatlands alone are estimated to store >50% of the world's soil carbon) (Gorham 1991, Eswaran et al. 1993, Roulet 2000, Tarnocai et al. 2009). The importance of wetlands as a global soil carbon sink is considerably offset by their production of the greenhouse gas CH_4 for which wetlands are the dominant global source, contributing 20 to 33% of global methane emissions (refer to Table 11.2; Bousquet et al. 2006, Bloom et al. 2010, Ringeval et al. 2010).

Wetlands support ideal conditions for the removal of reactive nitrogen by denitrification (Jordan et al. 2011) and for the sequestration of nitrogen and phosphorus in organic matter (Reddy et al. 1999). At the same time, wetlands can be important sources of dissolved organic matter (and organic nutrients) to downstream and coastal ecosystems (Schiff et al. 1998, Pellerin et al. 2004, Harrison et al. 2005). Despite the high potential for denitrification within wetlands, there are no good estimates of their contribution to global N_2O emission, and it has generally been assumed that N_2O production is much lower in wetlands than in upland soils (Bridgham et al. 2006, Mitsch and Gosselink 2007, Schlesinger 2009).

The heightened capacity for denitrification, CH_4 production, and soil carbon storage in wetlands are all a result of a lack of oxygen in wetland sediments. While aerobic oxidation ($CH_2O + O_2 \rightarrow CO_2 + H_2O$) dominates organic matter decomposition in most terrestrial ecosystems, microbes in flooded soils must use a variety of anaerobic metabolic pathways to obtain energy from organic matter. Microbial consumption of oxygen in wet soils and sediments frequently exceeds O_2 supply through diffusion. Without oxygen, microbes cannot use oxidative phosphorylation to decompose organic polymers to CO_2, and instead must rely on the alternate electron acceptors NO_3^-, Fe^{3+} and Mn^{4+}, SO_4^{2-}, or, in the most highly reducing environments, fermentation products such as acetate or CO_2 itself. Many of these primitive metabolic pathways evolved prior to the oxygenation of the Earth (Chapter 2), and continue to dominate the biogeochemistry of anoxic wetland sediments. These pathways yield less energy than aerobic respiration and the supply of alternate electron acceptors is often limiting, leading to far less efficient decomposition in wetlands and ultimately to large stores of organic matter. Over geologic time, the organic detritus accumulated in wetlands of the Carboniferous was buried and lithified to become modern coal deposits (Cross and Phillips 1990, McCabe 2009).

TYPES OF WETLANDS

The great variety of wetland types can be classified using hydrologic and physical properties that collectively constrain biogeochemical cycling and determine how wetlands influence the form, timing, and magnitude of chemical exports from catchments (Brinson 1993). Among the most important factors are water residence time, the degree of hydrologic connectivity between the wetland and regional rivers or groundwater, and the frequency, intensity, and duration of inundation. Wetlands may also be characterized by their soils and vegetation, which both respond to and exert control over wetland hydrology.

Wetland Hydrology

Water may enter a wetland by precipitation, tributary inflows, near-surface seepage, and exchange with deeper groundwater; water leaves wetlands through groundwater recharge, surface outflows, and evapotranspiration (Figure 7.2).

$$\text{Wetland volume } (V) = \text{Inputs } (P_n + S_i + G_i) - \text{Outputs } (ET - G_o - S_o). \quad (7.1)$$

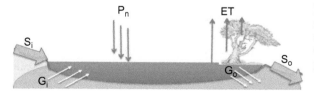

FIGURE 7.2   A wetland water budget. P_n represents precipitation inputs; ET represents evapotranspiration losses. S denotes surface water; G denotes groundwater. Subscript i indicates inputs; o indicates outputs.

The residence time of water within a wetland is calculated as the wetland volume (V) divided by the total inputs or outputs (MRT = V/Inputs or V/Outputs).[2] Wetlands in which precipitation and evapotranspiration are the only modes of water exchange usually have long water residence times and are sometimes referred to as closed systems because the rate of internal element turnover vastly exceeds the exchange of elements across ecosystem boundaries. In contrast, wetlands where runoff and outflow dominate the water budget are known as open systems, since water residence times are short and the flux of materials through the system may approach or exceed nutrient turnover within the ecosystem. The residence time of water ultimately limits the capacity of biota to change the composition of waters passing through a wetland.

The degree of hydrologic connectivity between a wetland and its catchment affects not only the source but the composition of biogeochemically important elements in the wetland, as well as the degree to which the wetland can affect biogeochemical patterns at larger, catchment scales. Comparisons of northern peatlands and riverine wetlands are instructive. The extensive peatlands at high latitudes tend to be hydrologically isolated wetlands, known as *ombrotrophic*[3] bogs, which receive all or most of their water from precipitation (Gorham 1957). In contrast, the wetlands that border many rivers experience episodic flooding and inundation accompanied by high rates of sediment deposition and erosion, and thus experience active biogeochemical exchange with adjacent surface waters.

Just as the amount and timing of annual precipitation is a key determinant of vegetation composition and productivity in terrestrial ecosystems, the annual hydroperiod of wetlands is a defining characteristic of their biological and biogeochemical properties (Brinson 1993). A wetland's hydroperiod describes the depth, duration, and frequency of inundation during a typical year (Figure 7.3). Some wetlands are permanently flooded, while others never have standing surface water. The periodicity and predictability of flooding in wetlands varies, with some wetlands flooded seasonally while other wetlands are only inundated following heavy rainfall. Wetlands subject to lunar tides will have predictable diel flooding frequencies, while wind tides lead to dynamic and unpredictable flooding in wetlands on low-lying coastal plains. The duration and intensity of flooding drives temporal variation in soil oxygen availability in wetland sediments, which places strong constraints on wetland vegetation. Flood pulses can also provide critical subsidies of material from upland and upstream ecosystems that fuel wetland productivity and food webs (Brinson et al. 1981, Megonigal et al. 1997).

Salinity merits special consideration. Wetlands that are hydrologically connected to the oceans or to inland salt lakes have intermittently or permanently high concentrations of seasalts. In salt marshes salinity may change from freshwater to full-strength seawater over the course of a single tide cycle, while more inland coastal wetlands may experience saltwater intrusion only during rare droughts or extreme wind tides associated with hurricanes. Only a few species of herbaceous vegetation have successfully adapted to life in full-strength seawater, since the high ionic strength of saltwater makes it difficult for plants to maintain osmotic balance.

---

[2] See Footnote 1, Chapter 3.

[3] Ombrotrophic, literally meaning "to feed on rain," is a term often used to describe vegetation dependent on atmospheric nutrients.

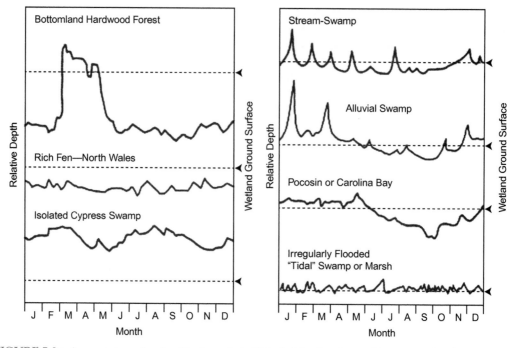

FIGURE 7.3   A comparison of wetland hydroperiods. Note that the duration and frequency of surface water vary greatly between wetland types. While the Bottomland Hardwood Forest and Carolina Bay are seasonally flooded, the hydrologically isolated cypress swamp is permanently flooded. In contrast, wetlands adjacent to rivers such as the stream-swamp and alluvial swamp shown here may undergo periodic inundation in conjunction with river floods. Tidal marshes may be flooded and drained on a daily basis, while some rich fens and bogs never have standing surface water. *Source: From Brinson (1993). Used with permission of Springer.*

Wetland Soils

Generally, wetland soils can be classified into three categories:

1. Soils permanently inundated with water above the soil surface
2. Saturated soils with the water table at or just below the soil surface
3. Soils where the water table depth is always below the surface

In saturated wetland soils, oxygen typically does not diffuse more than a few millimeters below the water table and reduced compounds and trace gases (N_2O, H_2S, CH_4) produced from anaerobic metabolic pathways can accumulate at high concentrations. Iron is a convenient indicator of anoxic conditions in the field because oxidized iron is easily recognized in soils by its red color, whereas reduced iron is grayish (Megonigal et al. 1993). Soil layers with reduced iron are called gley (Figure 7.4). In saturated wetland soils, the soil volume is generally 50% solids and 50% water, while in upland soils as much as 25% of the soil volume can consist of air-filled pore space. In upland soils only the interior of soil aggregates is typically anoxic (Chapter 6) and gases diffuse readily between the soil and the atmosphere. Wetland soils can

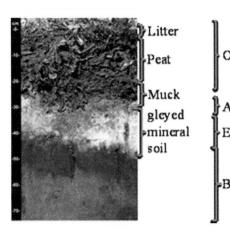

Litter

Peat

O

Muck

gleyed

mineral

soil

A

E

B

FIGURE 7.4 A hydric soil profile, with a thick dark layer of organic soil overlying a grey mineral soil characteristic of reduced iron. The traditional soil horizons for this Spodosol are indicated on the right. The organic (O) horizon overlies a mineral (A) horizon enriched in humic materials. The zone of eluviation (E) is characterized by a loss of silicate clays, iron, or aluminum and overlies the B horizon, or zone of illuviation. *Source: Image from NRCS 2010 Field Indicators of Hydric Soils in the United States; see* www.nwo.usace.army.mil/html/od-rwy/hydricsoils.pdf.

be converted to upland soils through drainage. The global loss of wetlands has largely resulted from efforts to drain wetlands so that formerly saturated sediments can support agriculture.

Wetland Vegetation

Having saturated sediments but sufficiently shallow surface water that allows vascular plants to dominate, wetlands occupy a special place along the terrestrial-to-aquatic continuum. The dominant autotrophs in wetlands have similar light and nutrient requirements to those of the plants of upland systems (Chapter 6) but must overcome the additional constraint of rooting in waterlogged soils with low O_2 concentration. Wetlands may be dominated by autotrophs ranging from sphagnum moss, sedges, reeds, and grasses to shrubs or trees—with the dominant vegetation both a function of and a control on wetland water balance and hydroperiod. Vascular plants growing in wetland ecosystems must cope with periodic or permanent saturation of their root tissues, a physiological challenge that prevents many plants from growing successfully in wetlands (Bailey-Serres and Voesenek 2008). A lack of oxygen in saturated soils directly interferes with root metabolism, creating root oxygen deficiency (Keeley 1979, Gibbs and Greenway 2003). In addition, anoxic sediments support microbial production of some metabolic products that are toxic to plants (e.g., H_2S from sulfate reduction; Eq. 2.12) (Lamers et al. 1998, Wang and Chapman 1999).

Wetland plants have developed a variety of morphological and physiological traits that allow them to persist in saturated sediments. Some wetland plants have the capacity to use anaerobic fermentation in their roots during periods of low or no oxygen (Keeley 1979, Gibbs and Greenway 2003, Greenway and Gibbs 2003); however, metabolism of organic compounds via fermentation is far less efficient than aerobic respiration. Many wetland plants have airspaces within their cortex (*aerenchyma*) that facilitate gas exchange between the atmosphere and the sediments surrounding their roots (Brix et al. 1992, Jackson and Armstrong 1999; Figure 7.5). Still other wetland plants, such as the bald cypress and coastal mangrove trees, have specialized aerial rooting structures (*pneumatophores*) that appear to facilitate gas exchange (Kurz and Demaree 1934, Scholander et al. 1955).

(a)   **(b)**

FIGURE 7.5   (a) Electron micrograph of a cross-sectioned stem of the aquatic macrophyte *Potamageton*; (b) photograph of Cypress pneumatophores. *Source: (a) From Jackson and Armstrong (1999). (b) From Wikipedia Commons.*

Facilitated gas exchange between the atmosphere and the rhizosphere, in addition to alleviating oxygen stress for plants, can also increase the oxygen content of wetland soils around plant roots (Wolf et al. 2007, Schmidt et al. 2010a), allowing aerobic metabolism by soil microbes in flooded soils. For many wetland plants the extent of aerynchymous tissues depends on the intensity or duration of inundation, suggesting that there is some physiological cost associated with these specialized tissues (Justin and Armstrong 1987). These adaptations allow wetland plants to persist in suboxic or anoxic conditions and to alter soil oxygen availability. Some plants are only found in wetland ecosystems (obligate wetland plants) whereas others are capable of growing across a broader range of hydrologic conditions (facultative wetland plants) and are merely more common in wetland ecosystems.

PRODUCTIVITY IN WETLAND ECOSYSTEMS

Emergent plants dominate the vegetation of most wetlands and net primary production is usually estimated using the harvest or eddy-covariance approaches outlined in Chapter 5. Net primary productivity varies widely across wetland ecosystems depending on nutrient supply (Brinson et al. 1981, Brown 1981). Unlike terrestrial ecosystems, where variation in vegetation type and stature is largely predictable from climate, the differences in wetland productivity is more strongly influenced by edaphic[4] factors (Brinson 1993). Variation in wetland hydroperiod has important consequences for productivity, because autotrophic respiration is less efficient in saturated soils (as discussed in the previous section) and because a high proportion of nutrients are sequestered in undecomposed soil organic matter, leaving low concentrations (and slow turnover) of available nutrients in the soil. Areas that are less frequently flooded tend to have higher productivity, since periodic soil drying allows for more rapid nutrient mineralization by aerobic microbes (Figure 7.6).

In contrast to upland terrestrial ecosystems, where numerous experiments have documented nutrient limitation of primary productivity, there have been far fewer experimental manipulations of nutrient supply in wetland ecosystems (Bedford et al. 1999). Venterink et al. (2001) reported that nearly half of 50 fertilization experiments in wetlands found

---

[4] Resulting from or influenced by the soil.

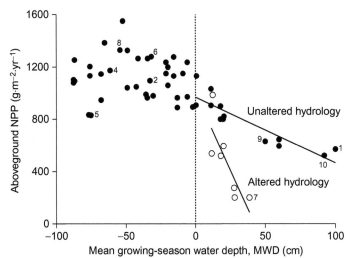

FIGURE 7.6 Water depth was negatively correlated with aboveground NPP for southern coastal wetlands. The effect of inundation was more pronounced when levees were built to maintain permanent flooding (plots shown in open circles). *Source: From Megonigal et al. 1997. Used with permission of Springer.*

significant N limitation of plant biomass, 8 experiments reported P limitation, and 13 reported colimitation by N with either P or K. In wetlands spanning gradients of atmospheric N deposition across Europe and Canada, N inputs appear to be correlated with increasing vascular plant biomass and reduced biomass of low stature mosses (Berendse et al. 2001, Turunen et al. 2004, Limpens et al. 2008). In a long-term (5-year) fertilization experiment in the Mer Bleue peatland in Ottawa, Canada, Bubier et al. (2007) saw no change in total aboveground plant biomass and lower net ecosystem production (NEP) in fertilized plots. They attributed this paradoxical finding to a nutrient stimulation of vascular plant growth that was accompanied by declines in the abundance of mosses, particularly *Sphagnum*, leading to a decline in organic matter accumulation. The loss of *Sphagnum* mosses due to N deposition, fertilization, or drainage-induced increases in N turnover could lead to substantial reductions in peat accumulation because replacement species typically produce higher-quality litter and have higher rates of evapotranspiration than *Sphagnum* (van Breemen 1995).

In closed wetland systems such as the extensive boreal peatlands at high latitudes in the Northern Hemisphere, nitrogen and phosphorus are typically both in short supply for plant growth and decomposition (Chapin et al. 1978, Damman 1988). In the tundra of Alaska, Chapin et al. (1978) found that the soil organic matter contained 64% of the total phosphorus in the ecosystem and had a mean residence time of 220 years, while available phosphorus in soil solution comprised 0.3% of the total phosphorus and had a residence time of 10 hours. Low temperatures and high water tables together limit nutrient mineralization in the tundra (Marion and Black 1987), and as a result of slow decomposition many boreal bogs show a net accumulation of nitrogen and phosphorus in peat (Hemond 1983, Damman 1988, Urban et al. 1989a). In a fertilization experiment, Shaver and Chapin (1986) found that the response of *Eriophorum vaginatum* in tussock tundra was greater for N than for P. Rates of nitrogen fixation within boreal wetlands can be very high (Barsdate and Alexander 1975, Waughman and Bellamy 1980, Schwintzer 1983). A variety of arctic plants are capable of assimilating low-molecular-weight organic nitrogen molecules (e.g., Chapin et al. 1993, Nasholm et al. 1998), which suggests that in isolated wetlands, nitrogen limitation is frequently severe.

Determining nutrient limitation of primary productivity in hydrologically open wetlands is more difficult because hydrologic losses complicate fertilization experiments. Wetlands receiving surface runoff can have high inputs of phosphorus and other elements derived from rock weathering (Mitsch et al. 1979, Waughman 1980, Frangi and Lugo 1985). In these ecosystems phosphorus and sulfur are retained on iron and aluminum minerals that are constituents of soil organic matter (Richardson 1985, Mowbray and Schlesinger 1988). With greater surface and groundwater inputs, net primary production is more likely to be limited by N than P (e.g., Tilton 1978) because large amounts of nitrogen can be lost through denitrification, while P tends to accumulate in soil organic material. Many wetlands receive high inorganic N loading from fertilizer, sewage-derived runoff, or N deposition, which can lead to substantial changes in plant composition (Bedford et al. 1999).

Net primary production (NPP) is highest in wetlands receiving nutrient enrichment or with high nutrient turnover. The degree, duration, and periodicity of flooding affect wetland productivity more than rainfall or temperature. Drainage can promote enhanced productivity by increasing nutrient mineralization. Tree growth and nitrogen content increase when northern wetlands are drained (Figure 7.7; Lieffers and Macdonald 1990, Westman and Laiho 2003, Choi et al. 2007, Turetsky et al. 2011). Flooding can enhance wetland productivity when it brings subsidies of nutrients from the contributing catchment, but can also stress wetland plants by suppressing organic matter mineralization and promoting the production of H_2S. This subsidy-stress relationship (*sensu* Odum et al. 1979) precludes a general relationship between water availability and NPP in wetland ecosystems. In a survey of temperate forested wetlands, Megonigal et al. (1997) found that intermittently flooded wetlands had higher litterfall and NPP than permanently flooded wetlands (Figure 7.8) and suggested that intermittent flooding allows soils to dry, which increases decomposition and promotes nutrient mineralization. In contrast, other studies have suggested that inundation by flowing water can deliver nutrients from upland areas to wetland forests (Conner and Day 1976, Conner et al. 2011). Several studies in forested floodplain wetlands found the highest litterfall in the wettest sites but little clear evidence that plant growth was affected by flooding regime (Clawson et al. 2001, Conner et al. 2011). The discrepancy in findings may be due to the type of inundation, with stagnant inundation suppressing nutrient mineralization and reducing productivity (e.g., Schlesinger 1978, Megonigal et al. 1997) and flowing water providing nutrient subsidies (e.g., Conner and Day 1976, Clawson et al. 2001).

ORGANIC MATTER STORAGE IN WETLANDS

Decomposition is impeded in flooded and saturated soils so that primary production in wetlands will often exceed decomposition, leading to a net accumulation of soil organic matter. As a result, over decadal to millennial timescales many wetlands have accumulated large standing stocks of soil organic matter (Table 7.2). If the plant remains are still recognizable, these organic materials are called peat. As decomposition removes carbon and the relative mineral fraction increases, oxidation of soil humus leads to a darker muck without recognizable plant tissues (see Figure 7.4). The rate of peat accumulation is determined by the rates of

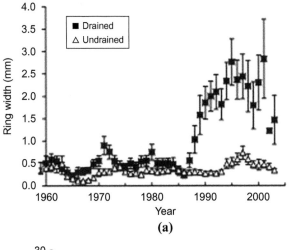

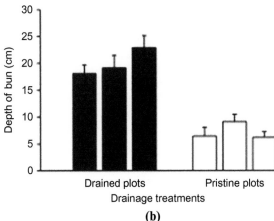

FIGURE 7.7 Drainage of a forested boreal fen in western Canada in 1986 doubled the rate of peat C accumulation through increases in tree biomass and detritus (indicated here by tree ring growth) (a), but also made the drained fen more susceptible to catastrophic losses of carbon in fire (b). In 2001, a wildfire burned ~450 years of accumulated peat in the drained portion while removing only ~58 years of accumulated peat in the undrained portions of the fen. *Source: Modified from Turetsky et al. (2011).*

decomposition in both the oxic upper level and the lower level of the deposit. Through time, older layers of organic material are buried and compacted beneath the weight of newly deposited plant detritus (Figure 7.7). The transport of solutes and diffusion of gases slow with depth as a result of water saturation and compaction. It is useful to differentiate between the biogeochemically active *acrotelm*, the surface layer of peat above the lowest water table elevation that experiences fluctuations between oxic and anoxic conditions, and the *catotelm*, or underlying layers that are permanently saturated.

Peatland ecosystems can be perceived as a special category of wetland wherein plants build the terrain through the deposition of litter into saturated soils. Accretion occurs in these low-energy environments through biogenic processes rather than by sediment deposition (Gosselink and Turner 1978, Brinson et al. 1981). Clymo (1984) proposed a model for peat accumulation which predicts that peatlands will eventually attain a steady state when the input of detritus from primary production at the peat surface is balanced by the loss of organic

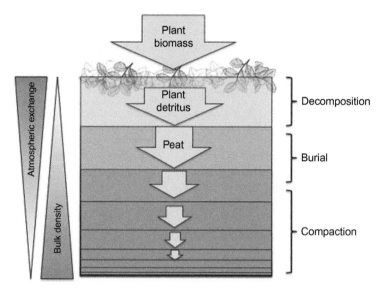

FIGURE 7.8   A model of peat accumulation and compaction over time. Fresh litter is deposited in the surface layers, where decomposition rates are highest due to oxygen diffusion and the supply of alternate electron acceptors. Organic matter that escapes decomposition is buried beneath new litter inputs and over time becomes compacted through the accumulation of overlying material. Models of peat accumulation predict that eventually peatlands reach a steady state where new biomass inputs are balanced by carbon losses through decomposition. *Source: Adapted from Clymo (1984).*

TABLE 7.2   Carbon Accumulation in Wetland Sediments: a Compilation of Reported Rates

Location	Wetland and/or vegetation type	Accumulation interval (yrs)	Accumulation rate (g C m^{-2} yr^{-1})	References
Peatlands			12–25	Malmer (1975)
Global wetlands			20–140	Mitra et al. (2005)
North America	Peatlands		29	Gorham (1991)
Boreal Wetlands			**8–80**	
Alaska and Canada	Peatlands		8–61	Ovenden (1990)
Alaska	*Picea* and *Sphagnum*	4790	11–61	Billings (1987)
Russia	Mires, bogs, and fens	3000–7000	12–80	Botch et al. (1995)
Manitoba	*Picea* and *Sphagnum*	2960–7939	13–26	Reader and Stewart (1972)
Western Canada	*Sphagnum* bogs	9000	13.6–34.9	Kuhry and Vitt (1996)
Sweden	Bogs		20–30	Armentano and Menges (1986)

TABLE 7.2 Cont'd

Location	Wetland and/or vegetation type	Accumulation interval (yrs)	Accumulation rate (g C m^{-2} yr^{-1})	References
Alaska	*Eriophorum vaginatum*	7000	27	Viereck (1966)
Ontario	*Sphagnum* bogs	5300	30–32	Belyea and Warner (1996)
Russia	Siberian mires	8000–10,000	12.1–23.7	Turunen et al. (2001)
Finland	Mires		18.5	Turunen et al. (2002)
Canada	Mer Bleue ombrotrophic bog	2700	10–25	Roulet et al. (2007)
Canada	23 ombrotrophic bogs	150	73 ± 17	Moore et al. (2005)
Finland	795 bogs and fens	5000	21	Clymo et al. (1998)
Sweden	Store Mosse mire	5000	14–72	Belyea and Malmer (2004)
Canada	Continental western Canadian peatlands	Current	19.4	Vitt et al. (2000)
Temperate Wetlands			**17–317**	
Georgia	Floodplain cypress gum forests	100	107	Craft and Casey (2000)
Georgia	Depressional wetlands	100	70	Craft and Casey (2000)
Wisconsin	*Sphagnum*	8260	17–38	Kratz and DeWitt (1986)
Massachusetts	Thoreau's bog		90	Hemond (1980)
North America	Protected prairie potholes		83	Euliss et al. (2006)
Ohio	Created marshes		180–190	Anderson and Mitsch (2006)
North America	Restored prairie potholes		305	Euliss et al. (2006)
Ohio	Depressional wetlands	42	317 ± 93	Bernal and Mitsch (2012)
Ohio	Riverine, flow-through	42	140 ± 16	Bernal and Mitsch (2012)
Eastern U.S.	Circumneutral freshwater peatlands	30	49 ± 11	Craft et al. (2008)
Eastern U.S.	Acidic freshwater peatlands	30	88 ± 20	Craft et al. (2008)
Subtropical Wetlands			**70–387**	
Louisiana	Salt marsh		200–300	Hatton et al. (1983)
Florida	*Cladium* swamp	25–30	70–105	Craft and Richardson (1993)

Continued

TABLE 7.2   Carbon Accumulation in Wetland Sediments: a Compilation of Reported Rates—Cont'd

Location	Wetland and/or vegetation type	Accumulation interval (yrs)	Accumulation rate (g C m^{-2} yr^{-1})	References
Florida Everglades	*Cladium sp.*		86–140	Reddy et al. (1993)
Florida Everglades	*Typha sp.*		163–387	Reddy et al. (1993)
Tropical Wetlands			**39–480**	
Amazon	Lowland peatlands	1700–2850	39–85	Lahteenoja et al. (2009)
Kenya	Papyrus wetlands		160	Jones and Humphries (2002)
Uganda	Papyrus wetlands		480	Saunders et al. (2007)
Costa Rica	Humid tropical wetland	42	255	Mitsch et al. (2010)
Mexico	Mangroves		100	Twilley et al. (1992)
Range of reported values			**8–480**	

matter by decomposition throughout the peat profile. The saturated soils of tundra and boreal forest region contain about 50% of the total storage of organic matter in soils of the world (Tarnocai et al. 2009, Frolking et al. 2011). Many of these ecosystems have accumulated soil carbon since the retreat of the last continental glaciers (Harden et al. 1992).

The unique aspect of wetland ecosystems is the dominance and diversity of anaerobic metabolic pathways employed by microbes for metabolism in the absence of oxygen. The drainage of wetland soils (through natural droughts or anthropogenic drainage) leads to rapid oxidation of their large stocks of organic matter by aerobic microbes (Armentano and Menges 1986, Turner 2004). The resulting decrease in soil elevation, or subsidence, has been notably documented for inland wetlands in England, Germany, and the Florida Everglades, where posts have been anchored in a stable subsurface layer and changing surface elevations are recorded relative to the immobile post. Soil elevation has declined more than 4 m in the past 130 and 150 years in the German and English sites, respectively (Heathwaite et al. 1990), and more than 3 m since 1924 at the Everglades site (Stephens and Stewart 1976). The rapid oxidation of soil organic matter in drained wetlands provides clear evidence that much of the organic material stored in wetland sediments is not inherently recalcitrant. Decomposition in flooded soils is extremely inefficient due to a lack of oxygen (Figure 7.9).

To understand how flooding constrains decomposition, we must compare the mechanisms and energy yield derived from anaerobic respiration and aerobic respiration. Both decomposition pathways are initiated by the cleavage of organic monomers from large complex organic polymers by extracellular enzymes. Aerobic respiration can

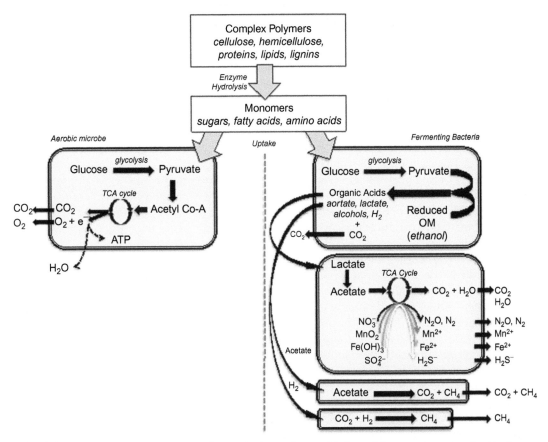

FIGURE 7.9   Contrasting the single aerobic respiration pathway with the multistage pathway involved in decomposition in the absence of oxygen. *Source: Figure drawn with inspiration from Megonigal et al. (2003) and Reddy and DeLaune (2008).*

completely degrade the resulting organic monomers to CO_2 using glycolysis followed by the Kreb's cycle (Figure 7.9). When oxygen is available, a single molecule of glucose yields 2 moles of ATP from glycolysis and a further 36 moles of ATP through the Kreb's cycle (Madigan and Martinko 2006). Without oxygen, this reaction sequence stops at pyruvate, and further degradation requires fermentative metabolism, which has a low energy yield (Figure 7.9). Aerobic respiration results in the complete degradation of monomers to CO_2, whereas fermentation results in the accumulation of a variety of organic acids and alcohols. The resulting fermentation products are subsequently further degraded to CO_2 by bacteria using NO_3^-, Mn^{4+}, Fe^{3+}, or SO_4^{2-} as alternative electron acceptors in place of O_2, or they may undergo additional fermentation steps to produce CH_4. These alternative respiratory pathways have lower energy yields, and thus support a smaller microbial biomass that in turn produces lower concentrations of extracellular enzymes (McLatchey and Reddy 1998).

There are two mechanistic explanations for the inefficient decomposition typical of wetlands. Until recently, decomposition was primarily assumed to be limited by the supply of oxygen and alternative electron acceptors necessary for the terminal steps in organic matter decomposition. Recent work has suggested additional enzyme-mediated constraints at earlier stages of the decomposition pathway (Limpens et al. 2008). The activity of phenol oxidase, a critical extracellular enzyme involved in the degradation of lignin and phenolics,[5] is substantially reduced under low-oxygen conditions, leading to an accumulation of phenolic compounds in wetland sediments (McLatchey and Reddy 1998, Freeman et al. 2001b). High concentrations of phenolic compounds can then further inhibit organic matter decomposition (Appel 1993, Freeman et al. 2001b, Ye et al. 2012).

When one or more alternate electron acceptors are abundant, the rate of soil organic matter decomposition will be limited by the pace of enzymatic hydrolysis or fermentation (Freeman et al. 2001b, Megonigal et al. 2003). In contrast, when alternate electron acceptors are in short supply, fermentation products may accumulate until sediments are resupplied with oxygen or alternative electron acceptors. Decomposition of soil organic matter in wetlands can be enhanced either by lowering the water table (allowing oxygen to penetrate to deeper soil layers) or by increasing the supply of alternate electron acceptors. Nitrogen deposition, amendments with oxidized Fe, and enhanced SO_4 availability resulting from acid rain or saltwater intrusion have all been shown to significantly enhance decomposition rates (Van Bodegom et al. 2005, Bragazza et al. 2006, Gauci and Chapman 2006, Weston et al. 2006). Decomposition in wetland sediments is typically highest at the wetland surface, where recently synthesized, more labile organic material comes into contact with the greatest potential supply of electron acceptors.

MICROBIAL METABOLISM IN SATURATED SEDIMENTS

In a closed aqueous system containing a large supply of organic material together with appreciable concentrations of oxidants (O_2, NO_3^-, Mn^{4+}, Fe^{3+}, and SO_4^{2-}), we can easily predict the order in which the oxidants will be depleted (Table 7.3). The exergonic (energy-yielding) oxidation of organic matter (A) would be paired first with oxygen respiration (B), then NO_3^- respiration (C), then Mn^{4+} (D), Fe^{3+} (E), and SO_4^{2-} (F) respiration would follow in sequence (Table 7.3). If organic matter remained after all of these oxidants were depleted, we might subsequently measure an accumulation of CH_4 in our closed vessel. This predictable sequence of biologically mediated chemical reactions occurs because there is a tendency for the highest energy yielding metabolic pathways to take precedence over lower energy yielding processes (Stumm and Morgan 1996). The same reaction sequence observed in a closed vessel can also be observed in wetland ecosystems examined through time following flooding or with depth in the soil profile (Figure 7.10).

Common reduction and oxidation half reactions are shown in Table 7.3 together with the standard electrical potential of each reaction. Standard electrical potentials are expressed per mol of electrons transferred; thus each reaction has been written to transfer one mol of

---

[5] Phenolics are a class of chemical compound consisting of a hydroxyl group bonded directly to an aromatic hydrocarbon. In wetlands, soluble humic acids make up a large fraction of phenolics.

TABLE 7.3 Common Reduction and Oxidation Half Reactions

Part A

Reduction	E° (V)	Oxidation	E° (V)
(A) $1/4O_2(g) + H^+ + e^- = 1/2H_2O$	+0.813	(L) $1/4CH_2O + 1/4H_2O = 1/4CO_2 + H^+ + e^-$	−0.485
(B) $1/5NO_3^- + 6/5H^+ + e^- = 1/10N_2 \ 3/5H_2O$	+0.749	(M) $1/2CH_4 + 1/2H_2O = 1/2CH_3OH + H^+ + e^-$	+0.170
(C) $1/2MnO_2(s) + 1/2HCO_3^- \ 3/2H^+ + e^- = 1/2MnCO_3 + H_2O$	+0.526	(N) $1/8HS^- + 1/2H_2O = 1/8SO_4^{2-} + 9/8H^+ + e^-$	−0.222
(D) $1/8NO_3^- + 5/4H^+ + e^- = 1/8NH_4^+ 3/8H_2O$	+0.363	(O) $FeCO_3(s) + 2H_2O = FeOOH(s) + HCO_3^-(10^{-3}) + 2H^+ + e^-$	−0.047
(E) $FeOOH(s) + HCO_3^-(10^{-3}) + 2H^+ + e^- = FeCO_3(s) + 2H_2O$	−0.047	(P) $1/8NH_4^+ 3/8H_2O = 1/8NO_3^- + 5/4H^+ + e^-$	+0.364
(F) $1/2CH_2O + H^+ + e^- = 1/2CH_3OH$	−0.178	(Q) $1/2MnCO^3(s) + H_2O = 1/2MnO_2(s) + 1/2HCO_3^-(10^{-3}) + 3/2H^+ + e^-$	+0.527
(G) $1/8SO_4^{2-} + 9/8H^+ + e^- = 1/8HS^- + 1/2H_2O$	−0.222		
(H) $1/8CO_2 + H^+ + e^- = 1/8CH_4 1/4H_2O$	−0.244		
(I) $1/6N_2 + 4/3H^+ + e^- = 1/3NH_4$	−0.277		

Part B

Examples	Combinations	$\Delta G°$ (W) pH=7 (kJ eq−1)
Aerobic respiration	A+L	−125
Denitrification	B+L	−119
Nitrate reduction to ammonium	D+L	−82
Fermentation	F+L	−27
Sulfate reduction	G+L	−25
Methane fermentation	H+L	−23
Methane oxidation	A+M	−62
Sulfide oxidation	A+N	−100
Nitrification	A+P	−43
Ferrous oxidation	A+O	−88
Mn(II) oxidation	A+Q	−30

Source: Modified from Stumm and Morgan (1996, p. 474).

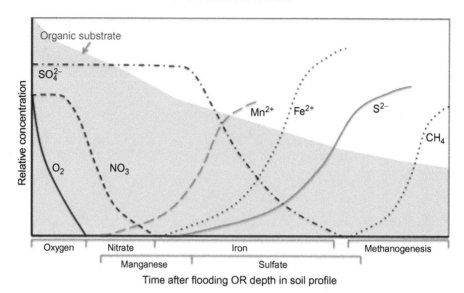

FIGURE 7.10   The concentrations of reactants and products of terminal decomposition pathways are shown for a wetland sediment over time following flooding. Rotating the figure 90° to the right shows the pattern of substrate concentrations (and the order of metabolic pathways) with depth in a soil profile.

electrons. Where $E° > 0$ the reaction will proceed spontaneously as written. Where $E° < 0$ the reaction will proceed in the opposite direction. The greater the difference in $E°$ between two half reactions, the greater the resulting free energy yield from their combination will be. In Part B the standard free energies of common redox couplets are shown. These are calculated from the $E°$ values in Part A using Equation 7.2. In the table, CH_2O represents an "average" organic substance. The actual free energy yield of different organic substances may differ from that given for CH_2O. This difference may be very large, particularly for anoxic processes involving carbon substrates with very different oxidation states than that assumed for CH_2O.

The terminal decomposition steps in wetlands are dominated by anaerobic metabolic pathways that yield a variety of reaction products in addition to the CO_2 and H_2O generated by aerobic oxidation. These pathways are responsible for the production of N_2, N_2O, and CH_4, the abundance of reduced H_2, Fe^{2+}, and H_2S, and the production of pyrite (FeS_2) in wetland soils. In any wetland the relative importance of these metabolic pathways to overall ecosystem carbon and nutrient cycling depends on the external supply or reoxidation of the various electron acceptors. The order of redox reactions results from the very limited supply of fermentation products. Since fermentation is slow and fermentation products are scarce, the metabolic pathways that maximize energy gain are highly favored. Successful metabolic strategies (and thus successful microbes) are those that garner the greatest energy given available substrates. The "redox ladder," the predictable sequence of reactions following flooding or with depth, thus arises from competitive interactions between microbes (Postma and Jakobsen 1996, Stumm and Morgan 1996). It seems at first paradoxical that in these carbon-rich systems we see such fierce competition for carbon substrates that the chemical reaction sequence is closely matched by an ecological succession of microorganisms (aerobic heterotrophs → denitrifiers → fermenters → sulfate reducers → methanogens). This paradox can be explained if we consider that the rate of decomposition is determined by fermentation while the relative dominance of terminal electron acceptor processes is predictable from their energy yield (Postma and Jakobsen 1996).

To understand and predict which microbial metabolisms will dominate at any given time or place in wetland sediments, we must understand how the possible reactions vary in the amount of energy generated (free energy yield). Thermodynamics allows us to predict the dominant metabolisms because particular microbial species, with different metabolic strategies, become competitively superior under different chemical conditions.

Free Energy Calculation

To calculate the energy yield from the oxidation of organic matter paired with the reduction of any electron acceptor, we calculate the standard Gibb's free energy yield ($\Delta G°$) for a redox couplet as:

$$\Delta G° = -nF\Delta E, \tag{7.2}$$

where n is the number of electrons, F is Faraday's constant (23.061 kcal volt^{-1}), and ΔE is the difference in electrical potential (V) between the oxidation and reduction reactions (Table 7.4). Reactions with a negative ΔG are energy yielding (exergonic) while reactions with a positive ΔG require energy (endergonic).

For aerobic respiration of a generic organic molecule (CH_2O), we can calculate the standard free energy yield $\Delta G°$. The standard free energy assumes that all substrates are available in abundance and that the reaction occurs at a standard temperature of 25°C. The free energy yield of each reaction is higher for carbon compounds that have more reduced chemical bonds and lower for more oxidized organic molecules. Here we will use a generic carbon molecule with 1/6 the free energy of glucose, which has six carbon atoms.

Oxidation 1/2 reaction:

$$[CH_2O] + H_2O \rightarrow CO_2(g) + 4\,H^+ + 4e^- \quad E° = -0.485 \text{ V} \tag{7.3}$$

Reduction 1/2 reaction:

$$O_2(g) + 4H^+(W) + 4e^- \rightarrow 2H_2O \quad E° = +0.813 \text{ V} \tag{7.4}$$

Joint reaction:

$$[CH_2O] + O_2 \rightarrow CO_2 + 2H_2O \quad \Delta E° = +1.30 \text{ V} \tag{7.5}$$

Thus,

$$\begin{aligned} \Delta G° &= -nF\Delta E \\ &= -(4)(23.061 \text{ kcal})(+1.30 \text{ V}) \\ &= -119.9 \text{ kcal per mol } CH_2O \end{aligned}$$

$$\text{or, since 1 kcal} = 4.184 \text{ kJ} = -502 \text{ kJ per mol } CH_2O$$
$$\text{or divide by 4 to determine} = -29.9 \text{ kcal per e}^{-1}$$
$$\text{or, since 1 kcal} = 4.184 \text{ kJ} = -125 \text{ kJ per e}^{-1}$$

I. PROCESSES AND REACTIONS

Note that the reduction step requires energy $(+E°)$ while the oxidation step yields energy $(-E°)$. We can express the energy yields for each equation per mole of carbon substrate or per mole of electrons transferred. The net energy yield of the paired reaction is ~ 125 kJ for every mole of electrons transferred. Expressing energy yield per mole of electrons is useful for comparing processes that oxidize inorganic reduced energy sources (e.g., sulfide oxidation, ferrous oxidation, or manganese oxidation) with those that oxidize organic matter.

If we pair the same organic matter oxidation reaction with the reduction of NO_3 as an alternative electron acceptor, we calculate a lower energy yield.

Oxidation 1/2 reaction:

$$[CH_2O] + H_2O \rightarrow CO_2(g) + 4H^+ + 4e^- \quad E° = -0.485 \text{ V} \tag{7.6}$$

Reduction 1/2 reaction:

$$0.8NO_3 + 4.8H^+ + 4e^- \rightarrow 0.4N_2 + 2.4H_2O \quad E° = +0.749 \text{ V} \tag{7.7}$$

Joint reaction:

$$[CH_2O] + 0.8NO_3 + 3.8H^+ \rightarrow CO_2 + 0.4N_2 + 1.4H_2O \quad \Delta E° = +1.23 \text{ V} \tag{7.8}$$

And since

$$\begin{aligned} \Delta G° &= -nF\Delta E \\ &= -(4)(23.061 \text{ kcal V}^{-1})(1.23 \text{ V}) \\ &= -113 \text{ kcal per mol } CH_2O \\ \text{or} &= -474 \text{ kJ per mol } CH_2O \\ \text{or} &= -28.5 \text{ kcal per e}^{-1} \\ \text{or} &= -119 \text{ kJ per e}^{-1} \end{aligned}$$

In comparing the $\Delta G°$ for these two reactions, we see that in denitrification nitrate respiration releases only 95% of the energy contained in our generic organic molecule (CH_2O) relative to aerobic respiration. Because of this difference in efficiency, whenever O_2 is available, heterotrophs utilizing aerobic respiration should outcompete denitrifiers for organic substrates.

Aerobic heterotrophs and denitrifiers often coexist in upland soils; indeed, many heterotrophic microbes are facultative denitrifiers that switch between aerobic respiration and denitrification depending on the supply of O_2 versus NO_3 (Carter et al. 1995; Chapter 6). In the previous comparisons we assume that substrates are not limiting. The "actual free energy yields" (ΔG) for these reactions, which take into account the concentrations of all reactants, indicate that under the conditions found in most oxic soils, aerobic respiration has a much higher ΔG than denitrification because oxygen is far more available than nitrate. In contrast, in wet soils where oxygen concentrations are low and nitrate concentrations are high (as in wet agricultural fields or wetlands receiving nitrogen-rich runoff), the two pathways may have nearly equivalent ΔG.

To calculate the actual free energy yield (ΔG) of a reaction we use the equation

$$\Delta G = \Delta G° + RT \ln Q, \tag{7.9}$$

where R is the universal gas constant (1.987×10^{-3} kcal K^{-1} mol^{-1}), T is temperature in °K, and Q represents the reaction quotient, or the concentration of reaction products relative to the concentration of reactants. For a generic reaction $^{a}Ox_1 + {}^{b}Red_2 \rightarrow {}^{c}Red_1 + {}^{d}Ox_2$, the reaction quotient would be calculated as

$$Q = [Red_1]^c [Ox_2]^d / [Ox_1]^a [Red_2]^b. \qquad (7.10)$$

Actual free energies thus modify our energy yield predictions by taking into account the relative abundance of reactants and products in the environment. This is a critical adjustment because the assumption of standard activities of all reactants inherent to the standard free energy calculations is rarely met in natural ecosystems. In most salt marsh ecosystems, for example, the dominant pathway for organic matter decomposition is sulfate reduction (Howarth 1984), a pathway that is not energetically favored according to standard free energy predictions (Table 7.4) but which becomes important in anoxic sediments where sulfate is abundant.

In a comparison of a freshwater and a brackish wetland in coastal Maryland, Neubauer et al. (2005) measured high rates of Fe^{3+} reduction in the early summer in both wetlands giving way later in the season to methanogenesis in the freshwater wetland (low SO_4^{2-}) and sulfate reduction in the brackish wetland receiving marine-derived SO_4^{2-} (Figure 7.11). We can use the actual free energy calculations to understand the environmental conditions under which we would

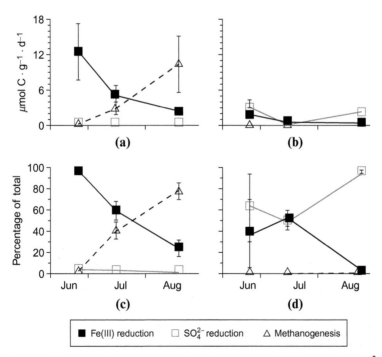

FIGURE 7.11   Seasonal changes in the rates and relative importance of Fe(III) reduction, SO_4^{2-} reduction, and methanogenesis during summer 2002 in (a) and (c) Jug Bay, a freshwater wetland, and (b) and (d) Jack Bay, a brackish wetland on the coastal plain of Maryland. *Source: From Neubauer et al. (2005). Used with permission of the Ecological Society of America.*

predict these shifts in the dominant metabolic pathways. Terminal electron acceptors that are extremely abundant will be likely to dominate decomposition pathways.

While the energy yield from sulfate reduction and methane fermentation is very low, the reduced products of sulfate reduction (HS^-) and methanogenesis (CH_4) are themselves energy-rich reduced substrates for other organisms. In the presence of oxygen, sulfide oxidation generates $100 \, kJ \, eq^{-1}$ of free energy, which, when combined with the free energy yield of sulfate reduction ($25 \, kJ \, eq^{-1}$), achieves the same total free energy yield as aerobic respiration of the same original carbon molecules (Table 7.4, Part B). This explains why very little sulfide gas or methane escapes from wetlands relative to the amounts produced in wetland sediments. There is energy to be extracted from these highly reduced gases.

Most wetlands undergo both flooding and drying cycles, and many wetlands are shallow enough to support rooted vegetation that can passively or actively transport gases between the atmosphere and sediments. Because of this variation in water level and in plant-facilitated gas exchange, oxygen depletion within wetlands is far from uniform in either time or space. As a result, wetland sediments are characterized by complex temporal and spatial gradients in the dominant metabolic pathways and rates of organic matter oxidation. Oxygen concentrations are typically high in shallow surface waters or drained surface sediments that are in direct contact with the atmosphere, but O_2 levels decline rapidly with depth in organic-rich sediments. However, the rhizospheres (root-associated sediments) of aerenchymous plants can remain well oxygenated and can support higher rates of mineralization that, in turn, may help alleviate nutrient limitation of wetland plant growth (Weiss et al. 2005, Laanbroek 2010, Schmidt et al. 2010b; Figure 7.12). Greater root biomass can thus be indirectly

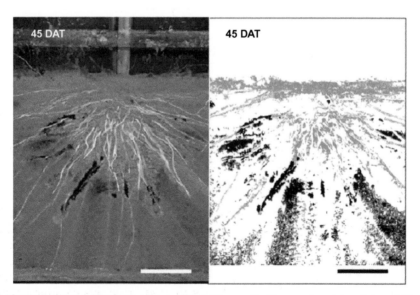

FIGURE 7.12   Schmidt et al. (2010) grew rice seedlings (*Oryza sativa*) in experimental containers (rhizotrons) with a clear plastic side. On the *left* is a photograph of the rhizosphere of a single rice seedling grown in a paddy soil for 45 days. On the *right* the same image is highlighted to show areas that are oxic in red and anoxic in black. *Source: From Schmidt et al. (2010b). Used with permission of Springer.*

associated with higher rates of organic matter mineralization due to rhizosphere oxidation (Wolf et al. 2007).

Measuring the Redox Potential of the Environment

We can measure the development of anoxic conditions in sediments by measuring redox potential (pe). Just as pH expresses the concentration of H^+ in solution, redox potential is used to express the tendency of an environmental sample (usually measured in situ) to either receive or supply electrons. Oxic environments have high redox potential because they have a high capacity to attract electrons (oxygen is the most powerful electron acceptor), while anoxic environments have a low redox potential (reducing conditions) because of an abundance of reduced compounds already replete with electrons. When a metal probe is inserted into a soil or sediment, the metal surface will begin to exchange electrons with its surroundings, and the net direction of the exchange will depend on both the reactivity of the metal and the relative abundance of electrons in the environment. To measure the direction and strength of this electron exchange, the metal probe can be connected to a reference electrode, with a voltmeter placed in between. The redox potential is then measured as the voltage necessary to stop the flow of electrons within the electrode.

In a laboratory setting, the redox potential of chemical mixtures is determined by connecting the redox probe to a standard hydrogen electrode. The relative abundance of electrons in the solution will alter the equilibrium constant for the exchange of electrons within the electrode, where electrons are shuttled between sulfuric acid and a hydrogen gas atmosphere:

$$H^+ + 4e^- \longleftrightarrow 2H_2 \; (g). \tag{7.11}$$

In the field it is not easy to maintain standard hydrogen electrodes, so investigators typically use either an Ag/AgCl electrode or a calomel reference electrode that has been calibrated against a hydrogen electrode (Fiedler 2004, Rabenhorst 2009). The Ag/AgCl electrode consists of a silver wire surrounded by AgCl salt that is contained within a concentrated KCl solution. The solid Ag exchanges electrons with the AgCl solution:

$$Ag \longleftrightarrow Ag^+ + e^-. \tag{7.12}$$

When the reference electrode is connected to the platinum probe and inserted into an oxic soil, oxygen will consume electrons along the platinum probe:

$$O_2 + 4e^- + 4H^+ \rightarrow 2H_2O. \quad E_h = +700 \text{ to } +400 \text{ mV}. \tag{7.13}$$

The reaction within the Ag/AgCl electrode (Eq. 7.12) will go to the right (Ag is oxidized) and the voltmeter will record a positive flow of electrons (Figure 7.13(a)). If instead the platinum probe is inserted into a highly reduced sediment, electrons will flow toward the reference electrode. The reaction (Eq. 7.12) will proceed to the left (Ag is reduced), and the voltmeter will record a negative flow of electrons (Figure 7.13(b)). Charge balance is maintained within the reference electrode by the diffusion of ions through a porous ceramic or membrane tip. Potassium ions (K^+) will be released through the ceramic tip when the redox potential is positive, and Cl^- ions will be released to the soil when the redox potential is negative (Figure 7.13)

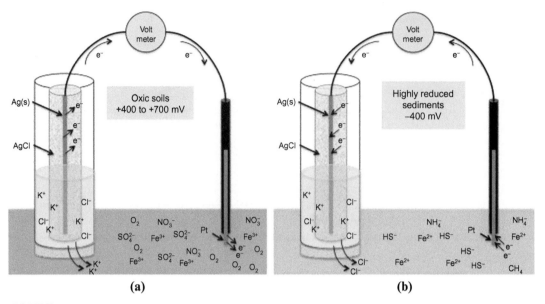

(a)   **(b)**

FIGURE 7.13   Schematic of an Ag/AgCl reference electrode and platinum electrode inserted into (a) an oxic soil and (b) a highly reduced sediment, demonstrating how the direction of electron flow depends on the availability of electron acceptors at the platinum electrode. Depending on the direction of e⁻ flow, either K^+ or Cl^- will be released through the ceramic or membrane tip of the reference electrode into the soil to maintain charge balance within the electrode.

The oxidizing or reducing potential of the soil is thus estimated relative to the reference electrode. Redox potential must be corrected when one uses Ag/AgCl rather than a standard hydrogen electrode using a correction factor (~200 mV for an Ag/AgCl electrode and ~250 mV for a calomel electrode). Typically, the redox potential in soils with any oxygen present varies only between +400 mV and +700 mV. As oxygen is depleted, other constituents, such as Fe^{3+} may accept electrons, but a lower voltage will be recorded since Fe^{3+} has a reduced capacity for attracting electrons relative to O_2.

$$Fe^{2+} + 3H_2O \leftrightarrow Fe(OH)_3 + 3H^+ + e^- \quad \text{and} \quad E_h = +100 \text{ to} -100 \text{ mV} \qquad (7.14)$$

In aqueous environments, redox potential can range from +400 mV to −800 mV (Figure 7.14). A negative sign means that the reducing environment has excess electrons, a situation in which excess electrons will form H_2 gas via Eq. 7.11.

The pH of the environment will affect the redox potential established by oxygen and other alternate electron acceptors (Figure 7.14). Aerobic oxidation, which generates H^+ ions (shown in Eq. 7.13), is more likely to proceed under neutral or alkaline conditions while anaerobic pathways that consume H^+ ions (shown in Eq. 7.14) are chemically more favorable in acidic environments (e.g., Weier and Gilliam 1986). Because of the pH sensitivity of redox reactions, it is often useful to express the redox potential of a reaction in units of pe, a constant that is derived from the equilibrium constant of the oxidation–reduction reaction (K) and which incorporates information on pH. For any reaction,

$$\text{Oxidized Species} + e^- + H^+ \longleftrightarrow \text{Reduced Species,} \qquad (7.15)$$

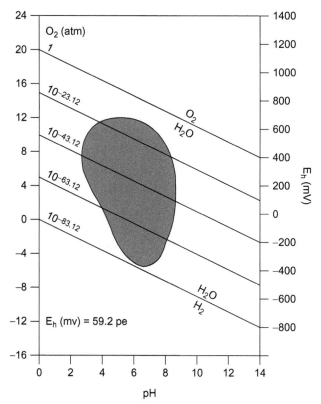

FIGURE 7.14 This stability diagram shows the relationship between pH and redox potential. Diagonal lines are the redox potential at different oxygen partial pressures. The *shaded zone* in the center of the figure shows the range of redox potentials that are found in natural aqueous environments. *Source: Modified from Lindsay (1979), which was based on the original compilation by Baas Becking et al. (1960). Used with permission of The University of Chicago Press.*

and the equilibrium constant can be determined by

$$\log K = \log[\text{reduced}] - \log[\text{oxidized}] - \log[e^-] - \log[H^+]. \qquad (7.16)$$

If we assume that the concentrations of oxidized and reduced species are equal, then

$$pe + pH = \log K. \qquad (7.17)$$

Here pe is the negative logarithm of the electron activity $(-\log[e^-])$, and it expresses the energy of electrons in the system (Bartlett 1986). Because the sum of pe and pH is constant, if one goes up, the other must decline. When a given reaction occurs at a lower pH, it will occur at a higher redox potential, expressed as pe. Measurements of redox potential that are expressed as voltage can be converted to pe following

$$pe = E_h/(RT/F)2.3, \qquad (7.18)$$

where R is the universal gas constant ($1.987\,\text{cal mol}^{-1}\,\text{K}^{-1}$), F is Faraday's constant ($23.06\,\text{kcal V}^{-1}\,\text{mol}^{-1}$), T is temperature in Kelvin, and 2.3 is a constant to convert natural to base-10 logarithms.

Environmental chemists use E_h-pH or pe-pH diagrams to predict the likely oxidation state of various constituents in natural environments based on these chemical relationships (e.g.,

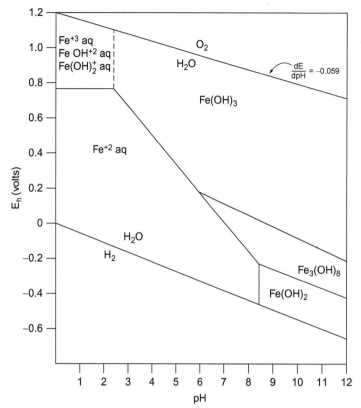

FIGURE 7.15 This stability diagram shows the expected form of iron in natural environments of varying pH and E_h. In interpreting such diagrams it is important to remember that E_h and pH are properties of the environment determined by the total suite of chemical species present. Thus, E_h predicts the forms of iron that will be present under a set of chemical conditions using a set of simplifying assumptions. Because iron is interacting with a variety of chemicals in solution, these predictions are not always accurate due to competing reactions. Important to biogeochemistry, E_h can be used to predict what modes of microbial activity are possible in a given environment, and E_h-pH predictions can be useful against which to compare field patterns. *Source: Modified from Lindsay (1979).*

Figure 7.15). Two lines bound all such diagrams. At any redox potential above the upper line, even water would be oxidized (Eq. 7.12 in reverse)—a condition not normally found on the surface of the Earth. Similarly any condition below the lower line would allow the reduction of water, again a condition rarely seen on Earth. These boundary redox conditions vary predictably with pH, with E_h declining by 59 mV with each unit of pH increase, reflecting that oxidation proceeds at a lower redox potential under more alkaline conditions.

In most cases organic matter contributes a large amount of "reducing power" that lowers the redox potential in flooded soils and sediments (Bartlett 1986). High concentrations of Fe^{2+} will be found in flooded low-redox potential environments where impeded decomposition leaves undecomposed organic matter in the soil and humic substances impart acidity to the soil solution. Where organic matter is sparse, iron may persist in its oxidized form (Fe^{3+}) even when the soils are flooded (e.g., Couto et al. 1985). Aeration and liming are each used as techniques for ameliorating acid mine drainage because iron tends to precipitate in oxidized forms at high redox potential or high pH.

Redox potential measurements do not actually measure the number of electrons in the environment, but instead provide a standard method for comparing their relative availability. Because individual anaerobic metabolic pathways vary predictably in their energy yield, they tend to dominate in a fairly narrow range of redox potentials so that a measure of field redox potential can provide a good prediction of the likely metabolic pathways at the point of measurement. The measure of a field redox potential is not equivalent to the measurement of the redox potential of a component chemical reaction (except when measurements are made in the lab in solutions where only one equilibrium reaction is possible). Instead, field redox potential provides a way of comparing the degree to which electron acceptor abundance and oxidizing efficiency vary in space and time.

Soils and sediments that resist change in their redox potential are said to be highly poised. Conceptually, poise is to redox as buffering capacity is to pH (Bartlett 1986). As long as soils are exposed to the atmosphere, they will appear to be highly poised, since O_2 will maintain a high redox potential under nearly all conditions. Soils with high concentrations of Mn^{4+} and Fe^{3+} are less likely to produce substantial amounts of H_2S or CH_4 during short-term flooding events because it is unlikely that microbes will be able to sufficiently deplete these electron acceptors and turn to the less efficient oxidizing constituents (SO_4, CO_2) (Lovley and Phillips 1987, Achtnich et al. 1995, Maynard et al. 2011). Such soils are also poised but at a lower redox potential than oxic soils. Because of the large differences in free energy yield for each anaerobic metabolic pathway, we can predict what metabolic products are likely to accumulate from a measure of redox potential (see Figure 7.14).

ANAEROBIC METABOLIC PATHWAYS

Prior to the oxidation of the Earth's atmosphere about 2.5 bya, anaerobic metabolism dominated the biosphere (Chapter 2). In the anoxic sediments of wetlands, lakes, rivers, and oceans, these metabolic pathways continue to dominate biogeochemical cycling today. Table 7.3 listed the more important overall reactions involved in anaerobic oxidation of organic matter in wetland sediments and allowed comparison among the resulting free energy yield from each reaction. These reactions have different biogeochemical consequences.

Fermentation

Fermentation of organic matter (Table 7.4) is a required precursor or accompaniment for each of the subsequent anaerobic metabolic pathways in Table 7.3 (Figure 7.9). Fermenting bacteria in wetlands are obligate anaerobes that use a variety of organic substrates, including

TABLE 7.4   Examples of Chemical Reactions and Associated Free Energy Yield (per mol of organic matter or sulfur compound) and Catalyzing Organisms Performing Fermentation

Reaction		$-\Delta G°$ (kJ mol^{-1})	Organisms catalyzing these reactions
(1) $3(CH_2O)$	$\rightarrow CO_2 + C_2H_6O$	23.4	For example, yeasts, *Sarcina ventriculi*, *Zymonas*, *Leuconostoc* sp., clostridia, *Thermonanaerobium brockii*, etc.
(2) $n(CH_2O)$	$\rightarrow mCO_2$ and/or fatty acids and/or alcohols and/or H_2	5–60	For example, yeasts, *clostridia, enterobacteria, lactobacilii, streptococci, propionibacteria*, and many others

Source: Adapted from Zehnder and Stumm (1988).

alcohols, sugars, and organic and amino acids, and convert them into CO_2 and various reduced fermentation products (predominantly C1–C18 acids and alcohols, molecular hydrogen, and CO_2) (Zehnder and Stumm 1988). Fermentation occurs inside microbial cells and does not require an external supply of electron acceptors. During fermentation ATP is produced by substrate-level phosphorylation. Fermentation pathways themselves have very low energy yield, but the low-molecular-weight organic acids, alcohols, and molecular hydrogen that are produced during fermentation ultimately determine the rate of terminal decomposition steps (Freeman et al. 2001b, Megonigal et al. 2003).

Although often ignored as an important pathway for the production of CO_2, a number of recent studies have suggested that fermentation (together with humic acid reduction; Lovley et al. 1996) can account for a significant fraction of anaerobic carbon mineralization in wetland sediments (Keller and Bridgham 2007). In addition to producing CO_2 or facilitating organic matter mineralization to CO_2 or CH_4, fermentation products can also accumulate as dissolved organic compounds (DOC) that are susceptible to leaching and hydrologic export.

Dissimilatory Nitrate Reduction

After O_2 is depleted by aerobic respiration, nitrate becomes the best alternative electron acceptor, and in anoxic sediments nitrate will be rapidly consumed whenever there are suitable organic substrates or reduced chemical compounds available for metabolism (Table 7.5). The term denitrification is typically used to refer to the process by which bacteria convert nitrate to gaseous N_2O or N_2 during the oxidation of organic matter (see Eqs. 1 and 2 in Table 7.5). These reactions are dissimilatory since the nitrogen used for denitrification is not incorporated into the biomass of denitrifying microbes. This type of denitrification is primarily performed by facultative anaerobes, organisms that switch from aerobic oxidation of organic matter when oxygen is present to denitrification when oxygen is depleted. The standard free energy yield of the reaction is only marginally lower than that of aerobic respiration

TABLE 7.5   Chemical Reactions and Associated Free Energy Yield (per mol of organic matter or sulfur compound) for Dissimilatory Nitrate Reduction Pathways

Reaction		$-\Delta G°$ (kJ mol^{-1})	Organisms catalyzing these reactions
(1) $2 NO_3^- + (CH_2O)$	$\rightarrow 2 NO_2^- + CO_2 + H_2O$	82.2	For example, members of the genus *Enterobacter*, *E. coli* and many others
(2) $4/5\ NO_2^- + (CH_2) + 4/5\ H^+$	$\rightarrow 2/5\ N_2 + CO_2 + 7/5\ H_2O$	112	For example, members of the genus *Pseudomonas*, *Bacillus lichenformis*, *Paracoccus denitrificans*, etc.
(3) $1/2\ NO_3^- + (CH_2O) + H^+$	$\rightarrow 1/2\ NH_4^+ + CO_2 + 1/2\ H_2O$	74	Members of the genus *Clostridium*
(4) $6/5\ NO_3^- + S^0 + 2/5\ H_2O$	$\rightarrow 3/5\ N_2 + SO_4^{2-} + 4/5\ H^+$	91.3	Member of the genus *Thiobacillus*
(5) $8/5\ NO_3^- + HS^- + 3/5\ H^+$	$\rightarrow 4/5\ N_2 + SO_4^{2-} + 4/5\ H_2O$	93	*Thoiosphaera pantotropha* and members of the genus *Thiobacillus*

Source: Adapted from Zehnder and Stumm (1988).

(refer to Table 7.3). Standard free energy calculations assume unlimited supplies of reactants; however, the NO_3^- in flooded soils is never as abundant as the O_2 in oxic soils. Thus even highly NO_3^--enriched wetland sediments are likely to have lower soil respiration rates once oxygen is depleted. In many flooded soils denitrification is limited by the availability of NO_3^-, a problem exacerbated by the fact that nitrification (Eqs. 2.17 and 2.18) cannot proceed without oxygen.

Although often discussed as a single process (e.g., Eq. 2.20), the denitrification process occurs in multiple steps whereby NO_3^- is converted sequentially to NO_2^-, NO, and N_2O (Figure 6.13). Particularly in nitrogen-enriched agricultural fields N_2O can be a significant fraction of the total gaseous nitrogen produced (Stehfest and Bouwman 2006) and dissolved nitrite (NO_2^-) can become a more important dissolved export than NH_4^+ (Stanley and Maxted 2008). It appears that the ratio of $N_2O:N_2$ production from denitrification in wetlands is typically lower than in upland soils (Schlesinger 2009), but there are concerns that continued nitrogen loading to wetlands from nitrogen deposition or fertilizer runoff may enhance N_2O yields since incomplete denitrification (with N_2O as a terminal product) becomes increasingly energetically favorable when the supply of NO_3^- is high (Verhoeven et al. 2006).

An alternative metabolic process to denitrification is dissimilatory nitrogen reduction to ammonium (usually referred to by the acronym DNRA), in which NO_3^- is converted to NH_4^+ through fermentation by obligate anaerobes (Zehnder and Stumm 1988, Megonigal et al. 2003; Eq. 3 in Table 7.5). DNRA appears to be a dominant process in some wet soils and wetland sediments (Scott et al. 2008, Dong et al. 2011), particularly in anoxic habitats where nitrate availability is very low and labile carbon supplies are high. Under these conditions, selection may favor microbes that retain fixed N over denitrifiers that further deplete limited N supplies (Tiedje 1988, Burgin and Hamilton 2007).

Anaerobic oxidation of NH_4^+ to N_2, or anammox, is another dissimilatory pathway, wherein NH_4^+ is oxidized by reaction with NO_2^- under anaerobic conditions to produce N_2. This process was only recently identified in wastewater treatment systems, but appears to be an important pathway for N_2 production in some coastal and marine sediments (Dalsgaard and Thamdrup 2002, Zehr and Ward 2002). Anammox appears to be competitively advantageous when carbon is highly limiting (Dalsgaard and Thamdrup 2002), but thus far there has been little research on its importance in freshwater ecosystems.

Nitrate may also be used in the oxidation of reduced sulfur, iron, or manganese compounds in anoxic sediments. Indeed, anaerobic sulfide oxidation using NO_3^- as an electron acceptor is an energetically favorable process accomplished by chemolithotrophic sulfur bacteria, which may occur in preference to denitrification or DNRA in situations where reduced sulfur compounds are abundant (refer to Table 7.5; Zehnder and Stumm 1988). Anaerobic oxidation of Fe^{2+} using NO_3^- by microbes is known to occur, but the ecosystem-level importance of this process is not currently well understood (Clement et al. 2005, Burgin and Hamilton 2007). Anaerobic Fe and Mn oxidation might both be expected to be important consumption pathways for NO_3^- when organic substrates are limiting and the concentrations of reduced Fe or Mn compounds are high.

Burgin and Hamilton (2007) proposed a useful conceptual model summarizing current understanding of the many alternative pathways for NO_3^- utilization under anoxic conditions (Figure 7.16). The relative importance of these processes is a function of the actual free

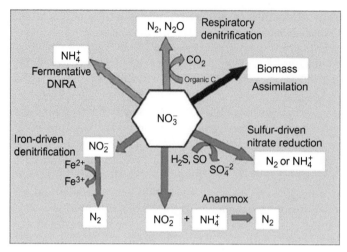

FIGURE 7.16 A conceptual diagram of important nitrate removal pathways in the absence of oxygen. *Blue arrows* denote autotrophic pathways and *dark pink* arrows denote heterotrophic pathways. In addition to using nitrate to acquire energy through dissimilatory reactions, all microbes require N assimilation into biomass (*black arrow*). *Source: From Burgin and Hamilton (2007). Used with permission of Ecological Society of America.*

energy yield of competing reactions resulting from the relative concentration of chemical substrates (e.g., heterotrophic pathways will likely dominate when suitable organic molecules are available).

Iron and Manganese Reduction

Manganese reduction, although thermodynamically favorable in many anoxic environments, is only locally important because Mn^{4+} is rarely found at high concentrations. The product of the reaction, soluble Mn^{2+}, is toxic to many plants and can affect productivity or species composition. In contrast, Fe reduction is a dominant metabolic pathway in many wetlands (e.g., Figure 7.11). In many cases there appears to be some overlap between the zone of denitrification and the zone of Mn reduction in sediments (e.g., Klinkhammer 1980, Kerner 1993; Figure 7.10), and most of the microbes in this zone are facultative anaerobes that can tolerate periods of oxic conditions (Chapter 6). In contrast, there is little overlap between the zone of Mn reduction and that of Fe reduction because soil bacteria show an enzymatic preference for Mn^{4+}, and Fe^{3+} reduction will not begin until Mn^{4+} is completely depleted (Lovley and Phillips 1988). Below the zone of Mn^{4+} reduction most redox reactions are performed by obligate anaerobes. Our earlier emphasis on the redox state of iron (refer to Figure 7.14) reflects the widespread use of Fe as an index of the transition from mildly oxidizing to strongly reducing conditions.

Certain types of bacteria (e.g., *Shewanella putrefaciens*) can couple the reduction of Mn and Fe directly to the oxidation of simple organic substances (Lovley and Phillips 1988, Lovley 1991, Caccavo et al. 1994), but usually these reactions are catalyzed by a suite of coexisting bacteria—with some species using fermentation to obtain metabolic energy (Eqs. 1 and 3 in Table 7.6), while others oxidize hydrogen, using Mn^{4+} and Fe^{3+} as electron acceptors (Eqs. 2 and 4 in Table 7.6) (Lovley and Phillips 1988, Weber et al. 2006). Below the depth of iron reduction, the redox potential progressively drops as sulfate reduction and then methanogenesis become the dominant terminal decomposition pathways (Lovley and Phillips 1987).

TABLE 7.6 Chemical Reactions and Associated Free Energy Yield (per mol of organic matter or hydrogen compound) for Fe and Mn Reduction Pathways

Reaction	$-\Delta G^\circ$ (kJ mol^{-1})	Organisms catalyzing these reaction
$2\,MnO_2 + (CH_2O) + 2\,H^+ \rightarrow MnCO_3 + Mn^{2+} + 2\,H_2O$	94.5	Members of the genus *Bacillus*, *Micrococcus*, and *Pseudomonas*
$2\,Mn^{3+} + H_2 \rightarrow 2\,Mn^{2+} + 2\,H^+$	285.3	Members of the genus *Shewanella*
$4\,FeOOH + (CH_2O) + 6\,H^+ \rightarrow FeCO_3 + 3\,Fe^{2+} + 6\,H_2O$	24.3	Members of the genus *Bacillus*
$2\,Fe^{3+} + H_2 \rightarrow 2\,Fe^{2+} + 2\,H^+$	148.5	Members of the genus *Pseudomonas* and *Shewanella*

Source: From Zehnder and Stumm (1988), Lovley (1991).

Sulfate Reduction

In Chapter 6 we examined the reduction of sulfate that accompanied the uptake, or assimilation, of sulfur by soil microbes and plants. In contrast, dissimilatory sulfate reduction in anaerobic soils is analogous to denitrification in which SO_4^{2-} acts as an electron acceptor during the oxidation of organic matter by bacteria (Table 7.7). This metabolic pathway evolved at least 2 bya (Chapter 2), and before widespread human air pollution the release of biogenic gases from wetlands was a dominant source of sulfur gases in the atmosphere (Chapter 13). Sulfate-reducing bacteria produce a variety of sulfur gases including hydrogen sulfide, H_2S; dimethylsulfide, $(CH_3)_2\,S$; and carbonyl sulfide, COS (Conrad, 1996). In brackish or saline waters, sulfate is typically the dominant electron acceptor and sulfate reduction can be an important fate for organic matter (e.g., Howarth 1984, Neubauer et al. 2005).

Although the production of reduced-sulfur gases in wetlands may be high, the escape of H_2S from wetland soils is often much less than the rate of sulfate reduction at depth as a result of reactions between H_2S and other soil constituents (e.g., NO_3, Table 7.5). Hydrogen sulfide

TABLE 7.7 Chemical Reactions and Associated Free Energy Yield (per mol of organic matter or H$_2$ compound) for Sulfate Reduction Pathways

Reaction	$-\Delta G^\circ$ (kJ mol^{-1})	Organisms catalyzing these reactions
$1/2\,SO_4^{2-} + (CH_2O) + 1/2\,H^+ \rightarrow 1/2\,HS^- + CO_2 + H_2O$	18.0	*Desulfobacter sp., Desulfovibrio sp., Desulfonema sp.*, etc.
$S^0 + (CH_2O) + H_2O \rightarrow HS^- + H^+$	12.0	*Desulfomonas acetoxidans, Campylobacter sp. Thermoproteus tenax, Pyrobaculum islandicum*
$S^0 + H_2 \rightarrow HS^- + H^+$	14.0	*Thermoproteus sp., Thermodiscus sp., Pyrodictum sp.*, various bacteria

Source: Adapted from Zehnder and Stumm (1988).

I. PROCESSES AND REACTIONS

can also react abiotically with Fe^{2+} to precipitate FeS, which gives the characteristic black color to anaerobic soils. FeS may be subsequently converted to pyrite in the reaction

$$FeS + H_2S \rightarrow FeS_2 + 2H^+ + 2e^-. \tag{7.19}$$

When H_2S diffuses upward through the zone of oxidized Fe^{3+}, pyrite (FeS_2) is precipitated following

$$2Fe(OH)_3 + 2H_2S + 2H^+ \rightarrow FeS_2 + 6H_2O + Fe^{2+}. \tag{7.20}$$

Thus, not all the reduced iron in wetland soils is formed directly by iron-reducing bacteria. In some cases the indirect pathways (Eqs. 7.19 and 7.20) may account for most of the total (Canfield 1989a, Jacobson 1994).

This effective sink for reduced sulfur initially led many researchers to believe that sulfate reduction was not a particularly important pathway for organic matter decomposition in wetland ecosystems since the emission of reduced sulfur gases from wetlands was too low to account for a large proportion of organic matter decomposition. However, Howarth and Teal (1980) used an elegant $^{34}SO_4^{2-}$ isotope tracer experiment to demonstrate that sulfate reduction was responsible for the majority of organic matter decomposition in the salt marsh surrounding Sapelo Island, Georgia, and that nearly all of the ^{34}S introduced to the marsh as sulfate was rapidly reduced and subsequently sequestered in the sediments as $Fe^{34}S_2$.

A low iron content limits the accumulation of iron sulfides in many wetland sediments (Rabenhorst and Haering 1989, Berner 1983, Giblin 1988). During periods of low water, specialized bacteria may reoxidize the iron sulfides (Ghiorse 1984), releasing SO_4^{2-} that can diffuse to the zone of sulfate-reducing bacteria. Thus, high rates of sulfate reduction may be maintained in soils and sediments that have relatively low SO_4^{2-} concentrations, owing to the recycling of sulfur between oxidized and reduced forms (Marnette et al. 1992, Urban et al. 1989b, Wieder et al. 1990).

Hydrogen sulfide also reacts with organic matter to form carbon-bonded sulfur that accumulates in peat (Casagrande et al. 1979, Anderson and Schiff 1987). In many areas, the majority of the sulfur in wetland soils is carbon-bonded and only small amounts are found in reduced inorganic forms—that is, H_2S, FeS, and FeS_2 (Spratt and Morgan 1990, Wieder and Lang 1988). Carbon-bonded forms—from the original plant debris, from the reaction of H_2S with organic matter, and from the direct immobilization of SO_4 by soil microbes—are relatively stable (Rudd et al. 1986, Wieder and Lang 1988). Carbon-bonded sulfur accounts for a large fraction of the sulfur in many coals (Casagrande and Siefert 1977, Altschuler et al. 1983) and thus for the sulfuric acid content of acid rain and coal-mining effluents. Organic sediments and coals containing carbon-bonded sulfur that is the result of dissimilatory sulfate reduction show negative value for $\delta^{34}S$ as a result of bacterial discrimination against the rare heavy isotope $^{34}SO_4^{2-}$ in favor of $^{32}SO_4^{2-}$ during sulfate reduction (Chambers and Trudinger 1979, Hackley and Anderson 1986).

Because H_2S can react with various soil constituents and is oxidized by sulfur bacteria as it passes through the overlying sediments and water (Eq. 2.13), many biogeochemists once believed that various organic sulfur gases might be the dominant forms escaping from wetland soils. Most studies, however, have found that H_2S accounts for a large fraction of the emission from wetland soils (Adams et al. 1981, Kelly 1990). Castro and Dierberg (1987) reported a flux of H_2S containing 1 to 110 mg S m^{-2} yr^{-1} from various wetlands in Florida. Nriagu et al. (1987) reported a total flux of sulfur gases ranging from 25 to 184 mg m^{-2} yr^{-1} from swamps in Ontario and found that the sulfate in rainfall in the surrounding region had a lower $\delta^{34}S$ value during the summer than during the winter. Presumably a portion of the SO_4^{2-} content in this ^{34}S-depleted rain is derived from the oxidation of sulfur gases released to the atmosphere by sulfate reduction in local wetlands.

Methanogenesis

Methanogenesis is the final step in anaerobic degradation of carbon. This is the metabolism that degrades carbon when all alternative electron acceptors have been exhausted. Despite its extremely low energy yield (refer to Table 7.3), methanogenesis is a dominant pathway for organic matter decomposition in many wetlands due to the lack of oxidants typical of water-logged soils. Methane can be produced from two different pathways in flooded sediments, both of which are accomplished by methanogens, a diverse group of strict anaerobes in the archaea (Zehnder and Stumm 1988). When organic matter fermentation produces organic acids at concentrations in excess of the availability of alternative electron donors (NO_3^-, Fe^{3+}, SO_4^{2-}), then methanogens can split acetate to produce methane in a process called acetoclastic methanogenesis or acetate fermentation (refer to Table 7.8, Reaction 1). The energy yield of acetoclastic methanogenesis is very low compared to other anaerobic metabolic path-ways (Table 7.4), and the product of this reaction produces a $\delta^{13}C$ of -50 to $-65‰$ in CH_4 (Woltemate et al. 1984, Whiticar et al. 1986, Cicerone et al. 1992). Acetoclastic methanogenesis is performed by only two genera of methanogens: the *Methanosarcina* and *Methanosaeta* (Megonigal et al. 2003).

When acetate is unavailable, a much wider variety of methanogens can perform hydrogen fermentation coupled to CO_2 reduction (Table 7.8, Reaction 2), in which the hydrogen serves as a source of electrons and energy while the CO_2 serves as the source of C and as an electron acceptor. Methanogenesis by CO_2 reduction accounts for the limited release of H_2 gas from wetland soil (Conrad 1996). This methane is even more highly depleted in ^{13}C than that produced from acetoclastic methanogenesis, with $\delta^{13}C$ of -60 to $-100‰$ (Whiticar et al. 1986).

TABLE 7.8   Chemical Reactions and Associated Free Energy Yield (per mol of organic matter or H_2) for the Methane-Producing Pathways of Acetate Splitting and CO_2 Reduction (or hydrogen fermentation)

Reaction		$-\Delta G°$ (kJ mol^{-1})	Organisms catalyzing these reactions
Acetate splitting **(1)** CH_3COOH	$\rightarrow CH_4 + CO_2$	28	Some methanogens (*M. barkeri*, *M. mazei*, *M. söhngenii*)
CO_2 reduction **(2)** $CO_2 + 4 H_2$	$\rightarrow CH_4 + H_2O$	17.4	Most methane bacteria

Source: Adapted from Zehnder and Stumm (1988).

Methanogenesis is often limited by the supply of fermentation products (H_2 or acetate), and can be stimulated through experimental additions of either organic matter or hydrogen (Coles and Yavitt 2002). Thus, methanogenesis generally declines with depth below the oxic-anoxic interface in wetland sediments (Megonigal and Schlesinger 1997), and with greater depth, methane production is increasingly from CO_2 reduction (Hornibrook et al. 2000). Methanogenic bacteria can use only certain organic substrates for acetate splitting and in many cases there is evidence that sulfate-reducing bacteria are more effective competitors for the same compounds (Kristjansson et al. 1982, Schonheit et al. 1982).

Sulfate-reducing bacteria also use H_2 as a source of electrons, and they are more efficient in the uptake of H_2 than methanogens engaging in CO_2 reduction (Kristjansson et al. 1982, Achtnich et al. 1995). Thus in most environments there is little to no overlap between the zone of methanogenesis and the zone of sulfate reduction in sediments (Lovley and Phillips 1986, Kuivila et al. 1989). Methanogenesis in marine sediments is inhibited by the high concentrations of SO_4 in seawater, and where methanogenesis occurs in marine environments CO_2 reduction is much more important than acetate splitting because acetate is entirely depleted within the zone of sulfate-reducing bacteria (Chapter 9). Because sulfate reduction provides a higher energy alternative, sulfate inputs to freshwater wetlands via acid deposition appear to suppress CH_4 flux (Dise and Verry 2001, Gauci et al. 2004, Gauci and Chapman 2006).

Methane flux varies widely across wetland ecosystem types, making global extrapolations challenging. Current estimates suggest that approximately 3% of wetlands' net annual ecosystem production is released to the atmosphere as CH_4 (an estimated 150 Tg CH_4 annually) (Dlugokencky et al. 2011; refer to Table 11.2). These scaling exercises also suggest that much of the interannual variation in global CH_4 fluxes could be explained by annual variation in wetland emissions (Bousquet et al. 2006). The flux of methane from wetland ecosystems results from the net effects of methanogenesis and methanotrophy. Methanogenesis is limited by the supply of labile organic matter (Bridgham and Richardson 1992, Cicerone et al. 1992, Valentine et al. 1994, Van der Gon Denier and Neue 1995), and CH_4 flux shows a direct correlation to net ecosystem production across a variety of wetland ecosystems (Whiting and Chanton 1993, Updegraff et al. 2001, Vann and Megonigal 2003; Figure 7.17).

The positive association between plant productivity and methane flux may be due at least in part to the facilitated gas exchange provided by many wetland plants (Sebacher et al. 1985, Chanton and Dacey 1991, Yavitt and Knapp 1995). Many wetland plants, including rice, have hollow stems composed of aerenchymous tissue, which allows O_2 to reach the roots. These hollow stems inadvertently act as a conduit for CH_4 transport to the surface (Kludze et al. 1993). When oxic soil layers overlay zones of highly reduced anoxic sediments, much of the CH_4 that diffuses to the sediment surface will be oxidized. Higher methane fluxes typically occur when water tables are higher and a greater proportion of sediment pore spaces are filled with water. Under these conditions obligate anaerobic methanogens can operate in shallower sediments, where high-quality organic matter is concentrated and there is less habitat available for methanotrophs and thus a greater chance for CH_4 escape to the atmosphere (Sebacher et al. 1986, Moore and Knowles 1989, Shannon and White 1994, von Fischer et al. 2010). In flooded soils CH_4 flux also increases with soil temperature (Roulet et al. 1992, Bartlett and Harriss 1993).

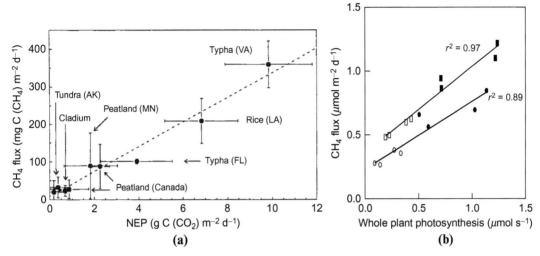

FIGURE 7.17 The relationship between wetland CH_4 emissions and various measures of primary productivity. (a) Emissions versus NEP in North American ecosystems ranging from the subtropics to the subarctic; here the slope is 0.033 g methane C/g CO_2. (b) Emissions versus whole-plant net photosynthesis in marsh microcosms planted with the emergent macrophyte *Orantium aquaticum* that were exposed to elevated and ambient concentrations of atmospheric CO_2. *Source: Figure (a) from Whiting and Chanton (1993); figure (b) from Vann and Megonigal (2003). Used with permission of Nature Publishing Group and Springer.*

Aerobic Oxidation of CH_4

Methanotrophs tend to outcompete nitrifiers for O_2 when CH_4 is abundant since more energy can be released from oxidizing methane than from oxidizing NH_4. Thus much of the methane produced at depth is oxidized to CO_2 before it is released from wetland sediments (Figure 7.18). When CH_4 concentrations are very high at depth—high enough to exceed the hydrostatic pressure of the overlying water—then CH_4-rich gas bubbles can escape to the surface in the process of *ebullition* (Figure 7.18). Effectively bypassing the methanotrophs, ebullition can account for a large fraction of the methane flux to the atmosphere (Fechner Levy and Hemond 1996, Neue et al. 1997, Baird et al. 2004, Goodrich et al. 2011, Comas and Wright 2012). Bubbles escaping by ebullition may be nearly pure CH_4, whereas bubbles emerging from vegetation are often diluted with N_2 from the atmosphere (Chanton and Dacey 1991). The effect of plant-facilitated gas exchange on methane transport and oxidation is complex because, by oxidizing their rhizosphere, aerenchymous plants create a larger volume of oxidized soil where methanotrophs can oxidize CH_4 at the same time that aerenchymous tissues can facilitate CH_4 escape through direct transport (Laanbroek 2010).

Anaerobic Oxidation of CH_4

It has been known for some time that in marine sediments anaerobic methane oxidation (AMO) by sulfate-reducing bacteria can be a major sink for methane (Reeburgh 1983, Henrichs and Reeburgh 1987, Blair and Aller 1995):

$$CH_4 + SO_4^{2-} \rightarrow HS^- + HCO_3^- + H_2O, \tag{7.21}$$

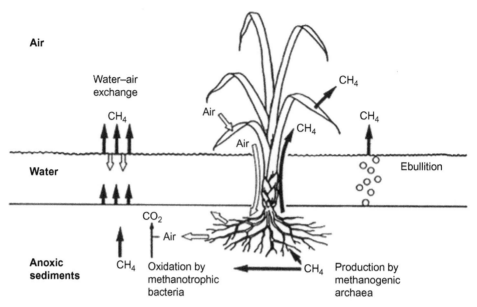

FIGURE 7.18   Processes of methane production, oxidation, and escape from wetland soils. *Source: From Schutz et al. (1991).*

and it has been suggested that AMO may be an important mechanism regulating methane emissions in freshwater wetlands as well (Smemo and Yavitt 2011). Either sulfate reduction or nitrate reduction may be coupled to anaerobic methane oxidation through close associations between methanogens and sulfate-reducing or denitrifying microbes (Conrad 1996, Boetius et al. 2000, Raghoebarsing et al. 2006).

Due to both anaerobic and aerobic methane oxidation, the net ecosystem fluxes of CH_4 from wetlands are often far lower than gross rates of methanogenesis. Reeburgh et al. (1993) have estimated that the global rate of methane production in wetlands is about 20% larger than the net release of methane from wetland soils. To understand the relative importance of methane consumption, investigators note that methane-oxidizing bacteria alter the $\delta^{13}C$ of CH_4 escaping to the atmosphere, and comparisons of the isotopic ratio of CH_4 in sediments and surface collections can indicate the importance of oxidation (Happell et al. 1993). In the Florida Everglades, for example, more than 90% of methane production is consumed by methanotrophs before it diffuses to the atmosphere (King et al. 1990).

The flux of methane from wetlands shows great spatial variability as a result of differences in soil properties, topography, and vegetation (Bartlett and Harriss 1993, Bubier 1995, Zou et al. 2005, Keller and Bridgham 2007, Levy et al. 2012, Morse et al. 2012), making global extrapolations difficult. Methane fluxes may increase with increasing temperatures if wetland sediments remain saturated (Christensen et al. 2003). In contrast, for wetlands that dry during the warmest months of the year, wetlands may shift from being net CH_4 sources during wet seasons to net CH_4 sinks as temperatures rise (Harriss et al. 1982). In an early comprehensive synthesis of wetland methane efflux studies, Bartlett and Harris (1993) reported that tropical wetlands were responsible for >60% of total global wetland CH_4 emissions while northern wetlands (north of 45° N) were responsible for nearly one-third of global emissions (Table 7.9).

TABLE 7.9   Compiled Estimates of Global Emissions of CH_4 from Natural Wetlands

Authors	Method	Global emissions from wetlands (Tg CH_4 yr^{-1})	Global natural emissions (Tg CH_4 yr^{-1})	Percent of natural sources
Matthews and Fung (1987)	Upscaling from field estimates	110	—	—
Aselmann and Crutzen (1989)	Upscaling from field estimates	40–160	—	—
Hein et al. (1997)	Global inverse modeling	231	—	—
Houweling et al. (2000b)	Global inverse modeling	163	222	73
Wuebbles and Hayhoe (2002)		100	145	69
Wang et al. (2004)	Global inverse modeling	176	200	88
Fletcher et al. (2004)	Global inverse modeling	231	260	89
Chen and Prinn (2006)	Global inverse modeling	145	168	86
Our synthesis: Table 11.2	Data synthesis	143	258	55

Although the range of measured rates of CH_4 emissions overlaps considerably across latitudes, the longer growing seasons and greater spatial extent of tropical wetlands explain their greater contribution to global atmospheric CH_4 (Figure 7.19; Bartlett and Harriss 1993). Net regional losses from wetlands are partially balanced by the consumption of atmospheric methane in adjacent upland soils, where it is consumed by methane-oxidizing bacteria (Whalen and Reeburgh 1990, Le Mer and Roger 2001; see also Chapter 11). Efforts over the last two decades to constrain the global estimates of CH_4 emissions from natural wetlands

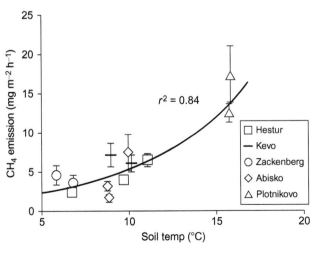

FIGURE 7.19   The relationship between mean seasonal soil temperature (at 5-cm depth) and mean seasonal CH_4 flux (measured at least 8 times throughout the growing season at each site) at the measurement site during all years. *Source: From Christensen et al. (2003). Used with permission of American Geophysical Union.*

have suggested that wetlands may produce 100 to 231 Tg CH_4 yr^{-1}, or 55 to 89% of the total annual flux of CH_4 from natural sources (refer to Table 11.2).

Microbial Consortia

The large differences in free energy yield between the various terminal decomposition pathways in wetlands can be used to predict the dominant metabolic processes under many environmental conditions. When NO_3^- or oxidized Fe or Mn is available in anoxic sediments, sulfate reduction is generally suppressed. Similarly, methanogenesis is suppressed by the provision of SO_4. Indeed, it has been suggested that sulfate delivered by acid rain could be suppressing current global CH_4 emissions by as much as 8% (Vile et al. 2003, Gauci et al. 2004, Weston et al. 2006). Most of our theoretical understanding of anaerobic metabolism is based on using the measured free energy yields from pure culture laboratory studies. A variety of recent studies have demonstrated that consortia of co-occurring microbial species can collectively perform chemical reactions that would not be predicted from studies of individual species (Lovley and Phillips 1988; Boetius et al. 2000; Raghoebarsing et al. 2005, 2006).

WETLANDS AND WATER QUALITY

Thus far we have primarily focused on the retention of C in solid organic matter within wetland sediments and the gaseous loss of C as CO_2 or CH_4. Hydrologic losses of dissolved organic carbon (DOC) from wetlands represent a major export term and an important source of organic carbon and nutrients for many rivers and coastal estuaries. Water in contact with peat leaches DOC from the peat matrix (Dalva and Moore 1991), and concentrations of DOC in peatland soils and overlying surface waters typically range from 20 to 100 mg C L^{-1}, with the vast majority of the DOC characterized as humic substances (Thurman 1985). Higher water tables tend to increase the rate of DOC production and, for wetlands that are hydrologically connected to downstream ecosystems or groundwaters, will increase rates of DOC loss (Blodau et al. 2004). Much of the variation in DOC fluxes between rivers can be explained by differences in the wetland area of their watersheds (Dalva and Moore 1991, Dillon and Molot 1997, Gergel et al. 1999, Pellerin et al. 2004, Johnston et al. 2008).

Without oxygen, decomposition is slow in saturated sediments, maintaining nutrient limitation of biomass growth in many wetlands and leading to the gradual buildup of soil organic matter. Wetlands can thus sequester large quantities of nutrients and trace elements delivered from their catchments, incorporating these elements into plant biomass and eventually into soil organic matter. In addition to retaining elements in tissues and soils, wetland sediments provide ideal conditions for denitrifers and possess a great capacity to remove excess NO_3 and convert inorganic phosphorus and other trace elements into organic molecules (e.g., Johnston et al. 1990, Emmett et al. 1994, Zedler and Kercher 2005, Fergus et al. 2011). In addition to sequestering trace elements, microbes in wetland sediments are responsible for the methylation of a wide variety of metals, some of which are toxic to biota and more rapidly assimilated in the methyl form (Chapter 13).

WETLANDS AND GLOBAL CHANGE

The areal extent of wetlands has declined through human history as wetlands have been drained, filled, or cultivated for agricultural or urban development. Although it is impossible to determine the global extent of wetlands that have been lost through direct human intervention, recent estimates suggest that 50% of all terrestrial wetlands have been destroyed by human activities in the United States, with higher rates for the developed regions of Canada, Europe, Australia, and Asia (Mitsch and Gosselink 2007).

Global Wetland Loss

Climate change is likely to accelerate the rate of wetland loss, as wetlands are particularly vulnerable to altered patterns of precipitation and evaporation on the world's continents (Chapter 10). It is difficult to anticipate the net effect of wetland loss on the global C cycle (Avis et al. 2011). Less frequent inundation is likely to foster oxidation of soil organic matter, but the loss of organic matter may be coupled with enhanced NPP (Megonigal et al. 1997, Choi et al. 2007) and reduced CH_4 emissions (Bousquet et al. 2006, Ringeval et al. 2010).

Sea Level Rise and Saltwater Intrusion

In some regions freshwater wetlands may become brackish or saline as a result of saltwater intrusion into coastal wetlands or through drought-induced evaporative concentration of salts. The resulting changes in ionic strength have been shown to drive "internal eutrophication," as PO_4^{3-} formerly bound to anion adsorption sites is displaced by Cl^- and SO_4^{2-} ions and reduced S compounds bind with Fe, reducing the efficiency of $Fe-PO_4$ binding (Caraco et al. 1989, Lamers et al. 1998, Beltman et al. 2000). The provision of the important electron acceptor SO_4^{2-}, a dominant constituent of seawater, alters the availability of electron acceptors in waterlogged soils, enhancing organic matter decomposition (through sulfate reduction), increasing concentrations of HS^-, and potentially suppressing methanogenesis and denitrification (Lamers et al. 1998, Weston et al. 2006, Sutton-Grier et al. 2011).

Rising Temperatures

Predicting the effect of rising temperatures on wetland biogeochemistry is difficult because it requires reconciling sometimes opposite direct effects and complex positive and negative feedbacks. For instance, higher temperatures will lead to a lower oxygen-holding capacity of water, perhaps further reducing the zones of aerobic respiration in many wetlands and slowing organic matter decomposition. At the same time, both methanogenesis and CO_2 production increase in wetland sediments at higher temperatures (Avery et al. 2002). The effect of warmer temperatures on plant evapotranspiration and the resulting changes in wetland hydroperiods will likely have the greatest impact on wetland processes globally.

Of particular concern is the potential for a loss of high-latitude wetlands due to permafrost thaw (Limpens et al. 2008, Schuur et al. 2008, Tarnocai et al. 2009, McGuire et al. 2010, Koven et al. 2011, Shaver et al. 2011). More than 50% of all wetlands are located at high latitudes (refer to Figure 7.1), and recent estimates suggest that as much as 1672 Pg of soil C is held within the northern permafrost region, of which about 88% occurs in perennially frozen soils (Tarnocai et al. 2009). The hydrology of many boreal wetlands is constrained by permafrost barriers to drainage; thus permafrost melting may lead to wetland drainage, with the potential for positive feedbacks to climate change through the decomposition of previously saturated organic matter (Avis et al. 2011) and the hydrologic export of large amounts of DOC (Guo et al. 2007, Olefeldt and Roulet 2012).

The abundance of stored soil organic matter in peatland ecosystems results from inefficient decomposition over long periods of time rather than high rates of organic matter production. Peat decomposition will increase as water tables drop (Yu et al. 2003, Fenner and Freeman 2011); thus drier arctic climates will lead to lower water table elevations and increased rates of organic matter oxidation. Changes in the flux of methane to the atmosphere may occur if global warming causes changes in the saturation of peatlands, particularly at northern latitudes. If these soils become warm and dry, the flux of methane may be lower while the flux of CO_2 to the atmosphere may increase (Freeman et al. 1993; Moore and Roulet 1993; Funk et al. 1994). The potential for catastrophic carbon losses from peatlands in a warming climate is of concern. Drainage of wetlands (either intentionally or as a result of a drier climate) can make wetlands more susceptible to wildfire and can support fires that burn more deeply into peat or spread over a greater spatial extent (Grosse et al. 2011). Recent peatland fires in Alaska (Mack et al. 2011) and Canada (Turetsky et al. 2011) each removed centuries of accumulated peat.

Elevated CO_2

Based on lab and field experiments, elevated CO_2 has less consistent effects on total aboveground plant biomass in wetlands than it does on upland ecosystems. Several experimental CO_2 enrichments in peatlands have shown that higher atmospheric CO_2 increases vascular plant biomass, but the effect is considerably offset by accompanying losses of peat-building bryophytes (especially *Sphagnum*) (Berendse et al. 2001, Freeman et al. 2001a). In brackish coastal marshes, Langley and Megonigal (2010) found that the vascular plant biomass response to CO_2 was constrained by low nutrient availability, and the low nutrient turnover in many wetlands may substantially limit their ability to sequester additional C in biomass in response to elevated CO_2. Both field and laboratory evidence suggests that elevated CO_2 leads to enhanced organic matter exudation by wetland plants that can exacerbate wetland DOC losses (Freeman et al. 2001a), promote organic matter oxidation (Wolf et al. 2007), and increase CH_4 production (Vann and Megonigal 2003, van Groenigen et al. 2011).

The flux of methane from natural wetlands accounts for $\sim$140 Tg of CH_4 annually, which is a large portion of the total global flux to the atmosphere (Chapter 11). With the atmospheric concentration increasing at about 1% yr^{-1}, various workers have asked whether changes in ecosystem processes within wetlands or changes in the spatial extent of wetlands might be responsible. Certainly, global methanogenesis has increased with the increasing cultivation of rice, which now accounts for $\sim$25% of the global production of methane from wetlands (Hein et al. 1997, Houweling et al. 2000a, Wuebbles and Hayhoe 2002, Fletcher et al. 2004, Wang et al. 2004). Because CH_4 production has been found to be highly correlated with NEP in a number of studies (refer to Figure 7.17), there is concern that as rising atmospheric

CO_2 stimulates the growth of wetland plants it may simultaneously provide a greater supply of organic substances for methane-producing bacteria (Megonigal and Schlesinger 1997, Vann and Megonigal 2003, Fenner et al. 2007). A meta-analysis of CO_2-enrichment experiments in 16 studies of natural wetlands and 21 studies in rice paddies found that increased CO_2 stimulated CH_4 emissions in wetlands by 13.2% and in rice paddies by 43.4% (van Groenigen et al. 2011), suggesting that enhanced production of CH_4 may offset C sequestration in wetlands exposed to elevated CO_2. Collectively, elevated CO_2 research in wetlands provides little evidence for increasing plant biomass in response to rising CO_2, and it raises concerns that increased CO_2 may stimulate enhanced rates of soil organic matter oxidation and CH_4 production.

SUMMARY

Wetlands occupy a small (and shrinking) proportion of the continental land surface, but play a critical role in global biogeochemistry by storing as much as 50% of soil organic matter, producing more than 20% of CH_4, and by substantially reducing the inorganic nutrient export to river networks and the sea. The unique role of wetlands in local to global biogeochemical cycling results from the tendency for wetlands to have waterlogged, anoxic soils for some or all of each year. The limited availability of oxygen to wetland heterotrophs impedes decomposition, slows nutrient turnover, and allows a variety of alternative metabolic pathways to dominate ecosystem C cycling and to link the cycling of C with that of nitrate, manganese, iron, sulfate, and hydrogen through microbial energetics.

Recommended Readings

Reddy, K.R., and R.D. DeLaune. 2008. *Biogeochemistry of Wetlands: Science and Applications*. Taylor and Francis.

Megonigal, J.P., M.E. Hines, and P.T. Visscher. 2003. Anaerobic metabolism: Linkages to trace gases and aerobic processes, pp. 317–324. In W.H. Schlesinger, editor. *Biogeochemistry*. Elsevier-Pergamon.

Stumm, W., and J.J. Morgan. 1996. *Aquatic Chemistry* (third ed.). Wiley.

Wetzel R.G. 2001. *Limnology* (third ed.). Academic Press.

Zehnder, A.J.B. 1988. *Biology of Anaerobic Microorganisms*. Wiley.

PROBLEMS

1. Calculate the oxygen concentration at which the actual free energy yield of denitrification is equivalent to aerobic respiration, assuming starting conditions of NO_3 of 0.1 mg N L^{-1}, DOC concentrations at 5 mg CL^{-1}, at a pH of 7, and standard atmospheric pressure. Now repeat your analysis, assuming that NO_3 concentrations elevated by fertilizer or acid rain inputs to be 5 mg L^{-1}.

2. Use the following equation to predict C losses from a 1-ha plot in the Siberian permafrost under two contrasting summer thaw conditions.

$$C_t = C_0(1 - e^{-kt}),$$

where C_t is the cumulative C release over time, C_0 is the proportion of C in the permafrost that is susceptible to microbial decomposition after thaw, k is the inverse of the turnover time for decomposition, and t is the length of the study period. Assume that bulk soil in the plot has a density of 1 g cm^{-3}, C content is 15 mg g^{-1}, of which 10% is susceptible to decomposition, Q_{10} is 1.9, and base respiration rate is 2.5 mg C gsoil^{-1}. Use this information to estimate the rate of C loss over 30 days in a cool growing season (average temperature of 5°C) versus a warm growing season (average temperature of 10°C). Assume that in the cooler year permafrost thaws to a depth of 10 cm and in the warmer year it thaws to a depth of 20 cm. Repeat your analyses for a longer and warmer growing season (average temperature of 10°C with a growing season of 45 days). Relevant numbers for this problem are derived from Dutta et al. (2006).

3. Both acid rain and saltwater intrusion are expected to reduce the production of methane from affected wetlands through the provision of SO_4^{2-} ions. Use free energy calculations to demonstrate how the free energy yield of sulfate reduction changes as you increase soil solution SO_4^{2-} concentrations from 10 to 2701 mg SO_4^{2-} per liter (freshwater to full-strength seawater). Graph your results. For comparison, show the free energy yield for methanogenesis and denitrification on the same plot. Assume that the sediments are anoxic, the pH $= 7$, and organic carbon is in excess of potential demand. Compare this free energy to methanogenesis and denitrification under the same conditions (assuming NO_3 of 1.0 mg N L^{-1}. At what sulfate concentration should sulfate reducers outcompete methanogens? Denitrifiers?

4. Use the relationship between plant photosynthesis and CH_4 flux from Figure 7.17(b) under ambient and elevated CO_2 treatments to calculate how much of the enhancement of plant C fixation into biomass resulting from elevated CO_2 is offset by CH_4 emissions. Assume that 50% of C fixed in photosynthesis is incorporated into biomass and that methane has a greenhouse warming potential 25× that of CO_2 (IPCC 2007). Does the offset differ between flooded and unflooded treatments? [*Hint:* To turn a graph into datapoints in a spreadsheet you can either use a ruler or, for a more elegant solution, go to the original source of the graph (Vann and Megonigal 2003), select and save the relevant figure as an image, and then use the free program "data thief" (*www.datathief.org*) to extract the datapoints.]

Inland Waters

INTRODUCTION

At any given time surface freshwater ecosystems (lakes and rivers) hold less than 0.02% of all water on Earth and occupy less than 3% of the land surface area (Figure 10.1; Wetzel 2001, Lehner and Doll 2004). The small footprint of surface waters on the terrestrial land surface belies their importance to human civilization and to global biogeochemistry. Because surface waters are constantly replenished and easily accessed, rivers and lakes provide the water supply for the vast majority of humans on Earth. In addition to providing the water necessary to grow most crops, inland waters support freshwater fisheries that currently provide around one-third of global annual fish production.[1] At the same time, surface waters are highly managed for the disposal of wastes, the transport of cargo, and the generation of electricity.

[1] From the Food and Agriculture Organization's "The State of World Fisheries and Aquaculture 2010" report, available from *www.fao.org/docrep/013/i1820e/i1820e.pdf*.

In Chapter 4 the role of rivers in moving elements from the continents to the oceans was discussed at length. Over geologic time the movement of water across the land surface has weathered igneous rock, producing sedimentary rocks and salty seas. However, the biota of freshwater aquatic ecosystems assimilate and transform elements many times during downstream transport, and large amounts of weathered material and terrestrially derived organic matter are stored for millennia in the sediments of lakes. In this chapter we describe the biogeochemistry within aquatic ecosystems, focusing particularly on their capacity to change the form, timing, and magnitude of downstream biogeochemical export from the terrestrial surface to the oceans.

Much of this chapter is devoted to the distinguishing features of lakes, rivers, and estuaries. Although traditional biogeochemical studies have focused on these as discrete systems, in reality streams, ponds, lakes, rivers, and estuaries exist along a continuum of freshwater habitats that stretch from the smallest headwaters to their confluence with the oceans or terminal lakes. Even water bodies that have no apparent surface water connection to rivers are hydrologically connected to a major river or its tributaries via groundwater or floodwaters at sufficiently long time scales. Examining satellite imagery of any landscape effectively demonstrates the network structure of rivers and the intersections between lotic (~*flowing water*) and lentic (~*standing water*) water habitats (Figure 8.1).

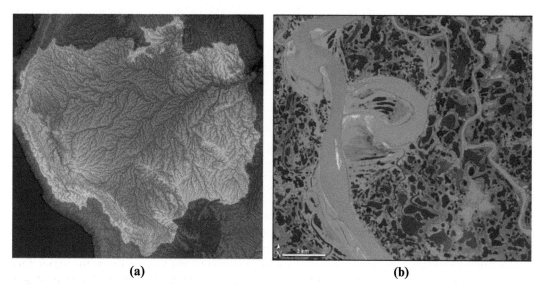

(a)   **(b)**

FIGURE 8.1   Two views of river networks: (a) the Amazon River Basin, which is currently estimated to be 6800 km long and drains a landscape that ranges in elevation from 4500 m (in white) to sea level (in dark green); (b) a segment of Canada's Mackenzie River Delta from August 4, 2005, when lakes throughout the Mackenzie's floodplain had thawed. *Source: Both images from NASA's Earth Observatory. (a) Created using Shuttle Radar Topography Mission (SRTM) data together with river data developed by the World Wildlife Fund's HydroSHEDS program (Lehner et al. 2008). (b) from NASA's Terra Satellite Advanced Spaceborne Thermal Emission and Reflection Radiometer (ASTER).*

Special Properties of Water

The unique physical and chemical properties of water exert significant control over the biogeochemistry of aquatic ecosystems. Light is attenuated as it passes through water so that the proportion of solar energy delivered to a freshwater ecosystem diminishes with depth. Typically surface waters are warmer and better lit than deep waters. In contrast to terrestrial ecosystems where the atmosphere is warmed from below (see Chapter 2), surface waters are warmed by light absorption at the surface. The density of liquid water varies with temperature, with freshwaters having their highest density at $4°C$. The surface delivery of solar energy can thus generate density gradients that physically separate surface and deepwater habitats with important consequences for the biogeochemistry of lakes, reservoirs, and large rivers.

Gas Diffusion and Solubility

In Chapter 7 we discussed how the slow diffusion of oxygen through water explains much of the difference between wetland and upland soil biogeochemistry. Surface waters readily exchange oxygen with the overlying atmosphere, but the oxygen content of deeper waters can become depleted when respiration exceeds the rate of O_2 diffusion. Therefore the sediments of aquatic ecosystems are often anoxic, and anaerobic metabolism dominates their biogeochemistry.

In contrast to terrestrial systems, aquatic biota are not bathed in an atmosphere of carbon dioxide. Instead, carbon dioxide dissolved in water is partitioned between dissolved CO_2 and bicarbonate and carbonate ions. Collectively these three forms are referred to as dissolved inorganic carbon, or DIC, and are often abbreviated as ΣCO_2. The relative proportion of forms depends on pH:

$$CO_2 + H_2O <=> H^+ + HCO_3^- <=> 2H^+ + CO_3^{2-}.$$

At pH <4.3 most carbon dioxide is found as a dissolved gas; between 4.3 and 8.3 bicarbonate (HCO_3^-) is the dominant form of dissolved inorganic carbon (DIC); and at pH >8.3 carbonate (CO_3^-) dominates. The majority of freshwater ecosystems have pHs between 5 and 8; thus bicarbonate dominates the inorganic C content of most freshwaters.

Terrestrial–Aquatic Linkages

Hydrologic Flowpaths

Precipitation exceeds evapotranspiration across most of the land surface, and surface waters flow downhill as a result of gravity acting on the excess water. When water is supplied to the land surface in excess of evapotranspiration, excess water may leach vertically into groundwater, may be transported laterally downslope through subsurface flowpaths, or, in cases where the rate of precipitation is higher than the rate of soil infiltration (or when soils have been compacted or replaced by pavements), water may flow over the land surface downslope. Water that makes its way to a receiving water body may thus enter as *groundwater flow, subsurface flow,* or *overland flow* (Figure 8.2). Permanent surface waters are found wherever the land surface is below the water table (Chapter 7). When water flow from the surrounding landscape is sufficient to erode

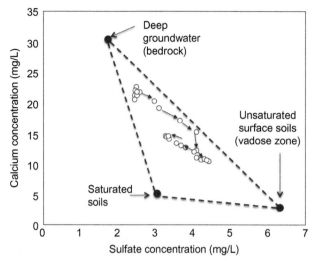

FIGURE 8.2   Using a three component model of hydrologic flowpaths in Walker Branch Watershed, changes in stream-water calcium and sulfate concentrations over the course of a March 1991 storm (*open symbols*) are shown relative to the average $[Ca^{2+}]$ and $[SO_4^{2-}]$ of the three contributing flowpaths. Note that as the storm progresses, stream water quickly becomes dominated by shallow vadose zone flowpaths and later by saturated soil flowpaths. *Source: Modified from Mulholland 1993.*

sediments and materials from the land surface, channels are formed (Montgomery and Dietrich 1988).

The route and the rate at which water moves along flowpaths will determine its chemical properties. For example, the subsurface flowpaths feeding a small stream in eastern Tennessee during summer low flow have high concentrations of Ca^{2+} because of their long residence time in carbonate bedrock, while stormwaters passing through the vadose zone[2] accumulate high concentrations of SO_4 deposited by acid deposition (Mulholland 1993). In contrast, water that flows through permanently saturated soils enters the stream with relatively low concentrations of both ions (Figure 8.2). Because each flowpath has a characteristic Ca^{2+}:SO_4^{2-} ratio, it is possible to determine the relative contribution of each source to stream flow by measuring stream water Ca^{2+}:SO_4^{2-} ratios over time.

Similar comparisons can distinguish the dominant sources of water for different aquatic ecosystems. In a comparison of stream and lake water chemistry across the Northern Highland Lake District of Wisconsin, Lottig et al. (2011) found that streams tended to have Ca^{2+} concentrations $\sim 4\times$ higher than lakes in the region, and suggested that stream flows in this region were primarily derived from groundwaters with long mineral–water contact times while lake inflows were dominated by precipitation and groundwater from short flowpaths with minimal water–mineral contact time.

As this example shows, for the majority of rivers the predominant source of water is delivery from the surrounding terrestrial landscape via groundwater and upstream headwaters, while precipitation is an important source of water to lakes, with its relative contribution increasing with lake size. At any given time, the volume of water (V) in a river or lake reflects the balance between inputs in precipitation, surface inflows, and subsurface exchange with soil waters and groundwater (as in Eq. 7.1). The residence time of water within an aquatic

[2] Unsaturated surface soils.

ecosystem (MRT) determines the extent to which organisms can affect the form, magnitude, and timing of element exports:

$$MRT = V/\Sigma \text{ Inputs.}$$

River inflows and precipitation are relatively easy inputs to measure; however, an accurate characterization of groundwater inputs is often difficult. Net groundwater inputs or exports are typically calculated by difference. The *benthos*, or bottom sediments, of many lakes and rivers are sites of active exchange between surface waters and groundwaters such that these net flux estimates may considerably underestimate the gross flux of groundwater exchange (Covino and McGlynn 2007, Poole et al. 2008).

The water delivered to surface waters and groundwaters from the surrounding terrestrial landscape carries with it the soluble and erodible elements characteristic of the watershed flowpaths through which it is routed. Because of this stream element fluxes have been used effectively to understand the biogeochemistry of terrestrial ecosystems (Chapters 4 and 6; Likens and Bormann 1995). For example, seasonal variation in stream NO_3 concentrations, with high concentrations outside of and very low concentrations during the growing season, is commonly used as an indicator that nitrogen uptake by vegetation drives watershed export (e.g., Bear Brook, NH; Bernhardt et al. 2005) (Figure 8.3). Yet this effect of terrestrial vegetation on stream nutrient concentrations will only be observed in watersheds where water residence times are relatively short (<1 year; Lutz et al. 2012).

In watersheds with deep soils and long groundwater residence times, seasonal variation in N uptake by terrestrial vegetation will be insufficient to change the concentration of the large groundwater volume, and patterns of N loading to streams may be relatively constant over time. In such watersheds stream nutrient concentrations will be reduced during periods of peak biological demand within the stream. Such is the case for Walker Branch, where NO_3 concentrations in groundwater seeps are constant year-round and the lowest annual NO_3^- concentrations are observed during spring algal blooms and following autumn litterfall (Figure 8.3; Mulholland 2004, Roberts et al. 2007). Despite draining mature deciduous forests

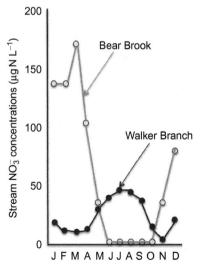

FIGURE 8.3 A comparison of average monthly stream NO_3 concentrations for Walker Branch, Tennessee (*black circles*) in the Oak Ridge National Lab and for Bear Brook, New Hampshire (*gray circles*) in the Hubbard Brook Experimental Forest. *Source: Drawn with data from Mulholland 2004 and Bernhardt et al. 2005.*

and having similarly steep, rocky channels, the difference in the annual pattern of NO_3 export from these two watersheds can be attributed primarily to differences in their hydrology.

Ion Chemistry

The ionic composition of surface waters is typically dominated by four major cations (Ca^{2+}, Mg^{2+}, Na^+, and K^+) and four major anions (HCO_3^-, CO_3^-, SO_4^{2-}, and Cl^-) with ionic forms of N, P, Fe, and other trace elements at lower concentrations (Livingstone 1963, Meybeck 1979) (Table 4.7). The concentration and composition of these ions within surface waters can vary considerably as a function of catchment geology, the chemistry of precipitation, and the extent of evaporative concentration (Gibbs 1970). Atmospheric deposition of marine salts and acid volatiles contributes substantial amounts of base cations (Na^+, Ca^{2+}, Mg^{2+}) or acid anions (SO_4^{2-}, NO_3^-) to catchments, many of which are transported to surface waters.

About two-thirds of the HCO_3^- in rivers is derived from the atmosphere, either directly from CO_2 or indirectly via organic decomposition and root respiration in contributing terrestrial ecosystems (Meybeck 1987, Jones and Mulholland 1998, Raymond and Cole 2003). River and lake waters are typically supersaturated with CO_2 and are a net source of CO_2 to the atmosphere (Table 8.1) (Cole et al. 1994, Quay et al. 1995, Richey et al. 2002, Mayorga

TABLE 8.1   Estimates of CO_2 Outgassing from Inland Waters

Zone-class	Area of inland waters (1000s km^2) min-max	pCO$_2$ (ppm) median	Gas exchange velocity (k$_{600}$ cm hr^{-1}) median	Areal outgassing (g C m^{-2}yr^{-1}) median	Zonal outgassing (Pg C yr^{-1}) median
Tropical (0°–25°)					
Lakes and reservoirs	1840–1840	1900	4.0	240	0.45
Rivers (>60–100 m wide)	146–146	3600	12.3	1600	0.23
Streams (>60–100 m wide)	60–60	4300	17.2	2720	0.16
Wetlands	3080–6170	2900	2.4	240	1.12
Temperate (25°–50°)					
Lakes and reservoirs	880–1050	900	4.0	80	0.08
Rivers (>60–100 m wide)	70–84	3200	6.0	720	0.05
Streams (<60–100 m wide)	29–34	3500	20.2	2630	0.08
Wetlands	880–3530	2500	2.4	210	0.47
Boreal and Arctic (50°–90°)					
Lakes and reservoirs	80–1650	1100	4.0	130	0.11
Rivers (>60–100 m wide)	7–131	1300	6.0	260	0.02
Streams (<60–100 m wide)	3–54	1300	13.1	560	0.02
Wetlands	280–5520	2000	2.4	170	0.49
Global		*Global land area*			
Lakes and reservoirs	2800–4540	2.1–3.4%			0.64
Rivers (>60–100 m wide)	220–360	0.2–0.3%			0.30
Streams (<60–100 m wide)	90–150	0.1–0.1%			0.26
Wetlands	4240–15 220	3.2–11.4%			2.08
All inland waters	**7350–20 260**	**5.5–15.2%**			**3.28**

Source: Aufdenkampe et al. 2011. Used with permission of the Ecological Society of America.

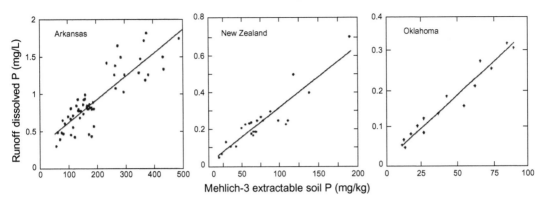

FIGURE 8.4   The nutrient content of surface waters reflect nutrient loading to their catchments. In A–C, the extractable soil phosphorus in agricultural watersheds is a good predictor of the concentrations of dissolved P in receiving streams. *Source: Sharpley et al. 1996. Used with permission of the Soil and Water Conservation Society.*

et al. 2005, Cole et al. 2007, Aufdenkampe et al. 2011). This is in part because of the decomposition of terrestrial organic matter within aquatic systems, but it is also due to the hydrologic transfer of terrestrially respired CO_2 through subsurface flowpaths into aquatic ecosystems. This displaced soil respiration represents a substantial fraction of CO_2 outgassing from many freshwater ecosystems.

Inorganic nutrients leached from terrestrial ecosystems become the nutrient supply to receiving waters. Just as water in excess of evapotranspiration results in surface runoff, nutrients supplied to the land surface in excess of biological demand and soil sorption capacity result in increasing nutrient loading to receiving surface waters. The concentrations of surface water N and P rise predictably with rates of watershed nutrient loading (Figures 8.4 and 8.5; Vollenweider 1976, Meybeck 1982, Peierls et al. 1991, Sharpley et al. 1996, Boyer et al. 2002, Howarth et al. 2012). The positive correlation between watershed nutrient loading and nutrient concentrations in receiving waters is enhanced by common watershed alterations (e.g., tile drains, stormwater pipes, soil compaction, and pavement) that reduce the residence time of water in terrestrial soils and the proportion of rainfall that is transmitted to groundwater.

Aquatic algae have body mass N:P ratios[3] of ~7.2 on average (although this varies between 3 and 20; Klausmeier et al. 2004), but runoff sources can vary much more widely. Runoff from unfertilized forests and fields typically has N:P mass ratios of 20 to 200 while many pollutant sources are enriched in both N and P—with raw sewage, urban stormwaters, and feedlot runoff each having N:P ratios between 1 and 10 (Downing and McCauley 1992). Both the absolute amount of N and P and their relative proportions are important drivers of freshwater biogeochemistry.

The total mass of dissolved ions in solution is its ionic strength. Ionic strength is typically reported in units of milliequivalents (mEq) of charge (Chapter 4) and is often measured in the

---

[3] Convert N:P mass ratios to N:P molar ratios by multiplying by 2.21. The mass ratio of 7.2 is equivalent to a molar N:P ratio of 16, and the range of molar N:P ratios reported by Klausmeier et al. 2004 is 6.6 to 44.2.

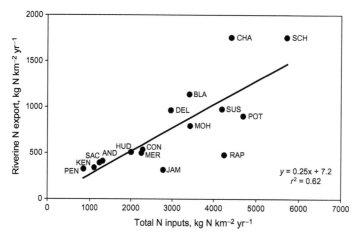

FIGURE 8.5 In an analysis of 16 large rivers in the northeastern U.S. nitrogen exports in streamflow were strongly related to the total new inputs of nitrogen to each catchment measured. From north to south, the catchments are: Penobscot (PEN), Kennebec (KEN), Androscoggin (AND), Saco (SAC), Merrimack (MER), Charles (CHA), Blackstone (BLA), Connecticut (CON), Hudson (HUD), Mohawk (MOH), Delaware (DEL), Schuylkill (SCH), Susquehanna (SUS), Potomac (POT), Rappahannock (RAP), and James (JAM). *Source: From Boyer et al. 2002. Used with permission of the Ecological Society of America.*

field as electrical conductivity.[4] The ionic composition of surface waters determines their alkalinity, which is a measure of the buffering capacity of the carbonate system in water. Alkalinity is defined as

$$\text{Alkalinity} = \left[2CO_3^{2-} + HCO_3^- + OH^-\right] - H^+, \tag{8.1}$$

which is roughly equivalent to the balance of cations and anions in water, where

$$\text{Alkalinity} = \left[2Ca^{2+} + 2Mg^{2+} + Na^+ + K^+ + NH_4^+\right] - \left[2SO_4^{2-} + NO_3^- + Cl^-\right]. \tag{8.2}$$

Generally, alkalinity is measured in milliequivalents per liter by titration of a water sample to a pH of 4.3. In addition to base cation exchange (as in Chapter 4), hydrogen ions (H^+) can be consumed through the protonation of organic acids (A^-) (Hedin et al. 1990) or through release of the acid cation Al^{3+} through dissolution or exchange processes (Eqs. 4.10–4.12) Thus, the titration of a water sample to a pH of 4.3 is often said to represent its *acid-neutralizing capacity* (ANC), a term used to denote the total suite of inorganic and organic constituents that allow a surface water to resist acidification (*sensu* Schindler 1988). *Alkalinity* and *ANC* are used interchangeably.

The easiest way to understand the acid-neutralizing capacity of a system is to examine its charge balance. Negatively and positively charged ions in solution must balance; thus as more acid anions (SO_4^{2-} and NO_3^-) are added to watersheds in acid rain or to surface waters as acid mine drainage, this charge must be balanced either by the accompanying protons (H^+) or by positively charged cations (Na^+, K^+, Ca^{2+}, Mg^{2+}, Al^{n+}) released through dissolution or ion exchange processes. In a well-buffered catchment receiving increasing deposition of H_2SO_4 and HNO_3, Hinderer et al. (1998) documented no changes in H^+ in a receiving lake. Acid rain derived increases in SO_4^{2-} and NO_3^- deposition were instead accompanied by large increases in base cation concentrations in lake inflows (Figure 8.6). It is by this mechanism that acid rain can

---

[4] The capacity of water to conduct electricity, typically reported in μScm^{-1}; conductivity increases with ionic strength.

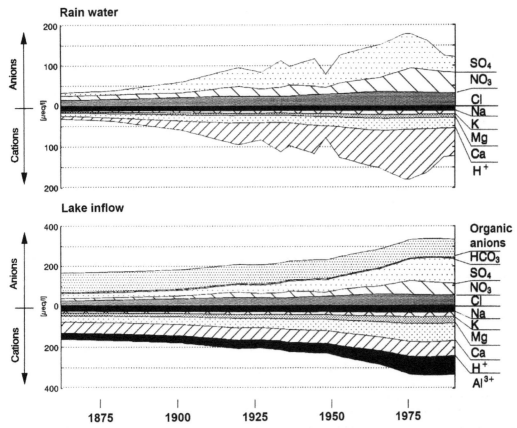

FIGURE 8. 6 A paleolimnological reconstruction of changes in the charge balance of rain and lake inflow over the period of the Industrial Revolution. Note that while rainwater became more acidic over the record (increasing contribution of H^+ and the acid anions SO_4^{2-} and NO_3^- through time) the interflow waters have seen little change in pH (H^+). Instead, large increases in the concentrations of sulfate and nitrate over time have been accompanied by increases in the base cations (K^+, Mg^{2+}, Ca^{2+}) and in soluble Al^{3+}. *Source: Hinderer et al. 1998.*

raise the alkalinity of receiving surface waters for watersheds dominated by carbonate rock (Kilham 1982, Lajewski et al. 2003).

In most freshwaters, ions delivered from the drainage basin are the dominant source of ANC, because the runoff of cations is usually balanced by HCO_3^- (Stoddard 1999). Land use change can contribute additional alkalinity through enhanced erosion of base cations from soils and enhanced weathering associated with cultivation (Edmondson and Lehman 1981, Renberg et al. 1993, Raymond and Cole 2003). In contrast, surface waters draining poorly buffered watersheds, where precipitation passes through soils of low cation exchange capacity and does not encounter carbonate rock, are more likely to become acidified in response to acid rain. Generally, the ANC of lakes is much lower than that of streams in the same landscape (Lottig et al. 2011), so lakes are more susceptible to acidification than streams.

Internal generation of alkalinity within lakes can become increasingly important as acid-ification progresses (Schindler 1986). Lake alkalinity increases as a result of processes that re-move SO_4^{2-} or NO_3^- from the water column, including sulfate reduction, sulfate adsorption on minerals, and denitrification (Schindler 1986, Kelly et al. 1987). It is because of this that, paradoxically, both nitrate and sulfate loading to lakes can enhance their ANC (Cook and Schindler 1983). In contrast, the production of organic carbon in photosynthesis and the deposition of calcite by phytoplankton reduce alkalinity internally by consuming HCO_3^-, converting it into biomass or sediment minerals.

Organic Subsidies

Organic C that is not stored within or respired from terrestrial ecosystems is exported hydrologically as soluble organic molecules (collectively DOM), OM attached to eroded soil minerals, or plant litter. These terrestrial carbon inputs subsidize, and often dominate, aquatic ecosystem metabolism. Allochthounous carbon inputs occur along a continuous gradient of particle sizes, with OM fractions often described by their particle size (Figure 8.7). For aquatic habitats with high edge:volume ratios (e.g., small streams, braided rivers, ponds), direct litterfall represents a substantial annual input of particulate organic material (POM), which varies greatly in particle size and availability to heterotrophs (Figure 8.7). Generally, POM is

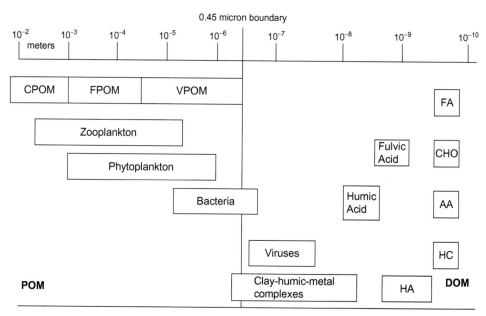

AA = amino acids; CHO = carbohydrates; CPOM = coarse particulate organic matter;
FA = fatty acids; FPOM = fine particulate organic matter; HA = hydrophilic acids;
HC = hydrocarbons; VPOM = very fine particulate organic matter

FIGURE 8.7   Size range of particulate and dissolved organic matter and carbon compounds in natural waters. The distinction between dissolved and particulate organic carbon is operationally defined, with investigators typically considering organic molecules that pass through a 0.45-mm filter dissolved. *Source: Reproduced from Hope et al. 1994.*

subdivided into two size classes of >1 mm coarse (CPOM) and 1 mm to 53 μm fine (FPOM) particulate organic matter (Hope et al. 1994). Particulate carbon inputs to freshwaters typically represent a small fraction (about 10%) of total allochthonous carbon exports from terrestrial watersheds (Schlesinger and Melack 1981) but can represent important seasonal or pulsed inputs to small forested streams (Fisher and Likens 1972, Wallace et al. 1997, Meyer and Eggert 1998).

In larger aquatic ecosystems, dissolved organic matter (DOM) and FPOM are the dominant forms of terrestrial carbon inputs. Riverine DOM is composed of a diverse and complex array of organic molecules, much of which is derived from the surrounding terrestrial watershed (Hope et al. 1994, Findlay and Sinsabaugh 1999, Kujawinski et al. 2004, Seitzinger et al. 2005). In general, the amount of DOM delivered to rivers increases with riverflow (Figure 8.8; Schlesinger and Melack 1981) and with catchment soil C:N ratios (Aitkenhead and McDowell 2000). As for POC, this DOM represents a small loss term for C and organic nutrients from terrestrial ecosystems (Hedin et al. 1995, Neff and Asner 2001) but is a primary energy source for receiving aquatic ecosystems (Wetzel 1992, Cole et al. 2007). The majority of freshwater DOM (50–75%) is comprised of fulvic and humic acids derived from terrestrial soils and upstream wetlands (Wetzel 1992, Hope et al. 1994). It is generally assumed that this terrestrial DOC (the C contained in DOM) is only exported because it is recalcitrant, yet

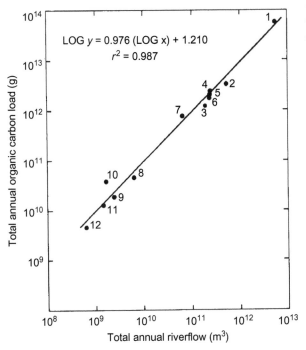

FIGURE 8.8 Total annual load of organic carbon shown as a logarithmic function of total annual riverflow for major rivers of the world. *Source: From Schlesinger and Melack 1981 with a revision of the data for the St. Lawrence derived from Pocklington and Tan (1987). Used with permission of the Ecological Society of America.*

Rivers 1–7 are among the 50 largest.
1 = Amazon; 2 = Mississippi; 3 = St. Lawrence;
4 = MacKenzie; 5 = Danube; 6 = Volga; 7 = Rhine

DOM is rapidly assimilated by freshwater biota (Brookshire et al. 2005, Kaushal and Lewis 2005, Bernhardt and McDowell 2008, Lutz et al. 2012).

There are several explanations for this seeming paradox. First, the proportion of the contributing watershed occupied by wetlands explains much of the variation in DOM across freshwater ecosystems (Chapter 7). The darkly stained organic rich waters of blackwater rivers contain high concentrations of humic and tannic acids. Recall, however, that the primary constraint on the decomposition of organic matter in wetlands is the absence of electron acceptors and the slow speed of fermentation (Chapter 7). Thus DOM that is unavailable to microbes in anoxic sediments becomes more available to freshwater microbes in the presence of oxygen, because aerobic respiration does not require these fermentation steps.

Second, freshwater ecosystems differ from terrestrial soils and wetland sediments in the availability of sunlight. Complex organic molecules within DOM can be degraded by UV exposure into a wide variety of photoproducts with enhanced lability (Wetzel 1992, Moran and Zepp 1997, Bertilsson and Tranvik 2000). Exposing field DOM samples to sunlight can substantially enhance their lability. For example, by simply exposing lake water DOM to natural sunlight prior to incubation studies, Lindell et al. (1996) were able to increase bacterial biomass growth by 83 to 175% above assays without prior sunlight exposure.

In many aquatic ecosystems the annual input of terrestrially derived C exceeds C fixation by aquatic algae and plants. In these "donor-controlled" aquatic systems, the annual rates of respiration, export, and storage of carbon may each be significantly larger than GPP, a condition that is rarely observed in terrestrial ecosystems (Delgiorgio and Peters 1994, Duarte and Agusti 1998, Cole et al. 2007, Battin et al. 2009). In addition to providing organic matter to aquatic heterotrophs, terrestrial DOM released to surface waters reduces the depth of light penetration and can substantially limit freshwater ecosystem GPP (Carpenter et al. 1998, Karlsson et al. 2009). Thus, increases in DOM loading drive aquatic ecosystems toward greater heterotrophy.

In remote watersheds with little direct human influence, organic matter not only is an important source of energy, but is also the dominant form of nutrient input. Dissolved organic nitrogen (DON) dominates annual nitrogen losses from unpolluted terrestrial ecosystems (Meybeck 1982, Hedin et al. 1995, Campbell et al. 2000, Perakis and Hedin 2002, Scott et al. 2007).

Unique Features of Aquatic Food Webs

On an annual basis, most freshwater ecosystems are net heterotrophic (NEP < 1), and they may be classified along a gradient of heterotrophy by their P:R ratio.[5] Generally, we expect the P:R ratio to increase as the size of lakes or rivers increases (Vannote 1980), because an aquatic ecosystem's receipt of solar energy is a function of its surface area and because the relative importance of terrestrial carbon inputs should decline with decreasing edge:volume ratios (Figure 8.9; Finlay 2001, Alin and Johnson 2007, Staehr et al. 2012a). There are many exceptions to this general rule because aquatic ecosystems vary in light attenuation with depth and because autochthonous production may become increasingly nutrient-limited in larger aquatic ecosystems with lower terrestrial subsidies (Figure 8.9).

[5] A widely used abbreviation for the ratio of GPP:R.

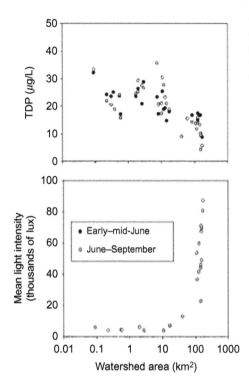

FIGURE 8.9   As the contributing watershed area increases, rivers grow wider and deeper. In a survey of streams in the Coast Range of northern California, Finlay et al. (2011) found that phosphorus availability declines and light availability increases with watershed (~stream) size. *Used with permission of the Ecological Society of America.*

In contrast to terrestrial food webs where autotrophs are large and the majority of grazers are small, the dominant autotrophs in most inland waters are algae, which are consumed by much larger zooplankton, grazing insects, and fish. Unlike woody plants, algae do not invest heavily in structural tissues and often the entire autotroph is consumed by herbivores. Because of this, and because terrestrial carbon inputs are a dominant energy source, the standing biomass of autotrophs in aquatic ecosystems often does not provide an effective index of ecosystem productivity as it does in terrestrial ecosystems. Rivers and lakes with low P:R ratios can still produce large numbers of insects and fish (Webster and Meyer 1997). Therefore, aquatic carbon budgets often focus instead on carbon fluxes into the biomass of higher trophic levels or may include estimates of ecosystem *secondary production.*[6]

The effort to calculate complete secondary production estimates for all heterotrophs within a system is usually cost and labor prohibitive; thus most secondary production estimates focus on the productivity of a limited subset of heterotrophs (e.g., bacterial, insect, or fish productivity). Stable isotope ratios (i.e., $\delta^{13}C$) are often used to estimate the relative contribution of autochthonous algae ($\delta^{13}C \sim$ atmospheric CO_2) compared to allochthonous inputs from land vegetation ($\delta^{13}C$ depleted relative to atmospheric CO_2) (Finlay 2001, Cole et al. 2002, McCutchan and Lewis 2002, Pace et al. 2004). Such studies have demonstrated that as aquatic ecosystems increase in P:R ratios, the $\delta^{13}C$ of herbivores becomes more similar to atmospheric CO_2 (indicative of autochthonous sources). These studies also show that allochthonous C

---

[6] The formation of a living mass of heterotrophs within an ecosystem (Benke and Huryn 2006).

remains an important energy source for the aquatic food webs of well-lit rivers and lakes (Bade et al. 2007, Pace et al. 2007).

LAKES

A lake is defined simply as a permanent body of water surrounded by land. Lakes can vary in size from small permanent ponds to the world's largest lake, Lake Baikal in southeastern Siberia, which is also the oldest (25 million years) and deepest (1700 m) lake in the world. Lake Baikal alone holds 20% of the total global surface freshwater on Earth ($\sim$23,000 km^3; Wetzel 2001). The vast majority of natural lakes are located in the northern temperate zone in formerly glaciated landscapes. Humans are increasing global lake area through the creation of manmade lakes. According to the international commission on large dams,[7] the global surface area of the impoundments created by 37,641 registered large reservoirs is $\sim$400,000 km^2. When smaller lakes and ponds are included, the estimate of the total global surface area of impoundments expands to $\sim$1,500,000 km^2 (St Louis et al. 2000). Lehner and Doll (2004) estimate the current global area of lakes and reservoirs at 3.2 million km^2, or 2.4% of the total global land surface.

Lake Water Budgets and Mixing

Lakes receive water from precipitation, surface inflows, and subsurface inputs, and they export water through evaporation, surface outflows, and export to groundwaters. The residence time of water within a lake varies as a function of both lake volume and watershed area. River and groundwater inflows increase with increasing catchment size; thus lakes that are small relative to their watersheds have shorter residence times. Small lake basins that are set within large river networks may have large river inputs relative to their total volume. These "open systems" effectively function as slow-moving pools within a network of rivers and the flux of elements through the lake may be much higher than the turnover of elements within the lake itself. Most reservoirs, for example, are typically built on large rivers and are small in size relative to their watersheds and have short MRT. In contrast, some terminal lakes, such as Utah's Great Salt Lake or Siberia's Aral Sea, have nearly closed water budgets, in which inputs from riverflows, and precipitation are balanced by evaporative losses. In these closed systems, elements without a gaseous phase accumulate over time. This is why most terminal lakes are saline.

Physical stratification can occur in water bodies of sufficient depth, where less dense warm waters float above colder bottom waters. In lakes, the warmer (low-density) surface layer is referred to as the *epilimnion*, which floats atop the darker, cooler (high-density) *hypolimnion*. The zone of rapid temperature change between these two water masses is known as the thermocline or *metalimnion* (Figure 8.10). The persistence of this density-driven separation of surface and bottom water masses depends on the extent of mixing by floods and winds, the bathymetery[8] of the water body, and the variation in external climate drivers. Many lakes

---

[7] ICOLD, available from *www.icold-cigb.org/*.

[8] The measurement of water depth at various places in a body of water—used to define the underwater topography of a freshwater system.

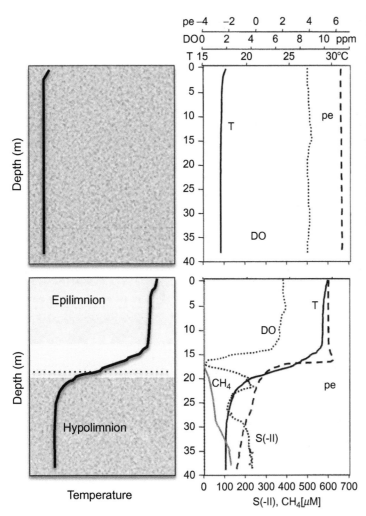

FIGURE 8.10 In the upper panels are shown a hypothetical and an actual lake temperature profile during winter (data from January). The lower panels show profiles during the period of summer stratification (data from July). The *dashed line* in the lower left panel indicates the lake thermocline. Depth profiles for temperature (T), dissolved oxygen (DO), redox potential (pe), total sulfide (S(-II)) and methane (CH_4) measured in the water column of Lake Kinneret in the Afro-Syrian rift valley during 1999. *Source: From Eckert and Conrad 2007. Used with permission of Springer.*

are seasonally stratified (Figure 8.10), and shallow lakes worldwide can have periods of stratification punctuated by storm-driven mixing events.

Very deep tropical lakes may be permanently stratified. Lake Nyos in Cameroon is a permanently stratified volcanic crater lake that receives high inputs of carbon dioxide from deep geothermal seeps. In 1986, the concentrations of CO_2 at depth became so high ($\sim$1–5 L of CO_2 per L of hypolimnetic water) that the lake explosively outgassed CO_2. As if the lake had been carbonated, the hypolimnetic and epilimnetic waters were rapidly mixed and a gas cloud was released from the lake surface. Because the nearly pure CO_2 was denser than air, the gas cloud flowed down the slopes of the ancient volcano, killing 1700 people and many livestock (Kling et al. 2005).

The depth of the epilimnion depends on air temperature, lake surface area (which determines the total capacity to absorb heat), and the extent of wind mixing (which acts to mix surface and deep waters and reduces the density gradient). The depth of the thermocline (h) for lakes less than 5000 meters in diameter can be predicted from

$$h \cong 2.0 \left[\tau/(g\Delta\rho) \right]^{1/2} L^{1/2}, \tag{8.3}$$

where τ is the wind stress associated with late summer storms, $\Delta\rho$ is the density contrast between the epilimnion and hypolimnion typical for lakes in the region of interest at the time of maximum heat content, g is gravity, and L is the square root of the surface area of the lake (Gorham and Boyce 1989). The mixing depth of the surface layer is an important determinant of lake productivity.

Trophic Status of Lakes

Lakes are often classified into trophic categories by the level of nutrient inputs relative to lake volume (Table 8.2). Trophic status is useful in distinguishing low-productivity *oligotrophic* lakes from high-productivity *eutrophic* lakes. Oligotrophic lakes are nutrient-poor and classified as having productivity $<300 \, mg \, C \, m^{-2} \, day^{-1}$ (Likens 1975). Oligotrophic lakes are often of relatively recent geologic origin (e.g., postglacial) and deep, with cold hypolimnetic waters. Such lakes often show a relatively large ratio between lake area and drainage area, and a long mean residence time for water (Dingman and Johnson 1971). In large lakes within small catchments, nutrient inputs are dominated by precipitation and terrestrial organic matter with a low nutrient content and a high N:P ratio.

Eutrophic lakes are dominated by inorganic nutrient inputs from the surrounding watershed, and they typically have lower N:P ratios as a result of higher P inputs from runoff, with the most eutrophic lakes having a TN:TP ratio less than 10. These nutrient-rich lakes are often shallow, with warm, highly productive waters. Sedimentation will eventually convert the physical state of many oligotrophic lakes to shallow, eutrophic conditions, so these concepts have also been used to describe lake age. However, in most cases, nutrient status remains the most useful criterion to distinguish between oligotrophic and eutrophic conditions (Table 8.2). Highly colored lakes are sometimes classified separately as dystrophic lakes, since as DOM increases, light penetration is reduced (Thienemann 1921), and algae in such lakes are likely to be light rather than nutrient limited (Karlsson et al. 2009).

Carbon Cycling in Lakes

The earliest efforts to understand carbon cycling in lakes involved carbon mass-balance studies in which investigators attempted to quantify all inputs and outputs of carbon over the course of a year (e.g., Richey et al. 1978). The idealized organic carbon budget for a lake can be expressed as

$$\Delta \text{Storage} = [\text{Inputs}] - [\text{Outputs}] \tag{8.4}$$

$$\Delta S = [P_W + P_B + A_I] - [R_W + R_B + B + H_O], \tag{8.5}$$

where ΔS=change in C storage within the lake, P_W=water column photosynthesis, P_B=benthic photosynthesis, A_I=allochthonous input of organic carbon, R_W=water column respiration, R_B=benthic respiration, B=permanent burial in sediments, and H_O = hydrologic loss of organic carbon from outflows.

TABLE 8.2 Lake Classification by Trophic Status

Trophic type	Mean primary productivity (mg C m^{-2} d^{-1})	Phytoplankton biomass (mg C m^{-3})	Chlorophyll a (mg m^{-3})	Light extinction coefficient (ηm^{-1})	Total Organic Carbon (mg L^{-1})	Total P (μg L^{-1})	Total N (μg L^{-1})
Ultra-oligotrophic	<50	<50	0.01–0.05	0.03–0.08		<1–5	<1–250
Oligotrophic	50–300	20–100	0.3–3	0.05–1	<1–3		
Oligomesotrophic						5–10	250–600
Mesotrophic	250–1000	100–300	2–15	0.1–2.0	<1–5		
Mesoeutrophic						10–30	500–1100
Eutrophic	>1000	>300	10–500	0.5–4.0	5–30		
Hypereutrophic						30–>5000	500–>15,000
Dystrophic	<50–500	<50–200	0.1–10	1.0–4.0	3–30	<1–10	<1–500

Source: Modified from Wetzel 2001 (Table 15.13, p. 389).

Studies of the production and fate of organic carbon are useful in understanding the overall biogeochemistry of lakes. Rich and Wetzel (1978) present a carbon budget for Lawrence Lake, a small shallow lake in Michigan in which rooted aquatic plants (macrophytes) contribute ~51.3% of annual autochthonous net primary production while phytoplankton contribute only 25.4% (Table 8.3). In contrast, Jordan and Likens (1975) showed that phytoplankton accounted for ~90% of annual NPP in Mirror Lake, a deep oligotrophic lake in New Hampshire. In Lawrence Lake, with its abundant macrophytes, NPP exceeds total ecosystem respiration (R), while in Mirror Lake NPP and R are equivalent. The more productive shallow lake has higher rates of carbon burial in sediments and exports more DOC (the C contained within DOM) in outflow than seen for Mirror Lake.

TABLE 8.3   Organic Matter Budgets for Lawrence Lake in Michigan and Mirror Lake in New Hampshire

| | Lawrence Lake, MI | | Mirror Lake, NH | |
Inputs	g C m^{-2} yr^{-1}	Inputs	g C m^{-2} yr^{-1}	Inputs
Net primary production (NPP)	191.4	88%	87.5	83%
POC				
Phytoplankton	43.3	20%	78.5	74%
Epiphytic algae	37.9	18%	2.2	2%
Epipelic algae	2	1%	–	–
Macrophytes	87.9	41%	2.8	3%
Bacterial CO_2 fixation	–	–	4	4%
DOC released by macrophytes				
Littoral	5.5	3%	–	–
Pelagic	14.7	7%	–	–
Imports	25.1	12%	17.93	17%
POC	4.1	2%	6.63	6%
DOC	21	10%	11.3	11%
Total available organic inputs	**216.5**		**105.43**	
Outputs	g C m^{-2} yr^{-1}	Outputs	g C m^{-2} yr^{-1}	Inputs
Respiration	159.7	74%	87.53	83%
Benthic	117.5	55%	43.13	41%
Water column	42.2	20%	44.4	42%
C Storage in Sediments	16.8	8%	7.6	7%
Exports	38.6	18%	10.2	10%
POC	2.8	1%	1.05	1%
DOC	35.8	17%	9.15	9%
Total removal of carbon	**215.1**		**105.33**	

Sources: Rich and Wetzel 1978; Jordan and Likens 1975.

Primary Production in Lakes

When air temperatures are cold or surface winds maintain a well-mixed water column (e.g., Figure 8.3a), phytoplankton, the free-floating algae that contribute most of the net production in large lakes, are frequently transported out of the well-lit surface waters (or photic zone). In deeper waters low light limits algal productivity. Light attenuates rapidly through the lake water column; the depth at which light levels are insufficient to support photosynthesis in excess of respiration is known as the *compensation depth*. During periods of stratification, phytoplankton are confined to the epilimnion. If the thermocline is above the compensation depth, phytoplankton will have sufficient light to meet their respiratory demands and productivity is likely to be limited by factors other than light.

During stratification, nutrients often limit productivity. Nutrients incorporated into biomass in the surface layers sink out of the epilimnion in the form of dead organisms and fecal pellets so that during prolonged periods of stratification surface waters become increasingly depleted in nutrients. This sinking organic matter is decomposed by heterotrophs in the hypolimnion, enriching bottom waters with inorganic nutrients and resulting in a depletion of oxygen and lower redox potential (Figure 8.10). Nutrients typically accumulate in the hypolimnion because heterotrophs in the dark, cool bottom waters are ultimately limited by the supply of fixed carbon from the lake surface. Periods of mixing bring these hypolimnetic nutrients back into contact with sunlight where they can stimulate phytoplankton growth. The spring algal blooms characteristic of many temperate lakes occur at the onset of seasonal stratification since at this time, nutrients are mixed throughout the lake water volume and thermal stratification constrains floating algae within the well-lit photic zone.

Measuring Primary Productivity

Rooted and floating aquatic plants in shallow lakes or along the margins of deep lakes contribute to lake productivity (Table 8.3), but their importance diminishes with lake size and depth. In deep lakes, the dominant primary producers are phytoplankton. Historically, methods for assessing net primary production in lakes ignored benthic productivity and used bottle assays to estimate lake NPP. Two different bottle assay approaches are still widely used to measure NPP or to assess algal nutrient limitation in lakes (Wetzel and Likens 2000). In the first method, lake water is placed in gas-tight glass bottles that are either clear or opaque to sunlight. Bottles are incubated (often in the lake itself) and researchers measure changes in oxygen concentration over time in both the light and dark bottles. In the light bottles, photosynthesis, photorespiration, and heterotrophic respiration co-occur. An increase in O_2 concentration over the course of the incubation is taken as the equivalent of net primary production—that is, photosynthesis in excess of respiration by the plankton. Over the same period, the reduction of O_2 in the dark bottles is taken as a measure of autotrophic and heterotrophic respiration. By summing the O_2 consumption of the dark bottles to the O_2 production in the light bottles, researchers can estimate gross primary production.

$$NPP = ([O_2]_{t2} - [O_2]_{t1})_{LIGHT} \tag{8.6}$$

$$GPP = ([O_2]_{t2} - [O_2]_{t1})_{LIGHT} - ([O_2]_{t2} - [O_2]_{t1})_{DARK}. \tag{8.7}$$

Although widely used and inexpensive, these bottle assays suffer from a series of experimental artifacts (reviewed by Peterson 1980). The sensitivity of most oxygen measurements is relatively low, so incubations must be of long duration in order to ensure measurable changes. Confining a small water sample in a bottle for long periods can exacerbate nutrient or CO_2 limitation of algal growth that may not occur in the well-mixed surface waters of a large lake. This method also makes a simplifying assumption that O_2 consumption in the light and dark bottles is equivalent.

A refinement of the classic light versus dark bottle approach uses ^{14}C-labeled DIC that is added to light bottles to measure the incorporation of ^{14}C into biomass over the course of short-term incubations (Wetzel and Likens 2000). Because ^{14}C can be measured with high precision, these assays can be conducted quickly and thus avoid many of the bottle artifacts mentioned earlier. Because the pH of most surface waters is in the range of 4.3 to 8.3, typically these assays add radiolabeled bicarbonate, frequently as $NaH^{14}CO_3$, to a plankton sample. After a short incubation period, the sample is filtered and the accumulation of ^{14}C in the solids collected on the filter is determined with a scintillation counter.

A major shortcoming of the ^{14}C method is that any DIC fixed and subsequently released from an algal cell as DOC (as exudates or cell contents lysed during zooplankton feeding) is not captured on the filter and thus not counted as part of carbon fixation. Filters may also not effectively capture very small picoplankton that can be an important component of productivity in some surface waters. When the two methods are compared side by side, the oxygen method typically gives higher values for net primary production than does the ^{14}C method. The ^{14}C method also provides no estimate of respiration, so it is not possible to use this method to estimate gross primary productivity. Each bottle method provides an instantaneous assessment of net primary production, but in order to determine annual production, or to make comparisons of productivity across lakes, it is necessary to repeat the assays many times over the course of a year.

With the development of field-deployable oxygen sensors, limnologists increasingly use diel variation in dissolved oxygen concentrations to calculate net ecosystem production at the ecosystem-scale (Staehr et al. 2010a, 2012b). In concept, such measurements are similar to estimates of NEP derived from eddy-flux tower measurements of terrestrial ecosystems (Chapter 5). To estimate lake NEP, an oxygen sensor is suspended from a floating buoy, usually in the center of a lake. Changes in oxygen concentration between measurements (often 5 to 30 minutes apart) are used to calculate instantaneous oxygen production (by photosynthesis) or consumption (by respiration) (Figure 8.11). At each time step, changes in oxygen concentration must be corrected for the physical processes of oxygen exchange with the overlying atmosphere:

$$NEP_{\Delta t} = \Delta O_2 - D/Z_{mix}, \tag{8.8}$$

where ΔO_2 (mmol O_2 m^{-3} time interval^{-1}) is change in oxygen concentration over the time interval, D is the physical exchange with the atmosphere over the time interval, and Z_{mix} is the mixing depth of the lake during the interval.

Atmospheric exchange (D) is calculated as

$$D = k \ (O_{2obs} - O_{2sat}), \tag{8.9}$$

where O_{2obs} is the actual oxygen concentration, O_{2sat} is the oxygen concentration in equilibrium with the atmosphere at the ambient temperature, and k is the oxygen exchange coefficient calculated for each time step. The gas exchange coefficient, k, can be estimated directly by measuring exchange of a gas tracer (e.g., sulfur hexafluoride or propane; Cole and

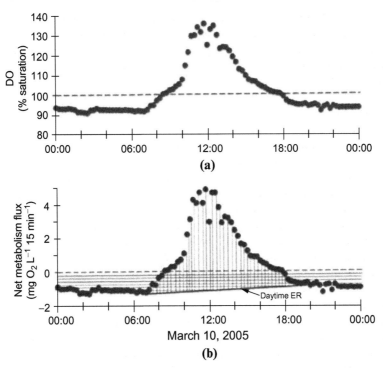

FIGURE 8.11 GPP and ER are typically derived from diel oxygen concentrations like the one shown here for a small stream in Tennessee. Diel profiles of (a) percent saturation of DO and (b) net metabolism flux showing the area representing gross primary production (GPP, *vertical lines*) and ecosystem respiration (ER, *horizontal lines*). *Dashed lines* indicate 100% saturation in (a) and a net metabolism flux of zero in (b). The *solid line* in (b) indicates the interpolated values of daytime ER. Data are from March 2005 in Walker Branch, Tennessee. *Source: From Roberts et al. 2007. Used with permission of Springer.*

Caraco 1998) and using the ratio of Schmidt numbers[9] between the tracer gas and oxygen to predict O_2 diffusion across the air–water interface (Jahne et al. 1987). The sum of oxygen produced over the course of daylight hours is used as an estimate of NEP. Using this method, nighttime oxygen consumption (from 1 hour past sunset to 1 hour prior to sunrise) is used to represent ecosystem respiration:

$$ER = \text{average nighttime R} \times 24 \text{ hours.} \tag{8.10}$$

The estimate of ER is added to daytime estimates of oxygen production to calculate a cumulative diel estimate of GPP. This technique makes the simplifying assumption that daytime respiration rates are similar to nighttime rates and is thus likely to provide a conservative estimate of GPP (Staehr et al. 2010a). By measuring the isotopic composition of atmospheric and dissolved O_2 and the rate of exchange of O_2 between air and water it is possible to gain more accurate estimates of daytime respiration (Quay et al. 1995) and thus better estimates of

[9] Each gas can be characterized by its Schmidt number (a ratio of momentum diffusivity to mass diffusivity). Exchange coefficients measured for inert tracer gases can be converted into exchange coefficients for oxygen using the ratio of Schmidt numbers for the two gases.

NPP and NEP. A major constraint on sensor-derived estimates of lake NPP is the difficulty of determining the volume of water from which the diel oxygen signal is derived, and thus the appropriate scaling term for extrapolation. The difficulty in attributing metabolism to the appropriate lake volume is similar to the challenge of attributing eddy-flux measurements to the appropriate area of terrestrial vegetation (Chapter 5).

Nutrient Limitation of Lake NPP

As for terrestrial systems, growing season length is an important constraint on lake NPP. While peak productivity in arctic lakes may reach that of temperate and tropical lakes, arctic lake phytoplankton have a far shorter growing season. In compilations of annual NPP estimates for low nutrient lakes, Wetzel (2001) and Alin and Johnson (2007) each found the lowest annual NPP in arctic, antarctic, and alpine lakes and the highest in tropical lakes (Figure 8.12). Lake size is also an important determinant of productivity since lake area is positively correlated with wind exposure and mixing depth and negatively correlated with nutrient and DOC inputs per water volume (Brylinski and Mann 1973, Duarte and Kalff 1989, Fee et al. 1994). The drainage ratio of a lake (the ratio of catchment area to lake area) is positively related to the external inputs of nutrients, DOC, and CO_2 (Gergel et al. 1999) and thus to increasing primary productivity (Alin and Johnson 2007). Put simply, the greater the edge:volume ratio, the larger the influence of terrestrial vegetation and soils on the biogeochemistry of a lake and thus the greater the supply of nutrients from the landscape. Larger lakes, with longer residence times and low edge:volume ratios, are more dependent on internal mineralization to supply the nutrients for new productivity.

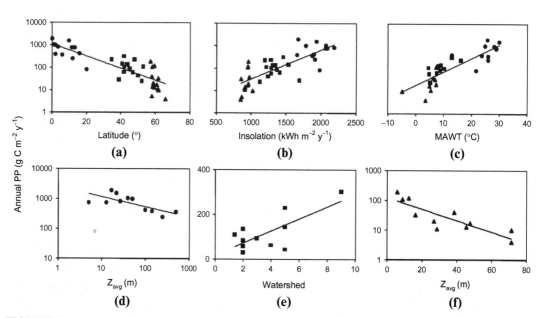

FIGURE 8.12 Global-scale relationships between annual primary production and the environmental variables: (a) latitude, (b) incident solar radiation, (c) mean annual water temperature, (d) depth, (e) watershed to lake area ratio, and (f) average lake depth (z). *Source: From Alin and Johnson 2007. Used with permission of the American Geophysical Union.*

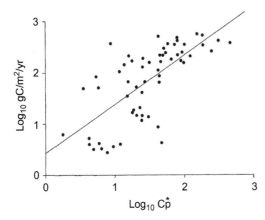

FIGURE 8.13 The relationship between net primary production and the phosphorus concentration of lakes of the world is fit by the line log $[P] = 0.83$ log NPP $+ 0.56$ ($r = 0.69$). Schindler excluded lakes with N:P ratios in inputs of $<5:1$ from this analysis. *Source: Adapted with permission from Schindler (1978).*

Within biomes, lakes receiving higher supplies of nitrogen and phosphorus (from urban or agricultural pollution) support greater phytoplankton NPP than do lakes with low nutrient loading (Wetzel 2001). In a series of influential studies in the 1970s, researchers consistently demonstrated that the variation in lake NPP or algal biomass across temperate lakes was closely associated with the lake concentration or annual loading rates of phosphorus (Schindler 1974, Vollenweider 1976, Smith 1983, Correll 1998; Figure 8.13). Phosphorus concentrations in the epilimnion are directly related to the total chlorophyll content of the water column, which is directly correlated to net primary productivity (Schindler 1978).

In 1974, David Schindler published the results of a whole-lake experiment in which a small hourglass-shaped lake in the Experimental Lakes Area of Canada was fertilized (Schindler 1974). The two basins of Lake 226 were split in half with a plastic curtain. Both sides of the lake received sucrose and nitrogen additions, but only one basin received phosphate addition. A large algal bloom erupted in the basin receiving C, N, and P, while the water remained clear and algal biomass low in the basin receiving only C and N (Schindler 1974). The aerial photo of this experimental effect (Figure 1.5) is undoubtedly one of the most influential environmental science photographs ever taken. This experimental demonstration and other data from lake ecosystems convinced regulators to remove phosphates from detergents, dramatically reducing municipal P loading to surface waters.

Despite the limited availability of phosphorus in surface waters, we might expect that, as for land vegetation, processes such as denitrification might limit the nitrogen supply in lakes. In a series of studies conducted over the last two decades, nitrogen limitation or N and P co-limitation of lake phytoplankton is frequently reported (Goldman 1988, Morris and Lewis 1988, Downing and McCauley 1992, Elser et al. 2009). Authors of several of these studies have suggested that the preponderance of evidence for P limitation of lake productivity from research throughout the 1970s may be a consequence of making such measurements in areas of North America and Canada that had experienced decades of excess atmospheric nitrogen deposition (Chapter 6) that over time increased the N:P ratio of nutrient inputs to lakes (Elser et al. 2009). Other authors argue that sewage and fertilizer inputs to lakes are typically phosphorus rich and thus may lower the N:P ratio of inputs and enhance internal stores of phosphorus, thereby increasing the potential for N limitation (Downing and McCauley 1992).

While lake phytoplankton productivity may be N, N + P, or P limited, there are several fundamental reasons to expect increases in phosphorus loading to play a disproportionate role in cultural eutrophication (Schindler 1977, Schindler et al. 2008). When phytoplankton grow in limited supplies of nitrogen, the prevalence of nitrogen-fixing algae, primarily cyanobacteria, typically increases, adding nitrogen through fixation and raising the N:P ratio, (compare to Figure 6.4). In a literature synthesis, Howarth et al. (1988) found that significant nitrogen fixation by lake phytoplankton occurred only when lake N:P ratios were below 16. A subsequent analysis by Smith (1990) suggested that total P loading, rather than N:P ratios, was the best predictor of N fixation rates across lakes globally. When phosphorus is added as a pollutant with low N:P, the algal community shifts to species of blue-green algae and primary productivity increases, with nitrogen inputs through fixation tending to maintain a phosphorus shortage for the growth and photosynthesis of phytoplankton (Smith 1982).

Håkanson et al. (2007) suggest that below N:P ratios of 15, the N:P ratio is a good predictor of cyanobacterial biomass but that in lakes with TN:TP ratios greater than 15, cyanobacterial biomass can be predicted from total phosphorus concentrations or loading alone (Figure 8.14). With high P loading, nitrogen fixation can supply up to 82% of the nitrogen input to the phytoplankton community (Howarth et al. 1988). Thus lake phytoplankton have a mechanism for acquiring additional inputs of nitrogen, but there is no equivalent biogeochemical process that can increase the supply of phosphorus in lakes when it is in short supply (Schindler 1977).

When the input of phosphorus to a lake ceases, blue-green algae typically decrease in importance and algal productivity declines (Edmondson and Lehman 1981, Schindler et al. 2008). A particularly influential demonstration of the potential for eutrophic lakes to recover following P reductions was recorded in Lake Washington (Edmondson and Lehman 1981).

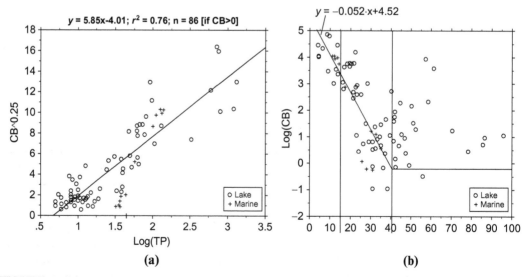

FIGURE 8.14   (a) The relationship between cyanobacterial biomass (CB) and total phosphorus concentrations across 86 lakes and coastal estuaries for which CB > 0. (b) The relationship between log CB and the TN:TP molar ratio. The regression equation given is for systems with TN:TP ratios <15. *Source: From Håkanson et al. 2007.*

Until 1967 the City of Seattle disposed of its sewage directly in Lake Washington; after that year sewage was diverted to the ocean instead of this closed basin. The resulting decline in noxious algal blooms in the lake provided strong evidence for P limitation of lake productivity. The time required for this recovery from eutrophication depends on the extent to which historic P inputs are mineralized and released from lake sediments (Genkai-Kato and Carpenter 2005, Mehner et al. 2008). Historic nitrogen loading is less problematic because significant quantities of nitrogen may be lost from lake sediments through denitrification (McCrackin and Elser 2012). Because of this difference in the persistence of N and P pollution in lakes, Schindler et al. (2008) argued that efforts to reduce N loading are likely to be less effective at reversing cultural eutrophication than reducing P inputs. In contrast, preventing future eutrophication in unpolluted lakes will require reductions in nitrogen as well as phosphorus loading (Conley et al. 2009b, Lewis et al. 2011).

Micronutrient Limitation

The potential for lake phytoplankton to respond to increasing N and P loading with enhanced NPP may be constrained by micronutrient availability. When phosphorus is added to nutrient-poor lakes, the growth of diatoms, single-celled algae with silicate cell walls, may reduce the concentrations of silica to low levels so that diatoms are competitively replaced by nonsiliceous green algae or cyanobacteria (Schelske and Stoermer 1971, Tilman et al. 1986). Similarly, enhanced loading of phosphorus may lead to Fe limitation in clearwater oligotrophic lakes (Sterner et al. 2004). All phytoplankton require Fe for photosynthesis (see Chapter 2) so Fe limitation may reduce whole ecosystem NPP. Because cyanobacteria typically have higher Fe requirements than eukaryotic algae (Morton and Lee 1974, Brand 1991), and N_2 fixation requires high Fe uptake (Murphy et al. 1976, Glass et al. 2009), increased competition for Fe could also lead to a decline in cyanobacteria or rates of N fixation (Molot et al. 2010). Changes in micronutrient loading or nutrient:micronutrient ratios may lead to more subtle changes in algal community structure. Titman (1976) showed that slight differences in the ratio of silica to phosphorus altered the outcome of competition between two dominant species of diatoms (*Asterionella* and *Cyclotella*). Other studies have shown that N fixation rates can be suppressed by the addition of trace micronutrients, such as Fe (Vrede and Tranvik 2006) or Cu (Horne and Goldman 1974).

Light Limitation of NPP

Much of the research on lake nutrient limitation has been conducted in clearwater lakes without high concentrations of dissolved organic carbon. In lakes stained with DOC, light is the primary limitation on productivity. Across a series of lakes in northern Sweden that ranged in DOC concentrations from about 10 to 100 mg L^{-1}, Ask et al. (2012) found that higher DOC was negatively correlated with GPP and positively correlated with ecosystem respiration (ER) (Figure 8.15). Similarly, Karlsson et al. (2009) found that phosphorus concentrations were negatively correlated with both primary productivity and the secondary production of fish in 12 nutrient-poor and DOC-rich Swedish lakes. Across these lakes, light availability was a much better predictor of secondary production (Figure 8.15). For dystrophic lakes (refer to Table 8.3), nutrient inputs are unlikely to stimulate algal blooms, but may instead enhance the rates of DOM decomposition and CO_2 release (Pace et al. 2007).

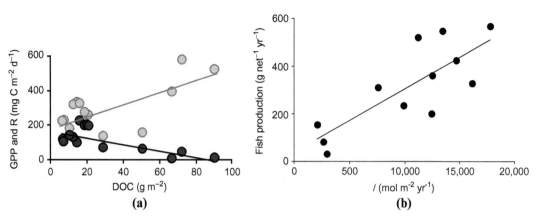

FIGURE 8.15   Light limitation of primary and secondary production in Swedish lakes. (a) Whole-lake gross primary production (GPP, *dark gray circles*) and respiration (R, *light gray circles*) for 15 lakes in northern Sweden. (b) Fish production as a function of the annual light climate (I, representing the mean PAR in the whole-lake volume during the ice-free period) for 12 lakes in northern Sweden ($r_2 = 0.63$, $p = 0.002$). *Sources: (a) From Ask et al. 2012, used with permission of American Geophysical Union; (b) from Karlsson et al. 2009, used with permission of Nature Publishing Group.*

Herbivore Control of NPP

Much of the algal biomass produced by lake phytoplankton is consumed by zooplankton, and changes in the food web can affect whole ecosystem NPP. More productive lakes typically produce higher fish biomass per unit volume (Melack 1976, Karlsson et al. 2009). This trophic linkage works in more than one direction: While nutrient and light availability may increase the productivity of higher trophic levels ("bottom-up" control), the abundance of top predators can influence the intensity of herbivory and alter NPP ("top-down" control; Carpenter et al. 1985). For example, in a series of whole-lake experiments, Carpenter et al. (2001) showed that adding pike to a lake reduced its NPP by a factor of 3 compared to a nearby lake that lacked piscivorous fish. Piscivores, by consuming zooplanktivorous fish, release herbivorous zooplankton from predation and allow them to substantially reduce standing stocks of phytoplankton. This effect has been termed the "trophic cascade."

In a formal meta-analysis of 54 experimental manipulations of top predators in lakes and ponds, Brett and Goldman (1996) found that nearly all such experimental additions of piscivores led to declines in zooplanktivore biomass and increases in phytoplankton. Not all lake manipulations of top predators, however, have led to changes in NPP (Elser et al. 1998, MacKay and Elser 1998) and it has been suggested that food web interactions are less likely to affect ecosystem productivity in ultraoligotrophic lakes where algae are extremely nutrient limited and in highly eutrophic lakes where algal growth rates are high and algal communities are dominated by unpalatable species (Kitchell and Carpenter 1993).

The Fate of Organic Carbon in Lakes

Organic carbon delivered to lakes from the surrounding catchment or fixed by lake autotrophs may be incorporated into aquatic consumer biomass, respired, sequestered in lake sediments, or transported downstream (Cole and Caraco 2001). Because the sediments of lake bottoms are frequently anoxic, organic material that reaches the lake bed will decompose slowly (Chapter 7) and lakes typically accumulate sediment carbon over time. In a literature synthesis Einsele et al. (2001) estimated that throughout the Holocene the amount of carbon stored in lake sediments globally is $\sim$820 Pg, with small lakes ($<$500 km^{2-}) containing about 70% of this total. Over geologic time lake basins fill with sediments and organic material. This can happen very quickly in manmade reservoirs where inputs of river sediment and organic materials are high (Downing et al. 2006).

Cole et al. (2007) compiled estimates of global lake C storage from the literature, which ranged from 0.03 to 0.07 Pg C yr^{-1} (midrange 0.05 Pg C yr^{-1}; Mulholland and Elwood 1982, Meybeck 1993, Dean and Gorham 1998, Stallard 1998, Einsele et al. 2001; Table 8.4). While this global rate is minuscule in comparison with the annual C storage in terrestrial vegetation and soils (about 1.3 Pg C yr^{-1}) or in the oceans (1.9 Pg C yr^{-1}; Sundquist 1993), it is roughly equivalent to the annual organic C storage in marine sediments ($\sim$0.12 Pg y; Sarmiento and Sundquist 1992) but occurs within lakes that occupy an area equivalent to only 2% of the ocean area. When lake carbon accumulation is corrected for area, long-term carbon burial rates range from 4.5 to 14 g C m^{-2} y^{-1} (Dean and Gorham 1998, Stallard 1998), which is higher than the long-term soil C accumulation rates estimated for most terrestrial soils (Schlesinger 1990; Figure 5.5).

Carbon burial in lake sediments is predicted to increase with eutrophication as a result of more constant and severe oxygen depletion of sediments coupled with high autochthonous production (Hutchinson 1938, Downing et al. 2008). Examining several lakes, Hutchinson (1938) suggested that the rate of depletion of O$_2$ in the hypolimnion during seasonal stratification was related to the productivity of the overlying waters. Highly productive waters should contribute large quantities of organic carbon for respiration in the hypolimnion, which is seasonally isolated from sources of oxygen.

The areal hypolimnetic oxygen deficit (abbreviated as AHOD) is a useful concept, but attempts to predict AHOD as a function of nutrient loading have been controversial. Cornett

TABLE 8.4   Carbon Burial Rates in Lakes and Impoundments

Environment	Mean or Median OC Burial (g m^{-2}yr^{-1})	Range
Eutrophic impoundments	2122	148–17,392
Impoundments (Asia)	980	20–3300
Impoundments (Central Europe)	465	14–1700
Impoundments (United States)	350	52–2000
Impoundments (Africa)	260	
Small mesoeutrophic lakes	94	11–198
Small oligotrophic lakes	27	3–128
Large mesoeutrophic lakes	18	10–30
Large oligotrophic lakes	6	2–9

Sources: Downing et al. 2008, Mulholland and Elwood 1982.

and Rigler (1979) concluded that "a simple proportionality between biomass in the epilimnion and area hypolimnetic oxygen deficit does not appear to exist." They suggested instead that the greatest O_2 consumption occurred in deep lakes with higher water temperatures and a thick hypolimnion (Cornett and Rigler 1979, 1980). It is logical that warmer water temperatures support higher rates of bacterial respiration in the hypolimnion. The relationship between AHOD and hypolimnion thickness, however, was unexpected because it suggests that the greatest deficits are found in deep lakes with large hypolimnetic volume. Their findings, while not without criticism (Chang and Moll 1980), suggest that the consumption of oxygen in the hypolimnion may be largely the result of respiration within the water column, which is greatest in deep lakes where the transit time for sinking detritus is long (Cole and Pace 1995).

Some of this controversy over whether nutrients, temperature, or lake depth drive hypolimnetic anoxia and carbon burial may arise because authors are attempting to find a single relationship that fits lakes of varying nutrient status. In a synthesis of carbon burial rates Downing et al. (2008) found extremely low carbon burial rates in large nutrient-poor lakes but very high rates of carbon burial in midsize eutrophic impoundments despite the fact that the impoundments were typically shallow (Table 8.4).

Carbon storage in lake sediments is increasing rapidly because of the rapid increase in manmade lakes. Rates of carbon burial in reservoirs are high (estimated at $\sim$400 g C $m^{-2} yr^{-1}$; Mulholland and Elwood 1982, Dean and Gorham 1998, Downing et al. 2008) and because of the rapid expansion of reservoirs, they bury more organic carbon than all natural lake basins combined. Extrapolating this areal rate to the estimated 1,500,000 km^2 of global reservoir area (St Louis et al. 2000) leads to a global reservoir C storage estimate of 0.6 Pg C y^{-1} (Cole et al. 2007). Given that reservoirs occupy only $\sim$1% of the continental land surface, this rate of C burial is impressive in comparison with the estimated 1 to 2 Pg C y^{-1} stored in terrestrial ecosystems and the 0.1 Pg C y^{-1} buried in ocean sediments (Chapter 11).

Carbon Export from Lakes

Carbon can be exported from lakes through the degassing of CO_2 or CH_4 or through the downstream export of particulate and dissolved organic matter. Respiration exceeds GPP in most oligotrophic lakes (Sand-Jensen and Staehr 2009, Staehr et al. 2010b, Ask et al. 2012, Jansson et al. 2012). The degree of net heterotrophy is, across many lakes, positively correlated with CO_2 supersaturation and lake DOC concentrations (Roehm et al. 2009, Ask et al. 2012). Globally, lakes may outgas 0.64 Pg C y^{-1} as CO_2 (Table 8.1; Cole et al. 2007, Aufdenkampe et al. 2011).

In low-sulfate, oligotrophic lakes, methanogenesis is often the dominant form of anaerobic metabolism, but rates of methanogenesis tend to be highly variable in both space and time (Rudd and Hamilton 1978, Kuivila et al. 1989, Zimov et al. 1997). As long as the hypolimnion remains oxygenated, most of the methane produced in hypolimnetic sediments is oxidized by methanotrophs during upward diffusion; however, in shallow sediments a much higher proportion of gross methanogenesis can escape to the atmosphere. Gaining an accurate estimate of lake CH_4 emissions is particularly problematic, since a dominant pathway for CH_4 releases from lakes is through ebullition (Bastviken et al. 2004). In a survey of Siberian thaw lakes, Walter et al. (2006) estimated that ebullition accounted for 95% of CH_4 emissions from these lakes. Although CH_4 emissions represent a small term within individual lake carbon budgets, the warming potential of CH_4 is $\sim 25\times$ higher than CO_2 such that CH_4 production may be an important link between lakes and climate change (Tranvik et al. 2009).

The construction of reservoirs is enhancing annual lake CH_4 fluxes by increasing lake area and providing ideal conditions for methanogenesis. Vegetation and organic soils are flooded as impoundments are created, providing a large source of organic matter to fuel methanogenesis in the early years following construction (Kelly et al. 1997, St Louis et al. 2000, Sobek et al. 2012, Teodoru et al. 2012). In some tropical hydroelectric reservoirs large amounts of CH_4 are released as deep methane-enriched waters are passed through turbines; it has been argued that for some reservoirs the resulting greenhouse gas emissions can be greater than that of fossil fuel alternatives these hydropower dams were intended to replace (Fearnside 1995, Abril et al. 2005). St Louis et al. (2000) made the first attempt to estimate the global CH_4 production from reservoirs, suggesting that 0.7×10^{14} g CH_4 per year, or 18% of anthropogenic emissions, could be produced in temperate and tropical reservoirs. Bastviken et al. (2011) arrived at a similar estimate of 0.92×10^{14} CH_4 yr^{-1} emissions from natural lakes and reservoirs combined. Refining these estimates will require more frequent estimates of CH_4 emissions from a larger number of lakes and reservoirs worldwide.

Nutrient Cycling in Lakes

Contemporary lake nutrient budgets are constructed by assessing the inputs of nutrients in precipitation, runoff, and N fixation and the losses of nutrients from lakes due to sedimentation, outflow, and the release of reduced gases. In many cases human impacts now dominate the nutrient budgets of lakes (Edmondson and Lehman 1981). For elements without a substantial gaseous phase (e.g., Fe, Si, P), patterns of element concentration in lake sediments can be used to reconstruct historical loading and retention patterns (Dillon and Evans 1993, Rippey and Anderson 1996). For example, we can use the sedimentary record to solve for changes in lake phosphorus loading over time as

$$L = TPz\rho + TPz\sigma \tag{8.11}$$

where L is total phosphorus loading to the lake (mg $m^{-2}yr^{-1}$), TP is the lake total phosphorus concentration (mg m^{-3}), z is lake depth (m), ρ is the hydraulic flushing coefficient ($\sim 1/$ MRT in yr^{-1}), and σ is the phosphorus sedimentation coefficient (yr^{-1}). Loading can thus be estimated for every layer in a lake sediment core for which investigators can determine the time period over which sediment accumulated (from ^{14}C dating), lake TP concentrations at the time of sedimentation (inferred from diatom composition), and total P content (Rippey and Anderson 1996). Lake sediments thus retain a record of changes in element exports from terrestrial catchments that can be used to understand how land use change or climate change influences both regional and in-lake nutrient cycling (Davis et al. 1985, Fritz 1996, Mackay et al. 1998, Riding 2000, Smol and Cumming 2000, Muller et al. 2005, Wagner et al. 2009).

Most lakes show a net retention of N, P and Si (Table 7.4; Cross and Rigler 1983; Muller et al. 2005; Harrison et al. 2009, 2012), although in lakes with a substantial annual turnover of water ("open systems"), the net storage of N and P may be relatively small during periods of high flow (Windolf et al. 1996). Iron derived from terrestrial runoff may also be sequestered in lake sediments (Dillon and Evans 2001). Net retention of Ca is seen in lakes where mollusk shells are accumulating in the sediments (Brown et al. 1992) and in some highly productive, alkaline lakes (pH $\sim$9) where calcite ($CaCO_3$) may precipitate directly as *marl* (Rosen et al. 1995, Hamilton et al. 2009):

$$Ca^{2+} + 2HCO_3^- \rightarrow CaCO_3\downarrow + H_2O + CO_2. \qquad (8.12)$$

These lakes will show a net retention of Ca and a relatively short mean residence time for Ca in the water column (Canfield et al. 1984). Calcite deposition is inhibited in lakes with high allocthonous DOC (Reynolds 1978, Hoch et al. 2000, Lin et al. 2005) and enhanced by high rates of photosynthesis that consume CO_2 (Hartley et al. 1995, Couradeau et al. 2012). In a phosphorus-enrichment experiment in a Michigan Lake, Hamilton et al. (2009) found that calcite deposition was enhanced by P fertilization and that the resulting biogenic calcite-phosphorus sedimentation was a substantial sink for the additional phosphorus. By this mechanism calcite precipitation has the potential to ameliorate eutrophication caused by nutrient loading (Koschel et al. 1983, Robertson et al. 2007).

In general, in-lake retention of nutrients increases with nutrient loading and water residence time (Seitzinger et al. 2006), and declines with lake depth. In deeper lakes, it takes longer for organic matter produced in the epilimnion to fall through the water column and be deposited in sediments. Falling organic matter or carbonates can be decomposed or mineralized during transport through the hypolimnion, reducing the absolute amount of material that reaches the sediments. It is only after sedimentation that elements have the potential to be permanently buried or, in the case of N, denitrified in anoxic sediments (Kelly et al. 1987, Dillon and Molot 1990, Molot and Dillon 1993, Windolf et al. 1996, Saunders and Kalff 2001).

Nitrogen Cycling in Lakes

Nitrogen-fixation rates in lakes range from 0.1 kg N ha^{-1} yr^{-1} to >90 kg N ha^{-1} yr^{-1} (Howarth et al. 1988), roughly spanning the range of nitrogen fixation reported for terrestrial ecosystems (Chapter 6). Lakes with high rates of nitrogen fixation can show large apparent accumulations of N (Horne and Galat 1985). Because N-fixation is competitively advantageous when N concentrations are low, rates of N fixation within individual lakes typically respond to changes in nitrogen availability. In many seasonally stratified lakes, a seasonal succession from eukaryotic algae to N fixing cyanobacteria is observed as epilimnetic nitrogen concentrations decline over the period of stratification (Sterner 1989) and N fixation rates are reduced when the supply of nitrogen is high (Doyle and Fisher 1994).

Fewer studies have assessed denitrification and other processes of gaseous loss in lakes. Denitrification can be studied by the application of acetylene-block techniques and ^{15}N isotopic labels as discussed in Chapter 6 (Seitzinger et al. 1993). Reported rates of denitrification in individual lakes ranges from 1.8 to 38 kg ha^{-1} yr^{-1} (from a literature synthesis by Piña-Ochoa and Alvarez-Cobelas 2006). The total loss of nitrogen by denitrification exceeds the input of nitrogen by fixation in almost all lakes where both processes have been measured simultaneously (Seitzinger 1988). In a synthesis of more than 100 individual studies, Harrison et al. (2009) found that lakes removed 0 to 99.7% of N inputs. Higher rates of N removal were found for lakes or reservoirs with high N-loading rates, high catchment:lake area ratios, and high settling velocities for N (Harrison et al. 2009).

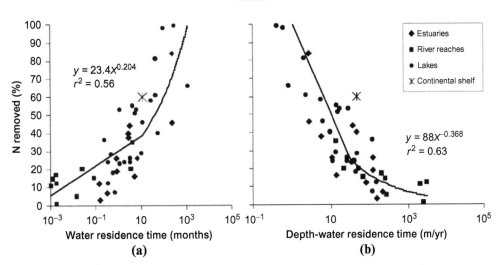

FIGURE 8.16 Relationship between the percentage of N removed (via burial or denitrification) and (a) water residence time (months) or (b) depth/water residence time (m/yr) for lakes, river reaches, estuaries, and continental shelves. *Source: From Seitzinger et al. 2006. Used with permission of the Ecological Society of America.*

Based on the statistical relationships developed from this dataset, Harrison et al. (2009) estimated that lakes and reservoirs collectively remove 19.7 Tg N yr^{-1}, with small lakes (<50 km^2) responsible for nearly 50% of the global total. Often denitrification appears limited by the production and availability of NO_3^- in the sediments (Rysgaard et al. 1994), and the proportion of N removed increases with water residence time (Figure 8.16; Yoh et al. 1983, 1988, Mengis et al. 1997, Seitzinger et al. 2006). Because atmospheric N_2O concentrations are so low, most lakes are likely to be supersaturated in N_2O (Whitfield et al. 2011); however, the loss of nitrogen from lakes as N_2 greatly exceeds the loss of N_2O (Seitzinger 1988, Beaulieu et al. 2011). As NO_3 loading to lakes increases (through atmospheric deposition or polluted surface runoff), higher N_2O production is likely to result (McCrackin and Elser 2011).

Lake Phosphorus Cycling

Because P is weathered slowly from rock and effectively bound in soils or assimilated by terrestrial vegetation, under most natural conditions the phosphorus inputs to lake ecosystems are relatively small (Ahl 1988, Reynolds and Davies 2001; Chapter 4). Much of the phosphorus entering lakes is carried with soil minerals, which are rapidly deposited in the sediments (Froelich 1988, Dillon and Evans 1993). The small amounts of inorganic phosphorus entering lakes undergo rapid abiotic precipitation with Fe, Ca, or Mn minerals that are insoluble in well-oxygenated waters, with the form depending on pH (Figure 4.10; Mortimer 1941, 1942; Blomqvist et al. 2004; Hamilton et al. 2009).

Analyses of lake water typically show that a large proportion of the phosphorus is contained in the plankton biomass and only a small portion is found in available forms (Lean 1973, Schindler 1977, Lewis and Wurtsbaugh 2008). Uptake of phosphorus by phytoplankton is an active process that shows a curvilinear relationship to increasing P concentration (Jansson 1993). Continued net primary production by phytoplankton depends on the rapid cycling of phosphorus between dissolved (e.g., HPO_4^{3-}) and organic forms in the epilimnion (Fee et al. 1994).

Studies of phosphorus cycling have shown that the turnover of phosphorus in the epilimnion is dominated by bacterial decomposition of organic material. Phytoplankton and bacteria excrete extracellular phosphatases to aid in the mineralization of P (Stewart and Wetzel 1982, Wetzel 1992), and planktonic bacteria may immobilize phosphorus when the C:P ratio of their substrate is high (Vadstein et al. 1993). Globally, the molar N:P ratio of freshwater phytoplankton ranges from 6 to 44 (Klausmeier et al. 2004) and net phosphorus mineralization begins at N:P < 16 (Tezuka 1990). Immobilization of N is less common because the C:N ratio of phytoplankton (8–20) is similar to that of bacterial biomass (Tezuka 1990, Downing and McCauley 1992, Elser et al. 2000b). Nutrient turnover in lakes is enhanced by the activities of grazing zooplankton (Porter 1976, Lehman 1980, Elser and Hassett 1994) and fish (Vanni 2002). Grazing zooplankton vary in N:P ratios, with the common cladoceran *Daphnia* having a low N:P ratio (~14:1) relative to most copepods (~30–50:1; Sterner et al. 1992); thus, changes in the identity of dominant grazers can alter the ratio as well as the rate of N and P turnover (Elser et al. 2000a).

During a period of stratification, the phosphorus pool in the surface waters is progressively depleted as phytoplankton and other organisms die and sink to the hypolimnion (Levine et al. 1986, Rippey and McSorley 2009). Baines and Pace (1994) found that 10 to 50% of NPP was exported to the hypolimnion of 12 lakes in the eastern United States, with a tendency for a greater fractional export in lakes of lower productivity (Figure 8.17). Higher rates of particle sinking are correlated with higher rates of bacterial respiration in the hypolimnion (Cole and Pace 1995). When fecal pellets and dead organisms sink through the thermocline, phosphorus mineralization continues in the lower water column and sediments (Gachter et al. 1988, Lehman 1988, Carignan and Lean 1991). Anoxic hypolimnetic waters often show high concentrations of P, which is returned to the surface during periods of seasonal mixing. Of course, the turnover of phosphorus through the biotic community is incomplete, so some phosphorus is permanently lost to the sediments.

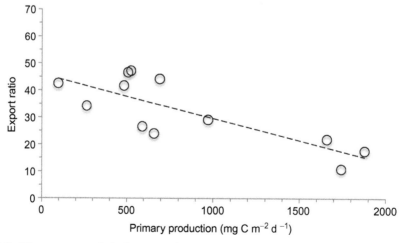

FIGURE 8.17   The percentage of planktonic productivity that sinks to the hypolimnion in lakes as a function of their net primary production. *Source: Modified from Baines and Pace 1994. Used with permission of NRC Research Press.*

As long as the hypolimnetic waters contain oxygen, a layer of Fe-oxide minerals at the sediment–water interface traps phosphorus that diffuses upward from bacterial decomposition of organic matter in the sediments or from the dissolution of Fe-P minerals at a lower redox potential at depth. However, when hypolimnetic waters become anoxic, this "iron trap for phosphorus" is lost as oxidized Fe minerals are reduced, releasing P to the overlying waters (Mortimer 1941, 1942; Caraco et al. 1990; Golterman 2001; Blomqvist et al. 2004). Interactions between elements may be important in determining the release of P from sediments. In most freshwaters the concentration of SO_4^{2-} is low, and P is strongly adsorbed by Fe minerals in the sediment. In the sea, concentrations of SO_4^{2-} are higher, and P limitation is less apparent (Chapter 9). Increasing concentrations of SO_4^{2-} in lakes due to acid rain or due to mining of pyrite-rich minerals may act through the anion exchange reactions to drive P into solution (Caraco et al. 1989, Wang and Chapman 1999).

In many cases the dissolution of Fe minerals is limited, and there is no regeneration of phosphorus from sediments (Davison et al. 1982, Levine et al. 1986, Caraco et al. 1990, Davison 1993, Golterman 1995). Sedimentary accumulations of undecomposed organic matter and Fe minerals contain P that is effectively lost from the ecosystem (Cross and Rigler 1983).

Sulfur Cycling in Lakes

In the sediments and low-oxygen bottom waters of lakes, as in the saturated sediments of wetlands, sulfur can play an important role in both carbon and nitrogen cycling. Sulfate concentrations in lakes are generally low; however, particularly in eutrophic lakes, where hypolimnetic oxygen concentrations are reduced, sulfate reduction is an important component of lake carbon cycling (Holmer and Storkholm 2001; Figure 8.18). In a highly eutrophic Michigan lake, Smith and Klug (1981) found that sulfate reduction was responsible for 7% of carbon mineralization.

In a review of the literature, Holmer and Storkholm (2001) concluded that sulfate reduction can account for a significant fraction (12–81%) of the total anaerobic carbon mineralization in lake sediments, although rates of sulfate reduction were typically lower than methanogenesis

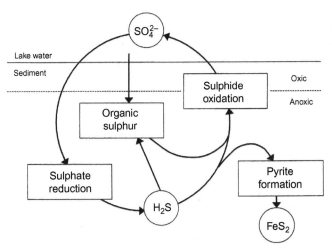

FIGURE 8.18 A simplified lake sediment sulfur cycle. *Source: From Holmer and Storkholm 2001.*

within the same lake. Sulfate reduction is enhanced in eutrophic lakes where more organic carbon is supplied to the sediments in deep water and shallow sediments are more likely to be anoxic. When reduced forms of sulfur are cycled through reoxidation pathways, high rates of SO_4^{2-} reduction can occur in lake sediments, despite low concentrations of SO_4^{2-} in lake water (Holmer and Storkholm 2001). Urban et al. (1994) found that rates of sulfide oxidation in the sediments of an oligotrophic lake in Wisconsin were nearly as rapid as sulfate reduction, indicating rapid fluxes of S despite the small pool sizes of S in the sediments.

Recent work has demonstrated that anaerobic sulfide oxidation during dissimilatory nitrate reduction or denitrification can be important sinks for NO_3^- in lake sediments, with sulfur bacteria gaining energy by oxidizing reduced sulfide to SO_4^{2-} and reducing NO_3^- either to N_2 or NH_4^+ (Chapter 7; Burgin and Hamilton 2008, Laverman et al. 2012). Although volatile losses of sulfur occur (Brinkman and Santos 1974), most H_2S appears to be reoxidized as it passes through the upper sediments (Dornblaser et al. 1994) or the water column (Mazumder and Dickman 1989), so little escapes to the atmosphere (Nriagu et al. 1987; Figure 8.10).

RIVERS

Rivers differ from lakes in several important ways that influence their biogeochemistry (Figure 8.19). The unidirectional flow of water in rivers maintains a constant supply of nutrients, water turbulence constantly mixes particulates into the water column, and frequent scouring flows limit the capacity for permanent burial of elements in sediments. In addition, the boundaries of rivers, both laterally and longitudinally, are extremely dynamic, with river flows expanding to encompass their floodplains and to include their ephemeral headwaters during periods of high runoff and contracting to a limited portion of their channels during periods of drought.

Like lakes, river biota can significantly alter the magnitude, timing, and form of chemical delivery to downstream waters (Meyer et al, 1988). Although "streams are the gutters down which flow the ruins of continents" (Leopold et al. 1964), stream channels are physically and

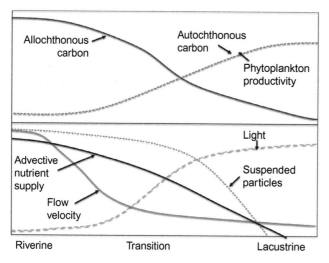

FIGURE 8.19 Commonly observed shifts in flow, light, nutrients, and sources of organic matter in the transition between rivers and lakes.

biologically complex and thus trap, delay, and attenuate the water, chemicals, and sediment pulses delivered from upslope and upstream (Bencala and Walters 1983). The communities of organisms within stream ecosystems largely survive by consuming and transforming terrestrial materials before passing them to the atmosphere or downstream, further altering the timing and quantity of chemical exports (Wallace and Webster 1996, Wallace et al. 1997).

Although materials accumulate in floodplains and large pools, rivers are not net aggrading systems and are less retentive of chemicals and solutes in comparison to hillslopes, wetlands, or lakes (Wagener et al. 1998, Essington and Carpenter 2000, Grimm et al. 2003). Because of the comparatively high velocities of flow in channels (i.e., limited residence time), chemicals and solutes that reach streams are routed much more rapidly to downstream receiving waters than would occur via subsurface flowpaths. Once introduced to streams, the only fate for elements that have no gaseous form (e.g., the limiting nutrient phosphorus, most trace metals) is transport to downstream floodplains, lakes, reservoirs, or coastal zones, although this transport may require anywhere from days to millennia.

For elements with a gaseous form at ambient conditions, substantial conversion and thus permanent export to the atmosphere can occur in rivers. In particular, denitrification can convert ~16 to 50% of nitrate (NO_3^-) inputs to N_2 (Galloway et al. 2008, Seitzinger et al. 2002, Mulholland et al. 2008) and >50% of the fixed carbon inputs are respired as CO_2 (Battin et al. 2009, Aufdenkampe et al. 2011).

Riverbeds provide ideal conditions for a variety of metabolic processes because of the large area where oxygenated waters flow over anoxic sediments; thus rivers typically have disproportionately high rates of nutrient transformations and decomposition (mass per unit area) compared to adjacent surrounding soils (Lohse et al. 2009). Anoxic habitats within streams and their associated riparian zones are often the primary locations where significant denitrification and methanogenesis occur within temperate and arid landscapes.

River Water Budgets and Mixing

Water may enter a segment of river channel as flow delivered from upstream segments, as direct precipitation delivered to the channel surface, or as lateral inflows that are delivered overland, through shallow subsurface pathways or via exchange with deep groundwater (Figure 8.20). In small headwater streams, lateral inputs can dominate the water balance, but as streams grow larger in size, inflow from upstream segments quickly becomes the

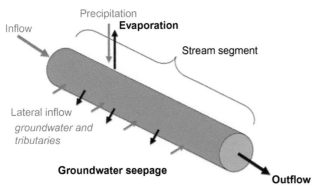

Precipitation
Inflow
Evaporation
Stream segment
Lateral inflow
groundwater and
tributaries
Groundwater seepage
Outflow

FIGURE 8.20   The water budget for a river segment.

dominant water source. Water may leave a channel through downstream flow, through evaporation, or through net losses to regional groundwater. Even when there is no net loss to groundwater, gross exchanges of water between the channel and groundwater can be quite large (Covino and McGlynn 2007, Poole et al. 2008). In some arid ecosystems the surface flow of large rivers is ultimately lost through net fluxes into regional groundwaters and during periods of high flow, rivers may overtop their banks and lose water to their floodplains. Under most conditions, however, the dominant route of water export is downstream flow.

The flow of water through terrestrial ecosystems to stream waters is often modeled using simplifying assumptions about the rate of plant uptake and the downward flow of water through the soil profile (Freeze 1974). Downward movement is assumed to occur during any period in which the percolation of water to a particular depth is in excess of the water-holding capacity of that depth and the rate of plant uptake during the interval. Water-holding capacity is commonly called *field capacity*, which is the water content that a soil can retain against the force of gravity. Excess water draining to the bottom of the soil profile will be transported to groundwater or receiving streams.

Long-term observations of streams show that hydrographs are affected by topography, vegetation, and soil characteristics, as well as the pattern and intensity of rainfall in individual storms (Ward 1967, Bosch and Hewlett 1982, McGuire et al. 2005). Stormflows tend to increase when vegetation is removed because bare soil allows a greater proportion of precipitation to leave via overland flow (Bosch and Hewlett 1982, Schlesinger et al. 2000; Figure 8.21). Stream hydrographs provide information about how rapidly rainwater or snowmelt is transported to channels, and how frequently channels are flooded or dry (Figure 8.22). Stream flow comprises surface runoff, which may carry organic debris and soil particles, and together with the drainage

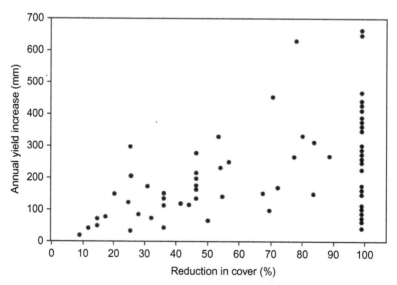

FIGURE 8.21   Water yield increases following changes in vegetation cover. *Source: Adapted from Bosch and Hewlett 1982.*

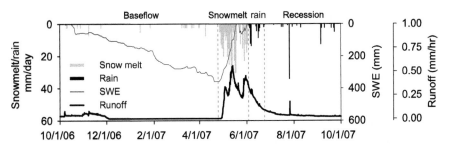

FIGURE 8.22   The annual hydrograph for a snowmelt-dominated river in Montana is shown in *black*. Stream flow reaches its peak during the late spring snowmelt period, but also floods following an intense summer rainstorm. The mass of snowpack is recorded as snow water equivalent (SWE) and is indicated by the *gray line*. Rain is shown with *black bars* and snowmelt is shown in *gray bars*. *Source: From Jensco et al. 2009. Used with permission of the American Geophysical Union.*

of permanently inundated soils or groundwater, the relative proportion changes both seasonally and on the scale of individual storms (refer to Figure 8.2; Bonell 1993, Sidle et al. 2000).

The relative importance of short flowpaths (~quickflow = overland flow + shallow soil flowpaths) versus baseflow (provided by groundwater and permanently saturated soil flowpaths) can vary greatly between biomes as a function of climate, and between watersheds as a function of soil depth and topography (McGuire et al. 2005, Lutz et al. 2012). Streams dominated by groundwater have consistent baseflows and tend to have stable channels with obvious headwaters. In contrast, streams with flows derived predominantly from precipitation, such as those in arid or urban watersheds, have dynamic and inconsistent flow origins (Stanley et al. 1997), such that the extent of headwater streams and the expanse of the channel network itself vary dramatically over time (Figure 8.23).

FIGURE 8.23   Changes in the longitudinal and lateral surface water area of Sycamore Creek, a desert stream in Arizona. *Source: Maps from Stanley et al. 1997. Used with permission of the University of California Press. Photos provided by Emily Stanley.*

Carbon Cycling in Rivers

While some fraction of the organic carbon in rivers is derived from internal productivity, much of the organic carbon entering a stream reach is DOC leached from the terrestrial landscape (Fisher and Likens 1973, Meyer 1981, Webster and Meyer 1997, Meyer et al. 1998, Mayorga et al. 2005). Dissolved organic carbon (DOC) becomes an increasingly important fraction of allochthonous organic carbon with increasing river size. The compounds of DOC include soluble carbohydrates and amino acids, which are leached from decomposing leaves and plant roots (Suberkropp et al. 1976), and humic and fulvic acids from soil organic matter (McDowell and Likens 1988, Qualls and Haines 1992; Chapter 5). A significant amount of the DOC in the Amazon River is of recent origin (Mayorga et al. 2005), but some fraction of the DOC exported to rivers is ancient. Using ^{14}C dating, Raymond and Bauer (2001) report DOC ages in the Susquehanna, Rappahannock, and Hudson rivers of 688, 736, and 1384 years bp.[10]

Although the age and composition of this terrestrial DOC would suggest that it is highly recalcitrant in soil, aquatic microbes assimilate and respire most terrestrial DOC during river transport (Wallace et al. 1999, Richey et al. 2002, Mayorga et al. 2005, Battin et al. 2009) and virtually no terrestrial DOC makes its way to the open ocean (Hedges et al. 1997). As discussed in the preceding section, DOC that was recalcitrant in dark soils or anoxic sediments may become labile in well-lit and well-oxygenated rivers.

Particulate matter dominates the carbon budget of many small streams. The leaves, needles, twigs, branches, and trunks of vegetation that fall into streams support diverse aquatic food webs as they are shredded and decomposed during downstream transport (Webster et al. 1999). Along river networks, the ratio of dissolved to particulate allochthonous carbon inputs generally increases as larger rivers have a smaller proportion of terrestrial "edge" from which to acquire particulate material and because CPOM delivered to tributaries is degraded rather than transported downstream (Vannote 1980, Webster and Meyer 1997, Webster et al. 1999). In narrow, shaded headwaters terrestrial inputs are the dominant energy source and R greatly exceeds NPP. Authochthonous production by benthic algae (periphyton) and aquatic macrophytes (vascular hydrophytes) becomes increasingly important as channels widen. This frequently observed pattern led Vannote et al. (1980) to predict increases in ecosystem productivity with stream size, or along the river continuum. The model predicts that in very large rivers, allochthonous inputs and heterotrophy will again dominate the carbon budget because large rivers are often too deep to support benthic autotrophs and too fast and turbid to support significant phytoplankton productivity.

New Inputs of C—Primary Productivity in Rivers

Where light is not limiting, primary production by benthic algae and aquatic macrophytes is an important carbon source to rivers. Biomass estimates or measures of chlorophyll are often misleading indicators of river productivity because scouring disturbances and variable light regimes drive substantial temporal variation in GPP (Grimm and Fisher

---

[10] Note that this ^{14}C dating provides a single age for the bulk pool of DOC, which will include DOC molecules that were fixed in photosynthesis seconds, hours, days, seasons, years, and millennia before the sample was collected.

1989, Hill 1996) and also because grazing invertebrates are capable of consuming a large fraction of GPP (Wallace and Webster 1996). Because algae and macrophyte tissues are more nutrient rich and palatable than terrestrial organic matter, autotrophic production often contributes disproportionately to secondary production in river food webs (McCutchan and Lewis 2002).

River NPP estimates are typically made using one of two approaches: respirometer chamber estimates or in situ dissolved oxygen change (Bott 2006). Respirometer chamber estimates are analogous to the light/dark bottle methods described above for lakes, and involve isolating stream sediments and water in closed containers and measuring changes in the dissolved oxygen concentration of the overlying water over time. Although chamber estimates are useful for comparative studies and experimental manipulations, NPP estimates derived from chambers are difficult to extrapolate to whole ecosystems for several reasons. First, enclosing stream sediments in a closed vessel dramatically alters the flow, nutrient supply, and gas exchange conditions typical of natural streams (Bott 2006). Second, because river sediments are typically very heterogeneous, scaling to the whole eocosystem requires adequate sampling of all benthic habitat types.

Finally, because chamber methods typically do not include subsurface sediments, they tend to considerably underestimate rates of ecosystem respiration because they do not measure oxygen consumption in the hyporheic zone[11] (Fellows et al. 2001). In general, chamber methods have indicated that primary production often exceeds respiration in well-lit streams (Minshall et al. 1983, Bott et al. 1985) while open channel methods are far more likely to find net heterotrophy for any 24-hour period (Mulholland et al. 2001, Hall and Tank 2003a, Bott et al. 2006, Bernot et al. 2010).

The open channel technique involves measuring diel patterns of stream-water oxygen, or less commonly CO_2 concentrations, and linking these diel changes in gas concentration to the processes of production, respiration, and groundwater or atmosphere exchange (Odum 1956; Figure 8.9). For any time step,

$$\Delta O_2 = GPP - R \pm E, \qquad (8.13)$$

where E is atmospheric exchange as estimated using a gas tracer (Wanninkhof et al. 1990). These data are analyzed as described in the lake metabolism section above. The chief difference is that, in rivers, flow turbulence is a more important driver of gas diffusion than is wind, and so gas-tracer–derived estimates of diffusion must be made for the same flows at which oxygen changes are measured. In two years of continuous monitoring of GPP in Walker Branch, Tennessee, Roberts et al. (2007) showed high day-to-day and seasonal variability in GPP that was largely predicted by light availability (Figure 8.24). In this stream, and probably many similar steams draining intact deciduous forests, the highest rates of stream metabolism are recorded outside of the terrestrial growing season.

---

[11] The hyporheic zone is the volume of subsurface sediments where groundwaters and surface waters are actively exchanged. Hyporheic zone volume varies widely across streams; there is no hyporheic zone in bedrock-constrained channels but the hyporheos may extend 10s to 100s of meters in coarse alluvial channels.

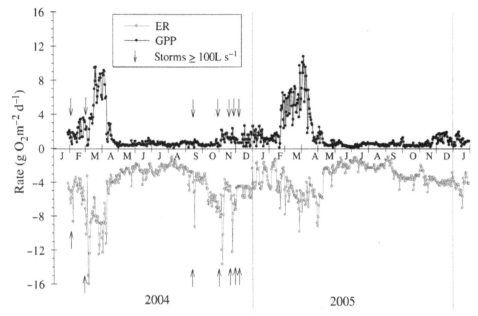

FIGURE 8.24   Daily estimates of gross primary production (GPP) and ecosystem respiration (ER) rates for Walker Branch, Tennessee, derived from open channel oxygen measurements over two years. *Arrows* indicate when large storms occurred. *Source: From Roberts et al. 2007. Used with permission of Springer.*

Limits to Autochthonous Production in Flowing Waters

In small streams light is often the primary factor limiting GPP (Hill 1996, Hall and Tank 2003b, Roberts et al. 2007) and surveys of algal standing stocks and stream nutrient concentrations typically fail to see the same strong positive correlations that have been observed in lakes (Biggs 1996, Francoeur 2001). The nutrient limitation status of benthic algae is often assessed using nutrient-diffusing substrates (NDS), which are agar-filled containers that slowly release nutrient solution through a porous membrane into the stream. The biomass of algae can be compared between substrates containing nutrient-enriched or nutrient-free agar (Pringle 1987).

In a comprehensive synthesis of 237 separate NDS experiments, Francoeur (2001) found that 43% of experiments measured no algal response to either N or P additions, while colimitation by both N and P (23%) was more common than limitation by either N (17%) or P (18%) alone. Results may vary seasonally according to light availability. For example, Bernhardt and Likens (2004) measured significant nutrient limitation of benthic algae for 10 streams of the Hubbard Brook experimental forest only in the spring before canopy leafout, with no nutrient-enrichment effects observed during summer or fall.

In well-lit streams, factors other than nutrient supply may constrain the response of river autotrophs to nutrient loading. Many rivers experience floods, which either scour or bury riverbeds with transported sediments. Thus the algae and macrophytes in many rivers are nearly always at some stage of recovery from the most recent flood (Grimm and Fisher 1989, Death and Winterbourn 1995). It is not surprising, then, that some of the most productive

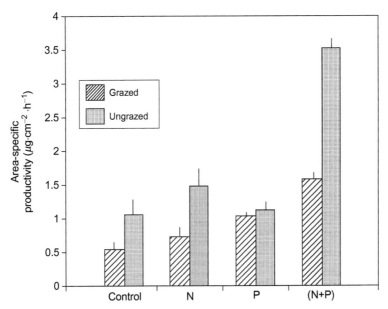

FIGURE 8.25 Effects of nutrient enrichment and grazing by snails on algal productivity in a series of streamside channels (measured by ^{14}C incorporation) in Walker Branch, Tennessee. *Source: From Rosemond et al. 1993. Used with permission of the Ecological Society of America.*

river systems on Earth are spring-fed rivers that never experience scouring flows and which are dominated by lush growths of aquatic macrophytes (Odum 1957). As in lakes, grazing invertebrates or fish can substantially reduce the standing crops and productivity of river autotrophs (Wallace and Webster 1996, Taylor et al. 2006), and can constrain the photosynthetic response to nutrient enrichment (Figure 8.25; Rosemond et al. 1993, Taylor et al. 2006).

Carbon Budgets for Rivers

Carbon budgets are often used to understand the relative importance of autochthonous and allochthonous carbon inputs to river systems. In the first such complete carbon budget, constructed for a small headwater stream in New Hampshire, Fisher and Likens (1972) were unable to find any algae in the sediments of Bear Brook. In these heavily shaded and acidic streams the only autotrophs were stream bryophytes, which contributed less than ~0.1% of annual C inputs (Table 8.5). In one year of monitoring surface and lateral carbon inputs and outputs, Fisher and Likens (1973) estimated that of the ~3260 kg of C delivered as leaves and branches to a segment of Bear Brook, ~2930 kg were respired by stream microbes. The energetics of this stream were almost completely dominated by allochthonous inputs with a P:R ratio <0.01.

In a compilation of subsequent stream organic matter budgets for 35 streams, Webster and Meyer (1997) documented a wide range in P:R ratios, from 0 (as in Bear Brook) to 1.7. Generally the P:R ratio increased with stream size; however, desert streams were far more productive, and the single blackwater river included in the study was far less productive than

TABLE 8.5   Annual Carbon Budget for Bear Brook, a Small New Hampshire Stream

Item	Kg—whole stream[a]	Kcal/m^2	Percentage
Inputs			
Litterfall			
Leaf	1990	1370	22.7
Branch	740	520	8.6
Miscellaneous	530	370	6.1
Wind transport			
Autumn	422	290	4.8
Spring	125	90	1.5
Throughfall	43	31	0.5
Fluvial transport			
CPOM	640	430	7.1
FPOM	155	128	2.1
DOM, surface	1580	1300	21.5
DOM, subsurface	1800	1500	24.8
Moss production	13	10	0.2
Input total	**8051**	**6039**	**99.9**
Outputs			
Fluvial transport			
CPOM	1370	930	15.0
FPOM	330	274	5.0
DOM	3380	2800	46.0
Respiration			
Macroconsumers	13	9	0.2
Microconsumers	2930	2026	34.0
Output total	**8020**	**6039**	**100.2**

[a] *Budget in kg does not balance because of different caloric equivalents of budgetary components.*
Source: Adapted from Fisher and Likens 1976.

predicted for their size (Figure 8.26; Webster and Meyer 1997). For desert streams, which drain a landscape of low-stature vegetation, stream size is not an effective indicator of light availability (Jones et al. 1997). For the blackwater Ogeechee River, as for dystrophic lakes, high concentrations of allochthonous DOC in rivers attenuated light availability (Meyer et al. 1997).

Much of the coarse particulate material provided to rivers from terrestrial vegetation is respired by microbes. While only a small fraction of terrestrial carbon is consumed by aquatic invertebrates (Fisher and Likens 1973), in shaded streams this is the primary resource for aquatic food webs. Terrestrial carbon thus contributes substantially to secondary production of aquatic insects. In a whole-ecosystem experiment, Wallace et al. (1997) excluded terrestrial litter inputs to a small mountain stream in North Carolina for four years and observed substantial declines in the biomass production of aquatic insects (Figure 8.27).

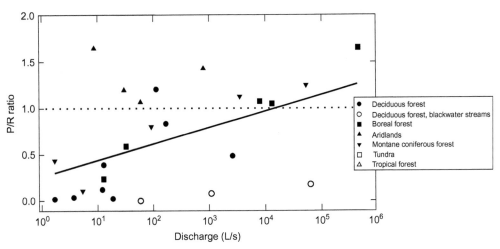

FIGURE 8.26   P:R ratios from a synthesis of 26 organic matter budgets reported in the literature. *Source: From Webster and Meyer 1997. Used with permission of the North American Benthological Society.*

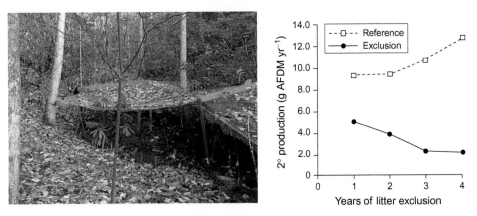

FIGURE 8.27   Effects of a multiyear leaf litter exclusion experiment on the secondary production (~total biomass of stream macroinvertebrates) of a small forested stream (Coweeta, North Carolina). *Source: From Wallace et al. 1997. Used with permission of the American Association for the Advancement of Science.*

Nutrient Spiraling in Rivers

An effective theory for nutrient cycling in stream ecosystems is the concept of nutrient spiraling, which recognizes that lotic nutrient cycles are constantly displaced downstream by advective flow (Figure 8.28; Webster and Patten 1979, Newbold et al. 1981, Newbold 1992, Webster and Ehrman 1996). During downstream transport, dissolved ions are accumulated by bacteria and other stream organisms and converted to organic forms. When these organisms die, they are degraded to inorganic forms that are returned to the water, only to be taken up again by organisms that are involved in the further degradation of organic

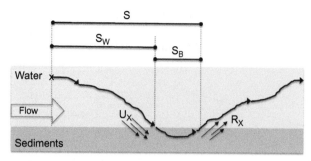

FIGURE 8.28   A nutrient spiral is a nutrient cycle that is displaced by advective flow. In this diagram the black line represents the path of an average nutrient molecule during downstream transport as it is moved downstream prior to uptake into sediments (U) and then transported yet further downstream prior to remineraliza- tion (R) and release to the water column. The spiraling length (S) is composed of the uptake length (S_w), which is the transport distance prior to removal from the water column, and the remi- neralization length (S_B), which is the down- stream distance transported within benthic sediments or biota. *Source: Modified from Newbold 1992.*

materials. The cycle between inorganic and organic forms may be completed many times before a nutrient atom is ultimately exported from the river.

Biogeochemists often compare the downstream transport of nutrients and conservative solutes (~elements available far in excess of biological demand and not effectively sorbed onto sediments) in order to understand the role of biota in determining element fate and transport. If a solution of conservative ions (e.g., Cl, Br) is injected into a river, their concen- tration will decline with distance downstream of the injection point as a function of dilution (Stream Solute Workshop 1990) or permanent loss to groundwater (Covino and McGlynn 2007, Poole et al. 2008). If the slope of the downstream decline in the concentration of a coinjected nutrient is steeper than measured for the conservative tracer, this indicates biolog- ical or chemical uptake in addition to dilution and mass loss. The uptake length (S_w) is thus calculated as the inverse slope of the line describing the downstream decline of tracer- corrected nutrient concentrations:

$$\ln A_x = \ln A_0 + kA_x, \tag{8.14}$$

where A_x is the corrected solute concentration x meters downstream from the point of addi- tion, A_0 is the corrected concentration of the solute at the addition site, and K_A is the per meter uptake rate of solute A. Solute concentrations are corrected for dilution by dividing them by the concentration of a coinjected conservative tracer (Tr) measured at the same location:

$$A_x = [N_x]/[Tr_x]. \tag{8.15}$$

From this and basic measures of stream flow and geomorphology, uptake lengths, velocities, and rates can be derived. The uptake length of a solute (S_w) represents the average down- stream distance that a dissolved nutrient molecule travels before it enters the particulate phase:

$$S_w = 1/k_A. \tag{8.16}$$

The nutrient spiral (between dissolved and particulate phases) is equivalent to a nutrient cycle (between organic and inorganic forms) for solutes that are not strongly sorbed onto min- eral particles, but it conflates both physical sorption and biological assimilation for ions such as NH_4^+ or PO_4^{2-}, which can also be removed from solution through abiotic sorption (see Chapter 4). Elements that are in greater biological demand will have shorter spiraling lengths. The spiraling length can be used to estimate whole ecosystem uptake rates (U) by

$$U = [Q \times C_{bkgrnd}]/[S_w \times w], \tag{8.17}$$

where Q is stream flow in $m^3 \ min^{-1}$, C_{bkgrnd} is the concentration of the solute of interest prior to the experimental enrichment, and w is the average width of the study reach in m. Spiraling length estimates can be made using either nutrient enrichments or isotopic tracers (^{15}N and ^{32}P) together with a conservative tracer (e.g., Cl^-, Br^-, or a fluorescing dye). Because uptake rates tend to increase with increasing concentration and demand can become saturated (Peterson et al. 2001, Bernhardt 2002, Mulholland et al. 2002, Covino et al. 2010), additions are likely to overestimate uptake lengths and underestimate uptake rates.

Across streams, nutrient uptake rates decline strongly with increasing flow velocities and increase with increasing sediment–water contact time indicating that hydrology ultimately constrains biological assimilation (Peterson et al. 2001, Hall et al. 2002, Webster et al. 2003, Wollheim et al. 2006). To allow comparisons of uptake efficiency across streams with very different flows, biogeochemists sometimes calculate uptake velocities, or mass transfer coefficients (V_f):

$$V_f = [Q/w]/S_w, \tag{8.18}$$

which measures the efficiency of benthic uptake accounting for differences in flow.

Modifications of the spiraling technique have made it possible to calculate from a single addition both the ambient uptake rate and the maximum potential uptake rate of a stream ecosystem (Covino et al. 2010). Coinjection of nutrient solutes (e.g., DOC together with NO_3, or inorganic N and P added simultaneously) can be compared to single nutrient injections to compare the relative strength of whole-ecosystem nutrient-limitation (Bernhardt and McDowell 2008, Lutz et al. 2012).

Most forms of land use change increase surface runoff and nitrogen (N) loading to streams. Receiving stream channels, regardless of their condition, will dampen these terrestrial signals through downstream dilution. Intact stream ecosystems will lead to more rapid attenuation of nutrient pulses than can be explained by dilution alone through biological assimilation and transformation (Alexander et al. 2000, Seitzinger et al. 2002, Bernhardt et al. 2003, Green et al. 2004).

Ecosystem nutrient demand can also be assessed using mass-balance approaches. Annual mass balances are very time consuming to construct and are rare in the literature (Table 8.6; Meyer et al. 1981, Triska et al. 1984). In a comprehensive annual N budget for a small stream in the U.S. Pacific Northwest, Triska et al. (1984) showed that one-third of the nitrogen delivered to the ~100-m study reach was effectively stored or converted into nitrogen gas or secondary production over the course of a year (Figure 8.29).

Several recent efforts have estimated the instantaneous mass flux difference between the upstream and downstream ends of a study reach. Roberts et al. (2007) compared upstream and downstream nitrogen fluxes (flux (mg/s=concentration (mg/L) × discharge (L/s)) two to three times per week for two years in Walker Branch, Tennessee. They used this information together with continuous estimates of stream GPP and R (refer to Figure 8.24) to show that variation in GPP could explain nearly 80% of instantaneous NO_3^- flux (Roberts et al. 2007). Since GPP in this stream is nearly entirely explained by light, they were able

TABLE 8.6   Yearly Fluxes of Organic Carbon, Nitrogen, and Phosphorus in Bear Brook

	Organic carbon (g/m²)	Nitrogen (g/m²)	Phosphorus (g/m²)	Atomic Ratio C:N:P
Inputs				
Total dissolved	200	56	0.39	1700:320:1
Total fine particulate	12	0.27	0.55	54:1:1
Total coarse particulate	340	8.2	0.7	1300:26:1
Total gaseous	1	<0.1	0	
Total inputs	**620**	**64**	**1.6**	**990:89:1**
Outputs				
Total dissolved	260	57	0.29	2300:440:1
Total fine particulate	25	0.43	1.1	59:0.9:1
Total coarse particulate	100	1.8	0.38	720:10:1
Total gaseous	230	?	0	
Total outputs	**620**	**59**	**1.8**	**890:72:1**

Source: Meyer et al. 1981. Used with permission of VG Wort, Germany.

to show that solar energy supply directly and rapidly drives nitrogen uptake capacity in the stream. Heffernan and Cohen (2010) provide a more extreme example of the tight linkages between light availability, stream GPP, and ecosystem scale NO_3^- uptake rates. Using oxygen and nitrate sensors recording every few minutes in a macrophyte-dominated spring system in south Florida, they estimated continuous metabolism and stream nitrate uptake (Heffernan and Cohen 2010). They documented dramatic diel covariation in GPP and NO_3^- uptake (Figure 8.30) that could be almost perfectly predicted by light availability.

River Nitrogen Cycling

Measurements and models of stream nitrogen cycling using spiraling approaches generally conclude that biological uptake within stream ecosystems significantly reduces the downstream flux of inorganic nitrogen (e.g., Alexander et al. 2000, Seitzinger et al. 2002, Bernhardt et al. 2002, Kemp and Dodds 2002b, Peterson et al. 2001, Mulholland et al. 2008, Wollheim et al. 2008). Across sites, much of the variation in nitrogen uptake lengths can be explained by differences in stream flow and depth (Peterson et al. 2001, Wollheim et al. 2008). Nutrients travel longer distances in streams with faster flow and reduced water–sediment contact.

In a cross biome survey of $^{15}NH_4$ uptake in streams of North America, Peterson et al. (2001) found that ammonium was typically removed from the water column within tens to hundreds of meters, while nitrate tended to travel much longer distances in the same streams. Rapid uptake of dissolved NH_4^+ by stream sediments is likely due to preferential uptake of NH_4^+ (Chapter 6) as well as cation exchange (Chapter 4) (Peterson et al. 2001). Much of the NH_4^+ that enters streams is immediately nitrified, with the proportion returning to the water column increasing with stream nitrate concentrations (Peterson et al. 2001, Bernhardt et al. 2002). Nitrification facilitates the downstream transport of nitrogen as nitrate, which

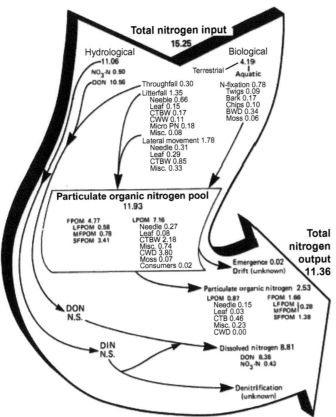

FIGURE 8.29 Nitrogen budget for the stream at watershed 10 of the H.J. Andrews Long Term Ecological Research site, indicating the source and magnitude of annual nitrogen inputs, mean pool size, and annual exports. DON and DIN as an instantaneous pool are not significant (NS). *Source: From Triska et al. 1984. Used with permission of the Ecological Society of America.*

DON = dissolved organic nitrogen; DIN = dissolved inorganic nitrogen; CTBW = cones, twigs, bark, and wood (>1.0 mm and <15.0 cm); CWD = coarse wood debris (>15.0 cm); BWD = wood blocks used to estimate the N fixation rate on woody debris; Micro PN = nitrogen input associated with microparticulate litterfall (<800 μm diameter); LPOM = large particulate organic matter (>1.0 mm diameter); FPOM = fine particulate organic matter (<1.0 mm diameter)

may exacerbate downstream nitrogen pollutant loading (Bernhardt et al. 2002). Nitrification may also be an important source of N_2O from stream ecosystems. There are few direct measures of nitrification-derived N_2O for streams, but most models assume, and limited field measurements support, the assumption that nitrification and denitrification generate low rates of N_2O production in streams as compared to terrestrial soils or wetlands (Beaulieu et al. 2011).

Nitrate spiraling is controlled almost entirely by biological processes because it is poorly sorbed. Generally, nitrate uptake rates tend to increase with nitrate loading, although this capacity can be easily saturated by high loading rates (Mulholland et al. 2008). Variations in

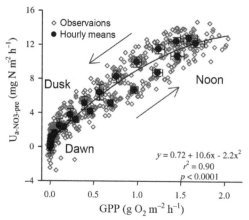

FIGURE 8.30 Nitrate uptake rate (U_{NO3}) varies as a function of GPP over the course of a single day in the macrophyte-dominated Ichetucknee River, Florida. The *small circles* represent individual data points, the *large circles* represent hourly averages. The *arrows* show the hysteresis of the relationship from predawn on day 1 to predawn on day 2. *Source: From Heffernan and Cohen 2010. Used with permission of the Association for the Sciences of Limnology and Oceanography. All Rights Reserved.*

carbon supply and carbon processing (metabolism) are good correlates of nitrogen uptake both within and among streams (Baker et al. 1999, Baker et al. 2000, Bernhardt and Likens 2002, Crenshaw et al. 2002, Hall and Tank 2003b, Roberts and Mulholland 2007). Tight linkages between carbon and nutrient processing are expected given stoichiometric and thermodynamic constraints on organisms, yet it can be difficult to conceive of small, forested streams as carbon limited. However, carbon can be strongly limiting even in streams with high organic matter standing stocks. Experimental additions of labile carbon to stream sediment samples have consistently been shown to decrease nitrification (Strauss and Lamberti 2000, Strauss and Lamberti 2002, Strauss et al. 2002). Organic carbon addition to hyporheic flowpaths have been shown to stimulate N assimilation and denitrification (Hedin et al. 1998, Baker et al. 1999, Sobczak et al. 2003, Crenshaw et al. 2002). During a two-month DOC enrichment to a stream in the Hubbard Brook Experimental Forest, Bernhardt and Likens (2002) were able to reduce inorganic nitrogen export from an entire watershed to below analytical detection throughout the period of C enrichment, indicating that the biological capacity for nitrate uptake was limited by labile carbon.

The flux of organic matter can strongly influence N cycling activity, often creating a patchy distribution of NO_3^- sources and sinks along a stream length (Triska and Oremland 1981, Holmes et al. 1996, Kemp and Dodds 2002a, McClain et al. 2003). Both organic matter storage and nitrogen uptake rates are strongly influenced by hydrologic factors such as stream size, water residence time, and the abundance of physical features to trap organic material (Bilby 1981, Triska 1989, Valett et al. 1996, Peterson et al. 2001, Wollheim et al. 2001, Hall et al. 2002, Webster et al. 2003); thus it is possible that a positive correlation between benthic C storage and N processing rates reflects similar hydrologic controls.

Since streams are not aggrading systems, a large portion of the net loss of N within stream ecosystems must be attributed to denitrification. In a series of $^{15}NO_3$-addition experiments in 72 North American streams, Mulholland et al. (2008) found that NO_3^- uptake rates varied widely (0.01–1000 mg N m^{-2} hr^{-1}). Across this series of isotopic enrichment experiments, anywhere from 0 to 100% of added $^{15}NO_3$ uptake was converted, within 24 hours, to $^{15}N_2$ (Mulholland et al. 2008, Mulholland et al. 2009). The median rate was 16% of NO_3 uptake converted to N_2 over the course of the single-day experiments. Given that these estimates do not include denitrification that occurs over longer time scales (NO_3 that is incorporated into biomass must first be mineralized and then nitrified before it can be converted to N_2), these direct measurements represent conservative estimates of denitrification rates. The N_2O yield from this series of experiments was very low, with only 0.04 to 5.6% of the total gaseous ^{15}N produced as N_2O (Beaulieu et al. 2011). Although the total N_2O produced increased with denitrification rates, Beaulieu et al. (2011) did not observe an increase in the proportional yield of N_2O with N loading, despite measuring gaseous end products across streams that varied widely in stream water NO_3^- concentrations. These observations suggest that streams provide ideal conditions for complete denitrification.

Hydrology constrains nitrogen assimilation and denitrification. Benthic algae and microbes are less able to assimilate N from the water column when flows (and thus fluxes) are high and scouring flows can remove most of algal standing stocks (Grimm and Fisher 1989). The capacity of river ecosystems to retain versus transport additional N depends not only on the biological capacity (set by either light or allochthonous carbon availability) but also on the timing of N delivery. Nitrogen that enters streams in stormflows will travel much farther downstream than nitrogen molecules that seep into the channel during baseflows (Shields et al. 2008).

River Phosphorus Cycling

Globally rivers transport $\sim 21 \times 10^{12}$ g of phosphorus to the oceans each year, with nearly all of this phosphorus in particulate form (Meybeck 1982, Ittekkot and Zhang 1989, Meybeck 1993). Only about 10% of the particulate phosphorus is biologically available; the rest is strongly bound to soil minerals (Meyer and Likens 1979, Ramirez and Rose 1992). For rivers without significant wastewater or fertilizer inputs, very little inorganic phosphorus is found in the water column. Phosphorus mineralized from organic matter is rapidly sorbed or assimilated, keeping P out of solution (Meyer 1979, Meyer and Likens 1979, Meyer 1980).

Because so little inorganic P remains in the water column in small streams, it is more difficult to measure its importance to ecosystem metabolism than in deepwater systems. P is less likely to be limiting to productivity and respiration in rivers than in deepwater habitats because P deposited in river sediments remains available to most river biota. Phosphorus spiraling is also more difficult to interpret than nitrogen spiraling, because a large proportion of inorganic phosphorus may be lost through physical sorption (Demars 2008, Stutter et al. 2010). The abundance of physical retention features (debris dams and slow-moving pools) in stream channels explained >90% of the variation in phosphorus uptake rates across streams of the Hubbard Brook Valley (Warren et al. 2007), consistent with previous observations that sediment sorption and trapping are the dominant drivers of P uptake in these streams (Meyer and Likens 1979). Changes in biological demand can affect P uptake rates. Grazing invertebrates reduced P uptake rates by suppressing algal productivity in laboratory streams (Mulholland et al. 1983), while autumn litterfall increased P uptake into coarse particulate organic matter in Walker Branch (Mulholland et al. 1985).

ESTUARIES

When large rivers reach sea level, their rate of flow slows, drastically reducing their ability to carry sediment. The load of suspended materials is deposited in the river channel and on the continental shelf. Rivers carrying large sediment loads, such as the Mississippi, may form

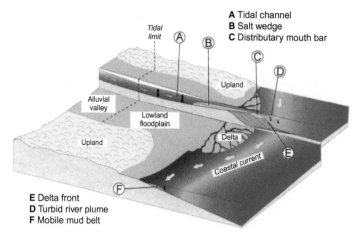

A Tidal channel
B Salt wedge
C Distributary mouth bar

FIGURE 8.31 Generic diagram of a river estuary. The estuary boundaries are defined as the upper limit of the tidal influence within the river inflow to the coastal boundary of freshwater influence. *Source: From Bianchi and Allison 2009. Used with permission of the National Academy of Sciences.*

E Delta front
D Turbid river plume
F Mobile mud belt

obvious deltas. The river channel is progressively confined and divided by deposited sediments, which may support broad, flat areas of salt marsh vegetation (Figure 8.31). The lower reaches of rivers and their salt marshes are subject to daily tidal inundation. An estuarine ecosystem consists of the river channel, to the maximum upstream extent of tidal influence, and the adjacent ocean waters, to the maximum seaward extent that they are affected by the addition of freshwater.

The estuary also includes any salt marshes that may develop along the shoreline. Estuaries are zones of mixing; within an estuary there is a strong gradient in salinity from land to sea. Estuaries are among the most challenging environments on Earth in which to study biogeochemistry, because, in addition to the underlying salinity gradient, turbulent mixing of fresh and saltwater within estuaries generates abrupt changes in temperature, salinity, pH, redox, and element concentrations with implications for biogeochemical cycling (see Table 8.7).

Estuarine Water Budgets and Mixing

The mixing of freshwater from rivers and salt water from the sea occurs in the central channel of an estuary. If the estuary is well mixed, the transition from freshwater to seawater is gradual and progressive as one moves downstream. In many cases, inflowing freshwater may extend over a "wedge" of denser saltwater, creating a sharp vertical gradient in salinity throughout much of the estuary (Figure 8.32). In either case, this transition zone is an arena of rapid biogeochemical transformations and high productivity (Burton 1988, Dagg et al. 2004). Seawater has high pH (about 8.1), redox potential ($>+200$ mV), and ionic strength (total dissolved ions), relative to freshwater (Figure 4.19; Table 9.1). The mixing of freshwater and seawater causes a rapid precipitation of the dissolved humic compounds carried by rivers. The cations in seawater replace H^+ on the exchange sites of the humic materials (Chapter 4), causing these materials to flocculate and sink to estuarine sediments (Sholkovitz 1976, Boyle et al. 1977).

Although humic acids make up only a small fraction of total riverine DOC, this flocculation is also responsible for the "salting out" of hydrocarbons and organometallic complexes which are precipitated in the estuary or within a short distance of the mouth of the river (Boyle et al.

TABLE 8.7   A Compilation of Literature Estimates of GPP, R, and NEP for Streams, Rivers, and Estuaries from Whole-Ecosystem Metabolism Estimates

Ecosystem	GPP (g Cm^{-2} d^{-1})	R (g Cm^{-2} d^{-1})	NEP (g Cm^{-2} d^{-1})	Global R (Pg Cy^{-1})	Global net heterotrophy (Pg Cy^{-1})
Streams (n = 62)	0.73±0.14 (0.02−5.62)	1.93±0.19 (0.29−8.16)	−1.20±0.15 (−5.86−2.51)	0.19	0.12
River (n = 37)	0.91±0.10 (0.06−2.28)	1.53±0.15 (0.20−3.54)	−0.66±0.11 (−2.06−1.60)	0.16	0.07
Estuaries (n = 31)	3.14±0.41 (0.72−10.4)	3.51±0.32 (0.83−7.58)	−0.39±0.21 (−2.98−2.86)	1.20	0.13

Note: Given is the mean standard error and the minimum and maximum in brackets. Ecosystems with the same superscript are not statistically different.
Source: Adapted from Battin et al. 2009.

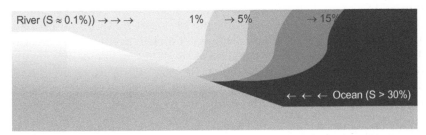

FIGURE 8.32   A diagram of a generic salinity gradient within a coastal estuary. River waters interacting with ocean waters lead to a gradient from fresh to full-strength ocean waters. The greater density of saltwater often leads to a salt wedge underlying a plume of less saline surface waters.

1974, Sholkovitz 1976, Jickells 1998, Turner and Millward 2002, Blair and Aller 2012). The flocculation of dissolved organic compounds and the deposition of larger plant debris account for a major portion of the organic carbon in estuarine sediments (Hedges et al. 1997, Blair and Aller 2012), and there is little evidence that organic matter from land contributes much to marine sediments beyond the continental shelf (Hedges and Parker 1976, Prahl et al. 1994, Hedges et al. 1997). As a result of the removal of terrestrial organic matter, the majority of the organic carbon in estuarine waters is composed of nonhumic substances, presumably resulting from net primary production in the estuary and its salt marshes (Fox 1983, Nixon et al. 1996).

Thermal stratification in estuaries is reinforced by salinity differences, such that thermohaline stratification separates more dilute, well-lit estuarine surface waters from darker and more saline bottom waters. Just as in lakes, this density gradient can lead to well-oxygenated and productive surface waters overlying deep waters where oxygen consumption can exceed diffusion. Together with enhanced nutrient loading, this density separation contributes to the widespread occurrence of coastal hypoxia (Diaz and Rosenberg 2008, Conley et al. 2009a).

Carbon Cycling in Estuaries

The inner reaches of estuaries receive large subsidies of organic matter from rivers. Despite having highly productive vegetation along their margins (mangroves, salt marshes) these regions are net heterotrophic because much of their metabolism is sustained by allochthonous carbon inputs from rivers, groundwaters, and, for many estuaries, urban wastewaters (Odum and Hoskin 1958, Odum and Wilson 1962, Heip et al. 1995, Kemp et al. 1997, Gattuso et al. 1998, Cai 2003, Gazeau et al. 2004, Wang and Cai 2004). As a result, most estuaries are supersaturated in CO_2. In a synthesis of literature estimates of air–water CO_2 fluxes from 32 estuaries from around the world, Chen and Borges (2009) found only one reported instance where the estuary was a net sink for CO_2, that is the air–water flux (FCO_2[12]) was 3.9 mol C m^{-2} yr^{-1} (from Kone and Borges 2008). For the other estuaries FCO_2 values ranged from -3.6 mol C m^{-2} yr^{-1} in Finland's Bothnian Bay (Algesten et al. 2004) to -76 mol C m^{-2} yr^{-1} in Portugal's Douro estuary (Frankignoulle et al. 1998). While acknowledging the wide variation in reported FCO_2 values and uncertainties in the spatial extent of inner estuaries, mangroves, and marshes (Borges 2005, Borges et al. 2005), Chen and Borges (2009) estimate that globally, estuaries emit CO_2 of ~ 0.50 Pg C yr^{-1} to the atmosphere (Figure 8.33).

Mounting evidence based on pCO_2 measurements and mass-balance calculations seems to indicate that the continental shelves are sinks for atmospheric CO_2 (Borges 2005, Chen and Borges 2009). Annual FCO_2 estimates on continental shelves average ~ 1.1 mol CO_2 m^{-2} yr^{-1}, which, scaled to a global continental shelf area of 26×10^6 km^2, yields an annual CO_2 uptake of about 0.35 Pg C yr^{-1} (Chen and Borges 2009). Because of their much greater spatial extent, coastal shelves more than offset the atmospheric CO_2 released from inner estuaries and their fringing marshes and mangroves (Table 8.8).

Primary Production in Estuaries

Nitrogen has been implicated as the nutrient responsible for eutrophication in the Chesapeake Bay (Cooper and Brush 1991, Bronk et al. 1998, Boesch et al. 2001), the Gulf of Mexico (Turner and Rabalais 1994, Rabalais et al. 2002), Narragansett Bay (Nixon et al. 1995, Howarth and Marino 2006), the Baltic Sea (Conley et al. 2007, Conley et al. 2009a), and many other estuaries in the developing world (Figure 8.34). Nitrogen fixation is often limited in estuarine systems because of the low availability of molybdenum (a critical cofactor in the nitrogenase enzyme) as well as sulfate interference with Mo uptake (Howarth and Cole 1985, Cole et al. 1993, Marino et al. 2003).

In addition, the higher turbulence of estuarine waters (Howarth et al. 1995a, Paerl 1996) and the abundance of zooplankton grazers (Marino et al. 2006) may together constrain the growth of filamentous cyanobacteria that are the common N fixers in freshwater lakes. At the same time that N fixation is hampered, the abundance of SO_4 found in saline waters limits Fe-P binding in estuarine sediments, such that organic P, once mineralized, is likely to be released to the water column (Figure 8.35; Caraco et al. 1990, Blomqvist et al. 2004, Jordan et al. 2008).

Many estuaries show a peak in net primary productivity at intermediate salinities, reflecting the zone of maximum nutrient availability and phytoplankton abundance (Anderson

---

[12] By convention, a positive FCO_2 indicates the ecosystem is a net sink for atmospheric CO_2 while a negative FCO_2 value indicates that the system is a net source of CO_2 to the atmosphere.

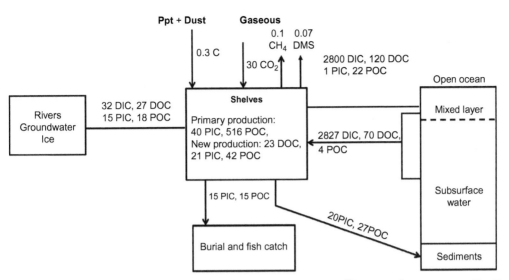

FIGURE 8.33 Mass balance of carbon in continental shelves (flows are in 10^{12} mol C yr^{-1}). *Source: From Chen-Tung and Borges 2009.*

TABLE 8.8 Air–Water CO_2 Flux in Open Oceanic Waters and Major Coastal Ecosystems (including inner estuaries and salt marshes)

	Surface (10^6 km^2)	FCO$_2$ Air–Water CO$_2$ flux (mol C m^{-2}yr^{-1})	Air–Water CO$_2$ flux (Pg C yr^{-1})
60°–90°			
Open oceanic waters	30.77	−0.75	−0.28
Inner estuaries	0.4	46	0.22
Open shelf	6.79	−1.88	−0.15
Subtotal	37.96	−0.46	−0.21
30°–60°			
Open oceanic waters	122.44	−1.4	−2.05
Inner estuaries	0.29	46	0.16
Non-estuarine salt marshes	0.14	23.45	0.04
Coastal upwelling systems	0.24	1.09	0.003
Open shelf	14.47	−1.74	−0.3
Subtotal	137.58	−1.3	−2.15
30°N–30°S			
Open oceanic waters	182.77	0.35	0.77
Inner estuaries	0.25	16.83	0.05
Coastal upwelling systems	1.25	1.09	0.02
Coral reefs	0.62	1.52	0.01
Mangroves	0.2	18.66	0.04
Open shelf	1.35	1.74	0.03
Subtotal	186.44	0.41	0.92
Coastal ocean	26	0.381	0.12
Open ocean	336	−0.388	−1.56
Global ocean	362	−0.331	−1.44

Source: Compiled by Borges 2005 (Table 3). Used with permission of Springer.

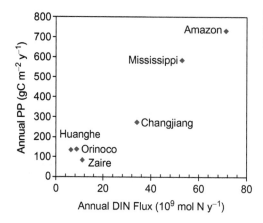

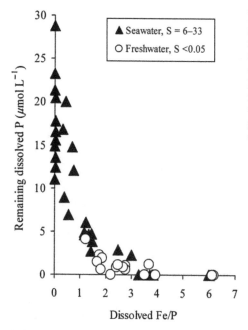

1986, Lohrenz et al. 1999, Dagg et al. 2004, Benner and Opsahl, 2001). In other cases, hydrologic mixing obscures any obvious relationship between net primary production and conservative properties, such as salinity, in the estuary (Powell et al. 1989).

Phytoplankton productivity and organic matter derived from the surrounding salt marshes fuel the high productivity of fish and shellfish in estuarine waters. For many years the large production of fish and shellfish in estuaries was attributed to an abundance of organic carbon flushing from salt marshes to the open water. Indeed, the losses of organic carbon from salt

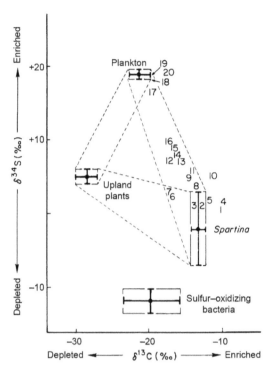

FIGURE 8.36 The isotope ratio for C and S in consumers is shown in relation to their ratios in upland plants, plankton, the salt marsh grass *Spartina*, and sulfur-oxidizing bacteria for Cape Cod's Great Sippewissett Salt Marsh in Massachusetts. The isotope ratios in sulfur-oxidizing bacteria are very different from those of consumers, indicating that sulfur oxidizers are not a major source of carbon for higher trophic levels in the estuary. Similarly, the C isotope ratio for terrestrial plants is considerably more depleted than in consumer biomass, suggesting that allochthonous C is less important than autochthonous C in this marsh. Consumers include shellfish, snails, shrimp, crabs, and fish. The values for each consumer represent pooled samples of 10 to 200 individuals except in the case of the flounder (9) and swordfish (19) *Source: From Peterson et al. 1986. Used with permission of the Ecological Society of America.*

marshes are usually $>100 \, \text{g C m}^{-2} \, \text{yr}^{-1}$, compared to values of 1 to $5 \, \text{g C m}^{-2} \, \text{yr}^{-1}$ from uplands (Nixon 1980, Schlesinger and Melack 1981). Haines (1977), however, suggested that this paradigm was questionable, because the isotopic ratio of carbon in estuarine animals did not match that of *Spartina*.

Using the natural abundance of stable isotopes of both sulfur and carbon, Peterson et al. (Peterson et al. 1985, 1986; Peterson and Howarth 1987) showed that the organic carbon in primary consumers within the Great Sippewissett Marsh in Massachusetts and the Sapelo Island marsh in Georgia was about equally derived from *Spartina* and from phytoplankton production in the open water (Figure 8.36). The shellfish, crabs, fish, and shrimp at the base of the marsh food web show isotopic ratios for C and S that are midway between these sources. Carbon from upland, terrestrial vegetation and carbon fixed by sulfur-oxidizing bacteria in salt marsh soils both play a minor role in supporting the abundant marine life of estuaries.

Nutrient Cycling in Estuaries

A great deal of effort has been directed toward understanding the nitrogen budget of estuaries. Most river waters do not contain large concentrations of available nitrogen (NO_3 and NH_4), and these forms are removed when the waters pass over coastal salt marshes. Indeed, the filtering action of land and marsh vegetation is so effective that inputs of nitrogen in rain can make a substantial contribution to the nitrogen budget of the central waters of estuaries (Correll and Ford 1982). However, as is the case for terrestrial ecosystems (Chapter 6), most of the nitrogen that supports estuarine productivity is not derived from new inputs but from mineralization and recycling of organic nitrogen within the estuary and its sediments (Stanley and Hobbie 1981).

As discussed previously, rates of N fixation in estuaries tend to be low and most nitrogen is supplied via atmospheric deposition and river runoff. At the pH and redox potential of seawater, nitrification occurs rapidly in estuarine waters (Billen 1975, Capone et al. 1990). Nitrification also occurs in the upper layers of sediment (Admiraal and Botermans 1989). Denitrification in the lower, anaerobic layers of sediment is primarily supported by nitrate diffusing down from the upper sediment (Seitzinger 1988, Kemp et al. 1990), although nitrate in the water column may also diffuse back into the sediments, where it is reduced (Simon 1988, Law et al. 1991b). In Narragansett Bay, Rhode Island, Seitzinger et al. (1980, 1984) found that denitrification removed about 50% of the available NO_3 entering in riverflow and about 35% of that derived from mineralization within the estuary. The major product of denitrification was N_2.

In Chesapeake Bay, denitrification leaves the nitrate in the lower water column enriched in $\delta^{15}N$ (Horrigan et al. 1990). When the nitrification rate in the sediments is low, available NO_3 may limit the rate of denitrification, and more NH_4^+ remains to support the growth of phytoplankton in the estuary (Kemp et al. 1990). Storms and tidal currents stir up the sediments in an estuary, releasing large quantities of NH_4^+ to the water column (Simon 1989). In oligotrophic estuaries where little carbon accumulates in the sediments, denitrification may be far less important. For example, Fulweiler et al. (2007) found that during periods of low productivity (and low organic matter deposition) the sediments of Narragansett Bay became N sources (through N fixation) rather than N sinks (through denitrification) to the water column.

Estuarine Phosphorus Cycling

Most river waters are supersaturated with dissolved CO_2, which is derived from the degradation of organic materials during downstream transport. High concentrations of dissolved CO_2 and humic materials cause river waters to be slightly acid. Under these conditions, phosphorus is bound within Fe-hydroxide minerals and is transported in the load of suspended sediment (Figure 4.4, Table 4.8; Eyre 1994). After mixing with the higher pH of seawater, phosphorus desorbs from these minerals and contributes to dissolved phosphorus in the estuary (Lebo 1991, Lebo and Sharp 1993, Berner and Rao 1994, Conley et al. 1995, Lin et al. 2012). Seitzinger (1991) found that an increase in pH in the Potomac River estuary caused a release of P from sediments, stimulating a bloom of N-fixing blue-green algae, a scenario that is analogous to the shifts in species dominance that are seen in P-polluted lakes (Chapter 7). Once phosphorus is released, it is less efficiently adsorbed in saltwater sediments. The iron trap that so effectively sequesters PO_4^{3-} in freshwater sediments is ineffective in saltwaters where Fe is rapidly bound to FeS_2 (Figure 8.35; Blomqvist et al. 2004). For both reasons, phosphorus is often more available in the waters of estuaries than in either freshwater or seawater. De Jonge and Villerius (1989) additionally suggest that the phosphorus bound to carbonate particles delivered to estuaries from the open ocean is released as seawater mixes with freshwater and the carbonates dissolve under the acidic conditions of the estuary.

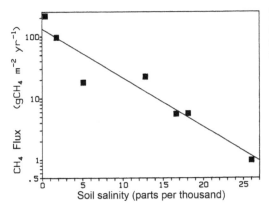

FIGURE 8.37   Annual methane flux as a function of average soil salinity across three southeastern United States salt marshes. *Source: From Bartlett et al. 1987. Used with permission of Springer.*

Anaerobic Metabolism in Estuarine Sediments

Estuarine sediments show high rates of sulfate reduction (Chapter 7), since they are rich in organic matter, flushed with high concentrations of SO_4^{2-} from seawater, and are frequently anaerobic. Although the exact magnitude of sulfate reduction is the subject of some controversy (Howes et al. 1984), various investigators have suggested that more than half of the CO_2 released during decomposition of organic matter in salt marshes and coastal marine sediments is associated with sulfate reduction (Jorgensen 1982, Howarth 1984, Henrichs and Reeburgh 1987, Skyring 1987, King 1988; see Chapter 7). Very little of the sulfide produced escapes to the atmosphere, so sulfide oxidation presumably contributes energy to sediment microbes. The importance of sulfate reduction depends on the concentration, and the rate of oxidation, of iron in estuarine sediments. When Fe(III) is abundant or when Fe(III) is continuously resupplied in oxidized rhizospheres or animal burrows, Fe reduction is often the most prevalent form of anaerobic metabolism (Gribsholt et al. 2003, Hyun et al. 2009, Attri et al. 2011, Kostka et al. 2012).

At a series of sites along the York River in the Chesapeake Bay estuary, Bartlett et al. (1987) found a gradient of decreasing methanogenesis with increasing salinity, as the SO_4^{2-} from seawater progressively inhibits methanogenesis (see Figure 8.37 above; see also Kelley et al. 1990). Howes et al. (1984) found that only about 0.3% of total carbon input to the sediments of Sippewissett Marsh in Massachusetts was lost through methanogenesis. Slightly higher rates have been reported for the Sapelo Island estuary in Georgia (King and Wiebe 1978), but globally the methane emission from saltwater marshes contributes little to the flux of CH_4 to the atmosphere (Chapter 11). Some salt marsh soils also appear to be a small source of phosphine (PH_3) gas to the atmosphere (Hou et al. 2011).

HUMAN IMPACTS ON INLAND WATERS

Through the creation of water infrastructure (e.g., levees, dams) and our dramatic alteration of the land surface (e.g., drainage of wetlands, stormwater pipes, pavements), humans have fundamentally changed the routing and the timing of hydrologic connections between terrestrial and aquatic ecosystems.

Water Infrastructure

Intensive land use by humans, whether for agriculture or for settlement, tends to dramatically reduce water residence time in upland soils while increasing water residence time in impoundments (e.g., Changnon and Demissie 1996, Walsh et al. 2005). Collectively, the vast increase in water control structures globally has tripled the average residence time of river water (Vörösmarty et al. 1997). Coinciding with this hydrologic rerouting, human activities have greatly increased nutrient and sediment loading to rivers and greatly reduced sediment export to coastal seas. There are no large rivers remaining in the world that are not directly impacted by human infrastructure and human wastes.

Humans now appropriate 17% of the global river volume (Postel 2000, Jackson et al. 2001). In many arid regions this proportion is much higher: In the American Southwest humans appropriate 76% of annual river flows (Sabo et al. 2010). Where annual water extraction exceeds annual runoff, groundwaters are depleted, small streams dry more frequently, river flows decline, and lakes are dewatered (Chapter 10). The most dramatic effect of human water extraction can be seen in the rapid dewatering of once large water bodies. Both the Aral Sea in Siberia and Lake Chad in northern Africa are examples of once great lakes that have shrunk to <20% of their former surface area as a result of water extraction for irrigation: these lakes have also become increasingly saline through evaporative concentration (Figure 8.38). The same process is occurring globally, as water extraction and catchment evapotranspiration increases and water inputs to rivers and lakes decline.

For thousands of years humans have constructed dams to support irrigation to grow crops and to ensure long-term water supply. Both the number and size of manmade dams have increased markedly since 1950, with more than 50,000 large dams (>15 m height) in operation worldwide (Figure 8.39; Berga et al. 2006, Lehner et al. 2011). Collectively the river segments converted to reservoirs by these dams are estimated to store 7000 to 8300 km^3 of water (Vörösmarty et al. 2003, Chao et al. 2008), a volume equivalent to 10% of the water stored

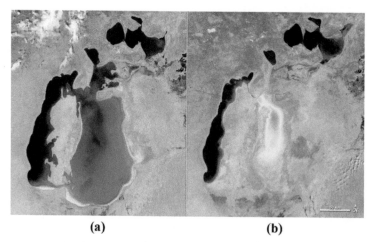

(a)   (b)

FIGURE 8.38   NASA satellite imagery shows the boundaries of the Aral Sea in 2000 (a) and in 2009 (b). Once the fourth largest lake in the world, the Aral Sea today is an example of a terminal lake that is both shrinking and becoming saltier as irrigation in the lake basin reduces annual river inflows below annual evaporative losses.

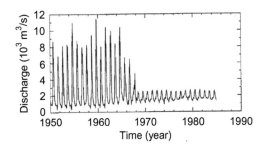

FIGURE 8.39 The effect of dam construction on river flows for the Nile River before and after construction of Aswan High Dam. Discharge is recorded just below the dam. The stabilization of flow is apparent, and it is not difficult to identify the time at which the dam was constructed and the Lake Nasser reservoir filled. The post-impoundment Nile shows reduced overall discharge, substantially truncated peak flows, higher low flows, and a many-month shift in the timing of the natural hydrograph. *Source: From Vörösmarty et al. 2004. Used with permission of the American Geophysical Union.*

in all natural freshwater lakes on Earth (Gleick 2000). Downing et al. (2006) estimate the current area of impounded surface waters at 258,570 km^2 (Table 8.9).

The magnitude of this large number of reservoirs on global C cycling is not yet certain, but qualitatively it is clear that reservoirs are C sources that are typically replacing terrestrial ecosystems that were likely carbon sinks. New reservoirs lead to the inundation of terrestrial vegetation and can have high rates of CO_2 and CH_4 emissions as large quantities of flooded organic matter are decomposed (St Louis et al. 2000, Downing et al. 2008). In contrast, reservoir construction appears to confer significant benefits with respect to the amelioration of nitrogen pollution as reservoirs have very high rates of N removal and provide conditions conducive to denitrification (anoxic, carbon-rich sediments receiving high concentrations of nitrogen in surface runoff (Harrison et al. 2005).

Reservoirs intercept more than 40% of global river discharge (Vörösmarty et al. 2003), and more than 50% of large river systems are affected by dams (Nilsson et al. 2005, Lehner et al. 2011). Dams lead to dramatic alterations in the timing and magnitude of riverflows. In many cases dam construction leads to reduced flows over the entire year (e.g., Figure 8.39), leading to less frequent and extensive hydrologic exchange between rivers and their

TABLE 8.9   Estimated Total Global Area of Small and Large Water Impoundments

A_{min} (km^2)	A_{max} (km^2)	Number of impoundments	Average impoundment area (km^2)	Total impoundment area (km^2)	d_L (impoundments per 10^6 km^2)
0.01	0.1	444,800	0.027	12,040	2965
0.1	1	60,740	0.271	16,430	405
1	10	8295	2.71	22,440	55.3
10	100	1133	27.1	30,640	7.55
100	1000	157	271	41,850	1.05
1000	10,000	21	2706	57,140	0.14
10,000	100,000	3	27,060	78,030	0.02
All impoundments		515,149	0.502	258,570	

Source: Downing et al. 2008. Used with permission of the American Geophysical Union.

surrounding floodplains. In the case of hydropower dams, dams may be operated to release peak flows several times per day during the hottest days of the summer in a practice known as "hydro-peaking"—with frequent, scouring flows reducing biological activity downstream.

Sediment delivery to many river deltas has been dramatically reduced as a result of flow regulation of rivers (Syvitski et al. 2005, Day et al. 2007, Syvitski and Saito 2007). Many of the major river deltas on Earth are now sinking at rates many times faster than global sea level is rising (Syvitski et al. 2009). Sinking deltas and rising seas are a bad combination because even in the absence of rising sea levels, storm surges can inundate increasingly large fractions of low deltas. Within estuaries, fringing salt marsh vegetation exists in a dynamic equilibrium between the rate of sediment accumulation and the rate of coastal subsidence or change in sea level (Kirwan and Murray 2007, Langley et al. 2009b, Kirwan and Blum 2011). As deposits accumulate, the rate of erosion and the oxidation of organic materials increase, slowing the rate of further accumulation.

Conversely, as sea level rises, deposits are inundated more frequently, leading to greater rates of sediment deposition and peat accumulation. Along the Gulf Coast of the United States, the rate of sedimentation has not kept pace with coastal subsidence, and substantial areas of marshland have been lost (DeLaune et al. 1983, Baumann et al. 1984). This loss of protective fringing wet-lands is considered to be a critical factor in the extensive flooding caused by Hurricane Katrina in 2005 (Tornqvist et al. 2008). Current models suggest that from 5 to 20% of coastal wetlands will be lost by the 2080s as a result of coastal subsidence and sea level rise (Nicholls 2004).

Reservoirs, because they trap and retain a large portion of the sediments transported by their contributing rivers, are very effective at retaining mineral elements. The construction of dams can significantly reduce the export of Fe, P, and Si from the continents. When the High Aswan Dam on the Nile River was built, for example, N and P exports to the Nile River estuary dropped precipitously (Table 8.10) and an 80% decline in fish and shrimp abundance was observed (Nixon 2003). The fisheries began to recover ~15 years later as the cities of Cairo and Alexandria grew and released larger fluxes of sewage N and P into the Nile (Table 8.10). Productivity in upstream reservoirs can sequester and trap silica in reservoir sediments, reducing Si inputs to coastal waters (Teodoru and Wehrli 2005, Humborg et al. 2006). In the post-dam period, diatoms were far less dominant in the Nile estuary, probably because urban wastewaters failed to replace the riverine Si flux. Across many coastal waters, reductions in Si coupled with coastal city pollutant inputs are leading to increases in N:Si or P:Si element ratios that favor the growth of nuisance algae over siliceous diatoms (Howarth et al. 2011).

Eutrophication

Humans are causing rapid "cultural eutrophication" of many inland and coastal waters by increasing the amount of nitrogen and phosphorus in the biosphere (Vitousek et al. 1997, Diaz and Rosenberg 2008, Galloway et al. 2008, Rabalais et al. 2009, Childers et al. 2011, Schipanski and Bennett 2012). While better sewage treatment and a ban on detergent phosphates have reduced P loading to many freshwaters, there is mixed evidence to suggest that cultural eutrophication can be reversed when pollution controls are implemented. Some lakes show rapid declines in algal productivity following reductions in nutrient loading (Edmondson and Lehman 1981, Jeppesen et al. 2005, Kronvang et al. 2005) while in other systems only limited responses are seen (Jeppesen et al. 2005, Kemp et al. 2009).

TABLE 8.10 Potential Release of P and N by the Urban Populations of Greater Cairo and Alexandria and the Total Urban Population of Egypt Compared with the Estimated Flux of Nutrients from the Nile

		10^3 tonnes yr^{-1}	
		P	N
The Nile			
Pre-Aswan High Dam			
Dissolved		3.2	6.7
On sediments		4–8	?
	Total	7–11	6.7
Post-High Dam			
Dissolved		0.03	0.2
On sediment		0	0
	Total	0.03	0.2
Human Waste			
Total generated in Cairo and Alexandria			
1965		4.4	21
1985		8.9	55
1995		12.6	87
Potential N and P in wastewater discharge			
Cairo and Alexandria[a]			
1965		1.1	5
1985		3.6	22
1995		9.5	65
Potential N and P in wasterwater discharge			
Total urban population[b]			
1965		2.4	12
1985		6.7	41
1995		15.8	108

[a] Assuming that the population connected to the sewers was 25% in 1965, 40% in 1985, and 75% in 1995. The 1965 estimate is very uncertain.
[b] Extrapolated from Cairo and Alexandria assuming that it accounted for 45% of the total urban population in 1965, 54% in 1985, and 65% in 1995 (see text).
Source: From Nixon 2003. Used with permission of Springer.

Many culturally eutrophied lakes contain large quantities of "legacy P" in their sediments, and the extent to which this P is susceptible to mineralization and mixing into the epilimnion appears to be a major constraint on reversing eutrophication (Martin et al. 2011). In addition, continued sulfate loading to lakes from acid rain or saltwater intrusion can counteract reductions in P loading through internal eutrophication (Caraco et al. 1989, Smolders et al. 2006). Under sulfur-rich reducing conditions, most Fe is present as FeS_x, leaving little oxidized Fe available to bind P in lake sediments; thus the "iron trap" for phosphorus becomes much less effective (Blomqvist et al. 2004).

Perhaps because primary productivity in streams and rivers is less likely to be nutrient limited (Dodds et al. 2002), the subject of river eutrophication is less well represented in the literature than lake or coastal eutrophication (Hilton et al. 2006). The well-recognized increases in nutrient loading to rivers (Green et al. 2004, Boyer et al. 2006, Alexander et al. 2008, Howarth et al. 2012) are typically linked to issues of coastal eutrophication. Yet the supply of anthropogenic nutrients to rivers can lead to algal blooms where there is sufficient light and limited flow disturbance (Peterson 1985, Hilton et al. 2006). In shaded streams, nutrient loading may speed the decomposition coarse particulate organic matter (Benstead et al. 2009, Woodward et al. 2012), and by lowering CPOM C:N and C:P ratios it may enhance C consumption by invertebrate consumers (Cross et al. 2007). Where allochthonous DOC concentrations are very high, nutrient loading is likely to directly stimulate heterotrophic activity which may exacerbate and expand problems of river hypoxia (Mallin et al. 2006).

The management of polluted estuaries is the subject of much controversy. Some workers argue that an improvement in estuarine conditions will be directly related to efforts to reduce nutrients in inflowing waters (Boesch 2002, Howarth and Marino 2006, Smith and Schindler 2009). Others suggest that the retention of prior inputs and the recirculation of nitrogen within the system mean that efforts to reduce new inputs will not necessarily produce immediate improvements in water quality (Kunishi 1988, Van Cappellen and Ingall 1994). Some have argued that the prevalence of nitrogen limitation in estuaries with long histories of human impacts occurs simply because of historic phosphorus loading.

This controversy over which element is most limiting, and thus what sorts of nutrient controls should be implemented, is similar to the ongoing debates about the relative importance of regulating N versus P inputs to lakes (Schindler et al. 2008, Conley et al. 2009b, Lewis et al. 2011). In fact, both elements can enhance eutrophication and exacerbate the duration and extent of anoxia. Anoxic conditions in turn, can enhance rates of phosphorus regeneration from sediments and provide a positive feedback to eutrophication (Van Cappellen and Ingall 1994, Vahtera et al. 2007). Phosphorus loading, because it cannot be easily removed, will likely have longer-term impacts than nitrogen loading—however, the impacts of enhanced N loads have more immediate effects on phytoplankton growth in estuaries (Conley et al. 2009b).

Global Climate Change

The extensive direct manipulation of inland waters by human activities makes it difficult to detect and predict the effects of climate change on freshwaters (Vörösmarty et al. 2000, Barnett et al. 2008, Milly et al. 2008, Arrigoni et al. 2010, Wang and Hejazi 2011). In the rare freshwaters where CO_2 concentrations are low and nutrients are abundant, rising atmospheric CO_2 may increase DIC and stimulate enhanced productivity (Schippers et al. 2004). The majority of lakes, rivers, and estuaries, however, are already supersaturated with CO_2. In these ecosystems rising atmospheric CO_2 is likely to enhance CO_2 evasion rates but is unlikely to fundamentally alter aquatic biogeochemistry. The global warming caused by rising atmospheric CO_2 is having much greater impacts on the water and nutrient budgets of freshwaters.

A warmer climate is predicted to generate a more rapid hydrologic cycle with higher evapotranspiration and rainfall across much of the planet but less certain consequences for soil moisture and surface runoff (see Chapter 10). Climate models suggest that climate warming will lead to a 10 to 40% increase in surface runoff by mid-century (Milly et al. 2005) and that in many regions a significant proportion of this increase will occur during extreme seasonal precipitation events (Milly et al. 2002, Palmer and Ralsanen 2002). Thus, somewhat paradoxically, climate change is expected to increase the severity of both droughts and floods in many regions because increases in stormflows will be rapidly transported off landscapes and toward terminal lakes and estuaries. In the equatorial to subtropical latitudes, intensification of Hadley cell circulation (Chapter 3) is expected to lead to a poleward expansion of the latitudinal bands of aridity (Held and Soden 2006, Lu et al. 2007), which will exacerbate water shortages for people in North America's desert southwest and throughout Europe's Mediterranean (Beniston et al. 2007, Seager et al. 2007).

Across the coterminous United States, a long-term increase in precipitation throughout the twentieth century and in stream flow since at least 1940 has been observed (Karl and Knight 1998, Lins and Slack 1999, McCabe and Wolock 2002, Groisman et al. 2004, Krakauer and Fung 2008). These patterns are primarily driven by data from the eastern United States, with evidence of declines or no change predominating for gauged streams of the Pacific Northwest (Luce and Holden 2009). These trends cannot be attributed solely to climate change; in some regions irrigation, damming, and urbanization are having much greater effects on runoff patterns than climate change (Arrigoni et al. 2010, Schilling et al. 2010). Generally, stream flows have increased in watersheds in proportion to their population density and the percent of land converted to urban areas or cropland. Streamflows have declined in proportion to reservoir volume and the area of irrigated land (Wang and Hejazi 2011). After accounting for these direct influences, the predominant effect of climate change on historical stream flows appears to be increases in annual stream flow (Figure 8.40; Wang and Hejazi 2011).

More frequent storm events are likely to contribute larger pollutant loads to rivers during peak flows when river biota have a limited capacity to assimilate excess nutrients (Kaushal et al. 2008). Increasing storm pulses of nutrients are thus likely to exacerbate problems of freshwater and coastal eutrophication (e.g., Paerl et al. 2001). Hotter temperatures combined with higher nutrient loading will likely contribute to already widespread problems with summer anoxia in rivers and estuaries.

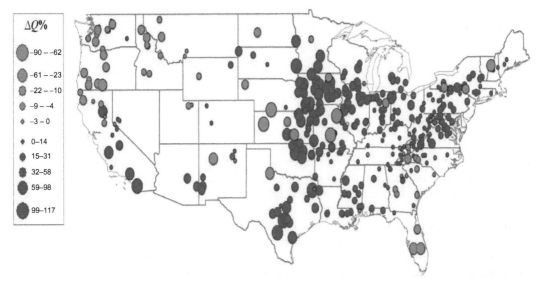

FIGURE 8.40   Spatial distribution of the change in stream flow (Q) recorded for 413 U.S. watersheds between the periods 1948–1970 and 1971–2003. *Source: From Wang and Hejazi 2011. Used with permission of the American Geophysical Union.*

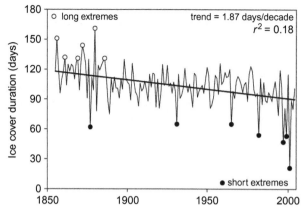

FIGURE 8.41   Graphic of the ice cover duration and extremes for the winters of 1855–1856
through 2004–2005 for Lake Mendota, Wisconsin.
Over the 150-year record the six most extreme short
and long ice duration winters (25-year events) are
marked as *closed* or *open circles. Source: From Benson
et al. 2012. Used with permission of Springer.*

Rising air temperatures are leading to later freeze dates and earlier thaw dates for
ice-covered lakes and rivers, earlier dates for snowmelt flows in rivers, and increasing
contributions of glacial and permafrost meltwater to rivers (Magnuson et al. 1997, Peterson
et al. 2002, Barnett et al. 2005). For many northern temperate lakes the period of ice cover has
declined over the last 50 years (Figure 8.41; Magnuson et al. 2000, Benson et al. 2012). The longer ice-free period and rising air temperatures are lengthening the period of thermal stratification in northern temperate lakes, a change that is expected to lead to accompanying
increases in lake productivity (Carpenter et al. 1992).

A major uncertainty in this prediction is how DOC loading to freshwaters may also change
in response to climate change or other anthropogenic forcing (Freeman et al. 2004) potentially
constraining productivity responses to increased growing season length. For the many arid-
land rivers that are fed primarily by snowmelt, earlier and smaller snowmelt flows may lead
to substantial declines in annual flows and the spatial extent of river networks. Finally, sea
level rise, declining coastal sediment accumulation, and drought-induced saltwater intrusion
are moving the saltwater–freshwater interface further inland (see Chapter 7).

SUMMARY

Freshwater ecosystems are intimately tied to biogeochemical reactions in the surrounding
terrestrial ecosystems. The rate of water delivery and the chemical properties of freshwater
are largely determined by the soil properties, vegetation, and hydrology of the contributing
watershed. Most inland water ecosystems are heterotrophic, showing an excess of respiration
over net primary production. During transport through freshwater, nutrients are removed
from the water column and sequestered in organic and inorganic forms in sediments or, in
the case of N, exported as gaseous products.

Because most inland waters are hydrologically connected, from the smallest headwater
streams, through rivers and lakes, all the way to estuaries or terminal lakes, their global importance really must be considered collectively. Although inland waters occupy only a small
portion of the terrestrial land surface and a small fraction of the total liquid water volume on
Earth, the relatively high rates of carbon and nutrient transformations in freshwater

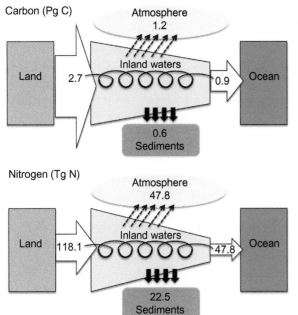

Carbon (Pg C)

Nitrogen (Tg N)

FIGURE 8.42 The cumulative effect of inland waters on global C and N cycling. Note that rivers and lakes deliver as much C and N to the atmosphere as to the ocean, indicating that biological processing of these elements within freshwaters is as important as their physical transport. *Source: Numbers for the C cycle from Cole et al. 2007 with modifications from Aufdenkampe et al. 2011. Numbers for the N cycle from Galloway et al. 2004.*

ecosystems makes them more important in global nutrient cycles than surface area alone would suggest. Collectively the biota of inland waters respire ~40% and store ~20% of the 2.7 Pg of allochthonous carbon, and denitrify or store ~60% of the 118 Tg of nitrogen they receive each year from terrestrial ecosystems (Cole et al. 2007, Galloway et al. 2004, Aufdenkampe et al. 2011; Figure 8.42). The construction of reservoirs has likely enhanced the storage and removal of both elements (St Louis et al. 2000, Downing et al. 2008, Harrison et al. 2009, Heathcote and Downing 2012).

Humans have had a dramatic impact on inland waters throughout the world, regulating the flow of water and altering the load of dissolved and suspended materials. The mixing of freshwater and seawater occurs in estuaries, located at the mouth of major rivers. In response to changes in pH, redox potential, and salinity, river waters feed estuaries with a rich solution of available N and P, and high rates of net primary production fuel a productive coastal marine ecosystem. Despite a temporary storage of nutrients in salt marshes and estuarine sediments, river waters are always a net source of nutrients to their estuary and the coastal ocean. As we shall see, rivers are a major source of nutrients in the global budgets of biogeochemical elements in the ocean.

Recommended Readings

Allan, J. D., and M. M. Castillo. 2007. *Stream Ecology* (second ed.). Springer.
Dodds, W. K. 2002. *Freshwater Ecology: Concepts and Environmental Applications.* Academic Press.
Hauer, R. H., and G. Lamberti. 2006. *Methods in Stream Ecology* (second ed.). Academic Press.
Stumm, W., and J. J. Morgan. 1996. *Aquatic Chemistry Chemical Equilibria and Rates in Natural Waters* (third ed.). Wiley.
Wetzel, R. G., and G. E. Likens. 2000. *Limnological Analyses* (third ed.). Springer-Verlag.
Wetzel, R. G. 2001. *Limnology* (third ed.). Academic Press.

PROBLEMS

1. Assume, as per Dean and Gorham (1998), that the average sedimentation rate of particulate material in reservoirs is 2 cm y^{-1}, that the average bulk density of lake sediments is 1 g cm^{-3}, and that the average C content of lake sediments is 2%. Under these assumptions, how much C might be stored annually in reservoir sediments if the total area of global reservoirs increased to twice their current land area of 1,500,000 km^3 (from St Louis et al. 2000)? Since methane efflux is also high from reservoirs, what average reservoir CH_4 flux would be necessary to completely offset this enhanced carbon sequestration (assume that CH_4 has a global warming potential 25× that of CO_2)? How does this calculated rate compare to estimates of reservoir CH_4 flux in the literature?

2. If global warming induces a 15% increase in DOC export from Scandinavian soils to receiving streams, what will the likely consequences be for GPP, ER, NEP, and pCO_2 in the 15 northern Swedish lakes shown in Figure 8.15 (from Ask et al. 2012)?

3. Go to the Hubbard Brook Ecosystem Study website and download the chemistry of stream water for experimental watersheds #6 from *www.hubbardbrook.org/data/dataset_search.php* and the instantaneous stream-flow data for watershed #6 from *www.hubbardbrook.org/data/dataset.php?id=1*. Match up weekly chemistry data with the discharge recorded for the sampling date. Prepare concentration x discharge (CxQ) relationships for NO_3, Ca^{2+}, and Cl for the entire record. Now compare these CxQ relationships for the first and last decades of the long-term record. Are the ranges of discharge or concentrations different between the two time periods? Now use these subsamples of the data to calculate volume-weighted concentrations for the two time periods. How have volume-weighted concentrations changed? For a more challenging assignment, take these volume-weighted concentrations and multiply them by annual flow to estimate the annual export of each solute.

4. Calculate the actual free energy yield for sulfate reduction in freshwater and in full-strength seawater. Assume pH=8, 1 atm of pressure, and unlimited CH_2O. Compare this free energy yield to methanogenesis under the same conditions.

The Oceans

INTRODUCTION

The Earth's waters constitute its hydrosphere. Only small quantities of freshwater contribute to the total; most water resides in the sea. In this chapter we examine the biogeochemistry of seawater and the contributions that oceans make to global biogeochemical cycles. We begin with a brief overview of the circulation of the oceans and the mass balance of the major elements that contribute to the salinity of seawater. Then we examine net primary productivity (NPP) in the surface waters and the fate of organic carbon in the sea. Net primary productivity in the oceans is related to the availability of essential nutrient elements, particularly nitrogen and phosphorus. Conversely, biotic processes strongly affect the chemistry of many elements in seawater, including N, P, Si, and a variety of trace metals. We examine the biogeochemical cycles of essential elements in the sea and the processes that lead to the exchange of gaseous components between the oceans and the atmosphere.

OCEAN CIRCULATION

In Chapter 3 we saw that the circulation of the atmosphere is driven by the receipt of solar energy which heats the atmosphere from the bottom, creating instability in the air column. Unlike the atmosphere, the oceans are heated from the top. Because warm water is less dense than cooler water, the receipt of solar energy conveys stability to the water column, preventing exchange between warm surface waters and cold deep waters over much of the ocean (Ledwell et al. 1993).

Global Patterns

At the surface, seawater is relatively well mixed by the wind (Thorpe 1985, Archer 1995) and by local changes in density that arise when the surface waters cool or increase in salinity. The mixed layers of the surface ocean range from 75 to 200 m in depth with a mean temperature of about 18°C. The surface temperature in some tropical seas may reach 30°C. The zone of rapid increase in density between the warm surface waters and the cold deep waters is known as the *pycnocline*. It roughly parallels the gradient in temperature, which is known as the thermocline (Chapter 8). The oceans' deep waters contain about 95% of the total volume with a mean temperature of 3°C.

Prevailing winds (Chapter 3) lead to the formation of surface currents in the oceans (Figure 9.1). These surface currents are now mapped at high resolution using the TOPEX/Poseidon satellite, which monitors small changes in the relative height of sea level (Ducet et al. 2000). Global wind patterns drive large gyres, circular rotating currents, that are observed at the surface of all ocean basins. These winds cause surface waters to converge in some areas and diverge in others. The horizontal pressure gradients that result from these areas of convergence and divergence drive motions in the ocean beneath the wind-driven flow at the surface.

These ocean currents are impacted by the Earth's rotation such that they circulate anticyclonically (clockwise in the Northern Hemisphere) around low-pressure centers (where surface waters have diverged). In the Northern Hemisphere, the currents are deflected to the right by the Coriolis force, which is observed when fluid motions occur in a rotating frame (see Figure 3.3). Thus, the Gulf Stream crosses the North Atlantic and delivers warm waters to northern Europe. In the Southern Hemisphere, the gyres move in a counter-clockwise direction and are deflected to the left.

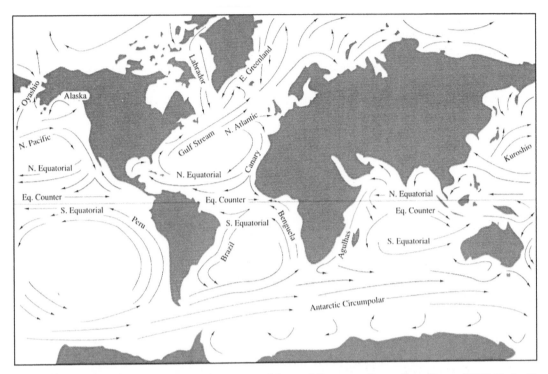

FIGURE 9.1 Major currents in the surface waters of the world's oceans. *Source: From Knauss (1978). Used with permission of Dr. John Knauss.*

The global circulation of the oceans transfers heat from the tropics to the polar regions of the Earth (Oort et al. 1994). About 10 to 20% of the net excess of solar energy received in the tropics is transferred to the poles by ocean circulation; the remainder is transferred through the atmosphere (Trenberth and Caron 2001). With the loss of heat at polar latitudes, the surface waters cool so that their density exceeds that of the underlying water. The resulting convective mixing permits exchange between the surface ocean and the deep waters. In contrast, deep convective mixing driven by winds is rarely observed (Wadhams et al. 2002, Gascard et al. 2002). During the winter in the Arctic and Antarctic oceans, the density of some polar waters also increases when freshwater is "frozen out" of seawater and added to the floating sea ice, leaving behind waters of greater salinity that sink to the deep ocean.

In contrast, during the summer, the polar oceans have lower surface salinity due to melting from the ice caps (Peterson et al. 2006). Because the seasonal downwelling of cold polar waters is driven by both temperature and salinity, it is known as *thermohaline circulation*. Global warming is anticipated to reduce the thermohaline circulation by increasing the density differences between surface and subsurface waters.

Our understanding of the circulation of the oceans has advanced with the deployment of sensors that are designed to follow currents at particular depths. Several thousand profiling floats compose the ARGO network, which transmits data to a satellite for subsequent relay to

FIGURE 9.2   Deployment of a RAFOS float in the North Atlantic Ocean. *Photograph courtesy of Susan Lozier, Duke University.*

ground stations. Among these, RAFOS floats[1] are subsurface floats that remain at depth, recording data for location, temperature, and pressure that are transmitted to a satellite when the floats rise to the sea surface after their mission is complete, typically after 2 years of deployment (Figure 9.2). These floats have documented an interior (mid-depth) circulation in the Atlantic, which returns polar waters to lower latitudes (Bower et al. 2009, Lozier 2010).

[1] ARGO derives from an ocean explorer in Greek mythology. RAFOS is SOFAR spelled backwards, reflecting that these floats receive signals from devices moored on the seafloor to record their position, whereas SOFAR (Sound Fixing and Ranging) devices emit signals that are received by moored devices.

Penetration of cold waters to the deep ocean at the poles, due to convective mixing, provides water for complete ocean mixing, or overturn. For example, North Atlantic deep water (NADW), which forms near Greenland, moves southward through the deep Atlantic, mixes with Antarctic water in the Southern Ocean, and eventually rounds the tip of Africa and enters the Indian and Pacific oceans (Lozier 2012; Figure 9.3). Major zones of upwelling are found in the Pacific Ocean and in the circumpolar Southern Ocean around 65°S latitude (Toggweiler and Samuels 1993). Deep waters are nutrient rich, so high levels of oceanic productivity are found in zones of upwelling. Upwelling along the western coast of South America yields high levels of net primary production that support the anchovy fishery of Peru. This global overturning circulation has traditionally been called the "conveyor belt" (Broecker 1991).

These patterns of ocean circulation have important implications for biogeochemistry. The overall mean residence time for seawater is 34,000 years with respect to river flow (i.e., total ocean volume/total annual river flow). In fact, most rivers mix only with the smaller volume of the surface ocean, which thus has a mean residence time of about 1700 years with respect to river water. If we account for the addition of rainwaters and upwelling waters to the surface ocean, the actual turnover time of the surface waters is even faster. For example, the mean residence time of surface waters in the North Pacific Ocean is about 9 to 15 years (Michel and Suess 1975). The surface water is also in rapid gaseous equilibrium with the atmosphere.

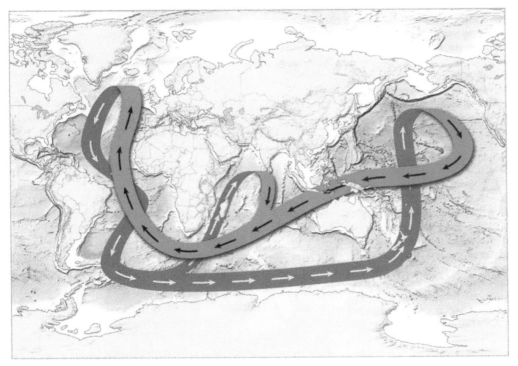

FIGURE 9.3 The global ocean thermohaline circulation forms a conveyor that moves water among the various ocean basins in surface (*red*) and deep-water (*blue*) currents. *Source: From Lozier (2010). Used with permission of the American Association for the Advancement of Science.*

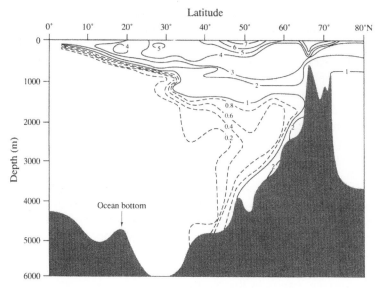

FIGURE 9.4   Penetration of bomb-derived tritium (^3H_2O) into the North Atlantic Ocean. Data are expressed as the ratio of $^3H/H \times 10^{-18}$ for samples collected in 1972. *Source: From Ostlund (1983).*

Mean residence time for CO_2 dissolved in the surface waters of the Atlantic Ocean is about 6 years (Stuiver 1980).

    Renewal or "ventilation" of the bottom waters is confined to the polar regions. Downward mixing of ^3H_2O and $^{14}CO_2$ produced from the testing of atomic bombs (Figure 9.4) and downward mixing of anthropogenic chemicals of recent origin (e.g., see Krysell and Wallace 1988) show the rate of entry of surface waters to the deep sea and the movement of deep water toward the equator. The downward transport in the North Atlantic is 15 Sv,[2] roughly 10 times greater than the total annual river flow to the oceans (Dickson and Brown 1994, Ganachaud 2003, Luo and Ku 2003). About 21 Sv sink in the Southern Ocean, including 4 to 5 Sv in the Weddell Sea, which is the origin of northward-flowing Antarctic bottom water (Hogg et al. 1982, Schmitz 1995). Because the volume of water entering the deep ocean is much greater than the total annual river flow to the sea, the mean residence time of the deep ocean is much less than 34,000 years. Estimates of the mean age of bottom waters using ^{14}C dating of total dissolved CO_2 range from 275 years for the Atlantic Ocean to 510 years for the Pacific (Stuiver et al. 1983), and the overall renewal of the oceans' bottom waters is normally assumed to occur in 500 to 1000 years. Thus, the deep waters maintain a historical record of the conditions of the surface ocean several centuries ago.

    Deep water currents also transfer seawater between the major ocean basins as a result of the Antarctic circumpolar current, which carries >130 Sv (Cunningham et al. 2003, Firing et al. 2011). In the Atlantic Ocean, evaporation exceeds the sum of river flow and precipitation, yielding higher seawater salinity than in the Pacific (Figure 9.5). The Atlantic receives a net

[2] As a unit to express the movement of large bodies of seawater, 1 Svedrup (Sv) $= 10^6$ m^3/sec $= 3.2 \times 10^{13}$ m^3/yr.

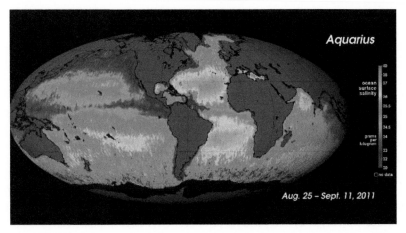

FIGURE 9.5   Salinity of the surface waters of the world's oceans. *Source: From NASA Aquarius (http://www.nasa.gov/ mission_pages/aquarius/multimedia/gallery/pia14786.html).*

inflow of less saline waters from the Pacific, via the Antarctic circumpolar current, to restore the water balance (Figure 9.3). To maintain a mass balance in the overall volume of the deep sea, any increase in the formation of deep waters in polar regions must be associated with an increase in upwelling in other regions (Marshall and Speer 2012).

Changes in ocean currents, particularly the formation of deep waters, may be associated with changes in global climate. At the end of the last glacial epoch, the concentration of atmospheric CO_2 rose from 200 ppm to about 280 ppm in the atmosphere (Figure 1.2). An increase in the rate of upwelling of deep waters in the Southern Ocean may have been associated with the release of CO_2 to the atmosphere (Burke and Robinson 2012). There is some indication that the formation of polar deep waters has slowed in recent years, perhaps indicative of global warming and a greater density stratification of the surface waters in the North Atlantic (Cunningham and Marsh 2010). Climate change is also likely to affect the pattern of surface currents. During the last glacial epoch, the Gulf Stream, which carries about 30 Sv, appears to have shifted southward, producing a humid climate in southern Europe (Keffer et al. 1988). During the Little Ice Age, 200 to 600 years ago, a weakened flow of the Gulf Stream appears to have been related to cold conditions in Europe (Lund et al. 2006).

El Niño

Ocean currents also show year-to-year variations that affect biogeochemistry and global climate. One of the best known variations in current occurs in the central Pacific Ocean. Under normal conditions, the trade winds drive warm surface waters to the western Pacific, allowing cold subsurface waters to upwell along the coast of Peru. Periodically, the surface transport breaks down in an event known as the El Niño–Southern Oscillation (ENSO). During El Niño years, the warm surface waters remain along the coast of Peru, reducing the upwelling of nutrient-rich water. Phytoplankton growth is limited and the normally productive anchovy fishery collapses (Glynn 1988).

Associated with these occasional warm surface waters in the eastern Pacific are changes in regional climate, for example, exceptionally warm winters and greater rainfall in western North America (Molles and Dahm 1990, Swetnam and Betancourt 1990, Redmond and Koch 1991). At the same time the absence of warm surface waters in the western Pacific reduces the intensity of the monsoon rainfalls in Southeast Asia and India. Working with atmospheric scientists, oceanographers now recognize that El Niño events are part of a cycle that yields opposite but equally extreme conditions during non-El Niño years, which are known as La Niña (Philander 1989). Although the switch from El Niño to La Niña is poorly understood, it is likely that the conditions at the beginning of each phase reinforce its development, with the cycle averaging between 3 and 5 years between El Niño events, which are recognized in sedimentary records extending to 5000 years ago (Rodbell et al. 1999). A similar but less powerful cyclic pattern of ocean circulation is seen in the Atlantic Ocean (Philander 1989).

The upwelling of cold, deep ocean waters during the La Niña years leads to lower atmospheric temperatures over much of the Northern Hemisphere. Thus, El Niño–La Niña cycles add variation to the global temperature record, complicating efforts to perceive atmospheric warming that may be due to the greenhouse effect. Moreover, the El Niño–La Niña cycle affects the concentrations of atmospheric CO_2, since the release of CO_2 from cold, upwelling waters is lower during years of El Niño (Bacastow 1976, Inoue and Sugimura 1992, Wong et al. 1993). During the 1991–1992 El Niño, the ocean released 0.3×10^{15} g C as CO_2 to the atmosphere, compared to its normal efflux of 1.0×10^{15} g C (Murray et al. 1994, 1995; Feely et al. 1999; compare Chavez et al. 1999 for the 1997–1998 El Niño), and the rate of CO_2 increase in the atmosphere slowed for several years (Keeling et al. 1995).

In addition to reducing marine net primary productivity in the eastern Pacific (Chavez et al. 1999, Behrenfeld et al. 2001, Turk et al. 2001), El Niño conditions can affect other aspects of biogeochemistry in the sea. Lower rates of denitrification in warm El Niño waters may decrease the total marine denitrification rate by as much as 25% over La Niña conditions (Codispoti et al. 1986, Cline et al. 1987).

THE COMPOSITION OF SEAWATER

Major Ions

Table 9.1 shows the concentration of major ions in seawater of average salinity, 35‰ (i.e., 35 g of salts per kilogram of seawater; Millero et al. 2008). Sodium and chloride dominate the mix, and in the surface waters the pH of seawater is close to 8.1. The mean residence time for each of these ions is much longer than the mean residence time for water in the oceans, allowing plenty of time for mixing. Although seawater varies slightly in salinity throughout the world (Figure 9.5), these ions are conservative in the sense that they maintain the same concentrations relative to one another in most ocean waters. For example, recent changes in the salinity of seawater, due to freshening of the North Atlantic by Greenland ice melt (Boyer et al. 2005, Curry and Mauritzen 2005), do not change the relationship between the concentration of

TABLE 9.1   Major Ion Composition of Seawater, Showing Relationships to Total Chloride
and Mean Residence Times for the Elements with Respect to Riverwater Inputs

Constituent	Concentration in seawater[a] (g kg^{-1})	Chlorinity ratio[a] (g kg^{-1})	Concentration in river water[b] (mg/kg)	Mean residence time[b] (10^6 yr)
Sodium	10.78145	0.556492	5.15	75
Magnesium	1.28372	0.066260	3.35	14
Calcium	0.41208	0.021270	13.4	1.1
Potassium	0.39910	0.020600	1.3	11
Strontium	0.00795	0.000410	0.03	12
Chloride	19.35271	0.998904	5.75	120
Sulfate	2.71235	0.140000	8.25	12
Bicarbonate	0.10481	0.005410	52	0.10
Bromide	0.06728	0.003473	0.02	100
Boron	0.02739	0.001413	0.01	10
Fluoride	0.00130	0.000067	0.10	0.05
Water	964.83496	49.800646		0.034

[a] Source: Millero et al. (2008).
[b] Source: Meybeck (1979) and Holland (1978).

various ions and Cl. Thus, a good estimate of total salinity can be calculated from the concentration of a single ion. Often chloride is used, and the relationship is

$$\text{Salinity} = (1.81) \times Cl^-, \tag{9.1}$$

with both values in g kg^{-1}. Table 9.1 shows the mean ratio between chloride and other major ions in standard seawater (Millero et al. 2008).

According to Table 9.1, the time for rivers to supply the elemental mass in the ocean, the mean residence time, varies from 120 million years for Cl to 1.1 million years for Ca. Biological processes, such as the deposition of calcium carbonate in the shells of animals, are responsible for the relatively rapid cycling of Ca. But even for Cl the mean residence time is much shorter than the age of the oceans; Cl has not simply accumulated in seawater through Earth's history.

Through geologic time, changes in the composition of seawater occurred when inputs and outputs of individual constituents were not in balance. In the face of continual inputs of ions in riverwater, the composition of seawater is maintained by processes that remove ions from the oceans. Processes controlling the composition of seawater must act differentially on the major ions, because their concentrations in seawater are much different from the concentrations in rivers. For example, whatever process removes Na from seawater must not be effective until the concentration of Na has built up to high levels (Drever 1988). On the other hand, Ca is the dominant cation in river water (Table 9.1), but the Ca concentration in seawater is relatively low. The Ca content of seawater has declined during the past 28 million years, as a result of changes in the relative importance of inputs from rock weathering and sedimentary losses to carbonate sediments (De La Rocha and DePaolo 2000, Griffith et al. 2008).

A number of processes act to remove the major elements from seawater. Earlier, we saw that wind blowing across the ocean surface produces sea spray and marine aerosols that contain the elements of seawater (Chapter 3). A significant portion of the river transport of Cl from land is derived directly from the sea (Figure 3.16). The atmospheric transport of these "cyclic salts" removes ions from the sea roughly in proportion to their concentration in seawater (Table 4.9).

During some periods of the Earth's history, vast deposits of minerals have formed when seawater evaporated from shallow, closed basins. Today, the extensive salt flats, or *sabkhas*, in the Persian Gulf region are the best examples. Although the area of such seas is limited, the formation of evaporite minerals has been an important mechanism for the removal of Na, Cl, and SO_4 from the oceans in the geologic past (Holland 1978). Huge deposits of salt laid down 400,000,000 years ago are now mined beneath Lake Erie, near Cleveland, Ohio.

Other mechanisms of loss occur in ocean sediments. Sediments are porous and the pores contain seawater. Burial of ocean sediments and their pore waters is significant in the removal of Na and Cl, which are the most concentrated ions in seawater. Biological processes are also involved in the burial of elements in sediments. As we will discuss in more detail in a later section Biogenic Carbonates (p. 363), the deposition of $CaCO_3$ by organisms is the major process removing Ca from seawater. Biological processes also cause the removal of SO_4, which is consumed in sulfate reduction and deposited as pyrite in ocean sediments (see Chapters 7 and 8).

Ions are removed from the oceans when the clays in the suspended sediments of rivers undergo ion exchange with seawater. In rivers, most of the cation exchange sites (Chapter 4) are occupied by Ca. When these clays are delivered to the sea, Ca is released and replaced by other cations, especially Na (Sayles and Mangelsdorf 1977, James and Palmer 2000). Most deep sea clays show higher concentrations of Na, K, and Mg than are found in the suspended matter of river water (Martin and Meybeck 1979). The clays eventually settle to the ocean floor, causing a net loss of these ions from ocean waters.

So far, the processes we have discussed for the removal of elements from seawater cannot explain the removal of much of the annual river flow of Mg and K to the sea. Marine geochemists have postulated several reactions of "reverse weathering," whereby silicate minerals were reconstituted (authigenic) in ocean sediments, removing Mg and other cations from the ocean (MacKenzie and Garrels 1966). Reverse weathering is a major sink for Li from seawater (Misra and Froelich 2012), and the formation of authigenic clay minerals is apparently a small sink for Mg and K (Kastner 1974, Sayles 1981). Michalopoulos and Aller (1995) found that aluminosilicate minerals were reconstituted in laboratory incubations of marine sediments from the Amazon River, suggesting that this mechanism sequesters as much as 10% of the annual flux of K to the sea (compare Hover et al. 2002).

In the late 1970s, Corliss et al. (1979) examined the emissions from hydrothermal (volcanic) vents in the sea. One of the best-known hydrothermal systems is found at a depth of 2500 m near the Galapagos Islands in the eastern Pacific Ocean. Hot fluids emanating from these vents are substantially depleted in Mg and SO_4 and enriched in Ca, Li, Rb, Si, and other elements compared to seawater (Elderfield and Schultz 1996, de Villiers and Nelson 1999). Globally the annual sink of Mg in hydrothermal vents, where it leads to the formation of Mg-rich silicate rocks, exceeds the delivery of Mg to the oceans in river water. The flux of

Ca to the oceans in rivers, 480×10^{12} g/yr, is incremented by an additional flux of up to 170×10^{12} g/yr from hydrothermal vents (Edmond et al. 1979). Changes in the Mg/Ca ratio in seawater through geologic time are a good index of the relative importance of hydrothermal activities (Horita et al. 2002, Coggon et al. 2010).

In sum, it appears that most Na and Cl are removed from the sea in pore water burial, sea spray, and evaporites. Magnesium is largely removed in hydrothermal exchange, and calcium and sulfate by the deposition of biogenic sediments. The mass balance of potassium is not well understood, but K appears to be removed by exchange with clay minerals, leading to the formation of illite, and by some reactions with basaltic sediments (Gieskes and Lawrence 1981). Whitfield and Turner (1979) show an indirect correlation between the mean residence time of elements in seawater and their tendency to incorporate into one or more sedimentary forms (Figure 9.6). Over long periods of time, ocean sediments are subducted to the Earth's mantle, where they are converted into primary silicate minerals, with a portion of the volatile components being released in volcanic gases (H_2O, CO_2, Cl_2, SO_2, etc.; Figure 1.3). The entire oceanic crust appears to circulate through this pathway in less than 300 million years (Muller et al. 2008), transferring sedimentary deposits to the mantle (Plank and Langmuir 1998).

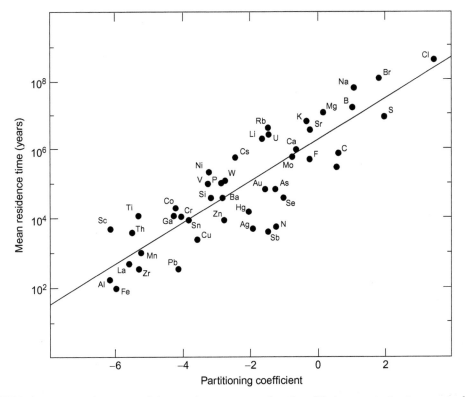

FIGURE 9.6 Mean residence time of elements in seawater as a function of their concentration in seawater divided by their mean concentration in the Earth's crust—with high values of the index indicating elements that are very soluble. *Source: From Whitfield and Turner (1979). Reprinted with permission from* Nature, *copyright 1979 Macmillan Magazines Limited.*

NET PRIMARY PRODUCTION

Measurement

Compared to massive forests, marine phytoplankton are easy to overlook, and much of their production of organic matter is rapidly decomposed in the surface waters. Thus, net primary production in the sea is often measured using modifications of the oxygen-bottle and ^{14}C techniques, as outlined for lake waters in Chapter 8. Controversy surrounding the exact magnitude of marine production derives from the tendency for O_2-bottle measurements of NPP to exceed those made using ^{14}C in the same waters (Peterson 1980). Part of the problem can be explained by the presence of a large biomass of picoplankton—small phytoplankton that pass through the filtration steps of the ^{14}C procedure. In the waters of the eastern tropical Pacific Ocean, Li et al. (1983) found that 25 to 90% of the photosynthetic biomass passes a 1-μm filter. Picoplankton are more important in warm, nutrient-poor waters than in the subpolar oceans (Agawin et al. 2000), and Stockner and Antia (1986) suggest that picoplankton may regularly account for up to 50% of ocean production. Marine phytoplankon also release large amounts of dissolved organic carbon to seawater (Baines and Pace 1991), and these compounds—technically a component of NPP—also pass through the filtration procedures of the ^{14}C method.

Acting to enhance estimates of marine NPP from ^{14}C methods, some ocean waters harbor anoxygenic photo-heterotrophic bacteria, which metabolize organic carbon when it is available and produce it by anoxygenic photosynthesis when it is not (Kolber et al. 2001). Photosynthesis by these bacteria constitutes 2 to 5% of total production in some waters, which would be overlooked by the O_2-bottle method (Kolber et al. 2000).

With either method, water samples isolated in bottles contain a mix of photosynthetic and heterotrophic organisms, and the net change in O_2 is often said to indicate *net community production* (NCP) during the incubation period. Some workers have tried to avoid the problems associated with bottles by measuring net increases in O_2 (Craig and Hayward 1987, Najjar and Keeling 1997, Emerson and Stump 2010) or decreases in HCO_3^- (Lee 2001) in the upper water column when photosynthesis is occurring. The concentration of O_2 or HCO_3^- is compared to the concentration expected as a result of the equilibrium dissolution of gases in seawater from the atmosphere. As with bottle methods, a supersaturation of O_2 or depletion of HCO_3^- is best taken as an index of NCP in the water column.[3]

Yet another, new technique involves the measurement of dissolved O_2 and $\delta^{17}O$ in waters (Luz and Barkan 2009). Oxygen is affected by the rate of photosynthesis and respiration, whereas $\delta^{17}O$ is solely affected by photosynthesis. Differential changes in these parameters can be used to estimate NCP in ocean waters (Juranek and Quay 2010). In side-by-side comparisons, this $\delta^{17}O$ method gave values more than twice as high as the ^{14}C method for measuring marine production (Quay et al. 2010).

Remote sensing offers significant potential for improved regional and global estimates of marine NPP without complications associated with community respiration. Where ocean waters contain little phytoplankton, there is limited absorption of incident radiation by

---

[3] Note that NCP is analogous to net ecosystem production (NEP) on land (Chapter 5), but NCP does not include deep-water and sediment respiration, which are supported by surface photosynthesis in the oceans.

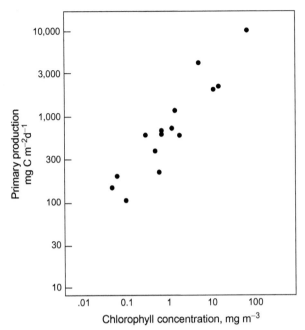

FIGURE 9.7 Net primary productivity as a function of surface chlorophyll in waters of coastal California. *Source: From Eppley et al. (1985),* Journal of Plankton Research. *Reprinted by permission of Oxford University Press.*

chlorophyll, and the reflected radiation is blue. Where chlorophyll and other pigments are abundant, the reflectance contains a greater proportion of green wavelengths (Prézelin and Boczar 1986). The reflected light is indicative of algal biomass in the upper 20 to 30% of the euphotic zone, where most NPP occurs (Balch et al. 1992). The reflectance data can be used to calculate the concentration of chlorophyll or organic carbon in the water column and hence production (Figure 9.7; Platt and Sathyendranath 1988; Najjar and Keeling 1997). With the deployment of the MODIS satellite (Chapter 5), multispectral images of the ocean are now available for worldwide estimates of marine NPP.

Global Patterns and Estimates

Differences in methodology account for much of the variation among published estimates of global marine production. Early estimates suggested that marine NPP was 23 to 27 $\times$ 10^{15} g C/yr (Berger 1989), but most workers now suggest that marine NPP may be 2$\times$ larger and roughly equivalent to NPP on land (Table 9.2). Behrenfeld and Falkowski (1997) used satellite measurements of pigment concentration in surface waters to estimate marine NPP at 43.5 $\times$ 10^{15} g C/yr. More recently, Friend et al. (2009) estimated marine NPP at 52.5 $\times$ 10^{15} g C/yr (Figure 9.8), with a spatial pattern similar to that for dissolved O_2 in seawater (Najjar and Keeling 1997). Net community production, yielding organic matter that can sink into the deep ocean, is about 15 to 20% of total NPP (Lee 2001, Laws et al. 2000, Falkowski 2005, Quay et al. 2010), with higher values in cold polar waters where bacterial respiration is lower.

TABLE 9.2   Estimates of Total Marine Primary Productivity and the Proportion That Is New Production

Province	% of ocean	Area $(10^{12}m^2)$	Mean production $(g\,C\,m^{-2}yr^{-1})$	Total global production $(10^{15}g\,C\,yr^{-1})$	New production[a] $(g\,C\,m^{-2}yr^{-1})$	Global new production $(10^{15}g\,C\,yr^{-1})$
Open ocean	90	326	130	42	18	5.9
Coastal zone	9.9	36	250	9.0	42	1.5
Upwelling area	0.1	0.36	420	0.15	85	0.03
Total		362		51		7.4

[a] *New productivity defined as C flux at 100 m.*
Source: From Knauer (1993). Used with permission of Springer-Verlag.

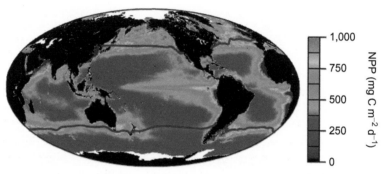

FIGURE 9.8   Global map of marine NPP. *Source: From Behrenfeld et al. (2006).*

The highest rates of NPP are measured in coastal regions, where nutrient-rich estuarine waters mix with seawater, and in regions of upwelling, where nutrient-rich deep water reaches the surface. NPP in beds of intertidal rockweed, *Ascophyllum nodosum,* in Cobscook Bay of eastern Maine is nearly 900 g C m^{-2} yr^{-1}, similar to that of a temperate forest (Table 5.2; Vadas et al. 2004). In contrast, among open-water environments, even the most productive areas show NPP of <300 g C m^{-2} yr^{-1}, roughly similar to the NPP of an arid woodland. As a result of their large area, the open oceans account for about 80% of the total marine NPP, with continental shelf areas accounting for the remainder (Table 9.2). Although massive beds of kelp are found along some coasts, such as the *Macrocystis* kelps of southern California, seaweed accounts for only about 0.1% of marine production globally (Smith 1981, Walsh 1984).

Dissolved Organic Matter

A large amount of dissolved organic matter is found in seawater (Martin and Fitzwater 1992). Based on its ^{14}C age and molecular properties, only a small fraction appears to derive from humic substances that have been delivered to the oceans by rivers (Opsahl and Benner

1997, Raymond and Bauer 2001, Hansell et al. 2004). Most marine DOC is derived from marine photosynthesis. It is well known that phytoplankton and bacteria leak dissolved organic compounds, and globally about 17% of net community production may leak from phytoplankton cells as dissolved organic carbon (DOC; Hansell and Carlson 1998b). DOC is also released following the death and lysis of phytoplankton cells (Agusti et al. 1998).

Most of this DOC is labile and rapidly decomposed in the surface ocean, where it supports bacterioplankton—free-floating bacteria (Kirchman et al. 1991, Druffel et al. 1992). However, a fraction of the DOC is relatively refractory, apparently composed of nitrogen compounds resynthesized by bacteria (Barber 1968, Ogawa et al. 2001, McCarthy et al. 1998, Jiao et al. 2010, Aluwihare et al. 2005). This DOC is entrained in downwelling waters and delivered to the deep sea (Aluwihare et al. 1997, Loh et al. 2004, Carlson et al. 1994). The mean residence time for DOC in the oceans, up to 6000 years, is longer than the time for deep-water renewal, implying that some DOC has made more than one cycle through the deep sea (Williams and Druffel 1987, Bauer et al. 1992).

Fate of Marine Net Primary Production

Most marine NPP is consumed by zooplankton and bacterioplankton in the surface waters. Bacterioplankton respire dissolved organic carbon and use extracellular enzymes to break down particulate organic carbon (POC) and colloids produced by phytoplankton (Druffel et al. 1992). Cho and Azam (1988) concluded that bacteria were more important than zooplankton in the consumption of POC in the North Pacific Ocean. Reviewing a large number of studies from marine and freshwater systems, Cole et al. (1988) found that net bacterial growth (production) is about twice that of zooplankton and accounted for the disappearance of 30% of NPP from the water column (compare del Giorgio and Cole 1998, Ducklow and Carlson 1992). In some areas, gross consumption by bacteria may reach 70% of NPP, especially when NPP is low (Biddanda et al. 1994).

Whereas zooplankton represent the first step in a trophic chain that eventually leads to large animals such as fish, bacteria are consumed by a large population of bacteriovores that mineralize nutrients and release CO_2 to the surface waters (Fuhrman and McManus 1984). Thus, when bacteria are abundant, a large fraction of the carbon fixed by NPP in the sea is not passed to higher trophic levels (Ducklow et al. 1986). In areas where bacterial growth is inhibited, such as in cold waters, more NPP is available to pass to higher trophic levels, including commercial fisheries (Pomeroy and Deibel 1986, Rivkin and Legendre 2001, Laws et al. 2000). Many of the world's most productive fisheries are found in cold, polar waters.

Fisheries' production is directly linked to the primary production in the sea (Iverson 1990, Ware and Thomson 2005). Already humans extract a large harvest of fish and shellfish from the oceans, which can be traced to the consumption of 8% of marine net primary productivity at the base of the food web (Pauly and Christensen 1995). Recent declines in the populations of important commercial fishes—up to 90% for some preferred species—suggest that it is unlikely that the current harvest is sustainable for future generations (Myers and Worm 2003). If warmer surface waters accompany climate change, we can also expect reductions in zooplankton and ultimately in the commercial harvest of fish (Vazquez-Dominquez et al. 2007).

There is general agreement among oceanographers that about 80 to 90% of the NPP is degraded to inorganic compounds (CO_2, NO_3, PO_4, etc.) in the surface waters, and the remainder sinks to the deep ocean. The sinking materials consist of particulate organic carbon (POC), including dead phytoplankton, fecal pellets, and organic aggregates, known as marine snow (Alldredge and Gotschalk 1990). The fraction (f_e) of net primary production that sinks out of the surface waters is known as *export production*—often called the *f*-ratio. Estimates of f_e are constrained, since high rates of sinking would remove unreasonably large quantities of nutrients from the surface ocean (Broecker 1974, Eppley and Peterson 1979).

The downward flux of organic matter varies seasonally depending on productivity in the surface water (Deuser et al. 1981, Asper et al. 1992, Sayles et al. 1994, Legendre 1998). Degradation continues as POC sinks through the water column of the deep ocean. With a mean sinking rate of about 350 m/day, the average particle spends about 10 days in transit to the bottom (Honjo et al. 1982). Bacterial respiration accounts for the consumption of O_2 and the production of CO_2 in the deep water.

As a result of progressive mineralization of the sinking debris, rates of oxygen consumption and bacterial activity show an exponential decline with depth below 200 m (Nagata et al. 2000, Andersson et al. 2004a). Honjo et al. (1982) found that respiration rates averaged 2.2 mg C m^{-2} day^{-1} in the deep ocean, where the rate of bacterial respiration is probably limited by cold temperatures. About 95% of the particulate carbon is degraded within a depth of 3000 m and only small quantities reach the sediments of the deep ocean (Suess 1980, Martin et al. 1987, Jahnke 1996, Hedges et al. 2001, Buesseler et al. 2007). Still, metabolically active bacteria are found at $\sim$11,000 m depth in the bottom waters of the Mariana Trench (Kato et al. 1998).

Significant rates of decomposition also continue at the surface of the sediments (Emerson et al. 1985, Cole et al. 1987, Bender et al. 1989, Smith 1992a), where the rate of decay is determined by the length of exposure of the organic matter to oxygen (Figure 9.9; Gelinas et al. 2001, Arnarson and Keil 2007). Where burrowing organisms stir or *bioturbate* the sediments, O_2 may penetrate to considerable depth (e.g., Ziebis et al. 1996, Lohrer et al. 2004), stimulating degradation of buried organic matter (Hulthe et al. 1998).

If the current, higher estimates of marine NPP are correct, then at least 7.4×10^{15} g C/yr (i.e., global $f_e = 0.15$) sinks to the deep waters of the ocean (Knauer 1993, Falkowski 2005). The export production includes both particulate organic carbon (5×10^{15} g C/yr; Henson et al. 2011) and dissolved organic carbon (Hopkinson and Vallino 2005). Seiter et al. (2005) suggest that 0.5×10^{15} g C pass 1000-m depth. From a compilation of data from sediment cores taken throughout the oceans, Berner (1982) estimated that the rate of incorporation of organic carbon in sediments is 0.157×10^{15} g C/yr. These values suggest that about 98% of the sinking organic materials are degraded in the deep sea (compare Martin et al. 1991). Degradation of organic carbon continues in marine sediments, and the ultimate rate of burial of organic carbon in the ocean is about 0.12×10^{15} g C/yr (Berner 1982, Seiter et al. 2005)—less than 1% of marine NPP.

Maps of the distribution of organic carbon in ocean sediments are similar to maps of the distribution of net primary production in the surface waters, except that a greater fraction of the total burial (83%) occurs on the continental shelf (Premuzic et al. 1982, Berner 1982). In most cases the delivery of materials to sediment (g C m^{-2} yr^{-1}) is correlated to measured rates of respiration in the sediment, linking accumulations of sedimentary carbon to surface productivity (Bertrand and Lailler-Verges 1993, Legendre 1998). Where this is not the case,

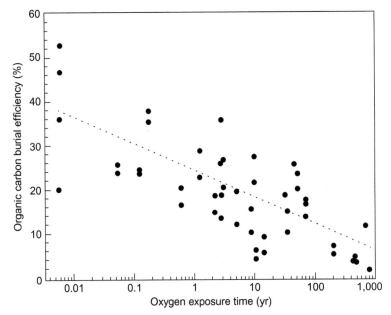

FIGURE 9.9   Organic carbon burial efficiency versus the time of its exposure to O_2 in sediments of the eastern North Pacific Ocean. *Source: From Hartnett et al. (1998).*

deposition from the episodic surface production or ancillary terrestrial sources is indicated (Smith et al. 1992a, 2001a). While near-shore sediments contain a large fraction of terrestrial material (Burdige 2005), isotopic analyses show that nearly all the sedimentary organic matter in the deep sea is derived from marine production and not from land (Hedges and Parker 1976, Prahl et al. 1994).

Indeed, most river-borne organic materials, both DOC and POC, must decompose in the ocean because the total burial of organic carbon in the ocean is less than the global delivery of organic matter in rivers, 0.4×10^{15} g C/yr (Schlesinger and Melack 1981). This has led to the curious suggestion that the ocean is a net heterotrophic system, because the ratio of total respiration to autochthonous production appears >1.0 (Smith and MacKenzie 1987, Serret et al. 2001, del Giorgio and Duarte 2002). However, net respiration is likely confined to near-shore regions and warm tropical waters, whereas photosynthesis and respiration are usually more closely balanced in the central ocean gyres (Williams 1998, Hoppe et al. 2002, Riser and Johnson 2008).

SEDIMENT DIAGENESIS

Organic Diagenesis

Slow metabolic rates characterize the inhabitants of sedimentary environments (D'Hondt et al. 2002, Røy et al. 2012), and archaea may be dominant in many of them (Lipp et al. 2008). Bacteria are found at 500-m depth in pelagic sediments (Parkes et al. 1994, Schippers et al.

2005, D'Hondt et al. 2004), and viable bacteria reported 1626 m below the seafloor near New-foundland, Canada, where ambient temperatures range from 60 to 100°C, currently represent the maximum known extent of the biosphere into the Earth's crust (Roussel et al. 2008). Given the age of some deep sediments (e.g., 111,000,000 years; Roussel et al. 2008), it is not surprising that all but the most refractory organic compounds have disappeared. Nevertheless, the live biosphere in sediments may contain as much as 100×10^{15} g C (Parkes et al. 1994, Lipp et al. 2008), equivalent to about 15% of the biomass on land (Chapter 5). Note that this living bio-mass represents only a small fraction of the total organic carbon contained in sedimentary environments (Hartgers et al. 1994; Table 2.3).

Change in the chemical composition of sediments after deposition is known as *diagenesis*. Many forms of diagenesis are the result of microbial activities that proceed following the or-der of redox reactions outlined in Chapter 7 (Thomson et al. 1993, D'Hondt et al. 2004). Within a few centimeters of the surface of sediments, NO_3^- and Mn^{4+} are exhausted by the anaerobic oxidation of organic matter. Supplementing the supply of NO_3^- that diffuses into the sediment from the overlying waters, some motile microbes appear to transport NO_3^- into the sediment, where it can be denitrified (Prokopenko et al. 2011). Sedimentary nitrogen can be oxidized anaerobically to NO_3^- (i.e., heterotrophic nitrification) in a coupling with the reduction of Mn^{4+} to Mn^{2+} (Hulth et al. 1999). In some cases, Mn^{3+} accumulates in Mn oxides in the sub-oxic zone, where it can act as an electron donor or an electron acceptor, depending on upward or downward shifts in redox potential (Anschutz et al. 2005, Trouwborst et al. 2006).

As in the case of freshwater sediments, the zone of Mn-reduction is underlain by zones of Fe-reduction and SO_4-reduction (Figure 9.10). Organic marine sediments undergo substantial diagenesis as a result of sulfate reduction (Froelich et al. 1979, Berner 1984). In marine

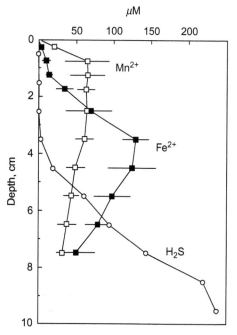

FIGURE 9.10 Pore water distribution of Mn^{2+}, Fe^{2+}, and H_2S in coastal sediments of Denmark, showing the approximate depth of Mn-reduction, Fe-reduction, and SO_4-reduction, respectively. *Source: From Thamdrup et al. (1994).*

environments, sulfate reduction leads to the release of reduced sulfur compounds (e.g., H_2S) and to the deposition of pyrite in sediments (Eqs. 7.19 and 7.20). The rate of pyrite formation is often limited by the amount of available iron (Boudreau and Westrich 1984, Morse et al. 1992), so only a small fraction of the sulfide is retained as pyrite and the remainder escapes to the upper layers of sediment where it is reoxidized (Jorgensen 1977, Thamdrup et al. 1994). When the rate of sulfate reduction is especially high, reduced gases may also escape to the water column. Other metabolic couplings, including anaerobic oxidation of pyrite by NO_3^-, Fe^{3+}, and Mn^{4+}, are seen when there is mixing of the sediment layers containing these substrates (Schippers and Jorgensen 2002).

Recently, oxidation/reduction reactions have been reported with electric currents transferring the electrons across local redox gradients in marine sediments (Nielsen et al. 2010). Mat-forming bacteria in the genus *Thiopioca* provide the spatial connection within their own cells, by capturing NO_3^- from the overlying water column and transferring it intracellularly to the sediments, where they oxidize sulfides produced by sulfate reduction at depth (Fossing et al. 1995). *Thiopioca* are related to the bacteria *Beggiatoa*, which oxidize upward diffusing hydrogen sulfide at the sediment surface.

In organic-rich sediments, sulfate reduction may begin within a few centimeters of the sediment surface where O_2 is depleted by aerobic respiration (e.g., Thamdrup et al. 1994). The various pathways of anaerobic metabolism are stimulated by increasing temperature with depth, which mobilizes acetate from buried organic matter (Wellsbury et al. 1997, Weston and Joye 2005). Simultaneously, adsorption of organic matter to clays and iron minerals tends to retard its degradation and lead to preservation (Keil et al. 1994, Kennedy et al. 2002a, Lalonde et al. 2012).

Globally, 9 to 14% of sedimentary organic carbon may be oxidized through anaerobic respiration, especially sulfate reduction (Lein 1984, Henrichs and Reeburgh 1987). The importance of sulfate reduction is much greater in organic-rich, near-shore sediments than in sediments of the open ocean (Skyring 1987, Canfield 1989b, 1991). Near-shore environments are characterized by high rates of NPP and a large delivery of organic particles to the sediment surface. Sulfate reduction generally increases with the overall rate of sedimentation, which is also greatest near the continents (Canfield 1989a, 1993).

Anoxic conditions develop rapidly as organic matter is buried in near-shore sediments. In a marine basin off the coast of North Carolina, Martens and Val Klump (1984) found that 149 moles C m^{-2} yr^{-1} were deposited, of which 35.6 moles were respired annually. The respiratory pathways included 27% in aerobic respiration, 57% in sulfate reduction leading to CO_2, and 16% in methanogenesis. In contrast, in pelagic sediments of the Pacific Ocean, net carbon burial was only 0.005 moles C m^{-2} yr^{-1} (D'Hondt et al. 2004). Near-shore environments promote the hydrogenation of sedimentary organic carbon, often by H_2S from sulfate-reducing bacteria (Hebting et al. 2006). These reduced organic residues are more resistant to decay and likely precursors to the formation of fossil petroleum (Gelinas et al. 2001).

The rate of burial of organic carbon depends strongly on the sedimentation rate (Figure 9.11; Muller and Suess 1979, Betts and Holland 1991). Greater preservation of organic matter in near-shore environments is likely to be due to the greater NPP in these regions (Bertrand and Lallier-Vergès 1993), rapid burial (Henrichs and Reeburgh 1987, Canfield 1991), and somewhat less efficient decomposition under anoxic conditions (Canfield

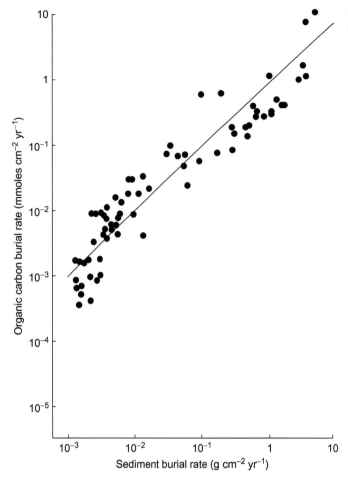

FIGURE 9.11 Burial of organic carbon in marine sediments as a function of the overall rate of sedimentation. *Source: From Berner and Canfield (1989). Reprinted by permission of* American Journal of Science.

1994, Kristensen et al. 1995). As seen in soils (Chapter 5), the long-term persistence of organic matter in marine sediments is also enhanced by association with mineral surfaces (Keil et al. 1994, Mayer 1994).

In contrast, pelagic areas have lower NPP, lower downward flux of organic particles, and lower overall rates of sedimentation. The sediments in these areas are generally oxic (Murray and Grundmanis 1980, Murray and Kuivila 1990), so aerobic respiration exceeds sulfate reduction by a large factor (Canfield 1989b, Hartnett and Devol 2003). Little organic matter remains to support sulfate reduction at depth (Berner 1984). Among near-shore and pelagic habitats, there is a strong positive correlation between the content of organic carbon and pyrite sulfur in sediments (Figure 9.12), but it is important to remember that the deposition of pyrite occurs at the expense of organic carbon, namely,

$$8FeO + 16CH_2O + 16SO_4^{2-} \rightarrow 15O_2 + 8FeS_2 \downarrow + 16HCO_3^- + 8H_2O. \qquad (9.2)$$

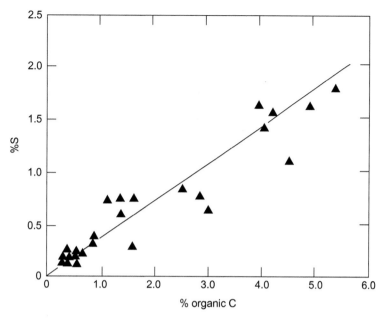

FIGURE 9.12 Pyrite sulfur content in marine sediments as a function of their organic carbon content. *Source: From Berner (1984).*

Thus, the net ecosystem production of marine environments is represented by the *total* of sedimentary organic carbon + sedimentary pyrite—with the latter resulting from the transformation of organic carbon to reduced sulfur (Eq. 9.2).

Permanent burial of reduced compounds (organic carbon and pyrite) accounts for the accumulation of O_2 in Earth's atmosphere (Chapter 3). The molar ratio is 1.0 for organic carbon, but the burial of 1 mole of reduced sulfur accounts for nearly 2.0 moles of O_2 (Raiswell and Berner 1986, Berner and Canfield 1989). The weight ratio of C/S in most marine shales is about 2.8—equivalent to a molar ratio of 7.5 (Raiswell and Berner 1986). Thus, through geologic time the deposition of reduced sulfur in pyrite may account for about 20% of the O_2 in the atmosphere. During periods of rapid continental uplift, erosion, and sedimentation, large amounts of organic substances and pyrite were buried and the oxygen content of the atmosphere increased (Des Marais et al. 1992). Rising atmospheric O_2 increases aerobic decomposition in marine sediments, consuming O_2 and limiting the further growth of O_2 in the atmosphere (Walker 1980).

In Chapter 7 we saw that redox potential controls the order of anaerobic metabolism by microbes in sediments. The zone of methanogenesis underlies the zone of sulfate reduction, because the sulfate-reducing bacteria are more effective competitors for reduced substrates. As a result of high concentrations of SO_4 in seawater, methanogenesis in ocean sediments is limited (Oremland and Taylor 1978, Lovley and Klug 1986, D'Hondt et al. 2002). Nearly all methanogenesis is the result of CO_2 reduction, because normally acetate is depleted before SO_4 is fully removed from the sediment (Sansone and Martens 1981, Crill and Martens

FIGURE 9.13   Methane volatilized from a frozen clathrate can be burned at the Earth's surface. *Source: Photo by Gary Klinkhammer, courtesy of NASA.*

1986, Whiticar et al. 1986). There is, however, some seasonal variation in the use of CO_2 and acetate that appears to be due to microbial response to temperature (Martens et al. 1986).

In some sediments, methane produced in cold, high-pressure conditions crystallizes with water to form methane hydrates or *clathrates*, which are unstable and volatilize CH_4 when brought to the surface of the Earth (Zhang et al. 2011b; Figure 9.13). There is great interest in clathrates as a commercial source of natural gas as well as great concern that a catastrophic degassing of clathrates in response to global warming might release vast quantities of methane to the atmosphere, exacerbating further warming (Archer et al. 2009). Evidence for large releases of methane from clathrates during climate warming at the end of the last glacial epoch is controversial (Kennett et al. 2000; compare Petrenko et al. 2009, Sowers 2006).

When methane produced at depth diffuses upward through the sediment, it is subject to anaerobic oxidation by methanotrophs (AOM) that use SO_4^{-2}, Mn^{4+}, and Fe^{3+} as alternative electron acceptors in the absence of O_2 (Reeburgh 2007, Beal et al. 2009). Some anaerobic methanotrophs are archaea that appear to coexist in consortia with sulfate-reducing bacteria (Hinrichs et al. 1999, Boetius et al. 2000, Michaelis et al. 2002). When sediments are low in organic matter, the rate of sulfate reduction may be determined solely by the upward flux of methane that provides an organic substrate for metabolism (Hensen et al. 2003, Sivan et al. 2007). Methane-consuming archaea are known to fix nitrogen to support their growth in deep sediments, where nitrate has been depleted by denitrification (Dekas et al. 2009).

Methane released from ocean sediments, natural seeps, and hydrothermal vents is easily oxidized by microbes before it reaches the surface (Iversen 1996). For example, a large amount of natural gas associated with the blow out of the Deep-Water Horizon oil well in the Gulf of Mexico was apparently oxidized before it reached the surface (Kessler et al. 2011).

Nevertheless, many areas of the surface ocean are supersaturated in CH_4 with respect to the atmosphere (Ward et al. 1987, Kelley and Jeffrey 2002). This methane appears to result from methanogenesis in decomposing, sinking particles (Scranton and Brewer 1977, Burke et al. 1983, Karl and Tilbrook 1994, Holmes et al. 2000). Methane is not highly soluble in seawater, but the global flux of CH_4 from the oceans to the atmosphere, $\sim 10 \times 10^{12}$ g/yr, is small compared to that from other sources (Liss and Slater 1974, Conrad and Seiler 1988, Reeburgh 2007; Table 11.2).

Biogenic Carbonates

A large number of marine organisms precipitate carbonate in their skeletal and protective tissues by the reaction:

$$Ca^{2+} + 2HCO_3^- \rightarrow CaCO_3 \downarrow + H_2O + CO_2. \tag{9.3}$$

Clams, oysters, and other commercial shellfish are the obvious examples, but a much larger quantity of $CaCO_3$ is produced by foraminifera, pteropods, and other small zooplankton that are found in the open ocean (Krumbein 1979, Simkiss and Wilbur 1989). Some carbonate formed in the guts of fish is carried to the deep sea in their fecal pellets (Wilson et al. 2009). Coccolithophores, a group of marine algae, are responsible for a large amount of $CaCO_3$ deposited on the seafloor of the open ocean.[4] The annual production of $CaCO_3$ by these organisms is much larger than what could be sustained by the supply of Ca to the oceans in river flow (Broecker 1974, Feely et al. 2004). However, not all of the $CaCO_3$ produced is stored permanently in the sediment.

Recall that CO_2 is produced in the deep ocean by the degradation of organic materials that sink from the surface waters. Deep ocean waters are supersaturated with CO_2 with respect to the atmosphere as a result of their long isolation from the surface and the progressive accumulation of respiratory CO_2. Carbon dioxide is also more soluble at the low temperatures and high pressures that are found in deep ocean water. (Note that CO_2 effervesces when the pressure of a warm soda bottle is released on opening). The accumulation of CO_2 makes the deep waters undersaturated with respect to $CaCO_3$, as a result of the formation of carbonic acid:

$$H_2O + CO_2 \rightarrow H^+ + HCO_3^- \rightarrow H_2CO_3. \tag{9.4}$$

When the skeletal remains of carbonate-producing organisms sink to the deep ocean, they dissolve:

$$CaCO_3 + H_2CO_3 \rightarrow Ca^{2+} + 2HCO_3^-. \tag{9.5}$$

Their dissolution increases the alkalinity, roughly equivalent to the concentration of HCO_3^-, in the deep ocean. Small particles may dissolve totally during transit to the bottom, while large particles may survive the journey, and their dissolution occurs as part of sediment diagenesis (Honjo et al. 1982, Berelson et al. 1990). Degradation of organic matter in marine sediments, by aerobic or anaerobic processes, generates acidity, furthering the rate of dissolution of biogenic

---

[4] Note that the precipitation of carbonate by phytoplankton supplies some of the CO_2 needed for photosynthesis, reducing the net uptake of CO_2 from seawater (Robertson et al. 1994).

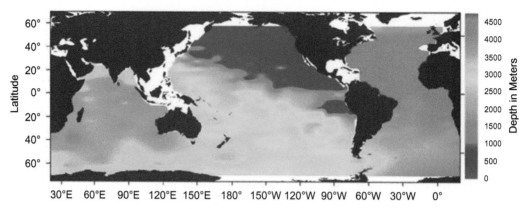

FIGURE 9.14   Calcite saturation depth in the world's oceans. *Source: Feely et al. (2004). Used with permission of the American Association for the Advancement of Science.*

carbonates that are coprecipitated (Jahnke et al. 1997, Ku et al. 1999, Berelson et al. 1990, Wenzhofer et al. 2001).

The depth at which the dissolution of $CaCO_3$ begins in the water column is called the carbonate lysocline and is an index of the *carbonate saturation depth* (CSD), the depth at which seawater is undersaturated with respect to $CaCO_3$. This depth is roughly 3000 m in the southern Pacific and 4000 to 4500 m in the Atlantic Ocean (Biscaye et al. 1976, Berger et al. 1976; Figure 9.14). Slightly deeper, the *carbonate compensation depth* (CCD) is the depth where the downward flux of carbonate balances the rate of dissolution, so there are no carbonate sediments (Kennett 1982). The tendency for a shallower saturation depth and CCD in the Pacific is the result of the older age of Pacific deep water, which allows a greater accumulation of respiratory CO_2 (Li et al. 1969). Dissolution of sinking $CaCO_3$ means that calcareous sediments are found only in shallow ocean basins, and no carbonate sediments are found over much of the pelagic area where the ocean is greater than 4500 m deep. About 10×10^{15} g/yr of $CaCO_3$ are produced in the surface water, and about 0.8×10^{15} g are preserved in deep-ocean sediments (Berelson et al. 2007, Feely et al. 2004).

Added to the estimated carbonate preservation in shallow-water sediments (2.2×10^{15} g; Milliman 1993), this estimate of total carbonate deposition consumes more than the estimated flux of Ca to the oceans, suggesting that the Ca budget of the oceans is not currently in steady state. Today, the preservation ratio of organic carbon to carbonate carbon in ocean sediments is about 0.20 by weight, close to the ratio in the Earth's sedimentary inventory (0.26 in Table 2.3).

Many studies of carbonate dissolution have employed sediment traps that are anchored at varying depths to capture sinking particles. In most areas, biogenic particles constitute most of the material caught in sediment traps, and most of the $CaCO_3$ is found in the form of calcite. Pteropods, however, deposit an alternative form of $CaCO_3$, known as aragonite, in their skeletal tissues. As much as 12% of the movement of biogenic carbonate to the deep ocean may occur as aragonite (Berner and Honjo 1981, Betzer et al. 1984). The downward movement of aragonite has been long overlooked because it is more easily dissolved than calcite and often disappears from sediment traps that are deployed for long periods. The carbonate lysocline for aragonite is found at 500- to 1000-m depth (Milliman et al. 1999, Feely et al. 2004).

Geochemists have long puzzled that dolomite—$(Ca, Mg)CO_3$—does not appear to be deposited abundantly in the modern oceans, despite the large concentration of Mg in seawater and the occurrence of massive dolomites in the geologic record. There are few organisms that precipitate Mg calcites in their skeletal carbonates, but thermodynamic considerations would predict that calcite should be converted to dolomite in marine sediments (e.g., Malone et al. 1994). Baker and Kastner (1981) show that the formation of dolomite is inhibited by SO_4^{2-}, but dolomite can form in organic-rich marine sediments in which HCO_3^- is enriched and SO_4^{2-} is depleted by sulfate reduction (Eq. 9.2; Baker and Burns 1985). Dolomite is precipitated in laboratory cultures of the sulfate-reducing bacterium *Desulfovibrio* (Vasconcelos et al. 1995). Thus, the precipitation of dolomite is directly linked to biogeochemical processes in marine sediments. Although dolomite has been a significant sink for marine Mg in the geologic past, its contribution to the removal of Mg from modern seawater is likely to be minor.

Rising CO_2 in Earth's atmosphere, as it dissolves in seawater, will raise the acidity of seawater (Eq. 9.4), potentially increasing the dissolution of biogenic carbonates (Eq. 9.5; Hoffman and Schellnhuber 2010). Already, large-scale observations suggest rising dissolved CO_2 and lower pH in the surface of the Pacific Ocean, amounting to a drop of 0.06 in pH during the past 15 years (Takahashi et al. 2006, Byrne et al. 2010; compare Dore et al. 2009). Losses of corals, which are largely built from aragonite, are likely to be the most immediate consequence (Kleypas et al. 1999, Hoegh-Guldberg et al. 2007, Feely et al. 2012), but higher seawater acidity could affect the ability of a wide variety of plankton to build carbonate skeletons (Riebesell et al. 2000, Orr et al. 2005). Aragonite-depositing species are likely to be the most severely affected (Gruber et al. 2012). Differential responses among species can be expected (Iglesias-Rodriguez et al. 2008), disrupting the current food web of marine ecosystems. Of course, the drop in seawater pH is buffered by the dissolution of carbonates (Eq. 9.5), but unregulated CO_2 emissions could lead to a decline of seawater pH of 0.7 units in the next several centuries—a greater change than any observed during the past 300,000,000 years (Zeebe et al. 2008, Hönisch et al. 2012).

THE BIOLOGICAL PUMP: A MODEL OF CARBON CYCLING IN THE OCEAN

CO_2 dissolves in seawater as a function of the concentration of CO_2 in the atmosphere. (Recall Henry's Law, Eq. 2.7.) The rate of dissolution increases with wind speed, which increases the turbulence of the surface waters[5] and the downward transport of bubbles (Watson et al. 1991, Wanninkhof 1992, Farmer et al. 1993). As it dissolves in water, CO_2 disassociates to form bicarbonate, following Eqs. 2.4 and 9.3 (Archer 1995). The solubility of CO_2 in seawater depends on temperature; CO_2 is about twice as soluble at 0°C as it is at 20°C (Broecker 1974).

The temperature of the upper 1 mm of the ocean's surface, the "skin" temperature, is critical to determining the atmosphere-to-ocean flux. Over much of the oceans' surface, the skin

[5] The term "piston velocity" is often used to describe the mixing of gases with seawater. A piston velocity of 5 m/day for CO_2 implies that atmosphere equilibrates with the amount of gas in the upper 5 m of seawater—as if pushed in by a piston—each day.

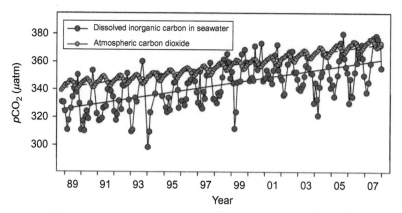

FIGURE 9.15   Dissolved carbon dioxide in seawater (dissolved inorganic carbon + total alkalinity) and pCO₂ in Earth's atmosphere at Mauna Loa, Hawaii, since 1989. *Source: From Dore et al. (2009). Used with permission of the National Academy of Sciences.*

temperature is about 0.3°C cooler than the underlying waters as a result of evaporation of water from the oceans' surface (Robertson and Watson 1992). Except in areas of upwelling, the surface waters of the oceans are undersaturated with respect to CO_2 in the atmosphere (Takahashi et al. 1997), allowing a net flux of CO_2 into the ocean. Indeed, surface waters show increasing concentrations of dissolved inorganic carbon during the past few decades presumably due to rising CO_2 in Earth's atmosphere (Figure 9.15; Takahashi et al. 2006, Inoue et al. 1995, Peng et al. 1998, Sabine et al. 2004, Dore et al. 2009).

Most CO_2 enters the deep oceans with the downward flux of cold water at polar latitudes. When cold waters form in equilibrium with an atmosphere of 400 ppm CO_2 (i.e., today), they carry more CO_2 than when they formed in equilibrium with an atmosphere of 280 ppm CO_2—the historical origin of most of today's deep waters that are 300 to 500 years old. Brewer et al. (1989) report that North Atlantic deep water now carries a *net* flux of 0.26×10^{15} g C/yr southward because of to the global rise in atmospheric CO_2 during this century.

In addition, the surface waters in many areas of the oceans are variably undersaturated in CO_2 as a result of photosynthesis. Sinking organic materials remove carbon from the surface ocean, and it is replaced by the dissolution of new CO_2 from the atmosphere. Taylor et al. (1992) found that during a 46-day period there was a net downward transport of carbon in the northeast Atlantic Ocean due to the sinking of live (2 g C/m²) and dead (17 g C/m²) cells and the downward mixing of living cells by turbulence (3 g C/m²). Thus, biotic processes act to convert inorganic carbon (CO_2) in the surface waters to organic carbon that is delivered to the deep waters of the ocean.

As we have seen, the storage of organic carbon in sediments accounts for <1% of marine NPP, so most of the organic carbon that sinks is liberated by bacterial respiration in the deep ocean and released back to the atmosphere centuries later as CO_2 in zones of upwelling. Nevertheless, in the absence of a marine biosphere, the atmospheric CO_2 concentration would be much higher than today—perhaps as high as 470 ppm (Broecker and Peng 1993). More marine NPP and a more active "biotic pump" are postulated explanations for the lower concentrations of atmospheric CO_2 during the last glacial epoch (Broecker 1982, Paytan et al. 1996, Kumar et al. 1995a).

The biotic pump delivers other forms of carbon to the deep sea. Recall the production and leakage of DOC from phytoplankton in the surface waters. When DOC is entrained in downwelling water, it is delivered to the deep sea, where some of it is respired. As a result of deepwater circulation, concentrations of DOC are greatest in the Atlantic Ocean and lower in the Indian and Pacific Oceans, in which the deep waters are older (Hansell and Carlson 1998a).

Finally, the production and sinking of $CaCO_3$ also delivers carbon to the deep ocean. Most of the Ca^{2+} is derived from weathering on land and is balanced in riverwater by $2HCO_3^-$ (Figure 1.3). Whether it is preserved in shallow-water calcareous sediments or sinks to the deep ocean, each molecule of $CaCO_3$ carries the equivalent of one CO_2 and leaves behind the equivalent of one CO_2 in the surface ocean (Eq. 9.3). Globally, the carbon sink in sedimentary $CaCO_3$ is about four times larger than the sink in organic sediments (Li 1972). Near-shore environments contain most of the sedimentary storage of $CaCO_3$ and organic carbon; $CaCO_3$ delivered to the deep sea dissolves by reaction with H^+, producing calcium and bicarbonate that return to the surface waters in zones of upwelling (Eq. 9.5; Berelson et al. 2007).

Thus, the biotic pump consists of three processes that remove carbon dioxide from the atmosphere—the fraction of NPP that sinks, downwelling of DOC, and sinking of carbonate skeletal debris (Figure 9.16). These *bio*geochemical processes are superimposed on a much larger, background flux of CO_2 that enters the oceans by dissolution of CO_2 in cold downwelling waters.[6] The biotic pump responds to changes in the activity of the biosphere, whereas net downwelling of dissolved CO_2 responds only to changes in the concentration of CO_2 in the atmosphere and the circulation of deep ocean waters.

Equilibrium with ocean waters controls the concentration of CO_2 in the atmosphere, but the equilibrium can be upset when changes in CO_2 in the atmosphere exceed the rate at which the ocean system can buffer the concentration. The seasonal cycle of terrestrial photosynthesis and the burning of fossil fuels are two processes that affect the concentration of atmospheric CO_2 more rapidly than the ocean can react. As a result we observe a seasonal oscillation of atmospheric CO_2 and an exponential increase in its concentration in the atmosphere (Figure 1.1). Given enough time, the oceans could take up nearly all of the CO_2 released from fossil fuels, and the atmosphere would once again show stable concentrations at only slightly higher levels than today (Laurmann 1979, Archer et al. 1998). As the oceans take up CO_2, they become more acid, with the acidity neutralized by the dissolution of marine carbonates.

A large number of models have been developed to explain the response of the ocean to higher concentrations of atmospheric CO_2 (Bacastow and Björkström 1981, Emanuel et al. 1985b). Most of these models are constructed to follow parcels of water as they circulate in a simplified ocean basin and to calculate the diffusion of CO_2 between layers that do not mix directly. Figure 9.17 on page 369 shows a multibox model in which the surface ocean is divided into cold polar waters and warmer waters. In this model, cold waters mix downward to eight layers of the deep ocean, while upwelling returns deep water to the surface, where it releases CO_2 to the atmosphere. The rate of mixing is calculated using oceanographic data for the rate at which ^{14}C and ^3H_2O from atomic bombs have entered the oceans (Killough and Emanuel 1981) and known constants for the dissolution of CO_2 in water as a function of

---

[6] In discussions of the biotic pump, some workers refer to the sinking of NPP as the "soft tissue pump" and the sinking of carbonates as the "hard tissue pump," both distinguished from the solubility pump, derived from the dissolution of CO_2 in downwelling waters.

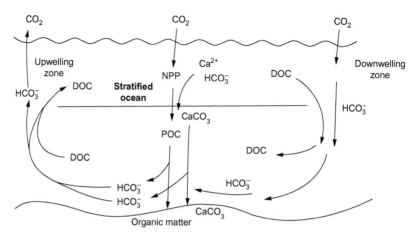

FIGURE 9.16 The marine biotic pump, showing the formation of organic matter (POC) and carbonate skeletons in the surface ocean and their downward transport and the downwelling of DOC and bicarbonate to the deep ocean.

temperature and pressure (Sundquist et al. 1979, Archer 1995). The models then adjust the chemistry of the water in each layer according to the carbonate equilibrium reactions given above.

As atmospheric carbon dioxide increases, we would expect an increased dissolution of CO_2 in the oceans, following Henry's Law. However, the surface ocean provides only a limited volume for CO_2 uptake, and the atmosphere is not in immediate contact with the much larger volume of the deep ocean. In the absence of large changes in NPP, it is the rate of formation of bottom waters in polar regions that limits the rate at which the oceans can take up CO_2. Reductions in the upwelling and CO_2 degassing of deep waters in the Southern Ocean may have had a large impact on the low concentrations of atmospheric CO_2 during the last glacial epoch (Francois et al. 1997, Kohfeld et al. 2005, Sigman et al. 2010; compare Kumar et al. 1995a). Conversely, greater upwelling may have led to rising CO_2 and warmer temperatures at the end of the last glacial period (Burke and Robinson 2012).

NUTRIENT CYCLING IN THE OCEAN

Net primary productivity in the sea is limited by a scarcity of nutrients. Production is highest in regions of high nutrient availability—the continental shelf and regions of upwelling (Figure 9.8)—and lower in the open ocean, where the concentrations of available N, P, Fe, and Si are normally very low. In most areas of the ocean, nitrate is not measurable in surface waters, and phytoplankton respond to nanomolar additions of nitrogen to seawater (Glover et al. 1988). Nutrients are continuously removed from the surface water by the downward sinking of dead organisms and fecal pellets. Working off the California coast, Shanks and Trent (1979) found that 4 to 22% of the nitrogen contained in particles (PON) was removed from the surface waters each day. The mean residence time of N, P, and Si in the surface ocean is much less than the mean residence time of the surface waters, and there are wide differences in the concentration of these elements between the surface and the deep ocean

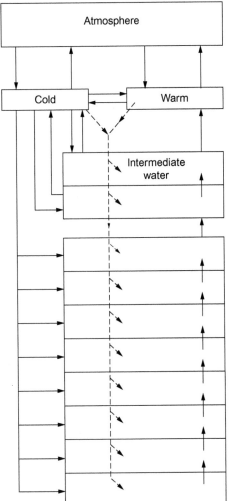

FIGURE 9.17 A box-diffusion model for the oceans, separating the surface oceans into cold polar waters and warmer waters at other latitudes. Cold polar waters mix with deeper waters as a result of downwelling. Other exchanges are by diffusion. *Source: From Emanuel et al. (1985a).*

(Figure 9.18). N, P, and Si are nonconservative elements in seawater; their behavior is strongly controlled by the presence of life.

Nutrients are regenerated in the deep ocean, where the concentrations are much higher than in surface waters. Recalling that the age of deep water in the Pacific Ocean is older than that in the Atlantic, we note that nutrient concentrations are higher in the deep Pacific Ocean (Figure 9.18), because its waters have had a longer time to receive sinking debris that are remineralized at depth. Similarly, in the Atlantic Ocean, nutrient concentrations increase progressively as North Atlantic deep water "ages" during its journey southward (Figure 9.19). Nutrients are also remineralized from DOC that mixes into the deep sea (Hopkinson and Vallino 2005). Nutrients are returned to the surface waters in the upwelling zones of the global thermohaline circulation (Figure 9.3; Sarmiento et al. 2004).

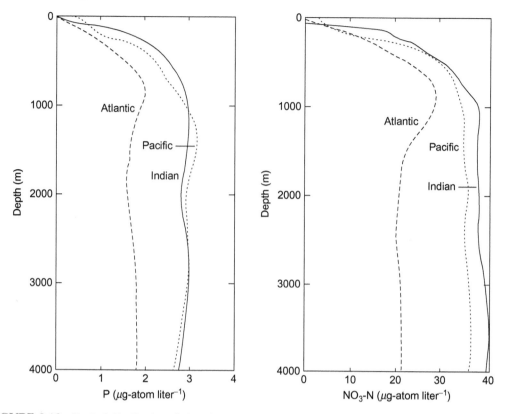

FIGURE 9.18   Vertical distribution of phosphate and nitrate in the world's oceans. *Source: From Svedrup et al. (1942).*

Internal Cycles

In 1958, Albert Redfield published a paper that has served as a focal point in marine bio-geochemistry for the past 50+ years. Redfield noted that marine phytoplankton contain N and P in a fairly constant molar ratio to the content of carbon, 106 C : 16 N : 1P (Redfield et al. 1963),[7] as a result of the incorporation of these elements in photosynthesis and growth:

$$106CO_2 + 16NO_3^- + HPO_4^{2-} + 122H_2O + 18H^+ \rightarrow$$
$$(CH_2O)_{106}(NH_3)_{16}(H_3PO_4) + 138O_2\uparrow. \tag{9.6}$$

The molar N/P ratio of 16 may reflect fundamental relationships between the requirements for protein and RNA synthesis common to all plants (Loladze and Elser 2011). Despite differences in nutrient concentration among the major oceans (Figure 9.18), upwelling waters contain available C, N, and P (i.e., HCO_3^-, NO_3^-, and HPO_4^{2-}) in the approximate ratio of

---

[7] Equivalent to a mass ratio of 40 C : 7 N : 1P. Thus, the N/P mass ratio in marine phytoplankton is lower than that in terrestrial plants ($\sim$15; Chapter 6).

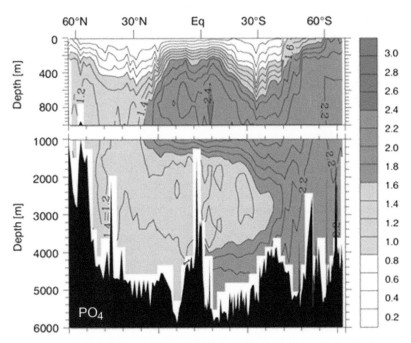

FIGURE 9.19 Phosphorus in the Atlantic Ocean, showing the increase in its concentration in deep waters as they travel from north to south. *Source: From Sarmiento and Gruber (2006). Used with permission of Princeton University Press.*

800 C : 16 N : 1P. Thus, even in the face of the high productivity found in upwelling waters, only about 10% of the HCO_3^- can be consumed by photosynthesis before the N and P are exhausted. Significantly, Redfield (1958) noted that the biota determined the relative concentrations of N and P in the deep sea, and that the biotic demand for N and P was closely matched to the availability of these elements in upwelling waters (Holland 1978).

Recognizing that the downward flux of biogenic particles carries $CaCO_3$ as well as organic carbon, Broecker (1974) recalculated Redfield's ratios to include $CaCO_3$. His modified Redfield ratio in sinking particles is 120 C : 15 N : 1P : 40Ca. The ratio in upwelling waters is 800 C : 15 N : 1P : 3200Ca. Based on these quantities, net production in the surface water could remove all the N and P but only 1.25% of the Ca in upwelling waters. Although biogenic $CaCO_3$ is the main sink for Ca in the ocean, the biota exerts only a tiny control on the availability of Ca in surface waters. Thus, calcium is a well-mixed and conservative element in seawater (Table 9.1).

It is important to remember that the Redfield ratio is an average value. The nutrient concentrations in individual plankton species may differ from the Redfield ratio depending on season and environmental conditions (Klausmeier et al. 2004, Weber and Deutsch 2010). Nevertheless, as an average value, the Redfield ratio allows us to compare the importance of riverflow, upward transport, and internal recycling for their contributions to the annual net primary production of the surface ocean. To sustain a global marine NPP of 50×10^{15} g C/yr (Table 9.2), phytoplankton must take up about 8.8×10^{15} g N and 1.2×10^{15} g P each year (Table 9.3). Rivers supply about 0.050×10^{15} g N/yr and 0.002×10^{15} g/yr of

TABLE 9.3   Calculation of the Sources of Nutrients That Would Sustain a Global
             Net Primary Productivity of 50×10^{15} g C/yr in the Surface
             Waters of the Oceans

Flux	Carbon (10^{12} g)	Nitrogen (10^{12} g)	Phosphorus (10^{12} g)
New primary production[a]	50,000	8800	1200
Amounts supplied			
By rivers[b]		50	2
By atmospheric deposition[c]		67	1
By N fixation[d]		150	—
By upwelling		700	100
Recycling (by difference)		7800	1100

Note: Values taken from Figures 9.21 and 9.22, with rounding. Based on an approach developed by Peterson (1981).
[a] *Assuming a Redfield atom ratio of 106:16:1.*
[b] *N from Galloway et al. (2004); P from Meybeck (1982).*
[c] *Duce et al. (2008).*
[d] *Deutsch et al. (2007).*

reactive P to the oceans (Chapters 8 and 12). However, the total nutrient supply from rivers, atmospheric inputs, and vertical movements (upwelling + diffusion + eddy convection) provides only a small fraction (11% for N and 9% for P) of the total nutrient requirement in the surface ocean, so nutrient recycling in the surface waters must supply the rest. Rapid turnover of nutrients is consistent with the rapid turnover of 80 to 90% of the organic carbon in the surface ocean.

In the face of nutrient-limited growth and efficient nutrient uptake, phytoplankton maintain very low concentrations of N and P in surface waters (Figure 9.18). McCarthy and Goldman (1979) showed that much of the nutrient cycling in the surface waters may occur in a small zone, perhaps in a nanoliter (10^{-9} liter) of seawater, which surrounds a dying phytoplankton cell. Growing phytoplankton in the immediate vicinity are able to assimilate the nitrogen as soon as it is released. Often it is difficult to study nutrient cycling on such a small scale, but various workers have applied isotopic tracers (e.g., $^{15}NH_4$ and $^{15}NO_3$) to measure nutrient uptake by phytoplankton and bacteria (Glibert et al. 1982, Goldman and Glibert 1982, Dickson and Wheeler 1995). Leakage of dissolved organic nitrogen compounds (DON) from phytoplankton may also account for a significant amount of the bacterial uptake and turnover of nitrogen in the surface waters (Kirchman et al. 1994, Bronk et al. 1994, Kroer et al. 1994). During decomposition of organic particles in the surface ocean, nitrogen is mineralized more rapidly than carbon, so that surviving particles carry C/N ratios that are somewhat greater than the Redfield ratio (Sambrotto et al. 1993) and that increase with depth (Honjo et al. 1982, Takahashi et al. 1985, Anderson and Sarmiento 1994, Alldredge 1998, Schneider et al. 2003).

Nutrient demand by phytoplankton is so great that it has been traditional to assume that little of the NH_4 released by mineralization remains for nitrification in the surface waters, and NH_4 dominates phytoplankton uptake of recycled N (Dugdale and Goering 1967; Harrison et al. 1992, 1996; but see Yool et al. 2007). In contrast, most of the nitrogen mineralized in the deep ocean is converted to NO_3—some by nitrifiers recently recognized as archaea (Konneke et al. 2005, Francis et al. 2005).[8] Nitrate also dominates the nitrogen supply in rivers, so oceanographers can use the fraction of NPP that derives from the uptake of NH_4 versus that derived from NO_3 to estimate the sources of nutrients that sustain NPP in the surface waters (Figure 9.20).

For example, Jenkins (1988) estimated that the upward flux of NO_3 from the deep ocean near Bermuda would support an NPP of about $36\,g\,C\,m^{-2}\,yr^{-1}$—about 38% of observed NPP (Michaels et al. 1994). The remaining production must depend on NH_4 supplied by recycling in the surface waters. The fraction of NPP that is sustained by nutrients delivered from atmospheric inputs (including N fixation), rivers, and upwelling is known as *new production*. Globally, new production is about 10 to 20% of total NPP, but the fraction, fn, is greatest in areas of cold, upwelling waters (Sathyendranath et al. 1991).

Rivers dominate the sources of N in many coastal waters, but in the pelagic oceans, the relative role of nutrient recycling in the surface waters and nutrient delivery by upwelling depends strongly on location. Outside of the major areas of thermohaline upwelling (Figure 9.3), nutrients are delivered to the surface waters by convection in eddies (McGillicuddy et al. 1998, Oschlies and Garcon 1998, Siegel et al. 1999, Johnson et al. 2010) and other processes of large-scale mixing (Uz et al. 2001) and advection (Palter et al. 2005). The vertical migration of diatoms is also known to transport nitrate to the surface (Villareal et al. 1999), and the vertical movements of whales transport nutrients from great depths, where whales feed, to the surface waters, where they defecate (Roman and McCarthy 2010).

Low, steady-state nutrient concentrations in the surface waters of the oceans—near the Redfield N/P ratio of 16—indicate that the sources of nutrients that sustain new production globally are about equal to the annual losses of nutrients in organic debris that sink through the thermocline to the deep ocean—that is, export production (f_e) (Eppley and Peterson 1979). However, it is now recognized that the traditional separation of new and recycled production based solely on the form of nitrogen taken up (NO_3^- vs. NH_4^+, respectively) is complicated by significant nitrogen fixation in the marine environment, atmospheric deposition of NH_4, and the presence of nitrifying bacteria and archaea that produce NO_3 in the surface waters (Yool et al. 2007, Martens-Habbena et al. 2009).

Air–Sea Exchange of Nitrogen

In many pelagic waters, nitrogen fixation provides inputs of nitrogen for new production. Observations of significant nitrogen fixation in the pelagic oceans are relatively new, stemming from the widespread occurrence of filamentous cyanobacteria, *Triochodesmium*, in warm, tropical waters (Carpenter and Romans 1991; Capone et al. 1997, 2005; Davis and McGillicuddy 2006) and unicellular cyanobacteria in other areas (Zehr et al. 2001, Montoya et al. 2004; Moisander et al. 2010).

---

[8] Nitrification in the deep sea is a form of chemoautotrophy (Eqs. 2.17 and 2.18) in which carbon is fixed in the dark. Chemoautotrophic nitrification and sulfide oxidation in the oceans is estimated to account for the addition of 0.77 Pg C/yr or 1.5% to NPP (Middelburg 2011).

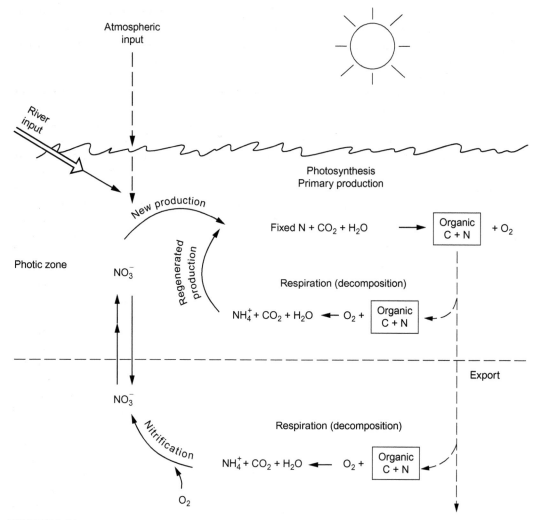

FIGURE 9.20   Links between the nitrogen and the carbon cycles in the surface ocean. Nitrogen regenerated in the surface waters is largely assimilated by phytoplankton as NH_4, while that diffusing and mixing up from the deep ocean is NO_3. When organic matter sinking to the deep ocean is mineralized, its nitrogen content is initially released as NH_4 and converted to nitrate by nitrifying bacteria. "New production" can be estimated as the fraction of net primary production that is derived from nitrate from rivers, atmospheric deposition, nitrogen fixation, and upwelling from the deep sea. *Source: From Jahnke (1990).*

External Inputs

Recalling that the nitrogen-fixing enzyme, nitrogenase, contains iron (Fe) and molybdenum (Mo) in its molecular structure (Chapter 2), the rate of nitrogen fixation in the oceans may be limited by the availability of these elements (Karl et al. 2002). Mo is a well-mixed and conservative element in seawater, but its uptake by phytoplankton appears inhibited by high concentrations of SO_4^{2-} (Howarth and Cole 1985). Fe is largely delivered to the pelagic oceans by the deposition of desert dusts, which also provide P and Si to surface waters (Jickells et al. 2005). Iron appears to limit nitrogen fixation over much of the pelagic oceans, where Fe inputs are low (Falkowski 1997, Berman-Frank et al. 2001a, Mills et al. 2004; compare Moore et al. 2009, Okin et al. 2011).

Paulsen et al. (1991) found that additions of carbohydrates to seawater stimulated N fixation and postulated that these compounds created local zones of active decomposition, where oxygen is depleted, so nitrogenase activity is possible. Natural aggregations of organic matter, forming "marine snow," create small microzones of anoxic conditions in seawater, in which a greater availability of trace micronutrients and low redox potentials can stimulate N fixation (Alldredge and Cohen 1987, Paerl and Carlton 1988). Anoxic microzones also develop in bundles of blue-green algae (Paerl and Bebout 1988) and in the endosymbiotic bacteria in diatoms (Martínez et al. 1983), both of which show significant N fixation in the sea. *Trichodesmium* shows spatial and temporal segregation of nitrogen fixation and photosynthesis within the cell to provide anoxic conditions for nitrogenase (Berman-Frank et al. 2001b).

Current estimates suggest global N fixation may account for about 150×10^{12} g N/yr added to the sea—roughly $10\times$ higher than estimates only a couple of decades ago (Mahaffey et al. 2005, Deutsch et al. 2007). As in terrestrial plants, nitrogen fixation in the phytoplankton is associated with $\delta^{15}N$ of $\sim 0‰$ in plankton biomass and in organic debris in sediments (Karl et al. 2002). This isotopic signature can be used to estimate nitrogen fixation in the water column (Mahaffey et al. 2003). Through glacial-interglacial cycles, the rate of nitrogen fixation in the oceans appears to relate to the distribution of desert dust in the atmosphere; high marine NPP during the last glacial may have derived from a greater deposition of Fe in the oceans from desert dust (Falkowski et al. 1998). The deposition of desert dust in the oceans links marine NPP to soil biogeochemistry of distant terrestrial ecosystems. Thus, in many areas primary productivity in the sea appears limited by available N, but the amount of N is largely determined by the deposition of Fe for the synthesis and activity of nitrogenase (Falkowski 1997, Wu et al. 2000, Moore and Doney 2007).

In the open ocean, direct atmospheric deposition of nitrogen in rainfall and dryfall is a small additional source of nutrients for new production, since these areas are distant from rivers and upwelling. Prospero and Savoie (1989) found that 40 to 70% of the nitrate in the atmosphere over the North Pacific Ocean was derived from soil dusts, presumably from the desert regions of China. Humans have increased the transport of reactive nitrogen from land to sea, largely as air pollutants (Duce et al. 2008). In some coastal regions downwind of pollution sources, an increased deposition of reactive nitrogen compounds from air pollution may cause higher marine NPP (Paerl 1995, Fanning 1989, Kim et al. 2011). Since reactive nitrogen is quickly removed from the atmosphere, its transport and deposition are more important in coastal areas than in the pelagic oceans.

Gaseous Losses of Nitrogen from the Sea

NPP in many ocean waters shows a tendency for limitation by available N (Howarth 1988, Falkowski 1997). What processes lead to N limitation in the sea? Denitrification in a zone of low O_2 concentration in the eastern Pacific Ocean may result in the loss of 50 to 60×10^{12} g N from the oceans each year (Codispoti and Christensen 1985, Deutsch et al. 2001). Other regions with suboxic conditions in the water column also provide local areas for marine denitrification. Often these are regions of high NPP, with oxygen depletion in the mid-water column due to the decomposition of sinking organic debris. The anoxic microzones created by flocculations of organic matter allow significant rates of denitrification in the oceans, despite the high redox potential of seawater (Alldredge and Cohen 1987).

As we saw in terrestrial ecosystems (Chapter 6), $^{14}NO_3$ is used preferentially as a substrate in the production of N_2 during denitrification. Denitrification results in a high content of ^{15}N in the residual nitrate in seawater (Liu and Kaplan 1989, Sigman et al. 2000, Voss et al. 2001). Denitrification is estimated from measures of $\delta^{15}N$ in the residual nitrate pool, as well as excess concentrations of dissolved N_2 gas in seawater (Chang et al. 2010). The pool of nitrate left in the ocean, which is taken up by phytoplankton, leaves an enriched ^{15}N signature in organic sediments. Recall we used this signature to ascertain the origin of denitrification in the geologic history of the Earth (Chapter 2).

Denitrification is also observed in ocean sediments, where it is performed by bacteria and a few specialized benthic eukaryotes (Piña-Ochoa et al. 2010). Christensen et al. (1987) estimate that over 50×10^{12} g N/yr may be lost from the sea by sedimentary denitrification in coastal regions. Devol (1991) found that nitrification occurring within the sediments supplied most of the nitrate for denitrification on the continental shelf of the western United States. Denitrification in sediments leaves the NO_3^- pool in the pore space enriched in ^{15}N (Lehmann et al. 2007). Including sediments, the overall rate of denitrification in the oceans is estimated between 270 and 482×10^{12} g N/yr (Brandes and Devol 2002, Codispoti et al. 2001, Galloway et al. 2004, Bianchi et al. 2012), which is slightly in excess of the current estimate of N inputs to the oceans. Changes in the balance of nitrogen fixation and denitrification have controlled the nitrogen content of the oceans through geologic time (Ganeshram et al. 1995, Ren et al. 2009).

Most of the gaseous nitrogen lost from marine environments by denitrification is N_2; however, seawater is supersaturated with N_2O in many regions (e.g., Walter et al. 2004b), and the oceans supply about 25% of the annual source of N_2O to the atmosphere (Table 12.4). Much of this N_2O may be derived from the oxidation of ammonia by archaea and other nitrifying species in the water column (Santoro et al. 2011).

Just a few years ago, several investigators reported an unusual microbial metabolism: the anaerobic oxidation of ammonium, using nitrite as an alternative electron acceptor in place of oxygen (Mulder et al. 1995, Strous et al. 1999, Kuypers et al. 2003, Schmidt et al. 2002). The reaction is performed by *Planctomyces* and a few other bacteria:

$$NH_4^+ + NO_2^- \rightarrow N_2\uparrow + 2H_2O, \tag{9.7}$$

and the mechanism of the reaction has recently been elucidated at the molecular level (Kartal et al. 2011). Known as *anammox*, the reaction produces N_2 gas, so it is difficult to separate anammox from denitrification. Both occur only in portions of the water column that are highly depleted in O_2. While significant in some high-productivity waters (Dalsgaard et al. 2003, Kuypers et al. 2005), the flux of N_2 from anammox is lower than that from denitrification in the suboxic waters of the Arabian Sea where both have been measured together (Ward et al. 2009, Bulow et al. 2010). In coastal sediments of the eastern United States and Sweden, Engstrom et al. (2005) found that 7 to 79% of the N_2 flux from sediments was due to anammox.

Finally, a small amount of ammonia is lost from the surface of the sea, where NH_4^+ is deprotonated to form gaseous NH_3 in the slightly alkaline conditions of seawater

(Eq. 6.4; Quinn et al. 1988, Jickells et al. 2003). This source of ammonia contributes about 15% to the annual global flux of NH_3 to the atmosphere (Chapter 12).

A Global Budget for Nitrogen in the Oceans

A global model for the N cycle of the oceans (Figure 9.21) offers a deceptive level of tidiness to our understanding of marine biogeochemistry, and the reader should realize that many fluxes, for example, nitrogen fixation, denitrification, and sedimentary preservation, are not known to better than a factor of two. Nevertheless, the model shows that most NPP is supported by nutrient recycling in the surface waters, and only small quantities of nutrients

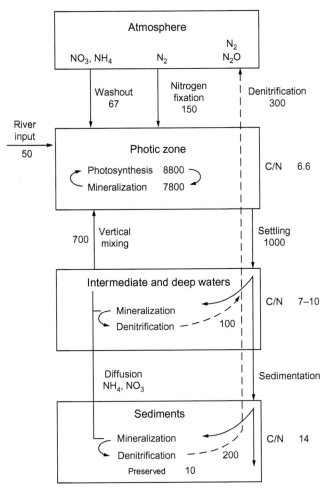

FIGURE 9.21   Nitrogen budget for the world's oceans, showing major fluxes in units of 10^{12} gN/yr. From an original conception by Wollast (1981), but with newer data added for atmospheric deposition (Duce et al. 2008), nitrogen fixation (Deutsch et al. 2007), riverflow (Galloway et al. 2004), denitrification (Brandes and Devol 2002), and nutrient regeneration in surface waters (compare Table 9.3). The global values have been rounded.

are lost to the deep ocean. Assuming that the total N pool in marine biota is ~500×10^{12} gN (Galloway et al. 2004), the mean residence time of the available nitrogen (inorganic and organic) in the surface ocean is about 125 days, whereas the mean residence time of organic N is about 20 days. Thus, each atom of N cycles through the biota many times.

In the absence of upwelling, the biotic pump would remove the pool of nutrients in the surface water in less than a year. After sinking and mineralization in the deep ocean, N enters pools with a mean residence time of about 500 years—largely controlled by the circulation of water through the deep ocean. The nitrogen cycle is dynamic; the overall mean residence time for N in the oceans is estimated to be ~2000 years, so the nitrogen cycle is responsive to global changes over relatively short periods (Brandes and Devol 2002).

Vertical mixing includes upwelling, upward convection, and diffusion from the deep ocean. Upwelling accounts for about half of the global upward flux, and it is centered in coastal areas where the resulting nutrient-rich waters yield high productivity. Away from areas of upwelling, diffusion and convection dominate the upward flux (Table 9.4), but diffusion rates are low (Ledwell et al. 1993), so the total supply of nutrients is limited in most of the open ocean (Lewis et al. 1986, Martin and Gordon 1988). Diffusion appears globally significant only as a result of the large area of pelagic ocean compared to the small area of upwellings.

Phosphorus

Phosphorus is nearly undetectable in most surface seawaters. Upwelling waters deliver N and P in amounts close to the Redfield ratio of 16, so it is reasonable to surmise that N and P would be depleted in tandem by the growth of phytoplankton. The long-standing debate over N versus P limitations to ocean NPP now appears settled in favor of N, as a result of the substantial rates of denitrification recently reported in marine environments. Nevertheless, phosphorus appears to limit phytoplankton activity in some regions (Wu et al. 2000), and

TABLE 9.4   Sources of Fe, PO_4, and NO_3 in Surface Waters of the North Pacific Ocean

Source	Fe	PO_4	NO_3
Concentration at 150 m (μmol m^{-3})	0.075	330	4300
Upwelling (μmol m^{-2} day^{-1})	0.00090	4.0	52
Net upward diffusion (μmol m^{-2} day^{-1})	0.0034	30	400
Atmospheric flux (μmol m^{-2} day^{-1})	0.16	0.102	26
Total fluxes (μmol m^{-2} day^{-1})	0.164	34	480
Percent from advective input	0.5	12	11
Percent from diffusive input	2	88	83
Percent from atmospheric input	98	0	5

Source: From Martin and Gordon (1988).

Fe and P appear to colimit the growth of nitrogen-fixing *Trichodesmium* in the Atlantic Ocean (Mills et al. 2004). *Trichodesmium* appears to have evolved specialized mechanisms to assimilate dissolved organic P compounds, known as phosphonates, from seawater (Dyhrman et al. 2006), and various phytoplankton appear to reduce the phospholipid content of cellular membranes when P is in particularly short supply (van Mooy et al. 2009).

As seen in Chapter 8, only a small portion of the total phosphorus transport in rivers (21×10^{12} g P/yr) is carried in dissolved forms; the remainder is adsorbed to Fe and Al oxide minerals that are carried as suspended particles. Some of the adsorbed P is released following the mixing of freshwater and seawater (Chase and Sayles 1980, Caraco et al. 1990), but most is probably buried with the deposition of terrigeneous sediments on the continental shelf (Filippelli 1997). The total flux of "bioreactive" P to the oceans is about 2.0×10^{12} g/yr (Ramirez and Rose 1992, Delaney 1998), giving an atom ratio of ~55 for N/P(reactive) in global riverflow.

Deposition of P on the ocean surface from the dust of deserts may play a special role in stimulating new production in areas of the open ocean that are distant from rivers and zones of upwelling (Wu et al. 2000; Mills et al. 2004). However, as seen for N, recycling of P in the surface waters accounts for the vast majority of the P uptake by phytoplankton (Figure 9.22). Most of the phosphorus pool in plankton is remineralized within a few days (Benitez-Nelson and Buesseler 1999). Phosphorus appears to be mineralized selectively from the phosphonate

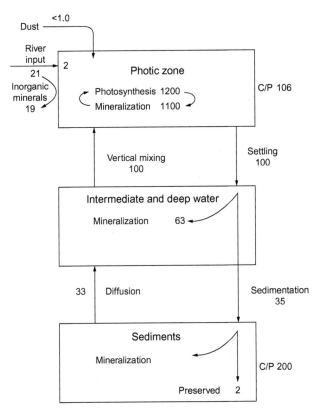

FIGURE 9.22 A phosphorus budget for the world's oceans, with important fluxes shown in units of 10^{12} g P/yr. From an original conception by Wollast (1981), but with newer data added for dust inputs (Graham and Duce 1979), riverflow (Meybeck 1982), sedimentary preservation (Wallman 2010), and nutrient regeneration in surface waters (compare Table 9.3). The global values have been rounded.

component of DOC (Clark et al. 1999). Each year a small amount of organic debris, with C/P ratios somewhat greater than the Redfield ratio, sinks through the thermocline to the deep ocean (Honjo et al. 1982). An average of 500 years later, mineralized P (i.e., HPO_4^{2-}) returns to the surface waters in upwelling.

The C/P ratio in organic matter that is buried in marine sediments is about 200 (Mach et al. 1987, Ingall and Van Cappellen 1990, Ramirez and Rose 1992), suggesting that P is mineralized more rapidly than C during the downward transport and sedimentary diagenesis of organic matter in the sea (Honjo et al. 1982, Froelich et al. 1979, Loh and Bauer 2000). Phosphorus release and C/P ratios are greatest in anoxic sediments (Ingall et al. 1993, Ingall and Jahnke 1997; Figure 9.23). Anoxic environments have lower concentrations of oxidized Fe minerals that can adsorb P as it is mineralized from organic matter (Krom and Berner 1981, Sundby et al. 1992, Berner and Rao 1994, Blomqvist et al. 2004). Both dissolved and inorganic P are adsorbed to iron oxide minerals, especially less crystalline forms (Ruttenberg and Sulak 2011; compare Chapter 4).

In contrast to N, there are no significant gaseous losses of P from the sea. At steady state, the inputs to the sea in river water must be balanced by the burial of phosphorus in ocean sediments. Most of the P carried in suspended sediments is probably deposited near the coast of continents, where P burial parallels overall rates of sedimentation (Filippelli 1997). Burial of biogenic P compounds in sediments of the open ocean is estimated between about 2.0×10^{12} g P/yr—similar to the delivery of bioreactive P in rivers (Howarth et al. 1995a, Delaney 1998, Wallman 2010). Burial occurs with the deposition of organic matter or $CaCO_2$ (Froelich et al. 1982), with at least a portion of the burial being in biogenic polyphosphates (Diaz et al. 2008). During sediment diagenesis, organic- and Fe-bound P are converted to phosphorites (authigenic apatite), which ultimately dominate the P storage in sediments (Ruttenberg 1993, Filippelli and Delaney 1996, Rasmussen 1996).

Phosphorite is formed when the PO_4^{3-} produced from the mineralization of organic P combines with Ca and F to form fluorapatite, in a mineral known as francolite (Ruttenberg and Berner 1993, Krajewski et al. 1994, Anderson et al. 2001). Organic P disappears from sediments in parallel with an increase in francolite formed within them (Filippelli and Delaney

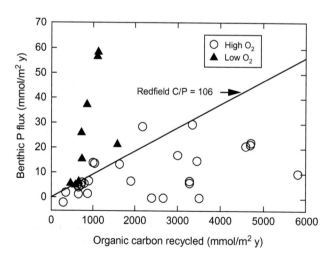

FIGURE 9.23 Flux of phosphorus from sediments to the water column as a function of the decomposition of organic carbon in areas of high and low O_2 in the overlying waters. *Source: From Ingall and Jahnke (1997).*

1996, Delaney 1998). In coastal California sediments, Kim et al. (1999) found that francolite sequestered about 30% of the P mineralized from organic compounds or adsorbed to Fe-oxides. The F is supplied by inward diffusion from seawater (Froelich et al. 1983, Schuffert et al. 1994). In some areas of the ocean, phosphorite nodules up to several centimeters in diameter accumulate on the seafloor. These nodules are an enigma; they remain on the surface of the sediment despite growing at rates slower than the rate of sediment accumulation (Burnett et al. 1982).

The mean residence time for reactive P in the oceans, relative to the input in rivers or the loss to sediments, is $>25,000$ years (Ruttenberg 1993, Filippelli and Delaney 1996, Delaney 1998). Thus, each atom of P that enters the sea may complete 50 cycles between the surface and the deep ocean before it is lost to sediments. The major sinks include the formation of authigenic phosphorites (francolite) and uptake at hydrothermal vents (Elderfield and Schultz 1996, Wheat et al. 2003). All forms of buried phosphorus complete a global biogeochemical cycle when geologic processes lift sedimentary rocks above sea level and weathering begins again. Thus, relative to N, the global cycle of P in the oceans turns very slowly (Chapter 12).

Human Perturbations of Marine Nutrient Cycling

The riverine input of N and P to the coastal oceans has increased in recent years, the result of direct human effluents and an increasing global use of nitrogen fertilizers (Howarth 1998, Boyer et al. 2006; Chapter 8). Fossil fuel pollutants and emissions of NH_3 from agricultural soils and livestock have also increased the atmospheric deposition of N on the ocean surface (Duce et al. 2008). These inputs have probably enhanced the productivity of coastal and estuarine ecosystems (Chapter 8) and perhaps the productivity of the entire ocean (Paerl 1995). Pahlow and Riebesell (2000) report a rising N/P ratio in export production from pelagic oceans in the Northern Hemisphere during the past 50 years, perhaps reflecting the widespread impacts of nitrogen deposition in the sea. Greater net primary production in the surface ocean should result in a greater transport of particulate carbon to the deep sea, potentially serving as a sink for increasing atmospheric CO_2.

Net primary production of $\sim 50 \times 10^{15}$ g C/yr is supported by nitrogen derived from a variety of sources (Table 9.3). If an additional 54×10^{12} g N/yr is deposited in the surface waters from atmospheric pollution (Duce et al. 2008), this "excess" nitrogen could result in an increase in the downward flux of organic carbon of about 0.30×10^{15} g/yr, assuming a Redfield atom ratio of 106 C/16 N in new production (Figure 9.24). Paerl et al. (1999) noted a significant stimulation of NPP in coastal and pelagic waters of the western Atlantic with the addition of synthetic rainfall containing small amounts of N.

Similar calculations using the "excess" flux of N in rivers (Schlesinger 2009) suggest an increased new production of $<0.20 \times 10^{15}$ g C/yr in coastal zones (Wollast 1991), although many studies indicate that much of the river flux of N is largely denitrified in coastal sediments (Seitzinger et al. 2006). For comparison, the current release of carbon dioxide to the atmosphere from fossil fuel combustion is about 9×10^{15} g C/yr (Chapter 11). This increased NPP in the oceans offers an additional sink for atmospheric CO_2 (Rabouille et al. 2001, Krishnamurthy et al. (2007), but it is associated with significant problems of coastal eutrophication (Diaz and Rosenberg 2008, Doney 2010; Chapter 8). Globally marine NPP has probably

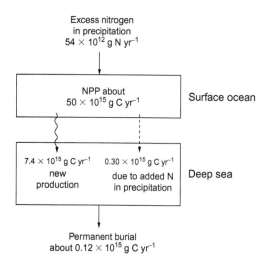

Excess nitrogen
in precipitation
54×10^{12} g N yr^{-1}

NPP about
50×10^{15} g C yr^{-1} Surface ocean

7.4×10^{15} g C yr^{-1} 0.30×10^{15} g C yr^{-1} Deep sea
new   due to added N
production   in precipitation

Permanent burial
about 0.12×10^{15} g C yr^{-1}

FIGURE 9.24   Estimated increase in the sedimentation of organic carbon that might be caused by human additions of nitrogen to the world's oceans by precipitation. Updated from an original conception by Peterson and Melillo (1985), based on current anthropogenic atmospheric inputs of Duce et al. (2008).

increased ~3% from anthropogenic nitrogen deposition from the atmosphere (Duce et al. 2008). The major ocean sink for anthropogenic CO_2, about 2.3×10^{15} g C/yr (Table 11.1), is found as a result of an increased dissolution of CO_2 in cold waters of the polar oceans (Shaffer 1993). As we discussed earlier, this inorganic sink for CO_2 is limited by the area of polar oceans and the amount of downwelling water.

Human perturbations of marine ecosystems are greatest in estuarine, coastal, and continental shelf waters (Chapter 8). These areas occupy only about 8% of the ocean's surface, but they account for about 18% of ocean productivity (refer to Table 9.2), and 83% of the carbon that is buried in sediments. Globally averaged models (e.g., Figures 9.21 and 9.22) mask the comparative importance of these regions to the overall biogeochemical cycles of the sea. For example, a significant amount of organic carbon may be transported from the continental shelf to the deep sea (Walsh 1991, Wollast 1993). If global climate change alters the rate of coastal upwelling (Bakun 1990), significant changes in the ocean's overall biogeochemistry can be expected (Walsh 1984).

Silicon, Iron, and Trace Metals

Phytoplankton acquire micronutrients from seawater, leaving an imprint on the distribution of trace elements in the sea. The concentrations of many essential elements, such as Si, Fe, Zn, Cu, Co, and Ni, are depleted in the surface waters and increase with depth (Bruland et al. 1991, Donat and Bruland 1995, Shelley et al. 2012). A few nonessential elements, such as Ti and Ba, which adsorb to sinking particles, also show this nonconservative behavior in seawater. Meanwhile, essential elements that are abundant in seawater (e.g., K, Mg, B, Sr, Mo, and SO_4) show conservative behavior, with long mean residence times and similar concentrations at all depths.

These patterns suggest that the geochemistry of many trace elements in seawater is controlled directly and indirectly by biota. Cherry et al. (1978) show that the mean residence time for 14 trace elements in seawater is inversely related to their concentration in sinking fecal

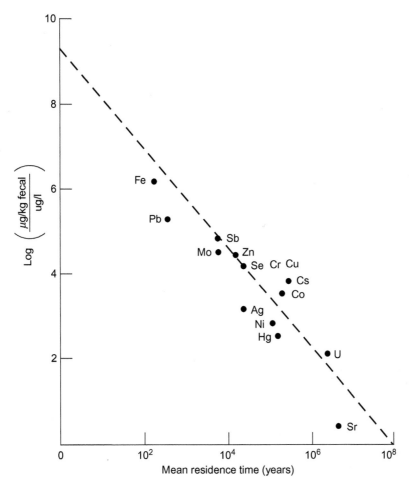

FIGURE 9.25 The ratio between the concentration of an element in sinking fecal pellets ($\mu g/kg$) and its concentration in seawater ($\mu g/l$), plotted as a function of its mean residence time in the ocean. *Source: From Cherry et al. (1978). Reprinted with permission from* Nature, *copyright 1978 Macmillan Magazines Limited.*

pellets (Figure 9.25). Some of these elements are remineralized by grazing zooplankton (Reinfelder and Fisher 1991) or by the degradation of POC by bacteria in the deep ocean. But the fate for many trace constituents is downward transport in organic particles and burial in the sediments of the deep sea (Turekian 1977, Lal 1977, Li 1981). Elements with less interaction with biota remain as the major constituents of seawater (refer to Table 9.1).

Diatoms compose a large proportion of the marine phytoplankton, and they require silicon (Si) as a constituent of their cell walls, where it is deposited as opal. As a result of biotic uptake, the concentration of dissolved Si in the surface waters is very low, usually $<2\ \mu M$, with the highest concentrations in the Southern and the North Pacific Oceans (Ragueneau et al. 2000). Globally, the annual uptake of Si by diatoms is about 6000×10^{12} g (Nelson et al.

1995). After the death of diatoms, a large fraction of the opal dissolves, and the Si is recycled in the surface waters, where bacterioplankton mediate the dissolution process (Bidle and Azam 1999).

Silicate concentrations generally increase with depth, but the dissolution of opal is dependent on temperature, so the rate of dissolution in the deep ocean is relatively slow (Honjo et al. 1982, Bidle et al. 2002, Van Cappellen et al. 2002). The average Si concentration in deep waters is about 70 μM. In sinking particles and in ocean sediments, Si/C ratios increase with depth, suggesting that C is mineralized more readily than Si (Nelson et al. 2002). Globally, the burial efficiency for opaline Si is about 3% of production—significantly greater than for organic carbon ($<1\%$) (DeMaster 2002). Silicon loss is retarded when dissolved silicon complexes with Al in the sediments (Dixit et al. 2001).

A mass-balance model for Si in the oceans shows that rivers (156×10^{12} g/yr), dust (14×10^{12} g/yr), and hydrothermal vents (17×10^{12} g/yr) are the main sources, and sedimentation of biogenic opal is the only significant sink (DeMaster 2002). The mean residence time for Si in the oceans is about 15,000 years, which is consistent with its nonconservative behavior in seawater. Most of the Si input is delivered by tropical rivers, as a result of high rates of rock weathering in tropical climates (Chapter 4). Sedimentation in the cold waters of the Antarctic Ocean accounts for 70% of the global sink (Ragueneau et al. 2000, DeMaster 2002), largely as a result of massive seasonal diatom blooms in the Southern Ocean near Antarctica (Nelson et al. 2002). About 10% of the sink is found in coastal regions, where the growth of diatoms in nutrient-enriched waters may be limited by silicon (Justic et al. 1995). Increasing deposition of desert dusts to the ocean surface, potentially stimulating diatom productivity and export production with a higher organic carbon content relative to $CaCO_3$, may explain periods with lower CO_2 in Earth's atmosphere (Harrison 2000).

Similar to the use of Si by diatoms, marine protists known as acantharians require strontium (Sr). These organisms precipitate celestite ($SrSO_4$) as a skeletal component. As a result of the uptake of Sr in surface waters and the sinking of acantharians to the deep sea, the Sr/Cl ratio in seawater varies from about 392 μg/g in surface waters to >405 μg/g with depth—relatively conservative behavior (Bernstein et al. 1987). The biotic demand for Sr leaves only a slight imprint on the large pool of Sr in the oceans; its overall mean residence time is about 12,000,000 years (Table 9.1).

All phytoplankton require a suite of micronutrients, for example, iron (Fe), copper (Cu), and zinc (Zn), in their biochemistry, where they are cofactors in metabolic enzymes (Morel and Price 2003). These elements are taken up in surface waters and mineralized when dead organisms sink to the deep ocean. Many of these elements are relatively insoluble in seawater, which typically has high redox potential (Chapter 7). Most metals are found at low concentrations in the surface waters, and concentrations increase with depth in the deep sea (e.g., Figure 9.26). In response to low concentrations of Fe, some bacterioplankton release organic compounds, known as siderophores, that chelate dissolved Fe and enhance its assimilation from seawater (Wilhelm et al. 1996, Butler 1998, Mendez et al. 2010). Organic complexes dominate the total concentrations of Fe, Zn, and Cu in surface waters (Bruland et al. 1991, Morel and Price 2003).

Near continents, the concentration of Fe in seawater is normally adequate to support phytoplankton growth, as a result of the inputs from rivers and local upwelling of waters (Hutchins et al. 1998, Firme et al. 2003, Elrod et al. 2004). In the central Pacific Ocean, however, Martin and Gordon (1988) found that internal sources of Fe could sustain only a small

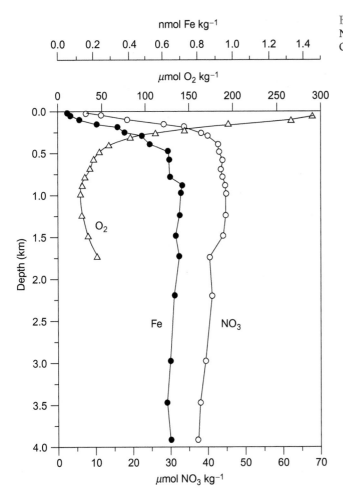

FIGURE 9.26   Vertical distribution of Fe, NO_3, and O_2 in the central North Pacific Ocean. *Source: From Martin et al. (1989).*

percentage of the observed NPP. They suggested that as much as 98% of the new production in this area is supported by Fe derived from dust deposited from the atmosphere (Table 9.4). Most of the dust deposited in this region is probably transported from the deserts of central China (Duce and Tindale 1991, Uematsu et al. 2003). Growth of phytoplankton appears to be limited by iron, so small quantities of NO_3 and PO_4 remain in surface seawaters even during periods of peak production (Figure 9.26). These waters are known as the HNLC—high nutrient, low chlorophyll—zones of the ocean.

During the last glacial epoch, arid environments were more widespread and more continental shelf area was exposed because of lower sea level. It is likely that there was greater wind erosion and atmospheric transport of Fe (Lambert et al. 2008). This added Fe deposition may have enhanced marine NPP (Kumar et al. 1995a, Martinez-Garcia et al. 2011). Simultaneously, bubbles of gas trapped in Antarctic ice 18,000 years ago show lower levels of atmospheric CO_2 (Figure 1.2). Several workers have suggested that fertilizing the oceans with Fe might be an effective way to stimulate marine NPP and export production, which would lower future levels of atmospheric CO_2. Export production is correlated to natural variations in dust deposition in the North Pacific (Bishop et al. 2002) and Southern oceans (Cessar et al. 2007), whereas the upwelling of Fe-rich waters controls NPP in other areas (Coale et al. 1996a, Blain et al. 2007).

Over a period of 10 years, a number of iron fertilization experiments were conducted in the Pacific (e.g., the IronEX experiment) and the Southern Ocean (SOFeX) to test the "iron hypothesis" (Boyd et al. 2007). In every case, the rate of net primary production by phytoplankton increased significantly—sometimes by a factor of 10—when Fe was added to the ocean surface (Martin et al. 1994; Coale et al. 1996b, 2004; Boyd et al. 2000), which lowered the concentration of CO_2 dissolved in the surface waters (Watson et al. 2000, Cooper et al. 1996). Iron increased the photosynthetic capacity of phytoplankton, in which it serves as an essential cofactor in several enzymes (Behrenfeld et al. 1996). Export production was calculated from the C/Fe ratio in POC, an extension of the Redfield ratio. For Fe-enrichment experiments in the Southern Ocean, this ratio was 3000, but increases in the export of POC to the deep ocean were modest (Buesseler et al. 2004). Smetacek et al. (2012) report increases in POC sinking below 1000-m depth during an Fe-enrichment experiment in the Antarctic ocean.

Fe in desert dust is normally found as Fe^{3+}, which is much less soluble than the organic complexes that normally dominate the pool of available Fe in seawater (Rue and Bruland 1997). Most Fe in the oceans is found in particulate matter (Johnson et al. 1997), which sinks quickly (Croot et al. 2004). The Fe added experimentally to the ocean surface rapidly disappeared from the upper water column, so the bloom of phytoplankton was short-lived (Boyd et al. 2007). In experiments of longer duration, the development of an active community of zooplankton might have regenerated Fe in surface waters through their grazing activities (Reinfelder and Fisher 1991, Hutchins et al. 1993). Heterotrophic bacteria accumulate Fe (Tortell et al. 1996) and are sometimes fed on by phytoplankton (Maranger et al. 1998).

Iron fertilization is probably not a cure for rising anthropogenic atmospheric CO_2 in Earth's atmosphere (Aumont and Bopp 2006, Zeebe and Archer 2005). In some iron-fertilization experiments, the flux of N_2O, another greenhouse gas, from the ocean surface increased significantly, potentially negating the uptake of CO_2 (Law and Ling 2001). Marine biologists warn of the disruptive effects of iron fertilization on the marine biosphere (Chisholm et al. 2001); by leading to other nutrient deficiencies, iron fertilization effects are likely to be short-lived, and the use of fossil fuels to mine, refine, and distribute Fe to the oceans may release more CO_2 than balanced by enhanced uptake in seawater.

Zinc (Zn) is an essential component of carbonic anhydrase—the enzyme that allows phytoplankton to convert HCO_3^- in seawater to CO_2 for photosynthesis (Morel et al. 1994). Low concentrations of Zn in surface waters can limit the growth of phytoplankton in marine environments (Brand et al. 1983, Sunda and Huntsman 1992). Zn is also an essential cofactor for alkaline phosphatase, which allows phytoplankton to extract P from dissolved organic forms (DOP) in low-phosphorus waters (Shaked et al. 2006). Like Fe, the concentrations of Zn increase with depth in the deep sea (Bruland 1989). Among samples of surface and deep waters, the concentrations of Fe and Zn are often well correlated to those of N, P, and Si, suggesting that biological processes control the distribution of all these elements in seawater. For example, Zn is correlated to Si in the northeast Pacific (Bruland et al. 1978a).

Uptake of trace metals also occurs for some nonessential, toxic metals, such as cadmium (Cd), which accumulates in phytoplankton. Cadmium appears to substitute for zinc in

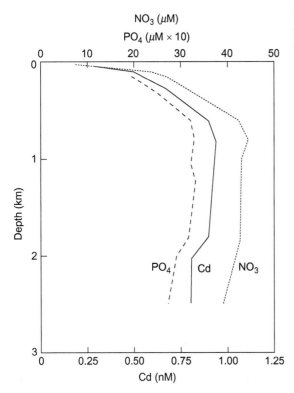

FIGURE 9.27   Depth distribution of nitrate, phosphate, and cadmium in the coastal waters of California. *Source: From Bruland et al. (1978b).*

biochemical molecules, allowing diatoms to maintain growth in zinc-deficient seawater (Price and Morel 1990, Lane et al. 2005, Park et al. 2008, Xu et al. 2008b). Cadmium is well correlated with available P in waters of the Pacific Ocean (Figure 9.27; Boyle et al. 1976, Abe 2002), and the concentration of Cd in marine sediments is sometimes used as an index of the availability of P in seawater in the geologic past (Hester and Boyle 1982, Elderfield and Rickaby 2000). When marine phosphate rock is used as a fertilizer, cadmium is often an undesirable trace contaminant (Smil 2000).

When nonessential elements (e.g., Al, Ti, Ba, and Cd) and essential elements (e.g., Si and P) show similar variations in concentration with depth, it is tempting to suggest that both are affected by biotic processes, but the correlation does not indicate whether the association is active or passive. Organisms actively accumulate essential micronutrients by enzymatic uptake, whereas other elements may show passive accumulations, as a result of coprecipitation or adsorption on dead, sinking particles. For instance, titanium (Ti), which is not essential for biochemistry, shows nonconservative behavior in seawater, with concentrations ranging from 10 μM at the surface to >200 μM at depth (Orians et al. 1990). The mean residence times for Ti, Ga (gallium), and Al (aluminum) in seawater range from 70 to 150 years (Orians et al. 1990).

Widespread observations of nonconservative behavior of barium (Ba) in seawater do not appear to result from direct biotic uptake (Sternberg et al. 2005). $BaSO_4$ precipitates on dead, sinking phytoplankton, especially diatoms and acantharians, as a result of the high concentrations of SO_4 that surround these organisms during decomposition (Bishop 1988, Bernstein and Byrne 2004). The precipitation of barite ($BaSO_4$) is an indication of the productivity of oceans in the past (Paytan et al. 1996).

In the Mediterranean Sea, Al (aluminum) shows a concentration minimum at a depth of 60 m, where Si and NO_3 are also depleted. MacKenzie et al. (1978) suggested that this distribution is the result of biotic activity, and active uptake has been confirmed in laboratory studies (Moran and Moore 1988). Other workers have found that organic particles carry Al to the deep ocean, but that the association is passive (Hydes 1979, Deuser et al. 1983). High Al in surface waters is due to atmospheric inputs of dust (Orians and Bruland 1986, Measures and Vink 2000, Kramer et al. 2004). Aluminum declines in concentration with depth as a result of scavenging by organic particles and sedimentation of mineral particles.

Manganese (Mn), an essential element for photosynthesis (Chapter 5), is found at higher concentrations in the surface waters (0.1 μg/liter) than in the deep waters (0.02 μg/liter) of the ocean. Calculating an Mn budget for the oceans, Bender et al. (1977) attribute the high surface concentrations to the input of dust to the ocean surface (Guieu et al. 1994, Shiller 1997, Mendez et al. 2010). Manganese appears less limiting than Fe and Zn for the growth of marine phytoplankton in surface waters (Brand et al. 1983). As in the case of Al, the deposition of Mn in dust in surface waters must exceed the rate of biotic uptake, downward transport, and remineralization of Mn in the deep sea.

The Mn budget of the ocean has long puzzled oceanographers, who recognized that the Mn concentration in ocean sediments greatly exceeds that found in the average continental rock (Broecker 1974, Martin and Meybeck 1979). Other sources of Mn are found in riverflow and in releases from hydrothermal vents (Edmond et al. 1979). Various deep-sea bacteria appear to concentrate Mn by oxidizing Mn^{2+} in seawater to Mn^{4+} that is deposited in sediment (Krumbein 1971; Ehrlich 1975, 1982). The most impressive sedimentary accumulations are seen in Mn nodules that range in diameter from 1 to 15 cm and cover large portions of the seafloor (Broecker 1974, McKelvey 1980). As we discussed for phosphorus nodules, the rate of growth of Mn nodules, about 1 to 300 mm per million years (Odada 1992), is slower than the mean rate of sediment accumulation (1000 mm per million years; Sadler 1981), yet they remain on the surface of the seafloor. Various hypotheses invoking sediment stirring by biota have been suggested to explain the enigma, but none is proven. In addition to a high concentration of Mn (15–25%), these nodules also contain high concentrations of Fe, Ni, Cu, and Co and are a potential economic mineral resource.

BIOGEOCHEMISTRY OF HYDROTHERMAL VENT COMMUNITIES

At a depth of 2500 m in the east Pacific Ocean, a remarkable community of organisms is found in association with hydrothermal vents. Discovered in 1977, this community consists of bacteria, tube worms, mollusks, and other organisms, many of which are recognized as new species (Corliss et al. 1979, Grassle 1985; Figure 9.28). Similar communities are found at hydrothermal vents in the Gulf of Mexico and other areas, including some located 15 km from mid-ocean rifts in the Atlantic Ocean (Kelley et al. 2005). In total darkness, these communities are supported by bacterial chemosynthesis, in which hydrogen sulfide (H_2S)

FIGURE 9.28 Medusa jellyfish at a hydrothermal vent at 2850-m depth on the East Pacific Rise, photographed from the ROV Jason II. *Photo courtesy of Emily M. Klein, chief scientist, Duke University.*

from the hydrothermal emissions is metabolized using O_2 and CO_2 from the deep sea waters to produce carbohydrates (Jannasch and Wirsen 1979, Jannasch and Mottl 1985):

$$O_2 + 4H_2S + CO_2 \rightarrow CH_2O + 4S \downarrow + 3H_2O. \tag{9.8}$$

Consumption of H_2S by chemosynthetic bacteria is correlated with declines in O_2 when seawater mixes with hydrothermal water (Johnson et al. 1986b). At first glance the reaction would appear to result in the production of organic matter without photosynthesis. We must remember, however, that the dependence of this reaction on O_2 links chemosynthesis in the deep sea to photosynthesis occurring in other locations on Earth. Other bacteria at hydrothermal vents employ chemosynthetic reactions based on methane, hydrogen, and reduced metals that are emitted in conjunction with H_2S (Jannasch and Mottl 1985, Petersen et al. 2011).

On the basis of the chemosynthetic reactions, bacterial growth feeds the higher organisms found in the hydrothermal communities (Grassle 1985, Levesque et al. 2005). Some of the bacteria are symbiotic in higher organisms. Symbiotic bacteria in the tube worm *Riftia* deposit elemental sulfur, leading to the rapid growth of tubular columns of sulfur up to 1.5 m long (Cavanaugh et al. 1981, Lutz et al. 1994). Filter-feeding clams up to 30 cm in diameter occur in dense mats near the vents. These communities are dynamic, and a particular vent may be active for only about 10 years. Because they are below the carbonate compensation depth, the clam shells slowly dissolve when the vent activity ceases (Grassle 1985). The offspring of these organisms must continually disperse to colonize new vent systems.

Various metallic elements and silicon are soluble in the hot, low-redox conditions of hydrothermal seawater. After mixing with seawater, the precipitation of metallic sulfides may remove as much as 100×10^{12} g S/yr from the ocean (Edmond et al. 1979, Jannasch 1989), although we have chosen a lower value (27×10^{12} g S/yr) in the sulfur cycle of Figure 9.29 (Elderfield and Schultz 1996). Mn and Fe are also deposited as insoluble oxides (MnO_2, FeO) and nodules on the seafloor. The iron oxides act to scavenge vanadium (V) and other elements from seawater and may remove 25% of the annual riverine input of V to the oceans each year (Trefry and Metz 1989).

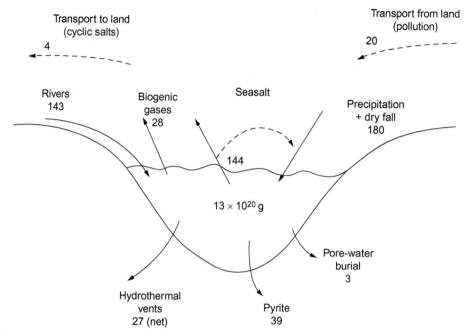

FIGURE 9.29 Sulfur budget for the world's oceans, showing important fluxes in units of 10^{12} g S/yr. (See also Figure 13.1.) *Sources: Riverflux from Meybeck (1979), gaseous output from Lana et al. (2011), hydrothermal flux from Elderfield and Schultz (1996), and pyrite deposition from Berner (1982).*

Hydrothermal vents attain global significance for their effect on the Ca, Mg, and SO_4 budgets of the oceans, and these bizarre chemosynthetic communities speak strongly for the potential for life to exist in unusual locations where oxidized and reduced substances are brought together by biogeochemical cycles. Persistent life in hydrothermal vent communities may be analogous to some of the earliest environments for the evolution of life on Earth and other planets (Chapter 2).

THE MARINE SULFUR CYCLE

Sulfur is the second most abundant anion in the oceans, where it is found overwhelmingly as SO_4^{2-} (refer to Table 9.1). Rivers and atmospheric deposition are the major sources of SO_4 in the sea (Figure 9.29), but most of the atmospheric deposition is derived from seasalt aerosols that are quickly redeposited on the ocean's surface (i.e., cyclic salt). Metallic sulfides precipitated at hydrothermal vents and biogenic pyrite forming in sediments are the major marine sinks. Sulfur is incorporated into protein by assimilatory reduction of SO_4 from seawater by marine phytoplankton and bacteria (Giordano et al. 2005). Sulfate shows a highly conservative behavior in seawater, with a mean residence time of about 10 million years relative to inputs from rivers.

Despite its high redox potential, seawater harbors various reduced sulfur compounds, presumably in anoxic microsites and suboxic waters. Rapid cycling of sulfur between reduced and oxidized forms is postulated, and gene sequences show the presence of chemoautotrophic sulfur-oxidizing pathways (Eq. 2.16; Canfield et al. 2010, Swan et al. 2011).

The oceans are recognized as a major source of dimethylsulfide—$(CH_3)_2$ S—to the atmosphere. Trace quantities of this gas contribute to the "odor of the sea" in coastal regions (Ishida 1968). Dimethylsulfide (DMS) is produced during the decomposition of dimethylsulfoniopropionate (DMSP) from dying phytoplankton cells (Kiene 1990). The reaction is mediated by the enzyme DMSP-lyase. Grazing by zooplankton seems to be important to the release of DMS to seawater (Dacey and Wakeham 1986, Wolfe et al. 1997). As a form of organic S, DMSP can account for 10% of the carbon content in phytoplankton (Stefels et al. 2007) and up to 15% of the carbon that flows through the grazing food chain (Kiene et al. 2000, Simo et al. 2002). However, not all DMSP is converted to DMS following release from a phytoplankton cell (Kiene and Linn 2000). Certain bacterioplankton appear to convert a substantial fraction of DMSP to other sulfur compounds (Howard et al. 2006), and some bacteria and phytoplankton can assimilate DMSP as a sulfur source (Kiene et al. 2000, Tripp et al. 2008, Vila-Costa et al. 2006, Reisch et al. 2011). Finally, only a small portion of the total production of DMS is lost to the atmosphere; the rest is degraded by microbes in the surface waters (Kiene and Bates 1990, del Valle et al. 2009). The mean residence time of DMS in seawater is about 2 days.

In an effort to balance the global sulfur cycle, DMS was first proposed as a major gaseous output of the sea by Lovelock et al. (1972), but it wasn't until 1977 that Maroulis and Bandy were able to measure DMS as an atmospheric constituent along the eastern coast of the United States. DMS is now widely recognized as a trace constituent in seawater and in the marine atmosphere, and the diffusion gradient of DMS across the sea–air interface indicates a global flux of 10 to 30×10^{12} g S/yr to the atmosphere (Lana et al. 2011; Figure 9.30). This is the largest natural emission of a sulfur gas to the atmosphere (Kjellstrom et al. 1999).

In contrast to terrestrial and freshwater wetland environments, where H_2S dominates the losses of gaseous sulfur, the oceans emit only small quantities of H_2S (Andreae et al. 1991, Shooter 1999). The oceans are also a source of carbonyl sulfide (COS) in the atmosphere, but the flux of COS is only a small component of the marine sulfur budget (about 0.04×10^{12} g S/yr; Chapter 13). Thus, dimethylsulfide is the major form of gaseous sulfur lost from the sea. Dimethylsulfide is also an important sulfur gas emitted from salt marshes (Steudler and Peterson 1985, Hines et al. 1993). Iverson et al. (1989) showed that the concentrations of DMS and its precursor DMSP increase as a function of increasing salinity in estuaries of the eastern United States.

In the atmosphere, DMS is rapidly oxidized by OH radicals, forming SO_2 and then sulfate aerosols that are deposited in precipitation (Shon et al. 2001, Faloona et al. 2009; Chapter 3). Nearly 80% of the nonseasalt sulfate (nss) in the atmosphere over the North Pacific Ocean appears to be derived from DMS, with the soil dust and pollution contributing the rest (Savoie and Prospero 1989). Marine DMS is estimated to contribute up to 10% of the atmospheric sulfur over industrial Europe (Tarrasón et al. 1995). The potential effects of DMS and SO_4^{2-} aerosols on climate are treated in detail in Chapter 13.

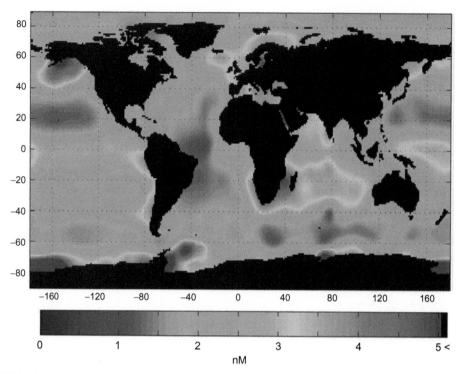

FIGURE 9.30 Mean annual dimethylsulfide concentration in the surface ocean (nM), showing zones of high concentrations in the high-latitude oceans. *Source: Redrawn from Lana et al. (2011). Used with permission of the American Geophysical Union. All rights reserved.*

THE SEDIMENTARY RECORD OF BIOGEOCHEMISTRY

Marine sediments contain a record of the conditions of the oceans through geologic time (Kastner 1999). Sediments and sedimentary rocks rich in $CaCO_3$ (calcareous ooze) show the past location of shallow, productive seas, where foraminifera and coccolithopores were abundant. Sediments deposited in the deep sea are dominated by silicate clay minerals, with high concentrations of Fe and Mn (red clays). Opal indicates the past environment of diatoms, whereas sediments with abundant organic carbon are associated with near-shore areas, where burial of organic materials is rapid (Figure 9.11). Changes in the species composition of preserved organisms have also been used to infer patterns of ocean climate, circulation, and productivity during the geologic past (Weyl 1978, Corliss et al. 1986). For instance, the ratio of germanium (Ge) to silicon (Si) in diatomaceous sediments has been used to infer variations in the past rates of continental weathering (Froelich et al. 1992).

Calcareous sediments contain a record of paleotemperature. When the continental ice caps grew during glacial periods, the water they contained was depleted in $H_2^{18}O$, relative to ocean water, because $H_2^{16}O$ evaporates more readily from seawater and subsequently contributes more to continental rainfall and snowfall. When large quantities of water were lost from the ocean and stored in ice, the waters that remained in the ocean were enriched in

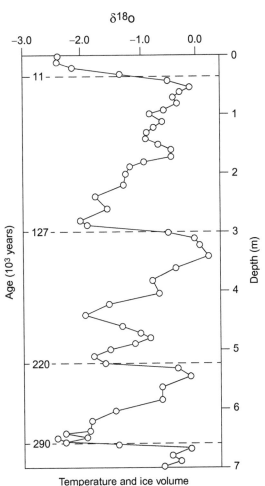

FIGURE 9.31 Changes in the $\delta^{18}O$ in sedimentary carbonates of the Caribbean Sea during 300,000 years. Enrichment of $\delta^{18}O$ during the last glacial epoch (20,000 years ago) is associated with lower sea levels and a greater proportion of $H_2^{18}O$ in seawater. *Source: From Broecker (1973).*

$H_2^{18}O$ compared to today. Carbonates precipitate in an equilibrium reaction with seawater (Eqs. 9.3–9.5), so an analysis of changes in the ^{18}O content of sedimentary carbonates is an indication of past changes in ocean volume and temperature (Figure 9.31).

The history of the Sr content of seawater is also of particular interest to geochemists because its isotopic ratio changes as a result of changes in the rate of rock weathering on land (Dia et al. 1992). Most strontium is ultimately removed from the oceans by coprecipitation with $CaCO_3$ (Kinsman 1969, Pingitore and Eastman 1986). During periods of extensive weathering, the ^{87}Sr content of seawater increases as a result of high content of that isotope in continental rocks. Thus, changes in the ^{87}Sr content of marine carbonate rocks offer an index of the relative rate of rock weathering over long periods (Richter et al. 1992). For calcium, changes in the rate of rock weathering versus the rate of carbonate sedimentation are reflected in the $\delta^{44}Ca$ ratio of calcium in carbonate sediments (De La Rocha and DePaolo 2000, Griffith et al. 2008). High rates of weathering are implicated for the Miocene, when atmospheric CO_2 levels were higher than today.

Carbonates are a small sink (20%) of boron in the oceans (Park and Schlesinger 2002), and the isotopic ratio of boron in carbonate varies as a function of seawater pH. The ratio measured in sedimentary foraminifera of the Miocene (21 mya) indicates that seawater pH was lower (7.4) than it is today (8.1), consistent with suggestions of higher atmospheric CO_2 during that period (Spivack et al. 1993, Pearson and Palmer 2000). Similarly, the boron isotope ratios of sedimentary carbonate indicate a higher seawater pH during the last glacial, when atmospheric CO_2 was low (Sanyal et al. 1995). As in all studies of sediments, the time resolution of the method is constrained by the mean residence time of the element in seawater, which for boron is >1,000,000 years (Park and Schlesinger 2002).

Sedimentary deposits of ^{13}C in organic matter and in $CaCO_3$ contain a record of the biotic productivity of Earth. Recall that photosynthesis discriminates against $^{13}CO_2$ relative to $^{12}CO_2$, slightly enriching plant materials in ^{12}C compared to the atmosphere (Chapter 5). When large amounts of organic matter are stored on land and in ocean sediments, $^{13}CO_2$ accumulates in the atmosphere and the ocean (i.e., $^{13}HCO_3$). Arthur et al. (1988) suggest that the relatively high ^{13}C content of marine carbonates during the late Cretaceous reflects a greater storage of organic carbon from photosynthesis. Similar changes are seen in the ^{13}C of coal age (Permian) brachiopods (Brand 1989). When the storage of organic carbon is greater, there is the potential for an increase in atmospheric O_2, as postulated for the Permian (Berner and Canfield 1989).

SUMMARY

Biogeochemistry in the oceans offers striking contrasts to that on land. The environment on land is spatially heterogeneous; within short distances there are great variations in soil characteristics, including redox potential and nutrient turnover. In contrast, the sea is relatively well mixed. Large, long-lived plants dominate the primary production on land, versus small, ephemeral phytoplankton in the sea. A fraction of the organic matter in the sea escapes decomposition and accumulates in sediments, whereas soils contain little permanent storage of organic matter. Terrestrial plants are rooted in the soil, which harbors most of the nutrient recycling by bacteria and fungi. The soil is sometimes dry, which limits NPP in many areas. In contrast, marine phytoplankton are bathed in the medium of nutrient cycling and never limited by water.

Through their buffering of atmospheric composition and temperature, the oceans exert enormous control over the climate of Earth. At a pH of 8.1 and a redox potential of +200 mV, seawater sets the conditions for biogeochemistry on the 71% of the Earth's surface that is covered by seawater. Most of the major ions in the oceans have long mean residence times and their concentration in seawater has been relatively constant for at least the past 1 million years or more. All of this reinforces the traditional, and unfortunate, view that the ocean is a body that offers nearly infinite dilution potential for the effluents of modern society. As we find high concentrations of mercury and other toxins in pelagic fish and birds, we realize that this is no longer true (Monteiro and Furness 1997, Vo et al. 2011).

Looking at the sedimentary record, we see that the ocean has been subject to large changes in volume, nutrients, and productivity, due to changes in global climate. Already, we have strong reason to suspect that the productivity of coastal waters is affected by human inputs of N and P (e.g., Beman et al. 2005). Changes in the temperature and productivity of the central ocean basins may well indicate that global changes are affecting the oceans as a whole (Behrenfeld et al. 2006, Polovina et al. 2008). Several studies report a decline in the oxygen content of the global ocean waters (Whitney et al. 2007, Helm et al. 2011). With warming climate, the overturning circulation of the oceans will decline, leading to lower NPP in the surface waters (Schmittner 2005, but see Lozier et al. 2011). The oceans of the future are likely to be warmer, more acidic, and less productive than those of today—just at a time when the growing human population will expect greater productivity from them.

Recommended Readings

Berner, E.K., and R.A. Berner. 2011. *Global Environment* (third ed.). Princeton University Press.
Broecker, W.S. 1974. *Chemical Oceanography*. Harcourt Brace Jovanovich.
Burdige, D.J. 2006. *Geochemistry of Marine Sediments*. Princeton University Press.
Drever, J.I. 1997. *The Geochemistry of Natural Waters* (third ed.). Prentice Hall.
Falkowski, P.G., and A.H. Knoll. 2007. *Evolution of Primary Producers in the Sea*. Academic Press/Elsevier.
Holland, H.D. 1978. *The Chemistry of the Atmosphere and Oceans*. Wiley.
Millero, F.J. 2005. *Chemical Oceanography*. CRC Press.
Sarmiento, J.L., and N. Gruber. 2006. *Ocean Biogeochemical Dynamics*. Princeton University Press.

PROBLEMS

1. What is the maximum concentration of dissolved inorganic phosphate that might be expected in seawater?
2. What is the recycling efficiency in the deep ocean for C, N, and P? Of the total annual inputs of each element to the ocean from the land and atmosphere, what fraction is deposited in the sediments?
3. For the major constituents in seawater (Table 9.1), plot the mean residence time in years versus the ratio of the concentration in seawater divided by the mean concentration in the Earth's crust (Table 4.1). How do you interpret your results?
4. Using Equation 8.2, estimate the alkalinity (meq/l) of seawater (Table 9.1). Then, referring to Equation 8.1, and remembering that the pH of seawater is about 8.1, what is the contribution of CO_3^{2-} to the total alkalinity?
5. The equilibrium between the concentration of CO_2 in the atmosphere and that dissolved in seawater $[H_2CO_3^*]$ is determined by Henry's Law, where

$$[H_2CO_3^*]/pCO_2 = 0.0347 \text{ mol liter}^{-1} \text{ atm}^{-1} \text{ at } 25°C \text{ and } 1 \text{ atmosphere pressure.}$$
($H_2CO_3^*$ refers to the sum of dissolved CO_2 plus carbonic acid, H_2CO_3.)

Carbonic acid disassociates in seawater to form HCO_3^-. Following:

$$H_2CO_3^* \rightarrow HCO_3^- + H^+, \text{ with a pK} = 6.54 - 0.0071 \text{ (T)}$$

and HCO_3^- dissociates to form CO_3^-, following:

$$HCO_3^- \rightarrow CO_3^- + H^+, \text{ with a pK} = 10.59 - 0.0102 \text{ (T)}.$$

If the atmospheric concentration of CO_2 increases from 400 ul/l to 560 ul/l during the next 50 years, how much CO_2 will enter the surface ocean (1–100 m)?

GLOBAL CYCLES

CHAPTER

10

The Global Water Cycle

INTRODUCTION

The annual circulation of water is the largest movement of a chemical substance at the surface of the Earth. Through evaporation and precipitation, water transfers much of the heat energy received by the Earth from the tropics to the poles, just as a steam-heating system transfers heat from the furnace to the rooms of a house. Movements of water vapor through the atmosphere determine the distribution of rainfall on Earth, and the annual availability of water on land is the single most important factor that determines the growth of plants (Kramer 1982). Where precipitation exceeds evapotranspiration on land, there is runoff. Runoff carries the products of mechanical and chemical weathering to the sea.

In this chapter, we examine a general outline of the global hydrologic cycle and then look briefly at some indications of past changes in the hydrologic cycle and global water balance. Finally, we look somewhat speculatively at changes in the water cycle that may accompany global climate change and other human impacts in the future. These changes will have direct effects on global patterns of plant growth, the height of sea level, and the movement of materials in biogeochemical cycles.

Too often, we forget that adequate freshwater is the most essential resource for human survival. Changes in the water cycle have strong implications for the future of agricultural productivity and for the social and economic well-being of human society (Vörösmarty et al. 2000). Archeologists find the ruins of elaborate water-transport infrastructure throughout human history (e.g., Bono and Boni 1996, Sandor et al. 2007). Widespread drought seems associated with the collapse of the early Mesopotamian civilization in the Middle East around 2200 B.C. (Weiss et al. 1993) and the disappearance of the Mayan civilization in Mexico around 900 A.D. (Hodell et al. 1995, Peterson and Haug 2005, Medina-Elizalde and Rohlijng 2012).

THE GLOBAL WATER CYCLE

The quantities of water in the global hydrologic cycle are so large that it is traditional to describe the pools and transfers in units of km^3 (Figure 10.1). Each km^3 contains 10^{12} liters and weighs 10^{15} g. The flux of water in the water cycle may also be expressed in units of average depth. For example, if all the rainfall on land were spread evenly over the surface, each weather station would record a depth of about 700 mm/yr. Units of depth can also be used to express runoff and evaporation (e.g., Figure 4.17). For instance, annual evaporation from the oceans removes the equivalent of 1000 mm of water each year from the surface area of the sea.

Not surprisingly, the oceans are the dominant pool in the global water cycle (Figure 10.1). Seawater composes more than 97% of all the water at the surface of the Earth. The mean depth of the oceans is 3500 m (Chapter 9). Water held in polar ice caps and in continental glaciers is the next largest contributor to the global pool. About 70% of the world's freshwater is frozen in Antarctica (Parkinson 2006). Discounting the water in the oceans and ice caps, less than 1% of the Earth's water is fresh and accessible in rivers, lakes, and groundwater.

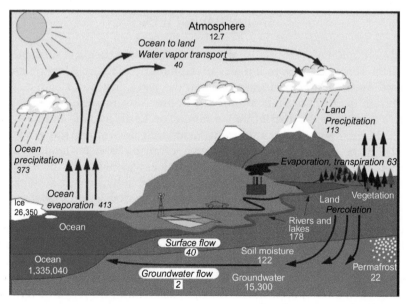

FIGURE 10.1   The global hydrologic cycle, with pools in units of 10^3 km^3 and flux in 10^3 km^3/yr. *Source: Modified from Trenberth (2007). Used with permission of the American Meteorological Society.*

Soils contain 121,800 km^3 of water, of which about 58,100 km^3 is within the rooting zone of plants (Webb et al. 1993). The large pool of freshwater below the unsaturated or vadose zone is known as groundwater (Chapter 7). Global estimates of the volume of groundwater are poorly constrained—4,200,000 to 23,400,000 km^3—but, except as a result of a few deep-rooted plants (e.g., Dawson and Ehleringer 1991) and the ingenuity of humans, groundwater is largely inaccessible to the biosphere. The pool of water in the atmosphere is tiny, equivalent to about 25 mm of rainfall at any given time (Eq. 3.4). Nevertheless, enormous quantities of water move through the atmosphere each year.

Evaporation removes about 413,000 km^3/yr of water from the world's oceans (compare Syed et al. 2010). Thus, the mean residence time of ocean water with respect to losses to the atmosphere is about 3100 years. Only about 373,000 km^3/yr of this water returns to the oceans in rainfall; the rest contributes to precipitation on land, which totals 113,000 km^3/yr. Plant transpiration and surface evaporation return 63,000 ± 13,100 km^3/yr from soils to the atmosphere (Ryu et al. 2011). The relative importance of plant transpiration (T) varies regionally (Table 10.1), but globally plants appear to be responsible for about 60% of total terrestrial evapotranspiration (ET) to the atmosphere (Choudhury et al. 1998, Alton et al. 2009), with the fraction T/ET increasing from 61% at 25% vegetation cover to 83% at 100% cover in greenhouse experiments (Wang et al. 2010a). With respect to precipitation inputs or evapotranspiration losses, the mean residence time of soil water is about 1 year.

When precipitation is below average, droughts develop relatively quickly, impacting a wide variety of ecosystem processes. During a 2005 drought, the rainforests of the Amazon Basin lost >1.2 × 10^{15} g C to the atmosphere (Phillips et al. 2009; compare Tian et al. 1998, Potter et al. 2011). Owing to the excess of precipitation over evapotranspiration on land, about 40,000 km^3/yr becomes runoff, derived from surface flow and groundwater. The runoff in rivers is supplemented by the subterranean flow of groundwater to the sea in coastal areas, which is potentially large but difficult to estimate (Moore 1996, 2010). One study estimates this flow as 2200 to 2400 km^3/yr (Zektser et al. 2007).

These global average values obscure large regional differences in the water cycle. Evaporation from the oceans is not uniform, but ranges from 4 mm/day in tropical latitudes to <1 mm/day at the poles (Mitchell 1983). Although much precipitation falls at tropical latitudes, an excess of evaporation over precipitation from the tropical oceans provides a net regional flux of water vapor to the atmosphere. Net evaporative loss accounts for the high salinity in tropical oceans (Figure 9.5), and the movement of water vapor in the atmosphere carries latent heat to polar regions (Trenberth and Caron 2001).

On land the relative balance of precipitation and evaporation differs strongly between regions. In tropical rainforests, precipitation may greatly exceed evapotranspiration. Shuttleworth (1988) calculates that 50% of the rainfall becomes runoff in the Amazon rainforests (Table 10.1). In desert regions, precipitation and evapotranspiration are essentially equal, so there is no runoff and only limited recharge of groundwater (Scanlon et al. 2006). As a global average, rivers carry about one-third of the precipitation from land to the sea (compare Alton et al. 2009). About 11% of precipitation becomes groundwater (12,666 km^3/yr; Doll and Fiedler 2008, Zektser and Loaiciga 1993), and the mean residence time of groundwater is about 1000 years (Slutsky and Yen 1997).

TABLE 10.1   Relative Importance of Pathways Leading to the Loss of Water from Terrestrial Ecosystems (except for T/ET, all data are shown in % of precipitation)

Vegetation	Precipitation (mm/yr)	Transpiration	Evaporation	Runoff and recharge	T/ET ×100	References
Tropical Rainforest						
	1623	45	56		45	Banerjee, cited in Galoux et al. 1981
	2000	49	26	26	65	Salati and Vose 1984
	2000	62	19	19	77	
		40	10	50	80	Shuttleworth 1988
	2209	56	11	32	84	Leopoldo et al. 1995
	2851	31	21		60	Calder et al. 1986
	3725	14	9		61	Frangi and Lugo 1985
Temperate Coniferous Forest						
	366	52	53		52	Pražák et al. 1994[a]
	595	59	36		55	Gash and Stewart 1977
	626	50	49		50	Lutzke and Simon, cited in Galoux et al. 1981
	627	40	48		41	Lutzke and Simon
	710	47	39		55	Rutter, cited in Galoux et al. 1981
	725	37	22	41	63	Tajchman 1972
	1085	49	15	41	77	Waring et al. 1981
	1127	8	12	80	39	Barbour et al. 2005[a]
	1225	49	15	38	77	McNulty et al. 1996
	2175	35	15	46	70	Waring et al. 1981
	2355	16	11	72	50	Waring et al. 1981
	2620	7	23		23	Hudson 1988
Temperate Deciduous Forest						
	349	39	58		40	Mitchell et al. 2009 (and personal communication)
	513	49	36	5	58	Molchanov, cited in Galoux et al. 1981
	549	54	9		86	Ladekarl 1998
	590	28-47				Gebauer et al. 2012[a]
	669	73	27		73	Paço et al. 2009
	763	33	15		69	Granier et al. 2000a[a]

TABLE 10.1   Cont'd

Vegetation	Precipitation (mm/yr)	Transpiration	Evaporation	Runoff and recharge	T/ET ×100	References
	1333	19	14		58	Wilson et al. 2001[a]
	2175	28	12	55	70	Waring et al. 1981
Boreal Forest						
	250	46	25		65	Grelle et al. 1997[a]
	271	51				Cienciala et al. 1997[a]
	872	45	8		85	Telmer and Veizer 2000
	502	39	35		53	Ten studies by Molchanov 1963, cited by Choudhury et al. 1998
	1237	19	26		42	Two studies by Delfs 1967, cited by Choudhury et al. 1998
Mediterranean Shrubland						
	475	60	40		60	Poole et al. 1981
	475	32	51	4	39	
	590	35	55	10	39	
Tropical Grassland						
	180	32	68		32	Hsieh et al. 1998[b]
	570	59	41		59	Chadwick et al. 2003
	1380	61	39		61	
	2500	72	28		72	
Temperate Grassland						
	365	65				Trlica and Biondini 1990[a]
	341	49	51		49	Lauenroth and Bradford 2006
		67	33	0	67	Massman 1992
	350	39	73		39	Hu et al. 2009
	477	37	67		37	
	580	56	83		56	
	580	39	49		44	
Steppe						
	144	45	55	0	45	Floret et al. 1982
	150	34	56	10	38	Paruelo and Sala 1995
	275	55	34		62	Huang et al. 2010[a]

Continued

TABLE 10.1   Relative Importance of Pathways Leading to the Loss of Water from Terrestrial Ecosystems (except for T/ET, all data are shown in % of precipitation)—Cont'd

Vegetation	Precipitation (mm/yr)	Transpiration	Evaporation	Runoff and recharge	T/ET × 100	References
Desert						
	150	35	65		35	Smith et al. 1995
	150	38	62		38	Liu et al. 2012
	165	27	73		27	Lane et al. 1984
	210	72	28		72	Schlesinger et al. 1987
	200	80	20		80	Liu et al. 1995
	260	21	36		37	Cavanaugh et al. 2011[a]
	212	21	27		44	

[a] *Growing season only*
[b] *Values are percentage of soil moisture lost at each site.*

The concept of *potential evapotranspiration* (PET), developed by hydrologists, expresses the maximum evapotranspiration that would be expected to occur under the climatic conditions of a particular site, assuming that water is always present in the soil and plant cover is 100%. Potential evapotranspiration is greater than the evaporation from an open pond, as a result of the plant uptake of water from the deep soil and a leaf area index >1.0 in many plant communities (Chapter 5). In tropical rainforests, PET and actual evapotranspiration (AET) are about equal (Vörösmarty et al. 1989). In deserts, PET greatly exceeds actual AET, owing to long periods when the soils are dry. In southern New Mexico, precipitation averages about 210 mm/year, but the receipt of solar energy could potentially evaporate over 2000 mm/yr from the soil (Phillips et al. 1988).

With higher values in warm, wet conditions, actual evapotranspiration (Figure 10.2) is often useful as a predictor of primary production (Figure 5.8), decomposition (Figure 5.17), and microbial activity (Figure 4.3). Changes in climate that affect rainfall and AET have a dramatic effect on the biosphere. Annual variability in AET is greatest in ecosystems with low AET, causing large year-to-year variations in net primary production related to annual precipitation in deserts (Frank and Inouye 1994, Prince et al. 1998). Actual evapotranspiration is more constant in tundra and boreal forest ecosystems, where wet soils do not constrain the supply of water to plants. The net primary productivity of land plants (60×10^{15} g C/yr) and the plant transpiration of water from land ($\sim$60% of 63×10^{18} g/yr) indicate that the global average water-use efficiency of vegetation is about 2.4 mmol of CO_2 fixed per mole of water lost (Eq. 5.3)—somewhat higher than the range measured by physiologists studying individual leaves (Chapter 5).

The sources of water contributing to precipitation also differ greatly among different regions of the Earth. Nearly all the rainfall over the oceans is derived from the oceans. On land, much of the rainfall in maritime and monsoonal climates is also derived from evaporation from the sea. Estimates of the percentage of rainfall derived from evapotranspiration from

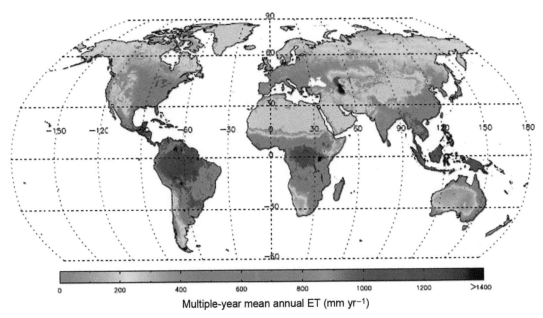

FIGURE 10.2 Global loss of water from the land surface by evapotranspiration. *Source: From Zhang et al. (2010). Used with permission of the American Geophysical Union.*

land are poorly constrained and regionally variable, with global estimates varying by 10 to 60% (Trenberth 1998, van der Ent 2010).[1]

Between 25 to 50% of the water falling in the Amazon Basin may be derived from evapotranspiration within the basin, with the rest derived from long-distance atmospheric transport from other regions (Salati and Vose 1984, Eltahir and Bras 1994). Evapotranspiration in Amazon forests is maximized by deep-rooted plants (Nepstad et al. 1994), and the regional importance of evapotranspiration in the Amazon basin speaks strongly for the long-term implications of forest destruction in that region. Using a general circulation model of the Earth's climate, Lean and Warrilow (1989) show that a replacement of the Amazon rainforest by a savanna would decrease regional evaporation and precipitation and increase surface temperatures (compare Shukla et al. 1990).

Regional cooling from plant transpiration is well known from comparisons of urban and rural areas (Juang et al. 2007). In semiarid regions, precipitation may decline as a result of the removal of vegetation, leading to soil warming (Balling 1989, Kurc and Small 2004, He et al. 2010) and increasing desertification (Schlesinger 1990, Chahine 1995, Koster et al. 2004). Thus, the transpiration of land plants is an important factor determining the movement of water in the hydrologic cycle and Earth's climate.

Estimates of global riverflow range from 33,500 to 47,000 km^3/yr (Lvovitch 1973, Speidel and Agnew 1982). A recent satellite-derived global estimate indicates 36,055 km^3 of annual

---

[1] Note the scale-dependence of such a statistic. At a local scale, nearly all the precipitation will be derived outside of the local area, whereas at larger regional scales an increased percentage will be derived within the area of interest (Trenberth 1998).

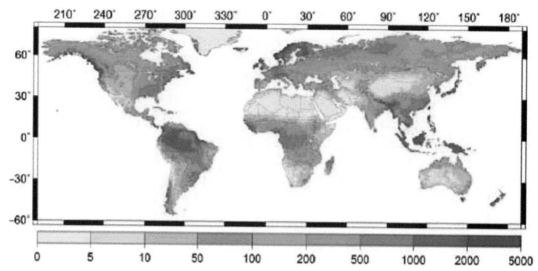

FIGURE 10.3   Annual runoff to the oceans, in mm/yr. *Source: From Oki and Kanae (2006). Used with permission of the American Association for the Advancement of Science.*

riverflow to the sea between 1994 and 2006 (Syed et al. 2010). For global models, many workers assume a value of about 40,000 km^3/yr (Figure 10.1). The distribution of flow among rivers is highly skewed. The 50 largest rivers carry about 43% of total riverflow; the Amazon alone carries ~20%. Reasonable estimates of the global transport of organic carbon, inorganic nutrients, and suspended sediments from land to sea can be based on data from a few large rivers. As a result of the positions of the continents and their surface topography, there are large regional differences in the delivery of runoff to the sea (Figure 10.3). The average runoff from North America is about 320 mm/yr, whereas the average runoff from Australia, which has a large area of internal drainage and deserts, is only 40 mm/yr (Tamrazyan 1989). Thus, the delivery of dissolved and suspended sediment to the oceans varies greatly between rivers draining different continents (Table 4.8).

In the Northern Hemisphere, 77% of the runoff is carried in rivers in which the flow is now regulated by dams and other human structures (Dynesius and Nilsson 1994, Nilsson et al. 2005), which strongly affect the sediment transport to the sea. The mean transit time of water to the sea has increased by an average of 60 days in rivers with significant impoundments and reservoirs (Vörösmarty and Sahagian 2000). Postel et al. (1996) calculate that humans now use 17% of the volume of rivers globally (~7000 km^2), converting a large portion of it to water vapor as a result of irrigated agriculture (Rost et al. 2008).

In some regions, such as the southwestern United States, human appropriation of surface runoff is as high as 76% (Sabo et al. 2010). Human extraction of groundwater, 145 to 283 km^3/yr (Wada et al. 2010, Konikow 2011), is less than 2% of groundwater recharge, but it is centered in arid and semiarid regions, which show large groundwater depletion (Rodell et al. 2009). Much of this groundwater is used for irrigation (Siebert et al. 2010, Wada et al. 2012). In the central valley of California, groundwater withdrawal of about 3 km^3/yr lowered the water table by ~20.3 mm/yr between 2003 and 2010 (Figure 10.4).

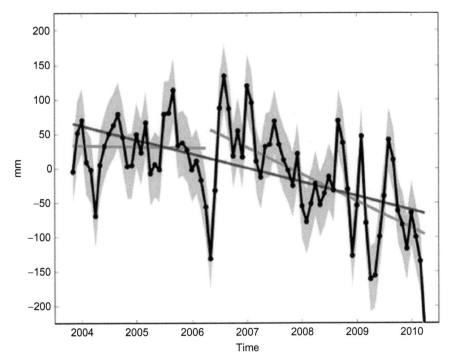

FIGURE 10.4 Groundwater levels in the Central Valley of California, showing changes in mm from a 2004 base-line value. *Source: From Famiglietti et al. (2011). Used with permission of the American Geophysical Union.*

The mean residence time of the oceans with respect to riverflow is about 34,000 years. The mean residence times differ among ocean basins. The mean residence time for water in the Pacific Ocean is 57,500 years—significantly longer than that for the Atlantic (18,000 years; Speidel and Agnew 1982, p. 30). This is consistent with the greater accumulation of nutrients in deep Pacific waters and a shallower carbonate compensation depth in the Pacific Ocean (Chapter 9). Despite the enormous riverflow of the Amazon, continental runoff to the Atlantic Ocean is less than the loss of water through evaporation. Thus, the Atlantic Ocean has a net water deficit, which accounts for its greater salinity (refer to Figure 9.5). Conversely, the Pacific Ocean receives a greater proportion of the total freshwater returning to the sea each year. Ocean currents carry water from the Pacific and Indian oceans to the Atlantic Ocean to restore the balance (refer to Figure 9.3)

MODELS OF THE HYDROLOGIC CYCLE

A variety of models have been developed to predict the movement of water in hydrologic cycles. Watershed models follow the fate of water received in precipitation and calculate runoff after subtraction of losses due to evaporation and plant uptake (Waring et al. 1981, Moorhead et al. 1989, Ostendorf and Reynolds 1993). In these models, the soil is considered as a collection of small boxes, in which the annual input and output of water must be equal.

Water entering the soil in excess of its water-holding capacity is routed to the next lower soil layer or to the next downslope soil unit via subsurface flow (Chapter 8). Models of water movement in the soil can be coupled to models of soil chemistry to predict the loss of elements in runoff (e.g., Nielsen et al. 1986, Knight et al. 1985, Furrer et al. 1990).

A major challenge in building these models is the calculation of plant uptake and transpiration loss. This flux is usually computed using a formulation of the basic diffusion law, in which the loss of water is determined by the gradient, or vapor pressure deficit, between plant leaves and the atmosphere. The loss is mediated by a resistance term, which includes stomatal conductance and wind speed (Chapter 5). In a model of forest hydrology, Running et al. (1989) assume that canopy conductance decreases to zero when air temperatures fall below 0°C or soil water potential declines below −1.6 MPa. Their model appears to give an accurate regional prediction of evapotranspiration and primary productivity for a variety of forest types in western Montana.

Larger-scale models have been developed to assess the contribution of continental land areas to the global hydrologic cycle. For example, Vörösmarty et al. (1989) divided South America into 5700 boxes, each 1/2° × 1/2° in size. Large-scale maps of each nation were used to characterize the vegetation and soils in each box, and data from local weather stations were used to characterize the climate. A model (Figure 10.5) was used to calculate the water balance

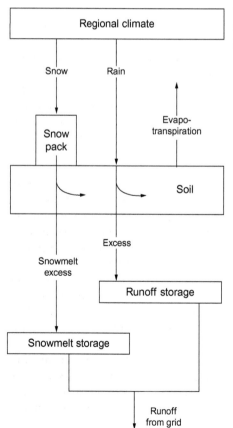

FIGURE 10.5   Components of a model for the hydrologic cycle of South America. *Source: From Vörösmarty et al. (1989). Used with permission of the American Geophysical Union.*

in each unit. During periods of rainfall, soil moisture storage is allowed to increase up to a maximum water-holding capacity determined by soil texture. During dry periods, water is lost to evapotranspiration, with the rate becoming a declining fraction of PET as the soil dries.

This type of model can be coupled to other models, including general circulation models of the Earth's climate (Chapter 3), to predict global biogeochemical phenomena. For example, a monthly prediction of soil moisture content for the South American continent can be used with known relationships between soil moisture and denitrification (Figure 6.17) to predict the loss of N_2O and the total loss of gaseous nitrogen from soils to the atmosphere (Potter et al. 1996). The excess water in the water-balance model is routed to stream channels, where it can be used to predict the flow of the major rivers draining the continent (Russell and Miller 1990, Milly et al. 2005). Changes in land use and the destruction of vegetation are easily added to these models, allowing a prediction of future changes in continental-scale hydrology and biogeochemistry.

THE HISTORY OF THE WATER CYCLE

As we learned in Chapter 2, water was delivered to the primitive Earth during its accretion from planetesimals, meteors, and comets. The accretion of the planets was largely complete by 4.5 billion years ago (bya). Then water was released from the Earth's mantle in volcanic eruptions (i.e., degassing), which continue to deliver water vapor to the atmosphere today—equivalent to $\sim 2.5 \ km^3/yr$ (Wallman 2001). As long as the Earth's temperature was greater than the boiling point of water, water vapor remained in the atmosphere. When the Earth cooled, nearly all the water condensed to form the oceans. Even then, a small amount of water vapor and CO_2 remained in the Earth's atmosphere—enough to maintain the temperature of the Earth above freezing via the greenhouse effect (Chapter 3). Today, water vapor and clouds account for 75% of the greenhouse effect on Earth (Lacis et al. 2010, Schmidt et al. 2010a). Without this greenhouse effect the surface of the Earth might be coated with a thick layer of ice, and biogeochemistry would be much less interesting.

There is good evidence of liquid oceans on Earth as early as 3.8 bya, and it is likely that the volume of water in the hydrologic cycle has not changed appreciably since that time. The total inventory of volatiles at the surface of the Earth (refer to Table 2.3) indicates that about 155×10^{22} g of water has been degassed from its crust. The difference between this value and the total of the pools of water shown in Figure 10.1 is largely contained in sedimentary rocks. Each year about 0.3 to 0.4 km^3 of water is carried to the Earth's mantle by subducted sediments (Wallace 2005, Wallman 2001, Parai and Mukhopadhyay 2012, Alt et al. 2012). This is less than the return of water vapor to the atmosphere in volcanic emissions since much of the water in subducted marine sediments is degassed before it reaches the lower mantle (Dixon et al. 2002, Green et al. 2010, van Keken et al. 2011, Alt et al. 2012). Nevertheless, a large volume of water, perhaps several times the current volume of the oceans, is retained in Earth's mantle (Marty 2012).

Owing to the low content of water vapor in the stratosphere, the Earth appears to have lost only a small amount of H_2O by photolysis, perhaps less than 35% of its total degassed water through Earth's history (Yung et al. 1989; Yung and DeMore 1999, p. 346; Pope et al. 2012). Much larger quantities appear to have been lost from Venus, where all water remains as vapor and exposed to ultraviolet light (Chapter 2). The loss of water from Venus is consistent with the high D/H ratio of its current inventory, reflecting the slower loss of its heavier isotope (Chapter 2). The accumulation of O_2 in the atmosphere and in oxidized minerals of the Earth's crust suggests that about 2% of Earth's water has been consumed by net photosynthesis through geologic time (refer to Table 2.3).

Throughout Earth's history, changes in relative sea level have accompanied periods of tectonic activity that increase (or decrease) the volume of submarine mountains. Changes in sea level also accompany changes in global temperature that lead to glaciations. The geologic record shows large changes in ocean volume during the 16 continental glaciations that occurred during the Pleistocene Epoch, extending to 2 million years ago (Bintanja et al. 2005, Dutton and Lambeck 2012). During the most recent glaciation, which reached a peak 22,000 years ago, an increment of $52,500 \times 10^3$ km^3 of seawater was sequestered in the polar and other glacial ice (Lambeck et al. 2002, Yokoyama et al. 2000). This represents nearly 4% of the ocean volume, and it lowered the sea level about 120 to 130 m from that of present day (Fairbanks 1989, Siddall et al. 2003). As we saw in Chapter 9, the Pleistocene glaciations are recorded in calcareous marine sediments. During periods of glaciation, the ocean was relatively rich in $H_2^{18}O$, which evaporates more slowly than $H_2^{16}O$. Calcium carbonate precipitated in these oceans shows higher values of $\delta^{18}O$, which can be used as an index of paleotemperature (refer to Figure 9.31).

Although many causes have been suggested, most climate scientists now believe that ice ages are related to small variations in the Earth's orbit around the Sun (Harrington 1987). These variations lead to differences in the receipt of solar energy, particularly in polar regions, and large changes in the Earth's hydrologic cycle. Once polar ice begins to accumulate, the cooling accelerates because snow has a high reflectivity or albedo to incoming solar radiation. Low concentrations of atmospheric CO_2 (refer to Figure 1.2) and high concentrations of sulfate aerosols (Legrand et al. 1991) and atmospheric dust (Lambert et al. 2008) during the last ice age were probably an effect, rather than a cause, of global cooling; however, these changes in the atmosphere may have reinforced the rate and onset of the cooling trend (Harvey 1988, Shakun et al. 2012).

Earth appears to enter each ice age relatively slowly ($\sim$50,000 years), whereas the warming to interglacial conditions occurs over a relatively short period (<1000 years; Figure 1.2). Some climate change is remarkably rapid; records of paleoclimate show periods when the mean annual temperature in Greenland rose as much as 9°C over a couple of decades (Taylor 1999, Severinghaus and Brook 1999). At the present time, the Earth is unusually warm; we are about halfway through an interglacial period, which should end about 12,000 A.D. There is substantial evidence that the Earth may have undergone several periods of frozen conditions in the Precambrian, and perhaps even more recently, yielding a "snowball" Earth, on which ice covered most of the planet (Hoffman et al. 1998, Kirschvink et al. 2000).

Continental glaciations represent a major disruption—a loss of steady-state conditions—in Earth's water cycle. Global cooling yields lower rates of evaporation, reducing the circulation of moisture through the atmosphere and reducing precipitation. One model of global climate suggests that during the last glacial epoch, total precipitation was 14% lower than that of today (Gates 1976). Throughout most of the world, the area of deserts expanded, and total net primary productivity and plant biomass on land were much lower (Chapter 5). Greater wind erosion of desert soils contributed to the accumulation of dust in ocean sediments, polar ice caps, and loess deposits (Yung et al. 1996, Lambert et al. 2008). The southwestern United States appears to have been an exception. Over most of this desert area, the climate of 18,000 years ago was wetter than today (Van Devender and Spaulding 1979, Wells 1983, Marion et al. 1985).

Changes in the rate of global riverflow produce changes in the delivery of dissolved and suspended matter to the sea. Broecker (1982) suggests that erosion of exposed continental shelf sediments during the glacial sea-level minimum may have led to a greater nutrient content of seawater and higher marine net primary productivity in glacial times. Worsley and Davies (1979) show that deep-sea sedimentation rates throughout geologic time have been greatest during periods of relatively low sea level, when a greater area of continents is displayed. These observations are consistent with the current imbalance in the ocean nitrogen cycle—the oceans may still be responding to large nitrogen inputs during the last glacial period (McElroy 1983; Figure 9.21).

THE WATER CYCLE AND CLIMATE CHANGE

Our ability to monitor ongoing changes in the global water cycle has improved dramatically with the deployment of the GRACE (Gravity Recovery and Climate Experiment) satellite system, which measures small changes in the distribution of Earth's gravity associated with changes in its mass, including that of water in soils, groundwater, and ice packs (Syed et al. 2008). For example, GRACE measurements show increasing water storage in the Amazon Basin during the seasonal period of heavy rains (Figure 10.6), and GRACE allows the calculation of evapotranspiration by subtracting changes in soil moisture from precipitation. The GRACE satellite has also allowed measurements of groundwater depletion in California (refer to Figure 10.4). Additionally, changes in sea level and ocean currents are accurately monitored by the TOPEX/Poseidon satellite (Chapter 9).

Rise in Sea Level

It is widely believed that global warming will cause a melting of the polar ice caps, leading to a rise in sea level and a flooding of coastal areas during the next century. Arctic rivers now carry a significantly enhanced flow of water compared to the early 1960s (Peterson et al. 2006,

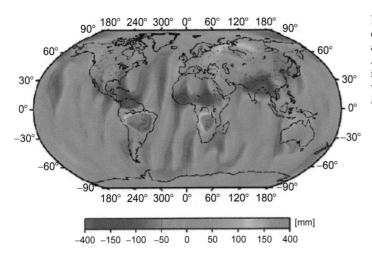

FIGURE 10.6 Changes in water content of the Earth's land surface, as measured by the GRACE satellite, April to August 2003. Note the large increase in water storage during the wet season in the Amazon. *Source: From Schmidt et al. (2006).*

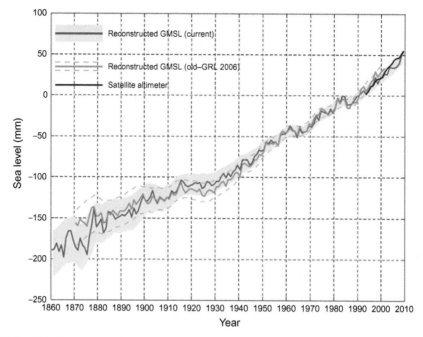

FIGURE 10.7   Global average sea level from 1860 to 2009. *Source: From Church and White (2011). Used with permission of Springer.*

McClelland et al. 2006). Deduced from a variety of methods, the long-term average rise in sea level has been between 1 to 2 mm/yr for the last 100 years (Church and White 2011, Merrifield et al. 2009, Kemp et al. 2011; Figure 10.7).

Observations of sea-level rise are complicated by the continuing isostatic adjustments of continental elevations in response to the melting of ice from the last continental glaciation (Sella et al. 2007). After removing this factor, Peltier and Tushingham (1989) found a rise of 2.4 mm/yr from 1920 to 1970, which they related to the effects of global warming on continental glaciers. Recent measurements by the TOPEX/Poseidon satellite suggest that relative sea level rose at a rate of 3.3 mm/yr for 1993 to 2009 (Cazenave and Llovel 2010), which is consistent with tide-gauge measurements during the same interval (Merrifield et al. 2009, Church and White 2011).

Sea-surface temperatures have also risen over the last 100 years (Levitus et al. 2001, 2005; Barnett et al. 2005), so some of the rise in sea level must be attributed to the thermal expansion of water at warmer temperatures (Miller and Douglas 2004, Antonov et al. 2005). Some rise in sea level may also stem from human activities, including the extraction of groundwater, which is then delivered to the sea by rivers (Sahagian et al. 1994, Konikow 2011, Pokhrel et al. 2012). The remaining rise in sea level is likely dominated by the melting of mountain glaciers throughout the world—an indication of a global warming trend (Oerlemans 2005, Meier et al. 2007, Jacob et al. 2012).

Loss of polar ice caps is associated with dramatic rise in sea level during past glacial cycles (Raymo and Mitrovica 2012, Deschamps et al. 2012). GRACE satellite measurements show that the ice pack on Greenland is in decline (van den Broeke et al. 2009), and the melting on Greenland is consistent with observations of lower salinity in North Atlantic waters (Curry and Mauritzer 2005, Dickson et al. 2002). The ice pack on Greenland shows the greatest decline at its margins and may actually be accumulating some mass in northern and interior regions (Krabill et al. 2000, Pritchard et al. 2009). However, the recent net loss of ice has accelerated to more than 250 km^3/yr (Velicogna and Wahr 2006a, Chen et al. 2006a, Rignot et al. 2011).

Overall, the massive ice cap on Antarctica shows smaller changes than in Greenland, although it also appears to be decreasing in volume (Velicogna and Wahr 2006a, Chen et al. 2009). Rapid melting on the Antarctic Peninsula may be balanced by slight accumulations of snowfall in East Antarctica (Rignot and Thomas 2002, Ramillien et al. 2006, Rignot et al. 2008). Although the record is not long, measurements from GRACE show an accelerating loss of ice mass from West Antarctica, now approaching 246 km^3/yr (Velicogna 2009, Rignot et al. 2011). The Ross Sea shows declining salinity that is consistent with these trends (Jacobs et al. 2002).

The measured losses of ice volume from Greenland, Antarctica, and continental glaciers yield a global rise in sea level of about 1.5 to 1.8 mm/yr (Meier et al. 2007, Jacob et al. 2012), about half of the observed rate. Presumably the thermal expansion of warmer seawater explains the difference. By 2100, losses from these ice packs may yield a total sea level rise >1 m, yielding widespread flooding of coastal environments and most of the world's major cities. Overall the Greenland ice pack contains the equivalent of 7 m of sea level rise, whereas Antarctica contains about 65 m. Thus, the expected melting during this century has released only a small fraction of the water in frozen ice packs.

Other human impacts on sea level include the extraction of groundwater, which adds water to the hydrologic cycle at the Earth's surface, and impoundments of water in reservoirs, which delays the delivery of water to the oceans. Impoundments now contain 8070 km^3 of water, and new dams are planned or under construction in many areas of Asia and South America (Lehner et al. 2011). Reservoirs and irrigation are thought to have reduced global riverflow by about 2% (930 km^3/yr; Biemans et al. 2011, Haddeland et al. 2006) and reduced the rise in sea level by 0.55 mm/yr (Chao et al. 2008). Groundwater extraction, on the other hand, may have increased the rise in sea level by 0.3 mm/yr (Sahagian et al. 1994, Konikow 2011).

Sea Ice

Just as the volume of a glass of water is not affected by ice cubes that may melt within it, sea level is not affected by changes in the area or volume of ice, known as *sea ice*, which floats on the ocean surface. Nevertheless, trends in sea ice are a useful index of Earth's climate. Dramatic losses of sea ice are seen in the Arctic, which may have ice-free summers within a few decades (Serreze et al. 2007, Comiso et al. 2008, Parkinson and Cavalieri 2008; Figure 10.8). The remaining ice has also lost thickness (Kwok and Rothrock 2009). In contrast, in the Southern Ocean, satellite measurements have shown relatively little change in the area of sea ice surrounding Antarctica during the past couple of decades (Zwally et al. 2002, Cavalieri and Parkinson 2008). Meanwhile, historical records of whaling near Antarctica show an

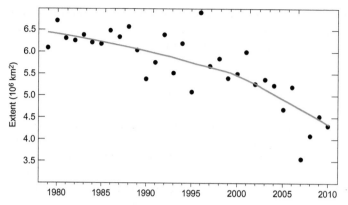

FIGURE 10.8 September sea ice extent, 1979 to 2010 in the Arctic. *Source: From Stroeve et al. (2012). Used with permission of Springer.*

increasing frequency of kills at extreme southern latitudes, consistent with a long-term decline in the extent of sea ice in the Southern Ocean (de la Mare 1997, Cotte and Guinet 2007). Loss of the natural reflectivity or albedo of snow cover and sea ice in the Arctic is estimated to have contributed 0.1 to 0.45 W m^{-2} to global warming in the interval from 1979 to 2008—reinforcing the ongoing climate change due to the accumulation of greenhouse gases in the atmosphere (Hudson 2011, Flanner et al. 2011).[2]

Terrestrial Water Balance

In response to future global warming, most climate models predict a more humid world, in which the movements of water in the hydrologic cycle through evaporation, precipitation, and runoff are enhanced (Loaiciga et al. 1996, Huntington 2006). Increased cloudiness may moderate the degree of warming, but a new steady state in Earth's temperature will be found at a higher value than that of today (Chapter 3). Not all areas of the land will be affected equally. Most of the anticipated temperature change is confined to high latitudes, and Manabe and Wetherald (1986) show that large areas of the central United States and Asia may experience a reduction in soil moisture, leading to more arid conditions. Due to the thermal buffering capacity of water, the oceans may warm more slowly than the land surface. Because most precipitation is generated from the oceans, land areas may experience severe drought during the transient period of global warming (Rind et al. 1990, Dirmeyer and Shulka 1996). Such changes in precipitation and temperature will lead to large-scale adjustments in the distribution of vegetation and global net primary production (Emanuel et al. 1985b, Smith et al. 1992b, Neilson and Marks 1994).

Analyzing the rainfall records of 1487 weather stations, Bradley et al. (1987) found an increase in precipitation over most of the midlatitudes in the Northern Hemisphere during

---

[2] Compare to the 2.3 W m^{-2} estimated to derive from the current accumulation of greenhouse gases from human activities (Chapter 3).

a 30- to 40-year period—consistent with changes expected for a warmer planet. Measured changes in global precipitation are small, since some areas with greater precipitation partially compensate for those reporting lesser amounts (Smith et al. 2006). Wentz et al. (2007) report a 2.8% increase in global precipitation between 1987 and 2006 (compare Dai et al. 1997).

In many areas, precipitation also seems to be becoming more variable; that is, floods and droughts are more frequent—consistent with the predictions of several general circulation models of future climate (Min et al. 2011). Over much of the world, the historical record of precipitation is scanty, and we must hope that global estimates of precipitation will improve dramatically with further application of satellite remote sensing (Petty 1995), especially the anticipated launch of the Global Precipitation Measurement (GPM) satellite in 2013.[3] Because water vapor absorbs microwave energy, the relative transmission of microwave radiation through the atmosphere is related to water vapor content and rainfall, and satellite remote sensing of the microwave emission from Earth can measure the rainfall and soil moisture over large areas (e.g., Weng et al. 1994).

One might expect that a warmer land surface would increase rates of evaporation and a warmer atmosphere might contain more water vapor. Water-balance studies of the terrestrial watersheds show an increase in evapotranspiration during the past 50 years (Walter et al. 2004a), correlated with an increased occurrence of drought (Dai et al. 2004), especially in the southwestern United States (Andreadis and Lettenmaier 2006, Seager et al. 2007). Analyses of long-term records show increasing evaporation in recent years (Szilagyi et al. 2001, Golubev et al. 2001, Brutsaert 2006). Global evapotranspiration has increased during most of the past 25 years, except during periods of drought associated with El Niño events (Jung et al. 2010). There are also indications of increases in humidity (Willett et al. 2007), particularly in regions where irrigated agriculture is widespread (Sorooshian et al. 2011).

Greater precipitation should lead to greater runoff from land (Miller and Russell 1992). Probst and Tardy (1987) found a 3% increase in stream flow in major world rivers over the last 65 years (compare Labat et al. 2004, McCabe and Wolock 2002, Syed et al. 2010). This increased stream flow may be an indication of global climate change, but it may also relate to the human destruction of vegetation leading to greater runoff (DeWalle et al. 2000, Brown et al. 2005, Rost et al. 2008, Peel et al. 2010). We might also speculate that greater stream flow is expected due to greater water-use efficiency by vegetation growing in a high-CO_2 atmosphere (Gedney et al. 2006, Betts et al. 2007; Chapter 5). In cold temperate regions, wintertime snowpack will be smaller and the spring runoff earlier in a warmer climate (Barnett et al. 2005, Burns et al. 2007).

The historical pattern of runoff for each continent and for the world as a whole shows a cyclic pattern (Probst and Tardy 1987). The cycles for the continents are not synchronous, so the trends in the global record are "damped," relative to those on each continent. Predictions of future runoff show differences between major regions of the world, with some

---

[3] http://pmm.nasa.gov/GPM/.

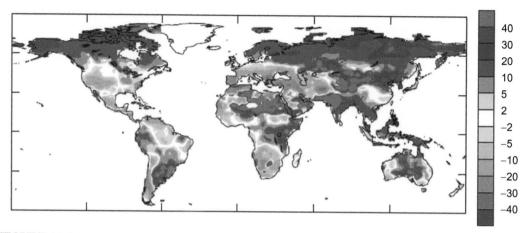

FIGURE 10.9   Projected percent changes in runoff from the Earth's land surface for 2041–2060, compared to the mean for 1900–1970. *Source: From Milly et al. (2005). Used with permission of the American Association for the Advancement of Science.*

showing increases and some showing extensive decreases (Figure 10.9). Impoundments of water in reservoirs may reduce regional impacts on the availability of water to humans.

In sum, the recent increases in precipitation, evaporation, and stream flow are consistent with predicted changes in the water cycle with global warming, but such observations must be evaluated in the context of long-term cycles in climate that have occurred through geologic time.

SUMMARY

Through evaporation and precipitation, the hydrologic cycle transfers water and heat throughout the global system. Receipt of water in precipitation is one of the primary factors controlling net primary production on land. Changes in the hydrologic cycle through geologic time are associated with changes in global temperature. All evidence suggests that movements in the hydrologic cycle were slower in glacial time and that they would be likely to increase with global warming. Movements of water on the surface of the Earth affect the rate of rock weathering and other biogeochemical phenomena.

Management of water resources will be critical to sustain the world's growing population. Data suggest that there is substantial room for improved management options. People use water very differently in different nations, with direct and indirect per capita consumption ranging from 700 m^3/yr in China to 2480 m^3/yr in the United States (Hoekstra and Chapagain 2007). Decreases in the per capita availability of water are likely to accompany changes in climate and human population during the next several decades, leaving substantial areas of the world with an increasing scarcity of water (Vörösmarty et al. 2000; Figure 10.10).

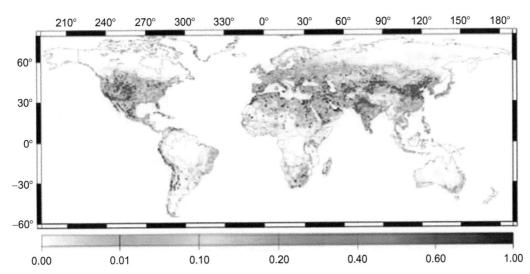

FIGURE 10.10 A water scarcity index for the Earth's land surface. Water scarcity is defined as the withdrawal from surface water divided by recycling, adjusted for regional supplements from desalinization. *Source: From Oki and Kanae (2006).*

Recommended Readings

Baumgartner, A., and E. Reichel. 1975. *The World Water Balance.* R. Olenburg.
Browning, K.A., and R.J. Gurney (Eds.). 1999. *Global Energy and Water Cycles.* Cambridge University Press.
National Academy of Sciences. 2012. *Challenges and Opportunities in the Hydrologic Sciences.* National Academy Press.
Oliver, H.R., and S.A. Oliver. 1995. *The Role of Water and the Hydrologic Cycle in Global Change.* Springer-Verlag.
Sumner, G. 1988. *Precipitation: Process and Analysis.* Wiley.
Van der Leeden, F., F.L. Troise, and D.K. Todd. 1990. *The Water Encyclopedia.* Lewis Publishers.
Ward, R.C. 1967. *The Principles of Hydrology.* McGraw-Hill.

PROBLEMS

1. If the 3% rise in global riverflow during the last 65 years (Probst and Tardy 1987) is all derived from a greater water-use efficiency of land vegetation in response to rising atmospheric CO_2, what has been the change in the water-use efficiency calculated on a global basis?
2. If the surface ocean (0–100 m) should increase in temperature by 3°C during the next century, how much will the thermal expansion of seawater contribute to a rise in sea level? (For this question, you will need to look up the equation for the thermal expansion of water as a function of temperature; try the *Handbook of Chemistry and Physics*).
3. How much larger would be the total inventory of water on Earth if some of it had not been consumed (i.e., split by photosynthesis) to produce molecular oxygen?
4. Calculate the amount of heat released to the atmosphere (calories) in the northward flow of water in the Gulf Stream (~31 Sv) as it leaves the tropical Atlantic at a temperature of 30°C and sinks to form North Atlantic Deep Water (NADW) at approximately 3°C. How does this compare to the heat that is released to Earth's atmosphere from the current annual combustion of fossil fuels?

The Global Carbon Cycle

INTRODUCTION

The carbon cycle is of central importance to biogeochemistry. Life is composed primarily of carbon, so estimates of the global production and destruction of organic carbon give us an overall index of the health of the biosphere—both past and present. Photosynthetic organisms capture sunlight energy in organic compounds that fuel the biosphere and account for the presence of molecular O_2 in our atmosphere. Thus, the carbon and oxygen cycles on Earth are inextricably linked, and the presence of O_2 in Earth's atmosphere sets the redox potential for organic metabolism in most habitats. Through oxidation and reduction, organisms transform the other important elements of life (e.g., N, P, and S) in reactions that capitalize on the presence of organic carbon and oxygen on Earth. When we understand the carbon cycle, we can make accurate first approximations of the movement of elements in other global cycles, recognizing the predictable stoichiometry of the chemical elements in organic matter. Finally, there is good evidence that through the burning of fossil fuels and other activities, humans have altered the global cycle of carbon, causing the atmospheric concentration of CO_2 to rise to levels that have never been experienced during the evolutionary history of most species that now occupy our planet (Pearson and Palmer 2000).

In this chapter, we consider a simple model for the carbon cycle on the Earth and for assessing human impacts on that cycle. We then consider the magnitude of past fluctuations in the carbon cycle to gain some perspective on the current human impact. We look briefly at the budgets of methane (CH_4) and carbon monoxide (CO) in the atmosphere. Because increasing concentrations of carbon dioxide and methane are associated with global warming through the greenhouse effect (Figure 3.2), the global carbon cycle is directly linked to considerations of global climate change and to international efforts to combat global warming. Finally, we examine the linkage of the carbon and oxygen cycles on Earth as a means of "cross-checking" our estimated budgets for these elements at the global level.

THE MODERN CARBON CYCLE

The Earth contains about 32×10^{23} g of carbon (Marty 2012). About 7×10^{22} g C is found in the upper mantle (Zhang and Zindler 1993). At the crust of the Earth, carbon is found in sedimentary rocks (Table 2.3), where it is held in organic compounds (1.56×10^{22} g C; Des Marais et al. 1992) and carbonates (5.4–6.5×10^{22} g C; Li 1972, Lecuyer et al. 2000; compare Table 2.3). Conventional fossil fuels are estimated to contain 4×10^{18} g C (Sundquist and Visser 2005). Live microbial biomass is a small component of the sedimentary organic carbon, perhaps containing up to 100×10^{15} g C (Lipp et al. 2008; compare to Whitman et al. 1998, who suggest up to 546×10^{15} g C).

The sum of the active pools of carbon at the Earth's surface is about 40×10^{18} g C (Figure 11.1). Dissolved inorganic carbon in the ocean is the largest near-surface pool, which has an enormous

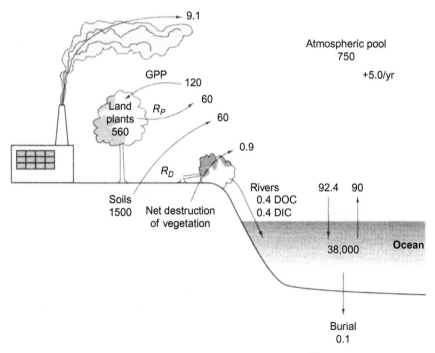

FIGURE 11.1   The global carbon cycle. All pools are expressed in units of 10^{15} g C and all annual fluxes in units of 10^{15} g C/yr, estimated for 2010. Values are taken from the text.

capacity to buffer changes in the atmosphere via Henry's Law (Eq. 2.7). At equilibrium, the sea contains about 50 times as much carbon as the atmosphere (Raven and Falkowski 1999). On land, the largest pool of carbon is contained in soils (Table 5.4). Even though atmospheric CO_2 is measured in parts per million, the atmosphere contains more carbon than all of the Earth's living vegetation (600×10^{15} g C; Table 5.3).

The largest fluxes of the global carbon cycle are those that link atmospheric carbon dioxide to land vegetation and to the sea (Figure 11.1). Global net primary production on land is estimated at 60×10^{15} g C/yr (Chapter 5), which is about half of recent estimates of gross primary production from the MODIS satellite (Beer et al. 2010, Ryu et al. 2011, Jung et al. 2011). Considering land vegetation alone, we find that each molecule of CO_2 in the atmosphere has the potential to be captured in net primary production in about 12.5 years. The annual exchange of CO_2 with the oceans is somewhat greater, so the overall mean residence time of CO_2 in the atmosphere is about 5 years.[1] Because this mean residence time is similar to the mixing time of the atmosphere, CO_2 shows small regional and seasonal variations that are superimposed on its global average concentration of nearly 400 ppm today (Figure 3.6).

The oscillations in the atmospheric content of CO_2 are the result of the seasonal uptake of CO_2 by photosynthesis in each hemisphere and seasonal differences in the use of fossil fuels and in the exchange of CO_2 with the oceans. The role of photosynthesis is indicated inasmuch as the oscillation in CO_2 is mirrored by slight oscillations in atmospheric O_2, which has a much longer mean residence time in the atmosphere and a much larger pool size (Figure 1.1). Globally, about two-thirds of the terrestrial vegetation occurs in regions with seasonal periods of growth, and the remainder occurs in the moist tropics, where growth occurs throughout the year (Box 1988). The seasonal effect of photosynthesis on atmospheric CO_2 is most pronounced in the Northern Hemisphere (Figure 3.6; Hammerling et al. 2012), which contains most of the world's land area. At high northern latitudes, vegetation accounts for about 50% of the seasonal variation in atmospheric CO_2 (D'Arrigo et al. 1987). In the Southern Hemisphere, smaller fluctuations in atmospheric CO_2 are seasonally reversed relative to the Northern Hemisphere, and they appear to be dominated by exchange with ocean waters (Keeling et al. 1984). The oscillation at Mauna Loa, Hawaii, located at 19° N latitude, is about 6 ppm/yr (Figure 1.1), equivalent to a transfer of about 13×10^{15} g C/yr to and from the atmosphere. This value is less than the annual net primary productivity of land plants (Table 5.3) because of the asynchrony of terrestrial photosynthesis and respiration throughout the globe and the buffering of atmospheric CO_2 concentrations by exchange with the oceans.

The release of CO_2 in fossil fuels, currently more than 9×10^{15} g C/yr, is one of the best-known values in the global carbon cycle.[2] If all of this CO_2 accumulated in the atmosphere, the annual increment would be $>1.0\%$/yr. In fact, the atmospheric increase is about 0.4%/yr (2 ppm) because only about 56% of the fossil fuel release remains in the atmosphere (Sabine

---

[1] This calculation is based on the traditional definition of mean residence time for a steady-state pool (i.e., mass/input) see footnote 1, Chapter 3. A recent report of the Intergovernmental Panel on Climate Change (Houghton et al. 1995; Table 3, p. 25) indicates a mean residence time of 50 to 200 years for atmospheric CO_2. This is actually the time it would take for the current human perturbation of the atmosphere to disappear into other pools at the Earth's surface (e.g., the ocean) if the use of fossil fuels were to cease. Thus, it would take several centuries to return steady-state conditions to the carbon cycle on Earth.

[2] *www.globalcarbonproject.org/carbonbudget/10/presentation.htm.*

et al. 2004). This is known as the "airborne fraction," which has increased slightly in recent years (Canadell et al. 2007, Le Quere et al. 2009). Where is the rest?

Oceanographers believe that about 32% ($\sim 2 \times 10^{15}$ g C/yr) of the CO_2 released from fossil fuels enters the oceans each year (Sabine et al. 2004, Khatiwala et al. 2009). This estimate is derived from measurements of the simultaneous increase in chloroflurocarbons, ^{3}H, ^{14}C, δ^{13}C, and total inorganic carbon in seawater as a result of human activities (e.g., McNeil et al. 2003, Sweeney et al. 2007, Quay et al. 2003, Takahashi et al. 1997). Thus, Figure 11.1 shows an annual uptake by the oceans (92×10^{15} g/yr) that is slightly greater than the return of CO_2 to the atmosphere (90×10^{15} g C/yr). Following Henry's Law (Eq. 2.7), the excess CO_2 dissolves in seawater, where it causes ocean acidification and dissolution of marine carbonates (Eqs. 9.3–9.5). The uptake of CO_2 by the oceans shows significant year-to-year variation, amounting to $\sim 20\%$ of the estimated net flux (Gruber et al. 2002, Watson et al. 2009). Most of the uptake occurs in the large areas of downwelling water in the North Atlantic and Southern oceans (Figure 11.2) and in continental shelf areas throughout the world (Thomas et al. 2004a). Owing to the low levels of nutrients in seawater, changes in marine NPP are believed to be relatively unimportant to the current oceanic uptake of anthropogenic CO_2, which is largely determined by the dissolution of CO_2 in the surface waters (Shaffer 1993).

Remembering that the exchange of CO_2 between the atmosphere and the oceans takes place only in the surface waters (Chapter 9), we can calculate the mean residence time of CO_2 in the surface ocean—about 10 years—by dividing the pool of carbon in surface waters

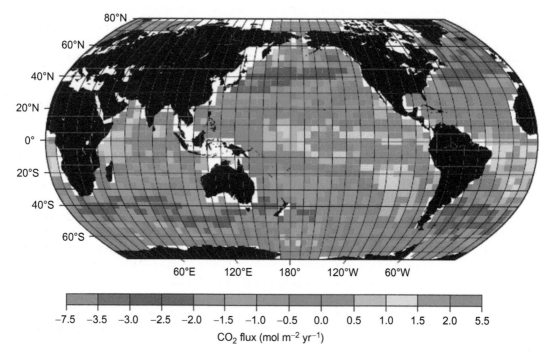

FIGURE 11.2   Estimates of the flux of CO_2 between the atmosphere and the oceans' surface for 1995. *Source: From Denman et al. (2007). Used with permission of Cambridge University Press.*

TABLE 11.1   Global Budget for Anthropogenic CO_2 in Earth's Atmosphere

	Fossil fuel		Biomass destruction[a]	=	Atmospheric increase		Ocean uptake		Terrestrial uptake	References
1990s	6.4	+	1.6	=	3.2	+	2.2	+	2.6	IPCC (2007)
2000–2007			1.1						2.3	Pan et al. (2011)
2010[b]	9.1		0.9		5.0		2.4		2.6	

Note: All data in 10^{15} g C/yr.
[a] Net biomass destruction in the tropics.
[b] Source: www.globalcarbonproject.org/carbonbudget/index.htm.

(921×10^{15} g C) by the rate of influx (92×10^{15} g C/yr). A similar mixing time is calculated from the distribution of ^{14}C in the surface ocean (Chapter 9). Turnover of carbon in the entire ocean is much slower, about 350 years—consistent with the age of deep ocean waters. Thus, the uptake of CO_2 by the oceans is constrained by mixing of surface and deep waters—not by the rate of dissolution of CO_2 across the surface (Chapter 9). Significantly, if the oceans were completely mixed, they might take up as much as 6×10^{15} g C/yr (Keeling 1983), indicating that it is the rate of release from fossil fuels relative to the rate at which the oceans can take up carbon that accounts for the current increase of CO_2 in the atmosphere. If the release of CO_2 from fossil fuels were curtailed, nearly all the CO_2 that has accumulated in the atmosphere would eventually dissolve in the oceans and the global carbon cycle would return to a steady state in a couple of hundred years (Laurmann 1979).

Taken alone, the atmospheric increase and oceanic uptake of CO_2 account for almost 90% of the annual emissions from fossil fuels (Sabine et al. 2004). Considering the errors associated with these global estimates, it would seem that we have a fairly tidy picture of the global carbon cycle. However, many terrestrial ecologists believe that there have also been substantial releases of CO_2 from terrestrial vegetation and soils, caused by the destruction of forest vegetation in favor of agriculture, especially in the tropics (Chapter 5). If their calculations are accurate, then the atmospheric budget is misbalanced, and a large amount of carbon dioxide that ought to be in the atmosphere is "missing" (Table 11.1).

The net release of CO_2 from vegetation is difficult to estimate globally (Chapter 5). At any time, some land is being cleared, while agriculture is abandoned in other areas that are allowed to regrow. For example, the Amazon Basin appears roughly in balance with respect to carbon flux, despite observations of widespread deforestation in the region (Houghton et al. 2000, Hansen et al. 2010). Estimates of carbon flux from the Amazonian rainforest are complicated by significant year-to-year variation, and the region is a net source of CO_2 during droughts (Tian et al. 1998, Phillips et al. 2009, Potter et al. 2011).

Historical changes in the ^{13}C and ^{14}C isotopic ratios in atmospheric CO_2 show unequivocal evidence of a net release from the biosphere early in this century (Wilson 1978).[3] Both are diluted by burning of fossil fuels, whereas only ^{13}C is reduced by burning vegetation.

[3] The changes in $^{14}CO_2$ in the atmosphere were only meaningful until the early 1960s, when the atmospheric signal was overwhelmed by the release of radiocarbon from atmospheric testing of nuclear weapons.

Until about 1960, the release from land clearing may have exceeded the release from fossil fuel combustion (Houghton et al. 1983). For the 1990s, estimates of the net CO_2 released from deforestation in the tropics range from 0.9 to 1.5 $\times$ 10^{15} g C (DeFries et al. 2002, Pan et al. 2011), but recent estimates of the *gross* release of carbon from tropical deforestation is much less—0.81 $\times$ 10^{15} g C/yr (Harris et al. 2012). The release of carbon from tropical deforestation is partially balanced by the carbon captured by the regrowth of forests in temperate and boreal regions, estimated as 0.65 to 1.17 $\times$ 10^{15} g C (Goodale et al. 2002, Myneni et al. 2001, Pan et al. 2011). Forests in the United States—mostly in New England—now appear to accumulate 0.1 to 0.2 $\times$ 10^{15} g C/yr (Birdsey et al. 1993, Turner et al. 1995, Woodbury et al. 2007, Thompson et al. 2011, Zhang et al. 2012a). Similar amounts of forest regrowth are estimated for Europe (0.165 $\times$ 10^{15} g C/yr; Peters et al. 2010), China (<0.26 $\times$ 10^{15} g C/yr; Piao et al. 2009), and Russia (<0.13 $\times$ 10^{15} g C/yr; Beer et al. 2006). Old-growth tropical forests also appear to provide net carbon storage from the atmosphere (Lewis et al. 2009, Davidson et al. 2012), so the global net sink for carbon in vegetation may be about 1.1 $\times$ 10^{15} g C/yr (Pan et al. 2011). Niwa et al. (2012) provide an independent estimate of the gross carbon sequestration in land vegetation (2.22 PgC/yr) using measurements of variations in atmospheric CO_2 obtained from passenger aircraft (compare Table 11.1). Global estimates of changes in the carbon held in vegetation and soils will improve with the application of remote sensing by satellites, which shows a significant flux of carbon from selective logging, not normally considered in estimates of deforestation (Asner et al. 2010). Any net release of carbon from land complicates our ability to balance a carbon dioxide budget for the atmosphere (Table 11.1).

We might reconcile the carbon dioxide budget of the atmosphere if we find evidence that the pool of carbon in land vegetation and soils has increased as a result of a global stimulation of plant growth by higher concentrations of atmospheric CO_2 (Chapter 5). Despite widespread forest destruction, enhanced uptake of CO_2 in areas of undisturbed vegetation could act as a sink for atmospheric CO_2 and add to the pool of carbon on land. The overall stimulation of terrestrial photosynthesis by human activities is informally known as the "beta" factor in models of the global carbon cycle. Beta is usually defined as the change in NPP that would derive from a doubling of atmospheric CO_2 concentration. In controlled experiments with tree seedlings, the beta factor usually lies in a range of 32 to 41% as a result of CO_2 fertilization (Poorter 1993, Curtis and Wang 1998, Wang et al. 2012). Free-Air CO_2 Enrichment (FACE) experiments (Chapter 5) show an average 18% stimulation of net primary production in forests grown at 1.5 times current levels of CO_2 (Norby et al. 2005). Most of the net carbon uptake is stored in woody biomass (McCarthy et al. 2010); changes in soils are less dramatic (van Groenigen et al. 2006, Lichter et al. 2008). Through a reallocation of photosynthate to roots, many trees appear to avoid the nutrient deficiencies that might be expected to develop as a result of faster growth at high CO_2 (Drake et al. 2011). However, soil nitrogen appears to constrain the long-term growth response at the FACE experiment at Oak Ridge, Tennessee (Norby et al. 2010, Garten et al. 2011).

The historical record of CO_2 in the atmosphere offers several indirect approaches for estimating changes in global net primary production and the potential for a significant, positive beta factor. For example, in the record of atmospheric CO_2 in the Northern Hemisphere, the seasonal decline each summer is largely due to photosynthesis, while the seasonal upswing in the autumn derives from decomposition. An increasing *amplitude* of the CO_2 oscillation, after the removal of fossil fuel and El Niño effects, implies a greater activity of the terrestrial

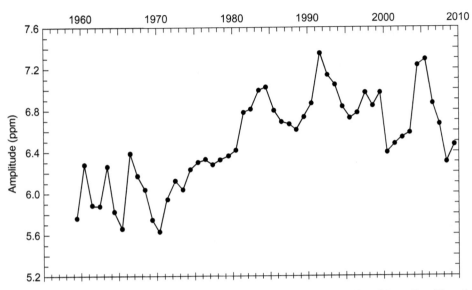

FIGURE 11.3   Increasing amplitude of the seasonal oscillations in atmospheric CO_2 at Mauna Loa, Hawaii. *Source: From Steve Piper, personal communication (2012).*

biosphere outside the tropics, where growth and decomposition occur year-round. Such a trend is evident in an analysis of the Mauna Loa record of CO_2 (Figure 11.3), in which the amplitude has increased by about 0.54%/yr from 1958 until the mid 1990s (Bacastow et al. 1985, Keeling 1993). The increase of amplitude at high northern latitudes has been about 0.66%/yr since 1960, perhaps as a result of climatic warming in this region over the same period (Keeling et al. 1996, Randerson et al. 1997). Although an increasing annual oscillation in the concentration of atmospheric CO_2 suggests that biospheric processes have been stimulated, we should not necessarily assume that a greater amount of carbon is being stored on land. Greater rates of decomposition may simply balance increased rates of photosynthesis (Houghton 1987, Keeling et al. 1996, Piao et al. 2008).

There is a small difference (about 4 ppm) in the concentration of atmospheric CO_2 between the Northern and the Southern Hemispheres, owing to the greater use of fossil fuel in the Northern Hemisphere (Keeling 1993). This observed latitudinal gradient in atmospheric CO_2 can be compared to the concentrations of CO_2 that would be expected due to the rate of mixing of the atmosphere, as calculated from general circulation models (Chapter 3). In fact, many of these models suggest that the concentration in the Northern Hemisphere, particularly at temperate latitudes, should be even greater than what is actually observed in the atmosphere, implying that there is significant uptake, perhaps by land plants (Tans et al. 1990, Denning et al. 1995, Fan et al. 1998b). The relative importance of land plants (versus uptake by the ocean) is also indicated by a latitudinal gradient in the isotopic ratio of atmospheric CO_2 (i.e., $\delta^{13}C$), which is fractionated by photosynthesis but not by dissolution in seawater (Ciais et al. 1995). These *inverse* models suggest a net uptake of carbon dioxide in vegetation worldwide—largely in forests, which contain most of the world's biomass.

Despite substantial theoretical and indirect evidence that it *should* occur, the direct evidence that an enhanced growth of land plants (e.g., changes in tree-ring thickness) currently acts as a large sink for atmospheric CO_2 is equivocal (Chapter 5). Our knowledge of the overall response of the terrestrial biosphere to future conditions is limited to a few studies that have examined changes in both CO_2 and temperature. Whereas plant uptake of CO_2 might increase as a result of longer growing seasons, one might expect substantial carbon losses from warmer soils. Oechel et al. (1994) found that wet tundra ecosystems in Alaska showed a complete physiological adjustment and no net carbon storage in response to a 3-year exposure to elevated CO_2. However, carbon storage in these ecosystems increased when both CO_2 and temperature were maintained at elevated levels. It is possible that the warmer soil temperatures enhanced decomposition, improving the supply of nutrients for plant growth (Van Cleve et al. 1990). Because the C/N ratio of soil organic matter (12–15) is lower than the C/N ratio of plant tissues (about 160; Table 6.5), a small amount of additional nitrogen mineralization in soils could yield a large enhancement of NPP and carbon sequestration in plants (Rastetter et al. 1992, McGuire et al. 1992). In a soil-warming experiment at the Harvard Forest in Massachusetts. Melillo et al. (2011) found that the mobilization of nitrogen from decomposing soil organic matter stimulated carbon uptake by the trees, so that changes in net carbon storage in the entire ecosystem were relatively minor.

Many areas of the world receive an excess atmospheric deposition of nitrogen derived from anthropogenic emissions of NO_x and NH_3 (Chapters 3 and 12). In some areas the N input is so extreme that symptoms of forest decline are observed (Chapter 6), but in other areas the added nitrogen has the potential to act as a fertilizer, stimulating plant growth (Townsend et al. 1996). If the global human production of fixed N ($\sim 100 \times 10^{12}$ g/yr; Chapter 12) were all stored in the woody tissues of plants with a C/N ratio of ~ 160 (Table 6.5), then as much as 16×10^{15} g C/yr might be stored in terrestrial ecosystems. Such a large storage is unlikely because not all nitrogen falls on forests and some nitrogen is removed from the land by runoff and denitrification (Schlesinger 2009). In some areas, the fate of the excess nitrogen deposited from the atmosphere is unclear—some may accumulate in soil organic matter, leading to greater carbon storage (Hyvonen et al. 2008, Nave et al. 2009, Liu and Greaver 2010; but see Mack et al. 2004). In European forests, the change in carbon storage per unit nitrogen deposition is about 40 ± 20 (Hogberg 2012). Using a slightly higher value, Thomas et al. (2010) indicate that forests may accumulate 0.31×10^{15} g C/yr as a result of inadvertent nitrogen fertilization worldwide.

Over longer periods of time, changes in the distribution of vegetation as a result of global climate change could also affect the concentration of atmospheric CO_2 (Chapter 5). Coupled to models of climate change, most models of the global carbon cycle suggest an increase in the carbon content of vegetation and soils when vegetation is in equilibrium with a warmer and wetter world of the future (Smith et al. 1992b). If the adjustment of vegetation to climate occurs over 100 years, these models suggest that the net uptake by the terrestrial biosphere could be as high as 1.8×10^{15} g C/yr—mostly in vegetation. However, other models suggest that changes in vegetation and soils during the transition in climate may yield the opposite effect. Smith and Shugart (1993) estimate large losses of carbon from vegetation during the transient period of drought that is likely to accompany global warming over the next century.

In sum, whole-ecosystem response will be determined by various factors—CO_2, nutrient availability, and global patterns of temperature and rainfall, which are all affected by human activities. Although it seems unlikely that enhanced growth by terrestrial vegetation will ultimately stem the rise of CO_2 that is derived from fossil fuels (see Idso and Kimball 1993), the response of the terrestrial biosphere could have a dramatic impact on the future composition of the atmosphere.

In our view of the global carbon cycle, it is important to recognize that the annual movements of carbon, rather than the amount stored in various reservoirs, are most important. Desert soil carbonates contain more carbon (930×10^{15} g) than land vegetation, but the exchange between desert soils and the atmosphere is tiny (0.023×10^{15} g C/yr), yielding a turnover time of 85,000 years in that pool (Schlesinger 1985). In the global carbon cycle, a flux that has not changed in recent times, no matter how large, is not likely to affect the concentration of atmospheric CO_2 (Houghton et al. 1983). For example, the release of CO_2 in forest fires is of no consequence to changes in atmospheric CO_2 unless the frequency or area of forest fires has changed in recent times (Adams et al. 1977, Auclair and Carter 1993, Kasischke et al. 1995). The carbon flux in rivers or sinking pteropods cannot serve as a net sink for anthropogenic CO_2 in the ocean, unless the flux in these pathways is greater as a result of human activities. Similarly, the storage of carbon in peatland soils is not a sink for fossil fuel CO_2, unless the rate of storage in these areas has increased significantly during the Industrial Revolution. These peatlands have accumulated carbon throughout the Holocene when atmospheric CO_2 has been relatively constant (Harden et al. 1992).[4] On the other hand, landfills (0.12×10^{15} g/yr; Barlaz 1998) and wood products (0.03×10^{15} g/yr for the United States; Pacala et al. 2001) are sinks for CO_2 released from fossil fuel combustion during the past century.

Relatively small changes in large pools of carbon can have a dramatic impact on the carbon dioxide content of the atmosphere, especially if they are not balanced by simultaneous changes in other components of the carbon cycle. A 1% increase in the rate of decomposition on land, as a result of global warming, would release nearly 0.6×10^{15} g C/yr to the atmosphere. Schimel et al. (1994) estimated that the soil carbon pool could lose 0.7% of its content (11×10^{15} g C) for every degree of global warming during the next century (compare to Kirschbaum 1995, Trumbore et al. 1996). Recent losses of carbon from soils throughout England are attributed to a warming climate (Bellamy et al. 2005). Losses of carbon from the large pool of soil organic matter frozen in permafrost have the potential to release large amounts of CO_2 to the atmosphere and exacerbate global warming (Dorrepaal et al. 2009, Schuur et al. 2009, Tarnocai et al. 2009, Ping et al. 2008, Schaefer et al. 2011).

Similarly, a 1%/yr increment to the biomass of carbon on land, as a result of a greater storage of NPP, could balance the CO_2 budget in the atmosphere and stem the rise of CO_2 (Table 11.1). We can speculate that this increment should be first realized in vegetation, which has a faster turnover time than soils; only a small percentage of NPP that enters the soil survives to become a component of soil organic matter (Schlesinger 1990; Chapter 5).

---

[4] Many who deny the potential impact of humans on climate argue that the flux from fossil fuels is so small compared to CO_2 from natural sources that it couldn't possibly have much effect on the concentration of atmospheric CO_2. This argument is easily dismissed, since the flux to and from the land and sea must have been balanced when CO_2 concentrations were relatively constant before the Industrial Revolution. The flux from fossil fuels is new to the carbon cycle and not balanced by other large human-induced sinks for carbon.

At the surface of the Earth, the largest global pool of carbon is found in sedimentary rocks, including fossil fuels (Table 2.3). Storage of organic carbon in these deposits accounts for the accumulation of O_2 in the atmosphere through geologic time (Chapter 2). In the absence of human perturbations, the exchange between the fossil pool and the atmosphere could be ignored in global models. Only a small amount of buried sedimentary organic matter is exposed to uplift, erosion, and oxidation each year—0.043 to 0.097×10^{15} g C/yr (Di-Giovanni et al. 2002, Copard et al. 2007). In extracting fossil fuels from the Earth's crust, humans affect the global system by creating a large biogeochemical flux, $\sim 9 \times 10^{15}$ g C/yr, where essentially none existed before. Carbon dioxide emission from fossil fuel combustion dwarfs the CO_2 exhaled by the human population worldwide, 0.6×10^{15} g C/yr (Prairie and Duarte 2006). The rise in atmospheric CO_2 is closely correlated with the rise in human population (Hofmann et al. 2009), economic activity, and the "carbon-intensity" of the modern economy (Canadell et al. 2007).

TEMPORAL PERSPECTIVES ON THE CARBON CYCLE

Studies of the biogeochemistry of carbon on Earth must begin with a consideration of the origin of carbon as an element and with theories that explain its differential abundance on the planets of our solar system (Chapter 2). During the early development of Earth, the carbon cycle was decidedly non-steady-state: The carbon content of the planet grew with the receipt of planetesimals and meteorites—especially carbonaceous chondrites—and the atmospheric content increased as volcanoes released CO_2 trapped in Earth's mantle. Today, the Earth's mantle appears to release about 0.03×10^{15} g C to the atmosphere each year (Kerrick 2001, Wallace 2005, Dasgupta and Hirschmann 2010). The oldest geologic sediments suggest that atmospheric CO_2 may have been as high as 3% on primitive Earth, contributing to a greenhouse effect during a time of low solar output (Walker 1985, Rye et al. 1995). Today small amounts of water vapor and CO_2 in our atmosphere maintain the surface temperature of the Earth above freezing—obviously an essential condition for the persistence of the biosphere (Ramanathan 1988, Lacis et al. 2010, Schmidt et al. 2010a).

As discussed in Chapter 4, CO_2 in the atmosphere interacts with the crust of the Earth, causing rock weathering (Eq. 4.3). Carbon dioxide is removed from the atmosphere and transferred via rivers to the oceans, where it is eventually deposited on the seafloor in carbonate rocks, adding to the Earth's crust (Figure 1.3). Along with carbonates, undecomposed organic materials also accumulate in marine sediments from photosynthesis on land and in the sea. As early as 1918, Arrhenius (1918, p. 177) speculated that the consumption of CO_2 by rock weathering might eventually cool the planet through a loss of its natural "greenhouse effect":

> As the crust grew thicker, the supply of this gas [CO_2] diminished and was further used up in the process of disintegration [weathering]. As a consequence the temperature slowly decreased, although decided fluctuations occurred with the changing volcanic activity during different periods. Supply and consumption of carbon dioxide fairly balanced as disintegration ran parallel with the proportion of this gas in the air.

Fortunately, CO_2 is returned to the atmosphere as a result of tectonic activity. In the complete geochemical cycle of carbon (Figure 1.3), subduction of the oceanic crust carries carbon deposited on the seafloor to the interior of the Earth, where CO_2 and other volatile elements are once again released by hydrothermal and volcanic emissions. Alt et al. (2012) estimate that 0.07×10^{15} g C/yr are subducted worldwide. Mixing of the subducted oceanic crust may extend to

1000-km depth in the mantle (Walter et al. 2011). Total volcanic emanation of volatile carbon from the Earth's surface, including both degassing of the mantle and of recycled sedimentary materials, is estimated at about 0.10×10^{15} g C/yr (Morner and Etiope 2002)—similar to estimates of carbon accumulation in ocean sediments. At these rates, the entire mass of carbon in the upper mantle, 7×10^{22} g, would recycle in less than a billion years (Dasgupta and Hirschmann 2010); thus, it is likely that much of the carbon in the mantle has spent some time in the biosphere in the geologic past. If this cycle were not complete, rock weathering would deplete CO_2 in the atmosphere and oceans in $\sim$500,000 years, and all carbon would be stored in sedimentary rocks.

On Earth, this geochemical cycle has helped to maintain the concentration of atmospheric CO_2 below 1% (10,000 ppm) for the last 100 million years (Berner and Lasaga 1989). On Mars, where this cycle has slowed or stopped, the atmosphere contains a small amount of CO_2, and the planet is very cold (Chapter 2). On Venus, which is too hot for CO_2 to react with crustal minerals, the atmosphere contains a large amount of CO_2, reinforcing its greenhouse effect (Walker 1977, Nozette and Lewis 1982). During periods of extensive volcanism, the atmospheric concentration of CO_2 on Earth may have been greater than today's, leading to warmer climates (Owen and Rea 1985); however, a continuous geologic record of liquid oceans on Earth indicates that CO_2 and other greenhouse gases have always remained at levels that produce relatively moderate surface temperatures. Deposition of carbon in ocean sediments by various biotic activities is one way that life may help promote long-term stability in Earth's climate, favorable to the persistence of life.

Despite their long-term significance in buffering atmospheric CO_2, the annual transfers of carbon in the geochemical cycle are relatively small. The massive quantities of CO_2 that are now tied up in the carbonate minerals of the Earth's crust are the result of a slow accumulation of these materials over long periods of Earth's history. Today rivers carry about 500×10^{12} g/yr of Ca^{2+} (Milliman 1993) and 0.40×10^{15} g/yr of carbon as HCO_3^- to the sea (Sarmiento and Sundquist 1992), with about 65% derived from the atmosphere and the rest from carbonate minerals (Gaillardet et al. 1999, Suchet et al. 2003, Hartmann et al. 2009). For seawater to maintain fairly constant concentrations of calcium, an equivalent amount of Ca must be deposited as $CaCO_3$ in ocean sediments, carrying 0.15×10^{15} g C/yr to the oceanic crust. Dividing the mass of carbonate rocks by their annual rate of formation, we find that each atom of carbon sequestered in marine carbonate spends more than 400 million years in that reservoir.

With the appearance of life, a *biogeochemical* cycle was added on top of the underlying geochemical cycle of carbon on Earth. Models of the modern biogeochemical cycle of carbon focus on the large annual transfer of CO_2 from the atmosphere to plants as a result of photosynthesis and the large return of CO_2 to the atmosphere as a result of decomposition (Figure 11.1). Today, the fluxes of carbon in the biogeochemical cycle of carbon, mostly expressed in units of 10^{15} to 10^{17} g C/yr, dwarf the fluxes of the underlying geochemical cycle of carbon, where the movements are typically 10^{13} to 10^{14} g C/yr.

During Earth's history, at times when the production of organic carbon by photosynthesis has exceeded its decomposition, organic carbon has accumulated in geologic sediments. The earliest organic carbon is present in rocks from 3.8 bya, with the pool increasing to 1.56×10^{22} g by about 540 mya (Des Marais et al. 1992). During that interval, about 20% of all the carbon buried in marine sediments was organic—similar to the ratio found in modern marine sediments (Li 1972; Holser et al. 1988; Dobrovolsky 1994, p. 163). During the Carboniferous Period (300 mya), large deposits of organic carbon were stored in freshwater environments, leading to modern economic deposits of coal.

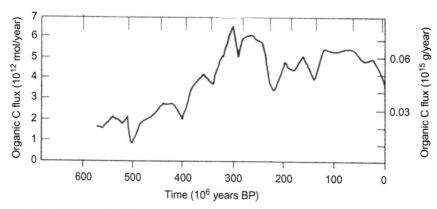

FIGURE 11.4   Burial of organic carbon on Earth during the past 600 million years. *Source: From Olson et al. (1985).*

During the Tertiary Epoch, the precursors to modern deposits of petroleum were added to marine sediments. Net storage of organic carbon in sediments has varied between about 0.04 and 0.07×10^{15} g C/yr during the last 300 million years (Figure 11.4; Berner and Raiswell 1983); a rate of 0.10×10^{15} g C/yr is estimated for the present (Chapter 9). Today, about 10 to 20% of the global organic burial occurs in the Bengal Fan as a result of rapid burial by Himalayan sediment (Galy et al. 2007, France-Lanord and Derry 1997).

Life also stimulated some of the reactions in the underlying geochemical cycle of carbon. Various marine organisms enhance the deposition of calcareous sediments, which now cover more than half of the oceans' seafloor (Kennett 1982). Land plants, by maintaining high concentrations of CO_2 in the soil pore space, raise the rate of carbonation weathering, speeding the reaction of CO_2 with the Earth's crust (Moulton et al. 2000; Chapter 4). Land plants and soil microbes also excrete a variety of organic compounds, byproducts of photosynthesis, which enhance rock weathering. Various models developed and summarized by Robert Berner of Yale University suggest that the atmospheric concentration of CO_2 declined precipitously as land plants gained dominance about 350 mya (Berner 1992, Berner and Kothavala 2001, Royer et al. 2001, Rothman 2002). The record of CO_2 in Earth's atmosphere suggests that concentrations have remained between 150 and 500 ppm for the past 24 million years (Pearson and Palmer 2000) and below 1500 ppm for the past 50 million years (Pagani et al. 2005, Zachos et al. 2008). Calcium levels in seawater were higher, atmospheric CO_2 levels were lower, and the Earth's climate was colder during the Miocene, 13,000,000 years ago (Griffith et al. 2008). Indeed, atmospheric CO_2 concentrations and global temperature are well correlated, especially for the last 20 million years (Came et al. 2007, Tripati et al. 2009).

Collections of gas trapped in ice cores from the Antarctic provide a historical record of atmospheric CO_2 for the last 800,000 years (Figure 1.2). The cyclic fluctuations of CO_2 in Earth's atmosphere imply non-steady-state conditions of the carbon cycle, with a periodicity of 100,000 years. Concentrations varied between 180 and 280 ppm, with the lowest values found in layers of ice that were deposited during glacial periods.[5] CO_2 concentrations and global

[5] The consistent minimum at ~200 ppm and maximum at ~280 ppm during glacial-interglacial intervals (Figure 1.2) is striking. Holland (1965) suggests that the minimum may be related to the onset of gypsum precipitation in ocean sediments, which raises the pH of seawater and leads to the precipitation of calcite (Lindsay 1979, p. 49).

temperatures have been well correlated for the past 160,000 years (Cuffey and Vimeux 2001). The lowest concentrations of CO_2 were likely to invoke significant physiological effects on land plants, reducing their photosynthesis (Gerhart and Ward 2010). Although the exact magnitude is controversial, the mass of carbon stored in vegetation and soils was also lower during the last glacial, as a result of the advance of continental ice sheets and widespread desertification of land habitats (Adams et al. 1990, Servant 1994, Bird et al. 1994, Beerling 1999). Thus, glacial conditions must have produced changes in the oceans that allowed a large uptake of CO_2 (Faure 1990, Sundquist 1993); this carbon dioxide was returned to the atmosphere at the end of the glacial epoch (Yu et al. 2010, Burke and Robinson 2012, Schmitt et al. 2012).

Increased marine NPP or an increased efficiency of the marine "biotic pump" (Chapter 9) seem unlikely mechanisms for CO_2 uptake during glacial conditions (Leuenberger et al. 1992), but a decrease in the amount of carbon stored as carbonates could lead to a greater retention of CO_2 in the oceans, following Eqs. 9.3 and 9.4. Most of the $CaCO_3$ dissolution in marine sediments is driven by CO_2 released during the decomposition of organic matter (Berelson et al. 1990), so Archer and Maier-Reimer (1994) suggest that an increase in the ratio of organic carbon to carbonate in sinking particles could lead to a greater dissolution of carbonate in marine sediments. Sanyal et al. (1995) report evidence for a higher pH of the glacial ocean, consistent with a greater dissolution of carbonates in the deep sea.

At the end of the last glacial, 17,000 years ago, atmospheric CO_2 rose to about 280 ppm, where it remained with minor variations until the beginning of the Industrial Revolution (Figure 11.5; Indermuhle et al. 1999, Meure et al. 2006, Alm et al. 2006). The increase in concentration from 280 ppm to today's value of about 400 ppm represents a global change of 43% in less than 200 years! Although the current level of CO_2 is not unprecedented in the geologic record, our concern is the speed at which a basic characteristic of the planet has changed to levels not previously experienced during human history or during the evolution of current ecosystems. Since global temperature and atmospheric CO_2 are related, we are destined for significant global warming during the near future (Shakun et al. 2012; Figure 1.2).

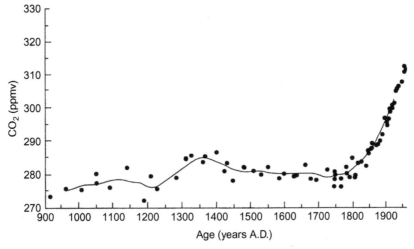

FIGURE 11.5 Concentrations of atmospheric CO_2 estimated from bubbles of gas trapped in ice cores from Antarctica. *Source: From Barnola et al. (1995).*

These perspectives on the global carbon cycle extend from processes that occur on a time scale of 10^9 years to those that occur annually. The global carbon cycle is composed of large, rapid transfers in the *bio*geochemical cycle superimposed on the underlying, small, slow transfers of the geologic cycle. Buffering of atmospheric CO_2 over geologic time involves small net changes in carbon storage that occur relatively slowly. Thus, an increase in the natural rate of rock weathering (which consumes $\sim 0.25 \times 10^{15}$ g C/yr) as a result of high CO_2 and rising global temperature is not likely to be an effective buffer to the rapid release of CO_2 from fossil fuels (9×10^{15} g C/yr). In contrast, the current net exchange of CO_2 between the atmosphere and the biosphere is about 150×10^{15} g C/yr, so the biosphere is more likely to buffer the rise of CO_2 due to human activities. The current increase in atmospheric CO_2 results from our ability to change the flux of CO_2 to the atmosphere by an amount that is significant relative to the biogeochemical reactions that buffer the system over long periods of time.

ATMOSPHERIC METHANE

At first glance, the annual flux of methane (CH_4) would seem to be only a minor component of the global carbon cycle. All sources of methane in the atmosphere are in the range of 10^{12} to 10^{14} g C/yr, which is several orders of magnitude lower than the flux for CO_2 shown in Figure 11.1. Globally, the atmospheric methane concentration is about 1.83 ppm, versus 400 ppm for CO_2 (Table 3.1). However, at the beginning of the Industrial Revolution, the concentration of methane in the atmosphere began to increase at an average rate of about 1%/yr (Figure 11.6), which was much faster than the rate of CO_2 increase over the same interval

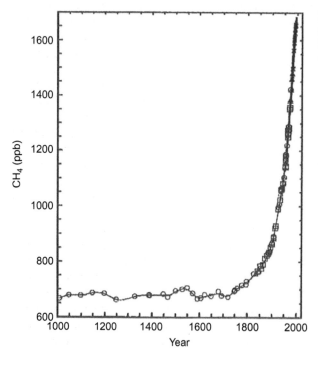

FIGURE 11.6   Concentrations of CH_4 in air extracted from Antarctic ice cores. *Source: From Etheridge et al. (1998). Used with permission of the American Geophysical Union.*

(Figure 11.5). The methane concentration in Earth's atmosphere has more than doubled since the beginning of the Industrial Revolution. Strangely, the rise in methane concentration slowed in the early 1990s, but an upward trend now seems to have commenced again (Dlugokencky et al. 2011, Terao et al. 2011). Over the next century each molecule of methane in the atmosphere has the potential to contribute 25 to $35\times$ or more to the human-induced greenhouse effect compared to each molecule of CO_2 (Lashof and Ahuja 1990, Shindell et al. 2009). Substantial progress in mitigating global warming could accompany a reduction in methane emissions (Montzka et al. 2011a, Shindell et al. 2012).

The cause of the increase in methane in Earth's atmosphere is not obvious, because a wide variety of sources contribute to the total annual production of about 600×10^{12} g/yr (Table 11.2). The sum of anthropogenic sources is about twice the sum of natural sources, so it is perhaps surprising that the annual increase of methane in the atmosphere is not larger. The estimate of total flux is fairly well constrained because it yields a mean residence time for atmospheric CH_4 of about 9 years, which is consistent with independent calculations based on methane consumption (Khalil and Rasmussen 1990, Prinn et al. 1995, Dentener et al. 2003) and with the spatial variation in CH_4 concentration in the atmosphere (Figure 3.5). The concentration of CH_4 is slightly higher in the Northern Hemisphere, suggesting that it is the location of major emissions (Figure 3.4).

Like CO_2, the concentration of methane in the atmosphere oscillates, showing a minimum concentration in midsummer in the Northern Hemisphere (Steele et al. 1987, Khalil et al. 1993a, Dlugokencky et al. 1994). While methane emissions from wetlands are greatest during warm periods, the summer is also the time of the most rapid destruction of atmospheric methane by OH radicals (Khalil et al. 1993b).

In 2006, do Carmo et al. reported anomalously high methane concentrations above the canopy of tropical rainforest in Brazil, also seen in satellite observations of atmospheric methane over tropical rainforests (Frankenberg et al. 2005). At about the same time, Keppler et al. (2006) suggested a large global flux of methane (62–236×10^{12} g/yr) from upland vegetation from an unknown, but aerobic, biochemical pathway. This conclusion has now been tempered by the failure of laboratory measurements to confirm the observation and by more cautious extrapolation to a global estimate (Kirschbaum et al. 2006, Megonigal and Guenther 2008, Dueck et al. 2007, Nisbet et al. 2009). Methane is produced in hollow trees (Covey et al. 2012), but a large flux of methane from upland vegetation cannot be accommodated in the current budget for atmospheric methane (refer to Table 11.2) without yielding a mean residence time that is incompatible with atmospheric measurements and conflicting with the observed midsummer minimum for methane concentrations in the atmosphere. The observations of methane flux from tropical rainforests are likely due to the extensive areas of flooded soils within them.

Methanogenesis in wetland habitats is widely acknowledged as the dominant natural source of atmospheric methane (Chapter 7). Matthews and Fung (1987) estimated that 110×10^{12} g/yr stem from anaerobic decomposition in natural wetlands globally. The rate of methane release would undoubtedly be much higher if it were not for high rates of methanotrophy by bacteria in wetland soils (Megonigal and Schlesinger 2002). The rate of production is higher in tropical wetlands than in boreal wetlands (Schütz et al. 1991, Bartlett and Harriss 1993, Cao et al. 1996), reflecting the positive relationship between temperature, net ecosystem production, and the rate of methanogenesis in many wetland

TABLE 11.2　Estimated Sources and Sinks of Methane in the Atmosphere in 2010

Natural sources	Flux (10^{12} g CH_4/yr)		References
Wetlands	143		Neef et al. 2010
Tropics		46	Bloom et al. 2010
Northern latitude		20	Christensen et al. 1996
Upland vegetation	10 (estimate)		Megonigal and Guenther 2008; Kirschbaum et al. 2006
Termites	19		Sanderson 1996
Oceans	10		Reeburgh 2007
Geological seepage[a]	33		Etiope et al. 2008
Anthropogenic sources			
Fossil fuel related			
Coal mines	30		Prather et al. 1995
Coal combustion	15		Prather et al. 1995
Oil and gas	72		Neef et al. 2010
Waste and waste management			
Landfills	18		Bogner and Matthews 2003
Animal waste	25		Prather et al. 1995
Sewage treatment	25		Prather et al. 1995
Ruminants	116		Neef et al. 2010
Reservoirs	70		St. Louis et al. 2000
Biomass burning	19		Kaiser et al. 2012
Rice cultivation	40		Sass and Fisher 1997, Bloom et al. 2010
Total sources	**645**		
Sinks			
Reaction with OH radicals	522		Neef et al. 2010
Removal in the stratosphere	34		Neef et al. 2010
Removal by soils	25		Curry 2007 Dutaur and Verchot 2007
Total sinks	**581**		
Atmospheric increase (2007)	23		Dlugokencky et al. 2009

Note: All data in 10^{12} g CH_4/yr from various sources as cited here and in the text.
[a] Total geological seepage less marine.

ecosystems (Chapter 7). Because tropical wetlands cover a large area of the world, they dominate the methane flux from wetlands globally (Aselmann and Crutzen 1989, Fung et al. 1991, Bartlett and Harriss 1993).

A large portion of the current increase in atmospheric methane may derive from an increase in the worldwide cultivation of rice (Sass and Fisher 1997). Because most rice paddies are found in warm climates, they often yield a large CH_4 flux, which is enhanced by the upward transport of CH_4 through the hollow stems of rice (Chapter 7). Matthews et al.

(1991) provide maps of the global distribution of CH_4 production from rice cultivation, which is likely to increase by more than 1%/yr during the next several decades (Anastasi et al. 1992). Better management of rice cultivation may be responsible for a reduced flux of CH_4 to the atmosphere in recent years (Kai et al. 2011).

Many grazing animals and termites maintain a population of anaerobic microbes that conduct fermentation at low redox potentials in their digestive tract. Digestion in these animals provides the functional equivalent of a mobile wetland soil! The belches[6] of grazing animals make a significant contribution to the global sources of methane (Table 11.2). In the early 1980s, about 78×10^{12} g/yr of CH_4 were derived from domestic and wild animals, with humans contributing 1×10^{12} g/yr (Crutzen et al. 1986, Lerner et al. 1988). Anastasi and Simpson (1993) suggest that larger herds of grazing animals may increase the global flux of methane from animals by about 1×10^{12} g/yr during the next several decades (compare to Table 11.2). Some termites and other insects also make a small but significant contribution to atmospheric methane as a result of anaerobic decomposition in their hindgut (Khalil et al. 1990, Brauman et al. 1992, Hackstein and Stumm 1994). It is not likely, however, that the flux of CH_4 from termites has increased significantly in recent years.

Forest fires produce methane as a product of incomplete combustion. We know little about the annual area of burning in the preindustrial world, but it is likely that the current release of CH_4 from forest fires has increased as a result of high, recent rates of biomass burning in the tropics (Andreae 1991). Kaufman et al. (1990) used remote sensing of fires in Brazil to calculate a loss of 7×10^{12} g CH_4/yr in that region in 1987, and Delmas et al. (1991) found a flux of 9.2×10^{12} g CH_4 from burning of African savannas. CH_4 typically accounts for 1% of the total carbon lost by fire (Levine et al. 1993), so the estimate of methane flux from forest fires (19 Tg CH_4/yr; Table 11.2) is compatible with a range of global estimates of the carbon released in forest fires (1.4 to 3.6×10^{15} g C/yr; Chapter 5).

Humans contribute directly to atmospheric methane during the production and use of fossil fuels and due to the disposal of wastes. The flux of methane from landfills increases linearly with the amount of material buried, which presumably decomposes under anoxic conditions (Thorneloe et al. 1993). The global flux from landfills is estimated at 16 to 20 Tg CH_4/yr (Bogner and Matthews 2003). Inadvertent releases of fossil CH_4 during the mining and use of coal and natural gas must account for about 20 to 30% of the total annual flux of CH_4 to the atmosphere, based on the ^{14}C age of atmospheric methane (Ehhalt 1974; Wahlen et al. 1989; Quay et al. 1991, 1999; Etiope et al. 2008).

Biomass burning and releases of natural gas appear to have increased the $\delta^{13}C$ of atmospheric methane from a preindustrial value of around $-50‰$ to the value of $-47‰$ that is observed today (Craig et al. 1988, Quay 1988). Increasing emissions from wetlands cannot be responsible, because methanogenesis by acetate or CO_2 reduction yields CH_4 that is more depleted in ^{13}C (Chapter 7). Indeed, an increased release of CH_4 by biomass burning would seem to be the only single source that is consistent with observed changes in both atmospheric $^{13}CH_4$ (Craig et al. 1988, Sowers et al. 2005) and $^{14}CH_4$ (Wahlen et al. 1989) in recent years.

---

[6] Although many believe that farts are the major pathway for methane release from cattle, about 90 to 95% of the methane is released from digestive burps from the rumen (Hayhoe and Farley 2009, p. 33). Wilkinson et al. (2012) suggest that similar methane emissions from dinosaurs may have warmed Earth in the Mesozoic.

However, a combination of changing contributions from various source and sink reactions is also possible (Whiticar 1993).

The major sink for atmospheric methane is reaction with hydroxyl radicals in the atmosphere (Chapter 3). Each year about 522×10^{12} g is removed from the troposphere by this process (Neef et al. 2010). As a result of its mean atmospheric lifetime of 9 years, a large portion of the tropospheric CH_4 mixes into the stratosphere, where about 30×10^{12} g is destroyed by reaction with OH, producing CO_2 and water vapor. Some workers have suggested that the rise in atmospheric methane during the Industrial Revolution was derived from a reduction in the sink strength offered by hydroxyl radicals, which react more rapidly with CO, which is also increasing in the atmosphere (Khalil and Rasmussen 1985). Although this mechanism cannot be dismissed, it is inconsistent with indirect observations that the concentration of hydroxyl radicals has not decreased, and may even have increased, in the atmosphere in recent years (Dentener et al. 2003, Prinn et al. 2005; Chapter 3).

A small amount of methane diffuses from the atmosphere into upland soils, where it is oxidized by methanotrophic bacteria (King 1992). In Mojave Desert soils, where the supply of labile organic matter is limited, soil bacteria consume an average of 0.66 mg CH_4 m^{-2} day^{-1}, with the greatest rates observed after rainstorms (Striegl et al. 1992). Consumption of CH_4 in temperate and tropical forest soils typically ranges from 1.0 to 5.0 mg CH_4 m^{-2} day^{-1} (Crill 1991, Adamsen and King 1993, Ishizuka et al. 2000, Smith et al. 2000, Price et al. 2004), with lower values after rainstorms, which tend to retard the diffusion of O_2 and CH_4 into clay-rich soils (Koschorreck and Conrad 1993; Castro et al. 1994, 1995). Methanotrophic bacteria remain active at extremely low CH_4 concentrations (Conrad 1994), so the global significance of soil methanotrophy appears limited by the rate of diffusion of methane into the soil (Born et al. 1990, King and Adamsen 1992).

Some of the methanotrophic activity in soils derives from the activities of nitrifying bacteria, which can use CH_4 as an alternative substrate to NH_4^+ (Jones and Morita 1983, Hyman and Wood 1983, Bédard and Knowles 1989). Steudler et al. (1989) suggested that the consumption of CH_4 by nitrifying bacteria may be lower in forests that currently receive a large atmospheric deposition of NH_4^+, because the NH_4^+/CH_4 ratio in soils has greatly increased in these regions. A number of workers reported reduced methane uptake when forest and grassland soils were fertilized with nitrogen, with the threshold being about 100 kg N ha^{-1} yr^{-1} (Aronson and Helliker 2010, Kim et al. 2012). Methane uptake by soils is also lower after land clearing, which stimulates nitrification (Hütsch et al. 1994, Keller and Reiners 1994). With fertilization or land clearing, ammonium oxidation produces small amounts of nitrite (NO_2), which may cause a persistent inhibition of methanotrophic bacteria in soils (King and Schnell 1994, Schnell and King 1994). Atmospheric deposition of nitrate is also known to reduce CH_4 uptake by forest soils (Steudler et al. 1984, Mochiguki et al. 2012).

Over large regions, the sink for methane in upland soils consumes only a small fraction of the production of methane in adjacent, wet lowland soils (e.g., Whalen et al. 1991, Delmas et al. 1992, Yavitt and Fahey 1993, Ullah and Moore 2011). The global estimate of the sink for atmospheric methane in soils is about 20 to 30×10^{12} g/yr (Curry 2007, Dutaur and Verchot 2007). Given this relatively small value, it is unlikely that changes in this process by human activities can account for the current increase in atmospheric CH_4 globally (e.g., Willison et al. 1995).

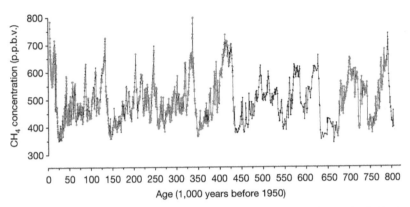

FIGURE 11.7   Concentrations of CH$_4$ in air that is extracted from ice cores in Antarctica dating to 800,000 years before present. *Source: From Loulergue et al. (2008). Compare to Figure 1.2.*

Ice-core records of atmospheric methane show that concentrations were about 400 ppb during the last glacial period, increasing abruptly to the preindustrial value of 700 ppb as the glaciers melted (Chappellaz et al. 1990, Loulergue et al. 2008; Figure 11.7). The increase during deglaciation seems to have occurred while many northern wetlands were still covered with ice, suggesting that changes in tropical wetlands may have caused the initial methane increase, which reinforced the global warming during deglaciation (Chappellaz et al. 1993). However, high-latitude, northern wetlands may have also contributed to the accumulation of methane in the atmosphere during the Holocene (Zimov et al. 1997, Walter et al. 2007, Smith et al. 2004).

Concentrations of atmospheric methane showed minor variation during the Holocene ($\pm15\%$; Blunier et al. 1995, Mitchell et al. 2011a), but beginning about 200 years ago the concentration began to increase rapidly (Figure 11.7). The atmospheric concentration of methane has doubled during this period. The rise in atmospheric methane is paralleled by a rise in formaldehyde—a methane oxidation product (Eq. 3.21)—in polar ice cores (Staffelbach et al. 1991). Recently measured increases in stratospheric water vapor are also linked to an increasing transport of CH$_4$ to the stratosphere, where it is oxidized (Thomas et al. 1989, Oltmans and Hofmann 1995).

Although the annual increase in methane in the atmosphere averaged about 1%/yr during the 1980s, the rate of increase slowed in the 1990s and was minimal during the early 2000s (Figure 11.8). A variety of explanations have been offered: Some workers suggest that less natural gas was leaking from gas fields of the former Soviet Union as a result of slower economic growth and of efforts to patch pipeline leaks (Law and Nisbet 1996, Aydin et al. 2011). Leakage was estimated at 40 Tg CH$_4$/yr in 1990 (Reshetnikov et al. 2000). Alternatively, Bekki et al. (1994) suggest that an increasing depletion of stratospheric ozone allowed greater amounts of uvB radiation to penetrate to the troposphere, where it produces OH radicals that oxidize CH$_4$. As the ozone hole has stabilized, so too has the production of OH radicals.

Future changes in the global methane budget as a result of rising CO$_2$ and global temperatures are difficult to predict. Warmer conditions may shift the balance between aerobic and anaerobic decomposition in wetlands and increase the ratio of CO$_2$ to CH$_4$ emitted from these ecosystems (Whalen and Reeburgh 1990, Funk et al. 1994, Moore and Dalva 1993). On the

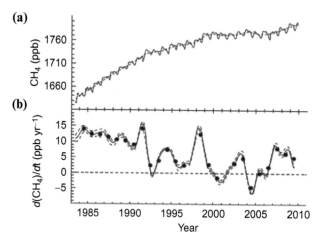

FIGURE 11.8   Concentration of methane (a) and its annual change (b) in Earth's atmosphere during the past three decades. *Source: From Dlugokencky et al. (2011). Used with permission of The Royal Society.*

other hand, methanogenic bacteria show a greater positive response to temperature than methane-oxidizing bacteria, suggesting that the flux of methane from wetland soils could increase with global warming (King and Adamsen 1992, Dunfield et al. 1993, Megonigal and Schlesinger 2002). An increasing flux of CH_4 may also accompany a CO_2-induced stimulation of wetland plants, which leak excess carbohydrate from their roots fueling methanogenesis in wetland soils (Dacey et al. 1994, Hutchin et al. 1995, Megonigal and Schlesinger 1997, van Groenigen et al. 2011). In contrast, methane consumption in upland soils appears to decline when forests are grown at high CO_2 (Phillips et al. 2001a, McLain et al. 2002).

Catastrophic release of methane from marine sediments, where it is held as methane hydrate (clathrate) could also yield a large increase in atmospheric methane and greenhouse warming in the future (MacDonald 1990). Burwicz et al. (2011) estimate the global pool in clathrates may be as large as 995×10^{15} g CH_4-C—more than 1000× current annual emissions (refer to Table 11.2). The geologic record indicates release of methane from hydrates in the past (Jahren et al. 2001, Katz et al. 1999), although evidence for such degassing at the end of the last glacial epoch is equivocal (Kennett et al. 2000, Sowers 2006, Petrenko et al. 2009). Given methane's potential as a greenhouse gas and indications that increasing concentrations of CH_4 may have preceded the global warming 10,000 years ago, a better understanding of the global methane budget is paramount if biogeochemists are to contribute to the development of effective international policy to combat global warming (Nisbet and Ingham 1995).

CARBON MONOXIDE

Carbon monoxide has a low concentration (45 to 250 ppb) and a short lifetime (2 months) in the atmosphere (Table 3.5). The short lifetime is consistent with wide regional and seasonal variations in its concentration (Figure 3.5); the concentration of CO in the Northern Hemisphere is typically three times larger than that in the Southern Hemisphere (Dianov-Klokov et al. 1989, Novelli et al. 2003). The budget for CO is dominated by anthropogenic sources (Table 11.3), especially fossil fuel combustion and biomass burning, which are

TABLE 11.3   Budget for Major Sources and Sinks for CO in the Atmosphere

Sources	Flux
Fossil fuel combustion	400
Biofuel combustion	160
Biomass burning	460[a]
Oxidation of methane	820
Oxidation of other volatile carbon compounds	521
Total	**2361**
Sinks	
Uptake by soils (Sanhueza et al. 1998)	115–230
Oxidation by OH reactions (Prather et al. 1995)	1400–2600
Stratospheric destruction	100
Total	**1615–3030**

Note: All units are 10^{12} g CO/yr; from Duncan et al. (2007) unless otherwise noted.
[a] *Kaiser et al. (2012), Jain (2007), and Mieville et al. (2010) give alternative estimates of 351, 372, and 500 × 10^{12} g CO/yr, respectively.*

concentrated in the Northern Hemisphere. Year-to-year variations in the occurrence of forest fires account for much of the variation in CO concentrations in downwind regions (Wotawa et al. 2001, Novelli et al. 2003, Vasileva et al. 2011). Variations in the concentration of CO and its isotopic composition (i.e., $\delta^{13}C$) in the Antarctic icepack have been used to trace biomass burning in the Southern Hemisphere for the past 650 years (Wang et al. 2010c).

Until recently, the concentration of CO was increasing at a rate of >1%/yr (Khalil and Rasmussen 1988, Dianov-Klokov et al. 1989), presumably as a result of increasing emissions from fossil fuel combustion and an increasing production of CO as a methane-oxidation product (Chapter 3). Surprisingly, CO concentrations began to decline slightly during the early 1990s (Novelli et al. 1994, Khalil and Rasmussen 1994), reflecting a greater emphasis on the control of air pollution in the United States and Europe (Novelli et al. 2003, Hudman et al. 2008). The decline in CO may also be related to the slower growth rate of atmospheric CH_4, inasmuch as methane oxidation to CO accounts for 28 to 35% of the inputs of CO to the troposphere (Granier et al. 2000b; Table 11.3).

A small amount of CO is taken up by vegetation and soils, but the dominant sink for CO is oxidation by hydroxyl radicals in the atmosphere (Eqs. 3.30–3.35). Because it is oxidized to CO_2 so rapidly, carbon monoxide is normally included as a component of the CO_2 flux in most accounts of the global carbon cycle (e.g., Figure 11.1). Actually, the direct release of CO may account for about 5% of the total carbon emitted during fossil fuel combustion and perhaps as much as 15% of the carbon released during biomass burning (Andreae 1991).

Carbon monoxide shows limited absorption of infrared radiation. Its main effect on the greenhouse warming of Earth is probably indirect—by slowing the destruction of methane in the atmosphere (Lashof and Ahuja 1990). More important, carbon monoxide plays a major role in atmospheric chemistry by controlling levels of tropospheric ozone (Chapter 3). High concentrations of atmospheric ozone over the tropical regions of South America and Africa appear to be related to the production of CO by forest burning (Figure 3.9), followed by the reaction of CO with OH radicals to produce ozone. About 2% of net primary production on land may be lost as CO or as volatile hydrocarbons that are oxidized to CO in the atmosphere (Chapter 5).

SYNTHESIS: LINKING THE CARBON AND OXYGEN CYCLES

Even on a lifeless Earth, photolysis of water vapor in the atmosphere would produce small amounts of O_2, as it has on planets such as Mars (Chapter 2). Early in its history, Earth may have lost up to 35% of the water degassed from the mantle by photolysis (Chapter 10). The process is now limited by cold temperatures and the ozone layer in the stratosphere, which minimize the amount of water vapor exposed to ultraviolet light. During the geologic history of Earth, significant amounts of atmospheric O_2 appeared only about 2.4 billion years ago, well after the advent of autotrophic photosynthesis. O_2 began to accumulate to its present level when the annual production exceeded the reaction of O_2 with reduced minerals in the Earth's crust.

The current atmospheric pool of O_2 is only a small fraction of the total O_2 produced over geologic time; most of the rest has been consumed in the oxidation of Fe and S (refer to Figure 2.8). The total net production of O_2 over geologic time is balanced stoichiometrically by the storage of reduced organic carbon (1.56×10^{22} g) and sedimentary pyrite (4.97×10^{21} gS; refer to Table 13.1) in the Earth's crust. Today, the content of oxygen in Earth's atmosphere is determined by the balance between the burial of organic carbon in sediments and the subsequent weathering of ancient sedimentary rocks, which have been uplifted to form the continents (Petsch et al. 2001). Oxygen is likely to have accumulated most rapidly during periods when large amounts of organic sediments were buried (Des Marais et al. 1992, France-Lanord and Derry 1997).

Today, the uplift and weathering of sedimentary rocks expose about 0.1 to 0.2×10^{15} g/yr of organic carbon to oxidation (Di-Giovanni et al. 2002; Figure 11.9). Much of the organic matter in exposed sedimentary rocks is labile (Schillawski and Petsch 2008, Galy et al. 2008). For instance, the biomass of microbes degrading organic matter in 365-million-year-old shales exposed to weathering in Kentucky is almost entirely derived from the shale (Petsch et al. 2001).

We have little evidence of historical variations in atmospheric O_2, but geochemical models suggest that the concentrations may have ranged from 15 to 35% during the past 500 million years (Berner and Canfield 1989, Berner 2001). The highest values would be expected in the Carboniferous and Permian periods, when a large amount of organic matter was buried in sediments (Berner et al. 2000; Figure 11.4). High concentrations of O_2 would have dramatic implications for the physiology, morphology, and evolution of most organisms (Graham et al. 1995). Giant insects are known to have been common during this period of high O_2 concentrations (Harrison et al. 2010). Fortunately, the pool of atmospheric O_2 is well buffered over geologic time because increases in O_2 expand the area and depth of aerobic respiration in marine sediments, leading to a greater consumption of O_2 and lower storage of organic carbon (Chapters 7 and 9). High levels of oxygen also reduce the rate of photosynthesis by causing increasing amounts of photorespiration (Tolbert et al. 1995; Chapter 5). The geologic record of sedimentary charcoal suggests that higher levels of O_2 may increase the occurrence

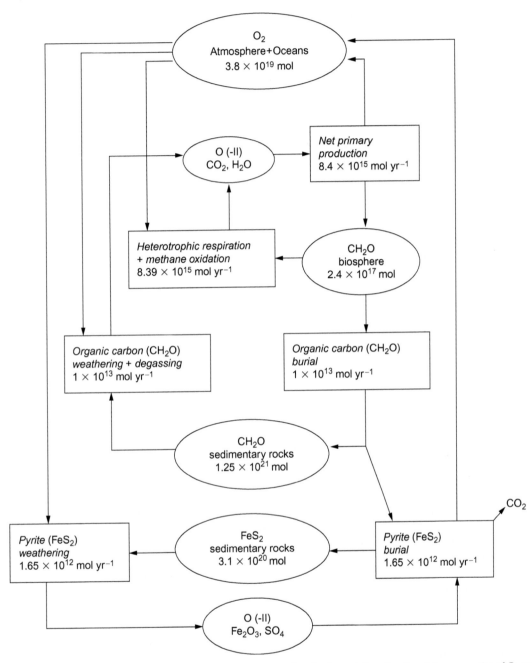

FIGURE 11.9 Linkage of the global carbon and oxygen cycles. Ovals contain estimates of the reservoirs of O_2 or the equivalent amount of reduced molecules that could be oxidized by O_2. Boxes indicate fluxes of O_2 or reduced molecules in moles/year. *Source: Modified from Lenton (2001).*

of forest fires (Scott and Glasspool 2006) and suppress the release of P by rock weathering (Lenton 2001). Finally, high dissolved O_2 in seawater would increase the adsorption of P to iron minerals in marine sediments, subsequently lowering nutrient availability and NPP in the sea (Van Cappellen and Ingall 1996). These interactions between the carbon, oxygen, and phosphorus cycles buffer the concentration of O_2 in Earth's atmosphere. Unlike geologic uplift and weathering, these processes are directly responsive to changes in the Earth's oxygen concentration.

Like the carbon cycle, the modern oxygen cycle is composed of a set of large, annual fluxes superimposed on the smaller, slow fluxes of the geologic cycle (Walker 1980). The current atmospheric pool of O_2 is maintained in a dynamic equilibrium between the production of O_2 by photosynthesis and its consumption in respiration, including fires (Figure 11.9). The annual fluctuation of O_2 in the atmosphere due to photosynthesis and respiration is about $\pm 0.0020\%$ in an average background concentration of 20.946% (Keeling and Shertz 1992). The mean residence time of O_2 in the atmosphere is about 4000 years—significantly shorter than what would be predicted merely by the reaction of O_2 with the Earth's crust, about 100,000,000 years (Figure 11.9).

Measured changes in the concentration of atmospheric O_2 provide a check on current estimates of CO_2 uptake by the terrestrial and marine biosphere (Bender et al. 1998). Photosynthesis on land produces O_2 and consumes CO_2 on an equal molar basis; the reverse is true when biomass and fossil fuels are burned. In contrast, uptake of CO_2 by the oceans, largely driven by Henry's Law, is not accompanied by a significant flux of O_2 to the atmosphere; the O_2 from marine photosynthesis remains dissolved in seawater and is consumed in the oceans during the degradation of organic matter. Thus, with fossil fuel combustion, the molar content of CO_2 should increase in Earth's atmosphere less rapidly than the molar decrease of O_2; the difference will reflect the dissolution of CO_2 in the oceans.

This method, developed by Ralph Keeling and Stephen Shertz (1992), in conjunction with precise measurements of changes in O_2 in the atmosphere, confirms a sink for about 1.7×10^{15} g C in the world's oceans and thus, an apparent net sink of 1.0×10^{15} g C in the terrestrial biosphere for the 1990s (compare to Bender et al. 2005; Table 11.1). This net sink for carbon in vegetation is relatively recent; the same methods show relatively little carbon accumulation on land during the 1980s (Langenfelds et al. 1999, Battle et al. 2000, Bopp et al. 2002). The sink for CO_2 in the oceans, increasing ocean acidification, is projected to exceed what is observed in sedimentary records for the past 300,000,000 years (Hönisch et al. 2012).

An examination of the isotopic composition of atmospheric O_2 (i.e., $\delta^{18}O$) also allows us to constrain the atmospheric CO_2 budget within certain limits. Photosynthesis does not discriminate among the oxygen isotopes of water—the O_2 released has an isotopic composition that is identical to that in the seawater or the soil water in which the plant is growing. Respiration discriminates among oxygen isotopes, consuming $^{16}O_2$ in preference to $^{18}O_2$; as a result $^{18}O_2$ is enriched in the atmosphere—known as the Dole effect (Luz and Barkan 2011). The $\delta^{18}O$ of atmospheric O_2 ($+23.5\permil$) suggests that gross primary production must be $>170 \times 10^{15}$ g C/yr on land and about 140×10^{15} g C/yr in the oceans (Bender et al. 1994).[7] Assuming that

---

[7] This calculation is based on the amount of GPP, evolving O_2 identical to that in soil or seawater (i.e., $\delta^{18}O = 0$), that is necessary to maintain a constant level of $\delta^{18}O$ in the atmosphere in the face of preferential respiratory consumption of the light isotope.

net primary production is one-half of gross primary production, in both cases, these values would seem to indicate that NPP is somewhat higher than what we have estimated independently for land (Table 5.3; Beer et al. 2010) and marine (Table 9.2) habitats globally. Nevertheless, these values offer an upper limit for NPP, which helps constrain our estimates for the global carbon cycle (Figure 11.1).

The oxygen cycle is directly linked to other biogeochemical cycles. For example, assuming that NO_3 accounts for about half the plant uptake of N on land (1200×10^{12} g) and about 10% of the N cycle in the oceans (8000×10^{12} g; Figure 9.21), then about 2% of the annual production of O_2 by photosynthesis is used to oxidize NH_4 in the nitrification reactions (compare to Ciais et al. 2007).

The formation and oxidation of sedimentary pyrite, through sulfate reduction, also affects the concentration of O_2 in the atmosphere. For every mole of pyrite-S oxidized, nearly 2 moles of O_2 are consumed from the atmosphere (Eq. 9.2). Currently, the annual burial of pyrite in marine sediments accounts for about 20% of the oxygen in our atmosphere (Chapter 9).

Methanogenesis in freshwater sediments returns CH_4 to the atmosphere, where it is oxidized (Henrichs and Reeburgh 1987). Methane oxidation in the atmosphere accounts for about 1% of the total consumption of atmospheric O_2 each year. In the absence of methanogenesis, the burial of organic carbon in freshwater sediments might be greater and the atmospheric content of O_2 might be slightly higher. Thus, methanogenesis acts as a negative feedback in the regulation of atmospheric O_2 (Watson et al. 1978, Kump and Garrels 1986).

It is perhaps entertaining to speculate whether the carbon cycle on Earth drives the oxygen cycle, or vice versa. Over geologic time, the answer is obvious: The conditions on our neighboring planets provide ample evidence that O_2 is derived from life. Now, however, the carbon and oxygen cycles are inextricably linked, and the metabolism of eukaryotic organisms, including humans, depends on the flow of electrons from reduced organic molecules to oxygen.

SUMMARY

Humans harvest about 20% of the annual production of organic carbon on land (i.e., NPP; Imhoff et al. 2004b, Haberl et al. 2007). In many areas we have destroyed land vegetation, while in other areas we have planted productive crops and forests and perhaps stimulated NPP by raising levels of atmospheric CO_2 and nitrogen deposition. At the moment, it appears that humans have created a net sink for carbon in the terrestrial biosphere, which mitigates some of the anticipated rise in atmospheric CO_2 from fossil fuel combustion (Table 11.1). With the use of fossil fuels, humans have, of course, supplemented the energy available to power modern society, including vast supplements to the agricultural systems that feed us. Dukes (2003) calculates that each year we burn the equivalent of the organic matter stored during 400 years of primary production in the geologic past. Accelerating use of fossil fuels is destined to lead to large changes in the Earth's conditions, which have otherwise been relatively stable during the 8000-year history of organized human society.

Recommended Readings

Archer, D. 2010. *The Global Carbon Cycle.* Princeton University Press.

Field, C.B., and M.R. Raupach (Eds.). 2004. *The Global Carbon Cycle: Integrating Humans, Climate, and the Natural World.* Island Press.

Solomon, S., D. Qin, M. Manning, M. Marquis, K. Averyt, M.M.B. Tignor, H.L. Miller, and Z. Chen. 2007. *Climate Change 2007: The Physical Science Basis.* Cambridge University Press.

PROBLEMS

1. How does the annual human extraction of fossil carbon (in grams) from the Earth's crust compare to the chemical and mechanical denudation of the continents (Chapter 4) by wind and water?

2. If wood has a density of 0.8 g/cm^3 and a carbon content of 50% by mass, what is the size of a cube of wood that contains 10^{15} g of carbon?

3. Calculate the pool of carbon in the global biomass of the human species, making and stating reasonable assumptions about the mass of individual humans and the abundance of their number globally.

4. Each molecule of methane contributes about $25\times$ as much to the Earth's greenhouse effect as each molecule of CO_2. If we were interested in preventing global warming, what would be the effect, versus reducing CO_2 emissions, if we were to eliminate all anthropogenic inputs of methane to the atmosphere (Table 11.2)?

The Global Cycles of Nitrogen and Phosphorus

INTRODUCTION

The availability of nitrogen and phosphorus controls many aspects of ecosystem function and global biogeochemistry. Nitrogen often limits the rate of net primary production on land and in the sea (Chapters 6 and 9). In living tissues, nitrogen is an integral part of enzymes, which mediate the biochemical reactions in which carbon is reduced (e.g., photosynthesis) or oxidized (e.g., respiration). Nearly all of the nitrogen in biomass is first assimilated by the attachment of an amine group ($-NH_2$) to the 5-carbon sugar oxoglutarate, linking the C and N cycles at the level of cellular biochemistry (Williams 1996, p. 158). Phosphorus is an essential component of DNA, ATP, and the phospholipid molecules of cell membranes. The ratio of N to P in plant tissues, about 16, finds its basis in the ratio of protein to RNA in protoplasm (Loladze and Elser 2011). Changes in the availability of N and P and their relative abundance are likely to have controlled the size and activity of the biosphere through geologic time.

A large number of biochemical transformations of nitrogen are possible, since nitrogen is found at valence states ranging from -3 (in NH_3) to $+5$ (in NO_3^-). A variety of microbes capitalize on the potential for transformations of N among these states and use the energy released by the changes in redox potential to maintain their life processes (Rosswall 1982).

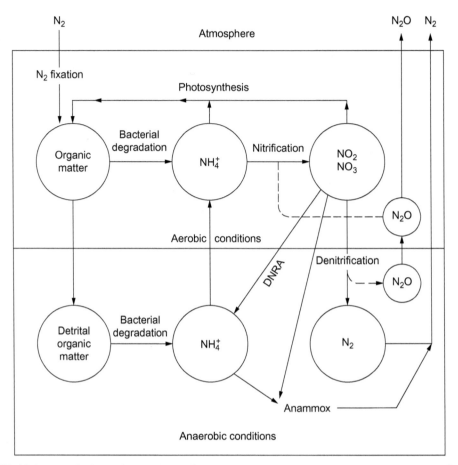

FIGURE 12.1   Microbial transformations in the nitrogen cycle (DNRA = Dissimilatory Nitrate Reduction to Ammonium). *Source: Modified from Wollast (1981).*

Collectively, these microbial reactions drive the cycle of nitrogen (Figure 12.1). In contrast, whether it occurs in soils or in biochemistry, phosphorus is almost always found in combination with oxygen (i.e., as PO_4^{3-}). Most metabolic activity is associated with the synthesis or destruction of high-energy bonds between a phosphate ion and various organic molecules, but in nearly all cases the phosphorus atom remains at a valence of $+5$ in these reactions.

    The most abundant form of nitrogen at the surface of the Earth, N_2, is the least reactive species. Nitrogen fixation converts atmospheric N_2 to one of the forms of reactive nitrogen ("fixed" or "odd"; Chapter 3) that can be used by biota. Nitrogen-fixing species are most abundant in nitrogen-poor habitats, where their activity increases the availability of nitrogen for the biosphere (Eq. 2.10). At the same time, denitrifying bacteria return N_2 to the atmosphere (Eq. 2.20), lowering the overall stock of nitrogen available for life on Earth.

    Rocks of the continental crust hold the reservoir of phosphorus that becomes available to the biosphere through rock weathering. Land plants can increase the rate of rock weathering in P-deficient habitats (Chapter 4), but in nearly all cases the phosphorus content of rocks is relatively low. Subsequent reactions between dissolved P and other minerals reduce the availability of P in soil solutions or seawater (Figure 4.10). Thus, in most habitats—both on land and in the sea—the availability of P is controlled by the degradation of organic forms of P (e.g., Figure 6.19). This *bio*geochemical cycle temporarily retains and recycles some P from the unrelenting flow of P from weathered rock to ocean sediments. The global P cycle is complete only when sedimentary rocks are lifted above sea level and the weathering begins again.

Because supplies of nitrogen and phosphorus often define soil fertility, humans have added enormous quantities of these elements to soils to enhance crop production. The production of fertilizer has more than doubled the supply of N and P on the land surface, altering biogeochemical cycling and leading to inadvertent enrichments of ecosystems downwind or downstream of the point of application.

In this chapter, we examine our current understanding of the global cycles of N and P. We attempt to balance N and P budgets for the world's land area and for the sea. For N, the balance between N fixation and denitrification through geologic time determines the nitrogen available to biota and the global nitrogen cycle. One of the byproducts of nitrification and denitrification is N_2O (nitrous oxide), which is both a greenhouse gas and a cause of ozone destruction in the stratosphere (Chapter 3). We present a budget for N_2O based on our current understanding of the sources of this gas in the atmosphere.

THE GLOBAL NITROGEN CYCLE

Land

Figure 12.2 presents the global nitrogen cycle, showing the linkage between the atmosphere, the land, and the oceans. The atmosphere contains the largest pool (3.9×10^{21} g N; Table 3.1). Relatively small amounts of N are found in terrestrial biomass (3.8×10^{15} g)[1] and in soil organic matter ($95–140 \times 10^{15}$ g to 1-m depth; Post et al. 1985, Batjes et al. 1996). The mean C/N ratios for terrestrial biomass and soil organic matter are about 160 and 12, respectively. At any time, the pool of inorganic nitrogen, NH_4^+ and NO_3^-, in soils is very small. The uptake of N by organisms is so rapid that little nitrogen remains in inorganic forms, despite the large annual flux through this pool (Chapter 6).[2] Since most N in soils is held in organic forms, soil organic matter is a good predictor of total nitrogen content in most circumstances (Glendining et al. 2011).

The nitrogen in the atmosphere is not available to most organisms because the great strength of the triple bond in N_2 makes this molecule practically inert.[3] All nitrogen that is

---

[1] This value is derived from an estimate of 600×10^{15} g C in terrestrial biomass (Table 5.3) and a C/N ratio of 160 in forest biomass (Table 6.4). Given the convergence of estimates for the carbon pool in biomass, higher estimates for the N pool in vegetation would require a lower estimate of the C/N ratio in biomass, which seems unlikely.

[2] Typically the C/N ratio in desert soils is very low (Post et al. 1985). In deserts, where soil biotic activity is limited, a large amount of nitrate may accumulate in the soil profile below the rooting zone of plants, perhaps amounting to $3–15 \times 10^{15}$ g N globally (Walvoord et al. 2003).

[3] The mean bond energy in N_2 is 226 kcal/mole, versus N-H (93), N-C (70), or N-O (48) (Davies 1972).

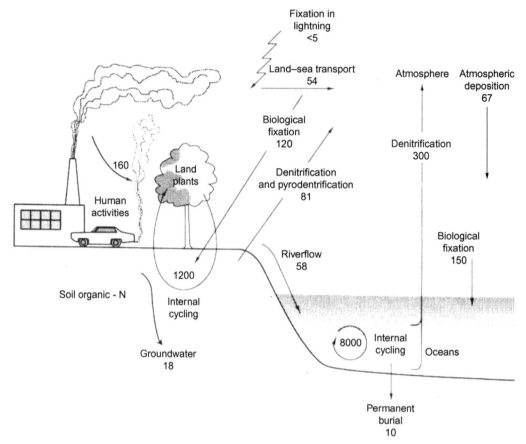

FIGURE 12.2   The global nitrogen cycle. Each flux is shown in units of 10^{12} g N/yr. Values as derived in the text. See also Table 12.3.

available to biota was originally derived from nitrogen fixation—either by lightning or by a few specialized species of microbes, which convert N_2 to forms of reactive nitrogen (Chapter 6). The rate of nitrogen fixation by lightning, which produces momentary conditions of high pressure and temperature allowing N_2 and O_2 to form NO_x, is relatively small. Most recent global estimates are in the range of 2 to 6×10^{12} g N/yr (Borucki and Chameides 1984, Kumar et al. 1995b, Ridley et al. 1996, Levy et al. 1996, Nesbitt et al. 2000, Martin et al. 2007, Schumann and Huntrieser 2007), and the estimated total annual deposition of oxidized N (NO_y) from the preindustrial atmosphere precludes an estimate higher than about 12×10^{12} g N/yr (Galloway et al. 2004). Assuming that lightning is distributed uniformly over land and sea, a liberal estimate for the deposition of N fixed by lightning over land would be 2×10^{12} g N/yr. The present-day deposition of oxidized nitrogen on land is about 25×10^{12} g/yr, owing to the additional NO_x that is emitted from soils, biomass burning, and human activities (Table 12.1).

TABLE 12.1   A Global Budget for Atmospheric NO_x (values are Tg N (10^{12} g N)/yr as NO)

Process	Annual Flux	References
Sources		
Fossil fuel combustion	25	Galloway et al. 2004
Net emissions from soils	12	Ganzeveld et al. 2002 (Gross flux $\sim$21 Tg N/yr; Davidson and Kingerlee 1997)
Biomass burning	9.6	Andreae and Merlet 2001, Kaiser et al. 2012 (compare 9.8 Tg N/yr, Mieville et al. 2010)
Lightning	5	See text references
NH_3 oxidation	1	Compare to Table 12.2 (Warneck 2000)
Aircraft	0.4	Prather et al. 1995
Transport from the stratosphere	0.6	For total NO_y (Prather et al. 1995)
Total sources	**53.6**	Compare 37 Tg N/yr from satellite measurements (Martin et al. 2003; 46 Tg N/yr (Galloway et al. 2004)
Sinks		
Deposition on land	24.8	Galloway et al. 2004
Deposition on the ocean surface	23.0	Duce et al. 2008, Dentener et al. 2006
Total sinks	**47.8**	

Biological nitrogen fixation is performed by several species of microbes, which are either free-living in lake waters, soils, and sediments or found in symbiotic association with the roots of plants (Chapter 6). Prior to widespread human activities, total biological nitrogen fixation on land is likely to have been 60 to 195 $\times$ 10^{12} g N/yr (Cleveland et al. 1999, Vitousek et al. in press). The current rate of biological nitrogen fixation is estimated at 120 to 180 $\times$ 10^{12} g N/yr for the sum of natural and agricultural ecosystems (Burns and Hardy 1975, Wang and Houlton 2009). As natural lands have given way to cultivation, the nitrogen fixation in agricultural systems has replaced (and even exceeded) the nitrogen fixation that was lost from natural ecosystems (Galloway et al. 2004). Estimates of N fixation in agriculture are as high as 50 to 70 $\times$ 10^{12} g N/yr, or nearly half of the land total (Herridge et al. 2008; compare to 21 $\times$ 10^{12} g N/yr, Liu et al. 2010).

These estimates of N fixation on land are equivalent to about 10 kg N/yr for each hectare of the Earth's land surface. Most studies of nitrogen fixation in free-living soil bacteria report values ranging from 1 to 5 kg ha^{-1} yr^{-1} (Chapter 6). A value of 3 kg N ha^{-1} yr^{-1} multiplied by the world's land area suggests that asymbiotic fixation contributes about one-third of the global total. The remainder is assumed to come from symbiotic association of bacteria with higher plants. This flux is not distributed uniformly among natural ecosystems; the greatest values are often found in tropical forests and in areas of disturbed or successional vegetation (Vitousek and Howarth 1991). In any case, biotic N fixation dwarfs abiotic fixation by lightning as the source of fixed N. The evolution of life and of nitrogen fixation on Earth greatly speeded the movement of nitrogen in a biogeochemical cycle. Taking all forms of N fixation as the only source, the mean residence time of nitrogen in the terrestrial biosphere is about 700 years (i.e., pool/input). This calculation is not altered significantly by considering the small amount of fixed nitrogen that is contained in sedimentary and metasedimentary rocks, which may contribute more than 20×10^{12} g N/yr to the terrestrial biosphere by chemical weathering (Holloway and Dahlgren 2002; Houlton, personal communication, 2011).

Assuming that the estimate of terrestrial net primary production, 60×10^{15} g C/yr, is roughly correct and that the mean C/N ratio of net primary production is about 50, the nitrogen requirement of land plants is about 1200×10^{12} g/yr (Chapter 6).[4] Thus, nitrogen fixation supplies only about 15% of the nitrogen that is assimilated by land plants each year. The remaining nitrogen must be derived from internal recycling and the decomposition of dead materials in the soil (Chapter 6). When the turnover in the soil is calculated with respect to the input of dead plant materials, the mean residence time of nitrogen in soil organic matter is >100 years. Thus, the mean residence time of N exceeds that of C in both land vegetation and soils (5 and 25 years, respectively; Chapter 5).

Humans have a dramatic impact on the global N cycle. In addition to planting N-fixing species for crops, humans produce nitrogen fertilizers through the Haber process; namely:

$$3CH_4 + 6H_2O \rightarrow 3CO_2 + 12H_2 \tag{12.1}$$

$$4N_2 + 12H_2 \rightarrow 8NH_3, \tag{12.2}$$

in which natural gas is burned to produce hydrogen, which is combined with N_2 to form ammonia under conditions of high temperature and pressure (Smil 2001). The industrial production of reactive N by these reactions supplies $>136 \times 10^{12}$ g N/yr for agricultural and chemical uses[5]—roughly matching the natural rate of nitrogen fixation on land.

A substantial fraction, perhaps half, of the annual application of nitrogen fertilizer to agricultural lands is lost to the atmosphere and to runoff waters (Erisman et al. 2007). Some NH_3 is volatilized directly to the atmosphere from cultivated soils (Figure 6.14), while some fertilizer N is lost indirectly from the excrement of domestic livestock that are fed forage crops (Table 12.2). The loss of NH_3 from agricultural lands to the atmosphere carries fixed N to adjacent natural ecosystems where it is deposited and enters biogeochemical cycles (Draaijers et al. 1989, Hesterberg et al. 1996).

Fossil fuel combustion also produces about 25×10^{12} g of fixed N (namely, NO_x) annually (Galloway et al. 2004). Some of this is derived from the organic nitrogen contained in fuels (Bowman 1991), but it is best regarded as a source of new, fixed N for the biosphere because in the absence of human activities, this N would remain inaccessible in the Earth's crust. NO_x also forms directly from N_2 and O_2 during the combustion of fossil fuels, especially in automobiles and coal-fired power plants (Bowman 1992, Davidson et al. 1998, Marufu et al. 2004,

---

[4] Most primary production consists of short-lived tissues with a C/N ratio that is much lower than that of wood (160), which composes most of the terrestrial biomass. Mineralization of $\sim$1000 $\times$ 10^{12} g N/yr is consistent with experimental observations suggesting 1 to 3% turnover of nitrogen in soils annually (Chapter 6). Raven et al. (1993) give an alternative estimate of 2338×10^{12} g/yr for the nitrogen uptake by land plants. Another recent global model for the N cycle assumes a C/N ratio of about 10 and uptake of 6207×10^{12} g N/yr by terrestrial NPP, which seems unlikely (Lin et al. 2000).

[5] *http://minerals.usgs.gov/minerals/pubs/commodity/nitrogen/mcs-2012-nitro.pdf.*

TABLE 12.2   A Global Budget for Atmospheric Ammonia

Process	Annual flux	References
Sources		
Domestic animals	18.5	Bouwman et al. 2002
Wild animals	0.1	
Sea surface	8.2	
Undisturbed soils	2.4	
Agricultural soils	3.6	
Fertilizers	9.0	
Biomass burning	7.7	Kaiser et al. 2012
Human excrement	2.6	
Coal combustion and industry	0.3	
Automobiles	0.2	Schlesinger and Hartley 1992
Total sources	**52.6**	Compare 58.2 Tg N/yr; Galloway et al. 2004
Sinks		
Deposition on land	38.7	
Deposition on the ocean surface	24.0	Duce et al. 2008; Dentener et al. 2006
Reaction with OH radicals	1.0	Schlesinger and Hartley 1992
Total sinks	**63.7**	

Note: Unless noted otherwise, sources are derived from Bouwman et al. (1997) and sinks from Galloway et al. (2004). All values are Tg N (10^{12} g N)/yr as NH_3 or NH_4^+ (in deposition).

Kim et al. 2006). In automobiles, some NO_x is converted to NH_3 by catalytic converters (Bhattacharyya and Das 1999, Emmenegger et al. 2004). Owing to the short residence time of NO_x and NH_3 in the atmosphere (Eq. 3.26; Table 3.5), most of this nitrogen is deposited by precipitation over land, where it enters biogeochemical cycles (Figure 12.3). A small portion of NO_x undergoes long-distance transport in the troposphere, accounting for excess nitrogen deposition in the oceans (Duce et al. 2008, Kim et al. 2011) and the rising levels of NO_3^- deposited in Greenland snow (Fischer et al. 1998, Burkhart et al. 2006; Figure 12.4).

Evidence for anthropogenic emissions of nitrogen to the atmosphere dates to about 1900 (Hastings et al. 2009, Holtgrieve et al. 2011). In some areas, the atmospheric deposition of N may fertilize plant growth (Chapter 6), leading to an increment in the sink for carbon in vegetation (Chapter 11). However, some high-elevation forests downwind of major population centers now receive enormous nitrogen inputs, both NH_4^+ and NO_3^-, which may be related to forest decline (Chapter 6). Various workers have tried to define the "critical load"—the level of nitrogen deposition that can be expected to cause changes in ecosystem properties—for example, increased nitrification and leaching. In many ecosystems, noticeable effects begin at nitrogen depositions of 1 kg ha^{-1} yr^{-1} (Pardo et al. 2011, Liu et al. 2011). As a

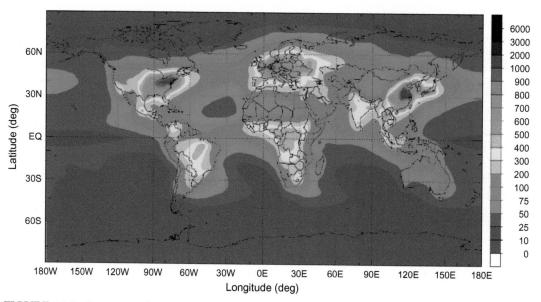

FIGURE 12.3   Deposition of NO$_y$ on Earth's surface. All values are mg N m^{-2} yr^{-1}. *Source: From Dentener et al. (2006).*

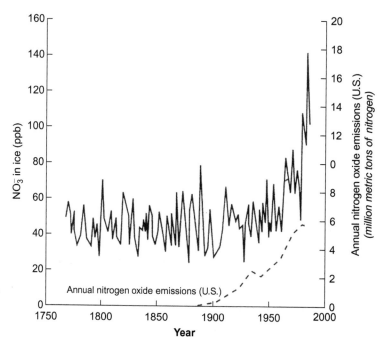

FIGURE 12.4   The 200-year record of nitrate in layers of the Greenland ice pack and the annual production of nitric oxides by fossil fuel combustion in the United States. *Source: Modified from Mayewski et al. (1990).*

result of air-pollution control legislation, deposition of nitrogen from the atmosphere is now declining in North America (Zbieranowski and Aherne 2011).

In total, about 300×10^{12} g of newly fixed N is delivered from the atmosphere to the Earth's land surface each year—28% from natural and 72% from human-derived sources (Table 12.3). Some of this nitrogen remains sequestered in the terrestrial biosphere, primarily in forests and in agricultural soils, which store about 30 to 40% of the applied nitrogen (Schindler and Knighton 1999, Fritschi et al. 2004, Stevens et al. 2005, Gardner and Drinkwater 2009). In the absence of processes removing nitrogen, a very large pool of nitrogen would be found on land in a relatively short time. About 23% of the nitrogen deposited on the land surface is lost to runoff (Howarth 1998, van Breemen et al. 2002), especially during periods of peak flow (Chapter 8). Runoff of nitrogen into Lake Michigan is well correlated to fertilizer and land-use changes in its basin (Han and Allan 2011). Globally, rivers carry nearly 60×10^{12} g N/yr from land to the sea (Boyer et al. 2006, Van Drecht et al. 2003), and humans may account for more than half of the present-day transport of N in rivers (Schlesinger 2009).

TABLE 12.3   Mass Balance for Nitrogen on the Earth's Land Surface

Inputs	Preindustrial	Human derived	Total
Biological N fixation	60[a]	60[b]	120
Lightning	5	0	5
Rock weathering	20[c]	0	20
Industrial N fixation	0	136[d]	136
Fossil fuel combustion	0	25	25
Total	**85**	**221**	**306**
Fates			
Biospheric increment	0	9	9
Soil accumulation	0	48	48
Riverflow	27	31	58
Groundwater	0	18	18
Denitrification	27[e]	17	44
Pyrodenitrification	25[f]	12	37
Atmospheric land–sea transport[g]	6	48	54
Total	**85**	**183**	**268**

Note: Updated from Schlesinger (2009), with permission from the National Academy of Sciences. Unless otherwise indicated, preindustrial values and human-derived inputs are from Galloway et al. (2004). Fates of anthropogenic nitrogen are derived in this chapter.

Note: All values are in Tg N/yr (=10^{12} g N/yr).

[a] *Vitousek et al. in press*
[b] *Herridge et al. (2008); value is the net from human activities.*
[c] *B.Z. Houlton, personal communication (2011).*
[d] *http://minerals.usgs.gov/minerals/pubs/commodity/nitrogen/mcs-2012-nitro.pdf.*
[e] *To balance.*
[f] *See derivation in the text.*
[g] *Duce et al. (2008).*

Human additions of fixed nitrogen to the terrestrial biosphere have also resulted in marked increases in the nitrogen content of groundwaters, especially in many agricultural areas (Spalding and Exner 1993, Rupert 2008, Kroeger and Charette 2008, Scanlon et al. 2010). For example, the loss of nitrate to groundwater was the largest single fate of nitrogen added to the fields of a dairy farm in Ontario (Barry et al. 1993). Nitrogen and other biochemical elements are also enriched in the drainage from cemeteries (Zychowski 2012). The global transport of N to groundwaters may approach 18×10^{12} g N/yr—calculated from an estimate of the annual flux of groundwater (12,666 km^3/yr; Doll and Fiedler 2008) and the median concentration of 1.9 mg N/liter in groundwaters of the United States (Nolan et al. 2002, Schlesinger 2009).

Despite these large transports, riverflow and groundwater cannot account for all of the nitrogen that is lost from land. The remaining nitrogen is assumed to be lost by denitrification and other gaseous pathways in terrestrial soils (Chapter 6), in wetlands (Chapter 7), and during forest fires (Chapter 6). Seitzinger et al. (2006) have compiled rates of denitrification showing that roughly 124×10^{12} g N/yr occur in soils and 110×10^{12} g/yr in freshwater environments (Table 12.4). If N fixation and denitrification were once in balance, then a terrestrial denitrification rate of $\sim 100 \times 10^{12}$ g N/yr was most likely in the preindustrial world (i.e., fixation minus riverflow; Schlesinger 2009). Most of the loss occurs as N_2, but the small fraction that is lost as N_2O during nitrification and denitrification (Chapter 6) contributes significantly to the global budget of this gas.

Indeed, the current rise in atmospheric N_2O can be used to estimate the overall increase in global denitrification as a result of human activities (Schlesinger 2009). If we assume that the $(N_2 + N_2O)/N_2O$ ratio for denitrification is about 4.0 and that the recent increase of N_2O in

TABLE 12.4   A Global Estimate of Denitrification of Nitrogen on or Applied to Land

System	Denitrification (Tg N/yr)
Terrestrial	
Soils	124 (65–175)
Freshwater	
Groundwater	44 (>0–138)
Lakes and reservoirs	31 (19–43)
Rivers	35 (20–35)
Subtotal	110 (39–216)
Marine	
Estuaries	8 (3–10)
Continental shelves	46 (>0–70)
Oxygen minimum zones	25 (>0–30?)
Subtotal	79 (3–145)

Source: From Seitzinger et al. (2006).

the atmosphere (nearly 4×10^{12} g N/yr) all derives from increased denitrification, then it is possible that the overall loss of N_2 from denitrification has increased by as much as 17×10^{12} g N/yr, helping to balance the present-day N budget on land (Figure 12.2).[6] Denitrification leaves soils enriched in $\delta^{15}N$ globally (Amundson et al. 2003, Houlton and Bai 2009); especially high rates of denitrification and high soil $\delta^{15}N$ are reported in the tropics (Houlton et al. 2006, Koba et al. 2012).

Nitrogen in biomass is volatilized as NH_3, NO_x, and N_2 during fires—the last constituting a form of *pyrodenitrification* (Chapter 6). About 30% of the nitrogen in fuel is converted to N_2, so globally biomass burning may return about 37×10^{12} g N/yr to the atmosphere as N_2 (Kuhlbusch et al. 1991). The rate of biomass burning has increased by about $1.5\times$ in recent years (Mouillot et al. 2006), so this form of denitrification may have increased from a preindustrial level of $\sim 25 \times 10^{12}$ g N/yr.

In balancing the terrestrial N cycle, we concentrate on processes that affect the net production or loss of fixed nitrogen (Table 12.3). We do not include processes that recycle N that was fixed at an earlier time. Thus, NH_3 volatilization from biomass burning (Table 12.2) and the natural emission of NO_x from soils (Table 12.1) can be ignored to the extent that these forms are redeposited on land in precipitation. Ammonia and NO_x have relatively short atmospheric lifetimes, so they are usually deposited in precipitation and dryfall near their point of origin (Chapter 3). Indeed, some have suggested that nitrogen "hop-scotches" across the landscape, where losses from one area result in increased deposition and local cycling in other areas. The cycle is complete only when nitrogen is returned to the atmosphere as N_2. By focusing on the new sources of available nitrogen from the atmosphere, we find that nearly all of the human perturbation of the global nitrogen cycle has occurred in the past 150 years.

Sea

The world's oceans receive about 60×10^{12} g N/yr in dissolved forms from rivers (Chapter 8), about 150×10^{12} g N/yr via biological N fixation (Chapter 9), and about 67×10^{12} g N in precipitation (Duce et al. 2008). Some of the precipitation flux is NH_4^+ that is derived from NH_3 volatilized from the sea (Quinn et al. 1988), but about 80% of the atmospheric deposition of N in the oceans derives from human activities on land (Duce et al. 2008). A significant fraction of the deposition of atmospheric nitrogen on the oceans is found in various types of organic compounds (Cornell et al. 2003). Through their various activities, humans have created a large flux of nitrogen through the atmosphere, from land to sea.

As we have shown for terrestrial ecosystems, most of the net primary production in the sea is supported by nitrogen recycling in the water column (Table 9.3). Nitrogen inputs from the atmosphere have the greatest influence in the open oceans, where the pool of inorganic nitrogen is very small. The riverflux of N assumes its greatest importance in coastal seas and estuaries. Here, runoff of excess nitrogen has caused significant changes in the productivity of coastal ecosystems, leading to eutrophication and hypoxia (Beman et al. 2005, Goolsby et al. 2001, Kim et al. 2011).

---

[6] The ratio ranges from 2.0 in upland soils to >12 in wetlands (Schlesinger 2009), so the total increase in denitrification caused by human inputs may range from 8 to 68×10^{12} g N/yr.

The deep ocean contains a large pool of inorganic nitrogen (720×10^{15} g N)[7] derived from the decomposition of sinking organic debris and mineralization from dissolved organic compounds. Permanent burial of organic nitrogen in sediments is small, so most of the nitrogen input to the oceans must be returned to the atmosphere as N_2 by denitrification and the anammox reaction (Figures 9.21 and 12.2). Important areas of denitrification are found in the anaerobic deep waters of the eastern tropical Pacific Ocean and the Arabian Sea (Chapter 9). Seitzinger et al. (2006) estimate denitrification of up to 145×10^{12} g N/yr in near-shore waters. Globally, coastal and marine denitrification may account for the return of 270 to 400×10^{12} g N/yr to the atmosphere as N_2, yielding an overall mean residence time for N in the oceans of < 2000 years (Brandes and Devol 2002, Codispoti 2007, Bianchi et al. 2012).

Although the estimates are subject to large uncertainty, the model of Figure 12.2 indicates a net loss of nitrogen from the oceans. The overall gaseous losses of nitrogen from the ocean exceed the inputs from rivers and the atmosphere, so that the oceans may be declining in nitrogen content (McElroy 1983, Codispoti 2007). Various workers have suggested that, in the absence of denitrification, higher concentrations of NO_3 would be found in the ocean and lower concentrations of N_2 in the atmosphere. The balance of nitrogen fixation and denitrification has probably controlled marine NPP through past glacial cycles (Ganeshram et al. 1995). Globally, suboxic waters, in which available nitrogen is depleted by denitrification, are also sites of N fixation, providing long-term self-regulation to the nitrogen cycle of the oceans (Deutsch et al. 2007).

TEMPORAL VARIATIONS IN THE GLOBAL NITROGEN CYCLE

The earliest atmosphere on Earth is thought to have been dominated by nitrogen, since N is abundant in volcanic emissions and only sparingly soluble in seawater (Chapter 2). The early rate of degassing of N_2 from the Earth's mantle must have been much greater than today. The present-day flux of N_2 from the mantle is about 78×10^9 g N/yr (Sano et al. 2001) to 123×10^9 g N/yr (Tajika 1998), which could not result in the observed accumulations of nitrogen at the Earth's surface (refer to Table 2.3), even after 4.5 billion years of Earth's history.[8]

Before the origin of life, nitrogen was fixed by lightning and in the shock waves of meteors, which create local conditions of high temperature and pressure in the atmosphere (Mancinelli and McKay 1988). The rate of N fixation was very low, perhaps about 6% of the present-day rate, because abiotic fixation in an atmosphere dominated by N_2 and CO_2 is much slower than in an atmosphere of N_2 and O_2 (Kasting and Walker 1981). The limited supply of fixed nitrogen in the primitive oceans on Earth is likely to have led to the early evolution of N fixation in marine biota (Chapter 2).

---

[7] Volume of the deep oceans ($0.95 \times 1.335 \times 10^{24}$ g; Figure 10.1) multiplied by the NO_3^- in the deep ocean (40 μmol NO_3/kg; Figures 9.18 and 9.27) multiplied by 14 g/mole.

[8] Some nitrogen degassed to the Earth's surface is carried back to the mantle by subduction of sediments (760×10^9 g N/yr; Busigny et al. 2003), with the contemporary net flux to the mantle perhaps amounting to 330×10^9 g/yr to 960×10^9 g N/yr (Goldblatt et al. 2009, Busigny et al. 2011). Thus the pool of nitrogen at the Earth's surface is decreasing slightly each year by net entrainment in the mantle.

The best estimates of abiotic N fixation on the primitive Earth suggest that it had a limited effect on the content of atmospheric nitrogen but provided a small though important supply of fixed nitrogen, largely NO_3^-, to the ocean's waters (Kasting and Walker 1981, Mancinelli and McKay 1988). Similar abiotic N fixation on Mars is postulated to have resulted in accumulations of nitrate or cyanide (HCN) on its surface (Segura and Navarro-Gonzalez 2005, Manning et al. 2008). Nitrate also accumulates in extremely arid soils on Earth, although here it is likely derived from distant biogenic sources (Michalski et al. 2004a).

With respect to N fixation by lightning, the mean residence time of N_2 in the atmosphere is about 1 billion years. The mean residence time of atmospheric nitrogen decreases to about 14,000,000 years when biological nitrogen fixation is included. This is much shorter than the history of life on Earth, and it speaks strongly to the importance of denitrification in returning N_2 to the atmosphere over geologic time. Denitrification closes the global biogeochemical cycle of nitrogen, but it also means that nitrogen remains in short supply for the biosphere. In the absence of denitrification, most nitrogen on Earth would be found as NO_3^- in seawater and the oceans would be quite acidic (Sillén 1966).

It is likely that denitrification appeared later than the other major metabolic pathways; denitrifying bacteria are facultative anaerobes, switching from simple heterotrophic respiration to NO_3 respiration under anaerobic conditions (Broda 1975, Betlach 1982). Denitrification enzymes are somewhat tolerant of low concentrations of O_2, allowing denitrifying bacteria to persist in environments with fluctuations in redox potential (Bonin et al. 1989, McKenney et al. 1994, Carter et al. 1995). The geologic record offers some insight regarding the origin of denitrification. Sedimentary rocks with enrichments of $\delta^{15}N$, indicating denitrification, date only to 2.7 bya—well after the origin of oxygenic photosynthesis (Beaumont and Robert 1999, Godfrey and Falkowski 2009, Thomazo et al. 2011).

Requiring oxygen as a reactant, nitrification clearly arose after photosynthesis and the development of an O_2-rich atmosphere (Eqs. 2.17 and 2.18). Some of the earliest nitrifying organisms may have been archaea, which are found in many soils (Leininger et al. 2006). Today, the rate of denitrification is controlled by the rate of nitrification, which supplies NO_3^- as a substrate (Eq. 2.20; Figure 6.13). In any case, the major microbial reactions in the nitrogen cycle (Figure 12.1) are all likely to have been in place at least 2 billion years ago.

Because NO_3^- is very soluble, there is little reliable record of changes in the content of NO_3^- in seawater through geologic time. Only changes in the deposition of organic nitrogen are recorded in sediments. Altabet and Curry (1989) show that the $^{15}N/^{14}N$ record in sedimentary foraminifera is useful in reconstructing the past record of ocean N chemistry. The isotope ratio in sedimentary organic matter increases when high rates of denitrification remove NO_3^- from the oceans, leaving the residual pool of nitrate enriched in ^{15}N (Altabet et al. 1995, Ganeshram et al. 1995).

Assuming a steady state in the ocean nitrogen cycle, the mean residence time for an atom of fixed N in the sea is < 2000 years. During this time, this atom will make several trips through the deep ocean, each lasting 200 to 500 years (Chapter 9). Because the turnover of N is much longer than the mixing time for ocean water, NO_3 shows a relatively uniform distribution in deep ocean water. The nitrogen budget in the oceans appears not in steady state; current estimates of the rate of denitrification exceed known inputs (compare to Figure 12.2). McElroy (1983) suggests that the oceans received a large input of nitrogen during the continental glaciation 20,000 years ago, and they have been recovering from this input ever since

(Christensen et al. 1987). This suggestion is consistent with sedimentary evidence of greater net primary production in the oceans during the last ice age (Broecker 1982), and with the low ratio of $^{15}N/^{14}N$ in sedimentary organic matter of glacial age implying low rates of denitrification (Ganeshram et al. 1995, Altabet et al. 2002, Gruber and Galloway 2008).

McElroy's paper should remind us to question the assumption of steady-state conditions when we construct global models of Earth's biogeochemistry, such as Figure 12.2. Earth has experienced large fluctuations in its biogeochemical function through geologic time. Changes in the global distribution and circulation of nitrogen may have accompanied climatic changes—just as the rise in CO_2 concentrations at the end of the last glacial epoch indicates a period of non-steady-state conditions in the global carbon cycle (Chapter 11).

At present, human activities have certainly disrupted the potential for steady-state conditions in the nitrogen cycle on land (Galloway et al. 2004, Liu et al. 2010). Humans have greatly accelerated the natural rate of N fixation; the production of N fertilizers and the cultivation of leguminous crops have increased dramatically since World War II, allowing higher crop yields to feed the world's growing human population (Figure 12.5). Archival collections show decreasing $\delta^{15}N$ in plant tissues, consistent with greater anthropogenic nitrogen inputs in recent years (Peñuelas and Filella 2001; compare to Hastings et al. 2009 and Holtgrieve et al. 2011).

Enrichments of nitrogen in terrestrial ecosystems, stimulating the rates of nitrification and denitrification, are likely to account for the rapid rise in the atmospheric content of N_2O, contributing to global climate change (Vitousek 1994). Changes in the global nitrogen cycle have important implications for human health (Townsend et al. 2003), carbon storage in the biosphere (Townsend et al. 1996), and the persistence of species diversity in nature (Bobbink et al. 2010, Stevens et al. 2010). Controls on N_2O emissions may be one of the most effective ways to reduce the threat of global climate change (Montzka et al. 2011a).

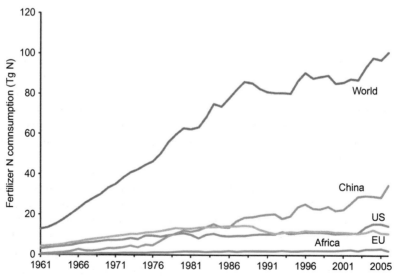

FIGURE 12.5   The production history of nitrogen fertilizer. *Source: From Robertson et al. (2009). Used with permission of the Annual Review.*

NITROUS OXIDE

Nitrous oxide, N_2O, has a mean concentration of 320 ppb in Earth's atmosphere, which indicates a global pool of 2.5×10^{15} g N_2O or 1.6×10^{15} g N (Table 3.1). The concentration of N_2O is increasing at a rate of 0.3%/yr (IPCC 2007). Each molecule of N_2O has the potential to contribute about $300\times$ to the greenhouse effect relative to each molecule of CO_2, so the current increase in the atmosphere has the potential to impact global climate over the next century (Shindell et al. 2009). Also, nitrous oxide is now the dominant human emission that causes ozone depletion in Earth's stratosphere (Ravishankara et al. 2009).

The only significant sink for N_2O—stratospheric destruction (Eqs. 3.47–3.49)—consumes about 12.2×10^{12} g N as N_2O per year (Minschwaner et al. 1993). A few soils also appear to consume N_2O, but the global sink in soils is probably very small (Chapuis-Lardy et al. 2007, Syakila and Kroeze 2011). The mean residence time for N_2O in the atmosphere is about 120 years, consistent with observations of a relatively uniform (320 ± 1 ppb) concentration of atmospheric N_2O around the world (Ishijima et al. 2009; see Figure 3.5). Nitrous oxide shows minor seasonal variation in the atmosphere, with greater amplitude at high northern latitudes (± 1.15 ppb) than at the South Pole (± 0.29 ppb; Jiang et al. 2007; compare to Ishijima et al. 2009). Unfortunately, estimates of sources—particularly sources that have changed greatly in recent years—are poorly constrained (Table 12.5).

The oceans appear to be a source of N_2O to the atmosphere as a result of nitrification in the deep sea (Cohen and Gordon 1979, Oudot et al. 1990). The isotopic content of N_2O from the marine environment is consistent with a source from nitrifying archaea in seawater (Santoro et al. 2011). Some of this N_2O may subsequently be *consumed* by denitrification as it passes upward through zones of low O_2 (Cohen and Gordon 1978, Kim and Craig 1990).

In many areas, surface waters are supersaturated in N_2O with respect to the atmosphere (Walter et al. 2004b). Specifically, the waters of the northwest Indian Ocean, a local zone of upwelling, may account for 20% of the total flux of N_2O from the oceans to the atmosphere (Law and Owens 1990). Based on the belief that the N_2O supersaturation of seawater was worldwide, calculated emissions from the ocean dominated the earliest global estimates of N_2O sources (Liss and Slater 1974, Hahn 1974). When more extensive sampling showed that the areas of supersaturation were regional, these workers substantially lowered their estimate of N_2O production in marine ecosystems. The most extensive survey of ocean waters suggests a flux of about 4 to 6×10^{12} g N/yr, emitted as N_2O to the atmosphere (Nevison et al. 1995, Bianchi et al. 2012). A large portion of this may derive from coastal waters, where increasing N_2O flux may derive from seawater enriched with NO_3^- from terrestrial runoff (Bange et al. 1996, Nevison et al. 2004, Naqvi et al. 2000).

Soil emissions from nitrification and denitrification (Chapter 6) are now thought to compose the largest global source of N_2O (Table 12.5). Particularly large emissions of N_2O are found from tropical soils (Bouwman et al. 1993, Kort et al. 2011). Conversion of tropical forests to cultivated lands and pasture results in greater N_2O emissions (Matson and Vitousek 1990, Keller and Reiners 1994), and the flux of N_2O increases when agricultural lands and forests are fertilized or manured (Chapter 6). Typically about 1% of the application of nitrogen fertilizer is lost to the atmosphere as N_2O (Bouwman et al. 2002b, Lesschen et al. 2011), and N_2O emissions have increased in parallel with fertilizer applications during the past century (Gao et al. 2011). Presumably the increased flux of N_2O from disturbed and fertilized soils stems

TABLE 12.5   A Global Budget for Nitrous Oxide (N_2O) in the Atmosphere (all values are Tg N/yr (10^{12} g/yr) nitrogen, as N_2O)

Natural sources	Annual flux	References
Soils	3.4 ± 1.3	Zhuang et al. 2012[a]
Ocean surface	6.2 ± 3.2	Bianchi et al. 2012
Total natural	9.6	
Anthropogenic sources		
Agricultural soils	2.8	Bouwman et al. 2002b[b]
Cattle and feed lots	2.8	Davidson 2009
Biomass burning	0.9	Kaiser et al. 2012
Industry and transportation	0.8	Davidson 2009
Human sewage	0.2	Mosier et al. 1998
Total anthropogenic	7.5	
Total sources	**17.1**	
Sinks		
Stratospheric destruction	12.3	Prather et al. 1995
Uptake by soils	<0.1	Syakila and Kroeze 2011
Atmospheric increase	4.0	IPCC 2007
Total identified sinks	**16.4**	

[a] *Alternative estimates for the flux of N_2O from natural soils includes 6.1 Tg N/yr (Potter et al. 1996) and 6.6 Tg N/yr (Bouwman et al. 1995).*
[b] *The sum of emissions from agriculture and domestic animals given here, 5.6 Tg N/yr, is in close agreement with the value of 5.0 Tg N/yr estimated by Syakila and Kroeze (2011). These estimates of N_2O flux from agricultural activities include emissions of N_2O from downstream ecosystems and groundwaters impacted by agricultural inputs in these regions.*

from higher rates of nitrification and greater availability of NO_3^- to denitrifying bacteria. Globally the flux of N_2O from agricultural soils is about 2.2 to 2.8×10^{12} g N/yr (Bouwman et al. 2002b, Davidson 2009), and the total production of N_2O from all soils is likely to be $<10 \times 10^{12}$ g N/yr (Table 12.5). In most ecosystems, soil CO_2 and N_2O losses are correlated, and a N_2O flux of 13.3×10^{12} g N/yr from soils is indicated from current estimates of CO_2 efflux (Xu et al. 2008b).

Downward leaching of fertilizer nitrate also has the potential to stimulate denitrification in groundwaters. Ronen et al. (1988) suggest that groundwater may be an important source of N_2O to the atmosphere—up to 1×10^{12} g N/yr, but most recent assessments suggest significantly lower values (Bottcher et al. 2011, Keuskamp et al. 2012). Excess nitrogen in surface runoff also causes a significant flux of N_2O from streams and rivers (Beaulieu et al. 2011), as does the disposal of human sewage (Kaplan et al. 1978, McElroy and Wang 2005).

Relatively small emissions of N_2O result from the combustion of fossil fuels or biomass (Table 12.5), but the industrial production of nylon and industrial chemicals results in

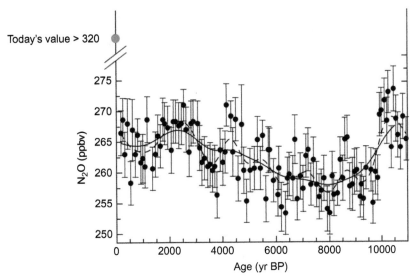

FIGURE 12.6   Nitrous oxide measurements from ice-core samples in Antarctica. *Source: From Flückiger et al. (2002).*

significant emissions of N_2O to the atmosphere (Thiemens and Trogler 1991). Total anthropogenic sources of N_2O are more than enough to explain its rate of increase in the atmosphere, and the total compilation of sources—both natural and anthropogenic—is slightly more than the known sinks, including the rate of N_2O accumulation in the atmosphere (Table 12.5).

Some constraints on the global budget for N_2O in Earth's atmosphere are set by studies of $\delta^{15}N$ and $\delta^{18}O$ in the gas. The flux of N_2O from soils is depleted in $\delta^{15}N$ and $\delta^{18}O$, since denitrifiers discriminate against the heavy isotopes in the pool of available NO_3^- (Chapter 6). Conversely, the "backflux" of N_2O from the stratosphere is *enriched* in both $\delta^{15}N$ and $\delta^{18}O$, since the photochemical destruction of N_2O in the stratosphere also discriminates against the heavy isotopes (Morgan et al. 2004). Estimates of changes in the sources of N_2O in the troposphere are constrained by its isotopic content, which is determined by N_2O from the stratosphere mixing with the weighted average of the isotopic content in various sources of N_2O at the Earth's surface (Kim and Craig 1993). In the past few decades, both $\delta^{15}N$ and $\delta^{18}O$ in atmospheric N_2O have declined—consistent with an increasing flux from soils to the atmosphere (Ishijima et al. 2007, Rockmann and Levin 2005, Park et al. 2012).[9]

Cores extracted from the Antarctic ice cap show that the concentration of N_2O was much lower (180 ppb) during the last glacial period (Leuenberger and Siegenthaler 1992, Sowers et al. 2003, Schilt et al. 2010). At the end of the Pleistocene, concentrations rose to ~265 ppb and remained fairly constant until the Industrial Revolution, when they increased to the present value of about 320 ppb (Figure 12.6; Flückiger et al. 2002, Sowers et al. 2003). Anticipating

---

[9] N_2O is a linear molecule (NNO). Analyses of the isotopic composition of the nitrogen atoms at the central (alpha) and end positions show promise to distinguish among various sources of N_2O, including nitrification and denitrification contributing to N_2O flux (Yoshida and Toyoda 2000, Sutka et al. 2006, Koba et al. 2009, Toyoda et al. 2011, Snider et al. 2012).

the future, field experiments show that rising CO_2 in Earth's atmosphere, excess N deposition in precipitation, and additions of Fe to ocean waters are all likely to increase the flux of N_2O to Earth's atmosphere and exacerbate global warming (Kammann et al. 2008, Liu and Greaver 2009, Law and Ling 2001, van Groenigen et al. 2011, Kim et al. 2012). Indeed, increasing N_2O emissions from soils may negate some of the benefits seen in the use of fertilizer to enhance crop growth for biofuels and soil carbon sequestration (Adler et al. 2007, Melillo et al. 2009).

THE GLOBAL PHOSPHORUS CYCLE

The global cycle of P is unique among the cycles of the major biogeochemical elements in having no significant gaseous component (Figure 12.7). The redox potential of most soils is too high to allow for the production of phosphine gas (PH_3), except under very specialized, local conditions (Bartlett 1986). Phosphine emissions are reported for sewage treatment ponds in Hungary (Dévai et al. 1988), marshes in Louisiana and Florida (Dévai and DeLaune 1995), and a lake in China (Geng et al. 2005). Phosphine is also found in the marine atmosphere over the Atlantic and Pacific oceans, where it may be derived from the impact of lightning on P-containing soil dusts (Glindemann et al. 2003, Zhu et al. 2007). The global flux of P in phosphine is probably $<0.04 \times 10^{12}$ g P/yr (Gassmann and Glindemann 1993).

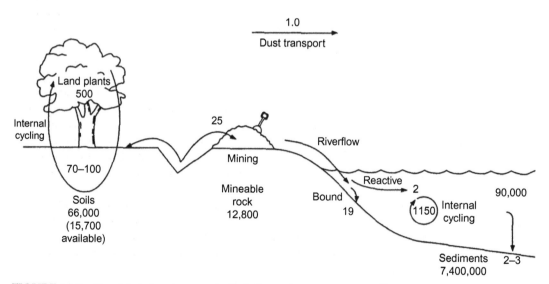

FIGURE 12.7   The global phosphorus cycle. Each flux is shown in units of 10^{12} g P/yr. Values for P production and reserves are taken from the U.S. Geological Survey. Estimate for sediments is from Van Cappellen et al. (1996), and estimates for other pools and flux are derived from the text.

The flux of P through the atmosphere in soil dust and seaspray (1×10^{12} g P/yr; Graham and Duce 1979, Mahowald et al. 2008) is also much smaller than other transfers in the global P cycle, which are largely derived from chemical weathering in soils. However, the atmospheric deposition of P is known to make a critical contribution to the supply of available P when it is deposited in some tropical forests on highly weathered soils (Swap et al. 1992, Chadwick et al. 1999, Okin et al. 2004) and in the open ocean (Talbot et al. 1986, Bristow et al. 2010). Newman (1995) reports P deposition from the atmosphere (0.07 to 1.7 kg ha^{-1} yr^{-1}) that rivals P derived from rock weathering (0.05 to 1.0 kg ha^{-1} yr^{-1}) in various terrestrial ecosystems. The Bodélé depression in Chad is thought to be the source of 0.12×10^{12} g P/yr in soil dusts that blow eastward over the Atlantic Ocean and into the Amazon Basin (Bristow et al. 2010, Ben-Ami et al. 2010).

Unlike transfers in the global nitrogen cycle, the major source of reactive P in the global P cycle is not provided by microbial reactions. Nearly all the phosphorus in terrestrial ecosystems is originally derived from the weathering of calcium phosphate minerals, especially apatite ($Ca_5(PO_4)_3$ OH; Eq. 4.7). The phosphorus content of most rocks is not large, and in most soils only a small fraction of the total P is available to biota (Chapter 4). Land vegetation appears to contain about 0.5×10^{15} g P (Smil 2000), while soils, to 50-cm depth, hold about 46×10^{15} g P, of which only 13.8×10^{15} g P is in labile or organic forms (X. Yang, personal communication, 2012). Estimated plant uptake on land, 70 to 100×10^{12} g P/yr (Smil 2000), implies a turnover of 0.5%/yr in soils. Root exudates and mycorrhizae may increase the rate of rock weathering on land (Chapter 4), but there is no process, equivalent to N fixation, that can produce dramatic increases in phosphorus availability for plants in P-deficient habitats. Thus, on both land and at sea, biota persist as a result of a well-developed recycling of phosphorus in organic forms (Figures 6.19 and 9.22).

The main flux of P in the global cycle is carried by rivers, which transport about 21×10^{12} g P/yr to the sea (Meybeck 1982, Smil 2000)—about twice as much as 300 years ago (Wallmann 2010, Liu et al. 2008). Only about 10% of this flux is potentially available to marine biota; the remainder is strongly bound to soil particles that are rapidly sedimented on the continental shelf (Chapter 9). The solubility product of apatite is only about 10^{-58} (Lindsay and Vlek 1977). At a seawater pH of 8.3, the phosphorus concentration in equilibrium with apatite is about 1.3×10^{-7} molar (~4 μg P/l; Atlas and Pytkowicz 1977; compare to Figure 4.10). In seawater, organic and colloidal forms maintain the concentration of P in excess of that in equilibrium with respect to apatite; the average content of P in deep ocean water is about 3×10^{-6} molar (~93 μg/l; compare to Figures 9.18 and 9.27). The concentration of PO_4^{3-} in the surface oceans is low, but the large volume of the deep sea accounts for a substantial pool of P (Figure 12.6). The overall mean residence time for reactive P in the sea is about 25,000 years (Chapter 9).

The turnover of P through the organic pools in the surface ocean occurs in a few days. Nearly 90% of the phosphorus taken up by marine biota is regenerated in the surface ocean, and most of the rest is mineralized in the deep sea (Figure 9.22). Eventually, however, phosphorus is deposited in ocean sediments, which contain the largest phosphorus pool near the surface of the Earth (Van Cappellen and Ingall 1996). About 2 to 3×10^{12} g P/yr are added to sediments of the open ocean—roughly equivalent to the delivery of reactive P to the oceans by rivers (Wallmann 2010, Baturin 2007; Figure 9.22). On a time scale of hundreds of millions of years, these sediments are uplifted and subject to rock weathering, completing the global cycle. Today, most of the phosphorus in rivers is derived from the weathering of sedimentary rocks, and it represents P that has made at least one complete journey through the global cycle (Griffith et al. 1977).

It is likely that PO_4 has always been the dominant form of P available to biota. Students of Earth's earliest life have speculated on mechanisms by which PO_4 could be polymerized and condensed so it could be incorporated into biochemical molecules. It is possible that lightning, volcanic eruptions, and other local high-energy environments may have been involved in the production of phosphite and polyphosphates (Yamagata et al. 1991, Pasek and Block 2009). Griffith et al. (1977) calculates that it took about 3 billion years for the weathering of igneous rocks to saturate seawater with PO_4, allowing the development of phosphorite skeletons in organisms (Cook and Shergold 1984) and the precipitation of authigenic apatites in sediments (Chapter 9). Today, these minerals are carried by subduction to the mantle, completing the global phosphorus cycle following tectonic uplift that returns rocks to the surface (Guidry et al. 2000, Buendia et al. 2010). Although the present-day burial of phosphorus appears similar to the rate during much of Earth's history (Filippelli and Delaney 1992), periods of massive uplift and erosion may have fueled high net primary productivity in the oceans (Filippelli and Delaney 1994, Filippelli 2008).

Van Cappellen and Ingall (1996) suggest a negative-feedback mechanism by which changes in the concentration of O_2 in Earth's atmosphere determine the availability of P in the deep sea, where the net mineralization of P is affected by its adsorption on Fe minerals in oxic sediments. This cycle stabilizes the level of O_2 in Earth's atmosphere through geologic time. For instance, if the concentration of O_2 in the atmosphere were to fall to low levels, available P would become more plentiful in waters upwelling from the deep sea, marine NPP would increase, and more O_2 would be released to the atmosphere from the oceans. Ocean productivity may have been first limited by P adsorption to Fe oxides shortly after the evolution of oxygenic photosynthesis, 3.5 to 3.8 bya (Bjerrum and Canfield 2002). Kump (1988) and Lenton (2001) offer analogous models centered on the terrestrial biosphere, in which the frequency of fire modulates changes in phosphorus availability, primary productivity, and changes in the levels of O_2 in the atmosphere.

In many areas, humans have enhanced the availability of P by mining phosphate rocks—25×10^{12} g P/yr—that can be used as fertilizer.[10] Most of the economic deposits of phosphate are found in sedimentary rocks of marine origin, so the mining activity directly enhances the turnover of the global P cycle. The largest deposits are found in Morocco (Cooper et al. 2011); in the United States, deposits of phosphate rock are found in Florida and North Carolina. In many areas, the flux of P in rivers is significantly higher than it was in prehistoric times as a result of erosion, pollution, and fertilizer runoff (Bennett et al. 2001, Liu et al. 2008, Yuan et al. 2011). By mining phosphate rock, humans impact the global P cycle at a rate that rivals their impact on the global N cycle. The ratio of industrial nitrogen fixation to the mining of phosphate rock is about 14.3 (molar), somewhat less than the ratio in the uptake of these elements by land plants (31.3) but similar to the Redfield ratio for the uptake by marine phytoplankton (16). Indeed, the extensive mining of phosphate rock, a nonrenewable resource, may lead to

---

[10] *http://minerals.usgs.gov/minerals/pubs/commodity/phosphate_rock/.*

phosphorus shortages for agriculture. At current rates of use, the global estimate of phosphorus reserves will last about 370 years (Cooper et al. 2011).

LINKING GLOBAL BIOGEOCHEMICAL CYCLES

The cycles of important biogeochemical elements are linked at many levels. Stock et al. (1990) describe how P is used to activate a transcriptional protein, stimulating nitrogen fixation in bacteria when nitrogen is in short supply. In this case, an understanding of the interaction between these elements is gained through the study of molecular biology. In Chapter 5, we saw that the photosynthetic rate of land plants is related to the N and P content of their leaves, linking the net production of organic carbon to the availability of these elements in plant cells. In marine ecosystems, net primary productivity is often calculated from the Redfield ratio of C:N:P in phytoplankton biomass (Chapter 9). The molar N/P ratio of 16 may reflect fundamental relationships between the requirements for protein and RNA synthesis common to all plants (Loladze and Elser 2011). Whatever our viewpoint—from molecules to whole ecosystems—the movements of N, P, and C are strongly linked in biogeochemistry (Reiners 1986).

Nitrogen fixation by free-living bacteria appears inversely related to the N/P ratio in soil (Figure 6.4), and the rate of accumulation of N is greatest in soils with high P content (Walker and Adams 1958). Similarly, N/P ratios <29 appear to stimulate N fixation in freshwater ecosystems (Chapter 7). One might speculate that the high demand for P by N-fixing organisms links the global cycles of N and P, with P being the ultimate limit on nitrogen availability and net primary production. Indeed, in many soils the accumulation of organic carbon is correlated to available P (Chapter 6). N fixation in ocean waters is stimulated by P and Fe deposition from soil dusts (Falkowski et al. 1998). Despite these theoretical arguments for a phosphorus limitation of the biosphere through geologic time, net primary production in most terrestrial and marine ecosystems usually shows an immediate response to additions of N (Figure 12.8). Denitrification appears to maintain small supplies of N in most ecosystems, and colimitation by N and P is commonplace (Harpole et al. 2011).

It is urgent that biogeochemists offer improved predictions of the response of the biosphere to ongoing changes in the availability of N and P, which may allow a sustained increase in plant growth in response to rising CO_2 in Earth's atmosphere. Theoretical considerations based on stoichiometry indicate that the sink for CO_2 in land plants will be limited (van Groenigen et al. 2006, Reay et al. 2008), but changes in soil nutrient turnover and plant uptake may provide enhanced nutrient supplies (Finzi et al. 2007, Drake et al. 2011, Dieleman et al. 2010). Similarly, changes in the supply of N and P to the surface oceans of the world, as determined by changes in circulation and nutrient turnover, may determine the long-term sink for carbon in the sea. Such changes in biogeochemistry are likely to have affected the concentration of CO_2 in the atmosphere and Earth's climate during glacial–interglacial cycles, implying that the marine biosphere could respond rapidly to future climate change (Falkowski et al. 1998, Altabet et al. 2002, Gruber and Galloway 2008).

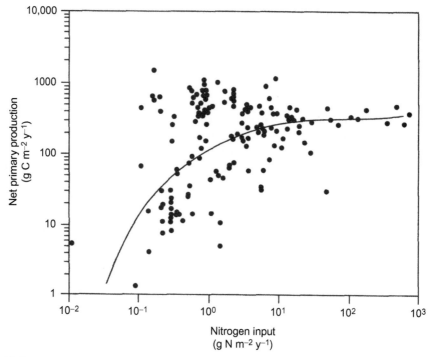

FIGURE 12.8 Net primary production versus nitrogen inputs to terrestrial, aquatic, and marine ecosystems. Net primary production increases in direct response to added nitrogen up to inputs of about $10 \, g \, N \, m^{-2} \, yr^{-1}$ (100 kg/ha). Inputs in excess of that level are rarely found in natural ecosystems, but are seen in polluted environments and agricultural soils *Source: Modified, with permission from Princeton University Press, from Levin (1989).*

SUMMARY

For both N and P, a small biogeochemical cycle with relatively rapid turnover is coupled to a large global pool with relatively slow turnover. For N, the major pool is found in the atmosphere. For P, the large pool is found in unweathered rock and soil.

The biogeochemical cycle of N begins with the fixation of atmospheric nitrogen, which transfers a small amount of inert N_2 to the biosphere. This transfer is balanced by denitrification, which returns N_2 to the atmosphere. The balance of these processes maintains a steady-state concentration of N_2 in the atmosphere with a turnover time of 10^7 years. In the absence of denitrification, most of the N inventory on Earth would eventually be sequestered in the ocean and in organic sediments. Denitrification closes the global nitrogen cycle, and it causes nitrogen to cycle more rapidly than phosphorus, which has no gaseous phase. The mean residence time of phosphorus in sedimentary rocks is measured in 10^8 yr, and the phosphorus cycle is complete only as a result of tectonic movements of the Earth's crust.

Once within the biosphere, the movements of N and P are more rapid than in their global cycles, showing turnover times ranging from hours (for soluble P in seawater) to hundreds of years (for N in biomass). In response to nutrient limitations, biotic recycling in terrestrial and marine habitats allows much greater rates of net primary production than rates of N fixation and rock weathering alone would otherwise support (Tables 6.1 and 9.3). The high efficiency of nutrient recycling may explain why, in the face of widespread nitrogen limitation, only about 2.5% of global net primary production is diverted to nitrogen fixation (Gutschick 1981).

Human perturbations of the global nitrogen and phosphorus cycles are widespread and dramatic. Through the production of fertilizers, humans have doubled the rate at which nitrogen enters the biogeochemical cycle on land. It is unclear how rapidly denitrification will respond to this global increase in nitrogen availability, but the rising concentrations of atmospheric N_2O are perhaps one indication of an ongoing biotic response (Vitousek 1994). Increasing nitrogen availability has led to the local extinction of species from polluted ecosystems and has shifted the limitation of net primary production in some systems from N to P (e.g., Mohren et al. 1986, Elser et al. 2007, Peñuelas et al. 2012). Increasing transport of N and P in rivers has shifted many estuarine and coastal ecosystems to a condition of Si deficiency (Justic et al. 1995). All of these changes indicate the effect of a single species—the human—in upsetting previous steady-state conditions in global nutrient cycling.

Recommended Readings

Melillo, J.M., C.B. Field, and B. Moldan. 2003. *Interactions of the Major Biogeochemical Cycles.* Island Press.
Smil, V. 2001. *Enriching the Earth.* MIT Press.
Schlesinger, W.H. (Ed.). 2004. *Biogeochemistry. Treatise on Geochemistry*, Volume 8. Elsevier.
Tiessen, H. (Ed.). 1995. *Phosphorus Cycling in Terrestrial and Aquatic Ecosystems.* Wiley.

PROBLEMS

1. Assuming that the concentration of P in crustal materials is 0.76 mg/g (Table 4.1), how much P is transported through the atmosphere by soil dust (Table 3.3)? How does this compare to the annual transport of P in rivers?
2. Assuming 100% efficiency in the industrial production of NH_3, compare the CO_2 "cost" in the production of ammonia fertilizer to the potential carbon sink that might be realized by higher NPP and soil organic matter in areas that receive this fertilizer.
3. Human pollution of the world's rivers has potentially doubled the transport of N and tripled the transport of P to the world's coastal seas. Assuming the Redfield ratio applies to NPP, what is the potential increase in NPP in coastal waters globally?

The Global Cycles of Sulfur and Mercury

INTRODUCTION

Sulfur is found in valence states ranging from $+6$ in SO_4^{2-} to -2 in sulfides. The original pool of sulfur on Earth was held in igneous rocks, largely as pyrite (FeS_2). Degassing of the mantle and, later, weathering of the crust under an atmosphere containing O_2 transferred a large amount of S to the oceans, where it is now found as SO_4^{2-} (Table 2.3). When SO_4^{2-} is assimilated by organisms, it is reduced and converted into organic sulfur, which is an essential component of protein. However, the live biosphere contains relatively little sulfur. Today, the major global pools of S are found in biogenic pyrite, seawater, and evaporites derived from ocean water (Table 13.1).

Like sulfur, the majority of the Earth's mercury (Hg) is found in the crust, and it is released to the atmosphere and oceans by volcanic eruptions, rock weathering, and human activities. Human and microbial activities yield a global biogeochemical cycle for this element, which has long been recognized as toxic to life.

As in the case of nitrogen, microbial transformations between valence states drive the global S and Hg cycles. Under anoxic conditions, SO_4 is a substrate for sulfate reduction, which may lead to the release of reduced gases to the atmosphere and to the deposition of biogenic pyrite in sediments (Chapters 7 and 9). Anoxic environments can also support sulfur-based photosynthesis, which is likely to have been one of the first forms of

TABLE 13.1   Active Reservoirs of Sulfur near the Surface of the Earth

Reservoir	10^{18} g S
Atmosphere	0.0000028
Seawater	1280
Sedimentary rocks	
Evaporites	2470
Shales	4970
Land plants	0.0085
Soil organic matter	0.0155
Total	**8720**

Sources: From Holser et al. (1989) and Dobrovolsky (1994).

photosynthesis on Earth (Eq. 2.12). In the presence of oxygen, reduced sulfur compounds are oxidized by microbes. In some cases, the oxidation of S is coupled to the reduction of CO_2 in the reactions of S-based chemosynthesis (Eq. 4.6).

Mercury is transformed among valence states that include dissolved ions (Hg^{2+}), elemental mercury (Hg°), and an especially toxic form known as methyl-mercury (CH_3Hg). Sulfate-reducing bacteria are implicated in the production of methyl-mercury in sediments, so the global cycles of S and Hg are linked by metabolism.

Understanding the biogeochemistry of S and Hg has enormous economic significance. Many metals are mined from sulfide minerals in hydrothermal ore deposits (Meyer 1985). Microbial reactions involving sulfur bacteria are used to concentrate metals from water (e.g., Zn, Labrenz et al. 2000; Cu, Sillitoe et al. 1996; and Au, Lengke and Southam 2006), and are used to remove metals from relatively low-grade ore (Lundgren and Silver 1980). Sulfur is an important constituent of both coal and oil. The organic S in coal is oxidized to sulfuric acid when coal is exposed to air. SO_2 is emitted to the atmosphere when coal and oil are burned.

A large amount of SO_2 is also emitted during the smelting of copper ores (Cullis and Hirschler 1980, Oppenheimer et al. 1985). An understanding of the relative importance of natural sulfur compounds in the atmosphere compared to anthropogenic SO_2 is essential to evaluate the causes of acid rain and the impact of acid rain on natural ecosystems. Similarly, understanding the natural and anthropogenic sources of mercury in the atmosphere allows rational policy decisions regarding the regulation of mercury emissions from power plants and other sources.

In this chapter, we compare the global cycles of S and Hg, since both are characterized by microbial reduction reactions that produce volatile forms. Assembling quantitative global cycles for these elements allows us to put the human emissions of S and Hg into a larger context. As we did for carbon (Chapter 11), nitrogen, and phosphorus (Chapter 12), we will attempt to establish a budget for S and Hg on land and in the atmosphere. Then we will couple these to marine budgets (Figure 9.29) to form an overall picture of the global cycles. The biogeochemical cycle of S has varied through Earth's history as a result of the appearance of new metabolic pathways and changes in their importance. We will review the history of the S cycle as it is told by sedimentary rocks.

THE GLOBAL SULFUR CYCLE

No sulfur gas is a long-lived or major constituent in the atmosphere. Thus, all attempts to model the global S cycle must explain the fate of the large annual input of sulfur compounds to the atmosphere. The short mean residence time for atmospheric sulfur compounds, as a result of their oxidation to SO_4, allows us to express all the fluxes in the global budget in units of 10^{12} g of S, regardless of the original form of emission. Despite the small atmospheric content of S compounds (totaling $\sim 4 \times 10^{12}$ g S at any moment; Rodhe 1999), the annual flux of S compounds through the atmosphere (about 300×10^{12} g S/yr) rivals the movements of N in the global nitrogen cycle (compare Figure 13.1 to Figure 12.2).

In 1960, Eriksson examined the potential origins of SO_4 in Swedish rainfall and hence, indirectly, sources of SO_4 in the atmosphere. He reasoned that if all the Cl^- in rainfall is

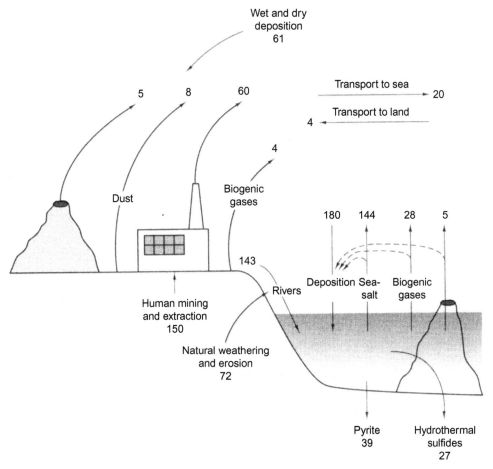

FIGURE 13.1 The global S cycle with annual flux shown in units of 10^{12} g S/yr. The derivation of most values is described in the text, with the marine values taken from Figure 9.22. The net flux from land to sea is extrapolated from Whelpdale and Galloway (1994).

derived from the ocean, then seaspray should also carry SO_4 roughly in proportion to the ratio of SO_4^{2-} to Cl^- in seawater. His calculation suggested that about 4×10^{12} g S/yr deposited on land must be derived from the sea. At about the same time, however, Junge (1960) was evaluating the SO_4 content of rainfall, and he estimated that about 73×10^{12} g S/yr was deposited on land globally. Clearly, there were other sources of SO_4 in the atmosphere and in rainfall. Junge's maps showed that SO_4 was abundant in the rainfall of industrial regions and in areas downwind of deserts (Figure 3.14a). Desert soils are a source of gypsum ($CaSO_4 \cdot 2H_2O$) in atmospheric dust (Reheis and Kihl 1995), and the burning of fossil fuels in industrial regions contributes SO_2 to air pollution (Langner et al. 1992, Spiro et al. 1992). By the mid 1970s, "acid rain" was clearly linked to coal-fired power plants, which release SO_2 (Likens and Bormann 1974). In the intervening years, new sources of S in the atmosphere have been recognized, and global flux estimates have been revised repeatedly. Our understanding of the global S cycle is much improved, but most of the estimates illustrated in Figure 13.1 remain subject to considerable uncertainty.

Episodic events, including volcanic eruptions and dust storms, contribute to the global biogeochemical cycle of S. Sulfur emissions from volcanoes are especially difficult and dangerous to measure. SO_2 dominates the volcanic release, but significant H_2S is reported in some eruptions (Aiuppa et al. 2005, Clarisse et al. 2011); both are oxidized to SO_4 in the atmosphere (compare to Eqs. 3.27–3.28). Legrand and Delmas (1987) used the deposition of SO_4 in the Antarctic ice pack to estimate the contribution of volcanoes to the global S cycle during the last 220 years. The Tambora eruption of 1815 was the largest, releasing 50×10^{12} g S to the atmosphere. Typically, major eruptions, such as that of Mt. Pinatubo (15 June 1991), release 5 to 10×10^{12} g S each (Bluth et al. 1993).

Following a major volcanic eruption, the Earth's climate is cooled for several years, as a result of SO_4 aerosols in the stratosphere (Briffa et al. 1998, McCormick et al. 1995, Rampino et al. 1988). When the volcanic emissions are averaged over many years, the annual global flux is about 7.5 to 10.5×10^{12} g S/yr (Halmer et al. 2002). About 70% of the sulfur gases leak passively from volcanoes and the remainder is derived from periodic, explosive events (Bluth et al. 1993, Allard et al. 1994). Huge eruptions in the Late Cretaceous (66 mya) may have released more than 1000 Tg S to the atmosphere (Self et al. 2008).

The movement of S in soil dust is also episodic and poorly understood. Many of the large particles are deposited locally, while smaller particles may undergo long-range transport in the atmosphere (Chapter 3). Savoie et al. (1987) found that dust from the deserts of the Middle East contributes SO_4 to the waters of the northwest Indian Ocean. Ivanov (1983) suggests a global flux of 8×10^{12} g S/yr owing to dust transport in the troposphere—about 10% of the fossil fuel release.

The total flux of biogenic sulfur gases from land is likely $<4 \times 10^{15}$ g S/yr (Watts 2000). The dominant sulfur gas emitted from freshwater wetlands and anoxic soils is H_2S, with dimethylsulfide and carbonyl sulfide (COS) playing lesser roles (Chapter 7). Emissions from plants are poorly understood and deserving of further study (Chapter 6). Forest fires emit an additional 2×10^{12} g S/yr (Andreae and Merlet 2001, Kaiser et al. 2012).

It seems certain that direct emissions from human industrial activities are the largest sources of S gases in the atmosphere. Ice cores from Greenland show a large increase in the deposition of SO_4 from the atmosphere at the beginning of the Industrial Revolution (Herron et al. 1977; Mayewski et al. 1986, 1990; Fischer et al. 1998). In Europe and the

United States, these emissions have been sharply curtailed by the control of air pollution in recent years (Likens et al. 2005, Stern 2006, Velders et al. 2011), and the SO_4 content in rainfall has fallen in parallel (Figure 13.2). Estimates of the recent global flux range from 50 to 70 × 10^{12} g S/yr (Smith et al. 2001a, Faloona 2009, Lee et al. 2011).

Owing to the reactivity of S gases in the atmosphere, most of the anthropogenic emission of SO_2 is deposited locally in precipitation and dryfall. Total deposition of S on land may be as high as 120 × 10^{12} g S/yr (Andreae and Jaeschke 1992), but a value of ~60 × 10^{12} g S/yr balances the global S cycle shown in Figure 13.1. Deposition in dryfall and the direct absorption of SO_2 are poorly understood, so this global estimate is subject to revision. The estimate of atmospheric deposition on land accounts for a large fraction of the total emissions from land. The remainder undergoes long-distance transport in the atmosphere and accounts for a net transfer of S from land to sea (Whelpdale and Galloway 1994).

A small fraction of the natural river load of SO_4 is derived from rainfall, which includes cyclic salts that are carried through the atmosphere from the ocean (4×10^{12} g S/yr in Figure 13.1). Weathering of pyrite and gypsum also contributes to the SO_4 content of river water (Table 4.9). The remainder of the global flux of S in rivers is derived from human activities. Berner (1971) suggested that at least 28% of the SO_4 content of modern rivers is derived from air pollution, mining, erosion, and other human activities. Meybeck (1979) estimates that the current river transport of about 143 × 10^{12} g S/yr is roughly double that of preindustrial conditions.

The marine portion of the global S cycle is largely taken from Figure 9.29. The ocean is a large source of seasalt aerosols that contain SO_4, but most of these are redeposited in the ocean in precipitation and dryfall. Dimethylsulfide—$(CH_3)_2$ S, or DMS—is the major biogenic gas

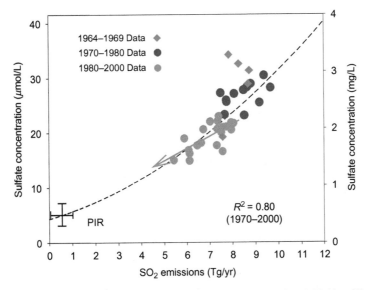

FIGURE 13.2 Sulfate concentration in wetfall precipitation at New Hampshire's Hubbard Brook Experimental Forest, as a function of SO_2 emissions in the estimated 24-hr source area. This shows the decline in both parameters as a result of the implementation of the Clean Air Act. (PRI shows the levels of the pre-industrial revolution.) *Source: From Likens et al. 2005. Used with permission of RSC Publishing, copyright Royal Society of Chemistry.*

emitted from the sea, with a global annual flux of 10 to 30×10^{12} g S/yr (Faloona 2009) and a recent best estimate of 28 Tg S/yr (Lana et al. 2011; Chapter 9). Thus, DMS is the largest natural source of sulfur gases in the atmosphere. The mean residence time of DMS in the atmosphere is <2 days (Table 3.5), as a result of its oxidation to SO_2. Much of the SO_2 is further oxidized to SO_4 (Faloona et al. 2009). Thus, most of the sulfur from DMS is also redeposited in the oceans.

In addition to helping balance the marine sulfur budget, dimethylsulfide attains global significance for its potential effects on climate (Shaw 1983). Charlson et al. (1987) postulated that the oxidation of DMS to sulfate aerosols would increase the scattering of solar radiation and the abundance of cloud condensation nuclei in the atmosphere, leading to greater cloudiness (Bates et al. 1987). Clouds over the sea reflect incoming sunlight, leading to global cooling (Chapter 3). SO_4 aerosols in the marine atmosphere, or deposited from it, that are not derived from seawater are known as non-seasalt sulfate (nss-sulfate), and are indicative of emissions of DMS, volcanoes, and other biogenic gases (Xie et al. 2002). A significant portion of the SO_4 in the Antarctic icepack and exposed soils is nss-sulfate, potentially from DMS (Legrand et al. 1991, Bao et al. 2000).

The hypothesis that DMS might provide a biotic regulation on global temperature is intriguing, for it may be responsible for the moderation of global climate throughout geologic time. If higher marine NPP is associated with warmer sea surface temperatures, then a greater flux of DMS would have the potential to act as a negative feedback on global warming through the greenhouse effect. Given the strong arguments pointing to global warming by increased atmospheric CO_2, the potential negative feedbacks of DMS are the subject of intense scientific scrutiny and debate. For instance, Quinn and Bates (2011) question the role of SO_4^{2-} aerosols derived from DMS versus the production of a variety of other aerosols in the marine atmosphere.

The production of DMS is often related to the growth of certain marine phytoplankton (Andreae et al. 1994a, Steinke et al. 2002). DMS flux is directly correlated to solar irradiance (Vallina and Simo 2007, Gali et al. 2011). The flux of DMS from the sea is greater in summer than in winter, as a result of greater sea surface temperature (Prospero et al. 1991, Tarrasón et al. 1995). The concentration of DMS in seawater is well correlated to that in the air over the North Pacific Ocean (Watanabe et al. 1995).

Cloud condensation nuclei also appear to be well correlated to the atmospheric burden of DMS in nonpolluted areas (Ayers and Gras 1991, Putaud et al. 1993, Andreae et al. 1995). The ice-core record of methanesulphonate (MSA)—a DMS degradation product—suggests higher concentrations during the last glacial epoch than today (Legrand et al. 1991, but see Castebrunet et al. 2006). Certainly the ocean's temperature was lower during the last glacial, but if, for other reasons (e.g., a greater deposition of iron-rich dust), marine NPP productivity was higher during the last glacial, an increased flux of DMS may have reinforced the global cooling of Earth's climate. Indeed, Turner et al. (1996) report more than a threefold increase in DMS flux to the atmosphere during an iron-enrichment experiment in the Pacific Ocean. If natural emissions of DMS increase in the future, it is possible that they will act to dampen the greenhouse effect during the next century.

The potential for atmospheric SO_4 aerosols to cool Earth's climate is clearly seen in observations of global temperature after major explosive volcanic events which inject S gases into the stratosphere. Schwartz (1988) argued that anthropogenic emissions of SO_2 should have the same effect on global climate as natural emissions of DMS, because SO_2 is also oxidized to produce condensation nuclei in the atmosphere. Most of the anthropogenic SO_2 emissions derive from land. Only about 4×10^{12} g S are emitted as SO_2 by international shipping (Corbett et al. 1999), and a very small additional amount by aircraft (Kjellstrom et al. 1999).

Using a general circulation model for global climate, Wigley (1989) suggested that climatic cooling by SO_2, largely from coal-fired power plants, may have offset some of the temperature change expected from the greenhouse effect, especially before the institution of air pollution controls in the United States and Europe. The Earth's albedo has apparently increased slightly in recent years, presumably due to particulate air pollutants (Pallé et al. 2009, Wang et al. 2009). Greater albedo reduces the radiation reaching the Earth's surface (i.e., global dimming). Reductions in man-made sulfur emissions might exacerbate global warming because the cooling effect of sulfur aerosols will be removed (Andreae et al. 2005). Because SO_2 has a short atmospheric lifetime (Chapter 3), its effect is regional and centered on areas of industry (Kiehl and Briegleb 1993, Falkowski et al. 1992, Langner et al. 1992). Some workers have even suggested injecting SO_4-forming aerosols into the stratosphere as a means of reducing Earth's temperature—a form of geo-engineering in response to the greenhouse warming of our planet (Crutzen 2006). This suggestion is not without controversy, given the potential for inadvertent impacts on the biosphere (Robock et al. 2009).

Temporal Perspectives on the Global Sulfur Cycle

During the accretion of the primordial Earth, sulfur was among the gases that were released by degassing of the Earth's mantle to form the secondary atmosphere (Chapter 2). Even today, volcanic emissions contain appreciable concentrations of SO_2 and H_2S (Table 2.2). When the oceans condensed on Earth, the atmosphere was essentially swept clear of S gases, owing to their high solubility in water (Eq. 2.6). On Venus, where there is no ocean, crustal degassing has resulted in a large concentration of SO_2 in the atmosphere (Oyama et al. 1979). The dominant form of S in the earliest seawater is likely to have been SO_4^{2-}; high concentrations of Fe^{2+} in the primitive ocean would have precipitated any sulfides, which are insoluble under anoxic conditions (Walker and Brimblecombe 1985). Nevertheless, the SO_4 content of the earliest oceans was low (Habicht et al. 2002) and increased in concert with the rise of O_2 in Earth's atmosphere (Canfield et al. 2000). The total inventory of S compounds on the surface of the Earth (nearly 10^{22} g S) represents the total crustal outgassing of S through geologic time (Table 2.3).

The ratio of ^{32}S to ^{34}S in the total S inventory on Earth is thought to be similar to the ratio of 22.22 measured in the Canyon Diablo Troilite (CDT), a meteorite collected in Arizona. The sulfur isotope ratio in this rock is accepted as an international standard and assigned a value of 0.00. In other samples, deviations from this ratio are expressed as δ^{34}S, with the units of parts per thousand parts (‰)—a convention that we also used for the isotopes of carbon (Chapter 5) and nitrogen (Chapter 6). Presumably the δ^{34}S isotope ratio in the earliest oceans was 0.00, because there is no reason to expect any discrimination between the isotopes of S during crustal degassing. When evaporite minerals precipitate from seawater, there is little differentiation among the isotopes of sulfur, so geologic deposits of gypsum ($CaSO_4 \cdot 2H_2O$) and barite ($BaSO_4$) carry a record of the isotopic composition of S in seawater. In the earliest sedimentary rocks, dating to 3.8 bya, δ^{34}S is close to 0.00 (Schidlowski et al. 1983).

Dissimilatory sulfate reduction by bacteria strongly differentiates among the isotopes of sulfur, as a result of a more rapid enzymatic reaction with $^{32}SO_4$. By itself, sulfate metabolism results in an isotopic depletion of $-18‰$ δ^{34}S in H_2S and sedimentary sulfides relative to the ratio in the

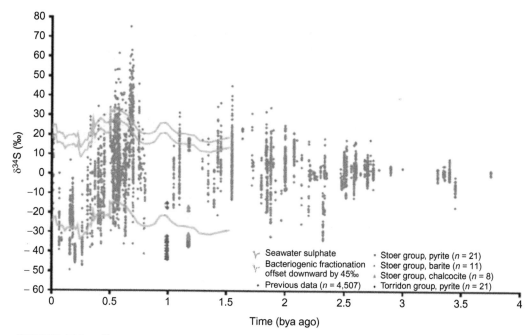

FIGURE 13.3 $\delta^{34}S$ in sedimentary pyrites through geologic time. *Source: Modified from Parnell et al. (2010).*

source reservoir (Canfield and Teske 1996). Stronger fractionations, sometimes as much as −66‰, are also reported (Sim et al. 2011). Strongly depleted sedimentary sulfides can result from repeated cycles of oxidation and reduction (Canfield and Teske 1996) and the reduction of thiosulfate ($S_2O_3^{2-}$) and sulfite ($S_2O_3^{2-}$) in sediments (Canfield and Thamdrup 1994, Habicht et al. 1998). The evolution of sulfate reduction dates to 2.4 to 2.7 bya, based on the first occurrence of sedimentary rocks with depletion of ^{34}S (Cameron 1982, Schidlowski et al. 1983, Habicht and Canfield 1996; Figure 13.3).[1] This also marks the time of the first substantial SO_4 concentrations in the primitive oceans. Today, the average $\delta^{34}S$ in sedimentary sulfides in all of the Earth's crust is about −10 to −12‰ (Holser and Kaplan 1966, Migdisov et al. 1983).

Figure 13.4 shows a three-box model for the S cycle, in which marine SO_4 and sedimentary sulfides are connected through microbial oxidation and reduction reactions, which discriminate between the sulfur isotopes (compare to Garrels and Lerman 1981). During periods of Earth's history when large amounts of sedimentary pyrite were formed as a result of sulfate reduction, seawater SO_4 became enriched in residual $^{34}SO_4$. Meanwhile, deposition or dissolution of evaporate deposits has no effect on the isotopic composition of seawater, but can strongly affect the concentration of SO_4 in seawater and thus the rate of sulfate-reduction and pyrite deposition (Wortmann and Paytan 2012).

Currently, about 50% of the pool of S near the surface of the Earth is found in reduced form (Li 1972, Holser et al. 1989), and the $\delta^{34}S$ of seawater is +21‰ (Kaplan 1975, Rees et al. 1978). Because there is little differentiation among isotopes of S during the precipitation of evaporites, the sedimentary record of evaporites indicates changes in the $\delta^{34}S$ of seawater over

[1] Scattered evidence is found for S-reduction as early as 3.4 bya (Ohmoto et al. 1993, Shen et al. 2001).

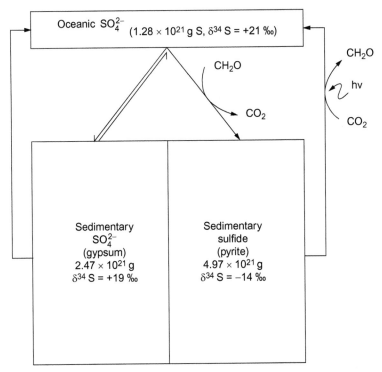

FIGURE 13.4   A model for the global sulfur cycle, showing the linkage and partitioning of S between oxidized and reduced pools near the surface of the Earth. Transfers of S from seawater to pyrite involve a major fractionation between ^{34}S and ^{32}S isotopes, whereas exchange between seawater SO_4 and sedimentary SO_4 (largely gypsum) involves only minor fractionation. The sum of all pools, nearly 10^{22} g S, represents the total outgassing of S from the mantle (compare to Table 2.3). About 15% now resides in the ocean. Estimates of the pool of S in sedimentary sulfides show a wide range of values; the value here, from Holser et al. (1989), is close to that estimated from the pool of sedimentary organic carbon (1.56×10^{22} g; Des Marais et al. 1992) divided by the mean C/S ratio in marine sediments (2.8; Raiswell and Berner 1986). Isotope ratios in seawater and gypsum are taken from Holser et al. (1989). The isotope rate of S in sedimentary sulfides is derived by mass balance to yield δ^{34}S of +4.2 in the global inventory.

geologic time. Changes in the isotopic composition of seawater indicate changes in the relative size of the reservoir of sedimentary pyrite, which occur as a result of changes in the net global balance between sulfate reduction and the oxidation of sedimentary sulfides ($\sim 100 \times 10^{12}$ g S/yr). By contrast, the annual uptake of S by marine phytoplankton and release through their decomposition ($\sim 1390 \times 10^{12}$ g S/yr; Dobrovolsky 1994, p. 183) have little effect on the isotopic composition of the major reservoirs of the global S cycle.

During the last 600,000,000 years, seawater SO_4 has varied between +10 and +30‰ in δ^{34}S (Figure 13.5), with an average value close to that of today. Seawater sulfate shows a marked positive excursion (+32‰) in δ^{34}S during the Cambrian (550 mya), when the deposition of pyrite must have been greatly in excess of the oxidation of sulfide minerals exposed on land. Seawater sulfate was less concentrated in ^{34}S, that is, δ^{34}S of +10‰, during the Carboniferous and Permian, when a large proportion of the Earth's net primary production occurred in freshwater swamps, where SO_4, sulfate reduction, and pyrite deposition are less important

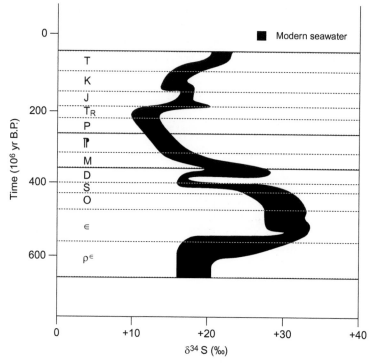

FIGURE 13.5   Variations in the isotopic composition of seawater SO_4 through geologic time. *Source: From Kaplan (1975). Used with permission from the Royal Society of London.*

(Berner 1984). Presumably the concentration of SO_4 in seawater was also greater during that interval, because the global rate of pyrite formation was depressed. The record of $\delta^{34}S$ in marine barite fluctuates in an inverse correlation with records of marine productivity during the past 130 million years (Paytan et al. 2004).

Although the sulfur cycle has shown shifts between net sulfur oxidation and net sulfur reduction in the geologic past, the current human impact is probably unprecedented in the geologic record. As we found for the carbon cycle, the present-day cycle of S is not in steady state. Human activities have added a large flux of gaseous sulfur to the atmosphere, some of which is transported globally. Humans are mining coal and extracting petroleum from the Earth's crust at a rate that mobilizes 150×10^{12} g S/yr, more than double the rate of 100 years ago (Brimblecombe et al. 1989). The net effect of these processes is to increase the pool of oxidized sulfur (SO_4) in the global cycle, at the expense of the storage of reduced sulfur in the Earth's crust. The current estimates of inputs to the ocean are slightly in excess of the estimate of total sinks, implying that the oceans are increasing in SO_4 by over 10^{13} g S/yr. Such an increase will be difficult to document because the content in the oceans is 1.28×10^{21} g S. As calculated in Chapter 9, the mean residence time for SO_4 in seawater is about 10,000,000 years with respect to the current inputs from rivers.

Various workers have attempted to use measurements of $\delta^{34}S$ to deduce the origin of the SO_4 in rainfall and the extent of human impact on the movement of S in the atmosphere (Grey and Jensen 1972, Nriagu et al. 1991, Mast et al. 2001, Puig et al. 2008). Unfortunately, the potential sources of SO_4 show a wide range of values for $\delta^{34}S$, making the identification of specific sources difficult (Nielsen 1974). For example, the sulfur in coal may be depleted in $\delta^{34}S$ if it is found as pyrite or enriched in $\delta^{34}S$ if it is derived from the sulfur that was originally assimilated by the plants forming coal (Hackley and Anderson 1986). Thus, coals show a wide range in $\delta^{34}S$. Similarly, petroleum shows a range of -10.0 to $+25‰$ in ^{34}S (Krouse and McCready 1979). Desert dusts containing SO_4 show a wide range in $\delta^{34}S$, averaging $+5.8‰$ (Bao and Reheis 2003). In the eastern United States, $\delta^{34}S$ of rainfall varies seasonally between $+6.4‰$ in winter and $+2.9‰$ in summer, consistent with any of these sources or a combination of them (Nriagu and Coker 1978). The lower values of summer are thought to reflect the influence of biogenic sulfur derived from sulfate reduction in wetlands (Nriagu et al. 1987).

When SO_2 is emitted as an air pollutant, it forms sulfuric acid through heterogeneous reactions with water in the atmosphere (Eqs. 3.27–3.30). As a strong acid that is completely dissociated in water, H_2SO_4 suppresses the disassociation of natural, weak acids in rainfall. For example, in the absence of strong acids, the dissolution of CO_2 in water will form a weak solution of carbonic acid, H_2CO_3, and rainfall pH will be about 5.6:

$$CO_2 + H_2O \rightarrow H^+ + HCO_3^-. \tag{13.1}$$

In the presence of strong acids that lower the pH below 4.3, this reaction moves to the left, and carbonic acid makes no contribution to free acidity. In many industrialized areas, free acidity in precipitation is almost wholly determined by the concentration of the strong acid anions, SO_4^{2-} and NO_3^- (Table 13.2). Rock weathering that was primarily driven by carbonation weathering in the preindustrial age is now driven by anthropogenic H^+ (Johnson et al. 1972).

TABLE 13.2 Sources of Acidity in Acid Rainfall Collected in Ithaca, New York, on July 11, 1975 (ambient pH 3.84)

Component	Concentration in precipitation (mg/liter)	Contribution to	
		Free acidity at pH 3.84 (μeq/liter)	Total acidity in a titration to pH 9.0 (μeq/liter)
H_2CO_3	0.62	0	20
Clay	5	0	5
NH_4^+	0.53	0	29
Dissolved Al	0.050	0	5
Dissolved Fe	0.040	0	2
Dissolved Mn	0.005	0	0.1
Total organic acids	0.43	2	5.7
HNO_3	2.80	40	40
H_2SO_4	5.60	102	103
Total		144	210

Source: From Galloway et al. (1976). Copyright 1976 by the American Association for the Advancement of Science. Used with permission.

It is interesting to estimate the global sources of acidity in the atmosphere. In this analysis, we are only interested in reactions that are net sources of H^+, so we can ignore the movements of soil dusts and seaspray because the strong-acid anions they contain, NO_3^- and SO_4^{2-}, are largely balanced by cations (especially Ca and Na) that are emitted at the same time. If the pH of all rainfall on Earth was 5.6 as a result of an equilibrium with atmospheric CO_2, the total deposition of H^+ ions would be 1.24×10^{12} moles/year. The production of NO by lightning produces additional acidity because NO dissolves in rainwater, forming HNO_3.

Globally, N fixation by lightning contributes 0.36×10^{12} moles of H^+/yr, and other natural sources of NO_x (soils and forest fires) contribute 1.57×10^{12} moles of H^+/yr. In the years following massive eruptions, SO_2 from volcanoes is distributed globally and dominates the atmospheric deposition of S (Mayewski et al. 1990, Langway et al. 1995). On average, however, volcanic emanations of SO_2 contribute $\sim 0.56 \times 10^{12}$ moles H^+/yr, and the oxidation of biogenic S gases produces 1.76×10^{12} moles H^+/yr. Thus, the total, natural production of H^+ in the atmosphere is normally about 4.25×10^{12} moles/yr. In contrast, the anthropogenic production of NO_x and SO_2 results in about 5.53×10^{12} moles of H^+/year—more than all natural sources of acidity combined.

The only net source of alkalinity in the atmosphere comes from the reaction of NH_3 with the strong acids H_2SO_4 and HNO_3 to form aerosols, $(NH_4)_2SO_4$ and NH_4NO_3 (Eq. 3.5). However, the "natural" global emission of NH_3, about 19×10^{12} g N/yr (Table 12.2), reduces the "natural" production of H^+ by only about 1.37×10^{12} moles/year, or 32% (compare to Savoie et al. 1993). Thus, even though the current acidity of the atmosphere is much higher as a result of human activities, the atmosphere has always acted as an acidic medium with respect to the Earth's crust throughout geologic time. Anthropogenic emissions of NH_3 neutralize 2.44×10^{12} moles H^+/yr, or 44% of the anthropogenic emissions of the acid-forming gases, NO_x and SO_2.

The Atmospheric Budget of Carbonyl Sulfide

Showing an average concentration of about 500 parts per trillion, carbonyl sulfide (COS; alternatively OCS) is the most abundant sulfur gas in the atmosphere (Table 3.1). The pool in the atmosphere contains about 2.8×10^{12} g S (Chin and Davis 1995). Based on the global budget shown in Table 13.3, the mean residence time for COS in the atmosphere is ~ 5 years. COS shows seasonal fluctuations in the atmosphere, parallel to CO_2 and likely to reflect plant activity (Montzka et al. 2004, 2007; Campbell et al. 2008; Geng and Mu 2006).

The global budget for COS in the atmosphere has been revised repeatedly during the last few decades (compare Khalil and Rasmussen 1984, Servant 1989, Chin and Davis 1995, Watts 2000, Kettle et al. 2002), and the current budget of COS is slightly imbalanced—showing an excess of sources over sinks. Several components of the budget, including net ocean uptake, are poorly constrained. The concentration of COS in the atmosphere has increased during the Industrial Revolution (Aydin et al. 2002, Montzka et al. 2004), but it declined slightly during the past couple of decades (Sturges et al. 2001, Rinsland et al. 2002).

The major source of COS in the atmosphere stems from the oxidation of carbon disulfide (CS_2) emitted from anoxic soils and industrial sources. The oceans are an indirect source of

TABLE 13.3 Global Budget for Carbonyl Sulfide in the Atmosphere

Sources	COS (10^{12} g S/yr)
Oceans	0.04[a]
Anoxic soils	0.03
Biomass burning	0.04[b]
Fossil fuels	0.06
Volcanoes	0.02
Oxidation of CS_2	0.25[c]
Oxidation of DMS	0.16
Total sources	**0.60**
Sinks	
Vegetation uptake	0.24
Soil uptake (oxic)	0.13
Oxidation by OH	0.09
Stratospheric photolysis	0.02
Total sinks	**0.48**

[a] *Net.*
[b] *Compare, 0.14 Tg S/yr; Andreae and Merlet (2001).*
[c] *44% from anthropogenic sources.*
Sources: From Kettle et al. (2002), except volcanic flux (Chin and Davis 1993).

COS, stemming from the oxidation of DMS. A small amount of COS is produced by the photochemical destruction of dissolved organic matter in seawater (Ferek and Andreae 1984). In the global budget (Table 13.3), the small direct source of COS from the oceans is a net value, recognizing that COS is also taken up by seawater (Weiss et al. 1995). Other sources of COS include biomass burning (Nguyen et al. 1995, Andreae and Merlet 2001) and direct industrial emissions. Wetland soils are a small source, but the global emission of COS from salt marshes is limited by the small extent of salt marsh vegetation (Aneja et al. 1979, Steudler and Peterson 1985, Carroll et al. 1986).

Some COS is oxidized in the troposphere via OH radicals, but the major tropospheric sink for COS, first reported by Goldan et al. (1988), appears to be uptake by vegetation and upland, oxic soils (Steinbacher et al. 2004, Kuhn et al. 1999, Simmons et al. 1999). Indeed, several workers have suggested that measurements of the uptake of COS through plant stomata may be useful as a measure of *gross primary production* (Stimler et al. 2010, Blonquist et al. 2011, Wohlfahrt et al. 2012). Kesselmeier and Merk (1993) found that a variety of crop plants take up COS whenever the ambient concentration is greater than 150 ppt. Uptake by vegetation is now believed to account for half of the total annual destruction of COS globally (Table 13.3).

A small amount of COS is mixed into the stratosphere, where it is destroyed by a photo-chemical reaction involving the OH radicals, producing SO_4 (Chapter 3). In fact, aside from the periodic eruptions of large volcanoes, COS appears to be the main source of SO_4 aerosols in the stratosphere (Hofmann and Rosen 1983, Servant 1986). There is some evidence that these aerosols have increased in recent years (Hofmann 1990). In the absence of trend toward increasing COS, some workers have suggested that SO_2 from high-altitude aircraft may be responsible (Hofmann 1991). These aerosols affect the amount of solar radiation entering the troposphere, and they are an important component of the radiation budget of the Earth (Turco et al. 1980). Through direct and indirect (CS_2) sources, humans appear to make large contributions to the budget of COS (Table 13.3), and any increase in COS-derived aerosols in the stratosphere has potential consequences for predictive models of future global warming (Hofmann and Rosen 1980). In return, global warming and an increasing flux of uvB radiation penetrating to the Earth's surface may enhance the production of COS in ocean waters (Najjar et al. 1995).

THE GLOBAL MERCURY CYCLE

Mercury is found as a trace metal in igneous rocks and is locally concentrated in economic deposits of a red mineral known as cinnabar (HgS) that is often associated with hydrother-mally altered rock. Volcanoes are a large natural source of mercury emissions to Earth's at-mosphere, in the form of elemental mercury, Hg^0—a volatile metal (Siegel and Siegel 1984, Engle et al. 2006, Pyle and Mather 2003, Bagnato et al. 2011; Figure 13.6). Hg^0 is oxidized

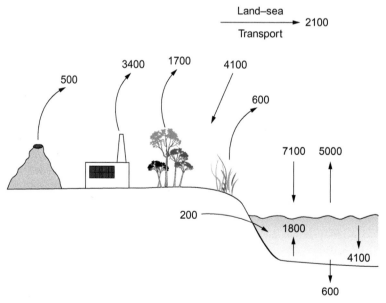

FIGURE 13.6　The global mercury cycle of the modern world. All values are 10^6 g Hg/yr. *Source: From Selin (2009).*

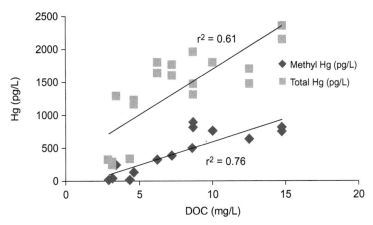

FIGURE 13.7   Concentration of total and methylmercury in stream waters draining into Lake Sunapee, New Hampshire as a function of the concentration of dissolved organic carbon. *Source: From Kathleen Weathers et al., unpublished.*

in the atmosphere to Hg^{2+}, which is the dominant form found in wet and dry deposition at the Earth's surface (Lin and Pehkonen 1999). The lifetime of Hg in the atmosphere ranges from about one year for Hg^0 to a few weeks for Hg^{2+}, so mercury shows regional patterns of deposition that reflect local emissions from natural sources and power plants (Engle et al. 2010, Prestbo and Gay 2009).

Rock weathering releases a small amount of mercury to runoff waters as Hg^{2+} (Figure 13.6). Humans have enhanced the flux of Hg from the Earth's crust by direct mining of mercury and others ores and by the combustion of coal (Pirrone et al. 2010, Gratz and Keeler 2011). As it circulates at the Earth's surface, mercury is rapidly converted among Hg^0, Hg^{2+}, and particulate forms. When mercury is deposited in forests, some is reduced to Hg^0 and revolatilized to the atmosphere (Graydon et al. 2012), while the remainder accumulates in soil organic matter (Demers et al. 2007, Gabriel et al. 2012) or binds to organic compounds that are transported in stream water (Dittman et al. 2010; Figure 13.7).

Similarly, some Hg^{2+} that is deposited in the oceans is photoreduced to elemental mercury that volatilizes from the ocean's surface, as a function of wind speed and Henry's Law for the dissolution of soluble gases in water (Andersson et al. 2008, Mason and Fitzgerald 1993, Kuss et al. 2011). Mercury tends to accumulate in polar ecosystems (Johnson et al. 2008b). Here the deposition of Hg is enhanced by a rapid oxidation to Hg^{2+}, probably mediated by atmospheric Br (Ariya et al. 2004, Jitaru et al. 2009).

Mercury has a long history of economic uses that led to the early human exploitation of ore deposits (Pacyna et al. 2006, Pirrone et al. 2010). Before its toxic properties were well known, mercury was widely used as a fungicide and as a preservative for leather, paint, and other products subject to microbial degradation.[2] Mercury has long been used to extract trace quantities of gold and other metals, and mining activities are often a source of mercury pollution

---

[2] See Act V, Scene 1 in Shakespeare's *Hamlet*, where the gravediggers comment that a hatmaker's corpse will decompose slowly, presumably because its content of mercury will impede microbial activity.

today (Pfeiffer et al. 1991, Artaxo et al. 2000). Mercury accumulates in organic sediments and coal, and is released to the atmosphere when these are burned (Billings and Matson 1972, Lee et al. 2006). Emissions of Hg rose to high values in the late 1800s, associated with the gold rush in the western United States, and again after World War II, associated with the proliferation of coal-fired power plants (Streets et al. 2011; compare Xu et al. 2011).

Lake sediments in New England and the Rocky Mountains show higher, recent levels of mercury accumulation than several centuries ago (Kamman and Engstrom 2002, Mast et al. 2010). Enhanced recent levels of mercury deposition are also recorded in peat bogs, where modern rates are elevated more than 15 times over those of the preindustrial period (Martinez-Cortizas et al. 1999, Roos-Barraclough et al. 2002, Shotyk et al. 2003). Accumulations are often greater in colder periods, when low temperatures reduce the revolatilization of Hg from sediments (Martinez-Cortizas et al. 1999).

An especially toxic form of mercury, methylmercury (CH_3Hg^+),[3] is produced by sulfate-reducing bacteria in anoxic sediments (Compeau and Bartha 1985; King et al. 2001; Gilmour et al. 1992, 2011; Schaefer and Morel 2009). The methylation process proceeds most rapidly with dissolved mercury (Hg^{2+}) and mercury associated with nanoparticles (Zhang et al. 2012b). Mercury accumulates in fishes as they age (Bache et al. 1971, Barber et al. 1972) and at progressively higher levels of food chains, so that large predatory fish often have the highest levels (Cabana and Rasmussen 1994). Higher concentrations of methylmercury are found in modern seabirds than in museum specimens of the same species collected a century ago (Monteiro and Furness 1997, Vo et al. 2011). Methylmercury is degraded by sunlight to Hg^0, which is revolatilized from the surface of lakes and ocean waters (Mason and Fitzgerald 1993, Sellers et al. 1996, Black et al. 2012).

Methylmercury is found in fishes in nearly all waters of the United States (Scudder et al. 2009). When toxic levels of mercury were found in many freshwater and saltwater fishes in the 1970s, environmental regulations were implemented to reduce airborne emissions, especially from coal-fired power plants. Concentrations of mercury in the atmosphere now show a declining trend (Butler et al. 2008, Slemr et al. 2011), which is expected to yield a dramatic decline of mercury deposition from the atmosphere and declining levels in fish (Harris et al. 2007).

Enormous controversy has accompanied the establishment of policies to regulate mercury emissions to the atmosphere. Biogeochemists have helped to elucidate how much of the deposition in various areas is due to local, regional, or global sources (Sigler and Lee 2006, Selin et al. 2008, Weiss-Penzias et al. 2011, Gratz and Keeler 2011). Similarly, biogeochemistry shows how methylmercury forms during sulfate reduction in sediments and explains why the fish in some lakes have relatively low concentrations while those in nearby lakes, often those with organic sediments, may have higher concentrations (Chen et al. 2005, Ward et al. 2010). Sulfate-reducing bacteria and SO_4-based acid rain are implicated in the methylation process (Gilmour and Henry 1991). The mercury content of river waters is controlled by chemical reactions in the water and by the types of environments that are a source of drainage (Burns et al. 2012). For Hg in the marine environment, we still have a poor understanding of the processes that produce methylmercury in seawater and contamination of marine fishes (Malcolm et al. 2010).

---

[3] Often abbreviated MeHg.

SUMMARY

The major pool of S in the global cycle is found in the crustal minerals gypsum and pyrite. Additional S is found dissolved in ocean water. Thus, with respect to pools, the global S cycle resembles the global cycle of phosphorus (Chapter 12). In contrast, the largest pool of the global N cycle is found in the atmosphere.

In other respects, however, there are strong similarities between the global cycles of N, S, and Hg. In all cases, the major annual movement of the element is through the atmosphere, and under natural conditions a large portion of the movement is through the production of reduced gases by biological activity. These gases return N, S, and Hg to the atmosphere, providing a closed global cycle with a relatively rapid turnover. In contrast, the ultimate fate for P is incorporation into ocean sediments, where the cycle of P is complete only as a result of long-term sedimentary uplift.

Biogeochemistry exerts a major influence on the global S cycle. The largest pool of S near the surface of the Earth is found in sedimentary pyrite, as a result of sulfate reduction. The sedimentary record shows that the relative extent of sulfate reduction has varied through geologic time. In the absence of sulfate-reducing bacteria, the concentration of SO_4 in seawater would be higher and O_2 in the atmosphere would be lower than present-day values (Eq. 9.2).

Current human perturbations of the sulfur and mercury cycles are extreme—roughly doubling the annual mobilization of these elements from the crust of the Earth. As a result of fossil fuel combustion, areas that are downwind of industrial regions now receive massive amounts of acidic deposition and mercury from the atmosphere. The excess acidity is likely to lead to changes in rock weathering (Chapter 4), forest growth (Chapter 6), and ocean productivity (Chapter 9). The deposition of mercury in aquatic ecosystems often leads to health advisories for fish consumption.

Recommended Reading

Amend, J.P., K.J. Edwards, and T. W. Lyons (Eds.). 2004. *Sulfur Biogeochemistry: Past and Present.* Geological Society of America.

Brimblecombe, P., and A.Y. Lein (Eds.). 1989. *Evolution of the Global Biogeochemical Sulphur Cycle.* Wiley.

Howarth, R.W., J.W.B. Stewart, and M.V. Ivanov (Eds.). 1992. *Sulphur Cycling on the Continents.* Wiley.

Ivanov, M.V., and J.R. Freney (Eds.). 1983. *The Global Biogeochemical Sulfur Cycle.* Wiley.

PROBLEMS

1. Today, there is approximately 3750×10^{18} g S found as sulfate in seawater and evaporite deposits. Assuming that this sulfur was originally delivered to the oceans in the form of sulfides emitted from hydrothermal emissions, how long would the oxidation of sulfides provide a sink for oxygen derived from the earliest photosynthetic organisms? (Assume that the net storage of organic carbon in sediments of the Precambrian ocean is roughly similar to that of today—0.1×10^{15} g C/yr.)

2. Volcanic gases contain Hg/CO_2 and Hg/H_2O in a ratio of 10^{-8} (Engle et al. 2006) and in a ratio of 10^{-5} with SO_2 (Bagnato et al. 2011). Calculate a global estimate of Hg emission from volcanoes that derives from the global cycles of C, H_2O, and S, as presented in this book.

3. When acid rain falls on carbonate-rich soils, CO_2 is released in the following reaction:

$$CaCO_3 + H_2SO_4 \rightarrow Ca^{2+} + SO_4^{2-} + H_2O + CO_2 \uparrow.$$

Provide a quantitative estimate of the maximum potential importance of this as a source of atmospheric CO_2 versus that derived directly from fossil fuel combustion.

14

Perspectives

Save for a few incoming meteors from space and volcanic emanations from the deep Earth, the surface of our planet, the arena of life, is a closed chemical system. The surface environment, including the atmosphere and the oceans, is a thin peel, roughly 50 km in thickness at the outside of the Earth's 6371-km radius. Certainly, the characteristics of Earth's surface have changed dramatically through geologic time, especially as Earth cooled and violent degassing of the mantle diminished. The most profound change appears due to life itself—the oxygenation of Earth's atmosphere about 2.5 billion years ago.

We study biogeochemistry with the recognition that many of the characteristics of the Earth's surface are determined by life. Earth is very different from its nearest neighbors, Mars and Venus, where pure geochemistry prevails. On Earth, rock weathering and the fluvial transport of materials to the sea are driven by plant-root and microbial activity in soils. The composition of seawater is determined by biotic processes that remove materials from the surface waters and deposit them in marine sediments. Many of the trace gases in the atmosphere are derived from the biosphere; they have short residence times in the atmosphere and must be restored to it by biotic activity. The constancy of Earth's atmospheric composition for the past 10,000 years is rather surprising, reflecting a close balance between biotic activities that produce and consume its gaseous constituents.

Today, when we study biogeochemistry, we see the pervasive influence of the human species on our planet's chemistry. We extract carbon from the Earth's crust at a rate that is more than 36× greater than the natural exposure of organic carbon in rocks at the Earth's surface; the mobilization of some metallic elements by mining is several-fold greater than their natural release by rock weathering (Table 14.1). The atmosphere's composition is changing rapidly as our expanding population taps into the Earth's resources in search of a better life for all peoples (Hofmann et al. 2009). A few decades ago, we missed the opportunity to institute worldwide programs for family planning that might have stabilized the Earth's human population at levels that would minimize our lasting impact on the chemical conditions of our planet. Our similar efforts today may trim only a billion or so from the population that is expected in 2050—9.3 billion neighbors for each of us (United Nations 2011).

With the increasing number of global citizens now expected, we must strive for a planet that will be habitable by many people for the foreseeable future. With the changes in Earth's chemistry that are now in progress, we are not on a sustainable course. The paradigm of growth, which has so long dominated economic theory, is obsolete. Preservation of natural

TABLE 14.1   Estimates of the Global Flux in the Biogeochemical Cycles of Certain Elements, Illustrating the Human Impact

Element	Juvenile flux[a] (1)	Chemical weathering (2)	Natural cycle[b] (3)	Biospheric recycling ratio[c] 3/(1+2)	Human mobilization[d] (4)	Human enhancement 4/(1+2)	Reference for global cycle
B	0.02	0.19	8.8	42	0.58	2.8	Park and Schlesinger (2002)
C	30	210	107,000	446	8700	36.3	Chapter 11
N	5	20[e]	9200[f]	368	221	8.8	Chapter 12
P	~0	2	1000	500	25	12.5	Chapter 12
S	10	70	450	5.6	130	1.6	Chapter 13
Cl	2	260	120	0.46	170	0.65	Figure 3.16
Ca	120	500	2300	3.7	65	0.10	Milliman et al. (1999), Caro et al. (2010)
Fe	6	1.5	40	5.3	1.1[g]	0.14	Muller et al. (2006)
Cu	0.05	0.056	2.5	23.6	1.5[g]	14.2	Rauch and Graedel (2007)
Hg	0.0005	0.0002	0.003	4.3	0.0023	3.3	Selin (2009)

Note: All data 10^{12} g/yr.

[a] *Degassing from the Earth's crust and mantle; sum of volcanic emissions to the atmosphere (subaerial) and net hydrothermal flux to the sea (Elderfield and Schultz 1996) and for N, fixation by lightning (Chapter 12).*

[b] *Annual biogeochemical cycle to/from the Earth's biota on land and in the oceans, in the absence of humans.*

[c] *Following Volk (1998).*

[d] *Direct and indirect mobilization by extraction and mining from the Earth's crust or (for N) industrial fixation. (Sources: From U.S. Geological Survey,* http://minerals.usgs.gov/minerals/pubs/commodity; *Klee and Graede, 2004.)*

[e] *B.Z. Holton (personal communication, 2011).*

[f] *Biological N fixation on land and in the oceans totals ~300 Tg N/yr (see Figure 12.2).*

[g] *Human enhancement in the atmosphere (Table 1.1) and rivers (Table 4.8).*

habitat is the foundation for the preservation of biodiversity, and night-time satellite photographs of the Earth's surface show that we have already left little of nature that is not fragmented, traversed, or converted to human use (Figure 14.1; Ellis et al. 2010, Hannah et al. 1995, Watts et al. 2007). As humans capture an increasing fraction of the planet's productive capacity (NPP), there will be a diminishing proportion left for the other species to persist with us (Haberl et al. 2007, Butchart et al. 2010).

And those species matter! Numerous studies show that sustainable levels of ecosystem function, on land and in the sea, depend on the rich diversity of life on Earth (Naeem et al. 1994; Tilman et al. 1996, 2001; Engelhardt and Ritchie 2001; Worm et al. 2006). With fewer species, ecosystems are less productive, more vulnerable to disturbance, and less likely to

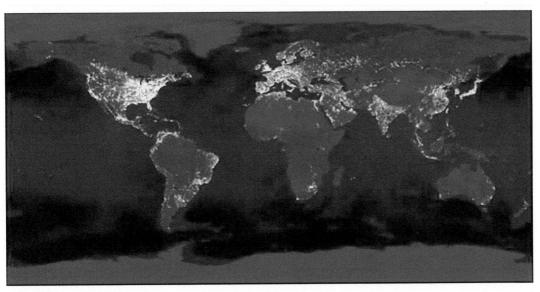

FIGURE 14.1   Night-time view of the world from satellite-derived measures of brightness. *Source: From http://eoimages.gsfc.nasa.gov/images/imagerecords/55000/55167/earth_lights.gif.*

TABLE 14.2   Annual Variation of NPP (% of 5-year mean) of Major Species and of Total Aboveground NPP in an Alaskan Tundra Ecosystem

	Production (% of average)					Coefficient of Variation (%)
	1968	1969	1970	1978	1981	
Eriophorum	77	58	148	101	116	35
Betula	30	52	55	248	121	88
Ledum	106	138	62	103	91	27
Vaccinium	135	172	96	28	71	56
Total production	93	110	106	84	107	11

Source: From Chapin et al. (1996). Used with permission from Cambridge University Press.

recover. Table 14.2 shows how a relatively stable total ecosystem NPP derives from the compensation of individual species in individual years. The stable conditions of our planet's chemistry are vulnerable to progressive biotic impoverishment. Often, increasing variability in ecosystem properties is the vanguard of impending collapse, analogous to similar observations in economics and medicine. (Scheffer et al. 2009, Carpenter et al. 2011). Biodiversity is the basis of stability in our life support system; maintaining biodiversity is more than a pastime for nature lovers.

If there are lessons in biogeochemistry that should guide the human practice of life, we should look for ways to transition to a zero-emission society, in which every product extracted from the Earth's surface is used and reused, without effluents delivered to the common pool of air and freshwater. Soil management is paramount, especially for elements such as phosphorus, for which there are limited supplies and no substitutes. We should look to ways to grow our food, without exogenous inputs, that will maximize yields while allowing us to maintain as much natural land as possible.

At root, almost every global environmental problem is the direct result of the exponential increase in the number of humans on Earth. As our population continues to rise, the material expectations of individuals will raise the cumulative impact on the planet's resources. Rising standards of living often reduce human fertility rates, but not fast enough to keep total resource consumption from increasing (Moses and Brown 2003). While the biosphere will no doubt survive of our species, without our stewardship, life will become increasingly unpleasant for most citizens, who will live on a hot, dirty, and crowded planet.

In every population of organisms studied to date, exponential population growth is followed by dramatic collapse (e.g., Klein 1968). In the history of our own species, wars and plagues have set us back temporarily along our inexorable trajectory toward the current world population peak (Turchin 2009, Zhang et al. 2011a). It is almost inevitable that water scarcity will grow worse as populations increase in arid regions, and that food security will become harder to achieve in many regions where climate grows hotter and rainfall becomes less predictable (Lobell et al. 2011). For those living in affluent countries, this will be unpleasant, while for those living in regions already suffering from food or water scarcity, this decline will be disastrous (Miranda et al. 2011).

Making meaningful changes will be inconvenient, but it is not too late to build a future where we waste much less energy, fertilizer, food, and water and in which we build innovative tools to better capture, store, and use the unlimited energy of the Sun to fuel our daily lives. For this to make a difference, we need to embrace a new culture of personal and corporate responsibility that values minimizing environmental impacts and maximizing long-term resource sustainability, over short-term economic gains and growth driven only by human numbers. The only other viable alternative to slow our impact on the biosphere would be to reverse our own trajectory of population growth. If we do not intentionally choose to use less, it is nearly certain that future droughts, famines, disease, and war will choose an alternative future for us.

Recommended Reading

Speth, J.G. 2008. *The Bridge at the End of the World*. Yale University Press.

References

Abbott, D., D. Sparks, C. Herzberg, W. Mooney, A. Nikishin, and Y. S. Zhang. 2000. Quantifying Precambrian crustal extraction: The root is the answer. Tectonophysics **322**: 163–190.

Abe, K. 2002. Preformed Cd and PO_4 and the relationship between the two elements in the northwestern Pacific and the Okhotsk Sea. Marine Chemistry **79**:27–36.

Aber, J., W. McDowell, K. Nadelhoffer, A. Magill, G. Berntson, M. Kamakea, S. McNulty, W. Currie, L. Rustad, and I. Fernandez. 1998. Nitrogen saturation in temperate forest ecosystems—Hypotheses revisited. BioScience **48**:921–934.

Aber, J. D., C. L. Goodale, S. V. Ollinger, M. L. Smith, A. H. Magill, M. E. Martin, R. A. Hallett, and J. L. Stoddard. 2003. Is nitrogen deposition altering the nitrogen status of northeastern forests? BioScience **53**:375–389.

Aber, J. D., and J. M. Melillo. 1980. Litter decomposition—Measuring relative contributions of organic matter and nitrogen to forest soils. Canadian Journal of Botany **58**:416–421.

Abramov, O., and S. J. Mojzsis. 2009. Microbial habitability of the Hadean Earth during the late heavy bombardment. Nature **459**:419–422.

Abril, G., F. Guerin, S. Richard, R. Delmas, C. Galy-Lacaux, P. Gosse, A. Tremblay, L. Varfalvy, M. A. Dos Santos, and B. Matvienko. 2005. Carbon dioxide and methane emissions and the carbon budget of a 10-year-old tropical reservoir (Petit Saut French Guiana). Global Biogeochemical Cycles **19**.

Accoe, F., P. Boeckx, J. Busschaert, G. Hofman, and O. Van Cleemput. 2004. Gross N transformation rates and net N mineralisation rates related to the C and N contents of soil organic matter fractions in grassland soils of different age. Soil Biology & Biochemistry **36**: 2075–2087.

Achard, F., H. D. Eva, H. J. Stibig, P. Mayaux, J. Gallego, T. Richards, and J. P. Malingreau. 2002. Determination of deforestation rates of the world's humid tropical forests. Science **297**:999–1002.

Achtnich, C., F. Bak, and R. Conrad. 1995. Competition for electron donors among nitrate reducers, ferric iron reducers, sulfate reducers, and methanogens in anoxic paddy soil. Biology and Fertility of Soils **19**:65–72.

Ackerman, S. A., and H. Chung. 1992. Radiative effects of airborne dust on regional energy budgets at the top of the atmosphere. Journal of Applied Meteorology **31**:223–233.

Adair, E. C., W. J. Parton, S. J. Del Grosso, W. L. Silver, M. E. Harmon, S. A. Hall, I. C. Burke, and S. C. Hart. 2008. Simple three-pool model accurately describes patterns of long-term litter decomposition in diverse climates. Global Change Biology **14**:2636–2660.

Adamec, L. 1997. Mineral nutrition of carnivorous plants: A review. Botanical Review **63**:273–299.

Adams, D. F., S. O. Farwell, E. Robinson, M. R. Pack, and W. L. Bamesberger. 1981. Biogenic sulfur source strengths. Environmental Science & Technology **15**: 1493–1498.

Adams, J.A.S., M.S.M. Mantovani, and L. L. Lundell. 1977. Wood versus fossil fuel as a source of excess carbon dioxide in the atmosphere–preliminary report. Science **196**:54–56.

Adams, J. M., H. Faure, L. Fauredenard, J. M. McGlade, and F. I. Woodward. 1990. Increases in terrestrial carbon storage from the Last Glacial maximum to the present. Nature **348**:711–714.

Adams, M. A., and P. M. Attiwill. 1982. Nitrate reductase activity and growth response of forest species to ammonium and nitrate sources of nitrogen. Plant and Soil **66**:373–381.

Adams, W. A., M. A. Raza, and L. J. Evans. 1980. Relationships between net redistribution of Al and Fe and extractable levels in Podzolic soils derived from Lower Paleozoic sedimentary rocks. Journal of Soil Science **31**:533–545.

Adamsen, A.P.S., and G. M. King. 1993. Methane consumption in temperate and sub-arctic forest soils—Rates, vertical zonation, and responses to water and nitrogen. Applied and Environmental Microbiology **59**:485–490.

Adler, P. R., S. J. Del Grosso, and W. J. Parton. 2007. Life-cycle assessment of net greenhouse-gas flux for bioenergy cropping systems. Ecological Applications **17**: 675–691.

Admiraal, W.I.M., and Y.J.H. Botermans. 1989. Comparison of nitrification rates in three branches of the Lower River Rhine. Biogeochemistry **8**:135–151.

Ae, N., J. Arihara, K. Okada, T. Yoshihara, and C. Johansen. 1990. Phosphorus uptake by pigeon pea and its role in cropping systems of the Indian subcontinent. Science **248**:477–480.

Aerts, R. 1996. Nutrient resorption from senescing leaves of perennials: Are there general patterns? Journal of Ecology 84:597–608.

Aerts, R., and F. S. Chapin. 2000. The mineral nutrition of wild plants revisited: A re-evaluation of processes and patterns. Advances in Ecological Research 30:1–67.

Agawin, N.S.R., C. M. Duarte, and S. Agusti. 2000. Nutrient and temperature control of the contribution of picoplankton to phytoplankton biomass and production. Limnology and Oceanography 45:591–600.

Agbenin, J. O. 2003. Extractable iron and aluminum effects on phosphate sorption in a savanna alfisol. Soil Science Society of America Journal 67:589–595.

Agee, C. B. 1990. A new look at differentiation of the Earth from melting experiments on the Allende Meteorite. Nature 346:834–837.

Aguilar, R., and R. D. Heil. 1988. Soil organic carbon, nitrogen, and phosphorus quantities in northern Great Plains rangeland. Soil Science Society of America Journal 52:1076–1081.

Agusti, S., M. P. Satta, M. P. Mura, and E. Benavent. 1998. Dissolved esterase activity as a tracer of phytoplankton lysis: Evidence of high phytoplankton lysis rates in the northwestern Mediterranean. Limnology and Oceanography 43:1836–1849.

Ahl, T. 1988. Background yield of phosphorus from drainage area and atmosphere—an empirical approach. Hydrobiologia 170:35–44.

Ahn, J., E. J. Brook, L. Mitchell, J. Rosen, J. R. McConnell, K. Taylor, D. Etheridge, and M. Rubino. 2012. Atmospheric CO_2 over the last 1000 years: A high-resolution record from the West Antarctic Ice Sheet (WAIS) Divide ice core. Global Biogeochemical Cycles 26.

Aitkenhead, J. A., and W. H. McDowell. 2000. Soil C:N ratio as a predictor of annual riverine DOC flux at local and global scales. Global Biogeochemical Cycles 14:127–138.

Aiuppa, A., S. Inguaggiato, A.J.S. McGonigle, M. O'Dwyer, C. Oppenheimer, M. J. Padgett, D. Rouwet, and M. Valenza. 2005. H_2S fluxes from Mt. Etna, Stromboli, and Vulcano (Italy) and implications for the sulfur budget at volcanoes. Geochimica et Cosmochimica Acta 69:1861–1871.

Alexander, C. O., G. Manhes, and C. Gopel. 2001. The early evolution of the inner solar systems: A meteoritic perspective. Science 293:64–68.

Alexander, E. B. 1988. Rates of soil formation—Implications for soil loss tolerance. Soil Science 145:37–45.

Alexander, R. B., R. A. Smith, and G. E. Schwarz. 2000. Effect of stream channel size on the delivery of nitrogen to the Gulf of Mexico. Nature 403:758–761.

Alexander, R. B., R. A. Smith, G. E. Schwarz, E. W. Boyer, J. V. Nolan, and J. W. Brakebill. 2008. Differences in phosphorus and nitrogen delivery to the Gulf of Mexico from the Mississippi River basin. Environmental Science & Technology 42:822–830.

Alexandre, A., J. D. Meunier, F. Colin, and J. M. Koud. 1997. Plant impact on the biogeochemical cycle of silicon and related weathering processes. Geochimica et Cosmochimica Acta 61:677–682.

Algesten, G., J. Wikner, S. Sobek, L. J. Tranvik, and M. Jansson. 2004. Seasonal variation of CO_2 saturation in the Gulf of Bothnia: Indications of marine net heterotrophy. Global Biogeochemical Cycles 18.

Alin, S. R., and T. C. Johnson. 2007. Carbon cycling in large lakes of the world: A synthesis of production, burial, and lake-atmosphere exchange estimates. Global Biogeochemical Cycles 21:GB3002.

Allan, C. J., and N. T. Roulet. 1994. Solid phase controls of dissolved aluminum within upland Precambrian Shield catchments. Biogeochemistry 26:85–114.

Allan, C. J., N. T. Roulet, and A. R. Hill. 1993. The biogeochemistry of pristine, headwater Precambrian Shield watersheds—An analysis of material transport within a heterogeneous landscape. Biogeochemistry 22:37–79.

Allard, P., J. Carbonnelle, N. Metrich, H. Loyer, and P. Zettwoog. 1994. Sulfur output and magma degassing budget of Stromboli volcano. Nature 368:326–330.

Alldredge, A. 1998. The carbon, nitrogen and mass content of marine snow as a function of aggregate size. Deep-Sea Research Part I—Oceanographic Research Papers 45:529-541.

Alldredge, A. L., and Y. Cohen. 1987. Can microscale chemical patches persist in the sea—Microelectrode study of marine snow fecal pellets. Science 235:689–691.

Alldredge, A. L., and C. C. Gotschalk. 1990. The relative contribution of marine snow of different origins to biological processes in coastal waters. Continental Shelf Research 10:41–58.

Allègre, C. J., G. Manhes, and C. Gopel. 1995. The age of the Earth. Geochimica et Cosmochimica Acta 59:1445–1456.

Allen, A. S., and W. H. Schlesinger. 2004. Nutrient limitations to soil microbial biomass and activity in loblolly pine forests. Soil Biology & Biochemistry 36:581–589.

Allen, M. F. 1992. Mycorrhizas: Plant-fungus Relationships. Chapman and Hall.

Allison, S. D. 2006. Brown ground: A soil carbon analogue for the green world hypothesis? American Naturalist 167:619–627.

Allison, S. D., M. D. Wallenstein, and M. A. Bradford. 2010. Soil-carbon response to warming dependent on microbial physiology. Nature Geoscience 3:336–340.

Alt, J. C., C. J. Garrido, W. C. Shanks, A. Turchyn, J. A. Padron-Navarta, V. L. Sanchez-Vizcaino,

M.T.G. Pugnaire, and C. Marchesi. 2012. Recycling of water, carbon and sulfur during subduction of serpentinites: A stable isotope study of Cerro de Almirez, Spain. Earth and Planetary Science Letters **327/328**:50–60.

Altabet, M. A., and W. B. Curry. 1989. Testing models of past ocean chemistry using foraminifera $^{15}N/^{14}N$. Global Biogeochemical Cycles 3:107–119.

Altabet, M. A., R. Francois, D. W. Murray, and W. L. Prell. 1995. Climate-related variations in denitrification in the Arabian Sea from sediment $^{15}N/^{14}N$ ratios. Nature **373**:506–509.

Altabet, M. A., M. J. Higginson, and D. W. Murray. 2002. The effect of millennial-scale changes in Arabian Sea denitrification on atmospheric CO_2. Nature **415**:159–162.

Alton, P., R. Fisher, S. Los, and M. Williams. 2009. Simulations of global evapotranspiration using semiempirical and mechanistic schemes of plant hydrology. Global Biogeochemical Cycles **23**.

Altschuler, Z. S., M. M. Schnepfe, C. C. Silber, and F. O. Simon. 1983. Sulfur diagenesis in Everglades peat and origin of pyrite in coal. Science **221**:221–227.

Aluwihare, L. I., D. J. Repeta, and R. F. Chen. 1997. A major biopolymeric component to dissolved organic carbon in surface sea water. Nature **387**:166–169.

Aluwihare, L. I., D. J. Repeta, S. Pantoja, and C. G. Johnson. 2005. Two chemically distinct pools of organic nitrogen accumulate in the ocean. Science **308**:1007–1010.

Alvarez, E., A. Martinez, and R. Calvo. 1992. Geochemical aspects of aluminum in forest soils in Galicia (NW Spain). Biogeochemistry **16**:167–180.

Ambrose, S. H., and N. E. Sikes. 1991. Soil carbon isotope evidence for Holocene habitat change in the Kenya Rift valley. Science **253**:1402–1405.

Ambus, P., and S. Christensen. 1994. Measurement of N_2O emission from a fertilized grassland—An analysis of spatial variability. Journal of Geophysical Research—Atmospheres **99**:16549–16555.

Amend, J. P., and E. L. Shock. 1998. Energetics of amino acid synthesis in hydrothermal ecosystems. Science **281**:1659–1662.

Ames, R. N., C.P.P. Reid, L. K. Porter, and C. Cambardella. 1983. Hyphal uptake and transport of nitrogen from two N-15 labeled sources by *Glomus mosseae*—a vesicular-arbuscular mycorrhizal fungus. New Phytologist **95**:381–396.

Amthor, J. S. 1984. The role of maintenance respiration in plant growth. Plant Cell and Environment 7:561–569.

Amthor, J. S. 1989. *Respiration and Crop Productivity*. Springer.

Amthor, J. S. 2010. From sunlight to phytomass: on the potential efficiency of converting solar radiation to phyto-energy. New Phytologist **188**:939–959.

Amundson, R. 2001. The carbon budget in soils. Annual Review of Earth and Planetary Sciences **29**:535–562.

Amundson, R., A. T. Austin, E.A.G. Schuur, K. Yoo, V. Matzek, C. Kendall, A. Uebersax, D. Brenner, and W. T. Baisden. 2003. Global patterns of the isotopic composition of soil and plant nitrogen. Global Biogeochemical Cycles **17**.

Amundson, R., and H. Jenny. 1991. The place of humans in the state factor theory of ecosystems and their soils. Soil Science **151**:99–109.

Amundson, R. G., O. A. Chadwick, and J. M. Sowers. 1989. A comparison of soil climate and biological activity along an elevation gradient in the eastern Mojave desert. Oecologia **80**:395–400.

Amundson, R. G., O. A. Chadwick, J. M. Sowers, and H. E. Doner. 1989b. Soil evolution along an altitudinal transect in the eastern Mojave desert of Nevada, USA. Geoderma **43**:349–371.

Amundson, R. G., and E. A. Davidson. 1990. Carbon dioxide and nitrogenous gases in the soil atmosphere. Journal of Geochemical Exploration **38**:13–41.

Anastasi, C., M. Dowding, and V. J. Simpson. 1992. Future CH_4 emissions from rice production. Journal of Geophysical Research—Atmospheres **97**:7521–7525.

Anastasi, C., and V. J. Simpson. 1993. Future methane emissions from animals. Journal of Geophysical Research—Atmospheres **98**:7181–7186.

Anbar, A. D., and A. H. Knoll. 2002. Proterozoic ocean chemistry and evolution: A bioinorganic bridge? Science **297**:1137–1142.

Anbar, A. D., Y. L. Yung, and F. P. Chavez. 1996. Methyl bromide: Ocean sources, ocean sinks, and climate sensitivity. Global Biogeochemical Cycles **10**:175–190.

Anbeek, C. 1993. The effect of natural weathering on dissolution rates. Geochimica et Cosmochimica Acta **57**:4963–4975.

Anders, E. 1989. Pre-biotic organic matter from coments and asteroids. Nature **342**:255–257.

Anders, E., and N. Grevesse. 1989. Abundances of the elements—Meteoritic and solar. Geochimica et Cosmochimica Acta **53**:197–214.

Anders, E., and T. Owen. 1977. Mars and Earth—Origin and abundance of volatiles. Science **198**:453–465.

Anderson, C. J., and W. J. Mitsch. 2006. Sediment, carbon, and nutrient accumulation at two 10-year-old created riverine marshes. Wetlands **26**:779–792.

Anderson, D. W., and E. A. Paul. 1984. Organo-mineral complexes and their study by radiocarbon dating. Soil Science Society of America Journal **48**:298–301.

Anderson, G. F. 1986. Silica, diatoms and a fresh water productivity maximum in Atlantic Coastal Plain estuaries, Chesapeake Bay. Estuarine Coastal and Shelf Science **22**:183–197.

Anderson, I. C., J. S. Levine, M. A. Poth, and P. J. Riggan. 1988. Enhanced biogenic emissions of nitric ocide and nitrous oxide following surface biomass burning. Journal of Geophysical Research—Atmospheres **93**:3893–3898.

Anderson, J. M., P. Ineson, and S. A. Huish. 1983. Nitrogen and cation mobilization by soil fauna feeding on leaf litter and soil organic matter from deciduous woodlands. Soil Biology & Biochemistry **15**:463–467.

Anderson, J.P.E., and K. H. Domsch. 1978. Physiological method for quantitative measurement of microbial biomass in soils. Soil Biology & Biochemistry **10**: 215–221.

Anderson, J.P.E., and K. H. Domsch. 1980. Quantities of plant nutrients in the microbial biomass of selected soils. Soil Science **130**:211–216.

Anderson, L. A., and J. L. Sarmiento. 1994. Redfield ratios of remineralization determined by nutrient data analysis. Global Biogeochemical Cycles **8**:65–80.

Anderson, L. D., M. L. Delaney, and K. L. Faul. 2001. Carbon to phosphorus ratios in sediments: Implications for nutrient cycling. Global Biogeochemical Cycles **15**:65–79.

Anderson, R. F., and S. L. Schiff. 1987. Alkalinity generation and the fate of sulfur in lake sediments. Canadian Journal of Fisheries and Aquatic Sciences **44**:188–193.

Anderson, S. P., W. E. Dietrich, and G. H. Brimhall. 2002. Weathering profiles, mass-balance analysis, and rates of solute loss: Linkages between weathering and erosion in a small, steep catchment. Geological Society of America Bulletin **114**:1143–1158.

Andersson, J. H., J.W.M. Wijsman, P.M.J. Herman, J. J. Middelburg, K. Soetaert, and C. Heip. 2004a. Respiration patterns in the deep ocean. Geophysical Research Letters **31**.

Andersson, M., A. Kjoller, and S. Struwe. 2004b. Microbial enzyme activities in leaf litter, humus and mineral soil layers of European forests. Soil Biology & Biochemistry **36**:1527–1537.

Andersson, M. E., K. Gardfeldt, I. Wangberg, and D. Stromberg. 2008. Determination of Henry's law constant for elemental mercury. Chemosphere **73**:587–592.

Andreadis, K. M., and D. P. Lettenmaier. 2006. Trends in 20th century drought over the continental United States. Geophysical Research Letters **33**.

Andreae, M. O. 1991. Biomass burning: Its history, use, and distribution and its impact on environmental quality and global climate. *In* J. S. Levine, editor. *Global Biomass Burning*, MIT Press, 3–21.

Andreae, M. O., B. E. Anderson, D. R. Blake, J. D. Bradshaw, J. E. Collins, G. L. Gregory, G. W. Sachse, and M. C. Shipham. 1994a. Influence of plumes from biomass burning on atmospheric chemistry over the equatorial and tropical South Atlantic during CITE-3. Journal of Geophysical Research—Atmospheres **99**: 12793–12808.

Andreae, M. O., and T. W. Andreae. 1988. The cycle of biogenic sulfur compounds over the Amazon basin. 1. Dry season. Journal of Geophysical Research—Atmospheres **93**:1487–1497.

Andreae, M. O., and W. R. Barnard. 1984. The marine chemistry of dimethylsulfide. Marine Chemistry **14**: 267–279.

Andreae, M. O., H. Berresheim, H. Bingemer, D. J. Jacob, B. L. Lewis, S. M. Li, and R. W. Talbot. 1990. The atmospheric sulfur cycle over the Amazon basin. 2. Wet season. Journal of Geophysical Research—Atmospheres **95**:16813–16824.

Andreae, M. O., R. J. Charlson, F. Bruynseels, H. Storms, R. Van Grieken, and W. Maenhaut. 1986. Internal mixture of sea salt, silicates, and excess sulfate in marine aerosols. Science **232**:1620–1623.

Andreae, M. O., W. Elbert, and S. J. Demora. 1995. Biogenic sulfur emissions and aerosols over the tropical South Atlantic. 3. Atmospheric dimethylsulfide, aerosols and cloud condensation nuclei. Journal of Geophysical Research—Atmospheres **100**:11335–11356.

Andreae, M. O., and W. A. Jaeschke. 1992. Exchange of sulphur between biosphere and atmosphere over temperate and tropical regions. *In* R. W. Howarth, J.W.B. Stewart, and M. V. Ivanov, editors. *Sulfur Cycling on the Continents*. Wiley, 27–61.

Andreae, M. O., C. D. Jones, and P. M. Cox. 2005. Strong present-day aerosol cooling implies a hot future. Nature **435**:1187–1190.

Andreae, M. O., and P. Merlet. 2001. Emission of trace gases and aerosols from biomass burning. Global Biogeochemical Cycles **15**:955–966.

Andreae, M. O., and D. Rosenfeld. 2008. Aerosol-cloud-precipitation interactions. Part 1. The nature and sources of cloud-active aerosols. Earth-Science Reviews **89**:13–41.

Andreae, T. W., M. O. Andreae, and G. Schebeske. 1994b. Biogenic sulfur emissions and aerosols over the tropical South Atlantic. 1. Dimethylsulfide in seawater and in the atmospheric boundary layer. Journal of Geophysical Research—Atmospheres **99**: 22819–22829.

Andreae, T. W., G. A. Cutter, N. Hussain, J. Radfordknoery, and M. O. Andreae. 1991. Hydrogen sulfide and radon in and over the western North Atlantic ocean. Journal of Geophysical Research—Atmospheres **96**:18753–18760.

Andreu-Hayles, L., O. Planells, E. Gutierrez, E. Muntan, G. Helle, K. J. Anchukaitis, and G. H. Schleser. 2011. Long tree-ring chronologies reveal 20th century increases in water-use efficiency but no enhancement of tree growth at five Iberian pine forests. Global Change Biology 17:2095–2112.

Andrews, J. A., K. G. Harrison, R. Matamala, and W. H. Schlesinger. 1999. Separation of root respiration from total soil respiration using carbon-13 labeling during Free-Air Carbon Dioxide Enrichment (FACE). Soil Science Society of America Journal 63:1429–1435.

Andrews, J. A., and W. H. Schlesinger. 2001. Soil CO_2 dynamics, acidification, and chemical weathering in a temperate forest with experimental CO_2 enrichment. Global Biogeochemical Cycles 15:149–162.

Andrews, M. 1986. The partitioning of nitrate assimilation betweeen root and shoot of higher plants. Plant Cell and Environment 9:511–519.

Aneja, V. P., D. S. Kim, M. Das, and B. E. Hartsell. 1996. Measurements and analysis of reactive nitrogen species in the rural troposphere of southeast United States: Southern oxidant study site SONIA. Atmospheric Environment 30:649–659.

Aneja, V. P., D. R. Nelson, P. A. Roelle, J. T. Walker, and W. Battye. 2003. Agricultural ammonia emissions and ammonium concentrations associated with aerosols and precipitation in the southeast United States. Journal of Geophysical Research—Atmospheres 108.

Aneja, V. P., J. H. Overton, L. T. Cupitt, J. L. Durham, and W. E. Wilson. 1979. Carbon disulfide and carbonyl sulfide from biogenic sources and their contributions to the global sulfur cycle. Nature 282:493–496.

Anenberg, S. C., L. W. Horowitz, D. Q. Tong, and J. J. West. 2010. An estimate of the global burden of anthropogenic ozone and fine particulate matter on premature human mortality using atmospheric modeling. Environmental Health Perspectives 118:1189–1195.

Anschutz, P., K. Dedieu, F. Desmazes, and G. Chaillou. 2005. Speciation, oxidation state, and reactivity of particulate manganese in marine sediments. Chemical Geology 218:265–279.

Antibus, R. K., J. G. Croxdale, O. K. Miller, and A. E. Linkins. 1981. Ectomycorrhizal fungi of *Salix rotundifolia*. 3. Resynthezied mycorrhizal complexes and their surface phosphatase activities. Canadian Journal of Botany 59:2458–2465.

Antonov, J. I., S. Levitus, and T. P. Boyer. 2005. Thermosteric sea level rise. Geophysical Research Letters 32: 1955–2003.

Antweiler, R. C., and J. I. Drever. 1983. The weathering of a late Tertiary volcanic ash—Importance of organic solutes. Geochimica et Cosmochimica Acta 47:623–629.

Appel, H. M. 1993. Phenolics in ecological interactions—The importance of oxidation. Journal of Chemical Ecology 19:1521–1552.

Appenzeller, C., and H. C. Davies. 1992. Structure of stratospheric intrusions into the troposphere. Nature 358:570–572.

April, R., and D. Keller. 1990. Mineralogy of the rhizosphere in forest soils of the eastern United States—Mineralogic studies of the rhizosphere. Biogeochemistry 9:1–18.

April, R., R. Newton, and L. T. Coles. 1986. Chemical weathering in two Adirondack watersheds—past and present-day rates. Geological Society of America Bulletin 97:1232–1238.

Arai, Y., and D. L. Sparks. 2007. Phosphate reaction dynamics in soils and soil minerals: A multiscale approach. Advances in Agronomy 94:135–179.

Archer, D. 1995. Upper ocean physics as relevant to ecosystem dynamics—A tutorial. Ecological Applications 5:724–739.

Archer, D., B. Buffett, and V. Brovkin. 2009. Ocean methane hydrates as a slow tipping point in the global carbon cycle. Proceedings of the National Academy of Sciences 106:20596–20601.

Archer, D., H. Kheshgi, and E. Maier-Reimer. 1998. Dynamics of fossil fuel—CO_2 neutralization by marine $CaCO_3$. Global Biogeochemical Cycles 12:259–276.

Archer, D., and E. Maier-Reimer. 1994. Effect of deep sea sedimentary calcite preservation on atmospheric CO_2 concentration. Nature 367:260–263.

Arianoutsou, M., and N. S. Margaris. 1981. Fire-induced nutrient losses in a phryganic (East Mediterranean) ecosystem. International Journal of Biometeorology 25:341–347.

Ariya, P. A., A. P. Dastoor, M. Amyot, W. H. Schroeder, L. Barrie, K. Anlauf, F. Raofie, A. Ryzhkov, D. Davignon, J. Lalonde, and A. Steffen. 2004. The Arctic: a sink for mercury. Tellus Series B—Chemical and Physical Meteorology 56:397–403.

Arkley, R. J. 1963. Calculation of carbonate and water movement in soil from climatic data. Soil Science 92: 239–248.

Armbruster, M., M. Abiy, and K. H. Feger. 2003. The biogeochemistry of two forested catchments in the Black Forest and the eastern Ore Mountains (Germany). Biogeochemistry 65:341–368.

Armentano, T. V., and E. S. Menges. 1986. Patterns of change in the carbon balance of organic-soil wetlands of the temeprate zone. Journal of Ecology 74:755–774.

Arnason, T. S., and R. G. Keil. 2007. Changes in organic matter-mineral interactions for marine sediments with varying oxygen exposure times. Geochimica et Cosmochimica Acta 71:3545–3556.

Arnold, F., and T. Bührke. 1983. New H_2SO_4 and HSO_3 vapor measurements in the stratosphere—Evidence for a volcanic influence. Nature 301:293–295.

Aronson, E. L., and B. R. Helliker. 2010. Methane flux in non-wetland soils in response to nitrogen addition: a meta-analysis. Ecology **91**:3242–3251.

Arrhenius, S. 1918. *The Destinies of the Stars*. G.P. Putnam's Sons, New york.

Arrigo, K. R. 2005. Marine microorganisms and global nutrient cycles. Nature **437**:349–355.

Arrigo, K. R., D. Lubin, G. L. van Dijken, O. Holm-Hansen, and E. Morrow. 2003. Impact of a deep ozone hole on Southern Ocean primary production. Journal of Geophysical Research—Oceans **108**.

Arrigoni, A. S., M. C. Greenwood, and J. N. Moore. 2010. Relative impact of anthropogenic modifications versus climate change on the natural flow regimes of rivers in the northern Rocky Mountains, United States. Water Resources Research **46**.

Art, H. W., F. H. Bormann, G. K. Voigt, and G. M. Woodwell. 1974. Barrier Island forest ecosystems: The role of meteorological inputs. Science **184**:60–62.

Artaxo, P., R. C. de Campos, E. T. Fernandes, J. V. Martins, Z. F. Xiao, O. Lindqvist, M. T. Fernandez-Jimenez, and W. Maenhaut. 2000. Large scale mercury and trace element measurements in the Amazon basin. Atmospheric Environment **34**:4085–4096.

Arthur, M. A., W. E. Dean, and L. M. Pratt. 1988. Geochemical and climatic effects of increased marine organic carbon burial at the Cenomanian-Turonian boundary. Nature **335**:714–717.

Aselmann, I., and P. J. Crutzen. 1989. Global distribution of natural fresh water wetlands and rice paddies, their net primary productivity, seasonality and possible methane emissions. Journal of Atmospheric Chemistry **8**:307–358.

Ask, J., J. Karlsson, and M. Jansson. 2012. Net ecosystem production in clear-water and brown-water lakes. Global Biogeochemical Cycles **26**:7.

Asman, W.A.H., and H. A. Vanjaarsveld. 1992. A variable-resolution transport model applied for NHx in Europe. Atmospheric Environment Part A-General Topics **26**:445–464.

Asner, G. P., G.V.N. Powell, J. Mascaro, D. E. Knapp, J. K. Clark, J. Jacobson, T. Kennedy-Bowdoin, A. Balaji, G. Paez-Acosta, E. Victoria, L. Secada, M. Valqui, and R. F. Hughes. 2010. High-resolution forest carbon stocks and emissions in the Amazon. Proceedings of the National Academy of Sciences **107**:16738–16742.

Asner, G. P., and P. M. Vitousek. 2005. Remote analysis of biological invasion and biogeochemical change. Proceedings of the National Academy of Sciences **102**:4383–4386.

Asper, V. L., W. G. Deuser, G. A. Knauer, and S.E. Lohrenz. 1992. Rapid coupling of sinking particle fluxes between surface and deep ocean waters. Nature **357**:670–672.

Atkinson, L. J., C. D. Campbell, J. Zaragoza-Castells, V. Hurry, and O. K. Atkin. 2010. Impact of growth temperature on scaling relationships linking photosynthetic metabolism to leaf functional traits. Functional Ecology **24**:1181–1191.

Atkinson, R. 2000. Atmospheric chemistry of VOCs and NOx. Atmospheric Environment **34**:2063–2101.

Atkinson, R., and J. Arey. 2003. Gas-phase tropospheric chemistry of biogenic volatile organic compounds: a review. Atmospheric Environment **37**:S197–S219.

Atlas, E., and R. M. Pytkowicz. 1977. Solubility behavior of apatites in seawater. Limnology and Oceanography **22**:290–300.

Attri, K., S. Kerkar, and P. A. LokaBharathi. 2011. Ambient iron concentration regulates the sulfate reducing activity in the mangrove swamps of Diwar, Goa, India. Estuarine Coastal and Shelf Science **95**:156–164.

Auclair, A.N.D., and T. B. Carter. 1993. Forest wildfires as a recent source of CO_2 at northern latitudes. Canadian Journal of Forest Research **23**:1528–1536.

Aufdenkampe, A. K., E. Mayorga, P. A. Raymond, J. M. Melack, S. C. Doney, S. R. Alin, R. E. Aalto, and K. Yoo. 2011. Riverine coupling of biogeochemical cycles between land, oceans, and atmosphere. Frontiers in Ecology and the Environment **9**:53–60.

Aumann, G. D. 1965. Microtine abundance and soil sodium levels. Journal of Mammalogy **46**:594–604.

Aumont, O., and L. Bopp. 2006. Globalizing results from ocean in situ iron fertilization studies. Global Biogeochemical Cycles **20**.

Austin, A. T., and L. Vivanco. 2006. Plant litter decomposition in a semi-arid ecosystem controlled by photodegradation. Nature **442**:555–558.

Autry, A., and J. W. Fitzgerald. 1993. Saturation potentials for sulfate adsorption by field-moist forest soils. Soil Biology & Biochemistry **25**:833–838.

Avery, G. G., R. D. Shannon, J. R. White, S. Christopher, M. J. Alperin, G. B. Avery, R. Jeffrey, and C. S. Martens. 2002. Controls on methane production in a tidal freshwater estuary and a peatland: methane production via acetate fermentation and CO_2 reduction. Biogeochemistry **62**:19–37.

Avila, A., M. Alarcon, and I. Queralt. 1998. The chemical composition of dust transported in red rains—Its contribution to the biogeochemical cycle of a Holm oak forest in Catalonia (Spain). Atmospheric Environment **32**:179–191.

Avis, C. A., A. J. Weaver, and K. J. Meissner. 2011. Reduction in areal extent of high-latitude wetlands in response to permafrost thaw. Nature Geoscience 4:444–448.

Avnimelech, Y., and J. R. McHenry. 1984. Enrichment of transported sediments with organic carbon, nutrients, and clay. Soil Science Society of America Journal 48:259–266.

Awasthi, O. P., E. Sharma, and L.M.S. Palni. 1995. Stemflow—A source of nutrients in some naturally growting epiphytic orchids of the Sikkim Himalaya. Annals of Botany 75:5–11.

Awramik, S. M., J. W. Schopf, and M. R. Walter. 1983. Filamentous fossil bacteria from the Archean of Western Australia. Precambrian Research 20:357–374.

Axelsson, B. 1981. Site differences in yield differences in biological production or in redistribution of carbon within trees. Department of Ecological and Environmental Research, Research Report 6. Swedish University of Agricultural Science, Uppsala.

Aydin, M., W. J. De Bruyn, and E. S. Saltzman. 2002. Pre-industrial atmospheric carbonyl sulfide (OCS) from an Antarctic ice core. Geophysical Research Letters 29.

Aydin, M., K. R. Verhulst, E. S. Saltzman, M. O. Battle, S. A. Montzka, D. R. Blake, Q. Tang, and M. J. Prather. 2011. Recent decreases in fossil-fuel emissions of ethane and methane derived from firn air. Nature 476:198–201.

Ayers, G. P., and J. L. Gras. 1991. Seasonal relationship between cloud condensation nuclei and aerosol methanesulfonate in marine air. Nature 353:834–835.

Ayers, G. P., S. A. Penkett, R. W. Gillett, B. Bandy, I. E. Galbally, C. P. Meyer, C. M. Elsworth, S. T. Bentley, and B. W. Forgan. 1992. Evidence for photochemical control of ozone concentrations in unpolluted marine air. Nature 360:446–449.

Baas Becking, L.G.M., I. R. Kaplan, and D. Moore. 1960. Limits of the natural environment in terms of pH and oxidation-reduction potentials. Journal of Geology 68:243–384.

Bacastow, R. B. 1976. Modulation of atmospheric carbon dioxide by the Southern Oscillation. Nature 261: 116–118.

Bacastow, R. B., and A. Björkström. 1981. Comparison of ocean models for the carbon cycle. In B. Bolin, editor. Carbon Cycle Modeling, Wiley, 29–79.

Bacastow, R. B., C. D. Keeling, and T. P. Whorf. 1985. Seasonal amplitude increase in atmospheric CO_2 concentration at Mauna Loa, Hawaii, 1959-1982. Journal of Geophysical Research—Atmospheres 90:10529–10540.

Bache, C. A., W. H. Gutenmann, and D. J. Lisk. 1971. Residues of total mercury and methyl-mercuric salts in lake trout as a function of age. Science 172:951–952.

Bache, C. W. 1984. The role of calcium in buffering soils. Plant, Cell and Environment 7:391–395.

Bachmann, P. A., P. L. Luisi, and J. Lang. 1992. Autocatalytic self-replicating micelles as models for prebiotic structures. Nature 357:57–59.

Bade, D. L., S. R. Carpenter, J. J. Cole, M. L. Pace, E. Kritzberg, M. C. Van de Bogert, R. M. Cory, and D. M. McKnight. 2007. Sources and fates of dissolved organic carbon in lakes as determined by whole-lake carbon isotope additions. Biogeochemistry 84:115–129.

Baethgen, W. E., and M. M. Alley. 1987. Nonexchangeable ammonium nitrogen contribution to plant available nitrogen. Soil Science Society of America Journal 51:110–115.

Bagnato, E., A. Aiuppa, F. Parello, P. Allard, H. Shinohara, M. Liuzzo, and G. Giudice. 2011. New clues on the contribution of Earth's volcanism to the global mercury cycle. Bulletin of Volcanology 73:497–510.

Bailey, J., A. Chrysostomou, J. H. Hough, T. M. Gledhill, A. McCall, S. Clark, F. Menard, and M. Tamura. 1998. Circular polarization in star-formation regions: Implications for biomolecular homochirality. Science 281:672–674.

Bailey, V. L., A. D. Peacock, J. L. Smith, and H. Bolton. 2002. Relationships between soil microbial biomass determined by chloroform fumigation-extraction, substrate-induced respiration, and phospholipid fatty acid analysis. Soil Biology & Biochemistry 34: 1385–1389.

Bailey-Serres, J., and L. A . C. J. Voesenek. 2008. Flooding stress: acclimations and genetic diversity. Annual review of plant biology 59:313–339.

Baines, S. B., and M. L. Pace. 1991. The production of dissolved organic matter by phytoplankton and its importance to bacteria—patterns across marine and freshwater systems. Limnology and Oceanography 36:1078–1090.

Baines, S. B., and M. L. Pace. 1994. Relationships between suspended particulate matter and sinking flux along a trophic gradient and implications for the fate of planktonic primary production. Canadian Journal of Fisheries and Aquatic Sciences 51:25–36.

Baird, A. J., C. W. Beckwith, S. Waldron, and J. M. Waddington. 2004. Ebullition of methane-containing gas bubbles from near-surface Sphagnum peat. Geophysical Research Letters 31.

Baker, J., M. Bizzarro, N. Wittig, J. Connelly, and H. Haack. 2005. Early planetesimal melting from an age of 4.5662 Gyr for differentiated meteorites. Nature 436:1127–1131.

Baker, L. A., D. Hope, Y. Xu, J. Edmonds, and L. Lauver. 2001. Nitrogen balance for the Central Arizona-Phoenix (CAP) ecosystem. Ecosystems 4:582–602.

Baker, M. A., C. N. Dahm, and H. M. Valett. 1999. Acetate retention and metabolism in the hyporheic zone of a mountain stream. Limnology and Oceanography 44:1530–1539.

Baker, M. A., H. M. Valett, and C. N. Dahm. 2000. Organic carbon supply and metabolism in a shallow groundwater ecosystem. Ecology 81:3133–3148.

Baker, P. A., and S. J. Burns. 1985. Occurrence and formation of dolomite in organic-rich continental margin sediments. Bulletin-American Association of Petroleum Geologists 69:1917–1930.

Baker, P. A., and M. Kastner. 1981. Constraints on the formation of sedimentary dolomite. Science 213: 214–216.

Baker, V. R. 1977. Stream-channel response to floods, with examples from central Texas. Geological Society of America Bulletin 88:1057–1071.

Bakun, A. 1990. Global climate change and intensification of coastal ocean upwelling. Science 247:198–201.

Balch, W., R. Evans, J. Brown, G. Feldman, C. McClain, and W. Esaias. 1992. The remote sensing of ocean primary productivity—Use of a new data compilation to test satellite algorithms. Journal of Geophysical Research—Oceans 97:2279–2293.

Baldocchi, D. D. 2003. Assessing the eddy covariance technique for evaluating carbon dioxide exchange rates of ecosystems: past, present and future. Global Change Biology 9:479–492.

Baldock, J. A., C. A. Masiello, Y. Gelinas, and J. I. Hedges. 2004. Cycling and composition of organic matter in terrestrial and marine ecosystems. Marine Chemistry 92:39–64.

Baldock, J. A., J. M. Oades, A. G. Waters, X. Peng, A. M. Vassallo, and M. A. Wilson. 1992. Aspects of the chemical structure of soil organic materials as revealed by solid-state C-13 NMR spectroscopy. Biogeochemistry 16:1–42.

Balling, R. C. 1989. The impact of summer rainfall on the temperature gradient along the United States-Mexico border. Journal of Applied Meteorology 28:304–308.

Bandfield, J. L., T. D. Glotch, and P. R. Christensen. 2003. Spectroscopic identification of carbonate minerals in the Martian dust. Science 301:1084–1087.

Banfield, J. F., W. W. Barker, S. A. Welch, and A. Taunton. 1999. Biological impact on mineral dissolution: Application of the lichen model to understanding mineral weathering in the rhizosphere. Proceedings of the National Academy of Sciences 96:3404–3411.

Bange, H. W., S. Rapsomanikis, and M. O. Andreae. 1996. Nitrous oxide in coastal waters. Global Biogeochemical Cycles 10:197–207.

Banin, A., and J. Navrot. 1975. Origin of life—Clues from relations between chemical compositions of living organisms and natural environments. Science 189:550–551.

Bao, H., D. A. Campbell, J. G. Bockheim, and M. H. Thiemens. 2000. Origins of sulphate in Antarctic dry-valley soils as deduced from anomalous ^{17}O compositions. Nature 407:499–502.

Bao, H. M., and M. C. Reheis. 2003. Multiple oxygen and sulfur isotopic analyses on water-soluble sulfate in bulk atmospheric deposition from the southwestern United States. Journal of Geophysical Research—Atmospheres 108.

Barber, R. T. 1968. Dissolved organic carbon from deep water resists microbial oxidation. Nature 220: 274–275.

Barber, R. T., A. Vijayakumar, and F. A. Cross. 1972. Mercury concentrations in recent and ninety-year old benthopelagic fish. Science 178:636–639.

Barber, S. A. 1962. A diffusion and mass-flow concept of soil nutrient availability. Soil Science 93:39–43.

Barber, V. A., G. P. Juday, and B. P. Finney. 2000. Reduced growth of Alaskan white spruce in the twentieth century from temperature-induced drought stress. Nature 405:668–673.

Barbour, M. M., J. E. Hunt, A. S. Walcroft, G.N.D. Rogers, T. M. McSeveny, and D. Whitehead. 2005. Components of ecosystem evaporation in a temperate coniferous rainforest, with canopy transpiration scaled using sapwood density. New Phytologist 165:549–558.

Barford, C. C., S. C. Wofsy, M. L. Goulden, J. W. Munger, E. H. Pyle, S. P. Urbanski, L. Hutyra, S. R. Saleska, D. Fitzjarrald, and K. Moore. 2001. Factors controlling long- and short-term sequestration of atmospheric CO_2 in a mid-latitude forest. Science 294: 1688–1691.

Barlaz, M. A. 1998. Carbon storage during biodegradation of municipal solid waste components in laboratory-scale landfills. Global Biogeochemical Cycles 12:373–380.

Barnett, T. P., D. W. Pierce, K. M. AchutaRao, P. J. Gleckler, B. D. Santer, J. M. Gregory, and W. M. Washington. 2005. Penetration of human-induced warming into the world's oceans. Science 309:284–287.

Barnett, T. P., D. W. Pierce, H. G. Hidalgo, C. Bonfils, B. D. Santer, T. Das, G. Bala, A. W. Wood, T. Nozawa, A. A. Mirin, D. R. Cayan, and M. D. Dettinger. 2008. Human-induced changes in the hydrology of the western United States. Science 319:1080–1083.

Barnola, J. M., M. Anklin, J. Porcheron, D. Raynaud, J. Schwander, and B. Stauffer. 1995. CO_2 evolution during the last millennium as recorded by Antarctic and Greenland ice. Tellus Series B—Chemical and Physical Meteorology 47:264–272.

Barre, P., G. Berger, and B. Velde. 2009. How element translocation by plants may stabilize illitic clays in the surface of temperate soils. Geoderma 151:22–30.

Barrett, L.R., and R. J. Schaetzl. 1992. An examination of podzolization near Lake Michigan using chronofunctions. Canadian Journal of Soil Science **72**:527–541.

Barron, A. R., N. Wurzburger, J. P. Bellenger, S. J. Wright, A.M.L. Kraepiel, and L. O. Hedin. 2009. Molybdenum limitation of asymbiotic nitrogen fixation in tropical forest soils. Nature Geoscience **2**:42–45.

Barry, D.A.J., D. Goorahoo, and M. J. Goss. 1993. Estimation of nitrate concentrations in groundwater using a whole-farm nitrogen budget. Journal of Environmental Quality **22**:767–775.

Barsdate, R. J., and V. Alexander. 1975. Nitrogen balance of arctic tundra—Pathways, rates, and environmental implications. Journal of Environmental Quality **4**:111–117.

Bartel-Ortiz, L. M., and M. B. David. 1988. Sulfur constituents and transfomations in upland and floodplain forest soils. Canadian Journal of Forest Research **18**: 1106–1112.

Bartlett, K. B., D. S. Bartlett, R. C. Harriss, and D. I. Sebacher. 1987. Methane emissions along a salt marsh salinity gradient. Biogeochemistry **4**:183–202.

Bartlett, K. B., and R. C. Harriss. 1993. Review and assessment of methane emissions from wetlands. Chemosphere **26**:261–320.

Bartlett, R. J. 1986. Soil redox behavior. *In* D. L. Sparks, editor. *Soil Physical Chemistry*, CRC Press, 179–207.

Barton, L., C.D.A. McLay, L. A. Schipper, and C. T. Smith. 1999. Annual denitrification rates in agricultural and forest soils: a review. Australian Journal of Soil Research **37**:1073–1093.

Bassirirad, H. 2000. Kinetics of nutrient uptake by roots: responses to global change. New Phytologist **147**: 155–169.

Bastviken, D., J. Cole, M. Pace, and L. Tranvik. 2004. Methane emissions from lakes: Dependence of lake characteristics, two regional assessments, and a global estimate. Global Biogeochemical Cycles **18**.

Bastviken, D., F. Thomsen, T. Svensson, S. Karlsson, P. Sanden, G. Shaw, M. Matucha, and G. Oberg. 2007. Chloride retention in forest soil by microbial uptake and by natural chlorination of organic matter. Geochimica et Cosmochimica Acta **71**:3182–3192.

Bastviken, D., L. J. Tranvik, J. A. Downing, P. M. Crill, and A. Enrich-Prast. 2011. Freshwater methane emissions offset the continental carbon sink. Science **331**:50.

Bates, T. R., and J. P. Lynch. 1996. Stimulation of root hair elongation in *Arabidopsis thaliana* by low phosphorus availability. Plant Cell and Environment **19**:529–538.

Bates, T. S., R. J. Charlson, and R. H. Gammon. 1987. Evidence for the climatic role of marine biogenic sulfur. Nature **329**:319–321.

Batjes, N. H. 1996. Total carbon and nitrogen in the soils of the world. European Journal of Soil Science **47**:151–163.

Bator, A., and J. L. Collett. 1997. Cloud chemistry varies with drop size. Journal of Geophysical Research—Atmospheres **102**:28071–28078.

Battin, T. J., S. Luyssaert, L. A. Kaplan, A. K. Aufdenkampe, A. Richter, and L. J. Tranvik. 2009. The boundless carbon cycle. Nature Geoscience **2**:598–600.

Battle, M., M. L. Bender, P. P. Tans, J.W.C. White, J. T. Ellis, T. Conway, and R. J. Francey. 2000. Global carbon sinks and their variability inferred from atmospheric O_2 and delta ^{13}C. Science **287**:2467–2470.

Baturin, G. N. 2007. Issue of the relationship between primary productivity of organic carbon in ocean and phosphate accumulation (Holocene-Late Jurassic). Lithol. Miner. Resourc. **42**:318–348.

Bauer, J. E., P. M. Williams, and E.R.M. Druffel. 1992. C-14 activity of dissolved organic carbon fractions in the North-central Pacific and Sargasso sea. Nature **357**:667–670.

Baumann, R. H., J. W. Day, and C. A. Miller. 1984. Mississippi deltaic wetland survival—sedimentation versus coastal submergence. Science **224**:1093–1095.

Baumgärtner, M., and R. Conrad. 1992. Effects of soil variables and season on the production and consumption of nitric oxide in oxic soils. Biology and Fertility of Soils **14**:166–174.

Bautista, D. M., P. Movahed, A. Hinman, H. E. Axelsson, O. Sterner, E. D. Hogestatt, D. Julius, S. E. Jordt, and P. M. Zygmunt. 2005. Pungent products from garlic activate the sensory ion channel TRPA1. Proceedings of the National Academy of Sciences **102**: 12248–12252.

Bazzaz, F. A. 1990. The response of natural ecosystems to the rising global CO_2 levels. Annual Review of Ecology and Systematics **21**:167–196.

Beadle, N.C.W. 1966. Soil phosphate and its role in molding segments of the Australian flora and vegetation, with special reference to xeromorphy and sclerophylly. Ecology **47**:992–1007.

Beal, E. J., C. H. House, and V. J. Orphan. 2009. Manganese- and iron-dependent marine methane oxidation. Science **325**:184–187.

Beaulieu, J. J., J. L. Tank, S. K. Hamilton, W. M. Wollheim, R. O. Hall, P. J. Mulholland, B. J. Peterson, L. R. Ashkenas, L. W. Cooper, C. N. Dahm, W. K. Dodds, N. B. Grimm, S. L. Johnson, W. H. McDowell, G. C. Poole, H. M. Valett, C. P. Arango, M. J. Bernot, A. J. Burgin, C. L. Crenshaw, A. M. Helton, L. T. Johnson, J. M. O'Brien, J. D. Potter, R. W. Sheibley, D. J. Sobota, and S. M. Thomas. 2011. Nitrous oxide emission from denitrification in stream and river networks. Proceedings of the National Academy of Sciences **108**:214–219.

Beaumont, V., and F. Robert. 1999. Nitrogen isotope ratios of kerogens in Precambrian cherts: a record of the evolution of atmosphere chemistry? Precambrian Research **96**:63–82.

Beckwith, S.V.W., and A. I. Sargent. 1996. Circumstellar disks and the search for neighbouring planetary systems. Nature **383**:139–144.

Bédard, C., and R. Knowles. 1989. Physiology, biochemistry, and specific inhibitors of CH_4, NH_4^+, and CO oxidation by methanotrophs and nitrifiers. Microbiological Reviews **53**:68–84.

Bedford, B. L., M. R. Walbridge, and A. Aldous. 1999. Patterns in nutrient availability and plant diversity of temperate North American wetlands. Ecology **80**: 2151–2169.

Bedison, J. E., and A. H. Johnson. 2010. Seventy four years of calcium loss from forest soils of the Adirondack mountains, New York. Soil Science Society of America Journal **74**:2187–2195.

Beer, C., W. Lucht, C. Schmullius, and A. Shvidenko. 2006. Small net carbon dioxide uptake by Russian forests during 1981-1999. Geophysical Research Letters **33**.

Beer, C., M. Reichstein, E. Tomelleri, P. Ciais, M. Jung, N. Carvalhais, C. Rodenbeck, M. A. Arain, D. Baldocchi, G. B. Bonan, A. Bondeau, A. Cescatti, G. Lasslop, A. Lindroth, M. Lomas, S. Luyssaert, H. Margolis, K. W. Oleson, O. Roupsard, E. Veenendaal, N. Viovy, C. Williams, F. I. Woodward, and D. Papale. 2010. Terrestrial gross carbon dioxide uptake: Global distribution and covariation with climate. Science **329**:834–838.

Beerling, D. J. 1999. New estimates of carbon transfer to terrestrial ecosystems between the last glacial maximum and the Holocene. Terra Nova **11**:162–167.

Beerling, D. J., and W. G. Chaloner. 1993a. Evolutionary responses of stomatal density to global CO_2 change. Biological Journal of the Linnean Society **48**:343–353.

Beerling, D. J., and W. G. Chaloner. 1993b. Stomatal density responses of Egyptian *Olea europaea* L. leaves to CO_2 change since 1327 BC. Annals of Botany **71**: 431–435.

Behera, S. N., and M. Sharma. 2011. Degradation of SO_2, NO_2 and NH_3 leading to formation of secondary inorganic aerosols: An environmental chamber study. Atmospheric Environment **45**:4015–4024.

Behrenfeld, M. J., A. J. Bale, Z. S. Kolber, J. Aiken, and P. G. Falkowski. 1996. Confirmation of iron limitation of phytoplankton photosynthesis in the equatorial Pacific Ocean. Nature **383**:508–511.

Behrenfeld, M. J., and P. G. Falkowski. 1997. Photosynthetic rates derived from satellite-based chlorophyll concentration. Limnology and Oceanography **42**: 1–20.

Behrenfeld, M. J., R. T. O'Malley, D. A. Siegel, C. R. McClain, J. L. Sarmiento, G. C. Feldman, A. J. Milligan,

P. G. Falkowski, R. M. Letelier, and E. S. Boss. 2006. Climate-driven trends in contemporary ocean productivity. Nature **444**:752–755.

Behrenfeld, M. J., J. T. Randerson, C. R. McClain, G. C. Feldman, S. O. Los, C. J. Tucker, P. G. Falkowski, C. B. Field, R. Frouin, W. E. Esaias, D. D. Kolber, and N. H. Pollack. 2001. Biospheric primary production during an ENSO transition. Science **291**:2594–2597.

Beilke, S., and D. Lamb. 1974. On the absorption of SO_2 in ocean water. Tellus **26**:268–271.

Bekker, A., H. D. Holland, P. L. Wang, D. Rumble, H. J. Stein, J. L. Hannah, L. L. Coetzee, and N. J. Beukes. 2004. Dating the rise of atmospheric oxygen. Nature **427**:117–120.

Bekki, S., K. S. Law, and J. A. Pyle. 1994. Effect of ozone depletion on atmospheric CH_4 and CO concentrations. Nature **371**:595–597.

Belcher, C. M., and J. C. McElwain. 2008. Limits for combustion in low O_2 redefine paleoatmospheric predictions for the mesozoic. Science **321**:1197–1200.

Belillas, C. M., and F. Rodà. 1991. Nutrient budgets in a dry heathland watershed in northeastern Spain. Biogeochemistry **13**:137–157.

Bell, D. R., and G. R. Rossman. 1992. Water in Earth's mantle—The role of nominally anhydrous minerals. Science **255**:1391–1397.

Bell, N., L. Hsu, D. J. Jacob, M. G. Schultz, D. R. Blake, J. H. Butler, D. B. King, J. M. Lobert, and E. Maier-Reimer. 2002. Methyl iodide: Atmospheric budget and use as a tracer of marine convection in global models. Journal of Geophysical Research—Atmospheres **107**.

Bell, R. A. 1993. Cryptoendolithic algae of hot semiarid lands and deserts. Journal of Phycology **29**:133–139.

Bellamy, P. H., P. J. Loveland, R. I. Bradley, R. M. Lark, and G.J.D. Kirk. 2005. Carbon losses from all soils across England and Wales 1978-2003. Nature **437**: 245–248.

Bellini, G., M. E. Sumner, D. E. Radcliffe, and N. P. Qafoku. 1996. Anion transport through columns of highly weathered acid soil: Adsorption and retardation. Soil Science Society of America Journal **60**:132–137.

Bellouin, N., O. Boucher, J. Haywood, and M. S. Reddy. 2005. Global estimate of aerosol direct radiative forcing from satellite measurements. Nature **438**:1138–1141.

Belser, L. W., and E. L. Mays. 1980. Specific inhibition of nitrite oxidation by chlorate and its use in assessing nitrification in soils and sediments. Applied and Environmental Microbiology **39**:505–510.

Beltman, B., T. G. Rouwenhorst, M. B. Van Kerkhoven, and T. Van der Krift. 2000. Internal eutrophication in peat soils through competition between chloride and sulphate with phosphate for binding sites. Biogeochemistry 50:183–194.

Belyea, L. R., and N. Malmer. 2004. Carbon sequestration in peatland: patterns and mechanisms of response to climate change. Global Change Biology 10: 1043–1052.

Belyea, L. R., and B. G. Warner. 1996. Temporal scale and the accumulation of peat in a Sphagnum bog. Canadian Journal of Botany 74:366–377.

Beman, J. M., K. R. Arrigo, and P. A. Matson. 2005. Agricultural runoff fuels large phytoplankton blooms in vulnerable areas of the ocean. Nature 434:211–214.

Ben-Ami, Y., I. Koren, Y. Rudich, P. Artaxo, S. T. Martin, and M. O. Andreae. 2010. Transport of North African dust from the Bodele depression to the Amazon Basin: a case study. Atmospheric Chemistry and Physics 10:7533–7544.

Ben-David, M., T. A. Hanley, and D. M. Schell. 1998. Fertilization of terrestrial vegetation by spawning Pacific salmon: the role of flooding and predator activity. Oikos 83:47–55.

Bencala, K. E., and R. A. Walters. 1983. Simulation of solute transport in a mountain pool and riffle stream—a transient storage model. Water Resources Research 19:718–724.

Bender, M., R. Jahnke, R. Weiss, W. Martin, D. T. Heggie, J. Orchardo, and T. Sowers. 1989. Organic carbon oxidation and benthic nitrogen and silica dynamics in San Clemente basin, a continental borderland site. Geochimica et Cosmochimica Acta 53:685–697.

Bender, M., T. Sowers, and L. Labeyrie. 1994. The Dole effect and its variations during the last 130,000 years as measured in the Vostock ice core. Global Biogeochemical Cycles 8:363–376.

Bender, M. L., M. Battle, and R. F. Keeling. 1998. The O_2 balance of the atmosphere: A tool for studying the fate of fossil-fuel CO_2. Annual Review of Energy and the Environment 23:207–223.

Bender, M. L., D. T. Ho, M. B. Hendricks, R. Mika, M. O. Battle, P. P. Tans, T. J. Conway, B. Sturtevant, and N. Cassar. 2005. Atmospheric O_2/N_2 changes, 1993-2002: Implications for the partitioning of fossil fuel CO_2 sequestration. Global Biogeochemical Cycles 19.

Bender, M. L., G. P. Klinkhammer, and D. W. Spencer. 1977. Manganese in seawater and marine manganese balance. Deep-Sea Research 24:799–812.

Beniston, M., D. B. Stephenson, O. B. Christensen, C.A.T. Ferro, C. Frei, S. Goyette, K. Halsnaes, T. Holt, K. Jylha, B. Koffi, J. Palutikof, R. Scholl, T. Semmler, and K. Woth. 2007. Future extreme events in European climate: an exploration of regional climate model projections. Climatic Change 81:71–95.

Benitez-Nelson, C. R., and K. O. Buesseler. 1999. Variability of inorganic and organic phosphorus turnover rates in the coastal ocean. Nature 398:502–505.

Benke, A. C., and A. D. Huryn. 2006. Secondary production of macroinvertebrates. In F. R. Hauer and G. A. Lam- berti, editors. Methods in Stream Ecology. Elsevier, 691–710.

Benner, R., and S. Opsahl. 2001. Molecular indicators of the sources and transformations of dissolved organic matter in the Mississippi River plume. Organic Geochemistry 32:597–611.

Bennett, E. M., S. R. Carpenter, and N. F. Caraco. 2001. Human impact on erodable phosphorus and eutrophication: A global perspective. BioScience 51: 227–234.

Bennett, P. C., M. E. Melcer, D. I. Siegel, and J. P. Hassett. 1988. The dissolution of quartz in dilute aqueous solutions of organic acids at 25° C. Geochimica et Cosmochimica Acta 52:1521–1530.

Benoit, G., and T. F. Rozan. 1999. The influence of size distribution on the particle concentration effect and trace metal partitioning in rivers. Geochimica et Cosmochimica Acta 63:113–127.

Ben-Shahar, R., and M. J. Coe. 1992. The relationships between soil factors, grass nutrients and the foraging behavior of wildebeest and zebra. Oecologia 90: 422–428.

Benson, B. J., J. J. Magnuson, O. P. Jensen, V. M. Card, G. Hodgkins, J. Korhonen, D. M. Livingstone, K. M. Stewart, G. A. Weyhenmeyer, and N. G. Granin. 2012. Extreme events, trends, and variability in Northern Hemisphere lake-ice phenology (1855-2005). Climatic Change 112:299–323.

Benstead, J. P., A. D. Rosemond, W. F. Cross, J. B. Wallace, S. L. Eggert, K. Suberkropp, V. Gulis, J. L. Greenwood, and C. J. Tant. 2009. Nutrient enrichment alters storage and fluxes of detritus in a headwater stream ecosystem. Ecology 90:2556–2566.

Benzing, D. H., and A. Renfrow. 1974. Mineral nutrition of Bromeliaceae. Botanical Gazette 135:281–288.

Berelson, W. M., W. M. Balch, R. Najjar, R. A. Feely, C. Sabine, and K. Lee. 2007. Relating estimates of $CaCO_3$ production, export, and dissolution in the water column to measurements of $CaCO_3$ rain into sediment traps and dissolution on the sea floor: A revised global carbonate budget. Global Biogeochemical Cycles 21.

Berelson, W. M., D. E. Hammond, and G. A. Cutter. 1990. In situ measurements of calcium carbonate dissolution rates in deep-sea sediments. Geochimica et Cosmochimica Acta 54:3013–3020.

Berendse, F., N. Van Breemen, H. Rydin, A. Buttler, M. Heijmans, M. R. Hoosbeek, J. A. Lee, E. Mitchell, T. Saarinen, H. Vasander, and B. Wallen. 2001. Raised atmospheric CO_2 levels and increased N deposition cause shifts in plant species composition and production in *Sphagnum* bogs. Global Change Biology 7: 591–598.

Berg, B. 1988. Dynamics of nitrogen (^{15}N) in decomposing Scots pine (*Pinus sylvestris*) needle litter-Long-term decompositon in a Scots pine forest—VI. Canadian Journal of Botany **66**:1539–1546.

Berg, B., A. Albrektson, M. P. Berg, J. Cortina, M. B. Johansson, A. Gallardo, M. Madeira, J. Pausas, W. Kratz, R. Vallejo, and C. McClaugherty. 1999. Amounts of litter fall in some pine forests in a European transect, in particular Scots pine. Annals of Forest Science **56**:625–639.

Berg, B., M. P. Berg, P. Bottner, E. Box, A. Breymeyer, R. C. Deanta, M. Couteaux, A. Escudero, A. Gallardo, W. Kratz, M. Madeira, E. Malkonen, C. McClaugherty, V. Meentemeyer, F. Munoz, P. Piussi, J. Remacle, and A. V. Desanto. 1993. Litter mass loss rates in pine forests of Europe and eastern United States—Some relationships with climate and litter quality. Biogeochemistry **20**:127–159.

Berg, P., L. Klemedtsson, and T. Rosswall. 1982. Inhibitory effect of low partial pressures of acetylene on nitrification. Soil Biology & Biochemistry **14**: 301–303.

Berga, L., J. M. Buil, E. Bofill, J. C. DeCea, J. A. Garcia Perez, G. Manueco, J. Polimon, A. Soriano, and J. Yaque. 2006. Dams and Reservoirs, Societies and Environment in the 21st Century. Taylor and Francis Group.

Berge, E., and H. A. Jakobsen. 1998. A regional scale multi-layer model for the calculation of long-term transport and deposition of air pollution in Europe. Tellus Series B—Chemical and Physical Meteorology **50**:205–223.

Berger, A. L. 1978. Long-term variations of caloric insolation resulting from Earth's orbital elements. Quaternary Research **9**:139–167.

Berger, T. W., and G. Glatzel. 1994. Deposition of atmospheric constituents and its impact on nutrient budgets of oak forests (*Quercus petraea* and *Quercus robur*) in lower Austria. Forest Ecology and Management **70**:183–193.

Berger, W. H. 1989. Global maps of ocean productivity. *In* W. H. Berger, V. S. Smetacek, and G. Wefer, editors. *Productivity of the Ocean: Present and Past*, Wiley, 429–455.

Berger, W. H., C. G. Adelseck, and L. A. Mayer. 1976. Distribution of carbonate in surface sediments of Pacific Ocean. Journal of Geophysical Research—Oceans and Atmospheres **81**:2617–2627.

Bergeron, O., and I. B. Strachan. 2011. CO_2 sources and sinks in urban and suburban areas of a northern mid-latitude city. Atmospheric Environment **45**: 1564–1573.

Berggren, D., and J. Mulder. 1995. The role of organic matter in controlling aluminum solubility in acidic mineral soil horizons. Geochimica et Cosmochimica Acta **59**:4167–4180.

Bergström, A. K., and M. Jansson. 2006. Atmospheric nitrogen deposition has caused nitrogen enrichment and eutrophication of lakes in the northern hemisphere. Global Change Biology **12**:635–643.

Berkner, L. V., and L. C. Marshall. 1965. On the origin and rise of oxygen concentration in the Earth's atmosphere. Journal of the Atmospheric Sciences **22**: 225–261.

Berliner, R., B. Jacoby, and E. Zamski. 1986. Absence of *Cistus incanus* from basaltic soils in Israel—Effect of mycorrhizae. Ecology **67**:1283–1288.

Berman-Frank, I., J. T. Cullen, Y. Shaked, R. M. Sherrell, and P. G. Falkowski. 2001a. Iron availability, cellular iron quotas, and nitrogen fixation in *Trichodesmium*. Limnology and Oceanography **46**: 1249–1260.

Berman-Frank, I., P. Lundgren, Y. B. Chen, H. Kupper, Z. Kolber, B. Bergman, and P. Falkowski. 2001b. Segregation of nitrogen fixation and oxygenic photosynthesis in the marine cyanobacterium *Trichodesmium*. Science **294**:1534–1537.

Bern, C. R., and A. R. Townsend. 2008. Accumulation of atmospheric sulfur in some Costa Rican soils. Journal of Geophysical Research—Biogeosciences **113**.

Bern, C. R., A. R. Townsend, and G. L. Farmer. 2005. Unexpected dominance of parent-material strontium in a tropical forest on highly weathered soils. Ecology **86**:626–632.

Bernal, B., and W. J. Mitsch. 2012. Comparing carbon sequestration in temperate freshwater wetland communities. Global Change Biology **18**: 1636–1647.

Berner, E. K., and R. A. Berner. 1996. *Global Environment: Water, Air, and Geochemical Cycles*. Prentice Hall.

Berner, R. A. 1971. Worldwide sulfur pollution of rivers. Journal of Geophysical Research **76**:6597–6600.

Berner, R. A. 1982. Burial of organic carbon and pyrite sulfur in the modern ocean—Its geochemical and environmental signfiicance. American Journal of Science **282**:451–473.

Berner, R. A. 1984. Sedimentary pyrite formation -an update. Geochimica et Cosmochimica Acta **48**: 605–615.

Berner, R. A. 1992. Weathering, plants, and the long-term carbon cycle. Geochimica et Cosmochimica Acta **56**:3225–3231.

Berner, R. A. 1997. Paleoclimate—The rise of plants and their effect on weathering and atmospheric CO_2. Science **276**:544–546.

Berner, R. A. 2001. Modeling atmospheric O_2 over Phanerozoic time. Geochimica et Cosmochimica Acta 65:685–694.

Berner, R. A., and D. E. Canfield. 1989. A new model for atmospheric oxygen over Phanerozoic time. American Journal of Science 289:333–361.

Berner, R. A., and S. Honjo. 1981. Pelagic sedimentation of aragonite—its geochemical significance. Science 211:940–942.

Berner, R. A., and Z. Kothavala. 2001. GEOCARB III: A revised model of atmospheric CO_2 over Phanerozoic time. American Journal of Science 301:182–204.

Berner, R. A., and A. C. Lasaga. 1989. Modeling the geochemical carbon cycle. Scientific American 260: 74–81.

Berner, R. A., S. T. Petsch, J. A. Lake, D. J. Beerling, B. N. Popp, R. S. Lane, E. A. Laws, M. B. Westley, N. Cassar, F. I. Woodward, and W. P. Quick. 2000. Isotope fractionation and atmospheric oxygen: Implications for Phanerozoic O_2 evolution. Science 287:1630–1633.

Berner, R. A., and R. Raiswell. 1983. Burial of organic carbon and pyrite sulfur in sediments over Phanerozoic time—a new theory. Geochimica et Cosmochimica Acta 47:855–862.

Berner, R. A., and J.-L. Rao. 1997. Alkalinity buildup during silicate weathering under a snow cover. Aquatic Geochemistry 2:301–312.

Berner, R. A., and J. L. Rao. 1994. Phosphorus in sediments of the Amazon River and estuary—Impilcations for the global flux of phosphorus to the sea. Geochimica et Cosmochimica Acta 58:2333–2339.

Bernhardt, E. S. 2002. Lessons from kinetic releases of ammonium in streams of the Hubbard Brook Experimental Forest (HBEF). Verhandlungen der Internationalen Vereingung fur theoretische und angewandte Limnologie 28:429–433.

Bernhardt, E. S., J. J. Barber, J. S. Pippen, L. Taneva, J. A. Andrews, and W. H. Schlesinger. 2006. Long-term effects of Free Air CO_2 Enrichment (FACE) on soil respiration. Biogeochemistry 77:91–116.

Bernhardt, E. S., R. O. Hall, and G. E. Likens. 2002. Whole-system estimates of nitrification and nitrate uptake in streams of the Hubbard Brook Experimental Forest. Ecosystems 5:419–430.

Bernhardt, E. S., and G. E. Likens. 2002. Dissolved organic carbon enrichment alters nitrogen dynamics in a forest stream. Ecology 83:1689–1700.

Bernhardt, E. S., and G. E. Likens. 2004. Controls on periphyton biomass in heterotrophic streams. Freshwater Biology 49:14–27.

Bernhardt, E. S., G. E. Likens, D. C. Buso, and C. T. Driscoll. 2003. In-stream uptake dampens effect of major forest disturbance on watershed nitrogen export. Proceedings of the National Academy of Science 100:10304–10308.

Bernhardt, E. S., G. E. Likens, R. O. Hall, D. C. Buso, S. G. Fisher, T. M. Burton, J. L. Meyer, M. H. McDowell, M. S. Mayer, W. B. Bowden, S.E.G. Findlay, K. H. Macneale, R. S. Stelzer, and W. H. Lowe. 2005. Can't see the forest for the stream?—In-stream processing and terrestrial nitrogen exports. BioScience 55:219–230.

Bernhardt, E. S., and W. H. McDowell. 2008. Twenty years apart: Comparisons of DOM uptake during leaf leachate releases to Hubbard Brook Valley streams in 1979 versus 2000. Journal of Geophysical Research—Biogeosciences 113:G03032.

Bernier, B., and M. Brazeau. 1988a. Foliar nutrient status in relation to sugar maple dieback and decline in the Quebec Appalachians. Canadian Journal of Forest Research 18:754–761.

Bernier, B., and M. Brazeau. 1988b. Magnesium-deficiency symptoms associated with sugar maple dieback in a lower Laurentians site in southeastern Quebec. Canadian Journal of Forest Research 18:1265–1269.

Bernot, M. J., D. J. Sobota, R. O. Hall, P. J. Mulholland, W. K. Dodds, J. R. Webster, J. L. Tank, L. R. Ashkenas, L. W. Cooper, C. N. Dahm, S. V. Gregory, N.B. Grimm, S. K. Hamilton, S. L. Johnson, W. H. McDowell, J. L. Meyer, B. Peterson, G. C. Poole, H. M. Valett, C. Arango, J. J. Beaulieu, A. J. Burgin, C. Crenshaw, A. M. Helton, L. Johnson, J. Merriam, B. R. Niederlehner, J. M. O'Brien, J. D. Potter, R. W. Sheibley, S. M. Thomas, and K. Wilson. 2010. Inter-regional comparison of land-use effects on stream metabolism. Freshwater Biology 55:1874–1890.

Bernstein, R. E., P. R. Betzer, R. A. Feely, R. H. Byrne, M. F. Lamb, and A. F. Michaels. 1987. Acantharian fluxes and strontium to chlorinity ratios in the North Pacific ocean. Science 237:1490–1494.

Bernstein, R. E., and R. H. Byrne. 2004. Acantharians and marine barite. Marine Chemistry 86:45–50.

Berresheim, H., and V. D. Vulcan. 1992. Vertical distributions of COS, CS_2, DMS and other sulfur compounds in a loblolly pine forest. Atmospheric Environment Part A—General Topics 26:2031–2036.

Berthrong, S. T., E. G. Jobbagy, and R. B. Jackson. 2009. A global meta-analysis of soil exchangeable cations, pH, carbon, and nitrogen with afforestation. Ecological Applications 19:2228–2241.

Bertilsson, S., and L. J. Tranvik. 2000. Photochemical transformation of dissolved organic matter in lakes. Limnology and Oceanography 45:753–762.

Bertine, K. K., and E. D. Goldberg. 1971. Fossil fuel combustion and the major sedimentary cycle. Science 173:233–235.

Bertram, T. H., A. E. Perring, P. J. Wooldridge, J. D. Crounse, A. J. Kwan, P. O. Wennberg, E. Scheuer, J. Dibb, M. Avery, G. Sachse, S. A. Vay, J. H. Crawford, C. S. - McNaughton, A. Clarke, K. E. Pickering, H. Fuelberg, G. Huey, D. R. Blake, H. B. Singh, S. R. Hall, R. E. Shetter, A. Fried, B. G. Heikes, and R. C. Cohen. 2007. Direct measurements of the convective recycling of the upper troposphere. Science 315:816–820.

Bertrand, P., and E. Lallier-Vergès. 1993. Past sedimentary organic matter accumulation and degradation controlled by productivity. Nature 364:786–788.

Betlach, M. R. 1982. Evolution of bacterial denitrification and denitrifier diversity. Antonie Van Leeuwenhoek Journal of Microbiology 48:585–607.

Betts, J. N., and H. D. Holland. 1991. The oxygen content of ocean bottom waters, the burial efficiency of organic carbon, and the regulation of atmospheric oxygen. Palaeogeography Palaeoclimatology Palaeoecology 97: 5–18.

Betts, R. A., O. Boucher, M. Collins, P. M. Cox, P. D. Falloon, N. Gedney, D. L. Hemming, C. Huntingford, C. D. Jones, D.M.H. Sexton, and M. J. Webb. 2007. Projected increase in continental runoff due to plant responses to increasing carbon dioxide. Nature 448: 1037–1041.

Betzer, P. R., R. H. Byrne, J. G. Acker, C. S. Lewis, R. R. Jolley, and R. A. Feely. 1984. The oceanic carbonate system—A reassessment of biogenic control. Science 226:1074–1077.

Beyer, L., H. R. Schulten, R. Fruend, and U. Irmler. 1993. Formation and properties of organic matter in a forest soil, as revealed by its biological activity, wet chemical analysis, CPMAS C-13 NMR spectroscopy and pyrolysis field ionization mass spectrometery. Soil Biology & Biochemistry 25:587–596.

Beyersdorf, A. J., D. R. Blake, A. Swanson, S. Meinardi, F. S. Rowland, and D. Davis. 2010. Abundances and variability of tropospheric volatile organic compounds at the South Pole and other Antarctic locations. Atmospheric Environment 44:4565–4574.

Bhattacharyya, S., and R. K. Das. 1999. Catalytic control of automotive NOx: A review. International Journal of Energy Research 23:351–369.

Bhatti, J. S., N. B. Comerford, and C. T. Johnston. 1998. Influence of oxalate and soil organic matter on sorption and desorption of phosphate onto a spodic horizon. Soil Science Society of America Journal 62:1089–1095.

Bianchi, D., J. P. Dunne, J. L. Sarmiento, and E. D. Galbraith. 2012. Data-based estimates of suboxia, denitrification, and N_2O production in the ocean and their sensitivities to dissolved O_2. Global Biogeochemical Cycles 26.

Bianchi, T. S., and M. A. Allison. 2009. Large-river delta-front estuaries as natural "recorders" of global environmental change. Proceedings of the National Academy of Sciences 106:8085–8092.

Biddanda, B., S. Opsahl, and R. Benner. 1994. Plankton respiration and carbon flux through bacterioplankton on the Louisiana shelf. Limnology and Oceanography 39:1259–1275.

Bidle, K. D., and F. Azam. 1999. Accelerated dissolution of diatom silica by marine bacterial assemblages. Nature 397:508–512.

Bidle, K. D., M. Manganelli, and F. Azam. 2002. Regulation of oceanic silicon and carbon preservation by temperature control on bacteria. Science 298: 1980–1984.

Biemann, K., T. Owen, D. R. Rushneck, A. L. Lafleur, and D. W. Howarth. 1976. Atmosphere of Mars' near-surface—Isotope ratios and upper limits on noble gases. Science 194:76–78.

Biemans, H., I. Haddeland, P. Kabat, F. Ludwig, R.W.A. Hutjes, J. Heinke, W. von Bloh, and D. Gerten. 2011. Impact of reservoirs on river discharge and irrigation water supply during the 20th century. Water Resources Research 47.

Biggs, B.J.F. 1996. Patterns of benthic algae in streams. In R. J. Stevenson, M. L. Bothwell and R. L. Lowe, editors. Algal Ecology, Academic Press, 31–56.

Bilby, R. E. 1981. Role of organic debris dams in regulating the export of dissolved and particulate matter from a forested watershed. Ecology 62:1234–1243.

Billen, G. 1975. Nitrification in Scheldt estuary (Belgium and Netherlands). Estuarine and Coastal Marine Science 3:79–89.

Billings, C. D., and W. R. Matson. 1972. Mercury emissions from coal combustion. Science 176: 1232–1233.

Billings, S. A., and D. D. Richter. 2006. Changes in stable isotopic signatures of soil nitrogen and carbon during 40 years of forest development. Oecologia 148:325–333.

Billings, S. A., S. M. Schaeffer, and R. D. Evans. 2003. Nitrogen fixation by biological soil crusts and heterotrophic bacteria in an intact Mojave Desert ecosystem with elevated CO_2 and added soil carbon. Soil Biology & Biochemistry 35:643–649.

Billings, W. D. 1987. Carbon balance of Alaskan tundra and taiga ecosystems—Past, present and future. Quaternary Science Reviews 6:165–177.

Billings, W. D., J. O. Luken, D. A. Mortensen, and K. M. Peterson. 1982. Arctic tundra—A source or sink for atmospheric carbon dioxide in a changing environment. Oecologia 53:7–11.

Binford, M. W., H. L. Gholz, G. Starr, and T. A. Martin. 2006. Regional carbon dynamics in the southeastern U.S. coastal plain: Balancing land cover type, timber harvesting, fire, and environmental variation. Journal of Geophysical Research—Atmospheres 111.

Binkley, D. 1986. *Forest Nutrition Management*. Wiley.

Binkley, D., and S. C. Hart. 1989. The components of nitrogen availability assessments in forest soils. Advances in Soil Science 10:57–112.

Binkley, D., and D. Richter. 1987. Nutrient cycles and H$^+$ budgets of forest ecosystems. Advances in Ecological Research 16:1–51.

Bintanja, R., R.S.W. van de Wal, and J. Oerlemans. 2005. Modelled atmospheric temperatures and global sea levels over the past million years. Nature 437: 125–128.

Bird, M. I., J. Lloyd, and G. D. Farquhar. 1994. Terrestrial carbon storage at the LGM. Nature 371:566.

Birdsey, R. A., A. J. Plantinga, and L. S. Heath. 1993. Past and prospective carbon storage in United States' forests. Forest Ecology and Management 58:33–40.

Birk, E. M., and P. M. Vitousek. 1986. Nitrogen availability and nitrogen-use efficiency in loblolly pine stands. Ecology 67:69–79.

Birkeland, P. W. 1984. *Soils and Geomorphology*, 2nd. edition. Oxford.

Biscaye, P. E., V. Kolla, and K. K. Turekian. 1976. Distribution of calcium carbonate in surface sediments of Atlantic ocean. Journal of Geophysical Research—Oceans and Atmospheres 81:2595–2603.

Bishop, J.K.B. 1988. The barite-opal-organic carbon association in oceanic particulate matter. Nature 332: 341–343.

Bishop, J.K.B., R. E. Davis, and J. T. Sherman. 2002. Robotic observations of dust storm enhancement of carbon biomass in the North Pacific. Science 298: 817–821.

Bjerrum, C. J., and D. E. Canfield. 2002. Ocean productivity before about 1.9 Gyr ago limited by phosphorus adsorption onto iron oxides. Nature 417:159–162.

Black, F. J., B. A. Poulin, and A. R. Flegal. 2012. Factors controlling the abiotic photo-degradatioon of monomethylmercury in surface waters. Geochimica et Cosmochimica Acta 84:492–507.

Black, R. A., J. H. Richards, and J. H. Manwaring. 1994. Nutrient uptake from enriched soil microsites by three Great Basin perennials. Ecology 75:110–122.

Blackard, J. A., M. V. Finco, E. H. Helmer, G. R. Holden, M. L. Hoppus, D. M. Jacobs, A. J. Lister, G. G. Moisen, M. D. Nelson, R. Riemann, B. Ruefenacht, D. Salajanu, D. L. Weyermann, K. C. Winterberger, T. J. Brandeis, R. L. Czaplewski, R. E. McRoberts, P. L. Patterson, and R. P. Tymcio. 2008. Mapping U.S. forest biomass using nationwide forest inventory data and moderate resolution information. Remote Sensing of Environment 112:1658–1677.

Blain, S., B. Queguiner, L. Armand, S. Belviso, B. Bombled, L. Bopp, A. Bowie, C. Brunet, C. Brussaard, F. Carlotti, U. Christaki, A. Corbiere, I. Durand, F. Ebersbach, J. L. Fuda, N. Garcia, L. Gerringa, B. Griffiths, C. Guigue, C. Guillerm, S. Jacquet, C. Jeandel, P. Laan, D. Lefevre, C. Lo Monaco, A. Malits, J. Mosseri, I. Obernosterer, Y. H. Park, M. Picheral, P. Pondaven, T. Remenyi, V. Sandroni, G. Sarthou, N. Savoye, L. Scouarnec, M. Souhaut, D. Thuiller, K. Timmermans, T. Trull, J. Uitz, P. van Beek, M. Veldhuis, D. Vincent, E. Viollier, L. Vong, and T. Wagener. 2007. Effect of natural iron fertilization on carbon sequestration in the Southern Ocean. Nature 446:1070–1074.

Blair, N. E., and R. C. Aller. 1995. Anaerobic methane oxidation on the Amazon shelf. Geochimica et Cosmochimica Acta 59:3707–3715.

Blair, N. E., and R. C. Aller. 2012. The fate of terrestrial organic carbon in the marine environment. Annual Review of Marine Science 4:401–423.

Blaise, T., and J. Garbaye. 1983. Effects of mineral fertilization on the mycorrhization of roots in a beech forest. Acta Oecologica-Oecologia Plantarum 4:165–169.

Blake, L., K.W.T. Goulding, C.J.B. Mott, and A. E. Johnston. 1999. Changes in soil chemistry accompanying acidification over more than 100 years under woodland and grass at Rothamsted Experimental Station, UK. European Journal of Soil Science 50:401–412.

Blake, R. E., S. J. Chang, and A. Lepland. 2010. Phosphate oxygen isotopic evidence for a temperate and biologically active Archaean ocean. Nature 464:1029–1032.

Blatt, H., and R. L. Jones. 1975. Proportions of exposed igneous, metamorphic, and sedimentary rocks. Geological Society of America Bulletin 86:1085–1088.

Blei, E., C. J. Hardacre, G. P. Mills, K. V. Heal, and M. R. Heal. 2010. Identification and quantification of methyl halide sources in a lowland tropical rainforest. Atmospheric Environment 44:1005–1010.

Blodau, C., N. Basiliko, and T. R. Moore. 2004. Carbon turnover in peatland mesocosms exposed to different water table levels. Biogeochemistry 67:331–351.

Blomqvist, S., A. Gunnars, and R. Elmgren. 2004. Why the limiting nutrient differs between temperate coastal seas and freshwater lakes: A matter of salt. Limnology and Oceanography 49:2236–2241.

Blonquist, J. M., Jr., S. A. Montzka, J. W. Munger, D. Yakir, A. R. Desai, D. Dragoni, T. J. Griffis, R. K. Monson, R. L. Scott, and D. R. Bowling. 2011. The potential of carbonyl sulfide as a proxy for gross primary production at flux tower sites. Journal of Geophysical Research—Biogeosciences **116**.

Bloom, A. A., P. I. Palmer, A. Fraser, D. S. Reay, and C. Frankenberg. 2010. Large-scale controls of methanogenesis inferred from methane and gravity spaceborne data. Science **327**:322–325.

Bloom, A. J., L. E. Jackson, and D. R. Smart. 1993. Root growth as a function of ammonium and nitrate in the root zone. Plant Cell and Environment **16**:199–206.

Bloom, A. J., S. S. Sukrapanna, and R. L. Warner. 1992. Root respiration associated with ammonium and nitrate absorption and assimilation by barley. Plant Physiology **99**:1294–1301.

Bloomfield, C. 1972. The oxidation of iron sulphides in soils in relation to the formation of acid sulphate soils, and of ochre deposits in field drains. Journal of Soil Science **23**:1–16.

Blum, J. D., A. Klaue, C. A. Nezat, C. T. Driscoll, C. E. Johnson, T. G. Siccama, C. Eagar, T. J. Fahey, and G. E. Likens. 2002. Mycorrhizal weathering of apatite as an important calcium source in base-poor forest ecosystems. Nature **417**:729–731.

Blunier, T., J. Chappellaz, J. Schwander, B. Stauffer, and D. Raynaud. 1995. Variations in atmospheric methane concentration during the Holocene epoch. Nature **374**:46–49.

Bluth, G.J.S., and L. R. Kump. 1994. Lithologic and climatologic controls of river chemistry. Geochimica et Cosmochimica Acta **58**:2341–2359.

Bluth, G.J.S., C. C. Schnetzler, A. J. Krueger, and L. S. Walter. 1993. The contribution of explosive volcanism to global atmospheric sulphur dioxide concentrations. Nature **366**:327–329.

Bobbink, R., K. Hicks, J. Galloway, T. Spranger, R. Alkemade, M. Ashmore, M. Bustamante, S. Cinderby, E. Davidson, F. Dentener, B. Emmett, J. W. Erisman, M. Fenn, F. Gilliam, A. Nordin, L. Pardo, and W. De Vries. 2010. Global assessment of nitrogen deposition effects on terrestrial plant diversity: a synthesis. Ecological Applications **20**:30–59.

Bockheim, J. G. 1980. Solution and use of chronofunctions in studying soil development. Geoderma **24**:71–85.

Boerner, R.E.J. 1984. Foliar nutrient dynamics and nutrient-use efficiency of four deciduous tree species in relation to site fertility. Journal of Applied Ecology **21**:1029–1040.

Boesch, D. F. 2002. Challenges and opportunities for science in reducing nutrient over-enrichment of coastal ecosystems. Estuaries **25**:886–900.

Boesch, D. F., R. B. Brinsfield, and R. E. Magnien. 2001. Chesapeake Bay eutrophication: Scientific understanding, ecosystem restoration, and challenges for agriculture. Journal of Environmental Quality **30**:303–320.

Boetius, A., K. Ravenschlag, C. J. Schubert, D. Rickert, F. Widdel, A. Gieseke, R. Amann, B. B. Jorgensen, U. Witte, and O. Pfannkuche. 2000. A marine microbial consortium apparently mediating anaerobic oxidation of methane. Nature **407**:623–626.

Boettcher, S. E., and P. J. Kalisz. 1990. Single-tree influence on soil properties in the mountains of eastern Kentucky. Ecology **71**:1365–1372.

Bogner, J., and E. Matthews. 2003. Global methane emissions from landfills: New methodology and annual estimates 1980-1996. Global Biogeochemical Cycles **17**.

Bolan, N. S. 1991. A critical review on the role of mycorrhizal fungi in the uptake of phosphorus by plants. Plant and Soil **134**:189–207.

Bolan, N. S., A. D. Robson, N. J. Barrow, and L.A.G. Aylmore. 1984. Specific activity of phosphorus in mycorrhizal and non-mycorrhizal plants in relation to the availability of phosphorus to plants. Soil Biology & Biochemistry **16**:299–304.

Bolan, N. S., S. Saggar, J. F. Luo, R. Bhandral, and J. Singh. 2004. Gaseous emissions of nitrogen from grazed pastures: Processes, measurements and modelling, environmental implications, and mitigation. Advances in Agronomy **84**:37–120.

Bollmann, A., and R. Conrad. 1998. Influence of O_2 availability on NO and N_2O release by nitrification and denitrification in soils. Global Change Biology **4**:387–396.

Bonanno, A., H. Schlattl, and L. Paterno. 2002. The age of the Sun and the relativistic corrections in the EOS. Astronomy & Astrophysics **390**:1115–1118.

Bond-Lamberty, B., S. D. Peckham, D. E. Ahl, and S. T. Gower. 2007. Fire as the dominant driver of central Canadian boreal forest carbon balance. Nature **450**:89–92.

Bond-Lamberty, B., and A. Thomson. 2010. Temperature-associated increases in the global soil respiration record. Nature **464**:579–582.

Bond-Lamberty, B., C. K. Wang, and S. T. Gower. 2004. A global relationship between the heterotrophic and autotrophic components of soil respiration? Global Change Biology **10**:1756–1766.

Bondarenko, N. V., J. W. Head, and M. A. Ivanov. 2010. Present-day volcanism on Venus: Evidence from microwave radiometry. Geophysical Research Letters **37**.

Bondietti, E. A., C. F. Baes, and S. B. McLaughlin. 1989. Radial trends in cation ratios in tree rings as indicators of the impact of atmospheric deposition on forests. Canadian Journal of Forest Research **19**:586–594.

Bonell, M. 1993. Progress in the understanding of runoff generation dynamics in forests. Journal of Hydrology **150**:217–275.

Bonin, P., M. Gilewicz, and J. C. Bertrand. 1989. Effects of oxygen on each step of denitrification on *Pseudomonas nautica*. Canadian Journal of Microbiology **35**:1061–1064.

Bono, P., and C. Boni. 1996. Water supply of Rome in antiquity and today. Environmental Geology **27**:126–134.

Booth, M. S., J. M. Stark, and E. Rastetter. 2005. Controls on nitrogen cycling in terrestrial ecosystems: A synthetic analysis of literature data. Ecological Monographs **75**:139–157.

Bopp, L., C. Le Quere, M. Heimann, A. C. Manning, and P. Monfray. 2002. Climate-induced oceanic oxygen fluxes: Implications for the contemporary carbon budget. Global Biogeochemical Cycles **16**.

Borges, A. V. 2005. Do we have enough pieces of the jigsaw to integrate CO_2 fluxes in the coastal ocean? Estuaries **28**:3–27.

Borges, A. V., B. Delille, and M. Frankignoulle. 2005. Budgeting sinks and sources of CO_2 in the coastal ocean: Diversity of ecosystems counts. Geophysical Research Letters **32**:4.

Boring, L. R., C. D. Monk, and W. T. Swank. 1981. Early regeneration of a clear-cut southern Appalachian forest. Ecology **62**:1244–1253.

Boring, L. R., W. T. Swank, J. B. Waide, and G. S. Henderson. 1988. Sources, fates, and impacts of nitrogen inputs to terrestrial ecosystems—Review and synthesis. Biogeochemistry **6**:119–159.

Bormann, B. T., and J. C. Gordon. 1984. Stand density effects in young red alder plantations—Productivity, photosynthate partitioning, and nitrogen fixation. Ecology **65**:394–402.

Bormann, F. H., G. E. Likens, T. G. Siccama, R. S. Pierce, and J. S. Eaton. 1974. The export of nutrients and recovery of stable conditions following deforestation at Hubbard Brook. Ecological Monographs **44**:255–277.

Born, M., H. Dorr, and I. Levin. 1990. Methane consumption in aerated soils of the temperate zone. Tellus **42B**:2–8.

Borucki, W. J., and W. L. Chameides. 1984. Lightning: Estimates of the rates of energy dissipation and nitrogen fixation. Reviews of Geophysics and Space Physics **22**:363–372.

Borucki, W. J., D. Koch, G. Basri, N. Batalha, T. Brown, D. Caldwell, J. Caldwell, J. Christensen-Dalsgaard, W. D. Cochran, E. DeVore, E. W. Dunham, A. K. Dupree, T. N. Gautier, J. C. Geary, R. Gilliland, A. Gould, S. B. Howell, J. M. Jenkins, Y. Kondo, D. W. Latham, G. W. Marcy, S. Meibom, H. Kjeldsen, J. J. Lissauer, D. G. Monet, D. Morrison, D. Sasselov, J. Tarter, A. Boss, D. Brownlee, T. Owen, D. Buzasi, D. Charbonneau, L. Doyle, J. Fortney, E. B. Ford, M. J. Holman, S. Seager, J. H. Steffen, W. F. Welsh, J. Rowe, H. Anderson, L. Buchhave, D. Ciardi, L. Walkowicz, W. Sherry, E. Horch, H. Isaacson, M. E. Everett, D. Fischer, G. Torres, J. A. Johnson, M. Endl, P. MacQueen, S. T. Bryson, J. Dotson, M. Haas, J. Kolodziejczak, J. Van Cleve, H. Chandrasekaran, J. D. Twicken, E. V. Quintana, B. D. Clarke, C. Allen, J. Li, H. Wu, P. Tenenbaum, E. Verner, F. Bruhweiler, J. Barnes, and A. Prsa. 2010. Kepler planet-detection mission: Introduction and first results. Science **327**:977–980.

Bosch, J. M., and J. D. Hewlett. 1982. A review of catchment experiments to determine the effect of vegetation changes on water yield and evapotranspiration. Journal of Hydrology **55**:3–23.

Boss, A. P. 1988. High temperatures in the early solar nebula. Science **241**:565–567.

Botch, M. S., K. I. Kobak, T. S. Vinson, and T. P. Kolchugina. 1995. Carbon pools and accumulation in peatlands of the former Soviet Union. Global Biogeochemical Cycles **9**:37–46.

Botkin, D. B., P. A. Jordan, A. S. Dominski, H. S. Lowendorf, and G. E. Hutchinson. 1973. Sodium dynamics in a northern ecosystem. Proceedings of the National Academy of Sciences, **70**:2745–2748.

Botkin, D. B., and C. R. Malone. 1968. Efficiency of net primary production based on light intercepted during the growing season. Ecology **49**:438–444.

Botkin, D. B., and L. G. Simpson. 1990. Biomass of the North American boreal forest—A step toward accurate global measures. Biogeochemistry **9**:161–174.

Botkin, D. B., L. G. Simpson, and R. A. Nisbet. 1993. Biomass and carbon storage of the North American deciduous forest. Biogeochemistry **20**:1–17.

Bott, T. L. 2006. Primary productivity and community respiration. *In* F. R. Hauer and G. A. Lamberti, editors. *Methods in Stream Ecology*. Elsevier, 663–690.

Bott, T. L., J. T. Brock, C. S. Dunn, R. J. Naimann, R. W. Ovink, and R. C. Peterson. 1985. Benthic community metabolism in four temperate stream systems: An inter-biome comparison and evaluation of the river continuum concept. Hydrobiologia **123**:3–45.

Bott, T. L., J. D. Newbold, and D. B. Arscott. 2006. Ecosystem metabolism in piedmont streams: Reach geomorphology modulates the influence of riparian vegetation. Ecosystems **9**:398–421.

Bottcher, J., D. Weymann, R. Well, C. von der Heide, A. Schwen, H. Flessa, and W.H.M. Duijnisveld. 2011. Emission of groundwater-derived nitrous oxide into the atmosphere: model simulations based on a ^{15}N-field experiment. European Journal of Soil Science **62**:216–225.

Bottke, W. F., D. Vokrouhlicky, D. Minton, D. Nesvorny, A. Morbidelli, R. Brasser, B. Simonson, and H. F. Levison. 2012. An Archaean heavy bombardment from a destabilized extension of the asteroid belt. Nature 485:78–81.

Bouchard, S. S., and K. A. Bjorndal. 2000. Sea turtles as biological transporters of nutrients and energy from marine to terrestrial ecosystems. Ecology 81: 2305–2313.

Boudreau, B. P., and J. T. Westrich. 1984. The dependence of bacterial sulfate reduction on sulfate concentration in marine sediments. Geochimica et Cosmochimica Acta 48:2503–2516.

Boudreau, J., R. F. Nelson, H. A. Margolis, A. Beaudoin, L. Guindon, and D. S. Kimes. 2008. Regional aboveground forest biomass using airborne and spaceborne LiDAR in Quebec. Remote Sensing of Environment 112:3876–3890.

Bousquet, P., P. Ciais, J. B. Miller, E. J. Dlugokencky, D. A. Hauglustaine, C. Prigent, G. R. Van der Werf, P. Peylin, E. G. Brunke, C. Carouge, R. L. Langenfelds, J. Lathiere, F. Papa, M. Ramonet, M. Schmidt, L. P. Steele, S. C. Tyler, and J. White. 2006. Contribution of anthropogenic and natural sources to atmospheric methane variability. Nature 443:439–443.

Boutron, C. F., J. P. Candelone, and S. M. Hong. 1994. Past and recent changes in the large-scale tropospheric cycles of lead and other heavy metals as documented in Antarctic and Greenland snow and ice—A review. Geochimica et Cosmochimica Acta 58:3217–3225.

Boutron, C. F., U. Gorlach, J. P. Candelone, M. A. Bolshov, and R. J. Delmas. 1991. Decrease in anthropogenic lead, cadmium and zinc in Greenland snows since the late 1960s. Nature 353:153–156.

Bouvier, A., and M. Wadhwa. 2010. The age of the Solar System redefined by the oldest Pb-Pb age of a meteoritic inclusion. Nature Geoscience 3:637–641.

Bouwman, A. F., L.J.M. Boumans, and N. H. Batjes. 2002a. Emissions of N$_2$O and NO from fertilized fields: Summary of available measurement data. Global Biogeochemical Cycles 16.

Bouwman, A. F., L.J.M. Boumans, and N. H. Batjes. 2002b. Estimation of global NH$_3$ volatilization loss from synthetic fertilizers and animal manure applied to arable lands and grasslands. Global Biogeochemical Cycles 16.

Bouwman, A. F., L.J.M. Boumans, and N. H. Batjes. 2002c. Modeling global annual N$_2$O and NO emissions from fertilized fields. Global Biogeochemical Cycles 16.

Bouwman, A. F., I. Fung, E. Matthews, and J. John. 1993. Global analysis of the potential for N$_2$O production in natural soils. Global Biogeochemical Cycles 7:557–597.

Bouwman, A. F., D. S. Lee, W.A.H. Asman, F. J. Dentener, K. W. VanderHoek, and J.G.J. Olivier. 1997. A global high-resolution emission inventory for ammonia. Global Biogeochemical Cycles 11:561–587.

Bouwman, A. F., K. W. Vanderhoek, and J.G.J. Olivier. 1995. Uncertainties in the global source distribution of nitrous oxide. Journal of Geophysical Research—Atmospheres 100:2785–2800.

Bowden, R. D., K. J. Nadelhoffer, R. D. Boone, J. M. Melillo, and J. B. Garrison. 1993. Contributions of aboveground litter, belowground litter, and root respiration to total soil respiration in a temperate mixed hardwood forest. Canadian Journal of Forest Research 23:1402–1407.

Bowden, W. B. 1986. Gaseous nitrogen emissions from undisturbed terrestrial ecosystems—An assessment of their impacts on local and global nitrogen budgets. Biogeochemistry 2:249–279.

Bowen, G. D., and S. E. Smith. 1981. The effects of mycorrhizas on nitrogen uptake by plants. In F. E. Clark and T. Rosswall, editors. Terrestrial Nitrogen Cycles. Swedish Natural Science Research Council, Stockholm, 237–247.

Bower, A. S., M. S. Lozier, S. F. Gary, and C. W. Boning. 2009. Interior pathways of the North Atlantic meridional overturning circulation. Nature 459:243–247.

Bowman, C. T. 1991. The chemistry of gaseous pollutant formation and destruction. In W. Bartok and A. F. Sarofim, editors. Fossil Fuel Combustion: A Source Book, Wiley, 215–260.

Bowman, C. T. 1992. Control of combustion-generated nitrogen oxide emissions. Technology driven by regulation. In Twenty-Fourth Symposium (International) on Combustion. The Combustion Institute, Pittsburgh, 859–878.

Bowman, D., J. K. Balch, P. Artaxo, W. J. Bond, J. M. Carlson, M. A. Cochrane, C. M. D'Antonio, R. S. DeFries, J. C. Doyle, S. P. Harrison, F. H. Johnston, J. E. Keeley, M. A. Krawchuk, C. A. Kull, J. B. Marston, M. A. Moritz, I. C. Prentice, C. I. Roos, A. C. Scott, T.W. Swetnam, G. R. van der Werf, and S. J. Pyne. 2009. Fire in the Earth System. Science 324: 481–484.

Bowring, S. A., and T. Housh. 1995. The Earth's early evolution. Science 269:1535–1540.

Box, E. O. 1988. Estimating the seasonal carbon source-sink geography of a natural, steady-state terrestrial biosphere. Journal of Applied Meteorology 27: 1109–1124.

Box, E. O., B. N. Holben, and V. Kalb. 1989. Accuracy of the AVHRR vegetation index as a predictor of biomass, primary productivity and net CO$_2$ flux. Vegetation 80:71–89.

Boyd, E. S., A. D. Anbar, S. Miller, T. L. Hamilton, M. Lavin, and J. W. Peters. 2011. A late methanogen origin for molybdenum-dependent nitrogenase. Geobiology 9:221–232.

Boyd, P. W., T. Jickells, C. S. Law, S. Blain, E. A. Boyle, K. O. Buesseler, K. H. Coale, J. J. Cullen, H.J.W. de Baar, M. Follows, M. Harvey, C. Lancelot, M. Levasseur, N.P.J. Owens, R. Pollard, R. B. Rivkin, J. Sarmiento, V. Schoemann, V. Smetacek, S. Takeda, A. Tsuda, S. Turner, and A. J. Watson. 2007. Mesoscale iron enrichment experiments 1993-2005: Synthesis and future directions. Science 315:612–617.

Boyd, P. W., A. J. Watson, C. S. Law, E. R. Abraham, T. Trull, R. Murdoch, D.C.E. Bakker, A. R. Bowie, K. O. Buesseler, H. Chang, M. Charette, P. Croot, K. Downing, R. Frew, M. Gall, M. Hadfield, J. Hall, M. Harvey, G. Jameson, J. LaRoche, M. Liddicoat, R. Ling, M. T. Maldonado, R. M. McKay, S. Nodder, S. Pickmere, R. Pridmore, S. Rintoul, K. Safi, P. Sutton, R. Strzepek, K. Tanneberger, S. Turner, A. Waite, and J. Zeldis. 2000. A mesoscale phytoplankton bloom in the polar Southern Ocean stimulated by iron fertilization. Nature 407: 695–702.

Boyer, E. W., C. L. Goodale, N. A. Jaworsk, and R. W. Howarth. 2002. Anthropogenic nitrogen sources and relationships to riverine nitrogen export in the northeastern United States. Biogeochemistry 57: 137–169.

Boyer, E. W., R. W. Howarth, J. N. Galloway, F. J. Dentener, P. A. Green, and C. J. Vörösmarty. 2006. Riverine nitrogen export from the continents to the coasts. Global Biogeochemical Cycles 20.

Boyer, T. P., S. Levitus, J. I. Antonov, R. A. Locarnini, and H. E. Garcia. 2005. Linear trends in salinity for the World Ocean, 1955-1998. Geophysical Research Letters 32.

Boyle, E., R. Collier, A. T. Dengler, J. M. Edmond, A. C. Ng, and R. F. Stallard. 1974. Chemical mass-balance in estuaries. Geochimica et Cosmochimica Acta 38:1719–1728.

Boyle, E. A., J. M. Edmond, and E. R. Sholkovitz. 1977. Mechanism of iron removal in estuaries. Geochimica et Cosmochimica Acta 41:1313–1324.

Boyle, E. A., F. Sclater, and J. M. Edmond. 1976. Marine geochemistry of cadmium. Nature 263:42–44.

Boyle, J. R., and G. K. Voigt. 1973. Biological weathering of silicate minerals. Implications for tree nutrition and soil genesis. Plant and Soil 38:191–201.

Boynton, W. V., D. W. Ming, S. P. Kounaves, S.M.M. Young, R. E. Arvidson, M. H. Hecht, J. Hoffman, P. B. Niles, D. K. Hamara, R. C. Quinn, P. H. Smith, B. Sutter, D. C. Catling, and R. V. Morris. 2009. Evidence for calcium carbonate at the Mars Phoenix landing site. Science 325:61–64.

Bradford, J. B., W. K. Lauenroth, and I. C. Burke. 2005. The impact of cropping on primary production in the US Great Plains. Ecology 86:1863–1872.

Bradley, R. S., H. F. Diaz, J. K. Eischeid, P. D. Jones, P. M. Kelly, and C. M. Goodess. 1987. Precipitation fluctuations over Northern Hemisphere land areas since the mid-19th century. Science 237:171–175.

Bragazza, L., C. Freeman, T. Jones, H. Rydin, J. Limpens, N. Fenner, T. Ellis, R. Gerdol, M. Hjek, T. Hjek, P. Iacumin, L. Kutnar, T. Tahvanainen, and H. Toberman. 2006. Atmospheric nitrogen deposition promotes carbon loss from peat bogs. Proceedings of the National Academy of Sciences 103:19386–19389.

Braissant, O., G. Cailleau, M. Aragno, and E. P. Verrecchia. 2004. Biologically induced mineralization in the tree Milicia excelsa (Moraceae): its causes and consequences to the environment. Geobiology 2: 59–66.

Bramley, R.G.V., and R. E. White. 1990. The variability of nitrifying activity in field soils. Plant and Soil 126: 203–208.

Brand, L. E. 1991. Minimum iron requirements of marine phytoplankton and the implications for the biogeochemical control of new production. Limnology and Oceanography 36:1756–1771.

Brand, L. E., W. G. Sunda, and R.R.L. Guillard. 1983. Limitation of marine phytoplankton reproductive rates by zinc, manganese, and iron. Limnology and Oceanography 28:1182–1198.

Brand, U. 1989. Biogeochemistry of Late Paleozoic North American brachiopods and secular variation of seawater composition. Biogeochemistry 7:159–193.

Brandes, J. A., and A. H. Devol. 2002. A global marine fixed-nitrogen isotopic budget: Implications for Holocene nitrogen cycling. Global Biogeochemical Cycles 16.

Brasier, M., O. Green, J. Lindsay, and A. Steele. 2004. Earth's oldest (~3.5 Ga) fossils and the 'Early Eden hypothesis': Questioning the evidence. Origins of Life and Evolution of the Biosphere 34:257–269.

Brasseur, G., and M. H. Hitchman. 1988. Stratospheric response to trace gas perturbations—Changes in ozone and temperature distributions. Science 240: 634–637.

Brauman, A., M. D. Kane, M. Labat, and J. A. Breznak. 1992. Genesis of acetate and methane by gut bacteria of nutritionally diverse termites. Science 257:1384–1387.

Bravard, S., and D. Righi. 1989. Geochemical differences in an Oxisol-Spodosol toposequence of Amazonia, Brazil. Geoderma 44:29–42.

Brenner, D. L., R. Amundson, W. T. Baisden, C. Kendall, and J. Harden. 2001. Soil N and N-15 variation with time in a California annual grassland ecosystem. Geochimica et Cosmochimica Acta 65:4171–4186.

Brett, M. T., and C. R. Goldman. 1996. A meta-analysis of the freshwater trophic cascade. Proceedings of the National Academy of Sciences 93:7723–7726.

Brewer, P. G., C. Goyet, and D. Dyrssen. 1989. Carbon dioxide transport by ocean currents at 25° N latitude in the Atlantic ocean. Science 246:477–479.

Breznak, J. A., W. J. Brill, J. W. Mertins, and H. C. Coppel. 1973. Nitrogen fixation in termites. Nature 244: 577–579.

Bridgham, S. D., J. P. Megonigal, J. K. Keller, N. B. Bliss, and C. Trettin. 2006. The carbon balance of North American wetlands. Wetlands 26:889–916.

Bridgham, S. D., and C. J. Richardson. 1992. Mechanisms controlling soil respiration (CO_2 and CH_4) in southern peatlands. Soil Biology & Biochemistry 24: 1089–1099.

Brierley, E.D.R., and M. Wood. 2001. Heterotrophic nitrification in an acid forest soil: isolation and characterisation of a nitrifying bacterium. Soil Biology & Biochemistry 33:1403–1409.

Briffa, K. R., P. D. Jones, F. H. Schweingruber, and T. J. Osborn. 1998. Influence of volcanic eruptions on Northern Hemisphere summer temperature over the past 600 years. Nature 393:450–455.

Brimblecombe, P., and G. A. Dawson. 1984. Wet removal of highly soluble gases. Journal of Atmospheric Chemistry 2:95–107.

Brimblecombe, P., C. Hammer, H. Rodhe, A. Ryaboshapko, and C. G. Boutron. 1989. Human influence on the sulphur cycle. In P. Brimblecombe and A. Y. Lein, editors. Evolution of the Global Biogeochemical Sulphur Cycle. Wiley, 77–121.

Brimhall, G. H., O. A. Chadwick, C. J. Lewis, W. Compston, I. S. Williams, K. J. Danti, W. E. Dietrich, M. E. Power, D. Hendricks, and J. Bratt. 1991. Deformational mass transport and invasive process in soil evolution. Science 255:695–702.

Brinkman, W. l., and U.D.M. Santos. 1974. Emission of biogenic hydrogen sulfide from Amazonian floodplain lakes. Tellus 26:261–267.

Brinson, M. M. 1993. Changes in the functioning of wetlands along environmental gradients. Wetlands 13: 65–74.

Brinson, M. M., A. E. Lugo, and J. Place. 1981. Primary productivity, decomposition and consumer activity in freshwater wetlands. Annual Review of Ecology and Systematics 12:123–161.

Bristow, C. S., K. A. Hudson-Edwards, and A. Chappell. 2010. Fertilizing the Amazon and equatorial Atlantic with West African dust. Geophysical Research Letters 37.

Brix, H., B. K. Sorrell, P. T. Orr, and T. Orr. 1992. Internal pressurization and convective gas flow in some emergent freshwater macrophytes. Limnology and Oceanography 37:1420–1433.

Brocks, J. J., G. A. Logan, R. Buick, and R. E. Summons. 1999. Archean molecular fossils and the early rise of eukaryotes. Science 285:1033–1036.

Broda, E. 1975. History of inorganic nitrogen in the biosphere. Journal of Molecular Evolution 7:87–100.

Broecker, W. S. 1973. Factors controlling CO_2 content in the oceans and atmosphere. In G. M. Woodwell and E. V. Pecan, editors. Carbon and the Biosphere, National Technical Information Service, Washington, D.C., 32–50.

Broecker, W. S. 1974. Chemical Oceanography. Hartcourt Brace Jovanovich.

Broecker, W. S. 1982. Ocean chemistry during glacial time. Geochimica et Cosmochimica Acta 46:1689–1705.

Broecker, W. S. 1985. How To Build a Habitable Planet. LaMont-Doherty Earth Observatory, Palisades.

Broecker, W. S. 1991. The great ocean conveyor. Oceanography 4:79–89.

Broecker, W. S., and T.-H. Peng. 1993. What caused the glacial to interglacial CO_2 change? In M. Heimann, editor. The Global Carbon Cycle. Springer, 95–115.

Bronk, D. A., P. M. Glibert, T. C. Malone, S. Banahan, and E. Sahlsten. 1998. Inorganic and organic nitrogen cycling in Chesapeake Bay: autotrophic versus heterotrophic processes and relationships to carbon flux. Aquatic Microbial Ecology 15:177–189.

Bronk, D. A., P. M. Glibert, and B. B. Ward. 1994. Nitrogen uptake, dissolved organic nitrogen release, and new production. Science 265:1843–1846.

Brook, G. A., M. E. Folkoff, and E. O. Box. 1983. A world model of soil carbon dioxide. Earth Surface Processes and Landforms 8:79–88.

Brookes, P. C., A. Landman, G. Pruden, and D. S. Jenkinson. 1985. Chloroform fumigation and the release of soil nitrogen—A rapid direct extraction method to measure microbial biomass nitrogen in soil. Soil Biology & Biochemistry 17:837–842.

Brookes, P. C., D. S. Powlson, and D. S. Jenkinson. 1982. Measurement of microbial biomass phosphorus in soil. Soil Biology & Biochemistry 14:319–329.

Brookes, P. C., D. S. Powlson, and D. S. Jenkinson. 1984. Phosphorus in the soil microbial biomass. Soil Biology & Biochemistry 16:169–175.

Brooks, R. R. 1973. Biogeochemical parameters and their significance for mineral exploraton. Journal of Applied Ecology 10:825–836.

Brookshire, E.N.J., S. Gerber, D.N.L. Menge, and L. O. Hedin. 2012. Large losses of inorganic nitrogen from tropical rainforests suggest a lack of nitrogen limitation. Ecology Letters 15:9–16.

Brookshire, E.N.J., H. M. Valett, S. A. Thomas, and J. R. Webster. 2005. Coupled cycling of dissolved organic nitrogen and carbon in a forest stream. Ecology 86:2487–2496.

Brown, A. E., L. Zhang, T. A. McMahon, A. W. Western, and R. A. Vertessy. 2005. A review of paired catchment studies for determining changes in water yield resulting from alterations in vegetation. Journal of Hydrology 310:28–61.

Brown, B. E., J. L. Fassbender, and R. Winkler. 1992. Carbonate production and sediment transport in a marl lake of southeastern Wisconsin. Limnology and Oceanography 37:184–191.

Brown, K., M. Tegoni, M. Prudencio, A. S. Pereira, S. Besson, J. J. Moura, I. Moura, and C. Cambillau. 2000. A novel type of catalytic copper cluster in nitrous oxide reductase. Nature Structural Biology 7:191–195.

Brown, R. H., R. N. Clark, B. J. Buratti, D. P. Cruikshank, J. W. Barnes, R.M.E. Mastrapa, J. Bauer, S. Newman, T. Momary, K. H. Baines, G. Bellucci, F. Capaccioni, P. Cerroni, M. Combes, A. Coradini, P. Drossart, V. Formisano, R. Jaumann, Y. Langevin, D. L. Matson, T. B. McCord, R. M. Nelson, P. D. Nicholson, B. Sicardy, and C. Sotin. 2006. Composition and physical properties of Enceladus' surface. Science 311:1425–1428.

Brown, S. 1981. A comparison of the structure, primary productivity, and transpiration of cypress ecosystems in Florida. Ecological Monographs 51:403–427.

Brown, S., and A. E. Lugo. 1982. The storage and production of organic matter in tropical forests and their role in the global carbon cycle. Biotropica 14:161–187.

Brown, S. S., T. B. Ryerson, A. G. Wollny, C. A. Brock, R. Peltier, A. P. Sullivan, R. J. Weber, W. P. Dube, M. Trainer, J. F. Meagher, F. C. Fehsenfeld, and A. R. Ravishankara. 2006. Variability in nocturnal nitrogen oxide processing and its role in regional air quality. Science 311:67–70.

Brown-Steiner, B., and P. Hess. 2011. Asian influence on surface ozone in the United States: A comparison of chemistry, seasonality, and transport mechanisms. Journal of Geophysical Research—Atmospheres 116.

Brownlee, D. E. 1992. The origin and evolution of the Earth. In S. S. Butcher, R. J. Charlson, G. H. Orians, and G. V. Wolfe, editors. Global Biogeochemical Cycles, Academic Press, 9–20.

Bruland, K. W. 1989. Complexation of zinc by natural organic ligands in the central North Pacific. Limnology and Oceanography 34:269–285.

Bruland, K. W., J. R. Donat, and D. A. Hutchins. 1991. Interactive influences of bioactive trace metals on biological production in oceanic waters. Limnology and Oceanography 36:1555–1577.

Bruland, K. W., G. A. Knauer, and J. H. Martin. 1978a. Cadmium in northeast Pacific waters. Limnology and Oceanography 23:618–625.

Bruland, K. W., G. A. Knauer, and J. H. Martin. 1978b. Zinc in northeast Pacific water. Nature 271:741–743.

Brumme, R., and F. Beese. 1992. Effects of liming and nitrogen-fertilization on emissions of CO_2 and N_2O from a temperate forest. Journal of Geophysical Research—Atmospheres 97:12851–12858.

Brunner, D., J. Staehelin, and D. Jeker. 1998. Large-scale nitrogen oxide plumes in the tropopause region and implications for ozone. Science 282:1305–1309.

Brutsaert, W. 2006. Indications of increasing land surface evaporation during the second half of the 20th century. Geophysical Research Letters 33.

Bryan, J. A., G. P. Berlyn, and J. C. Gordon. 1996. Toward a new concept of the evolution of symbiotic nitrogen fixation in the Leguminosae. Plant and Soil 186: 151–159.

Brylinski, M., and K. H. Mann. 1973. Analysis of factors governing productivity in lakes and reservoirs. Limnology and Oceanography 18:1–14.

Bubier, J. L. 1995. The relationship of vegetation to methane emission and hydrochemical gradients in northern peatlands. Journal of Ecology 83:403–420.

Bubier, J. L., T. R. Moore, and L. A. Bledzki. 2007. Effects of nutrient addition on vegetation and carbon cycling in an ombrotrophic bog. Global Change Biology 13: 1168–1186.

Bucher, M. 2007. Functional biology of plant phosphate uptake at root and mycorrhiza interfaces. New Phytologist 173:11–26.

Bucking, H., and Y. Shachar-Hill. 2005. Phosphate uptake, transport and transfer by the arbuscular mycorrhizal fungus Glomus intraradices is stimulated by increased carbohydrate availability. New Phytologist 165:899–912.

Buendia, C., A. Kleidon, and A. Porporato. 2010. The role of tectonic uplift, climate, and vegetation in the long-term terrestrial phosphorous cycle. Biogeosciences 7:2025–2038.

Buesseler, K. O., J. E. Andrews, S. M. Pike, and M. A. Charette. 2004. The effects of iron fertilization on carbon sequestration in the Southern Ocean. Science 304:414–417.

Buesseler, K. O., C. H. Lamborg, P. W. Boyd, P. J. Lam, T. W. Trull, R. R. Bidigare, J.K.B. Bishop, K. L. Casciotti, F. Dehairs, M. Elskens, M. Honda, D. M. Karl, D. A. Siegel, M. W. Silver, D. K. Steinberg, J. Valdes, B. Van Mooy, and S. Wilson. 2007. Revisiting carbon flux through the ocean's twilight zone. Science 316:567–570.

Bullen, T. D., and S. W. Bailey. 2005. Identifying calcium sources at an acid deposition-impacted spruce forest: a strontium isotope, alkaline earth element multitracer approach. Biogeochemistry 74:63–99.

Bulow, S. E., J. J. Rich, H. S. Naik, A. K. Pratihary, and B. B. Ward. 2010. Denitrification exceeds anammox as a nitrogen loss pathway in the Arabian Sea oxygen minimum zone. Deep-Sea Research Part I—Oceanographic Research Papers **57**:384-393.

Bundy, L. G., and J. M. Bremner. 1973. Inhibition of nitrification in soils. Soil Science Society of America Proceedings **37**:396–398.

Bunemann, E. K., P. Marschner, A. M. McNeill, and M. J. McLaughlin. 2007. Measuring rates of gross and net mineralisation of organic phosphorus in soils. Soil Biology & Biochemistry **39**:900–913.

Bunemann, E. K., A. Oberson, F. Liebisch, F. Keller, K. E. Anaheim, O. Huguenin-Elie, and E. Frossard. 2012. Rapid microbial phosphorus immobilization dominates gross phosphorus fluxes in a grassland soil with low inorganic phosphorus availability. Soil Biology and Biochemistry **51**:84–95.

Burbidge, E. M., G. E. Burbidge, W. A. Fowler, and F. Hoyle. 1957. Synthesis of the elements in stars. Reviews of Modern Physics **29**:547–650.

Burdige, D. J. 2005. Burial of terrestrial organic matter in marine sediments: A re-assessment. Global Biogeochemical Cycles **19**.

Burford, J. R., and J. M. Bremner. 1975. Relationships between denitrification capacities of soils and total, water-soluble and readily decomposable soil organic matter. Soil Biology & Biochemistry **7**:389–394.

Burgin, A. J., and P. M. Groffman. 2012. Soil O_2 controls denitrification rates and N_2O yield in a riparian wetland. Journal of Geophysical Research—Biogeosciences **117**.

Burgin, A. J., P. M. Groffman, and D. N. Lewis. 2010. Factors regulating denitrification in a riparian wetland. Soil Science Society of America Journal **74**:1826–1833.

Burgin, A. J., and S. K. Hamilton. 2007. Have we overemphasized the role of denitrification in aquatic ecosystems? A review of nitrate removal pathways. Frontiers in Ecology and the Environment **5**:89–96.

Burgin, A. J., and S. K. Hamilton. 2008. NO_3–driven SO_4^2- production in freshwater ecosystems: Implications for N and S cycling. Ecosystems **11**:908–922.

Burke, A., and L. F. Robinson. 2012. The Southern Ocean's role in carbon exchange during the last deglaciation. Science **335**:557–561.

Burke, M. K., and D. J. Raynal. 1994. Fine root growth phenology, production, and turnover in a northern hardwood forest ecosystem. Plant and Soil **162**:135–146.

Burke, R. A., D. F. Reid, J. M. Brooks, and D. M. Lavoie. 1983. Upper water column methane geochemistry in the eastern tropical North Pacific. Limnology and Oceanography **28**:19–32.

Burkhart, J. F., R. C. Bales, J. R. McConnell, and M.A. Hutterli. 2006. Influence of North Atlantic Oscillation on anthropogenic transport recorded in northwest Greenland ice cores. Journal of Geophysical Research—Atmospheres **111**.

Burnard, P., D. Graham, and G. Turner. 1997. Vesicle-specific noble gas analyses of "popping rock": Implications for primordial noble gases in earth. Science **276**:568–571.

Burnett, W. C., M. J. Beers, and K. K. Roe. 1982. Growth rates of phosphate nodules from the continental margin off Peru. Science **215**:1616–1618.

Burns, D. A., J. Klaus, and M. R. McHale. 2007. Recent climate trends and implications for water resources in the Catskill Mountain region, New York. Journal of Hydrology **336**:155–170.

Burns, D. A., and P. S. Murdoch. 2005. Effects of a clearcut on the net rates of nitrification and N mineralization in a northern hardwood forest, Catskill Mountains, New York. Biogeochemistry **72**:123–146.

Burns, D. A., K. Riva-Murray, P. M. Bradley, G. R. Aiken, and M. E. Brigham. 2012. Landscape controls on total and methyl Hg in the upper Hudson River basin, New York. Journal of Geophysical Research—Biogeosciences **117**.

Burns, R. C., and R.W.F. Hardy. 1975. *Nitrogen Fixation in Bacteria and Higher Plants.* Springer-Verlag, New York.

Burns, R. G. 1982. Enzyme activity in soil—Location and a possible role in microbial ecology. Soil Biology & Biochemistry **14**:423–427.

Burrows, A. 2000. Supernova explosions in the Universe. Nature **403**:727–733.

Burton, D. L., and E. C. Beauchamp. 1984. Field techniques using the acetylene blockage of nitrous oxide reduction to measure denitrification. Canadian Journal of Soil Science **64**:555–562.

Burton, J. D. 1988. Riverborne materials and the continent-ocean interface. *In* A. Lerman and M. Meybeck, editors. *Physical and Chemical Weathering in Geochemical Cycles,* Kluwer Academic Publishers, 299–321.

Burwicz, E. B., L. H. Ruepke, and K. Wallmann. 2011. Estimation of the global amount of submarine gas hydrates formed via microbial methane formation based on numerical reaction-transport modeling and a novel parameterization of Holocene sedimentation. Geochimica et Cosmochimica Acta **75**:4562–4576.

Busemann, H., A. F. Young, C.M.O. Alexander, P. Hoppe, S. Mukhopadhyay, and L. R. Nittler. 2006. Interstellar chemistry recorded in organic matter from primitive meteorites. Science **312**:727–730.

Busigny, V., P. Cartigny, and P. Philippot. 2011. Nitrogen isotopes in ophiolitic metagabbros: A re-evaluation of modern nitrogen fluxes in subduction zones and implication for the early Earth atmosphere. Geochimica et Cosmochimica Acta **75**:7502–7521.

Busigny, V., P. Cartigny, P. Philippot, M. Ader, and M. Javoy. 2003. Massive recycling of nitrogen and other fluid-mobile elements (K, Rb, Cs, H) in a cold slab environment: evidence from HP to UHP oceanic metasediments of the Schistes Lustres nappe (western Alps, Europe). Earth and Planetary Science Letters 215:27–42.

Butchart, S.H. M., M. Walpole, B. Collen, A. van Strien, J.P.W. Scharlemann, R.E.A. Almond, J.E.M. Baillie, B. Bomhard, C. Brown, J. Bruno, K. E. Carpenter, G. M. Carr, J. Chanson, A. M. Chenery, J. Csirke, N. C. Davidson, F. Dentener, M. Foster, A. Galli, J. N. Galloway, P. Genovesi, R. D. Gregory, M. Hockings, V. Kapos, J. F. Lamarque, F. Leverington, J. Loh, M. A. McGeoch, L. McRae, A. Minasyan, M. H. Morcillo, T.E.E. Oldfield, D. Pauly, S. Quader, C. Revenga, J. R. Sauer, B. Skolnik, D. Spear, D. Stanwell-Smith, S. N. Stuart, A. Symes, M. Tierney, T. D. Tyrrell, J. C. Vie, and R. Watson. 2010. Global biodiversity: Indicators of recent declines. Science 328: 1164–1168.

Butler, A. 1998. Acquisition and utilization of transition metal ions by marine organisms. Science 281:207–210.

Butler, J. H., M. Battle, M. L. Bender, S. A. Montzka, A. D. Clarke, E. S. Saltzman, C. M. Sucher, J. P. Severinghaus, and J. W. Elkins. 1999. A record of atmospheric halocarbons during the twentieth century from polar firn air. Nature 399:749–755.

Butler, J. H., D. B. King, J. M. Lobert, S. A. Montzka, S. A. Yvon-Lewis, B. D. Hall, N. J. Warwick, D. J. Mondeel, M. Aydin, and J. W. Elkins. 2007. Oceanic distributions and emissions of short-lived halocarbons. Global Biogeochemical Cycles 21.

Butler, T. J., M. D. Cohen, F. M. Vermeylen, G. E. Likens, D. Schmeltz, and R. S. Artz. 2008. Regional precipitation mercury trends in the eastern USA, 1998-2005: Declines in the Northeast and Midwest, no trend in the Southeast. Atmospheric Environment 42:1582–1592.

Butler, T. J., G. E. Likens, F. M. Vermeylen, and B.J.B. Stunder. 2003. The relation between NOx emissions and precipitation NO_3^- in the eastern USA. Atmospheric Environment 37:2093–2104.

Butler, T. J., G. E. Likens, F. M. Vermeylen, and B.J.B. Stunder. 2005. The impact of changing nitrogen oxide emissions on wet and dry nitrogen deposition in the northeastern USA. Atmospheric Environment 39:4851–4862.

Butterbach-Bahl, K., G. Willibald, and H. Papen. 2002. Soil core method for direct simultaneous determination of N_2 and N_2O emissions from forest soils. Plant and Soil 240:105–116.

Buyanovsky, G. A., M. Aslam, and G. H. Wagner. 1994. Carbon turnover in soil physical fractions. Soil Science Society of America Journal 58:1167–1173.

Buyanovsky, G. A., and G. H. Wagner. 1983. Annual cycles of carbon dioxide levels in soil air. Soil Science Society of America Journal 47:1139–1145.

Byrne, R. H., S. Mecking, R. A. Feely, and X. W. Liu. 2010. Direct observations of basin-wide acidification of the North Pacific Ocean. Geophysical Research Letters 37.

Cabana, G., and J. B. Rasmussen. 1994. Modeling food chain structure and contaminant bioaccumulation using stable nitrogen isotopes. Nature 372:255–257.

Caccavo, F., D. J. Lonergan, D. R. Lovley, M. Davis, J. F. Stolz, and M. J. McInerney. 1994. *Geobacter sulfurreducens* Spt.-Nov., A hydrogen-oxidizing and acetate-oxidizing dissimilatory metal-reducing microorganism. Applied and Environmental Microbiology 60: 3752–3759.

Cachier, H., M. P. Bremond, and P. Buatmenard. 1989. Carbonaceous aerosols from different tropical biomass burning sources. Nature 340:371–373.

Cachier, H., and J. Ducret. 1991. Influence of biomass burning on equatorial African rains. Nature 352: 228–230.

Cahoon, D. R., B. J. Stocks, J. S. Levine, W. R. Cofer, and K. P. Oneill. 1992. Seasonal distribution of African savanna fires. Nature 359:812–815.

Cai, T.E.B., L. B. Flanagan, and K. H. Syed. 2010. Warmer and drier conditions stimulate respiration more than photosynthesis in a boreal peatland ecosystem: Analysis of automatic chambers and eddy covariance measurements. Plant Cell and Environment 33: 394–407.

Cai, W. J. 2003. Riverine inorganic carbon flux and rate of biological uptake in the Mississippi River plume. Geophysical Research Letters 30.

Cailleau, G., O. Braissant, and E. P. Verrecchia. 2004. Biomineralization in plants as a long-term carbon sink. Naturwissenschaften 91:191–194.

Cairns, M. A., S. Brown, E. H. Helmer, and G. A. Baumgardner. 1997. Root biomass allocation in the world's upland forests. Oecologia 111:1–11.

Cairns-Smith, A. G. 1985. The first organisms. Scientific American 252(6), 90–100.

Caldeira, K., and J. F. Kasting. 1992. The life span of the biosphere revisited. Nature 360:721–723.

Calder, I. R., I. R. Wright, and D. Murdiyarso. 1986. A study of evaporation from tropical rain forest—West Java. Journal of Hydrology 89:13–31.

Caldwell, M. M., and S. D. Flint. 1994. Stratospheric ozone reduction, solar UvB radiation and terrestrial ecosystems. Climatic Change 28:375–394.

Callaway, R. M., E. H. Delucia, and W. H. Schlesinger. 1994. Biomass allocation of montane and desert ponderosa pine—An analog for response to climate change. Ecology 75:1474–1481.

Cama, J., and J. Ganor. 2006. The effects of organic acids on the dissolution of silicate minerals: A case study of oxalate catalysis of kaolinite dissolution. Geochimica et Cosmochimica Acta 70:2191–2209.

Cambardella, C. A., and E. T. Elliott. 1994. Carbon and nitrogen dynamics of soil organic matter fractions from cultivated grassland soils. Soil Science Society of America Journal 58:123–130.

Came, R. E., J. M. Eiler, J. Veizer, K. Azmy, U. Brand, and C. R. Weidman. 2007. Coupling of surface temperatures and atmospheric CO_2 concentrations during the Palaeozoic era. Nature 449:198–201.

Cameron, E. M. 1982. Sulfate and sulfate reduction in early Precambrian oceans. Nature 296:145–148.

Campbell, C. A., E. A. Paul, D. A. Rennie, and K. J. McCallum. 1967. Factors affecting the accuracy of the carbon-dating method in soil humus studies. Soil Science 104:81–85.

Campbell, J. E., G. R. Carmichael, T. Chai, M. Mena-Carrasco, Y. Tang, D. R. Blake, N. J. Blake, S. A. Vay, G. J. Collatz, I. Baker, J. A. Berry, S. A. Montzka, C. Sweeney, J. L. Schnoor, and C. O. Stanier. 2008. Photosynthetic control of atmospheric carbonyl sulfide during the growing season. Science 322:1085–1088.

Campbell, J. L., J. W. Hornbeck, W. H. McDowell, D. C. Buso, J. B. Shanley, and G. E. Likens. 2000. Dissolved organic nitrogen budgets for upland, forested ecosystems in New England. Biogeochemistry 49:123–142.

Campos, M., P. D. Nightingale, and T. D. Jickells. 1996. A comparison of methyl iodide emissions from seawater and wet depositional fluxes of iodine over the southern North Sea. Tellus Series B—Chemical and Physical Meteorology 48:106–114.

Canadell, J. G., C. Le Quere, M. R. Raupach, C. B. Field, E. T. Buitenhuis, P. Ciais, T. J. Conway, N. P. Gillett, R. A. Houghton, and G. Marland. 2007. Contributions to accelerating atmospheric CO_2 growth from economic activity, carbon intensity, and efficiency of natural sinks. Proceedings of the National Academy of Sciences 104:18866–18870.

Canfield, D. E. 1989a. Reactive iron in maine sediments. Geochimica et Cosmochimica Acta 53:619–632.

Canfield, D. E. 1989b. Sulfate reduction and oxic respiration in marine sediments—Implications for organic carbon preservation in euxinic environments. Deep-Sea Research Part A—Oceanographic Research Papers 36:121-138.

Canfield, D. E. 1991. Sulfate reduction in deep-sea sediments. American Journal of Science 291:177–188.

Canfield, D. E. 1993. Organic matter oxidation in marine sediments. In R. Wollast, F. T. MacKenzie and L. Chou, editors. Interactions of C, N, P and S in Biogeochemical Cycles and Global Change, Springer, 333–363.

Canfield, D. E. 1994. Factors influencing organic carbon preservation in marine sediments. Chemical Geology 114:315–329.

Canfield, D. E. 1997. The geochemistry of river particulates from the continental USA: Major elements. Geochimica et Cosmochimica Acta 61:3349–3365.

Canfield, D. E. 1998. A new model for Proterozoic ocean chemistry. Nature 396:450–453.

Canfield, D. E., W. J. Green, T. J. Gardner, and T. Ferdelman. 1984. Elemental residence times in Acton lake, Ohio. Archiv Fur Hydrobiologie 100:501–519.

Canfield, D. E., K. S. Habicht, and B. Thamdrup. 2000. The Archean sulfur cycle and the early history of atmospheric oxygen. Science 288:658–661.

Canfield, D. E., M. T. Rosing, and C. Bjerrum. 2006. Early anaerobic metabolisms. Philosophical Transactions of the Royal Society B—Biological Sciences 361:1819–1834.

Canfield, D. E., F. J. Stewart, B. Thamdrup, L. De Brabandere, T. Dalsgaard, E. F. Delong, N. P. Revsbech, and O. Ulloa. 2010. A cryptic sulfur cycle in oxygen-minimum-zone waters off the Chilean coast. Science 330:1375–1378.

Canfield, D. E., and A. Teske. 1996. Late Proterozoic rise in atmospheric oxygen concentration inferred from phylogenetic and sulphur-isotope studies. Nature 382:127–132.

Canfield, D. E., and B. Thamdrup. 1994. The production of ^{34}S-depleted sulfide during bacterial disproportionation of elemental sulfur. Science 266:1973–1975.

Cannon, H. L. 1960. Botanical prospecting for ore deposits. Science 132:591–598.

Cao, M. K., S. Marshall, and K. Gregson. 1996. Global carbon exchange and methane emissions from natural wetlands: Application of a process-based model. Journal of Geophysical Research—Atmospheres 101: 14399–14414.

Capo, R. C., and O. A. Chadwick. 1999. Sources of strontium and calcium in desert soil and calcrete. Earth and Planetary Science Letters 170:61–72.

Capone, D. G., J. A. Burns, J. P. Montoya, A. Subramaniam, C. Mahaffey, T. Gunderson, A. F. Michaels, and E. J. Carpenter. 2005. Nitrogen fixation by Trichodesmium spp.: An important source of new nitrogen to the tropical and subtropical North Atlantic Ocean. Global Biogeochemical Cycles 19.

Capone, D. G., S. G. Horrigan, S. E. Dunham, and J. Fowler. 1990. Direct determination of nitrification in marine waters by using the short-lived radioisotope of nitrogen, ^{13}N. Applied and Environmental Microbiology 56:1182–1184.

Capone, D. G., J. P. Zehr, H. W. Paerl, B. Bergman, and E. J. Carpenter. 1997. Trichodesmium, a globally significant marine cyanobacterium. Science 276:1221–1229.

Caraco, N., J. Cole, and G. E. Likens. 1990. A comparison of phosphorus immobilization in sediments of freshwater and coastal marine systems. Biogeochemistry 9:277–290.

Caraco, N. F., J. J. Cole, and G. E. Likens. 1989. Evidence for sulfate-controlled phosphorus release from sediments of aquatic ecosystems. Nature 341: 316–318.

Carignan, R., and D.R.S. Lean. 1991. Regeneration of dissolved substances in a seasonally anoxic lake—The relative importance of processes occurring in the wtaer column and in the sediments. Limnology and Oceanography 36:683–707.

Carlisle, A., A.H.F. Brown, and E. J. White. 1966. The organic matter and nutrient elements in the precipitation beneath a sessile oak (*Quercus patraea*) canopy. Journal of Ecology 54:87–98.

Carlson, C. A., H. W. Ducklow, and A. F. Michaels. 1994. Annual flux of dissolved organic carbon from the euphotic zone in the northwestern Sargasso sea. Nature 371:405–408.

Carlyle, J. C., and D. C. Malcolm. 1986. Larch litter and nitrogen availability in mixed larch-spruce stands. 1. Nutrient withdrawal, redistribution, and leaching loss from larch foliage at senescence. Canadian Journal of Forest Research 16:321–326.

Caro, G., D. A. Papanastassiou, and G. J. Wasserburg. 2010. $^{40}K/^{40}Ca$ isotopic constraints on the oceanic calcium cycle. Earth and Planetary Science Letters 296:124–132.

Carpenter, E. J., and K. Romans. 1991. Major role of the cyanobacterium *Trichodesmium* in nutrient cycling in the North Atlantic ocean. Science 254: 1356–1358.

Carpenter, S. R., J. J. Cole, J. R. Hodgson, J. F. Kitchell, M. L. Pace, D. Bade, K. L. Cottingham, T. E. Essington, J. N. Houser, and D. E. Schindler. 2001. Trophic cascades, nutrients, and lake productivity: whole-lake experiments. Ecological Monographs 71:163–186.

Carpenter, S. R., J. J. Cole, J. F. Kitchell, and M. L. Pace. 1998. Impact of dissolved organic carbon, phosphorus, and grazing on phytoplankton biomass and production in experimental lakes. Limnology and Oceanography 43:73–80.

Carpenter, S. R., J. J. Cole, M. L. Pace, R. Batt, W. A. Brock, T. Cline, J. Coloso, J. R. Hodgson, J. F. Kitchell, D. A. Seekell, L. Smith, and B. Weidel. 2011. Early warnings of regime shifts: A whole-ecosystem experiment. Science 332:1079–1082.

Carpenter, S. R., James F. Kitchell, and J. R. Hodgson. 1985. Cascading trophic interactions and lake productivity. BioScience 35:634–639.

Carpenter, S. R., S. G. Fisher, N. B. Grimm, and J. F. Kitchell. 1992. Global change and freshwater ecosystems. Annual Review of Ecology and Systematics 23:119–139.

Carr, J. S., and J. R. Najita. 2008. Organic molecules and water in the planet formation region of young circumstellar disks. Science 319:1504–1506.

Carr, M. H. 1987. Water on Mars. Nature 326:30–35.

Carr, M. H., M.J.S. Belton, C. R. Chapman, A. S. Davies, P. Geissler, R. Greenberg, A. S. McEwen, B. R. Tufts, R. Greeley, R. Sullivan, J. W. Head, R. T. Pappalardo, K. P. Klaasen, T. V. Johnson, J. Kaufman, D. Senske, J. Moore, G. Neukum, G. Schubert, J. A. Burns, P. Thomas, and J. Veverka. 1998. Evidence for a subsurface ocean on Europa. Nature 391:363–365.

Carroll, M. A., L. E. Heidt, R. J. Cicerone, and R. G. Prinn. 1986. OCS, H_2S, And CS_2 fluxes from a saltwater marsh. Journal of Atmospheric Chemistry 4:375–395.

Carter, D. O., D. Yellowlees, and M. Tibbett. 2007. Cadaver decomposition in terrestrial ecosystems. Naturwissenschaften 94:12–24.

Carter, J. P., Y. S. Hsiao, S. Spiro, and D. J. Richardson. 1995. Soil and sediment bacteria capable of aerobic nitrate respiration. Applied and Environmental Microbiology 61:2852–2858.

Casagrande, D., and K. Siefert. 1977. Origins of sulfur in coal—importance of ester sulfate content of peat. Science 195:675–676.

Casagrande, D. J., G. Idowu, A. Friedman, P. Rickert, K. Siefert, and D. Schlenz. 1979. H_2S Incorporation in coal precursors—origins of organic sulfur in coal. Nature 282:599–600.

Casey, W. H., H. R. Westrich, J. F. Banfield, G. Ferruzzi, and G. W. Arnold. 1993. Leaching and reconstruction at the surfaces of dissolving chain-silicate minerals. Nature 366:253–256.

Cassan, A., D. Kubas, J. P. Beaulieu, M. Dominik, K. Horne, J. Greenhill, J. Wambsganss, J. Menzies, A. Williams, U. G. Jorgensen, A. Udalski, D. P. Bennett, M. D. Albrow, V. Batista, S. Brillant, J.A. R. Caldwell, A. Cole, C. Coutures, K. H. Cook, S. Dieters, D. D. Prester, J. Donatowicz, P. Fouque, K. Hill, N. Kains, S. Kane, J. B. Marquette, R. Martin, K. R. Pollard, K. C. Sahu, C. Vinter, D. Warren, B. Watson, M. Zub, T. Sumi, M. K. Szymanski, M. Kubiak, R. Poleski, I. Soszynski, K. Ulaczyk, G. Pietrzynski, and L. Wyrzykowski. 2012. One or more bound planets per Milky Way star from microlensing observations. Nature 481:167–169.

Cassar, N., M. L. Bender, B. A. Barnett, S. Fan, W. J. Moxim, H. Levy, and B. Tilbrook. 2007. The Southern Ocean biological response to aeolian iron deposition. Science 317:1067–1070.

Castebrunet, H., C. Genthon, and P. Martinerie. 2006. Sulfur cycle at Last Glacial Maximum: Model results versus Antarctic ice core data. Geophysical Research Letters **33**.

Castelle, A. J., and J. N. Galloway. 1990. Carbon dioxide dynamics in acid forest soils in Shenandoah National Park, Virginia. Soil Science Society of America Journal **54**:252–257.

Castresana, J., and M. Saraste. 1995. Evolution of energetic metabolism—The respiration-early hypothesis. Trends in Biochemical Sciences **20**:443–448.

Castro, M. S., and F. E. Dierberg. 1987. Biogenic hydrogen sulfide emissions from selected Florida wetlands. Water Air and Soil Pollution **33**:1–13.

Castro, M. S., J. M. Melillo, P. A. Steudler, and J. W. Chapman. 1994. Soil moisture as a predictor of methane uptake by temperate forest soils. Canadian Journal of Forest Research **24**:1805–1810.

Castro, M. S., P. A. Steudler, J. M. Melillo, J. D. Aber, and R. D. Bowden. 1995. Factors controlling atmospheric methane consumption by temperate forest soils. Global Biogeochemical Cycles **9**:1–10.

Cates, R. G. 1975. Interface between slugs and wild ginger— Some evolutionary aspects. Ecology **56**:391–400.

Catling, D. C., K. J. Zahnle, and C. P. McKay. 2001. Biogenic methane, hydrogen escape, and the irreversible oxidation of early Earth. Science **293**:839–843.

Cavalieri, D. J., and C. L. Parkinson. 2008. Antarctic sea ice variability and trends. Journal of Geophysical Research—Oceans **113**:1979–2006.

Cavanaugh, C. M., S. L. Gardiner, M. L. Jones, H. W. Jannasch, and J. B. Waterbury. 1981. Prokaryotic cells in the hydrothermal vent tube worm *Riftia pachyptila* Jones—Possible chemoautotrophic symbionts. Science **213**:340–342.

Cavanaugh, M. L., S. A. Kurc, and R. L. Scott. 2011. Evapotranspiration partitioning in semiarid shrubland ecosystems: a two-site evaluation of soil moisture control on transpiration. Ecohydrology **4**: 671–681.

Cavigelli, M. A., and G. P. Robertson. 2000. The functional significance of denitrifier community composition in a terrestrial ecosystem. Ecology **81**:1402–1414.

Cayrel, R., V. Hill, T. C. Beers, B. Barbuy, M. Spite, F. Spite, B. Plez, J. Andersen, P. Bonifacio, P. Francois, P. Molaro, B. Nordstrom, and F. Primas. 2001. Measurement of stellar age from uranium decay. Nature **409**:691–692.

Cazenave, A., and W. Llovel. 2010. Contemporary sea level rise. Annual Review of Marine Science **2**:145–173.

Cebrián, J., and C. M. Duarte. 1995. Plant growth rate dependence of detrital carbon storage in ecosystems. Science **268**:1606–1608.

Cebrián, J., and J. Lartigue. 2004. Patterns of herbivory and decomposition in aquatic and terrestrial ecosystems. Ecological Monographs **74**:237–259.

Cejudo, F. J., A. Delatorre, and A. Paneque. 1984. Short-term ammonium inhibition of nitrogen fixation in *Azotobacter*. Biochemical and Biophysical Research Communications **123**:431–437.

Ceulemans, R., and M. Mousseau. 1994. Effects of elevated atmospheric CO_2 on woody plants. New Phytologist **127**:425–446.

Chacon, N., W. L. Silver, E. A. Dubinsky, and D. F. Cusack. 2006. Iron reduction and soil phosphorus solubilization in humid tropical forests soils: The roles of labile carbon pools and an electron shuttle compound. Biogeochemistry **78**:67–84.

Chadwick, O. A., L. A. Derry, P. M. Vitousek, B. J. Huebert, and L. O. Hedin. 1999. Changing sources of nutrients during four million years of ecosystem development. Nature **397**:491–497.

Chadwick, O. A., R. T. Gavenda, E. F. Kelly, K. Ziegler, C. G. Olson, W. C. Elliott, and D. M. Hendricks. 2003. The impact of climate on the biogeochemical functioning of volcanic soils. Chemical Geology **202**: 195–223.

Chadwick, O. A., E. F. Kelly, D. M. Merritts, and R. G. Amundson. 1994. Carbon dioxide consumption during soil development. Biogeochemistry **24**:115–127.

Chahine, M. T. 1995. Observation of local cloud and moisture feedbacks over high ocean and desert surface temperatures. Journal of Geophysical Research—Atmospheres **100**:8919–8927.

Chambers, L. A., and P. A. Trudinger. 1979. Thiosulfate formation and associated isotope effects during sulfite reduction by *Clostridium pasteurianum*. Canadian Journal of Microbiology **25**:719–721.

Chameides, W. L., and D. D. Davis. 1982. Chemistry in the troposphere. Chemical & Engineering News **60**:38–52.

Chameides, W. L., F. Fehsenfeld, M. O. Rodgers, C. Cardelino, J. Martinez, D. Parrish, W. Lonneman, D. R. Lawson, R. A. Rasmussen, P. Zimmerman, J. Greenberg, P. Middleton, and T. Wang. 1992. Ozone precursor relationships in the ambient atmosphere. Journal of Geophysical Research—Atmospheres **97**:6037–6055.

Chameides, W. L., P. S. Kasibhatla, J. Yienger, and H. Levy. 1994. Growth of continental-scale metro-agro-plexes, regional ozone pollution, and world food production. Science **264**:74–77.

Chameides, W. L., R. W. Lindsay, J. Richardson, and C. S. Kiang. 1988. The role of biogenic hydrocarbons in urban photochemical smog—Atlanta as a case study. Science **241**:1473–1475.

Chan, M. K., J. S. Kim, and D. C. Rees. 1993. The nitrogenase FeMo-cofactor and P-cluster pair—2.2-Angstrom resolution structures. Science 260:792–794.

Chandler, A. S., T. W. Choularton, G. J. Dollard, A.E.J. Eggleton, M. J. Gay, T. A. Hill, B.M.R. Jones, B. J. Tyler, B. J. Bandy, and S. A. Penkett. 1988. Measurements of H_2O_2 and SO_2 in clouds and estimates of their reaction rate. Nature 336:562–565.

Chang, B. X., A. H. Devol, and S. R. Emerson. 2010. Denitrification and the nitrogen gas excess in the eastern tropical South Pacific oxygen deficient zone. Deep-Sea Research Part I—Oceanographic Research Papers 57:1092-1101.

Chang, S., D. Des Marais, R. Mack, S. L. Miller, and G. E. Strathearn. 1983. Prebiotic organic synthesis and the origin of life. In J. W. Schopf, editor. Earth's Earliest Biosphere, Princeton University Press, 53–92.

Chang, W.Y.B., and R. A. Moll. 1980. Prediction of hypolimnetic oxygen deficits—Problems of interpretation. Science 209:721–722.

Changnon, S. A., and M. Demissie. 1996. Detection of changes in streamflow and floods resulting from climate fluctuations and land use-drainage changes. Climatic Change 32:411–421.

Chanton, J. P., and J.W.H. Dacey. 1991. Effects of vegetation on methane flux reservoirs and carbon isotopic composition. In T. D. Sharkey, E. A. Holland and H. A. Mooney, editors. Trace Gas Emissions by Plants. Academic Press, 65–92.

Chao, B. F., Y. H. Wu, and Y. S. Li. 2008. Impact of artificial reservoir water impound on global sea level. Science 320:212–214.

Chapelle, F. H., K. O'Neill, P. M. Bradley, B. A. Methe, S. A. Ciufo, L. L. Knobel, and D. R. Lovley. 2002. A hydrogen-based subsurface microbial community dominated by methanogens. Nature 415:312–315.

Chapin, D. M., L. C. Bliss, and L. J. Bledsoe. 1991. Environmental regulation of nitrogen fixation in a high arctic lowland ecosystem. Canadian Journal of Botany 69:2744–2755.

Chapin, F. S. 1974. Morphological and physiological mechanisms of temperature compensation in phosphate absorption along a latitudinal gradient. Ecology 55:1180–1198.

Chapin, F. S. 1980. The mineral nutrition of wild plants. Annual Review of Ecology and Systematics 11:233–260.

Chapin, F. S. 1988. Ecological aspects of plant mineral nutrition. Advances in Mineral Nutrition 3:161–191.

Chapin, F. S., R. J. Barsdate, and D. Barel. 1978. Phosphorus cycling in Alaskan coastal tundra—Hypothesis for the regulation of nutrient cycling. Oikos 31:189–199.

Chapin, F. S., and R. A. Kedrowski. 1983. Seasonal changes in nitrogen and phosphorus fractions and autumn retranslocation in evergreen and deciduous taiga trees. Ecology 64:376–391.

Chapin, F. S., and L. Moilanen. 1991. Nutritional controls over nitrogen and phosphorus resorption from Alaskan birch leaves. Ecology 72:709–715.

Chapin, F. S., L. Moilanen, and K. Kielland. 1993. Preferential use of organic nitrogen for growth by a nonmycorrhizal arctic sedge. Nature 361:150–153.

Chapin, F. S., and W. C. Oechel. 1983. Photosynthesis, respiration, and phosphate absorption by Carex aquatilis ecotypes along latitudinal and local environmental gradients. Ecology 64:743–751.

Chapin, F. S., H. L. Reynolds, C. M. D'Antonio, and V. M. Eckhart. 1996. The functional role of species in terrestrial ecosystems. In B. Walker and W. Steffen, editors. Global Change and Terrestrial Ecosystems. Cambridge University Press, 403–428.

Chapin, F. S., G. R. Shaver, and R. A. Kedrowski. 1986a. Environmental controls over carbon, nitrogen and phosphorus fractions in Eriophorum vaginatum in Alaskan tussock tundra. Journal of Ecology 74:167–195.

Chapin, F. S., P. M. Vitousek, and K. Van Cleve. 1986b. The nature of nutrient limitation in plant communities. American Naturalist 127:48–58.

Chapin, F. S., G. M. Woodwell, J. T. Randerson, E. B. Rastetter, G. M. Lovett, D. D. Baldocchi, D. A. Clark, M. E. Harmon, D. S. Schimel, R. Valentini, C. Wirth, J. D. Aber, J. J. Cole, M. L. Goulden, J. W. Harden, M. Heimann, R. W. Howarth, P. A. Matson, A. D. McGuire, J. M. Melillo, H. A. Mooney, J. C. Neff, R. A. Houghton, M. L. Pace, M. G. Ryan, S. W. Running, O. E. Sala, W. H. Schlesinger, and E. D. Schulze. 2006. Reconciling carbon-cycle concepts, terminology, and methods. Ecosystems 9:1041–1050.

Chapman, D. J., and J. W. Schopf. 1983. Biological and biochemical effects of the development of an aerobic environment. In J. W. Schopf, editor. Earth's Earliest Biosphere, Princeton University Press, 302–320.

Chappellaz, J., J. M. Barnola, D. Raynaud, Y. S. Korotkevich, and C. Lorius. 1990. Ice-core record of atmospheric methane over the past 160,000 years. Nature 345:127–131.

Chappellaz, J., T. Blunier, D. Raynaud, J. M. Barnola, J. Schwander, and B. Stauffer. 1993. Synchronous changes in atmospheric CH_4 and Greenland climate between 40-kyr and 8-kyr BP. Nature 366:443–445.

Chapuis-Lardy, L., N. Wrage, A. Metay, J. L. Chotte, and M. Bernoux. 2007. Soils, a sink for N_2O? A review. Global Change Biology 13:1–17.

Charlson, R. J., J. E. Lovelock, M. O. Andreae, and S. G. Warren. 1987. Oceanic phytoplankton, atmospheric sulfur, cloud albedo and climate. Nature **326**: 655–661.

Chase, E. M., and F. L. Sayles. 1980. Phosphorus in suspended sediments of the Amazon river. Estuarine and Coastal Marine Science **11**:383–391.

Chaussidon, M., and F. Robert. 1995. Nucleosynthesis of [11]B-rich boron in the pre-solar cloud recorded in meteoritic chondrules. Nature **374**:337–339.

Chavez, F. P., P. G. Strutton, C. E. Friederich, R. A. Feely, G. C. Feldman, D. C. Foley, and M. J. McPhaden. 1999. Biological and chemical response of the equatorial Pacific Ocean to the 1997–98 El Niño. Science **286**: 2126–2131.

Chen, C.T.A., and A. V. Borges. 2009. Reconciling opposing views on carbon cycling in the coastal ocean: Continental shelves as sinks and near-shore ecosystems as sources of atmospheric CO_2. Deep-Sea Research Part II—Topical Studies in Oceanography **56**:578–590.

Chen, C. Y., R. S. Stemberger, N. C. Kamman, B. M. Mayes, and C. L. Folt. 2005. Patterns of Hg bioaccumulation and transfer in aquatic food webs across multi-lake studies in the northeast US. Ecotoxicology **14**:135–147.

Chen, D. L., P. M. Chalk, and J. R. Freney. 1995. Distribution of reduced products of N-15-labelled nitrate in anaerobic soils. Soil Biology & Biochemistry **27**: 1539–1545.

Chen, J. L., C. R. Wilson, D. Blankenship, and B. D. Tapley. 2009. Accelerated Antarctic ice loss from satellite gravity measurements. Nature Geoscience **2**:859–862.

Chen, J. L., C. R. Wilson, and B. D. Tapley. 2006a. Satellite gravity measurements confirm accelerated melting of Greenland ice sheet. Science **313**:1958–1960.

Chen, J. M., B. Z. Chen, K. Higuchi, J. Liu, D. Chan, D. Worthy, P. Tans, and A. Black. 2006b. Boreal ecosystems sequestered more carbon in warmer years. Geophysical Research Letters **33**.

Chen, Y. H., and R. G. Prinn. 2006. Estimation of atmospheric methane emissions between 1996 and 2001 using a three-dimensional global chemical transport model. Journal of Geophysical Research—Atmospheres **111**.

Cheng, D. L., and K. J. Niklas. 2007. Above- and belowground biomass relationships across 1534 forested communities. Annals of Botany **99**:95–102.

Cherry, R. D., J.J.W. Higgo, and S. W. Fowler. 1978. Zooplankton fecal pellets and element residence times in the ocean. Nature **274**:246–248.

Chesworth, W., J. Dejou, and P. Larroque. 1981. The weathering of basalt and relative mobilities of the major elements at Belbex, France. Geochimica et Cosmochimica Acta **45**:1235–1243.

Chesworth, W., and F. Macias-Vasquez. 1985. pE, pH, and podzolization. American Journal of Science **285**: 128–146.

Chevalier, R. A., and C. L. Sarazin. 1987. Hot gas in the Universe. American Scientist **75**:609–618.

Childers, D. L., J. Corman, M. Edwards, and J. J. Elser. 2011. Sustainability challenges of phosphorus and food: Solutions from closing the human phosphorus cycle. BioScience **61**:117–124.

Chin, M., and D. D. Davis. 1995. A reanalysis of carbonyl sulfide as a source of stratospheric background sulfur aerosol. Journal of Geophysical Research **100**: 8993–9005.

Chin, M., and D. D. Davis. 1993. Global sources and sinks of OCS and CS_2 and their distributions. Global Biogeochemical Cycles **7**:321–337.

Chisholm, S. W., P. G. Falkowski, and J. J. Cullen. 2001. Oceans—Discrediting ocean fertilization. Science **294**: 309–310.

Chivian, D., E. L. Brodie, E. J. Alm, D. E. Culley, P. S. Dehal, T. Z. DeSantis, T. M. Gihring, A. Lapidus, L. H. Lin, S. R. Lowry, D. P. Moser, P. M. Richardson, G. Southam, G. Wanger, L. M. Pratt, G. L. Andersen, T. C. Hazen, F. J. Brockman, A. P. Arkin, and T. C. Onstott. 2008. Environmental genomics reveals a single-species ecosystem deep within earth. Science **322**:275–278.

Cho, B. C., and F. Azam. 1988. Major role of bacteria in biogeochemical fluxes in the ocean's interior. Nature **332**:441–443.

Cho, Y., C. T. Driscoll, C. E. Johnson, and T. G. Siccama. 2010. Chemical changes in soil and soil solution after calcium silicate addition to a northern hardwood forest. Biogeochemistry **100**:3–20.

Choi, W.-J., S. X. Chang, and J. S. Bhatti. 2007. Drainage affects tree growth and C and N dynamics in a minerotrophic peatland. Ecology **88**:443–453.

Chorover, J., M. K. Amistadi, and O. A. Chadwick. 2004. Surface charge evolution of mineral-organic complexes during pedogenesis in Hawaiian basalt. Geochimica et Cosmochimica Acta **68**:4859–4876.

Chorover, J., and G. Sposito. 1995. Surface-charge characteristics of kaolinitic tropical soils. Geochimica et Cosmochimica Acta **59**:875–884.

Chorover, J., P. M. Vitousek, D. A. Everson, A. M. Espesperanza, and D. Turner. 1994. Solution chemistry profiles of mixed-conifer forests before and after fire. Biogeochemistry **26**:115–144.

Choudhury, B. J., N. E. DiGirolamo, J. Susskind, W. L. Darnell, S. K. Gupta, and G. Asrar. 1998. A biophysical process-based estimate of global land surface evaporation using satellite and ancillary data—II. Regional and global patterns of seasonal and annual variations. Journal of Hydrology **205**:186–204.

Christ, M. J., C. T. Driscoll, and G. E. Likens. 1999. Watershed- and plot-scale tests of the mobile anion concept. Biogeochemistry **47**:335–353.

Christensen, J. P., J. W. Murray, A. H. Devol, and L. A. Codispoti. 1987. Denitrification in continental shelf sediments has major impact on the oceanic nitrogen budget. Global Biogeochemical Cycles **1**:97–116.

Christensen, N. L. 1977. Fire and soil-plant nutrient relations in a pine-wiregrass savanna on the coastal plain of North Carolina. Oecologia **31**:27–44.

Christensen, N. L., and T. MacAller. 1985. Soil mineral nitrogen transformations during succession in the piedmont of North Carolina. Soil Biology & Biochemistry **17**:675–681.

Christensen, P. R. 2003. Formation of recent martian gullies through melting of extensive water-rich snow deposits. Nature **422**:45–48.

Christensen, T. R., A. Ekberg, L. Strom, M. Mastepanov, N. Panikov, M. Oquist, B. H. Svensson, H. Nykanen, P. J. Martikainen, and H. Oskarsson. 2003. Factors controlling large scale variations in methane emissions from wetlands. Geophysical Research Letters **30**.

Christensen, T. R., I. C. Prentice, J. Kaplan, A. Haxeltine, and S. Sitch. 1996. Methane flux from northern wetlands and tundra—An ecosystem source modelling approach. Tellus Series B—Chemical and Physical Meteorology **48**:652–661.

Church, J. A., and N. J. White. 2011. Sea-level rise from the late 19th to the early 21st century. Surveys in Geophysics **32**:585–602.

Chyba, C., and C. Sagan. 1992. Endogenous production, exogenous delivery and impact-shock synthesis of organic molecules—An inventory for the origins of life. Nature **355**:125–132.

Chyba, C. F. 1990a. Impact delivery and erosion of planetary oceans in the early inner solar system. Nature **343**:129–133.

Chyba, C. F. 1990b. Meteoritics—Extraterrestrial amino acids and terrestrial life. Nature **348**:113–114.

Ciais, P., A. C. Manning, M. Reichstein, S. Zaehle, and L. Bopp. 2007. Nitrification amplifies the decreasing trends of atmospheric oxygen and implies a larger land carbon uptake. Global Biogeochemical Cycles **21**.

Ciais, P., M. Reichstein, N. Viovy, A. Granier, J. Ogee, V. Allard, M. Aubinet, N. Buchmann, C. Bernhofer, A. Carrara, F. Chevallier, N. De Noblet, A. D. Friend, P. Friedlingstein, T. Grunwald, B. Heinesch, P. Keronen, A. Knohl, G. Krinner, D. Loustau, G. Manca, G. Matteucci, F. Miglietta, J. M. Ourcival, D. Papale, K. Pilegaard, S. Rambal, G. Seufert, J. F. Soussana, M. J. Sanz, E. D. Schulze, T. Vesala, and R. Valentini. 2005. Europe-wide reduction in primary productivity caused by the heat and drought in 2003. Nature **437**:529–533.

Ciais, P., P. P. Tans, J.W.C. White, M. Trolier, R. J. Francey, J. A. Berry, D. R. Randall, P. J. Sellers, J. G. Collatz, and D. S. Schimel. 1995. Partitioning of ocean and land uptake of CO_2 as inferred by delta-C-13 measurements from the NOAA Climate Monitoring and Diagnostics Laboratory Global Air Sampling Network. Journal of Geophysical Research—Atmospheres **100**:5051–5070.

Cicerone, R. J. 1987. Changes in stratospheric ozone. Science **237**:35–42.

Cicerone, R. J., C. C. Delwiche, S. C. Tyler, and P. R. Zimmerman. 1992. Methane emissions from California rice paddies with varied treatments. Global Biogeochemical Cycles **6**:233–248.

Cicerone, R. J., and R. S. Oremland. 1988. Biogeochemical aspects of atmospheric methane. Global Biogeochemical Cycles **2**:299–327.

Cienciala, E., J. Kucera, A. Lindroth, J. Cermak, A. Grelle, and S. Halldin. 1997. Canopy transpiration from a boreal forest in Sweden during a dry year. Agricultural and Forest Meteorology **86**:157–167.

Ciesla, F. J., and S. A. Sandford. 2012. Organic synthesis via irradiation and warming of ice grains in the solar nebula. Science **336**:452–454.

Clarisse, L., C. Clerbaux, F. Dentener, D. Hurtmans, and P. F. Coheur. 2009. Global ammonia distribution derived from infrared satellite observations. Nature Geoscience **2**:479–483.

Clarisse, L., P.-F. Coheur, S. Chefdeville, J.-L. Lacour, D. Hurtmans, and C. Clerbaux. 2011. Infrared satellite observations of hydrogen sulfide in the volcanic plume of the August 2008 Kasatochi eruption. Geophysical Research Letters **38**.

Clark, D. A., S. Brown, D. W. Kicklighter, J. Q. Chambers, J. R. Thomlinson, and J. Ni. 2001. Measuring net primary production in forests: Concepts and field methods. Ecological Applications **11**:356–370.

Clark, J. D., and A. H. Johnson. 2011. Carbon and nitrogen accumulation in post-agricultural forest soils of western New England. Soil Science Society of America Journal **75**:1530–1542.

Clark, J. S. 1990. Fire and climate change during the last 750 yr in northwestern Minnesota. Ecological Monographs **60**:135–159.

Clark, J. S., and P. D. Royall. 1994. Preindustrial particulate emissions and carbon sequestration from biomass burning in North America. Biogeochemistry **24**:35–51.

Clark, J. S., B. J. Stocks, and P.J.H. Richard. 1996. Climate implications of biomass burning since the 19th century in eastern North America. Global Change Biology **2**:433–442.

Clark, K. L., H. L. Gholz, and M. S. Castro. 2004. Carbon dynamics along a chronosequence of slash pine plantations in north Florida. Ecological Applications 14:1154–1171.

Clark, K. L., N. M. Nadkarni, D. Schaefer, and H. L. Gholz. 1998. Cloud water and precipitation chemistry in a tropical montane forest, Monteverde, Costa Rica. Atmospheric Environment 32:1595–1603.

Clark, L. L., E. D. Ingall, and R. Benner. 1999. Marine organic phosphorus cycling: Novel insights from nuclear magnetic resonance. American Journal of Science 299:724–737.

Clark, M. L., D. A. Roberts, J. J. Ewel, and D. B. Clark. 2011. Estimation of tropical rain forest aboveground biomass with small-footprint lidar and hyperspectral sensors. Remote Sensing of Environment 115:2931–2942.

Clark, R. N. 2009. Detection of adsorbed water and hydroxyl on the Moon. Science 326:562–564.

Clarkson, D. T., and J. B. Hanson. 1980. The mineral nutrition of higher plants. Annual Review of Plant Physiology and Plant Molecular Biology 31:239–298.

Clawson, R. G., B. G. Lockaby, and B. Rummer. 2001. Changes in production and nutrient cycling across a wetness gradient within a floodplain forest. Ecosystems 4:126–138.

Clayton, H., J.R.M. Arah, and K. A. Smith. 1994. Measurement of nitrous oxide emissions from fertilized grassland using closed chambers. Journal of Geophysical Research—Atmospheres 99:16599–16607.

Clayton, J. L. 1976. Nutrient gains to adjacent ecosystems during a forest fire—Evaluation. Forest Science 22:162–166.

Cleland, E. E., and W. S. Harpole. 2010. Nitrogen enrichment and plant communities. Annals of the New York Academy of Sciences 1195:46–61.

Clement, J., J. Shrestha, J. Ehrenfeld, and P. Jaffe. 2005. Ammonium oxidation coupled to dissimilatory reduction of iron under anaerobic conditions in wetland soils. Soil Biology and Biochemistry 37:2323–2328.

Cleveland, C. C., B. Z. Houlton, C. Neill, S. C. Reed, A. R. Townsend, and Y. P. Wang. 2010. Using indirect methods to constrain symbiotic nitrogen fixation rates: a case study from an Amazonian rain forest. Biogeochemistry 99:1–13.

Cleveland, C. C., and D. Liptzin. 2007. C:N: P stoichiometry in soil: is there a "Redfield ratio" for the microbial biomass? Biogeochemistry 85:235–252.

Cleveland, C. C., and A. R. Townsend. 2006. Nutrient additions to a tropical rain forest drive substantial soil carbon dioxide losses to the atmosphere. Proceedings of the National Academy of Sciences 103:10316–10321.

Cleveland, C. C., A. R. Townsend, D. S. Schimel, H. Fisher, R. W. Howarth, L. O. Hedin, S. S. Perakis, E. F. Latty, J. C. Von Fischer, A. Elseroad, and M. F. Wasson. 1999. Global patterns of terrestrial biological nitrogen (N_2) fixation in natural ecosystems. Global Biogeochemical Cycles 13:623–645.

Cleveland, C. C., A. R. Townsend, and S. K. Schmidt. 2002. Phosphorus limitation of microbial processes in moist tropical forests: Evidence from short-term laboratory incubations and field studies. Ecosystems 5:680–691.

Cleveland, C. C., A. R. Townsend, P. Taylor, S. Alvarez-Clare, M.M.C. Bustamante, G. Chuyong, S. Z. Dobrowski, P. Grierson, K. E. Harms, B. Z. Houlton, A. Marklein, W. Parton, S. Porder, S. C. Reed, C. A. Sierra, W. L. Silver, E.V.J. Tanner, and W. R. Wieder. 2011. Relationships among net primary productivity, nutrients and climate in tropical rain forest: a pan-tropical analysis. Ecology Letters 14:939–947.

Cline, J. D., D. P. Wisegarver, and K. Kellyhansen. 1987. Nitrous oxide and vertical mixing in the equatorial Pacific during the 1982–1983 El Niño. Deep-Sea Research Part A—Oceanographic Research Papers 34:857–873.

Cloud, P. 1973. Paleoecological significance of the Banded Iron formation. Geology 68:1135–1145.

Clow, D. W., and M. A. Mast. 1999. Long-term trends in streamwater and precipitation chemistry at five headwater basins in the northeastern United States. Water Resources Research 35:541–554.

Clymo, R. S. 1984. The limits to peat bog growth. Philosophical Transactions of the Royal Society of London. Series B, Biological Sciences 303:605.

Clymo, R. S., J. Turunen, and K. Tolonen. 1998. Carbon accumulation in peatland. Oikos 81:368–388.

Coale, K. H., S. E. Fitzwater, R. M. Gordon, K. S. Johnson, and R. T. Barber. 1996. Control of community growth and export production by upwelled iron in the equatorial Pacific Ocean. Nature 379:621–624.

Coale, K. H., K. S. Johnson, S. E. Fitzwater, R. M. Gordon, S. Tanner, F. P. Chavez, L. Ferioli, C. Sakamoto, P. Rogers, F. Millero, P. Steinberg, P. Nightingale, D. Cooper, W. P. Cochlan, M. R. Landry, J. Constantinou, G. Rollwagen, A. Trasvina, and R. Kudela. 1996. A massive phytoplankton bloom induced by an ecosystem-scale iron fertilization experiment in the equatorial Pacific Ocean. Nature 383:495–501.

Codispoti, L. A. 2007. An oceanic fixed nitrogen sink exceeding 400 Tg N a^{-1} vs the concept of homeostasis in the fixed-nitrogen inventory. Biogeosciences 4:233–253.

Codispoti, L. A., J. A. Brandes, J. P. Christensen, A. H. Devol, S.W.A. Naqvi, H. W. Paerl, and T. Yoshinari. 2001. The oceanic fixed nitrogen and nitrous oxide budgets: Moving targets as we enter the anthropocene? Scientia Marina 65:85–105.

Codispoti, L. A., and J. P. Christensen. 1985. Nitrification, denitrification and nitrous oxide cycling in the eastern tropical South Pacific ocean. Marine Chemistry **16**:277–300.

Codispoti, L. A., G. E. Friederich, T. T. Packard, H. E. Glover, P. J. Kelly, R. W. Spinrad, R.T. Barber, J. W. Elkins, B. B. Ward, F. Lipschultz, and N. Lostaunau. 1986. High nitrite levels off northern Peru—A signal of instability in the marine denitrification rate. Science **233**:1200–1202.

Cody, G. D., N. Z. Boctor, T. R. Filley, R. M. Hazen, J. H. Scott, A. Sharma, and H. S. Yoder. 2000. Primordial carbonylated iron-sulfur compounds and the synthesis of pyruvate. Science **289**:1337–1340.

Cofer, W. R., J. S. Levine, E. L. Winstead, and B. J. Stocks. 1990. Gaseous emissions from Canadian boreal forest fires. Atmospheric Environment **24A**:1653–1659.

Cogbill, C. V., and G. E. Likens. 1974. Acid precipitation in the northeastern United States. Water Resources Research **10**:1133–1137.

Coggon, R. M., D.A.H. Teagle, C. E. Smith-Duque, J. C. Alt, and M. J. Cooper. 2010. Reconstructing past seawater Mg/Ca and Sr/Ca from mid-ocean ridge flank calcium carbonate veins. Science **327**:1114–1117.

Cohen, B. A., T. D. Swindle, and D. A. Kring. 2000. Support for the lunar cataclysm hypothesis from lunar meteorite impact melt ages. Science **290**:1754–1756.

Cohen, Y., and L. I. Gordon. 1978. Nitrous oxide in the oxygen minimum of eastern tropical North Pacific—Evidence for its consumption during denitrification and possible mechanisms for its preoduction. Deep-Sea Research **25**:509–524.

Cohen, Y., and L. I. Gordon. 1979. Nitrous oxide production in the ocean. Journal of Geophysical Research—Oceans and Atmospheres **84**:347–353.

Colaprete, A., P. Schultz, J. Heldmann, D. Wooden, M. Shirley, K. Ennico, B. Hermalyn, W. Marshall, A. Ricco, R. C. Elphic, D. Goldstein, D. Summy, G. D. Bart, E. Asphaug, D. Korycansky, D. Landis, and L. Sollitt. 2010. Detection of water in the LCROSS ejecta plume. Science **330**:463–468.

Cole, C. V., E. T. Elliott, H. W. Hunt, and D. C. Coleman. 1978. Trophic interactions in soils as they affect energy and nutrient dynamics. 5. Phosphorus transformations. Microbial Ecology **4**:381–387.

Cole, C. V., G. S. Innis, and J.W.B. Stewart. 1977. Simulation of phosphorus cycling in semiarid grasslands. Ecology **58**:1–15.

Cole, C. V., and S. R. Olsen. 1959. Phosphorus solubility in calcareous soils. I. Dicalcium phosphate activities in equilibrium solutions. Soil Science Society of America Proceedings **23**:116–118.

Cole, D. W., and M. Rapp. 1981. Element cycling in forest ecosystems. *In* D. E. Reichle, editor. *Dynamics Properties of Forest Ecosystems.* Cambridge University Press, 341–409.

Cole, J., and N. Caraco. 1998. Atmospheric exchange of carbon dioxide in a low-wind oligotrophic lake measured by the addition of SF_6. Limnology and Oceanography **43**:647–656.

Cole, J. J., and N. F. Caraco. 2001. Carbon in catchments: connecting terrestrial carbon losses with aquatic metabolism. Marine and Freshwater Research **52**:101–110.

Cole, J. J., N. F. Caraco, G. W. Kling, and T. K. Kratz. 1994. Carbon dioxide supersaturation in the surface waters of lakes. Science **265**:1568–1570.

Cole, J. J., S. R. Carpenter, J. F. Kitchell, and M. L. Pace. 2002. Pathways of organic carbon utilization in small lakes: Results from a whole-lake ^{13}C addition and coupled model. Limnology and Oceanography **47**: 1664–1675.

Cole, J. J., S. Findlay, and M. L. Pace. 1988. Bacterial production in fresh- and saltwater ecosystems—A cross-sysstem overview. Marine Ecology—Progress Series **43**:1–10.

Cole, J. J., S. Honjo, and J. Erez. 1987. Benthic decomposition of organic matter at a deep-water site in the Panama basin. Nature **327**:703–704.

Cole, J. J., J. M. Lane, R. Marino, and R. W. Howarth. 1993. Molybdenum assimilation by cyanobacteria and phytoplankton in freshwater and saltwater. Limnology and Oceanography **38**:25–35.

Cole, J. J., and M. L. Pace. 1995. Bacterial secondary production in oxic and anoxic freshwaters. Limnology and Oceanography **40**:1019–1027.

Cole, J. J., Y. T. Prairie, N. F. Caraco, W. H. McDowell, L. J. Tranvik, R. G. Striegl, C. M. Duarte, P. Kortelainen, J. A. Downing, J. J. Middelburg, and J. Melack. 2007. Plumbing the global carbon cycle: Integrating inland waters into the terrestrial carbon budget. Ecosystems **10**:171–184.

Coles, J.R.P., and J. B. Yavitt. 2002. Control of methane metabolism in a forested northern wetland, New York state, by aeration, substrates, and peat size fractions. Geomicrobiology Journal **19**:293–315.

Coley, P. D., J. P. Bryant, and F. S. Chapin. 1985. Resource availability and plant antiherbivore defense. Science **230**:895–899.

Collerson, K. D., and B. S. Kamber. 1999. Evolution of the continents and the atmosphere inferred from Th-U-Nb systematics of the depleted mantle. Science **283**: 1519–1522.

Colman, J. J., D. R. Blake, and F. S. Rowland. 1998. Atmospheric residence time of CH_3Br estimated from the Junge spatial variability relation. Science **281**:392–396.

Comas, X., and W. Wright. 2012. Heterogeneity of biogenic gas ebullition in subtropical peat soils is revealed using time-lapse cameras. Water Resources Research **48**.

Comiso, J. C., C. L. Parkinson, R. Gersten, and L. Stock. 2008. Accelerated decline in the Arctic Sea ice cover. Geophysical Research Letters **35**.

Compeau, G. C., and R. Bartha. 1985. Sulfate-reducing bacteria—Principal methylators of mercury in anoxic estuarine sediment. Applied and Environmental Microbiology 50:498–502.

Conant, R. T., M. G. Ryan, G. I. Agren, H. E. Birge, E. A. Davidson, P. E. Eliasson, S. E. Evans, S. D. Frey, C. P. Giardina, F. M. Hopkins, R. Hyvonen, M.U.F. Kirschbaum, J. M. Lavallee, J. Leifeld, W. J. Parton, J. M. Steinweg, M. D. Wallenstein, J.A.M. Wetterstedt, and M. A. Bradford. 2011. Temperature and soil organic matter decomposition rates—synthesis of current knowledge and a way forward. Global Change Biology 17:3392–3404.

Conley, D. J. 2002. Terrestrial ecosystems and the global biogeochemical silica cycle. Global Biogeochemical Cycles 16.

Conley, D. J., S. Bjorck, E. Bonsdorff, J. Carstensen, G. Destouni, B. G. Gustafsson, S. Hietanen, M. Kortekaas, K. Kuosa, H.E.M. Meier, B. Muller-Karulis, K. Nordberg, A. Norkko, G. Nurnberg, H. Pitkanen, N. N. Rabalais, R. Rosenberg, O. P. Savchuk, C. P. Slomp, M. Voss, F. Wulff, and L. Zillen. 2009a. Hypoxia-related processes in the Baltic Sea. Environmental Science & Technology 43:3412–3420.

Conley, D. J., J. Carstensen, G. Aertebjerg, P. B. Christensen, T. Dalsgaard, J.L.S. Hansen, and A. B. Josefson. 2007. Long-term changes and impacts of hypoxia in Danish coastal waters. Ecological Applications 17:S165–S184.

Conley, D. J., H. W. Paerl, R. W. Howarth, D. F. Boesch, S. P. Seitzinger, K. E. Havens, C. Lancelot, and G. E. Likens. 2009b. Controlling eutrophication: Nitrogen and phosphorus. Science 323:1014–1015.

Conley, D. J., W. M. Smith, J. C. Cornwell, and T. R. Fisher. 1995. Transformation of particle-bound phosphorus at the land-sea interface. Estuarine Coastal and Shelf Science 40:161–176.

Connell, M. J., R. J. Raison, and P. K. Khanna. 1995. Nitrogen mineralization in relation to site history and soil properties for a range of Australian forest soils. Biology and Fertility of Soils 20:213–220.

Conner, W. H., and J. W. Day Jr. 1976. Productivity and composition of a baldcypress-water tupelo site and a bottomland hardwood site in a Louisiana swamp. American Journal of Botany 63:1354–1364.

Conner, W. H., B. Song, T. M. Williams, and J. T. Vernon. 2011. Long-term tree productivity of a South Carolina coastal plain forest across a hydrology gradient. Journal of Plant Ecology-UK 4:67–76.

Conrad, R. 1994. Compensation concentration as critical variable for regulating the flux of trace gases between soil and atmosphere. Biogeochemistry 27:155–170.

Conrad, R. 1996. Soil microorganisms as controllers of atmospheric trace gases (H_2, CO, CH_4, OCS, N_2O and NO). Microbiological Reviews 60:609–640.

Conrad, R., and W. Seiler. 1988. Methane and hydrogen in seawater (Atlantic ocean). Deep-Sea Research Part A—Oceanographic Research Papers 35:1903-1917.

Conrad, R., W. Seiler, and G. Bunse. 1983. Factors influencing the loss of fertilizer nitrogen into the atmosphere as N_2O. Journal of Geophysical Research—Oceans and Atmospheres 88:6709–6718.

Conway, T. J., P. Tans, I. S. Waterman, K. W. Thoning, K. A. Masarie, and R. H. Gammon. 1988. Atmospheric carbon dioxide measurements in the remote global atmosphere, 1981–1984. Tellus 40B:81–115.

Cook, E. R., C. A. Woodhouse, C. M. Eakin, D. M. Meko, and D. W. Stahle. 2004. Long-term aridity changes in the western United States. Science 306:1015–1018.

Cook, P. J., and J. H. Shergold. 1984. Phosphorus, phosphorites and skeletal evolution at the Precambrian–Cambrian boundary. Nature 308:231–236.

Cook, R. B., and D. W. Schindler. 1983. The biogeochemistry of sulfur in an experimentally acidified lake. Ecological Bulletin 35:115–127.

Cooper, D. J., A. J. Watson, and P. D. Nightingale. 1996. Large decrease in ocean-surface CO_2 fugacity in response to in situ iron fertilization. Nature 383:511–513.

Cooper, G., N. Kimmich, W. Belisle, J. Sarinana, K. Brabham, and L. Garrel. 2001. Carbonaceous meteorites as a source of sugar-related organic compounds for the early Earth. Nature 414:879–883.

Cooper, J., R. Lombardi, D. Boardman, and C. Carliell-Marquet. 2011. The future distribution and production of global phosphate rock reserves. Resources Conservation and Recycling 57:78–86.

Cooper, O. R., D. D. Parrish, A. Stohl, M. Trainer, P. Nedelec, V. Thouret, J. P. Cammas, S. J. Oltmans, B. J. Johnson, D. Tarasick, T. Leblanc, I. S. McDermid, D. Jaffe, R. Gao, J. Stith, T. Ryerson, K. Aikin, T. Campos, A. Weinheimer, and M. A. Avery. 2010. Increasing springtime ozone mixing ratios in the free troposphere over western North America. Nature 463:344–348.

Cooper, S. R., and G. S. Brush. 1991. Long-term history of Chesapeake Bay anoxia. Science 254:992–996.

Copard, Y., P. Amiotte-Suchet, and C. Di-Giovanni. 2007. Storage and release of fossil organic carbon related to weathering of sedimentary rocks. Earth and Planetary Science Letters 258:345–357.

Copi, C. J., D. N. Schramm, and M. S. Turner. 1995. Big-bang nucleosynthesis and the baryon density of the Universe. Science 267:192–199.

Corbett, J. J., P. S. Fischbeck, and S. N. Pandis. 1999. Global nitrogen and sulfur inventories for ocean-going ships. Journal of Geophysical Research—Atmospheres 104:3457–3470.

Corliss, B. H., D. G. Martinson, and T. Keffer. 1986. Late Quaternary deep-ocean circulation. Geological Society of America Bulletin 97:1106–1121.

Corliss, J. B., J. Dymond, L. I. Gordon, J. M. Edmond, R.P.V. Herzen, R. D. Ballard, K. Green, D. Williams, A. Bainbridge, K. Crane, and T. H. Vanandel. 1979. Submarine thermal springs on the Galapagos rift. Science 203:1073–1083.

Cornelis, J. T., B. Delvaux, D. Cardinal, L. Andre, J. Ranger, and S. Opfergelt. 2010. Tracing mechanisms controlling the release of dissolved silicon in forest soil solutions using Si isotopes and Ge/Si ratios. Geochimica et Cosmochimica Acta 74:3913–3924.

Cornell, D. L., and D. Ford. 1982. Comparison of precipitation and land runoff as sources of estuarine nitrogen. Estuarine and Coastal Shelf Science 15:45–56.

Cornell, S. E., T. D. Jickells, J. N. Cape, A. P. Rowland, and R. A. Duce. 2003. Organic nitrogen deposition on land and coastal environments: A review of methods and data. Atmospheric Environment 37:2173–2191.

Cornett, R. J., and F. H. Rigler. 1979. Hypolimnetic oxygen deficits—Their prediction and interpretation. Science 205:580–581.

Cornett, R. J., and F. H. Rigler. 1980. The areal hypolimnetic oxygen deficit—An empirical test of the model. Limnology and Oceanography 25:672–679.

Corre, M. D., F. O. Beese, and R. Brumme. 2003. Soil nitrogen cycle in high nitrogen deposition forest: Changes under nitrogen saturation and liming. Ecological Applications 13:287–298.

Corre, M. D., and N. P. Lamersdorf. 2004. Reversal of nitrogen saturation after long-term deposition reduction: Impact on soil nitrogen cycling. Ecology 85:3090–3104.

Corre, M. D., E. Veldkamp, J. Arnold, and S. J. Wright. 2010. Impact of elevated N input on soil N cycling and losses in old-growth lowland and montane forests in Panama. Ecology 91:1715–1729.

Correll, D. L. 1998. The role of phosphorus in the eutrophication of receiving waters: A review. Journal of Environmental Quality 27:261–266.

Correll, D. L., C. O. Clark, B. Goldberg, V. R. Goodrich, D. R. Hayes, W. H. Klein, and W. D. Schecher. 1992. Spectral ultraviolet-B radiation fluxes at the Earth's surface—Long-term variations at 39°N, 77°W. Journal of Geophysical Research—Atmospheres 97:7579–7591.

Corsi, S. R., D. J. Graczyk, S. W. Geis, N. L. Booth, and K. D. Richards. 2010. A fresh look at road salt: Aquatic toxicity and water-quality impacts on local, regional, and national scales. Environmental Science & Technology 44:7376–7382.

Cosby, B. J., G. M. Hornberger, J. N. Galloway, and R. F. Wright. 1985. Modeling the effects of acid deposition—Assessment of a lumped parameter model of soil–water and streamwater chemistry. Water Resources Research 21:51–63.

Cosby, B. J., G. M. Hornberger, R. F. Wright, and J. N. Galloway. 1986. Modeling the effects of acid deposition: Control of long-term sulfate dynamics by soil sulfate adsorption. Water Resources Research 22:1283–1291.

Cotte, C., and C. Guinet. 2007. Historical whaling records reveal major regional retreat of Antarctic sea ice. Deep-Sea Research Part I—Oceanographic Research Papers 54:243–252.

Couradeau, E., K. Benzerara, E. Gerard, D. Moreira, S. Bernard, G. E. Brown Jr., and P. Lopez-Garcia. 2012. An early-branching microbialite cyanobacterium forms intracellular carbonates. Science 336:459–462.

Courchesne, F., and W. H. Hendershot. 1989. Sulfate retention in some podzolic soils of the southern Laurentians, Quebec. Canadian Journal of Soil Science 69:337–350.

Courty, P. E., M. Buee, A. G. Diedhiou, P. Frey-Klett, F. Le Tacon, F. Rineau, M. P. Turpault, S. Uroz, and J. Garbaye. 2010. The role of ectomycorrhizal communities in forest ecosystem processes: New perspectives and emerging concepts. Soil Biology & Biochemistry 42:679–698.

Couto, W., C. Sanzonowicz, and A.D.O. Barcellos. 1985. Factors affecting oxidation-reduction processes in an Oxisol with a seasonal water table. Soil Science Society of America Journal 49:1245–1248.

Covey, K. R., S. A. Wood, R. J. Warren, X. Lee, and M. A. Bradford. 2012. Elevated methane concentrations in trees of an upland forest. Geophysical Research Letters 39.

Covino, T. P., and B. L. McGlynn. 2007. Stream gains and losses across a mountain-to-valley transition: Impacts on watershed hydrology and stream water chemistry. Water Resources Research 43.

Covino, T. P., B. L. McGlynn, and R. A. McNamara. 2010. Tracer additions for spiraling curve characterization (TASCC): Quantifying stream nutrient uptake kinetics from ambient to saturation. Limnology and Oceanography—Methods 8:484–498.

Cowan, J. J., and C. Sneden. 2006. Heavy element synthesis in the oldest stars and the early Universe. Nature 440:1151–1156.

Cox, T. L., W. F. Harris, B. S. Ausmus, and N. T. Edwards. 1978. Role of roots in biogeochemical cycles in an eastern deciduous forest. Pedobiologia 18:264–271.

Craft, C., C. Washburn, and A. Parker. 2008. Latitudinal trends in organic carbon accumulation in temperate freshwater peatlands. In J. Vymagal, editor. Wastewater

Treatment, Plant Dynamics and Management in Constructed and Natural Wetlands. Springer, 23–31.

Craft, C. B., and W. P. Casey. 2000. Sediment and nutrient accumulation in floodplain and depressional freshwater wetlands of Georgia, USA. Wetlands 20: 323–332.

Craft, C. B., and C. J. Richardson. 1993. Peat accretion and N, P, and organic-C accumulation in nutrient-enriched and unenriched Everglades peatlands. Ecological Applications 3:446–458.

Craig, H., C. C. Chou, J. A. Welhan, C. M. Stevens, and A. Engelkemeir. 1988. The isotopic composition of methane in polar ice cores. Science 242:1535–1539.

Craig, H., and T. Hayward. 1987. Oxygen supersaturation in the ocean—Biological versus physical contributions. Science 235:199–202.

Crane, B. R., L. M. Siegel, and E. D. Getzoff. 1995. Sulfite reductase structure at 1.6 A—Evolution and catalysis for reduction of inorganic anions. Science 270:59–67.

Crawford, B., C.S.B. Grimmond, and A. Christen. 2011. Five years of carbon dioxide fluxes measurements in a highly vegetated suburban area. Atmospheric Environment 45:896–905.

Crenshaw, C. L., H. M. Valett, and J. R. Webster. 2002. Effects of augmentation of coarse particulate organic matter on metabolism and nutrient retention in hyporheic sediments. Freshwater Biology 47: 1820–1831.

Crews, T. E. 1999. The presence of nitrogen–fixing legumes in terrestrial communities: Evolutionary vs ecological considerations. Biogeochemistry 46:233–246.

Crews, T. E., K. Kitayama, J. H. Fownes, R. H. Riley, D. A. Herbert, D. Muellerdombois, and P. M. Vitousek. 1995. Changes in soil phosphorus fractions and ecosystem dynamics across a long chronosequence in Hawaii. Ecology 76:1407–1424.

Crill, P. M. 1991. Seasonal patterns of methane uptake and carbon dioxide release by a temperate woodland soil. Global Biogeochemical Cycles 5:319–334.

Crill, P. M., and C. S. Martens. 1986. Methane production from bicarbonate and acetate in an anoxic marine sediment. Geochimica et Cosmochimica Acta 50: 2089–2097.

Cromack, K., P. Sollins, W. C. Graustein, K. Speidel, A. W. Todd, G. Spycher, C. Y. Li, and R. L. Todd. 1979. Calcium-oxalate accumulation and soil weathering in mats of the hypogeous fungus *Hysterangium crassum.* Soil Biology & Biochemistry 11:463–468.

Cronan, C. S. 1980. Solution chemistry of a New Hampshire subalpine ecosystem—A biogeochemical analysis. Oikos 34:272–281.

Cronan, C. S., and G. R. Aiken. 1985. Chemistry and transport of soluble humic substances in forested watersheds of the Adirondack park, New York. Geochimica et Cosmochimica Acta 49:1697–1705.

Cronan, C. S., C. T. Driscoll, R. M. Newton, J. M. Kelly, C. L. Schofield, R. J. Bartlett, and R. April. 1990. A comparative analysis of aluminum biogeochemistry in a northeastern and a southeastern forested watershed. Water Resources Research 26:1413–1430.

Cronan, C. S., and D. F. Grigal. 1995. Use of calcium-aluminum ratios as indicators of stress in forest ecosystems. Journal of Environmental Quality 24: 209–226.

Croot, P. L., P. Streu, and A. R. Baker. 2004. Short residence time for iron in surface seawater impacted by atmospheric dry deposition from Saharan dust events. Geophysical Research Letters 31.

Cross, A. F., and W. H. Schlesinger. 1995. A literature review and evaluation of the Hedley fractionation—Applications to the biogeochemical cycle of soil phosphorus in natural ecosystems. Geoderma 64: 197–214.

Cross, A. T., and T. L. Phillips. 1990. Coal-forming plants through time in North America. International Journal of Coal Geology 16:1–46.

Cross, P. M., and F. H. Rigler. 1983. Phosphorus and iron retention in sediments measured by mass budget calculations and directly. Canadian Journal of Fisheries and Aquatic Sciences 40:1589–1597.

Cross, W. F., J. B. Wallace, and A. D. Rosemond. 2007. Nutrient enrichment reduces constraints on material flows in a detritus-based food web. Ecology 88: 2563–2575.

Crowley, T. J. 2000. Causes of climate change over the past 1000 years. Science 289:270–277.

Crutzen, P. J. 1983. Atmospheric interactions: Homogeneous gas reactions of C, N, and S containing compounds. *In* B. Bolin and R. B. Cook, editors. *The Major Biogeochemical Cycles and Their Interactions.* Wiley, 67–114.

Crutzen, P. J. 1988. Variability atmospheric chemical systems. *In* T. Rosswall, R. G. Woodmansee and P. G. Risser, editors. *Scales and Global Change.* Wiley, 81–108.

Crutzen, P. J. 2006. Albedo enhancement by stratospheric sulfur injections: A contribution to resolve a policy dilemma? Climatic Change 77:211–219.

Crutzen, P. J., and M. O. Andreae. 1990. Biomass burning in the tropics—Impact on atmospheric chemistry and biogeochemical cycles. Science 250:1669–1678.

Crutzen, P. J., I. Aselmann, and W. Seiler. 1986. Methane production by domestic animals, wild ruminants, and other herbivorous fauna, and humans. Tellus 38B:271–284.

Crutzen, P. J., A. C. Delany, J. Greenberg, P. Haagenson, L. Heidt, R. Lueb, W. Pollock, W. Seiler, A. Wartburg, and P. Zimmerman. 1985. Tropospheric chemical composition measurements in Brazil during the dry season. Journal of Atmospheric Chemistry 2: 233–256.

Crutzen, P. J., M. G. Lawrence, and U. Poschl. 1999. On the background photochemistry of tropospheric ozone. Tellus Series A—Dynamic Meteorology and Oceanography 51:123–146.

Crutzen, P. J., and P. H. Zimmermann. 1991. The changing photochemistry of the troposphere. Tellus Series A—Dynamic Meteorology and Oceanography 43: 136–151.

Cuevas, E., and E. Medina. 1986. Nutrient dynamics within Amazonian forest ecosystems. 1. Nutrient flux in fine litter fall and efficiency of nutrient utilization. Oecologia 68:466–472.

Cuevas, E., and E. Medina. 1988. Nutrient dynamics within Amazonian forests. 2. Fine root growth, nutrient availability and leaf litter decomposition. Oecologia 76:222–235.

Cuffey, K. M., and F. Vimeux. 2001. Covariation of carbon dioxide and temperature from the Vostok ice core after deuterium-excess correction. Nature 412: 523–527.

Cullis, C. F., and M. M. Hirschler. 1980. Atmospheric sulphur: Natural and man-made sources. Atmospheric Environment 14:1263–1278.

Cunningham, S. A., S. G. Alderson, B. A. King, and M. A. Brandon. 2003. Transport and variability of the Antarctic circumpolar current in Drake passage. Journal of Geophysical Research—Oceans 108.

Cunningham, S. A., and R. Marsh. 2010. Observing and modeling changes in the Atlantic MOC. Wiley Interdisciplinary Reviews-Climate Change 1: 180–191.

Curry, C. L. 2007. Modeling the soil consumption of atmospheric methane at the global scale. Global Biogeochemical Cycles 21.

Curry, R., and C. Mauritzen. 2005. Dilution of the northern North Atlantic Ocean in recent decades. Science 308:1772–1774.

Curtis, P. S., and X. Z. Wang. 1998. A meta-analysis of elevated CO_2 effects on woody plant mass, form, and physiology. Oecologia 113:299–313.

Cusack, D. F., W. L. Silver, M. S. Torn, and W. H. McDowell. 2011. Effects of nitrogen additions on above- and belowground carbon dynamics in two tropical forests. Biogeochemistry 104:203–225.

Cushon, G. H., and M. C. Feller. 1989. Asymbiotic nitrogen fixation and denitrification in a mature forest in coastal British Columbia. Canadian Journal of Forest Research 19:1194–1200.

Cyr, H., and M. L. Pace. 1993. Magnitude and patterns of herbivory in aquatic and terrestrial ecosystems. Nature 361:148–150.

Cziczo, D. J., D. S. Thomson, and D. M. Murphy. 2001. Ablation, flux, and atmospheric implications of meteors inferred from stratospheric aerosol. Science 291:1772–1775.

D'Arrigo, R., G. C. Jacoby, and I. Y. Fung. 1987. Boreal forests and atmosphere-biosphere exchange of carbon dioxide. Nature 329:321–323.

D'Hondt, S., B. B. Jorgensen, D. J. Miller, A. Batzke, R. Blake, B. A. Cragg, H. Cypionka, G. R. Dickens, T. Ferdelman, K. U. Hinrichs, N. G. Holm, R. Mitterer, A. Spivack, G. Z. Wang, B. Bekins, B. Engelen, K. Ford, G. Gettemy, S. D. Rutherford, H. Sass, C. G. Skilbeck, I. W. Aiello, G. Guerin, C. H. House, F. Inagaki, P. Meister, T. Naehr, S. Niitsuma, R. J. Parkes, A. Schippers, D. C. Smith, A. Teske, J. Wiegel, C. N. Padilla, and J.L.S. Acosta. 2004. Distributions of microbial activities in deep subseafloor sediments. Science 306:2216–2221.

D'Hondt, S., S. Rutherford, and A. J. Spivack. 2002. Metabolic activity of subsurface life in deep-sea sediments. Science 295:2067–2070.

Dacey, J.W.H., B. G. Drake, and M. J. Klug. 1994. Stimulation of methane emission by carbon dioxide enrichment of marsh vegetation. Nature 370:47–49.

Dacey, J.W.H., and S. G. Wakeham. 1986. Oceanic dimethylsulfide—Production during zooplankton grazing on phytoplankton. Science 233:1314–1316.

Dagg, M., R. Benner, S. Lohrenz, and D. Lawrence. 2004. Transformation of dissolved and particulate materials on continental shelves influenced by large rivers: plume processes. Continental Shelf Research 24:833–858.

Dahlgren, R. A., F. C. Ugolini, and W. H. Casey. 1999. Field weathering rates of Mt. St. Helens tephra. Geochimica et Cosmochimica Acta 63:587–598.

Dahlgren, R. A., and W. J. Walker. 1993. Aluminum release rates from selected Spodosol B_s horizons—Effect of pH and solid-phase aluminum pools. Geochimica et Cosmochimica Acta 57:57–66.

Dai, A., I. Y. Fung, and A. D. DelGenio. 1997. Surface observed global land precipitation variations during 1900-88. Journal of Climate 10:2943–2962.

Dai, A., K. E. Trenberth, and T. T. Qian. 2004. A global dataset of Palmer Drought Severity Index for 1870-2002: Relationship with soil moisture and effects of surface warming. Journal of Hydrometeorology 5: 1117–1130.

Dail, D. B., E. A. Davidson, and J. Chorover. 2001. Rapid abiotic transformation of nitrate in an acid forest soil. Biogeochemistry 54:131–146.

Dail, D. B., and J. W. Fitzgerald. 1999. S cycling in soil and stream sediment: influence of season and in situ concentrations of carbon, nitrogen and sulfur. Soil Biology & Biochemistry 31:1395–1404.

Dalsgaard, T., D. E. Canfield, J. Petersen, B. Thamdrup, and J. Acuna-Gonzalez. 2003. N_2 production by the anammox reaction in the anoxic water column of Golfo Dulce, Costa Rica. Nature 422:606–608.

Dalsgaard, T., and B. Thamdrup. 2002. Production of N_2 through anaerobic ammonium oxidation coupled to nitrate reduction in marine sediments. Applied and Environmental Microbiology 68:1312–1318.

Dalva, M., and T. R. Moore. 1991. Sources and sinks of dissolved organic carbon in a forested swamp catchment. Biogeochemistry 15:1–19.

Damman, A.W.H. 1988. Regulation of nitrogen removal and retention in *Sphagnum* bogs and other peatlands. Oikos 51:291–305.

Daniels, W. L., L. W. Zelazny, and C. J. Everett. 1987. Virgin hardwood foerst soils of the southern Appalachian mountains. 2. Weathering, mineralogy, and chemical properties. Soil Science Society of America Journal 51:730–738.

Dannenmann, M., K. Butterbach-Bahl, R. Gasche, G. Willibald, and H. Papen. 2008. Dinitrogen emissions and the N_2:N_2O emission ratio of a Rendzic Leptosol as influenced by pH and forest thinning. Soil Biology & Biochemistry 40:2317–2323.

Darling, M. S. 1976. Interpretation of global differences in plant calorific values—Significance of desert and arid woodland vegetation. Oecologia 23:127–139.

Dasgupta, R., and M. M. Hirschmann. 2010. The deep carbon cycle and melting in Earth's interior. Earth and Planetary Science Letters 298:1–13.

Dauphas, N. 2005. The U/Th production ratio and the age of the Milky Way from meteorites and Galactic halo stars. Nature 435:1203–1205.

Dauphas, N., M. van Zuilen, M. Wadhwa, A. M. Davis, B. Marty, and P. E. Janney. 2004. Clues from Fe isotope variations on the origin of early Archean BIFs from Greenland. Science 306:2077–2080.

David, M. B., M. J. Mitchell, and J. P. Nakas. 1982. Organic and inorganic sulfur constitutents of a forest soil and their relationship to microbial activity. Soil Science Society of America Journal 46:847–852.

David, M. B., J. O. Reuss, and P. M. Walthall. 1988. Use of a chemical-equilibrium model to understand soil chemical processes that influence soil solution and surface water alkalinity. Water Air and Soil Pollution 38:71–83.

David, M. B., G. F. Vance, and W. J. Fasth. 1991. Forest soil response to acid and salt additions of sulfate. 2. Aluminum and base cations. Soil Science 151:208–219.

Davidson, E. A. 2009. The contribution of manure and fertilizer nitrogen to atmospheric nitrous oxide since 1860. Nature Geoscience 2:659–662.

Davidson, E. A., J. Chorover, and D. B. Dail. 2003. A mechanism of abiotic immobilization of nitrate in forest ecosystems: the ferrous wheel hypothesis. Global Change Biology 9:228–236.

Davidson, E. A., A. C. de Araujo, P. Artaxo, J. K. Balch, I. F. Brown, M.M.C. Bustamante, M. T. Coe, R. S. DeFries, M. Keller, M. Longo, J. W. Munger, W. Schroeder, B. S. Soares, C. M. Souza, and S. C. Wofsy. 2012. The Amazon basin in transition. Nature 481:321–328.

Davidson, E. A., C.J.R. de Carvalho, A. M. Figueira, F. Y. Ishida, J. Ometto, G. B. Nardoto, R. T. Saba, S. N. Hayashi, E. C. Leal, I.C.G. Vieira, and L. A. Martinelli. 2007. Recuperation of nitrogen cycling in Amazonian forests following agricultural abandonment. Nature 447:995–998.

Davidson, E. A., S. C. Hart, and M. K. Firestone. 1992. Internal cycling of nitrate in soils of a mature coniferous forest. Ecology 73:1148–1156.

Davidson, E. A., S. C. Hart, C. A. Shanks, and M. K. Firestone. 1991a. Measuring gross nitrogen mineralization, immobilization, and nitrification by ^{15}N isotopic pool dilution in intact soil cores. Journal of Soil Science 42:335–349.

Davidson, E. A., M. Keller, H. E. Erickson, L. V. Verchot, and E. Veldkamp. 2000. Testing a conceptual model of soil emissions of nitrous and nitric oxides. BioScience 50:667–680.

Davidson, E. A., and W. Kingerlee. 1997. A global inventory of nitric oxide emissions from soils. Nutrient Cycling in Agroecosystems 48:37–50.

Davidson, E. A., P. A. Matson, P. M. Vitousek, R. Riley, K. Dunkin, G. Garciamendez, and J. M. Maass. 1993. Processes regulating soil emissions of NO and N_2O in a seasonally dry tropical forest. Ecology 74: 130–139.

Davidson, E. A., C. S. Potter, P. Schlesinger, and S. A. Klooster. 1998. Model estimates of regional nitric oxide emissions from soils of the southeastern United States. Ecological Applications 8:748–759.

Davidson, E. A., J. M. Stark, and M. K. Firestone. 1990. Microbial production and consumption of nitrate in an annual grassland. Ecology 71:1968–1975.

Davidson, E. A., and W. T. Swank. 1987. Factors limiting denitrification in soils from mature and disturbed southeastern hardwood forests. Forest Science 33:135–144.

Davidson, E. A., W. T. Swank, and T. O. Perry. 1986. Distinguishing between nitrification and denitrification as sources of gaseous nitrogen production in soil. Applied and Environmental Microbiology 52: 1280–1286.

Davidson, E. A., P. M. Vitousek, P. A. Matson, R. Riley, G. Garciamendez, and J. M. Maass. 1991b. Soil emissions of nitric oxide in a seasonally dry tropical forest of Mexico. Journal of Geophysical Research—Atmospheres 96:15439–15445.

Davies, W. G. 1972. *Introduction to Chemical Thermodynamics*. W.B. Saunders, Philadelphia.

Davis, C. S., and D. J. McGillicuddy. 2006. Transatlantic abundance of the N_2-fixing colonial cyanobacterium *Trichodesmium*. Science 312:1517–1520.

Davis, R. B., D. S. Anderson, and F. Berge. 1985. Paleolimnological evidence that lake acidification is accompanied by loss of organic matter. Nature 316: 436–438.

Davison, W. 1993. Iron and manganese in lakes. Earth-Science Reviews 34:119–163.

Davison, W., C. Woof, and E. Rigg. 1982. The dynamics of iron and manganese in a seasonally anoxic lake—Direct measurement of fluxes using sediment traps. Limnology and Oceanography 27:987–1003.

Dawson, T. E., and J. R. Ehleringer. 1991. Streamside trees that do not use stream water. Nature 350: 335–337.

Day, J. W., Jr., D. F. Boesch, E. J. Clairain, G. P. Kemp, S. B. Laska, W. J. Mitsch, K. Orth, H. Mashriqui, D. J. Reed, L. Shabman, C. A. Simenstad, B. J. Streever, R. R. Twilley, C. C. Watson, J. T. Wells, and D. F. Whigham. 2007. Restoration of the Mississippi Delta: Lessons from Hurricanes Katrina and Rita. Science 315:1679–1684.

Day, T. A., and P. J. Neale. 2002. Effects of UV-B radiation on terrestrial and aquatic primary producers. Annual Review of Ecology and Systematics 33: 371–396.

de Bergh, C. 1993. The D/H ratio and the evolution of water in the terrestrial planets. Origins of Life and Evolution of the Biosphere 23:11–21.

de Bergh, C., B. Bezard, T. Owen, D. Crisp, J. P. Maillard, and B. L. Lutz. 1991. Deuterium on Venus—Observations from Earth. Science 251:547–549.

de Duve, C. 1995. The beginnings of life on Earth. American Scientist 83:428–437.

De Jonge, V. N., and L. A. Villerius. 1989. Possible role of carbonate dissolution in estuarine phosphate dynamics. Limnology and Oceanography 34:332–340.

De Kimpe, C. R., and Y. A. Martel. 1976. Effects of vegetation on distribution of carbon, iron, and aluminum in B horizons of horthern Appalachian Spodosols. Soil Science Society of America Journal 40:77–80.

de la Mare, W. K. 1997. Abrupt mid-twentieth-century decline in Antarctic sea-ice extent from whaling records. Nature 389:57–60.

De La Rocha, C. L., and D. J. DePaolo. 2000. Isotopic evidence for variations in the marine calcium cycle over the Cenozoic. Science 289:1176–1178.

de Villiers, S., and B. K. Nelson. 1999. Detection of low-temperature hydrothermal fluxes by seawater Mg and Ca anomalies. Science 285:721–723.

Dean, J. M., and A. P. Smith. 1978. Behavioral and morphological adaptations of a tropical plant to high rainfall. Biotropica 10:152–154.

Dean, W. E., and E. Gorham. 1998. Magnitude and significance of carbon burial in lakes, reservoirs, and peatlands. Geology 26:535–538.

Death, R. G., and M. J. Winterbourn. 1995. Diversity patterns in stream benthic invertebrate communities—The influence of habitat stability. Ecology 76: 1446–1460.

DeBano, L. F., and C. E. Conrad. 1978. Effect of fire on nutrients in a chaparral ecosystem. Ecology 59:489–497.

DeBano, L. F., and J. M. Klopatek. 1988. Phosphorus dynamics of pinyon-juniper soils following simulated burning. Soil Science Society of America Journal 52:271–277.

DeBell, D. S., and C. W. Ralston. 1970. Release of nitrogen by burning light forest fuels. Soil Science Society of America Proceedings 34:936–938.

Deevey, E. S. 1970. Mineral cycles. Scientific American 223:148–158.

DeFries, R. S., R. A. Houghton, M. C. Hansen, C. B. Field, D. Skole, and J. Townshend. 2002. Carbon emissions from tropical deforestation and regrowth based on satellite observations for the 1980s and 1990s. Proceedings of the National Academy of Sciences 99: 14256–14261.

Dekas, A. E., R. S. Poretsky, and V. J. Orphan. 2009. Deep-sea archaea fix and share nitrogen in methane-consuming microbial consortia. Science 326:422–426.

del Arco, J. M., A. Escudero, and M. V. Garrido. 1991. Effects of site characteristics on nitrogen retranslocation from senescing leaves. Ecology 72:701–708.

del Giorgio, P. A., and J. J. Cole. 1998. Bacterial growth efficiency in natural aquatic systems. Annual Review of Ecology and Systematics 29:503–541.

del Giorgio, P. A., and C. M. Duarte. 2002. Respiration in the open ocean. Nature 420:379–384.

Del Grosso, S., W. Parton, T. Stohlgren, D. L. Zheng, D. Bachelet, S. Prince, K. Hibbard, and R. Olson. 2008. Global potential net primary production predicted from vegetation class, precipitation, and temperature. Ecology 89:2117–2126.

del Valle, D. A., D. J. Kieber, D. A. Toole, J. Brinkley, and R. P. Kiene. 2009. Biological consumption of dimethylsulfide (DMS) and its importance in DMS dynamics in the Ross Sea, Antarctica. Limnology and Oceanography 54:785–798.

Delaney, M. L. 1998. Phosphorus accumulation in marine sediments and the oceanic phosphorus cycle. Global Biogeochemical Cycles 12:563–572.

DeLaune, R. D., R. H. Baumann, and J. G. Gosselink. 1983. Relationships among vertical accretion, coastal submergence, and erosion in a Louisiana Gulf Coast marsh. Journal of Sedimentary Petrology 53:147–157.

Delcourt, H. R., and W. F. Harris. 1980. Carbon budget of the southeastern U.S. biota—Analysis of historical change in trend from source to sink. Science 210:321–323.

Delgiorgio, P. A., and R. H. Peters. 1994. Patterns in planktonic P-R ratios in lakes—Influence of lake trophy and dissolved organic carbon. Limnology and Oceanography 39:772–787.

Delmas, R., J. P. Lacaux, J. C. Menaut, L. Abbadie, X. Leroux, G. Helas, and J. Lobert. 1995. Nitrogen compound emission from biomass burning in tropical African savanna FOS/DECAFE-1991 Experiment (Lamto, Ivory Coast). Journal of Atmospheric Chemistry 22:175–193.

Delmas, R., and J. Servant. 1983. Atmospheric balance of sulfur above an equatorial forest. Tellus Series B-Chemical and Physical Meteorology 35:110–120.

Delmas, R. A., A. Marenco, J. P. Tathy, B. Cros, and J.G.R. Baudet. 1991. Sources and sinks of methane in the African savanna—CH_4 emissions from biomass burning. Journal of Geophysical Research—Atmospheres 96:7287–7299.

Delmas, R. A., J. P. Tathy, and B. Cros. 1992. Atmospheric methane budget in Africa. Journal of Atmospheric Chemistry 14:395–409.

DeLuca, T. H., and D. R. Keeney. 1993. Glucose-induced nitrate assimilation in prairie and cultivated soils. Biogeochemistry 21:167–176.

DeLuca, T. H., O. Zackrisson, M. C. Nilsson, and A. Sellstedt. 2002. Quantifying nitrogen-fixation in feather moss carpets of boreal forests. Nature 419:917–920.

DeLucia, E. H., R. M. Callaway, E. M. Thomas, and W. H. Schlesinger. 1997. Mechanisms of phosphorus acquisition for ponderosa pine seedlings under high CO_2 and temperature. Annals of Botany 79:111–120.

DeLucia, E. H., J. E. Drake, R. B. Thomas, and M. Gonzalez-Meler. 2007. Forest carbon use efficiency: is respiration a constant fraction of gross primary production? Global Change Biology 13:1157–1167.

DeLucia, E. H., and W. H. Schlesinger. 1991. Resource-use efficiency and drought tolerance in adjacent Great Basin and Sierran plants. Ecology 72:51–58.

DeLucia, E. H., and W. H. Schlesinger. 1995. Photosynthetic rates and nutrient-use efficiency among evergreen and deciduous shrubs in Okefenokee swamp. International Journal of Plant Sciences 156:19–28.

DeLucia, E. H., W. H. Schlesinger, and W. D. Billings. 1988. Water relations and the maintenance of Sierran conifers on hydrothermally-altered rock. Ecology 69:303–311.

Delwiche, C. C. 1970. The nitrogen cycle. Scientific American 223:136–146.

Demars, B.O.L. 2008. Whole-stream phosphorus cycling: Testing methods to assess the effect of saturation of sorption capacity on nutrient uptake length measurements. Water Research 42:2507–2516.

DeMaster, D. J. 2002. The accumulation and cycling of biogenic silica in the Southern Ocean: revisiting the marine silica budget. Deep-Sea Research Part I—Topical Studies in Oceanography 49:3155–3167.

Demers, J. D., C. T. Driscoll, T. J. Fahey, and J. B. Yavitt. 2007. Mercury cycling in litter and soil in different forest types in the Adirondack region, New York. Ecological Applications 17:1341–1351.

Denman, K. L., G. Brasseur, A. Chidthaisong, P. Ciais, P. M. Cox, R. E. Dickinson, D. Hauglustaine, C. Heinze, E. Holland, D. Jacob, U. Lohmann, S. Ramachandran, P. L. da Silva Dias, S. C. Wofsy, and X. Zhang. 2007. Couplings between changes in the climate system and biogeochemistry. Pages 499-587 in S. Solomon, D. Qin, M. Manning, Z. Chen, M. Marquis, K. B. Averyt, M. Tignor, and H. L. Miller, editors. Climate Change 2007: The Physical Science Basis. Contribution of Working Group I to the Fourth Assessment Report of the Intergovernmental Panel on Climate Change. Cambridge University Press.

Denning, A. S., I. Y. Fung, and D. Randall. 1995. Latitudinal gradient of atmospheric CO_2 due to seasonal exchange with land biota. Nature 376:240–243.

Dentener, F., J. Drevet, J. F. Lamarque, I. Bey, B. Eickhout, A. M. Fiore, D. Hauglustaine, L. W. Horowitz, M. Krol, U. C. Kulshrestha, M. Lawrence, C. Galy-Lacaux, S. Rast, D. Shindell, D. Stevenson, T. Van Noije, C. Atherton, N. Bell, D. Bergman, T. Butler, J. Cofala, B. Collins, R. Doherty, K. Ellingsen, J. Galloway, M. Gauss, V. Montanaro, J. F. Muller, G. Pitari, J. Rodriguez, M. Sanderson, F. Solmon, S. Strahan, M. Schultz, K. Sudo, S. Szopa, and O. Wild. 2006. Nitrogen and sulfur deposition on regional and global scales: A multimodel evaluation. Global Biogeochemical Cycles 20.

Dentener, F., W. Peters, M. Krol, M. van Weele, P. Bergamaschi, and J. Lelieveld. 2003. Interannual variability and trend of CH_2 lifetime as a measure for OH changes in the 1979-1993 time period. Journal of Geophysical Research—Atmospheres 108.

Derry, L. A., A. C. Kurtz, K. Ziegler, and O. A. Chadwick. 2005. Biological control of terrestrial silica cycling and export fluxes to watersheds. Nature 433: 728–731.

Deschamps, P., N. Durand, E. Bard, B. Hamelin, G. Camoin, A. L. Thomas, G. Henderson, J. Okuno, and Y. Yokoyama. 2012. Ice-sheet collapse and sea-level rise at the Bolling warming 14,600 years ago. Nature 483:559–564.

Des Marais, D.J.D., H. Strauss, R. E. Summons, and J. M. Hayes. 1992. Carbon isotope evidence for the stepwise oxidation of the Proterozoic environment. Nature 359:605–609.

Després, V. R., J. A. Huffman, S. M. Burrows, C. Hoose, A. S. Safatov, G. Buryak, J. Frohlich-Nowoisky, W. Elbert, M. O. Andreae, U. Poschl, and R. Jaenicke. 2012. Primary biological aerosol particles in the atmosphere: a review. Tellus Series B-Chemical and Physical Meteorology 64.

Dethier, D. P., S. B. Jones, T. P. Feist, and J. E. Ricker. 1988. Relations among sulfate, aluminum, iron, dissolved organic carbon and pH in upland forest soils of northwestern Massachusetts. Soil Science Society of America Journal 52:506–512.

Deuser, W. G., P. G. Brewer, T. D. Jickells, and R. F. Commeau. 1983. Biological control of the removal of abiogenic particles from the surface ocean. Science 219:388–391.

Deuser, W. G., E. H. Ross, and R. F. Anderson. 1981. Seasonality in the supply of sediment to the deep Sargasso sea and implications for the rapid transfer of matter to the deep ocean. Deep-Sea Research Part A—Oceanographic Research Papers 28:495-505.

Deutsch, C., N. Gruber, R. M. Key, J. L. Sarmiento, and A. Ganachaud. 2001. Denitrification and N_2 fixation in the Pacific Ocean. Global Biogeochemical Cycles 15:483–506.

Deutsch, C., J. L. Sarmiento, D. M. Sigman, N. Gruber, and J. P. Dunne. 2007. Spatial coupling of nitrogen inputs and losses in the ocean. Nature 445:163–167.

Dévai, I., and R. D. Delaune. 1995. Evidence for phosphine production and emission from Louisiana and Florida marsh soils. Organic Geochemistry 23:277–279.

Dévai, I., L. Felfoldy, I. Wittner, and S. Plosz. 1988. Detection of phosphine—New aspects of the phosphorus cycle in the hydrosphere. Nature 333: 343–345.

Devol, A. H. 1991. Direct measurement of nitrogen gas fluxes from continental shelf sediments. Nature 349:319–321.

DeWalle, D. R., B. R. Swistock, T. E. Johnson, and K. J. McGuire. 2000. Potential effects of climate change and urbanization on mean annual streamflow in the United States. Water Resources Research 36:2655–2664.

Dey, S., and L. Di Girolamo. 2011. A decade of change in aerosol properties over the Indian subcontinent. Geophysical Research Letters 38.

Dhamala, B. R., and M. J. Mitchell. 1995. Sulfur speciataion, vertical distribution, and seasonal variation in a northern hardwood forest soil, USA. Canadian Journal of Forest Research 25:234–243.

Di, H. J., K. C. Cameron, and R. G. McLaren. 2000. Isotopic dilution methods to determine the gross transformation rates of nitrogen, phosphorus, and sulfur in soil: a review of the theory, methodologies, and limitations. Australian Journal of Soil Research 38:213–230.

Di, H. J., K. C. Cameron, J. P. Shen, C. S. Winefield, M. O'Callaghan, S. Bowatte, and J. Z. He. 2009. Nitrification driven by bacteria and not archaea in nitrogen-rich grassland soils. Nature Geoscience 2:621–624.

Di, H. J., L. M. Condron, and E. Frossard. 1997. Isotope techniques to study phosphorus cycling in agricultural and forest soils: A review. Biology and Fertility of Soils 24:1–12.

Di-Giovanni, C., J. R. Disnar, and J. J. Macaire. 2002. Estimation of the annual yield of organic carbon released from carbonates and shales by chemical weathering. Global and Planetary Change 32:195–210.

Dia, A. N., A. S. Cohen, R. K. O'Nions, and N. J. Shackleton. 1992. Seawater Sr isotope variation over the past 300 kyr and influence of global climate cycles. Nature 356:786–788.

Dianov-Klokov, V. I., L. N. Yurganov, E. I. Grechko, and A. V. Dzhola. 1989. Spectroscopic measurements of atmospheric carbon monoxide and methane. 1. Latitudinal distribution. Journal of Atmospheric Chemistry 8:139–151.

Diaz, J., E. Ingall, C. Benitez-Nelson, D. Paterson, M. D. de Jonge, I. McNulty, and J. A. Brandes. 2008. Marine polyphosphate: A key player in geologic phosphorus sequestration. Science 320:652–655.

Diaz, R. J., and R. Rosenberg. 2008. Spreading dead zones and consequences for marine ecosystems. Science 321:926–929.

Diaz-Ravina, M., M. J. Acea, and T. Carballas. 1993. Microbial biomass and its contribution to nutrient concentrations in forest soils. Soil Biology & Biochemistry 25:25–31.

Dickerson, R. E. 1978. Chemical evolution and the origin of life. Scientific American 239:70–86.

Dickerson, R. R., G. J. Huffman, W. T. Luke, L. J. Nunnermacker, K. E. Pickering, A.C.D. Leslie, C. G. Lindsey, W.G.N. Slinn, T. J. Kelly, P. H. Daum, A. C. Delany, J. P. Greenberg, P. R. Zimmerman, J. F. Boatman, J. D. Ray, and D. H. Stedman. 1987. Thunderstorms—An important mechanism in the transport of air pollutants. Science 235:460–464.

Dickson, B., I. Yashayaev, J. Meincke, B. Turrell, S. Dye, and J. Holfort. 2002. Rapid freshening of the deep North Atlantic Ocean over the past four decades. Nature 416:832–837.

Dickson, M. L., and P. A. Wheeler. 1995. Nitrate uptake rates in a coastal upwelling regime—A comparison of pN-specific, absolute, and Chl A-specific rates. Limnology and Oceanography 40:533–543.

Dickson, R. R., and J. Brown. 1994. The production of North Atlantic deep water—Sources, rates, and pathways. Journal of Geophysical Research—Oceans **99**: 12319–12341.

Dieleman, W.I.J., S. Luyssaert, A. Rey, P. De Angelis, C.V.M. Barton, M.S.J. Broadmeadow, S. B. Broadmeadow, K. S. Chigwerewe, M. Crookshanks, E. Dufrene, P. G. Jarvis, A. Kasurinen, S. Kellomaki, V. Le Dantec, M. Liberloo, M. Marek, B. Medlyn, R. Pokorny, G. Scarascia-Mugnozza, V. M. Temperton, D. Tingey, O. Urban, R. Ceulemans, and I. A. Janssens. 2010. Soil N modulates soil C cycling in CO_2-fumigated tree stands: a meta-analysis. Plant Cell and Environment **33**:2001–2011.

Dillaway, D. N., and E. L. Kruger. 2010. Thermal acclimation of photosynthesis: a comparison of boreal and temperate tree species along a latitudinal transect. Plant Cell and Environment **33**:888–899.

Dillon, P. J., and H. E. Evans. 1993. A comparison of phosphorus retention in lakes determined from mass-balance and sediment core calculations. Water Research **27**:659–668.

Dillon, P. J., and H. E. Evans. 2001. Comparison of iron accumulation in lakes using sediment core and mass balance calculations. Science of the Total Environment **266**:211–219.

Dillon, P. J., and L. A. Molot. 1990. The role of ammonium and nitrate retention in the acidification of lakes and forested catchments. Biogeochemistry **11**:23–43.

Dillon, P. J., and L. A. Molot. 1997. Effect of landscape form on export of dissolved organic carbon, iron, and phosphorus from forested stream catchments. Water Resources Research **33**:2591–2600.

Dingman, S. L., and A. H. Johnson. 1971. Pollution potential of some New Hampshire lakes. Water Resources Research **7**:1208–1215.

Dinkelaker, B., and H. Marschner. 1992. *In vivo* demonstration of acid-phosphatase activity in the rhizosphere of soil-grown plants. Plant and Soil **144**:199–205.

Dirmeyer, P. A., and J. Shukla. 1996. The effect on regional and global climate of expansion of the world's deserts. Quarterly Journal of the Royal Meteorological Society **122**:451–482.

Dise, N. B., and E. S. Verry. 2001. Suppression of peatland methane emission by cumulative sulfate deposition in simulated acid rain. Biogeochemistry **53**:143–160.

Dittman, J. A., J. B. Shanley, C. T. Driscoll, G. R. Aiken, A. T. Chalmers, J. E. Towse, and P. Selvendiran. 2010. Mercury dynamics in relation to dissolved organic carbon concentration and quality during high flow events in three northeastern US streams. Water Resources Research **46**.

Dixit, S., and P. Van Cappellen. 2003. Predicting benthic fluxes of silicic acid from deep-sea sediments. Journal of Geophysical Research—Oceans **108**.

Dixit, S., P. Van Cappellen, and A. J. van Bennekom. 2001. Processes controlling solubility of biogenic silica and pore water build-up of silicic acid in marine sediments. Marine Chemistry **73**:333–352.

Dixon, J. E., L. Leist, C. Langmuir, and J. G. Schilling. 2002. Recycled dehydrated lithosphere observed in plume-influenced mid-ocean-ridge basalt. Nature **420**: 385–389.

Dixon, J. L., A. S. Hartshorn, A. M. Heimsath, R. A. DiBiase, and K. X. Whipple. 2012. Chemical weathering response to tectonic forcing: A soils perspective from the San Gabriel Mountains, California. Earth and Planetary Science Letters **323/324**:40–49.

Dixon, K. W., J. S. Pate, and W. J. Bailey. 1980. Nitrogen nutrition of the tuberous sundew *Drosera erythrorhiza* Lindl with special reference to catch of arthropod fauna by its glandular leaves. Australian Journal of Botany **28**:283–297.

Dlugokencky, E. J., L. Bruhwiler, J.W.C. White, L. K. Emmons, P. C. Novelli, S. A. Montzka, K. A. Masarie, P. M. Lang, A. M. Crotwell, J. B. Miller, and L. V. Gatti. 2009. Observational constraints on recent increases in the atmospheric CH_2 burden. Geophysical Research Letters **36**.

Dlugokencky, E. J., E. G. Nisbet, R. Fisher, and D. A. Lowry. 2011. Global atmospheric methane:budget,changes and dangers. Philosophical Transactions of the Royal Society of London. Series B, Biological Sciences 2058–2072.

Dlugokencky, E. J., L. P. Steele, P. M. Lang, and K. A. Masarie. 1994. The growth rate and distribution of atmospheric methane. Journal of Geophysical Research—Atmospheres **99**:17021–17043.

do Carmo, J. B., M. Keller, J. D. Dias, P. B. de Camargo, and P. Crill. 2006. A source of methane from upland forests in the Brazilian Amazon. Geophysical Research Letters **33**.

Dobrovolsky, V. V. 1994. *Biogeochemistry of the World's Land*. CRC Press.

Dodd, J. C., C. C. Burton, R. G. Burns, and P. Jeffries. 1987. Phosphatase activity associated with the roots and the rhizosphere of plants infected with vesicular-arbuscular mycorrhizal fungi. New Phytologist **107**:163–172.

Dodds, W. K., V. H. Smith, and K. Lohman. 2002. Nitrogen and phosphorus relationships to benthic algal biomass in temperate streams. Canadian Journal of Fisheries and Aquatic Sciences **59**: 865–874.

Doll, P., and K. Fiedler. 2008. Global-scale modeling of groundwater recharge. Hydrology and Earth System Sciences 12:863–885.

Don, A., J. Schumacher, and A. Freibauer. 2011. Impact of tropical land-use change on soil organic carbon stocks—a meta-analysis. Global Change Biology 17:1658–1670.

Donahue, T. M., J. H. Hoffman, R. R. Hodges, and A. J. Watson. 1982. Venus was wet—A measurement of the ratio of deuterium to hydrogen. Science 216: 630–633.

Donarummo, J., M. Ram, and E. F. Stoermer. 2003. Possible deposit of soil dust from the 1930's U.S. dust bowl identified in Greenland ice. Geophysical Research Letters 30.

Donat, J. R., and K. W. Bruland. 1995. Trace elements in the oceans. In B. Salbu and E. Steinnes, editors. Trace Elements in Natural Waters. CRC Press, 247–281.

Doney, S. C. 2010. The growing human footprint on coastal and open-ocean biogeochemistry. Science 328:1512–1516.

Dong, L. F., M. N. Sobey, C. J. Smith, I. Rusmana, W. Phillips, A. Stott, A. M. Osborn, and D. B. Nedwell. 2011. Dissimilatory reduction of nitrate to ammonium, not denitrification or anammox, dominates benthic nitrate reduction in tropical estuaries. Limnology and Oceanography 56:279–291.

Doolittle, R. F., D. F. Feng, S. Tsang, G. Cho, and E. Little. 1996. Determining divergence times of the major kingdoms of living organisms with a protein clock. Science 271:470–477.

Dore, J. E., R. Lukas, D. W. Sadler, M. J. Church, and D. M. Karl. 2009. Physical and biogeochemical modulation of ocean acidification in the central North Pacific. Proceedings of the National Academy of Sciences 106:12235–12240.

Dorn, H.-P., J. Callies, U. Platt, and D. H. Ehhalt. 1988. Measurement of tropospheric OH concentration by laser long-path absorption spectroscopy. Tellus 40B: 437–445.

Dornblaser, M., A. E. Giblin, B. Fry, and B. J. Peterson. 1994. Effects of sulfate concentration in the overlying water on sulfate reduction and sulfur storage in lake sediments. Biogeochemistry 24:129–144.

Dörr, H., and K. O. Münnich. 1989. Downward movement of soil organic matter and its influence on trace element transport (Pb-210, Cs-137) in the soil. Radiocarbon 31:655–663.

Dorrepaal, E., S. Toet, R.S.P. van Logtestijn, E. Swart, M. J. van de Weg, T. V. Callaghan, and R. Aerts. 2009. Carbon respiration from subsurface peat accelerated by climate warming in the subarctic. Nature 460:616–619.

Downing, J. A., J. J. Cole, J. J. Middelburg, R. G. Striegl, C. M. Duarte, P. Kortelainen, Y. T. Prairie, and K. A. Laube. 2008. Sediment organic carbon burial in agriculturally eutrophic impoundments over the last century. Global Biogeochemical Cycles 22.

Downing, J. A., and E. McCauley. 1992. The nitrogen-phosphorus relationship in lakes. Limnology and Oceanography 37:936–945.

Downing, J. A., Y. T. Prairie, J. J. Cole, C. M. Duarte, L. J. Tranvik, R. G. Striegl, W. H. McDowell, P. Kortelainen, N. F. Caraco, J. M. Melack, and J. J. Middelburg. 2006. The global abundance and size distribution of lakes, ponds, and impoundments. Limnology and Oceanography 51:2388–2397.

Downs, M. R., K. J. Nadelhoffer, J. M. Melillo, and J. D. Aber. 1996. Immobilization of a N-15-labeled nitrate addition by decomposing forest litter. Oecologia 105:141–150.

Doyle, R. D., and T. R. Fisher. 1994. Nitrogen fixation by periphyton and plankton on the Amazon floodplain at Lake Calado. Biogeochemistry 26:41–66.

Draaijers, G.P.J., W. Ivens, M. M. Bos, and W. Bleuten. 1989. The contribution of ammonia emissions from agriculture to the deposition of acidifying and eutrophying compounds onto forests. Environmental Pollution 60:55–66.

Drake, J. B., R. G. Knox, R. O. Dubayah, D. B. Clark, R. Condit, J. B. Blair, and M. Hofton. 2003. Aboveground biomass estimation in closed canopy Neotropical forests using lidar remote sensing: factors affecting the generality of relationships. Global Ecology and Biogeography 12:147–159.

Drake, J. E., A. Gallet-Budynek, K. S. Hofmockel, E. S. Bernhardt, S. A. Billings, R. B. Jackson, K. S. Johnsen, J. Lichter, H. R. McCarthy, M. L. McCormack, D.J.P. Moore, R. Oren, S. Palmroth, R. P. Phillips, J. S. Pippen, S. G. Pritchard, K. K. Treseder, W. H. Schlesinger, E. H. DeLucia, and A. C. Finzi. 2011. Increases in the flux of carbon belowground stimulate nitrogen uptake and sustain the long-term enhancement of forest productivity under elevated CO_2. Ecology Letters 14:349–357.

Dregne, H. 1976. The Soils of Arid Regions. Elsevier.

Drever, J. I. 1988. Geochemistry of Natural Waters. 2nd edition. Prentice-Hall.

Drever, J. I. 1994. The effect of land plants on weathering rates of silicate minerals. Geochimica et Cosmochimica Acta 58:2325–2332.

Drever, J. I., and C. L. Smith. 1978. Cyclic wetting and drying of the soil zone as an influence on chemistry of ground water in arid terrains. American Journal of Science 278:1448–1454.

Driscoll, C. T., and G. E. Likens. 1982. Hydrogen ion budget of an aggrading forested ecosystem. Tellus **34**: 283–292.

Driscoll, C. T., N. Vanbreemen, and J. Mulder. 1985. Aluminum chemistry in a forested Spodosol. Soil Science Society of America Journal **49**:437–444.

Drouet, T., J. Herbauts, W. Gruber, and D. Demaiffe. 2005. Strontium isotope composition as a tracer of calcium sources in two forest ecosystems in Belgium. Geoderma **126**:203–223.

Druffel, E.R.M., P. M. Williams, J. E. Bauer, and J. R. Ertel. 1992. Cycling of dissolved and particulate organic matter in the open ocean. Journal of Geophysical Research—Oceans **97**:15639–15659.

Drury, C. F., D. J. McKenney, and W. I. Findlay. 1992. Nitric oxide and nitrous oxide production from soil—Water and oxygen effects. Soil Science Society of America Journal **56**:766–770.

Drury, C. F., R. P. Voroney, and E. G. Beauchamp. 1991. Availability of NH_4^+-N to microorganisms and the soil internal N cycle. Soil Biology & Biochemistry **23**:165–169.

Duarte, C. M., and S. Agusti. 1998. The CO_2 balance of unproductive aquatic ecosystems. Science **281**: 234–236.

Duarte, C. M., and J. Kalff. 1989. The influence of catchment geology and lake depth on phytoplankton biomass. Archiv Fur Hydrobiologie **115**:27–40.

Dubayah, R. O., S. L. Sheldon, D. B. Clark, M. A. Hofton, J. B. Blair, G. C. Hurtt, and R. L. Chazdon. 2010. Estimation of tropical forest height and biomass dynamics using lidar remote sensing at La Selva, Costa Rica. Journal of Geophysical Research—Biogeosciences **115**.

Dubinsky, E. A., W. L. Silver, and M. K. Firestone. 2010. Tropical forest soil microbial communities couple iron and carbon biogeochemistry. Ecology **91**:2604–2612.

Duce, R. A., J. LaRoche, K. Altieri, K. R. Arrigo, A. R. Baker, D. G. Capone, S. Cornell, F. Dentener, J. Galloway, R. S. Ganeshram, R. J. Geider, T. Jickells, M. M. Kuypers, R. Langlois, P. S. Liss, S. M. Liu, J. J. Middelburg, C. M. Moore, S. Nickovic, A. Oschlies, T. Pedersen, J. Prospero, R. Schlitzer, S. Seitzinger, L. L. Sorensen, M. Uematsu, O. Ulloa, M. Voss, B. Ward, and L. Zamora. 2008. Impacts of atmospheric anthropogenic nitrogen on the open ocean. Science **320**:893–897.

Duce, R. A., P. S. Liss, J. T. Merrill, E. L. Atlas, P. Buat-Menard, B. B. Hicks, J. M. Miller, J. M. Prospero, R. Arimoto, T. M. Church, W. Ellis, J. N. Galloway, L. Hansen, T. D. Jickells, A. H. Knap, K. H. Reinhardt, B. Schneider, A. Soudine, J. J. Tokos, S. Tsunogai, R. Wollast, and M. Zhou. 1991. The atmospheric input of trace species to the world ocean. Global Biogeochemical Cycles **5**:193–259.

Duce, R. A., and N. W. Tindale. 1991. Atmospheric transport of iron and its deposition in the ocean. Limnology and Oceanography **36**:1715–1726.

Duce, R. A., C. K. Unni, B. J. Ray, J. M. Prospero, and J. T. Merrill. 1980. Long-range atmospheric transport of soil dust from Asia to the tropical North Pacific—Temporal variability. Science **209**:1522–1524.

Ducet, N., P. Y. Le Traon, and G. Reverdin. 2000. Global high-resolution mapping of ocean circulation from TOPEX/Poseidon and ERS-1 and-2. Journal of Geophysical Research—Oceans **105**:19477–19498.

Ducklow, H. W., and C. A. Carlson. 1992. Oceanic bacterial production. Advances in Microbial Ecology **12**:113–181.

Ducklow, H. W., D. A. Purdie, P.J.L. Williams, and J. M. Davies. 1986. Bacterioplankton—A sink for carbon in a coastal marine plankton community. Science **232**: 865–867.

Dueck, T. A., R. de Visser, H. Poorter, S. Persijn, A. Gorissen, W. de Visser, A. Schapendonk, J. Verhagen, J. Snel, F.J.M. Harren, A.K.Y. Ngai, F. Verstappen, H. Bouwmeester, L. Voesenek, and A. van der Werf. 2007. No evidence for substantial aerobic methane emission by terrestrial plants: a C-13-labelling approach. New Phytologist **175**:29–35.

Duff, S.M.G., G. Sarath, and W. C. Plaxton. 1994. The role of acid phosphatases in plant phosphorus metabolism. Physiologia Plantarum **90**:791–800.

Dugdale, R. C., and J. J. Goering. 1967. Uptake of new and regenerated forms of nitrogen in primary production. Limnology and Oceanography **12**:196–206.

Duggin, J. A., G. K. Voigt, and F. H. Bormann. 1991. Autotrophic and heterotrophic nitrification in response to clear cutting northern hardwood forest. Soil Biology & Biochemistry **23**:779–787.

Dukes, J. S. 2003. Burning buried sunshine: Human consumption of ancient solar energy. Climatic Change **61**:31–44.

Duncan, B. N., J. A. Logan, I. Bey, I. A. Megretskaia, R. M. Yantosca, P. C. Novelli, N. B. Jones, and C. P. Rinsland. 2007. Global budget of CO, 1988-1997: Source estimates and validation with a global model. Journal of Geophysical Research—Atmospheres **112**.

Dunfield, P., R. Knowles, R. Dumont, and T. R. Moore. 1993. Methane production and consumption in temperate and sub-arctic peat soils—response to temperature and pH. Soil Biology & Biochemistry **25**: 321–326.

Dunn, P. H., L. F. DeBano, and G. E. Eberlein. 1979. Effects of burning on chaparral soils. 2. Soil microbes and nitrogen mineralization. Soil Science Society of America Journal **43**:509–514.

Dunne, T., and L. B. Leopold. 1978. *Water in Environmental Planning*. Freeman, San Francisco.

Durka, W., E. D. Schulze, G. Gebauer, and S. Voerkelius. 1994. Effects of forest decline on uptake and leaching of deposited nitrate determined from ^{15}N and ^{18}O measurements. Nature 372:765–767.

Durr, H. H., M. Meybeck, and S. H. Durr. 2005. Lithologic composition of the Earth's continental surfaces derived from a new digital map emphasizing riverine material transfer. Global Biogeochemical Cycles 19.

Dutaur, L., and L. V. Verchot. 2007. A global inventory of the soil CH_4 sink. Global Biogeochemical Cycles 21.

Dutta, K., E.A.G. Schuur, J. C. Neff, and S. A. Zimov. 2006. Potential carbon release from permafrost soils of Northeastern Siberia. Global Change Biology 12:2336–2351.

Dutton, A., and K. Lambeck. 2012. Ice volume and sea level during the last interglacial. Science 337:216–218.

Duyzer, J., and D. Fowler. 1994. Modeling land-atmosphere exchange of gaseous oxides of nitrogen in Europe. Tellus Series B—Chemical and Physical Meteorology 46:353–372.

Dyhrman, S. T., P. D. Chappell, S. T. Haley, J. W. Moffett, E. D. Orchard, J. B. Waterbury, and E. A. Webb. 2006. Phosphonate utilization by the globally important marine diazotroph *Trichodesmium*. Nature 439:68–71.

Dynesius, M., and C. Nilsson. 1994. Fragmentation and flow regulation of river systems in the northern third of the world. Science 266:753–762.

Dyrness, C. T., K. Vancleve, and J. D. Levison. 1989. The effect of wildfire on soil chemistry in four forest types in interior Alaska. Canadian Journal of Forest Research 19:1389–1396.

Eckert, W., and R. Conrad. 2007. Sulfide and methane evolution in the hypolimnion of a subtropical lake: a three-year study. Biogeochemistry 82:67–76.

Eckstein, R. L., P. S. Karlsson, and M. Weih. 1999. Leaf life span and nutrient resorption as determinants of plant nutrient conservation in temperate-arctic regions. New Phytologist 143:177–189.

Edmond, J. M., C. Measures, R. E. McDuff, L. H. Chan, R. Collier, B. Grant, L. I. Gordon, and J. B. Corliss. 1979. Ridge crest hydrothermal activity and the balances of the major and minor elements in the ocean—Galapagos data. Earth and Planetary Science Letters 46:1–18.

Edmondson, W. T., and J. T. Lehman. 1981. The effect of changes in the nutrient income on the condition of Lake Washington. Limnology and Oceanography 26:1–29.

Edwards, D. P., J. F. Lamarque, J. L. Attie, L. K. Emmons, A. Richter, J. P. Cammas, J. C. Gille, G. L. Francis, M. N. Deeter, J. Warner, D. C. Ziskin, L. V. Lyjak, J. R. Drummond, and J. P. Burrows. 2003. Tropospheric ozone over the tropical Atlantic: A satellite perspective. Journal of Geophysical Research—Atmospheres 108.

Edwards, N. T. 1975. Effects of temperature and moisture on carbon dioxide evolution in a mixed deciduous forest floor. Soil Science Society of America Journal 39:361–365.

Edwards, N. T., and W. F. Harris. 1977. Carbon cycling in a mixed deciduous forest floor. Ecology 58:431–437.

Edwards, N. T., and P. Sollins. 1973. Continuous measurement of carbon dioxide evolution from partitioned forest floor components. Ecology 34:406–412.

Edwards, P. J. 1977. Studies of mineral cycling in a montane rain forest in New Guinea. 2. Production and disappearance of litter. Journal of Ecology 65:971–992.

Edwards, P. J. 1982. Studies of mineral cycling in a montane rain forest in New Guinea. 5. Rates of cycling in throughfall and litter fall. Journal of Ecology 70:807–827.

Edwards, P. J., and P. J. Grubb. 1982. Studies of mineral cycling in a montane rain forest in New Guinea. 4. Soil characteristics and the division of mineral elements between the vegetation and soil. Journal of Ecology 70:649–666.

Eghbal, M. K., R. J. Southard, and L. D. Whittig. 1989. Dynamics of evaporite distribution in soils on a fan-playa transect in the Carrizo Plain, California. Soil Science Society of America Journal 53:898–903.

Ehhalt, D. H. 1974. The atmospheric cycle of methane. Tellus 26:58–70.

Ehleringer, J. R., and R. K. Monson. 1993. Evolutionary and ecological apsects of photosynthetic pathway variation. Annual Review of Ecology and Systematics 24:411–439.

Ehlmann, B. L., J. F. Mustard, S. L. Murchie, F. Poulet, J. L. Bishop, A. J. Brown, W. M. Calvin, R. N. Clark, D. J. Des Marais, R. E. Milliken, L. H. Roach, T. L. Roush, G. A. Swayze, and J. J. Wray. 2008. Orbital identification of carbonate-bearing rocks on Mars. Science 322:1828–1832.

Ehrlich, H. L. 1975. Formation of ores in the sedimentary environment of the deep sea with microbial participation—Case for ferromanganese concretions. Soil Science 119:36–41.

Ehrlich, H. L. 1982. Enhanced removal of Mn^{2+} from seawater by marine sediments and clay minerals in the presence of bacteria. Canadian Journal of Microbiology 28:1389–1395.

Einsele, G., J. P. Yan, and M. Hinderer. 2001. Atmospheric carbon burial in modern lake basins and its significance for the global carbon budget. Global and Planetary Change 30:167–195.

Eisele, K. A., D. S. Schimel, L. A. Kapustka, and W. J. Parton. 1989. Effects of available P-ratio and N-P ratio on non-symbiotic dinitrogen fixation in tallgrass praire soils Oecologia 79:471–474.

Eissenstat, D. M., and R. D. Yanai. 1997. The ecology of root lifespan. Advances in Ecological Research 27: 1–60.

El Albani, A., S. Bengtson, D. E. Canfield, A. Bekker, R. Macchiarelli, A. Mazurier, E. U. Hammarlund, P. Boulvais, J. J. Dupuy, C. Fontaine, F. T. Fursich, F. Gauthier-Lafaye, P. Janvier, E. Javaux, F. O. Ossa, A. C. Pierson-Wickmann, A. Riboulleau, P. Sardini, D. Vachard, M. Whitehouse, and A. Meunier. 2010. Large colonial organisms with coordinated growth in oxygenated environments 2.1 Gyr ago. Nature 466: 100–104.

Elbert, W., M. R. Hoffmann, M. Kramer, G. Schmitt, and M. O. Andreae. 2000. Control of solute concentrations in cloud and fog water by liquid water content. Atmospheric Environment 34:1109–1122.

Elderfield, H., and R.E.M. Rickaby. 2000. Oceanic Cd/P ratio and nutrient utilization in the glacial Southern Ocean. Nature 405:305–310.

Elderfield, H., and A. Schultz. 1996. Mid-ocean ridge hydrothermal fluxes and the chemical composition of the ocean. Annual Review of Earth and Planetary Sciences 24:191–224.

Elkins, J. W., T. M. Thompson, T. H. Swanson, J. H. Butler, B. D. Hall, S. O. Cummings, D. A. Fisher, and A. G. Raffo. 1993. Decrease in the growth rates of atmospheric chlorofluorocarbon-11 and chlorofluorocarbon-12. Nature 364:780–783.

Elliott, E. T. 1986. Aggregate structure and carbon, nitrogen, and phosphorus in native and cultivated soils. Soil Science Society of America Journal 50:627–633.

Ellis, E. C., K. K. Goldewijk, S. Siebert, D. Lightman, and N. Ramankutty. 2010. Anthropogenic transformation of the biomes, 1700 to 2000. Global Ecology and Biogeography 19:589–606.

Elrod, V. A., W. M. Berelson, K. H. Coale, and K. S. Johnson. 2004. The flux of iron from continental shelf sediments: A missing source for global budgets. Geophysical Research Letters 31.

Elser, J. J., T. Andersen, J. S. Baron, A.-K. Bergstroem, M. Jansson, M. Kyle, K. R. Nydick, L. Steger, and D. O. Hessen. 2009. Shifts in lake N:P stoichiometry and nutrient limitation driven by atmospheric nitrogen deposition. Science 326:835–837.

Elser, J. J., M.E.S. Bracken, E. E. Cleland, D. S. Gruner, W. S. Harpole, H. Hillebrand, J. T. Ngai, E. W. Seabloom, J. B. Shurin, and J. E. Smith. 2007. Global analysis of nitrogen and phosphorus limitation of primary producers in freshwater, marine and terrestrial ecosystems. Ecology Letters 10:1135–1142.

Elser, J. J., T. H. Chrzanowski, R. W. Sterner, and K. H. Mills. 1998. Stoichiometric constraints on food-web dynamics: A whole-lake experiment on the Canadian shield. Ecosystems 1:120–136.

Elser, J. J., W. F. Fagan, R. F. Denno, D. R. Dobberfuhl, A. Folarin, A. Huberty, S. Interlandi, S. S. Kilham, E. McCauley, K. L. Schulz, E. H. Siemann, and R. W. Sterner. 2000a. Nutritional constraints in terrestrial and freshwater food webs. Nature 408:578–580.

Elser, J. J., W. F. Fagan, A. J. Kerkhoff, N. G. Swenson, and B. J. Enquist. 2010. Biological stoichiometry of plant production: metabolism, scaling and ecological response to global change. New Phytologist 186: 593–608.

Elser, J. J., and R. P. Hassett. 1994. A stoichiometric analysis of the zooplankton-phytoplankton interaction in marine and freshwater ecosystems. Nature 370:211–213.

Elser, J. J., R. W. Sterner, E. Gorokhova, W. F. Fagan, T. A. Markow, J. B. Cotner, J. F. Harrison, S. E. Hobbie, G. M. Odell, and L. J. Weider. 2000b. Biological stoichiometry from genes to ecosystems. Ecology Letters 3:540–550.

Elsgaard, L., M. F. Isaksen, B. B. Jorgensen, A. M. Alayse, and H. W. Jannasch. 1994. Microbial sulfate reduction in deep sea sediments at the Guaymas basin-hydrothermal vent area—Influence of temperature and substrates. Geochimica et Cosmochimica Acta 58:3335–3343.

Eltahir, E.A.B., and R. L. Bras. 1994. Precipitation recycling in the Amazon basin. Quarterly Journal of the Royal Meteorological Society 120:861–880.

Emanuel, W. R., I.Y.-S. Fung, G. G. Killough, B. Moore, and T.-H. Peng. 1985a. Modeling the global carbon cycle and changes in the atmospheric carbon dioxide. In J. R. Trabalka, editor. Atmospheric Carbon Dioxide and the Global Carbon Cycle. U.S. Department of Energy, Washington, D.C, 141–173.

Emanuel, W. R., H. H. Shugart, and M. P. Stevenson. 1985b. Climatic change and the broad-scale distribution of terrestrial ecosystem complexes. Climatic Change 7:29–43.

Emerson, S., K. Fischer, C. Reimers, and D. Heggie. 1985. Organic carbon dynamics and preservation in deep sea sediments. Deep-Sea Research Part A—Oceanographic Research Papers 32:1–21.

Emerson, S., and C. Stump. 2010. Net biological oxygen production in the ocean-II: Remote in situ measurements of O_2 and N_2 in subarctic Pacific surface waters. Deep-Sea Research Part I—Oceanographic Research Papers 57:1255–1265.

Emmenegger, L., J. Mohn, M. Sigrist, D. Marinov, U. Steinemann, F. Zumsteg, and M. Meier. 2004. Measurement of ammonia emissions using various techniques in a comparative tunnel study. International Journal of Environment and Pollution **22**:326–341.

Emmett, B. A., J. A. Hudson, P. A. Coward, and B. Reynolds. 1994. The impact of a riparian wetland on streamwater quality in a recently afforested upland catchment. Journal of Hydrology **162**:337–353.

Engel, M. H., and S. A. Macko. 1997. Isotopic evidence for extraterrestrial non-racemic amino acids in the Murchison meteorite. Nature **389**:265–268.

Engel, M. H., and S. A. Macko. 2001. The stereochemistry of amino acids in the Murchison meteorite. Precambrian Research **106**:35–45.

Engel, M. H., S. A. Macko, and J. A. Silfer. 1990. Carbon isotope composition of individual amino acids in the Murchison meteorite. Nature **348**:47–49.

Engelhardt, K.A.M., and M. E. Ritchie. 2001. Effects of macrophyte species richness on wetland ecosystem functioning and services. Nature **411**:687–689.

Engelstaedter, S., K. E. Kohfeld, I. Tegen, and S. P. Harrison. 2003. Controls of dust emissions by vegetation and topographic depressions: An evaluation using dust storm frequency data. Geophysical Research Letters **30**.

Engle, M. A., M. S. Gustin, F. Goff, D. A. Counce, C. J. Janik, D. Bergfeld, and J. J. Rytuba. 2006. Atmospheric mercury emissions from substrates and fumaroles associated with three hydrothermal systems in the western United States. Journal of Geophysical Research—Atmospheres **111**.

Engle, M. A., M. T. Tate, D. P. Krabbenhoft, J. J. Schauer, A. Kolker, J. B. Shanley, and M. H. Bothner. 2010. Comparison of atmospheric mercury speciation and deposition at nine sites across central and eastern North America. Journal of Geophysical Research—Atmospheres **115**.

Engstrom, P., T. Dalsgaard, S. Hulth, and R. C. Aller. 2005. Anaerobic ammonium oxidation by nitrite (anammox): Implications for N_2 production in coastal marine sediments. Geochimica et Cosmochimica Acta **69**:2057–2065.

Enoki, T., and H. Kawaguchi. 1999. Nitrogen resorption from needles of *Pinus thunbergii* Parl. growing along a topographic gradient of soil nutrient availability. Ecological Research **14**:1–8.

Enquist, B. J., A. J. Kerkhoff, S. C. Stark, N. G. Swenson, M. C. McCarthy, and C. A. Price. 2007. A general integrative model for scaling plant growth, carbon flux, and functional trait spectra. Nature **449**:218–222.

Enquist, B. J., and K. J. Niklas. 2002. Global allocation rules for patterns of biomass partitioning in seed plants. Science **295**:1517–1520.

Enriquez, S., C. M. Duarte, and K. Sandjensen. 1993. Patterns in decomposition rates among photosynthetic organisms—The importance of detritus C-N-P content. Oecologia **94**:457–471.

Ensign, S. H., and M. W. Doyle. 2006. Nutrient spiraling in streams and river networks. Journal of Geophysical Research **111**:G04009.

Eppley, R. W., and B. J. Peterson. 1979. Particulate organic matter flux and planktonic new production in the deep ocean. Nature **282**:677–680.

Eppley, R. W., E. Stewart, M. R. Abbott, and U. Heyman. 1985. Estimating ocean primary production from satellite chlorophyll—Introduction to regional differences and statistics for the southern California bight. Journal of Plankton Research **7**:57–70.

Epstein, C. B., and M. Oppenheimer. 1986. Empirical relation between sulfur dioxide emissions and acid deposition derived from monthly data. Nature **323**:245–247.

Epstein, H. E., I. C. Burke, and W. K. Lauenroth. 2002. Regional patterns of decomposition and primary production rates in the U.S. Great Plains. Ecology **83**:320–327.

Erb, T. J., P. Kiefer, B. Hattendorf, D. Gunther, and J. Vorholt. 2012. GFAJ-1 is an arsenate-resistant, phosphate-dependent organism. Science **337**:467–470.

Erickson, D. J., and R. A. Duce. 1988. On the global flux of atmospheric sea salt. Journal of Geophysical Research—Oceans **93**:14079–14088.

Eriksson, E. 1960. The yearly circulation of chloride and sulfur in nature; meteorological, geochemical and pedological implications. Part II. Tellus **12**:63–109.

Erisman, J. W., A. Bleeker, J. Galloway, and M. S. Sutton. 2007. Reduced nitrogen in ecology and the environment. Environmental Pollution **150**:140–149.

Escudero, A., J. M. Delarco, I. C. Sanz, and J. Ayala. 1992. Effects of leaf longevity and retranslocation efficiency on the retention time of nutrients in the leaf biomass of different woody species. Oecologia **90**:80–87.

Essington, T. E., and S. R. Carpenter. 2000. Nutrient cycling in lakes and streams: insights from a comparative analysis. Ecosystems **3**:131–143.

Eswaran, H., E. Berg, P. Reich, and E. Vandenberg. 1993. Organic carbon in soils of the world. Soil Science Society of America Journal **57**:192–194.

Etheridge, D. M., L. P. Steele, R. J. Francey, and R. L. Langenfelds. 1998. Atmospheric methane between 1000 AD and present: Evidence of anthropogenic emissions and climatic variability. Journal of Geophysical Research—Atmospheres **103**:15979–15993.

Etiope, G., K. R. Lassey, R. W. Klusman, and E. Boschi. 2008. Reappraisal of the fossil methane budget and related emission from geologic sources. Geophysical Research Letters 35.

Ettema, C. H., R. Lowrance, and D. C. Coleman. 1999. Riparian soil response to surface nitrogen input: temporal changes in denitrification, labile and microbial C and N pools, and bacterial and fungal respiration. Soil Biology & Biochemistry 31:1609–1624.

Euliss, N. H., R. A. Gleason, A. Olness, R. L. McDougal, H. R. Murkin, R. D. Robarts, R. A. Bourbonniere, and B. G. Warner. 2006. North American prairie wetlands are important nonforested land-based carbon storage sites. Science of the Total Environment 361:179–188.

Evans, J. R. 1989. Photosynthesis and nitrogen relationships in leaves of C-3 plants. Oecologia 78:9–19.

Evans, R. D., and J. R. Ehleringer. 1993. A break in the nitrogen cycle in aridlands—Evidence from delta-N-15 of soils. Oecologia 94:314–317.

Ewing, H. A., K. C. Weathers, P. H. Templer, T. E. Dawson, M. K. Firestone, A. M. Elliott, and V. K.S. Boukili. 2009. Fog water and ecosystem function: Heterogeneity in a California redwood forest. Ecosystems 12:417–433.

Ewing, S. A., B. Sutter, J. Owen, K. Nishiizumi, W. Sharp, S. S. Cliff, K. Perry, W. Dietrich, C. P. McKay, and R. Amundson. 2006. A threshold in soil formation at Earth's arid-hyperarid transition. Geochimica et Cosmochimica Acta 70:5293–5322.

Eyre, B. 1994. Nutrient biogeochemistry in the tropical Moresby River estuary system North Queensland, Australia. Estuarine Coastal and Shelf Science 39: 15–31.

Fahey, T. J. 1983. Nutrient dynamics of above-ground detritus in lodgepole pine (Pinus contorta ssp. latifolia) ecosystems, southeastern Wyoming. Ecological Monographs 53:51–72.

Fahey, T. J., and J. W. Hughes. 1994. Fine root dynamics in a northern hardwood forest ecosystem, Hubbard Brook Experimental Forest, NH. Journal of Ecology 82:533–548.

Fahey, T. J., G. L. Tierney, R. D. Fitzhugh, G. F. Wilson, and T. G. Siccama. 2005. Soil respiration and soil carbon balance in a northern hardwood forest ecosystem. Canadian Journal of Forest Research 35: 244–253.

Fahey, T. J., J. B. Yavitt, R. E. Sherman, P. M. Groffman, M. C. Fisk, and J. C. Maerz. 2011. Transport of carbon and nitrogen between litter and soil organic matter in a northern hardwood forest. Ecosystems 14:326–340.

Fairbanks, R. G. 1989. A 17,000-year glacio-eustatic sea-level record—Influence of glacial melting rates on the Younger Dryas event and deep ocean circulation. Nature 342:637–642.

Falkengren-Grerup, U. 1995. Interspecies differences in the preference of ammonium and nitrate in vascular plants. Oecologia 102:305–311.

Falkowski, P. G. 1997. Evolution of the nitrogen cycle and its influence on the biological sequestration of CO_2 in the ocean. Nature 387:272–275.

Falkowski, P. G. 2005. Biogeochemistry of primary production in the sea. In W. H. Schlesinger, editor. Biogeochemistry, Elsevier, 185–213.

Falkowski, P. G., R. T. Barber, and V. Smetacek. 1998. Biogeochemical controls and feedbacks on ocean primary production. Science 281:200–206.

Falkowski, P. G., T. Fenchel, and E. F. Delong. 2008. The microbial engines that drive Earth's biogeochemical cycles. Science 320:1034–1039.

Falkowski, P. G., Y. Kim, Z. Kolber, C. Wilson, C. Wirick, and R. Cess. 1992. Natural versus anthropogenic factors affecting low-level cloud albedo over the North Atlantic. Science 256:1311–1313.

Faloona, I. 2009. Sulfur processing in the marine atmospheric boundary layer: A review and critical assessment of modeling uncertainties. Atmospheric Environment 43:2841–2854.

Faloona, I., S. A. Conley, B. Blomquist, A. D. Clarke, V. Kapustin, S. Howell, D. H. Lenschow, and A. R. Bandy. 2009. Sulfur dioxide in the tropical marine boundary layer: dry deposition and heterogeneous oxidation observed during the Pacific Atmospheric Sulfur Experiment. Journal of Atmospheric Chemistry 63:13–32.

Famiglietti, J. S., M. Lo, S. L. Ho, J. Bethune, K. J. Anderson, T. H. Syed, S. C. Swenson, C. R. de Linage, and M. Rodell. 2011. Satellites measure recent rates of groundwater depletion in California's Central Valley. Geophysical Research Letters 38.

Fan, S., M. Gloor, J. Mahlman, S. Pacala, J. Sarmiento, T. Takahashi, and P. Tans. 1998a. A large terrestrial carbon sink in North America implied by atmospheric and oceanic carbon dioxide data and models. Science 282:442–446.

Fan, W. H., J. C. Randolph, and J. L. Ehman. 1998b. Regional estimation of nitrogen mineralization in forest ecosystems using geographic information systems. Ecological Applications 8:734–747.

Fanale, F. P. 1971. A case for catastrophic early degassing of the Earth. Chemical Geology 8:79–105.

Fang, J., A. Chen, C. Peng, S. Zhao, and L. Ci. 2001. Changes in forest biomass carbon storage in China between 1949 and 1998. Science 291:2320–2322.

Fang, Y. T., M. Yoh, K. Koba, W. X. Zhu, Y. Takebayashi, Y. H. Xiao, C. Y. Lei, J. M. Mo, W. Zhang, and

X. K. Lu. 2011. Nitrogen deposition and forest nitrogen cycling along an urban-rural transect in southern China. Global Change Biology 17:872–885.

Fanning, K. A. 1989. Influence of atmospheric pollution on nutrient limitation in the ocean. Nature 339: 460–463.

Farley, K. A., and E. Neroda. 1998. Noble gases in the Earth's mantle. Annual Review of Earth and Planetary Sciences 26:189–218.

Farman, J. C., B. G. Gardiner, and J. D. Shanklin. 1985. Large losses of total ozone in Antarctica reveal seasonal ClOx/NOx interaction. Nature 315:207–210.

Farmer, D. M., C. L. McNeil, and B. D. Johnson. 1993. Evidence for the importance of bubbles in increasing air-sea gas flux. Nature 361:620–623.

Farquhar, G. D., K. T. Hubick, A. G. Condon, and R. A. Richards. 1989. Carbon isotope fractionation and plant water-use efficiency. In P. W. Rundel, J. R. Ehleringer and K. A. Nagy, editors. Stable Isotopes in Ecological Research. Springer, 21–40.

Farquhar, J., H. M. Bao, and M. Thiemens. 2000. Atmospheric influence of Earth's earliest sulfur cycle. Science 289:756–758.

Farquhar, J., A. L. Zerkle, and A. Bekker. 2011. Geological constraints on the origin of oxygenic photosynthesis. Photosynthesis Research 107:11–36.

Farrar, J. F. 1985. The respiratory source of CO_2. Plant Cell and Environment 8:427–438.

Fassbinder, J.W.E., H. Stanjek, and H. Vali. 1990. Occurrence of magnetic bacteria in soil. Nature 343:161–163.

Faure, H. 1990. Changes in the global continental reservoir of carbon. Global and Planetary Change 82:47–52.

Fearnside, P. M. 1995. Hydroelectric dams in the Brazilian Amazon as sources of greenhouse gases. Environmental Conservation 22:7–19.

Fechner Levy, E. J., and H. F. Hemond. 1996. Trapped methane volume and potential effects on methane ebullition in a northern peatland. Limnology and Oceanography 41:1375–1383.

Federer, C. A. 1983. Nitrogen mineralization and nitrification—Depth variation in four New England forest soils. Soil Science Society of America Journal 47: 1008–1014.

Federer, C. A., and J. W. Hornbeck. 1985. The buffer capacity of forest soils in New England. Water Air and Soil Pollution 26:163–173.

Fedo, C. M., and M. J. Whitehouse. 2002. Metasomatic origin of quartz-pyroxene rock, Akilia, Greenland, and implications for Earth's earliest life. Science 296:1448–1452.

Fee, E. J., R. E. Hecky, G. W. Regehr, L. L. Hendzel, and P. Wilkinson. 1994. Effects of lake size on nutrient availability in the mixed layer during summer stratificaiton. Canadian Journal of Fisheries and Aquatic Sciences 51:2756–2768.

Feely, R. A., C. L. Sabine, R. H. Byrne, F. J. Millero, A. G. Dickson, R. Wanninkhof, A. Murata, L. A. Miller, and D. Greeley. 2012. Decadal changes in the aragonite and calcite saturation state of the Pacific Ocean. Global Biogeochemical Cycles 26.

Feely, R. A., C. L. Sabine, K. Lee, W. Berelson, J. Kleypas, V. J. Fabry, and F. J. Millero. 2004. Impact of anthropogenic CO_2 on the $CaCO_3$ system in the oceans. Science 305:362–366.

Feely, R. A., R. Wanninkhof, T. Takahashi, and P. Tans. 1999. Influence of El Niño on the equatorial Pacific contribution to atmospheric CO_2 accumulation. Nature 398:597–601.

Feller, M. C., and J. P. Kimmins. 1979. Chemical characteristics of small streams near Haney in southwestern British Columbia. Water Resources Research 15: 247–258.

Fellows, C. S., H. M. Valett, and C. N. Dahm. 2001. Whole-stream metabolism in two montane streams: Contribution of the hyporheic zone. Limnology and Oceanography 46:523–531.

Feng, X. 1999. Trends in intrinsic water-use efficiency of natural trees for the past 100-200 years: a response to atmospheric CO2 concentration. Geochimica et Cosmochimica Acta 63:1891–1903.

Fenn, M. E., M. A. Poth, J. D. Aber, J. S. Baron, B. T. Bormann, D. W. Johnson, A. D. Lemly, S. G. McNulty, D. E. Ryan, and R. Stottlemyer. 1998. Nitrogen excess in North American ecosystems: Predisposing factors, ecosystem responses, and management strategies. Ecological Applications 8:706–733.

Fenner, N., and C. Freeman. 2011. Drought-induced carbon loss in peatlands. Nature Geoscience 4:895–900.

Fenner, N., N. J. Ostle, N. McNamara, T. Sparks, H. Harmens, B. Reynolds, and C. Freeman. 2007. Elevated CO_2 effects on peatland plant community carbon dynamics and DOC production. Ecosystems 10: 635–647.

Ferek, R. J., and M. O. Andreae. 1984. Photochemical production of carbonyl sulfide in marine surface waters. Nature 307:148–150.

Fergus, C. E., P. A. Soranno, K. S. Cheruvelil, and M. T. Bremigan. 2011. Multiscale landscape and wetland drivers of lake total phosphorus and water color. Limnology and Oceanography 56:2127–2146.

Fernandez, I. J., L. E. Rustad, S. A. Norton, J. S. Kahl, and B. J. Cosby. 2003. Experimental acidification causes soil base-cation depletion at the Bear Brook watershed in Maine. Soil Science Society of America Journal 67:1909–1919.

Ferris, J. P., A. R. Hill, R. H. Liu, and L. E. Orgel. 1996. Synthesis of long prebiotic oligomers on mineral surfaces. Nature 381:59–61.

Fiedler, S. S. 2004. Water and redox conditions in wetland soils—Their influence on pedogenic oxides and morphology. Soil Science Society of America Journal 335:326–335.

Field, C., J. Merino, and H. A. Mooney. 1983. Compromises between water-use efficiency and nitrogen-use efficiency in five species of California evergreens. Oecologia 60:384–389.

Field, C. B., M. J. Behrenfeld, J. T. Randerson, and P. Falkowski. 1998. Primary production of the biosphere: Integrating terrestrial and oceanic components. Science 281:237–240.

Field, C. B., and H. A. Mooney. 1986. The photosynthesis-nitrogen relationship in wild plants. In T. J. Givnish, editor. On the Economy of Plant Form and Function. Cambridge University Press, 25–55.

Field, C. B., J. T. Randerson, and C. M. Malmstrom. 1995. Global net primary production—Combining ecology and remote sensing. Remote Sensing of Environment 51:74–88.

Fierer, N., M. A. Bradford, and R. B. Jackson. 2007. Toward an ecological classification of soil bacteria. Ecology 88:1354–1364.

Fife, D. N., and E.K.S. Nambiar. 1984. Movement of nutrients in radiata pine needles in relation to the growth of shoots. Annals of Botany 54:303–314.

Filippelli, G. M. 1997. Controls on phosphorus concentration and accumulation in oceanic sediments. Marine Geology 139:231–240.

Filippelli, G. M. 2008. The global phosphorus cycle: Past, present, and future. Elements 4:89–95.

Filippelli, G. M., and M. L. Delaney. 1992. Similar phosphorus fluxes in ancient phosphorite deposits and a modern phosphogenic environment. Geology 20:709–712.

Filippelli, G. M., and M. L. Delaney. 1994. The oceanic phosphorus cycle and continental weathering during the Neogene. Paleoceanography 9:643–652.

Filippelli, G. M., and M. L. Delaney. 1996. Phosphorus geochemistry of equatorial Pacific sediments. Geochimica et Cosmochimica Acta 60:1479–1495.

Filippelli, G. M., and C. Souch. 1999. Effects of climate and landscape development on the terrestrial phosphorus cycle. Geology 27:171–174.

Filoso, S., L. A. Martinelli, M. R. Williams, L. B. Lara, A. Krusche, V. Ballester, R. Victoria, and P. B. De Camargo. 2003. Land use and nitrogen export in the Piracicaba River basin, Southeast Brazil. Biogeochemistry 65:275–294.

Fimmen, R. L., D. D. Richter, D. Vasudevan, M. A. Williams, and L. T. West. 2008. Rhizogenic Fe-C redox cycling: a hypothetical biogeochemical mechanism that drives crustal weathering in upland soils. Biogeochemistry 87:127–141.

Findlay, S., and R. L. Sinsabaugh. 1999. Unravelling the sources and bioavailability of dissolved organic matter in lotic aquatic ecosystems. Marine and Freshwater Research 50:781–790.

Finlay, J. C. 2001. Stable-carbon-isotope ratios of river biota: Implications for energy flow in lotic food webs. Ecology 82:1052–1064.

Finlay, J. C., J. M. Hood, M. P. Limm, M. E. Power, J. D. Schade, and J. R. Welter. 2011. Light-mediated thresholds in stream-water nutrient composition in a river network. Ecology 92:140–150.

Finzi, A. C. 2009. Decades of atmospheric deposition have not resulted in widespread phosphorus limitation or saturation of tree demand for nitrogen in southern New England. Biogeochemistry 92: 217–229.

Finzi, A. C., and S. T. Berthrong. 2005. The uptake of amino acids by microbes and trees in three cold-temperate forests. Ecology 86:3345–3353.

Finzi, A. C., E. H. DeLucia, J. G. Hamilton, D. D. Richter, and W. H. Schlesinger. 2002. The nitrogen budget of a pine forest under free air CO_2 enrichment. Oecologia 132:567–578.

Finzi, A. C., D.J.P. Moore, E. H. DeLucia, J. Lichter, K. S. Hofmockel, R. B. Jackson, H. S. Kim, R. Matamala, H. R. McCarthy, R. Oren, J. S. Pippen, and W. H. Schlesinger. 2006. Progressive nitrogen limitation of ecosystem processes under elevated CO_2 in a warm-temperate forest. Ecology 87:15–25.

Finzi, A. C., R. J. Norby, C. Calfapietra, A. Gallet-Budynek, B. Gielen, W. E. Holmes, M. R. Hoosbeek, C. M. Iversen, R. B. Jackson, M. E. Kubiske, J. Ledford, M. Liberloo, R. Oren, A. Polle, S. Pritchard, D. R. Zak, W. H. Schlesinger, and R. Ceulemans. 2007. Increases in nitrogen uptake rather than nitrogen-use efficiency support higher rates of temperate forest productivity under elevated CO_2. Proceedings of the National Academy of Sciences 104:14014–14019.

Firestone, M. K. 1982. Biological denitrification. In F. J. Stevenson, editor. Nitrogen in Agricultural Soils. American Society of AgronomyMadison, Wisconsin, 289–326.

Firestone, M. K., and E. A. Davidson. 1989. Microbiological basis of NO and N_2O production and consumption in soil. In M. O. Andreae and D. S. Schimel, editors. Exchange of Trace Gases Between Terrestrial Ecosystems and the Atmosphere. Wiley, 7–21.

Firestone, M. K., R. B. Firestone, and J. M. Tiedje. 1980. Nitrous oxide from soil denitrification—Factors controlling its biological production. Science 208: 749–751.

Firing, Y. L., T. K. Chereskin, and M. R. Mazloff. 2011. Vertical structure and transport of the Antarctic Circumpolar Current in Drake Passage from direct velocity observations. Journal of Geophysical Research—Oceans 116.

Firme, G. F., E. L. Rue, D. A. Weeks, K. W. Bruland, and D. A. Hutchins. 2003. Spatial and temporal variability in phytoplankton iron limitation along the California coast and consequences for Si, N, and C biogeochemistry. Global Biogeochemical Cycles 17.

Firsching, B. M., and N. Claassen. 1996. Root phosphatase activity and soil organic phosphorus utilization by Norway spruce *Picea abies* (L) Karst. Soil Biology & Biochemistry 28:1417–1424.

Fischer, H., D. Wagenbach, and J. Kipfstuhl. 1998. Sulfate and nitrate firn concentrations on the Greenland ice sheet—2. Temporal anthropogenic deposition changes. Journal of Geophysical Research—Atmospheres 103:21935–21942.

Fisher, R. F. 1972. Spodosol development and nutrient distribution under *Hydnaceae* fungal mats. Soil Science Society of America Proceedings 36:492–495.

Fisher, R. F. 1977. Nitrogen and phosphorus mobilization by fairy ring fungus, *Marasmium oreades* (Bolt) Fr. Soil Biology & Biochemistry 9:239–241.

Fisher, S. G., and G. E. Likens. 1972. Stream ecosystem: organic energy budget. BioScience 22:33–35.

Fisher, S. G., and G. E. Likens. 1973. Energy flow in Bear Brook, New Hampshire—integrative approach to stream ecosystem metabolism. Ecological Monographs 43:421–439.

Fisher, S. G., and W. L. Minckley. 1978. Chemical characteristics of a desert stream in flash flood. Journal of Arid Environments 1:25–33.

Fishman, J., A. E. Wozniak, and J. K. Creilson. 2003. Global distribution of tropospheric ozone from satellite measurements using the empirically corrected tropospheric ozone residual technique: Identification of the regional aspects of air pollution. Atmospheric Chemistry and Physics 3:893–907.

Fisk, M. R., S. J. Giovannoni, and I. H. Thorseth. 1998. Alteration of oceanic volcanic glass: Textural evidence of microbial activity. Science 281:978–980.

Fissore, C., L. A. Baker, S. E. Hobbie, J. Y. King, J. P. McFadden, K. C. Nelson, and I. Jakobsdottir. 2011. Carbon, nitrogen, and phosphorus fluxes in household ecosystems in the Minneapolis-Saint Paul, Minnesota, urban region. Ecological Applications 21:619–639.

Fitzgerald, J. W., T. L. Andrew, and W. T. Swank. 1984. Availability of carbon-bonded sulfur for mineralization in forest soils. Canadian Journal of Forest Research 14:839–843.

Fitzgerald, J. W., T. C. Strickland, and J. T. Ash. 1985. Isolation and partial characterization of forest floor and soil organic sulfur. Biogeochemistry 1:155–167.

Fitzhugh, R. D., G. M. Lovett, and R. T. Venterea. 2003. Biotic and abiotic immobilization of ammonium, nitrite, and nitrate in soils developed under different tree species in the Catskill Mountains, New York, USA. Global Change Biology 9:1591–1601.

Flaig, W., H. Beutelspacher, and E. Rietz. 1975. Chemical composition and physical properties of humic substances. *In* J. E. Gieseking, editor. *Soil Components: Volume 1. Organic Components.* Springer, 1–211.

Flanner, M. G., K. M. Shell, M. Barlage, D. K. Perovich, and M. A. Tschudi. 2011. Radiative forcing and albedo feedback from the Northern Hemisphere cryosphere between 1979 and 2008. Nature Geoscience 4:151–155.

Fletcher, S.E.M., P. P. Tans, L. M. Bruhwiler, J. B. Miller, and M. Heimann. 2004. CH_4 sources estimated from atmospheric observations of CH_4 and its $^{12}C/^{12}C$-isotopic ratios: 1. Inverse modeling of source processes. Global Biogeochemical Cycles 18.

Fliegel, D., J. Kosler, N. McLoughlin, A. Simonetti, M. J. de Wit, R. Wirth, and H. Furnes. 2010. *In situ* dating of the Earth's oldest trace fossil at 3.34 Ga. Earth and Planetary Science Letters 299:290–298.

Flint, R. F. 1971. *Glacial and Quaternary Geology.* Wiley.

Floret, C., R. Pontanier, and S. Rambal. 1982. Measurement and modeling of primary production and water use in a South Tunisian steppe. Journal of Arid Environments 5:77–90.

Flückiger, J., E. Monnin, B. Stauffer, J. Schwander, T. F. Stocker, J. Chappellaz, D. Raynaud, and J. M. Barnola. 2002. High-resolution Holocene N_2O ice-core record and its relationship with CH_4 and CO_2. Global Biogeochemical Cycles 16.

Fontaine, S., S. Barot, P. Barre, N. Bdioui, B. Mary, and C. Rumpel. 2007. Stability of organic carbon in deep soil layers controlled by fresh carbon supply. Nature 450:277–280.

Forde, B., and H. Lorenzo. 2001. The nutritional control of root development. Plant and Soil 232:51–68.

Formisano, V., S. Atreya, T. Encrenaz, N. Ignatiev, and M. Giuranna. 2004. Detection of methane in the atmosphere of Mars. Science 306:1758–1761.

Fossing, H., V. A. Gallardo, B. B. Jorgensen, M. Huttel, L. P. Nielsen, H. Schulz, D. E. Canfield, S. Forster, R. N. Glud, J. K. Gundersen, J. Kuver, N. B. Ramsing, A. Teske, B. Thamdrup, and O. Ulloa. 1995. Concentration and transport of nitrate by the mat-forming sulfur bacterium *Thioploca*. Nature 374:713–715.

Foukal, P., C. Frohlich, H. Spruit, and T.M.L. Wigley. 2006. Variations in solar luminosity and their effect on the Earth's climate. Nature 443:161–166.

Fowells, H. A., and R. W. Krauss. 1959. The inorganic nutrition of loblolly pine and virginia pine with special reference to nitrogen and phosphorus. Forest Science 5:95–111.

Fowler, W. A. 1984. The quest for the origin of the elements. Science 226:922–935.

Fox, L. E. 1983. The removal of dissolved humic acid during estuarine mixing. Estuarine, Coastal and Shelf Science 16:431–440.

Fox, G. E., E. Stackebrandt, R. B. Hespell, J. Gibson, J. Maniloff, T. A. Dyer, R. S. Wolfe, W. E. Balch, R. S. Tanner, L. J. Magrum, L. B. Zablen, R. Blakemore, R. Gupta, L. Bonen, B. J. Lewis, D. A. Stahl, K. R. Luehrsen, K. N. Chen, and C. R. Woese. 1980. The phylogeny of prokaryotes. Science 209:457–463.

Fox, T. R., and N. B. Comerford. 1992a. Influence of oxalate loading on phosphorus and aluminum solubility in Spodosols. Soil Science Society of America Journal 56:290–294.

Fox, T. R., and N. B. Comerford. 1992b. Rhizosphere phosphatase activity and phosphatase hydrolyzable organic phosphorus in two forested Spodosols. Soil Biology & Biochemistry 24:579–583.

France-Lanord, C., and L. A. Derry. 1997. Organic carbon burial forcing of the carbon cycle from Himalayan erosion. Nature 390:65–67.

Francis, C. A., K. J. Roberts, J. M. Beman, A. E. Santoro, and B. B. Oakley. 2005. Ubiquity and diversity of ammonia-oxidizing archaea in water columns and sediments of the ocean. Proceedings of the National Academy of Sciences 102:14683–14688.

Francoeur, S. N. 2001. Meta-analysis of lotic nutrient amendment experiments: detecting and quantifying subtle responses. Journal of the North American Benthological Society 20:358–368.

Francois, R., M. A. Altabet, E. F. Yu, D. M. Sigman, M. P. Bacon, M. Frank, G. Bohrmann, G. Bareille, and L. D. Labeyrie. 1997. Contribution of Southern Ocean surface-water stratification to low atmospheric CO_2 concentrations during the last glacial period. Nature 389:929–935.

Frangi, J. L., and A. E. Lugo. 1985. Ecosystem dynamics of a subtropical floodplain forest. Ecological Monographs 55:351–369.

Frank, D. A., and R. D. Evans. 1997. Effects of native grazers on grassland N cycling in Yellowstone National Park. Ecology 78:2238–2248.

Frank, D. A., and P. M. Groffman. 1998. Ungulate vs. landscape control of soil C and N processes in grasslands of Yellowstone National Park. Ecology 79:2229–2241.

Frank, D. A., and R. S. Inouye. 1994. Temporal variation in actual evapotranspiration of terrestrial ecosystems—Patterns and ecological implications. Journal of Biogeography 21:401–411.

Frank, D. A., R. S. Inouye, N. Huntly, G. W. Minshall, and J. E. Anderson. 1994. The biogeochemistry of a north-temperate grassland with native ungulates—Nitrogen dynamics in Yellowstone National Park. Biogeochemistry 26:163–188.

Frank, D. A., A. W. Pontes, and K. J. McFarlane. 2012. Controls on soil organic carbon stocks and turnover among North American ecosystems. Ecosystems 15:604–615.

Frankenberg, C., J. B. Fisher, J. Worden, G. Badgley, S. S. Saatchi, J. E. Lee, G. C. Toon, A. Butz, M. Jung, A. Kuze, and T. Yokota. 2011. New global observations of the terrestrial carbon cycle from GOSAT: Patterns of plant fluorescence with gross primary productivity. Geophysical Research Letters 38.

Frankenberg, C., J. F. Meirink, M. van Weele, U. Platt, and T. Wagner. 2005. Assessing methane emissions from global space-borne observations. Science 308:1010–1014.

Frankignoulle, M., G. Abril, A. Borges, I. Bourge, C. Canon, B. DeLille, E. Libert, and J. M. Theate. 1998. Carbon dioxide emission from European estuaries. Science 282:434–436.

Frasier, R., S. Ullah, and T. R. Moore. 2010. Nitrous oxide consumption potentials of well-drained forest soils in southern Quebec, Canada. Geomicrobiology Journal 27:53–60.

Freedman, W. L., and B. F. Madore. 2010. The Hubble constant. Annual Review of Astronomy and Astrophysics 48:673–710.

Freeland, W. J., P. H. Calcott, and D. P. Geiss. 1985. Allelochemicals, minerals and herbivore population size. Biochemical Systematics and Ecology 13: 195–206.

Freeman, C., C. D. Evans, D. T. Monteith, B. Reynolds, and N. Fenner. 2001a. Export of organic carbon from peat soils. Nature 412:785.

Freeman, C., N. Fenner, N. J. Ostle, H. Kang, D. J. Dowrick, B. Reynolds, M. A. Lock, D. Sleep, S. Hughes, and J. Hudson. 2004. Export of dissolved organic carbon from peatlands under elevated carbon dioxide levels. Nature 430:195–198.

Freeman, C., J. Hawkins, M. A. Lock, and B. Reynolds. 1993. A laboratory perfusion system for the study of biogeochemical response of wetlands to climate change. In B. Gopal, A. Hillbricht-Ilowska and R. G. Wetzel, editors. Wetlands and Ecotones: Studies on Land-Water Interactions. National Institute of Ecology, New Delhi, 75–83.

Freeman, C., N. Ostle, and H. Kang. 2001b. An enzymic 'latch' on a global carbon store—A shortage of oxygen locks up carbon in peatlands by restraining a single enzyme. Nature 409:149.

Freeze, R. A. 1974. Streamflow generation. Reviews of Geophysics 12:627–647.

Freney, J. R., J. R. Simpson, and O. T. Denmead. 1983. Volatilization of ammonia. In J. R. Freney and J. R. Simpson, editors. Gaseous Loss of Nitrogen from Plant-Soil Systems. Martinus Nijhoff, 1–32.

Freschet, G. T., J.H.C. Cornelissen, R.S.P. van Logtestijn, and R. Aerts. 2010. Substantial nutrient resorption from leaves, stems and roots in a subarctic flora: what is the link with other resource economics traits? New Phytologist **186**:879–889.

Fridovich, I. 1975. Superoxide dismutases. Annual Review of Biochemistry **44**:147–159.

Friedland, A. J., and A. H. Johnson. 1985. Lead distribution and fluxes in a high-elevation forest in northern Vermont. Journal of Environmental Quality **14**: 332–336.

Friedlander, G., J. W. Kennedy, and J. M. Miller. 1964. *Nuclear and Radiochemistry*. John Wiley and Sons.

Friedmann, E. I. 1982. Endolithic microorganisms in the Antarctic cold desert. Science **215**:1045–1053.

Friend, A. D., R. J. Geider, M. J. Behrenfeld, and C. J. Still. 2009. Photosynthesis in global-scale models. *In* A. Laisk, L. Nedbal and Govindjee, editors. *Photosynthesis in silico: Understanding Complexity from Molecules to Ecosystems*. Springer, 465–497.

Fritschi, F. B., B. A. Roberts, D. W. Rains, R. L. Travis, and R. B. Hutmacher. 2004. Fate of nitrogen-15 applied to irrigated Acala and Pima cotton. Agronomy Journal **96**:646–655.

Fritz, S. C. 1996. Paleolimnological records of climatic change in North America. Limnology and Oceanography **41**:882–889.

Froelich, P. N. 1988. Kinetic control of dissolved phosphate in natural rivers and estuaries—A primer on the phosphate buffer mechanism. Limnology and Oceanography **33**:649–668.

Froelich, P. N., M. L. Bender, N. A. Luedtke, G. R. Heath, and T. Devries. 1982. The marine phosphorus cycle. American Journal of Science **282**:474–511.

Froelich, P. N., V. Blanc, R. A. Mortlock, S. N. Chillrud, W. Dunstan, A. Udomkit, and T.-H. Peng. 1992. River fluxes of dissolved silica to the ocean were higher during glacials: Ge/Si in diatoms, rivers, and oceans. Paleoceanography **7**:739–767.

Froelich, P. N., K. H. Kim, R. Jahnke, W. C. Burnett, A. Soutar, and M. Deakin. 1983. Pore-water fluoride in Peru continental margin sediments—Uptake from seawater. Geochimica et Cosmochimica Acta **47**: 1605–1612.

Froelich, P. N., G. P. Klinkhammer, M. L. Bender, N. A. Luedtke, G. R. Heath, D. Cullen, P. Dauphin, D. Hammond, B. Hartman, and V. Maynard. 1979. Early oxidation of organic matter in pelagic sediments of the eastern equatorial Atlantic—Suboxic diagenesis. Geochimica et Cosmochimica Acta **43**: 1075–1090.

Frolking, S., J. Talbot, M. C. Jones, C. C. Treat, J. B. Kauffman, E. S. Tuittila, and N. Roulet. 2011. Peatlands in the Earth's 21st century climate system. Environmental Reviews **19**:371–396.

Fruchter, J. S., D. E. Robertson, J. C. Evans, K. B. Olsen, E. A. Lepel, J. C. Laul, K. H. Abel, R. W. Sanders, P. O. Jackson, N. S. Wogman, R. W. Perkins, H. H. Vantuyl, R. H. Beauchamp, J. W. Shade, J. L. Daniel, R. L. Erikson, G. A. Sehmel, R. N. Lee, A. V. Robinson, O. R. Moss, J. K. Briant, and W. C. Cannon. 1980. Mount St Helens ash from the 18 May 1980 eruption—Chemical, physical, mineralogical, and biological properties. Science **209**: 1116–1125.

Fueglistaler, S., H. Wernli, and T. Peter. 2004. Tropical troposphere-to-stratosphere transport inferred from trajectory calculations. Journal of Geophysical Research—Atmospheres **109**.

Fuentes, J. D., M. Lerdau, R. Atkinson, D. Baldocchi, J. W. Bottenheim, P. Ciccioli, B. Lamb, C. Geron, L. Gu, A. Guenther, T. D. Sharkey, and W. Stockwell. 2000. Biogenic hydrocarbons in the atmospheric boundary layer: A review. Bulletin of the American Meteorological Society **81**: 1537–1575.

Fuhrman, J. A., and G. B. McManus. 1984. Do bacteria-sized marine eukaryotes consume significant bacterial production. Science **224**:1257–1260.

Fulweiler, R. W., S. W. Nixon, B. A. Buckley, and S. L. Granger. 2007. Reversal of the net dinitrogen gas flux in coastal marine sediments. Nature **448**: 180–182.

Fung, I., J. John, J. Lerner, E. Matthews, M. Prather, L. P. Steele, and P. J. Fraser. 1991. Three-dimensional model synthesis of the global methane cycle. Journal of Geophysical Research—Atmospheres **96**:13033–13065.

Funk, D. W., E. R. Pullman, K. M. Peterson, P. M. Crill, and W. D. Billings. 1994. Influence of water table on carbon dioxide, carbon monoxide and methane fluxes from taiga bog microcosms. Global Biogeochemical Cycles **8**:271–278.

Furrer, G., P. Sollins, and J. C. Westall. 1990. The study of soil chemistry through quasi-steady-state models. 2. Acidity of the soil solution. Geochimica et Cosmochimica Acta **54**:2363–2374.

Furrer, G., J. Westall, and P. Sollins. 1989. The study of soil chemistry through quasi-steady-state models. 1. Mathematical definition of the model. Geochimica et Cosmochimica Acta **53**:595–601.

Fuss, C. B., C. T. Driscoll, C. E. Johnson, R. J. Petras, and T. J. Fahey. 2011. Dynamics of oxidized and reduced iron in a northern hardwood forest. Biogeochemistry **104**:103–119.

Gabet, E. J., and S. M. Mudd. 2009. A theoretical model coupling chemical weathering rates with denudation rates. Geology **37**:151–154.

Gabriel, M., R. Kolka, T. Wickman, L. Woodruff, and E. Nater. 2012. Latent effect of soil organic matter oxidation on mercury cycling within a southern boreal ecosystem. Journal of Environmental Quality 41: 495–505.

Gachter, R., J. S. Meyer, and A. Mares. 1988. Contribution of bacteria to release and fixation of phosphorus in lake sediments. Limnology and Oceanography 33: 1542–1558.

Gaidos, E., N. Haghighipour, E. Agol, D. Latham, S. Raymond, and J. Rayner. 2007. New worlds on the horizon: Earth-sized planets close to other stars. Science 318:210–213.

Gaidos, E. J., K. H. Nealson, and J. L. Kirschvink. 1999. Biogeochemistry—Life in ice-covered oceans. Science 284:1631–1633.

Gaillardet, J., B. Dupre, P. Louvat, and C. J. Allegre. 1999. Global silicate weathering and CO_2 consumption rates deduced from the chemistry of large rivers. Chemical Geology 159:3–30.

Gale, P. M., and J. T. Gilmour. 1988. Net mineralization of carbon and nitrogen under aerobic and anaerobic conditions. Soil Science Society of America Journal 52:1006–1010.

Gali, M., V. Salo, R. Almeda, A. Calbet, and R. Simo. 2011. Stimulation of gross dimethylsulfide (DMS) production by solar radiation. Geophysical Research Letters 38.

Gallardo, A., and J. Merino. 1998. Soil nitrogen dynamics in response to carbon increase in a Mediterranean shrubland of SW Spain. Soil Biology & Biochemistry 30:1349–1358.

Gallardo, A., and R. Parama. 2007. Spatial variability of soil elements in two plant communities of NW Spain. Geoderma 139:199–208.

Gallardo, A., and W. H. Schlesinger. 1992. Carbon and nitrogen limitations of soil microbial biomass in desert ecosystems. Biogeochemistry 18:1–17.

Gallardo, A., and W. H. Schlesinger. 1994. Factors limiting microbial biomass in the mineral soil and forest floor of a warm-temperate forest. Soil Biology & Biochemistry 26:1409–1415.

Gallardo, A., and W. H. Schlesinger. 1996. Exclusion of Artemisia tridentata Nutt. from hydrothermally altered rock by low phosphorus availability. Madrono 43:292–298.

Gallo, M. E., A. Porras-Alfaro, K. J. Odenbach, and R. L. Sinsabaugh. 2009. Photoacceleration of plant litter decomposition in an arid environment. Soil Biology & Biochemistry 41:1433–1441.

Galloway, J. N., F. J. Dentener, D. G. Capone, E. W. Boyer, R. W. Howarth, S. P. Seitzinger, G. P. Asner, C. C. Cleveland, P. A. Green, E. A. Holland, D. M. Karl, A. F. Michaels, J. H. Porter, A. R. Townsend, and C. J. Vörösmarty. 2004. Nitrogen cycles: past, present, and future. Biogeochemistry 70:153–226.

Galloway, J. N., and G. E. Likens. 1979. Atmospheric enhancement of metal deposition in Adirondack lake sediments. Limnology and Oceanography 24:427–433.

Galloway, J. N., G. E. Likens, and E. S. Edgerton. 1976. Acid precipitation in northeastern United States—pH and acidity. Science 194:722–724.

Galloway, J. N. 2005. The global nitrogen cycle. In W. H. Schlesinger, editor. Treatise on Geochemistry. Volume 8. Biogeochemistry. Elsevier, 557–583.

Galloway, J. N., A. R. Townsend, J. W. Erisman, M. Bekunda, Z. Cai, J. R. Freney, L. A. Martinelli, S. P. Seitzinger, and M. A. Sutton. 2008. Transformation of the nitrogen cycle: Recent trends, questions, and potential solutions. Science 320:889–892.

Galloway, J. N., and D. M. Whelpdale. 1987. WATOX-86 overview and western North Atlantic Ocean S and N atmospheric budgets. Global Biogeochemical Cycles 1:261–281.

Galoux, A., P. Benecke, G. Gietl, H. Hager, C. Kayser, O. Kiese, K. R. Knoerr, C. E. Murphy, G. Schnock, and T. R. Sinclair. 1981. Radiation, heat, water and carbon dioxide balances. In D. E. Reichle, editor. Dynamic Properties of Forest Ecosystems. Cambridge University Press, 87–204.

Galy, V., O. Beyssac, C. France-Lanord, and T. Eglinton. 2008. Recycling of graphite during Himalayan erosion: A geological stabilization of carbon in the crust. Science 322:943–945.

Galy, V., C. France-Lanord, O. Beyssac, P. Faure, H. Kudrass, and F. Palhol. 2007. Efficient organic carbon burial in the Bengal fan sustained by the Himalayan erosional system. Nature 450:407–410.

Ganachaud, A. 2003. Large-scale mass transports, water mass formation, and diffusivities estimated from World Ocean Circulation Experiment (WOCE) hydrographic data. Journal of Geophysical Research—Oceans 108.

Ganeshram, R. S., T. F. Pedersen, S. E. Calvert, and J. W. Murray. 1995. Large changes in oceanic nutrient inventories from glacial to interglacial periods. Nature 376:755–758.

Ganzeveld, L. N., J. Lelieveld, F. J. Dentener, M. C. Krol, A. J. Bouwman, and G. J. Roelofs. 2002. Global soil-biogenic NOx emissions and the role of canopy processes. Journal of Geophysical Research—Atmospheres 107.

Gao, B., X. T. Ju, Q. Zhang, P. Christie, and F. S. Zhang. 2011. New estimates of direct N_2O emissions from Chinese croplands from 1980 to 2007 using localized emission factors. Biogeosciences 8: 3011–3024.

Garcia-Ruiz, J. M., S. T. Hyde, A. M. Carnerup, A. G. Christy, M. J. Van Kranendonk, and N. J. Welham. 2003. Self-assembled silica-carbonate structures and detection of ancient microfossils. Science **302**: 1194–1197.

Gardner, J. B., and L. E. Drinkwater. 2009. The fate of nitrogen in grain cropping systems: a meta-analysis of ^{15}N field experiments. Ecological Applications **19**:2167–2184.

Gardner, L. R. 1990. The role of rock weathering in the phosphorus budget of terrestrial watersheds. Biogeochemistry **11**:97–110.

Gardner, L. R., I. Kheoruenromne, and H. S. Chen. 1978. Isovolumetric geochemical investigation of a buried granite saprolite near Columbia, SC, USA. Geochimica et Cosmochimica Acta **42**:417–424.

Garrels, R. M., and A. Lerman. 1981. Phanerozoic cycles of sedimentary carbon and sulfur. Proceedings of the National Academy of Sciences—Physical Sciences **78**:4652–4656.

Garrels, R. M., and F. T. MacKenzie. 1971. *Evolution of Sedimentary Rocks*. W.W. Norton.

Garten, C. T. 1990. Foliar leaching, translocation, and biogenic emission of ^{35}S-radiolabeled loblolly pines. Ecology **71**:239–251.

Garten, C. T. 1993. Variation in foliar ^{15}N-abundance and the availability of soil nitrogen on Walker Branch watershed. Ecology **74**:2098–2113.

Garten, C. T., E. A. Bondietti, and R. D. Lomax. 1988. Contribution of foliar leaching and dry deposition to sulfate in net throughfall below deciduous trees. Atmospheric Environment **22**:1425–1432.

Garten, C. T., and P. J. Hanson. 1990. Foliar retention of ^{15}N-nitrate and ^{15}N-ammonium by red maple (*Acer rubrum*) and white oak (*Quercus alba*) leaves from simulated rain. Environmental and Experimental Botany **30**:333–342.

Garten, C. T., C. M. Iversen, and R. J. Norby. 2011. Litterfall ^{15}N abundance indicates declining soil nitrogen availability in a free-air CO_2 enrichment experiment. Ecology **92**:133–139.

Garten, C. T. 2011. Comparison of forest soil carbon dynamics at five sites along a latitudinal gradient. Geoderma **167–68**:30–40.

Garten, C. T., and H. Van Miegroet. 1994. Relationships between soil nitrogen dynamics and natural ^{15}N abundance in plant foliage from Great Smoky Mountains National Park. Canadian Journal of Forest Research **24**:1636–1645.

Garvin, J., R. Buick, A. D. Anbar, G. L. Arnold, and A. J. Kaufman. 2009. Isotopic evidence for an aerobic nitrogen cycle in the latest Archean. Science **323**: 1045–1048.

Gascard, J. C., A. J. Watson, M. J. Messias, K. A. Olsson, T. Johannessen, and K. Simonsen. 2002. Long-lived vortices as a mode of deep ventilation in the Greenland Sea. Nature **416**:525–527.

Gash, J.H.C., and J. B. Stewart. 1977. The evaporation from Thetford Forest during 1975. Journal of Hydrology **35**:385–396.

Gassmann, G., and D. Glindemann. 1993. Phosphane (PH_3) in the biosphere. Angewandte Chemie—International Edition in English **32**:761–763.

Gates, W. L. 1976. Modeling ice-age climate. Science **191**:1138–1144.

Gattuso, J. P., M. Frankignoulle, and R. Wollast. 1998. Carbon and carbonate metabolism in coastal aquatic ecosystems. Annual Review of Ecology and Systematics **29**:405–434.

Gatz, D. F., and A. N. Dingle. 1971. Trace substances in rain water: Concentration variations during convective rains, and their interpretation. Tellus **23**:14–27.

Gauci, V., and S. J. Chapman. 2006. Simultaneous inhibition of CH_4 efflux and stimulation of sulphate reduction in peat subject to simulated acid rain. Soil Biology & Biochemistry **38**:3506–3510.

Gauci, V., E. Matthews, N. Dise, B. Walter, D. Koch, G. Granberg, and M. Vile. 2004. Sulfur pollution suppression of the wetland methane source in the 20th and 21st centuries. Proceedings of the National Academy of Sciences **101**:12583–12587.

Gaudichet, A., F. Echalar, B. Chatenet, J. P. Quisefit, G. Malingre, H. Cachier, P. Buatmenard, P. Artaxo, and W. Maenhaut. 1995. Trace elements in tropical African savanna biomass burning aerosols. Journal of Atmospheric Chemistry **22**:19–39.

Gaudinski, J. B., M. S. Torn, W. J. Riley, T. E. Dawson, J. D. Joslin, and H. Majdi. 2010. Measuring and modeling the spectrum of fine-root turnover times in three forests using isotopes, minirhizotrons, and the Radix model. Global Biogeochemical Cycles **24**.

Gaudinski, J. B., S. E. Trumbore, E. A. Davidson, A. C. Cook, D. Markewitz, and D. D. Richter. 2001. The age of fine-root carbon in three forests of the eastern United States measured by radiocarbon. Oecologia **129**:420–429.

Gaudinski, J. B., S. E. Trumbore, E. A. Davidson, and S. H. Zheng. 2000. Soil carbon cycling in a temperate forest: radiocarbon-based estimates of residence times, sequestration rates and partitioning of fluxes. Biogeochemistry **51**:33–69.

Gazeau, F., S. V. Smith, B. Gentili, M. Frankignoulle, and J. P. Gattuso. 2004. The European coastal zone: characterization and first assessment of ecosystem metabolism. Estuarine Coastal and Shelf Science **60**: 673–694.

Gebauer, T., V. Homa, and C. Leuschner. 2012. Canopy transpiration of pure and mixed forest stands with variable abundance of European beech. Journal of Hydrology 442/443:2–14.

Gedalof, Z., and A. A. Berg. 2010. Tree-ring evidence for limited direct CO_2 fertilization of forests over the 20th century. Global Biogeochemical Cycles 24.

Gedney, N., P. M. Cox, R. A. Betts, O. Boucher, C. Huntingford, and P. A. Stott. 2006. Detection of a direct carbon dioxide effect in continental river run-off records. Nature 439:835–838.

Geisseler, D., W. R. Horwath, R. G. Joergensen, and B. Ludwig. 2010. Pathways of nitrogen utilization by soil microorganisms—A review. Soil Biology & Biochemistry 42:2058–2067.

Gelinas, Y., J. A. Baldock, and J. I. Hedges. 2001. Organic carbon composition of marine sediments: Effect of oxygen exposure on oil generation potential. Science 294:145–148.

Geng, C. M., and Y. J. Mu. 2006. Carbonyl sulfide and dimethyl sulfide exchange between trees and the atmosphere. Atmospheric Environment 40: 1373–1383.

Geng, J. J., X. J. Niu, X. C. Jin, X. R. Wang, X. H. Gu, M. Edwards, and D. Glindemann. 2005. Simultaneous monitoring of phosphine and of phosphorus species in Taihu Lake sediments and phosphine emission from lake sediments. Biogeochemistry 76: 283–298.

Genkai-Kato, M., and S. R. Carpenter. 2005. Eutrophication due to phosphorus recycling in relation to lake morphometry, temperature, and macrophytes. Ecology 86:210–219.

Gensel, P. G., and H. N. Andrews. 1987. The evolution of early land plants. American Scientist 75:478–489.

Georgiadis, M. M., H. Komiya, P. Chakrabarti, D. Woo, J. J. Kornuc, and D. C. Rees. 1992. Crystallographic structure of the nitrogenase iron protein from Azotobacter vinelandii. Science 257:1653–1659.

Georgii, H.-W., and D. Wötzel. 1970. On the relationship between drop size and concentration of trace elements in rainwater. Journal of Geophysical Research 75:1727–1731.

Gerard, F., K. U. Mayer, M. J. Hodson, and J. Ranger. 2008. Modelling the biogeochemical cycle of silicon in soils: Application to a temperate forest ecosystem. Geochimica et Cosmochimica Acta 72:741–758.

Gergel, S. E., M. G. Turner, and T. K. Kratz. 1999. Dissolved organic carbon as an indicator of the scale of watershed influence on lakes and rivers. Ecological Applications 9:1377–1390.

Gerhart, L. M., and J. K. Ward. 2010. Plant responses to low [CO_2] of the past. New Phytologist 188:674–695.

Geron, C., A. Guenther, J. Greenberg, T. Karl, and R. Rasmussen. 2006. Biogenic volatile organic compound emissions from desert vegetation of the southwestern US. Atmospheric Environment 40:1645–1660.

Gersper, P. L., and N. Holowaychuk. 1971. Some effects of stem flow from forest canopy trees on chemical properties of soils. Ecology 52:691–702.

Gessler, A., M. Rienks, and H. Rennenberg. 2000. NH_3 and NO_2 fluxes between beech trees and the atmosphere—correlation with climatic and physiological parameters. New Phytologist 147:539–560.

Ghiorse, W. C. 1984. Biology of iron-depositing and manganese-depositing bacteria. Annual Review of Microbiology 38:515–550.

Gholz, H. L. 1982. Environmental limits on aboveground net primary production, leaf-area, and biomass in vegetation zones of the Pacific Northwest. Ecology 63:469–481.

Gholz, H. L., R. F. Fisher, and W. L. Pritchett. 1985. Nutrient dynamics in slash pine plantation ecosystems. Ecology 66:647–659.

Gholz, H. L., D. A. Wedin, S. M. Smitherman, M. E. Harmon, and W. J. Parton. 2000. Long-term dynamics of pine and hardwood litter in contrasting environments: toward a global model of decomposition. Global Change Biology 6:751–765.

Ghude, S. D., D. M. Lal, G. Beig, R. van der A, and D. Sable. 2010. Rain-induced soil NOx emission from India during the onset of the summer monsoon: A satellite perspective. Journal of Geophysical Research—Atmospheres 115.

Giardina, C. P., R. L. Sanford, and I. C. Dockersmith. 2000. Changes in soil phosphorus and nitrogen during slash-and-burn clearing of a dry tropical forest. Soil Science Society of America Journal 64: 399–405.

Gibbs, J., and H. Greenway. 2003. Review: Mechanisms of anoxia tolerance in plants. I. Growth, survival and anaerobic catabolism. Functional Plant Biology 30:1–47.

Gibbs, R. J. 1970. Mechanisms controlling world water chemistry. Science 170:1088–1090.

Gibbs, R. J., D. M. Tshudy, L. Konwar, and J. M. Martin. 1989. Coagulation and transport of sediments in the Gironde estuary. Sedimentology 36:987–999.

Giblin, A. E. 1988. Pyrite formation in marshes during early diagenesis. Geomicrobiology Journal 6:77–97.

Gibson, D. G., J. I. Glass, C. Lartigue, V. N. Noskov, R. Y. Chuang, M. A. Algire, G. A. Benders, M. G. Montague, L. Ma, M. M. Moodie, C. Merryman, S. Vashee, R. Krishnakumar, N. Assad-Garcia, C. Andrews-Pfannkoch, E. A. Denisova, L. Young, Z. Q. Qi, T. H. Segall-Shapiro, C. H. Calvey, P. P. Parmar, C. A. Hutchison, H. O. Smith, and J. C. Venter.

2010. Creation of a bacterial cell controlled by a chemically synthesized genome. Science 329:52–56.

Gieskes, J. M., and J. R. Lawrence. 1981. Alteration of volcanic matter in deep-sea sediments—Evidence from the chemical composition of interstitial waters from deep-sea drilling cores. Geochimica et Cosmochimica Acta 45:1687–1703.

Gijsman, A. J. 1990. Nitrogen nutrition of douglas fir (*Pseudotsuga menziesii*) on strongly acid sandy soil. 1. Growth, nutrient uptake and ionic balance. Plant and Soil 126:53–61.

Gile, L. H., F. F. Peterson, and R. B. Grossman. 1966. Morphological and genetic sequences of carbonate accumulation in desert soils. Soil Science 101: 347–360.

Gill, R. A., and R. B. Jackson. 2000. Global patterns of root turnover for terrestrial ecosystems. New Phytologist 147:13–31.

Gillespie, A. R., and P. E. Pope. 1990. Rhizosphere acidification increases phosphorus recovery of black locust. 1. Induced acidification and soil response. Soil Science Society of America Journal 54:533–537.

Gillette, D. A., G. J. Stensland, A. L. Williams, W. Barnard, D. Gatz, P. C. Sinclair, and T. C. Johnson. 1992. Emissions of alkaline elements calcium, magnesium, potassium, and sodium from open sources in the contiguous United States. Global Biogeochemical Cycles 6:437–457.

Gilliland, A. B., K. W. Appel, R. W. Pinder, and R. L. Dennis. 2006. Seasonal NH_3 emissions for the continental United States: Inverse model estimation and evaluation. Atmospheric Environment 40:4986–4998.

Gillmann, C., P. Lognonne, and M. Moreira. 2011. Volatiles in the atmosphere of Mars: The effects of volcanism and escape constrained by isotopic data. Earth and Planetary Science Letters 303:299–309.

Gilmore, A. R., G. Z. Gertner, and G. L. Rolfe. 1984. Soil chemical changes associated with roosting birds. Soil Science 138:158–163.

Gilmour, C. C., D. A. Elias, A. M. Kucken, S. D. Brown, A. V. Palumbo, C. W. Schadt, and J. D. Wall. 2011. Sulfate-reducing bacterium *Desulfovibrio desulfuricans* ND132 as a model for understanding bacterial mercury methylation. Applied and Environmental Microbiology 77:3938–3951.

Gilmour, C. C., and E. A. Henry. 1991. Mercury methylation in aquatic systems affected by acid deposition. Environmental Pollution 71:131–169.

Gilmour, C. C., E. A. Henry, and R. Mitchell. 1992. Sulfate stimulation of mercury methylation in fresh-water sediments. Environmental Science & Technology 26:2281–2287.

Giordano, M., A. Norici, and R. Hell. 2005. Sulfur and phytoplankton: acquisition, metabolism and impact on the environment. New Phytologist 166:371–382.

Gislason, S. R., E. H. Oelkers, E. S. Eiriksdottir, M. I. Kardjilov, G. Gisladottir, B. Sigfusson, A. Snorrason, S. Elefsen, J. Hardardottir, P. Torssander, and N. Oskarsson. 2009. Direct evidence of the feedback between climate and weathering. Earth and Planetary Science Letters 277:213–222.

Glass, J. B., F. Wolfe-Simon, and A. D. Anbar. 2009. Co-evolution of metal availability and nitrogen assimilation in cyanobacteria and algae. Geobiology 7: 100–123.

Glass, S. J., and M. J. Matteson. 1973. Ion enrichment in aerosols dispersed from bursting bubbles in aqueous salt solutions. Tellus 25:272–280.

Gleick, P. H. 2000. The changing water paradigm—A look at twenty-first century water resources development. Water International 25:127–138.

Glendining, M. J., A. G. Dailey, D. S. Powlson, G. M. Richter, J. A. Catt, and A. P. Whitmore. 2011. Pedotransfer functions for estimating total soil nitrogen up to the global scale. European Journal of Soil Science 62:13–22.

Glibert, P. M., F. Lipschultz, J. J. McCarthy, and M. A. Altabet. 1982. Isotope-dilution models of uptake and remineralization of ammonium by marine plankton. Limnology and Oceanography 27:639–650.

Glindemann, D., M. Edwards, and P. Kuschk. 2003. Phosphine gas in the upper troposphere. Atmospheric Environment 37:2429–2433.

Glover, H. E., B. B. Prezelin, L. Campbell, M. Wyman, and C. Garside. 1988. A nitrate-dependent *Synechococcus* bloom in surface Sargasso sea-water. Nature 331:161–163.

Glynn, P. W. 1988. El-Nino Southern Oscillation 1982-1983—Nearshore population, community, and ecosystem responses. Annual Review of Ecology and Systematics 19:309–346.

Godbold, D. L., E. Fritz, and A. Huttermann. 1988. Aluminum toxicity and forest decline. Proceedings of the National Academy of Sciences 85:3888–3892.

Godden, J. W., S. Turley, D. C. Teller, E. T. Adman, M. Y. Liu, W. J. Payne, and J. Legall. 1991. The 2.3-Angstrom X-ray structure of nitrite reductase from *Achromobacter cycloclastes*. Science 253:438–442.

Godfrey, L. V., and P. G. Falkowski. 2009. The cycling and redox state of nitrogen in the Archaean ocean. Nature Geoscience 2:725–729.

Goldan, P. D., R. Fall, W. C. Kuster, and F. C. Fehsenfeld. 1988. Uptake of COS by growing vegetation—A major tropospheric sink. Journal of Geophysical Research—Atmospheres 93:14186–14192.

Goldan, P. D., W. C. Kuster, D. L. Albritton, and F. C. Fehsenfeld. 1987. The measurement of natural sulfur emissions from soils and vegetation: Three sites in the eastern United States revisited. Journal of Atmospheric Chemistry 5:439–467.

Goldberg, D. E. 1982. The distribution of evergreen and deciduous trees relative to soil type—An example from the Sierra-Madre, Mexico, and a general model. Ecology 63:942–951.

Goldberg, D. E. 1985. Effects of soil pH, competition, and seed predation on the distributions of tree species. Ecology 66:503–511.

Goldberg, S. D., and G. Gebauer. 2009. Drought turns a central European Norway spruce forest soil from an N_2O source to a transient N_2O sink. Global Change Biology 15:850–860.

Goldblatt, C., M. W. Claire, T. M. Lenton, A. J. Matthews, A. J. Watson, and K. J. Zahnle. 2009. Nitrogen-enhanced greenhouse warming on early Earth. Nature Geoscience 2:891–896.

Goldewijk, K. K., A. Beusen, and P. Janssen. 2010. Long-term dynamic modeling of global population and built-up area in a spatially explicit way: HYDE 3.1. Holocene 20:565–573.

Goldich, S. S. 1938. A study in rock weathering. Journal of Geology 46:17–58.

Goldman, C. R. 1988. Primary productivity, nutrients, and transparency during the early onset of eutrophication in ultra-oligotrophic Lake Tahoe, California-Nevada. Limnology and Oceanography 33: 1321–1333.

Goldman, J. C., and P. M. Glibert. 1982. Comparative rapid ammonium uptake by four species of marine phytoplankton. Limnology and Oceanography 27: 814–827.

Golley, F. B. 1972. Energy flux in ecosystems. In J. A. Wiens, editor. Ecosystem Structure and Function, Oregon State University Press, 69–90.

Golterman, H. L. 1995. The role of the ironhydroxide-phosphate-sulfide system in the phosphate exchange between sediments and overlying water. Hydrobiologia 297:43–54.

Golterman, H. L. 2001. Phosphate release from anoxic sediments or "What did Mortimer really write?" Hydrobiologia 450:99–106.

Golubev, V. S., J. H. Lawrimore, P. Y. Groisman, N. A. Speranskaya, S. A. Zhuravin, M. J. Menne, T. C. Peterson, and R. W. Malone. 2001. Evaporation changes over the contiguous United States and the former USSR: A reassessment. Geophysical Research Letters 28:2665–2668.

Golubiewski, N. E. 2006. Urbanization increases grassland carbon pools: Effects of landscaping in Colorado's front range. Ecological Applications 16:555–571.

Gong, S. L., L. A. Barrie, J. M. Prospero, D. L. Savoie, G. P. Ayers, J. P. Blanchet, and L. Spacek. 1997. Modeling sea-salt aerosols in the atmosphere.2. Atmospheric concentrations and fluxes. Journal of Geophysical Research—Atmospheres 102:3819–3830.

Goodale, C. L., M. J. Apps, R. A. Birdsey, C. B. Field, L. S. Heath, R. A. Houghton, J. C. Jenkins, G. H. Kohlmaier, W. Kurz, S. R. Liu, G. J. Nabuurs, S. Nilsson, and A. Z. Shvidenko. 2002. Forest carbon sinks in the Northern Hemisphere. Ecological Applications 12:891–899.

Goode, P. R., J. Qiu, V. Yurchyshyn, J. Hickey, M. C. Chu, E. Kolbe, C. T. Brown, and S. E. Koonin. 2001. Earthshine observations of the Earth's reflectance. Geophysical Research Letters 28:1671–1674.

Goodrich, J. P., R. K. Varner, S. Frolking, B. N. Duncan, and P. M. Crill. 2011. High-frequency measurements of methane ebullition over a growing season at a temperate peatland site. Geophysical Research Letters 38.

Goolsby, D. A., W. A. Battaglin, B. T. Aulenbach, and R. P. Hooper. 2001. Nitrogen input to the Gulf of Mexico. Journal of Environmental Quality 30: 329–336.

Goregues, C. M., V. D. Michotey, and P. C. Bonin. 2005. Molecular, biochemical, and physiological approaches for understanding the ecology of denitrification. Microbial Ecology 49:198–208.

Gorham, E. 1957. The development of peat lands. Quarterly Review of Biology 32:145–166.

Gorham, E. 1961. Factors influencing supply of major ions to inland waters, with special reference to the atmosphere. Geological Society of America Bulletin 72: 795–840.

Gorham, E. 1991. Northern peatlands—Role in the carbon cycle and probable responses to climatic warming. Ecological Applications 1:182–195.

Gorham, E., and F. M. Boyce. 1989. Influence of lake surface area and depth upon thermal stratification and the depth of the summer thermocline. Journal of Great Lakes Research 15:233–245.

Gorham, E., F. B. Martin, and J. T. Litzau. 1984. Acid rain—Ionic correlations in the eastern United States, 1980-1981. Science 225:407–409.

Gorham, E., P. M. Vitousek, and W. A. Reiners. 1979. Regulation of chemical budgets over the course of terrestrial ecosystem succession. Annual Review of Ecology and Systematics 10:53–84.

Gosselink, J. G., and R. E. Turner. 1978. The role of hydrology in freshwater wetland ecosystems. In R. E. Good, D. F. Whigham and R. L. Simpson, editors. Freshwater Wetlands: Ecological Processes and Management Potential. Academic Press, 63–78.

Gosz, J. R. 1981. Nitrogen cycling in coniferous ecosystems. *In* F. E. Clark and T. Rosswall, editors. *Terrestrial Nitrogen Cycles*. Swedish Natural Science Research Council, Stockholm, 405–426.

Gosz, J. R., R. T. Holmes, G. E. Likens, and F. H. Bormann. 1978. Flow of energy in a forest ecosystem. Scientific American **238**:92–102.

Gough, C. M., C. S. Vogel, H. P. Schmid, H. B. Su, and P. S. Curtis. 2008. Multi-year convergence of biometric and meteorological estimates of forest carbon storage. Agricultural and Forest Meteorology **148**: 158–170.

Govindarajulu, M., P. E. Pfeffer, H. R. Jin, J. Abubaker, D. D. Douds, J. W. Allen, H. Bucking, P. J. Lammers, and Y. Shachar-Hill. 2005. Nitrogen transfer in the arbuscular mycorrhizal symbiosis. Nature **435**: 819–823.

Gower, S. T., S. Pongracic, and J. J. Landsberg. 1996. A global trend in belowground carbon allocation: Can we use the relationship at smaller scales? Ecology **77**:1750–1755.

Gower, S. T., K. A. Vogt, and C. C. Grier. 1992. Carbon dynamics of Rocky Mountain douglas fir—Influence of water and nutrient availability. Ecological Monographs **62**:43–65.

Grace, J., C. Nichol, M. Disney, P. Lewis, T. Quaife, and P. Bowyer. 2007. Can we measure terrestrial photosynthesis from space directly, using spectral reflectance and fluorescence? Global Change Biology **13**:1484–1497.

Gradinger, R. 1995. Climate change and biological oceanography of the Arctic ocean. Philosophical Transactions of the Royal Society A—Mathematical Physical and Engineering Sciences **352**:277–286.

Graedel, T. E., and P. J. Crutzen. 1993. *Atmospheric Change*. W. H, Freeman.

Graedel, T. E., and W. C. Keene. 1995. Tropospheric budget of reactive chlorine. Global Biogeochemical Cycles **9**:47–77.

Graham, J. B., R. Dudley, N. M. Aguilar, and C. Gans. 1995. Implications of the late Paleozoic oxygen pulse for physiology and evolution. Nature **375**:117–120.

Graham, W. F., and R. A. Duce. 1979. Atmospheric pathways of the phosphorus cycle. Geochimica et Cosmochimica Acta **43**:1195–1208.

Grandstaff, D. E. 1986. The dissolution rate of forsteritic olivine from Hawaiian beach sand. *In* S. M. Colman and D. P. Dethier, editors. *Rates of Chemical Weathering of Rocks and Minerals*. Academic/Elsevier, 41–59.

Granhall, U. 1981. Biological nitrogen fixation in relation to environmental factors and functioning of natural ecosystems. *In* F. E. Clark and T. Rosswall, editors. *Terrestrial Nitrogen Cycles*. Swedish Natural Science Research Council, Stockholm, 131–144.

Granier, A., P. Biron, and D. Lemoine. 2000a. Water balance, transpiration and canopy conductance in two beech stands. Agricultural and Forest Meteorology **100**:291–308.

Granier, C., G. Petron, J. F. Muller, and G. Brasseur. 2000b. The impact of natural and anthropogenic hydrocarbons on the tropospheric budget of carbon monoxide. Atmospheric Environment **34**:5255–5270.

Grasman, B. T., and E. C. Hellgren. 1993. Phosphorus nutrition in white-tailed deer—Nutrient balance, physiological responses, and antler growth. Ecology **74**: 2279–2296.

Grassle, J. F. 1985. Hydrothermal vent animals—Distribution and biology. Science **229**:713–717.

Gratz, L. E., and G. J. Keeler. 2011. Sources of mercury in precipitation to Underhill, VT. Atmospheric Environment **45**:5440–5449.

Graustein, W. C., and R. L. Armstrong. 1983. The use of $^{87}Sr/^{86}Sr$ ratios to measure atmospheric transport into forested watersheds. Science **219**:289–292.

Graustein, W. C., K. Cromack, and P. Sollins. 1977. Calcium oxalate—Occurrence in soils and effect on nutrient and geochemical cycles. Science **198**:1252–1254.

Graveland, J., R. Vanderwal, J. H. Vanbalen, and A. J. Vannoordwijk. 1994. Poor reproduction in forest passerines from decline of snail abundance on acidified soils. Nature **368**:446–448.

Gray, J. T. 1982. Community structure and productivity in *Ceanothus* chaparral and coastal sage scrub of southern California. Ecological Monographs **52**:415–435.

Gray, J. T. 1983. Nutrient use by evergreen and deciduous shrubs in southern California. 1. Community nutrient cycling and nutrient-use efficiency. Journal of Ecology **71**:21–41.

Gray, J. T., and W. H. Schlesinger. 1981. Nutrient cycling in Mediterranean type ecosystems. *In* P. C. Miller, editor. *Resource Use by Chaparral and Matorral*. Springer, 259–285.

Graydon, J. A., V. L. St. Louis, S. E. Lindberg, K. A. Sandilands, J.W.M. Rudd, C. A. Kelly, R. Harris, M. T. Tate, D. P. Krabbenhoft, C. A. Emmerton, H. Asmath, and M. Richardson. 2012. The role of terrestrial vegetation in atmospheric Hg deposition: Pools and fluxes of spike and ambient Hg from the METAALICUS experiment. Global Biogeochemical Cycles **26**.

Greeley, R., and B. D. Schneid. 1991. Magma generation on Mars—Amounts, rates, and comparisons with Earth, Moon, and Venus. Science **254**:996–998.

Green, C. J., A. M. Blackmer, and N. C. Yang. 1994. Release of fixed ammonium during nitrification in soils. Soil Science Society of America Journal **58**: 1411–1415.

Green, H. W., W. P. Chen, and M. R. Brudzinski. 2010. Seismic evidence of negligible water carried below 400-km depth in subducting lithosphere. Nature **467**:828–831.

Green, P. A., C. J. Vörösmarty, M. Meybeck, J. N. Galloway, B. J. Peterson, and E. W. Boyer. 2004. Pre-industrial and contemporary fluxes of nitrogen through rivers: a global assessment based on typology. Biogeochemistry **68**:71–105.

Greenberg, J. P., and P. R. Zimmerman. 1984. Nonmethane hydrocarbons in remote tropical, continental, and marine atmospheres. Journal of Geophysical Research—Atmospheres **89**:4767–4778.

Greenland, D. J. 1971. Interactions between humic and fulvic acids and clays. Soil Science **111**:34–41.

Greenway, H., and J. Gibbs. 2003. Review: Mechanisms of anoxia tolerance in plants. II. Energy requirements for maintenance and energy distribution to essential processes. Functional Plant Biology **30**:999–1036.

Greenwood, J. P., S. Itoh, N. Sakamoto, E. P. Vicenzi, and H. Yurimoto. 2008. Hydrogen isotope evidence for loss of water from Mars through time. Geophysical Research Letters **35**.

Gregg, J. W., C. G. Jones, and T. E. Dawson. 2003. Urbanization effects on tree growth in the vicinity of New York City. Nature **424**:183–187.

Gregor, B. 1970. Denudation of the continents. Nature **228**:273–275.

Grelle, A., A. Lundberg, A. Lindroth, A. S. Moren, and E. Cienciala. 1997. Evaporation components of a boreal forest: Variations during the growing season. Journal of Hydrology **197**:70–87.

Gress, S. E., T. D. Nichols, C. C. Northcraft, and W. T. Peterjohn. 2007. Nutrient limitation in soils exhibiting differing nitrogen availabilities: What lies beyond nitrogen saturation? Ecology **88**:119–130.

Gressel, N., J. G. McColl, C. M. Preston, R. H. Newman, and R. F. Powers. 1996. Linkages between phosphorus transformations and carbon decomposition in a forest soil. Biogeochemistry **33**:97–123.

Grey, D. C., and M. L. Jensen. 1972. Bacteriogenic sulfur in air pollution. Science **177**:1099–1100.

Gribsholt, B., J. E. Kostka, and E. Kristensen. 2003. Impact of fiddler crabs and plant roots on sediment biogeochemistry in a Georgia saltmarsh. Marine Ecology—Progress Series **259**:237–251.

Grier, C. C., and S. W. Running. 1977. Leaf area of mature northwestern coniferous forests—Relation to site water balance. Ecology **58**:893–899.

Griffin, K. L., M. A. Bashkin, R. B. Thomas, and B. R. Strain. 1997. Interactive effects of soil nitrogen and atmospheric carbon dioxide on root/rhizosphere carbon dioxide efflux from loblolly and ponderosa pine seedlings. Plant and Soil **190**:11–18.

Griffin, M.P.A., M. L. Cole, K. D. Kroeger, and J. Cebrian. 1998. Dependence of herbivory on autotrophic nitrogen content and on net primary production across ecosystems. Biological Bulletin **195**:233–234.

Griffith, E. J., C. Ponnamperuma, and N. W. Gabel. 1977. Phosphorus: A key to life on the primitive Earth. Origins of Life **8**:71–85.

Griffith, E. M., A. Paytan, K. Caldeira, T. D. Bullen, and E. Thomas. 2008. A dynamic marine calcium cycle during the past 28 million years. Science **322**:1671–1674.

Griffiths, R. P., M. E. Harmon, B. A. Caldwell, and S. E. Carpenter. 1993. Acetylene reduction in conifer logs during early stages of decomposition. Plant and Soil **148**:53–61.

Grimm, N. B., and S. G. Fisher. 1989. Stability of periphyton and macroinvertebrates to disturbance by flash floods in a desert stream. Journal of the North American Benthological Society **8**:293–307.

Grimm, N. B., S. E. Gergel, W. H. McDowell, E. W. Boyer, C. L. Dent, P. Groffman, S. C. Hart, J. Harvey, C. Johnston, E. Mayorga, M. E. McClain, and G. Pinay. 2003. Merging aquatic and terrestrial perspectives of nutrient biogeochemistry. Oecologia **137**:485–501.

Groffman, P. M., and M. C. Fisk. 2011. Calcium constrains plant control over forest ecosystem nitrogen cycling. Ecology **92**:2035–2042.

Groffman, P. M., J. P. Hardy, M. C. Fisk, T. J. Fahey, and C. T. Driscoll. 2009. Climate variation and soil carbon and nitrogen cycling processes in a northern hardwood forest. Ecosystems **12**:927–943.

Groffman, P. M., N. L. Law, K. T. Belt, L. E. Band, and G. T. Fisher. 2004. Nitrogen fluxes and retention in urban watershed ecosystems. Ecosystems **7**:393–403.

Groffman, P. M., and J. M. Tiedje. 1989. Denitrification in north temperate forest soils—Spatial and temporal patterns at the landscape and seasonal scales. Soil Biology & Biochemistry **21**:613–620.

Groisman, P. Y., R. W. Knight, T. R. Karl, D. R. Easterling, B. M. Sun, and J. H. Lawrimore. 2004. Contemporary changes of the hydrological cycle over the contiguous United States: Trends derived from in situ observations. Journal of Hydrometeorology **5**:64–85.

Gross, M. G. 1977. *Oceanography: A View of the Earth.* Prentice-Hall.

Grosse, G., J. Harden, M. Turetsky, A. D. McGuire, P. Camill, C. Tarnocai, S. Frolking, E.A.G. Schuur, T. Jorgenson, S. Marchenko, V. Romanovsky, K. P. Wickickland, N. French, M. Waldrop, L. Bourgeau-Chavez, and R. G. Striegl. 2011. Vulnerability of high-latitude soil organic carbon in North America to disturbance. Journal of Geophysical Research—Biogeosciences **116**:23.

Grotzinger, J. P., and J. F. Kasting. 1993. New constraints on Precambrian ocean composition. Journal of Geology 101:235–243.

Gruber, N., and J. N. Galloway. 2008. An Earth-system perspective of the global nitrogen cycle. Nature 451: 293–296.

Gruber, N., C. Hauri, Z. Lachkar, D. Loher, T. L. Frolicher, and G.-K. Plattner. 2012. Rapid progression of ocean acidification in the California current system. Science 337:220–223.

Gruber, N., C. D. Keeling, and N. R. Bates. 2002. Interannual variability in the North Atlantic Ocean carbon sink. Science 298:2374–2378.

Gu, B. H., J. Schmitt, Z. Chen, L. Y. Liang, and J. F. McCarthy. 1995. Adsorption and desorption of different organic-matter fractions on iron oxide. Geochimica et Cosmochimica Acta 59:219–229.

Guadalix, M. E., and M. T. Pardo. 1991. Sulfate sorption by variable-charge soils. Journal of Soil Science 42:607–614.

Guenther, A. 2002. The contribution of reactive carbon emissions from vegetation to the carbon balance of terrestrial ecosystems. Chemosphere 49:837–844.

Guenther, A., C. Geron, T. Pierce, B. Lamb, P. Harley, and R. Fall. 2000. Natural emissions of non-methane volatile organic compounds; carbon monoxide, and oxides of nitrogen from North America. Atmospheric Environment 34:2205–2230.

Guidry, M. W., and F. T. Mackenzie. 2003. Experimental study of igneous and sedimentary apatite dissolution: Control of pH, distance from equilibrium, and temperature on dissolution rates. Geochimica et Cosmochimica Acta 67:2949–2963.

Guidry, M. W., F. T. MacKenzie, and R. S. Arvidson. 2000. Role of tectonics in phosphorus distribution and cycling. In C. R. Glenn, L. Prevot-Lucas and J. Lucas, editors. Marine Authigenesis: From Global to Microbial. Society of Sedimentary Geology, Tulsa, Oklahoma, 35–51.

Guieu, C., R. Duce, and R. Arimoto. 1994. Dissolved input of manganese to the ocean—Aerosol source. Journal of Geophysical Research—Atmospheres 99: 18789–18800.

Gunal, H., and M. D. Ransom. 2006. Clay illuviation and calcium carbonate accumulation along a precipitation gradient in Kansas. Catena 68:59–69.

Gunderson, C. A., K. H. O'Hara, C. M. Campion, A. V. Walker, and N. T. Edwards. 2010. Thermal plasticity of photosynthesis: the role of acclimation in forest responses to a warming climate. Global Change Biology 16:2272–2286.

Guo, J. H., X. J. Liu, Y. Zhang, J. L. Shen, W. X. Han, W. F. Zhang, P. Christie, K.W.T. Goulding, P. M. Vitousek, and F. S. Zhang. 2010. Significant acidification in major Chinese croplands. Science 327: 1008–1010.

Guo, L. B., and R. M. Gifford. 2002. Soil carbon stocks and land use change: a meta analysis. Global Change Biology 8:345–360.

Guo, L. D., C. L. Ping, and R. W. Macdonald. 2007. Mobilization pathways of organic carbon from permafrost to arctic rivers in a changing climate. Geophysical Research Letters 34:5.

Guo, Y. Y., R. Amundson, P. Gong, and Q. Yu. 2006. Quantity and spatial variability of soil carbon in the conterminous United States. Soil Science Society of America Journal 70:590–600.

Gurtz, M. E., G. R. Marzolf, K. T. Killingbeck, D. L. Smith, and J. V. McArthur. 1988. Hydrologic and riparian influences on the import and storage of coarse particulate organic matter in a prairie stream. Canadian Journal of Fisheries and Aquatic Sciences 45:655–665.

Gusewell, S. 2004. N:P ratios in terrestrial plants: variation and functional significance. New Phytologist 164:243–266.

Guskov, A., J. Kern, A. Gabdulkhakov, M. Broser, A. Zouni, and W. Saenger. 2009. Cyanobacterial photosystem II at 2.9-angstrom resolution and the role of quinones, lipids, channels and chloride. Nature Structural & Molecular Biology 16:334–342.

Gutschick, V. P. 1981. Evolved strategies in nitrogen acquisition by plants. American Naturalist 118: 607–637.

Haberl, H., K. H. Erb, F. Krausmann, V. Gaube, A. Bondeau, C. Plutzar, S. Gingrich, W. Lucht, and M. Fischer-Kowalski. 2007. Quantifying and mapping the human appropriation of net primary production in Earth's terrestrial ecosystems. Proceedings of the National Academy of Sciences 104:12942–12945.

Habicht, K. S., and D. E. Canfield. 1996. Sulphur isotope fractionation in modern microbial mats and the evolution of the sulphur cycle. Nature 382:342–343.

Habicht, K. S., D. E. Canfield, and J. Rethmeier. 1998. Sulfur isotope fractionation during bacterial reduction and disproportionation of thiosulfate and sulfite. Geochimica et Cosmochimica Acta 62: 2585–2595.

Habicht, K. S., M. Gade, B. Thamdrup, P. Berg, and D. E. Canfield. 2002. Calibration of sulfate levels in the Archean Ocean. Science 298:2372–2374.

Habing, H. J., C. Dominik, M. J. de Mulzon, M. F. Kessler, R. J. Laureijs, K. Leech, L. Metcalfe, A. Salama, R. Slebenmorgen, and N. Trams. 1999. Disappearance of stellar debris disks around main-sequence stars after 400 million years. Nature 401:456–458.

Hackley, K. C., and T. F. Anderson. 1986. Sulfur isotopic variations in low-sulfur coals from the Rocky Mountain region. Geochimica et Cosmochimica Acta 50: 1703–1713.

Hackstein, J.H.P., and C. K. Stumm. 1994. Methane production in terrestrial arthropods. Proceedings of the National Academy of Sciences 91:5441–5445.

Haddeland, I., T. Skaugen, and D. P. Lettenmaier. 2006. Anthropogenic impacts on continental surface water fluxes. Geophysical Research Letters 33.

Hagedorn, F., D. Spinnler, M. Bundt, P. Blaser, and R. Siegwolf. 2003. The input and fate of new C in two forest soils under elevated CO_2. Global Change Biology 9:862–872.

Hahn, J. 1974. The North Atlantic Ocean as a source of atmospheric N_2O. Tellus 26:160–168.

Hahn, J. 1980. Organic constituents of natural aerosols. Annals of the New York Academy of Sciences 338: 359–376.

Haines, B., M. Black, and C. Bayer. 1989. Sulfur emissions from roots of the rain forest tree Stryphnodendron excelsum. In E. S. Saltzman and W. J. Cooper, editors. Biogenic Sulfur in the Environment. American Chemical Society, 58–69.

Haines, E. B. 1977. Origins of detritus in Georgia salt marsh estuaries. Oikos 29:254–260.

Håkanson, L., A. C. Bryhn, and J. K. Hytteborn. 2007. On the issue of limiting nutrient and predictions of cyanobacteria in aquatic systems. Science of the Total Environment 379:89–108.

Hall, D. T., D. F. Strobel, P. D. Feldman, M. A. McGrath, and H. A. Weaver. 1995. Detection of an oxygen atmosphere on Jupiter's moon Europa. Nature 373:677–679.

Hall, R. O., E. S. Bernhardt, and G. E. Likens. 2002. Relating nutrient uptake with transient storage in forested mountain streams. Limnology and Oceanography 47: 255–265.

Hall, R. O., Jr., J. L. Tank, D. J. Sobota, P. J. Mulholland, J. M. O'Brien, W. K. Dodds, J. R. Webster, H. M. Valett, G. C. Poole, B. J. Peterson, J. L. Meyer, W. H. McDowell, S. L. Johnson, S. K. Hamilton, N. B. Grimm, S. V. Gregory, C. N. Dahm, L. W. Cooper, L. R. Ashkenas, S. M. Thomas, R. W. Sheibley, J. D. Potter, B. R. Niederlehner, L. T. Johnson, A. M. Helton, C. M. Crenshaw, A. J. Burgin, M. J. Bernot, J. J. Beaulieu, and C. P. Arango. 2009. Nitrate removal in stream ecosystems measured by ^{15}N-addition experiments: Total uptake. Limnology and Oceanography 54:653–665.

Hall, R. O., and J. L. Tank. 2003a. Ecosystem metabolism controls nitrogen uptake in streams in Grand Teton National Park, Wyoming. Limnology and Oceanography 48:1120–1128.

Hall, R. O., and J. L. Tank. 2003b. Ecosystem metabolism controls nitrogen uptake in streams in Grand Teton National Park, Wyoming. Limnology and Oceanography 48:1120–1128.

Hall, S. J., and P. A. Matson. 1999. Nitrogen oxide emissions after nitrogen additions in tropical forests. Nature 400:152–155.

Halmer, M. M., H.-U. Schmincke, and H.-F. Graf. 2002. The annual volcanic gas input into the atmosphere, in particular into the stratosphere: A global data set for the past 100 years. Journal of Volcanology and Geothermal Research 115:511–528.

Hames, R. S., K. V. Rosenberg, J. D. Lowe, S. E. Barker, and A. A. Dhondt. 2002. Adverse effects of acid rain on the distribution of the Wood Thrush Hylocichla mustelina in North America. Proceedings of the National Academy of Sciences 99:11235–11240.

Hamilton, J. G., E. H. DeLucia, K. George, S. L. Naidu, A. C. Finzi, and W. H. Schlesinger. 2002. Forest carbon balance under elevated CO_2. Oecologia 131: 250–260.

Hamilton, J.T.G., W. C. McRoberts, F. Keppler, R. M. Kalin, and D. B. Harper. 2003. Chloride methylation by plant pectin: An efficient environmentally significant process. Science 301:206–209.

Hamilton, S. K., D. A. Bruesewitz, G. P. Horst, D. B. Weed, and O. Sarnelle. 2009. Biogenic calcite-phosphorus precipitation as a negative feedback to lake eutrophication. Canadian Journal of Fisheries and Aquatic Sciences 66:343–350.

Hammerling, D. M., A. M. Michalak, C. O'Dell, and S. R. Kawa. 2012. Global CO_2 distributions over land from the Greenhouse Gases Observing Satellite (GOSAT). Geophysical Research Letters 39.

Han, H., and J. D. Allan. 2012. Uneven rise in N inputs to the Lake Michigan basin over the 20th century corresponds to agricultural and societal transitions. Biogeochemistry 109:175–187.

Han, T. M., and B. Runnegar. 1992. Megascopic eukaryotic algae from the 2.1-billion-year-old Negaunee iron formation, Michigan. Science 257:232–235.

Han, W. X., J. Y. Fang, D. L. Guo, and Y. Zhang. 2005. Leaf nitrogen and phosphorus stoichiometry across 753 terrestrial plant species in China. New Phytologist 168:377–385.

Hanczyc, M. M., S. M. Fujikawa, and J. W. Szostak. 2003. Experimental models of primitive cellular compartments: Encapsulation, growth, and division. Science 302:618–622.

Handley, L. L., and J. A. Raven. 1992. The use of natural abundance of nitrogen isotopes in plant physiology and ecology. Plant Cell and Environment 15:965–985.

Hanks, T. C., and D. L. Anderson. 1969. The early thermal history of the Earth. Physics of the Earth and Planetary Interiors 2:19–29.

Hannah, L., J. L. Carr, and A. Landerani. 1995. Human disturbance and natural habitat—A biome-level analysis of a global data set. Biodiversity and Conservation **4**:128–155.

Hannan, L. B., J. D. Roth, L. M. Ehrhart, and J. F. Weishampel. 2007. Dune vegetation fertilization by nesting sea turtles. Ecology **88**:1053–1058.

Hansell, D. A., and C. A. Carlson. 1998a. Deep-ocean gradients in the concentration of dissolved organic carbon. Nature **395**:263–266.

Hansell, D. A., and C. A. Carlson. 1998b. Net community production of dissolved organic carbon. Global Biogeochemical Cycles **12**:443–453.

Hansell, D. A., D. Kadko, and N. R. Bates. 2004. Degradation of terrigenous dissolved organic carbon in the western Arctic Ocean. Science **304**:858–861.

Hansen, M. C., S. V. Stehman, and P. V. Potapov. 2010. Quantification of global gross forest cover loss. Proceedings of the National Academy of Sciences **107**:8650–8655.

Hanson, P. J., N. T. Edwards, C. T. Garten, and J. A. Andrews. 2000. Separating root and soil microbial contributions to soil respiration: A review of methods and observations. Biogeochemistry **48**:115–146.

Happell, J. D., J. P. Chanton, G. J. Whiting, and W. J. Showers. 1993. Stable isotopes as tracers of methane dynamics in Everglades marshes with and without active populations of methane-oxidizing bacteria. Journal of Geophysical Research—Atmospheres **98**:14771–14782.

Harden, J. W. 1988. Genetic interpretations of elemental and chemical differences in a soil chronosequence, California. Geoderma **43**:179–193.

Harden, J. W., E. T. Sundquist, R. F. Stallard, and R. K. Mark. 1992. Dynamics of soil carbon during deglaciation of the Laurentide ice sheet. Science **258**: 1921–1924.

Harmon, M. E., J. F. Franklin, F. J. Swanson, P. Sollins, S. V. Gregory, J. D. Lattin, N. H. Anderson, S. P. Cline, N. G. Aumen, J. R. Sedell, G. W. Lienkaemper, K. Cromack, and K. W. Cummins. 1986. Ecology of coarse woody debris in temperate ecosystems. Advances in Ecological Research **15**:133–302.

Harmon, M. E., W. L. Silver, B. Fasth, H. Chen, I. C. Burke, W. J. Parton, S. C. Hart, W. S. Currie, and Lidet. 2009. Long-term patterns of mass loss during the decomposition of leaf and fine root litter: an intersite comparison. Global Change Biology **15**:1320–1338.

Harper, C. L., and S. B. Jacobsen. 1996. Noble gases and Earth's accretion. Science **273**:1814–1818.

Harpole, W. S., J. T. Ngai, E. E. Cleland, E. W. Seabloom, E. T. Borer, M.E.S. Bracken, J. J. Elser, D. S. Gruner, H. Hillebrand, J. B. Shurin, and J. E. Smith. 2011. Nutrient co-limitation of primary producer communities. Ecology Letters **14**:852–862.

Harries, J. E., H. E. Brindley, P. J. Sagoo, and R. J. Bantges. 2001. Increases in greenhouse forcing inferred from the outgoing longwave radiation spectra of the Earth in 1970 and 1997. Nature **410**:355–357.

Harrington, J. B. 1987. Climatic change—A Review of causes. Canadian Journal of Forest Research **17**: 1313–1339.

Harris, N. L., S. Brown, S. C. Hagen, S. S. Saatchi, S. Petrova, W. Salas, M. C. Hansen, P. V. Potapov, and A. Lotsch. 2012. Baseline map of carbon emissions from deforestation in tropical regions. Science **336**: 1573–1576.

Harris, R. C., J.W.M. Rudd, M. Amyot, C. L. Babiarz, K. G. Beaty, P. J. Blanchfield, R. A. Bodaly, B. A. Branfireun, C. C. Gilmour, J. A. Graydon, A. Heyes, H. Hintelmann, J. P. Hurley, C. A. Kelly, D. P. Krabbenhoft, S. E. Lindberg, R. P. Mason, M. J. Paterson, C. L. Podemski, A. Robinson, K. A. Sandilands, G. R. Southworth, V.L.S. Louis, and M. T. Tate. 2007. Whole-ecosystem study shows rapid fish-mercury response to changes in mercury deposition. Proceedings of the National Academy of Sciences **104**: 16586–16591.

Harris, W. G., K. A. Hollien, T. L. Yuan, S. R. Bates, and W. A. Acree. 1988. Nonexchangeable potassium associated with hydroxy-interlayered vermiculite from coastal plain soils. Soil Science Society of America Journal **52**:1486–1492.

Harrison, A. F. 1982. ^{32}P-method to compare rates of mineralization of labile organic phosphorus in woodland soils. Soil Biology & Biochemistry **14**: 337–341.

Harrison, J. A., N. Caraco, and S. P. Seitzinger. 2005. Global patterns and sources of dissolved organic matter export to the coastal zone: Results from a spatially explicit, global model. Global Biogeochemical Cycles **19**.

Harrison, J. A., P. J. Frings, and A.H.W. Beusen. et al., 2012. Global importance, patterns, and controls of dissolved silica retention in lakes and reservoirs. Global Biogeochemical Cycles **26**.

Harrison, J. A., R. J. Maranger, R. B. Alexander, A. E. Giblin, P.-A. Jacinthe, E. Mayorga, S. P. Seitzinger, D. J. Sobota, and W. M. Wollheim. 2009. The regional and global significance of nitrogen removal in lakes and reservoirs. Biogeochemistry **93**:143–157.

Harrison, J. F., A. Kaiser, and J. M. VandenBrooks. 2010. Atmospheric oxygen level and the evolution of insect body size. Proceedings of the Royal Society B-Biological Sciences **277**:1937–1946.

Harrison, K. A., R. Bol, and R. D. Bardgett. 2007. Preferences for different nitrogen forms by coexisting plant species and soil microbes. Ecology **88**:989–999.

Harrison, K. G. 2000. Role of increased marine silica input on paleo-pCO_2 levels. Paleoceanography **15**: 292–298.

Harrison, M. J., and M. L. van Buuren. 1995. A phosphate transporter from the mycorrhizal fungus *Glomus versiforme*. Nature **378**:626–629.

Harrison, R. B., D. W. Johnson, and D. E. Todd. 1989. Sulfate adsorption and desorption reversibility in a variety of forest soils. Journal of Environmental Quality **18**:419–426.

Harrison, W. G., L. R. Harris, and B. D. Irwin. 1996. The kinetics of nitrogen utilization in the oceanic mixed layer: Nitrate and ammonium interactions at nanomolar concentrations. Limnology and Oceanography **41**:16–32.

Harrison, W. G., L. R. Harris, D. M. Karl, G. A. Knauer, and D. G. Redalje. 1992. Nitrogen dynamics at the VERTEX time-series site. Deep-Sea Research Part A—Oceanographic Research Papers **39**: 1535–1552.

Harriss, R. C., D. I. Sebacher, and F. P. Day. 1982. Methane flux in the Great Dismal swamp. Nature **297**: 673–674.

Harte, J., S. Saleska, and T. Shih. 2006. Shifts in plant dominance control carbon-cycle responses to experimental warming and widespread drought. Environmental Research Letters **1**.

Hartgers, W. A., J.S.S. Damste, A. G. Requejo, J. Allan, J. M. Hayes, and J. W. Deleeuw. 1994. Evidence for only minor contributions from bacteria to sedimentary organic carbon. Nature **369**:224–227.

Hartley, A. E., and W. H. Schlesinger. 2000. Environmental controls on nitric oxide emission from northern Chihuahuan desert soils. Biogeochemistry **50**: 279–300.

Hartley, A. M., W. A. House, M. E. Callow, and B.S.C. Leadbeater. 1995. The role of a green alga in the precipitation of calcite and the coprecipitation of phosphate in freshwater. Internationale Revue Der Gesamten Hydrobiologie **80**:385–401.

Hartmann, J., N. Jansen, H. H. Durr, S. Kempe, and P. Kohler. 2009. Global CO_2-consumption by chemical weathering: What is the contribution of highly active weathering regions? Global and Planetary Change **69**:185–194.

Hartnett, H. E., and A. H. Devol. 2003. Role of a strong oxygen-deficient zone in the preservation and degradation of organic matter: A carbon budget for the continental margins of northwest Mexico and Washington State. Geochimica et Cosmochimica Acta **67**: 247–264.

Hartnett, H. E., R. G. Keil, J. I. Hedges, and A. H. Devol. 1998. Influence of oxygen exposure time on organic carbon preservation in continental margin sediments. Nature **391**:572–574.

Harvey, L.D.D. 1988. Climatic impact of ice-age aerosols. Nature **334**:333–335.

Hastings, M. G., J. C. Jarvis, and E. J. Steig. 2009. Anthropogenic impacts on nitrogen isotopes of ice-core nitrate. Science **324**:1288.

Hatch, D. J., S. C. Jarvis, and L. Philipps. 1990. Field measurement of nitrogen mineralization using soil core incubation and acetylene inhibition of nitrification. Plant and Soil **124**:97–107.

Hattenschwiler, S., F. Miglietta, A. Raschi, and C. Korner. 1997. Thirty years of *in situ* tree growth under elevated CO_2: a model for future forest responses? Global Change Biology **3**:463–471.

Hatton, R. S., R. D. Delaune, and W. H. Patrick. 1983. Sedimentation, accretion, and subsidence in marshes of Barataria basin, Louisiana. Limnology and Oceanography **28**:494–502.

Hawkesworth, C. J., and A.I.S. Kemp. 2006. Evolution of the continental crust. Nature **443**:811–817.

Hayhoe, K., and A. Farley. 2009. *A Climate for Change*. Faith Words.

Haynes, R. J., and K. M. Goh. 1978. Ammonium and nitrate nutrition of plants. Biological Reviews of the Cambridge Philosophical Society **53**:465–510.

He, H., T. M. Bleby, E. J. Veneklaas, and H. Lambers. 2011. Dinitrogen-fixing *Acacia* species from phosphorus-impoverished soils resorb leaf phosphorus efficiently. Plant, Cell & Environment **34**:2060–2070.

He, Y. F., P. D'Odorico, S.F.J. De Wekker, J. D. Fuentes, and M. Litvak. 2010. On the impact of shrub encroachment on microclimate conditions in the northern Chihuahuan desert. Journal of Geophysical Research—Atmospheres **115**.

Heathcote, A. J., and J. A. Downing. 2012. Impacts of eutrophication on carbon burial in freshwater lakes in an intensively agricultural landscape. Ecosystems **15**:60–70.

Heathwaite, A. L., R. Eggelsmann, K. H. Gottlich, and G. Haule. 1990. Ecohydrology, mire drainage and mire conservation. *In* A. L. Heathwaiter, editor. *Mires: Process, Exploitation and Conservation*. John Wiley & Sons, 417–484.

Hebting, Y., P. Schaeffer, A. Behrens, P. Adam, G. Schmitt, P. Schneckenburger, S. M. Bernasconi, and P. Albrecht. 2006. Biomarker evidence for a major preservation pathway of sedimentary organic carbon. Science **312**: 1627–1631.

Heckathorn, S. A., and E. H. Delucia. 1995. Ammonia volatilization during drought in perennial C-4 grasses of tallgrass prairie. Oecologia **101**:361–365.

Hedges, J. I., J. A. Baldock, Y. Gelinas, C. Lee, M. Peterson, and S. G. Wakeham. 2001. Evidence for non-selective preservation of organic matter in sinking marine particles. Nature **409**:801–804.

Hedges, J. I., R. G. Keil, and R. Benner. 1997. What happens to terrestrial organic matter in the ocean? Organic Geochemistry 27:195–212.

Hedges, J. I., and P. L. Parker. 1976. Land-derived organic matter in surface sediments from the Gulf of Mexico. Geochimica et Cosmochimica Acta 40:1019–1029.

Hedin, L. O., J. J. Armesto, and A. H. Johnson. 1995. Patterns of nutrient loss from unpolluted old-growth forests: Evaluation of biogeochemical theory. Ecology 76:493–509.

Hedin, L. O., G. E. Likens, K. M. Postek, and C. T. Driscoll. 1990. A field experiment to test whether organic acids buffer acid deposition. Nature 345:798–800.

Hedin, L. O., G. E. Likens, and F. H. Bormann. 1987. Decrease in precipitation acidity resulting from decreased SO_4^{2-} concentration. Nature 325:244–246.

Hedin, L. O., J. C. von Fischer, N. E. Ostrom, B. P. Kennedy, M. G. Brown, and G. P. Robertson. 1998. Thermodynamic constraints on the biogeochemical structure and transformation of nitrogen at terrestrial-lotic interfaces. Ecology 79:684–703.

Hedley, M. J., P. H. Nye, and R. E. White. 1982a. Plant-induced changes in the rhizosphere of rape (Brassica napus Var-Emerald) seedlings. 2. Origin of the pH change. New Phytologist 91:31–44.

Hedley, M. J., J.W.B. Stewart, and B. S. Chauhan. 1982b. Changes in inorganic and organic soil phosphorus fractions induced by cultivation practices and by laboratory incubations. Soil Science Society of America Journal 46:970–976.

Heffernan, J. B., and M. J. Cohen. 2010. Direct and indirect coupling of primary production and diel nitrate dynamics in a subtropical spring-fed river. Limnology and Oceanography 55:677–688.

Hein, R., P. J. Crutzen, and M. Heimann. 1997. An inverse modeling approach to investigate the global atmospheric methane cycle. Global Biogeochemical Cycles 11:43–76.

Heintzenberg, J. 1989. Fine particles in the global troposphere: A review. Tellus 41B:149–160.

Heip, C.H.R., N. K. Goosen, P.M.J. Herman, J. Kromkamp, J. J. Middelburg, and K. Soetaert. 1995. Production and consumption of biological particles in temperate tidal estuaries. In A. D. Ansell, R. N. Gibson, and M. Barnes, editors. Oceanography and Marine Biology, Taylor and Francis, 1–149.

Held, I. M., and B. J. Soden. 2006. Robust responses of the hydrological cycle to global warming. Journal of Climate 19:5686–5699.

Helfield, J. M., and R. J. Naiman. 2001. Effects of salmon-derived nitrogen on riparian forest growth and implications for stream productivity. Ecology 82:2403–2409.

Helm, K. P., N. L. Bindoff, and J. A. Church. 2011. Observed decreases in oxygen content of the global ocean. Geophysical Research Letters 38.

Hemond, H. F. 1980. Biogeochemistry of Thoreau's bog, Concord, Massachusetts. Ecological Monographs 50:507–526.

Hemond, H. F. 1983. The nitrogen budget of Thoreau's bog. Ecology 64:99–109.

Henderson, G. S., W. T. Swank, J. B. Waide, and C. C. Grier. 1978. Nutrient budgets of Appalachian and Cascade region watersheds—Comparison. Forest Science 24:385–397.

Henderson-Sellers, A., and K. McGuffie. 1987. A Climate Modeling Primer. Wiley.

Hendren, C. O., X. Mesnard, J. Droge, and M. R. Wiesner. 2011. Estimating production data for five engineered nanomaterials as a basis for exposure assessment. Environmental Science and Technology 45: 2562–2569.

Hendrey, G. R., D. S. Ellsworth, K. F. Lewin, and J. Nagy. 1999. A free-air enrichment system for exposing tall forest vegetation to elevated atmospheric CO_2. Global Change Biology 5:293–309.

Henrichs, S. M., and W. S. Reeburgh. 1987. Anaerobic mineralization of marine sediment organic matter—Rates and the role of anaerobic processes in the oceanic carbon economy. Geomicrobiology Journal 5:191–237.

Hensen, C., M. Zabel, K. Pfeifer, T. Schwenk, S. Kasten, N. Riedinger, H. D. Schulz, and A. Boettius. 2003. Control of sulfate pore-water profiles by sedimentary events and the significance of anaerobic oxidation of methane for the burial of sulfur in marine sediments. Geochimica et Cosmochimica Acta 67: 2631–2647.

Henson, S. A., R. Sanders, E. Madsen, P. J. Morris, F. Le Moigne, and G. D. Quartly. 2011. A reduced estimate of the strength of the ocean's biological carbon pump. Geophysical Research Letters 38.

Henze, D. K., and J. H. Seinfeld. 2006. Global secondary organic aerosol from isoprene oxidation. Geophysical Research Letters 33.

Herd, C.D.K., A. Blinova, D. N. Simkus, Y. Huang, R. Tarozo, C.M.O.D. Alexander, F. Gyngard, L. R. Nittler, G. D. Cody, M. L. Fogel, Y. Kebukawa, A.L.D. Kilcoyne, R. W. Hilts, G. F. Slater, D. P. Glavin, J. P. Dworkin, M. P. Callahan, J. E. Elsila, B. T. De Gregorio, and R. M. Stroud. 2011. Origin and evolution of prebiotic organic matter as inferred from the Tagish lake meteorite. Science 332: 1304–1307.

Herman, J. R. 2010. Global increase in UV irradiance during the past 30 years (1979-2008) estimated from satellite data. Journal of Geophysical Research—Atmospheres 115.

Herman, R. P., K. R. Provencio, R. J. Torrez, and G. M. Seager. 1993. Effect of water and nitrogen additions on free-living nitrogen fixer populations in desert grass root zones. Applied and Environmental Microbiology 59:3021–3026.

Herridge, D. F., M. B. Peoples, and R. M. Boddey. 2008. Global inputs of biological nitrogen fixation in agricultural systems. Plant and Soil 311:1–18.

Herron, M. M., C. C. Langway, H. W. Weiss, and J. H. Cragin. 1977. Atmospheric trace metals and sulfate in the Greenland ice sheet. Geochimica et Cosmochimica Acta 41:915–920.

Hess, S. L., R. M. Henry, C. B. Leovy, J. A. Ryan, J. E. Tillman, T. E. Chamberlain, H. L. Cole, R. G. Dutton, G. C. Greene, W. E. Simon, and J. L. Mitchell. 1976. Preliminary meteorological results on Mars from Viking-1 lander. Science 193:788–791.

Hester, K., and E. Boyle. 1982. Water chemistry control of cadmium content in recent benthic foraminifera. Nature 298:260–262.

Hesterberg, R., A. Blatter, M. Fahrni, M. Rosset, A. Neftel, W. Eugster, and H. Wanner. 1996. Deposition of nitrogen-containing compounds to an extensively managed grassland in central Switzerland. Environmental Pollution 91:21–34.

Hewitt, C. N., and R. M. Harrison. 1985. Tropospheric concentrations of the hydroxyl radical—A review. Atmospheric Environment 19:545–554.

Hibbard, K. A., B. E. Law, M. Reichstein, and J. Sulzman. 2005. An analysis of soil respiration across Northern Hemisphere temperate ecosystems. Biogeochemistry 73:29–70.

Hidy, G. M. 1970. Theory of diffusive and impactive scavenging. In R. J. Englemann and W.G.N. Slinn, editors. Precipitation Scavenging, 1970. U.S. Atomic Energy Commission, Division of Technical Information, Oak Ridge, Tennessee, 355–371.

Hietz, P., B. L. Turner, W. Wanek, A. Richter, C. A. Nock, and S. J. Wright. 2011. Long-term change in the nitrogen cycle of tropical forests. Science 334:664–666.

Hilderbrand, G. V., T. A. Hanley, C. T. Robbins, and C. C. Schwartz. 1999. Role of brown bears (Ursus arctos) in the flow of marine nitrogen into a terrestrial ecosystem. Oecologia 121:546–550.

Hilker, T., N. C. Coops, F. G. Hall, C. J. Nichol, A. Lyapustin, T. A. Black, M. A. Wulder, R. Leuning, A. Barr, D. Y. Hollinger, B. Munger, and C. J. Tucker. 2011. Inferring terrestrial photosynthetic light-use efficiency of temperate ecosystems from space. Journal of Geophysical Research—Biogeosciences 116.

Hill, W. R. 1996. Effects of light. In R. L. Stevenson, M. L. Bothwell and R. L. Lowe, editors. Algal Ecology. Academic Press, 121–148.

Hilley, G. E., C. P. Chamberlain, S. Moon, S. Porder, and S. D. Willett. 2010. Competition between erosion and reaction kinetics in controlling silicate-weathering rates. Earth and Planetary Science Letters 293:191–199.

Hilley, G. E., and S. Porder. 2008. A framework for predicting global silicate weathering and CO_2 drawdown rates over geologic time-scales. Proceedings of the National Academy of Sciences 105:16855–16859.

Hilton, J., M. O'Hare, M. J. Bowes, and J. I. Jones. 2006. How green is my river? A new paradigm of eutrophication in rivers. Science of the Total Environment 365:66–83.

Hinderer, M., I. Juttner, R. Winkler, C.E.W. Steinberg, and A. Kettrup. 1998. Comparing trends in lake acidification using hydrochemical modelling and paleolimnology: The case of the Herrenwieser See, Black Forest, Germany. Science of the Total Environment 218:113–121.

Hines, M. E., R. E. Pelletier, and P. M. Crill. 1993. Emissions of sulfur gases from marine and fresh-water wetlands of the Florida Everglades—Rates and extrapolation using remote sensing. Journal of Geophysical Research—Atmospheres 98:8991–8999.

Hingston, F. J., R. J. Atkinson, A. M. Posner, and J. P. Quirk. 1967. Specific adsorption of anions. Nature 215:1459–1461.

Hino, T., Y. Matsumoto, S. Nagano, H. Sugimoto, Y. Fukumori, T. Murata, S. Iwata, and Y. Shiro. 2010. Structural basis of biological N_2O generation by bacterial nitric oxide reductase. Science 330:1666–1670.

Hinrichs, K. U., J. M. Hayes, S. P. Sylva, P. G. Brewer, and E. F. DeLong. 1999. Methane-consuming archaebacteria in marine sediments. Nature 398:802–805.

Hirsch, A. I., J. W. Munger, D. J. Jacob, L. W. Horowitz, and A. H. Goldstein. 1996. Seasonal variation of the ozone production efficiency per unit NOx at Harvard Forest, Massachusetts. Journal of Geophysical Research—Atmospheres 101:12659–12666.

Hirsch, R. E., B. D. Lewis, E. P. Spalding, and M. R. Sussman. 1998. A role for the AKT1 potassium channel in plant nutrition. Science 280:918–921.

Hobbie, E. A. 2006. Carbon allocation to ectomycorrhizal fungi correlates with belowground allocation in culture studies. Ecology 87:563–569.

Hobbie, E. A., and A. P. Ouimette. 2009. Controls of nitrogen isotope patterns in soil profiles. Biogeochemistry 95:355–371.

Hobbie, J. E., and E. A. Hobbie. 2006. [15]N in symbiotic fungi and plants estimates nitrogen and carbon flux rates in arctic tundra. Ecology 87:816–822.

Hobbie, S. E. 1992. Effects of plant species on nutrient cycling. Trends in Ecology and Evolution 7:336–339.

Hoch, A. R., M. M. Reddy, and G. R. Aiken. 2000. Calcite crystal growth inhibition by humic substances with emphasis on hydrophobic acids from the Florida Everglades. Geochimica et Cosmochimica Acta **64**: 61–72.

Hocking, W. K., T. Carey-Smith, D. W. Tarasick, P. S. Argall, K. Strong, Y. Rochon, I. Zawadzki, and P. A. Taylor. 2007. Detection of stratospheric ozone intrusions by windprofiler radars. Nature **450**: 281–284.

Hodell, D. A., J. H. Curtis, and M. Brenner. 1995. Possible role of climate in the collapse of classic Maya civilization. Nature **375**:391–394.

Hodge, A., C. D. Campbell, and A. H. Fitter. 2001. An arbuscular mycorrhizal fungus accelerates decomposition and acquires nitrogen directly from organic material. Nature **413**:297–299.

Hodson, A., T. Heaton, H. Langford, and K. Newsham. 2010. Chemical weathering and solute export by meltwater in a maritime Antarctic glacier basin. Biogeochemistry **98**:9–27.

Hoegh-Guldberg, O., P. J. Mumby, A. J. Hooten, R. S. Steneck, P. Greenfield, E. Gomez, C. D. Harvell, P. F. Sale, A. J. Edwards, K. Caldeira, N. Knowlton, C. M. Eakin, R. Iglesias-Prieto, N. Muthiga, R. H. Bradbury, A. Dubi, and M. E. Hatziolos. 2007. Coral reefs under rapid climate change and ocean acidification. Science **318**:1737–1742.

Hoekstra, A. Y., and A. K. Chapagain. 2007. Water footprints of nations: Water use by people as a function of their consumption pattern. Water Resources Management **21**:35–48.

Hoff, T., B. M. Stummann, and K. W. Henningsen. 1992. Structure, function and regulation of nitrate reductase in higher plants. Physiologia Plantarum **84**:616–624.

Hoffman, P. F., A. J. Kaufman, G. P. Halverson, and D. P. Schrag. 1998. A Neoproterozoic snowball earth. Science **281**:1342–1346.

Hofmann, D. J. 1990. Increase in the stratospheric background sulfuric acid aerosol mass in the past 10 years. Science **248**:996–1000.

Hofmann, D. J. 1991. Aircraft sulfur emissions. Nature **349**:659.

Hofmann, D. J., J. H. Butler, and P. P. Tans. 2009. A new look at atmospheric carbon dioxide. Atmospheric Environment **43**:2084–2086.

Hofmann, D. J., and J. M. Rosen. 1980. Stratospheric sulfuric acid layer—Evidence for an anthropogenic component. Science **208**:1368–1370.

Hofmann, D. J., and J. M. Rosen. 1983. Sulfuric acid droplet formation and growth in the stratosphere after the 1982 eruption of El-Chichon. Science **222**: 325–327.

Hofmann, M., and H. J. Schellnhuber. 2010. Ocean acidification: a millennial challenge. Energy & Environmental Science **3**:1883–1896.

Hofmockel, K. S., N. Fierer, B. P. Colman, and R. B. Jackson. 2010. Amino acid abundance and proteolytic potential in North American soils. Oecologia **163**:1069–1078.

Hofzumahaus, A., F. Rohrer, K. D. Lu, B. Bohn, T. Brauers, C. C. Chang, H. Fuchs, F. Holland, K. Kita, Y. Kondo, X. Li, S. R. Lou, M. Shao, L. M. Zeng, A. Wahner, and Y. H. Zhang. 2009. Amplified trace gas removal in the troposphere. Science **324**: 1702–1704.

Hogberg, P. 1997. ^{15}N natural abundance in soil-plant systems. New Phytologist **137**:179–203.

Hogberg, P. 2012. What is the quantitative relation between nitrogen deposition and forest carbon sequestration? Global Change Biology **18**:1–2.

Hogberg, P., A. Nordgren, N. Buchmann, A.F.S. Taylor, A. Ekblad, M. N. Hogberg, G. Nyberg, M. Ottosson-Lofvenius, and D. J. Read. 2001. Large-scale forest girdling shows that current photosynthesis drives soil respiration. Nature **411**:789–792.

Hogg, N., P. Biscaye, W. Gardner, and W. J. Schmitz. 1982. On the transport and modification of Antarctic bottom water in the Vema Channel. Journal of Marine Research **40**:231–263.

Hojberg, O., N. P. Revsbech, and J. M. Tiedje. 1994. Denitrification in soil aggregates analyzed with microsensors for nitrous oxide and oxygen. Soil Science Society of America Journal **58**:1691–1698.

Hole, F. D. 1981. Effects of animals on soil. Geoderma **25**:75–112.

Holland, H. D. 1965. The history of ocean water and its effect on the chemistry of the atmosphere. Proceedings of the National Academy of Sciences **53**: 1173–1183.

Holland, H. D. 1978. *The Chemistry of the Atmosphere and Oceans.* Wiley.

Holland, H. D. 1984. *The Chemical Evolution of the Atmosphere and Oceans.* Princeton University Press.

Holland, H. D., C. R. Feakes, and E. A. Zbinden. 1989. The Flin Flon paleosol and the composition of the atmosphere 1.8 bybp. American Journal of Science **289**: 362–389.

Hollinger, D. Y. 1986. Herbivory and the cycling of nitrogen and phosphorus in isolated California oak trees. Oecologia **70**:291–297.

Hollinger, D. Y., J. Aber, B. Dail, E. A. Davidson, S. M. Goltz, H. Hughes, M. Y. Leclerc, J. T. Lee, A. D. Richardson, C. Rodrigues, N. A. Scott, D. Achuatavarier, and J. Walsh. 2004. Spatial and temporal variability in forest-atmosphere CO_2 exchange. Global Change Biology **10**:1689–1706.

Holloway, J. M., and R. A. Dahlgren. 2002. Nitrogen in rock: Occurrences and biogeochemical implications. Global Biogeochemical Cycles **16**.

Holloway, J. M., R. A. Dahlgren, B. Hansen, and W. H. Casey. 1998. Contribution of bedrock nitrogen to high nitrate concentrations in stream water. Nature **395**:785–788.

Holmer, M., and P. Storkholm. 2001. Sulphate reduction and sulphur cycling in lake sediments: a review. Freshwater Biology **46**:431–451.

Holmes, M. E., F. J. Sansone, T. M. Rust, and B. N. Popp. 2000. Methane production, consumption, and air-sea exchange in the open ocean: An evaluation based on carbon isotopic ratios. Global Biogeochemical Cycles **14**:1–10.

Holmes, R. M., J. B. Jones Jr., S. G. Fisher, and N. B. Grimm. 1996. Denitrification in a nitrogen-limited stream ecosystem. Biogeochemistry **33**:125–146.

Holser, W. T., and I. R. Kaplan. 1966. Isotope geochemistry of sedimentary sulfates. Chemical Geology **1**:93–135.

Holser, W. T., J. B. Maynard, and K. M. Cruikshank. 1989. Modelling the natural cycle of sulphur through Phanerozoic time. *In* P. Brimblecombe and A. Y. Lein, editors. *Evolution of the Global Biogeochemical Sulphur Cycle.* Wiley, 21–56.

Holser, W. T., M. Schidlowski, F. T. MacKenzie, and J. B. Maynard. 1988. Geochemical cycles of carbon and sulfur. *In* C. B. Gregor, R. M. Garrels, F. T. MacKenzie and J. B. Maynard, editors. *Chemical Cycles in the Evolution of the Earth.* Wiley, 105–173.

Holtgrieve, G. W., D. E. Schindler, W. O. Hobbs, P. R. Leavitt, E. J. Ward, L. Bunting, G. J. Chen, B. P. Finney, I. Gregory-Eaves, S. Holmgren, M. J. Lisac, P. J. Lisi, K. Nydick, L. A. Rogers, J. E. Saros, D. T. Selbie, M. D. Shapley, P. B. Walsh, and A. P. Wolfe. 2011. A coherent signature of anthropogenic nitrogen deposition to remote watersheds of the Northern Hemisphere. Science **334**:1545–1548.

Holton, J. R., P. H. Haynes, M. E. McIntyre, A. R. Douglass, R. B. Rood, and L. Pfister. 1995. Stratosphere-troposphere exchange. Reviews of Geophysics **33**:403–439.

Honeycutt, C. W., R. D. Heil, and C. V. Cole. 1990. Climatic and topographic relations of three Great-Plains soils. 1. Soil morphology. Soil Science Society of America Journal **54**:469–475.

Hong, J. I., Q. Feng, V. Rotello, and J. Rebek. 1992. Competition, cooperation, and mutation—Improving a synthetic replicator by light irradiation. Science **255**:848–850.

Hongoh, Y., V. K. Sharma, T. Prakash, S. Noda, H. Toh, T. D. Taylor, T. Kudo, Y. Sakaki, A. Toyoda, M. Hattori, and M. Ohkuma. 2008. Genome of an endosymbiont coupling N_2 fixation to cellulolysis within protist cells in termite gut. Science **322**:1108–1109.

Hönisch, B., A. Ridgwell, D. N. Schmidt, E. Thomas, S. J. Gibbs, A. Sluijs, R. Zeebe, L. Kump, R. C. Martindale, S. E. Greene, W. Kiessling, J. Ries, J. C. Zachos, D. L. Royer, S. Barker, T. M. Marchitto, R. Moyer, C. Pelejero, P. Ziveri, G. L. Foster, and B. Williams. 2012. The geological record of ocean acidification. Science **335**:1058–1063.

Honjo, S., S. J. Manganini, and J. J. Cole. 1982. Sedimentation of biogenic matter in the deep ocean. Deep-Sea Research Part A—Oceanographic Research Papers **29**:609-625.

Hooke, R. L. 1999. Spatial distribution of human geomorphic activity in the United States: Comparison with rivers. Earth Surface Processes and Landforms **24**:687–692.

Hooke, R. L. 2000. On the history of humans as geomorphic agents. Geology **28**:843–846.

Hooker, T. D., and J. E. Compton. 2003. Forest ecosystem carbon and nitrogen accumulation during the first century after agricultural abandonment. Ecological Applications **13**:299–313.

Hooper, P. R., I. W. Herrick, E. R. Laskowski, and C. R. Knowles. 1980. Composition of the Mount St Helens ashfall in the Moscow–Pullman area on 18 May 1980. Science **209**:1125–1126.

Hoosbeek, M. R., M. Lukac, D. van Dam, D. L. Godbold, E. J. Velthorst, F. A. Biondi, A. Peressotti, M. F. Cotrufo, P. de Angelis, and G. Scarascia-Mugnozza. 2004. More new carbon in the mineral soil of a poplar plantation under Free Air Carbon Enrichment (POPFACE): Cause of increased priming effect? Global Biogeochemical Cycles **18**.

Hope, D., M. F. Billett, and M. S. Cresser. 1994. A review of the export of carbon in river water: fluxes and processes. Environmental Pollution **84**:301–324.

Höpfner, M., M. Milz, S. Buehler, J. Orphal, and G. Stiller. 2012. The natural greenhouse effect of atmospheric oxygen (O_2) and nitrogen (N_2). Geophysical Research Letters **39**.

Hopkinson, C. S., and J. J. Vallino. 2005. Efficient export of carbon to the deep ocean through dissolved organic matter. Nature **433**:142–145.

Hoppe, H. G., K. Gocke, R. Koppe, and C. Begler. 2002. Bacterial growth and primary production along a north–south transect of the Atlantic Ocean. Nature **416**:168–171.

Horita, J., H. Zimmermann, and H. D. Holland. 2002. Chemical evolution of seawater during the

Phanerozoic: Implications from the record of marine evaporites. Geochimica et Cosmochimica Acta 66:3733–3756.

Hornberger, G. M., B. J. Cosby, and J. N. Galloway. 1986. Modeling the effects of acid deposition—Uncertainty and spatial variability in estimation of long-term sulfate dynamics in a region. Water Resources Research 22:1293–1302.

Horne, A. J., and D. L. Galat. 1985. Nitrogen fixation in an oligotrophic, saline desert lake—Pyramid Lake, Nevada. Limnology and Oceanography 30: 1229–1239.

Horne, A. J., and C. R. Goldman. 1974. Suppression of nitrogen fixation by blue-green algae in a eutrophic lake with trace additions of copper. Science 183: 409–411.

Hornibrook, E.R.C., F. J. Longstaffe, and W. S. Fyfe. 2000. Evolution of stable carbon isotope compositions for methane and carbon dioxide in freshwater wetlands and other anaerobic environments. Geochimica et Cosmochimica Acta 64:1013–1027.

Horodyski, R. J., and L. P. Knauth. 1994. Life on land in the Precambrian. Science 263:494–498.

Horrigan, S. G., J. P. Montoya, J. L. Nevins, and J. J. McCarthy. 1990. Natural isotopic composition of dissolved inorganic nitrogen in the Chesapeake Bay. Estuarine Coastal and Shelf Science 30:393–410.

Hosker, R. P., and S. E. Lindberg. 1982. Review—Atmospheric deposition and plant assimilation of gases and particles. Atmospheric Environment 16:889–910.

Hou, L. J., M. Liu, P. X. Ding, J. L. Zhou, Y. Yang, D. Zhao, and Y. L. Zheng. 2011. Influences of sediment dessication on phosphorus transformations in an intertidal marsh: Formation and release of phosphine. Chemosphere 83:917–924.

Hough, A. M., and R. G. Derwent. 1990. Changes in the global concentration of tropospheric ozone due to human activities. Nature 344:645–648.

Houghton, J. T. 1986. *The Physics of Atmospheres*. 2nd edition. Cambridge University Press.

Houghton, J. T., G. J. Jenkins, and J.J.E. Ephramus. and (eds). 1990. *Climate Change: The IPCC Scientific Assessment*. Cambridge University Press.

Houghton, J. T., L. G. Meira Filho, J. Bruce, H. Lee, B. A. Callander, E. Haites, N. Harris, and K. Maskell. 1995. *Climate Change 1994*. Cambridge University Press.

Houghton, R. A. 1987. Biotic changes consistent with the increased seasonal amplitude of atmospheric CO_2 concentrations. Journal of Geophysical Research—Atmospheres 92:4223–4230.

Houghton, R. A., J. E. Hobbie, J. M. Melillo, B. Moore, B. J. Peterson, G. R. Shaver, and G. M. Woodwell. 1983. Changes in the carbon content of terrestrial biota and soils between 1860 and 1980—A net release of CO_2 to the atmosphere. Ecological Monographs 53: 235–262.

Houghton, R. A., and D. L. Skole. 1990. Carbon. *In* W. C. Clark, B. L. Turner, R. W. Kates, J. F. Richards, J. T. Matthews, and W. B. Meyer, editors. *The Earth as Transformed by Human Action*. Cambridge University Press, 393–408.

Houghton, R. A., D. L. Skole, C. A. Nobre, J. L. Hackler, K. T. Lawrence, and W. H. Chomentowski. 2000. Annual fluxes of carbon from deforestation and regrowth in the Brazilian Amazon. Nature 403: 301–304.

Houle, D., and R. Carignan. 1992. Sulfur speciation and distribution in soils and aboveground biomass of a boreal coniferous forest. Biogeochemistry 16: 63–82.

Houle, D., R. Carignan, and R. Ouimet. 2001. Soil organic sulfur dynamics in a coniferous forest. Biogeochemistry 53:105–124.

Houlton, B. Z., and E. Bai. 2009. Imprint of denitrifying bacteria on the global terrestrial biosphere. Proceedings of the National Academy of Sciences 106: 21713–21716.

Houlton, B. Z., D. M. Sigman, and L. O. Hedin. 2006. Isotopic evidence for large gaseous nitrogen losses from tropical rainforests. Proceedings of the National Academy of Sciences 103:8745–8750.

Houweling, S., F. Dentener, and J. Lelieveld. 2000a. Simulation of preindustrial atmospheric methane to constrain the global source strength of natural wetlands. Journal of Geophysical Research—Atmospheres 105: 17243–17255.

Houweling, S., F. Dentener, J. Lelieveld, B. Walter, and E. Dlugokencky. 2000b. The modeling of tropospheric methane: How well can point measurements be reproduced by a global model? Journal of Geophysical Research—Atmospheres 105:8981–9002.

Hover, V. C., L. M. Walter, and D. R. Peacor. 2002. K uptake by modern estuarine sediments during early marine diagenesis, Mississippi Delta Plain, Louisiana. Journal of Sedimentary Research 72:775–792.

Howard, E. C., J. R. Henriksen, A. Buchan, C. R. Reisch, H. Buergmann, R. Welsh, W. Y. Ye, J. M. Gonzalex, K. Mace, S. B. Joye, R. P. Kiene, W. B. Whitman, and M. A. Moran. 2006. Bacterial taxa that limit sulfur flux from the ocean. Science 314:649–652.

Howard, J. A., and C. W. Mitchell. 1985. *Phytogeomorphology*. Wiley.

Howarth, R., F. Chan, D. J. Conley, J. Garnier, S. C. Doney, R. Marino, and G. Billen. 2011. Coupled biogeochemical cycles: eutrophication and hypoxia in temperate estuaries and coastal marine ecosystems. Frontiers in Ecology and the Environment 9: 18–26.

Howarth, R., D. Swaney, G. Billen, J. Garnier, B. G. Hong, C. Humborg, P. Johnes, C. M. Morth, and R. Marino. 2012. Nitrogen fluxes from the landscape are controlled by net anthropogenic nitrogen inputs and by climate. Frontiers in Ecology and the Environment 10:37–43.

Howarth, R. W. 1984. The ecological significance of sulfur in the energy dynamics of salt marsh and coastal marine sediments. Biogeochemistry 1:5–27.

Howarth, R. W. 1988. Nutrient limitation of net primary production in marine ecosystems. Annual Review of Ecology and Systematics 19:89–110.

Howarth, R. W. 1998. An assessment of human influences on fluxes of nitrogen from the terrestrial landscape to the estuaries and continental shelves of the North Atlantic Ocean. Nutrient Cycling in Agroecosystems 52:213–223.

Howarth, R. W., and J. J. Cole. 1985. Molybdenum availability, nitrogen limitation, and phytoplankton growth in natural waters. Science 229:653–655.

Howarth, R. W., H. S. Jensen, R. Marino, and H. Postma. 1995a. Transport to and processing of P in nearshore and oceanic waters. *In* H. Tiessen, editor. *Phosphorus in the Global Environment.* Wiley, 323–345.

Howarth, R. W., and R. Marino. 2006. Nitrogen as the limiting nutrient for eutrophication in coastal marine ecosystems: Evolving views over three decades. Limnology and Oceanography 51:364–376.

Howarth, R. W., R. Marino, and J. J. Cole. 1988. Nitrogen fixation in freshwater, estuarine, and marine ecosystems. 2. Biogeochemical controls. Limnology and Oceanography 33:688–701.

Howarth, R. W., D. Swaney, R. Marino, T. Butler, and C. R. Chu. 1995b. Turbulence does not prevent nitrogen fixation by plankton in estuaries and coastal seas—reply. Limnology and Oceanography 40: 639–643.

Howarth, R. W., and J. M. Teal. 1980. Energy flow in a salt marsh ecosystem: the role of reduced inorganic sulfur compounds. The American Naturalist 116:867–872.

Howell, D. G., and R. W. Murray. 1986. A budget for continental growth and denudation. Science 233: 446–449.

Howes, B. L., J.W.H. Dacey, and G. M. King. 1984. Carbon flow through oxygen and sulfate reduction pathways in salt marsh sediments. Limnology and Oceanography 29:1037–1051.

Hren, M. T., M. M. Tice, and C. P. Chamberlain. 2009. Oxygen and hydrogen isotope evidence for a temperate climate 3.42 billion years ago. Nature 462:205–208.

Hsieh, J.C.C., O. A. Chadwick, E. F. Kelly, and S. M. Savin. 1998. Oxygen isotopic composition of soil water: Quantifying evaporation and transpiration. Geoderma 82:269–293.

Hu, L., S. A. Yvon-Lewis, Y. Liu, J. E. Salisbury, and J. E. O'Hern. 2010. Coastal emissions of methyl bromide and methyl chloride along the eastern Gulf of Mexico and the east coast of the United States. Global Biogeochemical Cycles 24.

Hu, Z., G. Yu, Y. Zhou, X. Sun, Y. Li, P. Shi, Y. Wang, X. Song, Z. Zheng, L. Zhang, and S. Li. 2009. Partitioning of evapotranspiration and its controls in four grassland ecosystems: Application of a two-source model. Agricultural and Forest Meteorology 149: 1410–1420.

Huang, P. M. 1988. Ionic factors affecting aluminum transformations and the impact on soil and environmental sciences. Advances in Soil Science 8:1–78.

Huang, X., Y. Hao, Y. Wang, X. Cui, X. Mo, and X. Zhou. 2010. Partitioning of evapotranspiration and its relation to carbon dioxide fluxes in Inner Mongolia steppe. Journal of Arid Environments 74:1616–1623.

Huang, Y., J. Zou, X. Zheng, Y. Wang, and X. Xu. 2004. Nitrous oxide emissions as influenced by amendment of plant residues with different C:N ratios. Soil Biology & Biochemistry 36:973–981.

Huber, C., and G. Wächtershäuser. 2006. Alpha-hydroxy and alpha-amino acids under possible Hadean, volcanic origin-of-life conditions. Science 314:630–632.

Huber, C., and G. Wächtershäuser. 1998. Peptides by activation of amino acids with CO on (Ni, Fe)S surfaces: Implications for the origin of life. Science 281:670–672.

Huber, R., M. Kurr, H. W. Jannasch, and K. O. Stetter. 1989. A novel group of abyssal methanogenic archaebacteria (*Methanopyrus*) growing at 110° C. Nature 342:833–834.

Hudman, R. C., L. T. Murray, D. J. Jacob, D. B. Millet, S. Turquety, S. Wu, D. R. Blake, A. H. Goldstein, J. Holloway, and G. W. Sachse. 2008. Biogenic versus anthropogenic sources of CO in the United States. Geophysical Research Letters 35.

Hudson, B. D. 1995. Reassessment of the Polynov ion mobility series. Soil Science Society of America Journal 59:1101–1103.

Hudson, J. A. 1988. The contribution of soil moisture storage to the water balances of upland forested and grassland catchment. Hydrologic Sciences Journal 33:289–308.

Hudson, S. R. 2011. Estimating the global radiative impact of the sea ice-albedo feedback in the Arctic. Journal of Geophysical Research—Atmospheres 116.

Hue, N. V. 1991. Effects of organic acid anions on P sorption and phytoavailability in soils with different mineralogies. Soil Science 152:463–471.

Huemmrich, K. F., G. Kinoshita, J. A. Gamon, S. Houston, H. Kwon, and W. C. Oechel. 2010. Tundra carbon balance under varying temperature and moisture regimes. Journal of Geophysical Research—Biogeosciences 115.

Huenneke, L. F., S. P. Hamburg, R. Koide, H. A. Mooney, and P. M. Vitousek. 1990. Effects of soil resources on plant invasion and community structure in Californian serpentine grassland. Ecology 71:478–491.

Hueso, R., and A. Sanchez-Lavega. 2006. Methane storms on Saturn's moon Titan. Nature 442:428–431.

Hulett, L. D., A. J. Weinberger, K. J. Northcutt, and M. Ferguson. 1980. Chemical species in fly ash from coal-burning power plants. Science 210:1356–1358.

Hulth, S., R. C. Aller, and F. Gilbert. 1999. Coupled anoxic nitrification-manganese reduction in marine sediments. Geochimica et Cosmochimica Acta 63: 49–66.

Hulthe, G., S. Hulth, and P.O.J. Hall. 1998. Effect of oxygen on degradation rate of refractory and labile organic matter in continental margin sediments. Geochimica et Cosmochimica Acta 62:1319–1328.

Humborg, C., M. Pastuszak, J. Aigars, H. Siegmund, C. M. Morth, and V. Ittekkot. 2006. Decreased silica land-sea fluxes through damming in the Baltic Sea catchment—significance of particle trapping and hydrological alterations. Biogeochemistry 77:265–281.

Hungate, B. A., D. W. Johnson, P. Dijkstra, G. Hymus, P. Stiling, J. P. Megonigal, A. L. Pagel, J. L. Moan, F. Day, J. H. Li, C. R. Hinkle, and B. G. Drake. 2006. Nitrogen cycling during seven years of atmospheric CO$_2$ enrichment in a scrub oak woodland. Ecology 87:26–40.

Hungate, B. A., P. D. Stiling, P. Dijkstra, D. W. Johnson, M. E. Ketterer, G. J. Hymus, C. R. Hinkle, and B. G. Drake. 2004. CO$_2$ elicits long-term decline in nitrogen fixation. Science 304:1291.

Hungate, B. A., K. J. van Groenigen, J. Six, J. D. Jastrow, Y. Q. Lue, M. A. de Graaff, C. van Kessel, and C. W. Osenberg. 2009. Assessing the effect of elevated carbon dioxide on soil carbon: a comparison of four meta-analyses. Global Change Biology 15: 2020–2034.

Hunten, D. M. 1993. Atmospheric evolution of the terrestrial planets. Science 259:915–920.

Huntington, T. G. 2006. Evidence for intensification of the global water cycle: Review and synthesis. Journal of Hydrology 319:83–95.

Hurst, D. F., D.W.T. Griffith, and G. D. Cook. 1994. Trace gas emissions from biomass burning in tropical Australian savannas. Journal of Geophysical Research—Atmospheres 99:16441–16456.

Huss-Danell, K. 1986. Nitrogen in shoot litter, root litter and root exudates from nitrogen-fixing Alnus incana. Plant and Soil 91:43–49.

Huston, M. 1993. Biological diversity, soils, and economics. Science 262:1676–1680.

Hutchin, P. R., M. C. Press, J. A. Lee, and T. W. Ashenden. 1995. Elevated concentrations of CO$_2$ may double methane emissions from mires. Global Change Biology 1:125–128.

Hutchins, D. A., G. R. Ditullio, and K. W. Bruland. 1993. Iron and regenerated production: Evidence for biological iron recycling in two marine environments. Limnology and Oceanography 38:1242–1255.

Hutchins, D. A., G. R. DiTullio, Y. Zhang, and K. W. Bruland. 1998. An iron limitation mosaic in the California upwelling regime. Limnology and Oceanography 43:1037–1054.

Hutchins, K. S., and B. M. Jakosky. 1996. Evolution of Martian atmospheric argon: Implications for sources of volatiles. Journal of Geophysical Research—Planets 101:14933–14949.

Hutchinson, G. E. 1938. On the relation between oxygen deficit and the productivity and typology of lakes. Internationale Revue des Gesamten Hydrobiologie und Hydrographie 36:336–355.

Hutchinson, G. E. 1943. The biogeochemistry of aluminum and of certain related elements (concluded). Quarterly Review of Biology 18:331–363.

Hutchinson, G. L., W. D. Guenzi, and G. P. Livingston. 1993. Soil water controls on aerobic soil emission of gaseous nitrogen oxides. Soil Biology & Biochemistry 25:1–9.

Hutchinson, J. N. 1980. Record of peat wastage in the East Anglian fenlands at Holme Post, 1848-1978 AD. Journal of Ecology 68:229–249.

Hütsch, B. W., C. P. Webster, and D. S. Powlson. 1994. Methane oxidation in soil as affected by land use, soil pH and N fertilization. Soil Biology & Biochemistry 26:1613–1622.

Huxman, T. E., M. D. Smith, P. A. Fay, A. K. Knapp, M. R. Shaw, M. E. Loik, S. D. Smith, D. T. Tissue, J. C. Zak, J. F. Weltzin, W. T. Pockman, O. E. Sala, B. M. Haddad, J. Harte, G. W. Koch, S. Schwinning, E. E. Small, and D. G. Williams. 2004. Convergence across biomes to a common rain-use efficiency. Nature 429:651–654.

Hydes, D. J. 1979. Aluminum in seawater—Control by inorganic processes. Science 205:1260–1262.

Hyman, M. R., and P. M. Wood. 1983. Methane oxidtaion by Nitrosomonas europaea. Biochemical Journal 212: 31–37.

Hyun, J. H., J. S. Mok, H. Y. Cho, S. H. Kim, K. S. Lee, and J. E. Kostka. 2009. Rapid organic matter mineralization coupled to iron cycling in intertidal mud flats of the Han River estuary, Yellow Sea. Biogeochemistry 92:231–245.

Hyvonen, R., T. Persson, S. Andersson, B. Olsson, G. I. Agren, and S. Linder. 2008. Impact of long-term nitrogen addition on carbon stocks in trees and soils in northern Europe. Biogeochemistry 89:121–137.

Idso, C. D., S. B. Idso, and R. C. Balling. 2001. An intensive two-week study of an urban CO_2 dome in Phoenix, Arizona, USA. Atmospheric Environment 35:995–1000.

Idso, S. B., and B. A. Kimball. 1993. Tree growth in carbon dioxide enriched air and its implications for global carbon cycling and maximum levels of atmospheric CO_2. Global Biogeochemical Cycles 7:537–555.

Iglesias-Rodriguez, M. D., P. R. Halloran, R.E.M. Rickaby, I. R. Hall, E. Colmenero-Hidalgo, J. R. Gittins, D.R.H. Green, T. Tyrrell, S. J. Gibbs, P. von Dassow, E. Rehm, E. V. Armbrust, and K. P. Boessenkool. 2008. Phytoplankton calcification in a high-CO_2 world. Science 320:336–340.

Illmer, P., A. Barbato, and F. Schinner. 1995. Solubilization of hardly soluble $AlPO_4$ with P-solubilizing microorganisms. Soil Biology & Biochemistry 27:265–270.

Imhoff, M. L., L. Bounoua, R. DeFries, W. T. Lawrence, D. Stutzer, C. J. Tucker, and T. Ricketts. 2004a. The consequences of urban land transformation on net primary productivity in the United States. Remote Sensing of Environment 89:434–443.

Imhoff, M. L., L. Bounoua, T. Ricketts, C. Loucks, R. Harriss, and W. T. Lawrence. 2004b. Global patterns in human consumption of net primary production. Nature 429:870–873.

Indermuhle, A., T. F. Stocker, F. Joos, H. Fischer, H. J. Smith, M. Wahlen, B. Deck, D. Mastroianni, J. Tschumi, T. Blunier, R. Meyer, and B. Stauffer. 1999. Holocene carbon-cycle dynamics based on CO_2 trapped in ice at Taylor Dome, Antarctica. Nature 398:121–126.

Ingall, E., and R. Jahnke. 1997. Influence of water-column anoxia on the elemental fractionation of carbon and phosphorus during sediment diagenesis. Marine Geology 139:219–229.

Ingall, E. D., R. M. Bustin, and P. Van Cappellen. 1993. Influence of water column anoxia on the burial and preservation of carbon and phosphorus in marine shales. Geochimica et Cosmochimica Acta 57:303–316.

Ingall, E. D., and P. Van Cappellen. 1990. Relation between sedimentation rate and burial of organic phosphorus and organic carbon in marine sediments. Geochimica et Cosmochimica Acta 54:373–386.

Ingestad, T. 1979a. Mineral nutrient requirements of Pinus silvestris and Picea abies seedlings. Physiologia Plantarum 45:373–380.

Ingestad, T. 1979b. Nitrogen stress in birch seedlings. 2. N, K, P, Ca and Mg nutrition. Physiologia Plantarum 45:149–157.

Ingham, R. E., and J. K. Detling. 1990. Effects of root-feeding nematodes on above-ground net primary production in a North American grassland. Plant and Soil 121:279–281.

Inoue, H. Y., H. Matsueda, M. Ishii, K. Fushimi, M. Hirota, I. Asanuma, and Y. Takasugi. 1995. Long-term trend of the partial pressure of carbon dioxide (pCO_2) in surface waters of the western North Pacific, 1984-1993. Tellus Series B—Chemical and Physical Meteorology 47:391–413.

Inoue, H. Y., and Y. Sugimura. 1992. Variations and distributions of CO_2 in and over the equatorial Pacific during the period from the 1986/88 El Nino event to the 1988/89 La Nina event. Tellus Series B—Chemical and Physical Meteorology 44:1–22.

Insam, H. 1990. Are the soil microbial biomass and basal respiration governed by the cliimatic regime. Soil Biology & Biochemistry 22:525–532.

Inskeep, W. P., and P. R. Bloom. 1986. Kinetics of calcite precipitation in the presence of water-soluble organic ligands. Soil Science Society of America Journal 50: 1167–1172.

IPCC (Intergovernmental Panel on Climate Change) 2007. Climate Change 2007: The physical science basis. Cambridge University Press.

Irvine, W. M. 1998. Extraterrestrial organic matter: A review. Origins of Life and Evolution of Biospheres 28:365–383.

Irwin, J. G., and M. L. Williams. 1988. Acid rain—Chemistry and transport. Environmental Pollution 50:29–59.

Isaksen, I.S.A., and O. Hov. 1987. Calculation of trends in the tropospheric concentrations of O_3, OH, CO, CH_4, and NOx. Tellus 39B:271–285.

Ishida, Y. 1968. Physiological studies on evolution of dimethylsulfide. Mem. Coll. Agric. Kyoto Univ. 94: 47–82.

Ishijima, K., T. Nakazawa, and S. Aoki. 2009. Variations of atmospheric nitrous oxide concentration in the northern and western Pacific. Tellus Series B—Chemical and Physical Meteorology 61:408–415.

Ishijima, K., S. Sugawara, K. Kawamura, G. Hashida, S. Morimoto, S. Murayama, S. Aoki, and T. Nakazawa. 2007. Temporal variations of the atmospheric nitrous oxide concentration and its delta N-15 and delta O-18 for the latter half of the 20th century reconstructed from firn-air analyses. Journal of Geophysical Research—Atmospheres 112.

Ishizuka, S., T. Sakata, and K. Ishizuka. 2000. Methane oxidation in Japanese forest soils. Soil Biology & Biochemistry 32:769–777.

Ito, A. 2011. A historical meta-analysis of global terrestrial net primary productivity: are estimates converging? Global Change Biology 17:3161–3175.

Ittekkot, V., and R. Arain. 1986. Nature of particulate organic matter in the River Indus, Pakistan. Geochimica et Cosmochimica Acta 50:1643–1653.

Ittekkot, V., and S. Zhang. 1989. Pattern of particulate nitrogen transport in world rivers. Global Biogeochemical Cycles 3:383–392.

Ivanov, M. V. 1983. Major fluxes of the global biogeochemical cycle of sulphur. In M. V. Ivanov and J. R. Freney, editors. The Global Biogeochemical Sulphur Cycle. Wiley, 449–463.

Ivens, W., P. Kauppi, J. Alcamo, and M. Posch. 1990. Sulfur deposition onto European forests: Throughfall data and model estimates. Tellus 42B:294–303.

Iversen, N. 1996. Methane oxidation in coastal marine environments. In J. C. Murrell and D. P. Kelly, editors. Microbiology of Atmospheric Trace Gases. Springer, 51–68.

Iverson, R. L. 1990. Control of marine fish production. Limnology and Oceanography 35:1593–1604.

Iverson, R. L., F. L. Nearhoof, and M. O. Andreae. 1989. Production of dimethylsulfonium propionate and dimethylsulfide by phytoplankton in estuarine and coastal waters. Limnology and Oceanography 34:53–67.

Jackson, L. E., J. P. Schimel, and M. K. Firestone. 1989. Short-term partitioning of ammonium and nitrate between plants and microbes in an annual grassland. Soil Biology & Biochemistry 21:409–415.

Jackson, M. B., and W. Armstrong. 1999. Formation of aerenchyma and the processes of plant ventilation in relation to soil flooding and submergence. Plant Biology 1:274–287.

Jackson, M. G., R. W. Carlson, M. D. Kurz, P. D. Kempton, D. Francis, and J. Blusztajn. 2010. Evidence for the survival of the oldest terrestrial mantle reservoir. Nature 466:853–856.

Jackson, R. B., J. H. Manwaring, and M. M. Caldwell. 1990. Rapid physiological adjustment of roots to localized soil enrichment. Nature 344:58–60.

Jackson, R. B., S. R. Carpenter, C. N. Dahm, D. M. McKnight, R. J. Naiman, S. L. Postel, and S. W. Running. 2001. Water in a changing world. Ecological Applications 11:1027–1045.

Jackson, R. B., C. W. Cook, J. S. Pippen, and S. M. Palmer. 2009. Increased belowground biomass and soil CO_2 fluxes after a decade of carbon dioxide enrichment in a warm-temperate forest. Ecology 90:3352–3366.

Jackson, R. B., H. A. Mooney, and E. D. Schulze. 1997. A global budget for fine root biomass, surface area, and nutrient contents. Proceedings of the National Academy of Sciences 94:7362–7366.

Jacob, D. J., J. A. Logan, G. M. Gardner, R. M. Yevich, C. M. Spivakovsky, S. C. Wofsy, S. Sillman, and M. J. Prather. 1993. Factors regulating ozone over the United States and its export to the global atmosphere. Journal of Geophysical Research—Atmospheres 98:14817–14826.

Jacob, D. J., and S. C. Wofsy. 1990. Budgets of reactive nitrogen, hydrocarbons, and ozone over the Amazon forest during the wet season. Journal of Geophysical Research—Atmospheres 95:16737–16754.

Jacob, T., J. Wahr, W. T. Pfeffer, and S. Swenson. 2012. Recent contributions of glaciers and ice caps to sea level rise. Nature 482:514–518.

Jacobs, S. S., C. F. Giulivi, and P. A. Mele. 2002. Freshening of the Ross Sea during the late 20th century. Science 297:386–389.

Jacobson, M. E. 1994. Chemical and biological mobilization of Fe(III) in marsh sediments. Mobilization 25:41–60.

Jahne, B., O. Munnich, R. Bosinger, A. Dutzi, W. Huber, and P. Libner. 1987. On the parameters influencing air-water gas exchange. Journal of Geophysical Research 92:1937–1949.

Jahnke, R. A. 1990. Ocean flux studies—A status report. Reviews of Geophysics 28:381–398.

Jahnke, R. A. 1996. The global ocean flux of particulate organic carbon: Areal distribution and magnitude. Global Biogeochemical Cycles 10:71–88.

Jahnke, R. A., D. B. Craven, D. C. McCorkle, and C. E. Reimers. 1997. $CaCO_3$ dissolution in California continental margin sediments: The influence of organic matter remineralization. Geochimica et Cosmochimica Acta 61:3587–3604.

Jahren, A. H., N. C. Arens, G. Sarmiento, J. Guerrero, and R. Amundson. 2001. Terrestrial record of methane hydrate dissociation in the early Cretaceous. Geology 29:159–162.

Jain, A. K. 2007. Global estimation of CO emissions using three sets of satellite data for burned area. Atmospheric Environment 41:6931–6940.

Jain, A. K., Z. N. Tao, X. J. Yang, and C. Gillespie. 2006. Estimates of global biomass burning emissions for reactive greenhouse gases (CO, NMHCs, and NOx) and CO_2. Journal of Geophysical Research—Atmospheres 111.

James, B. R., and S. J. Riha. 1986. pH buffering in forest soil organic horizons—Relevance to acid precipitation. Journal of Environmental Quality 15:229–234.

James, P. B., H. H. Kieffer, and D. A. Paige. 1992. The seasonal cycle of carbon dioxide on Mars. In H. H. Kieffer, B. M. Jakosky, , C. W. Snyder and M. A. Matthews, editors. Mars, University of Arizona Press, 934–968.

James, R. H., and M. R. Palmer. 2000. Marine geochemical cycles of the alkali elements and boron: The role of sediments. Geochimica et Cosmochimica Acta 64: 3111–3122.

Jannasch, H. W. 1989. Sulphur emission and transformations at deep sea hydrothermal vents. *In* P. Brimblecombe and A. Y. Lein, editors. *Evolution of the Global Biogeochemical Sulphur Cycle.* Wiley, 181–190.

Jannasch, H. W., and M. J. Mottl. 1985. Geomicrobiology of deep sea hydrothermal vents. Science **229**: 717–725.

Jannasch, H. W., and C. O. Wirsen. 1979. Chemosynthetic primary production at east Pacific sea floor spreading centers. BioScience **29**:592–598.

Janos, D. P. 1980. Vesicular-arbuscular mycorrhizae affect lowland tropical rain forest plant growth. Ecology **61**:151–162.

Jansen, B., K.G.J. Nierop, and J. M. Verstraten. 2005. Mechanisms controlling the mobility of dissolved organic matter, aluminium and iron in podzol B horizons. European Journal of Soil Science **56**: 537–550.

Janssens, I. A., H. Lankreijer, G. Matteucci, A. S. Kowalski, N. Buchmann, D. Epron, K. Pilegaard, W. Kutsch, B. Longdoz, T. Grunwald, L. Montagnani, S. Dore, C. Rebmann, E. J. Moors, A. Grelle, U. Rannik, K. Morgenstern, S. Oltchev, R. Clement, J. Gudmundsson, S. Minerbi, P. Berbigier, A. Ibrom, J. Moncrieff, M. Aubinet, C. Bernhofer, N. O. Jensen, T. Vesala, A. Granier, E. D. Schulze, A. Lindroth, A. J. Dolman, P. G. Jarvis, R. Ceulemans, and R. Valentini. 2001. Productivity overshadows temperature in determining soil and ecosystem respiration across European forests. Global Change Biology **7**:269–278.

Jansson, M. 1993. Uptake, exchange, and excretion of orthophosphate in phosphate-starved *Scenedesmus quadricauda* and *Pseudomonas* K7. Limnology and Oceanography **38**:1162–1178.

Jansson, M., J. Karlsson, and A. Jonsson. 2012. Carbon dioxide supersaturation promotes primary production in lakes. Ecology Letters **15**:527–532.

Jaramillo, V. J., and J. K. Detling. 1988. Grazing history, defoliation, and competition—Effects on shortgrass production and nitrogen accumulation. Ecology **69**: 1599–1608.

Jarrell, W. M., and R. B. Beverly. 1981. The dilution effect in plant nutrition studies. Advances in Agronomy **34**:197–224.

Jastrow, J. D., R. M. Miller, R. Matamala, R. J. Norby, T. W. Boutton, C. W. Rice, and C. E. Owensby. 2005. Elevated atmospheric carbon dioxide increases soil carbon. Global Change Biology **11**:2057–2064.

Javaux, E. J. 2006. Extreme life on Earth—past, present and possibly beyond. Research in Microbiology **157**:37–48.

Javaux, E. J., C. P. Marshall, and A. Bekker. 2010. Organic-walled microfossils in 3.2-billion-year-old shallow-marine siliciclastic deposits. Nature **463**: 934–938.

Javoy, M. 1997. The major volatile elements of the Earth: Their origin, behavior, and fate. Geophysical Research Letters **24**:177–180.

Jencso, K. G., B. L. McGlynn, M. N. Gooseff, S. M. Wondzell, K. E. Bencala, and L. A. Marshall. 2009. Hydrologic connectivity between landscapes and streams: Transferring reach- and plot-scale understanding to the catchment scale. Water Resources Research **45**:1–16.

Jenkin, M. E., and K. C. Clemitshaw. 2000. Ozone and other secondary photochemical pollutants: chemical processes governing their formation in the planetary boundary layer. Atmospheric Environment **34**: 2499–2527.

Jenkins, J. C., R. A. Birdsey, and Y. Pan. 2001. Biomass and NPP estimation for the mid-Atlantic region (USA) using plot-level forest inventory data. Ecological Applications **11**:1174–1193.

Jenkins, W. J. 1988. Nitrate flux into the euphotic zone near Bermuda. Nature **331**:521–523.

Jenkinson, D. S., and D. S. Powlson. 1976. Effects of biocidal treatments on metabolism in soil. 1. Funigation with chloroform. Soil Biology & Biochemistry **8**:167–177.

Jenkinson, D. S., and J. H. Rayner. 1977. Turnover of soil organic matter in some of the Rothamstead classical experiments. Soil Science **123**:298–305.

Jennings, J. N. 1983. Karst landforms. American Scientist **71**:578–586.

Jenny, H. 1941. *Factors of Soil Formation.* Dover Publications.

Jenny, H. 1980. *The Soil Resource.* Springer.

Jeppesen, E., M. Sondergaard, J. P. Jensen, K. E. Havens, O. Anneville, L. Carvalho, M. F. Coveney, R. Deneke, M. T. Dokulil, B. Foy, D. Gerdeaux, S. E. Hampton, S. Hilt, K. Kangur, J. Kohler, E. Lammens, T. L. Lauridsen, M. Manca, M. R. Miracle, B. Moss, P. Noges, G. Persson, G. Phillips, R. Portielje, C. L. Schelske, D. Straile, I. Tatrai, E. Willen, and M. Winder. 2005. Lake responses to reduced nutrient loading—an analysis of contemporary long-term data from 35 case studies. Freshwater Biology **50**: 1747–1771.

Jersak, J., R. Amundson, and G. Brimhall. 1995. A mass-balance analysis of podzolization—Examples from the northeastern United States. Geoderma **66**:15–42.

Jiang, X., W. L. Ku, R. L. Shia, Q. B. Li, J. W. Elkins, R. G. Prinn, and Y. L. Yung. 2007. Seasonal cycle of N_2O: Analysis of data. Global Biogeochemical Cycles **21**.

Jiao, N., G. J. Herndl, D. A. Hansell, R. Benner, G. Kattner, S. W. Wilhelm, D. L. Kirchman, M. G. Weinbauer, T. W. Luo, F. Chen, and F. Azam. 2010. Microbial production of recalcitrant dissolved organic matter: long-term carbon storage in the global ocean. Nature Reviews Microbiology **8**: 593–599.

Jickells, T. D. 1998. Nutrient biogeochemistry of the coastal zone. Science 281:217–222.

Jickells, T. D., Z. S. An, K. K. Andersen, A. R. Baker, G. Bergametti, N. Brooks, J. J. Cao, P. W. Boyd, R. A. Duce, K. A. Hunter, H. Kawahata, N. Kubilay, J. laRoche, P. S. Liss, N. Mahowald, J. M. Prospero, A. J. Ridgwell, I. Tegen, and R. Torres. 2005. Global iron connections between desert dust, ocean biogeochemistry, and climate. Science 308:67–71.

Jickells, T. D., S. D. Kelly, A. R. Baker, K. Biswas, P. F. Dennis, L. J. Spokes, M. Witt, and S. G. Yeatman. 2003. Isotopic evidence for a marine ammonia source. Geophysical Research Letters 30.

Jimenez, J. L., M. R. Canagaratna, N. M. Donahue, A.S.H. Prevot, Q. Zhang, J. H. Kroll, P. F. DeCarlo, J. D. Allan, H. Coe, N. L. Ng, A. C. Aiken, K. S. Docherty, I. M. Ulbrich, A. P. Grieshop, A. L. Robinson, J. Duplissy, J. D. Smith, K. R. Wilson, V. A. Lanz, C. Hueglin, Y. L. Sun, J. Tian, A. Laaksonen, T. Raatikainen, J. Rautiainen, P. Vaattovaara, M. Ehn, M. Kulmala, J. M. Tomlinson, D. R. Collins, M. J. Cubison, E. J. Dunlea, J. A. Huffman, T. B. Onasch, M. R. Alfarra, P. I. Williams, K. Bower, Y. Kondo, J. Schneider, F. Drewnick, S. Borrmann, S. Weimer, K. Demerjian, D. Salcedo, L. Cottrell, R. Griffin, A. Takami, T. Miyoshi, S. Hatakeyama, A. Shimono, J. Y. Sun, Y. M. Zhang, K. Dzepina, J. R. Kimmel, D. Sueper, J. T. Jayne, S. C. Herndon, A. M. Trimborn, L. R. Williams, E. C. Wood, A. M. Middlebrook, C. E. Kolb, U. Baltensperger, and D. R. Worsnop. 2009. Evolution of organic aerosols in the atmosphere. Science 326:1525–1529.

Jin, V. L., and R. D. Evans. 2010. Elevated CO_2 increases plant uptake of organic and inorganic N in the desert shrub *Larrea tridentata*. Oecologia 163:257–266.

Jitaru, P., P. Gabrielli, A. Marteel, J.M.C. Plane, F.A.M. Planchon, P.-A. Gauchard, C. P. Ferrari, C. F. Boutron, F. C. Adams, S. Hong, P. Cescon, and C. Barbante. 2009. Atmospheric depletion of mercury over Antarctica during glacial periods. Nature Geoscience 2:505–508.

Jobbágy, E. G., and R. B. Jackson. 2000. The vertical distribution of soil organic carbon and its relation to climate and vegetation. Ecological Applications 10:423–436.

Jobbágy, E. G., and R. B. Jackson. 2001. The distribution of soil nutrients with depth: Global patterns and the imprint of plants. Biogeochemistry 53:51–77.

Jobbágy, E. G., and R. B. Jackson. 2004. The uplift of soil nutrients by plants: Biogeochemical consequences across scales. Ecology 85:2380–2389.

Jobson, B. T., S. A. McKeen, D. D. Parrish, F. C. Fehsenfeld, D. R. Blake, A. H. Goldstein, S. M. Schauffler, and J. C. Elkins. 1999. Trace gas mixing ratio variability versus lifetime in the troposphere and stratosphere: Observations. Journal of Geophysical Research—Atmospheres 104:16091–16113.

Joergensen, R. G. 1996. Quantification of the microbial biomass by determining ninhydrin-reactive N. Soil Biology & Biochemistry 28:301–306.

Joergensen, R. G., and T. Mueller. 1996. The fumigation-extraction method to estimate soil microbial biomass: Calibration of the k_{en} value. Soil Biology & Biochemistry 28:33–37.

Joergensen, R. G., J. S. Wu, and P. C. Brookes. 2011. Measuring soil microbial biomass using an automated procedure. Soil Biology & Biochemistry 43:873–876.

John, R., J. W. Dalling, K. E. Harms, J. B. Yavitt, R. F. Stallard, M. Mirabello, S. P. Hubbell, R. Valencia, H. Navarrete, M. Vallejo, and R. B. Foster. 2007. Soil nutrients influence spatial distributions of tropical tree species. Proceedings of the National Academy of Sciences 104:864–869.

Johnson, A. H., S. B. Andersen, and T. G. Siccama. 1994. Acid rain and soils of the Adirondacks. 1. Changes in pH and available calcium, 1930–1984. Canadian Journal of Forest Research 24:39–45.

Johnson, A. H., A. Moyer, J. E. Bedison, S. L. Richter, and S. A. Willig. 2008a. Seven decades of calcium depletion in organic horizons of Adirondack forest soils. Soil Science Society of America Journal 72:1824–1830.

Johnson, D. B., M. A. Ghauri, and M. F. Said. 1992. Isolation and characterization of an acidophilic, heterotrophic bacterium capable of oxidizing ferrous iron. Applied and Environmental Microbiology 58:1423–1428.

Johnson, D. W. 1984. Sulfur cycling in forests. Biogeochemistry 1:29–43.

Johnson, D. W. 2006. Progressive N limitation in forests: Review and implications for long-term responses to elevated CO_2. Ecology 87:64–75.

Johnson, D. W., W. Cheng, and I. C. Burke. 2000a. Biotic and abiotic nitrogen retention in a variety of forest soils. Soil Science Society of America Journal 64:1503–1514.

Johnson, D. W., and D. W. Cole. 1980. Anion mobility in soils: Relevance to nutrient transport from forest ecosystems. Environment International 3:79–90.

Johnson, D. W., D. W. Cole, S. P. Gessel, M. J. Singer, and R. V. Minden. 1977. Carbonic acid leaching in a tropical, temperate, subalpine, and northern forest soil. Arctic and Alpine Research 9:329–343.

Johnson, D. W., D. W. Cole, H. Vanmiegroet, and F. W. Horng. 1986a. Factors affecting anion movement and retention in four forest soils. Soil Science Society of America Journal 50:776–783.

Johnson, D. W., G. S. Henderson, D. D. Huff, S. E. Lindberg, D. D. Richter, D. S. Shriner, D. E. Todd, and J. Turner. 1982. Cycling of organic and inorganic sulfur in a chestnut oak forest. Oecologia 54:141–148.

Johnson, D. W., G. S. Henderson, and D. E. Todd. 1981a. Evidence of modern accumulations of adsorbed sulfate in an east Tennessee forested Ultisol. Soil Science 132:422–426.

Johnson, D. W., G. S. Henderson, and D. E. Todd. 1988. Changes in nutrient distribution in forests and soils of Walker Branch watershed, Tennessee, over an 11-year period. Biogeochemistry 5:275–293.

Johnson, D. W., R. B. Susfalk, R. A. Dahlgren, and J. M. Klopatek. 1998. Fire is more important than water for nitrogen fluxes in semi-arid forests. Environmental Science and Policy 1:79–86.

Johnson, D. W., and D. E. Todd. 1983. Relationships among iron, aluminum, carbon, and sulfate in a variety of forest soils. Soil Science Society of America Journal 47:792–800.

Johnson, D. W., D. E. Todd, and V. R. Tolbert. 2003. Changes in ecosystem carbon and nitrogen in a loblolly pine plantation over the first 18 years. Soil Science Society of America Journal 67:1594–1601.

Johnson, D. W., R. F. Walker, D. W. Glass, W. W. Miller, J. D. Murphy, and C. M. Stein. 2012. The effect of rock content on nutrients in a Sierra Nevada forest soil. Geoderma 173(174), 84–93.

Johnson, K. P., J. D. Blum, G. J. Keeler, and T. A. Douglas. 2008b. Investigation of the deposition and emission of mercury in arctic snow during an atmospheric mercury depletion event. Journal of Geophysical Research—Atmospheres 113.

Johnson, K. S., C. L. Beehler, C. M. Sakamotoarnold, and J. J. Childress. 1986b. In situ measurements of chemical distributions in a deep-sea hydrothermal vent field. Science 231:1139–1141.

Johnson, K. S., R. M. Gordon, and K. H. Coale. 1997. What controls dissolved iron concentrations in the world ocean? Marine Chemistry 57:137–161.

Johnson, K. S., S. C. Riser, and D. M. Karl. 2010. Nitrate supply from deep to near-surface waters of the North Pacific subtropical gyre. Nature 465:1062–1065.

Johnson, L. C., G. R. Shaver, D. H. Cades, E. Rastetter, K. Nadelhoffer, A. Giblin, J. Laundre, and A. Stanley. 2000b. Plant carbon-nutrient interactions control CO_2 exchange in Alaskan wet sedge tundra ecosystems. Ecology 81:453–469.

Johnson, N. M., C. T. Driscoll, J. S. Eaton, G. E. Likens, and W. H. McDowell. 1981b. Acid rain, dissolved aluminum and chemical weathering at the Hubbard Brook Experimental Forest, New Hampshire. Geochimica et Cosmochimica Acta 45:1421–1437.

Johnson, N. M., G. E. Likens, F. H. Bormann, D. W. Fisher, and R. S. Pierce. 1969. A working model for the variation in stream water chemistry at the Hubbard Brook Experimental Forest, New Hampshire. Water Resources Research 5:1353–1363.

Johnson, N. M., G. E. Likens, F. H. Bormann, and R. S. Pierce. 1968. Rate of chemical weathering of silicate minerals in New Hampshire. Geochimica et Cosmochimica Acta 32:531–545.

Johnson, N. M., R. C. Reynolds, and G. E. Likens. 1972. Atmospheric sulfur: Its effect on the chemical weathering of New England. Science 177:514–516.

Johnston, C. A., N. E. Detenbeck, and G. J. Niemi. 1990. The cumulative effect of wetlands on stream water quality and quantity—A landscape approach. Biogeochemistry 10:105–141.

Johnston, C. A., B. A. Shmagin, P. C. Frost, C. Cherrier, J. H. Larson, G. A. Lambert, and S. D. Bridgham. 2008. Wetland types and wetland maps differ in ability to predict dissolved organic carbon concentrations in streams. Science of the Total Environment 404:326–334.

Johnston, D. A. 1980. Volcanic contribution of chlorine to the stratosphere—More significant to ozone than previously estimated. Science 209:491–493.

Jonasson, S., J. P. Bryant, F. S. Chapin, and M. Andersson. 1986. Plant phenols and nutrients in relation to variations in climate and rodent grazing. American Naturalist 128:394–408.

Jones, D. L. 1998. Organic acids in the rhizosphere—a critical review. Plant and Soil 205:25–44.

Jones, J. B., J. D. Schade, S. G. Fisher, and N. B. Grimm. 1997. Organic matter dynamics in Sycamore Creek, a desert stream in Arizona. Journal of the North American Benthological Society 16:78–82.

Jones, J.B.J., and P. J. Mulholland. 1998. Influence of drainage basin topography and elevation on carbon dioxide and methane supersaturation of stream water. Biogeochemistry 40:57–72.

Jones, M. B., and S. W. Humphries. 2002. Impacts of the C-4 sedge Cyperus papyrus L. on carbon and water fluxes in an African wetland. Hydrobiologia 488:107–113.

Jones, M. J. 1973. The organic matter content of the savanna soils of west Africa. Journal of Soil Science 24:42–53.

Jones, R. D., and R. Y. Morita. 1983. Methane oxidation by Nitrosococcus oceanus and Nitrosomonas europaea. Applied and Environmental Microbiology 45:401–410.

Jones, R. L., and H. C. Hanson. 1985. Mineral Licks, Geophagy, and Biogeochemistry of North American Ungulates. Iowa State University Press.

Jordan, C. F. 1971. A world pattern in plant energetics. American Scientist 59:425–433.

Jordan, M., and G. E. Likens. 1975. An organic carbon budget for an oligotrophic lake in New Hampshire, U.S.A. Verhandlungen des Internationalen Verein Limnologie **19**:994–1003.

Jordan, S. J., J. Stoffer, and J. A. Nestlerode. 2011. Wetlands as sinks for reactive nitrogen at continental and global scales: A meta-analysis. Ecosystems **14**: 144–155.

Jordan, T. E., J. C. Cornwell, W. R. Boynton, and J. T. Anderson. 2008. Changes in phosphorus biogeochemistry along an estuarine salinity gradient: The iron conveyer belt. Limnology and Oceanography **53**: 172–184.

Jordan, T. E., D. L. Correll, and D. E. Weller. 1993. Nutrient interception by a riparian forest receiving inputs from adjacent cropland. Journal of Environmental Quality **22**:467–473.

Jordan, T. E., D. L. Correll, and D. E. Weller. 1997. Effects of agriculture on discharges of nutrients from coastal plain watersheds of Chesapeake Bay. Journal of Environmental Quality **26**:836–848.

Jorgensen, B. B. 1977. Sulfur cycle of a coastal marine sediment (Limfjorden, Denmark). Limnology and Oceanography **22**:814–832.

Jorgensen, B. B. 1982. Mineralization of organic matter in the sea bed—The role of sulfate reduction. Nature **296**:643–645.

Jorgensen, B. B., M. F. Isaksen, and H. W. Jannasch. 1992. Bacterial sulfate reduction above $100°$ C in deep-sea hydrothermal vent sediments. Science **258**:1756–1757.

Jorgensen, J. R., C. G. Wells, and L. J. Metz. 1980. Nutrient changes in decomposing loblolly pine forest floor. Soil Science Society of America Journal **44**:1307–1314.

Joslin, J. D., J. B. Gaudinski, M. S. Torn, W. J. Riley, and P. J. Hanson. 2006. Fine-root turnover patterns and their relationship to root diameter and soil depth in a C-14-labeled hardwood forest. New Phytologist **172**:523–535.

Ju, X. T., X. Lu, Z. L. Gao, X. P. Chen, F. Su, M. Kogge, V. Romheld, P. Christie, and F. S. Zhang. 2011. Processes and factors controlling N_2O production in an intensively managed low carbon calcareous soil under sub-humid monsoon conditions. Environmental Pollution **159**:1007–1016.

Juang, F.H.T., and N. M. Johnson. 1967. Cycling of chlorine through a forested watershed in New England. Journal of Geophysical Research **65**:227–237.

Juang, J. Y., G. Katul, M. Siqueira, P. Stoy, and K. Novick. 2007. Separating the effects of albedo from ecophysiological changes on surface temperature along a successional chronosequence in the southeastern United States. Geophysical Research Letters **34**.

Juang, J. Y., G. G. Katul, M.B.S. Siqueira, P. C. Stoy, S. Palmroth, H. R. McCarthy, H. S. Kim, and R. Oren. 2006. Modeling nighttime ecosystem respiration from measured CO_2 concentration and air temperature profiles using inverse methods. Journal of Geophysical Research—Atmospheres **111**.

Juice, S. M., T. J. Fahey, T. G. Siccama, C. T. Driscoll, E. G. Denny, C. Eagar, N. L. Cleavitt, R. Minocha, and A. D. Richardson. 2006. Response of sugar maple to calcium addition to northern hardwood forest. Ecology **87**:1267–1280.

Jung, M., M. Reichstein, P. Ciais, S. I. Seneviratne, J. Sheffield, M. L. Goulden, G. Bonan, A. Cescatti, J. Q. Chen, R. de Jeu, A. J. Dolman, W. Eugster, D. Gerten, D. Gianelle, N. Gobron, J. Heinke, J. Kimball, B. E. Law, L. Montagnani, Q. Z. Mu, B. Mueller, K. Oleson, D. Papale, A. D. Richardson, O. Roupsard, S. Running, E. Tomelleri, N. Viovy, U. Weber, C. Williams, E. Wood, S. Zaehle, and K. Zhang. 2010. Recent decline in the global land evapotranspiration trend due to limited moisture supply. Nature **467**:951–954.

Jung, M., M. Reichstein, H. A. Margolis, A. Cescatti, A. D. Richardson, M. A. Arain, A. Arneth, C. Bernhofer, D. Bonal, J. Q. Chen, D. Gianelle, N. Gobron, G. Kiely, W. Kutsch, G. Lasslop, B. E. Law, A. Lindroth, L. Merbold, L. Montagnani, E. J. Moors, D. Papale, M. Sottocornola, F. Vaccari, and C. Williams. 2011. Global patterns of land-atmosphere fluxes of carbon dioxide, latent heat, and sensible heat derived from eddy covariance, satellite, and meteorological observations. Journal of Geophysical Research—Biogeosciences **116**.

Junge, C. E. 1960. Sulfur in the atmosphere. Journal of Geophysical Research **65**:227–237.

Junge, C. E. 1974. Residence time and variability of tropospheric trace gases. Tellus **26**:477–488.

Junge, C. E., and R. T. Werby. 1958. The concentration of chloride, sodium, potassium, calcium, and sulfate in rain water over the United States. Journal of Meteorology **15**:417–425.

Juranek, L. W., and P. D. Quay. 2010. Basin-wide photosynthetic production rates in the subtropical and tropical Pacific Ocean determined from dissolved oxygen isotope ratio measurements. Global Biogeochemical Cycles **24**.

Jurinak, J. J., L. M. Dudley, M. F. Allen, and W. G. Knight. 1986. The role of calcium oxalate in the availability of phosphorus in soils of semiarid regions—A thermodynamic study. Soil Science **142**:255–261.

Justic, D., N. N. Rabalais, R. E. Turner, and Q. Dortch. 1995. Changes in nutrient structure of river-dominated coastal waters—Stoichiometric nutrient balance and its consequences. Estuarine Coastal and Shelf Science 40:339–356.

Justin, S. H . F. W., and W. Armstrong. 1987. The anatomical characteristics of roots and plant response to soil flooding. New Phytologist 106:465–495.

Kadeba, O. 1978. Organic matter status of some savanna soils of northern Nigeria. Soil Science 125:122–127.

Kah, L. C., T. W. Lyons, and T. D. Frank. 2004. Low marine sulphate and protracted oxygenation of the Proterozoic biosphere. Nature 431:834–838.

Kahn, R. 1985. The evolution of CO_2 on Mars. Icarus 62:175–190.

Kai, F. M., S. C. Tyler, J. T. Randerson, and D. R. Blake. 2011. Reduced methane growth rate explained by decreased Northern Hemisphere microbial sources. Nature 476:194–197.

Kaiser, C., L. Fuchslueger, M. Koranda, M. Gorfer, C. F. Stange, B. Kitzler, F. Rasche, J. Strauss, A. Sessitsch, S. Zechmeister-Boltenstern, and A. Richter. 2011. Plants control the seasonal dynamics of microbial N cycling in a beech forest soil by belowground C allocation. Ecology 92:1036–1051.

Kaiser, J. W., A. Heil, M. O. Andreae, A. Benedetti, N. Chubarova, L. Jones, J.-J. Morcrette, M. Razinger, M. G. Schultz, M. Suttie, and G. R. van der Werf. 2012. Biomass burning emissions estimated with a global fire assimilation system based on observed fire radiative power. Biogeosciences 9:527–554.

Kaiser, K., and G. Guggenberger. 2005. Dissolved organic sulphur in soil water under Pinus sylvestris L. and Fagus sylvatica L. stands in northeastern Bavaria, Germany—variations with seasons and soil depth. Biogeochemistry 72:337–364.

Kaiser, K., G. Guggenberger, and L. Haumaier. 2003. Organic phosphorus in soil water under a European beech (Fagus sylvatica L.) stand in northeastern Bavaria, Germany: seasonal variability and changes with soil depth. Biogeochemistry 66:287–310.

Kaiser, K., and W. Zech. 1998. Soil dissolved organic matter sorption as influenced by organic and sesquioxide coatings and sorbed sulfate. Soil Science Society of America Journal 62:129–136.

Kalbitz, K., S. Solinger, J. H. Park, B. Michalzik, and E. Matzner. 2000. Controls on the dynamics of dissolved organic matter in soils: A review. Soil Science 165:277–304.

Kamman, N. C., and D. R. Engstrom. 2002. Historical and present fluxes of mercury to Vermont and New Hampshire lakes inferred from Pb-210 dated sediment cores. Atmospheric Environment 36: 1599–1609.

Kammann, C., C. Muller, L. Grunhage, and H. J. Jager. 2008. Elevated CO_2 stimulates N_2O emissions in permanent grassland. Soil Biology & Biochemistry 40: 2194–2205.

Kang, D. W., V. P. Aneja, R. Mathur, and J. D. Ray. 2003. Nonmethane hydrocarbons and ozone in three rural southeast United States national parks: A model sensitivity analysis and comparison to measurements. Journal of Geophysical Research—Atmospheres 108.

Kaplan, I. R. 1975. Stable isotopes as a guide to biogeochemical processes. Proceedings of the Royal Society of London 189B:183–211.

Kaplan, W. A., J. W. Elkins, C. E. Kolb, M. B. McElroy, S. C. Wofsy, and A. P. Duran. 1978. Nitrous oxide in freshwater systems—Estimate for yield of atmospheric N_2O associated with disposal of human waste. Pure and Applied Geophysics 116:423–438.

Kaplan, W. A., S. C. Wofsy, M. Keller, and J. M. Dacosta. 1988. Emission of NO and deposition of O_3 in a tropical forest system. Journal of Geophysical Research—Atmospheres 93:1389–1395.

Kappler, A., C. Pasquero, K. O. Konhauser, and D. K. Newman. 2005. Deposition of banded iron formations by anoxygenic phototrophic Fe(II)-oxidizing bacteria. Geology 33:865–868.

Karberg, N. J., K. S. Pregitzer, J. S. King, A. L. Friend, and J. R. Wood. 2005. Soil carbon dioxide partial pressure and dissolved inorganic carbonate chemistry under elevated carbon dioxide and ozone. Oecologia 142: 296–306.

Karl, D., A. Michaels, B. Bergman, D. Capone, E. Carpenter, R. Letelier, F. Lipschultz, H. Paerl, D. Sigman, and L. Stal. 2002. Dinitrogen fixation in the world's oceans. Biogeochemistry 57/58:47–98.

Karl, D. M., and B. D. Tilbrook. 1994. Production and transport of methane in oceanic particulate organic matter. Nature 368:732–734.

Karl, T. R., and R. W. Knight. 1998. Secular trends of precipitation amount, frequency, and intensity in the United States. Bulletin of the American Meteorological Society 79:231–241.

Karlsson, J., P. Bystrom, J. Ask, P. Ask, L. Persson, and M. Jansson. 2009. Light limitation of nutrient-poor lake ecosystems. Nature 460:506–509.

Karltun, E., and J. P. Gustafsson. 1993. Interference by organic complexation of Fe and Al on the SO_4^{2-} adsorption in spodic B horizons in Sweden. Journal of Soil Science 44:625–632.

Kartal, B., W. J. Maalcke, N. M. de Almeida, I. Cirpus, J. Gloerich, W. Geerts, H. den Camp, H. R. Harhangi, E. M. Janssen-Megens, K. J. Francoijs, H. G. Stuntunnenberg, J. T. Keltjens, M.S.M. Jetten, and M. Strous. 2011. Molecular mechanism of anaerobic ammonium oxidation. Nature 479:127–130.

Kashefi, K., and D. R. Lovley. 2003. Extending the upper temperature limit for life. Science 301:934.

Kasischke, E. S., L. L. Bourgeauchavez, N. L. Christensen, and E. Haney. 1994. Observations on the sensitivity of ERS-1 SAR image intensity to changes in aboveground biomass in young loblolly pine forests. International Journal of Remote Sensing **15**:3–16.

Kasischke, E. S., N.H.F. French, L. L. Bourgeauchavez, and N. L. Christensen. 1995. Estimating release of carbon from the 1990 and 1991 forest fires in Alaska. Journal of Geophysical Research—Atmospheres **100**: 2941–2951.

Kaspari, M., S. P. Yanoviak, R. Dudley, M. Yuan, and N. A. Clay. 2009. Sodium shortage as a constraint on the carbon cycle in an inland tropical rainforest. Proceedings of the National Academy of Sciences **106**:19405–19409.

Kass, D. M., and Y. L. Yung. 1995. Loss of atmosphere from Mars due to solar wind-induced sputtering. Science **268**:697–699.

Kaste, J. M., A. J. Friedland, and E. K. Miller. 2005. Potentially mobile lead fractions in montane organic-rich soil horizons. Water Air and Soil Pollution **167**: 139–154.

Kasting, J. F., O. B. Toon, and J. B. Pollack. 1988. How climate evolved on the terrestrial planets. Scientific American **256**:90–97.

Kasting, J. F., and J.C.G. Walker. 1981. Limits on oxygen concentration in the prebiological atmosphere and the rate of abiotic fixation of nitrogen. Journal of Geophysical Research—Oceans and Atmospheres **86**: 1147–1158.

Kastner, M. 1974. The contribution of authigenic feldspars to the geochemical balance of alkalic metals. Geochimica et Cosmochimica Acta **38**:650–653.

Kastner, M. 1999. Oceanic minerals: Their origin, nature of their environment and significance. Proceedings of the National Academy of Sciences **96**: 3380–3387.

Kato, C., L. Li, Y. Nogi, Y. Nakamura, J. Tamaoka, and K. Horikoshi. 1998. Extremely barophilic bacteria isolated from the Mariana Trench, Challenger Deep, at a depth of 11,000 meters. Applied and Environmental Microbiology **64**:1510–1513.

Katterer, T., M. Reichstein, O. Andren, and A. Lomander. 1998. Temperature dependence of organic matter decomposition: a critical review using literature data analyzed with different models. Biology and Fertility of Soils **27**:258–262.

Katz, M. E., D. K. Pak, G. R. Dickens, and K. G. Miller. 1999. The source and fate of massive carbon input during the latest Paleocene thermal maximum. Science **286**:1531–1533.

Kauffman, J. B., R. L. Sanford, D. L. Cummings, I. H. Salcedo, and E. Sampaio. 1993. Biomass and nutrient dynamics asociated with slash fires in neotropical dry forests. Ecology **74**:140–151.

Kaufman, Y. J., D. Tanre, and O. Boucher. 2002. A satellite view of aerosols in the climate system. Nature **419**:215–223.

Kaufman, Y. J., C. J. Tucker, and I. Fung. 1990. Remote sensing of biomass burning in the tropics. Journal of Geophysical Research—Atmospheres **95**:9927–9939.

Kaul, M., G.M.J. Mohren, and V. K. Dadhwal. 2011. Phytomass carbon pool of trees and forests in India. Climatic Change **108**:243–259.

Kaushal, S. S., P. M. Groffman, L. E. Band, C. A. Shields, R. P. Morgan, M. A. Palmer, K. T. Belt, C. M. Swan, S.E.G. Findlay, and G. T. Fisher. 2008. Interaction between urbanization and climate variability amplifies watershed nitrate export in Maryland. Environmental Science & Technology **42**:5872–5878.

Kaushal, S. S., P. M. Groffman, G. E. Likens, K. T. Belt, W. P. Stack, V. R. Kelly, L. E. Band, and G. T. Fisher. 2005. Increased salinization of fresh water in the northeastern United States. Proceedings of the National Academy of Sciences **102**:13517–13520.

Kaushal, S. S., and W. M. Lewis. 2005. Fate and transport of organic nitrogen in minimally disturbed montane streams of Colorado, USA. Biogeochemistry **74**: 303–321.

Kavouras, I. G., N. Mihalopoulos, and E. G. Stephanou. 1998. Formation of atmospheric particles from organic acids produced by forests. Nature **395**: 683–686.

Keeley, J. E. 1979. Population differentiation along a flood frequency gradient: physiological adaptations to flooding in *Nyssa sylvatica*. Ecological Monographs **49**:89–108.

Keeley, J. E., and C. J. Fotheringham. 1997. Trace gas emissions and smoke-induced seed germination. Science **276**:1248–1250.

Keeling, C. D. 1983. The global carbon cycle: What we know and could know from atmospheric, biospheric, and oceanic observations. *Proceedings: Carbon dioxide research conference: Carbon Dioxide, Science and Consensus.* U.S. Department of EnergyWashington, D,C, II:3–62.

Keeling, C. D. 1993. Global observations of atmospheric CO_2. *In* M. Heimann, editor. *The Global Carbon Cycle*, Springer-Verlag, 1–29.

Keeling, C. D., A. F. Carter, and W. G. Mook. 1984. Seasonal, latitudinal, and secular variations in the abundance and isotopic ratios of atmospheric CO_2. 2. Results from oceanographic cruises in the tropical Pacific ocean. Journal of Geophysical Research—Atmospheres **89**:4615–4628.

Keeling, C. D., J.F.S. Chin, and T. P. Whorf. 1996. Increased activity of northern vegetation inferred from atmospheric CO_2 measurements. Nature **382**: 146–149.

Keeling, C. D., T. P. Whorf, M. Wahlen, and J. Vanderplicht. 1995. Interannual extremes in the rate of rise of atmospheric carbon dioxide since 1980. Nature 375:666–670.

Keeling, R. F., and S. R. Shertz. 1992. Seasonal and interannual variations in atmospheric oxygen and implications for the global carbon cycle. Nature 358: 723–727.

Keffer, T., D. G. Martinson, and B. H. Corliss. 1988. The position of the Gulf Stream during Quaternary glaciations. Science 241:440–442.

Keil, R. G., D. B. Montlucon, F. G. Prahl, and J. I. Hedges. 1994. Sorptive preservation of labile organic matter in marine sediments. Nature 370:549–552.

Keller, C. K., and D. H. Bacon. 1998. Soil respiration and georespiration distinguished by transport analyses of vadose CO_2, (CO_2)-^{13}C, and (CO_2)-^{14}C. Global Biogeochemical Cycles 12:361–372.

Keller, C. K., and B. D. Wood. 1993. Possibility of chemical weathering before the advent of vascular land plants. Nature 364:223–225.

Keller, J. K., and S. D. Bridgham. 2007. Pathways of anaerobic carbon cycling across an ombrotrophic-minerotrophic peatland gradient. Limnology and Oceanography 52:96–107.

Keller, M., W. A. Kaplan, S. C. Wofsy, and J. M. Dacosta. 1988. Emissions of N_2O from tropical forest soils—Response to fertilization with NH_4^+, NO_3^-, and PO_4^{3-}. Journal of Geophysical Research—Atmospheres 93:1600–1604.

Keller, M., and W. A. Reiners. 1994. Soil-atmosphere exchange of nitrous oxide, nitric oxide, and methane under secondary succession of pasture to forest in the Atlantic lowlands of Costa Rica. Global Biogeochemical Cycles 8:399–409.

Keller, M., E. Veldkamp, A. M. Weitz, and W. A. Reiners. 1993. Effect of pasture age on soil trace gas emissions from a deforested area of Costa Rica. Nature 365: 244–246.

Kelley, C. A., and W. H. Jeffrey. 2002. Dissolved methane concentration profiles and air-sea fluxes from 41° S to 27° N. Global Biogeochemical Cycles 16.

Kelley, C. A., C. S. Martens, and J. P. Chanton. 1990. Variations in sedimentary carbon remineralization rates in the White Oak river estuary, North Carolina. Limnology and Oceanography 35:372–383.

Kelley, D. S., J. A. Baross, and J. R. Delaney. 2002. Volcanoes, fluids, and life at mid-ocean ridge spreading centers. Annual Review of Earth and Planetary Sciences 30:385–491.

Kelley, D. S., J. A. Karson, G. L. Fruh-Green, D. R. Yoerger, T. M. Shank, D. A. Butterfield, J. M. Hayes, M. O. Schrenk, E. J. Olson, G. Proskurowski, M. Jakuba, A. Bradley, B. Larson, K. Ludwig, D. Glickson,

K. Buckman, A. S. Bradley, W. J. Brazelton, K. Roe, M. J. Elend, A. Delacour, S.M. Bernasconi, M. D. Lilley, J. A. Baross, R. T. Summons, and S. P. Sylva. 2005. A serpentinite-hosted ecosystem: The lost city hydrothermal field. Science 307:1428–1434.

Kellogg, L. E., S. D. Bridgham, and D. Lopez-Hernandez. 2006. A comparison of four methods of measuring gross phosphorus mineralization. Soil Science Society of America Journal 70:1349–1358.

Kellogg, W. W. 1992. Aerosols and global warming. Science 256:598.

Kelly, C. A., J.W.M. Rudd, R. A. Bodaly, N. P. Roulet, V. L. StLouis, A. Heyes, T. R. Moore, S. Schiff, R. Aravena, K. J. Scott, B. Dyck, R. Harris, B. Warner, and G. Edwards. 1997. Increases in fluxes of greenhouse gases and methyl mercury following flooding of an experimental reservoir. Environmental Science & Technology 31:1334–1344.

Kelly, C. A., J.W.M. Rudd, R. H. Hesslein, D. W. Schindler, P. J. Dillon, C. T. Driscoll, S. A. Gherini, and R. E. Hecky. 1987. Prediction of biological acid neutralization in acid-sensitive lakes. Biogeochemistry 3: 129–140.

Kelly, D. P. 1990. Organic sulfur compounds in the environment—biogeochemistry, microbiology, and ecological aspects. Advances in Microbial Ecology 11: 345–385.

Kelly, E. F., O. A. Chadwick, and T. E. Hilinski. 1998. The effect of plants on mineral weathering. Biogeochemistry 42:21–53.

Kelly, V. R., G. M. Lovett, K. C. Weathers, and G. E. Likens. 2002. Trends in atmospheric concentration and deposition compared to regional and local pollutant emissions at a rural site in southeastern New York. Atmospheric Environment 36: 1569–1575.

Kemp, A. C., B. P. Horton, J. P. Donnelly, M. E. Mann, M. Vermeer, and S. Rahmstorf. 2011. Climate related sea-level variations over the past two millennia. Proceedings of the National Academy of Sciences 108: 11017–11022.

Kemp, M. J., and W. K. Dodds. 2002a. Comparisons of nitrification and denitrification in prairie and agriculturally influenced streams. Ecological Applications 12:998–1009.

Kemp, M. J., and W. K. Dodds. 2002b. The influence of ammonium, nitrate, and dissolved oxygen concentrations on uptake, nitrification, and denitrification rates associated with prairie stream substrata. Limnology and Oceanography 47:1380–1393.

Kemp, W. M., P. Sampou, J. Caffrey, M. Mayer, K. Henriksen, and W. R. Boynton. 1990. Ammonium recycling versus denitrification in Chesapeake Bay sediments. Limnology and Oceanography 35: 1545–1563.

Kemp, W. M., E. M. Smith, M. MarvinDiPasquale, and W. R. Boynton. 1997. Organic carbon balance and net ecosystem metabolism in Chesapeake Bay. Marine Ecology-Progress Series 150:229–248.

Kemp, W. M., J. M. Testa, D. J. Conley, D. Gilbert, and J. D. Hagy. 2009. Temporal responses of coastal hypoxia to nutrient loading and physical controls. Biogeosciences 6:2985–3008.

Kennedy, M., M. Droser, L. M. Mayer, D. Pevear, and D. Mrofka. 2006. Late Precambrian oxygenation; Inception of the clay mineral factory. Science 311: 1446–1449.

Kennedy, M. J., O. A. Chadwick, P. M. Vitousek, L. A. Derry, and D. M. Hendricks. 1998. Changing sources of base cations during ecosystem development, Hawaiian Islands. Geology 26:1015–1018.

Kennedy, M. J., L. O. Hedin, and L. A. Derry. 2002a. Decoupling of unpolluted temperate forests from rock nutrient sources revealed by natural $^{87}Sr/^{86}Sr$ and ^{84}Sr tracer addition. Proceedings of the National Academy of Sciences 99:9639–9644.

Kennedy, M. J., D. R. Pevear, and R. J. Hill. 2002b. Mineral surface control of organic carbon in black shale. Science 295:657–660.

Kennett, J. 1982. Marine Geology. Prentice Hall.

Kennett, J. P., K. G. Cannariato, I. L. Hendy, and R. J. Behl. 2000. Carbon isotopic evidence for methane hydrate instability during Quaternary interstadials. Science 288:128–133.

Kenrick, P., and P. R. Crane. 1997. The origin and early evolution of plants on land. Nature 389:33–39.

Keppler, F., R. Eiden, V. Niedan, J. Pracht, and H. F. Scholer. 2000. Halocarbons produced by natural oxidation processes during degradation of organic matter. Nature 403:298–301.

Keppler, F., J.T.G. Hamilton, M. Brass, and T. Rockmann. 2006. Methane emissions from terrestrial plants under aerobic conditions. Nature 439:187–191.

Kerkhoff, A. J., B. J. Enquist, J. J. Elser, and W. F. Fagan. 2005. Plant allometry, stoichiometry and the temperature-dependence of primary productivity. Global Ecology and Biogeography 14:585–598.

Kerner, M. J. 1993. Coupling of microbial fermentation and respiration processes in an intertidal mudflat of the Elbe estuary. Limnology and Oceanography 38:314–330.

Kerr, J. B., and C. T. McElroy. 1993. Evidence for large upward trends of ultraviolet-B radiation linked to ozone depletion. Science 262:1032–1034.

Kerrick, D. M. 2001. Present and past nonanthropogenic CO_2 degassing from the solid Earth. Reviews of Geophysics 39:565–585.

Kerrick, D. M., and J.A.D. Connolly. 2001. Metamorphic devolatilization of subducted marine sediments and the transport of volatiles into the Earth's mantle. Nature 411:293–296.

Kesselmeier, J., P. Ciccioli, U. Kuhn, P. Stefani, T. Biesenthal, S. Rottenberger, A. Wolf, M. Vitullo, R. Valentini, A. Nobre, P. Kabat, and M. O. Andreae. 2002. Volatile organic compound emissions in relation to plant carbon fixation and the terrestrial carbon budget. Global Biogeochemical Cycles 16.

Kesselmeier, J., F. X. Meixner, U. Hofmann, A. L. Ajavon, S. Leimbach, and M. O. Andreae. 1993. Reduced-sulfur compound exchange between the atmosphere and tropical tree species in southern Cameroon. Biogeochemistry 23:23–45.

Kesselmeier, J., and L. Merk. 1993. Exchange of carbonyl sulfide (COS) between agricultural plants and the atmosphere—Studies on the deposition of COS to peas, corn and rapeseed. Biogeochemistry 23:47–59.

Kessler, J. D., D. L. Valentine, M. C. Redmond, M. R. Du, E. W. Chan, S. D. Mendes, E. W. Quiroz, C. J. Villanueva, S. S. Shusta, L. M. Werra, S. A. Yvon-Lewis, and T. C. Weber. 2011. A persistent oxygen anomaly reveals the fate of spilled methane in the deep Gulf of Mexico. Science 331:312–315.

Kettle, A. J., U. Kuhn, M. von Hobe, J. Kesselmeier, and M. O. Andreae. 2002. Global budget of atmospheric carbonyl sulfide: Temporal and spatial variations of the dominant sources and sinks. Journal of Geophysical Research—Atmospheres 107.

Keuskamp, J. A., G. van drecht, and A. F. Bouwman. 2012. European-scale modelling of groundwter denitrification and associated N_2O production. Environmental Pollution 165:67–78.

Khademi, S., J. O'Connell, J. Remis, Y. Robles-Colmenares, L.J.W. Miericke, and R. M. Stroud. 2004. Mechanism of ammonia transport by Amt/MEP/Rh: Structure of AmtB at 1.35 angstrom. Science 305:1587–1594.

Khalil, K., B. Mary, and P. Renault. 2004. Nitrous oxide production by nitrification and denitrification in soil aggregates as affected by O_2 concentration. Soil Biology & Biochemistry 36:687–699.

Khalil, M.A.K., and R. A. Rasmussen. 1984. Global sources, lifetimes and mass balances of carbonyl sulfide (OCS) and carbon disulfide (CS_2) in the Earth's atmosphere. Atmospheric Environment 18:1805–1813.

Khalil, M.A.K., and R. A. Rasmussen. 1985. Causes of increasing atmospheric methane—Depletion of hydroxyl radicals and the rise of emissions. Atmospheric Environment 19:397–407.

Khalil, M.A.K., and R. A. Rasmussen. 1988. Carbon monoxide in the Earth's atmosphere—Indications of a global increase. Nature 332:242–245.

Khalil, M.A.K., and R. A. Rasmussen. 1990. Constraints on the global sources of methane and an analysis of recent budgets. Tellus 42B:229–236.

Khalil, M.A.K., and R. A. Rasmussen. 1994. Global decrease in atmospheric carbon monoxide concentration. Nature 370:639–641.

Khalil, M.A.K., R. A. Rasmussen, J.R.J. French, and J. A. Holt. 1990. The influence of termites on atmospheric trace gases—CH_4, CO_2, $CHCl_3$, N_2O, CO, H_2, amd light hydrocarbons. Journal of Geophysical Research—Atmospheres 95:3619–3634.

Khalil, M.A.K., R. A. Rasmussen, and R. Gunawardena. 1993a. Atmospheric methyl bromide—Trends and global mass balance. Journal of Geophysical Research—Atmospheres 98:2887–2896.

Khalil, M.A.K., R. A. Rasmussen, and F. Moraes. 1993b. Atmospheric methane at Cape Meares—Analysis of a high-resolution data base and its environmental implications. Journal of Geophysical Research—Atmospheres 98:14753–14770.

Khalil, M.A.K., M. J. Shearer, and R. A. Rasmussen. 1993c. Methane sinks and distribution. In M.A.K. Khalil, editor. Atmospheric Methane: Sources, Sinks, and Role in Global Change, Springer-Verlag, 168–179.

Khan, S. A., R. L. Mulvaney, T. R. Ellsworth, and C. W. Boast. 2007. The myth of nitrogen fertilization for soil carbon sequestration. Journal of Environmental Quality 36:1821–1832.

Khatiwala, S., F. Primeau, and T. Hall. 2009. Reconstruction of the history of anthropogenic CO_2 concentrations in the ocean. Nature 462:346–349.

Kieffer, H. H. 1976. Soil and surface temperatures at Viking landing sites. Science 194:1344–1346.

Kiehl, J. T., and B. P. Briegleb. 1993. The relative roles of sulfate aerosols and greenhouse gases in climate forcing. Science 260:311–314.

Kielland, K. 1994. Amino acid absorption by arctic plants—Implications for plant nutrition and nitrogen cycling. Ecology 75:2373–2383.

Kiene, R. P. 1990. Dimethyl sulfide production from dimethylsulfoniopropionate in coastal seawater samples and bacterial cultures. Applied and Environmental Microbiology 56:3292–3297.

Kiene, R. P., and T. S. Bates. 1990. Biological removal of dimethyl sulfide from sea water. Nature 345:702–705.

Kiene, R. P., and L. J. Linn. 2000. The fate of dissolved dimethylsulfoniopropionate (DMSP) in seawater: Tracer studies using ^{35}S-DMSP. Geochimica et Cosmochimica Acta 64:2797–2810.

Kiene, R. P., L. J. Linn, and J. A. Bruton. 2000. New and important roles for DMSP in marine microbial communities. Journal of Sea Research 43:209–224.

Kilham, P. 1982. Acid precipitation—Its role in the alkalization of a lake in Michigan. Limnology and Oceanography 27:856–867.

Killingbeck, K. T. 1985. Autumnal resorption and accretion of trace metals in gallery forest trees. Ecology 66:283–286.

Killingbeck, K. T. 1996. Nutrients in senesced leaves: Keys to the search for potential resorption and resorption proficiency. Ecology 77:1716–1727.

Killough, G. G., and W. R. Emanuel. 1981. A comparison of several models of carbon turnover in the ocean with respect to their distributions of transit time and age, and responses to atmospheric CO_2 and [14]C. Tellus 33:274–290.

Kim, D., J. D. Schuffert, and M. Kastner. 1999. Francolite authigenesis in California continental slope sediments and its implications for the marine P cycle. Geochimica et Cosmochimica Acta 63:3477–3485.

Kim, J., and D. C. Rees. 1994. Nitrogenase and biological nitrogen fixation. Biochemistry 33:389–397.

Kim, J. S., and D. C. Rees. 1992. Structural models for the metal centers in the nitrogenase molybdenum-iron protein. Science 257:1677–1682.

Kim, K. R., and H. Craig. 1990. Two-isotope characterization of N_2O in the Pacific ocean and constraints on its origin in deep water. Nature 347:58–61.

Kim, K. R., and H. Craig. 1993. [15]N and [18]O characteristics of nitrous oxide—A global perspective. Science 262:1855–1857.

Kim, S. W., A. Heckel, S. A. McKeen, G. J. Frost, E. Y. Hsie, M. K. Trainer, A. Richter, J. P. Burrows, S. E. Peckham, and G. A. Grell. 2006. Satellite-observed US power plant NOx emission reductions and their impact on air quality. Geophysical Research Letters 33.

Kim, T. W., K. Lee, R. G. Najjar, H. D. Jeong, and H. J. Jeong. 2011. Increasing N abundance in the northwestern Pacific ocean due to atmospheric nitrogen deposition. Science 334:505–509.

Kim, Y. S., M. Imori, M. Watanabe, R. Hatano, M. J. Yi, and T. Koike. 2012. Simulated nitrogen inputs influence methane and nitrous oxide fluxes from a young larch plantation in northern Japan. Atmospheric Environment 46:36–44.

Kindler, R., J. Siemens, K. Kaiser, D. C. Walmsley, C. Bernhofer, N. Buchmann, P. Cellier, W. Eugster, G. Gleixner, T. Grunwald, A. Heim, A. Ibrom, S. K. Jones, M. Jones, K. Klumpp, W. Kutsch, K. S. Larsen, S. Lehuger,

B. Loubet, R. McKenzie, E. Moors, B. Osborne, K. Pilegaard, C. Rebmann, M. Saunders, M.W.I. Schmidt, M. Schrumpf, J. Seyfferth, U. Skiba, J.-F. Soussana, M. A. Sutton, C. Tefs, B. Vowinckel, M. J. Zeeman, and M. Kaupenjohann. 2011. Dissolved carbon leaching from soil is a crucial component of the net ecosystem carbon balance. Global Change Biology 17:1167–1185.

King, G. M. 1988. Patterns of sulfate reduction and the sulfur cycle in a South Carolina salt marsh. Limnology and Oceanography 33:376–390.

King, G. M. 1992. Ecological aspects of methane oxidation, a key determinant of global methane dynamics. Advances in Microbial Ecology 12:431–468.

King, G. M., and A.P.S. Adamsen. 1992. Effects of temperature on methane consumption in a forest soil and in pure cultures of the methanotroph *Methylomonas rubra*. Applied and Environmental Microbiology 58:2758–2763.

King, G. M., P. Roslev, and H. Skovgaard. 1990. Distribution and rate of methane oxidation in sediments of the Florida Everglades. Applied and Environmental Microbiology 56:2902–2911.

King, G. M., and S. Schnell. 1994. Effect of increasing atmospheric methane concentration on ammonium inhibition of soil methane consumption. Nature 370:282–284.

King, G. M., and W. J. Wiebe. 1978. Methane release from soils of a Georgia salt marsh. Geochimica et Cosmochimica Acta 42:343–348.

King, J. K., J. E. Kostka, M. E. Frischer, F. M. Saunders, and R. A. Jahnke. 2001. A quantitative relationship that demonstrates mercury methylation rates in marine sediments are based on the community composition and activity of sulfate-reducing bacteria. Environmental Science & Technology 35:2491–2496.

King, J. S., M. E. Kubiske, K. S. Pregitzer, G. R. Hendrey, E. P. McDonald, C. P. Giardina, V. S. Quinn, and D. F. Karnosky. 2005. Tropospheric O_3 compromises net primary production in young stands of trembling aspen, paper birch and sugar maple in response to elevated atmospheric CO_2. New Phytologist 168:623–635.

Kinsman, D.J.J. 1969. Interpretation of Sr^{2+} concentrations in carbonate minerals and rocks. Journal of Sedimentary Petrology 39:486–508.

Kira, T., and T. Shidei. 1967. Primary production and turnover of organic matter in different forest ecosystems of the western Pacific. Japanese Journal of Ecology 17:70–87.

Kirchman, D. L., H. W. Ducklow, J. J. McCarthy, and C. Garside. 1994. Biomass and nitrogen uptake by heterotrophic bacteria during the spring phytoplankton bloom in the North Atlantic ocean. Deep-Sea Research Part I—Oceanographic Research Papers 41:879–895.

Kirchman, D. L., Y. Suzuki, C. Garside, and H. W. Ducklow. 1991. High turnover rates of dissolved organic carbon during a spring phytoplankton bloom. Nature 352:612–614.

Kirk, G.J.D., P. H. Bellamy, and R. M. Lark. 2010. Changes in soil pH across England and Wales in response to decreased acid deposition. Global Change Biology 16:3111–3119.

Kirschbaum, M.U.F. 1995. The temperature dependence of soil organic matter decomposition, and the effect of global warming on soil organic-C storage. Soil Biology & Biochemistry 27:753–760.

Kirschbaum, M.U.F., D. Bruhn, D. M. Etheridge, J. R. Evans, G. D. Farquhar, R. M. Gifford, K. I. Paul, and A. J. Winters. 2006. A comment on the quantitative significance of aerobic methane release by plants. Functional Plant Biology 33:521–530.

Kirschvink, J. L., E. J. Gaidos, L. E. Bertani, N. J. Beukes, J. Gutzmer, L. N. Maepa, and R. E. Steinberger. 2000. Paleoproterozoic snowball Earth: Extreme climatic and geochemical global change and its biological consequences. Proceedings of the National Academy of Sciences 97:1400–1405.

Kirwan, M. L., and L. K. Blum. 2011. Enhanced decomposition offsets enhanced productivity and soil carbon accumulation in coastal wetlands responding to climate change. Biogeosciences 8:987–993.

Kirwan, M. L., and A. B. Murray. 2007. A coupled geomorphic and ecological model of tidal marsh evolution. Proceedings of the National Academy of Sciences 104:6118–6122.

Kitayama, K., N. Majalap-Lee, and S. Aiba. 2000. Soil phosphorus fractionation and phosphorus-use efficiencies of tropical rainforests along altitudinal gradients of Mount Kinabalu, Borneo. Oecologia 123:342–349.

Kitchell, J. F., and S. R. Carpenter. 1993. Cascading trophic interactions. *In* S. R. Carpenter and J. F. Kitchell, editors. *The Trophic Cascade in Lakes*. Cambridge University Press, 1–14.

Kivelson, M. G., K. K. Khurana, C. T. Russell, M. Volwerk, R. J. Walker, and C. Zimmer. 2000. Galileo magnetometer measurements: A stronger case for a subsurface ocean at Europa. Science 289:1340–1343.

Kjellstrom, E., J. Feichter, R. Sausen, and R. Hein. 1999. The contribution of aircraft emissions to the atmospheric sulfur budget. Atmospheric Environment 33:3455–3465.

Klappa, C. F. 1980. Rhizoliths in terrestrial carbonates—Classification, recognition, genesis and significance. Sedimentology 27:613–629.

Klausmeier, C. A., E. Litchman, T. Daufresne, and S. A. Levin. 2004. Optimal nitrogen-to-phosphorus stoichiometry of phytoplankton. Nature **429**:171–174.

Klee, R. J., and T. E. Graedel. 2004. Elemental cycles: A status report on human or natural dominance. Annual Review of Environment and Resources **29**:69–107.

Klein, D. R. 1968. The introduction, increase, and crash of reindeer on St. Matthew Island. Journal of Wildlife Management **32**:350–367.

Kleine, T., H. Palme, M. Mezger, and A. N. Halliday. 2005. Hf-W chronometry of lunar metals and the age and early differentiation of the Moon. Science **310**:1671–1674.

Kleinman, L., Y. N. Lee, S. R. Springston, L. Nunnermacker, X. L. Zhou, R. Brown, K. Hallock, P. Klotz, D. Leahy, J. H. Lee, and L. Newman. 1994. Ozone formation at a rural site in the southeastern United States. Journal of Geophysical Research—Atmospheres **99**:3469–3482.

Kleypas, J. A., R. W. Buddemeier, D. Archer, J. P. Gattuso, C. Langdon, and B. N. Opdyke. 1999. Geochemical consequences of increased atmospheric carbon dioxide on coral reefs. Science **284**:118–120.

Kling, G. W., W. C. Evans, G. Tanyileke, M. Kusakabe, T. Ohba, Y. Yoshida, and J. V. Hell. 2005. Degassing Lakes Nyos and Monoun: Defusing certain disaster. Proceedings of the National Academy of Sciences **102**:14185–14190.

Klinkhammer, G. P. 1980. Early diagenesis in sediments from the eastern equatorial Pacific II. Pore water metal results. Earth and Planetary Science Letters **49**:81–101.

Kludze, H. K., R. D. Delaune, and W. H. Patrick. 1993. Aerenchyma formation and methane and oxygen exchange in rice. Soil Science Society of America Journal **57**:386–391.

Kminek, G., and J. L. Bada. 2006. The effect of ionizing radiation on the preservation of amino acids on Mars. Earth and Planetary Science Letters **245**:1–5.

Knapp, A. K., and M. D. Smith. 2001. Variation among biomes in temporal dynamics of aboveground primary production. Science **291**:481–484.

Knauer, G. A. 1993. Productivity and new production of the oceanic systems. *In* R. Wollast, F. T. MacKenzie and L. Chou, editors. *Interactions of C, N, P and S Biogeochemical Cycles and Global Change*. Springer, 211–231.

Knauss, J. A. 1978. *Introduction to Physical Oceanography*. Prentice-Hall .

Knauth, L. P., and M. J. Kennedy. 2009. The late Precambrian greening of the Earth. Nature **460**:728–732.

Knight, D. H., T. J. Fahey, and S. W. Running. 1985. Water and nutrient outflow from contrasting lodgepole pine forests in Wyoming. Ecological Monographs **55**:29–48.

Knoll, A. H. 1992. The early evolution of eukaryotes—A geological perspective. Science **256**:622–627.

Knoll, A. H. 2003. *Life on a Young Planet*. Princeton University Press.

Knoll, M. A., and W. C. James. 1987. Effect of the advent and diversification of vascular land plants on mineral weathering through geological time. Geology **15**:1099–1102.

Knöller, K., C. Vogt, M. Haupt, S. Feisthauer, and H. H. Richnow. 2011. Experimental investigation of nitrogen and oxygen isotope fractionation in nitrate and nitrite during denitrification. Biogeochemistry **103**:371–384.

Knowles, R. 1982. Denitrification. Microbiological Reviews **46**:43–70.

Koba, K., Y. Fang, J. Mo, W. Zhang, X. Lu, L. Liu, T. Zhang, Y. Takebayashi, S. Toyoda, N. Yoshida, K. Suzuki, M. Yoh, and K. Senoo. 2012. The ^{15}N natural abundance of the N lost from an N-saturated subtropical forest in southern China. Journal of Geophysical Research—Biogeosciences **117**.

Koba, K., K. Osaka, Y. Tobari, S. Toyoda, N. Ohte, M. Katsuyama, N. Suzuki, M. Itoh, H. Yamagishi, M. Kawasaki, S. J. Kim, N. Yoshida, and T. Nakajima. 2009. Biogeochemistry of nitrous oxide in groundwater in a forested ecosystem elucidated by nitrous oxide isotopomer measurements. Geochimica et Cosmochimica Acta **73**: 3115–3133.

Koba, K., N. Tokuchi, T. Yoshioka, E. A. Hobbie, and G. Iwatsubo. 1998. Natural abundance of nitrogen-15 in a forest soil. Soil Science Society of America Journal **62**:778–781.

Kobe, R. K., C. A. Lepczyk, and M. Iyer. 2005. Resorption efficiency decreases with increasing green leaf nutrients in a global data set. Ecology **86**: 2780–2792.

Kodama, H., and M. Schnitzer. 1977. Effect of fulvic acid on crystallization of Fe (III) oxides. Geoderma **19**: 279–291.

Kodama, H., and M. Schnitzer. 1980. Effect of fulvic acid on the crystallization of aluminum hydroxides. Geoderma **24**:195–205.

Koehler, B., M. D. Corre, E. Veldkamp, H. Wullaert, and S. J. Wright. 2009. Immediate and long-term nitrogen oxide emissions from tropical forest soils exposed to elevated nitrogen input. Global Change Biology **15**: 2049–2066.

Koerselman, W., and A.F.M. Meuleman. 1996. The vegetation N:P ratio: A new tool to detect the nature of nutrient limitation. Journal of Applied Ecology **33**: 1441–1450.

Kohfeld, K. E., C. Le Quere, S. P. Harrison, and R. F. Anderson. 2005. Role of marine biology in glacial-interglacial CO_2 cycles. Science 308:74–78.

Kohler, I. H., P. R. Poulton, K. Auerswald, and H. Schnyder. 2010. Intrinsic water-use efficiency of temperate seminatural grassland has increased since 1857: an analysis of carbon isotope discrimination of herbage from the Park Grass Experiment. Global Change Biology 16:1531–1541.

Kohler, P., and H. Fischer. 2004. Simulating changes in the terrestrial biosphere during the last glacial/interglacial transition. Global and Planetary Change 43:33–55.

Koide, R. T. 1991. Nutrient supply, nutrient demand and plant response to mycorrhizal infection. New Phytologist 117:365–386.

Kokaly, R. F., G. P. Asner, S. V. Ollinger, M. E. Martin, and C. A. Wessman. 2009. Characterizing canopy biochemistry from imaging spectroscopy and its application to ecosystem studies. Remote Sensing of Environment 113:S78–S91.

Kolari, P., J. Pumpanen, U. Rannik, H. Ilvesniemi, P. Hari, and F. Berninger. 2004. Carbon balance of different aged Scots pine forests in southern Finland. Global Change Biology 10:1106–1119.

Kolber, Z. S., F. G. Plumley, A. S. Lang, J. T. Beatty, R. E. Blankenship, C. L. VanDover, C. Vetriani, M. Koblizek, C. Rathgeber, and P. G. Falkowski. 2001. Contribution of aerobic photoheterotrophic bacteria to the carbon cycle in the ocean. Science 292:2492–2495.

Kolber, Z. S., C. L. Van Dover, R. A. Niederman, and P. G. Falkowski. 2000. Bacterial photosynthesis in surface waters of the open ocean. Nature 407:177–179.

Kone, Y.J.M., and A. V. Borges. 2008. Dissolved inorganic carbon dynamics in the waters surrounding forested mangroves of the Ca Mau Province (Vietnam). Estuarine Coastal and Shelf Science 77:409–421.

Konhauser, K. O., S. V. Lalonde, L. Amskold, and H. D. Holland. 2007. Was there really an Archean phosphate crisis? Science 315:1234.

Konhauser, K. O., S. V. Lalonde, N. J. Planavsky, E. Pecoits, T. W. Lyons, S. J. Mojzsis, O. J. Rouxel, M. E. Barley, C. Rosiere, P. W. Fralick, L. R. Kump, and A. Bekker. 2011. Aerobic bacterial pyrite oxidation and acid rock drainage during the Great Oxidation Event. Nature 478:369–373.

Konhauser, K. O., E. Pecoits, S. V. Lalonde, D. Papineau, E. G. Nisbet, M. E. Barley, N. T. Arndt, K. Zahnle, and B. S. Kamber. 2009. Oceanic nickel depletion and a methanogen famine before the Great Oxidation Event. Nature 458:750–753.

Konikow, L. F. 2011. Contribution of global groundwater depletion since 1900 to sea-level rise. Geophysical Research Letters 38.

Konneke, M., A. E. Bernhard, J. R. de la Torre, C. B. Walker, J. B. Waterbury, and D. A. Stahl. 2005. Isolation of an autotrophic ammonia-oxidizing marine archaeon. Nature 437:543–546.

Kool, D. M., J. Dolfing, N. Wrage, and J. W. Van Groenigen. 2011. Nitrifier denitrification as a distinct and significant source of nitrous oxide from soil. Soil Biology & Biochemistry 43:174–178.

Körner, C. 1989. The nutritional status of plants from high altitudes—A worldwide comparison. Oecologia 81:379–391.

Kort, E. A., P. K. Patra, K. Ishijima, B. C. Daube, R. Jimenez, J. Elkins, D. Hurst, F. L. Moore, C. Sweeney, and S. C. Wofsy. 2011. Tropospheric distribution and variability of N_2O: Evidence for strong tropical emissions. Geophysical Research Letters 38.

Koschel, R., J. Benndorf, G. Proft, and F. Recknagel. 1983. Calcite precipitation as a natural control mechanism of eutrophication. Archiv Fur Hydrobiologie 98: 380–408.

Koschorreck, M., and R. Conrad. 1993. Oxidation of atmospheric methane in soil—Measurements in the field, in soil cores and in soil samples. Global Biogeochemical Cycles 7:109–121.

Koster, R. D., P. A. Dirmeyer, Z. C. Guo, G. Bonan, E. Chan, P. Cox, C. T. Gordon, S. Kanae, E. Kowalczyk, D. Lawrence, P. Liu, C. H. Lu, S. Malyshev, B. McAvaney, K. Mitchell, D. Mocko, T. Oki, K. Oleson, A. Pitman, Y. C. Sud, C. M. Taylor, D. Verseghy, R. Vasic, Y. K. Xue, T. Yamada, and G. Team. 2004. Regions of strong coupling between soil moisture and precipitation. Science 305:1138–1140.

Kostka, J. E., A. Roychoudhury, and P. Van Cappellen. 2012. Rates and controls of anaerobic microbial respiration across spatial and temporal gradients in salt-marsh sediments. Biogeochemistry 60:49–76.

Kostner, B., R. Schupp, E. D. Schulze, and H. Rennenberg. 1998. Organic and inorganic sulfur transport in the xylem sap and the sulfur budget of *Picea abies* trees. Tree Physiology 18:1–9.

Koven, C. D., B. Ringeval, P. Friedlingstein, P. Ciais, P. Cadule, D. Khvorostyanov, G. Krinner, and C. Tarnocai. 2011. Permafrost carbon-climate feedbacks accelerate global warming. Proceedings of the National Academy of Sciences 108:14769–14774.

Krabill, W., W. Abdalati, E. Frederick, S. Manizade, C. Martin, J. Sonntag, R. Swift, R. Thomas, W. Wright, and J. Yungel. 2000. Greenland ice sheet: High-elevation balance and peripheral thinning. Science 289:428–430.

Krajewski, K. P., P. Van Cappellen, J. Trichet, O. Kuhn, J. Lucas, A. Martinalgarra, L. Prevot, V. C. Tewari, L. Gaspar, R. I. Knight, and M. Lamboy. 1994.

Biological processes and apatite formation in sedimentary environments. Eclogae Geologicae Helvetiae **87**: 701–745.

Krakauer, N. Y., and I. Fung. 2008. Mapping and attribution of change in streamflow in the coterminous United States. Hydrology and Earth System Sciences **12**:1111–1120.

Kral, T. A., K. M. Brink, S. L. Miller, and C. P. McKay. 1998. Hydrogen consumptions by methanogens on the early Earth. Origins of Life and Evolution of the Biosphere **28**:311–319.

Kramer, J., P. Laan, G. Sarthou, K. R. Timmermans, and H.J.W. de Baar. 2004. Distribution of dissolved aluminium in the high atmospheric input region of the subtropical waters of the North Atlantic Ocean. Marine Chemistry **88**:85–101.

Kramer, M. G., P. Sollins, R. S. Sletten, and P. K. Swart. 2003. N isotope fractionation and measures of organic matter alteration during decomposition. Ecology **84**:2021–2025.

Kramer, P. J. 1982. Water and plant productivity of yield. *In* M. Rechcigl, editor. *Handbook of Agricultural Productivity*. CRC Press, 41–47.

Kramers, J. D. 2003. Volatile element abundance patterns and an early liquid water ocean on Earth. Precambrian Research **126**:379–394.

Krasnopolsky, V. A., and P. D. Feldman. 2001. Detection of molecular hydrogen in the atmosphere of Mars. Science **294**:1914–1917.

Kratz, T. R., and C. B. DeWitt. 1986. Internal factors controlling peatland-lake ecosystem development. Ecology **67**:100–107.

Krause, H. H. 1982. Nitrate formation and movement before and after clear cutting of a monitored watershed in central New Brunswick, Canada. Canadian Journal of Forest Research **12**:922–930.

Kreibich, H., J. Kern, P. B. de Camargo, M. Z. Moreira, R. L. Victoria, and D. Werner. 2006. Estimation of symbiotic N_2 fixation in an Amazon floodplain forest. Oecologia **147**:359–368.

Krishnamurthy, A., J. K. Moore, C. S. Zender, and C. Luo. 2007. Effects of atmospheric inorganic nitrogen deposition on ocean biogeochemistry. Journal of Geophysical Research—Biogeosciences **112**.

Kristensen, E., S. I. Ahmed, and A. H. Devol. 1995. Aerobic and anaerobic decomposition of organic matter in marine sediment: Which is fastest? Limnology and Oceanography **40**:1430–1437.

Kristjansson, J. K., P. Schonheit, and R. K. Thauer. 1982. Different Ks-values for hydrogen of methanogenic bacteria and sulfate-reducing bacteria—An explanation for the apparent inhibiton of methanogenesis by sulfate. Archives of Microbiology **131**:278–282.

Kroeger, K. D., and M. A. Charette. 2008. Nitrogen biogeochemistry of submarine groundwater discharge. Limnology and Oceanography **53**:1025–1039.

Kroehler, C. J., and A. E. Linkins. 1991. The absorption of inorganic phosphate from ^{32}P-labeled inositol hexaphosphate by *Eriophorum vaginatum*. Oecologia **85**: 424–428.

Kroer, N., N.O.G. Jorgensen, and R. B. Coffin. 1994. Utilization of dissolved nitrogen by heterotrophic bacterioplankton: A comparison of three ecosystems. Applied and Environmental Microbiology **60**:4116–4123.

Krom, M. D., and R. A. Berner. 1981. The diagenesis of phosphorus in a nearshore marine sediment. Geochimica et Cosmochimica Acta **45**:207–216.

Kronvang, B., E. Jeppesen, D. J. Conley, M. Sondergaard, S. E. Larsen, N. B. Ovesen, and J. Carstensen. 2005. Nutrient pressures and ecological responses to nutrient loading reductions in Danish streams, lakes and coastal waters. Journal of Hydrology **304**:274–288.

Kronzucker, H. J., M. Y. Siddiqi, and A.D.M. Glass. 1997. Conifer root discrimination against soil nitrate and the ecology of forest succession. Nature **385**:59–61.

Krouse, H. R., and R.G.L. McCready. 1979. Reductive reactions in the sulfur cycle. *In* P. A. Trudinger and D. J. Swaine, editors. *Biogeochemical Cycling of Mineral-Forming Elements*. Elsevier, 315–368.

Krumbein, W. E. 1971. Manganese-oxidizing fungi and bacteria in recent shelf sediments of the Bay of Biscay and the North Sea. Naturwissenschaften **58**: 56–57.

Krumbein, W. E. 1979. Calcification by bacteria and algae. *In* P. A. Trudinger and D. J. Swaine, editors. *Biogeochemical Cycling of Mineral-Forming Elements*. Elsevier, 47–68.

Krumholz, L. R., J. P. McKinley, F. A. Ulrich, and J. M. Suflita. 1997. Confined subsurface microbial communities in Cretaceous rock. Nature **386**:64–66.

Krysell, M., and D.W.R. Wallace. 1988. Arctic ocean ventilation studied with a suite of anthropogenic halocarbon tracers. Science **242**:746–749.

Ku, T.C.W., L. M. Walter, M. L. Coleman, R. E. Blake, and A. M. Martini. 1999. Coupling between sulfur recycling and syndepositional carbonate dissolution: Evidence from oxygen and sulfur isotope composition of pore water sulfate, South Florida Platform, USA. Geochimica et Cosmochimica Acta **63**:2529–2546.

Kuhlbrandt, W., D. N. Wang, and Y. Fujiyoshi. 1994. Atomic model of the plant light-harvesting complex by electron crystallography. Nature **367**:614–621.

Kuhlbusch, T. A., J. M. Lobert, P. J. Crutzen, and P. Warneck. 1991. Molecular nitrogen emissions from denitrificaiton during biomass burning. Nature **351**: 135–137.

Kuhn, U., C. Ammann, A. Wolf, F. X. Meixner, M. O. Andreae, and J. Kesselmeier. 1999. Carbonyl sulfide exchange on an ecosystem scale: soil represents a dominant sink for atmospheric COS. Atmospheric Environment 33:995–1008.

Kuhry, P., and D. H. Vitt. 1996. Fossil carbon/nitrogen ratios as a measure of peat decomposition. Ecology 77:271–275.

Kuivila, K. M., J. W. Murray, A. H. Devol, and P. C. Novelli. 1989. Methane production, sulfate reduction and competition for substrates in the sediments of Lake Washington. Geochimica et Cosmochimica Acta 53:409–416.

Kujawinski, E. B., R. Del Vecchio, N. V. Blough, G. C. Klein, and A. G. Marshall. 2004. Probing molecular-level transformations of dissolved organic matter: insights on photochemical degradation and protozoan modification of DOM from electrospray ionization Fourier transform ion cyclotron resonance mass spectrometry. Marine Chemistry 92:23–37.

Kulkarni, S. R. 1997. Brown dwarfs: A possible missing link between stars and planets. Science 276:1350–1354.

Kumar, N., R. F. Anderson, R. A. Mortlock, P. N. Froeroelich, P. Kubik, B. Dittrichhannen, and M. Suter. 1995a. Increased biological productivity and export production in the glacial Southern Ocean. Nature 378:675–680.

Kumar, P., A. Robins, S. Vardoulakis, and R. Britter. 2010. A review of the characteristics of nanoparticles in the urban atmosphere and the prospects for developing regulatory controls. Atmospheric Environment 44:5035–5052.

Kumar, P. P., G. K. Manohar, and S. S. Kandalgaonkar. 1995b. Global distribution of nitric oxide produced by lightning and its seasonal variation. Journal of Geophysical Research—Atmospheres 100:11203–11208.

Kump, L. R. 1988. Terrestrial feedback in atmospheric oxygen regulation by fire and phosphorus. Nature 335:152–154.

Kump, L. R., S. L. Brantley, and M. A. Arthur. 2000. Chemical, weathering, atmospheric CO_2 and climate. Annual Review of Earth and Planetary Sciences 28: 611–667.

Kump, L. R., and R. M. Garrels. 1986. Modeling atmospheric O_2 in the global sedimentary redox cycle. American Journal of Science 286:337–360.

Kump, L. R., J. F. Kasting, and R. G. Crane. 2010. *The Earth System*. Pearson/Prentice Hall.

Kump, L. R., C. Junium, M. A. Arthur, A. Brasier, A. Fallick, V. Melezhik, A. Lepland, A. E. Crne, and G. M. Luo. 2011. Isotopic evidence for massive oxidation of organic matter following the Great Oxidation Event. Science 334:1694–1696.

Kunishi, H. M. 1988. Sources of nitrogen and phosphorus in an estuary of the Chesapeake Bay. Journal of Environmental Quality 17:185–188.

Kunz, J., T. Staudacher, and C. J. Allegre. 1998. Plutonium-fission xenon found in Earth's mantle. Science 280:877–880.

Kurc, S. A., and E. E. Small. 2004. Dynamics of evapotranspiration in semiarid grassland and shrubland ecosystems during the summer monsoon season, central New Mexico. Water Resources Research 40.

Kurz, H., and D. Demaree. 1934. Cypress buttresses and knees in relation to water and air. Ecology 15:36–41.

Kurz, W. A., C. C. Dymond, G. Stinson, G. J. Rampley, E. T. Neilson, A. L. Carroll, T. Ebata, and L. Safranyik. 2008. Mountain pine beetle and forest carbon feedback to climate change. Nature 452:987–990.

Kuss, J., C. Zuelicke, C. Pohl, and B. Schneider. 2011. Atlantic mercury emission determined from continuous analysis of the elemental mercury sea–air concentration difference within transects between 50°N and 50°S. Global Biogeochemical Cycles 25.

Kuypers, M.M.M., G. Lavik, D. Woebken, M. Schmid, B. M. Fuchs, R. Amann, B. B. Jorgensen, and M.S.M. Jetten. 2005. Massive nitrogen loss from the Benguela upwelling system through anaerobic ammonium oxidation. Proceedings of the National Academy of Sciences 102:6478–6483.

Kuypers, M.M.M., A. O. Sliekers, G. Lavik, M. Schmid, B. B. Jorgensen, J. G. Kuenen, J.S.S. Damste, M. Strous, and M.S.M. Jetten. 2003. Anaerobic ammonium oxidation by anammox bacteria in the Black Sea. Nature 422:608–611.

Kvenvolden, K., J. Lawless, K. Pering, E. Peterson, J. Flores, C. Ponnamperuma, I. R. Kaplan, and C. Moore. 1970. Evidence for extraterrestrial amino acids and hydrocarbons in the Murchison meteorite. Nature 228:923–926.

Kwok, R., and D. A. Rothrock. 2009. Decline in Arctic sea ice thickness from submarine and ICESat records: 1958-2008. Geophysical Research Letters 36.

Laanbroek, H. J. 2010. Methane emission from natural wetlands: Interplay between emergent macrophytes and soil microbial processes. A mini-review. Annals of Botany 105:141–153.

Labat, D., Y. Godderis, J. L. Probst, and J. L. Guyot. 2004. Evidence for global runoff increase related to climate warming. Advances in Water Resources 27:631–642.

Labrenz, M., G. K. Druschel, T. Thomsen-Ebert, B. Gilbert, S. A. Welch, K. M. Kemner, G. A. Logan, R. E. Summons, G. De Stasio, P. L. Bond, B. Lai, S. D. Kelly, and J. F. Banfield. 2000. Formation of sphalerite (ZnS) deposits in natural biofilms of sulfate-reducing bacteria. Science 290:1744–1747.

Lacis, A. A., G. A. Schmidt, D. Rind, and R. A. Ruedy. 2010. Atmospheric CO_2: Principal control knob governing Earth's temperature. Science 330:356–359.

Laclau, J. P., J. P. Bouillet, J. Ranger, R. Joffre, R. Gouma, and A. Saya. 2001. Dynamics of nutrient translocation in stemwood across an age series of a eucalyptus hybrid. Annals of Botany 88:1079–1092.

Ladekarl, U. L. 1998. Estimation of the components of soil water balance in a Danish oak stand from measurements of soil moisture using TDR. Forest Ecology and Management 104:227–238.

Lagage, P. O., and E. Pantin. 1994. Dust depletion in the inner disk of Beta Pictoris as a possible indicator of planets. Nature 369:628–630.

Lagrange, A. M., M. Bonnefoy, G. Chauvin, D. Apai, D. Ehrenreich, A. Boccaletti, D. Gratadour, D. Rouan, D. Mouillet, S. Lacour, and M. Kasper. 2010. A giant planet imaged in the disk of the young star Beta Pictoris. Science 329:57–59.

Lahteenoja, O., K. Ruokolainen, L. Schulman, and M. Oinonen. 2009. Amazonian peatlands: an ignored C sink and potential source. Global Change Biology 15:2311–2320.

Lajewski, C. K., H. T. Mullins, W. P. Patterson, and C. W. Callinan. 2003. Historic calcite record from the Finger Lakes, New York: Impact of acid rain on a buffered terrain. Geological Society of America Bulletin 115:373–384.

Lajtha, K. 1987. Nutrient-reabsorption efficiency and the response to phosphorus fertilization in the desert shrub Larrea tridentata (DC) Cov. Biogeochemistry 4:265–276.

Lajtha, K., and S. H. Bloomer. 1988. Factors affecting phosphate sorption and phosphate retention in a desert ecosystem. Soil Science 146:160–167.

Lajtha, K., and W. H. Schlesinger. 1986. Plant response to variations in nitrogen availability in a desert shrubland community. Biogeochemistry 2:29–37.

Lajtha, K., and W. G. Whitford. 1989. The effect of water and nitrogen amendments on photosynthesis, leaf demography, and resource-use efficiency in Larrea tridentata, a desert evergreen shrub. Oecologia 80: 341–348.

Lal, D. 1977. Oceanic microcosm of particles. Science 198:997–1009.

Lal, R. 2004. Soil carbon sequestration in India. Climatic Change 65:277–296.

Lalonde, K., A. Mucci, A. Ouellet, and Y. Gelinas. 2012. Preservation of organic matter in sediments promoted by iron. Nature 483:198–200.

Lamb, B., H. Westberg, G. Allwine, L. Bamesberger, and A. Guenther. 1987. Measurement of biogenic sulfur emissions from soils and vegetation—Application of dynamic enclosure methods with Natusch filter and GC/FPD Analysis. Journal of Atmospheric Chemistry 5:469–491.

Lamb, D. 1985. The influence of insects on nutrient cycling in eucalypt forests—A beneficial role. Australian Journal of Ecology 10:1–5.

Lambeck, K., T. M. Esat, and E. K. Potter. 2002. Links between climate and sea levels for the past three million years. Nature 419:199–206.

Lambers, H., J. A. Raven, G. R. Shaver, and S. E. Smith. 2008. Plant nutrient-acquisition strategies change with soil age. Trends in Ecology & Evolution 23:95–103.

Lambert, F., B. Delmonte, J. R. Petit, M. Bigler, P. R. Kaufmann, M. A. Hutterli, T. F. Stocker, U. Ruth, J. P. Steffensen, and V. Maggi. 2008. Dust-climate couplings over the past 800,000 years from the EPICA Dome C ice core. Nature 452:616–619.

Lambert, R. L., G. E. Lang, and W. A. Reiners. 1980. Loss of mass and chemical change in decaying boles of a subalpine balsam fir forest. Ecology 61:1460–1473.

Lamers, L.P.M., H.B.M. Tomassen, and J.G.M. Roelofs. 1998. Sulfate-induced entrophication and phytotoxicity in freshwater wetlands. Environmental Science & Technology 32:199–205.

Lana, A., T. G. Bell, R. Simo, S. M. Vallina, J. Ballabrera-Poy, A. J. Kettle, J. Dachs, I. Bopp, E. S. Saltzman, J. Stefels, J. E. Johnson, and P. S. Liss. 2011. An updated climatology of surface dimethylsulfide concentrations and emission fluxes in the global ocean. Global Biogeochemical Cycles 25.

Landais, A., J. Lathiere, E. Barkan, and B. Luz. 2007. Reconsidering the change in global biosphere productivity between the Last Glacial Maximum and present day from the triple oxygen isotopic composition of air trapped in ice cores. Global Biogeochemical Cycles 21.

Lane, L. J., E. M. Romney, and T. E. Hakonson. 1984. Water-balance calculations and net production of perennial vegetation in the northern Mojave desert. Journal of Range Management 37:12–18.

Lane, T. W., M. A. Saito, G. N. George, I. J. Pickering, R. C. Prince, and F.M.M. Morel. 2005. A cadmium enzyme from a marine diatom. Nature 435:42.

Lang, G. E., and R. T. Forman. 1978. Detrital dynamics in a mature oak forest—Hutcheson Memorial forest, New Jersey. Ecology 59:580–595.

Lang, S. Q., D. A. Butterfield, M. Schulte, D. S. Kelley, and M. D. Lilley. 2010. Elevated concentrations of formate, acetate and dissolved organic carbon found at the Lost City hydrothermal field. Geochimica et Cosmochimica Acta 74:941–952.

Lange, O. L., E. D. Schulze, M. Evenari, L. Kappen, and U. Buschbom. 1974. The temperature-related photosynthetic capacity of plants under desert conditions. I. Seasonal changes of photosynthetic response to temperature. Oecologia 17:97–110.

Langenfelds, R. L., R. J. Francey, L. P. Steele, M. Battle, R. F. Keeling, and W. F. Budd. 1999. Partitioning of the global fossil CO_2 sink using a 19-year trend in atmospheric O_2. Geophysical Research Letters 26:1897–1900.

Langford, A. O., and F. C. Fehsenfeld. 1992. Natural vegetation as a source or sink for atmospheric ammonia—A case study. Science 255:581–583.

Langley, J. A., K. L. McKee, D. R. Cahoon, J. A. Cherry, and J. P. Megonigal. 2009a. Elevated CO_2 stimulates marsh elevation gain, counterbalancing sea-level rise. Proceedings of the National Academy of Sciences 106:6182–6186.

Langley, J. A., D. C. McKinley, A. A. Wolf, B. A. Hungate, B. G. Drake, and J. P. Megonigal. 2009b. Priming depletes soil carbon and releases nitrogen in a scrub-oak ecosystem exposed to elevated CO_2. Soil Biology & Biochemistry 41:54–60.

Langley, J. A., and J. P. Megonigal. 2010. Ecosystem response to elevated CO_2 levels limited by nitrogen-induced plant species shift. Nature 466:96–99.

Langley-Turnbaugh, S. J., and J. G. Bockheim. 1998. Mass balance of soil evolution on late Quaternary marine terraces in coastal Oregon. Geoderma 84:265–288.

Langmann, B., A. Folch, M. Hensch, and V. Matthias. 2012. Volcanic ash over Europe during the eruption of Eyjafjallajokull on Iceland, April-May 2010. Atmospheric Environment 48:1–8.

Langner, J., H. Rodhe, P. J. Crutzen, and P. Zimmermann. 1992. Anthropogenic influence on the distribution of tropospheric sulfate aerosol. Nature 359:712–716.

Langway, C. C., K. Osada, H. B. Clausen, C. U. Hammer, and H. Shoji. 1995. A 10-century comparison of prominent bipolar volcanic events in ice cores. Journal of Geophysical Research—Atmospheres 100:16241–16247.

Langway, C. C., K. Osada, H. B. Clausen, C. U. Hammer, H. Shoji, and A. Mitani. 1994. New chemical stratigraphy over the last millennium for Byrd Station, Antarctica. Tellus Series B—Chemical and Physical Meteorology 46:40–51.

Lantzy, R. J., and F. T. MacKenzie. 1979. Atmospheric trace metals: Global cycles and assessment of man's impact. Geochimica et Cosmochimica Acta 43:511–525.

Laothawornkitkul, J., J. E. Taylor, N. D. Paul, and C. N. Hewitt. 2009. Biogenic volatile organic compounds in the Earth system. New Phytologist 183:27–51.

Lapenis, A. G., G. B. Lawrence, A. A. Andreev, A. A. Bobrov, M. S. Torn, and J. W. Harden. 2004. Acidification of forest soil in Russia: From 1893 to present. Global Biogeochemical Cycles 18.

Lapeyrie, F., G. A. Chilvers, and C. A. Bhem. 1987. Oxalic acid synthesis by the mycorrhizal fungus Paxillus involutus (Batsch ex FR) FR. New Phytologist 106:139–146.

Laronne, J. B., and I. Reid. 1993. Very high rates of bedload sediment transport by ephemeral desert rivers. Nature 366:148–150.

Lashof, D. A., and D. R. Ahuja. 1990. Relative contributions of greenhouse gas emissions to global warming. Nature 344:529–531.

Laskowski, R., M. Niklinska, and M. Maryanski. 1995. The dynamics of chemical elements in forest litter. Ecology 76:1393–1406.

Latimer, J. S., and M. A. Charpentier. 2010. Nitrogen inputs to seventy-four southern New England estuaries: Application of a watershed nitrogen loading model. Estuarine Coastal and Shelf Science 89:125–136.

Laudelout, H., and M. Robert. 1994. Biogeochemistry of calcium in a broad-leaved forest ecosystem. Biogeochemistry 27:1–21.

Lauenroth, W. K., and J. B. Bradford. 2006. Ecohydrology and the partitioning AET between transpiration and evaporation in a semiarid steppe. Ecosystems 9:756–767.

Lauenroth, W. K., and W. C. Whitman. 1977. Dynamics of dry-matter production in a mixed-grass prairie in western North Dakota. Oecologia 27:339–351.

Laurmann, J. A. 1979. Market-penetration characteristics for energy production and atmospheric carbon dioxide growth. Science 205:896–898.

Laursen, K. K., P. V. Hobbs, L. F. Radke, and R. A. Rasmussen. 1992. Some trace gas emissions from North American biomass fires with an assessment of regional and global fluxes from biomass burning. Journal of Geophysical Research—Atmospheres 97:20687–20701.

Laverman, A. M., C. Pallud, J. Abell, and P. Van Cappellen. 2012. Comparative survey of potential nitrate and sulfate reduction rates in aquatic sediments. Geochimica et Cosmochimica Acta 77:474–488.

Law, B. E., E. Falge, L. Gu, D. D. Baldocchi, P. Bakwin, P. Berbigier, K. Davis, A. J. Dolman, M. Falk, J. D. Fuentes, A. Goldstein, A. Granier, A. Grelle, D. Hollinger, I. A. Janssens, P. Jarvis, N. O. Jensen, G. Katul, Y. Mahli, G. Matteucci, T. Meyers, R. Monson, W. Munger, W. Oechel, R. Olson, K. Pilegaard, K. T. Paw, H. Thorgeirsson, R. Valentini, S. Verma, T. Vesala, K. Wilson, and S. Wofsy. 2002. Environmental controls over carbon dioxide and water vapor exchange of terrestrial vegetation. Agricultural and Forest Meteorology 113:97–120.

Law, B. E., O. J. Sun, J. Campbell, S. Van Tuyl, and P. E. Thornton. 2003. Changes in carbon storage and fluxes in a chronosequence of ponderosa pine. Global Change Biology 9:510–524.

Law, B. E., M. Williams, P. M. Anthoni, D. D. Baldocchi, and M. H. Unsworth. 2000. Measuring and modelling seasonal variation of carbon dioxide and water vapour exchange of a *Pinus ponderosa* forest subject to soil water deficit. Global Change Biology 6: 613–630.

Law, C. S., and R. D. Ling. 2001. Nitrous oxide flux and response to increased iron availability in the Antarctic Circumpolar Current. Deep-Sea Research Part II—Topical Studies in Oceanography 48:2509–2527.

Law, C. S., and N.J.P. Owens. 1990. Significant flux of atmospheric nitrous oxide from the northwest Indian ocean. Nature 346:826–828.

Law, C. S., A. P. Rees, and N.J.P. Owens. 1991a. Temporal variability of denitrification in estuarine sediments. Estuarine Coastal and Shelf Science 33:37–56.

Law, K. R., H. W. Nesbitt, and F. J. Longstaffe. 1991b. Weathering of granitic tills and the genesis of a Podzol. American Journal of Science 291:940–976.

Law, K. S., and E. G. Nisbet. 1996. Sensitivity of the CH_4 growth rate to changes in CH_4 emissions from natural gas and coal. Journal of Geophysical Research—Atmospheres 101:14387–14397.

Lawless, J. G., and N. Levi. 1979. Role of metal ions in chemical evolution—Polymerization of alanine and glycine in a cation-exchanged clay environment. Journal of Molecular Evolution 13:281–286.

Lawrence, D., and W. H. Schlesinger. 2001. Changes in soil phosphorus during 200 years of shifting cultivation in Indonesia. Ecology 82:2769–2780.

Laws, E. A., P. G. Falkowski, W. O. Smith, H. Ducklow, and J. J. McCarthy. 2000. Temperature effects on export production in the open ocean. Global Biogeochemical Cycles 14:1231–1246.

Lawson, D. R., and J. W. Winchester. 1979. Standard crustal aerosol as a reference for elemental enrichment factors. Atmospheric Environment 13:925–930.

Le Mer, J., and P. Roger. 2001. Production, oxidation, emission and consumption of methane by soils: A review. European Journal of Soil Biology 37:25–50.

Le Quere, C., M. R. Raupach, J. G. Canadell, G. Marland, L. Bopp, P. Ciais, T. J. Conway, S. C. Doney, R. A. Feely, P. Foster, P. Friedlingstein, K. Gurney, R. A. Houghton, J. I. House, C. Huntingford, P. E. Levy, M. R. Lomas, J. Majkut, N. Metzl, J. P. Ometto, G. P. Peters, I. C. Prentice, J. T. Randerson, S. W. Running, J. L. Sarmiento, U. Schuster, S. Sitch, T. Takahashi, N. Viovy, G. R. van der Werf, and F. I. Woodward. 2009. Trends in the sources and sinks of carbon dioxide. Nature Geoscience 2:831–836.

Le Toan, T., S. Quegan, M.W.J. Davidson, H. Balzter, P. Paillou, K. Papathanassiou, S. Plummer, F. Rocca, S. Saatchi, H. Shugart, and L. Ulander. 2011. The BIOMASS mission: Mapping global forest biomass to better understand the terrestrial carbon cycle. Remote Sensing of Environment 115:2850–2860.

Leahey, A. 1947. Characteristics of soils adjacent to the MacKenzie River in the Northwest Territories of Canada. Soil Science Society of America Proceedings 12:458–461.

Lean, D.R.S. 1973. Phosphorus dynamics in lake water. Science 179:678–680.

Lean, J., and D. A. Warrilow. 1989. Simulation of the regional climatic impact of Amazon deforestation. Nature 342:411–413.

LeBauer, D. S., and K. K. Treseder. 2008. Nitrogen limitation of net primary productivity in terrestrial ecosystems is globally distributed. Ecology 89:371–379.

Lebo, M. E. 1991. Particle-bound phosphorus along an urbanized coastal plain estuary. Marine Chemistry 34: 225–246.

Lebo, M. E., and J. H. Sharp. 1993. Distribution of phosphorus along the Delaware, an urbanized coastal plain estuary. Estuaries 16:290–301.

Lechowicz, M. J., and G. Bell. 1991. The ecology and genetics of fitness in forest plants. 2. Microspatial heterogeneity of the edaphic environment. Journal of Ecology 79:687–696.

Leckie, S. E., C. E. Prescott, S. J. Grayston, J. D. Neufeld, and W. W. Mohn. 2004. Comparison of chloroform fumigation-extraction, phospholipid fatty acid, and DNA methods to determine microbial biomass in forest humus. Soil Biology & Biochemistry 36: 529–532.

Lecuyer, C., L. Simon, and F. Guyot. 2000. Comparison of carbon, nitrogen and water budgets on Venus and the Earth. Earth and Planetary Science Letters 181: 33–40.

Ledwell, J. R., A. J. Watson, and C. S. Law. 1993. Evidence for slow mixing across the pycnocline from an open-ocean tracer-release experiment. Nature 364:701–703.

Lee, C., R. V. Martin, A. van Donkelaar, H. Lee, R. R. Dickerson, J. C. Hains, N. Krotkov, A. Richter, K. Vinnikov, and J. J. Schwab. 2011. SO_2 emissions and lifetimes: Estimates from inverse modeling using *in situ* and global, space-based (SCIAMACHY and OMI) observations. Journal of Geophysical Research—Atmospheres 116.

Lee, D. C., A. N. Halliday, G. A. Snyder, and L. A. Taylor. 1997. Age and origin of the moon. Science 278: 1098–1103.

Lee, D. H., J. R. Granja, J. A. Martinez, K. Severin, and M. R. Ghadiri. 1996. A self-replicating peptide. Nature 382:525–528.

Lee, J. A., and G. R. Stewart. 1978. Ecological aspects of nitrogen assimilation. Advances in Botanical Research 6:1–43.

Lee, K. 2001. Global net community production estimated from the annual cycle of surface water total dissolved inorganic carbon. Limnology and Oceanography 46:1287–1297.

Lee, S. J., Y. C. Seo, H. N. Jang, K. S. Park, J. I. Baek, H. S. An, and K. C. Song. 2006. Speciation and mass distribution of mercury in a bituminous coal-fired power plant. Atmospheric Environment 40:2215–2224.

Legendre, L. 1998. Flux of particulate organic material from the euphotic zone of oceans: Estimation from phytoplankton biomass. Journal of Geophysical Research—Oceans 103:2897–2903.

Legrand, M., M. Deangelis, T. Staffelbach, A. Neftel, and B. Stauffer. 1992. Large perturbations of ammonium and organic acids content in the Summit Greenland ice core—Fingerprint from forest fires. Geophysical Research Letters 19:473–475.

Legrand, M., and R. J. Delmas. 1987. A 220-year continuous record of volcanic H_2SO_4 in the Antarctic ice sheet. Nature 327:671–676.

Legrand, M., C. Fenietsaigne, E. S. Saltzman, C. Germain, N. I. Barkov, and V. N. Petrov. 1991. Ice-core record of oceanic emissions of dimethylsulfide during the last climate cycle. Nature 350: 144–146.

Lehman, J. T. 1980. Lower food web dynamics. Cycling of P among phytoplankton, herbivorous zooplankton, and the lake water. Pages 49-68 Special Report of the Great Lakes Research Division. University of Michigan Great Lakes and Marine Waters Center.

Lehman, J. T. 1988. Hypolimnetic metabolism in Lake Washington—Relative effects of nutrinet load and food web structure on lake productivity. Limnology and Oceanography 33:1334–1347.

Lehmann, M. F., D. M. Sigman, D. C. McCorkle, J. Granger, S. Hoffmann, G. Cane, and B. G. Brunelle. 2007. The distribution of nitrate $^{15}N/^{14}N$ in marine sediments and the impact of benthic nitrogen loss on the isotopic composition of oceanic nitrate. Geochimica et Cosmochimica Acta 71:5384–5404.

Lehner, B., K. Verdin, and A. Jarvis. 2008. New global hydrography derived from spaceborne elevation data. EOS Transactions 89:93–94.

Lehner, B., and P. Doll. 2004. Development and validation of a global database of lakes, reservoirs and wetlands. Journal of Hydrology 296:1–22.

Lehner, B., C. R. Liermann, C. Revenga, C. Vörösmarty, B. Fekete, P. Crouzet, P. Doell, M. Endejan, K. Frenken, J. Magome, C. Nilsson, J. C. Robertson, R. Roedel, N. Sindorf, and D. Wisser. 2011. High-resolution mapping of the world's reservoirs and dams for sustainable river-flow management. Frontiers in Ecology and the Environment 9:494–502.

Leighton, R. B., and B. C. Murray. 1966. Behavior of carbon dioxide and other volatiles on Mars. Science 153: 136–144.

Lein, A. Y. 1984. Anaerobic consumption of organic matter in modern marine sediments. Nature 312:148–150.

Leininger, S., T. Urich, M. Schloter, L. Schwark, J. Qi, G. W. Nicol, J. I. Prosser, S. C. Schuster, and C. Schleper. 2006. Archaea predominate among ammonia-oxidizing prokaryotes in soils. Nature 442:806–809.

Lelieveld, J., and P. J. Crutzen. 1990. Influences of cloud photochemical processes on tropospheric ozone. Nature 343:227–233.

Lelieveld, J., P. J. Crutzen, V. Ramanathan, M. O. Andreae, C.A.M. Brenninkmeijer, T. Campos, G. R. Cass, R. R. Dickerson, H. Fischer, J. A. de Gouw, A. Hansel, A. Jefferson, D. Kley, A.T.J. de Laat, S. Lal, M. G. Lawrence, J. M. Lobert, O. L. Mayol-Bracero, A. P. Mitra, T. Novakov, S. J. Oltmans, K. A. Prather, T. Reiner, H. Rodhe, H. A. Scheeren, D. Sikka, and J. Williams. 2001. The Indian Ocean experiment: Widespread air pollution from south and southeast Asia. Science 291:1031–1036.

Lelieveld, J., J. van Aardenne, H. Fischer, M. de Reus, J. Williams, and P. Winkler. 2004. Increasing ozone over the Atlantic Ocean. Science 304:1483–1487.

Leman, L., L. Orgel, and M. R. Ghadiri. 2004. Carbonyl sulfide-mediated prebiotic formation of peptides. Science 306:283–286.

Lengke, M., and G. Southam. 2006. Bioaccumulation of gold by sulfate-reducing bacteria cultured in the presence of gold(I)-thio sulfate complex. Geochimica et Cosmochimica Acta 70:3646–3661.

Lenton, T. M. 1998. Gaia and natural selection. Nature 394:439–447.

Lenton, T. M. 2001. The role of land plants, phosphorus weathering and fire in the rise and regulation of atmospheric oxygen. Global Change Biology 7:613–629.

Leopold, L. B., M. G. Wolman, and J. R. Miller. 1964. Fluvial Processes in Geomorphology. W.H. Freeman.

Leopoldo, P. R., W. K. Franken, and N.A.V. Nova. 1995. Real evapotranspiration and transpiration through a tropical rain forest in central Amazonia as estimated by the water-balance method. Forest Ecology and Management 73:185–195.

Leri, A. C., and S.C.B. Myneni. 2010. Organochlorine turnover in forest ecosystems: The missing link in the terrestrial chlorine cycle. Global Biogeochemical Cycles 24.

Lerner, J., E. Matthews, and I. Fung. 1988. Methane emission from animals: A global high-resolution data base. Global Biogeochemical Cycles 2:139–156.

Lesack, L.F.W., and J. M. Melack. 1991. The deposition, composition, and potential sources of major ionic solutes in rain of the central Amazon basin. Water Resources Research 27:2953–2977.

Lesack, L.F.W., and J. M. Melack. 1996. Mass balance of major solutes in a rainforest catchment in the Central Amazon: Implications for nutrient budgets in tropical rainforests. Biogeochemistry 32:115–142.

Leschine, S. B., K. Holwell, and E. Canaleparola. 1988. Nitrogen fixation by anaerobic cellulolytic bacteria. Science 242:1157–1159.

Lesschen, J. P., G. L. Velthof, W. de Vries, and J. Kros. 2011. Differentiation of nitrous oxide emission factors for agricultural soils. Environmental Pollution 159:3215–3222.

Leuenberger, M., and U. Siegenthaler. 1992. Ice age atmospheric concentration of nitrous oxide from an Antarctic ice core. Nature 360:449–451.

Leuenberger, M., U. Siegenthaler, and C. C. Langway. 1992. Carbon isotope composition of atmospheric CO_2 during the last ice age from an Antarctic ice core. Nature 357:488–490.

Levesque, C., H. Limen, and S. K. Juniper. 2005. Origin, composition and nutritional quality of particulate matter at deep-sea hydrothermal vents on Axial Volcano, NE Pacific. Marine Ecology—Progress Series 289:43–52.

Levia, D. F., and E. E. Frost. 2003. A review and evaluation of stemflow literature in the hydrologic and biogeochemical cycles of forested and agricultural ecosystems. Journal of Hydrology 274:1–29.

Levin, S. A. 1989. Challenges in the development of a theory of community and ecosystem structure and function. In J. Roughgarden, R. M. May and S. A. Levin, editors. Perspectives in Ecological Theory. Princeton University Press, 242–255.

Levine, J. S., W. R. Cofe, D. L. Sebacher, E. L. Winstead, S. Sebacher, and P. J. Boston. 1988. The effects of fire on biogenic soil emissions of nitric oxide and nitrous oxide. Global Biogeochemical Cycles 2:445–449.

Levine, J. S., W. R. Cofer, and J. P. Pinto. 1993. Biomass burning. In M.A.K. Khalil, editor. Atmospheric Methane: Sources, Sinks and Role in Global Change. Springer, 299–313.

Levine, J. S., W. R. Cofer, D. I. Sebacher, R. P. Rhinehart, E. L. Winstead, S. Sebacher, C. R. Hinkle, P. A. Schmalzer, and A. M. Koller. 1990. The effects of fire on biogenic emissions of methane and nitric oxide from wetlands. Journal of Geophysical Research—Atmospheres 95:1853–1864.

Levine, S. N., M. P. Stainton, and D. W. Schindler. 1986. A radiotracer study of phosphorus cycling in a eutrophic Canadian shield lake, Lake-227, northwestern Ontario. Canadian Journal of Fisheries and Aquatic Sciences 43:366–378.

Levitus, S., J. Antonov, and T. Boyer. 2005. Warming of the world ocean, 1955–2003. Geophysical Research Letters 32.

Levitus, S., J. I. Antonov, J. L. Wang, T. L. Delworth, K. W. Dixon, and A. J. Broccoli. 2001. Anthropogenic warming of Earth's climate system. Science 292:267–270.

Levy, H., P. S. Kasibhatla, W. J. Moxim, A. A. Klonecki, A. I. Hirsch, S. J. Oltmans, and W. L. Chameides. 1997. The global impact of human activity on tropospheric ozone. Geophysical Research Letters 24:791–794.

Levy, H., W. J. Moxim, and P. S. Kasibhatla. 1996. A global three-dimensional time-dependent lightning source of tropospheric NOx. Journal of Geophysical Research—Atmospheres 101:22911–22922.

Levy, H., W. Moxim, A. Klonecki, and P. S. Kasibhatla. 1999. Simulated tropospheric NO_x: Its evaluation, global distribution and individual source contributions. Journal of Geophysical Research 104:26279–26306.

Levy, P. E., A. Burden, M.D.A. Cooper, K. J. Dinsmore, J. Drewer, C. Evans, D. Fowler, J. Gaiawyn, A. Gray, S. K. Jones, T. Jones, N. P. McNamara, R. Mills, N. Ostle, L. J. Sheppard, U. Skiba, A. Sowerby, S. E. Ward, and P. Zielinski. 2012. Methane emissions from soils: synthesis and analysis of a large UK data set. Global Change Biology 18:1657–1669.

Lewis, J. S., and R. G. Prinn. 1984. Planets and Their Atmospheres. Academic Press.

Lewis, M. R., W. G. Harrison, N. S. Oakey, D. Hebert, and T. Platt. 1986. Vertical nitrate fluxes in the oligotrophic ocean. Science 234:870–873.

Lewis, S. L., G. Lopez-Gonzalez, B. Sonke, K. Affum-Baffoe, T. R. Baker, L. O. Ojo, O. L. Phillips, J. M. Reitsma, L. White, J. A. Comiskey, M. N. Djuikouo, C.E.N. Ewango, T. R. Feldpausch, A. C. Hamilton, M. Gloor, T. Hart, A. Hladik, J. Lloyd, J. C. Lovett, J. R. Makana, Y. Malhi, F. M. Mbago, H. J. Ndangalasi, J. Peacock, K.S.H. Peh, D. Sheil, T. Sunderland, M. D. Swaine, J. Taplin, D. Taylor, S. C. Thomas, R. Votere, and H. Woll. 2009. Increasing carbon storage in intact African tropical forests. Nature 457:1003–1006.

Lewis, W. M. 1974. Effects of fire on nutrient movement in a South Carolina pine forest. Ecology 55:1120–1127.

Lewis, W. M. 1981. Precipitation chemistry and nutrient loading by precipitation in a tropical watershed. Water Resources Research 17:169–181.

Lewis, W. M. 1986. Nitrogen and phosphorus runoff losses from a nutrient-poor tropical moist forest. Ecology 67:1275–1282.

Lewis, W. M. 1988. Primary production in the Orinoco river. Ecology 69:679–692.

Lewis, W. M. 2002. Yield of nitrogen from minimally disturbed watersheds of the United States. Biogeochemistry 57:375–385.

Lewis, W. M., and M. C. Grant. 1979. Relationships between stream discharge and yield of dissolved substances from a Colorado mountain watershed. Soil Science 128:353–363.

Lewis, W. M., S. K. Hamilton, S. L. Jones, and D. D. Runnels. 1987. Major element chemistry, weathering and element yields for the Caura River drainage, Venezuela. Biogeochemistry 4:159–181.

Lewis, W. M., Jr., and W. A. Wurtsbaugh. 2008. Control of lacustrine phytoplankton by nutrients: Erosion of the phosphorus paradigm. International Review of Hydrobiology 93:446–465.

Lewis, W. M., W. A. Wurtsbaugh, and H. W. Paerl. 2011. Rationale for control of anthropogenic nitrogen and phosphorus to reduce eutrophication of inland waters. Environmental Science & Technology 45: 10300–10305.

Li, S. P., J. Matthews, and A. Sinha. 2008. Atmospheric hydroxyl radical production from electronically excited NO_2 and H_2O. Science 319:1657–1660.

Li, W.K.W., D.V.S. Rao, W. G. Harrison, J. C. Smith, J. J. Cullen, B. Irwin, and T. Platt. 1983. Autotrophic picoplankton in the tropical ocean. Science 219: 292–295.

Li, Y.-H. 1972. Geochemical mass balance among lithosphere, hydrosphere, and atmosphere. American Journal of Science 272:119–137.

Li, Y.-H., T. Takahashi, and W. S. Broecker. 1969. The degree of saturation of $CaCO_3$ in the oceans. Journal of Geophysical Research 74:5507–5525.

Li, Y. H. 1981. Geochemical cycles of elements and human perturbation. Geochimica et Cosmochimica Acta 45: 2073–2084.

Li, Z. P., F. X. Han, Y. Su, T. L. Zhang, B. Sun, D. L. Monts, and M. J. Plodinec. 2007. Assessment of soil organic and carbonate carbon storage in China. Geoderma 138:119–126.

Liang, J. Y., L. W. Horowitz, D. J. Jacob, Y. H. Wang, A. M. Fiore, J. A. Logan, G. M. Gardner, and J. W. Munger. 1998. Seasonal budgets of reactive nitrogen species and ozone over the United States, and export fluxes to the global atmosphere. Journal of Geophysical Research—Atmospheres 103:13435–13450.

Lichter, J. 1998. Rates of weathering and chemical depletion in soils across a chronosequence of Lake Michigan sand dunes. Geoderma 85:255–282.

Lichter, J., S. A. Billings, S. E. Ziegler, D. Gaindh, R. Ryals, A. C. Finzi, R. B. Jackson, E. A. Stemmler, and W. H. Schlesinger. 2008. Soil carbon sequestration in a pine forest after 9 years of atmospheric CO_2 enrichment. Global Change Biology 14:2910–2922.

Lieffers, V. J., and S. E. Macdonald. 1990. Growth and foliar nutrient status of black spruce and tamarack in relation to depth of water table in some Alberta peatlands. Canadian Journal of Forest Research 20:805–809.

Liengen, T. 1999. Conversion factor between acetylene reduction and nitrogen fixation in free-living cyanobacteria from high arctic habitats. Canadian Journal of Microbiology 45:223–229.

Lieth, H. 1975. Modeling the primary productivity of the world. In H. Lieth and R. H. Whittaker, editors. Primary Productivity of the Biosphere. Springer, 237–263.

Lightfoot, D. C., and W. G. Whitford. 1987. Variation in insect densities on desert creosotebush—Is nitrogen a factor? Ecology 68:547–557.

Likens, G. E. 1975. Primary production of inland water ecosystems. In H. Lieth and R. H. Whittaker, editors. Primary Productivity of the Biosphere. Springer-Verlag, 185–202.

Likens, G. E., and F. H. Bormann. 1974. Acid rain: A serious regional environmental problem. Science 184: 1176–1179.

Likens, G. E., and F. H. Bormann. 1995. Biogeochemistry of a Forested Ecosystem. 2nd edition. Springer.

Likens, G. E., F. H. Bormann, and N. M. Johnson. 1969. Nitrification—Importance to nutrient losses from a cutover forested ecosystem. Science 163: 1205–1206.

Likens, G. E., F. H. Bormann, and N. M. Johnson. 1981. Interactions between major biogeochemical cycles in terrestrial ecosystems. In G.E Likens, editor. Some Perspectives of the Major Biogeochemical Cycles, Wiley, 93–112.

Likens, G. E., F. H. Bormann, N. M. Johnson, D. W. Fisher, and R. S. Pierce. 1970a. Effects of forest cutting and herbicide treatment on nitrogen budgets in the Hubbard Brook watershed ecosystem. Ecological Monographs 40:23–47.

Likens, G. E., F. H. Bormann, R. S. Pierce, J. S. Eaton, and R. E. Munn. 1984. Long-term trends in precipitation chemistry at Hubbard brook, New Hampshire. Atmospheric Environment 18:2641–2647.

Likens, G. E., D. C. Buso, and T. J. Butler. 2005. Long-term relationships between SO_2 and NOx emissions and SO_4^{2-} and NO_3^- concentration in bulk deposition at the Hubbard Brook Experimental Forest, NH. Journal of Environmental Monitoring 7: 964–968.

Likens, G. E., C. T. Driscoll, and D. C. Buso. 1996. Long-term effects of acid rain: Response and recovery of a forest ecosystem. Science 272:244–246.

Likens, G. E., C. T. Driscoll, D. C. Buso, M. J. Mitchell, G. M. Lovett, S. W. Bailey, T. G. Siccama, W. A. Reiners, and C. Alewell. 2002. The biogeochemistry of sulfur at Hubbard Brook. Biogeochemistry 60:235–316.

Likens, G. E., C. T. Driscoll, D. C. Buso, T. G. Siccama, C. E. Johnson, G. M. Lovett, D. F. Ryan, T. Fahey, and W. A. Reiners. 1994. The biogeochemistry of potassium at Hubbard Brook. Biogeochemistry **25**:61–125.

Likens, G. E., F. H. Bormann, L. O. Hedin, C. T. Driscoll, and J. S. Eaton. 1990. Dry deposition of sulfur: A 23-year record for the Hubbard Brook forest ecosystem. Tellus **42B**:319–329.

Limpens, J., F. Berendse, C. Blodau, J. G. Canadell, C. Freeman, J. Holden, N. Roulet, H. Rydin, and G. Schaepman-Strub. 2008. Peatlands and the carbon cycle: from local processes to global implications – a synthesis. Biogeosciences Discussions **5**:1379–1419.

Lin, B. L., A. Sakoda, R. Shibasaki, N. Goto, and M. Suzuki. 2000. Modelling a global biogeochemical nitrogen cycle in terrestrial ecosystems. Ecological Modelling **135**:89–110.

Lin, C. J., and S. O. Pehkonen. 1999. The chemistry of atmospheric mercury: a review. Atmospheric Environment **33**:2067–2079.

Lin, L. H., P. L. Wang, D. Rumble, J. Lippmann-Pipke, E. Boice, L. M. Pratt, B. S. Lollar, E. L. Brodie, T. C. Hazen, G. L. Andersen, T. Z. DeSantis, D. P. Moser, D. Kershaw, and T. C. Onstott. 2006. Long-term sustainability of a high-energy, low-diversity crustal biome. Science **314**:479–482.

Lin, P., M. Chen, and L. Guo. 2012. Speciation and transformation of phosphorus and its mixing behavior in the Bay of St. Louis estuary in the northern Gulf of Mexico. Geochimica et Cosmochimica Acta **87**: 283–298.

Lin, Q., and P. C. Brookes. 1999. An evaluation of the substrate-induced respiration method. Soil Biology & Biochemistry **31**:1969–1983.

Lin, Y. P., P. C. Singer, and G. R. Aiken. 2005. Inhibition of calcite precipitation by natural organic material: Kinetics, mechanism, and thermodynamics. Environmental Science & Technology **39**:6420–6428.

Lindberg, S. E., and C. T. Garten. 1988. Sources of sulfur in forest canopy throughfall. Nature **336**:148–151.

Lindberg, S. E., and G. M. Lovett. 1985. Field measurements of particle dry deposition rates to foliage and inert surfaces in a forest canopy. Environmental Science & Technology **19**:238–244.

Lindberg, S. E., G. M. Lovett, D. D. Richter, and D. W. Johnson. 1986. Atmospheric deposition and canopy interactions of major ions in a forest. Science **231**:141–145.

Lindeboom, H. J. 1984. The nitrogen pathway in a penguin rookery. Ecology **65**:269–277.

Lindell, M. J., H. W. Graneli, and L. J. Tranvik. 1996. Effects of sunlight on bacterial growth in lakes of different humic content. Aquatic Microbial Ecology **11**:135–141.

Lindsay, W. L. 1979. *Chemical Equilibria in Soils*. Wiley.

Lindsay, W. L., and E. C. Moreno. 1960. Phosphate phase equilibria in soils. Soil Science Society of America Proceedings **24**:177–182.

Lindsay, W. L., and P.L.G. Vlek. 1977. Phosphate minerals. *In* J. B. Dixon and S. B. Weed, editors. *Minerals in Soil Environments*. Soil Science Society of America, Madison, Wisconsin, 639–672.

Linkins, A. E., R. L. Sinsabaugh, C. A. McClaugherty, and J. M. Melills. 1990. Cellulase activity on decomposing leaf litter in microcosms. Plant and Soil **123**:17–25.

Lins, H. F., and J. R. Slack. 1999. Streamflow trends in the United States. Geophysical Research Letters **26**: 227–230.

Lipp, J. S., Y. Morono, F. Inagaki, and K. U. Hinrichs. 2008. Significant contribution of archaea to extant biomass in marine subsurface sediments. Nature **454**: 991–994.

Liss, P. S., and P. G. Slater. 1974. Flux of gases across the air-sea interface. Nature **247**:181–184.

Lissauer, J. J., D. C. Fabrycky, E. B. Ford, W. J. Borucki, F. Fressin, G. W. Marcy, J. A. Orosz, J. F. Rowe, G. Torres, W. F. Welsh, N. M. Batalha, S. T. Bryson, L. A. Buchhave, D. A. Caldwell, J. A. Carter, D. Charbonneau, J. L. Christiansen, W. Cochran, J.-M. Desert, E. W. Dunham, M. N. Fanelli, J. J. Fortney, T. N. Gautier, III, J. C. Geary, R. L. Gilliland, M. R. Haas, J. R. Hall, M. J. Holman, D. G. Koch, D. W. Latham, E. Lopez, S. McCauliff, N. Miller, R. C. Morehead, E. V. Quintana, D. Ragozzine, D. Sasselov, D. R. Short, and J. H. Steffen. 2011. A closely packed system of low-mass, low-density planets transiting Kepler-11. Nature **470**:53–58.

Litton, C. M., J. W. Raich, and M. G. Ryan. 2007. Carbon allocation in forest ecosystems. Global Change Biology **13**:2089–2109.

Liu, B. L., F. Phillips, S. Hoines, A. R. Campbell, and P. Sharma. 1995. Water movement in desert soil traced by hydrogen and oxygen isotopes, chloride, and ^{36}Cl, southern Arizona. Journal of Hydrology **168**:91–110.

Liu, J. G., L. Z. You, M. Amini, M. Obersteiner, M. Herrero, A.J.B. Zehnder, and H. Yang. 2010. A high-resolution assessment on global nitrogen flows in cropland. Proceedings of the National Academy of Sciences **107**:8035–8040.

Liu, K. K., and I. R. Kaplan. 1989. The eastern tropical Pacific as a source of ^{15}N-enriched nitrate in seawater off southern California. Limnology and Oceanography **34**:820–830.

Liu, L., and T. L. Greaver. 2010. A global perspective on belowground carbon dynamics under nitrogen enrichment. Ecology Letters **13**:819–828.

Liu, L. L., and T. L. Greaver. 2009. A review of nitrogen enrichment effects on three biogenic GHGs: the CO_2 sink may be largely offset by stimulated N_2O and CH_4 emission. Ecology Letters **12**:1103–1117.

Liu, R., Y. Li, and Q.-X. Wang. 2012. Variations in water and CO_2 fluxes over a saline desert in western China. Hydrological Processes **26**:513–522.

Liu, T. S., X.F. Gu, Z.S. An, and Y.X. Fan. 1981. The dust fall in Beijing, China on April 18, 1980. *In* T. L. Pewe, editor. *Desert Dust: Origin, Characteristics, and Effects on Man.* Geological Society of America, 149–157. Special Paper 186, Boulder.

Liu, X. J., L. Duan, J. M. Mo, E. Z. Du, J. L. Shen, X. K. Lu, Y. Zhang, X. B. Zhou, C. N. He, and F. S. Zhang. 2011. Nitrogen deposition and its ecological impact in China: An overview. Environmental Pollution **159**: 2251–2264.

Liu, Y., G. Villalba, R. U. Ayres, and H. Schroder. 2008. Global phosphorus flows and environmental impacts from a consumption perspective. Journal of Industrial Ecology **12**:229–247.

Livingston, G. P., P. M. Vitousek, and P. A. Matson. 1988. Nitrous oxide flux and nitrogen transformations across a landscape gradient in Amazonia. Journal of Geophysical Research—Atmospheres **93**:1593–1599.

Livingstone, D. A. 1963. Chemical composition of rivers and lakes. *In* M. Fleischer, editor. *Data of Geochemistry.* USGS Prof. Paper 440-G. U.S. Geological Survey, Washington, D.C.

Llácer, J. L., I. Fita, and V. Rubio. 2008. Arginine and nitrogen storage. Current Opinion in Structural Biology **18**:673–681.

Llewellyn, M. 1975. The effects of the lime aphid (*Eucallipterus tiliae* L.) (Aphididae) on the growth of lime (*Tilia* X *vulgaris* Hayne). II. The primary production of saplings and mature trees, the energy drain imposed by the aphid population and revised standard deviations of aphid population energy budgets. Journal of Applied Ecology **12**:15–23.

Lloyd, J., and G. D. Farquhar. 1994. ^{13}C discrimination during CO_2 assimilation by the terrestrial biosphere. Oecologia **99**:201–215.

Loaiciga, H. A., J. B. Valdes, R. Vogel, J. Garvey, and H. Schwarz. 1996. Global warming and the hydrologic cycle. Journal of Hydrology **174**:83–127.

Lobell, D. B., W. Schlenker, and J. Costa-Roberts. 2011. Climate trends and global crop production since 1980. Science **333**:616–620.

Lobert, J. M., D. H. Scharffe, W. M. Hao, and P. J. Crutzen. 1990. Importance of biomass burning in the atmospheric budgets of nitrogen-containing gases. Nature **346**:552–554.

Logan, J. A. 1985. Tropospheric ozone—Seasonal behavior, trends, and anthropogenic influence. Journal of Geophysical Research—Atmospheres **90**:10463–10482.

Logan, J. A., M. J. Prather, S. C. Wofsy, and M. B. McElroy. 1981. Tropospheric chemistry—A global perspective. Journal of Geophysical Research—Oceans and Atmospheres **86**:7210–7254.

Loh, A. N., and J. E. Bauer. 2000. Distribution, partitioning and fluxes of dissolved and particulate organic C, N and P in the eastern North Pacific and Southern Oceans. Deep-Sea Research Part I—Oceanographic Research Papers **47**:2287-2316.

Loh, A. N., J. E. Bauer, and E.R.M. Druffel. 2004. Variable ageing and storage of dissolved organic components in the open ocean. Nature **430**:877–881.

Lohrenz, S. E., G. L. Fahnenstiel, D. G. Redalje, G. A. Lang, M. J. Dagg, T. E. Whitledge, and Q. Dortch. 1999. Nutrients, irradiance, and mixing as factors regulating primary production in coastal waters impacted by the Mississippi River plume. Continental Shelf Research **19**:1113–1141.

Lohrer, A. M., S. F. Thrush, and M. M. Gibbs. 2004. Bioturbators enhance ecosystem function through complex biogeochemical interactions. Nature **431**:1092–1095.

Lohrmann, R., and L. E. Orgel. 1973. Prebiotic activation processes. Nature **244**:418–420.

Lohse, K. A., P. D. Brooks, J. C. McIntosh, T. Meixner, and T. E. Huxman. 2009. Interactions between biogeochemistry and hydrologic systems. Annual Review of Environment and Resources **34**:65–96.

Loladze, I., and J. J. Elser. 2011. The origins of the Redfield nitrogen-to-phosphorus ratio are in a homoeostatic protein-to-rRNA ratio. Ecology Letters **14**: 244–250.

Lomstein, B. A., A. T. Langerhuus, S. D'Hondt, B. B. Jorgensen, and A. J. Spivack. 2012. Endospore abundance, microbial growth and necromass turnover in deep sub-seafloor sediment. Nature **484**: 101–104.

Long, S. P., E. A. Ainsworth, A.D.B. Leakey, J. Nosberger, and D. R. Ort. 2006. Food for thought: Lower-than-expected crop yield stimulation with rising CO_2 concentrations. Science **312**:1918–1921.

Longmore, M. E., B. M. Oleary, C. W. Rose, and A. L. Chandica. 1983. Mapping soil erosion and accumulation with the fallout isotope cesium-137. Australian Journal of Soil Research **21**:373–385.

Lonsdale, W. M. 1988. Predicting the amount of litterfall in forests of the world. Annals of Botany **61**:319–324.

Lopez-Hernandez, D., M. Brossard, and E. Frossard. 1998. P-isotopic exchange values in relation to Po mineralisation in soils with very low P-sorbing capacities. Soil Biology & Biochemistry **30**:1663–1670.

Lorenz, R. D., K. L. Mitchell, R. L. Kirk, A. G. Hayes, O. Aharonson, H. A. Zebker, P. Paillou, J. Radebaugh, J. I. Lunine, M. A. Janssen, S. D. Wall, R. M. Lopes, B. Stiles, S. Ostro, G. Mitri, and E. R. Stofan. 2008. Titan's inventory of organic surface materials. Geophysical Research Letters 35.

Lottig, N. R., E. H. Stanley, P. C. Hanson, and T. K. Kratz. 2011. Comparison of regional stream and lake chemistry: Differences, similarities, and potential drivers. Limnology and Oceanography 56:1551–1562.

Loulergue, L., A. Schilt, R. Spahni, V. Masson-Delmotte, T. Blunier, B. Lemieux, J. M. Barnola, D. Raynaud, T. F. Stocker, and J. Chappellaz. 2008. Orbital and millennial-scale features of atmospheric CH_4 over the past 800,000 years. Nature 453: 383–386.

Love, S. G., and D. E. Brownlee. 1993. A direct measurement of the terrestrial mass accretion rate of cosmic dust. Science 262:550–553.

Lovelock, J. E. 1979. Gaia: A New Look at Life on Earth. Oxford University Press.

Lovelock, J. E., R. J. Maggs, and R. A. Rasmussen. 1972. Atmospheric dimethyl sulphide and the natural sulphur cycle. Nature 237:452–453.

Lovelock, J. E., and M. Whitfield. 1982. Life span of the biosphere. Nature 296:561–563.

Lovett, G. M. 1994. Atmospheric deposition of nutrients and pollutants in North America—An ecological perspective. Ecological Applications 4:629–650.

Lovett, G. M., J. J. Cole, and M. L. Pace. 2006. Is net ecosystem production equal to ecosystem carbon accumulation? Ecosystems 9:152–155.

Lovett, G. M., and C. L. Goodale. 2011. A new conceptual model of nitrogen saturation based on experimental nitrogen addition to an oak forest. Ecosystems 14: 615–631.

Lovett, G. M., G. E. Likens, D. C. Buso, C. T. Driscoll, and S. W. Bailey. 2005. The biogeochemistry of chlorine at Hubbard Brook, New Hampshire. Biogeochemistry 72:191–232.

Lovett, G. M., and S. E. Lindberg. 1986. Dry deposition of nitrate to a deciduous forest. Biogeochemistry 2: 137–148.

Lovett, G. M., and S. E. Lindberg. 1993. Atmospheric deposition and canopy interactions of nitrogen in forests. Canadian Journal of Forest Research 23: 1603–1616.

Lovett, G. M., W. A. Reiners, and R. K. Olson. 1982. Cloud droplet deposition in subalpine balsam fir forests—Hydrological and chemical inputs. Science 218:1303–1304.

Lovett, G. M., K. C. Weathers, M. A. Arthur, and J. C. Schultz. 2004. Nitrogen cycling in a northern hardwood forest: Do species matter? Biogeochemistry 67:289–308.

Lovett, G. M., K. C. Weathers, and W. V. Sobczak. 2000. Nitrogen saturation and retention in forested watersheds of the Catskill Mountains, New York. Ecological Applications 10:73–84.

Lovley, D. R. 1991. Dissimilatory Fe(III) and Mn (IV) reduction. Microbiological Reviews 55: 259–287.

Lovley, D. R., J. D. Coates, E. L. Blunt-Harris, E.J.P. Phillips, and J. C. Woodward. 1996. Humic substances as electron acceptors for microbial respiration. Nature 382:445–448.

Lovley, D. R., and M. J. Klug. 1986. Model for the distribution of sulfate reduction and methanogenesis in freshwater sediments Geochimica et Cosmochimica Acta 50:11–18.

Lovley, D. R., and E.J.P. Phillips. 1988. Novel mode of microbial energy metabolism: organic carbon oxidation coupled to dissimilatory reduction of iron or manganese. Applied and Environmental Microbiology 54:1472–1480.

Lovley, D. R., and E.J.P. Phillips. 1986. Organic matter mineralization with reduction of ferric iron in anaerobic sediments. Applied and Environmental Microbiology 51:683–689.

Lovley, D. R., and E.J.P. Phillips. 1987. Competitive mechanisms for inhibition of sulfate reduction and methane production in the zone of ferric iron reduction in sediments. Applied and Environmental Microbiology 53:2636–2641.

Lowry, B., C. Hebant, and D. Lee. 1980. The origin of land plants—A new look at an old problem. Taxon 29:183–197.

Lozier, M. S. 2010. Deconstructing the conveyor belt. Science 328:1507–1511.

Lozier, M. S. 2012. Overturning in the North Atlantic. Annual Review of Marine Science 4:291–315.

Lozier, M. S., A. C. Dave, J. B. Palter, L. M. Gerber, and R. T. Barber. 2011. On the relationship between stratification and primary productivity in the North Atlantic. Geophysical Research Letters 38.

Lu, J., G. A. Vecchi, and T. Reichler. 2007. Expansion of the Hadley cell under global warming. Geophysical Research Letters 34.

Lu, M., Y. Yang, Y. Luo, C. Fang, X. Zhou, J. Chen, X. Yang, and B. Li. 2010. Responses of ecosystem nitrogen cycle to nitrogen addition: a meta-analysis. New Phytologist 189:1040–1050.

Lucas, Y., F. J. Luizao, A. Chauvel, J. Rouiller, and D. Nahon. 1993. The relation between biological activity of the rain forest and mineral composition of soils. Science 260:521–523.

Luce, C. H., and Z. A. Holden. 2009. Declining annual streamflow distributions in the Pacific Northwest United States, 1948-2006. Geophysical Research Letters 36.

Ludwig, J., F. X. Meixner, B. Vogel, and J. Forstner. 2001. Soil-air exchange of nitric oxide: An overview of processes, environmental factors, and modeling studies. Biogeochemistry 52:225–257.

Luke, W. T., R. R. Dickerson, W. F. Ryan, K. E. Pickering, and L. J. Nunnermacker. 1992. Tropospheric chemistry over the lower Great Plains of the United States. 2. Trace gas profiles and distributions. Journal of Geophysical Research—Atmospheres 97:20647–20670.

Lund, D. C., J. Lynch-Stieglitz, and W. B. Curry. 2006. Gulf Stream density structure and transport during the past millennium. Nature 444:601–604.

Lundgren, D. G., and M. Silver. 1980. Ore leaching by bacteria. Annual Review of Microbiology 34:263–283.

Lundström, U. S. 1993. The role of organic acids in the soil solution chemistry of a podzolized soil. Journal of Soil Science 44:121–133.

Lundström, U. S., N. van Breemen, and D. Bain. 2000a. The podzolization process. A review. Geoderma 94:91–107.

Lundström, U. S., N. van Breemen, D. C. Bain, P.A.W. van Hees, R. Giesler, J. P. Gustafsson, H. Ilvesniemi, E. Karltun, P. A. Melkerud, M. Olsson, G. Riise, O. Wahlberg, A. Bergelin, K. Bishop, R. Finlay, A. G. Jongmans, T. Magnusson, H. Mannerkoski, A. Nordgren, L. Nyberg, M. Starr, and L. T. Strand. 2000b. Advances in understanding the podzolization process resulting from a multidisciplinary study of three coniferous forest soils in the Nordic Countries. Geoderma 94:335–353.

Lunine, J. I. 1989. Origin and evolution of outer solar system atmospheres. Science 245:141–147.

Luo, S. D., and T. L. Ku. 2003. Constraints on deep-water formation from the oceanic distributions of ^{10}Be. Journal of Geophysical Research—Oceans 108.

Luo, Y., B. Su, W. S. Currie, J. S. Dukes, A. C. Finzi, U. Hartwig, B. Hungate, R. E. McMurtrie, R. Oren, W. J. Parton, D. E. Pataki, M. R. Shaw, D. R. Zak, and C. B. Field. 2004. Progressive nitrogen limitation of ecosystem responses to rising atmospheric carbon dioxide. BioScience 54:731–739.

Luo, Y. Q., and J. F. Reynolds. 1999. Validity of extrapolating field CO_2 experiments to predict carbon sequestration in natural ecosystems. Ecology 80:1568–1583.

Lupton, J. E., and H. Craig. 1981. A major ^{3}He source at 15°S on the east Pacific rise. Science 214:13–18.

Luterbacher, J., D. Dietrich, E. Xoplaki, M. Grosjean, and H. Wanner. 2004. European seasonal and annual temperature variability, trends, and extremes since 1500. Science 303:1499–1503.

Luthi, D., M. Le Floch, B. Bereiter, T. Blunier, J. M. Barnola, U. Siegenthaler, D. Raynaud, J. Jouzel, H. Fischer, K. Kawamura, and T. F. Stocker. 2008. High-resolution carbon dioxide concentration record 650,000-800,000 years before present. Nature 453:379–382.

Lutz, B. D., E. S. Bernhardt, B. J. Roberts, R. M. Cory, and P. J. Mulholland. 2012a. Distinguishing dynamics of dissolved organic matter components in a forested stream using kinetic enrichments. Limnology and Oceanography 57:76–89.

Lutz, B. D., E. S. Bernhardt, B. J. Roberts, and P. J. Mulholland. 2011. Examining the coupling of carbon and nitrogen cycles in Appalachian streams: the role of dissolved organic nitrogen. Ecology 92:720–732.

Lutz, B. D., P. J. Mulholland, and E. S. Bernhardt. 2012b. Long-term data reveal patterns and controls on streamwater chemistry in a forested stream: Walker Branch Tennessee, Ecological Monographs 82:367–387.

Lutz, R. A., T. M. Shank, D. J. Fornari, R. M. Haymon, M. D. Lilley, K. L. Vondamm, and D. Desbruyeres. 1994. Rapid growth at deep sea vents. Nature 371:663–664.

Luxmoore, R. J., T. Grizzard, and R. H. Strand. 1981. Nutrient translocation in the outer canopy and understory of an eastern deciduous forest. Forest Science 27:505–518.

Luyssaert, S., P. Ciais, S. L. Piao, E. D. Schulze, M. Jung, S. Zaehle, M. J. Schelhaas, M. Reichstein, G. Churkina, D. Papale, G. Abril, C. Beer, J. Grace, D. Loustau, G. Matteucci, F. Magnani, G. J. Nabuurs, H. Verbeeck, M. Sulkava, G. R. van der Werf, I. A. Janssens, and C.-I.S. Team. 2010. The European carbon balance. Part 3: Forests. Global Change Biology 16:1429–1450.

Luyssaert, S., I. Inglima, M. Jung, A. D. Richardson, M. Reichsteins, D. Papale, S. L. Piao, E. D. Schulzes, L. Wingate, G. Matteucci, L. Aragao, M. Aubinet, C. Beers, C. Bernhoffer, K. G. Black, D. Bonal, J. M. Bonnefond, J. Chambers, P. Ciais, B. Cook, K.J. Davis, A. J. Dolman, B. Gielen, M. Goulden, J. Grace, A. Granier, A. Grelle, T. Griffis, T. Grunwald, G. Guidolotti, P. J. Hanson, R. Harding, D. Y. Hollinger, L. R. Hutyra, P. Kolar, B. Kruijt, W. Kutsch, F. Lagergren, T. Laurila, B. E. Law, G. Le Maire, A. Lindroth, D. Loustau, Y. Malhi, J. Mateus, M. Migliavacca, L. Misson, L. Montagnani, J. Moncrieff, E. Moors, J. W. Munger, E. Nikinmaa, S. V. Ollinger, G. Pita, C. Rebmann, O. Roupsard, N. Saigusa, M. J. Sanz, G. Seufert, C. Sierra, M. L. Smith, J. Tang, R. Valentini, T. Vesala, and

I. A. Janssens. 2007. CO_2 balance of boreal, temperate, and tropical forests derived from a global database. Global Change Biology 13:2509–2537.

Luyssaert, S., M. Reichstein, E. D. Schulze, I. A. Janssens, B. E. Law, D. Papale, D. Dragoni, M. L. Goulden, A. Granier, W. L. Kutsch, S. Linder, G. Matteucci, E. Moors, J. W. Munger, K. Pilegaard, M. Saunders, and E. M. Falge. 2009. Toward a consistency cross-check of eddy covariance flux-based and biometric estimates of ecosystem carbon balance. Global Biogeochemical Cycles 23.

Luyssaert, S., E. D. Schulze, A. Borner, A. Knohl, D. Hessenmoller, B. E. Law, P. Ciais, and J. Grace. 2008. Old-growth forests as global carbon sinks. Nature 455:213–215.

Luz, B., and E. Barkan. 2009. Net and gross oxygen production from O_2/Ar, $^{17}O/^{16}O$ and $^{18}O/^{16}O$ ratios. Aquatic Microbial Ecology 56:133–145.

Luz, B., and E. Barkan. 2011. The isotopic composition of atmospheric oxygen. Global Biogeochemical Cycles 25.

Lvovitch, M. I. 1973. The global water balance. National Academy of Sciences, Washington, D.C.

Ma, Z., D. G. Bielenberg, K. M. Brown, and J. P. Lynch. 2001. Regulation of root hair density by phosphorus availability in Arabidopsis thaliana. Plant Cell and Environment 24:459–467.

Maathuis, F.J.M., and D. Sanders. 1994. Mechanism of high-affinity potassium uptake in roots of Arabidopsis thaliana. Proceedings of the National Academy of Sciences 91:9272–9276.

MacCracken, M. C. 1985. Carbon dioxide and climate change: Background and overview. In M. C. MacCraken and F. M. Luther, editors. Projecting the Climatic Effects of Increasing Carbon Dioxide. U.S. Department of Energy, 1–23.

Macdonald, G. J. 1990. Role of methane clathrates in past and future climates. Climatic Change 16:247–281.

Macdonald, N. W., and J. B. Hart. 1990. Relating sulfate adsorption to soil properties in Michigan forest soils. Soil Science Society of America Journal 54:238–245.

Mace, K. A., R. A. Duce, and N. W. Tindale. 2003. Organic nitrogen in rain and aerosol at Cape Grim. Tasmania, Australia. Journal of Geophysical Research—Atmospheres 108.

Mach, D. L., A. Ramirez, and H. D. Holland. 1987. Organic phosphorus and carbon in marine sediments. American Journal of Science 287:429–441.

Macia, E., M. V. Hernandez, and J. Oro. 1997. Primary sources of phosphorus and phosphates in chemical evolution. Origins of Life and Evolution of Biospheres 27:459–480.

MacIntyre, F. 1974. The top millimeter of the ocean. Scientific American 230:62–77.

Mack, M. C., M. S. Bret-Harte, T. N. Hollingsworth, R. R. Jandt, E.A.G. Schuur, G. R. Shaver, and D. L. Verbyla. 2011. Carbon loss from an unprecedented arctic tundra wildfire. Nature 475:489–492.

Mack, M. C., E.A.G. Schuur, M. S. Bret-Harte, G. R. Shaver, and F. S. Chapin. 2004. Ecosystem carbon storage in arctic tundra reduced by long-term nutrient fertilization. Nature 431:440–443.

Mackay, A. W., R. J. Flower, A. E. Kuzmina, L. Z. Granina, N. L. Rose, P. G. Appleby, J. F. Boyle, and R. W. Battarbee. 1998. Diatom succession trends in recent sediments from Lake Baikal and their relation to atmospheric pollution and to climate change. Philosophical Transactions of the Royal Society of London 353:1011–1055.

MacKay, N. A., and J. J. Elser. 1998. Factors potentially preventing trophic cascades: Food quality, invertebrate predation, and their interaction. Limnology and Oceanography 43:339–347.

MacKenzie, F. T., and R. M. Garrels. 1966. Chemical mass balance between rivers and oceans. American Journal of Science 264:507–525.

Mackenzie, F. T., M. Stoffyn, and R. Wollast. 1978. Aluminum in seawater—Control by biological activity. Science 199:680–682.

Mackney, D. 1961. A podzol development sequence in oakwoods and heath in central England. Journal of Soil Science 12:23–40.

Madigan, M., and J. Martinko. 2006. Brock Biology of Microorganisms, 11th edition. Pearson/Prentice Hall.

Madsen, H. B., and P. Nornberg. 1995. Mineralogy of four sandy soils developed under heather, oak, spruce and grass in the same fluvioglacial deposit in Denmark. Geoderma 64:233–256.

Maeda, M., H. Ihara, and T. Ota. 2008. Deep-soil adsorption of nitrate in a Japanese Andisol in response to different nitrogen sources. Soil Science Society of America Journal 72:702–710.

Magnani, F., M. Mencuccini, M. Borghetti, P. Berbigier, F. Berninger, S. Delzon, A. Grelle, P. Hari, P. G. Jarvis, P. Kolari, A. S. Kowalski, H. Lankreijer, B. E. Law, A. Lindroth, D. Loustau, G. Manca, J. B. Moncrieff, M. Rayment, V. Tedeschi, R. Valentini, and J. Grace. 2007. The human footprint in the carbon cycle of temperate and boreal forests. Nature 447:848–850.

Magnuson, J. J., D. M. Robertson, B. J. Benson, R. H. Wynne, D. M. Livingstone, T. Arai, R. A. Assel, R. G. Barry, V. Card, E. Kuusisto, N. G. Granin, T. D. Prowse, K. M. Stewart, and V. S. Vuglinski.

2000. Historical trends in lake and river ice cover in the Northern Hemisphere. Science 289: 1743–1746.

Magnuson, J. J., K. E. Webster, R. A. Assel, C. J. Bowser, P. J. Dillon, J. G. Eaton, H. E. Evans, E. J. Fee, R. I. Hall, L. R. Mortsch, D. W. Schindler, and F. H. Quinn. 1997. Potential effects of climate changes on aquatic systems: Laurentian Great Lakes and Precambrian Shield Region. Hydrological Processes 11:825–871.

Mahaffey, C., A. F. Michaels, and D. G. Capone. 2005. The conundrum of marine N_2 fixation. American Journal of Science 305:546–595.

Mahaffey, C., R. G. Williams, G. A. Wolff, N. Mahowald, W. Anderson, and M. Woodward. 2003. Biogeochemical signatures of nitrogen fixation in the eastern North Atlantic. Geophysical Research Letters 30.

Mahall, B. E., and W. H. Schlesinger. 1982. Effects of irradiance on growth, photosynthesis, and water-use efficiency of seedlings of the chaparral shrub, *Ceanothus megacarpus*. Oecologia 54:291–299.

Maher, K. 2010. The dependence of chemical weathering rates on fluid residence time. Earth and Planetary Science Letters 294:101–110.

Maher, K. 2011. The role of fluid residence time and topographic scales in determining chemical fluxes from landscapes. Earth and Planetary Science Letters 312:48–58.

Mahieu, N., D. S. Powlson, and E. W. Randall. 1999. Statistical analysis of published carbon-13 CPMAS NMR spectra of soil organic matter. Soil Science Society of America Journal 63:307–319.

Mahowald, N. 2011. Aerosol indirect effect on biogeochemical cycles and climate. Science 334:794–796.

Mahowald, N., T. D. Jickells, A. R. Baker, P. Artaxo, C. R. Benitez-Nelson, G. Bergametti, T. C. Bond, Y. Chen, D. D. Cohen, B. Herut, N. Kubilay, R. Losno, C. Luo, W. Maenhaut, K. A. McGee, G. S. Okin, R. L. Siefert, and S. Tsukuda. 2008. Global distribution of atmospheric phosphorus sources, concentrations and deposition rates, and anthropogenic impacts. Global Biogeochemical Cycles 22.

Mahowald, N. M., P. Artaxo, A. R. Baker, T. D. Jickells, G. S. Okin, J. T. Randerson, and A. R. Townsend. 2005a. Impacts of biomass burning emissions and land use change on Amazonian atmospheric phosphorus cycling and deposition. Global Biogeochemical Cycles 19.

Mahowald, N. M., A. R. Baker, G. Bergametti, N. Brooks, R. A. Duce, T. D. Jickells, N. Kubilay, J. M. Prospero, and I. Tegen. 2005b. Atmospheric global dust cycle and iron inputs to the ocean. Global Biogeochemical Cycles 19.

Maia, S.M.F., S. M. Ogle, C.E.P. Cerri, and C. C. Cerri. 2010. Soil organic carbon stock change due to land use activity along the agricultural frontier of the southwestern Amazon, Brazil, between 1970 and 2002. Global Change Biology 16:2775–2788.

Makkonen, K., and H. S. Helmisaari. 1999. Assessing fine-root biomass and production in a Scots pine stand—Comparison of soil core and root-ingrowth core methods. Plant and Soil 210:43–50.

Malaney, R. A., and W. A. Fowler. 1988. The transformation of matter after the Big Bang. American Scientist 76:472–477.

Malcolm, E. G., J. K. Schaefer, E. B. Ekstrom, C. B. Tuit, A. Jayakumar, H. Park, B. B. Ward, and F.M.M. Morel. 2010. Mercury methylation in oxygen deficient zones of the oceans: No evidence for the predominance of anaerobes. Marine Chemistry 122:11–19.

Malcolm, R. E. 1983. Assessment of phosphatase activity in soils. Soil Biology & Biochemistry 15:403–408.

Malin, M. C., and K. S. Edgett. 2000. Evidence for recent groundwater seepage and surface runoff on Mars. Science 288:2330–2335.

Malin, M. C., and K. S. Edgett. 2003. Evidence for persistent flow and aqueous sedimentation on early Mars. Science 302:1931–1934.

Malin, M. C., K. S. Edgett, L. V. Posiolova, S. M. McColley, and E.Z.N. Dobrea. 2006. Present-day impact cratering rate and contemporary gully activity on Mars. Science 314:1573–1577.

Mallin, M. A., V. L. Johnson, S. H. Ensign, and T. A. MacPherson. 2006. Factors contributing to hypoxia in rivers, lakes, and streams. Limnology and Oceanography 51:690–701.

Malmer, N. 1975. Development of bog mires. *In* A. D. Hassler, editor. *Coupling of Land and Water Systems*. Springer, 75–92.

Malone, M. J., P. A. Baker, and S. J. Burns. 1994. Recrystallization of dolomite- Evidence from the Monterey formation (Miocene), California. Sedimentology 41: 1223–1239.

Manabe, S., and R. T. Wetherald. 1980. Distribution of climate change resulting from an increase in the CO_2 content of the atmosphere. Journal of the Atmospheric Sciences 37:99–118.

Manabe, S., and R. T. Wetherald. 1986. Reduction in summer soil wetness induced by an increase in atmospheric carbon dioxide. Science 232:626–628.

Mancinelli, R. L., and C. P. McKay. 1988. The evolution of nitrogen cycling. Origins of Life and Evolution of the Biosphere 18:311–325.

Mankin, W. G., and M. T. Coffey. 1984. Increased stratospheric hydrogen chloride in the El Chichon cloud. Science 226:170–172.

Mankin, W. G., M. T. Coffey, and A. Goldman. 1992. Airborne observations of SO_2, HCl, and O_3 in the stratospheric plume of the Pinatubo volcano in July 1991. Geophysical Research Letters **19**:179–182.

Manley, S. L., N. Y. Wang, M. L. Walser, and R. J. Cicerone. 2006. Coastal salt marshes as global methyl halide sources from determinations of intrinsic production by marsh plants. Global Biogeochemical Cycles **20**.

Mann, M. E., R. S. Bradley, and M. K. Hughes. 1999. Northern hemisphere temperatures during the past millennium: Inferences, uncertainties, and limitations. Geophysical Research Letters **26**:759–762.

Manney, G. L., M. L. Santee, M. Rex, N. J. Livesey, M. C. Pitts, P. Veefkind, E. R. Nash, I. Wohltmann, R. Lehmann, L. Froidevaux, L. R. Poole, M. R. Schoeberl, D. P. Haffner, J. Davies, V. Dorokhov, H. Gernandt, B. Johnson, R. Kivi, E. Kyro, N. Larsen, P. F. Levelt, A. Makshtas, C. T. McElroy, H. Nakajima, M. C. Parrondo, D. W. Tarasick, P. von der Gathen, K. A. Walker, and N. S. Zinoviev. 2011. Unprecedented arctic ozone loss in 2011. Nature **478**:469–475.

Manning, C. V., C. P. McKay, and K. J. Zahnle. 2008. The nitrogen cycle on Mars: Impact decomposition of near-surface nitrates as a source for a nitrogen steady state. Icarus **197**:60–64.

Mansy, S. S., J. P. Schrum, M. Krishnamurthy, S. Tobe, D. A. Treco, and J. W. Szostak. 2008. Template-directed synthesis of a genetic polymer in a model protocell. Nature **454**:122–125.

Manzoni, S., R. B. Jackson, J. A. Trofymow, and A. Porporato. 2008. The global stoichiometry of litter nitrogen mineralization. Science **321**:684–686.

Manzoni, S., J. A. Trofymow, R. B. Jackson, and A. Porporato. 2010. Stoichiometric controls on carbon, nitrogen, and phosphorus dynamics in decomposing litter. Ecological Monographs **80**:89–106.

Maranger, R., D. F. Bird, and N. M. Price. 1998. Iron acquisition by photosynthetic marine phytoplankton from ingested bacteria. Nature **396**:248–251.

Marcano-Martinez, E., and M. B. McBridge. 1989. Calcium and sulfate retention by two Oxisols of the Brazilian cerrado. Soil Science Society of America Journal **53**:63–69.

Marchal, K., and J. Vanderleyden. 2000. The "oxygen paradox" of dinitrogen-fixing bacteria. Biology and Fertility of Soils **30**:363–373.

Marcq, E., D. Belyaev, F. Montmessin, A. Fedorova, J.-L. Bertaux, A. C. Vandaele, and E. Neefs. 2011. An investigation of the SO_2 content of the Venusian mesosphere using SPICAV-UV in nadir mode. Icarus **211**:58–69.

Marenco, A., H. Gouget, P. Nedelec, J. P. Pages, and F. Karcher. 1994. Evidence of a long-term increase in tropospheric ozone from PIC DU MIDI data series—

Consequences—Positive radiative forcing. Journal of Geophysical Research—Atmospheres **99**:16617–16632.

Marino, R., F. Chan, R. W. Howarth, M. L. Pace, and G. E. Likens. 2006. Ecological constraints on planktonic nitrogen fixation in saline estuaries. I. Nutrient and trophic controls. Marine Ecology—Progress Series **309**:25–39.

Marino, R., R. W. Howarth, F. Chan, J. J. Cole, and G. E. Likens. 2003. Sulfate inhibition of molybdenum-dependent nitrogen fixation by planktonic cyanobacteria under seawater conditions: a non-reversible effect. Hydrobiologia **500**:277–293.

Marion, G. M., and C. H. Black. 1987. The effect of time and temperature on nitrogen mineralization in Arctic tundra soils. Soil Science Society of America Journal **51**:1501–1508.

Marion, G. M., and C. H. Black. 1988. Potentially-available nitrogen and phosphorus along a chaparral fire cycle chronosequence. Soil Science Society of America Journal **52**:1155–1162.

Marion, G. M., C. H. Fritsen, H. Eicken, and M. C. Payne. 2003. The search for life on Europa: Limiting environmental factors, potential habitats, and Earth analogues. Astrobiology **3**:785–811.

Marion, G. M., W. H. Schlesinger, and P. J. Fonteyn. 1985. CALDEP—A regional model for soil $CaCO_3$ (caliche) deposition in southwestern deserts. Soil Science **139**: 468–481.

Marion, G. M., P.S.J. Verburg, B. Stevenson, and J. A. Arnone. 2008. Soluble element distributions in a Mojave desert soil. Soil Science Society of America Journal **72**:1815–1823.

Markewich, H. W., and M. J. Pavich. 1991. Soil chronosequence studies in temperate to subtropical, low-latitude, low-relief terrain with data from the eastern United States. Geoderma **51**:213–239.

Markewich, H. W., M. J. Pavich, M. J. Mausbach, R. G. Johnson, and V. M. Gonzalez. 1989. A guide for using soil and weathering profile data in chronosequence studies of the Coastal Plain of the eastern United States. U.S. Geological Survey.

Markewitz, D., and D. D. Richter. 1998. The *bio* in aluminum and silicon geochemistry. Biogeochemistry **42**:235–252.

Markewitz, D., D. D. Richter, H. L. Allen, and J. B. Urrego. 1998. Three decades of observed soil acidification in the Calhoun experimental forest: Has acid rain made a difference? Soil Science Society of America Journal **62**:1428–1439.

Marks, P. L., and F. H. Bormann. 1972. Revegetation following forest cutting: Mechanisms for return to steady-state nutrient cycling. Science **176**: 914–915.

Marnette, E. C., C. Hordijk, N. Van Breeman, and T. Cappenberg. 1992. Sulfate reduction and S-oxidation in a moorland pool sediment. Biogeochemistry 17:123–143.

Maron, J. L., J. A. Estes, D. A. Croll, E. M. Danner, S. C. Elmendorf, and S. L. Buckelew. 2006. An introduced predator alters Aleutian Island plant communities by thwarting nutrient subsidies. Ecological Monographs 76:3–24.

Maroulis, P. J., and A. R. Bandy. 1977. Estimate of the contribution of biologically-producced dimethyl sulfide to the global sulfur cycle. Science 196:647–648.

Marquis, R. J., and C. J. Whelan. 1994. Insectivorous birds increase growth of white oak through consumption of chewing insects. Ecology 75:2007–2014.

Marsh, A. S., J. A. Arnone, B. T. Bormann, and J. C. Gordon. 2000. The role of *Equisetum* in nutrient cycling in an Alaskan shrub wetland. Journal of Ecology 88:999–1011.

Marshall, J., and K. Speer. 2012. Closure of the meridional overturning circulation through Southern Ocean upwelling. Nature Geoscience 5:171–180.

Marteel, A., C. F. Boutron, C. Barbante, P. Gabrielli, G. Cozzi, V. Gaspari, P. Cescon, C. R. Ferrari, A. Dommergue, K. Rosman, S. M. Hong, and S. Do Hur. 2008. Changes in atmospheric heavy metals and metalloids in Dome C (East Antarctica) ice back to 672.0 kyr BP (Marine Isotopic Stages 16.2). Earth and Planetary Science Letters 272:579–590.

Martens, C. S., N. E. Blair, C. D. Green, and D. J. Desmarais. 1986. Seasonal variations in the stable carbon isotopic signature of biogenic methane in a coastal sediment. Science 233:1300–1303.

Martens, C. S., and J. Val Klump. 1984. Biogeochemical cycling in an organic-rich coastal marine basin. 4. An organic-carbon budget for sediments dominated by sulfate reduction and methanogenesis. Geochimica et Cosmochimica Acta 48:1987–2004.

Martens, R. 1995. Current methods for measuring microbial biomass-C in soil—Potentials and limitations. Biology and Fertility of Soils 19:87–99.

Martens-Habbena, W., P. M. Berube, H. Urakawa, J. R. de la Torre, and D. A. Stahl. 2009. Ammonia oxidation kinetics determine niche separation of nitrifying archaea and bacteria. Nature 461:976–979.

Martin, C. W., and R. D. Harr. 1988. Precipitation and streamwater chemistry from undisturbed watersheds in the Cascade mountains of Oregon. Water Air and Soil Pollution 42:203–219.

Martin, H. W., and D. L. Sparks. 1985. On the behavior of nonexchangeable potassium in soils. Communications in Soil Science and Plant Analysis 16:133–162.

Martin, J. H., K. H. Coale, K. S. Johnson, S. E. Fitzwater, R. M. Gordon, S. J. Tanner, C. N. Hunter, V. A. Elrod, J. L. Nowicki, T. L. Coley, R. T. Barber, S. Lindley, A.J. Watson, K. Vanscoy, C. S. Law, M. I. Liddicoat, R. Ling, T. Stanton, J. Stockel, C. Collins, A. Anderson, R. Bidigare, M. Ondrusek, M. Latasa, F. J. Millero, K. Lee, W. Yao, J. Z. Zhang, G. Friederich, C. Sakamoto, F. Chavez, K. Buck, Z. Kolber, R. Greene, P. Falkowski, S. W. Chisholm, F. Hoge, R. Swift, J. Yungel, S. Turner, P. Nightingale, A. Hatton, P. Liss, and N. W. Tindale. 1994. Testing the iron hypothesis in ecosystems of the equatorial Pacific ocean. Nature 371:123–129.

Martin, J. H., and S. E. Fitzwater. 1992. Dissolved organic carbon in the Atlantic, Southern and Pacific oceans. Nature 356:699–700.

Martin, J. H., and R. M. Gordon. 1988. Northeast Pacific iron distributions in relation to phytoplankton productivity. Deep-Sea Research Part A—Oceanographic Research Papers 35:177-196.

Martin, J. H., R. M. Gordon, S. Fitzwater, and W. W. Broenkow. 1989. VERTEX—Phytoplankton iron studies in the Gulf of Alaska. Deep-Sea Research Part A—Oceanographic Research Papers 36:649–680.

Martin, J. H., G. A. Knauer, D. M. Karl, and W. W. Broenkow. 1987. VERTEX—Carbon cycling in the northeast Pacific. Deep-Sea Research Part A—Oceanographic Research Papers 34:267–285.

Martin, J. M., and M. Meybeck. 1979. Elemental mass balance of material carried by major world rivers. Marine Chemistry 7:173–206.

Martin, M. E., and J. D. Aber. 1997. High spectral resolution remote sensing of forest canopy lignin, nitrogen, and ecosystem processes. Ecological Applications 7:431–443.

Martin, R. V. 2008. Satellite remote sensing of surface air quality. Atmospheric Environment 42:7823–7843.

Martin, R. V., D. J. Jacob, K. Chance, T. P. Kurosu, P. I. Palmer, and M. J. Evans. 2003. Global inventory of nitrogen oxide emissions constrained by space-based observations of NO_2 columns. Journal of Geophysical Research—Atmospheres 108.

Martin, R. V., B. Sauvage, I. Folkins, C. E. Sioris, C. Boone, P. Bernath, and J. Ziemke. 2007. Space-based constraints on the production of nitric oxide by lightning. Journal of Geophysical Research—Atmospheres 112.

Martin, S. L., D. B. Hayes, D. T. Rutledge, and D. W. Hyndman. 2011. The land-use legacy effect: Adding temporal context to lake chemistry. Limnology and Oceanography 56:2362–2370.

Martin, W. R., M. Bender, M. Leinen, and J. Orchardo. 1991. Benthic organic carbon degradation and biogenic silica dissolution in the central equatorial Pacific. Deep-Sea Research Part A—Oceanographic Research Papers 38:1481-1516.

Martinelli, L. A., M. C. Piccolo, A. R. Townsend, P. M. Vitousek, E. Cuevas, W. McDowell, G. P. Robertson, O. C. Santos, and K. Treseder. 1999. Nitrogen stable isotopic composition of leaves and soil: Tropical versus temperate forests. Biogeochemistry 46: 45–65.

Martínez, L., M. W. Silver, J. M. King, and A. L. Alldredge. 1983. Nitrogen fixation by floating diatom mats—A source of new nitrogen to oligotrophic ocean waters. Science 221:152–154.

Martinez-Cortizas, A., X. Pontevedra-Pombal, E. Garcia-Rodeja, J. C. Novoa-Munoz, and W. Shotyk. 1999. Mercury in a Spanish peat bog: Archive of climate change and atmospheric metal deposition. Science 284:939–942.

Martinez-Garcia, A., A. Rosell-Mele, S. L. Jaccard, W. Geibert, D. M. Sigman, and G. H. Haug. 2011. Southern Ocean dust-climate coupling over the past four million years. Nature 476:312–315.

Marty, B. 1995. Nitrogen content of the mantle inferred from N_2-Ar correlation in oceanic basalts. Nature 377:326–329.

Marty, B. 2012. The origins and concentrations of water, carbon, nitrogen and noble gases on Earth. Earth and Planetary Science Letters 313/314:56–66.

Marty, B., M. Chaussidon, R. C. Wiens, A.J.G. Jurewicz, and D. S. Burnett. 2011. A ^{15}N-poor isotopic composition for the solar system as shown by Genesis solar wind samples. Science 332:1533–1536.

Marufu, L. T., B. F. Taubman, B. Bloomer, C. A. Piety, B. G. Doddridge, J. W. Stehr, and R. R. Dickerson. 2004. The 2003 North American electrical blackout: An accidental experiment in atmospheric chemistry. Journal of Geophysical Reseach 31.

Marumoto, T., J.P.E. Anderson, and K. H. Domsch. 1982. Mineralization of nutrients from soil microbial biomass. Soil Biology & Biochemistry 14:469–475.

Marx, D. H., A. B. Hatch, and J. F. Mendicino. 1977. High soil fertility decreases surcrose content and susceptibility of loblolly pine roots to ectomycorrrhizal infection by *Pisolithus tinctorius*. Canadian Journal of Botany 55:1569–1574.

Masek, J. G., W. B. Cohen, D. Leckie, M. A. Wulder, R. Vargas, B. de Jong, S. Healey, B. Law, R. Birdsey, R. A. Houghton, D. Mildrexler, S. Goward, and W. B. Smith. 2011. Recent rates of forest harvest and conversion in North America. Journal of Geophysical Research—Biogeosciences 116.

Mason, R. P., and W. F. Fitzgerald. 1993. The distribution and biogeochemical cycling of mercury in the equatorial Pacific ocean. Deep-Sea Research Part I—Oceanographic Research Papers 40:1897-1924.

Massman, W. J. 1992. A surface-energy balance method for partitioning evapotranspiration data into plant and soil components for a surface with partical canopy cover. Water Resources Research 28: 1723–1732.

Mast, M. A., D. J. Manthorne, and D. A. Roth. 2010. Historical deposition of mercury and selected trace elements to high-elevation national parks in the western US inferred from lake-sediment cores. Atmospheric Environment 44:2577–2586.

Mast, M. A., J. T. Turk, G. P. Ingersoll, D. W. Clow, and C. L. Kester. 2001. Use of stable sulfur isotopes to identify sources of sulfate in Rocky Mountain snowpacks. Atmospheric Environment 35: 3303–3313.

Matamala, R., M. A. Gonzalez-Meler, J. D. Jastrow, R. J. Norby, and W. H. Schlesinger. 2003. Impacts of fine root turnover on forest NPP and soil C sequestration potential. Science 302:1385–1387.

Mathieu, O., J. Leveque, C. Henault, M. J. Milloux, F. Bizouard, and F. Andreux. 2006. Emissions and spatial variability of N_2O, N_2 and nitrous oxide mole fraction at the field scale, revealed with ^{15}N-isotopic techniques. Soil Biology & Biochemistry 38:941–951.

Matson, P. A., L. Johnson, C. Billow, J. Miller, and R. Pu. 1994. Seasonal patterns and remote spectral estimation of canopy chemistry across the Oregon transect. Ecological Applications 4:280–298.

Matson, P. A., and P. M. Vitousek. 1990. Ecosystem approach to a global nitrous oxide budget. BioScience 40:667–671.

Matson, P. A., P. M. Vitousek, J. J. Ewel, M. J. Mazzarino, and G. P. Robertson. 1987. Nitrogen transformations following tropical forest felling and burning on a volcanic soil. Ecology 68:491–502.

Matson, P. A., C. Volkmann, K. Coppinger, and W. A. Reiners. 1991. Annual nitrous oxide flux and soil nitrogen characteristics in sagebrush steppe ecosystems. Biogeochemistry 14:1–12.

Matthews, E. 1997. Global litter production, pools, and turnover times: Estimates from measurement data and regression models. Journal of Geophysical Research—Atmospheres 102:18771–18800.

Matthews, E., and I. Fung. 1987. Methane emission from natural wetlands: Global distribution, area, and environmental characteristics of sources. Global Biogeochemical Cycles 1:61–86.

Matthews, E., I. Fung, and J. Lerner. 1991. Methane emission from rice cultivation: Geographic and seasonal distribution of cultivated areas and emissions. Global Biogeochemical Cycles 5:3–24.

Mattson, W. J. 1980. Herbivory in relation to plant nitrogen content. Annual Review of Ecology and Systematics **11**:119–161.

Mattson, W. J., and N. D. Addy. 1975. Phytophagous insects as regulators of forest primary production. Science **190**:515–522.

Mayer, B., K. H. Feger, A. Giesemann, and H. J. Jager. 1995. Interpretation of sulfur cycling in two catchments in the Black Forest (Germany) using stable sulfur-and oxygen-isotope data. Biogeochemistry **30**:31–58.

Mayer, L. M. 1994. Relationships between mineral surfaces and organic carbon concentrations in soils and sediments. Chemical Geology **114**:347–363.

Mayewski, P. A., W. B. Lyons, M. J. Spencer, M. Twickler, W. Dansgaard, B. Koci, C. I. Davidson, and R. E. Honrath. 1986. Sulfate and nitrate concentrations from a south Greenland ice core. Science **232**:975–977.

Mayewski, P. A., W. B. Lyons, M. J. Spencer, M. S. Twickler, C. F. Buck, and S. Whitlow. 1990. An ice-core record of atmospheric response to anthropogenic sulfate and nitrate. Nature **346**:554–556.

Maynard, J. J., A. T. O'Geen, and R. A. Dahlgren. 2011. Sulfide induced mobilization of wetland phosphorus depends strongly on redox and iron geochemistry. Soil Science Society of America Journal **75**:1986.

Mayorga, E., A. K. Aufdenkampe, C. A. Masiello, A. V. Krusche, J. I. Hedges, P. D. Quay, J. E. Richey, and T. A. Brown. 2005. Young organic matter as a source of carbon dioxide outgassing from Amazonian rivers. Nature **436**:538–541.

Mazumder, A., and M. D. Dickman. 1989. Factors affecting the spatial and temporal distribution of phototrophic sulfur bacteria. Archiv fur Hydrobiologie **116**:209–226.

McCabe, G. J., and D. M. Wolock. 2002. A step increase in streamflow in the conterminous United States. Geophysical Research Letters **29**.

McCabe, P. J. 2009. Depositional environments of coal and coal-bearing strata. *In* R. A. Rahmani and R. M. Flores, editors. *Sedimentology of Coal and Coal-Bearing Sequences*. Blackwell Publishing Ltd, 11–42.

McCarthy, H. R., R. Oren, K. H. Johnsen, A. Gallet-Budynek, S. G. Pritchard, C. W. Cook, S. L. LaDeau, R. B. Jackson, and A. C. Finzi. 2010. Re-assessment of plant carbon dynamics at the Duke free-air CO_2 enrichment site: interactions of atmospheric CO_2 with nitrogen and water availability over stand development. New Phytologist **185**:514–528.

McCarthy, J. J., and J. C. Goldman. 1979. Nitrogenous nutrition of marine phytoplankton in nutrient-depleted waters. Science **203**:670–672.

McCarthy, M. D., J. I. Hedges, and R. Benner. 1998. Major bacterial contribution to marine dissolved organic nitrogen. Science **281**:231–234.

McCarty, G. W., J. J. Meisinger, and F.M.M. Jenniskens. 1995. Relationships between total N, biomass N and active N in soil under different tillage and N fertilizer treatments. Soil Biology & Biochemistry **27**:1245–1250.

McClain, M. E., E. W. Boyer, C. L. Dent, S. E. Gergel, N. B. Grimm, P. M. Groffman, S. C. Hart, J. W. Harvey, C. A. Johnston, E. Mayorga, W. H. McDowell, and G. Pinay. 2003. Biogeochemical hot spots and hot moments at the interface of terrestrial and aquatic ecosystems. Ecosystems **6**:301–312.

McClelland, J. W., S. J. Dery, B. J. Peterson, R. M. Holmes, and E. F. Wood. 2006. A pan-arctic evaluation of changes in river discharge during the latter half of the 20th century. Geophysical Research Letters **33**.

McColl, J. G., and D. F. Grigal. 1975. Forest fire—Effects on phosphorus movement to lakes. Science **188**:1109–1111.

McCormick, M. P., L. W. Thomason, and C. R. Trepte. 1995. Atmospheric effects of the Mt. Pinatubo eruption. Nature **373**:399–404.

McCrackin, M. L., and J. J. Elser. 2011. Greenhouse gas dynamics in lakes receiving atmospheric nitrogen deposition. Global Biogeochemical Cycles **25**.

McCrackin, M. L., and J. J. Elser. 2012. Denitrification kinetics and denitrifier abundances in sediments of lakes receiving atmospheric nitrogen deposition (Colorado, USA). Biogeochemistry **108**:39–54.

McCulloch, A., M. L. Aucott, C. M. Benkovitz, T. E. Graedel, G. Kleiman, P. M. Midgley, and Y. F. Li. 1999. Global emissions of hydrogen chloride and chloromethane from coal combustion, incineration and industrial activities: Reactive chlorine emissions inventory. Journal of Geophysical Research—Atmospheres **104**:8391–8403.

McCutchan, J. H., and W. M. Lewis. 2002. Relative importance of carbon sources for macroinvertebrates in a Rocky Mountain stream. Limnology and Oceanography **47**:742–752.

McDiffett, W. F., A. W. Beidler, T. F. Dominick, and K. D. McCrea. 1989. Nutrient concentration-stream discharge relationships during storm events in a first-order stream. Hydrobiologia **179**:97–102.

McDowell, W. H., and C. E. Asbury. 1994. Export of carbon, nitrogen, and major ions from three tropical montane watersheds. Limnology and Oceanography **39**:111–125.

McDowell, W. H., and G. E. Likens. 1988. Origin, composition, and flux of dissolved organic carbon in the Hubbard Brook valley. Ecological Monographs **58**:177–195.

McDowell, W. H., and T. Wood. 1984. Podzolization—Soil processes control dissolved organic carbon concentrations in stream water. Soil Science **137**:23–32.

McElroy, M. B. 1983. Marine biological controls on atmospheric CO_2 and climate. Nature 302:328–329.

McElroy, M. B. 2002. *The Atmospheric Environment.* Princeton University Press.

McElroy, M. B., M. J. Prather, and J. M. Rodriguez. 1982. Escape of hydrogen from Venus. Science 215: 1614–1615.

McElroy, M. B., and R. J. Salawitch. 1989. Changing composition of the global stratosphere. Science 243: 763–770.

McElroy, M. B., and Y.X.X. Wang. 2005. Human and animal wastes: Implications for atmospheric N_2O and NOx. Global Biogeochemical Cycles 19.

McElroy, M. B., Y. L. Yung, and A. O. Nier. 1976. Isotopic composition of nitrogen—Implications for past history of Mars' atmosphere. Science 194:70–72.

McEwen, A. S., L. Ojha, C. M. Dundas, S. S. Mattson, S. Byrne, J. J. Wray, S. C. Cull, S. L. Murchie, N. Thomas, and V. C. Gulick. 2011. Seasonal flows on warm Martian slopes. Science 333:740–743.

McFarland, J. W., R. W. Ruess, K. Kielland, K. Pregitzer, R. Hendrick, and M. Allen. 2010. Cross-ecosystem comparisons of *in situ* plant uptake of amino acid-N and NH_4^+. Ecosystems 13:177–193.

McGill, G. E., J. L. Warner, M. C. Malin, R. E. Arvidson, E. Eliason, S. Nozette, and R. D. Reasenberg. 1983. Topography, surface properties, and tectonic evolution. *In* D. M. Hunten, L. Colin, , T. M. Donahue and V. I. Moroz, editors. *Venus.* University of Arizona Press, 69–130.

McGill, W. B., and C. V. Cole. 1981. Comparative aspects of cycling of organic C, N, S and P through soil organic matter. Geoderma 26:267–286.

McGillicuddy, D. J., A. R. Robinson, D. A. Siegel, H. W. Jannasch, R. Johnson, T. Dickeys, J. McNeil, A. F. Michaels, and A. H. Knap. 1998. Influence of mesoscale eddies on new production in the Sargasso Sea. Nature 394:263–266.

McGroddy, M. E., T. Daufresne, and L. O. Hedin. 2004. Scaling of C:N:P stoichiometry in forests worldwide: Implications of terrestrial Redfield-type ratios. Ecology 85:2390–2401.

McGuire, A. D., R. W. Macdonald, E. A. Schuur, J. W. Harden, P. Kuhry, D. J. Hayes, T. R. Christensen, and M. Heimann. 2010. The carbon budget of the northern cryosphere region. Current Opinion in Environmental Sustainability 2:231–236.

McGuire, A. D., J. M. Melillo, L. A. Joyce, D. W. Kicklighter, A. L. Grace, B. Moore, and C. J. Vörösmarty. 1992. Interactions between carbon and nitrogen dynamics in estimating net primary production for potential vegetation in North America. Global Biogeochemical Cycles 6:101–124.

McGuire, K. J., J. J. McDonnell, M. Weiler, C. Kendall, B. L. McGlynn, J. M. Welker, and J. Seibert. 2005. The role of topography on catchment-scale water residence time. Water Resources Research 41:14.

McKay, C. P., and W. J. Borucki. 1997. Organic synthesis in experimental impact shocks. Science 276:390–392.

McKay, C. P., O. B. Toon, and J. F. Kasting. 1991. Making Mars habitable. Nature 352:489–496.

McKay, D. S., E. K. Gibson, K. L. ThomasKeprta, H. Vali, C. S. Romanek, S. J. Clemett, X.D.F. Chillier, C. R. Maechling, and R. N. Zare. 1996. Search for past life on Mars: Possible relic biogenic activity in Martian meteorite ALH84001. Science 273:924–930.

McKelvey, V. E. 1980. Seabed minerals and the law of the sea. Science 209:464–472.

McKenney, D. J., C. F. Drury, W. I. Findlay, B. Mutus, T. McDonnell, and C. Gajda. 1994. Kinetics of denitrification by *Pseudomonas fluorescens*—Oxygen effects. Soil Biology & Biochemistry 26:901–908.

McKenzie, R., B. Conner, and G. Bodeker. 1999. Increased summertime UV radiation in New Zealand in response to ozone loss. Science 285:1709–1711.

McLain, J.E.T., T. B. Kepler, and D. M. Ahmann. 2002. Belowground factors mediating changes in methane consumption in a forest soil under elevated CO_2. Global Biogeochemical Cycles 16.

McLatchey, G. P., and K. Reddy. 1998. Regulation of organic matter decomposition and nutrient release in a wetland soil. Journal of Environmental Quality 27: 1268–1274.

McLauchlan, K. K., S. E. Hobbie, and W. M. Post. 2006. Conversion from agriculture to grassland builds soil organic matter on decadal timescales. Ecological Applications 16:143–153.

McLaughlin, S. B., and G. E. Taylor. 1981. Relative humidity—Important modifier of pollutant uptake by plants. Science 211:167–169.

McNaughton, S. J. 1988. Mineral nutrition and spatial concentrations of African ungulates. Nature 334:343–345.

McNaughton, S. J. 1990. Mineral nutrition and seasonal movements of African migratory ungulates. Nature 345:613–615.

McNaughton, S. J., F. F. Banyikwa, and M. M. McNaughton. 1997. Promotion of the cycling of diet-enhancing nutrients by African grazers. Science 278:1798–1800.

McNaughton, S. J., and F. S. Chapin. 1985. Effects of phosphorus nutrition and defoliation on C-4 graminoids from the Serengeti plains. Ecology 66: 1617–1629.

McNaughton, S. J., M. Oesterheld, D. A. Frank, and K. J. Williams. 1989. Ecosystem-level patterns of primary productivity and herbivory in terrestrial habitats. Nature 341:142–144.

McNaughton, S. J., N.R.H. Stronach, and N. J. Georgiadis. 1998. Combustion in natural fires and global emissions budgets. Ecological Applications **8**: 464–468.

McNeil, B. I., R. J. Matear, R. M. Key, J. L. Bullister, and J. L. Sarmiento. 2003. Anthropogenic CO_2 uptake by the ocean based on the global chlorofluorocarbon data set. Science **299**:235–239.

McNulty, S. G., J. Boggs, J. D. Aber, L. Rustad, and A. Magill. 2005. Red spruce ecosystem level changes following 14 years of chronic N fertilization. Forest Ecology and Management **219**:279–291.

McNulty, S. G., J. M. Vose, and W. T. Swank. 1996. Loblolly pine hydrology and productivity across the southern United States. Forest Ecology and Management **86**:241–251.

McSween, H. Y. 1989. Chondritic meteorites and the formation of planets. American Scientist **77**:146–153.

Meade, C., J. A. Reffner, and E. Ito. 1994. Synchrotron-infrared absorbency measurements of hydrogen in $MgSiO_3$ perovskite. Science **264**:1558–1560.

Meade, R. H., T. Dunne, J. E. Richey, U. D. Santos, and E. Salati. 1985. Storage and remobilization of suspended sediment in the lower Amazon river of Brazil. Science **228**:488–490.

Measures, C. I., and S. Vink. 2000. On the use of dissolved aluminum in surface waters to estimate dust deposition to the ocean. Global Biogeochemical Cycles **14**:317–327.

Medina-Elizalde, M., and E. J. Rohling. 2012. Collapse of classic Maya civilization related to modest reduction in precipitation. Science **335**:956–959.

Meentemeyer, V. 1978a. Climatic regulation of decomposition rates of organic matter in terrestrial ecosystems. *In* D. C. Adriano and I. L. Brisbin, editors. *Environmental Chemistry and Cycling Processes*, National Technical Information Service, Springfield, Virginia, 779–789.

Meentemeyer, V. 1978b. Macroclimate and lignin control of litter decomposition rates. Ecology **59**:465–472.

Meentemeyer, V., E. O. Box, and R. Thompson. 1982. World patterns and amounts of terrestrial plant litter production. BioScience **32**:125–128.

Megonigal, J. P., W. H. Conner, S. Kroeger, and R. R. Sharitz. 1997. Aboveground production in southeastern floodplain forests: A test of the subsidy-stress hypothesis. Ecology **78**:370–384.

Megonigal, J. P., and A. B. Guenther. 2008. Methane emissions from upland forest soils and vegetation. Tree Physiology **28**:491–498.

Megonigal, J. P., M. E. Hines, and P. T. Visscher. 2003. Anaerobic metabolism: linkages to trace gases and aerobic processes. *In* W. H. Schlesinger, editor. *Biogeochemistry*, Elsevier-Pergamon, 317–324.

Megonigal, J. P., W. H. Patrick, and S. P. Faulkner. 1993. Wetland identification in seasonally flooded forest soils—Soil morphology and redox dyanmics. Soil Science Society of America Journal **57**:140–149.

Megonigal, J. P., and W. H. Schlesinger. 1997. Enhanced CH_4 emissions from a wetland soil exposed to elevated CO_2. Biogeochemistry **37**:77–88.

Megonigal, J. P., and W. H. Schlesinger. 2002. Methane-limited methanotrophy in tidal freshwater swamps. Global Biogeochemical Cycles **16**.

Mehner, T., M. Diekmann, T. Gonsiorczyk, P. Kasprzak, R. Koschel, L. Krienitz, M. Rumpf, M. Schulz, and G. Wauer. 2008. Rapid recovery from eutrophication of a stratified lake by disruption of internal nutrient load. Ecosystems **11**:1142–1156.

Meier, M. F., M. B. Dyurgerov, U. K. Rick, S. O'Neel, W. T. Pfeffer, R. S. Anderson, S. P. Anderson, and A. F. Glazovsky. 2007. Glaciers dominate eustatic sea-level rise in the 21st century. Science **317**:1064–1067.

Meier, R., T. C. Owen, H. E. Matthews, D. C. Jewitt, D. Bockelee-Morvan, N. Biver, J. Crovisier, and D. Gautier. 1998. A determination of the HDO/H_2O ratio in Comet C/1995 O1 (Hale-Bopp). Science **279**:842–844.

Melack, J. M. 1976. Primary productivity and fish yields in tropical lakes. Transactions of the American Fisheries Society **105**:575–580.

Melillo, J. M., J. D. Aber, and J. F. Muratore. 1982. Nitrogen and lignin control of hardwood leaf litter decomposition dynamics. Ecology **63**:621–626.

Melillo, J. M., J. D. Aber, P. A. Steudler, and J. P. Schimel. 1983. Denitrification potentials in a successional sequence of northern hardwood forest stands. *In* R. Hallberg, editor. *Environmental Biogeochemistry, Vol. 5*. Ecological Bulletins Stockholm, Sweden, **35**: 217–228.

Melillo, J. M., S. Butler, J. Johnson, J. Mohan, P. Steudler, H. Lux, E. Burrows, F. Bowles, R. Smith, L. Scott, C. Vario, T. Hill, A. Burton, Y. M. Zhou, and J. Tang. 2011. Soil warming, carbon-nitrogen interactions, and forest carbon budgets. Proceedings of the National Academy of Sciences **108**:9508–9512.

Melillo, J. M., A. D. McGuire, D. W. Kicklighter, B. Moore, C. J. Vörösmarty, and A. L. Schloss. 1993. Global climate change and terrestrial net primary production. Nature **363**:234–240.

Melillo, J. M., J. M. Reilly, D. W. Kicklighter, A. C. Gurgel, T. W. Cronin, S. Paltsev, B. S. Felzer, X. D. Wang, A. P. Sokolov, and C. A. Schlosser. 2009. Indirect emissions from biofuels: How important? Science **326**:1397–1399.

Melillo, J. M., P. A. Steudler, J. D. Aber, K. Newkirk, H. Lux, F. P. Bowles, C. Catricala, A. Magill,

T. Ahrens, and S. Morrisseau. 2002. Soil warming and carbon-cycle feedbacks to the climate system. Science **298**:2173–2176.

Melillo, J. M., P. A. Steudler, B. J. Feigl, C. Neill, D. Garcia, M. C. Piccolo, C. C. Cerri, and H. Tian. 2001. Nitrous oxide emissions from forests and pastures of various ages in the Brazilian Amazon. Journal of Geophysical Research—Atmospheres **106**:34179–34188.

Melosh, H. J., and A. M. Vickery. 1989. Impact erosion of the primordial atmosphere of Mars. Nature **338**: 487–489.

Mendez, J., C. Guieu, and J. Adkins. 2010. Atmospheric input of manganese and iron to the ocean: Seawater dissolution experiments with Saharan and North American dusts. Marine Chemistry **120**: 34–43.

Mengel, K., and H. W. Scherer. 1981. Release of nonexchangeable (fixed) soil ammonium under field conditions during the growing season. Soil Science **131**:226–232.

Mengis, M., R. Gachter, and B. Wehrli. 1997. Sources and sinks of nitrous oxide (N_2O) in deep lakes. Biogeochemistry **38**:281–301.

Mensing, S. A., J. Michaelsen, and R. Byrne. 1999. A 560-year record of Santa Ana fires reconstructed from charcoal deposited in the Santa Barbara Basin, California. Quaternary Research **51**:295–305.

Merrifield, M. A., S. T. Merrifield, and G. T. Mitchum. 2009. An anomalous recent acceleration of global sea level rise. Journal of Climate **22**:5772–5781.

Merrill, A. G., and D. R. Zak. 1992. Factors controlling denitrification rates in upland and swamp forests. Canadian Journal of Forest Research **22**: 1597–1604.

Metson, G. S., R. L. Hale, D. M. Iwaniec, E. M. Cook, J. R. Corman, C. S. Galletti, and D. L. Childers. 2012. Phosphorus in Phoenix: A budget and spatial representation of phosphorus in an urban ecosystem. Ecological Applications **22**:705–721.

Meure, C. M., D. Etheridge, C. Trudinger, P. Steele, R. Langenfelds, T. van Ommen, A. Smith, and J. Elkins. 2006. Law Dome CO_2, CH_4 and N_2O ice-core records extended to 2000 years BP. Geophysical Research Letters **33**.

Meybeck, M. 1977. Dissolved and suspended matter carried by rivers: Composition, time and space variations, and world balance. In H. L. Golterman, editor. Interactions between Sediments and Fresh Water. W. Junk, 25–32.

Meybeck, M. 1979. Major elements contents of river waters and dissolved inputs to the oceans. Revue de Geologie Dynamique et de Geographie Physique **21**: 215–246.

Meybeck, M. 1982. Carbon, nitrogen, and phosphorus transport by world rivers. American Journal of Science **282**:401–450.

Meybeck, M. 1987. Global chemical weathering of surficial rocks estimated from river dissolved loads. American Journal of Science **287**:401–428.

Meybeck, M. 1993. Riverine transport of atmospheric carbon—Sources, global typology and budget. Water Air and Soil Pollution **70**:443–463.

Meyer, C. 1985. Ore metals through geologic history. Science **227**:1421–1428.

Meyer, J. L. 1979. Role of sediments and bryophytes in phosphorus dynamics in a headwater stream ecosystem. Limnology and Oceanography **24**:365–375.

Meyer, J. L. 1980. Dynamics of phosphorus and organic matter during leaf decomposition in a forest stream. Oikos **34**:44–53.

Meyer, J. L., A. C. Benke, R. T. Edwards, and J. B. Wallace. 1997. Organic matter dynamics in the Ogeechee River, a blackwater river in Georgia. Journal of the North American Benthological Society **16**:82–87.

Meyer, J. L., G. E. Likens, and J. Sloane. 1981. Phosphorus, nitrogen and organic carbon flux in a headwater stream. Archiv fur Hydrobiologie **91**:28–44.

Meyer, J. L., J. B. Wallaca, and S. L. Eggert. 1998. Leaf litter as a source of dissolved organic carbon to streams. Ecosystems **1**:240–249.

Meyer, J. L., and G. E. Likens. 1979. Transport and transformation of phosphorus in a forest stream ecosystem. Ecology **60**:1255–1269.

Meyer, J. L., W. H. McDowell, T. L. Bott, J. W. Elwood, C. Ishizaki, J. M. Melack, B. L. Peckarsky, B. J. Peterson, and P. A. Rublee. 1988. Elemental dynamics in streams. Journal of the North American Benthological Society **7**:410–432.

Meyer, J. L., M. J. Paul, and W. K. Taulbee. In press. Ecosystem function in urban streams. Journal of the North American Benthological Society.

Meyer, O. 1994. Functional groups of microorganisms. In E.-D. Schulze and H. A. Mooney, editors. Biodiversity and Ecosystem Function, Springer-Verlag, 67–96.

Michaelis, W., R. Seifert, K. Nauhaus, T. Treude, V. Thiel, M. Blumenberg, K. Knittel, A. Gieseke, K. Peterknecht, T. Pape, A. Boetius, R. Amann, B. B. Jorgensen, F. Widdel, J. Peckmann, N. V. Pimenov, and M. B. Gulin. 2002. Microbial reefs in the Black Sea fueled by anaerobic oxidation of methane. Science **297**:1013–1015.

Michaels, A. F., N. R. Bates, K. O. Buesseler, C. A. Carlson, and A. H. Knap. 1994. Carbon cycle imbalances in the Sargasso sea. Nature **372**:537–540.

Michalopoulos, P., and R. C. Aller. 1995. Rapid clay mineral formation in Amazon delta sediments—Reverse weathering and oceanic elemental cycles. Science **270**: 614–617.

Michalski, G., J. K. Bohlke, and M. Thiemens. 2004a. Long term atmospheric deposition as the source of nitrate and other salts in the Atacama Desert, Chile: New evidence from mass-independent oxygen isotopic compositions. Geochimica et Cosmochimica Acta **68**:4023–4038.

Michalski, G., T. Meixner, M. Fenn, L. Hernandez, A. Sirulnik, E. Allen, and M. Thiemens. 2004b. Tracing atmospheric nitrate deposition in a complex semiarid ecosystem using delta ^{17}O. Environmental Science & Technology **38**:2175–2181.

Michel, R. L., and H. E. Suess. 1975. Bomb tritium in the Pacific ocean. Journal of Geophysical Research **80**: 4139–4152.

Middelburg, J. J. 2011. Chemoautotrophy in the ocean. Geophysical Research Letters **38**.

Middleton, K. R., and G. S. Smith. 1979. Comparison of ammoniacal and nitrate nutrition of prerennial ryegrass through a thermodynamic model. Plant and Soil **53**:487–504.

Mieville, A., C. Granier, C. Liousse, B. Guillaume, F. Mouillot, J. F. Lamarque, J. M. Gregoire, and G. Petron. 2010. Emissions of gases and particles from biomass burning during the 20th century using satellite data and an historical reconstruction. Atmospheric Environment **44**:1469–1477.

Migdisov, A. A., A. B. Ronov, and V. A. Grinenko. 1983. The sulphur cycle in the lithosphere. *In* M. V. Ivanov and J. R. Freney, editors. *The Global Biogeochemical Sulphur Cycle*. Wiley, 25–127.

Mikhailov, V. N. 2010. Water and sediment runoff at the Amazon River mouth. Water Resources **37**:145–159.

Mikutta, R., M. Kleber, M. S. Torn, and R. Jahn. 2006. Stabilization of soil organic matter: Association with minerals or chemical recalcitrance? Biogeochemistry **77**:25–56.

Milesi, C., C. D. Elvidge, R. R. Nemani, and S. W. Running. 2003. Assessing the impact of urban land development on net primary productivity in the southeastern United States. Remote Sensing of Environment **86**:401–410.

Miller, E. K., J. D. Blum, and A. J. Friedland. 1993. Determination of soil exchangeable cation loss and weathering rates using Sr isotopes. Nature **362**:438–441.

Miller, H. G., J. M. Cooper, and J. D. Miller. 1976. Effect of nitrogen supply on nutrients in litter fall and crown leaching in a stand of Corsican pine. Journal of Applied Ecology **13**:233–248.

Miller, J. R., and G. L. Russell. 1992. The impact of global warming on river runoff. Journal of Geophysical Research **97**:2757–2764.

Miller, L., and B. C. Douglas. 2004. Mass and volume contributions to twentieth-century global sea level rise. Nature **428**:406–409.

Miller, M. N., C. E. Dandie, B. J. Zebarth, D. L. Burton, C. Goyer, and J. T. Trevors. 2012. Influence of carbon amendments on soil denitrifier abundance in soil microcosms. Geoderma **170**:48–55.

Miller, S. L. 1953. A production of amino acids under possible primitive Earth conditions. Science **117**: 528–529.

Miller, S. L. 1957. The formation of organic compounds on the primitive Earth. Annals of the New York Academy of Sciences **69**:260–275.

Miller, W. R., and J. I. Drever. 1977. Chemical weathering and related controls on surface water chemistry in Absaroka mountains, Wyoming. Geochimica et Cosmochimica Acta **41**:1693–1702.

Millero, F. J., R. Feistel, D. G. Wright, and T. J. McDougall. 2008. The composition of standard seawater and the definition of the reference-composition salinity scale. Deep-Sea Research Part I—Oceanographic Research Papers **55**:50-72.

Millett, J., D. Godbold, A. R. Smith, and H. Grant. 2012. N_2 fixation and cycling in Alnus glutinosa, *Betula pendula* and *Fagus sylvatica* woodland exposed to free air CO_2 enrichment. Oecologia **169**:541–552.

Milliman, J. D. 1993. Production and accumulation of calcium carbonate in the ocean: Budget of a nonsteady state. Global Biogeochemical Cycles **7**: 927–957.

Milliman, J. D., and R. H. Meade. 1983. World-wide delivery of river sediment to the oceans. Journal of Geology **91**:1–21.

Milliman, J. D., and J.P.M. Syvitski. 1992. Geomorphic tectonic control of sediment discharge to the ocean—The importance of small mountainous rivers. Journal of Geology **100**:525–544.

Milliman, J. D., P. J. Troy, W. M. Balch, A. K. Adams, Y. H. Li, and F. T. Mackenzie. 1999. Biologically mediated dissolution of calcium carbonate above the chemical lysocline? Deep-Sea Research Part I—Oceanographic Research Papers **46**:1653-1669.

Mills, M. M., C. Ridame, M. Davey, J. La Roche, and R. J. Geider. 2004. Iron and phosphorus co-limit nitrogen fixation in the eastern tropical North Atlantic. Nature **429**:292–294.

Milly, P.C.D., J. Betancourt, M. Falkenmark, R. M. Hirsch, Z. W. Kundzewicz, D. P. Lettenmaier, and R. J. Stouffer. 2008. Climate change—Stationarity is dead: Whither water management? Science **319**: 573–574.

Milly, P.C.D., K. A. Dunne, and A. V. Vecchia. 2005. Global pattern of trends in streamflow and water availability in a changing climate. Nature **438**: 347–350.

Milly, P.C.D., R. T. Wetherald, K. A. Dunne, and T. L. Delworth. 2002. Increasing risk of great floods in a changing climate. Nature 415:514–517.

Min, S. K., X. B. Zhang, F. W. Zwiers, and G. C. Hegerl. 2011. Human contribution to more-intense precipitation extremes. Nature 470:376–379.

Minderman, G. 1968. Addition, decomposition and accumulation of organic matter in forests. Journal of Ecology 56:355–362.

Minoura, H., and Y. Iwasaka. 1996. Rapid change in nitrate and sulfate concentrations observed in early stage of precipitation and their deposition processes. Journal of Atmospheric Chemistry 24:39–55.

Minschwaner, K., R. J. Salawitch, and M. B. McElroy. 1993. Absorption of solar radiation by O_2—Implications for O_3 and lifetimes of N_2O, $CFCl_3$, and CF_2Cl_2. Journal of Geophysical Research—Atmospheres 98: 10543–10561.

Minshall, G. W., R. C. Petersen, K. W. Cummins, T. L. Bott, J. R. Sedell, C. E. Cushing, and R. L. Vannote. 1983. Interbiome comparison of stream ecosystem dynamics. Ecological Monographs 53:1–25.

Miranda, M. L., D. A. Hastings, J. E. Aldy, and W. H. Schlesinger. 2011. The environmental justice dimensions of climate change. Environmental Justice 4:17–25.

Mispagel, M. E. 1978. Ecology and bioenergetics of the acridid grasshopper, *Bootettix punctatus*, on creosotebush, *Larrea tridentata*, in the northern Mojave desert. Ecology 59:779–788.

Misra, S., and P. N. Froelich. 2012. Lithium isotope history of Cenozoic seawater: Changes in silicate weathering and reverse weathering. Science 335:818–823.

Mitchell, J.F.B. 1983. The hydrological cycle as simulated by an atmospheric general circulation model. *In* A. Street-Perrott and M. Beran, editors. *Variations in the Global Water Budget*. Reidel, 429–446.

Mitchell, J.F.B., T. C. Johns, J. M. Gregory, and S.F.B. Tett. 1995. Climate response to increasing levels of greenhouse gases and sulfate aerosols. Nature 376:501–504.

Mitchell, L. E., E. J. Brook, T. Sowers, J. R. McConnell, and K. Taylor. 2011a. Multidecadal variability of atmospheric methane, 1000-1800 CE. Journal of Geophysical Research—Biogeosciences 116.

Mitchell, M. J., M. B. David, D. G. Maynard, and S. A. Telang. 1986. Sulfur constituents in soils and streams of a watershed in the Rocky mountains of Alberta. Canadian Journal of Forest Research 16: 315–320.

Mitchell, M. J., C. T. Driscoll, R. D. Fuller, M. B. David, and G. E. Likens. 1989. Effect of whole-tree harvesting on the sulfur dynamics of a forest soil. Soil Science Society of America Journal 53:933–940.

Mitchell, M. J., N. W. Foster, J. P. Shepard, and I. K. Morrison. 1992. Nutrient cycling in Huntington forest and Turkey lakes deciduous stands—Nitrogen and sulfur. Canadian Journal of Forest Research 22: 457–464.

Mitchell, M. J., G. Lovett, S. Bailey, F. Beall, D. Burns, D. Buso, T. A. Clair, F. Courchesne, L. Duchesne, C. Eimers, I. Fernandez, D. Houle, D. S. Jeffries, G. E. Likens, M. D. Moran, C. Rogers, D. Schwede, J. Shanley, K. C. Weathers, and R. Vet. 2011b. Comparisons of watershed sulfur budgets in southeast Canada and northeast US: New approaches and implications. Biogeochemistry 103:181–207.

Mitchell, P. J., E. Veneklaas, H. Lambers, and S.S.O. Burgess. 2009. Partitioning of evapotranspiration in a semi-arid eucalypt woodland in south-western Australia. Agricultural and Forest Meteorology 149: 25–37.

Mitra, S., R. Wassmann, and P.L.G. Vlek. 2005. An appraisal of global wetland area and its organic carbon stock. Current Science 88:25–35.

Mitsch, W. J., C. L. Dorge, and J. R. Wiemhoff. 1979. Ecosystem dynamics and a phosphorus budget of an alluvial cypress swamp in southern Illinois. Ecology 60:1116–1124.

Mitsch, W. J., and J. G. Gosselink. 2007. *Wetlands*, 4th edition. John Wiley and Sons.

Mitsch, W. J., A. Nahlik, P. Wolski, B. Bernal, L. Zhang, and L. Ramberg. 2010. Tropical wetlands: Seasonal hydrologic pulsing, carbon sequestration, and methane emissions. Wetlands Ecology and Management 18:573–586.

Mizutani, H., H. Hasegawa, and E. Wada. 1986. High nitrogen isotope ratio for soils of seabird rookeries. Biogeochemistry 2:221–247.

Mizutani, H., and E. Wada. 1988. Nitrogen and carbon isotope ratios in seabird rookeries and their ecological implications. Ecology 69:340–349.

Mochizuki, Y., K. Koba, and Y. Muneoki. 2012. Strong inhibitory effect of nitrate on atmospheric methane oxidation in forest soils. Soil Biology & Biochemistry 50:164–166.

Mohren, G.M.J., J. Vandenburg, and F. W. Burger. 1986. Phosphorus deficiency induced by nitrogen input in douglas fir in the Netherlands. Plant and Soil 95: 191–200.

Moisander, P. H., R. A. Beinart, I. Hewson, A. E. White, K. S. Johnson, C. A. Carlson, J. P. Montoya, and J. P. Zehr. 2010. Unicellular cyanobacterial distributions broaden the oceanic N_2 fixation domain. Science 327:1512–1514.

Mojzsis, S. J., G. Arrhenius, K. D. McKeegan, T. M. Harrison, A. P. Nutman, and C.R.L. Friend. 1996. Evidence for life on Earth before 3,800 million years ago. Nature **384**:55–59.

Mojzsis, S. J., T. M. Harrison, and R. T. Pidgeon. 2001. Oxygen-isotope evidence from ancient zircons for liquid water at the Earth's surface 4,300 myr ago. Nature **409**:178–181.

Mokany, K., R. J. Raison, and A. S. Prokushkin. 2006. Critical analysis of root: shoot ratios in terrestrial biomes. Global Change Biology **12**:84–96.

Molina, M. J., and F. S. Rowland. 1974. Stratospheric sink for chlorofluoromethanes: chlorine atom-catalyzed destruction of ozone. Nature **249**:810–812.

Molina, M. J., T. L. Tso, L. T. Molina, and F.C.Y. Wang. 1987. Antarctic stratospheric chemistry of chlorine nitrate, hydrogen chloride and ice—Release of active chlorine. Science **238**:1253–1257.

Möller, D. 1990. The Na/Cl ratio in rainwater and the seasalt chloride cycle. Tellus **42B**:254–262.

Molles, M. C., and C. N. Dahm. 1990. A perspective on El Niño and La Nina—Global implications for stream ecology. Journal of the North American Benthological Society **9**:68–76.

Molot, L. A., and P. J. Dillon. 1993. Nitrogen mass balances and denitrification rates in central Ontario lakes. Biogeochemistry **20**:195–212.

Molot, L. A., G. Y. Li, D. L. Findlay, and S. B. Watson. 2010. Iron-mediated suppression of bloom-forming cyanobacteria by oxygen in a eutrophic lake. Freshwater Biology **55**:1102–1117.

Monger, H. C., L. A. Daugherty, W. C. Lindemann, and C. M. Liddell. 1991. Microbial precipitation of pedogenic calcite. Geology **19**:997–1000.

Monk, C. D. 1966. An ecological significance to evergreenness. Ecology **47**:504–505.

Monson, R. K., and E. A. Holland. 2001. Biospheric trace gas fluxes and their control over tropospheric chemistry. Annual Review of Ecology and Systematics **32**:547–576.

Monteiro, L. R., and R. W. Furness. 1997. Accelerated increase in mercury contamination in north Atlantic mesopelagic food chains as indicated by time series of seabird feathers. Environmental Toxicology and Chemistry **16**:2489–2493.

Montgomery, D. R., and W. E. Dietrich. 1988. Where do channels begin? Nature **336**:232–234.

Montoya, J. P., C. M. Holl, J. P. Zehr, A. Hansen, T. A. Villareal, and D. G. Capone. 2004. High rates of N_2 fixation by unicellular diazotrophs in the oligotrophic Pacific ocean. Nature **430**:1027–1031.

Montzka, S. A., M. Aydin, M. Battle, J. H. Butler, E. S. Saltzman, B. D. Hall, A. D. Clarke, D. Mondeel, and J. W. Elkins. 2004. A 350-year atmospheric history for carbonyl sulfide inferred from Antarctic firn air and air trapped in ice. Journal of Geophysical Research—Atmospheres **109**.

Montzka, S. A., J. H. Butler, R. C. Myers, T. M. Thompson, T. H. Swanson, A. D. Clarke, L. T. Lock, and J. W. Elkins. 1996. Decline in the tropospheric abundance of halogen from halocarbons: Implications for stratospheric ozone depletion. Science **272**:1318–1322.

Montzka, S. A., P. Calvert, B. D. Hall, J. W. Elkins, T. J. Conway, P. P. Tans, and C. Sweeney. 2007. On the global distribution, seasonality, and budget of atmospheric carbonyl sulfide (COS) and some similarities to CO_2. Journal of Geophysical Research—Atmospheres **112**.

Montzka, S. A., E. J. Dlugokencky, and J. H. Butler. 2011a. Non-CO_2 greenhouse gases and climate change. Nature **476**:43–50.

Montzka, S. A., M. Krol, E. Dlugokencky, B. Hall, P. Joeckel, and J. Lelieveld. 2011b. Small interannual variability of global atmospheric hydroxyl. Science **331**:67–69.

Mooney, H. A., and W. D. Billings. 1961. Comparative physiological ecology of arctic and alpine populations of *Oxyria digyna*. Ecological Monographs **31**:1–29.

Moore, C. M., M. M. Mills, E. P. Achterberg, R. J. Geider, J. LaRoche, M. I. Lucas, E. L. McDonagh, X. Pan, A. J. Poulton, M.J.A. Rijkenberg, D. J. Suggett, S. J. Ussher, and E.M.S. Woodward. 2009. Large-scale distribution of Atlantic nitrogen fixation controlled by iron availability. Nature Geoscience **2**:867–871.

Moore, J. K., and S. C. Doney. 2007. Iron availability limits the ocean nitrogen inventory stabilizing feedbacks between marine denitrification and nitrogen fixation. Global Biogeochemical Cycles **21**.

Moore, T., C. Blodau, J. Turunen, N. Roulet, and P.J.H. Richard. 2005. Patterns of nitrogen and sulfur accumulation and retention in ombrotrophic bogs, eastern Canada. Global Change Biology **11**:356–367.

Moore, T. R., and M. Dalva. 1993. The influence of temperature and water-table position on carbon dioxide and methane emissions from laboratory columns of peatland soils. Journal of Soil Science **44**:651–664.

Moore, T. R., and R. Knowles. 1989. The influence of water-table levels on methane and carbon dioxide emissions from peatland soils. Canadian Journal of Soil Science **69**:33–38.

Moore, T. R., and N. T. Roulet. 1993. Methane flux—Water table relations in northern wetlands. Geophysical Research Letters **20**:587–590.

Moore, W. S. 1996. Large groundwater inputs to coastal waters revealed by ^{226}Ra enrichments. Nature 380:612–614.

Moore, W. S. 2010. The effect of submarine groundwater discharge on the ocean. Annual Review of Marine Science 2:59–88.

Moorhead, D. L., J. F. Reynolds, and P. J. Fonteyn. 1989. Patterns of stratified soil-water loss in a Chihuahuan desert community. Soil Science 148:244–249.

Moosdorf, N., J. Hartmann, R. Lauerwald, B. Hagedorn, and S. Kempe. 2011. Atmospheric CO_2 consumption by chemical weathering in North America. Geochimica et Cosmochimica Acta 75:7829–7854.

Mora, C. I., S. G. Driese, and L. A. Colarusso. 1996. Middle to late Paleozoic atmospheric CO_2 levels from soil carbonate and organic matter. Science 271:1105–1107.

Moran, M. A., and R. G. Zepp. 1997. Role of photoreactions in the formation of biologically labile compounds from dissolved organic matter. Limnology and Oceanography 42:1307–1316.

Moran, S. B., and R. M. Moore. 1988. Evidence from mesocosm studies for biological removal of dissolved aluminum from seawater. Nature 335: 706–708.

Morel, F.M.M., and N. M. Price. 2003. The biogeochemical cycles of trace metals in the oceans. Science 300: 944–947.

Morel, F.M.M., J. R. Reinfelder, S. B. Roberts, C. P. Chamberlain, J. G. Lee, and D. Yee. 1994. Zinc and carbon co-limitation of marine phytoplankton. Nature 369: 740–742.

Morford, S. L., B. Z. Houlton, and R. A. Dahlgren. 2011. Increased forest ecosystem carbon and nitrogen storage from nitrogen-rich bedrock. Nature 477:78–81.

Morgan, C. G., M. Allen, M. C. Liang, R. L. Shia, G. A. Blake, and Y. L. Yung. 2004. Isotopic fractionation of nitrous oxide in the stratosphere: Comparison between model and observations. Journal of Geophysical Research—Atmospheres 109.

Morley, N., and E. M. Baggs. 2010. Carbon and oxygen controls on N_2O and N_2 production during nitrate reduction. Soil Biology & Biochemistry 42:1864–1871.

Morner, N. A., and G. Etiope. 2002. Carbon degassing from the lithosphere. Global and Planetary Change 33:185–203.

Morowitz, H. J. 1968. *Energy Flow in Biology: Biological Organization as a Problem in Thermal Physics*. Academic Press.

Morris, D. P., and W. M. Lewis. 1988. Phytoplankton nutrient limitation in Colorado mountain lakes. Freshwater Biology 20:315–327.

Morris, R. V., S. W. Ruff, R. Gellert, D. W. Ming, R. E. Arvidson, B. C. Clark, D. C. Golden, K. Siebach, G. Klingelhofer, C. Schroder, I. Fleischer, A. S. Yen, and S. W. Squyres. 2010. Identification of carbonate-rich outcrops on Mars by the Spirit rover. Science 329:421–424.

Morrow, P. A., and V. C. Lamarche. 1978. Tree-ring evidence for chronic insect suppression of productivity in sulalpine eucalyptus. Science 201: 1244–1246.

Morse, J. L., M. Ardon, and E. S. Bernhardt. 2012. Greenhouse gas fluxes in southeastern U.S. coastal plain wetlands under contrasting land uses. Ecological Applications 22:264–280.

Morse, J. W., J. C. Cornwell, T. Arakaki, S. Lin, and M. Huertadiaz. 1992. Iron sulfide and carbonate mineral diagenesis in Baffin bay, Texas. Journal of Sedimentary Petrology 62:671–680.

Morse, J. W., and F. T. Mackenzie. 1998. Hadean ocean carbonate geochemistry. Aquatic Geochemistry 4:301–319.

Mortimer, C. H. 1941. The exchange of dissolved substances between mud and water in lakes. Journal of Ecology 29:280–329.

Mortimer, C. H. 1942. The exchange of dissolved substances between mud and water in lakes. Journal of Ecology 30:147–201.

Morton, S. D., and T. H. Lee. 1974. Algal blooms—Possible effects of iron. Environmental Science & Technology 8:673–674.

Moses, M. E., and J. H. Brown. 2003. Allometry of human fertility and energy use. Ecology Letters 6: 295–300.

Mosier, A., C. Kroeze, C. Nevison, O. Oenema, S. Seitzinger, and O. van Cleemput. 1998. Closing the global N_2O budget: nitrous oxide emissions through the agricultural nitrogen cycle—OECD/IPCC/IEA phase II development of IPCC guidelines for national greenhouse gas inventory methodology. Nutrient Cycling in Agroecosystems 52:225–248.

Mosier, A., D. Schimel, D. Valentine, K. Bronson, and W. Parton. 1991. Methane and nitrous oxide fluxes in native, fertilized and cultivated grasslands. Nature 350:330–332.

Mosier, A. R., W. D. Guenzi, and E. E. Schweizer. 1986. Field denitrification estimation by ^{15}N and acetylene-inhibition techniques. Soil Science Society of America Journal 50:831–833.

Mouillot, F., and C. B. Field. 2005. Fire history and the global carbon budget: a 1° x 1° fire-history reconstruction for the 20th century. Global Change Biology 11:398–420.

Mouillot, F., A. Narasimha, Y. Balkanski, J. F. Lamarque, and C. B. Field. 2006. Global carbon emissions from biomass burning in the 20th century. Geophysical Research Letters 33.

Moulton, K. L., J. West, and R. A. Berner. 2000. Solute flux and mineral mass balance approaches to the quantification of plant effects on silicate weathering. American Journal of Science 300:539–570.

Mount, G. H. 1992. The measurement of tropospheric OH by long-path absorption. 1. Instrumentation. Journal of Geophysical Research—Atmospheres 97: 2427–2444.

Mount, G. H., J. W. Brault, P. V. Johnston, E. Marovich, R. O. Jakoubek, C. J. Volpe, J. Harder, and J. Olson. 1997. Measurement of tropospheric OH by long-path laser absorption at Fritz Peak observatory, Colorado, during the OH photochemistry experiment, fall 1993. Journal of Geophysical Research—Atmospheres 102: 6393–6413.

Mowbray, T., and W. H. Schlesinger. 1988. The buffer capacity of organic soils of the Bluff mountain fen, North Carolina. Soil Science 146:73–79.

Moyers, J. L., L. E. Ranweiler, S. B. Hopf, and N. E. Korte. 1977. Evaluation of particulate trace species in the southwest desert atmosphere. Environmental Science & Technology 11:789–795.

Muhs, D. R., E. A. Bettis, J. Been, and J. P. McGeehin. 2001. Impact of climate and parent material on chemical weathering in loess-derived soils of the Mississippi River valley. Soil Science Society of America Journal 65:1761–1777.

Mulder, A., A. A. Vandegraaf, L. A. Robertson, and J. G. Kuenen. 1995. Anaerobic ammonium oxidation discovered in a denitrifying fluidized-bed reactor. FEMS Microbiology Ecology 16:177–183.

Mulder, J., and A. Stein. 1994. The solubility of aluminum in acidic forest soils—Long-term changes due to acid deposition. Geochimica et Cosmochimica Acta 58:85–94.

Mulholland, P. J. 1993. Hydrometric and stream chemistry evidence of three storm flowpaths in Walker Branch Watershed. Journal of Hydrology 151:129–316.

Mulholland, P. J. 2004. The importance of in-stream uptake for regulating stream concentrations and outputs of N and P from a forested watershed: evidence from long-term chemistry records for Walker Branch Watershed. Biogeochemistry 70:403–426.

Mulholland, P. J., and J. W. Elwood. 1982. The role of lake and reservoir sediments as sinks in the perturbed global carbon cycle. Tellus 34:490–499.

Mulholland, P. J., C. S. Fellows, J. L. Tank, N. B. Grimm, J. R. Webster, S. K. Hamilton, E. Marti, L. Ashkenas, W. B. Bowden, W. K. Dodds, W. H. McDowell, M. J. Paul, and B. J. Peterson. 2001. Inter-biome comparison of factors controlling stream metabolism. Freshwater Biology 46:1503–1517.

Mulholland, P. J., R. O. Hall Jr., D. J. Sobota, W. K. Dodds, S. E.G. Findlay, N. B. Grimm, S. K. Hamilton, W. H. McDowell, J. M. O'Brien, J. L. Tank, L. R. Ashkenas, L. W. Cooper, C. N. Dahm, S. V. Gregory, S. L. Johnson, J. L. Meyer, B. J. Peterson, G. C. Poole, H. M. Valett, J. R. Webster, C. P. Arango, J. J. Beau-lieu, M. J. Bernot, A. J. Burgin, C. L. Crenshaw, A. M. Helton, L. T. Johnson, B. R. Niederlehner, J. D. Potter, R. W. Sheibley, and S. M. Thomas. 2009. Nitrate removal in stream ecosystems measured by ^{15}N-addition experiments: Denitrification. Limnology and Oceanography 54: 666–680.

Mulholland, P. J., A. M. Helton, G. C. Poole, R. O. Hall, S. K. Hamilton, B. J. Peterson, J. L. Tank, L. R. Ashkenas, L. W. Cooper, C. N. Dahm, W. K. Dodds, S.E.G. Findlay, S. V. Gregory, N. B. Grimm, S. L. Johnson, W. H. McDowell, J. L. Meyer, H. M. Valett, J. R. Webster, C. P. Arango, J. J. Beaulieu, M. J. Bernot, A. J. Burgin, C. L. Crenshaw, L. T. Johnson, B. R. Niederlehner, J. M. O'Brien, J. D. Potter, R. W. Sheibley, D. J. Sobota, and S. M. Thomas. 2008. Stream denitrification across biomes and its response to anthropogenic nitrate loading. Nature 452: 202–205.

Mulholland, P. J., J. D. Newbold, J. W. Elwood, and C. L. Hom. 1983. The effect of grazing intensity on phosphorus spiraling in autotrophic streams. Oecologia 58:358–366.

Mulholland, P. J., J. D. Newbold, J. W. Elwood, L. A. Ferren, and J. R. Webster. 1985. Phosphorus spiraling in a woodland stream—Seasonal variations. Ecology 66:1012–1023.

Mulholland, P. J., J. L. Tank, J. R. Webster, W. B. Bowden, W. K. Dodds, S. V. Gregory, N. B. Grimm, S. K. Hamilton, S. L. Johnson, E. Marti, W. H. McDowell, J. L. Merriam, J. L. Meyer, B. J. Peterson, H. M. Valett, and W. M. Wollheim. 2002. Can uptake length in streams be determined by nutrient addition experiments? Results from an interbiome comparison study. Journal of the North American Benthological Society 21:544–560.

Mulitza, S., D. Heslop, D. Pittauerova, H. W. Fischer, I. Meyer, J. B. Stuut, M. Zabel, G. Mollenhauer, J. A. Collins, H. Kuhnert, and M. Schulz. 2010. Increase in African dust flux at the onset of commercial agriculture in the Sahel region. Nature 466:226–228.

Muller, B., M. Maerki, M. Schmid, E. G. Vologina, B. Wehrli, A. Wuest, and M. Sturm. 2005. Internal carbon and nutrient cycling in Lake Baikal: sedimentation, upwelling, and early diagenesis. Global and Planetary Change 46:101–124.

Muller, D. B., T. Wang, B. Duval, and T. E. Graedel. 2006. Exploring the engine of anthropogenic iron cycles.

Proceedings of the National Academy of Sciences **103**:16111–16116.

Muller, P. J., and E. Suess. 1979. Productivity, sedimentation rate, and sedimentary organic matter in the oceans. 1. Organic carbon preservation. Deep-Sea Research Part A—Oceanographic Research Papers **26**:1347–1362.

Muller, R. D., M. Sdrolias, C. Gaina, and W. R. Roest. 2008. Age, spreading rates, and spreading asymmetry of the world's ocean crust. Geochemistry Geophysics Geosystems **9**.

Mumma, M. J., G. L. Villanueva, R. E. Novak, T. Hewagama, B. P. Bonev, M. A. DiSanti, A. M. Mandell, and M. D. Smith. 2009. Strong release of methane on Mars in northern summer 2003. Science **323**:1041–1045.

Mummey, D. L., J. L. Smith, and H. Bolton. 1994. Nitrous oxide flux from a shrub–steppe ecosystem—Sources and regulation. Soil Biology & Biochemistry **26**:279–286.

Munger, J. W., and S. J. Eisenreich. 1983. Continental-scale variations in precipitation chemistry. Environmental Science & Technology **17**:32–42.

Munger, J. W., S. M. Fan, P. S. Bakwin, M. L. Goulden, A. H. Goldstein, A. S. Colman, and S. C. Wofsy. 1998. Regional budgets for nitrogen oxides from continental sources: Variations of rates for oxidation and deposition with season and distance from source regions. Journal of Geophysical Research—Atmospheres **103**:8355–8368.

Murakami, M., K. Hirose, H. Yurimoto, S. Nakashima, and N. Takafuji. 2002. Water in Earth's lower mantle. Science **295**:1885–1887.

Murphy, D. M., D. S. Thomson, and T.M.J. Mahoney. 1998. *In situ* measurements of organics, meteoritic material, mercury, and other elements in aerosols at 5 to 19 kilometers. Science **282**:1664–1669.

Murphy, T. P., D.R.S. Lean, and C. Nalewajko. 1976. Blue-green algae—Their excretion of iron-selective chelators enables them to dominate other algae. Science **192**:900–902.

Murray, J. W., R. T. Barber, M. R. Roman, M. P. Bacon, and R. A. Feely. 1994. Physical and biological controls on carbon cycling in the equatorial Pacific. Science **266**:58–65.

Murray, J. W., and V. Grundmanis. 1980. Oxygen consumption in pelagic marine sediments. Science **209**:1527–1530.

Murray, J. W., E. Johnson, and C. Garside. 1995. A US JGOFS process study in the equatorial Pacific (EqPac)—Introduction. Deep-Sea Research Part II—Topical Studies in Oceanography **42**:275–293.

Murray, J. W., and K. M. Kuivila. 1990. Organic matter diagenesis in the northeast Pacific—Transition from aerobic red clay to suboxic hemipelagic sediments. Deep-Sea Research Part A—Oceanographic Research Papers **37**:59-80.

Murty, S.V.S., and R. K. Mohapatra. 1997. Nitrogen and heavy noble gases in ALH 84001: Signatures of ancient Martian atmosphere. Geochimica et Cosmochimica Acta **61**:5417–5428.

Mustard, J. F., C. D. Cooper, and M. K. Rifkin. 2001. Evidence for recent climate change on Mars from the identification of youthful near-surface ground ice. Nature **412**:411–414.

Myers, R. A., and B. Worm. 2003. Rapid worldwide depletion of predatory fish communities. Nature **423**:280–283.

Myneni, R. B., J. Dong, C. J. Tucker, R. K. Kaufmann, P. E. Kauppi, J. Liski, L. Zhou, V. Alexeyev, and M. K. Hughes. 2001. A large carbon sink in the woody biomass of northern forests. Proceedings of the National Academy of Sciences **98**:14784–14789.

Myneni, R. B., C. D. Keeling, C. J. Tucker, G. Asrar, and R. R. Nemani. 1997. Increased plant growth in the northern high latitudes from 1981 to 1991. Nature **386**:698–702.

Myneni, S.C.B., J. T. Brown, G. A. Martinez, and W. Meyer-Ilse. 1999. Imaging of humic substance macromolecular structures in water and soils. Science **286**:1335–1337.

Myrold, D. D., P. A. Matson, and D. L. Peterson. 1989. Relationships between soil microbial properties and above-ground stand characteristics of conifer forests in Oregon. Biogeochemistry **8**:265–281.

Nadelhoffer, K. F., and B. Fry. 1988. Controls on natural ^{15}N and ^{13}C abundances in forest soil organic matter. Soil Science Society of America Journal **52**:1633–1640.

Nadelhoffer, K. J., J. D. Aber, and J. M. Melillo. 1984. Seasonal patterns of ammonium and nitrate uptake in nine temperate forest ecosystems. Plant and Soil **80**:321–335.

Nadelhoffer, K. J., B. P. Colman, W. S. Currie, A. Magill, and J. D. Aber. 2004. Decadal-scale fates of ^{15}N tracers added to oak and pine stands under ambient and elevated N inputs at the Harvard Forest (USA). Forest Ecology and Management **196**:89–107.

Nadelhoffer, K. J., and J. W. Raich. 1992. Fine root production estimates and belowground carbon allocation in forest ecosystems. Ecology **73**:1139–1147.

Naeem, S., L. J. Thompson, S. P. Lawler, J. H. Lawton, and R. M. Woodfin. 1994. Declining biodiversity can alter the performance of ecosystems. Nature **368**:734–737.

Nagata, T., H. Fukuda, R. Fukuda, and I. Koike. 2000. Bacterioplankton distribution and production in deep Pacific waters: Large-scale geographic variations and possible coupling with sinking particle fluxes. Limnology and Oceanography **45**:426–435.

Najjar, R. G., D. J. Erickson, and S. Madronich. 1995. Modeling the air-sea fluxes of gases formed from the decomposition of dissolved organic matter: Carbonyl sulfide and carbon monoxide. *In* R. G. Zepp and C. Sonntag, editors. *The Role of Nonliving Organic Matter in the Earth's Carbon Cycle.* Wiley, 107–132.

Najjar, R. G., and R. F. Keeling. 1997. Analysis of the mean annual cycle of the dissolved oxygen anomaly in the world ocean. Journal of Marine Research **55**: 117–151.

Nance, J. D., P. V. Hobbs, and L. F. Radke. 1993. Airborne measurements of gases and particles from an Alaskan wildfire. Journal of Geophysical Research—Atmospheres **98**:14873–14882.

Naqvi, S.W.A., D. A. Jayakumar, P. V. Narvekar, H. Naik, V. Sarma, W. D'Souza, S. Joseph, and M. D. George. 2000. Increased marine production of N_2O due to intensifying anoxia on the Indian continental shelf. Nature **408**:346–349.

Nasholm, T., A. Ekblad, A. Nordin, R. Giesler, M. Hogberg, and P. Hogberg. 1998. Boreal forest plants take up organic nitrogen. Nature **392**: 914–916.

Navarre-Sitchler, A., and G. Thyne. 2007. Effects of carbon dioxide on mineral weathering rates at Earth-surface conditions. Chemical Geology **243**:53–63.

Nave, L. E., E. D. Vance, C. W. Swanston, and P. S. Curtis. 2009. Impacts of elevated N inputs on north temperate forest soil C storage, C/N, and net N-mineralization. Geoderma **153**:231–240.

Neary, A. J., and W. I. Gizyn. 1994. Throughfall and stemflow chemistry under deciduous and coniferous forest canopies in south-central Ontario. Canadian Journal of Forest Research **24**:1089–1100.

Neef, L., M. van Weele, and P. van Velthoven. 2010. Optimal estimation of the present-day global methane budget. Global Biogeochemical Cycles **24**.

Neff, J. C., and G. P. Asner. 2001. Dissolved organic carbon in terrestrial ecosystems: Synthesis and a model. Ecosystems **4**:2–48.

Neff, J. C., E. A. Holland, F. J. Dentener, W. H. McDowell, and K. M. Russell. 2002a. The origin, composition and rates of organic nitrogen deposition: A missing piece of the nitrogen cycle? Biogeochemistry **57**:99–136.

Neff, J. C., A. R. Townsend, G. Gleixner, S. J. Lehman, J. Turnbull, and W. D. Bowman. 2002b. Variable effects of nitrogen additions on the stability and turnover of soil carbon. Nature **419**:915–917.

Neill, C. 1992. Comparison of soil coring and ingrowth methods for measuring belowground production. Ecology **73**:1918–1921.

Neilson, R. P., and D. Marks. 1994. A global perspective of regional vegetation and hydrologic sensitivities from climatic change. Journal of Vegetation Science **5**:715–730.

Nelson, D. M., R. F. Anderson, R. T. Barber, M. A. Brzezinski, K. O. Buesseler, Z. Chase, R. W. Collier, M. L. Dickson, R. Francois, M. R. Hiscock, S. Honjo, J. Marra, W. R. Martin, R. N. Sambrotto, F. L. Sayles, and D. E. Sigmon. 2002. Vertical budgets for organic carbon and biogenic silica in the Pacific sector of the Southern Ocean, 1996-1998. Deep-Sea Research Part II-Topical Studies in Oceanography **49**:1645–1674.

Nelson, D. M., P. Treguer, M. A. Brzezinski, A. Leynaert, and B. Queguiner. 1995. Production and dissolution of biogenic silica in the ocean—Revised global estimates, comparison with regional data and relationship to biogenic sedimentation. Global Biogeochemical Cycles **9**:359–372.

Nelson, D. W. 1982. Gaseous losses of nitrogen other than through denitrification. *In* F. J. Stevenson, editor. *Nitrogen in Agricultural Soils.* American Society of Agronomy, Madison, Wisconsin, 327–363.

Nemani, R. R., C. D. Keeling, H. Hashimoto, W. M. Jolly, S. C. Piper, C. J. Tucker, R. B. Myneni, and S. W. Running. 2003. Climate-driven increases in global terrestrial net primary production from 1982 to 1999. Science **300**:1560–1563.

Nepstad, D. C., C. R. Decarvalho, E. A. Davidson, P. H. Jipp, P. A. Lefebvre, G. H. Negreiros, E. D. Dasilva, T. A. Stone, S. E. Trumbore, and S. Vieira. 1994. The role of deep roots in the hydrological and carbon cycles of Amazonian forests and pastures. Nature **372**:666–669.

Nesbitt, S. W., R. Y. Zhang, and R. E. Orville. 2000. Seasonal and global NOx production by lightning estimated from the Optical Transient Detector (OTD). Tellus Series B—Chemical and Physical Meteorology **52**:1206–1215.

Neubauer, S. C., K. Givler, S. K. Valentine, and J. P. Megonigal. 2005. Seasonal patterns and plant-mediated controls of subsurface wetland biogeochemistry. Ecology **86**:3334–3344.

Neue, H. U., R. Wassmann, H. K. Kludze, W. Bujun, and R. S. Lantin. 1997. Factors and processes controlling methane emissions from rice fields. Nutrient Cycling in Agroecosystems **49**:111–117.

Neukum, G. 1977. Lunar cratering. Philosophical Transactions of the Royal Society of London Series

A—Mathematical Physical and Engineering Sciences **285**:267–272.

Neukum, G., R. Jaumann, H. Hoffmann, E. Hauber, J. W. Head, A. T. Basilevsky, B. A. Ivanov, S. C. Werner, S. van Gasselt, J. B. Murray, T. McCord, and H.C.-I. Team. 2004. Recent and episodic volcanic and glacial activity on Mars revealed by the high resolution stereo camera. Nature **432**:971–979.

Nevison, C. D., T. J. Lueker, and R. F. Weiss. 2004. Quantifying the nitrous oxide source from coastal upwelling. Global Biogeochemical Cycles **18**.

Nevison, C. D., R. F. Weiss, and D. J. Erickson. 1995. Global oceanic emissions of nitrous oxide. Journal of Geophysical Research—Oceans **100**:15809–15820.

Newbold, J. D. 1992. Cycles and spirals of nutrients. *In* P. Calow and G. Petts, editors. *Rivers Handbook*, Blackwell, 379–408.

Newbold, J. D., J. W. Elwood, R. V. O'Neill, and W. VanWinkle. 1981. Measuring nutrient spiraling in streams. Candian Journal of Fisheries and Aquatic Science **38**:860–863.

Newman, D. K., and J. F. Banfield. 2002. Geomicrobiology: How molecular-scale interactions underpin biogeochemical systems. Science **296**:1071–1077.

Newman, E. I. 1995. Phosphorus inputs to terrestrial ecosystems. Journal of Ecology **83**:713–726.

Newman, E. I., and R. E. Andrews. 1973. Uptake of phosphorus and potassium in relation to root growth and density. Plant and Soil **38**:49–69.

Newman, P. A., E. R. Nash, S. R. Kawa, S. A. Montzka, and S. M. Schauffler. 2006. When will the Antarctic ozone hole recover? Geophysical Research Letters **33**.

Newsom, H. E., and K.W.W. Sims. 1991. Core formation during early accretion of the Earth. Science **252**:926–933.

Nezat, C. A., J. D. Blum, A. Klaue, C. E. Johnson, and T. G. Siccama. 2004. Influence of landscape position and vegetation on long-term weathering rates at the Hubbard Brook experimental forest, New Hampshire, USA. Geochimica et Cosmochimica Acta **68**: 3065–3078.

Nguyen, B. C., N. Mihalopoulos, J. P. Putaud, and B. Bonsang. 1995. Carbonyl sulfide emissions from biomass burning in the tropics. Journal of Atmospheric Chemistry **22**:55–65.

Nicholls, R. J. 2004. Coastal flooding and wetland loss in the 21st century: changes under the SRES climate and socio-economic scenarios. Global Environmental Change—Human and Policy Dimensions **14**:69–86.

Nielsen, D. R., M. T. Vangenuchten, and J. W. Biggar. 1986. Water flow and solute transport processes in the unsaturated zone. Water Resources Research **22**: S89–S108.

Nielsen, H. 1974. Isotopic composition of the major contributors to atmospheric sulfur. Tellus **26**:213–221.

Nielsen, L. P., N. Risgaard-Petersen, H. Fossing, P. B. Christensen, and M. Sayama. 2010. Electric currents couple spatially separated biogeochemical processes in marine sediment. Nature **463**:1071–1074.

Niemann, H. B., S. K. Atreya, S. J. Bauer, G. R. Carignan, J. E. Demick, R. L. Frost, D. Gautier, J. A. Haberman, D. N. Harpold, D. M. Hunten, G. Israel, J. I. Lunine, W. T. Kasprzak, T. C. Owen, M. Paulkovich, F. Raulin, E. Raaen, and S. H. Way. 2005. The abundances of constituents of Titan's atmosphere from the GCMS instrument on the Huygens probe. Nature **438**:779–784.

Niemann, H. B., S. K. Atreya, G. R. Carignan, T. M. Donahue, J. A. Haberman, D. N. Harpold, R. E. Hartle, D. M. Hunten, W. T. Kasprzak, P. R. Mahaffy, T. C. Owen, N. W. Spencer, and S. H. Way. 1996. The Galileo probe mass spectrometer: Composition of Jupiter's atmosphere. Science **272**:846–849.

Niklas, K. J., and B. J. Enquist. 2002. On the vegetative biomass partitioning of seed plant leaves, stems, and roots. American Naturalist **159**:482–497.

Niklinska, M., M. Maryanski, and R. Laskowski. 1999. Effect of temperature on humus respiration rate and nitrogen mineralization: Implications for global climate change. Biogeochemistry **44**:239–257.

Niles, P. B., W. V. Boynton, J. H. Hoffman, D. W. Ming, and D. Hamara. 2010. Stable isotope measurements of Martian atmospheric CO_2 at the Phoenix landing site. Science **329**:1334–1337.

Nilsson, C., C. A. Reidy, M. Dynesius, and C. Revenga. 2005. Fragmentation and flow regulation of the world's large river systems. Science **308**:405–408.

Nisbet, E. G., and B. Ingham. 1995. Methane output from natural and quasinatural sources: A review of the potential for change and for biotic and abiotic feedbacks. *In* G. M. Woodwell and F. T. MacKenzie, editors. *Biotic Feedbacks in the Global Climatic System*. Oxford University Press, 189–218.

Nisbet, R.E.R., R. Fisher, R. H. Nimmo, D. S. Bendall, P. M. Crill, A. V. Gallego-Sala, E.R.C. Hornibrook, E. Lopez-Juez, D. Lowry, P.B.R. Nisbet, E. F. Shuckburgh, S. Sriskantharajah, C. J. Howe, and E. G. Nisbet. 2009. Emission of methane from plants. Proceedings of the Royal Society B—Biological Sciences **276**: 1347–1354.

Niwa, Y., T. Machida, Y. Sawa, H. Matsueda, T. J. Schuck, C.A.M. Brenninkmeijer, R. Imasu, and M. Satoh. 2012. Imposing strong constraints on tropical terrestrial CO_2 fluxes using passenger aircraft-based measurements. Journal of Geophysical Research—Atmospheres **117**.

Nixon, S. W. 1980. Between coastal marshes and coastal waters—a review of twenty years of speculation and research on the role of salt marshes in estuarine productivity and water chemistry. *In* P. B. Hamilton and K. B. MacDonald, editors. *Estuarine and Wetland Processes*, Plenum, 437–525.

Nixon, S. W. 2003. Replacing the Nile: Are anthropogenic nutrients providing the fertility once brought to the Mediterranean by a great river? Ambio **32**: 30–39.

Nixon, S. W., J. W. Ammerman, L. P. Atkinson, V. M. Berounsky, G. Billen, W. C. Boicourt, W. R. Boynton, T. M. Church, D. M. Ditoro, R. Elmgren, J. H. Garber, A. E. Giblin, R. A. Jahnke, N.J. P. Owens, M.E.Q. Pilson, and S. P. Seitzinger. 1996. The fate of nitrogen and phosphorus at the land sea margin of the North Atlantic Ocean. Biogeochemistry **35**:141–180.

Nixon, S. W., S. L. Granger, and B. L. Nowicki. 1995. An assessment of the annual mass balance of carbon, nitrogen, and phosphorus in Narragansett Bay. Biogeochemistry **31**:15–61.

Nodvin, S. C., C. T. Driscoll, and G. E. Likens. 1988. Soil processes and sulfate loss at the Hubbard Brook experimental forest. Biogeochemistry **5**:185–199.

Nogueira, E. M., P. M. Fearnside, B. W. Nelson, R. I. Barbosa, and E.W.H. Keizer. 2008. Estimates of forest biomass in the Brazilian Amazon: New allometric equations and adjustments to biomass from wood-volume inventories. Forest Ecology and Management **256**:1853–1867.

Nolan, B. T., K. J. Hitt, and B. C. Ruddy. 2002. Probability of nitrate contamination of recently recharged groundwaters in the conterminous United States. Environmental Science & Technology **36**:2138–2145.

Norby, R. J., M. F. Cotrufo, P. Ineson, E. G. O'Neill, and J. G. Canadell. 2001. Elevated CO_2, litter chemistry, and decomposition: a synthesis. Oecologia **127**: 153–165.

Norby, R. J., E. H. DeLucia, B. Gielen, C. Calfapietra, C. P. Giardina, J. S. King, J. Ledford, H. R. McCarthy, D.J.P. Moore, R. Ceulemans, P. De Angelis, A. C. Finzi, D. F. Karnosky, M. E. Kubiske, M. Lukac, K. S. Pregitzer, G. E. Scarascia-Mugnozza, W. H. Schlesinger, and R. Oren. 2005. Forest response to elevated CO_2 is conserved across a broad range of productivity. Proceedings of the National Academy of Sciences **102**:18052–18056.

Norby, R. J., and C. M. Iversen. 2006. Nitrogen uptake, distribution, turnover, and efficiency of use in a CO_2-enriched sweetgum forest. Ecology **87**:5–14.

Norby, R. J., J. M. Warren, C. M. Iversen, B. E. Medlyn, and R. E. McMurtrie. 2010. CO_2 enhancement of forest productivity constrained by limited nitrogen availability. Proceedings of the National Academy of Sciences **107**:19368–19373.

Nordin, A., I. K. Schmidt, and G. R. Shaver. 2004. Nitrogen uptake by arctic soil microbes and plants in relation to soil nitrogen supply. Ecology **85**:955–962.

Norval, M., A. P. Cullen, F. R. de Gruijl, J. Longstreth, Y. Takizawa, R. M. Lucas, F. P. Noonan, and J. C. van der Leun. 2007. The effects on human health from stratospheric ozone depletion and its interactions with climate change. Photochemical & Photobiological Sciences **6**:232–251.

Notholt, J., Z. M. Kuang, C. P. Rinsland, G. C. Toon, M. Rex, N. Jones, T. Albrecht, H. Deckelmann, J. Krieg, C. Weinzierl, H. Bingemer, R. Weller, and O. Schrems. 2003. Enhanced upper tropical tropospheric COS: Impact on the stratospheric aerosol layer. Science **300**:307–310.

Novak, M., J. W. Kirchner, D. Fottova, E. Prechova, I. Jackova, P. Kram, and J. Hruska. 2005. Isotopic evidence for processes of sulfur retention/release in 13 forested catchments spanning a strong pollution gradient (Czech Republic, central Europe). Global Biogeochemical Cycles **19**.

Novelli, P. C., K. A. Masarie, P. M. Lang, B. D. Hall, R. C. Myers, and J. W. Elkins. 2003. Reanalysis of tropospheric CO trends: Effects of the 1997–1998 wildfires. Journal of Geophysical Research—Atmospheres **108**.

Novelli, P. C., K. A. Masarie, P. P. Tans, and P. M. Lang. 1994. Recent changes in atmospheric carbon monoxide. Science **263**:1587–1590.

Nowinski, N. S., L. Taneva, S. E. Trumbore, and J. M. Welker. 2010. Decomposition of old organic matter as a result of deeper active layers in a snow-depth manipulation experiment. Oecologia **163**: 785–792.

Nozette, S., and J. S. Lewis. 1982. Venus—Chemical weathering of igneous rocks and buffering of atmospheric composition. Science **216**:181–183.

Nriagu, J. O. 1989. A global assessment of natural sources of atmospheric trace metals. Nature **338**:47–49.

Nriagu, J. O., and R. D. Coker. 1978. Isotopic composition of sulfur in precipitation within the Great Lakes basin. Tellus **30**:365–375.

Nriagu, J. O., R. D. Coker, and L. A. Barrie. 1991. Origin of sulfur in Canadian arctic haze from isotope measurements. Nature **349**:142–145.

Nriagu, J. O., D. A. Holdway, and R. D. Coker. 1987. Biogenic sulfur and the acidity of rainfall in remote areas of Canada. Science **237**:1189–1192.

Nugent, M. A., S. L. Brantley, C. G. Pantano, and P. A. Maurice. 1998. The influence of natural mineral coatings on feldspar weathering. Nature **395**: 588–591.

Nye, P. H. 1977. The rate-limiting step in plant nutrient absorption from soil. Soil Science **123**:292–297.

Nye, P. H. 1981. Changes of pH across the rhizosphere induced by roots. Plant and Soil 61:7–26.

O'Brien, B. J., and J. D. Stout. 1978. Movement and turnover of soil organic matter as indicated by carbon isotope measurements. Soil Biology & Biochemistry 10:309–317.

O'Dowd, C. D., P. Aalto, K. Hameri, M. Kulmala, and T. Hoffmann. 2002. Aerosol formation—Atmospheric particles from organic vapours. Nature 416:497–498.

O'Hara, G. W., N. Boonkerd, and M. J. Dilworth. 1988. Mineral constraints to nitrogen fixation. Plant and Soil 108:93–110.

O'Leary, M. H. 1988. Carbon isotopes in photosynthesis. BioScience 38:328–336.

O'Neil, J., R. W. Carlson, D. Francis, and R. K. Stevenson. 2008. Neodymium-142 evidence for Hadean mafic crust. Science 321:1828–1831.

O'Neill, R. V., and D. L. De Angelis. 1981. Comparative productivity and biomass relations of forest ecosystems. In D. E. Reichle, editor. Dynamic Properties of Forest Ecosystems. Cambridge University Press, 411–449.

Oades, J. M. 1988. The retention of organic matter in soils. Biogeochemistry 5:35–70.

Oaks, A. 1992. A reevaluation of nitrogen assimilation in roots. BioScience 42:103–111.

Oaks, A. 1994. Primary nitrogen assimilation in higher plants and its regulation. Canadian Journal of Botany 72: 739–750.

Oberg, G., M. Holm, P. Sanden, T. Svensson, and M. Parikka. 2005. The role of organic-matter-bound chlorine in the chlorine cycle: a case study of the Stubbetorp catchment, Sweden. Biogeochemistry 75:241–269.

Oberhummer, H., A. Csoto, and H. Schlattl. 2000. Stellar production rates of carbon and its abundance in the universe. Science 289:88–90.

Odada, E. O. 1992. Growth rates of ferromanganese encrustations on rocks from the Romanche fracture zone, equatorial Atlantic. Deep-Sea Research Part A—Oceanographic Research Papers 39:235-&.

Odasz-Albrigtsen, A. M., H. Tommervik, and P. Murphy. 2000. Decreased photosynthetic efficiency in plant species exposed to multiple airborne pollutants along the Russian-Norwegian border. Canadian Journal of Botany 78:1021–1033.

Odum, E. P., J. T. Finn, and E. H. Franz. 1979. Subsidy-stress. BioScience 29:349–352.

Odum, H. T. 1956. Primary production in flowing waters. Limnology and Oceanography 1:102–117.

Odum, H. T. 1957. Trophic structure and productivity of Silver Springs, Florida. Ecological Monographs 27: 55–112.

Odum, H. T., and C. M. Hoskin. 1958. Comparative studies on the metabolism of marine waters. Publ Inst Marine Sci 5:16–46.

Odum, H. T., and R. F. Wilson. 1962. Further studies on reaeration and metabolism of Texas Bays, 1958-1960. Publ Inst Mar Sci 8:23–55.

Oechel, W. C., S. Cowles, N. Grulke, S. J. Hastings, B. Lawrence, T. Prudhomme, G. Riechers, B. Strain, D. Tissue, and G. Vourlitis. 1994. Transient nature of CO_2 fertilization in arctic tundra. Nature 371: 500–503.

Oechel, W. C., G. L. Vourlitis, S. J. Hastings, R. C. Zulueta, L. Hinzman, and D. Kane. 2000. Acclimation of ecosystem CO_2 exchange in the Alaskan arctic in response to decadal climate warming. Nature 406:978–981.

Oerlemans, J. 2005. Extracting a climate signal from 169 glacier records. Science 308:675–677.

Ogawa, H., Y. Amagai, I. Koike, K. Kaiser, and R. Benner. 2001. Production of refractory dissolved organic matter by bacteria. Science 292:917–920.

Oh, N. H., and D. D. Richter. 2005. Elemental translocation and loss from three highly weathered soil-bedrock profiles in the southeastern United States. Geoderma 126:5–25.

Ohmoto, H., T. Kakegawa, and D. R. Lowe. 1993. 3.4-billion-year-old biogenic pyrites from Barberton, South Africa—Sulfur isotope evidence. Science 262: 555–557.

Ohte, N., and N. Tokuchi. 1999. Geographical variation of the acid buffering of vegetated catchments: Factors determining the bicarbonate leaching. Global Biogeochemical Cycles 13:969–996.

Oki, T., and S. Kanae. 2006. Global hydrological cycles and world water resources. Science 313: 1068–1072.

Okin, G. S., A. R. Baker, I. Tegen, N. M. Mahowald, F. J. Dentener, R. A. Duce, J. N. Galloway, K. Hunter, M. Kanakidou, N. Kubilay, J. M. Prospero, M. Sarin, V. Surapipith, M. Uematsu, and T. Zhu. 2011. Impacts of atmospheric nutrient deposition on marine productivity: Roles of nitrogen, phosphorus, and iron. Global Biogeochemical Cycles 25.

Okin, G. S., N. Mahowald, O. A. Chadwick, and P. Artaxo. 2004. Impact of desert dust on the biogeochemistry of phosphorus in terrestrial ecosystems. Global Biogeochemical Cycles 18.

Olefeldt, D., and N. T. Roulet. 2012. Effects of permafrost and hydrology on the composition and transport of dissolved organic carbon in a subarctic peatland complex. Journal of Geophysical Research—Biogeosciences 117:15.

Olive, K. A., and D. N. Schramm. 1992. Astrophysical ^{7}Li as a product of Big Bang nucleosynthesis and galactic cosmic ray spallation. Nature 360:439–442.

Ollinger, S. V. 2011. Sources of variability in canopy reflectance and the convergent properties of plants. New Phytologist 189:375–394.

Ollinger, S. V., J. D. Aber, G. M. Lovett, S. E. Millham, R. G. Lathrop, and J. M. Ellis. 1993. A spatial model of atmospheric deposition for the northeastern United States. Ecological Applications 3:459–472.

Ollinger, S. V., A. D. Richardson, M. E. Martin, D. Y. Hollinger, S. E. Frolking, P. B. Reich, L. C. Plourde, G. G. Katul, J. W. Munger, R. Oren, M. L. Smith, K. T. Paw, P. V. Bolstad, B. D. Cook, M. C. Day, T. A. Martin, R. K. Monson, and H. P. Schmid. 2008. Canopy nitrogen, carbon assimilation, and albedo in temperate and boreal forests: Functional relations and potential climate feedbacks. Proceedings of the National Academy of Sciences 105: 19336–19341.

Ollinger, S. V., and M. L. Smith. 2005. Net primary production and canopy nitrogen in a temperate forest landscape: An analysis using imaging spectroscopy, modeling and field data. Ecosystems 8:760–778.

Ollinger, S. V., M. L. Smith, M. E. Martin, R. A. Hallett, C. L. Goodale, and J. D. Aber. 2002. Regional variation in foliar chemistry and N cycling among forests of diverse history and composition. Ecology 83: 339–355.

Olsen, S. R., C. V. Cole, F. S. Watanabe, and L. A. Dean. 1954. Estimation of available phosphorus in soils by extraction with sodium bicarbonate. U.S. Department of Agriculture.

Olson, J. S. 1963. Energy storage and the balance of producers and decomposers in ecological systems. Ecology 44:322–331.

Olson, J. S., R. M. Garrels, R. A. Berner, T. V. Armentano, M. I. Dyer, and D. H. Yaalon. 1985. The natural carbon cycle. In J. R. Trabalka, editor. Atmospheric Carbon Dioxide and the Global Carbon Cycle. U.S. Department of Energy, 175–213.

Olson, J. S., J. A. Watts, and L. J. Allison. 1983. Carbon in Live Vegetation of Major World Ecosystems. National Technical Information Service, Springfield, Virginia.

Olson, R. K., and W. A. Reiners. 1983. Nitrification in subalpine balsam fir soils—Tests for inhibitory factors. Soil Biology & Biochemistry 15:413–418.

Olsson, M., and P. A. Melkerud. 1989. Chemical and mineralogical changes during genesis of a podzol from till in southern Sweden. Geoderma 45:267–287.

Olsson, M. T., and P. A. Melkerud. 2000. Weathering in three podzolized pedons on glacial deposits in northern Sweden and central Finland. Geoderma 94: 149–161.

Oltmans, S. J., and D. J. Hofmann. 1995. Increase in lower stratospheric water vapor at a mid-latitude Northern Hemisphere site from 1981 to 1994. Nature 374: 146–149.

Oltmans, S. J., and H. Levy. 1992. Seasonal cycle of surface ozone over the western North Atlantic. Nature 358:392–394.

Oort, A. H. 1970. The energy cycle of the Earth. Scientific American 223(3), 54–63.

Oort, A. H., L. A. Anderson, and J. P. Peixoto. 1994. Estimates of the energy cycle of the oceans. Journal of Geophysical Research—Oceans 99:7665–7688.

Oppenheimer, M., C. B. Epstein, and R. E. Yuhnke. 1985. Acid deposition, smelter emissions, and the linearity issue in the western United States. Science 229: 859–862.

Opsahl, S., and R. Benner. 1997. Distribution and cycling of terrigenous dissolved organic matter in the ocean. Nature 386:480–482.

Ordonez, J. C., P. M. van Bodegom, J.P.M. Witte, I. J. Wright, P. B. Reich, and R. Aerts. 2009. A global study of relationships between leaf traits, climate and soil measures of nutrient fertility. Global Ecology and Biogeography 18:137–149.

Oremland, R. S., and B. F. Taylor. 1978. Sulfate reduction and methanogenesis in marine sediments. Geochimica et Cosmochimica Acta 42:209–214.

Oren, R., K. S. Werk, E. D. Schulze, J. Meyer, B. U. Schneider, and P. Schramel. 1988. Performance of two Picea abies (L) karst stands at different stages of decline. 6. Nutrient concentration. Oecologia 77:151–162.

Orgel, L. E. 1992. Molecular replication. Nature 358: 203–209.

Orgel, L. E. 1994. The origin of life on the Earth. Scientific American 271:77–83.

Orians, K. J., E. A. Boyle, and K. W. Bruland. 1990. Dissolved titanium in the open ocean. Nature 348: 322–325.

Orians, K. J., and K. W. Bruland. 1986. The biogeochemistry of aluminum in the Pacific ocean. Earth and Planetary Science Letters 78:397–410.

Orr, J. C., V. J. Fabry, O. Aumont, L. Bopp, S. C. Doney, R. A. Feely, A. Gnanadesikan, N. Gruber, A. Ishida, F. Joos, R. M. Key, K. Lindsay, E. Maier-Reimer, R. Matear, P. Monfray, A. Mouchet, R. G. Najjar, G. K. Plattner, K. B. Rodgers, C. L. Sabine, J. L. Sarmiento, R. Schlitzer, R. D. Slater, I. J. Totterdell, M. F. Weirig, Y. Yamanaka, and A. Yool. 2005. Anthropogenic ocean acidification over the twenty-first century and its impact on calcifying organisms. nature 437:681–686.

Oschlies, A., and V. Garcon. 1998. Eddy-induced enhancement of primary production in a model of the north Atlantic Ocean. Nature 394:266–269.

Osmond, C. B., K. Winter, and H. Ziegler. 1982. Functional significance of different pathways of CO_2 fixation in photosynthesis. In A. Person and M. H.

Zimmerman, editors. *Encyclopedia of Plant Physiology*. Springer, 479–547.

Ostendorf, B., and J. F. Reynolds. 1993. Relationships between a terrain-based hydrologic model and patch-scale vegetation patterns in an arctic tundra landscape. Landscape Ecology 8:229–237.

Ostlund, H. G. 1983. *Tritium and Radiocarbon, TTO Western North Atlantic Section, GEOSECS Reoccupation. Rosentiel School of Marine and Atmospheric Sciences*, University of Miami.

Ostman, N. L., and G. T. Weaver. 1982. Autumnal nutrient transfers by re-translocation, leaching, and litter fall in a chestnut oak forest in southern Illinois. Canadian Journal of Forest Research 12:40–51.

Oudot, C., C. Andrie, and Y. Montel. 1990. Nitrous oxide production in the tropical Atlantic ocean. Deep-Sea Research Part A—Oceanographic Research Papers 37:183-202.

Ovenden, L. 1990. Peat accumulation in northern wetlands. Quaternary Research 33:377–386.

Owen, D. F., and R. G. Wiegert. 1976. Do consumers maximize plant fitness? Oikos 27:488–492.

Owen, R. M., and D. K. Rea. 1985. Sea floor hydrothermal activity links climate to tectonics—The Eocene carbon dioxide greenhouse. Science 227:166–169.

Owen, T., and K. Biemann. 1976. Composition of the atmosphere at the surface of Mars—Detection of argon-36 and preliminary analysis. Science 193:801–803.

Owen, T., J. P. Maillard, C. Debergh, and B. L. Lutz. 1988. Deuterium on Mars—The abundance of HDO and the value of D/H. Science 240:1767–1770.

Oyama, V. I., G. C. Carle, F. Woeller, and J. B. Pollack. 1979. Venus lower atmospheric composition—Analysis by gas chromatography. Science 203:802–805.

Pacala, S. W., G. C. Hurtt, D. Baker, P. Peylin, R. A. Houghton, R. A. Birdsey, L. Heath, E. T. Sundquist, R. F. Stallard, P. Ciais, P. Moorcroft, J. P. Caspersen, E. Shevliakova, B. Moore, G. Kohlmaier, E. Holland, M. Gloor, M. E. Harmon, S. M. Fan, J. L. Sarmiento, C. L. Goodale, D. Schimel, and C. B. Field. 2001. Consistent land- and atmosphere-based U.S. carbon sink estimates. Science 292:2316–2320.

Pace, M. L., S. R. Carpenter, J. J. Cole, J. J. Coloso, J. F. Kitchell, J. R. Hodgson, J. J. Middelburg, N. D. Preston, C. T. Solomon, and B. C. Weidel. 2007. Does terrestrial organic carbon subsidize the planktonic food web in a clear-water lake? Limnology and Oceanography 52:2177–2189.

Pace, M. L., J. J. Cole, S. R. Carpenter, J. F. Kitchell, J. R. Hodgson, M. C. Van de Bogert, D. L. Bade, E. S. Kritzberg, and D. Bastviken. 2004. Whole-lake

carbon-13 additions reveal terrestrial support of aquatic food webs. Nature 427:240–243.

Paço, T. A., T. S. David, M. O. Henriques, J. S. Pereira, F. Valente, J. Banza, F. L. Pereira, C. Pinto, and J. S. David. 2009. Evapotranspiration from a Mediterranean evergreen oak savannah: The role of trees and pasture. Journal of Hydrology 369:98–106.

Pacyna, E. G., J. M. Pacyna, F. Steenhuisen, and S. Wilson. 2006. Global anthropogenic mercury emission inventory for 2000. Atmospheric Environment 40:4048–4063.

Paerl, H. W. 1995. Coastal eutrophication in relation to atmospheric nitrogen deposition—Current perspectives. Ophelia 41:237–259.

Paerl, H. W. 1996. A comparison of cyanobacterial bloom dynamics in freshwater, estuarine and marine environments. Phycologia 35:25–35.

Paerl, H. W., J. D. Bales, L. W. Ausley, C. P. Buzzelli, L. B. Crowder, L. A. Eby, J. M. Fear, M. Go, B. L. Peierls, T. L. Richardson, and J. S. Ramus. 2001. Ecosystem impacts of three sequential hurricanes (Dennis, Floyd, and Irene) on the United States' largest lagoonal estuary, Pamlico Sound, NC. Proceedings of the National Academy of Sciences 98:5655–5660.

Paerl, H. W., and B. M. Bebout. 1988. Direct measurement of O_2-depleted microzones in marine *Oscillatoria*—Relation to N_2 fixation. Science 241:442–445.

Paerl, H. W., and R. G. Carlton. 1988. Control of nitrogen fixation by oxygen depletion in surface-associated microzones. Nature 332:260–262.

Paerl, H. W., J. D. Willey, M. Go, B. L. Peierls, J. L. Pinckney, and M. L. Fogel. 1999. Rainfall stimulation of primary production in western Atlantic Ocean waters: roles of different nitrogen sources and co-limiting nutrients. Marine Ecology—Progress Series 176:205–214.

Pagani, M., J. C. Zachos, K. H. Freeman, B. Tipple, and S. Bohaty. 2005. Marked decline in atmospheric carbon dioxide concentrations during the Paleogene. Science 309:600–603.

Page, S. E., F. Siegert, J. O. Rieley, H.D.V. Boehm, A. Jaya, and S. Limin. 2002. The amount of carbon released from peat and forest fires in Indonesia during 1997. Nature 420:61–65.

Pagel, B.E.J. 1993. Abundances of light elements. Proceedings of the National Academy of Sciences 90:4789–4792.

Pahlow, M., and U. Riebesell. 2000. Temporal trends in deep ocean Redfield ratios. Science 287:831–833.

Paine, R. T. 1971. The measurement and application of the calorie to ecological problems. Annual Review of Ecology and Systematics 2:145–164.

Palit, S., A. Sharma, and G. Talukder. 1994. Effects of cobalt on plants. Botanical Review 60:149–181.

Pallé, E., P. R. Goode, and P. Montanes-Rodriguez. 2009. Interannual variations in Earth's reflectance 1999–2007. Journal of Geophysical Research—Atmospheres **114**.

Palmer, S. M., and C. T. Driscoll. 2002. Acidic deposition—Decline in mobilization of toxic aluminium. Nature **417**:242–243.

Palmer, S. M., C. T. Driscoll, and C. E. Johnson. 2004. Long-term trends in soil solution and stream water chemistry at the Hubbard Brook Experimental Forest: relationship with landscape position. Biogeochemistry **68**:51–70.

Palmer, T. N., and J. Ralsanen. 2002. Quantifying the risk of extreme seasonal precipitation events in a changing climate. Nature **415**:512–514.

Palter, J. B., M. S. Lozier, and R. T. Barber. 2005. The effect of advection on the nutrient reservoir in the North Atlantic subtropical gyre. Nature **437**:687–692.

Pan, Y., R. Birdsey, J. Hom, K. McCullough, and K. Clark. 2006. Improved estimates of net primary productivity from MODIS satellite data at regional and local scales. Ecological Applications **16**:125–132.

Pan, Y., R. A. Birdsey, J. Fang, R. Houghton, P. E. Kauppi, W. A. Kurz, O. L. Phillips, A. Shvidenko, S. L. Lewis, J. G. Canadell, P. Ciais, R. B. Jackson, S. W. Pacala, A. D. McGuire, S. Piao, A. Rautiainen, S. Sitch, and D. Hayes. 2011. A large and persistent carbon sink in the world's forests. Science **333**:988–993.

Paoli, G. D., L. M. Curran, and D. R. Zak. 2005. Phosphorus efficiency of Bornean rain forest productivity: Evidence against the unimodal efficiency hypothesis. Ecology **86**:1548–1561.

Paolini, J. 1995. Particulate organic carbon and nitrogen in the Orinoco river (Venezuela). Biogeochemistry **29**:59–70.

Papineau, D., S. J. Mojzsis, J. A. Karhu, and B. Marty. 2005. Nitrogen isotopic composition of ammoniated phyllosilicates: case studies from Precambrian metamorphosed sedimentary rocks. Chemical Geology **216**:37–58.

Parai, R., and S. Mukhopadhyay. 2012. How large is the subducted water flux? New constraints on mantle regassing rates. Earth and Planetary Science Letters **317**(318):396–406.

Pardo, L. H., M. E. Fenn, C. L. Goodale, L. H. Geiser, C. T. Driscoll, E. B. Allen, J. S. Baron, R. Bobbink, W. D. Bowman, C. M. Clark, B. Emmett, F. S. Gilliam, T. L. Greaver, S. J. Hall, E. A. Lilleskov, L. Liu, J. A. Lynch, K. J. Nadelhoffer, S. S. Perakis, M. J. Robin-Abbott, J. L. Stoddard, K. C. Weathers, and R. L. Dennis. 2011. Effects of nitrogen deposition and empirical nitrogen critical loads for ecoregions of the United States. Ecological Applications **21**:3049–3082.

Pardo, L. H., H. F. Hemond, J. P. Montoya, T. J. Fahey, and T. G. Siccama. 2002. Response of the natural abundance of N-15 in forest soils and foliage to high nitrate loss following clear cutting. Canadian Journal of Forest Research **32**:1126–1136.

Pardo, L. H., S. G. McNulty, J. L. Boggs, and S. Duke. 2007. Regional patterns in foliar ^{15}N across a gradient of nitrogen deposition in the northeastern US. Environmental Pollution **149**:293–302.

Parfitt, R. L., and R.S.C. Smart. 1978. Mechanism of sulfate adsorption on iron oxides. Soil Science Society of America Journal **42**:48–50.

Park, H., P. J. McGinn, and F.M.M. Morel. 2008. Expression of cadmium carbonic anhydrase of diatoms in seawater. Aquatic Microbial Ecology **51**:183–193.

Park, H., and W. H. Schlesinger. 2002. Global biogeochemical cycle of boron. Global Biogeochemical Cycles **16**.

Park, S., P. Croteau, K. A. Boering, D. M. Etheridge, D. Ferretti, P. J. Fraser, K.-R. Kim, P. B. Krummel, R. L. Langenfelds, T. D. van Ommen, L. P. Steele, and C. M. Trudinger. 2012. Trends and seasonal cycles in the isotopic composition of nitrous oxide since 1940. Nature Geoscience **5**:262–265.

Parker, G. G. 1983. Throughfall and stemflow in the forest nutrient cycle. Advances in Ecological Research **13**:57–133.

Parker, R. S., and B. M. Troutman. 1989. Frequency distribution for suspended sediment loads. Water Resources Research **25**:1567–1574.

Parkes, R. J., B. A. Cragg, S. J. Bale, J. M. Getliff, K. Goodman, P. A. Rochelle, J. C. Fry, A. J. Weightman, and S. M. Harvey. 1994. Deep bacterial biosphere in Pacific ocean sediments. Nature **371**:410–413.

Parkes, R. J., G. Webster, B. A. Cragg, A. J. Weightman, C. J. Newberry, T. G. Ferdelman, J. Kallmeyer, B. B. Jorgensen, I. W. Aiello, and J. C. Fry. 2005. Deep sub-seafloor prokaryotes stimulated at interfaces over geological time. Nature **436**:390–394.

Parkin, T. B. 1987. Soil microsites as a source of denitrification variability. Soil Science Society of America Journal **51**:1194–1199.

Parkin, T. B., A. J. Sexstone, and J. M. Tiedje. 1985. Comparison of field denitrification rates determined by acetylene-based soil core and ^{15}N methods. Soil Science Society of America Journal **49**:94–99.

Parkinson, C. L. 2006. Earth's cryosphere: Current state and recent changes. Annual Review of Environment and Resources **31**:33–60.

Parkinson, C. L., and D. J. Cavalieri. 2008. Arctic sea ice variability and trends, 1979–2006. Journal of Geophysical Research—Oceans **113**.

Parnell, J., A. J. Boyce, D. Mark, S. Bowden, and S. Spinks. 2010. Early oxygenation of the terrestrial environment during the Mesoproterozoic. Nature **468**:290–293.

Parrington, J. R., W. H. Zoller, and N. K. Aras. 1983. Asian dust—Seasonal transport to the Hawaiian islands. Science **220**:195–197.

Parrish, D. D., J. S. Holloway, M. Trainer, P. C. Murphy, G. L. Forbes, and F. C. Fehsenfeld. 1993. Export of North American ozone pollution to the North Atlantic ocean. Science **259**:1436–1439.

Parton, W., W. L. Silver, I. C. Burke, L. Grassens, M. E. Harmon, W. S. Currie, J. Y. King, E. C. Adair, L. A. Brandt, S. C. Hart, and B. Fasth. 2007. Global-scale similarities in nitrogen release patterns during long-term decomposition. Science **315**:361–364.

Parton, W. J., D. S. Schimel, C. V. Cole, and D. S. Ojima. 1987. Analysis of factors controlling soil organic matter levels in Great Plains grasslands. Soil Science Society of America Journal **51**:1173–1179.

Parton, W. J., J.W.B. Stewart, and C. V. Cole. 1988. Dynamics of C, N, P and S in grassland soils—A model. Biogeochemistry **5**:109–131.

Paruelo, J. M., and O. E. Sala. 1995. Water losses in the Patagonian steppe—A modeling approach. Ecology **76**:510–520.

Pasek, M., and K. Block. 2009. Lightning-induced reduction of phosphorus oxidation state. Nature Geoscience **2**:553–556.

Pastor, J., J. D. Aber, C. A. McClaugherty, and J. M. Melillo. 1984. Above-ground production and N and P cycling along a nitrogen mineralization gradient on Blackhawk island, Wisconsin. Ecology **65**:256–268.

Pastor, J., and S. D. Bridgham. 1999. Nutrient efficiency along nutrient availability gradients. Oecologia **118**:50–58.

Patterson, D. B., K. A. Farley, and M. D. Norman. 1999. [4]He as a tracer of continental dust: A 1.9-million-year record of aeolian flux to the west equatorial Pacific ocean. Geochimica et Cosmochimica Acta **63**:615–625.

Paulsen, D. M., H. W. Paerl, and P. E. Bishop. 1991. Evidence that molybdenum-dependent nitrogen fixation is not limited by high sulfate concentrations in marine environments. Limnology and Oceanography **36**:1325–1334.

Pauly, D., and V. Christensen. 1995. Primary production required to sustain global fisheries. Nature **374**:255–257.

Paytan, A., M. Kastner, D. Campbell, and M. H. Thiemens. 2004. Seawater sulfur isotope fluctuations in the Cretaceous. Science **304**:1663–1665.

Paytan, A., M. Kastner, and F. P. Chavez. 1996. Glacial to interglacial fluctuations in productivity in the equatorial Pacific as indicated by marine barite. Science **274**:1355–1357.

Pearson, J. A., D. H. Knight, and T. J. Fahey. 1987. Biomass and nutrient accumulation during stand development in Wyoming lodgepole pine forests. Ecology **68**:1966–1973.

Pearson, L. K., C. H. Hendy, D. P. Hamilton, and R. C. Pickett. 2010. Natural and anthropogenic lead in sediments of the Rotorua lakes, New Zealand. Earth and Planetary Science Letters **297**:536–544.

Pearson, P. N., and M. R. Palmer. 2000. Atmospheric carbon dioxide concentrations over the past 60 million years. Nature **406**:695–699.

Pedersen, H., K. A. Dunkin, and M. K. Firestone. 1999. The relative importance of autotrophic and heterotrophic nitrification in a conifer forest soil as measured by ^{15}N-tracer and pool-dilution techniques. Biogeochemistry **44**:135–150.

Pedro, G., M. Jamagne, and J. C. Begon. 1978. Two routes in genesis of strongly differentiated acid soils under humid, cool-temperate conditions. Geoderma **20**:173–189.

Peel, M. C., T. A. McMahon, and B. L. Finlayson. 2010. Vegetation impact on mean annual evapotranspiration at a global catchment scale. Water Resources Research **46**.

Peierls, B. L., N. F. Caraco, M. L. Pace, and J. J. Cole. 1991. Human influence on river nitrogen. Nature **350**:386–387.

Pellerin, B. A., W. M. Wollheim, C. S. Hopkinson, W. H. McDowell, M. R. Williams, C. J. Vörösmarty, and M. L. Daley. 2004. Role of wetlands and developed land use on dissolved organic nitrogen concentrations and DON/TDN in northeastern U.S. rivers and streams. Limnology and Oceanography **49**:910–918.

Peltier, W. R., and A. M. Tushingham. 1989. Global sea level rise and the greenhouse effect—Might they be connected? Science **244**:806–810.

Peng, T. H., R. Wanninkhof, J. L. Bullister, R. A. Feely, and T. Takahashi. 1998. Quantification of decadal anthropogenic CO_2 uptake in the ocean based on dissolved inorganic carbon measurements. Nature **396**:560–563.

Pennell, K. D., H. L. Allen, and W. A. Jackson. 1990. Phosphorus uptake capacity of a 14-year-old loblolly pine as indicated by a ^{32}P root bioassay. Forest Science **36**:358–366.

Penner, J. E., R. E. Dickinson, and C. A. Oneill. 1992. Effects of aerosols from biomass burning on the global radiation budget. Science **256**:1432–1434.

Peñuelas, J., and J. Azcón-Bieto. 1992. Changes in leaf delta ^{13}C of herbarium plant species during the last three centuries of CO_2 increase. Plant Cell and Environment 15:485–489.

Peñuelas, J., J. G. Canadell, and R. Ogaya. 2011. Increased water-use efficiency during the 20th century did not translate into enhanced tree growth. Global Ecology and Biogeography 20:597–608.

Peñuelas, J., and I. Filella. 2001. Herbaria century record of increasing eutrophication in Spanish terrestrial ecosystems. Global Change Biology 7:427–433.

Peñuelas, J., and R. Matamala. 1990. Changes in N and S leaf content, stomatal density and specific leaf area of 14 plant species during the last three centuries of CO_2 increase. Journal of Experimental Botany 41:1119–1124.

Peñuelas, J., J. Sardans, A. Rivas-Ubach, and I. A. Janssens. 2012. The human-induced imbalance between C, N and P in Earth's life system. Global Change Biology 18:3–6.

Penzias, A. A. 1979. Origin of the elements. Science 205:549–554.

Pepin, R. O. 2006. Atmospheres on the terrestrial planets: Clues to origin and evolution. Earth and Planetary Science Letters 252:1–14.

Perakis, S. S., and L. O. Hedin. 2002. Nitrogen loss from unpolluted South American forests mainly via dissolved organic compounds. Nature 415:416–419.

Perakis, S. S., and E. R. Sinkhorn. 2011. Biogeochemistry of a temperate forest nitrogen gradient. Ecology 92:1481–1491.

Perdue, E. M., K. C. Beck, and J. H. Reuter. 1976. Organic complexes of iron and aluminum in natural waters. Nature 260:418–420.

Perry, K. D., T. A. Cahill, R. A. Eldred, D. D. Dutcher, and T. E. Gill. 1997. Long-range transport of north African dust to the eastern United States. Journal of Geophysical Research—Atmospheres 102:11225–11238.

Peterjohn, W. T., M. B. Adams, and F. S. Gilliam. 1996. Symptoms of nitrogen saturation in two central Appalachian hardwood forest ecosystems. Biogeochemistry 35:507–522.

Peterjohn, W. T., and D. L. Correll. 1984. Nutrient dynamics in an agricultural watershed—Observations on the role of a riparian forest. Ecology 65:1466–1475.

Peterjohn, W. T., R. J. McGervey, A. J. Sexstone, M. J. Christ, C. J. Foster, and M. B. Adams. 1998. Nitrous oxide production in two forested watersheds exhibiting symptoms of nitrogen saturation. Canadian Journal of Forest Research 28:1723–1732.

Peterjohn, W. T., J. M. Melillo, P. A. Steudler, K. M. Newkirk, F. P. Bowles, and J. D. Aber. 1994. Responses of trace gas fluxes and N availability to experimentally-elevated soil temperatures. Ecological Applications 4:617–625.

Peterjohn, W. T., and W. H. Schlesinger. 1991. Factors controlling denitrification in a Chihuahuan desert ecosystem. Soil Science Society of America Journal 55:1694–1701.

Peters, W., M. C. Krol, G. R. van der Werf, S. Houweling, C. D. Jones, J. Hughes, K. Schaefer, K. A. Masarie, A. R. Jacobson, J. B. Miller, C. H. Cho, M. Ramonet, M. Schmidt, L. Ciattaglia, F. Apadula, D. Helta, F. Meinhardt, A. G. di Sarra, S. Piacentino, D. Sferlazzo, T. Aalto, J. Hatakka, J. Strom, L. Haszpra, H.A.J. Meijer, S. van der Laan, R.E.M. Neubert, A. Jordan, X. Rodo, J. A. Morgui, A. T. Vermeulen, E. Popa, K. Rozanski, M. Zimnoch, A. C. Manning, M. Leuenberger, C. Uglietti, A. J. Dolman, P. Ciais, M. Heimann, and P. P. Tans. 2010. Seven years of recent European net terrestrial carbon dioxide exchange constrained by atmospheric observations. Global Change Biology 16:1317–1337.

Petersen, J. M., F. U. Zielinski, T. Pape, R. Seifert, C. Moraru, R. Amann, S. Hourdez, P. R. Girguis, S. D. Wankel, V. Barbe, E. Pelletier, D. Fink, C. Borowski, W. Bach, and N. Dubilier. 2011. Hydrogen is an energy source for hydrothermal vent symbioses. Nature 476:176–180.

Peterson, B. J. 1980. Aquatic primary productivity and the ^{14}C-CO_2 method—A history of the productivity problem. Annual Review of Ecology and Systematics 11:359–385.

Peterson, B. J. 1981. Perspectives on the importance of the oceanic particulate flux in the global carbon cycle. Ocean Science and Engineering 6:71–108.

Peterson, B. J., R. M. Holmes, J. W. McClelland, C. J. Vörösmarty, R. B. Lammers, A. I. Shiklomanov, I. A. Shiklomanov, and S. Rahmstorf. 2002. Increasing river discharge to the Arctic Ocean. Science 298:2171–2173.

Peterson, B. J., and R. W. Howarth. 1987. Sulfur, carbon, and nitrogen isotopes used to trace organic matter flow in the salt marsh estuaries of Sapelo island, Georgia. Limnology and Oceanography 32:1195–1213.

Peterson, B. J., R. W. Howarth, and R. H. Garritt. 1985a. Multiple stable isotopes used to trace the flow of organic matter in estuarine food webs. Science 227:1361–1363.

Peterson, B. J., R. W. Howarth, and R. H. Garritt. 1986. Sulfur and carbon isotopes as tracers of salt marsh organic matter flow. Ecology 67:865–874.

Peterson, B. J., J. McClelland, R. Curry, R. M. Holmes, J. E. Walsh, and K. Aagaard. 2006. Trajectory shifts in the arctic and subarctic freshwater cycle. Science 313:1061–1066.

Peterson, B. J., and J. M. Melillo. 1985. The potential storage of carbon caused by eutrophication of the biosphere. Tellus Series B—Chemical and Physical Meteorology 37:117–127.

Peterson, B. J., W. M. Wollheim, P. J. Mulholland, J. R. Webster, J. L. Meyer, J. L. Tank, E. Marti, W. B. Bowden, H. M. Valett, A. E. Hershey, W. H. McDowell, W. K. Dodds, S. K. Hamilton, S. Gregory, and D. D. Morrall. 2001. Control of nitrogen export from watersheds by headwater streams. Science 292: 86–90.

Peterson, B. J., J. E., Hobbie, A. E. Hershey, M. A. Lock, T. E. Ford, J. R. Vestal, V. L. McKinley, M. J. Hullar, M. C. Miller, R. M. Ventullo, and G. S. Volk. 1985b. Transformation of a tundra river from heterotrophy to autotrophy by addition of phosphorus. Science 229:1383–1386.

Peterson, D. L., M. A. Spanner, S. W. Running, and K. B. Teuber. 1987. Relationship of Thematic Mapper simulator data to leaf-area index of temperate coniferous forests. Remote Sensing of Environment 22: 323–341.

Peterson, L. C., and G. H. Haug. 2005. Climate and the collapse of Maya civilization. American Scientist 93: 322–329.

Petrenko, V. V., A. M. Smith, E. J. Brook, D. Lowe, K. Riedel, G. Brailsford, Q. Hua, H. Schaefer, N. Reeh, R. F. Weiss, D. Etheridge, and J. P. Severinghaus. 2009. (CH_4)-^{14}C measurements in Greenland ice: Investigating last glacial termination CH_4 sources. Science 324:506–508.

Petroff, A., A. Mailliat, M. Amielh, and F. Anselmet. 2008. Aerosol dry deposition on vegetative canopies. Part I: Review of present knowledge. Atmospheric Environment 42:3625–3653.

Petsch, S. T., T. I. Eglinton, and K. J. Edwards. 2001. ^{14}C-dead living biomass: Evidence for microbial assimilation of ancient organic carbon during shale weathering. Science 292:1127–1131.

Petty, G. W. 1995. The status of satellite-based rainfall estimation over land. Remote Sensing of Environment 51:125–137.

Pfeiffer, W. C., O. Malm, C.M.M. Souza, L. D. Delacerda, E. G. Silveira, and W. R. Bastos. 1991. Mercury in the Madeira river ecosystem, Rondonia, Brazil. Forest Ecology and Management 38:239–245.

Philander, G. 1989. El Niño and La Niña. American Scientist 77:451–459.

Phillips, F. M., J. L. Mattick, T. A. Duval, D. Elmore, and P. W. Kubik. 1988. Chlorine-36 and tritium from nuclear weapons fallout as tracers for long-term liquid and vapor movement in desert soils. Water Resources Research 24:1877–1891.

Phillips, O. L., L.E.O.C. Aragao, S. L. Lewis, J. B. Fisher, J. Lloyd, G. Lopez-Gonzalez, Y. Malhi, A. Monteagudo, J. Peacock, C. A. Quesada, G. van der Heijden, S. Almeida, I. Amaral, L. Arroyo, G. Aymard, T. R. Baker, O. Banki, L. Blanc, D. Bonal, P. Brando, J. Chave, A. C. Alves de Oliveira, N. D. Cardozo, C. I. Czimczik, T. R. Feldpausch, M. A. Freitas, E. Gloor, N. Higuchi, E. Jimenez, G. Lloyd, P. Meir, C. Mendoza, A. Morel, D. A. Neill, D. Nepstad, S. Patino, M. Cristina Penuela, A. Prieto, F. Ramirez, M. Schwarz, J. Silva, M. Silveira, A. S. Thomas, H. ter Steege, J. Stropp, R. Vasquez, P. Zelazowski, E. Alvarez Davila, S. Andelman, A. Andrade, K.-J. Chao, T. Erwin, A. Di Fiore, E. Honorio C, H. Keeling, T. J. Killeen, W. F. Laurance, A. Pena Cruz, N.C.A. Pitman, P. Nunez Vargas, H. Ramirez-Angulo, A. Rudas, R. Salamao, N. Silva, J. Terborgh, and A. Torres-Lezama. 2009. Drought sensitivity of the Amazon rainforest. Science 323:1344–1347.

Phillips, R. J., B. J. Davis, K. L. Tanaka, S. Byrne, M. T. Mellon, N. E. Putzig, R. M. Haberle, M. A. Kahre, B. A. Campbell, L. M. Carter, I. B. Smith, J. W. Holt, S. E. Smrekar, D. C. Nunes, J. J. Plaut, A. F. Egan, T. N. Titus, and R. Seu. 2011. Massive CO_2 ice deposits sequestered in the south polar layered deposits of Mars. Science 332:838–841.

Phillips, R. L., S. C. Whalen, and W. H. Schlesinger. 2001. Influence of atmospheric CO_2 enrichment on methane consumption in a temperate forest soil. Global Change Biology 7:557–563.

Phillips, R. P., A. C. Finzi, and E. S. Bernhardt. 2011. Enhanced root exudation induces microbial feedbacks to N cycling in a pine forest under long-term CO_2 fumigation. Ecology Letters 14:187–194.

Phoenix, V. R., P. C. Bennett, A. S. Engel, S. W. Tyler, and F. G. Ferris. 2006. Chilean high-altitude hot-spring sinters: a model system for UV screening mechanisms by early Precambrian cyanobacteria. Geobiology 4:15–28.

Piao, S. L., P. Ciais, P. Friedlingstein, P. Peylin, M. Reichstein, S. Luyssaert, H. Margolis, J. Y. Fang, A. Barr, A. P. Chen, A. Grelle, D. Y. Hollinger, T. Laurila, A. Lindroth, A. D. Richardson, and T. Vesala. 2008. Net carbon dioxide losses of northern ecosystems in response to autumn warming. Nature 451:49–52.

Piao, S. L., J. Y. Fang, P. Ciais, P. Peylin, Y. Huang, S. Sitch, and T. Wang. 2009. The carbon balance of terrestrial ecosystems in China. Nature 458:1009–1013.

Piao, S. L., S. Luyssaert, P. Ciais, I. A. Janssens, A. P. Chen, C. Cao, J. Y. Fang, P. Friedlingstein, Y. Q. Luo, and S. P. Wang. 2010. Forest annual carbon cost: a global-scale analysis of autotrophic respiration. Ecology 91:652–661.

Piao, S. L., X. H. Wang, P. Ciais, B. Zhu, T. Wang, and J. Liu. 2011. Changes in satellite-derived vegetation growth trend in temperate and boreal Eurasia from 1982 to 2006. Global Change Biology 17: 3228–3239.

Piccolo, M. C., C. Neill, J. M. Melillo, C. C. Cerri, and P. A. Steudler. 1996. [15]N natural abundance in forest and pasture soils of the Brazilian Amazon basin. Plant and Soil 182:249–258.

Piccot, S. D., J. J. Watson, and J. W. Jones. 1992. A global inventory of volatile organic compound emissions from anthropogenic sources. Journal of Geophysical Research—Atmospheres 97:9897–9912.

Pierce, R. S., J. W. Hornbeck, G. E. Likens, and F. H. Bormann. 1970. Effects of elimination of vegetation on stream water quantity and quality. Proceedings of the International Association of Scientific Hydrology, Wellington, New Zealand, 311–328.

Pietruczuk, A., J. W. Krzyscin, J. Jaroslawski, J. Podgorski, P. Sobolewski, and J. Wink. 2010. Eyjafjallajokull volcano ash observed over Belsk (52°N, 21°E), Poland, in April 2010. International Journal of Remote Sensing 31:3981–3986.

Pigott, C. D., and K. Taylor. 1964. The distribution of some woodland herbs in relation to the supply of nitrogen and phosphorus in the soil. Journal of Ecology 52(Supplm.):175–185.

Piña-Ochoa, E., and M. Alvarez-Cobelas. 2006. Denitrification in aquatic environments: A cross-system analysis. Biogeochemistry 81:111–130.

Piña-Ochoa, E., S. Hogslund, E. Geslin, T. Cedhagen, N. P. Revsbech, L. P. Nielsen, M. Schweizer, F. Jorissen, S. Rysgaard, and N. Risgaard-Petersen. 2010. Widespread occurrence of nitrate storage and denitrification among Foraminifera and Gromiida. Proceedings of the National Academy of Sciences 107:1148–1153.

Pinay, G., L. Roques, and A. Fabre. 1993. Spatial and temporal patterns of denitrification in a riparian forest. Journal of Applied Ecology 30:581–591.

Ping, C. L., G. J. Michaelson, M. T. Jorgenson, J. M. Kimble, H. Epstein, V. E. Romanovsky, and D. A. Walker. 2008. High stocks of soil organic carbon in the North American arctic region. Nature Geoscience 1:615–619.

Pingitore, N. E., and M. P. Eastman. 1986. The coprecipitation of Sr^{2+} with calcite at 25°C and 1-atm. Geochimica et Cosmochimica Acta 50:2195–2203.

Pinker, R. T., B. Zhang, and E. G. Dutton. 2005. Do satellites detect trends in surface solar radiation? Science 308:850–854.

Pinol, J., J. M. Alcaniz, and F. Roda. 1995. Carbon dioxide efflux and pCO_2 in soils of three Quercus-Ilex montane forests. Biogeochemistry 30:191–215.

Pinto, J. P., G. R. Gladstone, and Y. L. Yung. 1980. Photochemical production of formaldehyde in Earth's primitive atmosphere. Science 210:183–184.

Pinto-Tomas, A. A., M. A. Anderson, G. Suen, D. M. Stevenson, F.S.T. Chu, W. W. Cleland, P. J. Weimer, and C. R. Currie. 2009. Symbiotic nitrogen fixation in the fungus gardens of leaf-cutter ants. Science 326: 1120–1123.

Pirozynski, K. A., and D. W. Malloch. 1975. Origin of land plants—A matter of mycotropism. Biosystems 6:153–164.

Pirrone, N., S. Cinnirella, X. Feng, R. B. Finkelman, H. R. Friedli, J. Leaner, R. Mason, A. B. Mukherjee, G. B. Stracher, D. G. Streets, and K. Telmer. 2010. Global mercury emissions to the atmosphere from anthropogenic and natural sources. Atmospheric Chemistry and Physics 10:5951–5964.

Pizzarello, S., Y. S. Huang, L. Becker, R. J. Poreda, R. A. Nieman, G. Cooper, and M. Williams. 2001. The organic content of the Tagish Lake meteorite. Science 293:2236–2239.

Pizzarello, S., and A. L. Weber. 2004. Prebiotic amino acids as asymmetric catalysts. Science 303:1151.

Planavsky, N. J., P. McGoldrick, C. T. Scott, C. Li, C. T. Reinhard, A. E. Kelly, X. Chu, A. Bekker, G. D. Love, and T. W. Lyons. 2011. Widespread iron-rich conditions in the mid-Proterozoic ocean. Nature 477: 448–451.

Plank, T., and C. H. Langmuir. 1998. The chemical composition of subducting sediment and its consequences for the crust and mantle. Chemical Geology 145:325–394.

Platt, T., and S. Sathyendranath. 1988. Oceanic primary production—Estimation by remote sensing at local and regional scales. Science 241:1613–1620.

Platt, U., M. Rateike, W. Junkermann, J. Rudolph, and D. H. Ehhalt. 1988. New tropospheric OH measurements. Journal of Geophysical Research—Atmospheres 93:5159–5166.

Pletscher, D. H., F. H. Bormann, and R. S. Miller. 1989. Importance of deer compared to other vertebrates in nutrient cycling and energy flow in a northern hardwood ecosystem. American Midland Naturalist 121:302–311.

Pocklington, R., and F. C. Tan. 1987. Seasonal and annual variations in the organic matter contributed by the St Lawrence River to the Gulf of St Lawrence. Geochimica et Cosmochimica Acta 51:2579–2586.

Pöschl, U., S. T. Martin, B. Sinha, Q. Chen, S. S. Gunthe, J. A. Huffman, S. Borrmann, D. K. Farmer, R. M. Garland, G. Helas, J. L. Jimenez, S. M. King, A. Manzi, E. Mikhailov, T. Pauliquevis, M. D. Petters, A. J. Prenni, P. Roldin, D. Rose, J. Schneider, H. Su, S. R. Zorn, P. Artaxo, and M. O. Andreae. 2010.

Rainforest aerosols as biogenic nuclei of clouds and precipitation in the Amazon. Science **329**:1513–1516.

Pokhrel, Y. N., N. Hanasaki, P.J.-F. Yeh, T. J. Yamada, S. Kanae, and T. Oki. 2012. Model estimates of sealevel change due to anthropogenic impacts on terrestrial water storage. Nature Geoscience **5**:389–392.

Polglase, P. J., P. M. Attiwill, and M. A. Adams. 1992. Nitrogen and phosphorus cycling in relation to stand age of *Eucalyptus regnans* F-Muell. 3. Labile inorganic and organic P, phosphatase activity and P availability. Plant and Soil **142**:177–185.

Pollack, J. B., and D. C. Black. 1982. Noble gases in planetary atmospheres—Implications for the origin and evolution of atmospheres. Icarus **51**:169–198.

Pollack, J. B., J. F. Kasting, S. M. Richardson, and K. Poliakoff. 1987. The case for a wet, warm climate on early Mars. Icarus **71**:203–224.

Polovina, J. J., E. A. Howell, and M. Abecassis. 2008. Ocean's least productive waters are expanding. Geophysical Research Letters **35**.

Polubesova, T. A., J. Chorover, and G. Sposito. 1995. Surface-charge characteristics of podzolized soil. Soil Science Society of America Journal **59**:772–777.

Pomeroy, L. R., and D. Deibel. 1986. Temperature regulation of bacterial activity during the spring bloom in Newfoundland coastal waters. Science **233**:359–361.

Pomowski, A., W. G. Zumft, P.M.H. Kroneck, and O. Einsle. 2011. N_2O binding at a 4Cu:2S copper–sulphur cluster in nitrous oxide reductase. Nature **477**:234–U143.

Ponnamperuma, F. N., E. M. Tianco, and T. Loy. 1967. Redox equilibria in flooded soils: I. The iron hydroxide systems. Soil Science **103**:374–382.

Poole, D. K., S. W. Roberts, and P. C. Miller. 1981. Water utilisation. *In* P. C. Miller, editor. *Resource Use by Chaparral and Matorral*. Springer-Verlag, 123–149.

Poole, G. C., S. J. O'Daniel, K. L. Jones, W. W. Woessner, E. S. Bernhardt, A. M. Helton, J. A. Stanford, B. R. Boer, and T. J. Beechie. 2008. Hydrologic spiralling: The role of multiple interactive flow paths in stream ecosystems. River Research and Applications **24**:1018–1031.

Poorter, H. 1993. Interspecific variation in the growth response of plants to an elevated ambient CO_2 concentration. Vegetatio **104**:77–97.

Poorter, H., K. J. Niklas, P. B. Reich, J. Oleksyn, P. Poot, and L. Mommer. 2012. Biomass allocation to leaves, stems and roots: meta-analyses of interspecific variation and environmental control. New Phytologist **193**:30–50.

Poorter, H., and M. Perez-Soba. 2001. The growth response of plants to elevated CO_2 under non-optimal environmental conditions. Oecologia **129**:1–20.

Pope, C. A., III, M. Ezzati, and D. W. Dockery. 2009. Fineparticulate air pollution and life expectancy in the United States. New England Journal of Medicine **360**:376–386.

Pope, E. C., D. K. Bird, and M. T. Rosing. 2012. Isotope composition and volume of Earth's early oceans. Proceedings of the National Academy of Sciences **109**:4371–4376.

Porder, S., and O. A. Chadwick. 2009. Climate and soil-age constraints on nutrient uplift and retention by plants. Ecology **90**:623–636.

Porder, S., D. A. Clark, and P. M. Vitousek. 2006. Persistence of rock-derived nutrients in the wet tropical forests of La Selva, Costa Rica. Ecology **87**:594–602.

Porter, K. G. 1976. Enhancement of algal growth and productivity by grazing zooplankton. Science **192**:1332–1334.

Post, W. M., and K. C. Kwon. 2000. Soil carbon sequestration and land-use change: processes and potential. Global Change Biology **6**:317–327.

Post, W. M., J. Pastor, P. J. Zinke, and A. G. Stangenberger. 1985. Global patterns of soil nitrogen storage. Nature **317**:613–616.

Postberg, F., J. Schmidt, J. Hillier, S. Kempf, and R. Srama. 2011. A salt-water reservoir as the source of a compositionally stratified plume on Enceladus. Nature **474**:620–622.

Postel, S. L. 2000. Entering an era of water scarcity: The challenges ahead. Ecological Applications **10**:941–948.

Postel, S. L., G. C. Daily, and P. R. Ehrlich. 1996. Human appropriation of renewable fresh water. Science **271**:785–788.

Postma, D., and R. Jakobsen. 1996. Redox zonation: Equilibrium constraints on the Fe (III)/SO_4-reduction interface. Geochimica et Cosmochimica Acta **60**:3169–3175.

Potter, C., S. Klooster, C. Hiatt, V. Genovese, and J. C. Castilla-Rubio. 2011. Changes in the carbon cycle of Amazon ecosystems during the 2010 drought. Environmental Research Letters **6**.

Potter, C. S., P. A. Matson, P. M. Vitousek, and E. A. Davidson. 1996. Process modeling of controls on nitrogen trace gas emissions from soils worldwide. Journal of Geophysical Research—Atmospheres **101**:1361–1377.

Potts, M. J. 1978. Deposition of air-borne salt on *Pinus radiata* and underlying soil. Journal of Applied Ecology **15**:543–550.

Pouyat, R. V., I. D. Yesilonis, and N. E. Golubiewski. 2009. A comparison of soil organic stocks between residential turf grass and native soil. Urban Ecosystems **12**:49–62.

Powell, T. M., J. E. Cloern, and L. M. Huzzey. 1989. Spatial and temporal variability in south San Francisco Bay (USA). 1. Horizontal distributions of salinity, suspended sediments, and phytoplankton biomass and productivity. Estuarine Coastal and Shelf Science 28:583–597.

Powers, J. S., R. A. Montgomery, E. C. Adair, F. Q. Brearley, S. J. DeWalt, C. T. Castanho, J. Chave, E. Deinert, J. U. Ganzhorn, M. E. Gilbert, J. A. Gonzalez-Iturbe, S. Bunyavejchewin, H. R. Grau, K. E. Harms, A. Hiremath, S. Iriarte-Vivar, E. Manzane, A. A. de Oliveira, L. Poorter, J. B. Ramanamanjato, C. Salk, A. Varela, G. D. Weiblen, and M. T. Lerdau. 2009. Decomposition in tropical forests: a pan-tropical study of the effects of litter type, litter placement and mesofaunal exclusion across a precipitation gradient. Journal of Ecology 97:801–811.

Powers, J. S., K. K. Treseder, and M. T. Lerdau. 2005. Fine roots, arbuscular mycorrhizal hyphae and soil nutrients in four neotropical rain forests: patterns across large geographic distances. New Phytologist 165:913–921.

Powers, J. S., and E. Veldkamp. 2005. Regional variation in soil carbon and delta ^{13}C in forests and pastures of northeastern Costa Rica. Biogeochemistry 72:315–336.

Powner, M. W., B. Gerland, and J. D. Sutherland. 2009. Synthesis of activated pyrimidine ribonucleotides in prebiotically plausible conditions. Nature 459:239–242.

Prahl, F. G., J. R. Ertel, M. A. Goni, M. A. Sparrow, and B. Eversmeyer. 1994. Terrestrial organic carbon contributions to sediments on the Washington margin. Geochimica et Cosmochimica Acta 58:3035–3048.

Prairie, Y. T., and C. M. Duarte. 2006. Direct and indirect metabolic CO_2 release by humanity. Biogeosciences Discussion 3:1781–1789.

Prather, M. J. 1985. Continental sources of halocarbons and nitrous oxide. Nature 317:221–225.

Prather, M. J., R. Derwent, D. Ehhalt, P. Fraser, E. Sanhueza, and X. Zhou. 1995. Other trace gases and atmospheric chemistry. In J. T. Houghton, L. G. Meira Filho, J. Bruce, H. Lee, B. A. Callender, E. Haites, N. Harris, and K. Maskell, editors. Climate Change 1994. Cambridge University Press, 73–126.

Prather, M. J., and J. Hsu. 2008. NF_3, the greenhouse gas missing from Kyoto. Geophysical Research Letters 35.

Pražák, J., M. Sir, and M. Tesar. 1994. Estimation of plant transpiration from meteorological data under conditions of sufficient soil moisture. Journal of Hydrology 162:409–427.

Pregitzer, K. S., A. J. Burton, D. R. Zak, and A. F. Talhelm. 2008. Simulated chronic nitrogen deposition increases carbon storage in northern temperate forests. Global Change Biology 14:142–153.

Pregitzer, K. S., and E. S. Euskirchen. 2004. Carbon cycling and storage in world forests: biome patterns related to forest age. Global Change Biology 10:2052–2077.

Pregitzer, K. S., D. R. Zak, A. F. Talhelm, A. J. Burton, and J. R. Eikenberry. 2010. Nitrogen turnover in the leaf litter and fine roots of sugar maple. Ecology 91:3456–3462.

Premuzic, E. T., C. M. Benkovitz, J. S. Gaffney, and J. J. Walsh. 1982. The nature and distribution of organic matter in the surface sediments of the world's oceans and seas. Organic Geochemistry 4:63–77.

Press, F., and R. Siever. 1986. Earth. W.H. Freeman.

Prestbo, E. M., and D. A. Gay. 2009. Wet deposition of mercury in the US and Canada, 1996–2005: Results and analysis of the NADP mercury deposition network (MDN). Atmospheric Environment 43:4223–4233.

Preunkert, S., M. Legrand, and D. Wagenbach. 2001. Sulfate trends in a Col du Dome (French Alps) ice core: A record of anthropogenic sulfate levels in the European midtroposphere over the twentieth century. Journal of Geophysical Research—Atmospheres 106:31991–32004.

Preunkert, S., D. Wagenbach, and M. Legrand. 2003. A seasonally resolved alpine ice core record of nitrate: Comparison with anthropogenic inventories and estimation of preindustrial emissions of NO in Europe. Journal of Geophysical Research—Atmospheres 108.

Prézelin, B. B., and B. A. Boczar. 1986. Molecular basis of cell absorption and fluorescence in phytoplankton: Potential applications to studies in optical oceanography. Progress in Phycological Research 4:349–464.

Price, N. M., and F.M.M. Morel. 1990. Cadmium and cobalt substitution for zinc in a marine diatom. Nature 344:658–660.

Price, S. J., R. R. Sherlock, F. M. Kelliher, T. M. McSeveny, K. R. Tate, and L. M. Condron. 2004. Pristine New Zealand forest soil is a strong methane sink. Global Change Biology 10:16–26.

Prince, S. D., E. B. De Colstoun, and L. L. Kravitz. 1998. Evidence from rain-use efficiencies does not indicate extensive Sahelian desertification. Global Change Biology 4:359–374.

Pringle, C. M. 1987. Effects of water and substratum nutrient supplies on lotic periphyton growwth: an integrated bioassay. Canadian Journal of Fisheries and Aquatic Science 44:619–629.

Prinn, R. G. 2003. The cleansing capacity of the atmosphere. Annual Review of Environment and Resources 28:29–57.

Prinn, R. G., J. Huang, R. F. Weiss, D. M. Cunnold, P. J. Fraser, P. G. Simmonds, A. McCulloch, C. Harth, S. Reimann, P. Salameh, S. O'Doherty, R.H.J. Wang, L. W. Porter, B. R. Miller, and P. B. Krummel. 2005. Evidence for variability of atmospheric hydroxyl radicals over the past quarter century. Geophysical Research Letters 32.

Prinn, R. G., R. F. Weiss, B. R. Miller, J. Huang, F. N. Alyea, D. M. Cunnold, P. J. Fraser, D. E. Hartley, and P. G. Simmonds. 1995. Atmospheric trends and lifetime of CH_3CCl_3 and global OH concentrations. Science 269:187–192.

Priscu, J. C., C. H. Fritsen, E. E. Adams, S. J. Giovannoni, H. W. Paerl, C. P. McKay, P. T. Doran, D. A. Gordon, B. D. Lanoil, and J. L. Pinckney. 1998. Perennial Antarctic lake ice: An oasis for life in a polar desert. Science 280:2095–2098.

Pritchard, H. D., R. J. Arthern, D. G. Vaughan, and L. A. Edwards. 2009. Extensive dynamic thinning on the margins of the Greenland and Antarctic ice sheets. Nature 461:971–975.

Pritchard, S. G., A. E. Strand, M. L. McCormack, M. A. Davis, A. C. Finzi, R. B. Jackson, R. Matamala, H. H. Rogers, and R. Oren. 2008a. Fine root dynamics in a loblolly pine forest are influenced by free-air-CO_2-enrichment: a six-year-minirhizotron study. Global Change Biology 14:588–602.

Pritchard, S. G., A. E. Strand, M. L. McCormack, M. A. Davis, and R. Oren. 2008b. Mycorrhizal and rhizomorph dynamics in a loblolly pine forest during 5 years of free-air-CO_2-enrichment. Global Change Biology 14:1252–1264.

Probst, J. L., and Y. Tardy. 1987. Long range streamflow and world continental runoff fluctuations since the beginning of this century. Journal of Hydrology 94: 289–311.

Prokopenko, M. G., D. M. Sigman, W. M. Berelson, D. E. Hammond, B. Barnett, L. Chong, and A. Townsend-Small. 2011. Denitrification in anoxic sediments supported by biological nitrate transport. Geochimica et Cosmochimica Acta 75:7180–7199.

Prospero, J. M., and D. L. Savoie. 1989. Effect of continental sources on nitrate concentrations over the Pacific ocean. Nature 339:687–689.

Prospero, J. M., D. L. Savoie, E. S. Saltzman, and R. Larsen. 1991. Impact of oceanic sources of biogenic sulfur on sulfate aerosol concentrations at Mawson, Antarctica. Nature 350:221–223.

Protz, R., G. J. Ross, I. P. Martini, and J. Terasmae. 1984. Rate of podzolic soil formation near Hudson Bay, Ontario. Canadian Journal of Soil Science 64: 31–49.

Protz, R., G. J. Ross, M. J. Shipitalo, and J. Terasmae. 1988. Podzolic soil development in the southern James Bay lowlands, Ontario. Canadian Journal of Soil Science 68:287–305.

Pryor, S. C., R. J. Barthelmie, L. L. Sorensen, and B. Jensen. 2001. Ammonia concentrations and fluxes over a forest in the midwestern USA. Atmospheric Environment 35:5645–5656.

Pugnaire, F. I., and F. S. Chapin. 1993. Controls over nutrient resorption from leaves of evergreen Mediterranean species. Ecology 74:124–129.

Puig, R., A. Avila, and A. Soler. 2008. Sulphur isotopes as tracers of the influence of a coal-fired power plant on a Scots pine forest in Catalonia (NE Spain). Atmospheric Environment 42:733–745.

Putaud, J. P., S. Belviso, B. C. Nguyen, and N. Mihalopoulos. 1993. Dimethylsulfide, aerosols, and condensation nuclei over the tropical northeastern Atlantic ocean. Journal of Geophysical Research—Atmospheres 98:14863–14871.

Pye, K. 1987. *Eolian Dust and Dust Deposits*. Academic Press, London.

Pyle, D. M., and T. A. Mather. 2003. The importance of volcanic emissions for the global atmospheric mercury cycle. Atmospheric Environment 37:5115–5124.

Quack, B., and D.W.R. Wallace. 2003. Air-sea flux of bromoform: Controls, rates, and implications. Global Biogeochemical Cycles 17.

Quade, J., T. E. Cerling, and J. R. Bowman. 1989a. Development of Asian monsoon revealed by marked ecological shift during the latest Miocene in northern Pakistan. Nature 342:163–166.

Quade, J., T. E. Cerling, and J. R. Bowman. 1989b. Systematic variations in the carbon and oxygen isotopic composition of pedogenic carbonate along elevation transects in the southern Great Basin, United States. Geological Society of America Bulletin 101:464–475.

Qualls, R. G., and B. L. Haines. 1991. Fluxes of dissolved organic nutrients and humic substances in a deciduous forest. Ecology 72:254–266.

Qualls, R. G., and B. L. Haines. 1992. Biodegradability of dissolved organic matter in forest throughfall, soil solution, and stream water. Soil Science Society of America Journal 56:578–586.

Quay, P., R. Sonnerup, T. Westby, J. Stutsman, and A. McNichol. 2003. Changes in the $^{13}C/^{12}C$ of dissolved inorganic carbon in the ocean as a tracer of anthropogenic CO_2 uptake. Global Biogeochemical Cycles 17.

Quay, P., J. Stutsman, D. Wilbur, A. Snover, E. Dlugokencky, and T. Brown. 1999. The isotopic composition of atmospheric methane. Global Biogeochemical Cycles 13:445–461.

Quay, P., D. Wilbur, J. Richey, and A. Devol. 1995. The $^{18}O:^{16}O$ of dissolved oxygen in rivers and lakes in the Amazon Basin: Determining the ratio of respiration to photosynthesis rates in freshwaters. Limnology and Oceanography 40:718–729.

Quay, P. D. 1988. Isotopic composition of methane released from wetlands: Implications for the increase in atmospheric methane. Global Biogeochemical Cycles 2:385–397.

Quay, P. D., S. L. King, J. Stutsman, D. O. Wilbur, L. P. Steele, I. Fung, R. H. Gammon, T. A. Brown, F. W. Farrell, P. M. Gootes, and F. H. Schmidt. 1991. Carbon isotopic composition of atmospheric CH_4: Fossil and biomass burning source strengths. Global Biogeochemical Cycles 5:25–47.

Quay, P. D., C. Peacock, K. Bjorkman, and D. M. Karl. 2010. Measuring primary production rates in the ocean: Enigmatic results between incubation and non-incubation methods at Station ALOHA. Global Biogeochemical Cycles 24.

Quinn, J. M., and M. J. Stroud. 2002. Water quality and sediment and nutrient export from New Zealand hill-land catchments of contrasting land use. New Zealand Journal of Marine and Freshwater Research 36:409–429.

Quinn, P. K., R. J. Charlson, and T. S. Bates. 1988. Simultaneous observations of ammonia in the atmosphere and ocean. Nature 335:336–338.

Quinn, P.K. and T. S. Bates. 2011. The case against climate regulation via oceanic phytoplankton sulphur emissions. Nature 480:51–56.

Quinn, P. K., R. J. Charlson, and W. H. Zoller. 1987. Ammonia, the dominant base in the remote marine atmosphere. Tellus 39B:413–425.

Quist, M. E., T. Nasholm, J. Lindeberg, C. Johannisson, L. Hogbom, and P. Hogberg. 1999. Responses of a nitrogen-saturated forest to a sharp decrease in nitrogen input. Journal of Environmental Quality 28:1970–1977.

Raaimakers, D., R.G.A. Boot, P. Dijkstra, S. Pot, and T. Pons. 1995. Photosynthetic rates in relation to leaf phosphorus content in pioneer versus climax tropical rain forest trees. Oecologia 102:120–125.

Rabalais, N. N., R. E. Turner, R. J. Diaz, and D. Justic. 2009. Global change and eutrophication of coastal waters. Ices Journal of Marine Science 66:1528–1537.

Rabalais, N. N., R. E. Turner, and W. J. Wiseman. 2002. Gulf of Mexico hypoxia, aka "The dead zone." Annual Review of Ecology and Systematics 33:235–263.

Rabenhorst, A. C. 2009. Making soil oxidation-reduction potential measurements using multimeters. Soil Science Society of America Journal 73:2198–2201.

Rabenhorst, M., and K. Haering. 1989. Soil micromorphology of a Chesapeake Bay tidal marsh—Implications for sulfur accumulation. Soil Science 147:339–347.

Rabinowitz, J., J. Flores, R. Krebsbach, and G. Rogers. 1969. Peptide formation in the presence of linear or cyclic polyphosphates. Nature 224:795–796.

Rabouille, C., F. T. Mackenzie, and L. M. Ver. 2001. Influence of the human perturbation on carbon, nitrogen, and oxygen biogeochemical cycles in the global coastal ocean. Geochimica et Cosmochimica Acta 65:3615–3641.

Rachmilevitch, S., A. B. Cousins, and A. J. Bloom. 2004. Nitrate assimilation in plant shoots depends on photorespiration. Proceedings of the National Academy of Sciences 101:11506–11510.

Raciti, S. M., P. M. Groffman, J. C. Jenkins, R. V. Pouyat, T. J. Fahey, S.T.A. Pickett, and M. L. Cadenasso. 2011. Accumulation of carbon and nitrogen in residential soils with different land-use histories. Ecosystems 14:287–297.

Raes, F., R. Van Dingenen, E. Vignati, J. Wilson, J. P. Putaud, J. H. Seinfeld, and P. Adams. 2000. Formation and cycling of aerosols in the global troposphere. Atmospheric Environment 34: 4215–4240.

Raghoebarsing, A. A., A. Pol, K. T. van de Pas-Schoonen, A.J.P. Smolders, K. F. Ettwig, W.I.C. Rijpstra, S. Schouten, J.S.S. Damste, H.J.M. Op den Camp, M.S.M. Jetten, and M. Strous. 2006. A microbial consortium couples anaerobic methane oxidation to denitrification. Nature 440:918–921.

Raghoebarsing, A. A., A.J.P. Smolders, M. C. Schmid, W.I.C. Rijpstra, M. Wolters-Arts, J. Derksen, M.S.M. Jetten, S. Schouten, J. S. Sinninghe Damsté, L.P.M. Lamers, J.G.M. Roelofs, H.J.M. Op den Camp, and M. Strous. 2005. Methanotrophic symbionts provide carbon for photosynthesis in peat bogs. Nature 436:1153–1156.

Raghubanshi, A. S. 1992. Effect of topography on selected soil properties and nitrogen mineralization in a dry tropical forest. Soil Biology & Biochemistry 24:145–150.

Ragueneau, O., P. Treguer, A. Leynaert, R. F. Anderson, M. A. Brzezinski, D. J. DeMaster, R. C. Dugdale, J. Dymond, G. Fischer, R. Francois, C. Heinze, E. Maier-Reimer, V. Martin-Jezequel, D. M. Nelson, and B. Queguiner. 2000. A review of the Si cycle in the modem ocean: recent progress and missing gaps in the application of biogenic opal as a paleoproductivity proxy. Global and Planetary Change 26:317–365.

Rahn, K. A., and D. H. Lowenthal. 1984. Elemental tracers of distant regional pollution aerosols. Science 223:132–139.

Raich, J. W., and K. J. Nadelhoffer. 1989. Belowground carbon allocation in forest ecosystems—Global trends. Ecology 70:1346–1354.

Raich, J. W., C. S. Potter, and D. Bhagawati. 2002. Inter-annual variability in global soil respiration, 1980-94. Global Change Biology 8:800–812.

Raich, J. W., and W. H. Schlesinger. 1992. The global carbon dioxide flux in soil respiration and its relationship to vegetation and climate. Tellus Series B—Chemical and Physical Meteorology 44:81–99.

Raich, J. W., and A. Tufekcioglu. 2000. Vegetation and soil respiration: Correlations and controls. Biogeochemistry 48:71–90.

Raison, R. J. 1979. Modification of the soil environment by vegetation fires, with particular reference to nitrogen transformations—Review. Plant and Soil 51:73–108.

Raison, R. J., M. J. Connell, and P. K. Khanna. 1987. Methodology for studying fluxes of soil mineral N *in situ*. Soil Biology & Biochemistry 19:521–530.

Raison, R. J., P. K. Khanna, and P. V. Woods. 1985. Mechanisms of element transfer to the atmosphere during vegetation fires. Canadian Journal of Forest Research 15:132–140.

Raiswell, R., and R. A. Berner. 1986. Pyrite and organic matter in Phanerozoic normal marine shales. Geochimica et Cosmochimica Acta 50:1967–1976.

Ralph, B. J. 1979. Oxidative reactions in the sulfur cycle. *In* P. A. Trudinger and D. J. Swaine, editors. *Biogeochemical Cycling of Mineral-Forming Elements*. Elsevier, 369–400.

Ramanathan, V. 1988. The greenhouse theory of climate change—A test by an inadvertent global experiment. Science 240:293–299.

Ramankutty, N., and J. A. Foley. 1998. Characterizing patterns of global land use: An analysis of global croplands data. Global Biogeochemical Cycles 12:667–685.

Ramillien, G., A. Lombard, A. Cazenave, E. R. Ivins, M. Llubes, F. Remy, and R. Biancale. 2006. Interannual variations of the mass balance of the Antarctica and Greenland ice sheets from GRACE. Global and Planetary Change 53:198–208.

Ramirez, A. J., and A. W. Rose. 1992. Analytical geochemistry of organic phosphorus and its correlation with organic carbon in marine and fluvial sediments and soils. American Journal of Science 292:421–454.

Rampino, M. R., S. Self, and R. B. Stothers. 1988. Volcanic winters. Annual Review of Earth and Planetary Sciences 16:73–99.

Randel, W. J., M. Park, L. Emmons, D. Kinnison, P. Bernath, K. A. Walker, C. Boone, and H. Pumphrey. 2010. Asian monsoon transport of pollution to the stratosphere. Science 328:611–613.

Randerson, J. T., F. S. Chapin, J. W. Harden, J. C. Neff, and M. E. Harmon. 2002. Net ecosystem production: A comprehensive measure of net carbon accumulation by ecosystems. Ecological Applications 12:937–947.

Randerson, J. T., M. V. Thompson, T. J. Conway, I. Y. Fung, and C. B. Field. 1997. The contribution of terrestrial sources and sinks to trends in the seasonal cycle of atmospheric carbon dioxide. Global Biogeochemical Cycles 11:535–560.

Randlett, D. L., D. R. Zak, and N. W. Macdonald. 1992. Sulfate adsorption and microbial immobilization in northern hardwood forests along an atmospheric deposition gradient. Canadian Journal of Forest Research 22:1843–1850.

Raper, C. D., D. L. Osmond, M. Wann, and W. W. Weeks. 1978. Interdependence of root and shoot activities in determining nitrogen uptake rate of roots. Botanical Gazette 139:289–294.

Rasmussen, B. 1996. Early-diagenetic REE-phosphate minerals (florencite, gorceixite, crandallite, and xenotime) in marine sandstones: A major sink for oceanic phosphorus. American Journal of Science 296:601–632.

Rasmussen, B. 2000. Filamentous microfossils in a 3,235-million-year-old volcanogenic massive sulphide deposit. Nature 405:676–679.

Rasmussen, B., I. R. Fletcher, J. J. Brocks, and M. R. Kilburn. 2008. Reassessing the first appearance of eukaryotes and cyanobacteria. Nature 455:1101–1104.

Rasmussen, C., R. J. Southard, and W. R. Horwath. 2006. Mineral control of organic carbon mineralization in a range of temperate conifer forest soils. Global Change Biology 12:834–847.

Rastetter, E. B., R. B. McKane, G. R. Shaver, and J. M. Melillo. 1992. Changes in C-storage by terrestrial ecosystems—How C-N interactions restrict responses to CO_2 and temperature. Water Air and Soil Pollution 64:327–344.

Rauch, J. N., and T. E. Graedel. 2007. Earth's anthrobiogeochemical copper cycle. Global Biogeochemical Cycles 21.

Raval, A., and V. Ramanathan. 1989. Observational determination of the greenhouse effect. Nature 342:758–761.

Raven, J. A., and C. S. Cockell. 2006. Influence on photosynthesis of starlight, moonlight, planetlight, and light pollution (reflections on photosynthetically active radiation in the Universe). Astrobiology 6:668–675.

Raven, J. A., and P. G. Falkowski. 1999. Oceanic sinks for atmospheric CO_2. Plant Cell and Environment 22:741–755.

Raven, J. A., A. A. Franco, E. L. Dejesus, and J. Jacobneto. 1990. H^+ extrusion and organic acid synthesis in

N_2-fixing symbioses involving vascular plants. New Phytologist **114**:369–389.

Raven, J. A., B. Wollenweber, and L. L. Handley. 1992. A comparison of ammonium and nitrate as nitrogen sources for photolithotrophs. New Phytologist **121**: 19–32.

Raven, J. A., B. Wollenweber, and L. L. Handley. 1993. The quantitative role of ammonia/ammonium transport and metabolism by plants in the global nitrogen cycle. Physiologia Plantarum **89**:512–518.

Ravi, S., P. D'Odorico, D. D. Breshears, J. P. Field, A. S. Goudie, T. E. Huxman, J. Li, G. S. Okin, R. J. Swap, A. D. Thomas, S. Van Pelt, J. J. Whicker, and T. M. Zobeck. 2011. Aeolian processes and the biosphere. Reviews of Geophysics **49**.

Ravishankara, A. R. 1997. Heterogeneous and multiphase chemistry in the troposphere. Science **276**: 1058–1065.

Ravishankara, A. R., J. S. Daniel, and R. W. Portmann. 2009. Nitrous oxide (N_2O): The dominant ozone-depleting substance emitted in the 21st century. Science **326**:123–125.

Raymo, M. E., and J. X. Mitrovica. 2012. Collapse of polar ice sheets during the stage 11 interglacial. Nature **483**:453–456.

Raymond, J., and D. Segre. 2006. The effect of oxygen on biochemical networks and the evolution of complex life. Science **311**:1764–1767.

Raymond, P. A., and J. E. Bauer. 2001. Riverine export of aged terrestrial organic matter to the North Atlantic Ocean. Nature **409**:497–500.

Raymond, P. A., and J. J. Cole. 2003. Increase in the export of alkalinity from North America's largest river. Science **301**:88–91.

Raymond, P. A., N. H. Oh, R. E. Turner, and W. Broussard. 2008. Anthropogenically enhanced fluxes of water and carbon from the Mississippi River. Nature **451**:449–452.

Reader, R. J., and J. M. Stewart. 1972. Relationship between net primary production and accumulation for a peatland in southeastern Manitoba. Ecology **53**: 1024–1037.

Reay, D. S., F. Dentener, P. Smith, J. Grace, and R. A. Feely. 2008. Global nitrogen deposition and carbon sinks. Nature Geoscience **1**:430–437.

Reddy, K. J., W. L. Lindsay, S. M. Workman, and J. I. Drever. 1990. Measurement of calcite ion activity products in soils. Soil Science Society of America Journal **54**:67–71.

Reddy, K. R., and R. D. DeLaune. 2008. *Biogeochemistry of Wetlands: Science and applications.* Taylor and Francis.

Reddy, K. R., R. D. DeLaune, W. F. Debusk, and M. S. Koch. 1993. Long-term nutrient accumulation rates in the Everglades. Soil Science Society of America Journal **57**:1147–1155.

Reddy, K. R., R. H. Kadlec, E. Flaig, and P. M. Gale. 1999. Phosphorus retention in streams and wetlands: A review. Critical Reviews in Environmental Science and Technology **29**:83–146.

Redfield, A. C. 1958. The biological control of chemical factors in the environment. American Scientist **46**: 205–221.

Redfield, A. C. 1963. The influence of organisms on the composition of sea-water. *In* M. N. Hill, editor. *The Sea.* Wiley, 26–77.

Redmond, K. T., and R. W. Koch. 1991. Surface climate and streamflow variability in the western United States and their relationship to large-scale circulation indexes. Water Resources Research **27**:2381–2399.

Reeburgh, W. S. 1983. Rates of biogeochemical processes in anoxic sediments. Annual Review of Earth and Planetary Sciences **11**:269–298.

Reeburgh, W. S. 2007. Oceanic methane biogeochemistry. Chemical Reviews **107**:486–513.

Reeburgh, W. S., S. C. Whalen, and M. J. Alperin. 1993. The role of microbially-mediated oxidation in the global CH_4 budget. *In* J. C. Murrell and D. P. Kelley, editors. *Microbiology of C1 Compounds.* Intercept, 1–14.

Reed, S. C., C. C. Cleveland, and A. R. Townsend. 2011. Functional ecology of free-living nitrogen fixation: A contemporary perspective. Annual Review of Ecology, Evolution and Systematics **42**:489–512.

Reed, S. C., A. R. Townsend, C. C. Cleveland, and D. R. Nemergut. 2010. Microbial community shifts influence patterns in tropical forest nitrogen fixation. Oecologia **164**:521–531.

Rees, C. E., W. J. Jenkins, and J. Monster. 1978. Sulfur isotopic composition of ocean water sulfate. Geochimica et Cosmochimica Acta **42**:377–381.

Reeves, H. 1994. On the origin of the light elements ($Z < 6$). Reviews of Modern Physics **66**:193–216.

Reheis, M. C., and R. Kihl. 1995. Dust deposition in southern Nevada and California, 1984–1989—Relations to climate, source area, and source lithology. Journal of Geophysical Research—Atmospheres **100**:8893–8918.

Reich, P. B., D. S. Ellsworth, M. B. Walters, J. M. Vose, C. Gresham, J. C. Volin, and W. D. Bowman. 1999. Generality of leaf trait relationships: A test across six biomes. Ecology **80**:1955–1969.

Reich, P. B., D. F. Grigal, J. D. Aber, and S. T. Gower. 1997. Nitrogen mineralization and productivity in 50 hardwood and conifer stands on diverse soils. Ecology **78**:335–347.

Reich, P. B., B. D. Kloeppel, D. S. Ellsworth, and M. B. Walters. 1995. Different photosynthesis-nitrogen relations in deciduous hardwood and evergreen coniferous tree species. Oecologia 104:24–30.

Reich, P. B., and J. Oleksyn. 2004. Global patterns of plant leaf N and P in relation to temperature and latitude. Proceedings of the National Academy of Sciences 101:11001–11006.

Reich, P. B., J. Oleksyn, J. Modrzynski, P. Mrozinski, S. E. Hobbie, D. M. Eissenstat, J. Chorover, O. A. Chadwick, C. M. Hale, and M. G. Tjoelker. 2005. Linking litter calcium, earthworms and soil properties: a common garden test with 14 tree species. Ecology Letters 8:811–818.

Reich, P. B., J. Oleksyn, and I. J. Wright. 2009. Leaf phosphorus influences the photosynthesis-nitrogen relation: a cross-biome analysis of 314 species. Oecologia 160:207–212.

Reich, P. B., J. Oleksyn, I. J. Wright, K. J. Niklas, L. Hedin, and J. J. Elser. 2010. Evidence of a general 2/3-power law of scaling leaf nitrogen to phosphorus among major plant groups and biomes. Proceedings of the Royal Society B—Biological Sciences 277: 877–883.

Reich, P. B., and A. W. Schoettle. 1988. Role of phosphorus and nitrogen in photosynthetic and whole-plant carbon gain and nutrient-use efficiency in eastern white pine. Oecologia 77:25–33.

Reich, P. B., M. G. Tjoelker, J. L. Machado, and J. Oleksyn. 2006. Universal scaling of respiratory metabolism, size and nitrogen in plants. Nature 439:457–461.

Reich, P. B., M. B. Walters, and D. S. Ellsworth. 1992. Leaf life span in relation to leaf, plant, and stand characteristics among diverse ecosystems. Ecological Monographs 62:365–392.

Reich, P. B., M. B. Walters, D. S. Ellsworth, and C. Uhl. 1994. Photosynthesis-nitrogen relations in Amazonian tree species. 1. Patterns among species and communities. Oecologia 97:62–72.

Reichle, D. E., B. E. Dinger, N. T. Edwards, W. F. Harris, and P. Sollins. 1973a. Carbon flow and storage in a forest ecosystem. In G. M. Woodwell and E. V. Pecan, editors. Carbon and the Biosphere. National Technical Information Service, 345–365.

Reichle, D. E., R. A. Goldstein, R. I. Van Hook, and C. J. Dodson. 1973b. Analysis of insect consumption in a forest canopy. Ecology 54:1076–1084.

Reid, C. D., and E. L. Fiscus. 1998. Effects of elevated CO_2 and/or ozone on limitations to CO_2 assimilation in soybean (Glycine max). Journal of Experimental Botany 49:885–895.

Reiners, P. W., T. A. Ehlers, S. G. Mitchell, and D. R. Montgomery. 2003. Coupled spatial variations in precipitation and long-term erosion rates across the Washington Cascades. Nature 426:645–647.

Reiners, W. A. 1972. Structure and energetics of three Minnesota forests. Ecological Monographs 42:71–94.

Reiners, W. A. 1986. Complementary models for ecosystems. American Naturalist 127:59–73.

Reiners, W. A. 1992. Twenty years of ecosystem reorganization following experimental deforestation and regrowth suppression. Ecological Monographs 62: 503–523.

Reinfelder, J. R., and N. S. Fisher. 1991. The assimilation of elements ingested by marine copepods. Science 251:794–796.

Reinhard, C. T., R. Raiswell, C. Scott, A. D. Anbar, and T. W. Lyons. 2009. A late Archaen sulfidic sea simulated by early oxidative weathering of the continents. Science 326:713–716.

Reisch, C. R., M. J. Stoudemayer, V. A. Varaljay, I. J. Amster, M. A. Moran, and W. B. Whitman. 2011. Novel pathway for assimilation of dimethylsulphoniopropionate widespread in marine bacteria. Nature 473:208–211.

Remde, A., and R. Conrad. 1991. Role of nitrification and denitrifciation for NO metabolism in soil. Biogeochemistry 12:189–205.

Ren, H., D. M. Sigman, A. N. Meckler, B. Plessen, R. S. Robinson, Y. Rosenthal, and G. H. Haug. 2009. Foraminiferal isotope evidence of reduced nitrogen fixation in the ice-age Atlantic ocean. Science 323: 244–248.

Renberg, I., and M. Wik. 1984. Dating recent lake sediments by soot particle counting. Verhandlungen Internationale Vereiniguer Theoretische und Angewandte Limnologie 22:712–718.

Renberg, I., T. Korsman, and H.J.B. Birks. 1993. Prehistoric increases in the pH of acid-sensitive Swedish lakes caused by land use changes. Nature 362:824–827.

Rennenberg, H. 1991. The significance of higher plants in the emission of sulfur compounds from terrestrial ecosystems. In T. D. Sharkey, E. A. Holland and H. A. Mooney, editors. Trace Gas Emissions by Plants. Academic Press, 217–260.

Reshetnikov, A. I., N. N. Paramonova, and A. A. Shashkov. 2000. An evaluation of historical methane emissions from the Soviet gas industry. Journal of Geophysical Research—Atmospheres 105:3517–3529.

Retallack, G. J. 1997. Early forest soils and their role in Devonian global change. Science 276:583–585.

Reuss, J. O. 1980. Simulation of soil nutrient losses resulting from rainfall acidity. Ecological Modelling 11: 15–38.

Reuss, J. O., B. J. Cosby, and R. F. Wright. 1987. Chemical processes governing soil and water acidification. Nature 329:27–32.

Reuss, J. O., and D. W. Johnson. 1986. *Acid Deposition and the Acidification of Soils and Waters*. Springer.

Reynolds, C. S., and P. S. Davies. 2001. Sources and bioavailability of phosphorus fractions in freshwaters: a British perspective. Biological Reviews **76**:27–64.

Reynolds, R. C. 1978. Polyphenol inhibition of calcite precipitation in Lake Powell. Limnology and Oceanography **23**:585–597.

Reynolds-Vargas, J. S., D. D. Richter, and E. Bornemisza. 1994. Environmental impacts of nitrification and nitrate adsorption in fertilized Andisols in the Valle Central of Costa Rica. Soil Science **157**:289–299.

Rhew, R. C., B. R. Miller, and R. F. Weiss. 2000. Natural methyl bromide and methyl chloride emissions from coastal salt marshes. Nature **403**:292–295.

Rich, P. H., and R. G. Wetzel. 1978. Detritus in the lake ecosystem. American Naturalist **112**:57–71.

Richardson, C. J. 1985. Mechanisms controlling phosphorus retention capacity in freshwater wetlands. Science **228**:1424–1427.

Richardson, C. J., T. W. Sasek, and E. A. Fendick. 1992. Implications of physiological responses to chronic air pollution for forest decline in the southeastern United States. Environmental Toxicology and Chemistry **11**:1105–1114.

Richardson, S. J., D. A. Peltzer, R. B. Allen, M. S. McGlone, and R. L. Parfitt. 2004. Rapid development of phosphorus limitation in temperate rainforest along the Franz Josef soil chronosequence. Oecologia **139**:267–276.

Richey, J. E., J. M. Melack, A. K. Aufdenkampe, V. M. Ballester, and L. L. Hess. 2002. Outgassing from Amazonian rivers and wetlands as a large tropical source of atmospheric CO_2. Nature **416**:617–620.

Richey, J. E., R. C. Wissmar, A. H. Devol, G. E. Likens, J. S. Eaton, R. G. Wetzel, W. E. Odum, N. M. Johnson, O. L. Loucks, R. T. Prentki, and P. H. Rich. 1978. Carbon flow in four lake ecosystems—Structural approach. Science **202**:1183–1186.

Richter, A., J. P. Burrows, H. Nuss, C. Granier, and U. Niemeier. 2005. Increase in tropospheric nitrogen dioxide over China observed from space. Nature **437**:129–132.

Richter, D. D. 2007. Humanity's transformation of Earth's soil: Pedology's new frontier. Soil Science **172**:957–967.

Richter, D. D., H. L. Allen, J. W. Li, D. Markewitz, and J. Raikes. 2006. Bioavailability of slowly cycling soil phosphorus: major restructuring of soil P fractions over four decades in an aggrading forest. Oecologia **150**:259–271.

Richter, D. D., and L. I. Babbar. 1991. Soil diversity in the tropics. Advances in Ecological Research **21**:315–389.

Richter, D. D., and D. Markewitz. 1995. How deep is soil? BioScience **45**:600–609.

Richter, D. D., D. Markewitz, S. E. Trumbore, and C. G. Wells. 1999. Rapid accumulation and turnover of soil carbon in a re-establishing forest. Nature **400**:56–58.

Richter, D. D., D. Markewitz, C. G. Wells, H. L. Allen, R. April, P. R. Heine, and B. Urrego. 1994. Soil chemical change during three decades in an old-field loblolly pine (*Pinus taeda* L.) ecosystem. Ecology **75**:1463–1473.

Richter, D. D., C. W. Ralston, and W. R. Harms. 1982. Prescribed fire—Effects on water quality and forest nutrient cycling. Science **215**:661–663.

Richter, F. M., D. B. Rowley, and D. J. Depaolo. 1992. Sr-isotope evolution of seawater—The role of tectonics. Earth and Planetary Science Letters **109**:11–23.

Riding, R. 2000. Microbial carbonates: the geological record of calcified bacterial-algal mats and biofilms. Sedimentology **47**:179–214.

Ridley, B. A., J. E. Dye, J. G. Walega, J. Zheng, F. E. Grahek, and W. Rison. 1996. On the production of active nitrogen by thunderstorms over New Mexico. Journal of Geophysical Research—Atmospheres **101**: 20985–21005.

Riebe, C. S., J. W. Kirchner, and R. C. Finkel. 2003. Long-term rates of chemical weathering and physical erosion from cosmogenic nuclides and geochemical mass balance. Geochimica et Cosmochimica Acta **67**:4411–4427.

Riebesell, U., I. Zondervan, B. Rost, P. D. Tortell, R. E. Zeebe, and F.M.M. Morel. 2000. Reduced calcification of marine plankton in response to increased atmospheric CO_2. Nature **407**:364–367.

Rignot, E., J. L. Bamber, M. R. Van Den Broeke, C. Davis, Y. H. Li, W. J. Van De Berg, and E. Van Meijgaard. 2008. Recent Antarctic ice mass loss from radar interferometry and regional climate modelling. Nature Geoscience **1**:106–110.

Rignot, E., and R. H. Thomas. 2002. Mass balance of polar ice sheets. Science **297**:1502–1506.

Rignot, E., I. Velicogna, M. R. van den Broeke, A. Monaghan, and J. Lenaerts. 2011. Acceleration of the contribution of the Greenland and Antarctic ice sheets to sea level rise. Geophysical Research Letters **38**.

Riley, W. J., J. B. Gaudinski, M. S. Torn, J. D. Joslin, and P. J. Hanson. 2009. Fine-root mortality rates in a temperate forest: estimates using radiocarbon data and numerical modeling. New Phytologist **184**:387–398.

Rind, D., E. W. Chiou, W. Chu, J. Larsen, S. Oltmans, J. Lerner, M. P. McCormick, and L. McMaster. 1991. Positive water-vapor feedback in climate models confirmed by satellite data. Nature **349**:500–503.

Rind, D., R. Goldberg, J. Hansen, C. Rosenzweig, and R. Ruedy. 1990. Potential evapotranspiration and the likelihood of future drought. Journal of Geophysical Research—Atmospheres 95:9983–10004.

Ringeval, B., N. de Noblet-Ducoudre, P. Ciais, P. Bousquet, C. Prigent, F. Papa, and W. B. Rossow. 2010. An attempt to quantify the impact of changes in wetland extent on methane emissions on the seasonal and interannual time scales. Global Biogeochemical Cycles 24.

Rinne, J., H. Hakola, T. Laurila, and U. Rannik. 2000. Canopy-scale monoterpene emissions of *Pinus sylvestris* dominated forests. Atmospheric Environment 34:1099–1107.

Rinsland, C. P., A. Goldman, E. Mahieu, R. Zander, J. Notholt, N. B. Jones, D.W.T. Griffith, T. M. Stephen, and L. S. Chiou. 2002. Ground-based infrared spectroscopic measurements of carbonyl sulfide: Free tropospheric trends from a 24-year time series of solar absorption measurements. Journal of Geophysical Research—Atmospheres 107.

Rippey, B., and N. J. Anderson. 1996. Reconstruction of lake phosphorus loading and dynamics using the sedimentary record. Environmental Science & Technology 30:1786–1788.

Rippey, B., and C. McSorley. 2009. Oxygen depletion in lake hypolimnia. Limnology and Oceanography 54:905–916.

Riser, S. C., and K. S. Johnson. 2008. Net production of oxygen in the subtropical ocean. Nature 451:323–325.

Risley, L. S., and D. A. Crossley. 1988. Herbivore-caused greenfall in the southern Appalachians. Ecology 69:1118–1127.

Rivkin, R. B., and L. Legendre. 2001. Biogenic carbon cycling in the upper ocean: Effects of microbial respiration. Science 291:2398–2400.

Roberts, B. J., and P. J. Mulholland. 2007. In-stream biotic control on nutrient biogeochemistry in a forested stream, West Fork of Walker Branch. JGR Biogeosciences 112.

Roberts, B. J., P. J. Mulholland, and W. R. Hill. 2007. Multiple scales of temporal variability in ecosystem metabolism rates: Results from 2 years of continuous monitoring in a forested headwater stream. Ecosystems 10:588–606.

Roberts, T. L., J.W.B. Stewart, and J. R. Bettany. 1985. The influence of topography on the distribution of organic and inorganic soil phosphorus across a narrow environmental gradient. Canadian Journal of Soil Science 65:651–665.

Robertson, D. M., H. S. Garn, and W. J. Rose. 2007. Response of calcareous Nagawicka Lake, Wisconsin, to changes in phosphorus loading. Lake and Reservoir Management 23:298–312.

Robertson, G. P. 1982a. Factors regulating nitrification in primary and secondary succession. Ecology 63:1561–1573.

Robertson, G. P. 1982b. Nitrification in forested ecosystems. Philosophical Transactions of the Royal Society of London Series B—Biological Sciences 296:445–457.

Robertson, G. P., M. A. Huston, F. C. Evans, and J. M. Tiedje. 1988. Spatial variability in a successional plant community—Patterns of nitrogen availability. Ecology 69:1517–1524.

Robertson, G. P., and P. M. Vitousek. 1981. Nitrification potentials in primary and secondary succession. Ecology 62:376–386.

Robertson, G. P., and P. M. Vitousek. 2009. Nitrogen in agriculture: Balancing the cost of an essential resource. Annual Review of Environment and Resources 34:97–125.

Robertson, J. E., C. Robinson, D. R. Turner, P. Holligan, A. J. Watson, P. Boyd, E. Fernandez, and M. Finch. 1994. The impact of a coccolithophore bloom on oceanic carbon uptake in the northeast Atlantic during summer 1991. Deep-Sea Research Part I—Oceanographic Research Papers 41:297-314.

Robertson, J. E., and A. J. Watson. 1992. Thermal skin effect on the surface ocean and its implications for CO_2 uptake. Nature 358:738–740.

Robertson, M. P., and S. L. Miller. 1995. An efficient prebiotic synthesis of cytosine and uracil. Nature 375:772–774.

Robinson, D. 1986. Limits to nutrient inflow rates in roots and root systems. Physiologia Plantarum 68:551–559.

Robinson, D. 1994. The responses of plants to nonuniform supplies of nutrients. New Phytologist 127:635–674.

Robinson, D. 2001. Delta ^{15}N as an integrator of the nitrogen cycle. Trends in Ecology & Evolution 16:153–162.

Robock, A., A. Marquardt, B. Kravitz, and G. Stenchikov. 2009. Benefits, risks, and costs of stratospheric geoengineering. Geophysical Research Letters 36.

Rockmann, T., and I. Levin. 2005. High-precision determination of the changing isotopic composition of atmospheric N_2O from 1990 to 2002. Journal of Geophysical Research—Atmospheres 110.

Rockstrom, J., W. Steffen, K. Noone, A. Persson, F. S. Chapin, E. F. Lambin, T. M. Lenton, M. Scheffer, C. Folke, H. J. Schellnhuber, B. Nykvist, C. A. de Wit, T. Hughes, S. van der Leeuw, H. Rodhe, S. Sorlin, P. K. Snyder, R. Costanza, U. Svedin, M. Falkenmark, L. Karlberg, R. W. Corell, V. J. Fabry, J. Hansen, B. Walker, D. Liverman, K. Richardson, P. Crutzen, and J. A. Foley. 2009.

A safe operating space for humanity. Nature **461**: 472–475.

Rodbell, D. T., G. O. Seltzer, D. M. Anderson, M. B. Abbott, D. B. Enfield, and J. H. Newman. 1999. An ~15,000-year record of El Niño–driven alluviation in southwestern Ecuador. Science **283**:516–520.

Rodell, M., I. Velicogna, and J. S. Famiglietti. 2009. Satellite-based estimates of groundwater depletion in India. Nature **460**:999–1002.

Rodhe, H. 1981. Formation of sulfuric and nitric acid in the atmosphere during long-range transport. Tellus **33**:132–141.

Rodhe, H. 1999. Human impact on the atmospheric sulfur balance. Tellus Series A—Dynamic Meteorology and Oceanography **51**:110–122.

Rodriguez, A., J. Duran, F. Covelo, J. Maria Fernandez-Palacios, and A. Gallardo. 2011. Spatial pattern and variability in soil N and P availability under the influence of two dominant species in a pine forest. Plant and Soil **345**:211–221.

Roehm, C. L., Y. T. Prairie, and P. A. del Giorgio. 2009. The pCO_2 dynamics in lakes in the boreal region of northern Quebec, Canada. Global Biogeochemical Cycles, Canada **23**.

Roelandt, C., B. van Wesemael, and M. Rounsevell. 2005. Estimating annual N_2O emissions from agricultural soils in temperate climates. Global Change Biology **11**:1701–1711.

Roelle, P., V. P. Aneja, J. O'Connor, W. Robarge, D. S. Kim, and J. S. Levine. 1999. Measurement of nitrogen oxide emissions from an agricultural soil with a dynamic chamber system. Journal of Geophysical Research—Atmospheres **104**:1609–1619.

Rogers, H. H., G. B. Runion, and S. V. Krupa. 1994. Plant responses to atmospheric CO_2 enrichment with emphasis on roots and the rhizosphere. Environmental Pollution **83**:155–189.

Rogers, J. R., P. C. Bennett, and W. J. Choi. 1998. Feldspars as a source of nutrients for microorganisms. American Mineralogist **83**:1532–1540.

Rohrer, F., and H. Berresheim. 2006. Strong correlation between levels of tropospheric hydroxyl radicals and solar ultraviolet radiation. Nature **442**:184–187.

Roman, J., and J. J. McCarthy. 2010. The whale pump: Marine mammals enhance primary productivity in a coastal basin. PLoS One 5.

Romell, L. G. 1935. Ecological problems in the humus layer in the forest. Cornell University, Agricultural Experiment Station.

Rondón, A., and L. Granat. 1994. Studies on the dry deposition of NO_2 to coniferous species at low NO_2 concentrations. Tellus Series B—Chemical and Physical Meteorology **46**:339–352.

Ronen, D., M. Magaritz, and E. Almon. 1988. Contaminated aquifers are a forgotten component of the global N_2O budget. Nature **335**:57–59.

Roos-Barraclough, F., A. Martinez-Cortizas, E. Garcia-Rodeja, and W. Shotyk. 2002. A 14,500-year record of the accumulation of atmospheric mercury in peat: volcanic signals, anthropogenic influences and a correlation to bromine accumulation. Earth and Planetary Science Letters **202**:435–451.

Rose, S. L., and C. T. Youngberg. 1981. Tripartite associations in snowbrush (*Ceanothus velutinus*)—Effect of vesicular-arbuscular mycorrhizae on growth, nodulation, and nitrogen fixation. Canadian Journal of Botany **59**:34–39.

Rosemond, A. D., P. J. Mulholland, and J. W. Elwood. 1993. Top-down and bottom-up control of stream periphyton: Effects of nutrients and herbivores. Ecology **74**:1264–1280.

Rosen, M. R., J. V. Turner, L. Coshell, and V. Gailitis. 1995. The effects of water temperature, stratification, and biological activity on the stable isotopic composition and timing of carbonate precipitation in a hypersaline lake. Geochimica et Cosmochimica Acta **59**:979–990.

Rosenzweig, M. L. 1968. Net primary productivity of terrestrial communities: Prediction from climatological data. American Naturalist **102**:67–74.

Rosing, M. T. 1999. ^{13}C-depleted carbon microparticles in >3700-Ma sea-floor sedimentary rocks from west Greenland. Science **283**:674–676.

Roskoski, J. P. 1980. Nitrogen fixation in hardwood forests of the northeastern United States. Plant and Soil **54**:33–44.

Ross, D. S., and R. J. Bartlett. 1996. Field-extracted Spodosol solutions and soils: Aluminum, organic carbon, and pH interrelationships. Soil Science Society of America Journal **60**:589–595.

Ross, J. E., and L. H. Aller. 1976. The chemical composition of the Sun. Science **191**:1223–1229.

Rosswall, T. 1982. Microbiological regulation of the biogeochemical nitrogen cycle. Plant and Soil **67**:15–34.

Rost, S., D. Gerten, A. Bondeau, W. Lucht, J. Rohwer, and S. Schaphoff. 2008. Agricultural green and blue water consumption and its influence on the global water system. Water Resources Research 44.

Rothman, D. H. 2002. Atmospheric carbon dioxide levels for the last 500 million years. Proceedings of the National Academy of Sciences **99**:4167–4171.

Rothman, J. M., P. J. Van Soest, and A. N. Pell. 2006. Decaying wood is a sodium source for mountain gorillas. Biology Letters **2**:321–324.

Rothschild, L. J., and R. L. Mancinelli. 2001. Life in extreme environments. Nature **409**:1092–1101.

Roulet, N., T. Moore, J. Bubier, and P. Lafleur. 1992. Northern fens—Methane flux and climatic change. Tellus Series B—Chemical and Physical Meteorology **44**:100–105.

Roulet, N. T. 2000. Peatlands, carbon storage, greenhouse gases, and the Kyoto Protocol: Prospects and significance for Canada. Wetlands **20**:605–615.

Roulet, N. T., P. M. Lafleur, P.J.H. Richard, T. R. Moore, E. R. Humphreys, and J. Bubier. 2007. Contemporary carbon balance and late Holocene carbon accumulation in a northern peatland. Global Change Biology **13**:397–411.

Roussel, E. G., M. A. Cambon-Bonavita, J. Querellou, B. A. Cragg, G. Webster, D. Prieur, and R. J. Parkes. 2008. Extending the sub-sea-floor biosphere. Science **320**:1046.

Routson, R. C., R. E. Wildung, and T. R. Garland. 1977. Mineral weathering in an arid watershed containing soil developed from mixed basaltic-felsic parent materials. Soil Science **124**:303–308.

Rowland, F. S. 1989. Chlorofluorocarbons and the depletion of stratospheric ozone. American Scientist **77**: 36–45.

Rowland, F. S. 1991. Stratospheric ozone depletion. Annual Review of Physical Chemistry **42**:731–768.

Røy, H., J. Kallmeyer, R. R. Adhikari, R. Pockalny, B. B. Jorgensen, and S. D'Hondt. 2012. Aerobic microbial respiration in 86-million-year-old deep-sea red clay. Science **336**:922–925.

Royer, D. L., R. A. Berner, and D. J. Beerling. 2001. Phanerozoic atmospheric CO_2 change: evaluating geochemical and paleobiological approaches. Earth-Science Reviews **54**:349–392.

Rudaz, A. O., E. A. Davidson, and M. K. Firestone. 1991. Sources of nitrous oxide production following wetting of dry soil. FEMS Microbiology Ecology **85**: 117–124.

Rudd, J.W.M., and R. D. Hamilton. 1978. Methane cycling in a eutrophic shield lake and its effects on whole lake metabolism. Limnology and Oceanography **23**:337–348.

Rudd, J.W.M., C. A. Kelly, and A. Furutani. 1986. The role of sulfate reduction in long-term accumulation of organic and inorganic sulfur in lake sediments. Limnology and Oceanography **31**:1281–1291.

Rue, E. L., and K. W. Bruland. 1997. The role of organic complexation on ambient iron chemistry in the equatorial Pacific Ocean and the response of a mesoscale iron addition experiment. Limnology and Oceanography **42**:901–910.

Ruess, R. W., and S. W. Seagle. 1994. Landscape patterns in soil microbial processes in the Serengeti national park, Tanzania. Ecology **75**:892–904.

Ruhe, R. V. 1984. The soil-climate system across the prairies in the midwestern USA. Geoderma **34**:201–219.

Rumpel, C., and I. Kogel-Knabner. 2011. Deep soil organic matter-a key but poorly understood component of terrestrial C cycle. Plant and Soil **338**:143–158.

Running, S. W., R. R. Nemani, F. A. Heinsch, M. S. Zhao, M. Reeves, and H. Hashimoto. 2004. A continuous satellite-derived measure of global terrestrial primary production. BioScience **54**:547–560.

Running, S. W., R. R. Nemani, D. L. Peterson, L. E. Band, D. F. Potts, L. L. Pierce, and M. A. Spanner. 1989. Mapping regional forest evapotranspiration and photosynthesis by coupling satellite data with ecosystem simulation. Ecology **70**:1090–1101.

Runyon, J., R. H. Waring, S. N. Goward, and J. M. Welles. 1994. Environmental limits on net primary production and light-use efficiency across the Oregon transect. Ecological Applications **4**:226–237.

Rupert, M. G. 2008. Decadal-scale changes of nitrate in ground water of the United States, 1988–2004. Journal of Environmental Quality **37**:S240–S248.

Russell, A. E., D. A. Laird, T. B. Parkin, and A. P. Mallarino. 2005. Impact of nitrogen fertilization and cropping system on carbon sequestration in midwestern Mollisols. Soil Science Society of America Journal **69**:413–422.

Russell, G. L., and J. R. Miller. 1990. Global river runoff calculated from a global atmospheric general-circulation model. Journal of Hydrology **117**: 241–254.

Russell, J. M., M. Z. Luo, R. J. Cicerone, and L. E. Deaver. 1996. Satellite confirmation of the dominance of chlorofluorocarbons in the global stratospheric chlorine budget. Nature **379**:526–529.

Russell, M. 2006. First life. American Scientist **94**:32–39.

Rustad, L. E. 1994. Element dynamics along a decay continuum in a red spruce ecosystem in Maine, USA. Ecology **75**:867–879.

Rustad, L. E., J. L. Campbell, G. M. Marion, R. J. Norby, M. J. Mitchell, A. E. Hartley, J.H.C. Cornelissen, J. Gurevitch, and N. Gcte. 2001. A meta-analysis of the response of soil respiration, net nitrogen mineralization, and aboveground plant growth to experimental ecosystem warming. Oecologia **126**: 543–562.

Ruttenberg, K. C. 1993. Reassessment of the oceanic residence time of phosphorus. Chemical Geology **107**: 405–409.

Ruttenberg, K. C., and R. A. Berner. 1993. Authigenic apatite formation and burial in sediments from non-upwelling, continental marine environments. Geochimica et Cosmochimica Acta **57**:991–1007.

Ruttenberg, K. C., and D. J. Sulak. 2011. Sorption and desorption of dissolved organic phosphorus onto iron (oxyhydr)oxides in seawater. Geochimica et Cosmochimica Acta **75**:4095–4112.

Rütting, T., D. Huygens, C. Müller, O. Cleemput, R. Godoy, and P. Boeckx. 2008. Functional role of DNRA and nitrite reduction in a pristine south Chilean *Nothofagus* forest. Biogeochemistry **90**:243–258.

Ryan, D. F., and F. H. Bormann. 1982. Nutrient resorption in northern hardwood forests. BioScience **32**:29–32.

Ryan, M. C., G. R. Graham, and D. L. Rudolph. 2001. Contrasting nitrate adsorption in Andisols of two coffee plantations in Costa Rica. Journal of Environmental Quality **30**:1848–1852.

Ryan, M. G. 1991. Effects of climate change on plant respiration. Ecological Applications **1**:157–167.

Ryan, M. G. 1995. Foliar maintenance respiration of subalpine and boreal trees and shrubs in relation to nitrogen content. Plant Cell and Environment **18**:765–772.

Ryan, M. G., S. T. Gower, R. M. Hubbard, R. H. Waring, H. L. Gholz, W. P. Cropper, and S. W. Running. 1995. Woody tissue maintenance respiration of four conifers in contrasting climates. Oecologia **101**:133–140.

Ryan, M. G., R. M. Hubbard, D. A. Clark, and R. L. Sanford. 1994. Woody tissue respiration for *Simarouba amara* and *Minqyartia guianensis*, two tropical wet forest trees with different growth habits. Oecologia **100**:213–220.

Ryan, M. H., M. Tibbett, T. Edmonds-Tibbett, L.D.B. Suriyagoda, H. Lambers, G. R. Cawthray, and J. Pang. 2012. Carbon trading for phosphorus gain: the balance between rhizosphere carboxylates and arbuscular mycorrhizal symbiosis in plant phosphorus acquisition. Plant, Cell & Environment **35**: in press.

Rychert, R., J. Skujins, D. Sorensen, and D. Porcella. 1978. Nitrogen fixation by lichens and free-living microorganisms in deserts. *In* N. E. West and J. Skujins, editors. *Nitrogen in Desert Ecosystems*. Dowden, Hutchinson and Ross, 20–30.

Rye, R., P. H. Kuo, and H. D. Holland. 1995. Atmospheric carbon dioxide concentrations before 2.2-billion years ago. Nature **378**:603–605.

Rygiewicz, P. T., and C. P. Andersen. 1994. Mycorrhizae alter quality and quantity of carbon allocated below ground. Nature **369**:58–60.

Rysgaard, S., N. Risgaardpetersen, N. P. Sloth, K. Jensen, and L. P. Nielsen. 1994. Oxygen regulation of nitrification and denitrification in sediments. Limnology and Oceanography **39**:1643–1652.

Ryu, Y., D. D. Baldocchi, H. Kobayashi, C. van Ingen, J. Li, T. A. Black, J. Beringer, E. van Gorsel, A. Knohl, B. E. Law, and O. Roupsard. 2011. Integration of MODIS land and atmosphere products with a coupled-process model to estimate gross primary productivity and evapotranspiration from 1 km to global scales. Global Biogeochemical Cycles **25**.

Saa, A., M. C. Trasarcepeda, F. Gilsotres, and T. Carballas. 1993. Changes in soil phosphorus and acid-phosphatase activity immediately following forest fires. Soil Biology & Biochemistry **25**:1223–1230.

Saa, A., M. C. Trasarcepeda, B. Soto, F. Gilsotres, and F. Diazfierros. 1994. Forms of phosphorus in sediments eroded from burnt soils. Journal of Environmental Quality **23**:739–746.

Sabine, C. L., R. A. Feely, N. Gruber, R. M. Key, K. Lee, J. L. Bullister, R. Wanninkhof, C. S. Wong, D.W.R. Wallace, B. Tilbrook, F. J. Millero, T. H. Peng, A. Kozyr, T. Ono, and A. F. Rios. 2004. The oceanic sink for anthropogenic CO_2. Science **305**:367–371.

Sabo, J. L., T. Sinha, L. C. Bowling, G.H.W. Schoups, W. W. Wallender, M. E. Campana, K. A. Cherkauer, P. L. Fuller, W. L. Graf, J. W. Hopmans, J. S. Kominoski, C. Taylor, S. W. Trimble, R. H. Webb, and E. E. Wohl. 2010. Reclaiming freshwater sustainability in the Cadillac Desert. Proceedings of the National Academy of Sciences **107**: 21263–21270.

Sadler, P. M. 1981. Sediment accumulation rates and the completeness of stratigraphic sections. Journal of Geology **89**:569–584.

Sagan, C., W. R. Thompson, R. Carlson, D. Gurnett, and C. Hord. 1993. A search for life on Earth from the Galileo spaccecraft. Nature **365**:715–721.

Saggar, S., J. R. Bettany, and J.W.B. Stewart. 1981. Sulfur transformations in relation to carbon and nitrogen in incubated soils. Soil Biology & Biochemistry **13**: 499–511.

Sahagian, D. L., F. W. Schwartz, and D. K. Jacobs. 1994. Direct anthropogenic contributions to sea-level rise in the twentieth century. Nature **367**:54–57.

Sahai, N. 2000. Estimating adsorption enthalpies and affinity sequences of monovalent electrolyte ions on oxide surfaces in aqueous solution. Geochimica et Cosmochimica Acta **64**:3629–3641.

Saito, T., Y. Yokouchi, Y. Kosugi, M. Tani, E. Philip, and T. Okuda. 2008. Methyl chloride and isoprene emissions from tropical rain forest in southeast Asia. Geophysical Research Letters **35**.

Sala, O. E., W. J. Parton, L. A. Joyce, and W. K. Lauenroth. 1988. Primary production of the central grassland region of the United States. Ecology **69**:40–45.

Salati, E., and P. B. Vose. 1984. Amazon Basin—A system in equilibrium. Science **225**:129–138.

Saleska, S. R., J. Harte, and M. S. Torn. 1999. The effect of experimental ecosystem warming on CO_2 fluxes in a montane meadow. Global Change Biology **5**:125–141.

Salifu, K. F., and V. R. Timmer. 2003. Nitrogen retranslocation response of young *Picea mariana* to nitrogen-15 supply. Soil Science Society of America Journal **67**: 309–317.

Saltzman, E. S., M. Aydin, C. Tatum, and M. B. Williams. 2008. 2,000-year record of atmospheric methyl bromide from a South Pole ice core. Journal of Geophysical Research—Atmospheres **113**.

Saltzman, E. S., M. Aydin, M. B. Williams, K. R. Verhulst, and B. Gun. 2009. Methyl chloride in a deep ice core from Siple Dome, Antarctica, Geophysical Research Letters **36**.

Sambrotto, R. N., G. Savidge, C. Robinson, P. Boyd, T. Takahashi, D. M. Karl, C. Langdon, D. Chipman, J. Marrs, and L. Codispoti. 1993. Elevated consumption of carbon relative to nitrogen in the surface ocean. Nature **363**:248–250.

Samet, J. M., F. Dominici, F. C. Curriero, I. Coursac, and S. L. Zeger. 2000. Fine particulate air pollution and mortality in 20 U.S. Cities, 1987–1994. New England Journal of Medicine **343**:1742–1749.

Sanborn, P. T., and T. M. Ballard. 1991. Combustion losses of sulfur from conifer foliage—Implications of chemical form and soil nitrogen status. Biogeochemistry **12**:129–134.

Sanchez, P. A., D. E. Bandy, J. H. Villachica, and J. J. Nicholaides. 1982a. Amazon basin soils—Management for continuous crop production. Science **216**:821–827.

Sanchez, P. A., M. P. Gichuru, and L. B. Katz. 1982b. Organic matter in major soils of the tropical and temperate regions. International Congress of Soil Science **12**: 99–114.

SanClements, M. D., I. J. Fernandez, and S. A. Norton. 2010. Phosphorus in soils of temperate forests: Linkages to acidity and aluminum. Soil Science Society of America Journal **74**:2175–2186.

Sand-Jensen, K., and P. A. Staehr. 2009. Net heterotrophy in small Danish lakes: A widespread feature over gradients in trophic status and land cover. Ecosystems **12**:336–348.

Sanderson, M. G. 1996. Biomass of termites and their emissions of methane and carbon dioxide: A global database. Global Biogeochemical Cycles **10**:543–557.

Sandor, J. A., J. B. Norton, J. A. Homburg, D. A. Muenchrath, C. S. White, S. E. Williams, C. I. Havener, and P. D. Stahl. 2007. Biogeochemical studies of a native American runoff agroecosystem. Geoarchaeology-an International Journal **22**:359–386.

Sanhueza, E., L. Cardenas, L. Donoso, and M. Santana. 1994. Effect of plowing on CO_2, CO, CH_4, N_2O, and NO fluxes from tropical savanna soils. Journal of Geophysical Research—Atmospheres **99**:16429–16434.

Sanhueza, E., Y. Dong, D. Scharffe, J. M. Lobert, and P. J. Crutzen. 1998. Carbon monoxide uptake by temperate forest soils: the effects of leaves and humus layers. Tellus Series B—Chemical and Physical Meteorology **50**:51–58.

Sano, Y., N. Takahata, Y. Nishio, T. P. Fischer, and S. N. Williams. 2001. Volcanic flux of nitrogen from the Earth. Chemical Geology **171**:263–271.

Sansone, F. J., and C. S. Martens. 1981. Methane production from acetate and associated methane fluxes from anoxic coastal sediments. Science **211**:707–709.

Santoro, A. E., C. Buchwald, M. R. McIlvin, and K. L. Casciotti. 2011. Isotopic signature of N_2O produced by marine ammonia-oxidizing archaea. Science **333**:1282–1285.

Sanyal, A., N. G. Hemming, G. N. Hanson, and W. S. Broecker. 1995. Evidence for a higher pH in the glacial ocean from boron isotopes in foraminifera. Nature **373**:234–236.

Sarbu, S. M., T. C. Kane, and B. K. Kinkle. 1996. A chemoautotrophically based cave ecosystem. Science **272**: 1953–1955.

Sarmiento, J. L., and N. Gruber. 2006. *Ocean Biogeochemical Dynamics*. Princeton University Press.

Sarmiento, J. L., N. Gruber, M. A. Brzezinski, and J. P. Dunne. 2004. High-latitude controls of thermocline nutrients and low latitude biological productivity. Nature **427**:56–60.

Sarmiento, J. L., and E. T. Sundquist. 1992. Revised budget for the oceanic uptake of anthropogenic carbon dioxide. Nature **356**:589–593.

Sass, R. L., and F. M. Fisher. 1997. Methane emissions from rice paddies: a process study summary. Nutrient Cycling in Agroecosystems **49**:119–127.

Sathyendranath, S., T. Platt, E.P.W. Horne, W. G. Harrison, O. Ulloa, R. Outerbridge, and N. Hoepffner. 1991. Estimation of new production in the ocean by compound remote sensing. Nature **353**:129–133.

Saugier, B., J. Roy, and H. A. Mooney. 2001. Estimations of global terrestrial productivity: converging toward a single number? In J. Roy, B. Saugier and H. A. Mooney, editors. *Terrestrial Global Productivity*. Academic Press, 543–557.

Saunders, D. L., and J. Kalff. 2001. Nitrogen retention in wetlands, lakes and rivers. Hydrobiologia **443**: 205–212.

Saunders, M. J., M. B. Jones, and F. Kansiime. 2007. Carbon and water cycles in tropical papyrus wetlands. Wetlands Ecology and Management **15**:489–498.

Saurer, M., R.T.W. Siegwolf, and F. H. Schweingruber. 2004. Carbon isotope discrimination indicates improving water-use efficiency of trees in northern Eurasia over the last 100 years. Global Change Biology **10**:2109–2120.

Savarino, J., and M. Legrand. 1998. High northern latitude forest fires and vegetation emissions over the last millennium inferred from the chemistry of a central Greenland ice core. Journal of Geophysical Research—Atmospheres 103:8267–8279.

Savoie, D. I., J. M. Prospero, R. J. Larsen, F. Huang, M. A. Izaguirre, T. Huang, T. H. Snowdon, L. Custals, and C. G. Sanderson. 1993. Nitrogen and sulfur species in Antarctic aerosols at Mawson, Palmer station, and Marsh (King George Island). Journal of Atmospheric Chemistry 17:95–122.

Savoie, D. L., and J. M. Prospero. 1989. Comparison of oceanic and continental sources of non-sea-salt sulfate over the Pacific ocean. Nature 339:685–687.

Savoie, D. L., J. M. Prospero, and R. T. Nees. 1987. Nitrate, non-sea-salt sulfate, and mineral aerosol over the northwestern Indian ocean. Journal of Geophysical Research 92:933–942.

Savva, Y., and F. Berninger. 2010. Sulphur deposition causes a large-scale growth decline in boreal forests in Eurasia. Global Biogeochemical Cycles 24.

Sayles, F. L. 1981. The composition and diagenesis of interstitial solutions. 2. Fluxes and diagenesis at the water-sediment interface in the high latitude north and south Atlantic. Geochimica et Cosmochimica Acta 45:1061–1086.

Sayles, F. L., and P. C. Mangelsdorf. 1977. Equilibration of clay minerals with seawater—Exchange reactions. Geochimica et Cosmochimica Acta 41:951–960.

Sayles, F. L., W. R. Martin, and W. G. Deuser. 1994. Response of benthic oxygen demand to particulate organic carbon supply in the deep sea near Bermuda. Nature 371:686–689.

Scanlon, B. R., J. B. Gates, R. C. Reedy, W. A. Jackson, and J. P. Bordovsky. 2010. Effects of irrigated agroecosystems: 2. Quality of soil water and groundwater in the southern High Plains, Texas. Water Resources Research 46.

Scanlon, B. R., K. E. Keese, A. L. Flint, L. E. Flint, C. B. Gaye, W. M. Edmunds, and I. Simmers. 2006. Global synthesis of groundwater recharge in semi-arid and arid regions. Hydrological Processes 20: 3335–3370.

Schaefer, D. A., and W. A. Reiners. 1989. Throughfall chemistry and canopy processing mechanisms. In S. E. Lindberg, A. L. Page and S. A. Norton, editors. Acid Precipitation: Sources, Deposition and Canopy Interactions. Springer-Verlag, 241–284.

Schaefer, D. A., and W. G. Whitford. 1981. Nutrient cycling by the subterranean termite Gnathamitermes tubiformans in a Chihuahuan desert ecosystem. Oecologia 48:277–283.

Schaefer, J. K., and F.M.M. Morel. 2009. High methylation rates of mercury bound to cysteine by Geobacter sulfurreducens. Nature Geoscience 2:123–126.

Schaefer, K., T. Zhang, L. Bruhwiler, and A. P. Barrett. 2011. Amount and timing of permafrost carbon release in response to climate warming. Tellus Series B—Chemical and Physical Meteorology 63: 165–180.

Schaefer, L., and B. Fegley. 2010. Chemistry of atmospheres formed during accretion of the Earth and other terrestrial planets. Icarus 208:438–448.

Schaefer, M. 1990. The soil fauna of a beech forest on limestone—Trophic structure and energy budget. Oecologia 82:128–136.

Schaller, M., J. D. Blum, S. P. Hamburg, and M. A. Vadeboncoeur. 2010. Spatial variability of long-term chemical weathering rates in the White Mountains, New Hampshire. Geoderma 154:294–301.

Scheffer, M., J. Bascompte, W. A. Brock, V. Brovkin, S. R. Carpenter, V. Dakos, H. Held, E. H. van Nes, M. Rietkerk, and G. Sugihara. 2009. Early-warning signals for critical transitions. Nature 461:53–59.

Schelske, C. L., and E. F. Stoermer. 1971. Eutrophication, silica depletion, and predicted changes in algal quality in Lake Michigan. Science 173:423–424.

Schidlowski, M. 1980. The atmosphere. In O. Hutzinger, editor. The Handbook of Environmental Chemistry. Volume 1, Part A. The Natural Environment and the Biogeochemical Cycles. Springer-Verlag, 1–16.

Schidlowski, M. 1983. Evolution of photoautotrophy and early atmospheric oxygen levels. Precambrian Research 20:319–335.

Schidlowski, M. 2001. Carbon isotopes as biogeochemical recorders of life over 3.8 Ga of Earth history: evolution of a concept. Precambrian Research 106: 117–134.

Schiff, S., R. Aravena, E. Mewhinney, R. Elgood, B. Warner, P. Dillon, and S. Trumbore. 1998. Precambrian shield wetlands: Hydrologic control of the sources and export of dissolved organic matter. Climatic Change 40:167–188.

Schillawski, S., and S. Petsch. 2008. Release of biodegradable dissolved organic matter from ancient sedimentary rocks. Global Biogeochemical Cycles 22.

Schilling, K. E., K.-S. Chan, H. Liu, and Y.-K. Zhang. 2010. Quantifying the effect of land use-land cover change on increasing discharge in the Upper Mississippi River. Journal of Hydrology 387:343–345.

Schilt, A., M. Baumgartner, J. Schwander, D. Buiron, E. Capron, J. Chappellaz, L. Loulergue, S. Schupbach, R. Spahni, H. Fischer, and T. F. Stocker. 2010. Atmospheric nitrous oxide during the last 140,000 years. Earth and Planetary Science Letters 300:33–43.

Schimel, D., M. A. Stillwell, and R. G. Woodmansee. 1985. Biogeochemistry of C, N, and P in a soil catena of the shortgrass steppe. Ecology **66**:276–282.

Schimel, D. S., B. H. Braswell, E. A. Holland, R. Mc-Keown, D. S. Ojima, T. H. Painter, W. J. Parton, and A. R. Townsend. 1994. Climatic, edaphic, and biotic controls over storage and turnover of carbon in soils. Global Biogeochemical Cycles **8**: 279–293.

Schimel, D. S., J. I. House, K. A. Hibbard, P. Bousquet, P. Ciais, P. Peylin, B. H. Braswell, M. J. Apps, D. Baker, A. Bondeau, J. Canadell, G. Churkina, W. Cramer, A. S. Denning, C. B. Field, P. Friedlingstein, C. Goodale, M. Heimann, R. A. Houghton, J. M. Melillo, B. Moore, D. Murdiyarso, I. Noble, S. W. Pacala, I. C. Prentice, M. R. Raupach, P. J. Rayner, R. J. Scholes, W. L. Steffen, and C. Wirth. 2001. Recent patterns and mechanisms of carbon exchange by terrestrial ecosystems. Nature **414**:169–172.

Schimel, J. P., and J. Bennett. 2004. Nitrogen mineralization: Challenges of a changing paradigm. Ecology **85**:591–602.

Schimel, J. P., and F. S. Chapin. 1996. Tundra plant uptake of amino acid and NH_4^+ nitrogen *in situ*: Plants compete well for amino acid N. Ecology **77**: 2142–2147.

Schimel, J. P., and M. K. Firestone. 1989. Nitrogen incorporation and flow through a coniferous forest soil profile. Soil Science Society of America Journal **53**: 779–784.

Schimel, J. P., M. K. Firestone, and K. S. Killham. 1984. Identification of heterotrophic nitrification in a Sierran forest soil. Applied and Environmental Microbiology **48**:802–806.

Schindler, D. 1977. Evolution of phosphorus limitation in lakes. Science **195**:260–262.

Schindler, D. W. 1974. Eutrophication and recovery in experimental lakes: Implications for lake management. Science **184**:897–899.

Schindler, D. W. 1978. Factors regulating phytoplankton production and standing crop in the world's freshwaters. Limnology and Oceanography **23**: 478–486.

Schindler, D. W. 1986. The significance of in-lake production of alkalinity. Water Air and Soil Pollution **30**:931–944.

Schindler, D. W. 1988. Confusion over the origin of alkalinity in lakes. Limnology and Oceanography **33**: 1637–1640.

Schindler, D. W., R. E. Hecky, D. L. Findlay, M. P. Stainton, B. R. Parker, M. J. Paterson, K. G. Beaty, M. Lyng, and S.E.M. Kasian. 2008. Eutrophication of lakes cannot be controlled by reducing nitrogen input: Results of a 37-year whole-ecosystem experiment.

Proceedings of the National Academy of Sciences **105**:11254–11258.

Schindler, F. V., and R. E. Knighton. 1999. Fate of fertilizer nitrogen applied to corn as estimated by the isotopic and difference methods. Soil Science Society of America Journal **63**:1734–1740.

Schindler, S. C., M. J. Mitchell, T. J. Scott, R. D. Fuller, and C. T. Driscoll. 1986. Incorporation of ^{35}S-sulfate into inorganic and organic constituents of two forest soils. Soil Science Society of America Journal **50**: 457–462.

Schipanski, M. E., and E. M. Bennett. 2012. The influence of agricultural trade and livestock production on the global phosphorus cycle. Ecosystems **15**:256–268.

Schipper, L. A., A. B. Cooper, C. G. Harfoot, and W. J. Dyck. 1993. Regulators of denitrification in an organic riparian soil. Soil Biology & Biochemistry **25**: 925–933.

Schippers, A., and B. B. Jorgensen. 2002. Biogeochemistry of pyrite and iron sulfide oxidation in marine sediments. Geochimica et Cosmochimica Acta **66**: 85–92.

Schippers, A., L. N. Neretin, J. Kallmeyer, T. G. Ferdelman, B. A. Cragg, R. J. Parkes, and B. B. Jorgensen. 2005. Prokaryotic cells of the deep sub-seafloor biosphere identified as living bacteria. Nature **433**: 861–864.

Schippers, P., M. Lurling, and M. Scheffer. 2004. Increase of atmospheric CO_2 promotes phytoplankton productivity. Ecology Letters **7**:446–451.

Schlesinger, W. H. 1977. Carbon balance in terrestrial detritus. Annual Review of Ecology and Systematics **8**: 51–81.

Schlesinger, W. H. 1978. Community structure, dyanamics and nutrient cycling in Okefenokee cypress swamp forest. Ecological Monographs **48**:43–65.

Schlesinger, W. H. 111-127, 1984. Soil organic matter: A source of atmospheric CO_2. *In* G. M. Woodwell, editor. *The Role Of Terrestrial Vegetation in the Global Carbon Cycle*. Wiley, 111–127.

Schlesinger, W. H. 1985. The formation of caliche in soils of the Mojave desert, California. Geochimica et Cosmochimica Acta **49**:57–66.

Schlesinger, W. H. 1986. Changes in soil carbon storage and associated properties with disturbance and recovery. *In* J. R. Trabalka and D. E. Reichle, editors. *The Changing Carbon Cycle: A Global Analysis*. Springer, 194–220.

Schlesinger, W. H. 1990. Evidence from chronosequence studies for a low carbon-storage potential of soils. Nature **348**:232–234.

Schlesinger, W. H. 2000. Carbon sequestration in soils: some cautions amidst optimism. Agriculture Ecosystems & Environment **82**:121–127.

Schlesinger, W. H. 2009. On the fate of anthropogenic nitrogen. Proceedings of the National Academy of Sciences 106:203–208.

Schlesinger, W. H., L. A. Bruijnzeel, M. B. Bush, E. M. Klein, K. A. Mace, J. A. Raikes, and R. J. Whittaker. 1998. The biogeochemistry of phosphorus after the first century of soil development on Rakata Island, Krakatau, Indonesia. Biogeochemistry 40:37–55.

Schlesinger, W. H., J. J. Cole, A. C. Finzi, and E. A. Holland. 2011. Introduction to coupled biogeochemical cycles. Frontiers in Ecology and the Environment 9:5–8.

Schlesinger, W. H., E. H. Delucia, and W. D. Billings. 1989. Nutrient-use efficiency of woody plants on contrasting soils in the western Great Basin, Nevada. Ecology 70:105–113.

Schlesinger, W. H., P. J. Fonteyn, and G. M. Marion. 1987. Soil moisture content and plant transpiration in the Chihuahuan desert of New Mexico. Journal of Arid Environments 12:119–126.

Schlesinger, W. H., J. T. Gray, D. S. Gill, and B. E. Mahall. 1982a. *Ceanothus megacarpus* chaparral—A synthesis of ecosystem processes during development and annual growth. Botanical Review 48:71–117.

Schlesinger, W. H., J. T. Gray, and F. S. Gilliam. 1982b. Atmospheric deposition processes and their importance as sources of nutrients in a chaparral ecosystem of southern California. Water Resources Research 18:623–629.

Schlesinger, W. H., and A. E. Hartley. 1992. A global budget for atmospheric NH_3. Biogeochemistry 15:191–211.

Schlesinger, W. H., and J. M. Melack. 1981. Transport of organic carbon in the world's rivers. Tellus 33:172–187.

Schlesinger, W. H., J. S. Pippen, M. D. Wallenstein, K. S. Hofmockel, D. M. Klepeis, and B. E. Mahall. 2003. Community composition and photosynthesis by photoautotrophs under quartz pebbles, southern Mojave Desert. Ecology 84:3222–3231.

Schlesinger, W. H., J. A. Raikes, A. E. Hartley, and A. E. Cross. 1996. On the spatial pattern of soil nutrients in desert ecosystems. Ecology 77:364–374.

Schlesinger, W. H., S. L. Tartowski, and S. M. Schmidt. 2006. Nutrient cycling within an arid ecosystem. *In* K. M. Havstad, L. F. Huenneke and W. H. Schlesinger, editors. *Structure and Function of a Chihuahuan Desert Ecosystem.* Oxford University Press, 133–149.

Schlesinger, W. H., T. J. Ward, and J. Anderson. 2000. Nutrient losses in runoff from grassland and shrubland habitats in southern New Mexico: II. Field plots. Biogeochemistry 49:69–86.

Schmidt, G. A., R. A. Ruedy, R. L. Miller, and A. A. Lacis. 2010a. Attribution of the present-day total greenhouse effect. Journal of Geophysical Research—Atmospheres 115.

Schmidt, H., T. Eickhorst, and R. Tippkötter. 2010b. Monitoring of root growth and redox conditions in paddy soil rhizotrons by redox electrodes and image analysis. Plant and Soil 341:221–232.

Schmidt, I., O. Sliekers, M. Schmid, I. Cirpus, M. Strous, E. Bock, J. G. Kuenen, and M.S.M. Jetten. 2002. Aerobic and anaerobic ammonia oxidizing bacteria competitors or natural partners? FEMS Microbiology Ecology 39:175–181.

Schmidt, M.W.I., M. S. Torn, S. Abiven, T. Dittmar, G. Guggenberger, I. A. Janssens, M. Kleber, I. Kogel-Knabner, J. Lehmann, D.A.C. Manning, P. Nanni-pieri, D. P. Rasse, S. Weiner, and S. E. Trumbore. 2011. Persistence of soil organic matter as an ecosystem property. Nature 478:49–56.

Schmidt, R., P. Schwintzer, F. Flechtner, C. Reigber, A. Guntner, P. Doll, G. Ramillien, A. Cazenave, S. Petrovic, H. Jochmann, and J. Wunsch. 2006. GRACE observations of changes in continental water storage. Global and Planetary Change 50:112–126.

Schmitt, J., R. Schneider, J. Elsig, D. Leuenberger, A. Lourantou, J. Chappellaz, P. Koehler, F. Joos, T. F. Stocker, M. Leuenberger, and H. Fischer. 2012. Carbon isotope constraints on the deglacial CO_2 rise from ice cores. Science 336:711–714.

Schmittner, A. 2005. Decline of the marine ecosystem caused by a reduction in the Atlantic overturning circulation. Nature 434:628–633.

Schmitz, O. J., D. Hawlena, and G. C. Trussell. 2010. Predator control of ecosystem nutrient dynamics. Ecology Letters 13:1199–1209.

Schmitz, W. J. 1995. On the interbasin-scale thermohaline circulation. Reviews of Geophysics 33:151–173.

Schneider, B., R. Schlitzer, G. Fischer, and E. M. Nothig. 2003. Depth-dependent elemental compositions of particulate organic matter (POM) in the ocean. Global Biogeochemical Cycles 17.

Schnell, S., and G. M. King. 1994. Mechanistic analysis of ammonium inhibition of atmospheric methane consumption in forest soils. Applied and Environmental Microbiology 60:3514–3521.

Schnitzer, M., and H. R. Schulten. 1992. The analysis of soil organic matter by pyrolysis field-ionization mass spectrometry. Soil Science Society of America Journal 56:1811–1817.

Schoenau, J. J., and J. R. Bettany. 1987. Organic matter leaching as a component of carbon, nitrogen, phosphorus, and sulfur cycles in a forest, grassland, and gleyed soil. Soil Science Society of America Journal 51:646–651.

Scholander, P. F., L. Van Dam, and S. I. Scholander. 1955. Gas exchange in the roots of mangroves. American Journal of Botany 42:92–98.

Scholefield, D., J.M.B. Hawkins, and S. M. Jackson. 1997. Development of a helium atmosphere soil incubation technique for direct measurement of nitrous oxide and dinitrogen fluxes during denitrification. Soil Biology & Biochemistry 29:1345–1352.

Schönbächler, M., R. W. Carlson, M. F. Horan, T. D. Mock, and E. H. Hauri. 2010. Heterogeneous accretion and the moderately volatile element budget of Earth. Science 328:884–887.

Schonheit, P., J. K. Kristjansson, and R. K. Thauer. 1982. Kinetic mechanism for the ability of sulfate reducers to out compete methanogens for acetate. Archives of Microbiology 132:285–288.

Schopf, J. W., A. B. Kudryavtsev, D. G. Agresti, T. J. Wdowiak, and A. D. Czaja. 2002. Laser-Raman imagery of Earth's earliest fossils. Nature 416:73–76.

Schöttelndreier, M., and U. Falkengren-Grerup. 1999. Plant induced alteration in the rhizosphere and the utilisation of soil heterogeneity. Plant and Soil 209:297–309.

Schrenk, M. O., K. J. Edwards, R. M. Goodman, R. J. Hamers, and J. F. Banfield. 1998. Distribution of *Thiobacillus ferrooxidans* and *Leptospirillum ferrooxidans*: Implications for generation of acid mine drainage. Science 279:1519–1522.

Schroth, A. W., B. C. Bostick, M. Graham, J. M. Kaste, M. J. Mitchell, and A. J. Friedland. 2007. Sulfur species behavior in soil organic matter during decomposition. Journal of Geophysical Research-Biogeosciences 112:1–10.

Schuffert, J. D., R. A. Jahnke, M. Kastner, J. Leather, A. Sturz, and M. R. Wing. 1994. Rates of formation of modern phosphorite off western Mexico. Geochimica et Cosmochimica Acta 58:5001–5010.

Schulten, H. R., and M. Schnitzer. 1993. A state-of-the-art structural concept for humic substances. Naturwissenschaften 80:29–30.

Schulten, H. R., and M. Schnitzer. 1997. Chemical model structures for soil organic matter and soils. Soil Science 162:115–130.

Schulthess, C. P., and C. P. Huang. 1991. Humic and fulvic acid adsorption by silicon and aluminum oxide surfaces on clay minerals. Soil Science Society of America Journal 55:34–42.

Schultz, M. G., A. Heil, J. J. Hoelzemann, A. Spessa, K. Thonicke, J. G. Goldammer, A. C. Held, J.M.C. Pereira, and M. van het Bolscher. 2008. Global wildland fire emissions from 1960 to 2000. Global Biogeochemical Cycles 22.

Schultz, R. C., P. P. Kormanik, W. C. Bryan, and G. H. Brister. 1979. Vesicular-arbuscular mycorrhiza influence growth but not mineral concentrations in seedlings of eight sweetgum families. Canadian Journal of Forest Research 9:218–223.

Schulze, E.-D., P. Ciais, S. Luyssaert, M. Schrumpf, I. A. Janssens, B. Thiruchittampalam, J. Theloke, M. Saurat, S. Bringezu, J. Lelieveld, A. Lohila, C. Rebmann, M. Jung, D. Bastviken, G. Abril, G. Grassi, A. Leip, A. Freibauer, W. Kutsch, A. Don, J. Nieschulze, A. Borner, J. H. Gash, and A. J. Dolman. 2010. The European carbon balance. Part 4. Integration of carbon and other trace-gas fluxes. Global Change Biology 16:1451–1469.

Schumann, U., and H. Huntrieser. 2007. The global lightning-induced nitrogen oxides source. Atmospheric Chemistry and Physics 7:3823–3907.

Schutz, H., P. Schroder, and H. Rennenberg. 1991. Role of plants in regulating the methane flux to the atmosphere. *In* T. D. Sharkey, E. A. Holland and H. A. Mooney, editors. *Trace Gas Emissions by Plants*, Academic Press, 29–63.

Schuur, E.A.G. 2003. Productivity and global climate revisited: The sensitivity of tropical forest growth to precipitation. Ecology 84:1165–1170.

Schuur, E.A.G., J. Bockheim, J. G. Canadell, E. Euskirchen, C. B. Field, S. V. Goryachkin, S. Hagemann, P. Kuhry, P. M. Lafleur, H. Lee, G. Mazhitova, F. E. Nelson, A. Rinke, V. E. Romanovsky, N. Shiklomanov, C. Tarnocai, S. Venevsky, J. G. Vogel, and S. A. Zimov. 2008. Vulnerability of permafrost carbon to climate change: Implications for the global carbon cycle. BioScience 58:701–714.

Schuur, E.A.G., J. G. Vogel, K. G. Crummer, H. Lee, J. O. Sickman, and T. E. Osterkamp. 2009. The effect of permafrost thaw on old carbon release and net carbon exchange from tundra. Nature 459:556–559.

Schwartz, S. E. 1988. Are global cloud albedo and climate controlled by marine phytoplankton? Nature 336:441–445.

Schwartz, S. E. 1989. Acid deposition—Unraveling a regional phenomenon. Science 243:753–763.

Schwartzman, D. W., and T. Volk. 1989. Biotic enhancement of weathering and the habitability of Earth. Nature 340:457–460.

Schwertmann, U. 1966. Inhibitory effect of soil organic matter on the crystallization of amorphous ferric hydroxide. Nature 212:645–646.

Schwintzer, C. R. 1983. Non-symbiotic and symbiotic nitrogen fixation in a weakly minerotrophic peatland. American Journal of Botany 70:1071–1078.

Schwintzer, C. R., and J. D. Tjepkema. 1994. Factors affecting the acetylene to $^{15}N_2$-conversion ratio in root nodules of *Myrica gale* L. Plant Physiology **106**: 1041–1047.

Scott, A. C., and I. J. Glasspool. 2006. The diversification of Paleozoic fire systems and fluctuations in atmospheric oxygen concentration. Proceedings of the National Academy of Sciences **103**:10861–10865.

Scott, D., J. Harvey, R. Alexander, and G. Schwarz. 2007. Dominance of organic nitrogen from headwater streams to large rivers across the conterminous United States. Global Biogeochemical Cycles **21**.

Scott, J. T., M. J. McCarthy, W. S. Gardner, and R. D. Doyle. 2008. Denitrification, dissimilatory nitrate reduction to ammonium, and nitrogen fixation along a nitrate concentration gradient in a created freshwater wetland. Biogeochemistry **87**:99–111.

Scott, N. A., and D. Binkley. 1997. Foliage litter quality and annual net N mineralization: Comparison across North American forest sites. Oecologia **111**: 151–159.

Scott, R. M. 1962. Exchangeable bases of mature, well-drained soils in relation to rainfall in East Africa. Journal of Soil Science **13**:1–9.

Scranton, M. I., and P. G. Brewer. 1977. Occurrence of methane in near-surface waters of western subtropical north Atlantic. Deep-Sea Research **24**:127–138.

Scriber, J. M. 1977. Limiting effects of low leaf water content on nitrogen-utilization, energy budget, and larval growth of *Hyalophora cecropia* (Lepidoptera Saturniidae). Oecologia **28**:269–287.

Scudder, B. C., L. C. Chasar, D. A. Wentz, N. J. Bauch, M. E. Brigham, P. W. Moran, and D. P. Krabbenhoft. 2009. Mercury in fish, bed sediment, and water from streams across the United States, 1998–2005. U.S. Geological Survey, 74.

Scurlock, J.M.O., and R. J. Olson. 2002. Terrestrial net primary productivity—a brief history and a new worldwide database. Environmental Research **10**: 91–109.

Seager, R., M. Ting, I. Held, Y. Kushnir, J. Lu, G. Vecchi, H.-P. Huang, N. Harnik, A. Leetmaa, N.-C. Lau, C. Li, J. Velez, and N. Naik. 2007. Model projections of an imminent transition to a more arid climate in southwestern North America. Science **316**: 1181–1184.

Sears, S. O., and D. Langmuir. 1982. Sorption and mineral equilibria controls on moisture chemistry in a C-horizon soil. Journal of Hydrology **56**:287–308.

Seastedt, T. R. 1985. Maximization of primary and secondary productivity by grazers. American Naturalist **126**:559–564.

Seastedt, T. R. 1988. Mass, nitrogen, and phosphorus dynamics in foliage and root detritus of tallgrass prairie. Ecology **69**:59–65.

Seastedt, T. R., and D. A. Crossley. 1980. Effects of microarthropods on the seasonal dynamics of nutrients in forest litter. Soil Biology & Biochemistry **12**:337–342.

Seastedt, T. R., and D. C. Hayes. 1988. Factors influencing nitrogen concentrations in soil water in a North American tallgrass prairie. Soil Biology & Biochemistry **20**:725–729.

Seastedt, T. R., and C. M. Tate. 1981. Decomposition rates and nutrient contents of arthropod remains in forest litter. Ecology **62**:13–19.

Sebacher, D. I., R. C. Harriss, and K. B. Bartlett. 1985. Methane emissions to the atmosphere through aquatic plants. Journal of Environmental Quality **14**:40–46.

Sebacher, D. I., R. C. Harriss, K. B. Bartlett, S. M. Sebacher, and S. S. Grice. 1986. Atmospheric methane sources: Alaskan tundra bogs, an alpine fen, and a subarctic boreal marsh. Tellus Series B—Chemical and Physical Meteorology **38**:1–10.

Seemann, J. R., T. D. Sharkey, J. L. Wang, and C. B. Osmond. 1987. Environmental effects on photosynthesis, nitrogen-use efficiency, and metabolite pools in leaves of sun and shade plants. Plant Physiology **84**:796–802.

Segura, A., and R. Navarro-Gonzalez. 2005. Nitrogen fixation on early Mars by volcanic lightning and other sources. Geophysical Research Letters **32**.

Sehmel, G. A. 1980. Particle and gas dry deposition—A review. Atmospheric Environment **14**:983–1011.

Seiler, W., and R. Conrad. 1987. Contribution of tropical ecosystems to the global budgets of trace gases, especially CH_4 H_2, CO, and N_2O. *In* R. E. Dickinson, editor. *The Geophysiology of Amazonia*. Wiley, 133–162.

Seinfeld, J. H. 1989. Urban air pollution—State of the science. Science **243**:745–752.

Seiter, K., C. Hensen, and M. Zabel. 2005. Benthic carbon mineralization on a global scale. Global Biogeochemical Cycles **19**.

Seitzinger, S., J. A. Harrison, J. K. Bohlke, A. F. Bouwman, R. Lowrance, B. Peterson, C. Tobias, and G. Van Drecht. 2006. Denitrification across landscapes and waterscapes: A synthesis. Ecological Applications **16**:2064–2090.

Seitzinger, S., S. Nixon, M.E.Q. Pilson, and S. Burke. 1980. Denitrification and N_2O production in nearshore marine sediments. Geochimica et Cosmochimica Acta **44**:1853–1860.

Seitzinger, S. P. 1988. Denitrification in freshwater and coastal marine ecosystems: Ecological and geochemical significance. Limnology and Oceanography **33**: 702–724.

Seitzinger, S. P. 1991. The effect of pH on the release of phosphorus from Potomac estuary sediments—Implications for blue–green algal blooms. Estuarine Coastal and Shelf Science 33:409–418.

Seitzinger, S. P., H. Hartnett, R. Lauck, M. Mazurek, T. Minegishi, G. Spyres, and R. Styles. 2005. Molecular-level chemical characterization and bio-availability of dissolved organic matter in stream water using electrospray-ionization mass spectrometry. Limnology And Oceanography 50:1–12.

Seitzinger, S. P., E. Mayorga, A. F. Bouwman, C. Kroeze, A.H.W. Beusen, G. Billen, G. Van Drecht, E. Dumont, B. M. Fekete, J. Garnier, and J. A. Harrison. 2010. Global river nutrient export: A scenario analysis of past and future trends. Global Biogeochemical Cycles 24.

Seitzinger, S. P., L. P. Nielsen, J. Caffrey, and P. B. Christensen. 1993. Denitrification measurements in aquatic sediments—A comparison of three methods. Biogeochemistry 23:147–167.

Seitzinger, S. P., S. W. Nixon, and M.E.Q. Pilson. 1984. Denitrification and nitrous oxide production in a coastal marine ecosystem. Limnology and Oceanography 29:73–83.

Seitzinger, S. P., R. V. Styles, E. W. Boyer, R. B. Alexander, G. Billen, R. W. Howarth, B. Mayer, and N. Van Breemen. 2002. Nitrogen retention in rivers: model development and application to watersheds in the northeastern United States. Biogeochemistry 57: 199–237.

Self, S., S. Blake, K. Sharma, M. Widdowson, and S. Sephton. 2008. Sulfur and chlorine in Late Cretaceous Deccan magmas and eruptive gas release. Science 319:1654–1657.

Selin, N. E. 2009. Global biogeochemical cycling of mercury: A review. Annual Review of Environment and Resources 34:43–63.

Selin, N. E., D. J. Jacob, R. M. Yantosca, S. Strode, L. Jaegle, and E. M. Sunderland. 2008. Global 3-D land-ocean-atmosphere model for mercury: Present-day versus preindustrial cycles and anthropogenic enrichment factors for deposition. Global Biogeochemical Cycles 22.

Sella, G. F., S. Stein, T. H. Dixon, M. Craymer, T. S. James, S. Mazzotti, and R. K. Dokka. 2007. Observation of glacial isostatic adjustment in "stable" North America with GPS. Geophysical Research Letters 34.

Sellers, P., C. A. Kelly, J.W.M. Rudd, and A. R. MacHutchon. 1996. Photodegradation of methylmercury in lakes. Nature 380:694–697.

Selmants, P. C., and S. C. Hart. 2010. Phosphorus and soil development: Does the Walker and Syers model apply to semiarid ecosystems? Ecology 91: 474–484.

Seo, K. H., and K. P. Bowman. 2002. Lagrangian estimate of global stratosphere-troposphere mass exchange. Journal of Geophysical Research—Atmospheres 107.

Sequeira, R. 1993. On the large-scale impact of arid dust on precipitation chemistry of the continental Northern Hemisphere. Atmospheric Environment Part A—General Topics 27:1553–1565.

Serca, D., A. Guenther, L. Klinger, D. Helmig, D. Hereid, and P. Zimmerman. 1998. Methyl bromide deposition to soils. Atmospheric Environment 32:1581–1586.

Serrasolsas, I., and P. K. Khanna. 1995. Changes in heated and autoclaved forest soils of SE Australia. 2. Phosphorus and phosphatase activity. Biogeochemistry 29:25–41.

Serret, P., C. Robinson, E. Fernandez, E. Teira, and G. Tilstone. 2001. Latitudinal variation of the balance between plankton photosynthesis and respiration in the eastern Atlantic ocean. Limnology and Oceanography 46:1642–1652.

Serreze, M. C., M. M. Holland, and J. Stroeve. 2007. Perspectives on the arctic's shrinking sea-ice cover. Science 315:1533–1536.

Servant, J. 1986. The burden of the sulphate layer of the stratosphere during volcanic "quiescent" periods. Tellus 38B:74–79.

Servant, J. 1989. Les sources et les puits d'oxysulfure de carbone (COS) a l'echelle mondiale. Atmospheric Research 23:105–116.

Servant, J. 1994. The continental carbon cycle during the last glacial maximum. Atmospheric Research 31: 253–268.

Sessions, A. L., D. M. Doughty, P. V. Welander, R. E. Summons, and D. K. Newman. 2009. The continuing puzzle of the Great Oxidation event. Current Biology 19:R567–R574.

Severinghaus, J. P., and E. J. Brook. 1999. Abrupt climate change at the end of the last glacial period inferred from trapped air in polar ice. Science 286:930–934.

Sexstone, A. J., T. B. Parkin, and J. M. Tiedje. 1985a. Temporal response of soil denitrification rates to rainfall and irrigation. Soil Science Society of America Journal 49:99–103.

Sexstone, A. J., N. P. Revsbech, T. B. Parkin, and J. M. Tiedje. 1985b. Direct measurement of oxygen profiles and denitrification rates in soil aggregates Soil Science Society of America Journal 49:645–651.

Shaffer, G. 1993. Effects of the marine biota on global carbon cycling. In M. Heimann, editor. The Global Carbon Cycle. Springer-Verlag, 431–455.

Shaked, Y., Y. Xu, K. Leblanc, and F.M.M. Morel. 2006. Zinc availability and alkaline phosphatase activity in Emiliania huxleyi: Implications for Zn-P co-limitation in the ocean. Limnology and Oceanography 51: 299–309.

Shakun, J. D., P. U. Clark, F. He, S. A. Marcott, A. C. Mix, Z. Liu, B. Otto-Bliesner, A. Schmittner, and E. Bard. 2012. Global warming preceded by increasing carbon dioxide concentrations during the last deglaciation. Nature 484:49–54.

Shanks, A. L., and J. D. Trent. 1979. Marine snow—Microscale nutrient patches. Limnology and Oceanography 24:850–854.

Shannon, R. D., and J. R. White. 1994. Three-year study of controls on methane emissions from two Michigan peatlands. Biogeochemistry 27:35–60.

Sharkey, T. D. 1985. Photosynthesis in intact leaves of C-3 plants—Physics, physiology and rate limitations. Botanical Review 51:53–105.

Sharkey, T. D. 1988. Estimating the rate of photorespiration in leaves. Physiologia Plantarum 73:147–152.

Sharpley, A., T. C. Daniel, J. T. Sims, and D. H. Pote. 1996. Determining environmentally sound soil phosphorus levels. Journal of Soil and Water Conservation 51:160–166.

Sharpley, A. N. 1985. The selective erosion of plant nutrients in runoff. Soil Science Society of America Journal 49:1527–1534.

Sharpley, A. N., H. Tiessen, and C. V. Cole. 1987. Soil phosphorus forms extracted in soil tests as a function of pedogenesis. Soil Science Society of America Journal 51:362–365.

Shaver, G. R., W. D. Billings, F. S. Chapin, A. E. Giblin, K. J. Nadelhoffer, W. C. Oechel, and E. B. Rastetter. 2011. Global change and carbon balance of arctic ecosystems: Changes in global terrestrial carbon cycling. BioScience 42:433–441.

Shaver, G. R., and F. S. Chapin. 1986. Effect of fertilizer on production and biomass of tussock tundra, Alaska, USA. Arctic and Alpine Research 18:261–268.

Shaver, G. R., A. E. Giblin, K. J. Nadelhoffer, K. K. Thieler, M. R. Downs, J. A. Laundre, and E. B. Rastetter. 2006. Carbon turnover in Alaskan tundra soils: effects of organic matter quality, temperature, moisture and fertilizer. Journal of Ecology 94:740–753.

Shaver, G. R., and J. M. Melillo. 1984. Nutrient budgets of marsh plants—Efficiency concepts and relation to availability. Ecology 65:1491–1510.

Shaw, G. E. 1983. Bio-controlled thermostasis involving the sulfur cycle. Climatic Change 5:297–303.

Shaw, M. R., and J. Harte. 2001. Response of nitrogen cycling to simulated climate change: differential responses along a subalpine ecotone. Global Change Biology 7:193–210.

Shaw, R. W. 1987. Air pollution by particles. Scientific American 257:96–103.

Shear, C. R., H. L. Crane, and A. T. Myers. 1946. Nutrient-element balance: A fundamental concept in plant nutrition. Proceedings of the American Society for Horticultural Science 47:239–248.

Shearer, G., and D. H. Kohl. 1988. Natural ^{15}N abundance as a method of estimating the contribution of biologically-fixed nitrogen to N_2-fixing systems—Potential for non-legumes. Plant and Soil 110:317–327.

Shearer, G., and D. H. Kohl. 1989. Estimates of N_2 fixation in ecosystems: The need for and basis of the ^{15}N natural abundance method. In P. W. Rundel, J. R. Ehleringer and K. A. Nagy, editors. Stable Isotopes in Ecological Research, Springer, 342–374.

Shearer, G., D. H. Kohl, R. A. Virginia, B. A. Bryan, J. L. Skeeters, E. T. Nilsen, M. R. Sharifi, and P. W. Rundel. 1983. Estimates of N_2-fixation from variation in the natural abundance of ^{15}N in Sonoran desert ecosystems. Oecologia 56:365–373.

Shelley, R. U., P. N. Sedwick, T. S. Bibby, P. Cabedo-Sanz, T. M. Church, R. J. Johnson, A. I. Macey, C. M. Marsay, E. R. Sholkovitz, S. J. Ussher, P. J. Worsfold, and M. C. Lohan. 2012. Controls on dissolved cobalt in surface waters of the Sargasso Sea: Comparisons with iron and aluminum. Global Biogeochemical Cycles 26.

Shen, S. M., G. Pruden, and D. S. Jenkinson. 1984. Mineralization and immobilization of nitrogen in fumigated soil and the measurement of microbial biomass nitrogen. Soil Biology & Biochemistry 16:437–444.

Shen, Y. A., R. Buick, and D. E. Canfield. 2001. Isotopic evidence for microbial sulphate reduction in the early Archaean era. Nature 410:77–81.

Shepherd, M. F., S. Barzetti, and D. R. Hastie. 1991. The production of atmospheric Nox and N_2O from a fertilized agricultural soil. Atmospheric Environment Part A—General Topics 25:1961–1969.

Shields, C. A., L. E. Band, N. Law, P. M. Groffman, S. S. Kaushal, K. Savvas, G. T. Fisher, and K. T. Belt. 2008. Streamflow distribution of non-point source nitrogen export from urban-rural catchments in the Chesapeake Bay watershed. Water Resources Research 44.

Shiller, A. M. 1997. Manganese in surface waters of the Atlantic Ocean. Geophysical Research Letters 24:1495–1498.

Shimshock, J. P., and R. G. de Pena. 1989. Below-cloud scavenging of tropospheric ammonia. Tellus 41B:296–304.

Shindell, D., J.C.I. Kuylenstierna, E. Vignati, R. van Dingenen, M. Amann, Z. Klimont, S. C. Anenberg, N. Muller, G. Janssens-Maenhout, F. Raes, J. Schwartz, G. Faluvegi, L. Pozzoli, K. Kupiainen, L. Hoglund-Isaksson, L. Emberson, D. Streets, V. Ramanathan, K. Hicks, N.T.K. Oanh, G. Milly, M. Williams, V. Demkine, and D. Fowler. 2012. Simultaneously mitigating near-term climate change and improving human health and food security. Science **335**:183–189.

Shindell, D. T., G. Faluvegi, D. M. Koch, G. A. Schmidt, N. Unger, and S. E. Bauer. 2009. Improved attribution of climate forcing to emissions. Science **326**:716–718.

Shoji, S., M. Nanzyo, Y. Shirato, and T. Ito. 1993. Chemical kinetics of weathering in young Andisols from northeastern Japan using soil age normalized to 10°C. Soil Science **155**:53–60.

Sholkovitz, E. R. 1976. Flocculation of dissolved organic and inorganic matter during mixing of river water and seawater. Geochimica et Cosmochimica Acta **40**: 831–845.

Shon, Z. H., D. Davis, G. Chen, G. Grodzinsky, A. Bandy, D. Thornton, S. Sandholm, J. Bradshaw, R. Stickel, W. Chameides, G. Kok, L. Russell, L. Mauldin, D. Tanner, and F. Eisele. 2001. Evaluation of the DMS flux and its conversion to SO_2 over the Southern Ocean. Atmospheric Environment **35**:159–172.

Shooter, D. 1999. Sources and sinks of oceanic hydrogen sulfide—An overview. Atmospheric Environment **33**:3467–3472.

Shorter, J. H., C. E. Kolb, P. M. Crill, R. A. Kerwin, R. W. Talbot, M. E. Hines, and R. C. Harriss. 1995. Rapid degradation of atmospheric methyl bromide in soils. Nature **377**:717–719.

Shortle, W. C., and K. T. Smith. 1988. Aluminum-induced calcium deficiency syndrome in declining red spruce. Science **240**:1017–1018.

Shotyk, W., M. E. Goodsite, F. Roos-Barraclough, R. Frei, J. Heinemeier, G. Asmund, C. Lohse, and T. S. Hansen. 2003. Anthropogenic contributions to atmospheric Hg, Pb and As accumulation recorded by peat cores from southern Greenland and Denmark, dated using the ^{14}C "bomb pulse curve" Geochimica et Cosmochimica Acta **67**:3991–4011.

Shugart, H. H., S. Saatchi, and F. G. Hall. 2010. Importance of structure and its measurement in quantifying function of forest ecosystems. Journal of Geophysical Research—Biogeosciences **115**.

Shukla, J., and Y. Mintz. 1982. Influence of land-surface evapotranspiration on the Earth's climate. Science **215**:1498–1501.

Shukla, J., C. Nobre, and P. Sellers. 1990. Amazon deforestation and climate change. Science **247**:1322–1325.

Shukla, S. P., and M. Sharma. 2010. Neutralization of rainwater acidity at Kanpur, India. Tellus Series B-Chemical and Physical Meteorology **62**:172–180.

Shuttleworth, W. J. 1988. Evaporation from Anazonian rainforest. Proceedings of the Royal Society of London Series B—Biological Sciences **233**:321–346.

Siddall, M., E. J. Rohling, A. Almogi-Labin, C. Hemleben, D. Meischner, I. Schmelzer, and D. A. Smeed. 2003. Sea-level fluctuations during the last glacial cycle. Nature **423**:853–858.

Sidle, R. C., Y. Tsuboyama, S. Noguchi, I. Hosoda, M. Fujieda, and T. Shimizu. 2000. Stormflow generation in steep forested headwaters: a linked hydrogeomorphic paradigm. Hydrological Processes **14**: 369–385.

Siebert, S., J. Burke, J. M. Faures, K. Frenken, J. Hoogeveen, P. Doell, and F. T. Portmann. 2010. Groundwater use for irrigation—A global inventory. Hydrology and Earth System Sciences **14**:1863–1880.

Siegel, D. A., D. J. McGillicuddy, and E. A. Fields. 1999. Mesoscale eddies, satellite altimetry, and new production in the Sargasso Sea. Journal of Geophysical Research—Oceans **104**:13359–13379.

Siegel, S. M., and B. Z. Siegel. 1984. First estimate of annual mercury flux at the Kilauea main vent. Nature **309**:146–147.

Siever, R. 1974. The steady state of the Earth's crust, atmosphere, and oceans. Scientific American **230**:72–79.

Sigg, A., and A. Neftel. 1991. Evidence for a 50-percent increase in H_2O_2 over the past 200 years from a Greenland ice core. Nature **351**:557–559.

Sigler, J. M., and X. Lee. 2006. Recent trends in anthropogenic mercury emission in the northeast United States. Journal of Geophysical Research—Atmospheres **111**.

Sigman, D. M., M. A. Altabet, D. C. McCorkle, R. Francois, and G. Fischer. 2000. The delta 15N of nitrate in the Southern Ocean: Nitrogen cycling and circulation in the ocean interior. Journal of Geophysical Research—Oceans **105**: 19599–19614.

Sigman, D. M., P. J. DiFiore, M. P. Hain, C. Deutsch, Y. Wang, D. M. Karl, A. N. Knapp, M. F. Lehmann, and S. Pantoja. 2009. The dual isotopes of deep nitrate as a constraint on the cycle and budget of oceanic fixed nitrogen. Deep-Sea Research Part I—Oceanographic Research Papers **56**:1419–1439.

Sigman, D. M., M. P. Hain, and G. H. Haug. 2010. The polar ocean and glacial cycles in atmospheric CO_2 concentration. Nature **466**:47–55.

Sillén, L. G. 1966. Regulation of O_2, N_2 and CO_2 in the atmosphere: Thoughts of a laboratory chemist. Tellus **18**:198–206.

Sillitoe, R. H., R. L. Folk, and N. Saric. 1996. Bacteria as mediators of copper sulfide enrichment during weathering. Science **272**:1153–1155.

Sillman, S. 1999. The relation between ozone, NOx and hydrocarbons in urban and polluted rural environments. Atmospheric Environment **33**:1821–1845.

Silver, W. L. 1994. Is nutrient availability related to plant nutrient use in humid tropical forests? Oecologia **98**:336–343.

Silver, W. L., D. J. Herman, and M. K. Firestone. 2001. Dissimilatory nitrate reduction to ammonium in upland tropical forest soils. Ecology 82:2410–2416.

Silver, W. L., and R. K. Miya. 2001. Global patterns in root decomposition: comparisons of climate and litter quality effects. Oecologia 129:407–419.

Silvester, W. B. 1989. Molybdenum limitation of asymbiotic nitrogen fixation in forests of Pacific northwest America. Soil Biology & Biochemistry 21:283–289.

Silvester, W. B., P. Sollins, T. Verhoeven, and S. P. Cline. 1982. Nitrogen fixation and acetylene reduction in decaying conifer boles—Effects of incubation time, aeration, and moisture content. Canadian Journal of Forest Research 12:646–652.

Sim, M. S., T. Bosak, and S. Ono. 2011. Large sulfur isotope fractionation does not require disproportionation. Science 333:74–77.

Simas, F.N.B., C. Schaefer, V. F. Melo, M. R. Albuquerque-Filho, R.F.M. Michel, V. V. Pereira, M.R.M. Gomes, and L. M. da Costa. 2007. Ornithogenic Cryosols from maritime Antarctica: Phosphatization as a soil-forming process. Geoderma 138:191–203.

Simkiss, K., and K. M. Wilbur. 1989. *Biomineralization: Cell Biology and Mineral Deposition*. Academic Press.

Simmons, J. S., L. Klemedtsson, H. Hultberg, and M. E. Hines. 1999. Consumption of atmospheric carbonyl sulfide by coniferous boreal forest soils. Journal of Geophysical Research—Atmospheres 104:11569–11576.

Simo, R., S. D. Archer, C. Pedros-Alio, L. Gilpin, and C. E. Stelfox-Widdicombe. 2002. Coupled dynamics of dimethylsulfoniopropionate and dimethylsulfide cycling and the microbial food web in surface waters of the North Atlantic. Limnology and Oceanography 47:53–61.

Simon, L., J. Bousquet, R. C. Levesque, and M. Lalonde. 1993. Origin and diversification of endomycorrhizal fungi and coincidence with vascular land plants. Nature 363:67–69.

Simon, N. S. 1988. Nitrogen cycling between sediment and the shallow water column in the transition zone of the Potomac river and estuary. 1. Nitrate and ammonium fluxes. Estuarine Coastal and Shelf Science 26:483–497.

Simon, N. S. 1989. Nitrogen cycling between sediment and the shallow water column in the transition zone of the Potomac river and estuary. 2. The role of wind-driven resuspension and adsorbed ammonium. Estuarine Coastal and Shelf Science 28:531–547.

Simonich, S. L., and R. A. Hites. 1994. Importance of vegetation in removing polycyclic aromatic hydrocarbons from the atmosphere. Nature 370:49–51.

Simonson, R. W. 1995. Airborne dust and its significance to soils. Geoderma 65:1–43.

Singer, A. 1989. Illite in the hot aridic soil environment. Soil Science 147:126–133.

Singh, H., Y. Chen, A. Staudt, D. Jacob, D. Blake, B. Heikes, and J. Snow. 2001. Evidence from the Pacific troposphere for large global sources of oxygenated organic compounds. Nature 410:1078–1081.

Singh, J. S., and S. R. Gupta. 1977. Plant decomposition and soil respiraton in terrestrial ecosystems. Botanical Review 43:499–528.

Singh, J. S., W. K. Lauenroth, and R. K. Steinhorst. 1975. Review and assessment of various techniques for estimating net primary production in grasslands from harvest data. Botanical Review 41:181–232.

Singleton, G. A., and L. M. Lavkulich. 1987a. A soil chronosequence on beach sands, Vancouver island, British Columbia. Canadian Journal of Soil Science 67:795–810.

Singleton, G. A., and L. M. Lavkulich. 1987b. Phosphorus transformations in a soil chronosequence, Vancouver island, British Columbia. Canadian Journal of Soil Science 67:787–793.

Singsaas, E. L., D. R. Ort, and E. H. DeLucia. 2001. Variation in measured values of photosynthetic quantum yield in ecophysiological studies. Oecologia 128:15–23.

Sinsabaugh, R. L., R. K. Antibus, A. E. Linkins, C. A. McClaugherty, L. Rayburn, D. Repert, and T. Weiland. 1993. Wood decomposition—Nitrogen and phosphorus dynamics in relation to extracellular enzyme activity. Ecology 74:1586–1593.

Sinsabaugh, R. L., M. M. Carreiro, and D. A. Repert. 2002. Allocation of extracellular enzymatic activity in relation to litter composition, N deposition, and mass loss. Biogeochemistry 60:1–24.

Sinsabaugh, R. L., C. L. Lauber, M. N. Weintraub, B. Ahmed, S. D. Allison, C. Crenshaw, A. R. Contosta, D. Cusack, S. Frey, M. E. Gallo, T. B. Gartner, S. E. Hobbie, K. Holland, B. L. Keeler, J. S. Powers, M. Stursova, C. Takacs-Vesbach, M. P. Waldrop, M. D. Wallenstein, D. R. Zak, and L. H. Zeglin. 2008. Stoichiometry of soil enzyme activity at global scale. Ecology Letters 11:1252–1264.

Sivan, O., D. P. Schrag, and R. W. Murray. 2007. Rates of methanogenesis and methanotrophy in deep-sea sediments. Geobiology 5:141–151.

Sjöberg, A. 1976. Phosphate analysis of anthropic soils. Journal of Field Archaeology 3:447–454.

Sjöström, J., and U. Qvarfort. 1992. Long-term changes of soil chemistry in central Sweden. Soil Science 154:450–457.

Skiba, U., K. A. Smith, and D. Fowler. 1993. Nitrification and denitrificaiton as sources of nitric oxide and nitrous oxide in a sandy loam soil. Soil Biology & Biochemistry 25:1527–1536.

Skyring, G. W. 1987. Sulfate reduction in coastal ecosystems. Geomicrobiology Journal 5:295–374.

Sleep, N. H., K. J. Zahnle, J. F. Kasting, and H. J. Morowitz. 1989. Annihilation of ecosystems by large asteroid impacts on the early Earth. Nature 342:139–142.

Slemr, F., E. G. Brunke, R. Ebinghaus, and J. Kuss. 2011. Worldwide trend of atmospheric mercury since 1995. Atmospheric Chemistry and Physics 11:4779–4787.

Slemr, F., and W. Seiler. 1991. Field study of environmental variables controlling the NO emissions from soil and the NO compensation point. Journal of Geophysical Research—Atmospheres 96:13017–13031.

Slinn, W.G.N. 1988. A simple model for Junge's relationship between concentration fluctuations and residence times for tropospheric trace gases. Tellus 40B:229–232.

Sloan, G. C., M. Matsuura, A. A. Zijlstra, E. Lagadec, M.A.T. Groenewegen, P. R. Wood, C. Szyszka, J. Bernard-Salas, and J. T. van Loon. 2009. Dust formation in a galaxy with primitive abundances. Science 323:353–355.

Slutsky, A. H., and B. C. Yen. 1997. A macro-scale natural hydrologic cycle water availability model. Journal of Hydrology 201:329–347.

Smeck, N. E. 1985. Phosphorus dynamics in soils and landscapes. Geoderma 36:185–199.

Smedley, S. R., and T. Eisner. 1995. Sodium uptake by puddling in a moth. Science 270:1816–1818.

Smemo, K. A., and J. B. Yavitt. 2011. Anaerobic oxidation of methane: an underappreciated aspect of methane cycling in peatland ecosystems? Biogeosciences 8:779–793.

Smetacek, V., C. Klaas, V. H. Strass, P. Assmy, M. Montresor, B. Cisewski, N. Savoye, A. Webb, F. dOvidio, J. M. Arrieta, U. Bathmann, R. Bellerby, G. M. Berg, P. Croot, S. Gonzalez, J. Henjes, G. J. Herndl, L. J. Hoffmann, H. Leach, M. Losch, M. M. Mills, C. Neill, I. Peeken, R. Rottgers, O. Sachs, E. Sauter, M. M. Schmidt, J. Schwarz, A. Terbruggen, D. Wolf-Gladrow. 2012. Deep carbon export from a Southern Ocean iron-fertilized diatom bloom. Nature 487:313–319.

Smil, V. 2000. Phosphorus in the environment: Natural flows and human interferences. Annual Review of Energy and the Environment 25:53–88.

Smil, V. 2001. Enriching the Earth. MIT Press.

Smirnoff, N., P. Todd, and G. R. Stewart. 1984. The occurrence of nitrate reduction in the leaves of woody plants. Annals of Botany 54:363–374.

Smith, K. A., K. E. Dobbie, B. C. Ball, L. R. Bakken, B. K. Sitaula, S. Hansen, R. Brumme, W. Borken, S. Christensen, A. Prieme, D. Fowler, J. A. Macdonald, U. Skiba, L. Klemedtsson, A. Kasimir-Klemedtsson, A. Degorska, and P. Orlanski. 2000. Oxidation of atmospheric methane in northern European soils, comparison with other ecosystems, and uncertainties in the global terrestrial sink. Global Change Biology 6:791–803.

Smith, K. L. 1992a. Benthic boundary layer communities and carbon cycling at abyssal depths in the central north Pacific. Limnology and Oceanography 37:1034–1056.

Smith, K. L., R. J. Baldwin, and P. M. Williams. 1992a. Reconciling particulate organic carbon flux and sediment community oxygen consumption in the deep north Pacific. Nature 359:313–316.

Smith, K. L., R. S. Kaufmann, R. J. Baldwin, and A. F. Carlucci. 2001a. Pelagic-benthic coupling in the abyssal eastern North Pacific: An 8-year time-series study of food supply and demand. Limnology and Oceanography 46:543–556.

Smith, L. C., G. M. MacDonald, A. A. Velichko, D. W. Beilman, O. K. Borisova, K. E. Frey, K. V. Kremenetski, and Y. Sheng. 2004. Siberian peatlands a net carbon sink and global methane source since the early Holocene. Science 303:353–356.

Smith, M. L., S. V. Ollinger, M. E. Martin, J. D. Aber, R. A. Hallett, and C. L. Goodale. 2002. Direct estimation of aboveground forest productivity through hyperspectral remote sensing of canopy nitrogen. Ecological Applications 12:1286–1302.

Smith, M. S., and J. M. Tiedje. 1979. Phases of denitrification following oxygen depletion in soil. Soil Biology & Biochemistry 11:261–267.

Smith, P. H., L. K. Tamppari, R. E. Arvidson, D. Bass, D. Blaney, W. V. Boynton, A. Carswell, D. C. Catling, B. C. Clark, T. Duck, E. DeJong, D. Fisher, W. Goetz, H. P. Gunnlaugsson, M. H. Hecht, V. Hipkin, J. Hoffman, S. F. Hviid, H. U. Keller, S. P. Kounaves, C. F. Lange, M. T. Lemmon, M. B. Madsen, W. J. Markiewicz, J. Marshall, C. P. McKay, M. T. Mellon, D. W. Ming, R. V. Morris, W. T. Pike, N. Renno, U. Staufer, C. Stoker, P. Taylor, J. A. Whiteway, and A. P. Zent. 2009. H$_2$O at the Phoenix landing site. Science 325:58–61.

Smith, R. A., R. B. Alexander, and M. G. Wolman. 1987. Water-quality trends in the nation's rivers. Science 235:1607–1615.

Smith, R. B., R. H. Waring, and D. A. Perry. 1981. Interpreting foliar analyses from douglas fir as weight per unit of leaf area. Canadian Journal of Forest Research 11:593–598.

Smith, R. C., B. B. Prezelin, K. S. Baker, R. R. Bidigare, N. P. Boucher, T. Coley, D. Karentz, S. Macintyre, H. A. Matlick, D. Menzies, M. Ondrusek, Z. Wan, and K. J. Waters. 1992b. Ozone depletion—Ultraviolet radiation and phytoplankton biology in Antarctic waters. Science **255**:952–959.

Smith, R. L., and M. J. Klug. 1981. Reduction of sulfur compounds in the sediments of a eutrophic lake basin. Applied and Environmental Microbiology **41**:1230–1237.

Smith, S. D., C. A. Herr, K. L. Leary, and J. M. Piorkowski. 1995. Soil-plant water relations in a Mojave desert mixed shrub community: a comparison of three geomorphic surfaces. Journal of Arid Environments **29**:339–351.

Smith, S. J., H. Pitcher, and T.M.L. Wigley. 2001b. Global and regional anthropogenic sulfur dioxide emissions. Global and Planetary Change **29**:99–119.

Smith, S. J., J. F. Power, and W. D. Kemper. 1994. Fixed ammonium and nitrogen availability indexes. Soil Science **158**:132–140.

Smith, S. V. 1981. Marine macrophytes as a global carbon sink. Science **211**:838–840.

Smith, S. V., and F. T. MacKenzie. 1987. The ocean as a net heterotrophic system: Implications from the carbon biogeochemical cycle. Global Biogeochemical Cycles **1**:187–198.

Smith, T. M., R. Leemans, and H. H. Shugart. 1992c. Sensitivity of terrestrial carbon storage to CO_2-induced climate change: Comparison of four scenarios based on general-circulation models. Climatic Change **21**:367–384.

Smith, T. M., and H. H. Shugart. 1993. The transient response of terrestrial carbon storage to a perturbed climate. Nature **361**:523–526.

Smith, T. M., X. G. Yin, and A. Gruber. 2006. Variations in annual global precipitation (1979-2004), based on the Global Precipitation Climatology Project 2.5 degrees analysis. Geophysical Research Letters **33**.

Smith, V. H. 1982. The nitrogen and phosphorus dependence of algal biomass in lakes—An empirical and theoretical analysis. Limnology and Oceanography **27**:1101–1112.

Smith, V. H. 1983. Low nitrogen to phosphorus ratios favor dominance by blue-green algae in lake phytoplankton. Science **221**:669–671.

Smith, V. H. 1992b. Effects of nitrogen-phosphorus supply ratios on nitrogen fixation in agricultural and pastoral ecosystems. Biogeochemistry **18**:19–35.

Smith, V. H., and D. W. Schindler. 2009. Eutrophication science: where do we go from here? Trends in Ecology & Evolution **24**:201–207.

Smith, W. H. 1976. Character and significance of forest tree root exudates. Ecology **57**:324–331.

Smith, W. H. 1990. The atmosphere and the rhizosphere: linkages with potential significance for forest tree health. *In* A. A. Lucier and S. G. Haines, editors. *Mechanisms of Forest Response to Acidic Deposition.* Springer-Verlag, 188–241.

Smith, W. H., and T. G. Siccama. 1981. The Hubbard Brook ecosystem study—Biogeochemistry of lead in the northern hardwood forest. Journal of Environmental Quality **10**:323–332.

Smol, J. P., and B. F. Cumming. 2000. Tracking long-term changes in climate using algal indicators in lake sediments. Journal of Phycology **36**:986–1011.

Smolders, A.J.P., L.P.M. Lamers, E.C.H.E.T. Lucassen, G. Van Der Velde, and J.G.M. Roelofs. 2006. Internal eutrophication: How it works and what to do about it—a review. Chemistry and Ecology **22**:93–111.

Smrekar, S. E., E. R. Stofan, N. Mueller, A. Treiman, L. Elkins-Tanton, J. Helbert, G. Piccioni, and P. Drossart. 2010. Recent hotspot volcanism on Venus from VIRTIS emissivity data. Science **328**:605–608.

Snaydon, R. W. 1962. Micro-distribution of *Trifolium repens* L. and its relation to soil factors. Journal of Ecology **50**:133–143.

Snider, D. M., S. L. Schiff, and J. Spoelstra. 2009. $^{15}N/^{14}N$ and $^{18}O/^{16}O$ stable isotope ratios of nitrous oxide produced during denitrification in temperate forest soils. Geochimica et Cosmochimica Acta **73**:877–888.

Snider, D. M., J. J. Venkiteswaran, S. L. Schiff, and J. Spoelstra. 2012. Deciphering the oxygen isotope composition of nitrous oxide produced by nitrification. Global Change Biology **18**:356–370.

Sobczak, W. V., S. Findlay, and S. Dye. 2003. Relationships between DOC bioavailability and nitrate removal in an upland stream: An experimental approach. Biogeochemistry **62**:309–327.

Sobek, S., T. DelSontro, N. Wongfun, and B. Wehrli. 2012. Extreme organic carbon burial fuels intense methane bubbling in a temperate reservoir. Geophysical Research Letters **39**:4.

Soden, B. J., D. L. Jackson, V. Ramaswamy, M. D. Schwarzkopf, and X. L. Huang. 2005. The radiative signature of upper tropospheric moistening. Science **310**:841–844.

Sofiev, M., J. Soares, M. Prank, G. de Leeuw, and J. Kukkonen. 2011. A regional-to-global model of emission and transport of sea salt particles in the atmosphere. Journal of Geophysical Research—Atmospheres **116**.

Sollins, P., G. P. Robertson, and G. Uehara. 1988. Nutrient mobility in variable-charge and permanent-charge soils. Biogeochemistry **6**:181–199.

Sollins, P., G. Spycher, and C. A. Glassman. 1984. Net nitrogen mineralization from light- and heavy-fraction forest soil organic matter. Soil Biology & Biochemistry **16**:31–37.

Solomon, D. K., and T. E. Cerling. 1987. The annual carbon dioxide cycle in a montane soil—Observations, modeling, and implications for weathering. Water Resources Research 23:2257–2265.

Solomon, P., J. Barrett, T. Mooney, B. Connor, A. Parrish, and D. E. Siskind. 2006. Rise and decline of active chlorine in the stratosphere. Geophysical Research Letters 33.

Solomon, S. 1990. Progress towards a quantitative understanding of Antarctic ozone depletion. Nature 347:347–354.

Solomon, S., R. R. Garcia, F. S. Rowland, and D. J. Wuebbles. 1986. On the depletion of Antarctic ozone. Nature 321:755–758.

Solomon, S., R. W. Portmann, and D.W.J. Thompson. 2007. Contrasts between Antarctic and Arctic ozone depletion. Proceedings of the National Academy of Sciences 104:445–449.

Solomon, S. C., O. Aharonson, J. M. Aurnou, W. B. Banerdt, M. H. Carr, A. J. Dombard, H. V. Frey, M. P. Golombek, S. A. Hauck, J. W. Head, B. M. Jakosky, C. L. Johnson, P. J. McGovern, G. A. Neumann, R. J. Phillips, D. E. Smith, and M. T. Zuber. 2005. New perspectives on ancient Mars. Science 307:1214–1220.

Son, Y. 2001. Non-symbiotic nitrogen fixation in forest ecosystems. Ecological Research 16:183–196.

Sorooshian, S., J. L. Li, K. L. Hsu, and X. G. Gao. 2011. How significant is the impact of irrigation on the local hydroclimate in California's Central Valley? Comparison of model results with ground and remote-sensing data. Journal of Geophysical Research—Atmospheres 116.

Soule, P. T., and P. A. Knapp. 2006. Radial growth rate increases in naturally occurring ponderosa pine trees: A late-20th century CO_2 fertilization effect? New Phytologist 171:379–390.

Sowers, T. 2006. Late Quaternary atmospheric CH_4-isotope record suggests marine clathrates are stable. Science 311:838–840.

Sowers, T., R. B. Alley, and J. Jubenville. 2003. Ice core records of atmospheric N_2O covering the last 106,000 years. Science 301:945–948.

Sowers, T., S. Bernard, O. Aballain, J. Chappellaz, J.-M. Barnola, and T. Marik. 2005. Records of the $\delta^{13}C$ of atmospheric CH_4 over the last two centuries as recorded in Antarctic snow and ice. Global Biogeochemical Cycles 19.

Sowers, T., and M. Bender. 1995. Climate records covering the last deglaciation. Science 269:210–214.

Spalding, R. F., and M. E. Exner. 1993. Occurrence of nitrate in groundwater—A review. Journal of Environmental Quality 22:392–402.

Sparks, D. L. 2002. Environmental Soil Chemistry (2nd ed.). Academic/Elsevier.

Sparks, J. P., J. M. Roberts, and R. K. Monson. 2003. The uptake of gaseous organic nitrogen by leaves: A significant global nitrogen transfer process. Geophysical Research Letters 30.

Speidel, D. H., and A. F. Agnew. 1982. The Natural Geochemistry of Our Environment. Westview Press.

Spichtinger, N., M. Wenig, P. James, T. Wagner, U. Platt, and A. Stohl. 2001. Satellite detection of a continental-scale plume of nitrogen oxides from boreal forest fires. Geophysical Research Letters 28:4579–4582.

Spiro, P. A., D. J. Jacob, and J. A. Logan. 1992. Global inventory of sulfur emissions with 1° x 1° resolution. Journal of Geophysical Research 97:6023–6036.

Spivack, A. J., C. F. You, and H. J. Smith. 1993. Foraminiferal boron isotope ratios as a proxy for surface ocean pH over the past 21-myr. Nature 363:149–151.

Spoelstra, J., S. L. Schiff, P. W. Hazlett, D. S. Jeffries, and R. G. Semkin. 2007. The isotopic composition of nitrate produced from nitrification in a hardwood forest floor. Geochimica et Cosmochimica Acta 71:3757–3771.

Sposito, G. 1989. The Surface Chemistry of Natural Particles. Oxford University Press.

Spratt, H. G., and M. D. Morgan. 1990. Sulfur cycling in a cedar-dominated, freshwater wetland. Limnology and Oceanography 35:1586–1593.

Springer, C. J., E. H. DeLucia, and R. B. Thomas. 2005. Relationships between net photosynthesis and foliar nitrogen concentrations in a loblolly pine forest ecosystem grown in elevated atmospheric carbon dioxide. Tree Physiology 25:385–394.

Spycher, G., P. Sollins, and S. Rose. 1983. Carbon and nitrogen in the light fraction of a forest soil—Vertical distribution and seasonal patterns. Soil Science 135:79–87.

Squyres, S. W., R. E. Arvidson, S. Ruff, R. Gellert, R. V. Morris, D. W. Ming, L. Crumpler, J. D. Farmer, D. J. Des Marais, A. Yen, S. M. McLennan, W. Calvin, J. F. Bell, B. C. Clark, A. Wang, T. J. McCoy, M. E. Schmidt, and P. A. de Souza. 2008. Detection of silica-rich deposits on Mars. Science 320:1063–1067.

Squyres, S. W., J. P. Grotzinger, R. E. Arvidson, J. F. Bell, W. Calvin, P. R. Christensen, B. C. Clark, J. A. Crisp, W. H. Farrand, K. E. Herkenhoff, J. R. Johnson, G. Klingelhofer, A. H. Knoll, S. M. McLennan, H. Y. McSween, R. V. Morris, J. W. Rice, R. Rieder, and L. A. Soderblom. 2004. In situ evidence for an ancient aqueous environment at Meridiani Planum, Mars. Science 306:1709–1714.

Srinivasan, G., J. N. Goswami, and N. Bhandari. 1999. ^{26}Al in eucrite Piplia Kalan: Plausible heat source and formation chronology. Science 284:1348–1350.

Srivastava, S. C., and J. S. Singh. 1988. Carbon and phosphorus in the soil biomass of some tropical soils of India. Soil Biology & Biochemistry 20:743–747.

St Louis, V. L., C. A. Kelly, E. Duchemin, J.W.M. Rudd, and D. M. Rosenberg. 2000. Reservoir surfaces as sources of greenhouse gases to the atmosphere: A global estimate. BioScience 50:766–775.

Staaf, H., and B. Berg. 1982. Accumulation and release of plant nutrients in decomposing scots pine needle litter—Long-term decomposition in a scots pine forest. 2. Canadian Journal of Botany 60:1561–1568.

Staehr, P., L. Baastrup-Spohr, K. Sand-Jensen, and C. Stedmon. 2012a. Lake metabolism scales with lake morphometry and catchment conditions. Aquatic Sciences—Research Across Boundaries 74:155–169.

Staehr, P. A., D. Bade, M. C. Van de Bogert, G. R. Koch, C. Williamson, P. Hanson, J. J. Cole, and T. Kratz. 2010a. Lake metabolism and the diel oxygen technique: State of the science. Limnology and Oceanography-Methods 8:628–644.

Staehr, P. A., K. Sand-Jensen, A. L. Raun, B. Nilsson, and J. Kidmose. 2010b. Drivers of metabolism and net heterotrophy in contrasting lakes. Limnology and Oceanography 55:817–830.

Staehr, P. A., J. M. Testa, W. M. Kemp, J. J. Cole, K. Sand-Jensen, and S. V. Smith. 2012b. The metabolism of aquatic ecosystems: history, applications, and future challenges. Aquatic Sciences 74:15–29.

Staffelbach, T., A. Neftel, B. Stauffer, and D. Jacob. 1991. A record of the atmospheric methane sink from formaldehyde in polar ice cores. Nature 349:603–605.

Stallard, R. F. 1998. Terrestrial sedimentation and the carbon cycle: Coupling weathering and erosion to carbon burial. Global Biogeochemical Cycles 12:231–257.

Stallard, R. F., and J. M. Edmond. 1981. Geochemistry of the Amazon. 1. Precipitation chemistry and the marine contribution to the dissolved load at the time of peak discharge. Journal of Geophysical Research—Oceans and Atmospheres 86:9844–9858.

Stallard, R. F., and J. M. Edmond. 1983. Geochemistry of the Amazon. 2. The influence of geology and weathering environment on the dissolved load. Journal of Geophysical Research—Oceans and Atmospheres 88:9671–9688.

Stanley, D. W., and J. E. Hobbie. 1981. Nitrogen recycling in a North Carolina coastal river. Limnology and Oceanography 26:30–42.

Stanley, E. H., S. G. Fisher, and N. B. Grimm. 1997. Ecosystem expansion and contraction in streams. BioScience 47:427–435.

Stanley, E. H., and J. T. Maxted. 2008. Changes in the dissolved nitrogen pool across land cover gradients in Wisconsin streams. Ecological Applications 18:1579–1590.

Stanley, S. R., and E. J. Ciolkosz. 1981. Classification and genesis of Spodosols in the central Appalachians. Soil Science Society of America Journal 45:912–917.

Stark, J. M., and S. C. Hart. 1997. High rates of nitrification and nitrate turnover in undisturbed coniferous forests. Nature 385:61–64.

Starr, G., and S. F. Oberbauer. 2003. Photosynthesis of arctic evergreens under snow: Implications for tundra ecosystem carbon balance. Ecology 84:1415–1420.

Staubes, R., H.-W. Georgii, and G. Ockelmann. 1989. Flux of COS, DMS, and CS_2 from various soils in Germany. Tellus 41B:305–313.

Steele, L. P., E. J. Dlugokencky, P. M. Lang, P. P. Tans, R. C. Martin, and K. A. Masarie. 1992. Slowing down of the global accumulation of atmospheric methane during the 1980s. Nature 358:313–316.

Steele, L. P., P. J. Fraser, R. A. Rasmussen, M.A.K. Khalil, T. J. Conway, A. J. Crawford, R. H. Gammon, K. A. Masarie, and K. W. Thoning. 1987. The global distribution of methane in the troposphere. Journal of Atmospheric Chemistry 5:125–171.

Stefels, J., M. Steinke, S. Turner, G. Malin, and S. Belviso. 2007. Environmental constraints on the production and removal of the climatically active gas dimethylsulphide (DMS) and implications for ecosystem modelling. Biogeochemistry 83:245–275.

Stehfest, E., and L. Bouwman. 2006. N_2O and NO emission from agricultural fields and soils under natural vegetation: Summarizing available measurement data and modeling of global annual emissions. Nutrient Cycling in Agroecosystems 74:207–228.

Steinbacher, M., H. G. Bingemer, and U. Schmidt. 2004. Measurements of the exchange of carbonyl sulfide (OCS) and carbon disulfide (CS_2) between soil and atmosphere in a spruce forest in central Germany. Atmospheric Environment 38:6043–6052.

Steinke, M., G. Malin, S. D. Archer, P. H. Burkill, and P. S. Liss. 2002. DMS production in a coccolithophorid bloom: evidence for the importance of dinoflagellate DMSP lyases. Aquatic Microbial Ecology 26: 259–270.

Stephens, J. C., and E. H. Stewart. 1976. Effect of climate on organic soil subsidence. In Proceedings of the 2nd International Symposium on Land Subsidence, Publication 121, International Association of Hydrological Sciences, 649–655.

Sterling, S., and A. Ducharne. 2008. Comprehensive data set of global land cover change for land surface model applications. Global Biogeochemical Cycles 22.

Stern, D. I. 2006. Reversal of the trend in global anthropogenic sulfur emissions. Global Environmental Change-Human and Policy Dimensions 16:207–220.

Sternberg, E., D. G. Tang, T. Y. Ho, C. Jeandel, and F.M.M. Morel. 2005. Barium uptake and adsorption in diatoms. Geochimica et Cosmochimica Acta 69: 2745–2752.

Sterner, R. W. 1989. Resource competition during seasonal succession toward dominance by cyanobacteria. Ecology **70**:229–245.

Sterner, R. W., and J. J. Elser. 2002. *Ecological Stoichiometry.* Princeton University Press.

Sterner, R. W., J. J. Elser, and D. O. Hessen. 1992. Stoichiometric relationships among producers, consumers and nutrient cycling in pelagic ecosystems. Biogeochemistry **17**:49–67.

Sterner, R. W., T. M. Smutka, R.M.L. McKay, X. M. Qin, E. T. Brown, and R. M. Sherrell. 2004. Phosphorus and trace metal limitation of algae and bacteria in Lake Superior. Limnology and Oceanography **49**:495–507.

Stetter, K. O., G. Lauerer, M. Thomm, and A. Neuner. 1987. Isolation of extremely thermophilic sulfate reducers—Evidence for a novel branch of archaebacteria. Science **236**:822–824.

Steudler, P. A., R. D. Bowden, J. M. Melillo, and J. D. Aber. 1989. Influence of nitrogen fertilization on methane uptake in temperate forest soils. Nature **341**:314–316.

Steudler, P. A., and B. J. Peterson. 1985. Annual cycle of gaseous sulfur emissions from a New England *Spartina alterniflora* marsh. Atmospheric Environment **19**:1411–1416.

Stevens, C. J., N. B. Dise, D.J.G. Gowing, and J. O. Mountford. 2006. Loss of forb diversity in relation to nitrogen deposition in the UK: regional trends and potential controls. Global Change Biology **12**:1823–1833.

Stevens, C. J., C. Dupre, E. Dorland, C. Gaudnik, D.J.G. Gowing, A. Bleeker, M. Diekmann, D. Alard, R. Bobbink, D. Fowler, E. Corcket, J. O. Mountford, V. Vandvik, P. A. Aarrestad, S. Muller, and N. B. Dise. 2010. Nitrogen deposition threatens species richness of grasslands across Europe. Environmental Pollution **158**:2940–2945.

Stevens, C. J., P. Manning, L.J.L. van den Berg, M.C.C. de Graaf, G.W.W. Wamelink, A. W. Boxman, A. Bleeker, P. Vergeer, M. Arroniz-Crespo, J. Limpens, L.P.M. Lamers, R. Bobbink, and E. Dorland. 2011. Ecosystem responses to reduced and oxidised nitrogen inputs in European terrestrial habitats. Environmental Pollution **159**:665–676.

Stevens, T. O., and J. P. McKinley. 1995. Lithoautotrophic microbial ecosystems in deep basalt aquifers. Science **270**:450–454.

Stevens, W. B., R. G. Hoeft, and R. L. Mulvaney. 2005. Fate of nitirogen-15 in a long-term nitrogen rate study: II. Nitrogen uptake efficiency. Agronomy Journal **97**:1046–1053.

Stevenson, D. J. 1983. The nature of the Earth prior to the oldest known rock record: The Hadean Earth. *In* J. E. Schopf, editor. *Earth's Earliest Biosphere.* Princeton University Press, 32–40.

Stevenson, D. J. 2008. A planetary perspective on the deep Earth. Nature **451**:261–265.

Stevenson, F. J. 1982. Origin and distribution of nitrogen in soil. *In* F. J. Stevenson, editor. *Nitrogen in Agricultural Soils,* Soil Science Society of America, 1–42.

Stevenson, F. J. 1986. *Cycles of Soil.* Wiley.

Stewart, A. J., and R. G. Wetzel. 1982. Influence of dissolved humic materials on carbon assimilation and alkaline-phosphatase activity in natural algal bacterial assemblages. Freshwater Biology **12**:369–380.

Still, C. J., J. A. Berry, G. J. Collatz, and R. S. DeFries. 2003. Global distribution of C-3 and C-4 vegetation: Carbon cycle implications. Global Biogeochemical Cycles **17**.

Stimler, K., S. A. Montzka, J. A. Berry, Y. Rudich, and D. Yakir. 2010. Relationships between carbonyl sulfide (COS) and CO_2 during leaf gas exchange. New Phytologist **186**:869–878.

Stock, J. B., A. M. Stock, and J. M. Mottonen. 1990. Signal transduction in bacteria. Nature **344**:395–400.

Stockner, J. G., and N. J. Antia. 1986. Algal picoplankton from marine and freshwater ecosystems—A multidisciplinary perspective. Canadian Journal of Fisheries and Aquatic Sciences **43**:2472–2503.

Stoddard, J. L., D. S. Jeffries, A. Lükewille, T. A. Clair, P. J. Dillon, C. T. Driscoll, M. Forsius, M. Johannessen, J. S. Kahl, J. H. Kellogg, A. Kemp, J. Mannio, D. T. Monteith, P. S. Murdoch, S. Patrick, A. Rebsdorf, B. L. Skjelkvåle, M. P. Stainton, T. Traaen, H. van Dam, K. E. Webster, J. Wieting, and A. Wilander. 1999. Regional trends in aquatic recovery from acidification in North America and Europe. Nature **401**:575–576.

Stofan, E. R., C. Elachi, J. I. Lunine, R. D. Lorenz, B. Stiles, K. L. Mitchell, S. Ostro, L. Soderblom, C. Wood, H. Zebker, S. Wall, M. Janssen, R. Kirk, R. Lopes, F. Paganelli, J. Radebaugh, L. Wye, Y. Anderson, M. Allison, R. Boehmer, P. Callahan, P. Encrenaz, E. Flamini, G. Francescetti, Y. Gim, G. Hamilton, S. Hensley, W.T.K. Johnson, K. Kelleher, D. Muhleman, P. Paillou, G. Picardi, F. Posa, L. Roth, R. Seu, S. Shaffer, S. Vetrella, and R. West. 2007. The lakes of Titan. Nature **445**:61–64.

Stolt, M. H., J. C. Baker, and T. W. Simpson. 1992. Characterization and genesis of saprolite derived from gneissic rocks of Virginia. Soil Science Society of America Journal **56**:531–539.

Stone, E. L., and S. Boonkird. 1963. Calcium accumulation in the bark of *Terminalia spp.* in Thailand. Ecology **44**:586–588.

Stone, E. L., and R. Kszystyniak. 1977. Conservation of potassium in a *Pinus resinosa* ecosystem. Science **198**:192–194.

Stott, P. A., S.F.B. Tett, G. S. Jones, M. R. Allen, J.F.B. Mitchell, and G. J. Jenkins. 2000. External control of 20th century temperature by natural and anthropogenic forcings. Science **290**:2133–2137.

Strahm, B. D., and R. B. Harrison. 2006. Nitrate sorption in a variable-charge forest soil of the Pacific Northwest. Soil Science **171**:313–321.

Strahm, B. D., and R. B. Harrison. 2007. Mineral and organic matter controls on the sorption of macronutrient anions in variable-charge soils. Soil Science Society of America Journal **71**:1926–1933.

Strand, A. E., S. G. Pritchard, M. L. McCormack, M. A. Davis, and R. Oren. 2008. Irreconcilable differences: Fine-root life spans and soil carbon persistence. Science **319**:456–458.

Strauss, E. A., and G. A. Lamberti. 2000. Regulation of nitrification in aquatic sediments by organic carbon. Limnology and Oceanography **45**:1854–1859.

Strauss, E. A., and G. A. Lamberti. 2002. Effect of dissolved organic carbon quality on microbial decomposition and nitrification rates in stream sediments. Freshwater Biology **47**:65–74.

Strauss, E. A., N. L. Mitchell, and G. A. Lamberti. 2002. Factors regulating nitrification in aquatic sediments: effects of organic carbon, nitrogen availability, and pH. Canadian Journal of Fisheries and Aquatic Sciences **59**:554–563.

Streets, D. G., M. K. Devane, Z. Lu, T. C. Bond, E. M. Sunderland, and D. J. Jacob. 2011. All-time releases of mercury to the atmosphere from human activities. Environmental Science & Technology **45**: 10485–10491.

Streets, D. G., C. Yu, Y. Wu, M. Chin, Z. Zhao, T. Hayasaka, and G. Shi. 2008. Aerosol trends over China, 1980-2000. Atmospheric Research **88**:174–182.

Striegl, R. G., T. A. McConnaughey, D. C. Thorstenson, E. P. Weeks, and J. C. Woodward. 1992. Consumption of atmospheric methane by desert soils. Nature **357**:145–147.

Strobel, B. W. 2001. Influence of vegetation on low-molecular-weight carboxylic acids in soil solution—a review. Geoderma **99**:169–198.

Stroeve, J. C., M. C. Serreze, M. M. Holland, J. E. Kay, J. Malanik, and A. P. Barrett. 2012. The Arctic's rapidly shrinking sea ice cover: a research synthesis. Climatic Change **110**:1005–1027.

Strojan, C. L., D. C. Randall, and F. B. Turner. 1987. Relationship of leaf litter decomposition rates to rainfall in the Mojave desert. Ecology **68**:741–744.

Strother, P. K., L. Battison, M. D. Brasier, and C. H. Wellman. 2011. Earth's earliest non-marine eukaryotes. Nature **473**:505–509.

Strous, M., J. A. Fuerst, E.H.M. Kramer, S. Logemann, G. Muyzer, K. T. van de Pas-Schoonen, R. Webb, J. G. Kuenen, and M.S.M. Jetten. 1999. Missing lithotroph identified as new planctomycete. Nature **400**:446–449.

Stuiver, M. 1980. ^{14}C distribution in the Atlantic ocean. Journal of Geophysical Research—Oceans and Atmospheres **85**:2711–2718.

Stuiver, M., P. D. Quay, and H. G. Ostlund. 1983. Abyssal water ^{14}C distribution and the age of the world oceans. Science **219**:849–851.

Stumm, W., and J. J. Morgan. 1996. *Aquatic Chemistry Chemical Equilibria and Rates in Natural Waters*, 3rd edition. Wiley.

Sturges, W. T., S. A. Penkett, J. M. Barnola, J. Chappellaz, E. Atlas, and V. Stroud. 2001. A long-term record of carbonyl sulfide (COS) in two hemispheres from firn air measurements. Geophysical Research Letters **28**: 4095–4098.

Stutter, M. I., B.O.L. Demars, and S. J. Langan. 2010. River phosphorus cycling: Separating biotic and abiotic uptake during short-term changes in sewage effluent loading. Water Research **44**:4425–4436.

Su, H., Y. Cheng, R. Oswald, T. Behrendt, I. Trebs, F. X. Meixner, M. O. Andreae, P. Cheng, Y. Zhang, and U. Poeschl. 2011. Soil nitrite as a source of atmospheric HONO and OH radicals. Science **333**: 1616–1618.

Suarez, D. L., J. D. Wood, and I. Ibrahim. 1992. Reevaluation of calcite supersaturation in soils. Soil Science Society of America Journal **56**:1776–1784.

Subbarao, G. V., K. Nakahara, M. P. Hurtado, H. Ono, D. E. Moreta, A. F. Salcedo, A. T. Yoshihashi, T. Ishikawa, M. Ishitani, M. Ohnishi-Kameyama, M. Yoshida, M. Rondon, I. M. Rao, C. E. Lascano, W. L. Berry, and O. Ito. 2009. Evidence for biological nitrification inhibition in *Brachiaria* pastures. Proceedings of the National Academy of Sciences **106**:17302–17307.

Suberkropp, K., G. L. Godshalk, and M. J. Klug. 1976. Changes in the chemical composition of leaves during processing in a woodland stream. Ecology **57**: 720–727.

Subke, J. A., I. Inglima, and M. F. Cotrufo. 2006. Trends and methodological impacts in soil CO_2 efflux partitioning: A metaanalytical review. Global Change Biology **12**:921–943.

Subke, J. A., N. R. Voke, V. Leronni, M. H. Garnett, and P. Ineson. 2011. Dynamics and pathways of autotrophic and heterotrophic soil CO_2 efflux revealed by forest girdling. Journal of Ecology **99**:186–193.

Suchet, P. A., J. L. Probst, and W. Ludwig. 2003. Worldwide distribution of continental rock lithology: Implications for the atmospheric/soil CO_2 uptake by continental weathering and alkalinity river transport to the oceans. Global Biogeochemical Cycles **17**.

Suess, E. 1980. Particulate organic carbon flux in the oceans—Surface productivity and oxygen utilization. Nature **288**:260–263.

Summons, R. E., L. L. Jahnke, J. M. Hope, and G. A. Logan. 1999. 2-Methylhopanoids as biomarkers for cyanobacterial oxygenic photosynthesis. Nature **400**:554–557.

Sunda, W. G., and S. A. Huntsman. 1992. Feedback interactions between zinc and phytoplankton in seawater. Limnology and Oceanography **37**:25–40.

Sundby, B., C. Gobeil, N. Silverberg, and A. Mucci. 1992. The phosphorus cycle in coastal marine sediments. Limnology and Oceanography **37**:1129–1145.

Sundquist, E. T. 1993. The global carbon dioxide budget. Science **259**:934–941.

Sundquist, E. T., L. N. Plummer, and T.M.L. Wigley. 1979. Carbon dioxide in the ocean surface—Homogeneous buffer factor. Science **204**:1203–1205.

Sundquist, E. T., and K. Visser. 2005. The geologic history of the carbon cycle. *In* W. H. Schlesinger, editor. *Biogeochemistry*. Elsevier, 425–472.

Sunshine, J. M., T. L. Farnham, L. M. Feaga, O. Groussin, F. Merlin, R. E. Milliken, and M. F. A'Hearn. 2009. Temporal and spatial variability of Lunar hydration as observed by the Deep Impact spacecraft. Science **326**:565–568.

Sutherland, R. A., C. Vankessel, R. E. Farrell, and D. J. Pennock. 1993. Landscape-scale variations in plant and soil ^{15}N natural abundance. Soil Science Society of America Journal **57**:169–178.

Sutka, R. L., N. E. Ostrom, P. H. Ostrom, J. A. Breznak, H. Gandhi, A. J. Pitt, and F. Li. 2006. Distinguishing nitrous oxide production from nitrification and denitrification on the basis of isotopomer abundances. Applied and Environmental Microbiology **72**: 638–644.

Sutton, M. A., C.E.R. Pitcairn, and D. Fowler. 1993. The exchange of ammonia between the atmosphere and plant communities. Advances in Ecological Research **24**:301–393.

Sutton, M. A., D. Simpson, P. E. Levy, R. I. Smith, S. Reis, M. van Oijen, and W. de Vries. 2008. Uncertainties in the relationship between atmospheric nitrogen deposition and forest carbon sequestration. Global Change Biology **14**:2057–2063.

Sutton-Grier, A. E., J. K. Keller, R. Koch, C. Gilmour, and J. P. Megonigal. 2011. Electron donors and acceptors influence anaerobic soil organic matter mineralization in tidal marshes. Soil Biology & Biochemistry **43**:1576–1583.

Svedhem, H., D. V. Titov, F. W. Taylor, and O. Witasse. 2007. Venus as a more Earth-like planet. Nature **450**: 629–632.

Svedrup, H. U., M. W. Johnson, and R. H. Fleming. 1942. *The Oceans*. Prentice Hall.

Svensson, T., G. M. Lovett, and G. E. Likens. 2012. Is chloride a conservative ion in forest ecosystems? Biogeochemistry **107**:125–134.

Swain, E. B., D. R. Engstrom, M. E. Brigham, T. A. Henning, and P. L. Brezonik. 1992. Increasing rates of atmospheric mercury deposition in midcontinental North America. Science **257**:784–787.

Swain, M. R., G. Vasisht, and G. Tinetti. 2008. The presence of methane in the atmosphere of an extrasolar planet. Nature **452**:329–331.

Swan, B. K., M. Martinez-Garcia, C. M. Preston, A. Sczyrba, T. Woyke, D. Lamy, T. Reinthaler, N. J. Poulton, E.D.P. Masland, M. L. Gomez, M. E. Sieracki, E. F. DeLong, G. J. Herndl, and R. Stepanauskas. 2011. Potential for chemolithoautotrophy among ubiquitous bacteria lineages in the dark ocean. Science **333**:1296–1300.

Swank, W. T., and J. E. Douglass. 1977. Nutrient budgets for undisturbed and manipulated hardwood forest ecosystems in the mountains of North Carolina. *In* D. L. Correll, editor. *Watershed Research in Eastern North America*. Smithsonian Institution, 343–362.

Swank, W. T., J. W. Fitzgerald, and J. T. Ash. 1984. Microbial transformation of sulfate in forest soils. Science **223**:182–184.

Swank, W. T., and G. S. Henderson. 1976. Atmospheric input of some cations and anions to forest ecosystems in North Carolina and Tennessee. Water Resources Research **12**:541–546.

Swank, W. T., J. B. Waide, D. A. Crossley, and R. L. Todd. 1981. Insect defoliation enhances nitrate export from forest ecosystems. Oecologia **51**:297–299.

Swanson, F. J., F. L. Fredriksen, and F. M. McCorison. 1982. Material transfer in a western Oregon forested watershed. *In* R. L. Edmonds, editor. *Analysis of Coniferous Forest Ecosystems in the Western United States*. Dowden, Hutchinson and Ross, 233–266.

Swap, R., M. Garstang, S. Greco, R. Talbot, and P. Kallberg. 1992. Saharan dust in the Amazon basin. Tellus Series B—Chemical and Physical Meteorology **44**:133–149.

Swap, R., S. Ulanski, M. Cobbett, and M. Garstang. 1996. Temporal and spatial characteristics of Saharan dust outbreaks. Journal of Geophysical Research—Atmospheres **101**:4205–4220.

Sweeney, C., E. Gloor, A. R. Jacobson, R. M. Key, G. McKinley, J. L. Sarmiento, and R. Wanninkhof. 2007. Constraining global air-sea gas exchange for CO_2 with recent bomb ^{14}C measurements. Global Biogeochemical Cycles **21**.

Swetnam, T. W., and J. L. Betancourt. 1990. Fire—Southern Oscillation relations in the southwestern United States. Science **249**:1017–1020.

Swift, M. J., O. W. Heal, and J. M. Anderson. 1979. *Decomposition in Terrestrial Ecosystems*. University of California Press.

Syakila, A., and C. Kroeze. 2011. The global nitrous oxide budget revisited. Greenhouse Gas Measurement and Management 1:17–26.

Syed, T. H., J. S. Famiglietti, D. P. Chambers, J. K. Willis, and K. Hilburn. 2010. Satellite-based global-ocean mass balance estimates of interannual variability and emerging trends in continental freshwater discharge. Proceedings of the National Academy of Sciences 107:17916–17921.

Syed, T. H., J. S. Famiglietti, M. Rodell, J. Chen, and C. R. Wilson. 2008. Analysis of terrestrial water storage changes from GRACE and GLDAS. Water Resources Research 44.

Syvitski, J.P.M., A. J. Kettner, I. Overeem, E.W.H. Hutton, M. T. Hannon, G. R. Brakenridge, J. Day, C. J. Vörösmarty, Y. Saito, L. Giosan, and R. J. Nicholls. 2009. Sinking deltas due to human activities. Nature Geoscience 2:681–686.

Syvitski, J.P.M., and Y. Saito. 2007. Morphodynamics of deltas under the influence of humans. Global and Planetary Change 57:261–282.

Syvitski, J.P.M., C. J. Vörösmarty, A. J. Kettner, and P. Green. 2005. Impact of humans on the flux of terrestrial sediment to the global coastal ocean. Science 308:376–380.

Szikszay, M., A. A. Kimmelmann, R. Hypolito, R. M. Figueira, and R. H. Sameshima. 1990. Evolution of the chemical composition of water passing through the unsaturated zone to groundwater at an experimental site at the University of Sao Paulo, Brazil. Journal of Hydrology 118:175–190.

Szilagyi, J., G. G. Katul, and M. B. Parlange. 2001. Evapotranspiration intensifies over the conterminous United States. Journal of Water Resources Planning and Management-Asce 127:354–362.

Tabatabai, M. A., and W. A. Dick. 1979. Distribution and stability of pyrophosphatase in soils. Soil Biology & Biochemistry 11:655–659.

Tabazadeh, A., and R. P. Turco. 1993. Stratospheric chlorine injection by volcanic eruptions—HCl scavenging and implications for ozone. Science 260:1082–1086.

Tajchman, S. J. 1972. The radiation and energy balance of coniferous and deciduous forests. Journal of Applied Ecology 9:359–375.

Tajika, E. 1998. Mantle degassing of major and minor volatile elements during the Earth's history. Geophysical Research Letters 25:3991–3994.

Takahashi, T., W. S. Broecker, and S. Langer. 1985. Redfield ratio based on chemical data from isopycnal surfaces. Journal of Geophysical Research—Oceans 90:6907–6924.

Takahashi, T., R. A. Feely, R. F. Weiss, R. H. Wanninkhof, D. W. Chipman, S. C. Sutherland, and T. T. Takahashi. 1997. Global air-sea flux of CO_2: An estimate based on measurements of sea-air pCO_2 difference. Proceedings of the National Academy of Sciences 94:8292–8299.

Takahashi, T., S. C. Sutherland, R. A. Feely, and R. Wanninkhof. 2006. Decadal change of the surface water pCO_2 in the North Pacific: A synthesis of 35 years of observations. Journal of Geophysical Research—Oceans 111.

Talbot, R. W., R. C. Harriss, E. V. Browell, G. L. Gregory, D. I. Sebacher, and S. M. Beck. 1986. Distribution and geochemistry of aerosols in the tropical north Atlantic troposphere—Relationship to Saharan dust. Journal of Geophysical Research—Atmospheres 91:5173–5182.

Talukdar, R. K., A. Mellouki, A. M. Schmoltner, T. Watson, S. Montzka, and A. R. Ravishankara. 1992. Kinetics of the OH reaction with methyl chloroform and its atmospheric implications. Science 257:227–230.

Tamm, C. O., and L. Hällbäcken. 1986. Changes in soil pH over a 50-year period under different forest canopies in SW Sweden. Water Air and Soil Pollution 31:337–341.

Tamrazyan, G. P. 1989. Global peculiarities and tendencies in river discharge and wash-down of the suspended sediments—The Earth as a whole. Journal of Hydrology 107:113–131.

Tan, K. H. 1980. The release of silicon, aluminum, and potassium during decomposition of soil minerals by humic acid. Soil Science 129:5–11.

Tan, K. H., and P. S. Troth. 1982. Silica-sesquioxide ratios as aids in characterization of some temperate region and tropical soil clays. Soil Science Society of America Journal 46:1109–1114.

Tanaka, N., and K. K. Turekian. 1991. Use of cosmogenic ^{35}S to determine the rates of removal of atmospheric SO_2. Nature 352:226–228.

Tanaka, N., and K. K. Turekian. 1995. Determination of the dry deposition flux of SO_2 using cosmogenic ^{35}S and ^7Be measurements. Journal of Geophysical Research—Atmospheres 100:2841–2848.

Tank, J. L., E. J. Rosi-Marshall, M. A. Baker, and R. O. Hall. 2008. Are rivers just big streams? A pulse method to quantify nitrogen demand in a large river. Ecology 89:2935–2945.

Tanre, D., F. M. Breon, J. L. Deuze, M. Herman, P. Goloub, F. Nadal, and A. Marchand. 2001. Global observation of anthropogenic aerosols from satellite. Geophysical Research Letters 28:4555–4558.

Tans, P. P., I. Y. Fung, and T. Takahashi. 1990. Observational constraints on the global atmospheric CO_2 budget. Science 247:1431–1438.

Tapia-Hernandez, A., M. R. Bustillos-Cristales, T. Jimenez-Salgado, J. Caballero-Mellado, and L. E. Fuentes-Ramirez. 2000. Natural endophytic occurrence of *Acetobacter diazotrophicus* in pineapple plants. Microbial Ecology 39:49–55.

Tarafdar, J. C., and N. Claassen. 1988. Organic phosphorus compounds as a phosphorus source for higher plants through the activity of phosphatases produced by plant roots and microorganisms. Biology and Fertility of Soils 5:308–312.

Taran, Y. A., J. W. Hedenquist, M. A. Korzhinsky, S. I. Tkachenko, and K. I. Shmulovich. 1995. Geochemistry of magmatic gases from Kudryavy volcano, Iturup, Kuril islands. Geochimica et Cosmochimica Acta 59:1749–1761.

Tarnocai, C., J. G. Canadell, E.A.G. Schuur, P. Kuhry, G. Mazhitova, and S. Zimov. 2009. Soil organic carbon pools in the northern circumpolar permafrost region. Global Biogeochemical Cycles 23.

Tarrasón, L., S. Turner, and I. Floisand. 1995. Estimation of seasonal dimethyl sulfide fluxes over the north Atlantic ocean and their contribution to European pollution levels. Journal of Geophysical Research—Atmospheres 100:11623–11639.

Tavares, P., A. S. Pereira, J.J.G. Moura, and I. Moura. 2006. Metalloenzymes of the denitrification pathway. Journal of Inorganic Biochemistry 100:2087–2100.

Taylor, A. B., and M. A. Velbel. 1991. Geochemical mass balances and weathering rates in forested watersheds of the southern Blue Ridge. 2. Effects of botanical uptake terms. Geoderma 51:29–50.

Taylor, A. H., A. J. Watson, and J. E. Robertson. 1992. The influence of the spring phytoplankton bloom on carbon dioxide and oxygen concentrations in the surface waters of the northeast Atlantic during 1989. Deep-Sea Research Part A—Oceanographic Research Papers 39:137–152.

Taylor, B. R., D. Parkinson, and W.F.J. Parsons. 1989. Nitrogen and lignin content as predictors of litter decay rates—A microcosm test. Ecology 70:97–104.

Taylor, B. W., A. S. Flecker, and R. O. Hall. 2006. Loss of a harvested fish species disrupts carbon flow in a diverse tropical river. Science 313:833–836.

Taylor, K. 1999. Rapid climate change. American Scientist 87:320–327.

Taylor, S., J. H. Lever, and R. P. Harvey. 1998. Accretion rate of cosmic spherules measured at the South Pole. Nature 392:899–903.

Tegen, I., M. Werner, S. P. Harrison, and K. E. Kohfeld. 2004. Relative importance of climate and land use in determining present and future global soil dust emission. Geophysical Research Letters 31.

Telmer, K., and J. Veizer. 2000. Isotopic constraints on the transpiration, evaporation, energy, and gross primary production budgets of a large boreal watershed: Ottawa River basin, Canada. Global Biogeochemical Cycles 14:149–165.

Temple, K. L., and A. R. Colmer. 1951. The autotrophic oxidation of iron by a new bacterium: *Thiobacillus ferrooxidans*. Journal of Bacteriology 62:605–611.

Templer, P. H., M. A. Arthur, G. M. Lovett, and K. C. Weathers. 2007. Plant and soil natural abundance delta [15]N: indicators of relative rates of nitrogen cycling in temperate forest ecosystems. Oecologia 153:399–406.

Templer, P. H., W. L. Silver, J. Pett-Ridge, K. M. DeAngelis, and M. K. Firestone. 2008. Plant and microbial controls on nitrogen retention and loss in a humid tropical forest. Ecology 89:3030–3040.

Teodoru, C., and B. Wehrli. 2005. Retention of sediments and nutrients in the Iron Gate I Reservoir on the Danube River. Biogeochemistry 76:539–565.

Teodoru, C. R., J. Bastien, M. C. Bonneville, P. A. del Giorgio, M. Demarty, M. Garneau, J. F. Helie, L. Pelletier, Y. T. Prairie, N. T. Roulet, I. B. Strachan, and A. Tremblay. 2012. The net carbon footprint of a newly created boreal hydroelectric reservoir. Global Biogeochemical Cycles 26.

Terao, Y., H. Mukai, Y. Nojiri, T. Machida, Y. Tohjima, T. Saeki, and S. Maksyutov. 2011. Interannual variability and trends in atmospheric methane over the western Pacific from 1994 to 2010. Journal of Geophysical Research—Atmospheres 116.

Terman, G. L. 1977. Quantitative relationships among nutrients leached from soils. Soil Science Society of America Journal 41:935–940.

Terman, G. L. 1979. Volatilization losses of nitrogen as ammonia from surface-applied fertilizers, organic amendments, and crop residues. Advances in Agronomy 31:189–223.

Textor, C., H. F. Graf, M. Herzog, and J. M. Oberhuber. 2003. Injection of gases into the stratosphere by explosive volcanic eruptions. Journal of Geophysical Research—Atmospheres 108.

Tezuka, Y. 1990. Bacterial regeneration of ammonium and phosphate as affected by the carbon-nitrogen-phosphorus ratio of organic substrates. Microbial Ecology 19:227–238.

Thamdrup, B., H. Fossing, and B. B. Jorgensen. 1994. Manganese, iron, and sulfur cycling in a coastal marine sediment, Aarhus bay, Denmark. Geochimica et Cosmochimica Acta 58:5115–5129.

Theobald, M. R., P. D. Crittenden, A. P. Hunt, Y. S. Tang, U. Dragosits, and M. A. Sutton. 2006. Ammonia emissions from a Cape fur seal colony, Cape Cross Namibia. Geophysical Research Letters **33**.

Thiemens, M. H., and W. C. Trogler. 1991. Nylon production—An unknown source of atmospheric nitrous oxide. Science **251**:932–934.

Thienemann, A. 1921. Seetypen. Naturwissenschaften **18**:1–3.

Thomas, C. D., A. Cameron, R. E. Green, M. Bakkenes, L. J. Beaumont, Y. C. Collingham, B.F.N. Erasmus, M. F. de Siqueira, A. Grainger, L. Hannah, L. Hughes, B. Huntley, A. S. van Jaarsveld, G. F. Midgley, L. Miles, M. A. Ortega-Huerta, A. T. Peterson, O. L. Phillips, and S. E. Williams. 2004a. Extinction risk from climate change. Nature **427**:145–148.

Thomas, G. E., J. J. Olivero, E. J. Jensen, W. Schroeder, and O. B. Toon. 1989. Relation between increasing methane and the presence of ice clouds at the mesopause. Nature **338**:490–492.

Thomas, H., Y. Bozec, K. Elkalay, and H.J.W. de Baar. 2004b. Enhanced open ocean storage of CO_2 from shelf sea pumping. Science **304**:1005–1008.

Thomas, R. B., J. D. Lewis, and B. R. Strain. 1994. Effects of leaf nutrient status on photosynthetic capacity in loblolly pine (*Pinus taeda* L) seedlings grown in elevated atmospheric CO_2. Tree Physiology **14**:947–960.

Thomas, R. Q., C. D. Canham, K. C. Weathers, and C. L. Goodale. 2010. Increased tree carbon storage in response to nitrogen deposition in the US. Nature Geoscience **3**:13–17.

Thomazo, C., M. Ader, and P. Philippot. 2011. Extreme ^{15}N-enrichments in 2.72-gyr-old sediments: evidence for a turning point in the nitrogen cycle. Geobiology **9**:107–120.

Thompson, A. E., R. S. Anderson, J. Rudolph, and L. Huang. 2002. Stable carbon isotope signatures of background tropospheric chloromethane and CFC113. Biogeochemistry **60**:191–211.

Thompson, A. M. 1992. The oxidizing capacity of the Earth's atmosphere—Probable past and future changes. Science **256**:1157–1165.

Thompson, J. R., D. R. Foster, R. Scheller, and D. Kittredge. 2011. The influence of land use and climate change on forest biomass and composition in Massachusetts, USA. Ecological Applications **21**:2425–2444.

Thompson, L. G., T. Yao, E. Mosley-Thompson, M. E. Davis, K. A. Henderson, and P. N. Lin. 2000. A high-resolution millennial record of the south Asian monsoon from Himalayan ice cores. Science **289**:1916–1919.

Thomson, J., N. C. Higgs, I. W. Croudace, S. Colley, and D. J. Hydes. 1993. Redox zonation of elements at an oxic post-oxic boundary in deep sea sediments. Geochimica et Cosmochimica Acta **57**:579–595.

Thorneloe, S. A., M. A. Barlaz, R. Peer, L. C. Huff, L. Davis, and J. Mangino. 1993. Waste management. *In* M.A.K. Khalil, editor. *Atmospheric Methane: Sources, Sinks and Role in Global Change*. Springer, 362–398.

Thorpe, S. A. 1985. Small-scale processes in the upper ocean boundary layer. Nature **318**:519–522.

Thurman, E. M. 1985. *Organic Geochemistry of Natural Waters*. Junk Inc., Dordrecht.

Ti, C., J. Pan, Y. Xia, and X. Yan. 2012. A nitrogen budget of mainland China with spatial and temporal variation. Biogeochemistry **108**:381–394.

Tian, F., O. B. Toon, A. A. Pavlov, and H. De Sterck. 2005. A hydrogen-rich early Earth atmosphere. Science **308**:1014–1017.

Tian, H. Q., J. M. Melillo, D. W. Kicklighter, A. D. McGuire, J.V.K. Helfrich, B. Moore, and C. J. Vörösmarty. 1998. Effect of interannual climate variability on carbon storage in Amazonian ecosystems. Nature **396**:664–667.

Tice, M. M., and D. R. Lowe. 2004. Photosynthetic microbial mats in the 3,416-myr-old ocean. Nature **431**:549–552.

Tice, M. M., and D. R. Lowe. 2006. Hydrogen-based carbon fixation in the earliest known photosynthetic organisms. Geology **34**:37–40.

Tiedje, J. M. 1988. Ecology of denitrification and dissimilatory nitrate reduction to ammonium. *In* J. B. Zehnder, editor. *Biology of Anaerobic Microorganisms*. Wiley.

Tiedje, J. M., A. J. Sexstone, T. B. Parkin, N. P. Revsbech, and D. R. Shelton. 1984. Anaerobic processes in soil. Plant and Soil **76**:197–212.

Tiedje, J. M., S. Simkins, and P. M. Groffman. 1989. Perspectives on measurement of denitrification in the field including recommended protocols for acetylene-based methods. Plant and Soil **115**:261–284.

Tiessen, H., and J.W.B. Stewart. 1983. Particle-size fractions and their use in studies of soil organic matter. 2. Cultivation effects on organic matter composition in size fractions. Soil Science Society of America Journal **47**:509–514.

Tiessen, H., and J.W.B. Stewart. 1988. Light and electron microscopy of stained microaggregates-the role of organic matter and microbes in soil aggregation. Biogeochemistry **5**:312–322.

Tiessen, H., J.W.B. Stewart, and J. R. Bettany. 1982. Cultivation effects on the amounts and concentration of carbon, nitrogen, and phosphorus in grassland soils. Agronomy Journal **74**:831–835.

Tiessen, H., J.W.B. Stewart, and C. V. Cole. 1984. Pathways of phosphorus transformations in soils of differing pedogenesis. Soil Science Society of America Journal **48**:853–858.

Tilman, D. 1985. The resource-ratio hypothesis of plant succession. American Naturalist **125**:827–852.

Tilman, D., R. Kiesling, R. Sterner, S. S. Kilham, and F. A. Johnson. 1986. Green, blue-green and diatom algae—Taxonomic differences in competitive ability for phosphorus, silicaon and nitrogen. Archiv fur Hydrobiologie **106**:473–485.

Tilman, D., P. B. Reich, J. Knops, D. Wedin, T. Mielke, and C. Lehman. 2001. Diversity and productivity in a long-term grassland experiment. Science **294**:843–845.

Tilman, D., D. Wedin, and J. Knops. 1996. Productivity and sustainability influenced by biodiversity in grassland ecosystems. Nature **379**:718–720.

Tilton, D. L. 1978. Comparative growth and foliar element concentrations of *Larix laricina* over a range of wetland types in Minnesota. Journal of Ecology **66**: 499–512.

Timmer, V. R., and E. L. Stone. 1978. Comparative foliar analysis of young balsam fir fertilized with nitrogen, phosphorus, potassium, and lime. Soil Science Society of America Journal **42**:125–130.

Tipping, E., P. M. Chamberlain, C. L. Bryant, and S. Buckingham. 2010. Soil organic matter turnover in British deciduous woodlands, quantified with radiocarbon. Geoderma **155**:10–18.

Tischner, R. 2000. Nitrate uptake and reduction in higher and lower plants. Plant Cell and Environment **23**:1005–1024.

Tisdall, J. M., and J. M. Oades. 1982. Organic matter and water-stable aggregates in soils. Journal of Soil Science **33**:141–163.

Tissue, D. T., J. P. Megonigal, and R. B. Thomas. 1997. Nitrogenase activity and N_2 fixation are stimulated by elevated CO_2 in a tropical N_2-fixing tree. Oecologia **109**:28–33.

Titman, D. 1976. Ecological competition between algae—Experimental confirmation of resource-based competition theory. Science **192**:463–465.

Titus, J. H., R. S. Nowak, and S. D. Smith. 2002. Soil resource heterogeneity in the Mojave Desert. Journal of Arid Environments **52**:269–292.

Titus, T. N., H. H. Kieffer, and P. R. Christensen. 2003. Exposed water ice discovered near the south pole of Mars. Science **299**:1048–1051.

Toggweiler, J. R., and B. Samuels. 1993. New radiocarbon constraints on the upwelling of abyssal water to the ocean's surface. *In* M. Heimann, editor. *The Global Carbon Cycle*. Springer, 333–366.

Tolbert, N. E., C. Benker, and E. Beck. 1995. The oxygen and carbon dioxide compensation points of C-3 plants—Possible role in regulating atmospheric oxygen. Proceedings of the National Academy of Sciences **92**:11230–11233.

Toon, O. B., J. F. Kasting, R. P. Turco, and M. S. Liu. 1987. The sulfur cycle in the marine atmosphere. Journal of Geophysical Research—Atmospheres **92**:943–963.

Torn, M. S., A. G. Lapenis, A. Timofeev, M. L. Fischer, B. V. Babikov, and J. W. Harden. 2002. Organic carbon and carbon isotopes in modern and 100-year-old-soil archives of the Russian steppe. Global Change Biology **8**:941–953.

Torn, M. S., S. E. Trumbore, O. A. Chadwick, P. M. Vitousek, and D. M. Hendricks. 1997. Mineral control of soil organic carbon storage and turnover. Nature **389**:170–173.

Tornqvist, T. E., D. J. Wallace, J.E.A. Storms, J. Wallinga, R. L. Van Dam, M. Blaauw, M. S. Derksen, C.J.W. Klerks, C. Meijneken, and E.M.A. Snijders. 2008. Mississippi Delta subsidence primarily caused by compaction of Holocene strata. Nature Geoscience **1**:173–176.

Tortell, P. D., M. T. Maldonado, and N. M. Price. 1996. The role of heterotrophic bacteria in iron-limited ocean ecosystems. Nature **383**:330–332.

Touboul, M., T. Kleine, B. Bourdon, H. Palme, and R. Wieler. 2007. Late formation and prolonged differentiation of the Moon inferred from W isotopes in lunar metals. Nature **450**:1206–1209.

Towe, K. M. 1990. Aerobic respiration in the Archean. Nature **348**:54–56.

Townsend, A. R., B. H. Braswell, E. A. Holland, and J. E. Penner. 1996. Spatial and temporal patterns in terrestrial carbon storage due to deposition of fossil fuel nitrogen. Ecological Applications **6**:806–814.

Townsend, A.R., R. W. Howarth, F. A. Bazzaz, M.S. Booth, C. C. Cleveland, S. K. Collinge, A. P. Dobson, P. R. Epstein, D. R. Keeney, M. A. Mallin, C. A. Rogers, P. Wayne, and A. H. Wolfe. 2003. Human health effects of a changing global nitrogen cycle. Frontiers in Ecology and the Environment **1**:240–246.

Townsend-Small, A., and C. I. Czimczik. 2010. Carbon sequestration and greenhouse gas emissions in urban turf. Geophysical Research Letters **37**.

Toyoda, S., M. Yano, S. Nishimura, H. Akiyama, A. Hayakawa, K. Koba, S. Sudo, K. Yagi, A. Makabe, Y. Tobari, N. O. Ogawa, N. Ohkouchi, K. Yamada, and N. Yoshida. 2011. Characterization and production and consumption processes of N_2O emitted from temperate agricultural soils determined via isotopomer ratio analysis. Global Biogeochemical Cycles **25**.

Trail, D., E. B. Watson, and N. D. Tailby. 2011. The oxidation state of Hadean magmas and implications for early Earth's atmosphere. Nature **480**:79–82.

Tranvik, L. J., J. A. Downing, J. B. Cotner, S. A. Loiselle, R. G. Striegl, T. J. Ballatore, P. Dillon, K. Finlay, K. Fortino, L. B. Knoll, P. L. Kortelainen, T. Kutser,

S. Larsen, I. Laurion, D. M. Leech, S. L. McCallister, D. M. McKnight, J. M. Melack, E. Overholt, J. A. Porter, Y. Prairie, W. H. Renwick, F. Roland, B. S. Sherman, D. W. Schindler, S. Sobek, A. Tremblay, M. J. Vanni, A. M. Verschoor, E. von Wachenfeldt, and G. A. Weyhenmeyer. 2009. Lakes and reservoirs as regulators of carbon cycling and climate. Limnology and Oceanography 54:2298–2314.

Travis, C. C., and E. L. Etnier. 1981. A survey of sorption relationships for reactive solutes in soil. Journal of Environmental Quality 10:8–17.

Trefry, J. H., and S. Metz. 1989. Role of hydrothermal precipitates in the geochemical cycling of vanadium. Nature 342:531–533.

Trefry, J. H., S. Metz, R. P. Trocine, and T. A. Nelsen. 1985. A decline in lead transport by the Mississippi river. Science 230:439–441.

Trenberth, K. E. 1998. Atmospheric moisture residence times and cycling implications from rainfall rate and climate change. Climatc Change 39:667–694.

Trenberth, K. E., and J. M. Caron. 2001. Estimates of meridional atmosphere and ocean heat transports. Journal of Climate 14:3433–3443.

Trenberth, K. E., and C. J. Guillemot. 1994. The total mass of the atmosphere. Journal of Geophysical Research—Atmospheres 99:23079–23088.

Trenberth, K. E., L. Smith, T. T. Qian, A. Dai, and J. Fasullo. 2007. Estimates of the global water budget and its annual cycle using observational and model data. Journal of Hydrometeorology 8:758–769.

Treseder, K. K. 2008. Nitrogen additions and microbial biomass: a meta-analysis of ecosystem studies. Ecology Letters 11:1111–1120.

Treseder, K. K., and P. M. Vitousek. 2001. Effects of soil nutrient availability on investment in acquisition of N and P in Hawaiian rain forests. Ecology 82:946–954.

Treuhaft, R. N., F. G. Goncalves, J. B. Drake, B. D. Chapman, J. R. dos Santos, L. V. Dutra, P. Graca, and G. H. Purcell. 2010. Biomass estimation in a tropical wet forest using Fourier transforms of profiles from lidar or interferometric SAR. Geophysical Research Letters 37.

Treuhaft, R. N., B. E. Law, and G. P. Asner. 2004. Forest attributes from radar interferometric structure and its fusion with optical remote sensing. BioScience 54:561–571.

Trimble, S. W. 1977. Fallacy of stream equilibrium in contemporary denudation studies. American Journal of Science 277:876–887.

Trimble, V. 1997. Origin of the biologically important elements. Origins of Life and Evolution of the Biosphere 27:3–21.

Tripati, A. K., C. D. Roberts, and R. A. Eagle. 2009. Coupling of CO_2 and ice sheet stability over major climate transitions of the last 20 million years. Science 326:1394–1397.

Tripp, H. J., J. B. Kitner, M. S. Schwalbach, J.W.H. Dacey, L. J. Wilhelm, and S. J. Giovannoni. 2008. SAR11 marine bacteria require exogenous reduced sulphur for growth. Nature 452:741–744.

Triska, F. J. 1989. Retention and transport of nutrients in a third-order stream: channel processes. Ecology 70:1877–1892.

Triska, F. J., James R. Sedell, Kermit Cromack Jr., Stan V. Gregory, and F. M. McCorison. 1984. Nitrogen budget for a small coniferous forest stream. Ecological Monographs 54:119–140.

Triska, F. J., and R. S. Oremland. 1981. Denitrification associated with periphyton communities. Applied and Environmental Microbiology 42:745–748.

Trlica, M. J., and M. E. Biondini. 1990. Soil-water dynamics, transpiration, and water losses in a crested wheatgrass and native shortgrass ecosystem. Plant and Soil 126:187–201.

Trouwborst, R. E., B. G. Clement, B. M. Tebo, B. T. Glazer, and G. W. Luther. 2006. Soluble Mn(III) in suboxic zones. Science 313:1955–1957.

Trouwborst, R. E., A. Johnston, G. Koch, G. W. Luther, and B. K. Pierson. 2007. Biogeochemistry of Fe(II) oxidation in a photosynthetic microbial mat: Implications for Precambrian Fe(II) oxidation. Geochimica et Cosmochimica Acta 71:4629–4643.

Trumbore, S. E. 1993. Comparison of carbon dynamics in tropical and temperate soils uding radiocarbon measurements. Global Biogeochemical Cycles 7:275–290.

Trumbore, S. E. 1997. Potential responses of soil organic carbon to global environmental change. Proceedings of the National Academy of Sciences 94:8284–8291.

Trumbore, S. E., O. A. Chadwick, and R. Amundson. 1996. Rapid exchange between soil carbon and atmospheric carbon dioxide driven by temperature change. Science 272:393–396.

Tukey, H. B. 1970. The leaching of substances from plants. Annual Review of Plant Physiology 21:305–324.

Turchin, P. 2009. Long-term population cycles in human societies. Annals of the New York Academy of Science. 1162:1–17.

Turco, R. P., R. C. Whitten, O. B. Toon, J. B. Pollack, and P. Hamill. 1980. OCS, stratospheric aerosols and climate. Nature 283:283–286.

Turekian, K. K. 1977. Fate of metals in the oceans. Geochimica et Cosmochimica Acta 41:1139–1144.

Turetsky, M. R., W. F. Donahue, and B. W. Benscoter. 2011. Experimental drying intensifies burning and carbon losses in a northern peatland. Nature Communications 2:514.

Turk, D., M. J. McPhaden, A. J. Busalacchi, and M. R. Lewis. 2001. Remotely sensed biological production in the equatorial Pacific. Science 293:471–474.

Turner, A., and G. E. Millward. 2002. Suspended particles: Their role in estuarine biogeochemical cycles. Estuarine Coastal and Shelf Science 55:857–883.

Turner, B. L., L. M. Condron, S. J. Richardson, D. A. Peltzer, and V. J. Allison. 2007. Soil organic phosphorus transformations during pedogenesis. Ecosystems 10:1166–1181.

Turner, B. L., and B.M.J. Engelbrecht. 2011. Soil organic phosphorus in lowland tropical rain forests. Biogeochemistry 103:297–315.

Turner, B. L., and P. M. Haygarth. 2000. Phosphorus forms and concentrations in leachate under four grassland soil types. Soil Science Society of America Journal 64:1090–1099.

Turner, D. P., G. J. Koerper, M. E. Harmon, and J. J. Lee. 1995. A carbon budget for forests of the conterminous United States. Ecological Applications 5:421–436.

Turner, J. 1982. The mass-flow component of nutrient supply in three western Washington forest types. Acta Oecologica-Oecologia Plantarum 3:323–329.

Turner, J., D. W. Johnson, and M. J. Lambert. 1980. Sulfur cycling in a douglas fir forest and its modification by nitrogen applications. Acta Oecologica-Oecologia Plantarum 1:27–35.

Turner, J., and P. R. Olson. 1976. Nitrogen relations in a douglas fir plantation. Annals of Botany 40:1185–1193.

Turner, R. E. 2004. Coastal wetland subsidence arising from local hydrologic manipulations. Estuaries 27:266–272.

Turner, R. E., and N. N. Rabalais. 1994. Coastal eutrophication near the Mississippi river delta. Nature 368:619–621.

Turner, S. M., G. Malin, P. S. Liss, D. S. Harbour, and P. M. Holligan. 1988. The seasonal variation of dimethyl sulfide and dimethylsulfoniopropionate concentrations in nearshore waters. Limnology and Oceanography 33:364–375.

Turner, S. M., P. D. Nightingale, L. J. Spokes, M. I. Liddicoat, and P. S. Liss. 1996. Increased dimethyl sulphide concentrations in sea water from in situ iron enrichment. Nature 383:513–517.

Turnipseed, A. A., L. G. Huey, E. Nemitz, R. Stickel, J. Higgs, D. J. Tanner, D. L. Slusher, J. P. Sparks, F. Flocke, and A. Guenther. 2006. Eddy covariance fluxes of peroxyacetyl nitrates (PANs) and NOy to a coniferous forest. Journal of Geophysical Research—Atmospheres 111.

Turtle, E. P., J. E. Perry, A. G. Hayes, R. D. Lorenz, J. W. Barnes, A. S. McEwen, R. A. West, A. D. Del Genio, J. M. Barbara, J. I. Lunine, E. L. Schaller, T. L. Ray, R.M.C. Lopes, and E. R. Stofan. 2011. Rapid and extensive surface changes near Titan's equator: Evidence of April showers. Science 331:1414–1417.

Turunen, J., N. T. Roulet, and T. R. Moore. 2004. Nitrogen deposition and increased carbon accumulation in ombrotrophic peatlands in eastern Canada. Global Biogeochemical Cycles 18:1–12.

Turunen, J., T. Tahvanainen, K. Tolonen, and A. Pitkanen. 2001. Carbon accumulation in West Siberian mires, Russia. Global Biogeochemical Cycles 15:285–296.

Turunen, J., E. Tomppo, K. Tolonen, and A. Reinikainen. 2002. Estimating carbon accumulation rates of undrained mires in Finland—Application to boreal and subarctic regions. Holocene 12:69–80.

Twilley, R. R., R. H. Chen, and T. Hargis. 1992. Carbon sinks in mangroves and their implications to the carbon budget of tropical coastal ecosystems. Water Air and Soil Pollution 64:265–288.

Tyler, G. 1994. A new approach to understanding the calcifuge habit of plants. Annals of Botany 73:327–330.

Tyler, G., and L. Ström. 1995. Differing organic acid exudation pattern explains calcifuge and acidifuge behavior of plants. Annals of Botany 75:75–78.

Uehara, G., and G. Gillman. 1981. The Mineralogy, Chemistry, and Physics of Tropical Soils with Variable Charge Clays. Westview Press.

Uematsu, M., Z. F. Wang, and I. Uno. 2003. Atmospheric input of mineral dust to the western North Pacific region based on direct measurements and a regional chemical transport model. Geophysical Research Letters 30.

Ueno, Y., K. Yamada, N. Yoshida, S. Maruyama, and Y. Isozaki. 2006. Evidence from fluid inclusions for microbial methanogenesis in the early Archaean era. Nature 440:516–519.

Ugolini, F. C., R. Minden, H. Dawson, and J. Zachara. 1977. Example of soil processes in the Abies amabilis zone of the central Cascades, Washington. Soil Science 124:291–302.

Ugolini, F. C., and R. S. Sletten. 1991. The role of proton donors in pedogenesis as revealed by soil solution studies. Soil Science 151:59–75.

Ugolini, F. C., M. G. Stoner, and D. J. Marrett. 1987. Arctic pedogenesis. 1. Evidence for contemporary podzolization. Soil Science 144:90–100.

Uhl, C., and C. F. Jordan. 1984. Succession and nutrient dynamics following forest cutting and burning in Amazonia. Ecology 65:1476–1490.

Ullah, S., and T. R. Moore. 2011. Biogeochemical controls on methane, nitrous oxide, and carbon dioxide fluxes from deciduous forest soils in eastern Canada. Journal of Geophysical Research—Biogeosciences 116.

Umena, Y., K. Kawakami, J. R. Shen, and N. Kamiya. 2011. Crystal structure of oxygen-evolving photosystem II at a resolution of 1.9 angstrom. Nature 473: 55–60.

United Nations. 2011. 2010 Revision of World Population Prospects, New York.

Uno, I., K. Eguchi, K. Yumimoto, T. Takemura, A. Shimizu, M. Uematsu, Z. Y. Liu, Z. F. Wang, Y. Hara, and N. Sugimoto. 2009. Asian dust transported one full circuit around the globe. Nature Geoscience 2:557–560.

Unrau, P. J., and D. P. Bartel. 1998. RNA-catalysed nucleotide synthesis. Nature 395:260–263.

Updegraff, K., S. D. Bridgham, J. Pastor, P. Weishampel, and C. Harth. 2001. Response of CO_2 and CH_4 emissions from peatlands to warming and water table manipulation. Ecological Applications 11:311–326.

Urban, N. R., S. E. Bayley, and S. J. Eisenreich. 1989a. Export of dissolved organic carbon and acidity from peatlands. Water Resources Research 25:1619–1628.

Urban, N. R., P. L. Brezonik, L. A. Baker, and L. A. Sherman. 1994. Sulfate reduction and diffusion in sediments of Little Rock lake, Wisconsin. Limnology and Oceanography 39:797–815.

Urban, N. R., S. J. Eisenreich, and D. F. Grigal. 1989b. Sulfur cycling in a forested Sphagnum bog in northern Minnesota. Biogeochemistry 7:81–109.

Uz, B. M., J. A. Yoder, and V. Osychny. 2001. Pumping of nutrients to ocean surface waters by the action of propagating planetary waves. Nature 409:597–600.

Vadas, R. L., W. A. Wright, and B. F. Beal. 2004. Biomass and productivity of intertidal rockweeds (Ascophyllum nodosum LeJolis) in Cobscook Bay. Northeastern Naturalist 11:123–142.

Vadstein, O., Y. Olsen, H. Reinertsen, and A. Jensen. 1993. The role of planktonic bacteria in phosphorus cycling in lakes—Sink and link. Limnology and Oceanography 38:1539–1544.

Vahtera, E., D. J. Conley, B. G. Gustafsson, H. Kuosa, H. Pitkanen, O. P. Savchuk, T. Tamminen, M. Viitasalo, M. Voss, N. Wasmund, and F. Wulff. 2007. Internal ecosystem feedbacks enhance nitrogen-fixing cyanobacteria blooms and complicate management in the Baltic Sea. Ambio 36:186–194.

Valentine, D. W., E. A. Holland, and D. S. Schimel. 1994. Ecosystem and physiological controls over methane production in northern wetlands. Journal of Geophysical Research—Atmospheres 99: 1563–1571.

Valentini, R., G. Matteucci, A. J. Dolman, E. D. Schulze, C. Rebmann, E. J. Moors, A. Granier, P. Gross, N. O. Jensen, K. Pilegaard, A. Lindroth, A. Grelle, C. Bernhofer, T. Grunwald, M. Aubinet, R. Ceulemans, A. S. Kowalski, T. Vesala, U. Rannik, P. Berbigier, D. Loustau, J. Guomundsson, H. Thorgeirsson, A. Ibrom, K. Morgenstern, R. Clement, J. Moncrieff, L. Montagnani, S. Minerbi, and P. G. Jarvis. 2000. Respiration as the main determinant of carbon balance in European forests. Nature 404:861–865.

Valett, H. M., J. A. Morrice, C. N. Dahm, and M. E. Campana. 1996. Parent lithology, surface-groundwater exchange and nitrate retention in headwater streams. Limnology and Oceanography 41:333–345.

Vallina, S. M., and R. Simo. 2007. Strong relationship between DMS and the solar radiation dose over the global surface ocean. Science 315:506–508.

Van Bodegom, P. M., R. Broekman, J. Van Dijk, C. Bakker, and R. Aerts. 2005. Ferrous iron stimulates phenol oxidase activity and organic matter decomposition in waterlogged wetlands. Biogeochemistry 76:69–83.

van Breemen, N. 1995. How Sphagnum bogs down other plants. Trends in Ecology & Evolution 10.

van Breemen, N., E. W. Boyer, C. L. Goodale, N. A. Jaworski, K. Paustian, S. P. Seitzinger, K. Lajtha, B. Mayer, D. Van Dam, R. W. Howarth, K. J. Nadelhoffer, M. Eve, and G. Billen. 2002. Where did all the nitrogen go? Fate of nitrogen inputs to large watersheds in the northeastern USA. Biogeochemistry 57:267–293.

van Breemen, N., P. A. Burrough, E. J. Velthorst, H. F. Vandobben, T. Dewit, T. B. Ridder, and H.F.R. Reijnders. 1982. Soil acidification from atmospheric ammonium sulfate in forest canopy throughfall. Nature 299:548–550.

van Breemen, N., R. Finlay, U. Lundstrom, A. G. Jongmans, R. Giesler, and M. Olsson. 2000. Mycorrhizal weathering: A true case of mineral plant nutrition? Biogeochemistry 49:53–67.

Van Cappellen, P., S. Dixit, and J. van Beusekom. 2002. Biogenic silica dissolution in the oceans: Reconciling experimental and field-based dissolution rates. Global Biogeochemical Cycles 16.

Van Cappellen, P., and E. D. Ingall. 1994. Benthic phosphorus regeneration, net primary production, and ocean anoxia—A model of the coupled marine biogeochemical cycles of carbon and phosphorus. Paleoceanography 9:677–692.

Van Cappellen, P., and E. D. Ingall. 1996. Redox stabilization of the atmosphere and oceans by phosphorus-limited marine productivity. Science 271:493–496.

Van Cleve, K., R. Barney, and R. Schlentner. 1981. Evidence of temperature control of production and nutrient cycling in two interior Alaska black spruce ecosystems. Canadian Journal of Forest Research 11:258–273.

Van Cleve, K., W. C. Oechel, and J. L. Hom. 1990. Response of black spruce (*Picea mariana*) ecosystems to soil temperature modification in interior Alaska. Canadian Journal of Forest Research 20:1530–1535.

Van Cleve, K., and R. White. 1980. Forest-floor nitrogen dynamics in a 60-year old paper birch ecosystem in interior Alaska. Plant and Soil 54:359-381.

Van de Water, P. K., S. W. Leavitt, and J. L. Betancourt. 1994. Trends in stomatal density and $^{13}C/^{12}C$ ratios of *Pinus flexilis* needles during last glacial-interglacial cycle. Science 264:239–243.

van den Broeke, M., J. Bamber, J. Ettema, E. Rignot, E. Schrama, W. J. van de Berg, E. van Meijgaard, I. Velicogna, and B. Wouters. 2009. Partitioning recent Greenland mass loss. Science 326:984–986.

van den Driessche, R. 1974. Prediction of mineral nutrient status of trees by foliar analysis. Botanical Review 40:347–394.

van der Ent, R. J., H.H.G. Savenije, R. Schaefli, and D. Steele S. C. 2010. Origin and fate of atmospheric moisture over continents. Water Resources Research 46.

Van der Gon Denier, H.A.C., and H. U. Neue. 1995. Influence of organic matter incorporation on the methane emission from a wetland rice field. Global Biogeochemical Cycles 9:11–22.

Van der Hoven, S. J., and J. Quade. 2002. Tracing spatial and temporal variations in the sources of calcium in pedogenic carbonates in a semiarid environment. Geoderma 108:259–276.

van der Lee, G.E.M., B. de Winder, W. Bouten, and A. Tietema. 1999. Anoxic microsites in douglas fir litter. Soil Biology & Biochemistry 31:1295–1301.

van der Werf, G. R., J. T. Randerson, G. J. Collatz, and L. Giglio. 2003. Carbon emissions from fires in tropical and subtropical ecosystems. Global Change Biology 9:547–562.

Van Devender, T. R., and W. G. Spaulding. 1979. Development of vegetation and climate in the southwestern United States. Science 204:701–710.

Van Dijk, A., and A. J. Dolman. 2004. Estimates of CO_2 uptake and release among European forests based on eddy covariance data. Global Change Biology 10:1445–1459.

Van Dijk, S. M., and J. H. Duyzer. 1999. Nitric oxide emissions from forest soils. Journal of Geophysical Research—Atmospheres 104:15955–15961.

Van Drecht, G., A. F. Bouwman, J. M. Knoop, A.H.W. Beusen, and C. R. Meinardi. 2003. Global modeling of the fate of nitrogen from point and nonpoint sources in soils, groundwater, and surface water. Global Biogeochemical Cycles 17.

van Groenigen, K. J., C. W. Osenberg, and B. A. Hungate. 2011. Increased soil emissions of potent greenhouse gases under increased atmospheric CO_2. Nature 475:214–216.

van Groenigen, K. J., J. Six, B. A. Hungate, M. A. de Graaff, N. van Breemen, and C. van Kessel. 2006. Element interactions limit soil carbon storage. Proceedings of the National Academy of Sciences 103:6571–6574.

van Hees, P.A.W., D. L. Jones, R. Finlay, D. L. Godbold, and U. S. Lundstomd. 2005. The carbon we do not see—the impact of low molecular weight compounds on carbon dynamics and respiration in forest soils: a review. Soil Biology & Biochemistry 37:1–13.

Van Houten, F. B. 1973. Origin of red beds: A review. Annual Review of Earth and Planetary Sciences 1:39–61.

van Keken, P. E., B. R. Hacker, E. M. Syracuse, and G. A. Abers. 2011. Subduction factory: 4. Depth-dependent flux of H_2O from subducting slabs worldwide. Journal of Geophysical Research—Solid Earth 116.

van Kessel, C., R. E. Farrell, J. P. Roskoski, and K. M. Keane. 1994. Recycling of the naturally occurring ^{15}N in an established stand of *Leucaena leucocephala*. Soil Biology & Biochemistry 26:757–762.

van Kessel, C., J. Nitschelm, W. R. Horwath, D. Harris, F. Walley, A. Luscher, and U. Hartwig. 2000. Carbon-13 input and turn-over in a pasture soil exposed to long-term elevated atmospheric CO_2. Global Change Biology 6:123–135.

Van Mooy, B.A.S., H. F. Fredricks, B. E. Pedler, S. T. Dyhrman, D. M. Karl, M. Koblizek, M. W. Lomas, T. J. Mincer, L. R. Moore, T. Moutin, M. S. Rappe, and E. A. Webb. 2009. Phytoplankton in the ocean use non-phosphorus lipids in response to phosphorus scarcity. Nature 458:69–72.

Van Oost, K., T. A. Quine, G. Govers, S. De Gryze, J. Six, J. W. Harden, J. C. Ritchie, G. W. McCarty, G. Heckrath, C. Kosmas, J. V. Giraldez, J.R.M. da Silva, and R. Merckx. 2007. The impact of agricultural soil erosion on the global carbon cycle. Science 318:626–629.

van Scholl, L., T. W. Kuyper, M. M. Smits, R. Landeweert, E. Hoffland, and N. van Breemen. 2008. Rock-eating mycorrhizas: their role in plant nutrition and biogeochemical cycles. Plant and Soil 303:35–47.

Van Sickle, J. 1981. Long-term distributions of annual sediment yields from small watersheds. Water Resources Research 17:659–663.

Van Trump, J. E., and S. L. Miller. 1972. Prebiotic synthesis of methionine. Science 178:859–860.

van Zuilen, M. A., A. Lepland, and G. Arrhenius. 2002. Reassessing the evidence for the earliest traces of life. Nature **418**:627–630.

Vance, G. F., and M. B. David. 1991. Chemical characteristics and acidity of soluble organic substances from a northern hardwood forest floor, central Maine, USA. Geochimica et Cosmochimica Acta **55**:3611–3625.

Vandal, G. M., W. F. Fitzgerald, C. F. Boutron, and J. P. Candelone. 1993. Variations in mercury deposition to Antarctica over the past 34,000 years. Nature **362**:621–623.

Vandenberg, J. J., and K. R. Knoerr. 1985. Comparison of surrogate surface techniques for estimation of sulfate dry deposition. Atmospheric Environment **19**:627–635.

Vann, C. D., and J. P. Megonigal. 2003. Elevated CO_2 and water-depth regulation of methane emissions:Comparison of woody and non-woody wetland plant species. Biogeochemistry **63**:117–134.

Vanni, M. J. 2002. Nutrient cycling by animals in freshwater ecosystems. Annual Review of Ecology and Systematics **33**:341–370.

Vannote, R. L., G. W. Minshall, K. W. Cummins, J. R. Sedell, and C. E. Cushing. 1980. The river continuum concept. Canadian Journal of Fisheries and Aquatic Science **37**:130–137.

Vasconcelos, C., J. A. McKenzie, S. Bernasconi, D. Grujic, and A. J. Tien. 1995. Microbial mediation as a possible mechanism for natural dolomite formation at low temperatures. Nature **377**:220–222.

Vasileva, A. V., K. B. Moiseenko, J. C. Mayer, N. Jurgens, A. Panov, M. Heimann, and M. O. Andreae. 2011. Assessment of the regional atmospheric impact of wildfire emissions based on CO observations at the ZOTTO tall tower station in central Siberia. Journal of Geophysical Research—Atmospheres **116**.

Vazquez-Dominguez, E., D. Vaque, and J. M. Gasol. 2007. Ocean warming enhances respiration and carbon demand of coastal microbial plankton. Global Change Biology **13**:1327–1334.

Velbel, M. A. 1990. Mechanisms of saprolitization, isovolumetric weathering, and pseudomorphous replacement during rock weathering—A review. Chemical Geology **84**:17–18.

Velbel, M. A. 1992. Geochemical mass balances and weathering rates in forested watersheds of the southern Blue Ridge. 3. Cation budgets and the weathering rate of amphibole. American Journal of Science **292**:58–78.

Velders, G.J.M., A. Snijder, and R. Hoogerbrugge. 2011. Recent decreases in observed atmospheric concentrations of SO_2 in the Netherlands in line with emission reductions. Atmospheric Environment **45**:5647–5651.

Velicogna, I. 2009. Increasing rates of ice mass loss from the Greenland and Antarctic ice sheets revealed by GRACE. Geophysical Research Letters **36**.

Velicogna, I., and J. Wahr. 2006a. Acceleration of Greenland ice mass loss in spring 2004. Nature **443**: 329–331.

Velicogna, I., and J. Wahr. 2006b. Measurements of time-variable gravity show mass loss in Antarctica. Science **311**:1754–1756.

Venterea, R. T., P. M. Groffman, L. V. Verchot, A. H. Magill, J. D. Aber, and P. A. Steudler. 2003. Nitrogen oxide gas emissions from temperate forest soils receiving long-term nitrogen inputs. Global Change Biology **9**:346–357.

Venterea, R. T., and D. E. Rolston. 2000. Mechanisms and kinetics of nitric and nitrous oxide production during nitrification in agricultural soil. Global Change Biology **6**:303–316.

Venterink, H. O. 2011. Legumes have a higher root phosphatase activity than other forbs, particularly under low inorganic P and N supply. Plant and Soil **347**: 137–146.

Venterink, H. O., R. E. Van der Vliet, and M. J. Wassen. 2001. Nutrient limitation along a productivity gradient in wet meadows. Plant and Soil **234**:171–179.

Verchot, L. V., E. A. Davidson, J. H. Cattanio, I. L. Ackerman, H. E. Erickson, and M. Keller. 1999. Land use change and biogeochemical controls of nitrogen oxide emissions from soils in eastern Amazonia. Global Biogeochemical Cycles **13**:31–46.

Vergutz, L., S. Manzoni, A. Porporato, R. F. Novais, and R. B. Jackson. 2012. Global resorption efficiencies and concentrations of carbon and nutrients in leaves of terrestrial plants. Ecological Monographs **82**: 205–220.

Verhoeven, J.T.A., B. Arheimer, C. Q. Yin, and M. M. Hefting. 2006. Regional and global concerns over wetlands and water quality. Trends in Ecology & Evolution **21**:96–103.

Vernadsky, V. 1998. *The Biosphere*. Copernicus.

Verstraten, J. M., J.C.R. Dopheide, J. Duysings, A. Tietema, and W. Bouten. 1990. The proton cycle of a deciduous forest ecosystem in the Netherlands and its implications for soil acidification. Plant and Soil **127**:61–69.

Viereck, L. A. 1966. Plant succession and soil development on gravel outwash of the Muldrow glacier, Alaska. Ecological Monographs **36**:181–199.

Vila-Costa, M., R. Simo, H. Harada, J. M. Gasol, D. Slezak, and R. P. Kiene. 2006. Dimethylsulfoniopropionate uptake by marine phytoplankton. Science **314**:652–654.

Vile, M. A., S. D. Bridgham, R. K. Wieder, and M. Novak. 2003. Atmospheric sulfur deposition alters pathways of gaseous carbon production in peatlands. Global Biogeochemical Cycles 17:1–7.

Villareal, T. A., C. Pilskaln, M. Brzezinski, F. Lipschultz, M. Dennett, and G. B. Gardner. 1999. Upward transport of oceanic nitrate by migrating diatom mats. Nature 397:423–425.

Virginia, R. A., and C. C. Delwiche. 1982. Natural ^{15}N abundance of presumed N_2-fixing and non-N_2-fixing plants from selected ecosystems. Oecologia 54:317–325.

Virginia, R. A., and W. M. Jarrell. 1983. Soil properties in a mesquite-dominated Sonoran desert ecosystem. Soil Science Society of America Journal 47:138–144.

Virginia, R. A., W. M. Jarrell, and E. Francovizcaino. 1982. Direct measurement of denitrification in a Prosopis (mesquite)-dominated Sororan desert ecosystem. Oecologia 53:120–122.

Vitousek, P. M. 1977. Regulation of element concentrations in mountain streams in the northeastern United States. Ecological Monographs 47:65–87.

Vitousek, P. M. 1994. Beyond global warming—Ecology and global change. Ecology 75:1861–1876.

Vitousek, P. M. 2004. Nutrient Cycling and Limitation. Princeton University Press.

Vitousek, P. M., J. D. Aber, R. W. Howarth, G. E. Likens, P. A. Matson, D. W. Schindler, W. H. Schlesinger, and D. G. Tilman. 1997. Human alteration of the global nitrogen cycle: Sources and consequences. Ecological Applications 7:737–750.

Vitousek, P. M., K. Cassman, C. Cleveland, T. Crews, C. B. Field, N. B. Grimm, R. W. Howarth, R. Marino, L. Martinelli, E. B. Rastetter, and J. I. Sprent. 2002. Towards an ecological understanding of biological nitrogen fixation. Biogeochemistry 57:1–45.

Vitousek, P. M., P. R. Ehrlich, A. H. Ehrlich, and P. A. Matson. 1986. Human appropriation of the products of photosynthesis. BioScience 36:368–373.

Vitousek, P. M., T. Fahey, D. W. Johnson, and M. J. Swift. 1988. Element interactions in forest ecosystems—Succession, allometry and input–output budgets. Biogeochemistry 5:7–34.

Vitousek, P. M., J. R. Gosz, C. C. Grier, J. M. Melillo, and W. A. Reiners. 1982. A comparative analysis of potential nitrification and nitrate mobility in forest ecosystems. Ecological Monographs 52:155–177.

Vitousek, P. M., and R. W. Howarth. 1991. Nitrogen limitation on land and in the sea—How can it occur? Biogeochemistry 13:87–115.

Vitousek, P. M., T. N. Ladefoged, P. V. Kirch, A. S. Hartshorn, M. W. Graves, S. C. Hotchkiss, S. Tuljapurkar, and O. A. Chadwick. 2004. Soils, agriculture, and society in precontact Hawai. Science 304:1665–1669.

Vitousek, P. M., and P. A. Matson. 1984. Mechanisms of nitrogen retention in forest ecosystems—A field experiment. Science 225:51–52.

Vitousek, P. M., and P. A. Matson. 1988. Nitrogen transformations in a range of tropical forest soils. Soil Biology & Biochemistry 20:361–367.

Vitousek, P. M., and J. M. Melillo. 1979. Nitrate losses from disturbed forests—Patterns and mechanisms. Forest Science 25:605–619.

Vitousek, P. M., D.N.L. Menge, S. C. Reed, and C. C. Cleveland. in press. Biological nitrogen fixation: rates, patterns, and ecological controls in terrestrial ecosystems. Phil Trans Royal Soc. B.

Vitousek, P. M., and W. A. Reiners. 1975. Ecosystem succession and nutrient retention—Hypothesis. BioScience 25:376–381.

Vitousek, P. M., and R. L. Sanford. 1986. Nutrient cycling in moist tropical forest. Annual Review of Ecology and Systematics 17:137–167.

Vitousek, P. M., L. R. Walker, L. D. Whiteaker, D. Muellerdombois, and P. A. Matson. 1987. Biological invasion by Myrica faya alters ecosystem development in Hawaii. Science 238:802–804.

Vitt, D. H., L. A. Halsey, I. E. Bauer, and C. Campbell. 2000. Spatial and temporal trends in carbon storage of peatlands of continental western Canada through the Holocene. Canadian Journal of Earth Sciences 37:683–693.

Vo, A.T.E., M. S. Bank, J. P. Shine, and S. V. Edwards. 2011. Temporal increase in organic mercury in an endangered pelagic seabird assessed by century-old museum specimens. Proceedings of the National Academy of Sciences 108:7466–7471.

Vogt, K. A., C. C. Grier, C. E. Meier, and R. L. Edmonds. 1982. Mycorrhizal role in net primary production and nutrient cycling in Abies amabilis ecosystems in western Washington. Ecology 63:370–380.

Vogt, K. A., C. C. Grier, C. E. Meier, and M. R. Keyes. 1983. Organic matter and nutrient dynamics in forest floors of young and mature Abies amabilis stands in western Washington, as affected by fine root input. Ecological Monographs 53:139–157.

Vogt, K. A., C. C. Grier, and D. J. Vogt. 1986. Production, turnover, and nutrient dynamics of aboveground and belowground detritus of world forests. Advances in Ecological Research 15:303–377.

Vogt, K. A., D. J. Vogt, and J. Bloomfield. 1998. Analysis of some direct and indirect methods for estimating root biomass and production of forests at an ecosystem level. Plant and Soil 200:71–89.

Vogt, K. A., D. J. Vogt, P. A. Palmiotto, P. Boon, J. Ohara, and H. Asbjornsen. 1996. Review of root dynamics in forest ecosystems grouped by climate, climatic forest type and species. Plant and Soil 187:159–219.

Volk, T. 1998. *Gaia's Body*. Springer/Copernicus.

Vollenweider, R. A. 1976. Advances in defining critical loading levels for phosphorus in lake eutrophication. Memorie 1st Ital. Idriobiol. **33**:53–83.

Volz, A., and D. Kley. 1988. Evaluation of the Montsouris series of ozone measurements made in the 19th-century. Nature **332**:240–242.

von Fischer, J. C., R. C. Rhew, G. M. Ames, B. K. Fosdick, and P. E. von Fischer. 2010. Vegetation height and other controls of spatial variability in methane emissions from the Arctic coastal tundra at Barrow, Alaska. Journal of Geophysical Research—Biogeosciences **115**:11.

Voroney, R. P., and E. A. Paul. 1984. Determination of k_c and K_n *in situ* for calibration of the chloroform fumigation-incubation method. Soil Biology & Biochemistry **16**:9–14.

Vörösmarty, C. J., P. Green, J. Salisbury, and R. B. Lammers. 2000. Global water resources: Vulnerability from climate change acid population growth. Science **289**:284–288.

Vörösmarty, C. J., D. Lettenmaier, C. Levecque, M. Meybeck, C. Pahl-Wostl, J. Alcamo, W. Cosgrove, H. Grassl, H. Hoff, P. Kabat, F. Lansigan, R. Lawford, and R. Naiman. 2004. Humans transforming the global water system. EOS, Transactions, American Geophysical Union **85**:509–520.

Vörösmarty, C. J., M. Meybeck, B. Fekete, K. Sharma, P. Green, and J.P.M. Syvitski. 2003. Anthropogenic sediment retention: major global impact from registered river impoundments. Global and Planetary Change **39**:169–190.

Vörösmarty, C. J., B. Moore, A. L. Grace, M. P. Gildea, J. M. Melillo, B. J. Peterson, E. B. Rastetter, and P. A. Steudler. 1989. Continental scale models of water balance and fluvial transport: An application to South America. Global Biogeochemical Cycles **3**: 241–265.

Vörösmarty, C. J., and D. Sahagian. 2000. Anthropogenic disturbance of the terrestrial water cycle. BioScience **50**:753–765.

Vörösmarty, C. J., K. P. Sharma, B. M. Fekete, A. H. Copeland, J. Holden, J. Marble, and J. A. Lough. 1997. The storage and aging of continental runoff in large reservoir systems of the world. Ambio **26**:210–219.

Vose, J. M., and M. G. Ryan. 2002. Seasonal respiration of foliage, fine roots, and woody tissues in relation to growth, tissue N, and photosynthesis. Global Change Biology **8**:182–193.

Voss, M., J. W. Dippner, and J. P. Montoya. 2001. Nitrogen isotope patterns in the oxygen-deficient waters of the eastern tropical north Pacific ocean.

Deep-Sea Research Part I—Oceanographic Research Papers **48**:1905-1921.

Vossbrinck, C. R., D. C. Coleman, and T. A. Woolley. 1979. Abiotic and biotic factors in litter decomposition in a semiarid grassland. Ecology **60**:265–271.

Vrede, T., and L. J. Tranvik. 2006. Iron constraints on planktonic primary production in oligotrophic lakes. Ecosystems **9**:1094–1105.

Vuichard, N., P. Ciais, L. Belelli, P. Smith, and R. Valentini. 2008. Carbon sequestration due to the abandonment of agriculture in the former USSR since 1990. Global Biogeochemical Cycles **22**.

Wacey, D., M. R. Kilburn, M. Saunders, J. Cliff, and M. D. Brasier. 2011. Microfossils of sulphur-metabolizing cells in 3.4-billion-year-old rocks of western Australia. Nature Geoscience **4**:698–702.

Wackett, L. P., A. G. Dodge, and L.B.M. Ellis. 2004. Microbial genomics and the periodic table. Applied and Environmental Microbiology **70**:647–655.

Wada, Y., L.P.H. van Beek, and M.F.P. Bierkens. 2012. Nonsustainable groundwater sustaining irrigation: A global assessment. Water Resources Research **48**.

Wada, Y., L.P.H. van Beek, C. M. van Kempen, J. Reckman, S. Vasak, and M.F.P. Bierkens. 2010. Global depletion of groundwater resources. Geophysical Research Letters **37**.

Wadhams, P., J. Holfort, E. Hansen, and J. P. Wilkinson. 2002. A deep convective chimney in the winter Greenland Sea. Geophysical Research Letters **29**.

Wagener, S. M., M. W. Oswood, and J. P. Schimel. 1998. Rivers and soils: Parallels in carbon and nutrient processing. BioScience **48**:104–108.

Wagner, B., A. F. Lotter, N. Nowaczyk, J. M. Reed, A. Schwalb, R. Sulpizio, V. Valsecchi, M. Wessels, and G. Zanchetta. 2009. A 40,000-year record of environmental change from ancient Lake Ohrid (Albania and Macedonia). Journal of Paleolimnology **41**:407–430.

Wagner, C., A. Griesshammer, and H. L. Drake. 1996. Acetogenic capacities and the anaerobic turnover of carbon in a Kansas prairie soil. Applied and Environmental Microbiology **62**:494–500.

Wahlen, M., N. Tanaka, R. Henry, B. Deck, J. Zeglen, J. S. Vogel, J. Southon, A. Shemesh, R. Fairbanks, and W. Broecker. 1989. ^{14}C in methane sources and in atmospheric methane—The contribution from fossil carbon. Science **245**:286–290.

Waite, J. H., M. R. Combi, W. H. Ip, T. E. Cravens, R. L. McNutt, W. Kasprzak, R. Yelle, J. Luhmann, H. Niemann, D. Gell, B. Magee, G. Fletcher, J. Lunine, and W. L. Tseng. 2006. Cassini ion and neutral mass spectrometer: Enceladus plume composition and structure. Science **311**:1419–1422.

Wakatsuki, T., and A. Rasyidin. 1992. Rates of weathering and soil formation. Geoderma 52:251–263.

Wakefield, A. E., N. J. Gotelli, S. E. Wittman, and A. M. Ellison. 2005. Prey addition alters nutrient stoichiometry of the carnivorous plant *Sarracenia purpurea*. Ecology 86:1737–1743.

Walbridge, M. R., C. J. Richardson, and W. T. Swank. 1991. Vertical-distribution of biological and geochemical phosphorus subcycles in two southern Appalachian forest soils. Biogeochemistry 13:61–85.

Walbridge, M. R., and P. M. Vitousek. 1987. Phosphorus mineralization potentials in acid organic soils—Processes affecting $PO_4^3–^{32}P$ isotope-dilution measurements. Soil Biology & Biochemistry 19:709–717.

Waldman, J. M., J. W. Munger, D. J. Jacob, R. C. Flagan, J. J. Morgan, and M. R. Hoffmann. 1982. Chemical composition of acid fog. Science 218:677–680.

Waldman, J. M., J. W. Munger, D. J. Jacob, and M. R. Hoffmann. 1985. Chemical characterization of stratus cloudwater and its role as a vector for pollutant deposition in a Los Angeles pine forest. Tellus Series B—Chemical and Physical Meteorology 37:91–108.

Waldrop, M. P., K. P. Wickland, R. White, A. A. Berhe, J. W. Harden, and V. E. Romanovsky. 2010. Molecular investigations into a globally important carbon pool: permafrost-protected carbon in Alaskan soils. Global Change Biology 16:2543–2554.

Walker, J.C.G. 1977. *Evolution of the Atmosphere*. Macmillan.

Walker, J.C.G. 1980. The oxygen cycle. *In* O. Hutzinger, editor. *Handbook of Environmental Chemistry. The Natural Environment and the Biogeochemical Cycles, Vol. 1, Part A*. Springer-Verlag, 87–104.

Walker, J.C.G. 1983. Possible limits on the composition of the Archean ocean. Nature 302:518–520.

Walker, J.C.G. 1984. How life affects the atmosphere. BioScience 34:486–491.

Walker, J.C.G. 1985. Carbon dioxide on the early Earth. Origins of Life and Evolution of the Biosphere 16:117–127.

Walker, J.C.G., and P. Brimblecombe. 1985. Iron and sulfur in the pre-biologic ocean. Precambrian Research 28:205–222.

Walker, J.C.G., C. Klein, S.M.J.W. Schopf, D. J. Stevenson, and M. R. Walter. 1983. Environmental evolution of the Archean-early Proterozoic Earth. *In* J. W. Schopf, editor. *Earth's Earliest Biosphere*. Princeton University Press, 260–290.

Walker, T. W., and A.F.R. Adams. 1958. Studies on soil organic matter: I. Influence of phosphorus content of parent materials on accumulations of carbon, nitrogen, sulfur, and organic phosphorus in grassland soils. Soil Science 85:307–318.

Walker, T. W., and J. K. Syers. 1976. Fate of phosphorus during pedogenesis. Geoderma 15:1–19.

Wallace, A., S. A. Bamburg, and J. W. Cha. 1974. Quantitative studies of roots of perennial plants in the Mojave desert. Ecology 55:1160–1162.

Wallace, J. B., S. L. Eggert, J. L. Meyer, and J. R. Webster. 1997. Multiple trophic levels of a forest stream linked to terrestrial litter inputs. Science 277:102–104.

Wallace, J. B., S. L. Eggert, J. L. Meyer, and J. R. Webster. 1999. Effects of resource limitation on a detrital-based ecosystem. Ecological Monographs 69:409–442.

Wallace, J. B., and J. R. Webster. 1996. The role of macroinvertebrates in stream ecosystem function. Annual Review of Entomology 41:115–139.

Wallace, P. J. 2005. Volatiles in subduction zone magmas: Concentrations and fluxes based on melt inclusion and volcanic gas data. Journal of Volcanology and Geothermal Research 140:217–240.

Wallenstein, M. D., S. McNulty, I. J. Fernandez, J. Boggs, and W. H. Schlesinger. 2006. Nitrogen fertilization decreases forest soil fungal and bacterial biomass in three long-term experiments. Forest Ecology and Management 222:459–468.

Wallerstein, G. 1988. Mixing in stars. Science 240:1743–1750.

Walling, D. E. 2006. Human impact on land-ocean sediment transfer by the world's rivers. Geomorphology 79:192–216.

Walling, D. E., and D. Fang. 2003. Recent trends in the suspended sediment loads of the world's rivers. Global and Planetary Change 39:111–126.

Wallmann, K. 2001. The geological water cycle and the evolution of marine delta ^{18}O values. Geochimica et Cosmochimica Acta 65:2469–2485.

Wallmann, K. 2010. Phosphorus imbalance in the global ocean? Global Biogeochemical Cycles 24.

Walsh, C. J., A. H. Roy, J. W. Feminella, P. D. Cottingham, P. M. Groffman, and R. P. Morgan. 2005. The urban stream syndrome: current knowledge and the search for a cure. Journal of the North American Benthological Society 24:706–723.

Walsh, J. J. 1984. The role of ocean biota in accelerated ecological cycles—A temporal view. BioScience 34:499–507.

Walsh, J. J. 1991. Importance of continental margins in the marine biogeochemical cycling of carbon and nitrogen. Nature 350:53–55.

Walter, K. M., M. E. Edwards, G. Grosse, S. A. Zimov, and F. S. Chapin. 2007. Thermokarst lakes as a source of atmospheric CH_4 during the last deglaciation. Science 318:633–636.

Walter, K. M., S. A. Zimov, J. P. Chanton, D. Verbyla, and F. S. Chapin. 2006. Methane bubbling from Siberian thaw lakes as a positive feedback to climate warming. Nature 443:71–75.

Walter, M. J., S. C. Kohn, D. Araujo, G. P. Bulanova, C. B. Smith, E. Gaillou, J. Wang, A. Steele, and S. B. Shirey. 2011. Deep mantle cycling of oceanic crust: Evidence from diamonds and their mineral inclusions. Science 333:54–57.

Walter, M. T., D. S. Wilks, J. Y. Parlange, and R. L. Schneider. 2004a. Increasing evapotranspiration from the conterminous United States. Journal of Hydrometeorology 5:405–408.

Walter, S., H. W. Bange, and D.W.R. Wallace. 2004b. Nitrous oxide in the surface layer of the tropical North Atlantic Ocean along a west to east transect. Geophysical Research Letters 31.

Walvoord, M. A., F. M. Phillips, D. A. Stonestrom, R. D. Evans, P. C. Hartsough, B. D. Newman, and R. G. Striegl. 2003. A reservoir of nitrate beneath desert soils. Science 302:1021–1024.

Wan, S. Q., D. F. Hui, and Y. Q. Luo. 2001. Fire effects on nitrogen pools and dynamics in terrestrial ecosystems: A meta-analysis. Ecological Applications 11:1349–1365.

Wanek, W., M. Mooshammer, A. Bloechl, A. Hanreich, and A. Richter. 2010. Determination of gross rates of amino acid production and immobilization in decomposing leaf litter by a novel ^{15}N-isotope pool dilution technique. Soil Biology & Biochemistry 42: 1293–1302.

Wanek, W., and G. Zotz. 2011. Are vascular epiphytes nitrogen or phosphorus limited? A study of plant ^{15}N fractionation and foliar N:P stoichiometry with the tank bromeliad *Vriesea sanguinolenta*. New Phytologist 192:462–470.

Wang, D., S. A. Heckathorn, X. Wang, and S. M. Philpott. 2012. A meta-analysis of plant physiological and growth responses to temperature and elevated CO_2. Oecologia 169:1–13.

Wang, D. B., and M. Hejazi. 2011. Quantifying the relative contribution of the climate and direct human impacts on mean annual streamflow in the contiguous United States. Water Resources Research 47:16.

Wang, F. Y., and P. M. Chapman. 1999. Biological implications of sulfide in sediment—A review focusing on sediment toxicity. Environmental Toxicology and Chemistry 18:2526–2532.

Wang, J. S., J. A. Logan, M. B. McElroy, B. N. Duncan, I. A. Megretskaia, and R. M. Yantosca. 2004. A 3-D model analysis of the slowdown and interannual variability in the methane growth rate from 1988 to 1997. Global Biogeochemical Cycles 18.

Wang, K. C., R. E. Dickinson, and S. L. Liang. 2009. Clear sky visibility has decreased over land globally from 1973 to 2007. Science 323:1468–1470.

Wang, L., K. K. Caylor, J. C. Villegas, G. A. Barron-Gafford, D. D. Breshears, and T. E. Huxman. 2010a.

Partitioning evapotranspiration across gradients of woody plant cover: Assessment of a stable isotope technique. Geophysical Research Letters 37.

Wang, L. X., and S. A. Macko. 2011. Constrained preferences in nitrogen uptake across plant species and environments. Plant Cell and Environment 34: 525–534.

Wang, W. Q., A. Kohl, and D. Stammer. 2010b. Estimates of global ocean volume transports during 1960 through 2001. Geophysical Research Letters 37.

Wang, Y., Y. He, H. Zhang, J. Schroder, C. Li, and D. Zhou. 2008. Phosphate mobilization by citric, tartaric, and oxalic acids in a clay loam Ultisol. Soil Science Society of America Journal 72:1263–1268.

Wang, Y. P., and B. Z. Houlton. 2009. Nitrogen constraints on terrestrial carbon uptake: Implications for the global carbon-climate feedback. Geophysical Research Letters 36.

Wang, Z., J. Chappellaz, K. Park, and J. E. Mak. 2010c. Large variations in Southern Hemisphere biomass burning during the last 650 years. Science 330:1663–1666.

Wang, Z.H.A., and W. J. Cai. 2004. Carbon dioxide degassing and inorganic carbon export from a marsh-dominated estuary (the Duplin River): A marsh CO_2 pump. Limnology and Oceanography 49: 341–354.

Wanninkhof, R. 1992. Relationship between wind speed and gas exchange over the ocean. Journal of Geophysical Research—Oceans 97:7373–7382.

Wanninkhof, R., P. J. Mulholland, and J. W. Elwood. 1990. Gas-exchange rates for a 1st-order stream determined with deliberate and natural tracers. Water Resources Research 26:1621–1630.

Warby, R.A.F., C. E. Johnson, and C. T. Driscoll. 2009. Continuing acidification of organic soils across the northeastern USA: 1984–2001. Soil Science Society of America Journal 73:274–284.

Ward, B. B., A. H. Devol, J. J. Rich, B. X. Chang, S. E. Bulow, H. Naik, A. Pratihary, and A. Jayakumar. 2009. Denitrification as the dominant nitrogen loss process in the Arabian Sea. Nature 461:78–81.

Ward, B. B., K. A. Kilpatrick, P. C. Novelli, and M. I. Scranton. 1987. Methane oxidation and methane fluxes in the ocean surface layer and deep anoxic waters. Nature 327:226–229.

Ward, D. M., K. H. Nislow, and C. L. Folt. 2010. Bioaccumulation syndrome: identifying factors that make some stream food webs prone to elevated mercury bioaccumulation. *Year in Ecology and Conservation Biology 2010* 1195:62–83.

Ward, R. C. 1967. *Principles of Hydrology*. McGraw-Hill.

Wardle, D. A. 1992. A comparative assessment of factors which influence microbial biomass carbon and nitrogen levels in soil. Biological Reviews of the Cambridge Philosophical Society **67**:321–358.

Wardle, D. A., L. R. Walker, and R. D. Bardgett. 2004. Ecosystem properties and forest decline in contrasting long-term chronosequences. Science **305**:509–513.

Ware, D. M., and R. E. Thomson. 2005. Bottom-up ecosystem trophic dynamics determine fish production in the northeast Pacific. Science **308**:1280–1284.

Warembourg, F. R., and E. A. Paul. 1977. Seasonal transfers of assimilated ^{14}C in grassland—Plant production and turnover, soil and plant respiration. Soil Biology & Biochemistry **9**:295–301.

Waring, R. H., J. J. Landsberg, and M. Williams. 1998. Net primary production of forests: a constant fraction of gross primary production? Tree Physiology **18**:129–134.

Waring, R. H., J. J. Rogers, and W. T. Swank. 1981. Water relations and hydrologic cycles. *In* D. E. Reichle, editor. *Dynamic Properties of Forest Ecosystems*, Cambridge University Press, 205–264.

Waring, R. H., and W. H. Schlesinger. 1985. *Forest Ecosystems*. Academic.

Warneck, P. 2000. *Chemistry of the Natural Atmosphere*, 2nd edition. Academic Press.

Warren, D. R., E. S. Bernhardt, R. O. Hall, and G. E. Likens. 2007. Forest age, wood and nutrient dynamics in headwater streams of the Hubbard Brook Experimental Forest, NH. Earth Surface Processes and Landforms **32**:1154–1163.

Watanabe, S., H. Yamamoto, and S. Tsunogai. 1995. Relation between the concentrations of DMS in surface seawater and air in the temperate north Pacific region. Journal of Atmospheric Chemistry **22**:271–283.

Watkins, N. D., R.S.J. Sparks, H. Sigurdsson, T. C. Huang, A. Federman, S. Carey, and D. Ninkovich. 1978. Volume and extent of Minoan tephra from Santorini volcano—New evidence from deep sea sediment cores. Nature **271**:122–126.

Watmough, S. A., R. McNeely, and P. M. Lafleur. 2001. Changes in wood and foliar delta ^{13}C in sugar maple at Gatineau park, Quebec, Canada. Global Change Biology **7**:955–960.

Watson, A., J. E. Lovelock, and L. Margulis. 1978. Methanogenesis, fires and the regulation of atmospheric oxygen. Biosystems **10**:293–298.

Watson, A. J., D.C.E. Bakker, A. J. Ridgwell, P. W. Boyd, and C. S. Law. 2000. Effect of iron supply on Southern Ocean CO_2 uptake and implications for glacial atmospheric CO_2. Nature **407**:730–733.

Watson, A. J., and J. E. Lovelock. 1983. Biological homeostasis of the global environment—The parable of Daisyworld. Tellus Series B—Chemical and Physical Meteorology **35**:284–289.

Watson, A. J., C. Robinson, J. E. Robinson, P.J.L. Williams, and M.J.R. Fasham. 1991. Spatial variability in the sink for atmospheric carbon dioxide in the north Atlantic. Nature **350**:50–53.

Watson, A. J., U. Schuster, D.C.E. Bakker, N. R. Bates, A. Corbiere, M. Gonzalez-Davila, T. Friedrich, J. Hauck, C. Heinze, T. Johannessen, A. Kortzinger, N. Metzl, J. Olafsson, A. Olsen, A. Oschlies, X. A. Padin, B. Pfeil, J. M. Santana-Casiano, T. Steinhoff, M. Telszewski, A. F. Rios, D.W.R. Wallace, and R. Wanninkhof. 2009. Tracking the variable north Atlantic sink for atmospheric CO_2. Science **326**:1391–1393.

Watson, E. B., and T. M. Harrison. 2005. Zircon thermometer reveals minimum melting conditions on earliest Earth. Science **308**:841–844.

Watts, R. D., R. W. Compton, J. H. McCammon, C. L. Rich, S. M. Wright, T. Owens, and D. S. Ouren. 2007. Roadless space of the conterminous United States. Science **316**:736–738.

Watts, S. F. 2000. The mass budgets of carbonyl sulfide, dimethyl sulfide, carbon disulfide, and hydrogen sulfide. Atmospheric Environment **34**:761–779.

Watwood, M. E., and J. W. Fitzgerald. 1988. Sulfur transformations in forest litter and soil—Results of laboratory and field incubations. Soil Science Society of America Journal **52**:1478–1483.

Watwood, M. E., J. W. Fitzgerald, W. T. Swank, and E. R. Blood. 1988. Factors inovlved in potential sulfur accumulation in litter and soil from a coastal pine forest. Biogeochemistry **6**:3–19.

Waugh, D. W., and T. M. Hall. 2002. Age of stratospheric air: Theory, observations, and models. Reviews of Geophysics **40**.

Waughman, G. J. 1980. Chemical aspects of the ecology of some south German peatlands. Journal of Ecology **68**:1025–1046.

Waughman, G. J., and D. J. Bellamy. 1980. Nitrogen fixation and the nitrogen balance in peatland ecosystems. Ecology **61**:1185–1198.

Wayne, R. P. 1991. *Chemistry of Atmospheres: An Introduction to the Chemistry of the Atmospheres of Earth, the Planets, and their Satellites*, 2nd. edition. Oxford University Press.

Weand, M. P., M. A. Arthur, G. M. Lovett, F. Sikora, and K. C. Weathers. 2010. The phosphorus status of northern hardwoods differs by species but is unaffected by nitrogen fertilization. Biogeochemistry **97**:159–181.

Weathers, K. C., G. E. Likens, F. H. Bormann, J. S. Eaton, W. B. Bowden, J. L. Andersen, D. A. Cass, J. N. Galloway, W. C. Keene, K. D. Kimball, P. Huth, and D. Smiley. 1986. A regional acidic cloud fog water event in the eastern United States. Nature **319**:657–658.

Weathers, K. C., G. M. Lovett, G. E. Likens, and R. Lathrop. 2000. The effect of landscape features on deposition to Hunter Mountain, Catskill Mountains, New York. Ecological Applications **10**:528–540.

Webb, R. S., C. E. Rosenzweig, and E. R. Levine. 1993. Specifying land surface characteristics in general circulation models—Soil profile data set and derived water-holding capacities. Global Biogeochemical Cycles **7**:97–108.

Webb, W., S. Szarek, W. Lauenroth, R. Kinerson, and M. Smith. 1978. Primary productivity and water-use in native forest, grassland, and desert ecosystems. Ecology **59**:1239–1247.

Webb, W. L., W. K. Lauenroth, S. R. Szarek, and R. S. Kinerson. 1983. Primary production and abiotic controls in forests, grasslands, and desert ecosystems in the United States. Ecology **64**:134–151.

Weber, K. A., M. M. Urrutia, P. F. Churchill, R. K. Kukkadapu, and E. E. Roden. 2006. Anaerobic redox cycling of iron by freshwater sediment microorganisms. Environmental Microbiology **8**:100–113.

Weber, T. S., and C. Deutsch. 2010. Ocean nutrient ratios governed by plankton biogeography. Nature **467**:550–554.

Webster, J. R., E. F. Benfield, T. P. Ehrman, M. A. Schaeffer, J. L. Tank, J. J. Hutchens, and D. J. D'Angelo. 1999. What happens to allochthonous material that falls into streams? A synthesis of new and published information from Coweeta. Freshwater Biology **41**:687–705.

Webster, J. R., and T. P. Ehrman. 1996. Solute dynamics. In F. R. Hauer and G. A. Lamberti, editors. *Methods in Stream Ecology*. Elsevier, 145–160.

Webster, J. R., and J. L. Meyer. 1997. Organic matter budgets for streams: A synthesis. Journal of the North American Benthological Society **16**:141–161.

Webster, J. R., P. J. Mulholland, J. L. Tank, H. M. Valett, W. K. Dodds, B. J. Peterson, W. B. Bowden, C. N. Dahm, S. Findlay, S. V. Gregory, N. B. Grimm, S. K. Hamilton, S. L. Johnson, E. Marti, W. H. McDowell, J. L. Meyer, D. D. Morrall, S. A. Thomas, and W. M. Wollheim. 2003. Factors affecting ammonium uptake in streams—An inter-biome perspective. Freshwater Biology **48**:1329–1352.

Webster, J. R., and B. C. Patten. 1979. Effects of watershed perturbation on stream potassium and calcium dynamics. Ecological Monographs **19**:51–52.

Wedepohl, K. H. 1995. The composition of the continental crust. Geochimica et Cosmochimica Acta **59**:1217–1232.

Wedin, D. A., and D. Tilman. 1996. Influence of nitrogen loading and species composition on the carbon balance of grasslands. Science **274**:1720–1723.

Weier, K. L., J. W. Doran, J. F. Power, and D. T. Walters. 1993. Denitrification and the dinitrogen/nitrous oxide ratio as affected by soil water, available carbon, and nitrate. Soil Science Society of America Journal **57**:66–72.

Weier, K. L., and J. W. Gilliam. 1986. Effect of acidity on denitrification and nitrous oxide evolution from Atlantic coastal plain soils. Soil Science Society of America Journal **50**:1202–1206.

Weir, J. S. 1972. Spatial distribution of elephants in an African national park in relation to environmental sodium. Oikos **23**:1–13.

Weiss, H., M. A. Courty, W. Wetterstrom, F. Guichard, L. Senior, R. Meadow, and A. Curnow. 1993. The genesis and collapse of third millennium north Mesopotamian civilization. Science **261**:995–1004.

Weiss, H. V., M. Koide, and E. D. Goldberg. 1971. Mercury in a Greeland ice sheet: Evidence of recent input by man. Science **174**:692–694.

Weiss, J. V., D. Emerson, and J. P. Megonigal. 2005. Rhizosphere iron(III) deposition and reduction in a *Juncus effusus* L.-dominated wetland. Soil Science Society of America Journal **69**:1861.

Weiss, P. S., J. E. Johnson, R. H. Gammon, and T. S. Bates. 1995. Reevaluation of the open ocean source of carbonyl sulfide to the atmosphere. Journal of Geophysical Research—Atmospheres **100**:23083–23092.

Weiss, R. F., J. Muehle, P. K. Salameh, and C. M. Harth. 2008. Nitrogen trifluoride in the global atmosphere. Geophysical Research Letters **35**.

Weiss-Penzias, P. S., M. S. Gustin, and S. N. Lyman. 2011. Sources of gaseous oxidized mercury and mercury dry deposition at two southeastern U.S. sites. Atmospheric Environment **45**:4569–4579.

Weitz, A. M., E. Veldkamp, M. Keller, J. Neff, and P. M. Crill. 1998. Nitrous oxide, nitric oxide, and methane fluxes from soils following clearing and burning of tropical secondary forest. Journal of Geophysical Research—Atmospheres **103**:28047–28058.

Welch, S. A., A. E. Taunton, and J. F. Banfield. 2002. Effect of microorganisms and microbial metabolites on apatite dissolution. Geomicrobiology Journal **19**:343–367.

Welch, S. A., and W. J. Ullman. 1993. The effect of organic acids on plagioclase dissolution rates and stoichiometry. Geochimica et Cosmochimica Acta **57**:2725–2736.

Wells, P. V. 1983. Paleobiogeography of montane islands in the Great Basin since the last glaciopluvial. Ecological Monographs **53**:341–382.

Wellsbury, P., K. Goodman, T. Barth, B. A. Cragg, S. P. Barnes, and R. J. Parkes. 1997. Deep marine biosphere fuelled by increasing organic matter availability during burial and heating. Nature **388**:573–576.

Weng, F. Z., R. R. Ferraro, and N. C. Grody. 1994. Global precipitation estimations using Defense Meteorological Satellite Program F10 and F11 special sensor microwave imager data. Journal of Geophysical Research—Atmospheres 99:14493–14502.

Wennberg, P. O., R. C. Cohen, R. M. Stimpfle, J. P. Koplow, J. G. Anderson, R. J. Salawitch, D. W. Fahey, E. L. Woodbridge, E. R. Keim, R. S. Gao, C. R. Webster, R. D. May, D. W. Toohey, L. M. Avallone, M. H. Proffitt, M. Loewenstein, J. R. Podolske, K. R. Chan, and S. C. Wofsy. 1994. Removal of stratospheric O_3 by radicals—*in situ* measurements of OH, HO_2, NO, NO_2, ClO, AND BrO. Science 266: 398–404.

Wentz, F. J., L. Ricciardulli, K. Hilburn, and C. Mears. 2007. How much more rain will global warming bring? Science 317:233–235.

Wenzhofer, F., M. Adler, O. Kohls, C. Hensen, B. Strotmann, S. Boehme, and H. D. Schulz. 2001. Calcite dissolution driven by benthic mineralization in the deep-sea: *In situ* measurements of Ca^{2+}, pH, pCO_2 and O_2. Geochimica et Cosmochimica Acta 65: 2677–2690.

Wesely, M. L., and B. B. Hicks. 2000. A review of the current status of knowledge on dry deposition. Atmospheric Environment 34:2261–2282.

Wessman, C. A., J. D. Aber, D. L. Peterson, and J. M. Melillo. 1988a. Foliar analysis using near-infrared reflectance spectroscopy. Canadian Journal of Forest Research 18:6–11.

Wessman, C. A., J. D. Aber, D. L. Peterson, and J. M. Melillo. 1988b. Remote sensing of canopy chemistry and nitrogen cycling in temperate forest ecosystems. Nature 335:154–156.

West, A. J., A. Galy, and M. Bickle. 2005. Tectonic and climatic controls on silicate weathering. Earth and Planetary Science Letters 235:211–228.

West, T. O., C. C. Brandt, L. M. Baskaran, C. M. Hellwinckel, R. Mueller, C. J. Bernacchi, V. Bandaru, B. Yang, B. S. Wilson, G. Marland, R. G. Nelson, D.G.D. Ugarte, and W. M. Post. 2010. Cropland carbon fluxes in the United States: increasing geospatial resolution of inventory-based carbon accounting. Ecological Applications 20:1074–1086.

West, T. O., and G. Marland. 2002. A synthesis of carbon sequestration, carbon emissions, and net carbon flux in agriculture: comparing tillage practices in the United States. Agriculture Ecosystems & Environment 91:217–232.

Westheimer, F. H. 1987. Why nature chose phosphates. Science 235:1173–1178.

Westman, C. J., and R. Laiho. 2003. Nutrient dynamics of drained peatland forests. Biogeochemistry 63: 269–298.

Weston, N. B., R. E. Dixon, and S. B. Joye. 2006. Ramifications of increased salinity in tidal freshwater sediments: Geochemistry and microbial pathways of organic matter mineralization. Journal of Geophysical Research 111:1–14.

Weston, N. B., and S. B. Joye. 2005. Temperature-driven decoupling of key phases of organic matter degradation in marine sediments. Proceedings of the National Academy of Sciences 102:17036–17040.

Wetherill, G. W. 1985. Occurrence of giant impacts during the growth of the terrestrial planets. Science 228: 877–879.

Wetherill, G. W. 1994. Provenance of the terrestrial planets. Geochimica et Cosmochimica Acta 58:4513–4520.

Wetselaar, R. 1968. Soil organic nitrogen mineralization as affected by low soil water potentials. Plant and Soil 29:9–17.

Wetzel, R. G. 1992. Gradient-dominated ecosystems—Sources and regulatory functions of dissolved organic matter in freshwater ecosystems. Hydrobiologia 229:181–198.

Wetzel, R. G. 2001. *Limnology*, 3rd edition. Academic Press.

Wetzel, R. G., and G. E. Likens. 2000. *Limnological Analyses*, 3rd edition. Springer-Verlag.

Weyl, P. K. 1978. Micro-paleontology and ocean surface climate. Science 202:475–481.

Whalen, S. C., and W. S. Reeburgh. 1990. Consumption of atmospheric methane by tundra soils. Nature 346: 160–162.

Whalen, S. C., W. S. Reeburgh, and K. S. Kizer. 1991. Methane consumption and emission by taiga. Global Biogeochemical Cycles 5:261–273.

Wheat, C. G., J. McManus, M. J. Mottl, and E. Giambalvo. 2003. Oceanic phosphorus imbalance: Magnitude of the mid-ocean ridge flank hydrothermal sink. Geophysical Research Letters 30.

Whelpdale, D. M., and J. N. Galloway. 1994. Sulfur and reactive nitrogen oxide fluxes in the North Atlantic atmosphere. Global Biogeochemical Cycles 8: 481–493.

White, A. F., and A. E. Blum. 1995. Effects of climate on chemical weathering in watersheds. Geochimica et Cosmochimica Acta 59:1729–1747.

White, A. F., A. E. Blum, M. S. Schulz, T. D. Bullen, J. W. Harden, and M. L. Peterson. 1996. Chemical weathering rates of a soil chronosequence on granitic alluvium. 1. Quantification of mineralogical and surface area changes and calculation of primary silicate reaction rates. Geochimica et Cosmochimica Acta 60: 2533–2550.

White, A. F., A. E. Blum, M. S. Schulz, D. V. Vivit, D. A. Stonestrom, M. Larsen, S. F. Murphy, and D. Eberl. 1998. Chemical weathering in a tropical watershed, Luquillo mountains, Puerto Rico: I. Long-term versus short-term weathering fluxes. Geochimica et Cosmochimica Acta **62**:209–226.

White, A. F., T. D. Bullen, M. S. Schulz, A. E. Blum, T. G. Huntington, and N. E. Peters. 2001. Differential rates of feldspar weathering in granitic regoliths. Geochimica et Cosmochimica Acta **65**:847–869.

White, C. S. 1988. Nitrification inhibition by monoterpenoids—Theoretical mode of action based on molecular structures. Ecology **69**:1631–1633.

White, E. J., and F. Turner. 1970. A method of estimating income of nutrients in catch of airborne particles by a woodland canopy. Journal of Applied Ecology **7**:441–461.

White, P. J., and M. R. Broadley. 2001. Chloride in soils and its uptake and movement within the plant: A review. Annals of Botany **88**:967–988.

White, T.C.R. 1984. The abundance of invertebrate herbivores in relation to the availability of nitrogen in stressed food plants. Oecologia **63**:90–105.

Whitehead, D. C., D. R. Lockyer, and N. Raistrick. 1988. The volatilization of ammonia from perennial ryegrass during decomposition, drying and induced senescence. Annals of Botany **61**:567–571.

Whiteside, M. D., K. K. Treseder, and P. R. Atsatt. 2009. The brighter side of soils: Quantum dots track organic nitrogen through fungi and plants. Ecology **90**:100–108.

Whitfield, C. J., J. Aherne, and H. M. Baulch. 2011. Controls on greenhouse gas concentrations in polymictic headwater lakes in Ireland. Science of the Total Environment **410**:217–225.

Whitfield, M., and D. R. Turner. 1979. Water-rock partition coefficients and the composition of seawater and river water. Nature **278**:132–137.

Whitfield, P. H., and H. Schreier. 1981. Hysteresis in relationships between discharge and water chemistry in the Fraser river basin, British Columbia. Limnology and Oceanography **26**:1179–1182.

Whiticar, M. J. 1993. Stable isotopes and global budgets. In M.A.K. Khalil, editor. Atmospheric Methane: Sources, Sinks and Role in Global Change. Springer, 138–167.

Whiticar, M. J., E. Faber, and M. Schoell. 1986. Biogenic methane formation in marine and freshwater environments—CO_2 reduction vs. acetate fermentation isotope evidence. Geochimica et Cosmochimica Acta **50**:693–709.

Whiting, G. J., and J. P. Chanton. 1993. Primary production control of methane emission from wetlands. Nature **364**:794–795.

Whitlow, S., P. Mayewski, J. Dibb, G. Holdsworth, and M. Twickler. 1994. An ice-core based record of biomass burning in the arctic and subarctic, 1750-1980. Tellus Series B—Chemical and Physical Meteorology **46**:234–242.

Whitman, W. B., D. C. Coleman, and W. J. Wiebe. 1998. Prokaryotes: The unseen majority. Proceedings of the National Academy of Sciences **95**:6578–6583.

Whitney, F. A., H. J. Freeland, and M. Robert. 2007. Persistently declining oxygen levels in the interior waters of the eastern subarctic Pacific. Progress in Oceanography **75**:179–199.

Whittaker, R. H. 1970. Communities and Ecosystems. Macmillan.

Whittaker, R. H., F. H. Bormann, G. E. Likens, and T. G. Siccama. 1974. The Hubbard Brook ecosystem study: Forest biomass and production. Ecological Monographs **44**:233–254.

Whittaker, R. H. 1975. Communities and Ecosystems, 2nd edition. Macmillian.

Whittaker, R. H., and G. E. Likens. 1973. Carbon in the biota. In G. M. Woodwell and E. V. Pecan, editors. Carbon and the Biosphere. National Technical Information Service, 281–302.

Whittaker, R. H., and P. L. Marks. 1975. Methods of assessing terrestrial productivity. In H. Lieth and R. H. Whittaker, editors. Primary Productivity of the Biosphere, Springer, 55–118.

Whittaker, R. H., and W. A. Niering. 1975. Vegetation of the Santa Catalina Mountains, Arizona. V. Biomass, production and diversity along the elevation gradient. Ecology **56**:771–790.

Whittaker, R. H., and G. M. Woodwell. 1968. Dimension and production relations of trees and shrubs in the Brookhaven forest. Journal of Ecology **56**:1–25.

Wickham, J. D., T. G. Wade, K. H. Riitters, R. V. O'Neill, J. H. Smith, E. R. Smith, K. B. Jones, and A. C. Neale. 2003. Upstream-to-downstream changes in nutrient export risk. Landscape Ecology **18**:195–208.

Widdel, F., S. Schnell, S. Heising, A. Ehrenreich, B. Assmus, and B. Schink. 1993. Ferrous iron oxidation by anoxygenic phototrophic bacteria. Nature **362**:834–836.

Widmer, F., B. T. Shaffer, L. A. Porteous, and R. J. Seidler. 1999. Analysis of nifH gene pool complexity in soil and litter at a douglas fir forest site in the Oregon Cascade mountain range. Applied and Environmental Microbiology **65**:374–380.

Wieder, R. K., and G. E. Lang. 1988. Cycling of inorganic and organic sulfur in peat from Big Run bog, West Virginia. Biogeochemistry **5**:221–242.

Wieder, R. K., J. B. Yavitt, and G. E. Lang. 1990. Methane production and sulfate reduction in two Appalachian peatlands. Biogeochemistry **10**:81–104.

Wieder, W. R., C. C. Cleveland, and A. R. Townsend. 2009. Controls over leaf litter decomposition in wet tropical forests. Ecology **90**:3333–3341.

Wiegert, R. G., and F. C. Evans. 1964. Primary production and the disappearance of dead vegetation on an old field in southeastern Michigan. Ecology 45:49–63.

Wigley, T.M.L. 1989. Possible climate change due to SO_2-derived cloud condensation nuclei. Nature 339: 365–367.

Wilde, S. A., J. W. Valley, W. H. Peck, and C. M. Graham. 2001. Evidence from detrital zircons for the existence of continental crust and oceans on the Earth 4.4 Gyr ago. Nature 409:175–178.

Wilhelm, S. W., D. P. Maxwell, and C. G. Trick. 1996. Growth, iron requirements, and siderophore production in iron-limited *Synechococcus* PCC 7002. Limnology and Oceanography 41:89–97.

Wilkinson, B. H., and B. J. McElroy. 2007. The impact of humans on continental erosion and sedimentation. Geological Society of America Bulletin 119:140–156.

Wilkinson, B. H., B. J. McElroy, S. E. Kesler, S. E. Peters, and E. D. Rothman. 2009. Global geologic maps are tectonic speedometers—Rates of rock cycling from area-age frequencies. Geological Society of America Bulletin 121:760–779.

Wilkinson, D. M., E. G. Nisbet, and G. D. Ruxton. 2012. Could methane produced by sauropod dinosaurs have helped drive Mesozoic climate warmth? Current Biology 22:292–293.

Willbold, M., T. Elliott, and S. Moorbath. 2011. The tungsten isotopic composition of the Earth's mantle before the terminal bombardment. Nature 477:195–198.

Willett, K. M., N. P. Gillett, P. D. Jones, and P. W. Thorne. 2007. Attribution of observed surface humidity changes to human influence. Nature 449:710–712.

Williams, E. J., G. L. Hutchinson, and F. C. Fehsenfeld. 1992. NOx and N_2O emissions from soil. Global Biogeochemical Cycles 6:351–388.

Williams, E. L., L. M. Walter, T.C.W. Ku, G. W. Kling, and D. R. Zak. 2003. Effects of CO_2 and nutrient availability on mineral weathering in controlled tree growth experiments. Global Biogeochemical Cycles 17.

Williams, G. R. 1996. *The Molecular Biology of Gaia.* Columbia University Press.

Williams, L. E., and A. J. Miller. 2001. Transporters responsible for the uptake and partitioning of nitrogenous solutes. Annual Review of Plant Physiology and Plant Molecular Biology 52:659–688.

Williams, M., C. Hopkinson, E. Rastetter, and J. Vallino. 2004. N budgets and aquatic uptake in the Ipswich River basin, northeastern Massachusetts. Water Resources Research 40.

Williams, M., C. Hopkinson, E. Rastetter, J. Vallino, and L. Claessens. 2005. Relationships of land use and stream solute concentrations in the Ipswich River basin, northeastern Massachusetts. Water Air and Soil Pollution 161:55–74.

Williams, P.J.L. 1998. The balance of plankton respiration and photosynthesis in the open oceans. Nature 394:55–57.

Williams, P. M., and E.R.M. Druffel. 1987. Radiocarbon in dissolved organic matter in the central north Pacific ocean. Nature 330:246–248.

Williams, R.J.P., and J.J.R. Fraústo da Silva. 1996. *The Natural Selection of the Chemical Elements: The Environment and Life's Chemistry.* Clarendon.

Williams, S. N., R. E. Stoiber, N. Garcia, A. Londono, J. B. Gemmell, D. R. Lowe, and C. B. Connor. 1986. Eruption of the Nevado del Ruiz volcano, Colombia, on 13 November 1985—Gas flux and fluid geochemistry. Science 233:964–967.

Willison, T. W., K.W.T. Goulding, and D. S. Powlson. 1995. Effect of land-use change and methane mixing ratio on methane uptake from United Kingdom soil. Global Change Biology 1:209–212.

Wilson, A. T. 1978. Pioneer agriculture explosion and CO_2 levels in the atmosphere. Nature 273:40–41.

Wilson, K. B., P. J. Hanson, P. J. Mulholland, D. D. Baldocchi, and S. D. Wullschleger. 2001. A comparison of methods for determining forest evapotranspiration and its components: sap-flow, soil water budget, eddy covariance and catchment water balance. Agricultural and Forest Meteorology 106: 153–168.

Wilson, R. W., F. J. Millero, J. R. Taylor, P. J. Walsh, V. Christensen, S. Jennings, and M. Grosell. 2009. Contribution of fish to the marine inorganic carbon cycle. Science 323:359–362.

Windolf, J., E. Jeppesen, J. P. Jensen, and P. Kristensen. 1996. Modelling of seasonal variation in nitrogen retention and in-lake concentration: A four-year mass balance study in 16 shallow Danish lakes. Biogeochemistry 33:25–44.

Winner, W. E., C. L. Smith, G. W. Koch, H. A. Mooney, J. D. Bewley, and H. R. Krouse. 1981. Rates of emission of H_2S from plants and patterns of stable sulfur isotope fractionation. Nature 289:672–673.

Wittmann, H., F. von Blanckenburg, L. Maurice, J. L. Guyot, N. Filizola, and P. W. Kubik. 2011. Sediment production and delivery in the Amazon river basin quantified by *in situ*-produced cosmogenic nuclides and recent river loads. Geological Society of America Bulletin 123:934–950.

Wogelius, R. A., and J. V. Walther. 1991. Olivine dissolution at 25 °C—Effects of pH, CO_2, and organic acids. Geochimica et Cosmochimica Acta 55:943–954.

Wohlfahrt, G., F. Brilli, L. Hortnagl, X. Xu, H. Bingemer, A. Hansel, and F. Loreto. 2012. Carbonyl sulfide

(COS) as a tracer for canopy photosynthesis, transpiration and stomatal conductance: potential and limitations. Plant Cell and Environment **35**:657–667.

Wolf, A. A., B. G. Drake, J. E. Erickson, and J. P. Megonigal. 2007. An oxygen-mediated positive feedback between elevated carbon dioxide and soil organic matter decomposition in a simulated anaerobic wetland. Global Change Biology **13**:2036–2044.

Wolf, I., and R. Russow. 2000. Different pathways of formation of N_2O, N_2 and NO in black earth soil. Soil Biology & Biochemistry **32**:229–239.

Wolfe, D. W. 2001. *Tales from the Underground*. Perseus.

Wolfe, G. V., M. Steinke, and G. O. Kirst. 1997. Grazing-activated chemical defence in a unicellular marine alga. Nature **387**:894–897.

Wolfe-Simon, F., J. S. Blum, T. R. Kulp, G. W. Gordon, S. E. Hoeft, J. Pett-Ridge, J. F. Stolz, S. M. Webb, P. K. Weber, P.C.W. Davies, A. D. Anbar, and R. S. Oremland. 2011. A bacterium that can grow by using arsenic instead of phosphorus. Science **332**:1163–1166.

Wolin, M. J., and T. L. Miller. 1987. Bioconversion of organic carbon to CH_4 and CO_2. Geomicrobiology Journal **5**:239–259.

Wollast, R. 1981. Interactions between major biogeochemical cycles in marine ecosystems. *In* G. E. Likens, editor. *Some Perspectives of the Major Biogeochemical Cycles*. Wiley, 125–142.

Wollast, R. 1991. The coastal organic carbon cycle: Fluxes, sources, and sinks. *In* R.F.C. Mantoura, J.-M. Martin and R. Wollast, editors. *Ocean Margin Proceses in Global Change*, Wiley, 365–381.

Wollast, R. 1993. Interactions of carbon and nitrogen cycles in the coastal zone. *In* R. Wollast, F. T. MacKenzie and L. Chou, editors. *Interactions of C, N, P, and S Biogeochemical Cycles and Global Change*. Springer, 195–210.

Wollheim, W. M., B. J. Peterson, L. A. Deegan, J. E. Hobbie, B. Hooker, W. B. Bowden, K. J. Edwardson, D. B. Arscott, and A. E. Hershey. 2001. Influence of stream size on ammonium and suspended particulate nitrogen processing. Limnology and Oceanography **46**:1–13.

Wollheim, W. M., C. J. Vörösmarty, B. J. Peterson, S. P. Seitzinger, and C. S. Hopkinson. 2006. Relationship between river size and nutrient removal. Geophysical Research Letters **33**.

Wollheim, W. M., C. J. Vörösmarty, A. F. Bouwman, P. Green, J. Harrison, E. Linder, B. J. Peterson, S. P. Seitzinger, and J.P.M. Syvitski. 2008. Global N removal by freshwater aquatic systems using a spatially distributed, within-basin approach. Global Biogeochemical Cycles **22**.

Woltemate, I., M. J. Whiticar, and M. Schoell. 1984. Carbon and hydrogen isotopic composition of bacterial methane in a shallow freshwater lake. Limnology and Oceanography **29**:985–992.

Wong, C. S., Y. H. Chan, J. S. Page, G. E. Smith, and R. D. Bellegay. 1993. Changes in equatorial CO_2 flux and new production estimated from CO_2 and nutrient levels in Pacific surface waters during the 1986–87 El Niño. Tellus Series B—Chemical and Physical Meteorology **45**:64–79.

Wong, M.T.F., R. Hughes, and D. L. Rowell. 1990. Retarded leaching of nitrate in acid soils from the tropics—Measurement of the effective anion exchange capacity. Journal of Soil Science **41**:655–663.

Wood, B. J., M. J. Walter, and J. Wade. 2006. Accretion of the Earth and segregation of its core. Nature **441**:825–833.

Wood, T., F. H. Bormann, and G. K. Voigt. 1984. Phosphorus cycling in a northern hardwood forest—Bioloigcal and chemical control. Science **223**:391–393.

Wood, W. W., and M. J. Petraitis. 1984. Origin and distribution of carbon dioxide in the unsaturated zone of the southern High Plains of Texas. Water Resources Research **20**:1193–1208.

Woodbury, P. B., J. E. Smith, and L. S. Heath. 2007. Carbon sequestration in the US forest sector from 1990 to 2010. Forest Ecology and Management **241**:14–27.

Woodmansee, R. G. 1978. Additions and losses of nitrogen in grassland ecosystems. BioScience **28**:448–453.

Woodmansee, R. G., J. L. Dodd, R. A. Bowman, F. E. Clark, and C. E. Dickinson. 1978. Nitrogen budget of a shortgrass prairie ecosystem. Oecologia **34**:363–376.

Woodmansee, R. G., and L. S. Wallach. 1981. Effects of fire regimes on biogeochemical cycles. *In* F. E. Clark and T. Rosswall, editors. *Terrestrial Nitrogen Cycles*. Swedish Natural Science Research Council, 649–669.

Woodward, F. I. 1987. Stomatal numbers are sensitive to increases in CO_2 from preindustrial levels. Nature **327**:617–618.

Woodward, F. I. 1993. Plant responses to past concentrations of CO_2. Vegetation **104**:145–155.

Woodward, G., M. O. Gessner, P. S. Giller, V. Gulis, S. Hladyz, A. Lecerf, B. Malmqvist, B. G. McKie, S. D. Tiegs, H. Cariss, M. Dobson, A. Elosegi, V. Ferreira, M.A.S. Graca, T. Fleituch, J. O. Lacoursiere, M. Nistorescu, J. Pozo, G. Risnoveanu, M. Schindler, A. Vadineanu, L.B.M. Vought, and E. Chauvet. 2012. Continental-scale effects of nutrient pollution on stream ecosystem functioning. Science **336**:1438–1440.

Woodwell, G. M. 1974. Variation in the nutrient content of leaves of *Quercus alba, Quercus coccinea,* and *Pinus rigida* in the Brookhaven forest from bud-break to abscission. American Journal of Botany **61**:749–753.

Woolf, D., J. E. Amonette, F. A. Street-Perrott, J. Lehmann, and S. Joseph. 2010. Sustainable biochar to mitigate global climate change. Nature Communications **1**.

Woosley, S. E. 1986. Nucleosynthesis and stellar evolution. *In* J. Audouze, C. Chiosi and S. E. Woosley, editors. *Nucleosynthesis and Chemical Evolution.* The Geneva Observatory, 1–195.

Woosley, S. E., and M. M. Phillips. 1988. Supernova 1987A. Science **240**:750–759.

Workshop, S. S. 1990. Concepts and methods for assessing solute dynamics in stream ecosystems. Journal of the North American Bentholocial Society **9**:95–119.

Worm, B., E. B. Barbier, N. Beaumont, J. E. Duffy, C. Folke, B. S. Halpern, J.B.C. Jackson, H. K. Lotze, F. Micheli, S. R. Palumbi, E. Sala, K. A. Selkoe, J. J. Stachowicz, and R. Watson. 2006. Impacts of biodiversity loss on ocean ecosystem services. Science **314**:787–790.

Worsley, T. R., and T. A. Davies. 1979. Sea level fluctuations and deep sea sedimentation rates. Science **203**:445–446.

Wortmann, U. G., and A. Paytan. 2012. Rapid variability of seawater chemistry over the past 130 million years. Science **337**:334–336.

Wotawa, G., P. C. Novelli, M. Trainer, and C. Granier. 2001. Inter-annual variability of summertime CO concentrations in the Northern Hemisphere explained by boreal forest fires in North America and Russia. Geophysical Research Letters **28**:4575–4578.

Wotawa, G., and M. Trainer. 2000. The influence of Canadian forest fires on pollutant concentrations in the United States. Science **288**:324–328.

Wrage, N., G. L. Velthof, M. L. van Beusichem, and O. Oenema. 2001. Role of nitrifier denitrification in the production of nitrous oxide. Soil Biology & Biochemistry **33**:1723–1732.

Wright, J. R., A. Leahey, and H. M. Rice. 1959. Chemical, morphological and mineralogical characteristics of a chronosequence of soils on alluvial deposits in the Northwest Territories. Canadian Journal of Soil Science **39**:32–43.

Wright, R. F. 1976. Impact of forest fire on nutrient influxes to small lakes in northeastern Minnesota. Ecology **57**:649–663.

Wright, R. F., E. Lotse, and A. Semb. 1994. Experimental acidification of alpine catchments at Sogndal, Norway—Results after 8 years. Water Air and Soil Pollution **72**:297–315.

Wu, J. 1981. Evidence of sea spray produced by bursting bubbles. Science **212**:324–326.

Wu, J., A. G. Odonnell, and J. K. Syers. 1995. Influences of glucose, nitrogen and plant residues on the immobilization of sulfate-S in soil. Soil Biology & Biochemistry **27**:1363–1370.

Wu, J. F., W. Sunda, E. A. Boyle, and D. M. Karl. 2000. Phosphate depletion in the western North Atlantic Ocean. Science **289**:759–762.

Wu, Z. T., P. Dijkstra, G. W. Koch, J. Penuelas, and B. A. Hungate. 2011. Responses of terrestrial ecosystems to temperature and precipitation change: a meta-analysis of experimental manipulation. Global Change Biology **17**:927–942.

Wuebbles, D. J., and K. Hayhoe. 2002. Atmospheric methane and global change. Earth-Science Reviews **57**:177–210.

Wuebbles, D. J., and J. S. Tamaresis. 1993. The role of methane in the global environment. *In* M.A.K. Khalil, editor. *Atmospheric Methane: Sources, Sinks, and Role in Global Climate.* Springer, 469–513.

Wullstein, L. H., and S. A. Pratt. 1981. Scanning electron microscopy of rhizosheaths of *Oryzopsis hymenoides.* American Journal of Botany **68**:408–419.

Xia, J., and S. Wan. 2008. Global response patterns of terrestrial plant species to nitrogen addition. New Phytologist **179**:428–439.

Xiao, J. F., Q. L. Zhuang, B. E. Law, J. Q. Chen, D. D. Baldocchi, D. R. Cook, R. Oren, A. D. Richardson, S. Wharton, S. Y. Ma, T. A. Martini, S. B. Verma, A. E. Suyker, R. L. Scott, R. K. Monson, M. Litvak, D. Y. Hollinger, G. Sun, K. J. Davis, P. V. Bolstad, S. P. Burns, P. S. Curtis, B. G. Drake, M. Falk, M. L. Fischer, D. R. Foster, L. H. Gu, J. L. Hadley, G. G. Katul, Y. Roser, S. McNulty, T. P. Meyers, J. W. Munger, A. Noormets, W. C. Oechel, K. T. Paw, H. P. Schmid, G. Starr, M. S. Torn, and S. C. Wofsy. 2010. A continuous measure of gross primary production for the conterminous United States derived from MODIS and AmeriFlux data. Remote Sensing of Environment **114**:576–591.

Xie, Z.-Q., L.-G. Sun, J.-J. Wang, and B.-Z. Liu. 2002. A potential source of atmospheric sulfur from penguin colony emissions. Journal of Geophysical Research **107**.

Xiong, J., W. M. Fischer, K. Inoue, M. Nakahara, and C. E. Bauer. 2000. Molecular evidence for the early evolution of photosynthesis. Science **289**:1724–1730.

Xu, L.-Q., X.-D. Liu, L.-g. Sun, Q.-Q. Chen, H. Yan, Y. Liu, Y.-H. Luo, and J. Huang. 2011. A 700-year record of mercury in avian eggshells of Guangjin Island, South China Sea. Environmental Pollution **159**:889–896.

Xu, X. F., H. Q. Tian, and D. F. Hui. 2008a. Convergence in the relationship of CO_2 and N_2O exchanges between soil and atmosphere within terrestrial ecosystems. Global Change Biology 14:1651–1660.

Xu, Y., L. Feng, P. D. Jeffrey, Y. G. Shi, and F.M.M. Morel. 2008b. Structure and metal exchange in the cadmium carbonic anhydrase of marine diatoms. Nature 452:56–61.

Yaalon, D. H. 1965. Downward movement and distribution of anions in soil profiles with limited wetting. *In* E. D. Hallsworth and D. V. Crawford, editors. *Experimental Pedology*. Butterworth, 157–164.

Yagi, K., J. Williams, N. Y. Wang, and R. J. Cicerone. 1995. Atmospheric methyl bromide (CH_3Br) from agricultural soil fumigations. Science 267:1979–1981.

Yamada, A., T. Inoue, D. Wiwatwitaya, M. Ohkuma, T. Kudo, and A. Sugimoto. 2006. Nitrogen fixation by termites in tropical forests, Thailand. Ecosystems 9:75–83.

Yamagata, Y., H. Watanabe, M. Saitoh, and T. Namba. 1991. Volcanic production of polyphosphates and its relevance to prebiotic evolution. Nature 352:516–519.

Yanagawa, H., Y. Ogawa, K. Kojima, and M. Ito. 1988. Construction of protocellular structures under stimulated primitive Earth conditions. Origins of Life and Evolution of the Biosphere 18:179–207.

Yanai, R. D. 1992. Phosphorus budget of a 70-year-old northern hardwood forest. Biogeochemistry 17:1–22.

Yang, E. S., D. M. Cunnold, M. J. Newchurch, R. J. Salawitch, M. P. McCormick, J. M. Russell, J. M. Zawodny, and S. J. Oltmans. 2008. First stage of Antarctic ozone recovery. Journal of Geophysical Research—Atmospheres 113.

Yang, E. S., P. Gupta, and S. A. Christopher. 2009. Net radiative effect of dust aerosols from satellite measurements over Sahara. Geophysical Research Letters 36.

Yang, L. H. 2004. Periodical cicadas as resource pulses in North American forests. Science 306:1565–1567.

Yang, X., and W. M. Post. 2011. Phosphorus transformations as a function of pedogenesis: A synthesis of soil phosphorus data using Hedley fractionation method. Biogeosciences 8:2907–2916.

Yano, J., J. Kern, K. Sauer, M. J. Latimer, Y. Pushkar, J. Biesiadka, B. Loll, W. Saenger, J. Messinger, A. Zouni, and V. K. Yachandra. 2006. Where water is oxidized to dioxygen: Structure of the photosynthetic Mn4Ca cluster. Science 314:821–825.

Yavitt, J. B., and T. J. Fahey. 1993. Production of methane and nitrous oxide by organic soils within a northern hardwood forest ecosystem. *In* R. S. Oremland, editor. *Biogeochemistry and Global Change*. Chapman and Hall, 261–277.

Yavitt, J. B., and A. K. Knapp. 1995. Methane emission to the atmosphere through emergent cattail (*Typha*

latifolia L) plants. Tellus Series B—Chemical and Physical Meteorology 47:521–534.

Ye, R., Q. Jin, B. Bohannan, J. K. Keller, S. A. McAllister, and S. D. Bridgham. 2012. pH controls over anaerobic carbon mineralization, the efficiency of methane production, and methanogenic pathways in peatlands across an ombrotrophic-minerotrophic gradient. Soil Biology & Biochemistry 55: in press.

Ye, R. W., B. A. Averill, and J. M. Tiedje. 1994. Denitrification: Production and consumption of nitric oxide. Applied and Environmental Microbiology 60:1053–1058.

Yi, Z. G., X. M. Wang, M. G. Ouyang, D. Q. Zhang, and G. Y. Zhou. 2010. Air-soil exchange of dimethyl sulfide, carbon disulfide, and dimethyl disulfide in three subtropical forests in south China. Journal of Geophysical Research—Atmospheres 115.

Yin, Q. Z., S. B. Jacobsen, K. Yamashita, J. Blichert-Toft, P. Telouk, and F. Albarede. 2002. A short timescale for terrestrial planet formation from Hf-W chronometry of meteorites. Nature 418:949–952.

Yin, X. W. 1993. Variation in foliar nitrogen concentration by forest type and climatic gradients in North America. Canadian Journal of Forest Research 23: 1587–1602.

Yoh, M., H. Terai, and Y. Saijo. 1983. Accumulation of nitrous oxide in the oxygen-deficient layer of freshwater lakes. Nature 301:327–329.

Yoh, M., H. Terai, and Y. Saijo. 1988. Nitrous oxide in freshwater lakes. Archiv fur Hydrobiologie 113:273–294.

Yokouchi, Y., M. Ikeda, Y. Inuzuka, and T. Yukawa. 2002a. Strong emission of methyl chloride from tropical plants. Nature 416:163–165.

Yokouchi, Y., K. Osada, M. Wada, F. Hasebe, M. Agama, R. Murakami, H. Mukai, Y. Nojiri, Y. Inuzuka, D. Toom-Sauntry, and P. Fraser. 2008. Global distribution and seasonal concentration change of methyl iodide in the atmosphere. Journal of Geophysical Research—Atmospheres 113.

Yokouchi, Y., D. Toom-Sauntry, K. Yazawa, T. Inagaki, and T. Tamaru. 2002b. Recent decline of methyl bromide in the troposphere. Atmospheric Environment 36:4985–4989.

Yokoyama, Y., K. Lambeck, P. De Deckker, P. Johnston, and L. K. Fifield. 2000. Timing of the Last Glacial Maximum from observed sea-level minima. Nature 406:713–716.

Yoneyama, T., T. Muraoka, T. Murakami, and N. Boonkerd. 1993. Natural abundance of ^{15}N in tropical plants with emphasis on tree legumes. Plant and Soil 153:295–304.

Yool, A., A. P. Martin, C. Fernandez, and D. R. Clark. 2007. The significance of nitrification for oceanic new production. Nature 447:999–1002.

Yoshida, N., and S. Toyoda. 2000. Constraining the atmospheric N_2O budget from intramolecular site preference in N_2O isotopomers. Nature **405**:330–334.

Yoshida, Y., Y. H. Wang, C. Shim, D. Cunnold, D. R. Blake, and G. S. Dutton. 2006. Inverse modeling of the global methyl chloride sources. Journal of Geophysical Research—Atmospheres **111**.

Yoshinari, T., and R. Knowles. 1976. Acetylene inhibition of nitrous oxide reduction by denitrifying bacteria. Biochemical and Biophysical Research Communications **69**:705–710.

Young, J. R., E. C. Ellis, and G. M. Hidy. 1988. Deposition of air-borne acidifiers in the western environment. Journal of Environmental Quality **17**:1–26.

Youngberg, C. T., and A. G. Wollum. 1976. Nitrogen accretion in developing *Ceanothus velutinus* stands. Soil Science Society of America Journal **40**:109–112.

Yu, D. S., X. Z. Shi, H. Wang, W. X. Sun, J. M. Chen, Q. H. Liu, and Y. C. Zhao. 2007. Regional patterns of soil organic carbon stocks in China. Journal of Environmental Management **85**:680–689.

Yu, J. M., W. S. Broecker, H. Elderfield, Z. D. Jin, J. McManus, and F. Zhang. 2010. Loss of carbon from the deep sea since the Last Glacial Maximum. Science **330**:1084–1087.

Yu, Z., Q. Zhang, T.E.C. Kraus, R. A. Dahlgren, C. Anastasio, and R. J. Zasoski. 2002. Contribution of amino compounds to dissolved organic nitrogen in forest soils. Biogeochemistry **61**:173–198.

Yu, Z. C., I. D. Campbell, C. Campbell, D. H. Vitt, G. C. Bond, and M. J. Apps. 2003. Carbon sequestration in western Canadian peat highly sensitive to Holocene wet–dry climate cycles at millennial timescales. Holocene **13**:801–808.

Yuan, G., and L. M. Lavkulich. 1994. Phosphate sorption in relation to extractable iron and aluminum in Spodosols. Soil Science Society of America Journal **58**:343–346.

Yuan, T. L., N. Gammon, and R. G. Leighty. 1967. Relative contribution of organic and clay fractions to cation-exchange capacity of sandy soils from several groups. Soil Science **104**:123–128.

Yuan, X. L., S. H. Xiao, and T. N. Taylor. 2005. Lichen-like symbiosis 600 million years ago. Science **308**:1017–1020.

Yuan, Z. W., L. Sun, J. Bi, H. J. Wu, and L. Zhang. 2011. Phosphorus flow analysis of the socioeconomic ecosystem of Shucheng County, China. Ecological Applications **21**:2822–2832.

Yuji, A., and D. L. Sparks. 2007. Phosphate reaction dynamics in soils and soil components: A multiscale approach. Advances in Agronomy **94**:135–179.

Yung, Y. L., and W. B. DeMore. 1999. *Photochemistry of Planetary Atmospheres*. Oxford University Press.

Yung, Y. L., T. Lee, C. H. Wang, and Y. T. Shieh. 1996. Dust: A diagnostic of the hydrologic cycle during the last glacial maximum. Science **271**:962–963.

Yung, Y. L., and M. B. McElroy. 1979. Fixation of nitrogen in the prebiotic atmosphere. Science **203**:1002–1004.

Yung, Y. L., J. S. Wen, J. I. Moses, B. M. Landry, M. Allen, and K. J. Hsu. 1989. Hydrogen and deuterium loss from the terrestrial atmosphere—A quantitative assessment of nonthermal escape fluxes. Journal of Geophysical Research—Atmospheres **94**:14971–14989.

Yvon-Lewis, S. A., E. S. Saltzman, and S. A. Montzka. 2009. Recent trends in atmospheric methyl bromide: Analysis of post-Montreal Protocol variability. Atmospheric Chemistry and Physics **9**:5963–5974.

Zachos, J. C., G. R. Dickens, and R. E. Zeebe. 2008. An early Cenozoic perspective on greenhouse warming and carbon-cycle dynamics. Nature **451**:279–283.

Zagal, E., and J. Persson. 1994. Immobilization and remineralization of nitrate during glucose decomposition at four rates of nitrogen addition. Soil Biology & Biochemistry **26**:1313–1321.

Zak, D. R., G. E. Host, and K. S. Pregitzer. 1989. Regional variability in nitrogen mineralization, nitrification, and overstory biomass in northern Lower Michigan. Canadian Journal of Forest Research **19**:1521–1526.

Zak, D. R., D. Tilman, R. R. Parmenter, C. W. Rice, F. M. Fisher, J. Vose, D. Milchunas, and C. W. Martin. 1994. Plant production and soil microorganisms in late successional ecosystems—A continental-scale study. Ecology **75**:2333–2347.

Zbieranowski, A. L., and J. Aherne. 2011. Long-term trends in atmospheric reactive nitrogen across Canada: 1988–2007. Atmospheric Environment **45**:5853–5862.

Zechmeister-Boltenstern, S., and H. Kinzel. 1990. Non-symbiotic nitrogen fixation associated with temperate soils in relation to soil properties and vegetation. Soil Biology & Biochemistry **22**:1075–1084.

Zedler, J. B., and S. Kercher. 2005. Wetland resources: Status, trends, ecosystem services, and restorability. Annual Review of Environment and Resources **30**:39–74.

Zeebe, R. E., and D. Archer. 2005. Feasibility of ocean fertilization and its impact on future atmospheric CO_2 levels. Geophysical Research Letters **32**.

Zeebe, R. E., J. C. Zachos, K. Caldeira, and T. Tyrrell. 2008. Oceans—Carbon emissions and acidification. Science **321**:51–52.

Zehetner, F. 2010. Does organic carbon sequestration in volcanic soils offset volcanic CO_2 emissions? Quaternary Science Reviews **29**:1313–1316.

Zehnder, A. J., and W. Stumm. 1988. Geochemistry and biogeochemistry of anaerobic habitats. *In* A. J.

Zehnder, editor. *Biology of Anaerobic Microorganisms.* Wiley, 1–38.

Zehr, J. P., and B. B. Ward. 2002. Nitrogen cycling in the ocean: New perspectives on processes and paradigms. Applied and Environmental Microbiology **68**:1015–1024.

Zehr, J. P., J. B. Waterbury, P. J. Turner, J. P. Montoya, E. Omoregie, G. F. Steward, A. Hansen, and D. M. Karl. 2001. Unicellular cyanobacteria fix N_2 in the subtropical North Pacific Ocean. Nature **412**: 635–638.

Zektser, I. S., L. G. Everett, and R. G. Dzhamalov. 2007. *Submarine Groundwater.* CRC Press.

Zektser, I. S., and H. A. Loaiciga. 1993. Groundwater fluxes in the global hydrologic cycle—Past, present and future. Journal of Hydrology **144**:405–427.

Zender, C. S., R. L. Miller, and I. Tegen. 2004. Supplemental material to "Quantifying mineral dust mass budgets: Systematic terminology, constraints, and current estimates." Eos Trans. AGU **85**:509.

Zerihun, A., B. A. McKenzie, and J. D. Morton. 1998. Photosynthate costs associated with the utilization of different nitrogen-forms: influence on the carbon balance of plants and shoot-root biomass partitioning. New Phytologist **138**:1–11.

Zhang, D. D., H. F. Lee, C. Wang, B. Li, J. Zhang, Q. Pei, and J. Chen. 2011a. Climate change and large-scale human population collapses in the pre-industrial era. Global Ecology and Biogeography **20**:520–531.

Zhang, F., J. M. Chen, V. D. Pan, R. A. Birdsey, S. H. Shen, W. M. Ju, and L. M. He. 2012a. Attributing carbon changes in conterminous U.S. forests to disturbance and non-disturbance factors from 1901 to 2010. Journal of Geophysical Research—Biogeosciences **117**.

Zhang, H. L., and P. R. Bloom. 1999. Dissolution kinetics of hornblende in organic acid solutions. Soil Science Society of America Journal **63**:815–822.

Zhang, H. M., and B. G. Forde. 1998. An Arabidopsis MADS box gene that controls nutrient-induced changes in root architecture. Science **279**:407–409.

Zhang, J. B., Z. C. Cai, Y. Cheng, and T. B. Zhu. 2009. Denitrification and total nitrogen gas production from forest soils of eastern China. Soil Biology & Biochemistry **41**:2551–2557.

Zhang, K., J. S. Kimball, R. R. Nemani, and S. W. Running. 2010. A continuous satellite-derived global record of land surface evapotranspiration from 1983 to 2006. Water Resources Research **46**.

Zhang, P. Z., P. Molnar, and W. R. Downs. 2001. Increased sedimentation rates and grain sizes 2-4 Myr ago due to the influence of climate change on erosion rates. Nature **410**:891–897.

Zhang, T., B. Kim, C. Levard, B. C. Reinsch, G. V. Lowry, M. A. Deshusses, and H. Hsu-Kim. 2012. Methylation of mercury by bacteria exposed to dissolved, nanoparticulate and microparticulate mercuric sulfides. Environmental Science and Technology **46**: 6950–6958.

Zhang, X., K. C. Hester, W. Ussler, P. M. Walz, E. T. Peltzer, and P. G. Brewer. 2011b. *In situ* Raman-based measurements of high dissolved methane concentrations in hydrate-rich ocean sediments. Geophysical Research Letters **38**.

Zhang, X. B., F. W. Zwiers, G. C. Hegerl, F. H. Lambert, N. P. Gillett, S. Solomon, P. A. Stott, and T. Nozawa. 2007. Detection of human influence on twentieth-century precipitation trends. Nature **448**:461–465.

Zhang, X. Y., and S. Kondragunta. 2006. Estimating forest biomass in the USA using generalized allometric models and MODIS land products. Geophysical Research Letters **33**.

Zhang, Y., L. Song, X. J. Liu, W. Q. Li, S. H. Lu, L. X. Zheng, Z. C. Bai, G. Y. Cai, and F. S. Zhang. 2012. Atmospheric organic nitrogen deposition in China. Atmospheric Environment **46**:195–204.

Zhang, Y. X., and A. Zindler. 1993. Distribution and evolution of carbon and nitrogen in Earth. Earth and Planetary Science Letters **117**:331–345.

Zhao, F. J., J. Lehmann, D. Solomon, M. A. Fox, and S. P. McGrath. 2006. Sulphur speciation and turnover in soils: evidence from sulphur K-edge XANES spectroscopy and isotope dilution studies. Soil Biology & Biochemistry **38**:1000–1007.

Zhao, M. S., and S. W. Running. 2010. Drought-induced reduction in global terrestrial net primary production from 2000 through 2009. Science **329**: 940–943.

Zhou, G., S. Liu, Z. Li, D. Zhang, X. Tang, C. Zhou, J. Yan, and J. Mo. 2006. Old-growth forests can accumulate carbon in soils. Science **314**:1417.

Zhou, T., and Y. Q. Luo. 2008. Spatial patterns of ecosystem carbon residence time and NPP-driven carbon uptake in the conterminous United States. Global Biogeochemical Cycles **22**.

Zhu, R. B., D. Glindemann, D. M. Kong, L. G. Sun, J. J. Geng, and X. R. Wang. 2007. Phosphine in the marine atmosphere along a hemispheric course from China to Antarctica. Atmospheric Environment **41**: 1567–1573.

Zhuang, Q., Y. Lu, and M. Chen. 2012. An inventory of global N_2O emissions from the soils of natural terrestrial ecosystems. Atmospheric Environment **47**: 66–75.

Ziebis, W., S. Forster, M. Huettel, and B. B. Jorgensen. 1996. Complex burrows of the mud shrimp

Callianassa truncata and their geochemical impact in the sea bed. Nature 382:619–622.

Zierl, B. 2001. A water balance model to simulate drought in forested ecosystems and its application to the entire forested area in Switzerland. Journal of Hydrology 242:115–136.

Zimmerman, P. R., J. P. Greenberg, and C. E. Westberg. 1988. Measurements of atmospheric hydrocarbons and biogenic emission fluxes in the Amazon boundary layer. Journal of Geophysical Research—Atmospheres 93:1407–1416.

Zimov, S. A., S. P. Davydov, G. M. Zimova, A. I. Davydova, E.A.G. Schuur, K. Dutta, and F. S. Chapin. 2006. Permafrost carbon: Stock and decomposability of a globally significant carbon pool. Geophysical Research Letters 33.

Zimov, S. A., Y. V. Voropaev, I. P. Semiletov, S. P. Davidov, S. F. Prosiannikov, F. S. Chapin, M. C. Chapin, S. Trumbore, and S. Tyler. 1997. North Siberian lakes: A methane source fueled by Pleistocene carbon. Science 277:800–802.

Zinke, P. J. 1980. Influence of chronic air pollution on mineral cycling in forests. Pacific Southwest Forest and Range Experiment Station, General Technical Report 43:88-99, U.S. Forest Service.

Zobel, D. B., and J. A. Antos. 1991. 1980 tephra from Mount St Helens—Spatial and temporal variation beneath forest canopies. Biology and Fertility of Soils 12:60–66.

Zou, J. W., Y. Huang, J. Y. Jiang, X. H. Zheng, and R. L. Sass. 2005. A 3-year field measurement of methane and nitrous oxide emissions from rice paddies in China: Effects of water regime, crop residue, and fertilizer application. Global Biogeochemical Cycles 19:9.

Zuber, M. T., J. W. Head, D. E. Smith, G. A. Neumann, E. Mazarico, M. H. Torrence, O. Aharonson, A. R. Dye, C. L. Fassett, M. A. Rosenburg, and H. J. Melosh. 2012. Constraints on the volatile distribution within Shackleton crater at the lunar south pole. Nature 486:378–381.

Zwally, H. J., J. C. Comiso, C. L. Parkinson, D. J. Cavalieri, and P. Gloersen. 2002. Variability of Antarctic sea ice 1979–1998. Journal of Geophysical Research—Oceans 107.

Zychowski, J. 2012. Impact of cemeteries on groundwater chemistry: A review. Catena 93:29–37.

Zysset, M., P. Blaser, J. Luster, and A. U. Gehring. 1999. Aluminum solubility control in different horizons of a Podzol. Soil Science Society of America Journal 63:1106–1115.

Index

Note: Page numbers followed by *f* indicate figures, *t* indicate tables, and *np* indicate footnotes.

1	2	3	4	5	6	7	8	9	10	11	12	13	14	15	16	17	18
H 1 Hydrogen 1.0																	**He** 2 Helium 4.0
Li 3 Lithium 6.9	**Be** 4 Beryllium 9.0											**B** 5 Boron 10.8	**C** 6 Carbon 12.0	**N** 7 Nitrogen 14.0	**O** 8 Oxygen 16.0	**F** 9 Fluorine 19.0	**Ne** 10 Neon 20.2
Na 11 Sodium 23.0	**Mg** 12 Magnesium 9.0											**Al** 13 Aluminum 27.0	**Si** 14 Silicon 28.1	**P** 15 Phosphorus 31.0	**S** 16 Sulfur 32.1	**Cl** 17 Chlorine 35.5	**Ar** 18 Argon 40.0
K 19 Potassium 39.1	**Ca** 20 Calcium 40.2	**Sc** 21 Scandium 45.0	**Ti** 22 Titanium 47.9	**V** 23 Vanadium 50.9	**Cr** 24 Chromium 52.0	**Mn** 25 Manganese 54.9	**Fe** 26 Iron 55.9	**Co** 27 Cobalt 58.9	**Ni** 28 Nickel 58.7	**Cu** 29 Copper 63.5	**Zn** 30 Zinc 65.4	**Ga** 31 Gallium 69.7	**Ge** 32 Germanium 72.6	**As** 33 Arsenic 74.9	**Se** 34 Selenium 79.0	**Br** 35 Bromine 79.9	**Kr** 36 Krypton 83.8
Rb 37 Rubidium 85.5	**Sr** 38 Strontium 87.6	**Y** 39 Yttrium 88.9	**Zr** 40 Zirconium 91.2	**Nb** 41 Niobium 92.9	**Mo** 42 Molybdenum 95.9	**Tc** 43 Technetium 99	**Ru** 44 Ruthenium 101.0	**Rh** 45 Rhodium 102.9	**Pd** 46 Palladium 106.4	**Ag** 47 Silver 107.9	**Cd** 48 Cadmium 112.4	**In** 49 Indium 114.8	**Sn** 50 Tin 118.7	**Sb** 51 Antimony 121.8	**Te** 52 Tellurium 127.6	**I** 53 Iodine 126.9	**Xe** 54 Xenon 131.3
Cs 55 Caesium 132.9	**Ba** 56 Barium 137.4	**La** 57 Lanthanum 138.9	**Hf** 72 Hafnium 178.5	**Ta** 73 Tantalum 181.0	**W** 74 Tungsten 183.9	**Re** 75 Rhenium 186.2	**Os** 76 Osmium 190.2	**Ir** 77 Iridium 192.2	**Pt** 78 Platinum 195.1	**Au** 79 Gold 197.0	**Hg** 80 Mercury 200.6	**Tl** 81 Thallium 204.4	**Pb** 82 Lead 207.2	**Bi** 83 Bismuth 209.0	**Po** 84 Polonium 210.0	**At** 85 Astatine 210.0	**Rn** 86 Radon 222.0
Ra 88 Radium 226.0		**Ac** 89 Actinium 132.9	**Th** 90 Thorium 232.0	**Pa** 91 Protactinium 231.0	**U** 92 Uranium 238.0												

Geologic Time Scale

Era	System & Period	Series & Epoch	Some Distinctive Features	Years Before Present
CENOZOIC	Quaternary	Recent	Modern man.	11,000
		Pleistocene	Early man; northern glaciation.	1/2 to 2 million
	Tertiary	Pliocene	Large carnivores.	13 + 1 million
		Miocene	First abundant grazing mammals.	25 + 1 million
		Oligocene	Large running mammals.	36 + 2 million
		Eocene	Many modern types of mammals.	58 + 2 million
		Paleocene	First placental mammals.	63 + 2 million
MESOZOIC	Cretaceous		First flowering plants; climax of dinosaurs and ammonites, followed by Cretaceous-Tertiary extinction.	135 + 5 million
	Jurassic		First birds, first mammals dinosaurs and ammonites abundant.	181 + 5 million
	Triassic		First dinosaurs. Abundant cycads and conifers.	230 + 10 million
PALEOZOIC	Permian		Extinction of most kinds of marine animals, including trilobites. Southern glaciation.	280 + 10 million
	Carboniferous	Pennsylvanian	Great coal forests, conifers. First reptiles.	310 + 10 million
		Mississippian	Sharks and amphibians abundant. Large and numerous scale trees and seed ferns.	345 + 10 million
	Devonian		First amphibians; ammonites; fishes abundant.	405 + 10 million
	Silurian		First terrestrial plants and animals.	425 + 10 million
	Ordovician		First fishes; invertebrates dominant.	500 + 10 million
	Cambrian		First abundant record of marine life; trilobites dominant.	600 + 50 million
	Precambrian		Fossils extremely rare, consisting of primitive aquatic plants. Evidence of glaciation. Oldest dated algae, over 2,600 million years; oldest dated meteorites 4,500 million years.	